智能与绿色建筑文集 2

MEMOIRS ON INTELLIGENT BUILDING & GREEN BUILDING 2

——第二届国际智能、绿色建筑与建筑节能大会

——The 2nd International Conference on Intelligent Green and Energy Efficient Building

中华人民共和国建设部科学技术司
《智能与绿色建筑文集 2》编委会　编

中国建筑工业出版社

图书在版编目（CIP）数据

智能与绿色建筑文集 2——第二届国际智能、绿色建筑与建筑节能大会/中华人民共和国建设部科学技术司、《智能与绿色建筑文集 2》编委会编.—北京：中国建筑工业出版社，2006

ISBN 7-112-08095-9

Ⅰ.智… Ⅱ.①中…②智… Ⅲ.①智能建筑—国际会议—文集②建筑工程—无污染技术—文集 Ⅳ.①TU243-53②TU712-53

中国版本图书馆 CIP 数据核字(2006)第 011495 号

本文集 2 介绍智能、节能与绿色建筑的发展以及关键技术的应用和研究成果。内容包括 5 部分：A 绿色建筑设计理论、方法和实践，B 建筑智能化与绿色建筑、C 建筑节能与绿色建筑，D 建筑生态、材料与绿色建筑，E 房地产发展与绿色建筑等。内容新颖，技术先进，图文并茂，颇具参考价值，可供建筑设计和科研人员、施工技术和管理人员、建材技术和产品生产人员以及土建类大专院校师生参考。

* * *

责任编辑：蒋协炳
责任设计：郑秋菊
责任校对：孙 爽 王雪竹

智能与绿色建筑文集 2

MEMOIRS ON INTELLIGENT BUILDING & GREEN BUILDING 2

——第二届国际智能、绿色建筑与建筑节能大会

——The 2nd International Conference on Intelligent Green and Energy Efficient Building

中华人民共和国建设部科学技术司
《智能与绿色建筑文集 2》编委会 编

*

中国建筑工业出版社出版、发行(北京西郊百万庄)
新 华 书 店 经 销
北京天成排版公司制版
北京中科印刷有限公司印刷

*

开本：787×1092 毫米 1/16 印张：60¾ 字数：1513 千字
2006 年 3 月第一版 2006 年 3 月第一次印刷
印数：1—3000 册 定价：**138.00** 元
ISBN 7-112-08095-9
(14049)

本社网址：http：//www.cabp.com.cn
网上书店：http：//www.china-building.com.cn

《智能与绿色建筑文集 2》编委会名单

序

自工业革命以来，人类社会进入到一个前所未有的高速发展时期。机器的出现使得劳动生产率得到大幅度提高，人类因而有能力最大化地将自然资源转化为供人类生产和消费的物质产品，满足人类日益增长的物质需要。但是，在人类的这种需要得到极大满足的同时，经济发展对资源环境的影响逐渐显露出来，甚至成为严重制约经济发展的瓶颈，引起了人们深刻的反思。我们的发展无论是在本国、还是在其他国家或地区，无论是目前还是未来，都不能以无节制地消耗资源(尤其是能源)和生态环境的日趋恶化为代价，只有如此人类及其社会才能可持续发展下去。因此，人类现在对生存环境的关注远远超过了以往任何时期：开始对传统发展模式进行深刻反省，重新审视自己的生产方式和生活方式。在建筑领域逐渐发展出绿色建筑的理念，促进资源节约与环境保护，将各种先进适用技术应用于建筑物，开拓一条人与自然和谐共处、全社会整体可持续发展的新的建筑发展道路。

建筑作为人类改造自然的方式之一，为我们提供了居住和活动的场所，但同时，建筑也对社会、经济及人类赖以生存的生态环境产生了相当大的负面影响：能源、资源耗费巨大，用于人类建筑活动的自然资源大约占到全球总消耗量的40%，发电产生的温室气体占全球温室气体排放的37%，而其中48%直接用于建筑的能量供应；水资源短缺，水体污染频频发生等等，建筑产生的相关空气污染、电磁污染占环境污染总量的34%，已经威胁到了人类自身的生存与未来发展。

从绿色建筑的理念出发，要求人们关注建筑的全生命周期，尽一切可能建造节约资源、能源的建筑，尽量减少建筑在照明、采暖制冷方面的能耗，减低CO_2的排放，减弱全球性温室效应，提高水资源的利用效率，从而减少资源、能源的不断紧缺对人们工作、生活产生的危机，使人类能够长久地生存在舒适的环境中。

中国作为世界上最大的发展中国家，经过长期努力已经取得了巨大的经济与社会发展成就。但随着城镇化速度的加快，年建筑量不断快速增长，土地、水、能源以及环保等问题已日益加剧。面临如此严峻形势，仅仅靠简单的限制资源、能源的使用和加大开采量是无法扭转和克服的，必须尽快找到解决问题的途径。因此，大力发展绿色建筑是中国建筑业的发展方向，是迎接挑战的根本策略。

世界各国和地区必须联合起来，加快推进绿色建筑的发展，才能为拯救我们共同的星球做出贡献。必须按照科学发展观的思想，可持续发展的理念，加强政策性宏观调控，提高建筑行业行为规范与自律责任，节约能源资源，保护环境，以人为本，推动绿色建筑发展；修复自然生态系统与人居生态系统之间形成的破损、阻断关系，坚持资源保护与高效利用原则，发展可再生、可循环的资源利用技术系统；带动绿色建筑相关生态和绿色产业发展，调整健全社会发展与社会进步中以建筑科技、文化、艺术为载体的先进人文理念、社会价值观，发展人居系统与自然系统的和谐结构模式；完善建筑学科理论体系，实现理论与社会实践相结合的科技生产力转化，科学与法制的管理，建立保障社会发展与进步的生态规划体系与绿色建筑标准体系；鼓励绿色建筑科技推广与创新，促进资源节约型和环境友好型的宜居城市建设，城乡一体化统筹发展绿色建筑，改善和提高村镇建筑和生态环境质量和水平。

智能系统在绿色建筑中可以控制、记录、分析能源、水资源在使用过程中的消耗，监控环境，为信息共享、交换提供了便捷的平台，提供语音通讯、数据和图像传输、办公服务的平台，为绿色建筑及使用者提供安全保障。因此智能系统不是用高科技手段对绿色建筑进行的点缀，而是管理整个绿色建筑的运行指挥中心。如果没有智能系统，在建筑使用过程中实现资源、能源的有效利用、环境监控、提供便利和保障安全，实现绿色建筑的目标就很困难。智能系统必将完全融入绿色建筑之中，智能技术的进步也将推进绿色建筑的发展。

发展绿色建筑可以在环境、经济、社会三方面获得很好的效果。对于环境，可增强和保护生态系统和生物多样性，提高空气和水质量，减少固体污染，保护自然资源；对于经济，可降低运行成本，增加资产价值和收益，提高员工劳动生产率和工作满意度，优化生命周期经济效益；对于社会，可提高空气环境、热环境和声音环境，增强居住者舒适度和健康水平，保障安全，最大限度降低对基础设施的影响，提高整体生活质量。

为了进一步加强智能与绿色建筑技术的交流与合作，巩固自“首届国际智能与绿色建筑技术研讨会”以来取得的成效，总结和探讨应用经验，推广先进技术，探索中国的智能与绿色建筑的未来发展道路，展示中国发展智能与绿色建筑的实力，在出版了“首届国际智能与绿色建筑技术研讨会”论文集《智能与绿色建筑文集》之后，配合2006年3月28～29日在北京召开的“第二届国际智能、绿色建筑与建筑节能大会”，建设部科技司和大会组委会向国际、国内的各知名专家、学者、研究人员征集了大量的论文，并从中精选了118篇送交中国建筑工业出版社编辑出版，刊登在本《智能与绿色建筑文集2》上。与首届大会的论文征集情况相比(首届论文征集从2004年6月到2004年12月，收到论文240篇，后经专家评审选取收录了155篇，其中英文22篇)，《智能绿色建筑文集2》增加了专家约稿的数量，并相对增大了建筑节能相关论文的比重和分量。另外国外论文的征集量比首届有较大增长，学术委员会的相关外国专家也对论文进行了严格把关和筛选。论文的征集和评审更加严格，论文主题与大会的各议题切合性更强，论文质量整体比首届有了较大的提高。希望智能与绿色建筑系列文集的出版对中国智能与绿色建筑的发展起到指导作用。

中华人民共和国建设部科技司
《智能与绿色建筑文集2》编委会
2006年2月22日

PREFACE

Human society has entered into an unprecedented high-speed development period since the Industrial Revolution. Labor productivity was largely improved by the appearing of machine. Hence people could furthest transform natural resources into physical products for their production or consumption to meet progressively improved physical needs of mankind. However, just at the moment that these needs was satisfied tremendously, the effects of economic development to resources and environment were emerging gradually, and even became a bottleneck seriously conditioning economic development, which lead people to a deep reconsideration. No matter in our homeland or other countries or regions, no matter at present or in the future, we cannot develop at the price of resources (especially energy) consuming in excess and ecological environment deteriorating gradually, and only by taking this way, human and their society can develop sustainably. Therefore, human pays much more attention to the environment they live in than any other days: they began to reconsider the traditional developing model, and rethink their Production mode and living style. We gradually form the concept of green building in the building field, meanwhile, people have gradually evolved a concept of green building in the architecture field, promote resources saving and environmental protection, and apply different kinds of advanced technologies into buildings, so as to find a new developing road to coexist in harmony between human and nature, and to achieve sustainable development in whole society. Building, as one of the ways that human innovate nature, provide them the sites to reside and work. However, at the same time, buildings also have quite a few negative influences on society、economy and ecological environment that human depend on to live: energy and resources consumption is huge, and the natural resources used to construction activities is up to 40% of the world's total consumption; 37% of the world's total greenhouse gas(GHG) comes from electricity generation, 48% of which are used to building energy supply; water shortage, water pollution and so on happens frequently; air pollution, electromagnetism pollution produced by buildings account for 34% of the total environmental pollution, all of which have seriously threatened human's survival and development in the future.

The concept of green building requires people pay attention to building lifecycle, make every effort to construct resource-and-energy-saving buildings, reduce building energy consumption in lighting, heating and refrigerating to the best, decrease CO_2 emission and global Green House Gas effects, increase water efficiency, in order to lessen crisis in people's working and living caused by resources and energy shortage, and further to make people live in a comfortable environment for long.

As the largest developing country in the world, China has gained great achievements in economic and social development by the long-term efforts, but accompanying with the rapid urbanization, the number of new buildings grows fast, and problems like land, water, energy and environmental protection are becoming worse and worse, which can not be turned or solved by simply restricting resources and energy use and increasing exploitation amounts. Thus, to develop green building is the direction for China' s building industry and the ultimate strategy to meet the challenges.

We could not save our own planet unless countries and regions in the world unite together to quicken the green building development. We must adhere to the scientific development view and sustainable development concept, enhance policy-based macro-regulation, improve the code of conduct and self-responsibilities in the building industry, save energy and resources, protect environment and promote the development of green building based on the people-oriented principle; reset the damaged and interdicted relationship between natural ecological system and residential system, stick to the principle of resource-protection and high-efficient utilization, and develop the technology system on using reusable and recyclable resources; bring along the development of ecological and green industries related to green building, and adjust and enrich the advanced humanities and social views carried by building technologies, culture, arts and etc, so as to establish a harmony mode between residential and natural systems; perfect the theoretical system of building study, realize the scientific and legitimate management and the transfer of scientific and technological achievements by the integration between theory and practice, and set up ecological planning system and green building criterion system to protect social developments and advancements; encourage the promotion and innovation of green building technologies, advance the construction of resource-saving, environmental-friendly and residential-preferable cities, develop green building by the integration of urban and rural areas, and improve the quality of buildings and ecological environment in rural areas.

The intelligent system in green building could control, record, analysis energy and water resources consumption in the using process, monitor environment, provide a convenient platform for sharing and exchanging the information, provide the platform for voice communication, data and image transmission, work service, and provide safeguard security for green building and its user. Therefore, the intelligent system is not a high-tech ornament, but an operation management control center for entire green building. If there is no intelligent system, it will be difficult to realize the efficient use of resources and energy, the environment monitoring in the building using process, and provide the convenient safeguard security. The intelligent system will certainly be integrated with green building, and the intelligent technology progress will also promote green building development .

The development of green building can obtain very good effects in three aspects——

environment, economy, and society. Regarding the environment, it can strengthen and protect ecosystem and the biodiversity, improve air and the water quality, reduce solid pollution, and protect natural resource. Regarding the economy, it can reduce operational cost, increase property value and income, enhance staff productivity and degree of satisfaction, and optimize the economic efficiency in life cycle . Regarding the society, it may enhance the air, heat and sound environment, strengthen occupants' comfort and health level, safeguard the security, reduce to the infrastructure influence in maximum, and improve the overall life quality.

In order to further strengthen the communication and cooperation of intelligent and green building technologies, consolidate the fruits obtained since "The First International Intelligent and Green Building Technologies and Products Conference & Expo", summary and discuss application experience, promote advanced technology, explore the development way of intelligent and green building in the future, demonstrate China's abilities to develop the intelligent and green building, after the first publication of *Memoir of the Intelligent & Green Building*, to respond "the 2nd International Conference on Intelligent Green and Energy Efficient Building", the Department of Science and Technology of M. O. C. and the Conference Organizing Committee have collected the massive papers from the international and domestic well-known experts, scholars and researchers, and selected 118 articles delivered to the China Architecture & Building Press to edit and print, which will be published on *Memoir of the Intelligent & Green Building II.*

Compared with the paper collection in the first conference(totally received 240 papers from July to December 2004, and finally selected 155 papers including 22 in English), *Memoir of the Intelligent & Green Building II* solicited more papers from the experts, and also increase the papers on energy efficiency. Moreover, the international papers which are also selected by the relevant international experts strictly are more than in the first conference. The collection and selection are more strict, the themes are more relevant to the conference topics, and the overall quality are much more improved, so we hope, the series of *Memoir of the Intelligent & Green Building* will guide the development of intelligent and green building in China.

Department of Science and Technology, Ministry of Construction, P. R. C.

Editor Committee of *Memoir of the Intelligent & Green Building II*

February 22, 2006

目　录

A　绿色建筑设计理论、方法和实践

B 建筑智能化与绿色建筑

C　建筑节能与绿色建筑

D 建筑生态，材料与绿色建筑

E 房地产发展与绿色建筑

建设部　科技部印发

2005 年 10 月

绿色建筑技术导则

主编单位：中国建筑科学研究院

参编单位：清华大学

城市建设研究院

中国建筑材料科学研究院

绿色建筑技术导则

目　　录

绿色建筑技术导则

前言

推进绿色建筑是发展节能省地型住宅和公共建筑的具体实践。党的“十六大”报告指出我国要实现“可持续发展能力不断增强，生态环境得到改善，资源利用效率显著提高，促进人与自然的和谐，推动整个社会走上生产发展、生活富裕、生态良好的文明发展道路”。发展绿色建筑必须牢固树立和认真落实科学发展观，必须从建筑全寿命周期的角度，全面审视建筑活动对生态环境和住区环境的影响，采取综合措施，实现建筑业的可持续发展。为引导、促进和规范绿色建筑的发展，特制定《绿色建筑技术导则》（下简称导则）。

1 总则

1.1 我国正处于经济快速发展阶段，作为大量消耗能源和资源的建筑业，必须发展绿色建筑，改变当前高投入、高消耗、高污染、低效率的模式，承担起可持续发展的社会责任和义务。

1.2 本导则所称的绿色建筑是指在建筑的全寿命周期内，最大限度地节约资源（节能、节地、节水、节材）、保护环境和减少污染，为人们提供健康、适用和高效的使用空间，与自然和谐共生的建筑。

1.3 发展绿色建筑，应倡导城乡统筹、循环经济的理念和紧凑型城市空间的发展模式；全社会参与，挖掘建筑节能、节地、节水、节材的潜力；正确处理节能、节地、节水、节材、环保及满足建筑功能之间的辩证关系。

1.4 发展绿色建筑，应坚持技术创新，走科技含量高、资源消耗低与环境污染少的新型工业化道路。

1.5 发展绿色建筑，应注重经济性，从建筑的全寿命周期综合核算效益和成本，引导市场发展需求，适应地方经济状况，提倡朴实简约，反对浮华铺张。

1.6 发展绿色建筑，应注重地域性，尊重民族习俗，依据当地自然资源条件、经济状况、气候特点等，因地制宜地创造出具有时代特点和地域特征的绿色建筑。

1.7 发展绿色建筑，应注重历史性和文化特色，要尊重历史，加强对已建成环境和历史文脉的保护和再利用。

1.8 绿色建筑的建设必须符合国家的法律法规与相关的标准规范，实现经济效益、社会效益和环境效益的统一。

2 适用范围

本导则用于指导绿色建筑(主要指民用建筑)的建设，适用于建设单位、规划设计单位、施工与监理单位、建筑产品研发企业和有关管理部门等。

3 绿色建筑应遵循的原则

3.1 绿色建筑应坚持“可持续发展”的建筑理念。理性的设计思维方式和科学程序的把握，是提高绿色建筑环境效益、社会效益和经济效益的基本保证。

3.2 绿色建筑除满足传统建筑的一般要求外，尚应遵循以下基本原则：

3.2.1 关注建筑的全寿命周期

建筑从最初的规划设计到随后的施工建设、运营管理及最终的拆除，形成了一个全寿命周期。关注建筑的全寿命周期，意味着不仅在规划设计阶段充分考虑并利用环境因素，而且确保施工过程中对环境的影响最低，运营管理阶段能为人们提供健康、舒适、低耗、无害空间，拆除后又对环境危害降到最低，并使拆除材料尽可能再循环利用。

3.2.2 适应自然条件，保护自然环境

(1) 充分利用建筑场地周边的自然条件，尽量保留和合理利用现有适宜的地形、地貌、植被和自然水系；

(2) 在建筑的选址、朝向、布局、形态等方面，充分考虑当地气候特征和生态环境；

(3) 建筑风格与规模和周围环境保持协调，保持历史文化与景观的连续性；

(4) 尽可能减少对自然环境的负面影响，如减少有害气体和废弃物的排放，减少对生态环境的破坏。

3.2.3 创建适用与健康的环境

(1) 绿色建筑应优先考虑使用者的适度需求，努力创造优美和谐的环境；

(2) 保障使用的安全，降低环境污染，改善室内环境质量；

(3) 满足人们生理和心理的需求，同时为人们提高工作效率创造条件。

3.2.4 加强资源节约与综合利用，减轻环境负荷

(1) 通过优良的设计和管理，优化生产工艺，采用适用技术、材料和产品；

(2) 合理利用和优化资源配置，改变消费方式，减少对资源的占有和消耗；

(3) 因地制宜，最大限度利用本地材料与资源；

(4) 最大限度地提高资源的利用效率，积极促进资源的综合循环利用；

(5) 增强耐久性能及适应性，延长建筑物的整体使用寿命；

(6) 尽可能使用可再生的、清洁的资源和能源。

4 绿色建筑指标体系

绿色建筑指标体系是按定义，对绿色建筑性能的一种完整的表述，它可用于评估实体建筑物与按定义表述的绿色建筑相比在性能上的差异。绿色建筑指标体系由节地与室外环

境、节能与能源利用、节水与水资源利用、节材与材料资源、室内环境质量和运营管理六类指标组成。这六类指标涵盖了绿色建筑的基本要素，包含了建筑物全寿命周期内的规划设计、施工、运营管理及回收各阶段的评定指标的子系统。

图 1 为绿色建筑指标体系框图，表 1 为绿色建筑的分项指标与重点应用阶段汇总。

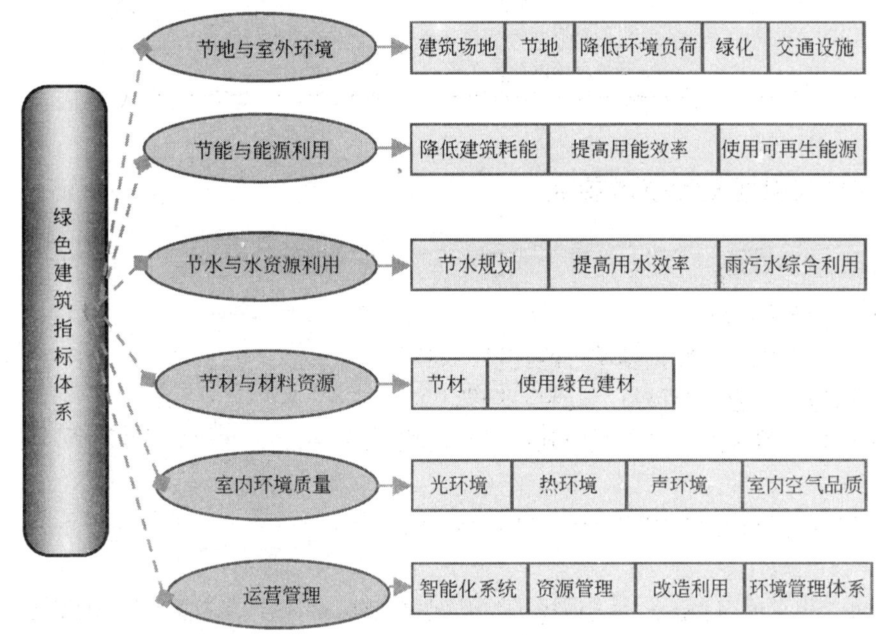

图 1 绿色建筑指标体系框图

绿色建筑分项指标与重点应用阶段汇总表 **表 1**

项目	分项指标	重点应用阶段
节地与室外环境	建筑场地	规划、施工
	节地	规划、设计
	降低环境负荷	全寿命周期
	绿化	全寿命周期
	交通设施	规划、设计、运营管理
节能与能源利用	降低建筑能耗	全寿命周期
	提高用能效率	设计、施工、运营管理
	使用可再生能源	规划、设计、运营管理
节水与水资源利用	节水规划	规划
	提高用水效率	设计、运营管理
	雨污水综合利用	规划、设计、运营管理
节材与材料资源	节材	设计、施工、运营管理
	使用绿色建材	设计、施工、运营管理
室内环境质量	光环境	规划、设计
	热环境	设计、运营管理
	声环境	设计、运营管理
	室内空气品质	设计、运营管理

续表

项　目	分项指标	重点应用阶段
运营管理	智能化系统	规划、设计、运营管理
	资源管理	运营管理
	改造利用	设计、运营管理
	环境管理体系	运营管理

5　绿色建筑规划设计技术要点

5.1　节地与室外环境

5.1.1　建筑场地

（1）优先选用已开发且具城市改造潜力的用地；

（2）场地环境应安全可靠，远离污染源，并对自然灾害有充分的抵御能力；

（3）保护自然生态环境，充分利用原有场地上的自然生态条件，注重建筑与自然生态环境的协调；

（4）避免建筑行为造成水土流失或其他灾害。

5.1.2　节地

（1）建筑用地适度密集，适当提高公共建筑的建筑密度，住宅建筑立足创造宜居环境确定建筑密度和容积率；

（2）强调土地的集约化利用，充分利用周边的配套公共建筑设施，合理规划用地；

（3）高效利用土地，如开发利用地下空间，采用新型结构体系与高强轻质结构材料，提高建筑空间的使用率。

5.1.3　降低环境负荷

（1）建筑活动对环境的负面影响应控制在国家相关标准规定的允许范围内；

（2）减少建筑产生的废水、废气、废物的排放；

（3）利用园林绿化和建筑外部设计以减少热岛效应；

（4）减少建筑外立面和室外照明引起的光污染；

（5）采用雨水回渗措施，维持土壤水生态系统的平衡。

5.1.4　绿化

（1）优先种植乡土植物，采用少维护、耐候性强的植物，减少日常维护的费用；

（2）采用生态绿地、墙体绿化、屋顶绿化等多样化的绿化方式，应对乔木、灌木和攀缘植物进行合理配置，构成多层次的复合生态结构，达到人工配置的植物群落自然和谐，并起到遮阳、降低能耗的作用；

（3）绿地配置合理，达到局部环境内保持水土、调节气候、降低污染和隔绝噪声的目的。

5.1.5　交通

（1）充分利用公共交通网络；

（2）合理组织交通，减少人车干扰；

(3) 地面停车场采用透水地面，并结合绿化为车辆遮荫。

5.2 节能与能源利用

5.2.1 降低能耗

(1) 利用场地自然条件，合理考虑建筑朝向和楼距，充分利用自然通风和天然采光，减少使用空调和人工照明；

(2) 提高建筑围护结构的保温隔热性能，采用由高效保温材料制成的复合墙体和屋面及密封保温隔热性能好的门窗，采用有效的遮阳措施；

(3) 采用用能调控和计量系统。

5.2.2 提高用能效率

(1) 采用高效建筑供能、用能系统和设备

1) 合理选择用能设备，使设备在高效区工作；

2) 根据建筑物用能负荷动态变化，采用合理的调控措施。

(2) 优化用能系统，采用能源回收技术

1) 考虑部分空间、部分负荷下运营时的节能措施；

2) 有条件时宜采用热、电、冷联供形式，提高能源利用效率；

3) 采用能量回收系统，如采用热回收技术；

4) 针对不同能源结构，实现能源梯级利用。

5.2.3 使用可再生能源

充分利用场地的自然资源条件，开发利用可再生能源，如太阳能、水能、风能、地热能、海洋能、生物质能、潮汐能以及通过热泵等先进技术取自自然环境(如大气、地表水、污水、浅层地下水、土壤等)的能量。可再生能源的使用不应造成对环境和原生态系统的破坏以及对自然资源的污染。可再生能源的应用可参考表 2。

可再生能源的应用 **表 2**

可再生能源	利用方式
太阳能	太阳能发电
	太阳能供暖与热水
	太阳能利用(不含采光)于干燥、炊事等较高温用途热量的供给
	太阳能制冷
地热(100%回灌)	地热发电+梯级利用
	地热梯级利用技术(地热直接供暖—热泵供暖联合利用)
	地热供暖技术
风能	风能发电技术
生物质能	生物质能发电
	生物质能转换热利用
其他	地源热泵技术
	污水和废水热泵技术
	地表水水源热泵技术
	浅层地下水热泵技术(100%回灌)
	浅层地下水直接供冷技术(100%回灌)
	地道风空调

5.2.4　确定节能指标

(1) 各分项节能指标；

(2) 综合节能指标。

5.3　节水与水资源利用

5.3.1　节水规划

根据当地水资源状况，因地制宜地制定节水规划方案，如中水、雨水回用等，保证方案的经济性和可实施性。

5.3.2　提高用水效率

(1) 按高质高用、低质低用的原则，生活用水、景观用水和绿化用水等按用水水质要求分别提供、梯级处理回用；

(2) 采用节水系统、节水器具和设备，如采取有效措施，避免管网漏损，空调冷却水和游泳池用水采用循环水处理系统，卫生间采用低水量冲洗便器、感应出水龙头或缓闭冲洗阀等，提倡使用免冲厕技术等；

(3) 采用节水的景观和绿化浇灌设计，如景观用水不使用市政自来水，尽量利用河湖水、收集的雨水或再生水，绿化浇灌采用微灌、滴灌等节水措施。

5.3.3　雨污水综合利用

(1) 采用雨水、污水分流系统，有利于污水处理和雨水的回收再利用；

(2) 在水资源短缺地区，通过技术经济比较，合理采用雨水和中水回用系统；

(3) 合理规划地表与屋顶雨水径流途径，最大程度降低地表径流，采用多种渗透措施增加雨水的渗透量。

5.3.4　确定节水指标

(1) 各分项节水指标；

(2) 综合节水指标。

5.4　节材与材料资源

5.4.1　节材

(1) 采用高性能、低材耗、耐久性好的新型建筑体系；

(2) 选用可循环、可回用和可再生的建材；

(3) 采用工业化生产的成品，减少现场作业；

(4) 遵循模数协调原则，减少施工废料；

(5) 减少不可再生资源的使用。

5.4.2　使用绿色建材

(1) 选用蕴能低、高性能、高耐久性和本地建材，减少建材在全寿命周期中的能源消耗；

(2) 选用可降解、对环境污染少的建材；

(3) 使用原料消耗量少和采用废弃物生产的建材；

(4) 使用可节能的功能性建材。

5.5　室内环境质量

5.5.1　光环境

(1) 设计采光性能最佳的建筑朝向，发挥天井、庭院、中庭的采光作用，使天然光线能照亮人员经常停留的室内空间；

（2）采用自然光调控设施，如采用反光板、反光镜、集光装置等，改善室内的自然光分布；

（3）办公和居住空间，开窗能有良好的视野；

（4）室内照明尽量利用自然光，如不具备自然采光条件，可利用光导纤维引导照明，以充分利用阳光，减少白天对人工照明的依赖；

（5）照明系统采用分区控制、场景设置等技术措施，有效避免过度使用和浪费；

（6）分级设计一般照明和局部照明，满足低标准的一般照明与符合工作面照度要求的局部照明相结合；

（7）局部照明可调节，以有利使用者的健康和照明节能；

（8）采用高效、节能的光源、灯具和电器附件。

5.5.2　热环境

（1）优化建筑外围护结构的热工性能，防止因外围护结构内表面温度过高过低、透过玻璃进入室内的太阳辐射热等引起的不舒适感；

（2）设置室内温度和湿度调控系统，使室内的热舒适度能得到有效的调控，建筑物内的加湿和除湿系统能得到有效调节；

（3）根据使用要求合理设计温度可调区域的大小，满足不同个体对热舒适性的要求。

5.5.3　声环境

（1）采取动静分区的原则进行建筑的平面布置和空间划分，如办公、居住空间不与空调机房、电梯间等设备用房相邻，减少对有安静要求房间的噪声干扰；

（2）合理选用建筑围护结构构件，采取有效的隔声、减噪措施，保证室内噪声级和隔声性能符合《民用建筑隔声设计规范》(GBJ 118)的要求；

（3）综合控制机电系统和设备的运行噪声，如选用低噪声设备，在系统、设备、管道(风道)和机房采用有效的减振、减噪、消声措施，控制噪声的产生和传播。

5.5.4　室内空气品质

（1）对有自然通风要求的建筑，人员经常停留的工作和居住空间应能自然通风；可结合建筑设计提高自然通风效率，如采用可开启窗扇自然通风、利用穿堂风、竖向拔风作用通风等；

（2）合理设置风口位置，有效组织气流，采取有效措施防止串气、泛味，采用全部和局部换气相结合，避免厨房、卫生间、吸烟室等处的受污染空气循环使用；

（3）室内装饰、装修材料对空气质量的影响应符合《民用建筑室内环境污染控制规范》GB 50325 的要求；

（4）使用可改善室内空气质量的新型装饰装修材料；

（5）设集中空调的建筑，宜设置室内空气质量监测系统，维护用户的健康和舒适；

（6）采取有效措施防止结露和滋生霉菌。

6　绿色建筑施工技术要点

6.1　场地环境

6.1.1　施工场地

(1) 通过合理布置，减少施工对场地及场地周边环境的扰动和破坏；

(2) 设置专门场地堆置弃土，土方尽量原地回填利用，并采取防止土壤流失的措施；

(3) 采取保护表层土壤、稳定斜坡、植被覆盖等措施；

(4) 使用淤泥栅栏、沉淀池等措施控制沉淀物。

6.1.2 降低环境负荷

(1) 施工废弃物分类处理，且符合国家及地方法律法规的要求；

(2) 避免或减少排放污染物对土壤的污染，如：仓库、油库、化粪池、垃圾站等处应采取防漏防渗措施，防止危险品、化学品、污染物、固体废物中有害物质的泄漏；

(3) 施工结束后应恢复施工活动中被破坏的植被(一般指临时占地内)，补偿施工活动中人为破坏植被和地貌造成的土壤侵蚀等损失。

6.1.3 保护水文环境

(1) 岩土工程勘察和基础工程施工前应采取避免对地下水污染的对策；

(2) 保护场地内及周围的地下水与自然水体，减少施工活动对其水质、水量的负面影响；

(3) 优化施工降水方案，减少地下水抽取，且保证回灌水水质。

6.2 节能

6.2.1 降低能耗

(1) 通过改善能源使用结构，有效地控制施工过程中的能耗；

(2) 根据具体情况合理组织施工，积极推广节能新技术、新工艺。

6.2.2 提高用能效率

(1) 制定合理施工能耗指标，提高施工能源利用率；

(2) 确保施工设备满负荷运转，减少无用功，禁止不合格临时设施用电，以免造成损失。

6.3 节水

6.3.1 提高用水效率

(1) 采用施工节水工艺、节水设备和设施；

(2) 加强节水管理，施工用水进行定额计量。

6.4 节材与材料资源

6.4.1 节材

(1) 临时设施充分利用旧料和现场拆迁回收材料，使用装配方便、可循环利用的材料；

(2) 周转材料、循环使用材料和机具应耐用、维护与拆卸方便、易于回收和再利用；

(3) 采用工业化的成品，减少现场作业与废料；

(4) 减少建筑垃圾，充分利用废弃物。

6.4.2 使用绿色建材

(1) 施工单位应按照国家、行业或地方管理部门对绿色建材做出的法律、法规及评价方法，选择建筑材料；

(2) 就地取材，充分利用本地资源进行施工，减少运输对环境造成的影响。

7 绿色建筑的智能技术要点

7.1 智能技术

应用以智能技术为支撑的系统与产品，提高绿色建筑性能。发展节能与节水控制系统与产品、利用可再生能源的智能系统与产品、室内环境综合控制系统与产品等。可采用综合性智能采光控制、地热与协同控制、外遮阳自动控制、能源消耗与水资源消耗自动统计与管理、空调与新风综合控制、中水雨水利用综合控制等技术。

7.2 智能化系统

7.2.1 功能效益

(1) 定位正确，满足用户功能性、安全性、舒适性和高效率的需求；

(2) 采用的技术适用先进、系统可扩充性强、具有前瞻性，能满足较长时间的应用需求。

7.2.2 功能质量

(1) 智能化系统中的子系统，如：通信网络子系统、信息网络子系统、建筑设备监控子系统、火灾自动报警及消防联动子系统、安全防范子系统、综合布线子系统、智能化系统集成等的功能质量满足设计要求，且先进、可靠与实用；

(2) 住宅小区智能化系统中的子系统，如：安全防范子系统、管理与设备监控子系统、信息网络子系统、智能化系统集成等的功能质量满足设计要求，且先进、可靠与实用；

(3) 能源消耗与水资源消耗自动统计与管理体系。

8 绿色建筑运营管理技术要点

8.1 管理网络

(1) 建立运营管理的网络平台，加强对节能、节水的管理和环境质量的监视，提高物业管理水平和服务质量；

(2) 建立必要的预警机制和突发事件的应急处理系统。

8.2 资源管理

8.2.1 节能与节水管理

(1) 建立节能与节水的管理机制；

(2) 实现分户、分类计量与收费；

(3) 节能与节水的指标达到设计要求；

(4) 对绿化用水进行计量，建立并完善节水型灌溉系统。

8.2.2 耗材管理

(1) 建立建筑、设备与系统的维护制度，减少因维修带来的材料消耗；

(2) 建立物业耗材管理制度，选用绿色材料。

8.2.3 绿化管理

(1) 建立绿化管理制度；

(2) 采用无公害病虫害技术，规范杀虫剂、除草剂、化肥、农药等化学药品的使用，有效避免对土壤和地下水环境的损害。

8.2.4 垃圾管理

(1) 建筑装修及维修期间，对建筑垃圾实行容器化收集，避免或减少建筑垃圾遗撒；

(2) 建立垃圾管理制度，对垃圾流向进行有效控制，防止无序倾倒和二次污染；

(3) 生活垃圾分类收集、回收和资源化利用。

8.3 改造利用

(1) 通过经济技术分析，采用加固、改造延长建筑物的使用年限；

(2) 通过改善建筑空间布局和空间划分，满足新增的建筑功能需求；

(3) 设备、管道的设置合理、耐久性好，方便改造和更换。

8.4 环境管理体系

加强环境管理，建立 ISO 14000 环境管理体系，达到保护环境，节约资源，改善环境质量的目的。

9 推进绿色建筑技术产业化

9.1 绿色建筑技术产业化

绿色建筑技术产业化应以政府引导下的市场需求为导向，构建绿色建筑的技术保障体系、建筑结构体系、部品与构配件体系和质量控制体系；开展绿色建筑技术产业化基地示范工程；将绿色建筑的研究、开发、设计、施工、部品与构配件的生产、销售和服务等诸环节联结为一个完整的产业系统。实现绿色建筑技术的标准化、系列化、工业化、工程化与集约化。

9.2 发展绿色建筑的新技术、新产品、新材料与新工艺

9.2.1 发展适合绿色建筑的资源利用与环境保护技术，如新型结构体系、围护结构体系、室内环境污染防治与改善技术、废弃物收集处理与回用技术、计算机模拟分析、太阳能利用与建筑一体化技术、分质供水技术与成套设备、污水收集、处理与回用成套技术、节水器具与设施等。先发展量大面广、可推广应用、见效快、产业化前景好的技术项目，如太阳能利用、地源热泵、垃圾处理、污水处理、节能型空调等新技术。

9.2.2 加强信息技术应用，如规划设计中应用 GIS(地理信息系统)技术、虚拟仿真技术等工具，建立三维地表模型，对场地的自然属性及生态环境等进行量化分析，辅助规划设计；在建筑设计与施工中采用 CAD(计算机辅助设计)、CAC(计算机辅助施工)技术和基于网络的协同设计与建造等技术；建立新型的运营管理方式，实现传统物业管理模式向数字化物业管理模式的提升等。通过应用信息技术，进行精密规划、设计、精心建造和优化集成，实现与提高绿色建筑的各项指标。

9.2.3 发展新型绿色建筑材料，加强材料性能、环境等指标的检测，及时淘汰落后产品，加速新型绿色建材的推广应用。

9.3 绿色建筑评价和认定

9.3.1 绿色建筑的评价和认定应在本导则的指导下，通过开展试点和示范工程，不断总结完善，逐步建立完整系统的绿色建筑评价和认证体系，包括等级划分、评价指标、认证方法与工作流程和认证机构等。

9.3.2 绿色建筑创新奖是建设部促进绿色建筑发展的重要奖项。本导则提供了绿色建筑创新奖评奖的评定指标体系。

Published by
Ministry of Construction
Ministry of Science and Technology
October 2005

Technical Guidelines for Green Building

Editor in Chief: China Academy of Building Research
Editors: Tsinghua University
China Urban Construction Design & Research Institute
China Building Materials Academy

TECHNICAL GUIDELINES FOR GREEN BUILDING

CONTENTS

TECHNICAL GUIDELINES FOR GREEN BUILDING

Preface

Promoting green building is the concrete practice of developing energy-efficient-and-land-saving housing and public buildings. It is advocated in the report of the 16th Party Congress that

"The capability of sustainable development will be steadily enhanced. The ecological environment will be improved. The efficiency of using resources will be increased significantly. We will enhance harmony between man and nature to push the whole society onto a path to civilized development featuring the growth of production, an affluent life and a sound ecosystem."

In the development of green building, it is very necessary to establish and adhere to the scientific view of development, to take into full consideration the influence of construction activities to the ecological and residential environment from the view of the life cycle of building, and to adopt comprehensive measures to realize the sustainable development of construction industry. The Technical Guidelines for Green Building (TGGB) is established specially to lead, promote and regulate the green building development.

1 General Principles

1.1 China is at the stage of rapid economic development, so as an industry consuming a large amount of energy and resources, the construction industry must develop green building, to change the present mode that is high investment, high consumption, heavy pollution and low profit, and take the social responsibilities and obligations to achieve sustainable development.

1.2 Green building in the Guideline refers to the building, within its life cycle that can save resources (including energy, land, water and materials) at the maximum, protect environment and reduce pollution, and provide the human beings a healthy, applicable and efficient space, to achieve the harmony between nature and human beings.

1.3 To develop green building, it is necessary to advocate urban and rural harmonious development, circular economy concept and the mode of compact urban space, to tap latent potential in the efficiency of energy, land, water and materials with the involvement of the whole society, and to properly handle the dialectic relations between the efficiency of energy, land, water and materials, environmental protection and architectural functions.

1.4 To develop green building, it is necessary to stick to technological innovation, take a new road to industrialization featuring high scientific and technological content, low resources consumption, little environmental pollution.

1.5 To develop green building, it is necessary to pay attention to economical efficiency, to check and calculate benefits and costs comprehensively by taking the life cycle of building into full consideration, to guide market development and demand, to fit local economic situations, and to advocate the simple and unadorned spirit, but not the flashy and lavish.

1.6 To develop green building, it is necessary to pay attention to the regional features and show respect ethnical conventions, and to create green building with times and regional features based on its natural, economic and climatic conditions.

1.7 To develop green building, it is necessary to pay attention to and show respect to the history and culture, and enhance the protection and reuse to the existing environment and the historical and cultural context.

1.8 Green building construction shall be in conformity with the state laws, rules and regulations and the related standards and codes, to bring about harmony of economic returns and contribution to society and environmental protection.

2 Scope of Application

The Guideline is used to guide the construction work of green building (mainly referring to civil building) done by construction units, planning and design units, builders, supervisors, enterprises developing building products and the related management departments.

3 Principles for Green Building

3.1 Green building should stick to the concept of sustainable development. Rational design and thinking methods and scientific procedures are the basic guarantee for improving the environmental, social and economic efficiency of green building.

3.2 Except from the principles for traditional building (general building), green building also needs to follow these ones:

3.2.1 Regard to life cycle of building

It forms a life cycle of building that from the primary planning and design, to the following construction and operational management, till the final demolition. Paying regard to the life cycle of building means that the planning and design should fully consider and use environmental factors, the construction process should make sure the effects to environment be minimized, the operational management can provide people healthy, comfortable, low-consumptive and harmless space, the demolition can cause the least damage to environment, and the materials demolished can be recyclable.

3.2.2 Adapt to natural conditions and protect natural environment

(1) Make full use of the natural conditions around construction site, and try to preserve and properly use the present fitting landform, physiognomy, vegetation and natural water system;

(2) Make full consideration of the local climatic and ecological environment on building location, orientation, layout, form and etc;

(3) Building style and scale should be in harmony with the surrounding environment, and be in community with history, culture and landscape;

(4) Reduce negative influence to the natural environment with all efforts, such as reduce the emission of harmful gas and rejectamenta, and the damage to ecological environment.

3.2.3 Create applicable and healthy environment

(1) Green building should consider the user's moderate requirements in priority, and create beautiful and harmonious environment with great efforts;

(2) Ensure safe utilization, reduce environmental pollution and improve indoor environmental quality;

(3) Satisfy people's physical and psychological requirements, and at the same time create conditions to increase their work efficiency.

3.2.4 Enhance the efficiency and comprehensive utilization of resources, and reduce environmental loading

(1) Optimize manufacturing techniques, and adopt applicable technologies, materials and products by the excellent design and management;

(2) Utilize and optimize resource allocation reasonably, change consuming patterns, and reduce the possession and consumption of resources;

(3) Maximize the utilization of local materials and resources in accordance with local conditions;

(4) Increase resource efficiency in maximum, and actively promote comprehensive and recyclable utilization of resources;

(5) Enhance the endurance and adaptability of building, and prolong the overall length of building;

(6) Utilize renewable clear resources and energy as possible as it can be.

4 Green Building Index System

Green building index system is an integrated system for the performance of green building according to definition, which can be used to evaluate performance differences between substantial building and the defined green building, including six indicators: land efficiency and outdoor environment, energy efficiency and energy utilization, water efficiency and water resource efficiency, materials efficiency and material resources, indoor environmental quality and operational management, all of which have covered the basic elements of green building. And contained the respective subsystems of assessment index in the phase of planning, design, construction, operational management and recycle, within a life cycle of building.

Chart 1 is green building index system; form 1 is a collection of each sub-indicator and its key applicable phase.

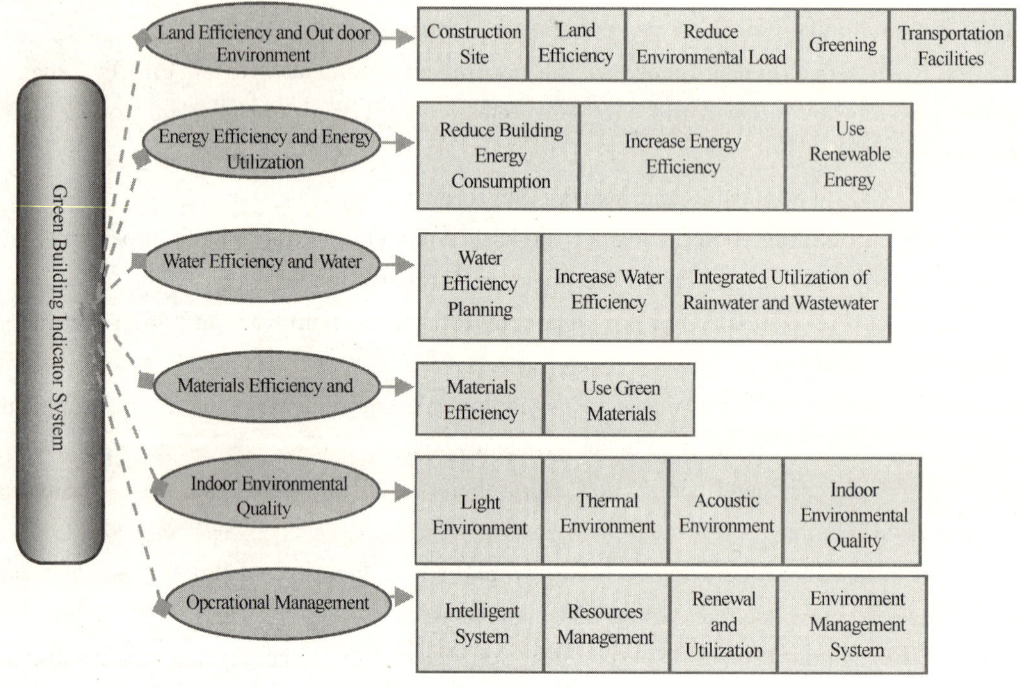

Chart 1 Green Building Index System

A Collection of Each Sub-indicator and its Key Applicable Phase **Form 1**

Item	Sub-indicator	Key Applicable Phase
Land Efficiency and Outdoor Environment	Construction Site	Planning, Construction
	Land Efficiency	Planning, Design
	Reduce Environmental Load	Life Cycle of Building
	Greening	Life Cycle of Building
	Transportation Facilities	Planning, Design, Operational Management
Energy Efficiency and Energy Utilization	Reduce Building Energy Consumption	Life Cycle of Building
	Increase Energy Efficiency	Design, Construction, Operational Management
	Use Renewable Energy	Planning, Design, Operational Management

续表

Item	Sub-indicator	Key Applicable Phase
Water Efficiency and Water Resource Utilization	Water Efficiency Planning	Planning
	Increase Water Efficiency	Design, Operational Management
	Integrated Utilization of Rainwater and Wastewater	Planning, Design, Operational Management
Materials Efficiency and Materials Resource	Materials Efficiency	Design, Construction, Operational Management
	Use Green Materials	Design, Construction, Operational Management
Indoor Environmental Quality	Light Environment	Planning, Design
	Thermal Environment	Design, Operational Management
	Acoustic Environment	Design, Operational Management
	Indoor Environmental Quality	Design, Operational Management
Operational Management	Intelligent System	Planning, Design, Operational Management
	Resources Management	Operational Management
	Renewal and Utilization	Planning, Operational Management
	Environment Management System	Operational Management

5 Technical Points for Green Building Planning and Design

5.1 Land Efficiency and Outdoor Environment

5.1.1 Construction site

(1) Choose the land developed and with urban renewal potential in priority;

(2) Site environment should be safe, reliable, away from pollution source and with enough ability to resist natural disasters;

(3) Protect the natural ecological environment, take full advantage of the original natural ecological conditions, and pay attention to the harmony between buildings and natural ecological environment;

(4) Avoid soil and water losses or other disasters caused by building.

5.1.2 Land efficiency

(1) Make construction land in moderate density. The density for public buildings should be raised, and the density and capacity for residential buildings should be determined under the condition of providing residence-preferred environment for people;

(2) Emphasize intensive land utilization, take full advantage of public building facilities around, and plan and utilize land effectively;

(3) Utilize land high-efficiently, for example, develop and utilize the underground space, adopt new-style structural system and high-strong light structural materials, to improve the building space utilization rate.

5.1.3 Reduce environmental load

(1) The negative effects to environment caused by construction activities should not exceed the permission stated by the related state standards and regulations;

(2) Reduce waste gas emission and wastes and wastewater discharge produced by construction;

(3) Use landscape greening and external design of building to reduce heat island effects;

(4) Reduce the light pollution caused by building external elevation and outdoor lighting;

(5) Adopt rainwater infiltration measures to maintain the balance of soil and water ecosystem.

5.1.4 Vegetation

(1) Grow local plants in priority, and adopt strong weatherability plants remanding less maintenance to reduce daily maintenance;

(2) Adopt multiple vegetation measures, such as ecological greenbelt, walls planting, roof planting and etc, and make good allocation to arbors, shrubs and climbers to create a multilayer composite ecological structure, to achieve the harmony of artificial allocated flora, to provide shade and to reduce energy consumption;

(3) Good greenbelt allocation to achieve the goals of soil and water conservation, climate adjustment, and pollution and noise reduction in local environment.

5.1.5 Transportation

(1) Take full advantage of public transportation network;

(2) Appropriately organize transportation to reduce the disturbance between passengers and vehicles;

(3) Ground parking lots should use infiltration pavement and provide shade for vehicles by landscape or vegetation design.

5.2 Energy Efficiency and Energy Utilization

5.2.1 Reduce energy consumption

(1) Use natural conditions of the site, give an appropriate consideration of building orientations and distance between buildings, in order to reduce the use of air conditionings and artificial lighting;

(2) Improve heat insulation and preservation performance of building envelope, adopt composed roofs and sides made by high efficient heat-preserving materials, windows and doors with excellent performance on sealing, heat insulation and preservation, and effective shading measures;

(3) Adopt energy control and measure system.

5.2.2 Improve energy efficiency

(1) Adopt high-efficient energy supply and use system and equipments:

1) Choose good energy use equipments and facilities to work in high-efficient zone;

2) Adopt corresponding control measures according to the dynamic variety of building energy load.

(2) Optimize energy use system and adopt energy recycle technologies:

1) Take into account the energy efficient measures under operation of partial space or load;

2) When applicable, use heat, electricity, and cool cogeneration to improve energy efficiency;

3) Adopt energy recycle system, such as thermal recycle technologies;

4) Adopt energy stepwise utilization for different energy structure.

5.2.3 Utilize renewable resources

Take full advantage of local natural resources and conditions, develop and utilize renewable resources, such as solar energy, water energy, wind energy, geothermal energy, ocean energy, bio-energy, tidal energy and other energy from natural environment (such as atmosphere, surface water, sewage, shallow groundwater, soil and etc.) obtained by heat pumps and other advanced technologies. And the utilization of renewable energy should not cause any damage to environment and original ecological system or pollution to natural resources. The applicable renewable energy can see form 2 for reference:

The application of renewable energy **Form 2**

Renewable Energy	Utilization Methods
Solar Energy	Electricity Generation
	Heat and Hot Water Supply
	Use solar energy (except daylighting) to support high temperature demand works such as drying and cooking
	Cooling
Geothermal Energy (100% water recirculation)	Geothermal Electricity Generation+Stepwise Utilization
	Geothermal Stepwise utilization technology (collaboration of geothermal direct heating and heat pump)
	Geothermal Heating Supply Technologies
Wind Energy	Electricity Generation Technologies
Bio-energy	Electricity Generation Technologies
	Heat Utilization of Biomass Energy Transition
Others	Ground Source Heat Pumps Technologies
	Sewage and Wastewater Source Heat Pumps Technologies
	Surface Water Heat Pumps Technologies
	Shallow Groundwater Heat Pumps Technologies (100% water recirculation)
	Shallow Groundwater Direct Cooling Technologies (100% water recirculation)
	Underground Tunnel Air Conditioning

5. 2. 4 Set a target for energy efficiency

(1) Sub-target for energy efficiency;

(2) Integrated target for energy efficiency.

5. 3 Water Efficiency and Water Resource Utilization

5. 3. 1 Water efficiency planning

Make water efficiency scheme in accordance with local conditions, such as reusing rainwater and grey water, and ensure its economical efficiency and applicability.

5. 3. 2 Increase water efficiency

(1) Based on the principle of higher quality higher utilization and lower quality lower utilization, water for resident, scenery and landscaping should be supplied respectively according to water quality required and be treated and stepwise recycled;

(2) Adopt water-efficient system, equipments and facilities, for example, avoid pipe network leakage by effective measures, adopt cycle water treatment system for cooling water for air-conditioning and swimming pool water, use toilet required low-quantity water, sensor faucet or slow-closure flushing valve in the toilet, advocate rinse-free technologies and etc;

(3) Adopt water-efficient irrigating design for landscape and vegetation, for example, use river water, collected rainwater or grey water for landscape, but not municipal water, and adopt micro-irrigation, drop irrigation and other water-saving measures for vegetation.

5. 3. 3 Integrated utilization of rainwater and wastewater

(1) Adopt separate system for rainwater and wastewater in favor of wastewater treatment and rainwater collect and reuse;

(2) In water-shortage region, adopt rainwater and grey water reuse system reasonably by comparison technical and economical aspects;

(3) Plan the surface and rooftop pathways for rainwater reasonably to reduce surface runoffs in maximum and increase rainwater infiltration by adopting multiple measures.

5. 3. 4 Set a target for water efficiency

(1) Sub-target for water efficiency;

(2) Integrated target for water efficiency.

5. 4 Material Efficiency and Materials Resource

5. 4. 1 Material efficiency

(1) Adopt high-performance, low-consumption, well-endurance and new-tech building system;

(2) Choose recyclable, reusable and renewable building materials;

(3) Adopt industrial manufactured products and reduce sitemade work;

(4) Follow the principle of module coordination and reduce construction waste;

(5) Reduce the utilization of unrenewable resources.

5. 4. 2 Use green materials

(1) Choose local materials with low embodied energy, high performance and well-

endurance, in order to reduce energy consumption of building materials within the life cycle of building;

(2) Choose biodegradable and low-pollution materials;

(3) Use materials with low raw materials consumption, and adopt materials produced by wastes;

(4) Use energy-efficient and functional materials.

5.5 Indoor Environmental Quality

5.5.1 Light Environment

(1) Building should be designed with excellent orientation to optimize interior daylight, through skylight, courtyard, or atrium to make daylight illuminating the regularly occupied areas of the building;

(2) Improve indoor daylight distribution through daylight control devices, such as daylight reflectors and daylight gathering facilities;

(3) Windows possess good views in office and residential space;

(4) Maximize daylight utilization in the indoor illumination. If daylight is not available, strategy to be considered could be optical fiber, which can introduce daylight in order to reduce the dependence on artificial illumination during day time;

(5) Use strategies such as sub area control or scene setting in lighting system to effectively avoid overspend and waste;

(6) Classify general and local lighting in design, and combine general lighting, which meets low criteria, with local lighting, which meets luminous intensity of working face;

(7) Make local lighting controllable, in favor of occupants' health and energy efficiency;

(8) Use light sources, lamps, and electronic accessories with high efficiency and low energy consumption.

5.5.2 Thermal Environment

(1) Optimize thermal efficiency of building envelope, prevent discomfort caused by over heat or over cold on the internal surface of building envelope, or interior sunlight emission through glazing;

(2) Establish interior temperature and humidity control system to effectively adjust the indoor thermal comfort and the HVAC and dehumidify system inside a building;

(3) Design the extent of temperature controllable range according to use requirement, in order to meet the thermal comfort demands of different individuals.

5.5.3 Acoustic Environment

(1) Divide noisy and quiet zone in floor plan layout and spatial partition. For instance, disjoin working and living area away from air conditioning machine room and elevator chamber to reduce noise disturbance;

(2) Reasonably choose building envelope materials, take effective sound insulation

measures to ensure indoor noise level and sound insulation performance comply with the requirements of "Code for Sound Insulation Design of Civil Buildings" (GBJ 118);

(3) Integrate running noise control of electro mechanical system and equipment, for example, select low-noise equipment, take efficient vibration reduction and noise elimination measures in system, equipment, conduit and machine room, and control noise generation and transmission.

5.5.4 Indoor Air Quality

(1) For buildings with naturally ventilated requirement, natural airflow should be provided to working and living spaces used frequently by occupants. Architecture design, such as operable window, cross ventilation, buoyancy ventilation, could be a way to improve natural ventilation efficiency;

(2) Reasonably locate air outlet, efficiently organize air flow, take measures to prevent emission of stale air, and combine complete and local air exchange to avoid recycle of polluted air in kitchen, toilet, and smoking room;

(3) Indoor Air Quality (IAQ) impact of interior decoration materials should comply with "Indoor Environmental Pollution Control Regulation of Residential Buildings" (GB 50325);

(4) Use innovative decoration materials, which have the potential to improve IAQ;

(5) For the buildings with central air conditioning, IAQ monitoring system should be established a healthy and comfortable spaces for occupants;

(6) Use effective strategies to prevent dew and mildew.

6 Main Points of Green Building Construction Techniques

6.1 Site Environment

6.1.1 Construction Site

(1) Reduce construction disturbance and damage to the site and its surrounding environment through rational disposal;

(2) Arrange special area to pile waste soil, maximize earthwork reuse by onsite backfill, and take measures to prevent soil from running off;

(3) Take measures, such as protecting surface soil, stabilizing slope, and covering vegetation;

(4) Use mud barrier and sedimentation tank to control sediment.

6.1.2 Lessen Environmental Load

(1) Classify construction wastes, and comply with national and regional laws and regulations;

(2) Avoid or decrease soil pollution of discharged pollutants, for example: anti seepage should be applied in places such as storage, oil depot, septic tank, and dump area to prevent leaking from dangerous articles, chemicals, pollutants, and harmful

substances of solid wastes;

(3) Damaged vegetation should be renovated after construction (commonly in temporally occupied land), to compensate damages, such as soil erosion arising from vegetation and landform demolishment caused by human activities during construction.

6.1.3 Hydrological Environment Protection

(1) Take measures to avoid ground water pollution before geotechnical engineering reconnaissance and understructure construction;

(2) Protect ground water and natural water bodies on-site and in the surroundings, weaken construction negative impacts to water quality and quantity;

(3) Optimize construction plan for underground-water-level droping, reduce underground water pumping and ensure the quality of recharged underground water.

6.2 Energy Efficiency

6.2.1 Reduce Energy Consumption

(1) Through improving energy utilization structure, effectively control energy consumption during the construction process;

(2) According to the specific situation, rationally organize construction, and actively promote new technologies and techniques for energy efficiency.

6.2.2 Improve the efficiency of operation energy

(1) Establish rational target for construction energy consumption, and improve energy utilization efficiency during construction;

(2) Ensure the full load running of construction equipments, reduce inefficiency, and prohibit the use of incompetent temporary electricity facilities to avoid loss.

6.3 Water Efficiency

6.3.1 Improve the efficiency of water usage

(1) Use water efficiency techniques, equipments and facilities in construction;

(2) Enhance water efficiency management, meter construction water usage with rations.

6.4 Material Efficiency and Resources

6.4.1 Material Efficiency

(1) Maximize the use of old materials and recycled materials from on-site demolition for temporary equipments, and use portable and recyclable materials;

(2) Reusable, recyclable materials and facilities should be durable, and be convenient to maintain, disassemble, recycle, and reuse;

(3) Use industrialized products to reduce on-site works and wastes;

(4) Decrease construction rubbish, and use rejectamenta materials sufficiently.

6.4.2 Use Green Construction Materials

(1) Builders should comply with green material laws, regulations, and assessment methods established by state, industry, or regional management authorities in selecting construction materials;

(2) Obtain materials from local sources to the utmost extent, to reduce transport impacts on environment.

7 Key points of Intelligent Technology for Green Building

7.1 Intelligent technology

Use the systems and equipments supported by intelligent technologies to promote the performance of green building. Develop the systems and equipments for energy-efficiency and water-efficiency, intelligent systems and equipments for using renewable energy, combination control systems and equipments for indoor environment and etc. Use technologies on integrated intelligent daylighting control, geothermal and the related control, auto control of external sun shading, auto statistic and management for energy and water consumption, integrated control of air conditioning and fresh air system, integrated control of using grey water and rain water.

7.2 Intelligent System

7.2.1 Functionality and Benefit

(1) Choose rightly, and meet the requirements of user on functionality, security, comfort and high efficiency.

(2) The technologies being uesed must be applicable, advanced, and the system be easy to expand, forward-looking, and be suitable for long-term utilization.

7.2.2 Functionality and Quality

(1) The functionality and quality of subsystems in intelligent system, like communication network system, information network system, building automation system , fire automation system, security automation system, generic cabling system, intelligent systems integration should accord with the design requirements, and be advanced, reliable and practical.

(2) The functionality and quality of the subsystems in intelligent system for dwellings, like security automation system, information network system, intelligent systems integration, should accord with the design requirement, and be advanced, reliable and practical.

(3) Auto statistic and management system for energy and water consumption.

8 Technical Points of Green Building Management

8.1 Management Networking

(1) Establish a platform for operational management networking, reinforce energy and water efficiency management and environmental quality surveillance, and improve property management level and service quality;

(2) Establish essential pre-warning mechanism and emergency treatment system.

8.2 Resource Management

8.2.1 Energy and Water Efficiency Management

(1) Establish energy and water efficiency management mechanism;

(2) Realize household based, classified metering and fee charging;

(3) Energy and water efficiency target reach design requirements;

(4) Meter planting water, establish and improve water efficiency irrigation system.

8.2.2 Consumable Material Management

(1) Establish maintenance scheme for building, equipment and system, and reduce material consumption due to servicing;

(2) Establish management scheme for property consumable materials, choose green materials.

8.2.3 Planting Management

(1) Establish planting management scheme;

(2) Use harmless and plant-diseases and insect-pests prevention techniques, standardize the usage of chemicals such as pesticide, herbicide, fertilizer and etc., to avoid negative impacts on soil and ground water environment.

8.2.4 Waste Management

(1) During building decoration and renovation, collect construction wastes in dustbin, to avoid or reduce garbage leakage;

(2) Establish waste management scheme, to efficiently control rubbish flow, and prevent unorderly dumpage and secondary pollution;

(3) Classify, recycle and reuse household wastes.

8.3 Reuse

(1) Through cost and technique analysis, reinforce and renovate building to prolong its lifecycle;

(2) Optimize building spatial layout to meet its new functional requirements;

(3) Setup of equipment and conduit should be reasonable, durable, and easy to renovate and alteration.

8.4 Environmental Management System

Reinforce environment management, establish ISO 14000 Environmental Management System, to achieve the purpose of protecting environment, saving resources, and improving environmental quality.

9 Boost Green Building Technology Industrialization

9.1 Green Building Technology Industrialization

Green building technology industrialization should be guided by the government-orientated market demand, to establish green building technology guarantee system, building structure system, products and components system and quality control system; develop demonstration projects of green building technology industrialization bases; join together green building research, development, design, construction, as well as production, distribution, and service of products and components to an integrated industrial system, so as to achieve standardization, serialization, industrialization, and intensification.

9.2 Develop New Technologies, Products, Materials, and Techniques for Green Building

9.2.1 Develop resource usage and environmental protection technologies that adapt to green buildings, such as new structure system, envelope structure system, prevention and cure technologies for indoor environmental pollution, collection, disposal and reuse technologies for wastes, computer based simulation and analysis, solar use and building incorporate technologies, water differentiated supply technology and whole set equipment, sewage collection, treatment and reuse whole set technologies, water-efficient equipments and facilities and etc.. The projects that could be extensively used, easy to be popularized and applied, with fast payback period and good prospect in industrialization should be developed in priority, which might include new technologies such as solar energy, soil heat pump, refuse treatment, sewage disposal, energy - efficient air conditioning and etc..

9.2.2 Reinforce information technology's application. For instance, in urban planning, use GIS (Geographical Information System) and simulation technologies to establish 3D model of the earth surface, and analysis natural feature and ecological environment of the site to aid planning and designing; in architecture design and construction, use technologies such as CAD (Computer Aided Design), CAC (Computer Aided Construction), and collaborative design and construction technologies based on network; establish new management mode to update traditional property management to digital management. Through the application of information technology, plan, design, and construct precisely and integrally, to implement and improve green building targets.

9.2.3 Develop new green building materials, reinforce monitoring products' capacity and environment index, eliminate behindhand products in time, and accelerate the popularization and application of new green building products.

9.3 Green Building Assessment and Validation

9.3.1 Green building assessment and validation should be oriented by this guideline. Through the development of pilot and demonstration projects, and experience accumulation, establish an integrated green building assessment and validation system, which should include rating scheme, assessment guideline, certification methods and flow sheet, and certification organizations.

9.3.2 Green Building Innovation Award is an important award initiated by the Ministry of Construction to promote green building development. This guideline provides assessment indicator system for Green Building Innovation Award.

A 绿色建筑设计理论、方法和实践

基于生物气候条件的绿色城市设计生态策略*

王建国　徐小东

（东南大学建筑学院，南京　210000）

摘　要：本文基于不同生物气候区域的地理特征，探讨了湿热地区、干热地区、冬冷夏热地区和寒冷地区的人居环境类型及其绿色城市设计生态策略和方法应用要点。

关键词：城市设计　可持续发展　生态策略

1　城市设计学术领域中绿色思想的缘起

20世纪是人类创造空前繁荣的物质文明的时期，同时也是一个滥用自然资源、对地球生态环境和人居环境产生严重破坏的时期。近20年来，人、城市建筑与环境之间的矛盾日益严峻和尖锐，并对人类的生存和发展构成严峻挑战。从1970年起，全球性环境问题、能源问题开始从自然领域逐渐扩展到政治舞台，一系列高层次的国际会议围绕这一主题而召开，并形成一批国际性的行动纲领和文件。一个时期以来，与绿色设计、生态环保等相关的各种概念和思想在国际建筑界层出不穷，未来采用何种规划设计和发展运作模式才能使得我们的城市获得持续性的健康发展成为学术界广泛关注的焦点。

客观上，20世纪是西方世界建立工业哲学的时代。很多人相信，大量工业化的生产是经济增长的基础。在这个时代里，工业技术的发展和更进一步的知识体系的建立被认为是社会进步的先决条件，城市发展和建设亦然。城市总是被占有主导地位的思想基础所组织和营造。在20世纪上半叶，现代主义那种注重“功能”和“效率”的城市规划设计理念深刻影响了世界城市的建设发展，同等化和标准化成为一个完整的体系。城市和环境建设通常也被打上了工业技术进步和革命的烙印，并被理性的规划赋予不同的特征。

1973年世界发生的第一次石油危机，又一次引发了人们对利用太阳能进行城镇建设的关注，1992年全球环境发展峰会签署的“里约宣言”，第一次以官方联盟的方式提出“可持续发展”准则。从此，人们开始真正告别以工业哲学为思想基础的现代主义时代，进入了国际交流合作、经济全球化和可持续发展的新纪元。总体而言，世纪之交的世界城市和建筑发展正处于十字路口，需要我们对前人的城市发展建设的思想和实践进行检视和反思，并在此基础上去探索和开拓新的城市发展思想和城市设计途径，寻找可持续发展的物质和技术途径。

1974年，E. R. 舒马赫的论著《小的是美好的》为自足性设计提供了完整的哲学理

---

*　国家自然科学基金资助项目：50125820，江苏省科技发展计划：JS 200202。

论，他提倡设计师应更多地关注和利用地方性适用技术和可再生能源，如风能、水能和太阳能等。

1975年，英国著名建筑师罗杰斯等人在美国加利福尼亚州发起成立了相关的城市生态组织，并在其编著的《小小地球上的城市》中系统诠释了他对21世纪世界城市发展的看法，其核心是“重建与自然平衡的城市”。

在城市设计的方法和技术层面，I. L. 麦克哈格(1969)的“设计结合自然”思想在城市规划设计的自然生态基础及其与自然环境的整合方面，为城市设计建立了一个新的基准，他的设计目的只有两个：生存和成功，也即为人们营造健康舒适的城市环境。J. O. 西蒙兹则为绿色景观规划设计提出富有实际操作意义的建议和主张。

荷夫(M. Hough)在《城市形态及其自然过程》(1984)一书中指出：必须重新检讨目前城市形态构成的基础，用生态学的视角去重新发掘我们日常生活场所的内在品质和特性是十分重要的。他认为，廉价石油的到来，使得无生物结构的规划理论坚持认为，一个良好的户外气候的创造与城市发展无关，这一看法如今无疑已受到大多数人的质疑和反对。

此后，雅涅斯基(O. Yanitsky，1987)甚至提出了完整的“生态城”设想，它包含自然—地理层、社会—功能层和文化—意识层三个层次以及基础研究、应用研究与发展、规划设计、建设实施和城市有机组织结构的形成等五个层次，表达了生态城市所追求的城市与环境、城市环境与人类意识共同进化的思想。

1996年3月，30位来自欧洲11个国家的著名建筑师共同签署了《在建筑和城市规划中应用太阳能的欧洲宪章》，为城市规划师和建筑师指明了在未来人类社会发展和建设工作中应具有的社会责任和价值标准。

1997年，笔者在中国《建筑学报》发表论文，提出“绿色城市设计”的概念，认为城市设计在经典的美学、经济、人文基础上，还要加上基于“生态优先”的设计准则。同时，我们又从价值取向角度，将城市设计发展大致划分成三个阶段：1920年以前主要基于建筑美学及视觉有序准则的第一代城市设计，20世纪20～60年代遵循技术美学和经济性准则的第二代城市设计，和70年代以来崛起的第三代城市设计，亦即笔者所概括并提出的“绿色城市设计”。同时，论文还论及宏观、中观和微观三个不同层次的绿色城市设计要点(王建国，1997)。本文拟在此基础上，进一步展开对不同生物气候条件下的绿色城市设计策略的探讨。

特定地域的生物气候条件是城市形态最重要的决定因素之一，它不仅造化了自然界本身，还是人类行为和地域文化特征的重要成因。

世界各地的气候条件错综复杂，划分因素和标准也很多。英国人斯欧克莱在《建筑环境科学手册》中根据空气温度、湿度和太阳辐射等因素，将地球上的地域大致分为四种不同的气候类型区：湿热气候区、干热气候区、温和气候区和寒冷气候区。尽管这种分法比较感性、主观和粗略，但在研究城市、建筑与气候关系时常采用的就是这种分类法。考虑到我国已有的分类习惯，本文将之改为湿热地区、干热地区、冬冷夏热地区和寒冷地区，并针对其不同特征探讨城市设计策略问题。

2 湿热地区的城市设计生态策略与方法

目前全球大约有40%的人口居住在湿热型气候区域，且比例还在增长(Landsberg,

1984)。湿热地区有着共同的气候特征，它主要包括两种气候类型：一类是夏季湿热，但有短暂寒冬的次湿热气候区(包括我国南方地区、长江流域局部地区)；另一类是典型的赤道气候类型区(沿赤道两侧的狭长区域，纬度0°～15°之间)和热带海洋气候区(南非、澳大利亚的东北部)，其主要气候特征是年平均温度较高且湿度相对稳定。高温潮湿的气候，除影响人类的舒适性之外，还促进了霉菌和真菌的增长、建筑材料的腐蚀以及各种虫害的滋生。

从城市和建筑设计来看，湿热地区的气候具有以下显著特点：首先，该地区夏天的气候“并不在于其单纯的热，而在于是高温高湿组合的热湿”，通常较难通过设计来改善。这是因为，随着温度的升高，从植物和潮湿土壤蒸发的水汽升高，会导致更高的温度和太阳辐射，致使当地居民感觉十分不适，也影响了一些被动式冷却系统的实用性。其次，该地区常受到具有强烈破坏作用的飓风和洪水的影响。这是由于信风经过辽阔的海洋之后常会聚于赤道地带，造成潮湿空气对流加剧，并导致该地区午后降雨并伴有雷暴这一有规律的现象。再次，该地区温度几乎没有季节性的变化，除受太阳直接辐射外，云层的漫射影响很大，仅靠截取直接太阳辐射的遮阳措施往往效果不佳。此外，相对于其他气候区，对适于该气候特征的建筑和城市设计所作的系统性研究还很少，再加上当地所住的多为穷人，大多数人支付不起昂贵的空调费用。因此，应通过低成本的、适宜的城市和建筑设计细节来从根本上减少热应力对人体健康和生产的影响。

从湿热地区的上述气候特征我们不难发现该地区城市设计的重点所在。简要概括即为：最大限度地提供良好的自然通风条件，提高环境的热舒适性并降低制冷所需的能源消耗；同时，最大限度地减轻热带风暴和洪水的危害。

2.1 基地选择原则

应选择那些温度较低、通风良好以及周边地形特征适于自然排水的地方，并避免将密集的住区或商业街区建造在洪水易发地段。

(1) 选择通风良好的区域

良好的通风对湿热地区居民的舒适性而言是至关重要的。决定湿热地区热应力水平的一个主要因素即为通风能力，除了积极利用自然风外，也应依靠地形地貌变化产生的局地风。在无风的夜晚，山谷的坡度可使气流向下运动产生谷地风，沿海或滨水地区则可受益于白天及夜间生成的水陆风的作用。

(2) 避免洪涝危害

湿热地区的城市建设，应选择有利的地形，避免位于江河两侧或江河出口附近的平坦地区，这些地方常是雨水汇集之处，水位较高。另外，也应避免紧邻沿海区域，这些地区易受飓风和暴雨袭击，海平面上升和大风引发的海浪也会造成严重破坏。

(3) 满足城市防洪的要求

城市洪涝灾害主要由来自远方的客水以及市区汇集的超量雨水所引发。前者需从整体区域的层面解决，别无它法；后者则可以通过采用适当的城市设计细节来减少暴雨的径流量，延迟径流，延长雨水排入排水管道的时间。为此，可采取以下具体措施：

1) 尽可能保持自然植被和具有渗透性的表层土；

2) 用具有渗透性的路面铺装(如植草砖等)替代原先密实的人行道、广场和停车场；

3) 设立屋顶蓄水池(既能蓄水，又能减轻太阳热辐射)或在市区开挖人工湖；

4）保留土地结构的自然灌溉特征，如交错的谷地等。

2.2 城市结构和建筑物密度

在湿热地区，过高的密度常会导致市区维持在较高的热应力水平。现代城市比传统城市具有更高的密度，只有采取适宜的城市设计策略才能将环境质量的恶化减到最小。

总体看，湿热地区的城市应尽量采用分散式结构，尽端开敞以有利于通风。建筑物密度应维持在相对较低的水平，鼓励不同高度的建筑交错布置，并使建筑物的长边与主导风向成一微小角度，以增强城市通风性能。在城市建设时，尽量避免采用高层板式建筑，以免阻碍风在街道和建筑间的流动；或避免与主导风向垂直布置的具有相同高度的建筑群，以免产生“风影区”。积极利用高层建筑对风环境的改变原理，鼓励建造高层塔楼，并将之相互远离，以促使高空的气流与地面附近的空气混合，将高处相对较冷的空气带到地面，增加近地风速，提高行人的舒适性。此外，高层塔楼能使更多的人居住到地面以上更高海拔的区域，其温度和湿度会随高度上升而降低，从而提高了这部分人的居住质量。

2.3 街道网络的规划设计

湿热地区的街道网络规划设计的首要目标是提高步行环境的舒适性，并为沿路的建筑提供良好的通风能力，尽量降低城市密集区的“热岛效应”。可采用尽端开放式和分散式布局，或将不同高度的建筑物相互结合，或通过街道的宽阔开口都可提高湿热气候条件下的城市通风性能。

在建筑物低密度区，风能够在建筑及其周边地区自由流动，街道对风运动所起的阻碍较小；而在高密度区，这种影响就非常重要。当街道与风向平行时，街道能获得最好的通风效果，但沿街建筑内部的通风潜力受到影响；当街道与风向垂直时，沿街建有长条形建筑的街道会妨碍整个地区的通风能力，这在湿热地区是十分不利的。因而，从湿热地区城市通风性能来看，一个良好的街道规划应使宽阔的林荫道与主导风向成大约30°的倾斜角，能使风顺利穿过街道到达市区，也有利于沿街建筑前后面的空气产生压差，增强建筑自然通风潜力。具体处理时，应进一步挖掘湿热地区传统建筑形式中那些适应气候的有效手段。例如，骑楼就是结合南方气候条件和商业经营需要而发展起来的，夏日遮阳，雨天挡雨，方便穿行，在现代城市设计时应加以考虑和利用。目前我国南方地区，如广州、福州等地常采用住宅底层架空的做法，一方面有利于防潮、通风和日照，方便居民活动；另一方面，也有利于改善局地气候，通过底层架空绿化形成室内外环境的相互渗透，丰富了住区环境，实践证明是适应南方地域气候的空间形式。

2.4 开放空间设计

湿热地区的建筑通风和遮阳都需考虑，但相比之下，通风比遮阳更为重要。在开放空间设计时，应使城市绿地系统形成良好的网络，贯穿整个建成区域，并注意与主导风向结合，成为城市的通风走廊，缓减市区热岛效应。我们在中国海口市开展总体城市设计时，注意保留了连续的有树荫遮蔽的开敞绿地，并与大海、河流以及整个城市的绿地系统相连。此举不仅为城市提供了“风廊”，有利于降低夏日炎热的温度，也为行人提供了良好的步行和骑车通道，保护了海鸟的栖息环境。

湿热地区植物与局地气候有时也存在矛盾，树木提供的阴影在任何时候都是受欢迎的，但是树木又阻碍了风和空气的流通，再加上植物的蒸发作用增加了空气湿度，这些无疑更加重了原先的不适感，尤其是在市区通风能力本来就较弱的时候。在湿热气候条件下，当通风与遮荫起冲突时，通风更为重要，这时就要避免树木对风的阻碍作用。当地最好的植被类型是草坪、花圃和高大树木的结合，避免种植高耸的灌木丛。

2.5　案例研究

20 多年来，杨经文一直在维护一种广义的职业责任。“设计每一个作品的时候，他总是思考着未来，思考着通过怎样的设计，获得一份充分考虑资源恢复和补充的经济性，而不是掠夺和耗尽环境资源的经济性”。杨经文建立起一整套“生物气候”设计理论以及对环境知识体系的完整理解，他总结了在湿热气候条件下适应气候的城市设计策略，归纳起来有以下几个方面：

(1) 城市的绿化系统要贯穿整个城市和市内的建筑，要注意主导风向，使风能进入城市内部，避免市区“热岛”的形成；

(2) 要鼓励和引导市民到室外公共空间进行户外活动，不要让他们总是处于室内空调环境中；

(3) 在建筑密集的市中心地区要减少汽车流量，以降低污染、降低热量；

(4) 在市区内均匀布置公共活动场所(总面积占 10%～20%)，这些场地应为露天的并有绿化或由架子、罩篷遮挡的半封闭空间；

(5) 兼顾平面和垂直绿化，使植物同建筑构成一体来反映一种绿色的形象；

(6) 道路交通规划应尽量减少人们对汽车的依赖，并鼓励在一定的范围内使用步行道，楼群之间不必让汽车穿越；

(7) 人行道要设计成半封闭或不封闭形式，并形成系统；

(8) 要尽量设计可渗透地面，避免雨水从地面上流失；

(9) 要保护好风景性水面用来蒸发降温。

3　干热地区的城市设计生态策略与方法

干热地区大约在赤道南北 15°～30°之间的亚热带纬度范围内，包括亚洲的中部和西部地区、中东、非洲、美洲北部和南部以及澳大利亚中部和西北部地区。我国新疆吐鲁番盆地一带，以及川西攀枝花地区、川东长江谷地、云南元江谷地以及海南岛西部的部分地区也属于这一气候区。由于西北及东南信风在经过干热地区上空时带走了大量水分，使得空气十分干燥，该地区总体气候特征主要表现为干旱、高盐碱化、大面积高温和强烈的太阳辐射，通常在中午和下午风很强，但在夜间却较弱(局部干热地区夜间也有强风)，从而引发当地下午共同的气候特征——沙尘暴，这也是导致不适和麻烦的主要因素之一。

干热地区的夏季气候条件最为严酷，而冬季通常较为舒适。干热气候给人们生活带来巨大压力，主要包括由高温和强烈太阳辐射引发的热应力、刺眼的光线、盛行的沙尘暴以及一些地区冬季的严寒。因此，该地区建筑物和街道设计应以确保城镇建筑环境夏季的热舒适性为最高目标，这是因为在通常情况下，“如能满足人在夏天的热舒适条件，也就等于满足了冬天的热舒适条件。”

3.1　基地选择和布局原则

干热地区城市设计的主要目标是如何减轻恶劣气候给人们室外活动带来的压力，尽可能提高单体建筑物的节能性能，并综合利用地形变化来获取良好的通风条件，比如：选在山的迎风坡或较高海拔位置，可比下风坡获得更好的通风条件和更适宜的气温。

针对夏天的燥热，在具体布局时，首先，使居住区与工作场所能够通过快速、便捷的交通系统连接起来，并将社会公共服务设施置于适宜的服务半径内，以减少路途上的体力消耗，节约能源。其次，通过选择方位和遮蔽方式，使城镇建筑环境免受过多的太阳辐射，并利用自然通风促进热量扩散。在傍晚和夜间，建筑通风是十分必需的，它可以加快室内降温速度，改善居住环境。第三，采取地表保护措施，加快白天所积累的辐射热的扩散。人行道应尽可能不暴露在阳光下，尤其是在白天室外活动的聚集场所，以确保通风良好、减少沙尘和减弱刺眼的光线为目的。

3.2 城市结构和建筑物密度

干热地区的城市通常呈现为高密集型、紧凑式的结构形态，这对改善室内外环境的局地气候有着积极影响。

(1) 干热地区城市建筑物密度与城市气温

对干热地区的城市而言，狭窄的街道和密集的建筑是比较适宜的形式，与宽阔的街道相比，会产生更多的阴影。对于同高度的建筑物来说，宽阔的街道与狭窄的街道相比在白天会产生更大的温度波动。Dhaka通过对班格拉德西(Bangladesh，北纬23°)研究后发现：在夏季白天，高宽比为1∶1的街道温度会比高宽比为3∶1的街道高出4℃。究其原因，主要因为在白天，城市街道上空会比狭窄的街道地面温度更高，这是因为它们比地面受到更多的太阳辐射；在夜晚，与此相反，由于只有一个狭窄开口敞向天空，街道地面会比街道上空温度下降慢些，到后半夜，街道上空的冷空气就会沉降，从而改善了地面的热环境。

街道两侧建筑墙体围合的空间形式通常称为“城市峡谷”，东西向的“峡谷”，随着其间距的减小，在东西墙上形成的阴影面积将会增大，有利于墙壁的相互遮掩从而减少了阳光的直接辐射量，并利于风从建筑物之间的间隙吹过；对南北向的“峡谷”而言，虽然有利于建筑物投射出最大的阴影，但当改变南北方向建筑物之间的距离时，在冬天，其形成的阴影会降低建筑物所吸收到的太阳光和热，效果截然不同。

同样高度的建筑，如果增加了城市建筑物密集度，也就意味着减少了建筑周围的露天空间面积，而它的减少对城市气候的影响还在很大程度上取决于建筑物的方向、色彩尤其是屋顶的色彩。此时，如能综合利用城市密度、建筑物高度和城市中建筑元件的平均反射率，将屋顶、墙壁涂成白色(干热地区的“浅色化”措施)，能明显减少城市建筑物对太阳辐射的吸收。无论是从对街道及其周边环境的理论分析，还是从实际测量中都可以得出，在干热地区设计高密集型的居住环境从总体上来说是有利于降低建筑物白天的温度的。然而，狭窄的街道可能会造成较为严重的空气污染，更多的噪声和较低的风速。

(2) 干热地区城市建筑物密度与通风潜力

干热地区通常白天风较强而晚上较弱，令人感到意外的是一般该地区白天不需通风(避让沙尘)，而夜间为确保室内比较舒适通风又成为必需，所以我们主要关注的是如何提高城市及其建筑物夜间的通风性能。

在高密度且层数接近的住区，街道很窄，建筑物间距较小，风几乎全部从屋顶掠过，建筑底部风很小，这时应结合建筑设计充分利用屋顶空间作为夜间休息的场所，其他楼层则可利用“风斗”采集从楼顶吹来的风并向下引导用以改善底部建筑的通风条件。此外，也应保证建筑物之间有合理的间距，并要求板式建筑在与主导风向垂直的方向上不能有“墙”，这样风能在建筑物之间“应付自如”而获得良好的自然通风能力。在塔楼林立的街道，高层建筑产生的局地风也可以改善地面的风环境。

（3）干热地区建筑物密度与城市开放空间

干热地区还受地方性沙尘暴和沙尘“波”的影响，裸露的空地通常是沙尘的源泉，而植被覆盖的土地有助于滤出空气中的沙尘。但雨水的缺乏和异地引水的高成本限制了城市美化露天环境的能力。这时较为合理的城市设计政策就是限制建筑物之间的距离(按规则退让)至个体居民能够绿化的尺度，这也使得干热地区的城市密度高于其他气候类型区。

3.3 街道网络的规划设计

在干热地区，对通风的关注主要是保证夜间建筑物的通风能力，其重要性是占第二位的。街道的宽度决定了街道两侧建筑物的间距，能够影响通风能力和太阳能潜力。干热地区的街道布局原则是在夏季尽量为行人提供阴凉，并减少建筑物的暴晒。在当地常见的沙化地带，与风向平行的宽阔街道会从总体上恶化城市的沙尘问题。由于干热地区的风主要从西向吹来，所以与街道方向相关的阳光暴晒问题与沙尘问题之间存在矛盾，需要通过整个城市的“抑沙”措施加以解决。例如，通过街道铺地设计来改善地表覆盖率和近地风速；或者保持场地的自然状况，种植沙漠植物以限制沙尘的形成，降低空气沙尘污染。

不同方向的街道提供的遮荫模式是不一样的，南北向街道与东西向街道相比，夏天具有更好的阴影条件。在“对角线”方向上的街道栅格从阳光暴晒角度来看，东北—西南方向和西北—东南方向更好一些，能在夏季提供更多的阴凉而在冬季获得更多的日晒（Knowles，1981）。在干热气候下，通常不考虑利用穿堂风来降温，建筑布局非常紧凑，就像突尼斯城，街道很窄，建筑相互遮蔽。哈桑·法赛为埃及 Bariz 新城所做的规划，利用狭窄的南北向街道以增大清晨和午后的阴影。

3.4 开放空间设计

干热地区的贫瘠土地上水汽蒸发率较高，从覆盖植物的土壤中蒸发的水汽可以降低气温、提高湿度。因此，城市开放空间包括私人绿地、公共绿地、公园等对该地区的气候影响最为显著。从舒适性角度考虑，在绿化区附近的户外活动要比在混凝土建筑附近更舒适。大面积的浅色屋顶与树木种植的有机结合，将会提高空气湿度，并明显降低该地区的室外温度。在炎热的气候下，人行道应尽可能地不暴露在阳光下，尤其是在白天户外活动的聚集场所。那些狭窄且具有较好遮阳设施的街道对于步行人流、室外活动以及购物环境来说都是更为舒适宜人的。

当地居民在与恶劣气候长期斗争的过程中积累了丰富的经验，一些地方性降温措施常令人拍案叫绝。在有条件的地区，水体的引入无疑具有直接而显著的作用。如西班牙阿尔罕布拉宫采用了凉快的、有遮阳的庭院和喷水池，它不只是华丽的景致，而是对最深刻、最丰饶的建筑创造力的源泉——对格拉纳达干热气候挑战的直接应对，其基本原理为：在干热气候下，尽量堵住温热空气并使之湿润，再利用水汽蒸腾降低温度。几个世纪以来，这种应用已经取得惊人成就，诸如伊朗和巴基斯坦能巧妙引风的房子，以及印度的莫卧儿城堡——在这有围墙的花园里，淙淙水渠将一座座凉爽的大理石凉亭连接起来。再如，我国新疆维吾尔族居民在各家院子里用葡萄棚遮阳，下面常有一条小溪缓缓流淌以增加空气湿度，从而将葡萄种植与改善室内外局地气候很好地结合起来；而在吐鲁番市中心地区，竟有一条长达一公里多的葡萄棚步行街，能在夏日提供阴凉，颇具地方特色。

3.5 案例研究

由柯克设计的美国凤凰城太阳绿洲是一项蕴含着生物气候设计思想和技术手段的综合性城市设计。它将建筑、生态、机械设计融于一体，其主要目的是想改善这个位于世界上最热的城市中心地下停车场顶部的外部空间，使之成为一处真正向市民开放的、人人能共享的交流场所。

方案采用一个呈对角线的、不对称的张拉膜结构帐篷覆盖在广场上，方便行人穿越，一方面，可使冬日温暖的朝阳自由洒落，而在其他时间又能遮蔽阳光；另一方面，其富有动态的形体能使热气流从顶部逸出，而雨水恰好可从底部排出。最为特殊的是，该方案还运用生态技术手段，在绿洲的东侧布置了两排冷却塔，对酷暑的沙漠气流加以过滤和弱化。10个18m高的冷却塔，在顶部配有可反转的烟囱罩以蒸发顶部用于被动制冷的水。由于相对密度大的湿冷气流下沉，在底部形成一股冷气流，从而确保夏日广场上1.7m高的接近人体水平面上的空气每隔20秒被冷却和更换一次。

4 冬冷夏热地区的城市设计生态策略与方法

冬天寒冷而夏天炎热的地区集中在30°～40°纬度之间，主要位于各大洲的东部，如中国、日本和美国等。其特点是夏季比较炎热干燥，白天的温度为30～35℃，最高可达37～39℃，甚至40℃；冬季较为寒冷，气温一般在－10～5℃之间；其相对湿度变化较大，白天为30％～40％，夜晚则达80％。这种气候区夏季需要空调，冬季需要采暖，只有春秋季可通过自然通风获得较为理想的热舒适性，总体而言，能耗较大，需要特殊的节能设计。我国长江中下游地区大量人口均位于该气候区，由于不能完全依靠空调，因而舒适的热环境主要依靠科学的城市规划和建筑设计策略。

4.1 基地选择原则

在冬冷夏热气候区，夏天常高温、高湿、多雨，而冬天则非常寒冷，常在0℃以下。更为重要的是，这个地区冬、夏两季的主导风向经常是不同的。如在我国东部地区，冬天的风主要来自北方，夏天则主要来自东南方向。因此，该地区基地选择一方面要保证冬季日照良好，夏季通风流畅，在东南方向没有大的地形起伏、遮挡；另一方面能防止夏季高温辐射，又能阻断冬季寒流侵袭，在西北方向最好有高大地形或成片防护林阻隔。

4.2 城市结构和密度的综合考虑

冬冷夏热地区，城市结构布局首先应鼓励夏季风(我国为东南风)尽可能穿越城市空间，它要求建筑适当地分散布置；而在冬天，为了最大可能地节省采暖费用，需要拥有最小暴露、紧凑布局的建筑。因而，冬冷夏热地区要求我们通过特殊的城市规划和设计细节，建造一种由各种建筑类型混合排列的“夏天暴露分散，而冬天紧凑”的城市结构模式。

在我国大部分地区，应该依靠建筑群体形态设计尽可能地使南向、东南向的夏季风得到强化，而阻挡冬季寒冷的西北风。为了达到这个目的，可合理安排不同长度和高度的建筑物，使它们尽可能地面对主导风向逐级布置，首先尽量将一些体量最小的独立住宅布置在最南边，然后依次是一些低矮的建筑类型，而在用地的北部边界则建造最高和最长的建筑。这样，整个地区就由高层板式公寓楼、多层方形公寓楼、两三层的联排住宅、双拼或独立式别墅组成，形成了迎合夏季东南风的“凹口”状态，同时能阻挡冬季北向来风。这些建筑类型的混合与那些由单一类型建筑组成的地区来比，城市居住区的总体密度更高一些，也具有相对更好的环境质量和热舒适性。

4.3 街道网络的规划设计

街道方位对城市通风有直接影响，应尽可能通过适当的布局来适应全年的风向变化。当街道与风向垂直时，应避免目前通常的沿街长条形的建筑布局，该模式对城市通风具有最大的阻碍作用，将大大减弱屋顶上方的气流和地面的风速；平行于风向或与风向大约成45°倾斜角的街道，将有利于产生无障碍的“风道”，诱导风穿越市区。此外，街道方位在一定程度上还影响沿街建筑交叉通风的潜能。当街道与风向平行时，大多数的建筑处于风力的“真空”地带；而当街道与风向成30°～60°角时，对建筑内部的自然通风较为有利。因此，通过对城市空间的总体通风条件和建筑物的自然通风进行综合考虑，比较理想的街道方位应与主导风向成30°～60°，这样能产生较好的综合通风效果。

在我国冬冷夏热地区，东西走向的街道在冬天与主导风向(北风)垂直，而在夏天与主导风向(东南风)成45°斜角，这种街道方位和布局将有利于冬天最大限度地减少北风的影响，而在夏天则能增进街道和沿街建筑的通风。同时，这种布局对于加强冬日沿街建筑的日照也是一个好的方位选择，但对于人行道上的行人来说不甚理想。因此，从冬日街道自身的环境质量来看，在防风保护和阳光照射的考虑上存在冲突。但总体而言，上述推荐的街道方位在季节更替中已经能够提供比较适宜的生活环境。

4.4 开放空间设计

冬冷夏热地区的夏日需要凉风习习，浓荫蔽日，冬天则需远离寒风，阳光普照。舒适的环境总由这样一系列矛盾的参数控制着，它要求我们在城市开放空间的设计过程中，充分考虑冬冷夏热地区城市特定的地域生态条件和气候特征，通过双极控制原则积极加以调适。例如，作为行道树的法国梧桐，夏日树叶茂密，给行人提供了舒适的阴凉世界；冬天树叶凋零，又将阳光还于行人，这是自然法则所提供的最好的生物气候策略。

冬冷夏热地区室外活动较为频繁，城市开放空间非常重要，我国在这方面做得远远不够。以南京为例，由于气候原因，在炎热的夏季，街道和一些广场、街头小游园缺乏基本的遮荫设施，一到午后酷热难耐，居民难以外出活动，再加上一些公共场所不定时限电，基本丧失了吸引力，导致市民的出行明显减少；而在寒冷的冬季，近年来，随着沿街高层建筑的不断增多，冬季寒风形成的“峡谷风”、下沉湍流给行人造成很大的不便，在一定程度上导致夏季白天和冬季夜晚南京的城市开放空间缺乏活力。

5 寒冷地区的城市设计生态策略与方法

寒冷地区是指夏天凉爽舒适，冬天(11月～第二年3月)平均温度低于0℃的地区，主要分布在高纬度区域(纬度40°以上)。世界上至少有30个国家位于地球北半部，6亿以上的人口有着生活在严寒气候的经历。从气候特征而言，它主要分布在冰岛、格陵兰岛、瑞典、俄罗斯、加拿大、美国、中国以及阿富汗、伊朗等北半球高纬度区域。由于地理位置特殊，自然条件严峻，该地区的城市一年中很长时间总与严寒、黑暗、寒风和冰雪相伴，气候条件对其发展来说无疑会产生很大影响，甚至会成为制约经济发展的瓶颈，导致一些人群因生活无法适应而远赴它乡。

寒冷地区夏季的舒适性一般仅需良好的通风就能保证室内的舒适。因而，该地区的城市设计应以保护和改善城市生态环境、减少冬季热能损耗以及降低由于室外寒冷、降雪和刮风对人体造成的不适作为一切设计的出发点。与其相适应，该地区城市规划设计有着独特的标准和要求，为此，亟需制定一些最基本的目标：通过适当的城市形态和阳光通道政策，鼓励土地混合使用，综合开发；通过减少到户外活动场所、停车场、学校和娱乐中心的距离使路径最优；通过拱廊、走廊、穿越街区的通道以及相互连接的中庭和地下走道的一体化设计来保护行人；季节性的使用公共领域空间(Pressman，1989)。此外，寒冷地区的城市冬季严寒，景观单调，令人感到沉重和压抑，此时适宜的城市色彩和夜景观设计有利于调节生活在漫长冬季里居民的视觉、心理感受。城市色彩应遵循“明快、含蓄、温暖、和谐”的原则，基本以色彩明快的暖色调为主；在夜景设计时，采用以暖色钠灯为主，可在夜间给人带来温暖、舒适的感觉。

为了克服气候因素的制约，发挥自身优势，包括前苏联、日本和加拿大等国家都制定了针对严寒地域特点的城市规划。前苏联在 1985 年制定了“北域 2005”计划，旨在提高约占国土面积一半的西伯利亚地区的城镇建设水平。世界上降雪最多的城市札幌制定了包括中远期综合规划、城市规划以及 5 年建设规划，都充分考虑到结合自身地域特点，严格控制发展规模，保持紧凑的用地布局模式，并制定冬季节能、防雪的特殊计划。加拿大许多城市也针对寒冷地区的特点在规划设计中采取充分措施，如在为圣约翰斯制定的“寒地城市设计导则”中就包括了一些适应气候的策略，如保持日照、防风、防雪处理等，并在街道、公园和开放空间、住宅和商业建筑以及停车场、绿化配置、色彩、材质和照明等方面的设计导则。

5.1 基地选择原则

对于严寒地区的城市或街区选址，那些受庇护、有阳光的地点会为居民提供更舒适的环境。因此，一方面要考虑风的来向，另一方面要充分利用地形、地貌对气候的有利“修正”。中国传统风水理论提供的“负阴抱阳”的理想模式，它遵循某种“‘全息同构’的准则，是环境内各项自然地理要素的有机协调”，有着一定的科学性，非常适用于寒冷地区。该模式周围的地形能够阻挡冬日的寒风，其凹向开口又能很好地接受太阳辐射；在中国，南向总是比北向能够得到更充足的阳光。

5.2 城市结构和密度

一般认为，寒冷地区的城市应为集中紧凑的城市形态，在确保建筑享有充分日照的前提下，合理地提高建筑密度，这样有利于减少交通需求和节约能源。这是因为，高密度意味着城市土地的密集使用，寒冷地区的城市居住区、商业区和服务部门的高密集性，可以节省步行和乘车的距离，减少交通需求和建筑取暖能耗。如果再结合一些特殊的构造方法，如在一定区域布置高度相同的建筑物，使冬天的冷风越过屋顶而不影响室外活动空间，能起到很好的防风效果。此外，与小型建筑相比，大型的高密度多用户使用的建筑减少了建筑表面积，可以减少热损失。

对冬天寒风进行有效屏蔽也是一项很重要的策略。在建筑布局和形体设计时，“一个弯曲的凸面，或一栋宽的、V 字形的、长条形的、东南朝向的建筑可遮挡北风，从而在它的南侧产生一个受庇护的区域。一系列这样的建筑能保护一个建造了较低建筑物的大片地区”。板式高层建筑可以为其院落内部的开放空间、公共设施、儿童游戏场地以及其他低层建筑抵御北向的寒风提供有效屏蔽，这对改善冬季居住环境，增加户外活动大有裨益。

5.3 街道网络的规划设计

寒冷地区街道网络规划设计的一项重要任务就是街道风的预防。街道风受走向、宽度与两旁建筑物、绿化的影响很大，也与街区所在的地理位置以及该地段常年气候条件有关。在规划设计时，应合理确定建筑物间距，科学布置树木、灌木丛、廊架等防风隔断，留出足够的风道，给风多一些“自由”空间，这样就可以大大削弱街道风的危害。以下措施将会有助于上述目标的实现：

（1）曲折型道路系统：主要道路应尽量与主导风向垂直；沿街连续的板式建筑会降低路面的风速，而弯曲的或有角度的街道比相同方位垂直的街道具有更低的风速，当狭窄的街道走向与风向平行时，这个特征尤其显著。

（2）桥式连接：结合过街交通，在宽阔的道路上用一些建筑化的人行天桥、大型横幅等横跨在街道上，可有效降低街道的整体风速。

（3）玻璃街道：在城市商业区，带有玻璃屋顶的“街道”可为行人提供防风防雪保护。加拿大卡尔加里市以“＋5m”人行天桥步行系统而闻名。市民可在离地 5m 高的封闭天桥内行走，不仅有效实行了人车分流，还在冬季为人们提供气候防护。

（4）带顶的街道：带“顶”的高速公路通常建于地下，它可连接多个城市以及市内主要网点，丝毫不受风雪影响。在挪威的奥森市，一条新建的三车道地下公路已经实现了无雪汽车交通，也减轻了城市中心区地面交通的压力，并可结合地下交通组织商业、银行、酒店、办公等不受季节影响的地下公共空间。

（5）防护墙或防护林：在寒冷地区提高室外温度是困难的，由于热量很快散失到周围环境中，尤其是有风的时候，防护墙或防护林可以用来保护建筑物及其外部区域不受冷风侵袭。研究表明，逆风的墙体可以降低市区流动的风速，密植的防风林也能起到同样的功效。

5.4 开放空间设计

寒冷地区的城市开放空间设计，需充分考虑特定地域的生态条件和气候特征，减少开放空间在冬夏利用率上的差异，增强它在冬天的活力。首先考虑的是避免将它建造在阴暗区和可能频繁产生近地高速风的地段，能获取阳光和免受寒风侵袭。其次，应积极利用绿化植物来获取舒适的外部环境。北半球的冬天一般吹北风，这时应在寒风来源的北方密植高大的常绿树木，可以防风而且不会遮挡阳光照射，同时需沿着树木种植常绿灌木林带以防止寒风从树冠往下渗透。在寒冷的冬天为公园提供防风设施尤为重要，公园休息场地、运动场地的南侧应多种草坪，北侧多植大树，这样冬天可挡风，夏日可遮阳，因而无论什么时候都会受到欢迎。

寒冷地区的城市规划设计面临两难选择，人们是躲在封闭的空间里“逃避”冬天，还是在开放空间中享受冬季户外运动的乐趣？这就需要制定长期的“冬季自觉”的城市开发计划，它要求积极拓展滑雪、冰上运动等“冬季文化”内容，发挥寒冷地区城市冰雪资源的独特魅力，深入研究在冬季利用公园、广场、街道和河流开展冰雪活动的方式，从总体上对城市冰雪景观作统一规划，为老人和小孩建设专门的设施；鼓励清除积雪的计划和设施，确保交通畅达，提高城市吸引力。以哈尔滨市为例，其城市总体设计就充分结合松花江沿江绿带以及流经市区的马家沟河生态廊道规划建设城市冰雪观光走廊，以创造四季皆宜的城市景观。

5.5 案例研究

在居住区气候防护方面，最广为人知的要数英国的寒地城市设计专家拉尔夫·厄斯金(Ralph·Erskin)所提出的“风屏蔽”设计策略，即在场地北部建造环绕的板式多层建筑，为居住院落内的开放空间、公共设施、儿童游戏场地以及其他层数较低的住宅抵御北向寒风提供有效屏蔽。由他主持设计的英国纽卡斯尔城的贝克(Byker)地段再开发项目，位于一片朝向西南的斜坡上，在此可以鸟瞰到整个城中心。他使用连续的“薄墙型”建筑环绕整个基地的北侧边界，从而可以有效阻止北海吹来的寒风，并可以成功隔绝来自铁路、公路的噪声。

厄斯金认为：住宅和城市应该像鲜花一样向着春夏的太阳开放，并背向阴影和寒冷的北风，同时对平台、花园和街道提供阳光的温暖和寒风的防护。他提出的巨构建筑形式的亚寒带城市模式，完全是从气候角度出发作出的理性判断。出于对阳光的极度关注，他将整个城市置于群山环抱之中，坐落在向阳山坡上，这样既能最大限度利用宝贵的太阳能资源，又有利于躲避严寒的侵袭。城市的步行交通系统也根据气候和季节被精心设计成彼此独立又互为补充的两套系统，以确保在寒冷季节各种活动的正常进行。

6 几点认识

生物气候条件是城市建设时首先面临的自然挑战，它直接关系到一个城市的布局形态模式和人居环境的舒适性。在极端气候环境中，它甚至在很大程度上决定了一个城市的结构形态、街道建筑布局、开放空间设计等。对特定生物气候条件的把握是绿色城市设计的重要内容，正因其特殊所以就更需要设计来反映它、表达它。因此，无论是现在还是将来，“形式追随气候”应像“形式追随功能”一样，成为城市设计的重要原则。

绿色城市设计遵循生态学的适应与补偿原理，重点关注自然条件制约与城市和建筑形式应变的内在机理，提倡因具体时空位置和生物气候条件的不同而具有不同的结构、形态和建筑特征，处理好城镇建筑环境的规划、建设与地方生物气候条件的结合，并促使传统文化特色和技术手段得到继承与发展。这对于改变目前千篇一律、“放之四海而皆准”的城市和建筑模式是很有意义的。未来的城市建设应根植于地方生物气候条件，因殊途而呈现出多样的发展，只有这样，世界才能呈现出多元、丰富多彩的特征。

Ecological Strategies for Green Urban Design Based on Biological and Climatic Conditions

Wang Jianguo　Xu Xiaodong
(School of Architecture, Southeast University, Nanjing　210000　China)

Abstract: Based on the geographic characteristics in different bio-climate regions, the paper analyses the different types of human living environment, in which, including the humid-hot, dry-hot, cold winter and warm summer and cold climate region, and expounds the green urban design ecological strategies and application principles concerned.

参考文献

[1] 王建国. 生态原则和绿色城市设计，建筑学报，1997(7)

[2] 徐小东. 基于生物气候条件的绿色城市设计生态策略研究(博士学位论文). 南京：东南大学建筑学院，2005

[3] 王建国，徐小东. 绿色城市设计与城市可持续发展. 中国(厦门)国际城市绿色环保博览会组委会编. 呼唤绿色新世纪. 厦门：厦门大学出版社，2001

[4] 王建国. 城市设计生态理念初探 [J]. 规划师，2002(4)

[5] 董卫，王建国. 可持续发展的城市与建筑，南京：东南大学出版社，1999

[6] Brenda, Robert Vale. Green Architecture—Design for a Sustainable Future [M]. London: Thames and Hudson, 1996

[7] Michael Hough. City Form and Natural Process: Towards a New Urban Vernacular [M]. VNR Company, 1984

[8] Baruch Givani. Climate Consideration in Building and Urban Design [M]. A Division of International Thomson Publishing Inc, 1998

[9] Gary, O. Robinette. Landscape Planning for Energy Conservation [M]. New York: Van Nostrand Reinhold Company Inc. 1983

上海世博会生态评价标准体系研究

吴志强　车　乐
（同济大学建筑与城市规划学院，上海　200092）

摘　要：2010 年世博会以“和谐城市”为理念提出建设“生态世博”，上海世博会生态评价标准体系研究在国内外经验研究基础上，首创性以规划控制为主线，以建设空间为载体，针对世博会自然和经济生态内容，因地制宜确立三维生态评价标准体系框架。
关键词：上海世博会　生态评价标准体系

1　研究背景

1.1　2002 年 12 月 3 日摩纳哥蒙特卡洛国际展览局（BIE）宣布中国上海获得 2010 年世博会主办权。

纵观世博会 150 余年的发展历史，对生态的观点日益关注，从 20 世纪 90 年代开始，可持续发展已成为世博会的主要内容。1992 年西班牙塞维利亚世界博览会的“发现的时代”、1998 年葡萄牙里斯本海洋博览会的“海洋——未来的财富”、2000 年德国汉诺威世界博览会的“人类、自然、技术：一个诞生中的世界”，从这几届世博会的主题中不难看出“可持续发展”概念的重要性。日本爱知世博会更进一步被定位为以人为本，让环保世博成为永久主题，实现将日本从目前的循环型社会向可持续型社会的转变，其主题即为“自然的睿智”，对整个场地进行了环境影响评价并发表了《生态宣言》。

构建和谐社会的内涵之一是人与自然的和谐相处。针对 2010 年上海世博会这一大事件，本文作者、世博会总规划师吴志强进一步提出以“和谐城市”为核心理念，构筑了“三大和谐”的基本思想，包含了“人与自然的和谐”、“人与社会的和谐”、“历史与未来的和谐”三个向度的内涵。同时，上海世博会提出了“科技世博、人文世博、生态世博”的口号，如何使园区建设与管理真正能够实现生态化的内涵，建成可持续的园区与环境，是能否真正实现 2010 年“生态世博”的关键。围绕生态世博这个目标，上海世博会举办了“绿色世博研讨会”；围绕“2010 年上海世博会与可持续发展”这一主题，第三届 2010 年上海世博会论坛在爱知世博园友爱之家开幕（2005.9.23）。一系列的政策与研讨都表明着决策者对生态、对可持续的强烈关注度。

1.2　进入 21 世纪，上海提出了建设世界级城市的战略目标。从当代世界城市的研究进展来看，上海建设世界级城市的努力需要考虑三个维度的发展[1]：一是在经济发展方面，要

---

[1] 世界银行．绿色工业．北京：中国财政经济出版社，2001。

努力建设成为在全球经济网络中具有重要影响作用的枢纽型城市，二是在社会发展方面，要努力建设成为具有高人类发展指数的适宜人居住的城市，三是在资源环境方面，要努力建设成为具有减物质化发展特征的循环经济型城市。期冀将上海建设成为具有减物质化特征的循环经济型国际大都市，就是要从资源环境方面把上海未来的经济增长与世界的经济发展趋势进行比较研究的结果。城市的经济、社会功能固然要求很高，然而，生态环境子系统的协调和支持能力在整个城市综合竞争力中的重大作用日益凸现。

世博会作为上海战略性提升的全新阶段一个最富有挑战性和上海向更高层次发展的推动力的大项目，第一次把“城市”这个字眼体现在了世博会的主题中。该大事件的运作，既有申办时的先发效应，又有筹备举办时的直接效应，更有结束后较长时段内的后续效应，对产业发展、城市规划、基础设施、生态环境、社会服务、市民素质等各个方面都会产生广泛影响。对于这样一个重大事件，最为直接的就是建立一个生态评价的标准和框架来控制和规范它，并为上海生态城市和循环经济大都市建设提供参考范本。

回顾相似城市大事件的开发运作，悉尼奥运会的绿色建设中澳大利亚绿色建筑评估体系 NABERS 功不可没，2008 北京绿色奥运建设亦进行了绿色奥运建筑评估体系研究，上海世博会建设，需要通过建立中观层面严格的、可操作的建设全过程监督管理机制，落实到选址、招标、设计、施工、调试及运营管理的每个环节。通过中观层面的规划标准控制，将宏观层面的可持续发展思想和微观层面的绿色生态技术有机结合起来，对于构建完整的城市生态建设体系具有一定的理论和现实意义。

2 上海世博会生态评价标准体系框架设计

2.1 指标框架构建经验方法

生态评价指标框架是一种方法、结构和工具，用来组织和构建生态评价的各种指标。对于开发和选择指标体系而言，框架意义重大。对于生态评价这一复杂的巨系统，任何一个框架对于组织和表达其复杂性及相互联系都是不够完备的。对于某一特定的框架，必须针对性满足用户的需要和优先权。以适用于特定的地区的专家、社会团体及决策者可以用这些指标来监测和管理特定的开发建设活动。由于本文实质研究为可持续发展体系中的环境可持续发展方面，故先将可持续发展体系的设计框架类型简单回顾。

可持续发展指标体系的设计框架包含不同的类型，这些框架相互之间都是有联系的。

2.1.1 单项指标框架

该框架是一系列不同类型的指数集成，指标体系简洁明了。如生态需求指标❶(ER)、联合国人类发展指数(HDI)❷、国内生产总值、可持续压力指数、可持续经济福利指数(WMDS)❸等。

2.1.2 压力—状态—响应 PSR 框架

该框架 1970 年由加拿大统计学者安史尼·弗雷德提出，被联合国经济合作与发展组

---

❶ 生态需求指标(ER)1971 年由美国麻省理工学院提出，旨在定量测算经济增长对于资源环境的压力。

❷ 联合国人类发展指数(HDI)1990 年由联合国开发计划署(UNDP)提出。

❸ 可持续经济福利指数(WMDS)1990 年由资深经济学家戴利(Daly)和科布(Cobb)提出。

织(OECD)采用并应用。该框架基于人类活动会对环境产生压力，同时会改变自然资源的数量和状态，社会通过环境、经济和部门政策对这些变化进行响应。压力指标可以描绘人类活动对环境(包括自然资源的数量和质量)产生的影响；状态指标可以描绘出环境质量和自然资源的数量和质量；响应指标可以描绘出社会对环境变化的响应和关注程度。

2.1.3　驱动力—压力—状态—影响—响应 DPSIR 框架

该框架是由欧洲环境组织(EEA)提出并首先使用。“驱动力”是指社会、人口统计学和经济学的发展及相应的生活方式和消费、生产形式的改变(如能源生产、交通、农业、旅游业等)；“压力”描述的是有关化学和生物机构的运营和排放，以及土地和其他资源的使用状况等信息(如道路交通产生的有毒物质排放)；“状态”描述了特定区域和特定时间内的物理、生物和化学现象的数量和质量；(如全球温度升高)；“影响”指标反映环境变化产生的影响(如农产品减产、飓风、洪水等)；“响应”指标涉及政府、制度、人群和个人为防止、减少、减轻和适应环境变化而采取的对策(如研究太阳能、制定行动计划)。

2.1.4　三维结构框架

该框架是与社会、经济和环境参数相应的一种测度和显示协同性行为的框架。该类指标与可持续发展的经济、环境和社会三维指标有直接的关系。

2.1.5　基于 21 世纪议程的框架

在这个框架中，指标的内容与 21 世纪议程中某些章节的内容直接相关，基于某地有代表性的主题或问题构架的种类或问题框架。

2.1.6　目标—指标—联系框架

在这个框架中，每个指标的内容都与国家、地区和地方的可持续发展目标相关。

2.2　世博会生态评价指标体系框架设计特殊性

2.2.1　选择目标分层框架

世博会规划设计首要目标在于满足世博会各种活动运作的需求，保证世博会“顺利、成功、精彩”。这一目标的追求具有一定的阶段性；而同时，近年的世博会越来越关注会后展区的利用与再开发，因此世博会规划设计也必须满足周边地区甚至城市整体的发展，从这个角度来看，规划设计的目标又超越了事件本身，变得更为复杂。由此可见，相比一般规划设计活动而言世博会规划设计的目标具有阶段性与复杂性的特点。对于这样一个多因素的复杂事件，本研究将指标拟定采用目标分层框架。

2.2.2　类目指标的增设

作为特定的国际性会展事件的空间承载，世博会规划设计围绕世博会活动的一系列运作要求产生的建筑环境、设施、道路等构成要素的具体形式和空间上的分布关系，包括会展活动的组织方式，构成了规划设计内容最主要的部分。世博会规划设计的特殊性还在于大部分场馆和设施只在世博会期间作为临时性使用，会后要进行重新功能置换，而保留的场馆也要充分考虑会后的再利用。同时，为了保证世博会的成功举办，规划设计中还必须增加一系列特设的运营保障体系，例如安全防恐设施、信息指示系统等。世博会规划设计内容之特殊性决定其类目指标较一般规划设计活动有较大区别。

2.2.3　整体运营监测指标

世博会规划设计是一个主题突出的规划设计工作，必须在给定的时间段里，实现满足给定活动需求的设施搭建和空间布置。功能要求上的限定与时间上的限定，决定了此类规划设计的运作实施具备十分显著的具体性和针对性特征，其在短期内，消耗材料之巨、建设活动之密集、对城市基础设施改善之大、对周边居民扰动之强，都是一般建设活动所难比拟的，有必要对其运营实施过程设立监测指标，以保证其良性生态循环。

2.2.4　后续利用指标

相对于一般的大型城市设计，世博会规划设计的后续影响更久远，并且在社会、经济、文化等各个层面发挥作用。利用世博会的契机对城市基础设施和环境的改善是世博会规划设计实践的直接影响，而世博园区的后续利用则提升了园区及周边土地价值；除了物质空间发展的影响，成功的世博会将在很大程度上提高城市的知名度，提升城市文化和城市精神，并为当地带来前所未有的自信心和荣誉感。

2.3　上海世博会生态评价指标体系框架设计

2010年上海世博会是世博会历史上首次以"城市"为主题，上海世博会的园区位于相对靠近城市中心的区段，其开发建设必然与上海城市整体建设密切关联，并以满足上海市未来整体发展要求为目标。同时，上海世博会还是世博会历史上占地规模最大、预期参观人数最多的一届综合博览会。世博园规划范围内现状各类用地包括居住、工业、历史街区等；生态组成要素包括黄浦江、白莲泾等水系，滨江湿地、城市绿化等；规划建设内容包括世博会场馆、广场绿化、配套服务设施、大型交通和市政基础设施等。

上海世博会生态评价指标体系框架设计在经验方法的基础上，根据世博会生态评价的特点和上海世博会的实际情况，创新的以城市规划实施控制的过程为主线，对世博园区的各空间要素进行自然和经济生态评价和引导，在以往一维或者二维指标体系框架基础上构建了全新的三维体系框架岛(图1)。

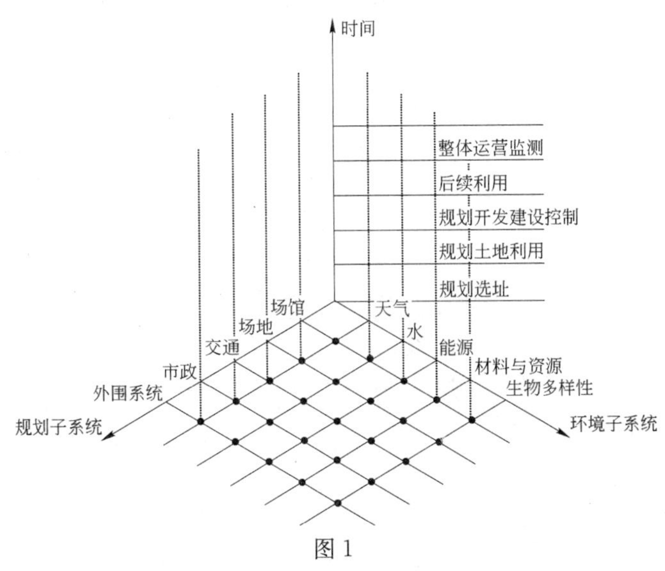

图1

首先确定上海世博会可持续发展的阶段目标，即目标层。阶段目标层按时间轴，即规划阶段划分，可分为规划选址、规划土地利用、规划开发建设控制、后续利用和整体运营监测五阶段目标。

在阶段目标层下，根据各阶段规划工作的要求，建立一个或数个较为具体的分目标，即准则(类目指标)。规划选址阶段下分5条准则，分别是上层次规划要求、建设用地、场地环境质量、现有交通和市政公用设施和外围系统。规划土地利用阶段下分 6 条准则，分别是用地生态布局、场馆用地、交通体系、市政设施体系、防灾设施体系和非建设用地。规划开发建设控制阶段下分 5 条准则，分别是场地、建筑、市政基础设施、道路与交通设施和综合防灾设施。后续利用阶段下分 4 条准则，分别为场址、建筑、交通和市政基础设施。整体运营监测阶段下分 2 条准则，即自然环境监测和人工环境监测。

准则层由具体指标组成，具体指标是环境生态和经济生态与规划主线的交集，包括水环境、空气环境、声环境、电磁环境、土地生态和经济、生物多样性和生态景观质量、建筑经济、节约资源等方面。

最后对具体指标进行指标释义。

上海世博会生态评价指标体系强调的是建立一个动态和开放的发展过程，不仅对于二级、三级指标可以进行修改、完善、增删等工作，而且对于一级指标、二级指标的权重设定也应该结合城市的发展现状和目标而应当有相应的改变和体现。根据世博会建设的周期和上海城市发展的政策等而做出相应的调整和不同侧重。

3 世博会生态评价体系指标选用原则

城市生态的评价是一个复杂的系统问题。子系统之间既有相互作用，又有相互间的输入与输出。某些因子的改变可能会影响整个系统的优劣变化，要在众多指标中选取具有前瞻性、灵敏性、内涵型的核心指标，确实不是一件易事。本研究参考国内外可持续发展指标之选用原则与方法为此设定了指标的选用原则(表 1)。

国内外可持续发展指标选用原则对照表　　表 1

指标选取原则	Liverman❶	可持续西雅图❷	联合国可持续发展委员会 UNCSD❸	英国环境、交通与区域部 DETR❹	台北可持续发展指标研究❺
系统性	■				
前瞻性	■	■		■	
完备性				■	
代表性			■	■	■
平易性		■	■	■	■
可操作性	■	■	■	■	
一致性		■			■
简明性	■	■	■	■	■
适应性	■		■	■	■
敏感性	■			■	
重要性			■		
国际相容性			■	■	■

❶ Liverman(1988)等人于《Global Sustainability》提出选择可持续发展指标是必须考虑的准则。

❷ 在《可持续的西雅图 1993》中良好指标的选取原则。

❸ 联合国可持续发展委员会(United Nation Center for Sustainable Development)采用的可持续指标建立原则。

❹ 英国环境、交通与区域部(Department of Environment, Transport and Regions)建立可持续指标准则。

❺ 黄书礼(1996)在台北市都市可持续发展指标与策略中指标选取原则。

3.1 适用性原则

针对指标体系的案例区域——上海世博会园区，这套城市生态环境指标体系会体现出其独特性。由于区域之间的差异是绝对的，相似是相对的，各城市生态建设发展的状况不同，面临问题的侧重点不同，故而这些也需要在评价区域的指标体系中体现对具体区域、具体问题的适用性。

3.2 简明性原则

选用指标的目的，具体含义，使用发放都应该体现简明易懂的原则，以方便该生态评价指标体系的实际使用和推广。

3.3 可操作性原则

由于上海世博会生态评价标准体系直接以服务上海世博会建设和运营为目的，其选用指标必须具备在指标测度和指标控制方面体现可操作性。

3.4 国际相容性原则

作为在国际相关经验基础上建立，并具有可持续发展弹性的应用性评价标准体系，其指标选用应与既有国际相关体系保持内在的关系，从而方便研究者和应用者理解和研究发展。

3.5 系统性原则

各选用指标之间应在体系构建框架思想的指导下成一系统，体现系统的整体性、层次性和开放性。

4 上海世博会生态评价标准体系初步构建

根据以上对生态评价标准体系框架方法的研究和指标选用原则的基本确立，上海世博会生态评价标准体系初步构建如表2。

表2

目　标	主　题	准　则	具体指标
合理选择规划用地，避免用地周边环境对建设项目可能产生的不良影响，并减少建设用地选址给周边环境及整个城市造成负面影响	规划选址	上层次规划要求	用地性质 人口容量 历史保护范围 区域发展计划
		建设用地	地基承载力 水资源承载力 拆迁用地比重 占用林地及生态湿地比重 对褐地开发利用比重 用地安全指数
		场地环境质量	大气质量 声质量 水质量 电磁辐射污染水平 土壤化学污染程度 生物多样性
		现有交通和市政公用设施	交通设施容量 市政设施容量
		外围系统	用地性质相容性 生态容度

续表

目　标	主　题	准　则	具 体 指 标
在总体设计阶段，预测和评价世博会项目建设的环境影响，以及环境对建设项目的影响，根据不同情况提出控制对策，使建设项目在满足功能要求的同时对环境的影响和对经济的消耗减至最小	规划土地利用	用地生态布局	土地利用的生态等级分区指标 用地平衡指标 土地利用效率指标
		场馆用地	各类场馆占地面积和人口容量
		交通体系	环境低负荷的交通组织 交通网络 多模式的交通转换系数
		市政设施体系	与城市能源供应体系的衔接 能源利用和转换效率 可再生能源与新能源的使用 能源循环利用率 水循环利用率和用水效率 分质供水 污/废/雨水处理与回用 垃圾、固体废弃物处置 环境外部性
		防灾设施体系	防灾设施布点服务范围 防灾设施体系智能化程度 防灾设施体系整合度
		非建设用地	绿容率指标 生态景观指标 生态保育能力指标
在详细设计和开发建设阶段，对绿色原则和要求进行衔接和细化。保证建筑设计施工、设施配置运营、场地开发建设贯彻可持续发展的设计理念和技术策略	规划开发建设控制	场地	侵蚀和沉积控制 减少对场地的扰动 减少热岛效应 减少光污染 空地率 生物走廊 对周边区域的安全影响
		建筑	对既有设施的利用 建筑规模、容积与面积控制 建筑全生命周期节能 建筑结构材料选择 建筑功能性与可适应性设计 室内热环境设计 建筑日照与自然采光、通风、隔声与噪声控制
		市政基础设施	负荷服务能力 相邻外部性
		道路与交通设施	公共交通设施 广场及停车场配置 无污染、轻噪声的交通工具 人流组织 交通导引标识系统 交通智能化程度
		综合防灾设施	各类防灾设施负荷度

续表

目　　标	主　　题	准　　则	具　体　指　标
在世博会后的后续利用阶段，提出符合生态要求的改造和再利用建议对策	后续利用	场址	功能符合度、场址生态恢复度
		建筑	建筑重复利用率 建筑改造利用率 建筑使用周期 临时设施 建筑功能转换时节能技术使用
		交通	交通设施和交通线路保留使用率 交通设施和交通线路转换使用率
		市政基础设施	基础设施管网及设施保留利用率 基础设施管网及设施转换利用率 基础设施管网及设施整合利用率
在整个开发运营过程中，提出运营监测的生态指标，保证整体运营过程的可持续性	整体运营监测	自然环境监测	大气环境质量 水环境质量 声环境质量 绿化环境质量 生物多样性监测指标
		人工环境监测	空间形态与自然的结合状况 用地功能布局整合程度 历史地段环境监测 建筑的物理环境质量 人工环境的防灾与安全性

5　结语

进入 21 世纪，上海提出了建设成为世界级城市的战略目标。作为上海战略性提升的全新阶段一个最富有挑战性和上海向更高层次发展的推动力的大项目，上海世博会生态评价标准体系的研究对建立“生态世博”标准的科学内核，为上海和其他城市的类似大型开发建设活动的生态建设提供经验，具有重要的理论和实践意义。并可通过中观层面的规划标准控制，将宏观层面的可持续发展思想和微观层面的绿色生态技术有机结合起来，对于构建完整的城市生态建设体系具有一定的理论和现实意义。该研究目前正在完善过程中，拟采用德尔非法对各项指标权重进一步研究确定。并对具体指标进行指标释义与评分标准确定。

Study on the eco-assessment standard system of Shanghai EXPO

Wu Zhiqiang　Che le

(College of Architecture&Urban Planning，Tong Ji University，Shanghai　200092　China)

Abstract: “Ecology EXPO” is one of the prime guidance of 2010 Shanghai EXPO, according to the idea of “harmonious city”, based on the research of sustainable development standards in China and abroad, the paper establish three dimension eco-assessment standard system in the angle of plan view.

参考文献

[1]　吴志强. 论上海世界博览会 EXPO 2010 主题. 现代城市研究，2003 [1]

[2]　吴志强. 21 世纪城市规划师宣言(草案). 城市规划 1998 [4]

[3]　吴志强. 21 世纪的城市建筑走向三大和谐. 城市规划 1999 [10]

[4]　同济大学主编. 上海规划院. 德国 AS & P 规划建筑事务所. 日本日建规划设计株式会社协编. 中国 2010 年上海世博会规划设计方案. 2004 年 11 月

[5]　《The Green City. sustainable homes，sustainable suburbs.》2004

[6]　干靓硕士论文，作为大型城市设计实践的上海世博会规划设计过程研究，2004

[7]　《绿色尺度. Green Dimensions.》[英] 克利夫 芒福汀著. 陈贞，高文艳译. 北京：中国建筑工业出版社，2004

[8]　深圳市城市规划研究院. 可持续的住区. 规划师设计师发展商的指南. 2000

[9]　英国城乡规划协会. 可持续环境的规划对策. 1993

[10]　“绿色世博”国际环保研讨会论文集. 2005

[11]　中国 2010 年上海世博会国际论坛文集. 上海：上海译文出版社，2004

[12]　智能与绿色建筑文集. 北京：中国建筑工业出版社，2005

迈向绿色建筑的不寻常之路

徐一骐
（浙江省建设厅科技与勘察设计处，杭州　310025）

摘　要：社会生态学与建筑秩序的融合，已越来越成为可持续发展的主流议题。实现生态社会，需要我们抛弃现有环境和生产体系的紧身衣的束缚；但道路依然崎岖，未来生态图景的实现依然存在着不确定因素。本文从文化批判的视角，剖析了“发展主义”语境下被认同的世界经济观对人与自然的破坏性影响。文章通过广泛和确切的信息，对“可持续性”到底意味着什么进行了必要的阐释。作者还指出了绿色建筑和生态城市建设，对于超越那些过去一直引导我们生活方式的基本立足点所具有的人文含义，对理想价值观念的重塑提出了期望。本文系《绿色建筑与文化塑造》一文的片断。

关键词：绿色建筑　发展主义批判　可持续性　城市生态　环境保护　价值观　文化塑造

在进入 21 世纪之际，我们看到了一种普遍对地球创伤的觉醒。从建筑师对“轻触这片土地”❶ 的朴素解读中，从致力于温和的绿色建筑的设计机构身上，我们可以观察到这一点。在建筑界，有一些人对建筑的生态适切性给予了诙谐表达，另外一些人则正向我们解释和叙说着建成环境可见世界的个性。

绿色建筑虽然得到了越来越多的重视，但其却又是难以用常见的技术语言来加以定义的。不同的价值观也会产生相异的观点。然而就“绿色”而言，这种观点却也会在不同的时间序列中发生变化。

迄今为止，科技被认为是将自然界中获取的原材料转化成有用产品的手段，可是在科技发展的同时对环境造成的破坏却熟视无睹的现象倒未加表达。但这种暗讽的正确性尚需要加以验证。福斯特、杨经文、卡普里茨基、爱德华兹等建筑师使我们看到，随着可持续发展成为主导，随着科学技术与社会需要逐渐联系起来，随着自然成为设计过程的新工具，以上这些问题将转化为积极的因素。例如在英国，高科技这一现代化的产物到 20 世纪 90 年代与可持续发展的矛盾已经得到缓解。20 世纪即将结束之时，我们欣喜地看到建筑设计集合了严谨的工程学、计算机技术以及生态学而称为生态科技或生态酷儿❷。建筑这一固定的重要实体变得轻质、富有弹性、可伸缩、部分可活动。倘若我们能学会区分开以可持续方式来解决问题的技术与对地球产生危害的技术，即对生态科学的本质就有了新

---

❶ 此处引用了有关澳大利亚建筑师格伦·莫屈特(Glenn Murcutt)作品的一句评语。由他早期创作的一组建筑代表作，因具有以瓦楞铁为材料的“澳洲民间建筑风格”而备受推崇。详见布莱恩·爱德华兹主编，朱玲等译．绿色建筑．沈阳：辽宁科学技术出版社，2005 年 5 月，P53。

❷ ［英］布莱恩·爱德华兹主编．绿色建筑．沈阳：辽宁科学技术出版社，2005 年 5 月，P9。

的理解。生态科学意味着人们以一种可更新的方式，在不会给自然界和环境增加负担的情况下，将废弃物提炼加工成有用的产品。

关于可持续的辩论使得可持续发展理论成为人们研究的主要议题之一，它不是单纯地适用于某一类建筑，而是丰富和完善了世界范围的建筑规划。很多著作和杂志列举了绿色建筑案例，在城市建筑领域，理查德·雷吉斯特(Richard Register)在资助四个以此为主题的国际会议中做出了杰出贡献。他的《生态城伯克利》(Eco-City Berkeley)一书，指出了一些实际有效的可能性。另外一些人，如彼得·伯格(Peter. Berg)则在他的书中提出了新的城市生活的倡议和指南。巴西南部的大城市库里提巴(Curitiba)，在进入21世纪时，恢复了它作为一个社区的整体性，成为大城市最佳现代范式之一。

在中国，关于城市的理论及规划实践虽然还处在探求之中，而且还仍然存在着误区，对生态城市建设的长期性、综合性、全面性等认识不足❶；但有统计数据表明：我国自20世纪80年代以来，已有许多城市提出建设生态城市的设想。据不完全统计，至2003年底，约有135座城市提出了建设生态型城市的目标❷。第五届生态城市国际会议在我国深圳市的召开，更表明生态城市规划建设已成为当今中国城市发展的主流之一。

在过去的10年间，我们关注全球变暖，从对相关的国际协定(《京都议定书》)的关注，转移到对建筑设计、城市状况、环境及生态健康的更广泛的关注上。这种转移是可持续发展观点的中心。就影响环境形成而言，所有的发展规划都涉及到生态可持续性设计的理念，虽然建筑师在设计场所谈论“环境设计”、“生态设计”、“可持续设计”，抑或“绿色设计”，但所有这些努力都是希望人类的自然性栖居保持在对其他生命形式的尊重、对生态环境的创造性理解之中。

正如理查德·罗杰斯1996年在他的演讲中所提到的，我们目前研究重点正从建筑设计转向城市规划，从单一的学科(如能量)转向如生态学这样更复杂的领域。人口的增长给环境造成的压力愈大，对建筑师追求生态设计思想精神实质的要求也愈高。当我们将建筑设计的想法放置在与地球的亲密关系来研究时，可持续设计的创作便立刻获得了新的审美体验；当可持续设计所包含的时间片断融入到建筑环境之中时，仿佛就得到了大地的认可，也获得了“未来”的允诺。在这方面，也许建筑师玛丽·古佐夫斯基所说的更能概括善待地球的一种共识，她说：“我们有责任通过时间去保存、维护和培养环境”❸。这是因为，在环境问题上最有希望的是“温和”地支配传统而非绝对地支配传统。我们可以铲除菜园里的草，但这仅仅是对那些多余的植物进行的清理；我们可以砍伐森林，但这种砍伐要以不破坏大地的美观、有用性为前提；我们可以建造房屋，但这种建造首先要尊重大地、生态、环境和物种多样性，而不是无限制地扩大地盘，任意地切割、划分土地和抄作房地产。用帕斯莫尔的话说，就是“托管人精神和协助自然”的理念❹。人不是地球的支配者，而只是地球的管理人，人只有“协助自然”和“帮助自然完善”的权利，也就是帮助自然完善有缺陷的可能形态，把它改造成为更易为人类理性所理解，更符合人的审美愿

❶ [美] 理查德·瑞杰斯特. 生态城市伯克利：为一个健康的未来建设城市，沈清基等译. 译者序. 北京：中国建筑工业出版社，2005年，P7。

❷ 黄兴宇. “生态城市研究回顾与展望”，21世纪城市发展国际研讨会论文集，2004年11月。

❸ [美] 玛丽·古佐夫斯基. 可持续建筑的自然光运用，汪芳等译. 北京：中国建筑工业出版社，2004年，导言xxi。

❹ 韩立新. 环境价值论——环境伦理：一场真正的道德革命. 昆明：云南人民出版社，2005年4月，P45。

望……

所有这些生态思想家、环境论者、生态城市规划师、景观设计师和建筑师的社会理想和实践，都基于这样一个事实：地球脆弱生态系统的斑斑伤痕早已暴露出来。至今我们还可以看到一种屠杀生物、毁灭这颗行星生命系统、破坏它的地质的倾向，人类的工业和商业活动已使蒙受其害的大自然和人类本身处于极不协调的对立关系之中，也造成了文明的缺憾。

跟保存、维护及培养环境的温和传统不同，绝对支配自然的传统可以追溯到工业革命早期，但其思想和文化渊源，则受到《圣经》中包含征服、统治自然的观念的影响，尽管《圣经》中同样也包含了不少生态思想，包括保护濒临灭绝物种的思想。在西方文明过程的深处，存在着其他的一些历史的和文化的原因，这些其他原因既可以在利益最大化和资本最大化的经济学逻辑的世俗化过程中找到，又可以从“发展主义”文化的历史环境中找到。这种人对自然绝对支配的传统观念是建立在人对人的支配、西方文化对非西方文化的支配、发达经济对欠发达经济支配的观念之上的。这是一种建立在注重个人财富基础之上的价值观念，而由不同传统文化根深蒂固地维系着的世界观所引发的伦理争论，都表明了所谓“宏大叙事”的“发展主义”的不平等性。

哈佛大学现任校长萨默斯(Lawrence H. summers)就曾经发表过如此荒谬的高论：“我认为在低收入国家堆置有毒废弃物，其经济上的逻辑是没有错的，我们应当面对这个事实。”❶ 经济学逻辑、事实、理性——一个理论，发展到如此炉火纯青的地步，听上去真的有些让人沮丧。西方社会的富裕时代还尽在眼中，但我们的平民式理想却怎样也无法接受这样的逻辑和事实。

巴西生态思想家和环境保护者卢岑贝格对此话有一番评论：“这样有影响力的现代经济学家，竟然提出要把环境污染的工业转移到非洲等地区，理由是地球上的这些区域还可以继续承受污染。让我们设想一下，一个肿瘤专家，一个在攻克癌症方面颇有建树的医生，会对脑瘤扩散和转移到身体其他健康的部位持赞成态度……究竟是哪一种观念使萨默斯先生站到了他曾经执教的那所著名大学的对立面呢?”❷ 倘若我们将卢岑贝格的形象比喻“脑瘤”两字，看成是现代工业文明的价值观，以及由此引发的当今世界消费主义文化盛行的“故事主线”，那是再贴切不过的了。夸富式的军备扩张，社会—经济政治系统过度的“交易成本”❸，由于追求竞争力而促动的生产能力的“创造性毁灭”，对设施和所谓“耐用消费品”过快的更新换代，都是这一故事的离奇情节，而其主人公——各种工业、商业财团和跨国公司、金融企业，它们在矛盾中演绎着自己的角色，既担负追求自己金融利益的任务，又担负推动人类事务的使命，依赖榨取地球非再生物而建立的生活方式，直到把故事情节推向更为复杂、扑朔迷离的方方面面，如由各种社会现象所引发的“异化”和弊端：盲目的消费主义、苛刻的劳动条件、克隆、疯牛病危机、迷恋毒品、自杀等等……人类历史上没有哪一种价值观像“发展主义”那样更具有冒险、扩张、改造世界的

❶ ［美］贾丁斯. 环境伦理学—环境哲学导论，林官明等译. 北京：北京大学出版社，2002年，P271～272。

❷ ［巴西］卢岑贝格. 自然不可改良：经济全球化与环境科学，黄凤祝译. 生活、读书、新知三联书店，1999年，P67。

❸ 许多概念被提出以解释此类成本。马克思谈到过资本主义生产的“虚假费用”，Georges Bataille称之为“可诅咒的部分”。

雄心了。但技术文明在给人类社会带来物质进步和繁荣的同时，也在现行的经济制度一国之内和国家之间造就了不平等，也对这颗行星的生态环境和资源的合理利用造成了莫大的威胁。

“发展”是无限的吗?❶ 有人用人的欲望是无限的来合法化“发展”的无限性。但是，他们忘记了，欲望实际上是被体制和体制下的机构(例如商业财团和广告公司)制造和再生产出来的。而“发展”之所以不可能无限制地进行下去，因为可利用资源不是无限的，我们“只有一个地球”，它在“发展主义”的驱使下，正给我们带来一个又一个“寂静的春天”❷。

审视、研究和争论是必然的。在这令人目眩的物质文明的发展背后，隐藏着多少深刻的忧虑与不安啊：人口急剧膨胀，自然资源枯竭，生态环境恶化……这些问题严重地制约着许多国家，特别是发展中国家，它迫使人们不得不追问：现代化的目标选择和道路取向究竟是什么？是继续采取以牺牲自然资源和生态环境为代价来谋求一时经济繁荣的传统发展模式？还是选择经济、社会、人口、资源与环境协调发展？

这就解释了为什么政治上的颠覆反而成就了可持续发展。因为可持续发展超越了以往任何一项环境运动：文明迫使我们重新思考两个多世纪以来的西方工业哲学❸。当下，世界各地每一个领域都在兴起这一种新意识。自从1992年“可持续发展”一词被里约热内卢召开的联合国环境与发展大会用作主题以后，这个短语几乎已经成为一个最具有意义的中心词。这一短语是如此流行，人类行为关注环境的愿望和要求是如此广泛，以至于现在更深层的问题是如何把这些要求和愿望变成现实。

可以这样说：我们需要的是一种可持续的生活方式。我们面临的困难极其巨大，我们面对的未来极其危险。我们需要慎重考虑这样的未来：当前的人口与所剩一半的资源❹。可持续发展也是目前摆在建筑师们面前的压力最大、最复杂和最具挑战性的课题。全球城市人口无时无刻不在膨胀着。这意味着在过去的10年间，问题的焦点已经从全球变暖转移到更广泛的领域，环境与生态的健康发展正受到威胁。

城市人口的增长使资源压力和废弃物排放的问题更加严峻。今天对于全球的生态学家来讲，更大的问题是资源匮乏和污染问题是否会阻碍人类作为一个物种而继续向前发展。“我们在挖掘更多的能源方面已经显得无能为力，成功地控制废物的增长是否能更有效?”建筑师布莱恩·爱德华兹如此不无怆然地诘问到。❺ 城市化过程对发展中国家资源和环境

---

❶ 雅克·鲍多特等. 与地球重新签约——哥本哈根社会发展论坛文选之一，黄平选编. 详见编者前言. 北京：人民文学出版社，2003年7月，P3～7。

❷ 在美国中部的一片整齐、繁荣的农场中坐落着一个小镇，这里到处绿树掩映，草木蓁蓁，繁花似锦，小溪汩汩流淌；池塘里鳟鱼跳跃，树枝上小鸟歌唱；春华秋实，生机无限……直到许多年后的某一天：神秘莫测的疾病袭来，引起大批牛羊家禽的死亡；农民们诉说着不幸，兽医们一筹莫展。昔日这里清晨荡漾着乌鸦、鸥鸟、鸽子的合唱消失了，田野、树林、沼泽和小路两旁的植物相继焦黄、枯萎，只有失去了生命的小溪。被生命抛弃了的地方只有寂静……这就是卡森在《寂静的春天》中所讲述的可怕的关于“明天的寓言”。诚如作者所言，上述城镇虽然是虚构的，但在美国和世界其他地方都可以容易地找到上千个这种城乡的翻版。

❸ [英] 布莱恩·爱德华兹主编. 绿色建筑. 沈阳：辽宁科学技术出版社，朱玲等译，2005年5月，P14。

❹ [美] 托马斯·贝里. 伟大的事业：人类未来之路. 生活、读书、新知三联书店，曹静译，2005年5月，P135。

❺ [英] 布莱恩·爱德华兹主编. 绿色建筑. 沈阳：辽宁科学技术出版社，朱玲等译，2005年5月，P21。

所造成的压力尤为巨大。在中国，近年来每年约有1500万农民进入城市工作、定居，2005年有1亿多农村人口外出打工，这种影响最初已在城市显现出来。尽管各大城市还在不断改进、扩建铁路、汽车客运站，一再扩建民航楼站，不断设计、延伸城市交通网络，扩大城市版图，但由疾速增长的流动人口所引起的生存空间的紧张压力和污染已更多地聚集在城市这一更小更集中的地理区域范围之内。

无可置疑地，城市建设越来越重要。2000年是人类历史上第一次城市人口超过农村人口，也就是说，在全球60亿人口中，更多的人口住在城市。在今日世界，随着城市人口快速增长，也导致了城市规划和建设规模不断扩大，到处都是忙碌、繁荣的建设工地，仿佛是一台失去了刹车制动系统的隆隆列车，无止境地奔跑着，令大地震撼着。然而，随着人类城市化水平的不断提高，我们的消费更多，空调、车辆以及各种能耗设备也更多，污染也不断地加剧。莱斯特·布朗(Lester Brown)在他的《致命迹象》(Vital Signs)一书中，以对世界状态的概览为开篇，告诉我们："今天的世界比任何时候都更热，更拥挤，更城市化，经济上更富裕，环境上更贫穷。"❶ 可以说，随着城镇的发展，健康已成为一个更重要的主题，可持续发展是人类定居地球后的又一次挑战。

然而，值得注意的是，可持续发展的思想在很大程度上被人们误解甚至有意曲解了❷，一些国家或地区现行的发展战略虽然也美其名曰可持续发展，但实际上并没有去考虑确保生态可持续性和后代人的基本生存条件为前提，依然以发展为中心、为根本、为压倒一切的原则，依然在走高成本、高消耗、非循环的老路，既造成不可再生资源的巨大浪费，又造成环境的严重污染，而在环境治理、开发替代资源等方面的投入却不容乐观。

说真的，这样一种发展因为冠上了"可持续"的美名作掩饰，是更具有危害性的，因为它能够轻易地使人们丧失忧患意识、危机意识、理性判断以及对地球上生态万物的责任感。贾丁斯说得好："可持续的说法太普遍了"，"让我们先停下来"，先问"可持续什么?""很显然，可持续经常被认为是'持续目前的生活方式和消费水平'。这样的可持续生活模式只是持续现在的状态。但是目前的消费方式，尤其是在消费驱动的工业经济中的消费方式，正是导致环境恶化的元凶，现在的消费情形正是要改变的东西。……我们要警惕，不要只是简单地把可持续发展当作时髦词来谈论经济和消费的继续增长"❸。

我们确实要警惕。城市建设理应考虑历史、地理和地方文化所提供的有价值的经验。生态意义上的建筑形式总是和文化的个性联系在一起的。事实上，现实生活中我们遇到"不可持续"的威胁，我们关于建筑的现状和价值的意义已在很大程度上陷入到工业过程所控制的标准之中，以至于置身于设计场所的建筑师很难看到建筑个性作为一种观念受到威胁的严重性。从消费观念上讲，我们的生存方式、设计观念，已越来越被纳入到商业—工业秩序的控制之下，仿佛我们非得走工业化国家的老路，汽车、高速公路、停车场、购物中心、高档商务楼、别墅，仿佛这些东西都是人类幸福不可缺少的。

西方社会消费汽车的习性已成为我们中许多人向往的东西。被美国生态城市规划师理

---

❶ Brown, Lester, et al., eds. *Vital Signs*. New York: W. W. Norton. An annual publication of the World Watch Institute since 1992, P15。

❷ 王诺. 欧美生态文学. 北京：北京大学出版社，2003年8月，P78~79。

❸ [美] 贾丁斯. 环境伦理学——环境哲学导论. 林官明等译，北京：北京大学出版社，2002年，P96。

查德·雷吉斯特讥之为瓦解社会、破坏生态、浪费能源、使人和环境受到毒害甚至谋害人命的元凶——私家车，已成为工商界拉动需求的操作引擎，“小汽车蔓延综合症”（Auto Sprawl Syndrome）这一西方国家司空见惯的社会现象也已屡屡在我们日常生活周围发生。在我国东南沿海发达地区，土地本来就很紧张，开发商在青山绿水之间兴建 $2000m^2$ 的豪宅，一掷千金，绝非传闻。这正好应验了《温和建筑》一书的作者马尔科姆·韦尔斯所说的一句话：“在前行的路上，我们中的大多数正在无情地毁坏着地球，却赢得了所有从大肆消费和无情浪费中获益的人的喝彩”❶。

在广告、影视商业文化强大攻势之下，人的消费节奏越来越快，标新立异，时尚名牌，用过东西就扔的坏脾气，在一些青少年中间滋生、蔓延。商业的消费之网，铺天盖地，无孔不入。过度消费常常表现在建筑的过度浮夸，室内装修穿金戴银、珠光宝气、矫揉造作的仿古和仿西洋建筑；故作姿态地夸张应用高科技，以及对震撼空间和舒适度的过度追求。“建筑成了流行的附属品——商业和私人生活舞台的背景幕。在家中，我们为了一些附加的小玩意而营造空间，并不考虑人的使用要求。我们的办公室里摆满了电脑，飞机场里挤满了免税店。”❷ 这种无止境的商品生产，这种填不满的消费欲求，不知是为了追求更多时尚和舒适，还是为了相互攀比，满足虚荣心，还是真的物质世界太丰富了，抑或真正是一种精神状态的麻木？

在日常生活中，上班、下班，手表和时钟以及各种时间表控制着我们的生存，但在忙忙碌碌中我们却忘记了自己是谁，忘记了应有的责任。我们受过教育，会用方程式和数学模型的方式来表达建筑结构或经济现象，以及它们的相互关系。但我们很少用同样的方式来表达对生态关系的关注和对物种多样性应有的尊重。当一只野生的白鹭飞临一座城市，不会留恋充斥了水泥森林、汽车的噪声和光污染的街道、小区。让我们问问自己，一个节约、和谐的社会、家庭和个人，究竟需要多少消费？

当年在瓦尔登湖畔，梭罗在这里为自己建造了一座小木屋，一座不透风、有木板、粉刷的房子。屋里有一个阁楼，一间盥洗室，每一边有扇大窗户，两个活板天窗，一头还有一扇大门，门对面是一座砖头砌的壁炉。整个建造过程都是由梭罗独自完成。建造这座房子需要多少开销？抛开人工，只计算买材料的费用，梭罗留下了一份详细的清单，总计仅28.125美元。这是一座没有故意装点的简单木屋和农舍，是生态保护者向往的地方。在梭罗看来，房子就是居民的外壳，但是导致这些住宅很别致的，不是它的外表，而是房子里面的居民生活。他所领略到的建筑学的美，是从内向外扩展的，是从房屋的居住者（他们是真正的建造者）的需求和人性中逐渐发展起来的；假如房子造起来，居住在里面的人没有精神，那么只可以叫“坟墓建筑学”。在湖畔筑造这么一座简朴的木屋，可以在里面自由地呼吸，这房子因为造价低廉、简朴而显得体面，让人快乐。梭罗写道：“在这件事上，我问心无愧，这对灵魂和身体是一个很大的安慰，我已经决定，永远不会无耻地去为魔鬼辩护，我要竭尽全力地赞美真理”❸。读《瓦尔登湖》，梭罗的语言就和他建造的房子

---

❶ ［美］玛丽·古佐夫斯基．可持续建筑的自然光运用．汪芳等译，北京：中国建筑工业出版社，2005年8月，前言 Pxviii。

❷ ［英］布莱恩·爱德华兹主编．绿色建筑．沈阳：辽宁科学技术出版社，朱玲、郑志宇译，2005年，P13。

❸ ［美］亨利·梭罗．瓦尔登湖．哈尔滨：哈尔滨出版社，张知遥译，2003年8月，P44～46。

一样本真朴实，平淡中蕴涵着深邃的思想，有如同朋友间亲切地交谈，聆听过后，真是“胜读十年书”！

虽然工业文明给人类带来了丰硕的物质成果，但人类又不满意和无法忍受生活中诗意的丧失。为什么人们一般都更留连或怀念居住在遍布世界各地的旧城区？因为那里保存着朴素的场所精神：古朴整洁的庭园、屋前小径、一把修复后的摇椅、抑或改建过的生态村落，带给居住者的不是物质生活的满足，而是人与建筑环境的自然联系和人文习惯、思维方式的传承。相比之下，虽然某些新建筑最终能赢得与旧建筑相等的传统地位，但现代建筑很少能获得与旧建筑同等的情感亲和力。

青石板砌筑成的洗衣台板，代代相传的古家具，破损却不失优雅沉静的陶器，烧炭用火盆和铜制茶壶内滋滋冒出的缕缕青烟，都是一段无法割舍的情结，都是人的对往昔岁月的怀念。当我们谈论旧村庄时，同时也论及它们所拥有的物品，我们谈论它们，不是这些物品的机械性质，而是用亲切的语言去保护它们的存在。

有一种现象，即人类关注其他生命形式并期望融入自然生命系统的天性，被称为亲生命性(biophilia)。大地、树林、花草、水塘、动物，都是我们每个人天生都希愿为之亲近的。我们喜爱绿色家园这个词，有时只是为了那绿的意念，想听一句幽静的语言，仿佛就会呼吸进清醒的芬芳，仿佛绿色的茎管催动意念的力，在我们灵魂中起作用。

在我国，有着幅员辽阔的大地，在广阔的农村、牧区、农场、林区，到处都可以找到亲生命性的建筑物。“走在中国的大地上，你要找一个村庄，那你首先不要刻意去寻找建筑物。……你应该先找树，哪儿有一片绿林，那绿树下就可能掩映着一个村庄”，《绿色建筑》一书的作者如此生动地描写道❶。各种乡土民居与自然环境有着很强的适应性和共融性，东北暖居、华北四合院、西南山居、黄土窑洞、新疆的土拱民居、“木朵木房”、“阿以旺”、江南水乡、干栏建筑……都真实地反映了本土的气候特征和地域文化，所有这些民居所蕴涵着的丰富、朴素的绿色思想，都给我们这样一个最重要的启示：凡是适宜的技术都是根据生态原则筛选出来的，而不是用生硬的技术来打造环境。这里面包含了传统手艺、建筑技能中对本土自然条件的知晓和娴熟，而正是这些知识，告诉我们，它比大规模商业性房地产项目内涵更丰富而更少破坏性。我们看到传统建筑所用的建材，主要有石材、木材、土砖、竹、草等，这些基本上都属于天然建材，它们在制造与使用过程中，对地球环境负荷相对最小，而其中的木、竹、草完全是生物材料，在生态系统的物质循环中，它们都是绿色生命的产生者，而且会被分解掉，还原于自然。这些材料在建筑物中的过程只不过是其生物循环的一个环节，人们充分利用了它们，却没有对环境造成危害。

笔者有一次去拜访戴复东先生，在请教建筑节能的事情时，戴先生从橱窗里搬出一件他新完成的劳作：一块普通的砖——当你细细打量它时，很难想像，这块轻盈、坚实、棱角分明而又色彩均匀的砖块，竟是采用向日葵直立的茎杆——用这种喜爱阳光的农作物纤维制成的！在新奇之余，你开始慢慢思量那些在朴素思想引导下的绿色求索所具有的意义。一块砖，包含了一位建筑师对土地亲密关系的认识，而这些认识是朴素的、沉静的。这种求索远离了人们的视线，需要专注和耐心。时下那些争相追逐为富有的委托人设计豪宅大厦的明星建筑师却……，记得路易斯·康曾经用近似神圣的语气，问一块砖，它最想

❶ 西安建筑科技大学绿色建筑研究中心编．绿色建筑．北京：中国计划出版社，1999年，P175。

成为什么，它回答："一个拱"。如果我们问一片土地它最想成为什么，它将不会回答："一座城市"。如果我们问一片洪水泛滥的平原，它将不会回答"一片住宅区"。在今天，当我们中许多人在持续地放纵自己，把地球看成是经济目标的舞台背景时，建筑的社会责任是什么呢?

路易斯·康讲的是一个寓言：是关于建筑本体的生命意象，人与自然关系失衡所造成的灾难和大地伦理对人类行为所发出的忠告和期望。我们确实需要超越理性——追寻一个在各个层面充实生活、有生态意义的建筑。

有必要驳斥这样一种观点：即认为城市建设中，出于"改善人居环境"的需要，即便牺牲城郊大片良田多少都是为了"造福于民"，然而，当人与土地的亲密关系在我们视野中消失之后，大自然再也不可能从铺满沥青的马路和建筑物周围长出野生植物和农作物了。尽管以下这段话，出现在关于美洲地理学的论文中，但巴里·洛佩斯(Barry Lopez)所说的，对我国城市规划和建设中的误区同样具有警示作用："一个社会关于它所占据土地真正维度的知识越是肤浅，土地对于开发、对于为短期利益而进行的操纵来说就越是脆弱。面对政治和商业现实，土地最终发现自己没有保卫者。它发现自己与不可缺少的具体理解失去了亲密关系"[1]。

土地的不幸遭遇是它发现自己没有保卫者，但它遭到的漠视态度和缺乏绿色意义上的理解，则是人的认识上的悲哀。土地只会为万物生灵默默地提供栖息的场所，为人提供耕地、割草、放牧、护林，但却不会诉说它与生物多样性失去亲密关系的感觉。而我们人对大地的视觉、听觉与知觉却因为"人类中心论"的自在性活动而遮蔽起来了。

这种遮蔽，使人类置生态衰败的严重状况而不顾，既显示了人类智力某些部分的麻痹或瘫痪，也显示了人类敏感性的一种阻断。[1]我们需要抛弃文化固着、发展成瘾的思维方式，需要抛弃现有环境和生产体系的紧身衣的束缚，我们需要"从美开始"，对"地球读写能力"(Earth Literacy)的培养作出承诺，在建造活动中及生活方式的选择上，我们不能不考虑基本的价值准则，不考虑未来和长远的影响。如果这些问题不能得以迅速解答的话，大好机会将不复存在。

建筑活动不能接受那种独断的假定，即自然界在那里存在，建造居住区和城市就是为了人的利益而供人类无限制地占有和使用土地。一个有成效的干预机制(政府、舆论)应该催动"紧凑型"城镇规划的研究和实施，而决不是放任商业权贵用金钱去竞买更多珍贵的土地资源和生态空间。

毕竟，我们还没有失去诚实。因为，在失去人文精神的时候，物质文明有可能代表"时代"。倘若那样的话，我们离主体异化、价值崩溃、意义失落便不远了。

作为人类的一分子，每个人都有相应的社会责任、生态责任。责任使人面向自己真正的生活成为可能。正如华兹华斯所讴歌的，责任出自良知，是"指路的明灯"，是"防范或惩罚过错的荆条"。我们应该增加对环境、社会和精神关系之间的理解。

在我国，"五四"运动以来，每一代人的文化努力中，都有着特定的基本价值观，在那时，是对封建旧礼教的抗争，对新文化的向往、呐喊，是德先生和赛先生，而现在，是

[1] [美] 托马斯·贝里. 伟大的事业——人类未来之路. 生活、读书、新知三联书店，曹静译，2005年8月，P115、P135。

社会经济、政治和文化的积极变革，是热情的介入、绿色思维的汇集、生态意识的季动，是对简朴生活发出的吁请！在教育及建筑师职业中，尤其应当把保存、维护、培养生态环境看作是现实和价值的终极意义上的向导，并对设计活动提供解释。尤其是建筑教育，应当在青年中唤醒一种对他们将生活于其中的世界的意识，懂得克服那些游离于道德感召力之外的市场商业习性，懂得个体价值取决于自然世界的精神知觉，懂得创作与社群环境、文化风景的朴素对话。当建筑设计的理想与生态平衡、公共艺术、残疾人的自由出行等结合在一起时，广场、街道、绿地、林荫道、街旁休憩区就和人的需求形成了共生的环境，可以说，建筑师对美的需求因此也有了一种归属感。

中国是个农业大国，在我们广袤的祖国大地上，农民最懂得“栽种”能源，而不仅仅是粮食作物。玉米、高粱、向日葵、收割下来的麦穗和稻秆，无一不是会生长的能源，或许绿色建筑的智慧正向这些取之不尽的能源敞开着生态的大门。

An Unusual Road to Green Architecture

Xu Yiqi

(Department of Science and Technology & Investigation Design, Zhejiang Provincial Administration of Construction, Hangzhou 310025 China)

Abstract: the integration of social ecology and construction regulation has become a mainstream subject for sustainable development. To build an eco-society, we need to get rid of the staitjacket of the current environment and production system. However, the road to the ideal future is full of twists and turns and uncertainties do exist for the realization of the prospects of the ecological society. From the perspective of cultural criticism, this paper analyzes the destructive impacts on man and nature of the global economic view in the context of "developmentalism"; the paper also provides necessary interpretations on the meaning of "sustainability" based on extensive and definite information. The author highlights the cultural significance of green architecture and ecological urban development, which may help us surpass the fundamental beliefs that have been guiding us in our life style. The author also looks forward to reshaping the value so that it can become ideal. This paper is an excerption of *Green Architecture and Cultural Shaping*.

绿色住宅技术集成与示范
——上海生态住宅示范楼

汪　维　韩继红　安　宇

（上海市建筑科学研究院有限公司，上海　200032）

摘　要：绿色建筑遵循“节约能源、节省资源、保护环境、以人为本”的基本理念，是最能体现可持续发展能力的建筑模式。本文重点介绍上海市科委重大科研攻关项目主要成果之一——“上海生态住宅示范楼”的“零(低)建筑能耗、资源高效循环利用、智能高品质居住环境”等先进技术集成体系，为大力发展节能省地型绿色住宅、推广“四节一环保”等关键技术提供示范借鉴。

关键词：绿色住宅　示范楼　技术集成

1　前言

绿色建筑是21世纪国际建筑业发展的主要趋势。基于对国内外绿色建筑发展的经验探索，其理念可以概括为“节约能源、节省资源、保护环境、以人为本”十六个字。它是指为人们提供健康、舒适、安全的居住、工作和活动的空间，同时在建筑全生命周期(材料生产、建筑规划、设计、施工、运营维护和拆除过程)中实现高效率的利用资源(能源、土地、水资源、材料等)，最低限度地影响环境的建筑物。绿色建筑也称为生态建筑。

发展绿色建筑是建筑业贯彻落实科学发展观、实现可持续发展的具体体现，是大力发展循环经济，建设资源节约、环境友好型社会的必然要求。

绿色建筑在我国的发展尚处于起步阶段。通过开展广泛的国际交流，学习和借鉴发达国家绿色建筑先进理念和技术体系，结合中国特色、地域特点和时代特征，上海在全国率先开展绿色建筑技术集成及其示范研究。2004年9月，由上海市建筑科学研究院总体负责的上海市科委重大科研攻关项目“生态建筑关键技术研究与系统集成”取得阶段性成果，汇集了上海交通大学、上海电力学院、宏润建设集团、上海植物园和上海格爱绿色环保科技有限公司以及国内外60多家企业的先进技术研究成果的“上海生态办公示范楼”（图1）在上海市建筑科学研究院莘庄科技园区落成，为我国绿色建筑关键技术的研究

图1　上海生态办公示范楼

和建筑一体化集成提供了示范和实验平台。该工程已获得 2005 年建设部授予的首届“全国绿色建筑创新奖”一等奖，研究成果已汇编在中国建筑工业出版社出版的《上海生态建筑示范工程——生态办公示范楼》一书中，不少先进适用技术已经在许多民用建筑设计中予以推广应用。2005 年 9 月，该重大科研攻关项目的又一重要成果“上海生态住宅示范楼”(图 2)也在同一场址相继落成。本文重点介绍其“零(低)建筑能耗、资源高效循环利用、智能高品质居住环境”等先进技术集成体系，为大力发展节能省地型绿色住宅、推广“四节一环保”等关键技术提供示范借鉴。

图 2　上海生态住宅示范楼

2　示范楼概况

上海生态住宅示范楼由一幢代表联排小住宅的一个单元(一户)的“零能耗”独立住宅(图 3)和一幢代表多层公寓的低能耗生态公寓组成(图 4)。独立住宅建筑面积 238m²，为二层砖混凝土结构，在这一幢建筑中全面集成了当今国内外的绿色住宅先进技术，力求达到“零建筑能耗”的目标，许多技术将能在五至十年内得到广泛推广应用。生态公寓将多层建筑一梯两户型单体两套和木结构轻质屋顶加层合为一体，总建筑面积 402m²，以技术推广为目的，在该建筑中集成了大量一两年内即可在住宅建设中推广应用的绿色住宅实用

(*a*) 示范楼实景图

(*b*) 复原后的效果图

图 3　联排小住宅之一个单元的独立住宅

技术，增加的建筑造价控制在 500 元/m^2 以内。

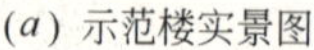

(*a*) 示范楼实景图

(*b*) 复原后的效果图

图 4　多层建筑中两套单体公寓和木结构屋顶加层

整体示范楼实现“零(低)建筑能耗、资源高效循环利用、智能高品质居住环境”等先进技术集成目标。

3　示范楼先进技术集成体系

3.1　“零(低)能耗”建筑节能技术

“零能耗”建筑是通过最优化的建筑节能设计，采用先进的节能材料和节能技术使建筑物的建筑能耗降低到尽可能低的水平，再采用可再生能源补充剩余所需能量，以实现建筑物采暖空调不使用常规能源即“零能耗”的目标。显然，“零能耗”是降低建筑能耗的理想状态。

建筑节能的技术措施不仅包括提高围护结构的保温隔热效率，还包括降低系统设备的运行能耗、提高设备运行效率和充分利用取之不尽的可再生能源。降低建筑能耗是以上三方面的综合效果。

为实现“零(低)能耗”建筑的目标，独立住宅和生态公寓建设中主要集成了以下节能技术措施。

3.1.1　超低能耗围护结构

“零能耗”独立住宅围护结构：采用高效外墙外保温系统，结构形式为非水泥基 EPS100 外墙外保温体系与砂加气 200 填充墙，平均传热系数为 0.32W/(m^2·K)(图 5)；采用高效的节能门窗系统，外窗采用真空低辐射中空塑钢窗，玻璃传热系数为 1.2W/(m^2·K)，外窗平均传热系数为 1.5W/(m^2·K)，遮阳系数 0.72；天窗采用夹胶钢化低辐射中空塑钢窗，玻璃传热系数 1.8W/(m^2·K)，

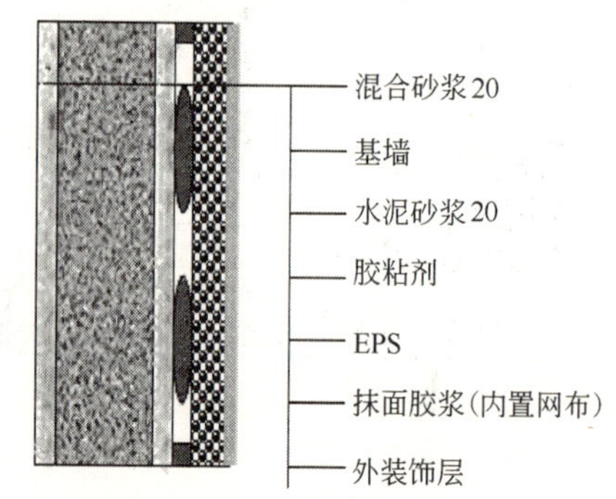

图 5　高效外墙外保温系统

外窗平均传热系数为 2.5 W/(m^2·K)，遮阳系数 0.69；采用倒置式保温与种植屋面相结合的屋面保温体系，屋面采用 XPS100 保温，种植屋面传热系数设计为 0.24W/(m^2·K)，坡屋面传热系数设计为 0.31W/(m^2·K)。

根据建筑设计风格和日照规律，采用多种高效智能遮阳系统，包括户外可调铝合金百叶帘(图 6)、可伸缩外遮阳篷(图 7)、户外天窗遮阳帘(图 8)、户外卷闸百叶帘(图 9)、户内百叶帘等内外遮阳系列产品，其中南窗、东窗和天窗采用外遮阳方式，北窗采用内遮阳方式；应用并展示了日光增强型百叶帘、太阳能驱动卷闸帘、太阳能驱动风光感应及无线控制器、无线遥控及编程控制器、户外 24V 安全性遮阳帘等一些具世界领先水平的遮阳技术产品，通过固定开关、无线遥控发射器、风光感应控制器共同实现对全部遮阳帘的控制，提高其工作效率和安全性。使外窗的综合遮阳系数达到 0.4，天窗遮阳系数为 0.2。

图 6　户外可调铝合金百叶帘

图 7　可伸缩外遮阳篷

图 8　户外天窗遮阳帘

图 9　户外卷闸百叶帘

“低能耗公寓”围护结构：外墙采用混凝土空心砌块与 XPS 外保温体系，平均传热系数为 0.81W/(m^2·K)，坡屋面采用木龙骨与 OSD 板保温体系，平均传热系数为 0.16W/(m^2·K)，窗采用中空 Low-e 塑钢窗，传热系数 1.8W/(m^2·K)。采用南向铝合金遮阳百叶。

3.1.2　地源热泵空调系统

在独立住宅中应用了地源热泵空调系统，比常规空调系统节能 20%～40%，具有较高的室内热舒适性，无吹风感和噪声。组成共计三部分：冷热源部分为土壤热泵机组加地下埋管换热器系统，末端系统为毛细管辐射，加独立除湿新风系统。土壤热泵系统的地下换热器采用垂直埋管形式(图 10)，通过地下埋管管内的介质循环与土壤进行闭式热交换达到供冷供热目的。辐射末端均选用直接采用特制砂浆粘贴在顶棚上的毛细管席 KS15 系列来供冷及供暖(图 11)。

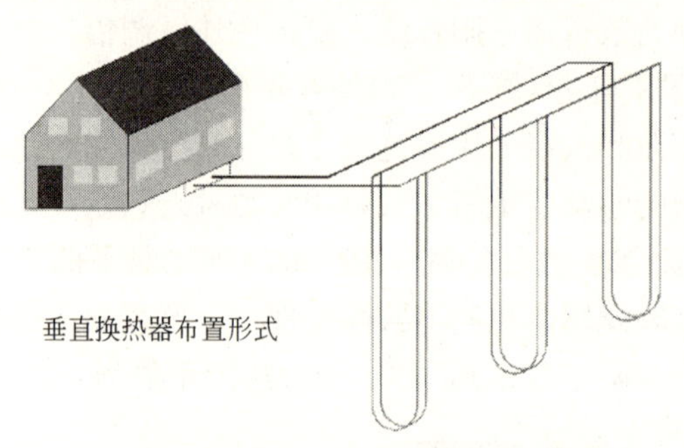

图 10 地源热泵地下换热器示意图

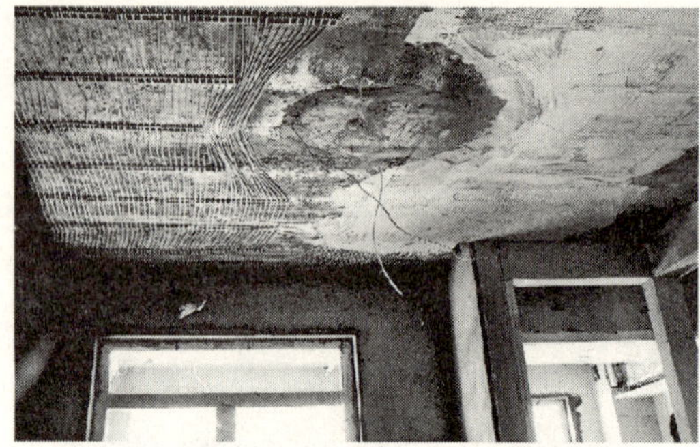

图 11 毛细管辐射末端

3.1.3 太阳能光伏发电

独立住宅采用 3kW 太阳能光伏发电和并网技术，光伏电池每块 200W，是目前世界上最大的单块光伏电池，并与屋面结构浑然一体(图 12)。园区采用太阳能庭院灯、草坪灯和风光互补路灯。在理想光照强度下，充电 4 小时即可保证景观灯 3～5 天的正常工作。据计算一台太阳能路灯一般每年可节约电费 1000 元。

图 12 3kW 太阳能 BIPV 系统和太阳能庭院灯

3.1.4 太阳能集热器与建筑一体化

太阳能利用与建筑一体化已成为太阳能大量推广应用的技术关键。生态住宅示范楼注重实现了太阳能集热器与建筑一体化设计：生态公寓的二、三层阳台分别安装 2.7m^2 和 4.2m^2 的阳台护栏悬挂式太阳能热水系统(图 13)；独立住宅上方安装了 4.6m^2 的遮阳屋檐悬挂型太阳能建筑一体化热水系统(图 14)。该新颖设计不仅为建筑增添特有的美学建筑风格，而且为多层住宅建筑中规模化利用太阳能提供了工程示范和应用可行性。

3.1.5 风力发电系统

独立住宅采用了一套性能优异的涡轮式小型风力发电机(图 15)。该系统寿命长达几十年，额定功率 140W，启动风速仅 2m/s，额定风速 15m/s，扫掠面积仅 0.3m^2，比常规风力发电机可多发 50%的电力，与建筑实现一体化设计，具有高效、美观、长寿和无噪声等优点。

图 13　阳台护栏悬挂式太阳能集热器

图 14　遮阳屋檐悬挂型太阳能集热器

3.1.6　空气源热泵热水系统

独立住宅还应用了空气源热泵热水器，将空气能与热泵节能技术有机地结合起来，采用逆卡诺循环原理，以极少的电能通过热泵工质把空气中的低温热能吸收起来，其用电量仅仅是电热水器的四分之一，燃气热水器的三分之一，是继燃气热水器、电热水器、太阳能热水器之后的第四种热水器。环境温度为 5℃时，系统能效比 COP 达到 2.93，在环境温度为 25℃时，系统能效比 COP 达到 4.52。

空气源热泵热水器还可充分利用低谷时低价电，节约开支。

3.1.7　相变储能材料

生态公寓中采用纳米石墨相变储能材料制成的蓄能罐安放在吊顶层，用作空调相变储能装置(图 16)。夏季在电力低谷时段开启空调器制冷功能，冷量便直接传入相变蓄能罐中蓄冷，待相变材料相变完全后，空调器停止运转，在电力需求高峰时段，再需要制冷时，仅需启动风机，利用空气循环换热，将蓄能罐中的冷量逐步释放到室内空间；冬季相变材料可以发挥蓄热功能；从而实现电力调峰和节省电费支出的目的。

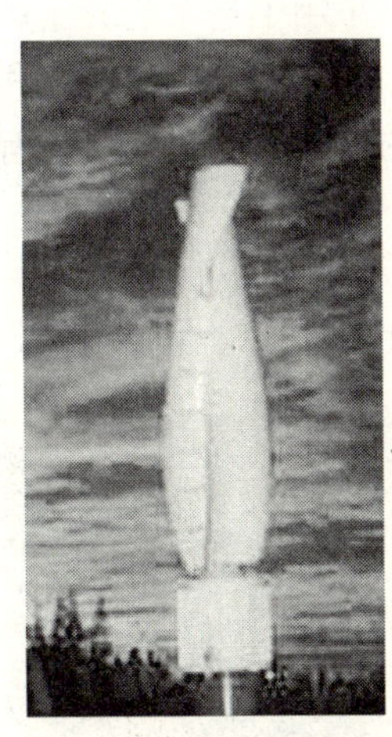

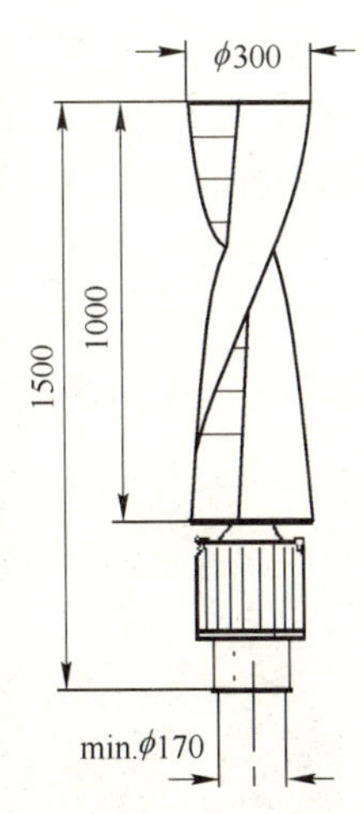

图 15　涡轮式小型风力发电机

图 16　吊顶层内的相变材料蓄能罐

3.1.8　家用燃气空调

生态公寓采用最新开发的家用燃气中央空调，以天然气作为动力直接使用初级能源，可以减少城市用电负荷，优化能源结构，减少城市污染。该系统可制冷、制热并供热水，与同类空调相比节能效果明显。

经节能效果分析，独立住宅全年采暖空调耗电量约 3100kWh(中庭不采用空调时)，3kW 的光伏发电系统，在上海一年可发电 3300kWh，再加上风力发电机的发电量，可满足建筑全年采暖空调的耗电量，实现了“零”能耗住宅示范楼节能设计目标。而生态公寓通过采取不同节能措施，分别实现了 50%和 65%的节能目标。

3.2 资源高效循环利用

资源节约的核心是占用资源少、环境负荷小、可循环率大，上海生态住宅示范楼采用了资源高效循环利用的三方面技术措施。

3.2.1 3R(Reduce Reuse Recycle)环保材料

3R 材料是指可减量使用、能重复使用和可再生使用的建筑环保材料。

示范楼选用了大量 3R 环保材料，混凝土采用高性能混凝土，提高混凝土的强度和耐久性，减少建筑中混凝土的使用量；采用低水泥用量混凝土，利用粉煤灰和矿渣粉等工业废料取代水泥作为掺合料，变废为宝，减少自然资源、能源的消耗和二氧化碳的排放，改善环境质量；采用再生骨料混凝土，将废旧混凝土和建筑垃圾加工制成的骨料取代天然碎石，节约天然石材资源，使混凝土成为可再生利用的材料。砌筑、抹灰和地面砂浆均采用了再生骨料、粉煤灰等制成的商品砂浆，减少天然用砂 25%、水泥用量 15%，内外墙均采用由废渣、废骨料制成的加气混凝土或砌块砌筑而成。土建和装饰工程中精心选用了旧木料、废石材、废渣、旧砖碎瓦等废旧和可再生的环保材料。

3.2.2 节水综合技术

建筑节水技术途径主要有三方面：一是采用节水器具；二是采用雨污水收集回用措施；三是应用就地保水措施。

示范楼全部采用经主管部门评定的 3L、6L 两段型节水型坐便器和节水型龙头；采用雨污水处理装置回收处理雨水和生活污水，出水水质采用自动在线监测，再生水回用于绿化浇水、景观补充用水和冲厕用水；步行道路和附近的停车场全部采用了透水混凝土就地保水(图 17)，其透水率为 800～1000mL/min，强度达到 C25，能提高地表的透水性和透气性，雨水能渗入地下平衡地表含水率，同时可增强地表与空气的热量、水分交换，有利于调节住宅周围的微气候。

3.2.3 中央吸尘和垃圾真空分类收集

将生活垃圾分类收集可防止垃圾在运输和储存过程中对建筑物和环境造成的二次污

图 17 透水混凝土路面

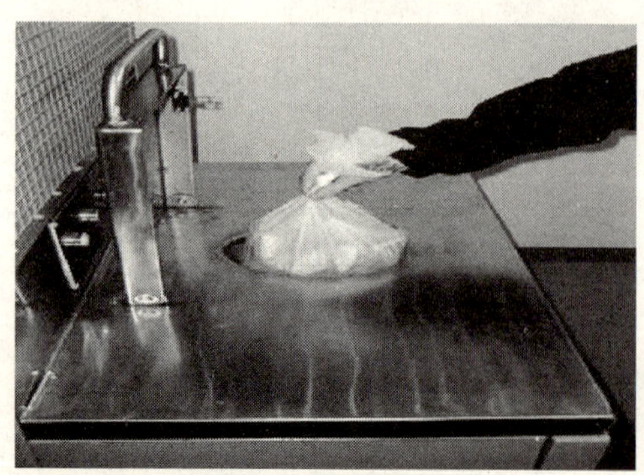

图 18 中央垃圾分类收集处理

染，为减少粉尘污染，独立住宅除设计有中央吸尘外，还采用中央垃圾分类收集处理和传输控制系统对生活垃圾进行收集和处理利用(图 18)。

3.2.4　轻质木结构加层

为了发挥现代木框架结构技术坚固、安全、舒适、资源利用高效的特点，公寓顶层进行了混凝土多层结构与木结构整合的尝试(图 19)，为上海在旧房改造中推广应用轻质木结构进行了有益的探索。

(*a*) 建成后的木结构加层

(*b*) 木结构内部构造

图 19　结构加层

3.3　智能高品质居住环境

3.3.1　智能家居控制系统(i-HOME)

示范楼采用先进的 i-HOME 家庭网络智能控制系统，实现以下功能：

- 家电智能控制系统：

设置多种模式，如离家模式、回家模式、就餐模式、睡眠模式等，并可对控制模块进行个性化控制：对不同房间中的照明灯、窗帘、电视机、厨房中的电器和遮阳篷等分别进行控制，或对多个设备的联动控制。

- 可视对讲系统：

借助住宅内的电视机或网络视频技术随时查看住宅门口和来访者的情况。

- 家庭安防和门禁系统：

布设了燃气泄漏报警、红外幕帘入侵报警、红外动感入侵报警、紧急求助按钮等，可紧急触发短信功能；掌形门禁系统进行住户身份识别。

- 家庭信息系统

家庭服务网是家庭宽带网的入口，支持 ADSL、CableModem、FTTB+LAN 等多种宽带接入方式。

- 远程视频监控

用户可以随时在 Internet 网上查看网络摄像机实时拍摄的画面，从而可以了解到家中的状况。

- 家庭留言功能

当用户不在家想给家人留言时，可上网进入系统中的“家庭留言”功能，选择预先已编写好的留言。室内可视对讲主机即可接收并显示短信内容。

3.3.2 自然通风和天然采光

利用 CFD 模拟技术辅助设计，模拟建筑周边的风环境(图 20)。独立住宅设计了可遥控开启关闭的 $3m^2$ 天窗和通透明亮的中庭，辅以南高北低的建筑结构合理的南向窗墙比 0.45～0.5。经模拟计算，在东南主导风向、室外平均标准高度风速为 3m/s 时，室内主要人体活动的区域范围内各层居住高度(1m)的风速，基本处于 0.5～1m/s 的范围之内，达到自然通风的设计要求，从而达到利用自然通风减少全年空调使用时间、提高换气次数和空气质量的目的。

图 20 室外风场模拟结果

采用光学模拟软件优化设计方案，并对建筑实际采光效果进行测试评价。模拟结果显示底楼中庭区域的照度相对较弱，而直接立面开窗口附近区域则照度很高，形成一定的眩光，经对设计方案进行了相应的优化，最终通过测试采光满足设计要求(图 21)。

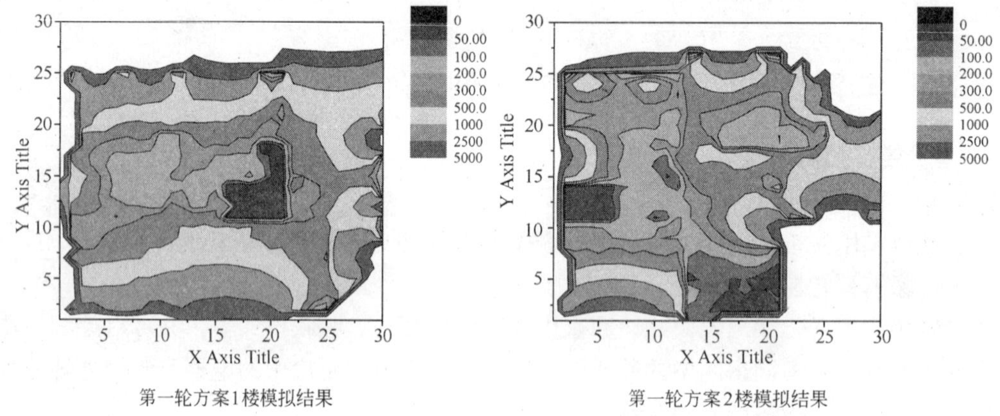

图 21 独立别墅示范楼采光模拟结果(9 月 21 日 12 时，阴)

3.3.3 通风隔声窗

在公寓中安装的双层机械通风隔声窗包含隔声系统、消声系统和通风系统。具有微小夹角的双层窗将吻合低谷降低，从而整体提高窗体的隔声性能；风道被设计成阻抗综合式消声器，在满足通风速率的前提下，隔离室外噪声并消减气流噪声；通风系统采用 $135m^3/h$ 的低噪声风机，可以分档控制风量供给，这种排风式通风隔声窗有助于改善室内空气质量(图 22)。

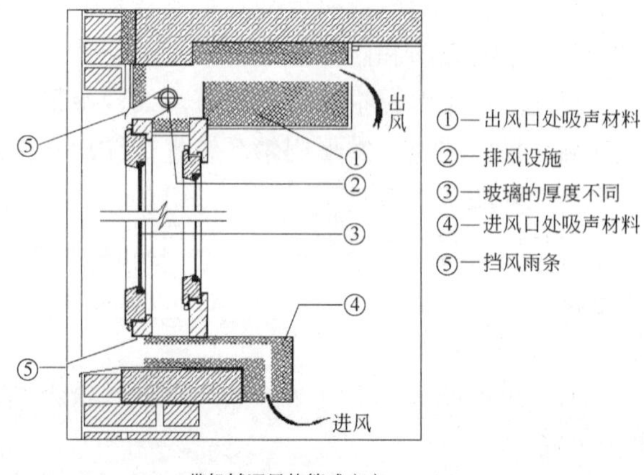

图 22 通风隔声窗

3.3.4 生态绿化

独立住宅 $90m^2$ 平屋面选用耐寒性、慢生常绿草坪做屋顶绿化，既容易人工保养维护，又能提高屋面保温隔热效果和储水功能，能将 50%的屋面降水保留在屋面上，然后再通过植物蒸发掉，从而改善了气候环境。与没有屋顶绿化的同类建筑相比，夏季酷热的白天室内温度可降低 3～4℃，冬天取暖费可节约 1/3。

生态公寓则采用成本低、易于维护的窗台开槽绿化形式，营造美观视觉环境。

示范楼东西外墙采用爬藤等垂直绿化，既减弱日晒，又美化环境，提升居住品质(图 23)。

(*a*) 屋顶绿化　　(*b*) 窗台开槽绿化　　(*c*) 垂直绿化

图 23　三种形式生态绿化

4　示范楼生态性能评估

在住宅竣工运行一段时间后，为了评价独立住宅的绿色综合性能，分别应用美国和英国的生态住宅绿色性能评估工具 LEED-H(V1.4)和 Eco-homes 对独立住宅进行了绿色性能评估。

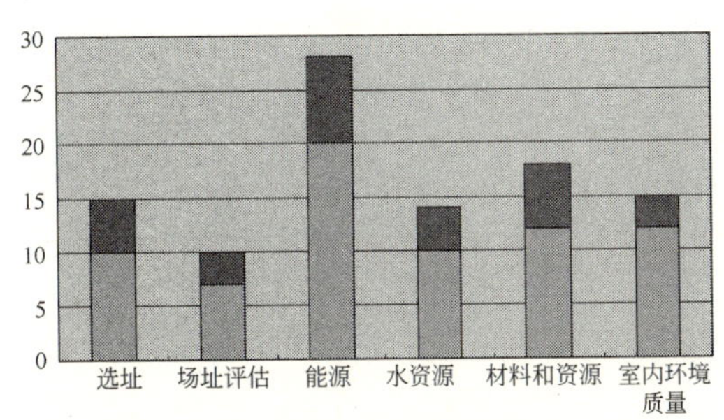

图 24　LEED-H 评估结果

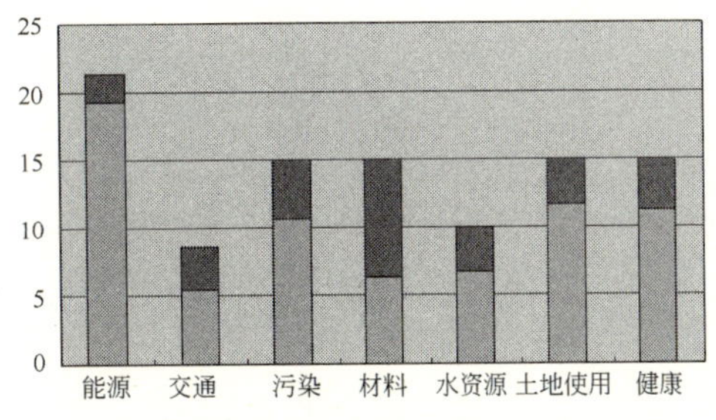

图 25　Eco-home 评估结果

应用 LEED-H 对独立住宅的绿色性能评价结果(图 24)，总分为 71 分，评估等级属“金”等级。同时应用英国 Eco-home 评估工具对独立住宅的评价结果(图 25)，总分达到 71.23 分，属“最好”等级，其中能源的评估分数相对较高，这得益于独立住宅采用的先进节能技术体系。

5　结束语

上海生态住宅示范楼的建成，为全面展示绿色建筑理念和集成技术体系、引导我国绿色建筑的研究和推广应用提供了示范平台。目前对该示范楼的运营研究工作已全面开展。通过跟踪实测评价其绿色技术集成体系效果，并开展绿色新技术、新产品的应用研究，形成适宜推广的绿色技术集成体系，为房产商建设绿色建筑提供技术支撑；为我国的绿色建筑设计和建造及上海2010年世博建筑示范区的建设提供可行的技术借鉴。

Integrated Technologies Study on Green House Demonstration

——Green House Demonstration in Shanghai

Wang Wei　Han Jihong　An Yu

(Shanghai Research Institute of Building Sciences, Shanghai　200032　China)

Abstract: Based on the idea of energy-efficiency, resource-saving, environmental protection and people-oriented, green building is the representative of sustainable development in construction industry. This paper mainly introduces the integrated green house technologies concerned "zero" energy consumption, high resource recycle efficiency, and smart living environment quality, which is from the key research project of Shanghai Science and Technology Committee (SSTC)- "Shanghai Green House Demonstration Project". It will give the reference for promoting the green house development and popularizing the integrated technologies in China.

绿色建筑标准及评估初探
——《浙江省绿色建筑标准》编制

刘晓天　刘　宏　任　涛　周和进
（依柯尔绿色建筑技术研究发展（北京）有限公司，北京　100835）

摘　要：本文介绍了绿色建筑发展的背景，国内外现有绿色建筑评估体系的发展现状，着重讲述浙江省绿色建筑标准和评估体系的主要内容及其特点，最后对绿色建筑的发展进行了展望。

关键词：绿色建筑　标准　评估体系

1　背景

随着人类文明的不断发展，人类对环境的影响越来越大，并日益扩展至全球范围，如臭氧层破坏、温室效应、热带雨林破坏、土地荒漠化等；再加上能源危机、水资源短缺，人类赖以生存的地球正在面对着各种危机，这给全人类敲响了警钟。进入新世纪，如何解决日益紧迫的人口、资源、环境与工业化和城镇化加快、经济快速增长的矛盾，是全人类共同面临的严峻挑战。

建筑，这个与人类生活息息相关的完全人工的产物，在为人们提供基本活动场所的同时，大量地消耗了资源、能源并对环境造成了严重的污染。在建筑的建造和使用过程中，如何节约能源资源，提高使用效率，努力缓解能源资源短缺的矛盾，保护和改善环境，各国都根据自己的国情进行了不懈的研究与实践。

我国现有建筑总面积 400 多亿 m^2，预计到 2020 年还将新增建筑面积约 300 亿 m^2。建筑需用大量土地，在建造和使用过程中直接消耗的能源占全社会总能耗的 30%，使用的建材的生产能耗占 16.7%，用水占城市用水的 47%，使用钢材占全国用钢量的 30%，水泥占 25%，建材生产、建筑活动造成的污染约占全部污染的 34%。为此，2004 年中央经济工作会议上，胡锦涛同志明确指出，要大力发展“节能省地型”住宅，全面推广和普及节能技术，制定并强制推行更严格的节能节材节水标准。温家宝同志也指出，大力抓好能源、资源节约，加快发展循环经济。要充分认识节约能源、资源的重要性和紧迫性，增强危机感和责任感。2005 年政府工作报告中又明确提出，鼓励发展“节能省地型”住宅和公共建筑。建设部为了贯彻落实党中央、国务院的要求，高度重视发展“节能省地型”住宅和公共建筑，及时提出了“绿色建筑”的新理念。

绿色建筑指在建筑的全生命周期内，在适宜条件下，最大限度地节约资源（能源、土地、水资源、材料等）、保护环境和减少污染，为人们提供安全、健康和适用的使用空间，与自然和谐共生的建筑。

绿色建筑是我国经济社会不断发展、城镇化进程不断加快、产业结构调整和居住舒适度提高的产物，是城乡经济、社会、环境可持续发展的必然，是生态环境、生存环境的客观要求。

2 国内外评估体系的发展现状

围绕推广和规范绿色建筑的目标，近年来许多国家出台了各自的绿色建筑评估体系。例如，美国LEED绿色建筑评估体系、德国的生态建筑导则LNB、英国的BREEM评估体系、澳大利亚的建筑环境评价体系NABERS、加拿大的GBTool、挪威的Eco Profile、法国的ESCALE、日本的CASBEE等。这些评估体系，基本上都是以健康、舒适的居住环境、节约能源和资源、减少对自然环境污染为目标，从场地规划与土地利用、自然和生态环境影响、节约能源和可再生能源利用、节约资源和资源再利用、室内外环境质量等诸方面，制定了相应的评分体系。这些评估体系的制定及推广对各个国家在城市建设中倡导"绿色"概念，引导建造者注重绿色和可持续发展起到了重要的作用。

与欧美各国的实践相比，中国面临更严峻的挑战，包括：我国人口三大高峰的压力；能源和自然资源的超常规开发；生态环境的整体恶化；区域间发展的不平衡；实施城市化战略的困扰等。这些绿色建筑的评估体系，大多是由发达国家根据其自身特点制定的，以他们的技术水平比较容易达到，而我国除管理因素外，或者由于技术水平落后，或者由于经济实力不足，在客观上有许多标准和要求也是难以达到的，了解国外的理论和实践时，不能机械地照搬。

因此，中国的绿色建筑标准必须具有主权和中国特色。根据中国的现实情况，在广大地区由于不同的生态环境、生态条件、生态功能以及经济能力与文化特色，一统式的绿色建筑标准缺少可操作性和科学引导与管理作用。所以，绿色建筑需要根据不同类型进行分级标准制定，即中国绿色建筑分级标准体系。标准编制包括：行政等级、行业等级、技术类型等级分类。如：省级、市级、地区级；又如：行业等级、企业等级和项目等级。

我国绿色建筑评价工作还处于起步阶段，但是发展迅速，目前已经推出的有中国的生态住宅技术评估体系、绿色建筑导则，绿色建筑评价体系以及绿色奥运建筑评估体系等。随着首届国际智能与绿色建筑技术研讨会在北京的召开，绿色建筑在中国的土地上如雨后春笋般生根、发芽。

3 浙江省绿色建筑标准及评估体系

为了规范和引导浙江省绿色建筑行业的发展，依柯尔绿色建筑技术研究发展(北京)有限公司受浙江省建设厅委托，与温州市建筑联合设计院共同编制《浙江省绿色建筑标准》和《浙江省绿色建筑评估体系》。编制组深入地研究国内外绿色建筑评估体系，广泛征询专家意见，收集基础资料，并结合建设部示范项目——杭州市"绿建·北秀蓝湾住宅小区"的实践经验，对内容进行反复推敲和不断修订，最后完成定稿。标准编制进度见表1。

标准编制进度表 **表 1**

2005.7	2005.10.31	2005.11.30	2006.1.8	2006.1.18
标准大纲	初　稿	征询意见稿	送 审 稿	报 批 稿

该标准和评估体系结合我国绿色建筑发展的现状，依照浙江省经济、技术、资源等实际情况，吸收了发达国家和地区相关绿色建筑评估体系的经验，遵循了绿色建筑目标切实可行、适度超前的原则。以下文中将《浙江省绿色建筑标准》简称为《标准》，而《浙江省绿色建筑评估体系》简称为《评估》。

3.1 整体思路和框架

浙江省绿色建筑技术体系的思路是：充分考虑浙江的具体特点，加大建筑节能和省地的力度，并使绿色建筑能在浙江省顺利地推广施行。

浙江省绿色建筑技术体系是建立在已有的建筑技术体系上的，具体的框架如图 1 所示。

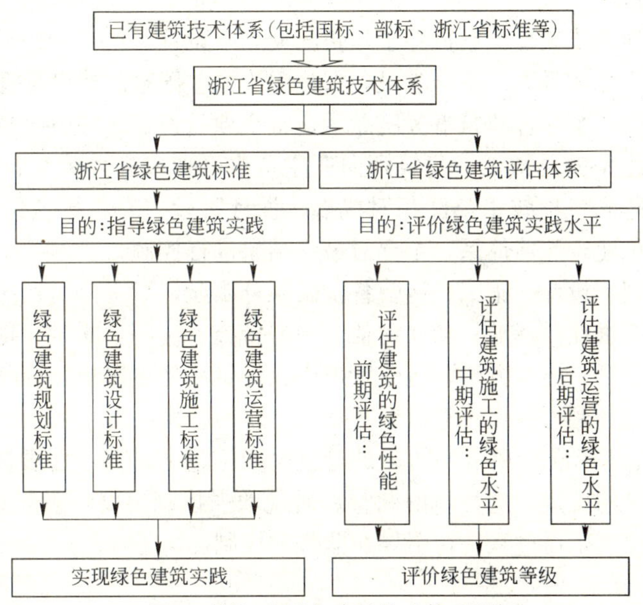

图 1　浙江省绿色建筑技术体系框架

3.2 主要内容

1. 关注建筑的全生命周期：从最初的规划、设计到随后的施工、运营管理到最终的拆除：

——规划和设计阶段充分考虑和利用环境因素；

——确保施工过程中对环境的影响最低；

——运营管理阶段为人们提供健康、舒适、安全、低耗、无害空间；

——拆除后又对环境危害降到最低，并使拆除材料尽可能再循环利用。

2. 最大限度地节约资源(能源、土地、水资源、材料等)：

——优化规划设计方案，减少对各种资源的占用和消耗；

——优化生产工艺，采用适当的技术、材料和产品；

——合理利用现有资源及本地资源；

——最大限度提高土地、能源、水资源及建筑材料的利用效率；

——合理延长建筑物的使用年限。

3. 保护环境、减少污染：

——充分利用建筑场地周边的自然条件；

——在建筑的选址、布局、朝向、形态等方面，充分考虑本地气候特征和生态环境；

——建筑风格与规模和周围环境保持协调，尽可能减少对自然环境的负面影响；

——减少建筑对周围环境声污染、光污染、水污染、大气污染等。

4. 以人为本，创造健康、适用和高效的使用空间：

——提高室内健康舒适度，改善室内环境质量；

——充分考虑使用者的适度需求，努力创造优美和谐的环境；

——保障使用的安全和高效。

3.3 《标准》和《评估》的特点

3.3.1 划分阶段

绿色建筑是一项系统工程，它贯穿于建筑的规划、设计、施工、运营管理以及最后拆除的全过程。国外评估体系大都不分阶段，那是因为这些国家的建设管理和运作模式与我国有很大不同，是由一个综合建筑事务所总承包整个项目设计，并由建筑师负责项目策划设计及施工配合过程中的总协调，建筑师向业主负责，形成开发策划、设计、施工一体化。在施工和运营阶段各项技术都已相对成熟，设计阶段占有主导地位，权利和责任重大。因此国外的绿色建筑评估体系，主要对最终结果进行控制。

但在我国，规划、设计、施工、运营各阶段划分明显，职责分担到各单位，而各单位又相互独立，这就要求在《标准》和《评估》中进行统一协调。同时，我国建筑界对绿色建筑的概念理解，设计、施工方法，运营管理模式等仍然处在探索的阶段，在建筑各个阶段也还存在着诸多的问题。

基于以上原因，《标准》和《评估》是对建设实施过程的各个主要阶段都进行控制，以及时获得工程信息的反馈、进行纠偏调整。通过增强控制力度，对建筑的规划、设计、施工、运营直至拆除全生命过程进行引导、监督和控制。

3.3.2 确定指标

《标准》和《评估》分为场地与环境、建筑、能源、水资源、材料与资源、智能化监控和管理这六个指标。专业分工明确，权利和义务清晰，简化了绿色建筑全生命周期内的实施过程。

“场地与环境”、“能源”、“水资源”、“材料与资源”：这四个指标主要从四节的角度出发，兼顾环保和舒适性。

“建筑”：这里的“建筑”是指单体建筑物本身。它主要体现的是“以人为本”的理念，即建筑的节约不应以牺牲人类的舒适度为代价。“建筑”主要关注人居的舒适以及与自然环境的协调。

“智能化监控与管理”：将“智能”部分列为单项指标，是《标准》和《评估》的一大特点。绿色建筑必然是节能建筑，而节能建筑如果要真正实现其全部目标，就必须采用智能化监控、管理，利用计算机技术和自动控制技术，监控建筑设备运行的全过程，使得办公、居住环境更加安全、健康、舒适，最大限度地降低资源的消耗，最大限度地降低对环境的影响。智能化是绿色建筑的技术保障，智能化系统可以为绿色建筑提供各种信息，提高建筑性能、增加建筑价值，智能化系统的技术水平影响着绿色建筑的整体功效。

《标准》和《评估》中的“智能”部分主要是对系统中设备运行情况进行实时测量、监控、记录，并进行分析处理，其中包括：空调系统，送风、排风系统，外窗遮阳系统，制冷站，锅炉、地源—热泵、换热器系统，生活给水系统，排水、水处理系统、喷泉系统及绿化浇灌系统，供配电系统，照明系统，室内、室外环境系统。

然而许多国外评估体系都未将智能作为单项指标，那是因为智能技术在国外的发展已经相对成熟。在我国，智能建筑技术虽然发展迅速，但与发达国家相比仍然比较落后，为了将智能技术在绿色建筑中推广和应用，就需要政府的鼓励和正确引导。所以，在《标准》中将其列为单项指标。

当然，推广智能建筑技术也要量力而行。在标准和评估中，对该项指标不作强制性要求，使标准扩大了适用范围，真正起到约束和引导的作用。

各个指标相对独立，但又相互联系。单项指标中又可能包含四节，环境保护及舒适性。所以要真正落实绿色建筑理念，就需要提倡一种在项目前期就有多个工种、多个责任方参与的“整体设计”或者“参与设计”理念。

3.3.3　设置权重

国外的评估体系根据各自特点，对各指标得分采用了不同的计算方法。如英国BREEM分权重，而美国LEED则不分权重。LEED是按各项子标得分直接相加，简化了操作过程。但是，我国地域辽阔，各地区气候、人文、经济等不尽相同，即使在浙江省，地理环境、自然资源、城乡发展和经济发展、生活水平和社会习俗都存在着巨大的差异。不分权重将导致不同重要性的内容得分却相同，失去了公平性和公正性。这就要求我们因地制宜，实事求是，在《评估》中引入权重。权重可根据具体地区情况进行灵活的调整，也可以通过调节权重方便地进行条款增减，有针对性地改变评价内容。同时，权重系统的存在也使得基准值的改变不必改变分值，不会对系统的主体框架造成较大影响。

然而，目前我国绿色建筑系统研究还不太成熟，基础数据较为缺乏，只有通过大量的调查，统计，利用数学方法得出权重。层次分析法可以合理地将定性与定量的决策结合起来，按照思维、心理的规律把决策过程层次化、数量化。通过层次分析法，对数据进行一致性判断，可以得出一组与理想结果较为接近的数据。针对各指标重要性，我们咨询了专家和学者、设计单位、施工单位、政府部门等。根据气候、经济、社会、人文等方面因素的不同得出各指标之间的重要性比度，通过层次分析法，得出了权重。

《评估》中按照公建、住宅及地区的不同进行了分类，分别给出了权重(见表2)。当然，分类还是比较笼统，甚至也可以说是“粗线条”的，这将随着科研工作的不断深入以及工程实践的广泛推广而不断改进。

3.3.4　简单适用

1.《评估》是依据《标准》制定的，内容相互对照，通过《评估》的内容加强《标准》内涵。

2.《评估》中各项指标做到了充分的细化，各子指标含义明确，确定子指标值简单，不易产生歧义。

权 重 表 表 2

地区分类 \ 指标	场地与环境	建 筑	能 源	水资源	材料与资源	智 能
Ⅰ类城区						
Ⅱ类城区						
郊区						
自然保护区或风景区						

3.《评估》中各项子指标尽量以定量指标表达，最大限度摆脱了评价的主观性。

4. 对各种子指标均提出了优、中、差档的确定方法，便于实际操作。

5. 整个评价体系为层次化结构，既可以对总目标评价，又可以对各指标或各阶段进行评价。不仅可以比较整个建筑全生命周期的绿色度，又可以比较各个阶段或各个指标完成的情况(参见表 3、表 4、表 5)。

总得分统计与评价表 表 3

指 标	场地与环境	建 筑	能 源	水资源	材料与资源	智 能
指标得分						
总 得 分						
总绿色度						
绿色建筑等级						

指标得分统计与评价表 表 4

指标 \ 阶段	规 划	设 计	施 工	运 营	各指标得分	指标绿色度
场地与环境						
建 筑						
能 源						
水 资 源						
材料与资源						
智 能						
评 价						

阶段得分统计与评价表 表 5

阶段 \ 指标	场地与环境	建 筑	能 源	水资源	材料与资源	智 能	各阶段得分	阶段绿色度
规 划								
设 计								
施 工								
运 营								
评 价								

6. 在建筑竣工验收后，可先对规划设计阶段和施工阶段进行绿色建筑初步评价，并给出初评结果证明；运营管理阶段应在通过初评的建筑运营管理一年后评定，并结合初评结果给出最后的绿色建筑评估结果并定级，之后每3年重新评定一次，对建筑的全生命周期做了全程跟踪。

4 结语

21世纪人类共同的主题是可持续发展。党的十六届五中全会明确提出，要加快建设资源节约型、环境友好型社会，促进经济发展与人口、资源、环境相协调，归根结底是为了可持续发展。而绿色建筑正是我国在建设领域贯彻落实可持续科学发展观，促进经济结构调整，转变经济增长方式的有效途径。

发展是人类生存的基本要求。建筑是人类改变和适应周围环境的一种开发行为。建筑行为包含了以不同形式大量消耗、改变和转化自然资源，显然这些行为在各方面对环境造成了影响，也将影响到人类的可持续发展。“绿色”是自然、生态、生命和活力的象征，它代表了人类与自然和谐共处、协调发展的文化，贴切而直观地表达了可持续发展的概念和内涵。绿色建筑的产生和发展是将绿色思想引入建筑领域的结果，这是国际建筑界对人类可持续发展战略所采取的积极回应，也必将成为未来建筑的主导趋势。

然而，在我国，由于绿色建筑的发展还处于起步阶段，相关体制还不够完善，公民环保意识有待提高，法律法规须要完善，技术水平仍待提高，基础数据和研究尚需积累等种种严峻现实，如何真正推广绿色建筑而不仅仅是概念的炒作，是一项十分艰巨的任务。

《标准》和《评估》的出台对促进浙江省绿色建筑的发展，在全国推广绿色建筑具有开创性的意义。虽然标准和评估中还存在着不足，部分需要定量评价的内容难以科学地界定。但是在现阶段，允许《标准》和《评估》定性规定多一些，即侧重于考察是否具有了绿色建筑的概念和基本特征，是否尽心采取适宜的生态技术措施等。当然，随着建筑体系和经济、技术状况的不断发展，人们对绿色建筑的理解的不断深入，《标准》和《评估》也将适时做出相应的修订，为绿色建筑在我国的发展贡献微薄之力。

最后感谢浙江省建设厅、北京21世纪城市生态研究院、浙江大学、上海建研院、浙江省建筑设计研究院等单位及个人对我们工作给予的支持和帮助。

The Primary Exploration for Standard and Assessment System for Green Building

——The Compilation of “Standard for Green Building of ZheJiang Province”

Liu Xiaotian　Liu Hong　Ren Tao　Zhou Hejin
(Eco Green Building Technologies Research and Development Co., Ltd (Beijing), Beijing　100835　China)

Abstract: The paper mainly introduces green building development background, and the current development situation of the available green building assessment system in China and overseas, with the emphasis on main content and features of green building standard and assessment system of Zhejiang Province, and anticipates the prospect for the green building development in conclusion.

参考文献

[1] 绿色奥运建筑课题组. 绿色奥运建筑评估体系. 北京：中国建筑工业出版社. 2003

[2] (美)戴维·纪森编，林耕，刘宪，姚小琴译. 大且绿——走向21世纪的可持续性建筑. 天津：天津科技翻译出版公司. 2005.1

[3] 中华人民共和国建设部科学技术司和《智能与绿色建筑文集》编委会编. 智能与绿色建筑文集. 北京：中国建筑工业出版社. 2005.3

[4] 中国建筑科学研究院编. 绿色建筑技术文集. 北京：中国建筑工业出版社. 2005.3

新疆生态住居模式构建研究*

姜曙光　李　刚　李俊锋　邢海峰
（石河子大学水利建筑工程学院，新疆　石河子　832003）

摘　要：分析了新疆住居的现状，指出其存在环境污染严重、聚落和单体缺乏地域特色，能耗大等问题。结合新疆地域特点，提出构建新疆生态住居模式的 6 个组成模块：即生态保护屏障——“3＋2”绿地系统模块；“家园式”住宅聚落模块；“内向型”住宅单体模块；绿色建材模块；绿色能源模块；地域特色文化模块。阐明该模式建设的必要性、紧迫性和可行性。

关键词：新疆　生态住居　模式　构建

1　新疆住居现状以及存在的问题

1.1　住居环境

新疆以独立的封闭流域为自然生态单元，流域的上游是绿洲，下游是荒漠。绿洲是新疆干旱区人类活动的载体，绿洲的中心通常是城镇。570 条河流形成了相互分隔的近 800 多个绿洲，面积仅占 8.6%，散布在茫茫荒漠中。生态环境脆弱及区域条件严酷，使新疆的空间联系弱，区域活动具有明显的内向性和封闭性，城市化进程缓慢[1]。

近些年，新疆绿洲生态林建设中，树种单一、树相单调、树龄短等问题很突出。在绿洲边缘地带居住的农牧民，习惯了以砍伐荒漠植被梭梭、野生胡杨、红柳等作为烹饪、取暖的燃料。这些植物都具有不同程度的防风固沙、压盐制碱的生态功能。荒漠生态的破坏，再加上绿洲本身缺乏坚实可靠的生态保护屏障，结果导致了沙尘在夏、秋两季向农田、城镇频繁的侵袭和肆虐。新疆维吾尔自治区环保局发布的《2004 年环境状况公报》显示，2004 年全疆发生了 8 次区域性沙尘暴天气，绿洲内发生局部沙尘天气 130 次。与上年相比，2004 年新疆沙尘污染天气增加，影响范围扩大，影响强度增强[2]。

新疆以天山为界，南疆属于寒冷地区，采暖期 5 个月，北疆属于严寒地区，采暖期长达 6 个月，居住建筑采暖能耗巨大。由于当地煤炭资源非常丰富，埋深浅，挖掘成本很低，故城镇、农村靠燃煤取暖的占绝大多数，由此造成的冬、春两季大气污染也非常严重。新疆维吾尔自治区环保局发布的近几年《环境状况公报》显示，新疆城市空气环境质量总体水平在逐渐下降。2004 年全区城市空气质量达到一、二级的天数与 2003 年相比减少了 5.8 个百分点。新疆目前约 90%的 SO_2 和氮氧化物排放来自化石能源的生产和消费。

---

* 国家自然科学基金资助项目（批准号：50568004）；兵团科技攻关项目（05GG32）；石河子大学工科振兴科技行动专项（GKZX 200511）。

大气污染物造成的酸雨、呼吸道疾病等已经严重威胁到当地经济发展和人类健康[2]。

1.2　住宅聚落

新疆城市住宅聚落形式单一，通常采用一字形组合的行列式，连续多排平行布置，既不利于冬季避风、日照，又不利于夏季防尘、遮阳；同时还缺乏平面错落和空间错落的景致美感。新疆小城镇地多人稀，住宅聚落布置散乱，缺乏集中成片的统一规划和优化，既浪费了土地，又加大了城镇管网、道路等基础设施建设费用。住宅聚落普遍缺乏地域文化的渗透，能够反映多民族地方文化的住宅聚落，仅在乌鲁木齐、喀什、伊犁等少数城市可以看到。

1.3　住宅单体

新疆大、中城市从 2003 年开始强制性执行新建居住建筑节能达标 50%的规定，2006 年 1 月开始，小城镇也将执行节能 50%的标准。近几年，新疆地区受到许多国家“康居示范工程”的影响，开发商为迎合居民追求时尚的心理需求，不顾寒区地域的客观条件，大量使用南方地区常用的落地窗和外凸飘窗等流行元素，使住宅空透、开敞，窗墙面积比严重超标，外窗成为住宅中最大的“热桥”。住宅也失去了北方地区应有的厚重、封闭的个性。

2　新疆生态住居模式构建

2.1　绿洲地区生态保护屏障——“3+2”绿地系统模块

如图 1 所示，绿洲地区生态保护屏障——“3+2”绿地系统模块，即 3 道密集型封闭式生态防护林带，2 类居民聚集带绿地(农村林网化农田绿地和城区绿地)。新疆干旱少雨、风沙大，3 道防护林要从生态系统的需要出发，注重本土特色抗风沙、耐干旱树种的选择，如胡杨、沙枣树、榆树等，在城区多采用乡土树种中抗病性强、对污染物有吸收净化作用的树种，以多品种多层次的群植、丛植为主，乔、灌、花、草合理组合，突出立体绿化，形成高低错落、疏密有致的立体植物群落。

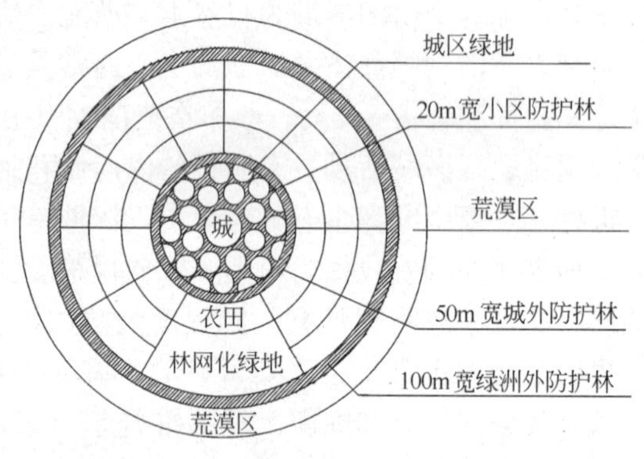

图 1　“3+2”绿地系统模块

采用组团、小区、城镇、绿洲外围多层次绿化保护措施，加强绿洲大型防风基干林与农田林网化配套建设，保护荒漠植被，阻止风沙向绿洲的侵袭，建立可靠的小区域气候防护单元。

2.2　“家园式”住宅聚落模块

被茫茫荒漠包围的绿洲上，人类在恶劣的自然环境中，渴望群体聚集在一起的安全感，聚落组合上反映出较强的凝聚性。如图 2 所示，新疆第一代军垦人居住的“兵营式”单层土坯房聚落，每组两栋房屋，相距仅 10m 左右，入口相对，组成一共用院落，每个院落由 10～16 家组成，共用院落扩大了人们交往的机率和范围。

现代聚落设计在满足居民安全感需要的同时，应该突出聚落的归宿感，强化聚落的家园性，使其与严酷的自然气候条件形成反差，以安抚人们的心灵。如图3所示，“家园式”聚落比“兵营式”聚落具有更强的围合性、封闭性、更大的内院，更多的交往对象。另外，还应该结合当地的自然环境因素、气候特征，按照冬季能为聚落阻挡西北风的要求，尽量靠宽厚的绿化带封闭聚落的西北向，使得聚落的组合避风节能，以形成优化微气候的良好界面，建立微气候防护单元。

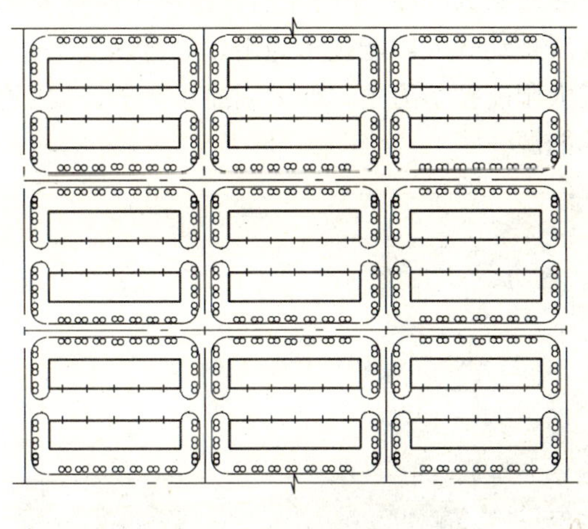

图2 “兵营式”聚落

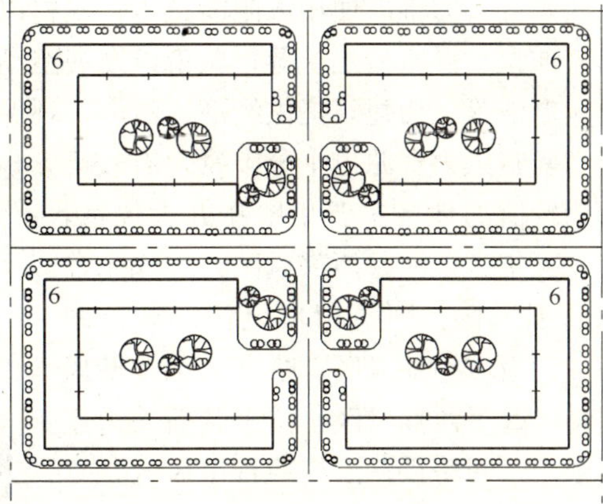

图3 “家园式”聚落

新疆是以农业为主的城镇体系，城市化进程缓慢，住居宜发展低、多层的形式，体现地域文化的内涵，体现以人为本的思想，与地域有机的融合。

2.3 “内向型”住宅单体模块

新疆既是严寒地区，又是地震多发区，住宅单体设计应该充分考虑当地的自然环境因素、气候特征、住宅的功能、居民的行为活动特点等，尽量减少住宅形体的凸凹，体形系数宜控制在0.3以内。尽量少设或不设错层、跃层，合理确定窗墙面积比，这样既有利于节约能源，又有利于结构抗震。我们认为该地区外围窗户应以密闭的固定窗为主，附带设置少量的、内平开的小窗扇。这样一方面能更好地达到采光、保温、节能和防风沙的目的，另一方面，还缩小了材料断面，减少许多配件，可以节省大量的材料和费用。封闭、内向是严寒地区住居应有的特征，把注重地域气候和环境因素作为建筑设计的原点，是现代可持续发展建筑理论的一个重要思想，设计人员和开发商有责任为居民提供一种健康的居住理念和居住环境[3]。

2.4 绿色建材模块

新疆是我国棉花的重要生产基地。该地区70%以上农作物是棉花，每年秋收后大量棉秆作为当地农民的薪柴收割或烧荒，造成严重的环境污染和资源浪费。利用废弃物棉秆为主要原材料，配以胶凝料，采用现代科技手段经特殊工艺加工而制成的免烧型秸秆砌块、秸秆板材等，国内已有多项专利技术，生产该类产品，可为绿色住宅建设就近提供环保墙材，以取代传统的粘土砖，成本低、性能好；既能够使该地区大量的秸秆得到充分利用、土地资源得到有效保护，又能提高住宅墙体的热工性能，具有广阔的市场应用前景。另外，南疆库尔勒地区蕴藏有我国最大的蛭石矿产，可开发出性能优良的墙体保温砌块、墙体复合板材和屋面保温板材。这些具有就地取材，造价低廉，绿色环保等特点的地方材料，是建设生态住居的重要基石。

2.5 绿色能源模块

新疆属于经济欠发达而能源资源非常丰富的地区，煤炭、石油、天然气储量均居全国前列，另外，可再生能源太阳能、风能也非常丰富，但目前当地主要使用的能源仍然是污染严重、能效低下的煤炭。

新疆具有得天独厚的太阳能资源，全年日照时数为3000小时左右，仅次于西藏地区，在新疆应大力推广使用太阳能技术，农村以发展被动式太阳能技术为主，如蓄热墙型、蓄热屋面型、温室型建筑等；城市以发展主动式太阳能技术为主，如太阳能集热器、太阳能光伏电池等，为建筑提供经济、清洁、环保的能源。

新疆风力资源也十分丰富，有较高利用价值的地区主要集中在九大风区，即：额尔齐斯河谷西部风区；准噶尔盆地西部风区；阿拉山口风区；乌鲁木齐—达板城谷地风区；哈密北部戈壁风区；吐鲁番西部风区；位于兰新铁道辽墩至十三间房铁路沿线的百里风区；哈密南部戈壁风区和罗布泊风区。目前仅有达板城地区的风力得到有效开发，新疆风能公司在达坂城地区建起了大型风力发电站(图4)，每年为主电网系统提供5000万kWh的电量。国家已把新疆列入风能利用的重点推广地区。

图4 达坂城风力发电站

依照可持续发展的理论，新疆住宅能源结构亟待改变，最终目标是以太阳能、风能取代化石燃料。在可再生能源能够形成商业化应用规模之前，推广使用储量大、成本低、排污量几乎为零的天然气是最优方案。据探测，新疆全区天然气资源总量约为13万亿m^3。准噶尔盆地预测油气资源量159亿t，天然气具有清洁、使用方便、燃烧效率高、热当量的价格低等优点。采用天燃气壁挂炉分户采暖，比起集中供暖，可以省掉锅炉房、煤炭堆场、废渣堆场以及长距离输送管线的热能损失等，新疆地区的空气质量将得到全面提高。建筑将具有显著的节能效果；居民热费也将大大降低。

以丰富的资源为依托，实施优势资源转换战略，促进能源结构的“绿色化”，实现能源与环境的和谐发展，对于新疆地区是紧迫的，也是可行的。

2.6 地域特色文化模块

新疆是多民族省区，拥有绚丽多彩的民族文化，浓郁的民族风情，乌鲁木齐市国际大巴扎建筑群(图5)就是依斯兰文化的杰出代表作品，具有独特的塞外风情。挖掘展现地域特点的民族文化，灵活运用到现代住居中，将使当地的住居展现出的异域风格，充满无限的生机。

“军垦文化”是新疆地域文化中不可缺少的重要组成部分，以“艰苦奋斗、开拓进取、无私奉献”为主要内涵的“军垦文化”，激励着三代兵团人恪守着屯垦戍边的特殊使命，创造了“荒漠变绿洲”神话。近几年，国家陆续投入了大量资金，用于新疆边境团场

图 5　国际大巴扎建筑群

的基本建设，即“金边工程”，注意保留最能反映军垦战士艰苦创业艰难历程的见证物，最具团场个性特色的内容，尊重军垦历史，建设具有“军垦文化”特色的“金边工程”是和谐社会的需要。

3　结束语

新疆的生态环境极其脆弱，住居建设按生态模式发展，是必须的，也是紧迫的；新疆具有丰富的地域优势资源，建设生态住居，从物质基础方面考虑，也是可行的。建设符合区域特征的具有高效、活力、健康、和谐的生态住居模式，对边疆的稳定、地方经济的可持续发展具有重大的现实意义。

Study of Constructing Ecological Inhabitancy Model in Xinjiang

Jiang Shuguang　Li Gang　Li Junfeng　Xing Haifeng

(College of water conservancy and architecture,
Shihezi university, Shihezi　Xinjiang　832003　China)

Abstract: Study on structure and building of ecological residence model in Xinjiang, Based on the analysis of present residence condition in Xinjiang, this paper proposes some existing problems to be solved, such as severe environmental pollution, lack of local color in single and group building and large energy waste. Combined with regional color in Xinjiang, six component model blocks for ecological residence module are proposed, they are as follows: ecological protection cordon, “3+2” green ground system, homestead building group, introvert single building, healthy construction material, conserve energy and local colorful culture. In the end, the necessity, urgency and feasibility of this construction model are explained and proved.

参考文献

[1] 张军民. 新疆绿洲经济活动的非连续性及封闭性研究 [J]. 石河子大学学报(哲社版)，2004，(4)：11～12

[2] 新疆维吾尔自治区环保局.《2004 年环境状况公报》. 中国环境报，2005.6.10

[3] 姜曙光，唐艳娟，马玉薇. 新疆地震多发区节能住宅的几点思考. 低温建筑技术，2005，(2)：100～101

绿色建筑设计中的CAD技术
——建筑设计中BIM技术与能耗分析相结合的应用研究

曾旭东　赵　昂
（重庆大学建筑城规学院，重庆　400045）

摘　要：本文针对我国目前绿色建筑设计中CAD技术的应用现状和国内外建筑CAD技术的发展，根据当前传统的能耗分析方法无法满足绿色建筑设计需要的现状，提出将BIM（建筑信息模型）技术与建筑能耗分析相结合，进行绿色建筑设计这一新的设计方法。运用BIM技术创建的虚拟建筑模型包含了建筑的所有信息，将这个虚拟建筑模型导入建筑能耗分析软件中，可以自动地识别、转换并分析模型中包含的大量建筑数据信息，从而方便快捷的得到建筑能耗分析结果。

关键词：绿色建筑设计　CAD技术　BIM（建筑信息模型）　能耗分析

我国正处于经济建设的高速增长时期，每年新建建筑80%以上为高耗能建筑，既有建筑95%以上是高能耗建筑。目前我国单位建筑面积能耗是发达国家的2～3倍以上。严峻的事实表明，中国要走可持续发展道路，发展节能与绿色建筑刻不容缓。绿色建筑能为人们提供健康、舒适、安全的居住、工作和活动的空间，同时在建筑全生命周期（物料生产、建筑规划、设计、施工、运营维护及拆除、回用过程）中实现高效率地利用资源（能源、土地、水资源、材料）、最低限度地影响环境。[1]

建筑的可持续发展，不仅对建筑环境工程师、建筑设备工程师提出挑战，更重要的是对建筑师的挑战。绿色建筑的一个重要方面就是节约能源，降低建筑能耗。在决定建筑能量性能的各种因素中，建筑的体形、方位及围护结构形式起着决定性作用，直接的影响包括建筑物与外环境的换热量、自然通风状况和自然采光水平。而这三方面涉及的内容将构成70%以上的建筑采暖通风空调能耗。[2]因此建筑设计对建筑的能量性能起着主导作用。不同的建筑设计方案，在能耗方面会有巨大的差别。单凭经验或者手工计算，很难正确判断建筑设计的优劣。目前凭借先进的计算机技术进行复杂的数据计算和实时的动态模拟是实现绿色建筑设计的科学性和合理性的重要保障。

1　绿色建筑设计中的CAD技术的应用现状与发展方向

在利用CAD技术进行绿色建筑设计，特别是分析和模拟建筑能量性能这一领域，国外的研究开始于20世纪60年代，特别是20世纪70年代世界石油危机后，在建筑能源领域成为研究热点。各个国家都在建筑能耗分析与模拟方面做了大量的研究，基于各自的节能标准开发了相当多的能量分析软件。如美国的DOE 2、Energy plus，加拿大的Hot

2000，英国的 ESP-r，日本的 HASP，我国的 DeST 等等。

中国的建筑节能工作自 1980 年以来，取得了很大成绩，但与发达国家相比，建筑能效设计意识尚很缺乏。目前，中国的建筑设计更多的是侧重于建筑的外观和功能，而忽视建筑的能效设计——即通过被动设计的方法，在建筑设计的初始阶段，就开始进行建筑能效设计，从而选择最优化的设计方案。

在 AutoCAD 占据统治地位的 2D CAD 时代，只能通过手工输入的方式将建筑设计的相关数据输入到相关软件中，才能进行能量分析。而操作和使用这些软件不仅需要具备复杂的能量分析基础知识，而且本身也是专业化的程序，必须由专业人士经过专业培训才能进行操作。大量的专业数据、繁琐的输入工作使得能量分析与模拟对于建筑师来说是可望而不可及的。这是建筑师无法在设计过程中直接对设计方案进行建筑能量分析的一个重要原因。因此在传统的 2D 设计模式中，能效计算通常安排在设计的最终阶段，由专业人士进行操作和分析、模拟，但此时建筑设计方案已经很难改变。于是，能量分析就成为一种象征性的姿态，对绿色建筑设计起不到什么作用。

在我国，《公共建筑节能设计标准》已于 2005 年 7 月 1 日起开始实施。节能设计成为强制性的设计标准，而传统的建筑能量分析方法却无法满足广大建筑师和建筑设计的需求。根据调查分析，能效设计概念难以在实践中实施的原因是多方面的，其中主要原因之一是缺乏能为建筑师所接受和使用的、在建筑设计的任意阶段，特别是初始阶段进行的能效设计方法和软件。如果能充分运用先进的计算机技术，使建筑师能够随意的在设计的任意阶段，特别是在设计的初期阶段毫不费力的对整个建筑的能源消耗和生态效应进行准确的计算和评估，得到关于建筑能量性能的宝贵的反馈信息，那么建筑师就能及时准确地对设计方案进行调整和优化，从而满足建筑节能的要求，最大限度地减少不可再生能源的消耗，真正实现绿色建筑设计的自动化、数字化。这就需要一个包含建筑全部信息的数字化模型，和一个能识别这些信息的建筑性能分析工具。

2 虚拟建筑设计方法——建筑信息模型(Building Information Modeling，简称 BIM)技术

在过去 20 多年里，我国绝大多数设计师一直在使用以 AutoCAD 为代表的 2D CAD 软件进行计算机辅助建筑设计。确切的说，AutoCAD 是典型的“计算机辅助绘图”软件。随着科技的发展和时代的需要，2D CAD 软件的局限性也越来越突出，不仅禁锢了建筑师的想像力和创造力，将大量宝贵的时间花费在 2D 图形的线条表达上——图纸和建筑空间是分离的，也在某种程度上大大制约了建筑设计水平的提高。

随着计算机软硬件技术突飞猛进的发展，一类新的建筑设计方法——“虚拟建筑”(Virtual Building)设计方法正逐渐被世界各地的建筑师广泛应用。“虚拟建筑”设计方法的核心是 BIM 技术。

BIM 即“建筑信息模型”(Building Information Modeling)，是基于对象的建模技术(Object-based Modeling)。[3]运用 BIM 技术，建筑师作设计的过程，就是建造一个真实建筑的过程。这个虚拟的建筑模型，包含了大量建筑材料和建筑构件特征等信息，是一个包含了建筑全部信息的综合电子数据库。在这样一个真实的智能的建筑模型中，建筑师可以

任意地输出平面、剖面、立面，以及各种细部详图、建筑材料、门窗表，还可以输出预算报表、施工进度等等。这种设计方法我们称之为“虚拟建筑”设计方法，它将建筑师从传统的2D线条中解放出来，在“虚拟建筑”中做设计。[4]

随着数字化、信息化和智能化技术的发展，以BIM技术为核心的多种建筑3D CAD软件日趋完善和成熟，在提高设计质量、缩短时间、节约成本等方面，有着2D CAD软件无法比拟的优越性。目前国际上采用BIM技术的主流建筑CAD软件主要有Graphisoft公司的ArchiCAD，Autodesk公司的Revit，Bentley公司的Microstation Triforma，Nemetschek North America公司的VectorWorks，Nemetschek的Allplan，Advanced Relational Technology公司的Chief Architect，SoftPlan Systems公司的SoftPlan以及Gehry Technologies公司的Digital Project等等。

从2D CAD过渡到基于BIM技术的3D CAD，是未来计算机辅助建筑设计的发展趋势。越来越多的世界知名建筑师和事务所开始使用BIM软件进行建筑设计。我国的一些设计院、事务所也开始关注这一世界上最先进的建筑CAD技术。这就为绿色建筑设计中能量分析的自动化、智能化提供了基础和平台。

3 建筑设计中BIM技术与能耗分析相结合的应用研究

利用BIM技术，建筑师在设计过程中创建的虚拟建筑模型已经包含了大量设计信息，包括几何信息、材料性能、构件属性等等，只要将模型导入相关的能量分析软件，就可以得到相应的能量分析结果。原本需要专业人士花费大量时间输入大量专业数据这个过程，如今利用先进的计算机技术就可以自动完成。建筑师不需要额外花费精力，运用CAD技术进行绿色建筑设计也就变得得心应手。

目前并不是所有的建筑性能分析软件都能识别和提取以BIM技术创建的虚拟建筑模型中包含的信息。为了适应目前CAD技术从2D向BIM转变这一潮流，许多能量分析软件都开始进行研究开发以适应BIM技术。而大多数主流BIM软件都支持一种通用的、得到广泛认可的IFC数据格式，这种数据格式为虚拟建筑模型与能量分析软件之间的数据交换提供了理论基础。

3.1 利用BIM技术进行绿色建筑设计的理论基础——IFC数据标准

IFC(Industry Foundation Classes)标准是由国际协同工作联盟IAI(International Alliance for Interoperability)制定的国际建筑业工程数据交换标准，并且是惟一一个得到ISO(国际标准化组织)认可的建筑信息标准。IFC的目标就是将CAD以及非CAD软件中在整个业界不断共享的信息应用到建筑的生命周期当中。

通过IFC标准，建筑设计过程中得到的“虚拟建筑”模型可以用于结构工程、HVAC、电力设备、管道消防、电梯系统、评估等各个专业。例如结构设计软件Tekla Structure、流体分析软件Airpak、模型检查软件Solibri、评估软件Timberline's Precision Estimating等，都可以通过IFC文件格式工作，如图1所示。正是在IFC这个通用的数据标准基础上，BIM技术与建筑能量分析软件相结合进行绿色建筑设计才具有理论上的可行性。

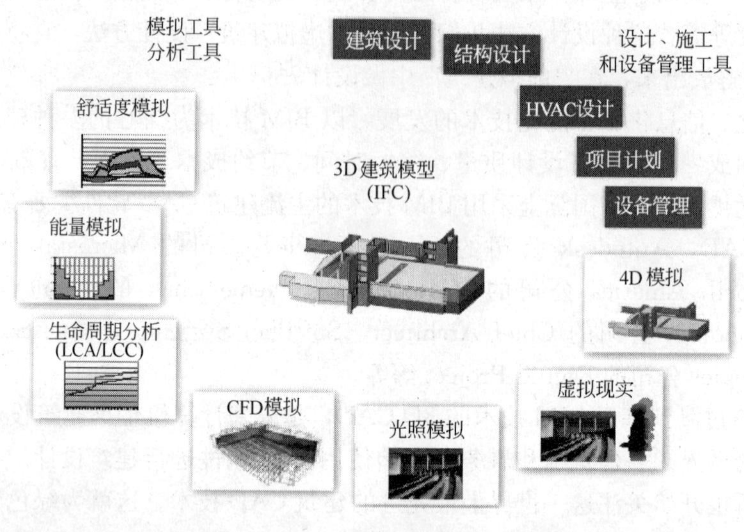

图 1

3.2 建筑设计中 BIM 技术与能耗分析相结合的应用

3.2.1 方案设计阶段的能量分析

在建筑设计的方案阶段，能充分利用建筑信息模型和能量分析工具，简化能量分析的操作过程，是建筑师进行绿色建筑设计迫切需要解决的问题。目前，美国的 Green Building Studio 可以满足建筑师的这一需求。

Green Building Studio(简称 GBS)是一个免费的基于网络服务的能量分析工具，可以在初期概念设计阶段，提供专业的、详细的能量分析结果。目前 GBS 支持 ArchiCAD、Revit 和 Microstation 等 BIM 软件创建的虚拟建筑模型。

GBS 直接从 BIM 软件中导入建筑模型，利用其中包含的大量建筑信息来建立一个准确的热模型(其中包括合理的分区和方位)，并将其转换成 XML 格式(gbXML 是一种开放的 XML 格式，已被 HVAC 软件业界迅速接受，成为其数据交换标准)。并根据当地建筑标准和法规，对不同的建筑空间类型进行智能化的假定。最后结合当地典型的气候数据，采用 DOE2.2 模拟引擎(一个被广泛接受的建筑分析程序)进行逐时模拟。每年能量消耗、费用以及一系列建筑采暖制冷负荷、系统(诸如照明、HVAC、空间供暖的主要电力和天然气的能源使用)数据都能立刻展现出来。[5] 而整个过程中，建筑师只需在 Green Building Studio 中手动的输入建筑类型和地理位置即可。

除了给出详尽的能量分析报告外，GBS 还能输出 gbXML、3D VRML、DOE2.2 等文件格式，可以利用其他工具诸如 Trane 的 Trace 700，或 eQuest、Energy Plus 等对建筑能效进行进一步的分析，如图 2 所示。

目前 GBS 中只提供美国各地的气象数据，仅适用于美国。国内的建筑设计方案要想用 GBS 进行能量分析，需要选择与建筑所在地气候条件相似的美国地区。2005 年 GBS 公司已经开始整合中国各地的气象数据。预计 2006 年，中国部分地区将可以直接通过网络使用 GBS 进行建筑能量分析。

另一个主要用于设计初期的能量分析软件是 ECOTECT。在建筑分析领域里，它是惟

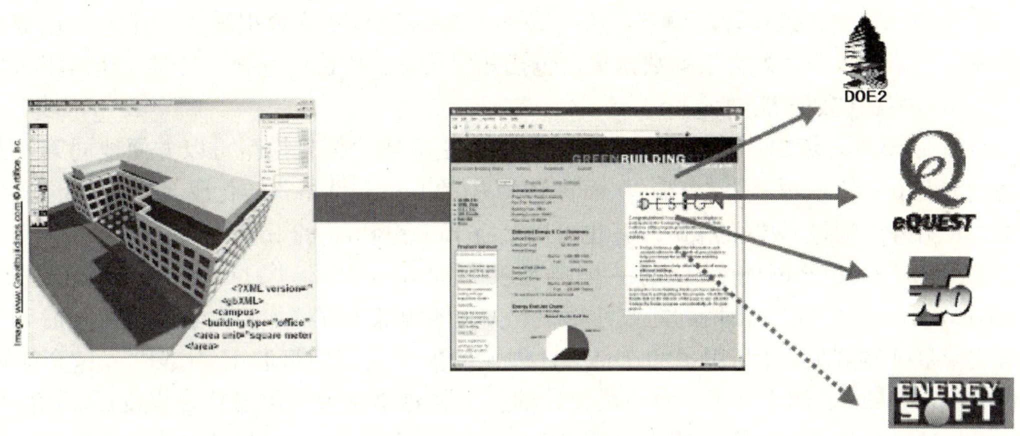

图 2

一由建筑师研发，主要为建筑师使用的软件。ECOTECT 的建模和分析功能可以处理任何复杂的几何形体模型，可以对包括太阳能、热量分析、照明分析、声学分析、混响时间、主导风向、资源消耗以及材料造价等方面进行分析。它的主要优点在于能在建筑设计早期阶段就能够得到非常直观的反馈信息。该软件目前尚不能直接导入 BIM 模型，然而即将发布的 ECOTECT 5.5 版本将支持 IFC 数据格式。

3.2.2　精确的能量分析与模拟

在建筑设计基本完成之后，需要对建筑物的能效性能进行准确的计算、分析与模拟。在这方面，美国的 Energy Plus 软件是其中的佼佼者。EnergyPlus 是一个建筑全能耗分析软件(whole building energy analysis tool)，[6]是个独立的没有图形界面的模拟软件，包含上百个子程序，可以模拟整个建筑的热性能和能量流、计算暖通空调设备负荷等，并可以对整个建筑的能量消耗进行分析，如图 3 所示。

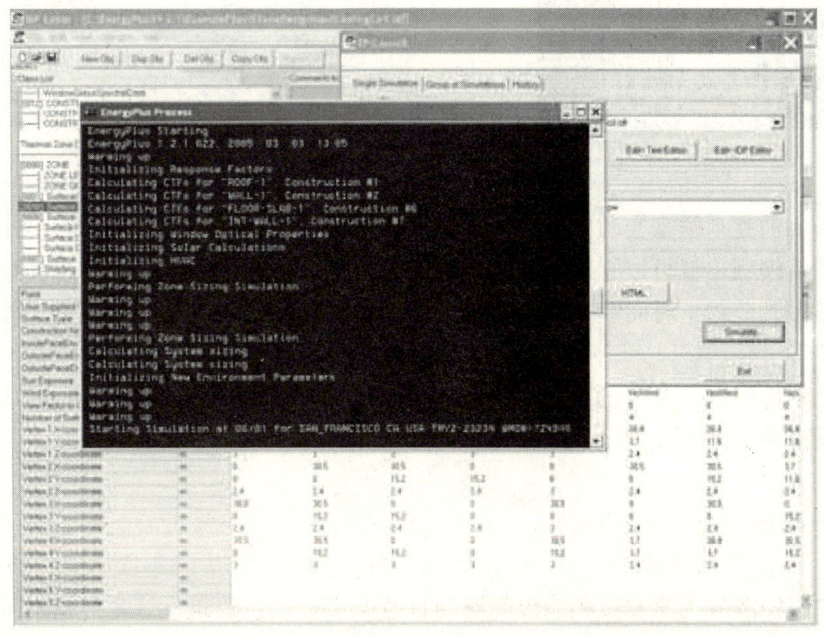

图 3　EnergyPlus 结合后台的两个辅助软件进行模拟的过程

在 2D CAD 的建筑设计环境下，运行 EnergyPlus 进行精确模拟需要专业人士花费大量时间，手工输入一系列大量的数据集，包括几何信息、构造、场地、气候、建筑用途以及 HVAC 的描述数据等。

然而在 BIM 环境中，建筑师在设计过程中创建的建筑信息模型可以方便地同第三方设备例如 Bspro Com 服务器结合，从而将 BIM 中的 IFC 文件格式转化成 EnergyPlus 的数据格式。另外，通过 GBS 的 gbXML 也可以获得 EnergyPlus 的 IDF 格式。

BIM 与 EnergyPlus 相结合的一个典型实例是位于纽约“911”遗址上的自由塔(Freedom Tower)。在自由塔的能效计算中，美国能源部主管的加州大学劳伦斯·伯克利国家实验室(LBNL)充分利用了 ArchiCAD 创建的虚拟建筑模型和 EnergyPlus 这个能量分析软件。如图 4、图 5 所示，自由塔设计的一大特点是精致的褶皱状外表皮。LBNL 利用 ArchiCAD 软件将这个高而扭曲的建筑物的中间(办公区)部分建模，将外表几何形状非常复杂的模型导入了 EnergyPlus，模拟了选择不同外表皮时的建筑性能，并且运用 EnergyPlus 来确定最佳的日照设计和整个建筑物的能量性能，最后建筑师根据模拟结果来选择最优化的设计方案。

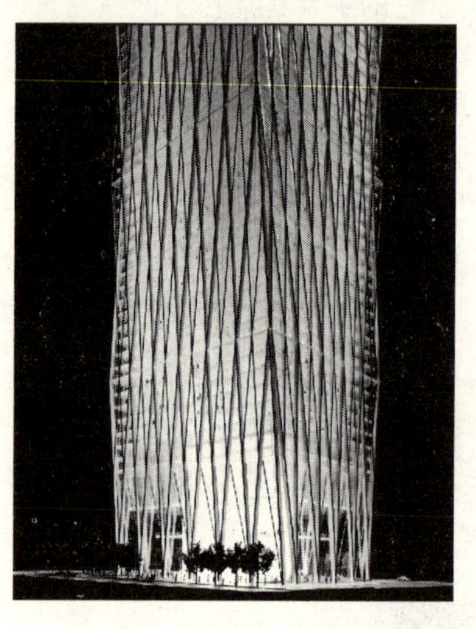

图 4

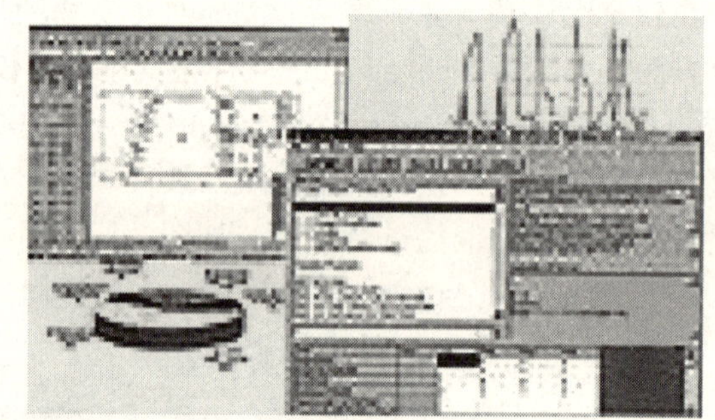

图 5

劳伦斯·伯克利国家实验室 (LBNL)的 Vladimir Bazjanac 说：“我们用 ArchiCAD 建造出虚拟建筑模型，就可以立即对建筑进行能量性能模拟，而不需要为了模拟再重新创建模型。这样做既节约了时间和费用，并且立即就可以得到结果，同时还有助于设计可选方案的表达。”[7]

除以上软件以外，芬兰的 Riuska 软件，CFD 软件 Airpak 等等，都支持 IFC 数据标准，可以直接导入 BIM 模型，方便快捷的得到能量分析结果。

4 小结与展望

随着我国经济的飞速发展和能源问题的日益严重，绿色建筑设计变得越来越重要。从 2D CAD 向基于 BIM 技术的 3D CAD 转换，是计算机辅助建筑设计的发展趋势。利用

BIM 技术，建筑师可以在设计过程中创建包含建筑全部信息的虚拟建筑模型，在设计的任意阶段将其直接导入相关的能量分析软件中，就可以方便快捷的得到能量分析结果，从而及时调整设计方案，优化建筑性能，充分利用 CAD 技术来保障绿色建筑设计的科学性和合理性。

目前在采用 BIM 技术的 3D 建筑 CAD 软件与建筑能耗分析软件的结合方面，仍然存在许多尚未解决的问题，如一些主流能量分析软件无法直接导入虚拟建筑模型，不支持 IFC 数据标准等。但越来越多的软件厂商、开发者、使用者已经开始关注这些问题并在进行研究和开发。相信在不久的将来，综合利用 BIM 和建筑能耗分析进行绿色建筑设计的技术，将越来越完善和成熟。

CAD Technology in Green Building Design

——Application Study of Integration of BIM Technology and Energy Efficiency Analysis in Building Design

Zeng Xudong Zhao Ang

(Faculty of Architecture and Urban Planning
Chongqing University, Chongqing 400045 China)

Abstract: Aimed at the situation of green building design and CAD technology development, this article put forward a new green building design approach, integration of BIM technology and building energy efficiency analysis. The virtual building model created with BIM technology is a electronic database including all of the building's information. Importing this virtual building model to the building energy efficiency analysis application, we can get the energy analysis results quickly and easily.

参考文献

[1] 建设部副部长仇保兴在国务院新闻办新闻发布会上的讲话

[2] 薛志峰等. 超低能耗建筑技术及应用. 北京：中国建筑工业出版社，2005

[3] G. Aound, A. Lee and S. Wu, The Utilisation of Building Information Models In nD Modeling: A Study of Data Interfacing and Adoption Barriers, http://www.itcon.org/2005/08/, April 2005

[4] Dianne Davis, BIM (Building Information Modeling) Update, http://www.aia.org

[5] http://www.greenbuildingstudio.com

[6] Jacobs P, Henderson H. State-of-the-art review of whole building, building envelope, and HVAC component and system simulation and design tools. http://www.arti-21cr.org/research

[7] Graphisoft, Energy Analysis with ArchiCAD9 _ whitepaper10

长江三角洲地域绿色住居评价体系研究

王　竹　王　杉　裘晓莲
（浙江大学建筑系，浙江杭州　310027）

摘　要：随着“可持续发展”战略日益成为人类共识，人们对绿色住居的认识与理解也不断深化。浙江大学建筑系在课题《长江三角洲绿色住居可持续发展评价体系》中提出了运用神经元网络概念，融入模糊思想，建立绿色住居综合评价目标体系，并以它为主要理论基础，实现评估模型与评估软件的开发。

关键词：绿色住居　评价体系　软件设计

1　国内外绿色住居评价体系研究进展

1.1　国内外绿色住居评价体系概况

随着“可持续发展”战略日益成为人类共识，人们对绿色住居的认识与理解也不断深化，许多国家纷纷开始了与之有关的评价体系研究工作。绿色住居评价体系工程复杂而庞大，它不仅需要各方各面的专业人士在可持续发展的思想指导下合作进行，还需要城市管理者和普通大众共同参与构建工程的始末，并达成共识。

发达国家和地区早在20世纪60年代就开始探索“绿色建筑/可持续建筑的发展战略与技术”，研究并制定了相应的评价指标体系。近年来，国外一些相关评价体系已被相继采纳或运用，主要是检测建筑物在建造和使用中相对于以前的改进程度，其中英国BREEAM、美国LEED™、加拿大GBC等是实施的比较成功的评价体系，其他还有荷兰ECO Quantum，德国ECO-PRO，法国EQUER，日本《环境共生住宅A-Z》等，各自都有不同的特点。这些评价体系对当地建筑是否达到节能、环保的性能标准给出了完整的分析与评估方法，并设计了各类图表及电脑软件，以供设计者或使用者评估。

我国绿色住居评价体系的研究工作现还处于初期研究开发阶段，缺乏实践经验的积累，许多相关的技术研究领域还是空白。但是近年来围绕着节约能源和减少污染等问题，国家相关部门陆续颁布了一些单项的技术法规，并且研制开发出不少节能新技术，这对推动绿色住居评价体系工作是十分有利的。如：台湾《绿建筑解说与评估手册》、香港建筑环境评估法(HK BEAM)、《绿色生态住宅小区建设要点与技术导则》、《现代房地产绿色开发与评价》以及《中国生态住宅技术评估手册》等等。

1.2　对国内外各类型绿色住居评价体系的比较分析

有关国内外各类型评价体系的内容与比较，见表1。

国内外评价体系主要特点　　表 1

国际	BREEAM	① 针对英国本国 ② 采用树桩分支的多层次结构形式 ③ 过程：输入数据—综合评分—确定等级
	LEED	① 针对美国本国 ② 辅助设计 ③ 采用树桩分支的多层次结构形式 ④ 过程：输入数据—综合评分—确定等级
	GBC	① 适用于不同国家和地区 ② 辅助设计 ③ 制定基准指标作为评定建筑的评定参照 ④ 采用树桩分支的多层次结构形式 ⑤ 建立在 EXCEL 平台上，结果为直方图形式
	《环境共生住宅 A-Z》	① 规划设计的综合评价(定性) ② 基本性能评价(定量) ③ 建筑全寿命周期环境冲击评价 ④ 事后验证
国内	《绿建筑解说与评估手册》	选择有科学化计算标准的指标
	建筑环境评估法 HK BEAM	① 针对香港地区 ② 采用树桩分支的多层次结构形式 ③ 过程：输入数据—综合评分—确定等级
	《绿色生态住宅小区建设要点与技术导则》	① 设定了详细的评价指标和评价标准 ② 采用了树桩分支的多层次结构形式
	《现代房地产绿色开发与评价》	① 设定了详细的评价指标和评价标准 ② 采用了树桩分支的多层次结构形式
	《中国生态住宅技术评估手册》	① 体系从 5 个子项对小区进行评估，各子项均为 100 分 ② 被认定的生态住宅需满足指标体系中的基本要求(得分≥60 分以上)；在此基础上，分体系得分在 80 分以上者可进行生态住宅单项认定 ③ 对于与生态住宅基本原则密切相关的条款，作为必备条件，若子项满足所有必备条件要求，则该子项可得基本分 40 分；在此基础上，附加分另算，该子项最后得分为基本分与附加分之和。若违背必备条件，则计分为 0 分

通过对国内外各类型评价体系的比较与分析，从评价内容到评价机制等各方面都对我们有很多的启发，但并不能将它们直接引用到我们的实践中来。原因主要有以下几点：

- 操作界面复杂，操作方法不易于理解与实施；
- 评价需要较多的取样数据，给评价带来一定难度；
- 软件不易在不同的评价层次进行操作；
- 数据繁多、复杂，过于依赖专家群体，不能给予非专业人士以及普通使用者明确的结论和指导依据。

2 长江三角洲地域绿色住居评价体系研究

2.1 课题研究背景

长江三角洲地区四季分明、气候温和、江河密布、土地肥沃，属于典型的夏热冬冷地区。该地区在我国国民经济建设和人居环境可持续发展建设中将发挥越来越重要的作用，因此研究该地域绿色住居建设是中国人居环境研究的重要组成部分。2001 年我们获得了国家自然科学基金资助项目《长江三角洲城镇基本住居单位可持续发展适宜性模式研究》(50178062)。在其子课题《长江三角洲绿色住居可持续发展评价体系》中提出了运用神经元网络概念，融入模糊思想，建立绿色住居综合评价目标体系，并且实现评估模型与评估软件。

2.2　方法论与技术支持

《长江三角洲地域绿色住居可持续发展评价体系》(以下简称《评价体系》)在可持续性指标产生的机理和方法论上提出建立可持续性指标的方法论框架：

明确可持续发展目标——确定问题——选择基本指标——根据可持续性原理扩充相关指标——考虑边界问题——考虑不确定性——可持续性指标的评价。

其主要技术支持包括以下三个部分：

(1) 无量纲法

《评价体系》选用正规化转换方案进行原始指标的转换。由于无量纲方法是将不同的物理量，以一定的标准和方式以百分数或比例等表示，将物理量量纲不同的不可比指标变成无量纲的可比指标，不同领域的信息可集成考虑，对系统整体状况的研究比较适合。

(2) 神经元网络

《评价体系》在已有的研究基础上创造性地引入了人工神经元 BP 网络，并讨论应用最广的前向人工神经网络模型的回想评价功能。这个模型是由大量的、同时也是很简单的处理单元(也称神经元)广泛地互相连接而形成的复杂网络系统，把纷繁复杂的大量基础数据通过多维向一维的映射最终得到少量或者是惟一的目标指标。

(3) 模糊思维

《评价体系》引入了模糊综合评价的概念，它是指对多种模糊因素所影响的事物或现象进行总的评价。由于地域住区指标体系是一个多因素耦合的巨复杂系统，表现出极大的不确定性和随机性，因此，引入模糊思维的概念是符合评价的客观要求的。

3　评价软件设计指导思想

3.1　评价软件设计理念——设计体现易用性

如果一个评价软件的功能强大但却很难使用，那么用户会拒绝该程序。复杂的东西不一定是高科技的东西，而傻瓜相机因为有技术含量而变得更加简单。因此，在创建评价软件的环境中，我们将软件使用者——而不是评价体系的构建——摆在设计过程的核心，从一开始就将使用者关心的问题考虑在内，并且将软件使用者的需要摆在首位。简而言之，我们将软件的易用性作为我们最主要的设计理念，使评价软件“傻瓜化”。

易用性可以细化为五个方面：

(1) 易理解性：要求软件研制过程中形成的所有文档语言简练、前后一致、易于理解以及语句无歧义；

(2) 高效性：使用有效，能很快实现预期目标；

(3) 易学习性：要求提供的用户文档内容详细、结构清晰、语言准确；

（4）易操作性：软件的人机界面友好、界面设计科学合理以及操作简单等；

（5）用户满意度：用户对产品表示满意并乐于使用它。

3.2 评价软件需求分析

评价软件的易用性，体现在人与计算机的交互过程中，其前提条件就是对人机关系的深入了解。用户分析与角色设计是了解用户的科学方法。

笔者对用户的状况进行了讨论，并抽象出三组典型的用户，见表2。

评价软件用户状况 **表2**

	典型用户	职　业	特　征	使用动机	设计对策
	住宅研究专家、学者、设计师	管理机构人员、研究学者、规划设计师等	专业知识丰富，对国家政策、规范以及设计有全面的认识，认知能力强	监督并规范绿色住居设计，对已建成项目进行评测、分级	提供专业性的评价细则、资料
	房地产开发商	房地产开发	专业知识较为丰富，对国家政策、规范以及设计有一定的理解，在商业利益与产品品质之间游离，认知能力较强	指导并完善项目开发，提高产品品质	提供专业性的建议及指导
	普通用户	各类型职业	缺乏专业经验，对国家政策、规范以及设计理解片面，缺少专业指导，具有强烈的认知欲望	对绿色生态住居形成全面、专业普及的认识	评价过程操作简单，体现人文关怀

4 评价体系框架

4.1 评价方法——混沌神经网络评价法

神经网络和混沌之间具有共同的特性，即系统的非线性。其主要研究目标是建立含有混沌动力学的神经网络模型，并用之于智能信息处理。混沌神经网络能恰当地描述我们复杂的绿色住居体系。

（1）该评价法能恰当地描述系统的非均匀性。在绿色住居系统里，基本单元分布及相互之间的作用具有非均匀性，因子所具有的权重相差较大，在体系中的影响力是不同的。

（2）该评价法能恰当地描述系统的非线性。由于基本单元之间的相互作用具有非均匀性，从而导致系统在各种条件下可能存在无序（混沌）和有序（规则）的解，也就引发了无序与有序之间转化的问题。

（3）该评价法能恰当地描述系统的自适应性。由于绿色住居系统的开放性，它必定与其他系统及元素发生作用，也应有能力对外界做出相应的反应。

(4) 该评价法能恰当地描述系统的网络性。绿色住居系统在结构上具有网络性，其所构成网络是由确定和随机两种因素决定的。

4.2 评价体系的层次结构

绿色住居评价体系的层次结构是人工神经网络的多层次体系，需要考虑计算机算法统一的可行性，因此评价体系中的同一层次必须平等，且针对每一个层次有固定的评价规范。同一层次的元素具备同等级的属性，但并非均匀地分布，这个层次的元素成为其上一级的评价的基本单元。依次类推，由下至上，形成了递推型的评价体系结构。图1即是这种结构的理想模型，其中 A 代表绿色住居体系；B_1、$B_2 \cdots B_a$ 代表组成这个体系的 a 个因素，它们有可能是居住环境因子、居住建筑因子等；C 层级内的因子 C_1、$C_2 \cdots C_{ad}$ 代表分别组成 B_1 至 B_a 的各个元素，例如，组成 B_1(居住环境因子)有可能是能源使用、资源利用、环境污染等；组成 B_2(居住建筑因子)有可能是室外环境、室内环境、活化地域等……；其他层级以此类推。

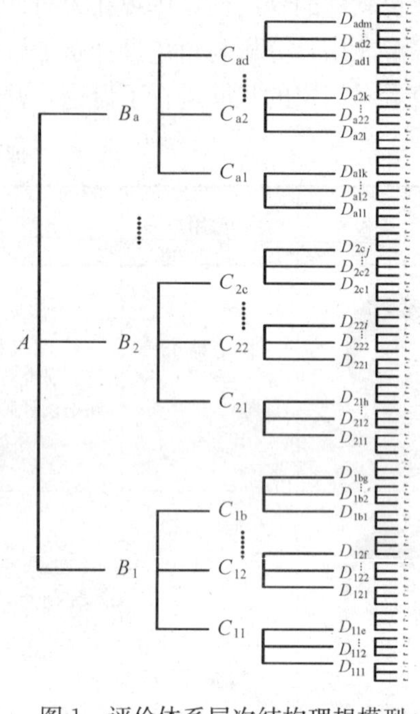

图1 评价体系层次结构理想模型

4.3 评价体系的指标选择

指标是反映总体现象的特定概念和具体数值。因此，评估指标的选择较为复杂和困难。因此，在本评价体系中，将整个系统的评价步骤分解为基于多个子系统的评价，并依据各个元素的重要性，给每个级别的元素分配一定的权值，以合理的体现项目开发或评价的重点。根据评价的目的、细节要求及范围限制，有选择性地在某一级对相关元素进行独立的评价。

采用混沌神经网络评价法模型表达长江三角洲地区绿色住居评价指标体系，共分为六层：总目标层，因子层，亚因子层，准则层，基本指标层以及附加细则层。总目标层至准则层如表3所示。

评价体系总目标层至准则层 **表3**

总目标层	因子层	亚因子层	准则层
长江三角洲地域绿色住居评价体系	环境	资源消耗	能源消耗
			用地情况
			建材消耗
			水的消耗
		环境负荷	三废处理
			施工全过程
			电磁污染
			对周边环境的影响

续表

<table>
<tr><th>总目标层</th><th>因子层</th><th>亚因子层</th><th>准则层</th></tr>
<tr><td rowspan="20">长江三角洲地域绿色住居评价体系</td><td rowspan="8">社　会</td><td rowspan="3">人文环境</td><td>继承历史</td></tr>
<tr><td>融入地域</td></tr>
<tr><td>活化地域</td></tr>
<tr><td rowspan="5">服务管理</td><td>建筑物的适应性</td></tr>
<tr><td>设备控制</td></tr>
<tr><td>维护与管理</td></tr>
<tr><td>私密性与视觉景观</td></tr>
<tr><td>配套公共建筑</td></tr>
<tr><td rowspan="10">建　筑</td><td rowspan="5">室外环境</td><td>小区绿化</td></tr>
<tr><td>空气质量</td></tr>
<tr><td>噪声控制</td></tr>
<tr><td>建筑间距</td></tr>
<tr><td>小区微环境</td></tr>
<tr><td rowspan="5">室内环境</td><td>空间环境</td></tr>
<tr><td>空气质量</td></tr>
<tr><td>热环境</td></tr>
<tr><td>光环境</td></tr>
<tr><td>声环境</td></tr>
</table>

4.4　评价体系指标的权重

由于评价指标体系中的各个指标对评价目标的影响是不同的，对于被评判事物，从不同的着眼因素可以得到决然不同的结论，而且在诸多着眼因素对总评价的影响程度不一，存在着模糊择优因素。因此在模型建立的下一步是设想最适当的计算方法。

选择的BP算法是目前神经网络应用最为广泛的算法，其基本思想：对一个输入样本，经过神经元网络的正向推理得到一个输出，然后该输出与期望的样本输出层开始向回传播，调整连接权值和阈值，使网络输出的误差在允许的范围之内。因此这样的计算方法切合了评价系统的特点。可得到对评价对象的评价结果，再现评价专家的经验、知识、主观判断及对指标的重要性倾向，实现定性与定量的有效结合，保证评价的客观性与一致性。

4.5　评价体系的基准

为了能够体现项目整体品质，接近实际，将权重乘以100，化成百分制计分标准。由于将理想点作为假设点，因此评价对象不可能在每一个指标中都获得满分，只可能尽量靠近。所以将评价等级可以划分为“很差”、“较差”、“及格”、“较好”以及“很好”五个等级，能更真实地反映项目某些方面的“极差”或者“极优”对整体品质的影响。如：某方面太差，不但不能获得评分，甚至减分；某方面很优秀，有可能获得更高评分，弥补其他的不足。

5　评价体系程序化实现

该部分内容涉及：①软件功能的实现(题库结构的架构、文字和自动生成图表相结合、多人评价结果汇总、强调局部性评价、评价结果保存、矩阵矢量计算评分)；②软件构成

（评价体系的数据库、软件的操作模块及 UIBU 布局、软件的可扩展性、运行环境要求）；③软件演示（评价输入演示、结果分析演示、评分输出演示）等。

5.3.1 评价输入演示（图 2～图 5）

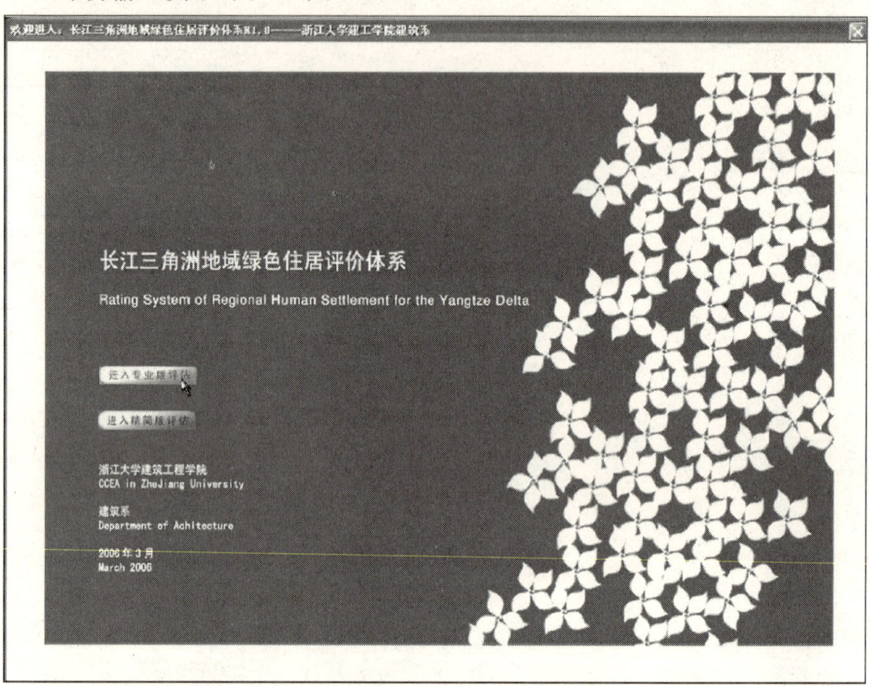

图 2 评价输入——进入界面

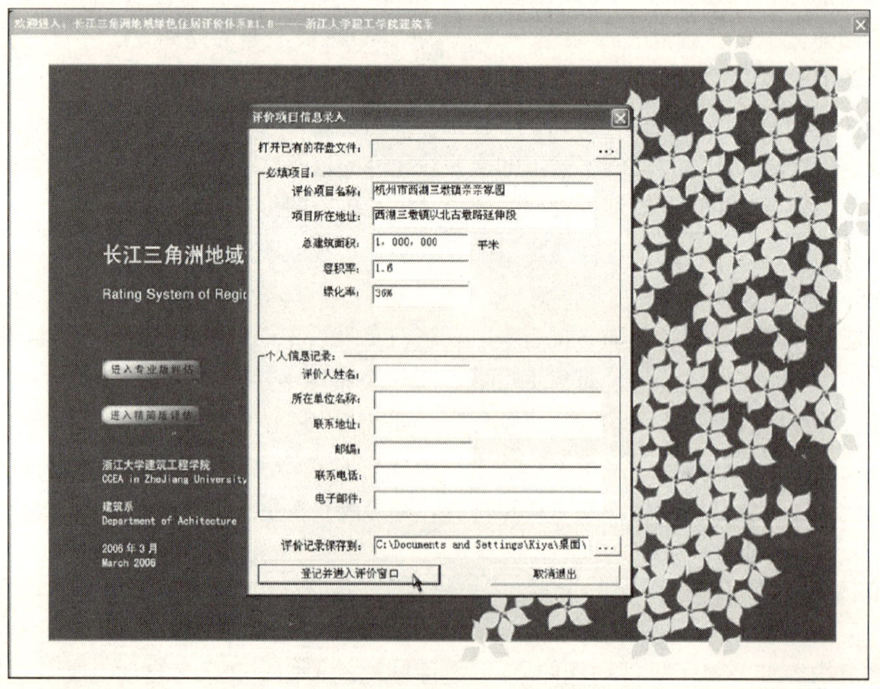

图 3 评价输入——登录信息

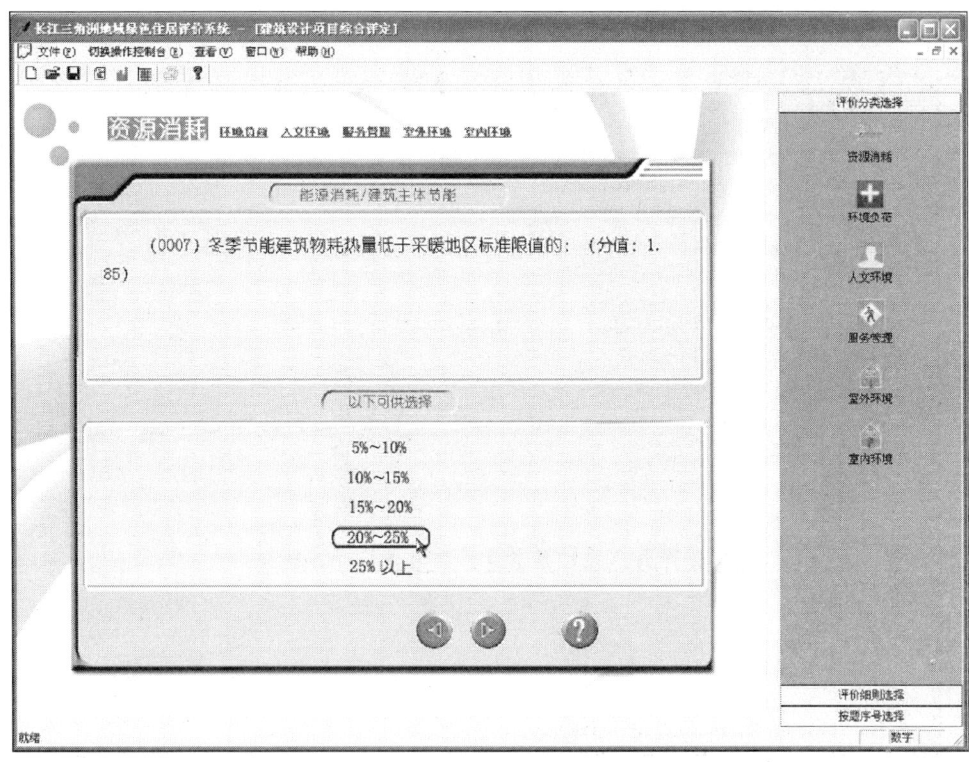

图 4　评价输入——评价答题

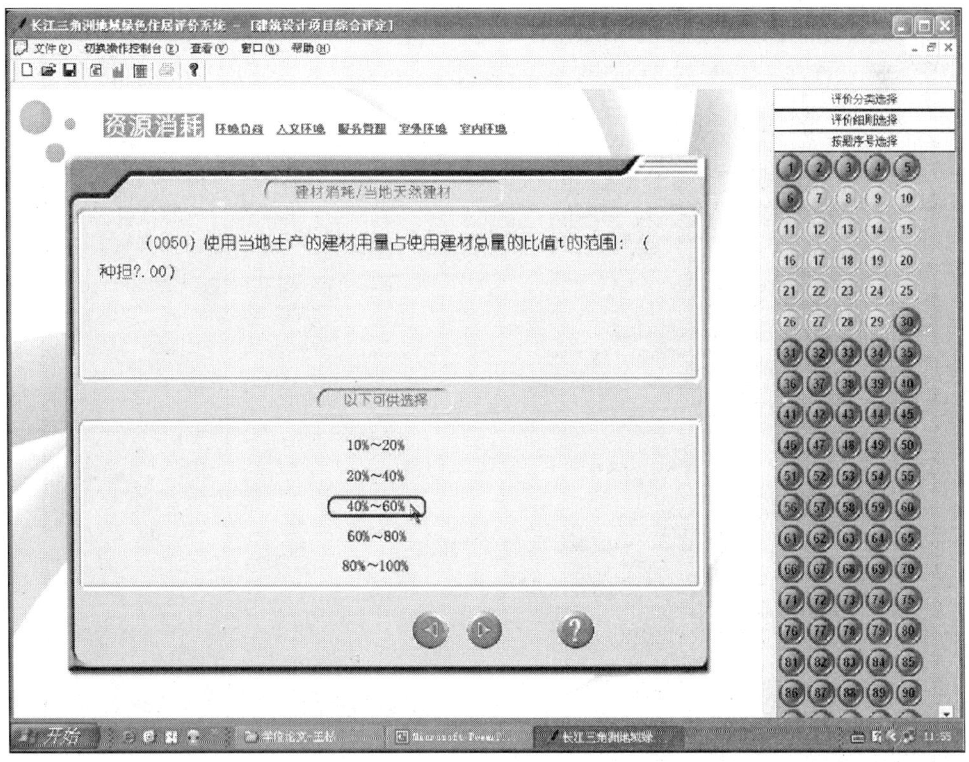

图 5　评价输入——按题序号选择

5.3.2　结果分析演示(图 6)

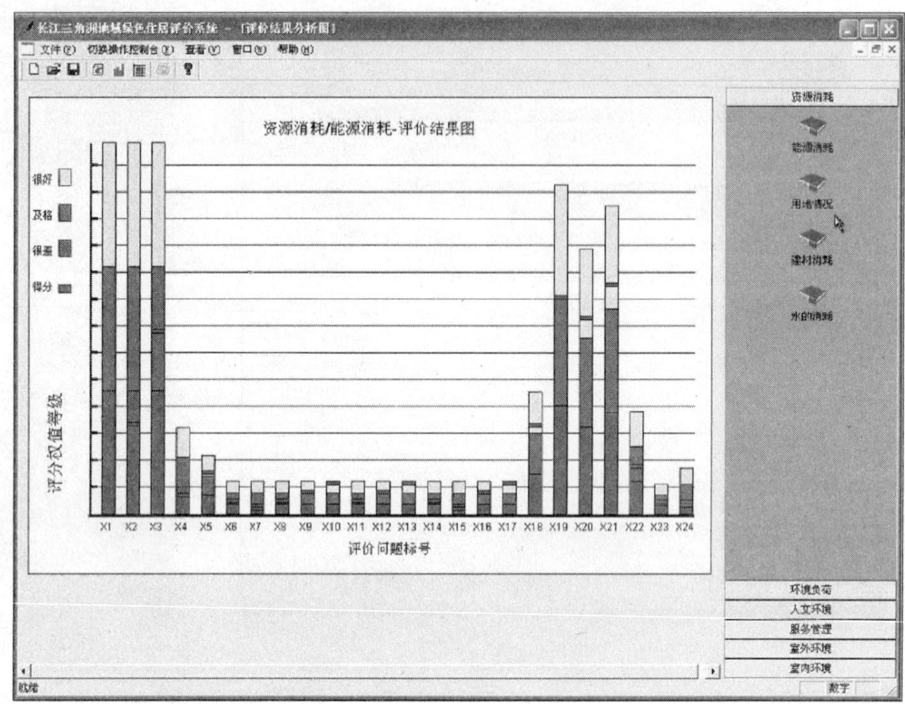

图 6　结果分析——柱图显示

5.3.3　评分输出演示(图 7～图 9)

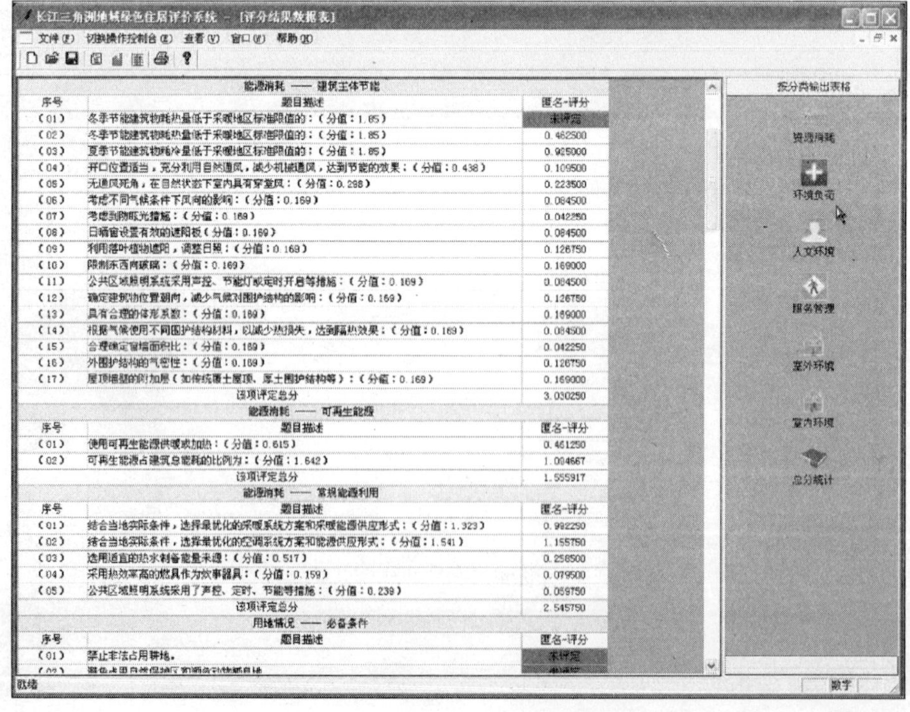

序号	题目描述	匿名-评分
	能源消耗 —— 建筑主体节能	
(01)	冬季节能建筑物耗热量低于采暖地区标准限值的：(分值：1.85)	未评定
(02)	冬季节能建筑物耗热量低于采暖地区标准限值的：(分值：1.85)	0.462500
(03)	夏季节能建筑物耗冷量低于采暖地区标准限值的：(分值：1.85)	0.925000
(04)	开口位置适当，充分利用自然通风，减少机械通风，达到节能的效果：(分值：0.438)	0.109500
(05)	无通风死角，在自然状态下室内具有穿堂风：(分值：0.298)	0.223500
(06)	考虑不同气候条件下风向的影响：(分值：0.169)	0.084500
(07)	考虑到防眩光措施：(分值：0.169)	0.042250
(08)	日晒窗设置有效的遮阳板(分值：0.169)	0.084500
(09)	利用落叶植物遮阳，调整日照：(分值：0.169)	0.126750
(10)	限制东西向玻璃：(分值：0.169)	0.169000
(11)	公共区域照明系统采用声控、节能灯或定时开启等措施：(分值：0.169)	0.084500
(12)	确定建筑物位置朝向，减少气候对围护结构的影响：(分值：0.169)	0.126750
(13)	具有合理的体形系数：(分值：0.169)	0.169000
(14)	根据气候使用不同围护结构材料，以减少热损失，达到隔热效果：(分值：0.169)	0.084500
(15)	合理确定窗墙面积比：(分值：0.169)	0.042250
(16)	外围护结构的气密性：(分值：0.169)	0.126750
(17)	屋顶墙壁的附加层(如传统覆土屋顶、厚土围护结构等)：(分值：0.169)	0.169000
	该项评定总分	3.030250
	能源消耗 —— 可再生能源	
序号	题目描述	匿名-评分
(01)	使用可再生能源供暖或加热：(分值：0.615)	0.461250
(02)	可再生能源占建筑总能耗的比例为：(分值：1.642)	1.094667
	该项评定总分	1.555917
	能源消耗 —— 常规能源利用	
序号	题目描述	匿名-评分
(01)	结合当地实际条件，选择最优化的采暖系统方案和采暖能源供应形式：(分值：1.323)	0.992250
(02)	结合当地实际条件，选择最优化的空调系统方案和能源供应形式：(分值：1.541)	1.155750
(03)	选用适宜的热水制备能量来源：(分值：0.517)	0.258500
(04)	采用热效率高的燃具作为炊事器具：(分值：0.159)	0.079500
(05)	公共区域照明系统采用了声控、定时、节能等措施：(分值：0.239)	0.059750
	该项评定总分	2.545750
	用地情况 —— 必备条件	
序号	题目描述	匿名-评分
(01)	禁止非法占用耕地。	未评定

图 7　评价输出——单人评价输出

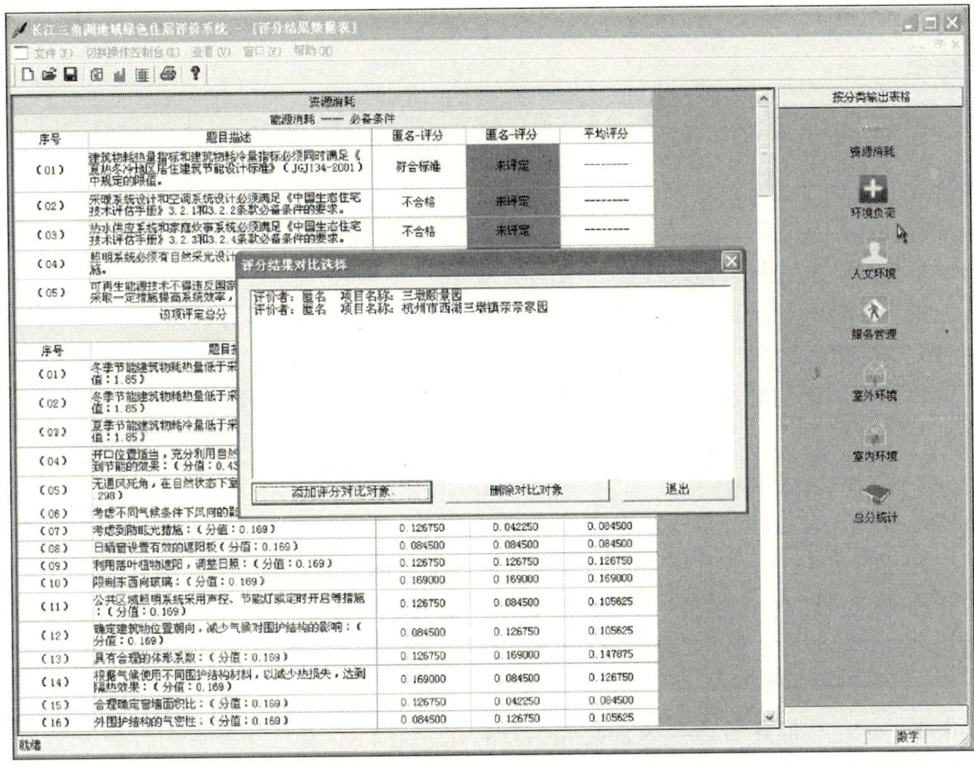

图 8　评价输出——添加对比评价

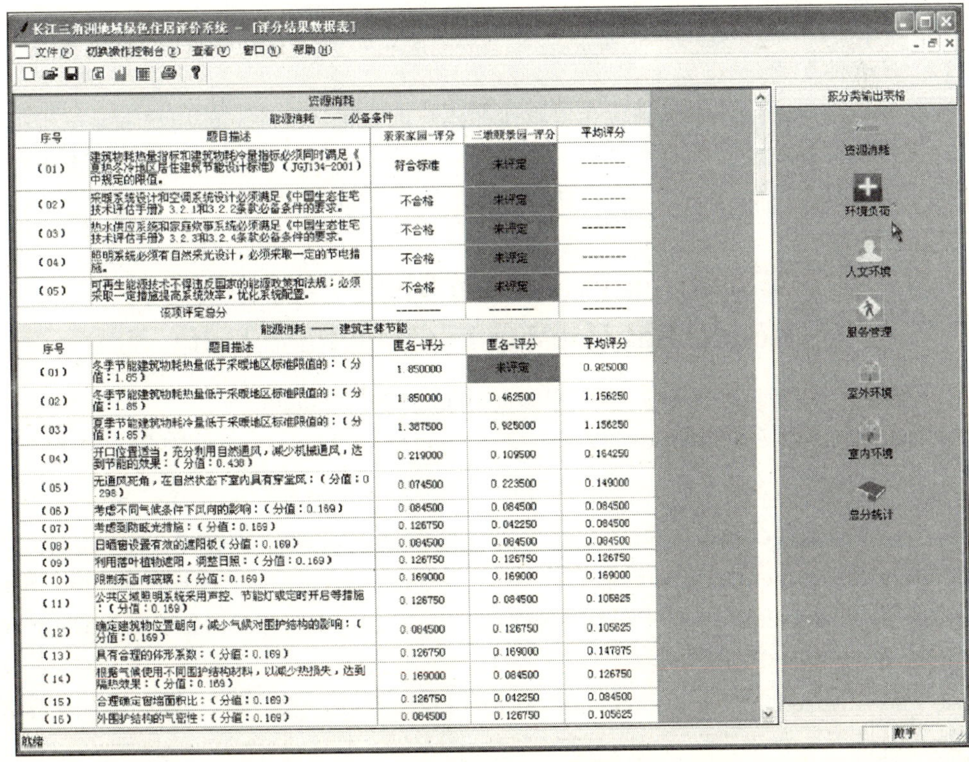

图 9　评价输出——多人评价输出

6 结语

长江三角洲绿色住居评价体系软件在已经应用的过程中效果较理想。但是，由于本研究所涉及的方面十分宽广，加上时间与精力的限制，目前评价软件的运行性能还不是完全令人满意。相信随着对评价体系的不断完善，地域绿色住居评价软件的性能会更加优越，并推动绿色、生态住区环境的发展。

未来绿色住居评价体系软件功能应该具有如下特点：

（1）支持多种题库

可以对不同类型的后台数据库进行直接的修改、编辑操作，实现评价数据在不同系统之间的传递与转换。

（2）结构模块化

软件系统在设计和开发过程中要保证各子系统、子系统中的各项功能甚至每一个应用程序的高度模块化。只有这样才可以实现对系统的自由剪裁和重新配置，如对各子系统内部的评价内容、权重等功能的取舍，以此达到根据不同类型用户的不同需求配置系统的目的。

（3）分布式应用系统

未来的绿色住居评价体系软件是超大规模的，它将不再是部署在一台服务器上的系统，因此支持分布式应用和分布式数据库是未来绿色住居评价体系软件的一个特征。

（4）支持 Internet 应用

随着网络的发展，绿色住居评价体系软件应支持 Internet 上的信息获取的实现。

（5）多语种支持

一套绿色住居评价体系软件应当可以按照用户的设定，在不同的用户端显示出不同语种的应用界面。由此还可以引申出另一种功能，即可以由用户来改变应用系统输出界面上使用的术语，这样不同行业的用户可能会感到应用系统的专业性更强。

A Study on Evaluation System for Green Residential Housing in Changjiang Delta

Wang zhu　Wang shan　Qiu xiaolian

（Department of Architecture，Zhejiang University，Hangzhou　310027　China）

Abstract: As the statement of sustainable development turns into the global common consensus，people get more and more ideas of things about the green settlement. The architecture department in Zhejiang University brought up the idea that using the theories of neural networks and fussy rules to construct the mode of sustainable development index system in the subject "the index system of the green human settlement in Yangtze River Delta" and focused attention on the evaluation software development to deepen the subject.

参考文献

[1] 王竹，贺勇，于文波. 长江三角洲绿色住居机理与适宜性模式研究，国家自然科学基金资助项目(50178062)，浙江大学

[2] 王竹，贺勇，魏秦，王玲. 关于绿色建筑评价的思考. 浙江大学学报(工学版)，2002

[3] 裘晓莲. 长江三角洲地域绿色住居可持续发展评价方法探讨研究. 浙江大学，2004

[4] 徐子苹，刘少瑜. 英国建筑研究所环境评估法 BREEAM 引介. 新建筑，2002，(1)：55

[5] 周双海. 美国能源及环境先导计划 LEED 引介. 中外建筑，2003，(2)：33

[6] 建设部科技发展促进中心. 中国生态住宅技术评估手册(2002 版). 北京：中国建筑工业出版社，2002

[7] 吴良镛，人居环境科学导论. 北京：中国建筑工业出版社，2001

[8] 西安建筑科技大学绿色研究中心. 绿色建筑. 北京：中国计划出版社，1999

[9] 绿色建筑技术手册. 北京：中国建筑工业出版社，1996

[10] 吴祥兴，陈忠. 混沌学导论. 上海：上海科学技术文献出版社，1996

[11] 朱怿. 多元视野下的建筑美观. 建筑学报，2005

[12] 董军，胡上序. 混沌神经网络研究进展与展望. 信息与控制. 1997，(5)：359

[13] 袁琦. 浅析软件用户界面易用性的设计概念. 电脑知识与技术. 2005，(12)：15

[14] 夏热冬冷地区居住建筑节能设计标准(JGJ 134—2001)

[15] 齐康. 长江三角洲地区发展与整合. 城市研究，2000.1

德国节能建筑设计方法与实践以及如何与我国国情相结合

曲翠松

（同济大学建筑城规学院建筑系，上海　200092）

摘　要： 20世纪中期西方工业化国家进入信息社会以来，新的建筑施工技术、新型建筑材料的应用，使得建筑以前所未有的姿态出现。以建筑环保和可持续发展为原则，德国建筑与工业长期以来受到政府的支持，进行了大量的试验和测试，其节能建筑的设计水平领先于世界。本文着重介绍被动式建筑以及先进的混凝土核心激活技术、地热资源利用、太阳能作为建筑能源等均已经发展成熟的技术。最后通过一个国内项目设计的实践来探讨他山之石的借鉴与运用。

关键词： 被动式建筑　混凝土核心激活　地热资源利用　太阳能

1　德国节能建筑设计方法与实践

1.1　德国节能建筑发展概况

20世纪中期西方工业化国家进入信息社会以来，新的建筑施工技术、新型建筑材料的应用，使得建筑以前所未有的姿态出现。以建筑环保和可持续发展为原则，德国建筑与工业长期以来受到政府的支持，进行了大量的试验和测试，其节能建筑的设计水平领先于世界。在政策法规方面制定了具有法律效应的最低标准，从1976年制定的《节能法》、1984年出台的《热保温规范》WSVO 1984到1995年的《热保温规范》WSVO′95、1998年的《暖气设备规范》HeizAnlV′98被《节能建筑规范》2002 EnEV 2002所取代，经过2004年的补充修订出台的《节能建筑规范》EnEV 2004将95规范中的低能耗建筑列为标准，也就是说所有新建和改建建筑必须满足节能规范的要求，其用于取暖的能源消耗在95规范基础上降低30%，并且对于热源的来源——是否是可再生能源也做了具体的指标规定，见图1。

1.2　被动式建筑(住宅)

在符合《节能建筑规范》EnEV 2004的基础上将能耗的指标进一步降低，同时提升使用者的舒适度就得到了被动式建筑(Passive House)。一幢被动式建筑的特征具体体现在：低能耗——建筑取暖能耗低于15kWh/(m^2 · a)，也就是说建筑每平方米每年取暖消耗燃油低于1.5L，所以被动式建筑又叫1.5L建筑。目前在德国、瑞士以及奥地利地区这样的建筑已建成3000多幢，而且数量还在不断增加。与低能耗建筑相比节省能耗80%，而与传统的节能建筑相比节省值达90%。要达到如此高节能效率仅有两个不可忽视的要

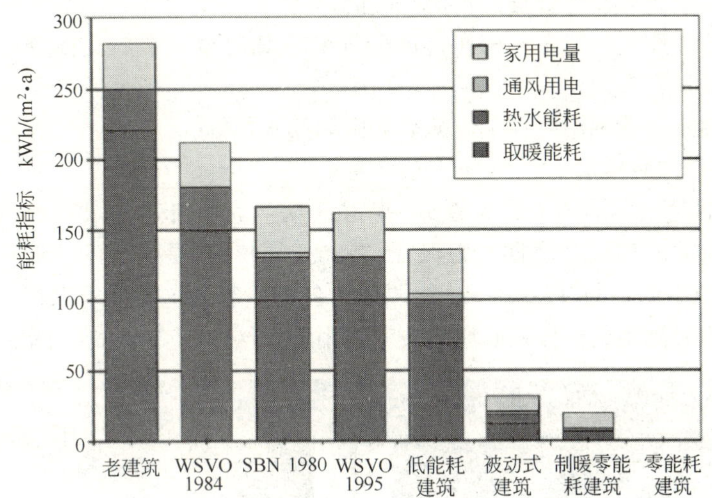

图 1 德国不同时期节能建筑规范所规定的建筑能耗指标

（其中 SBN 1980 为瑞典热保温规范）

（来源：PASSIVHAUS INSTITUT，Darmstadt）

点：1. 尽量避免建筑通过围护结构的热损失。2. 最大争取自由热量的利用。体现在具体措施上有以下几点(见图 2)：

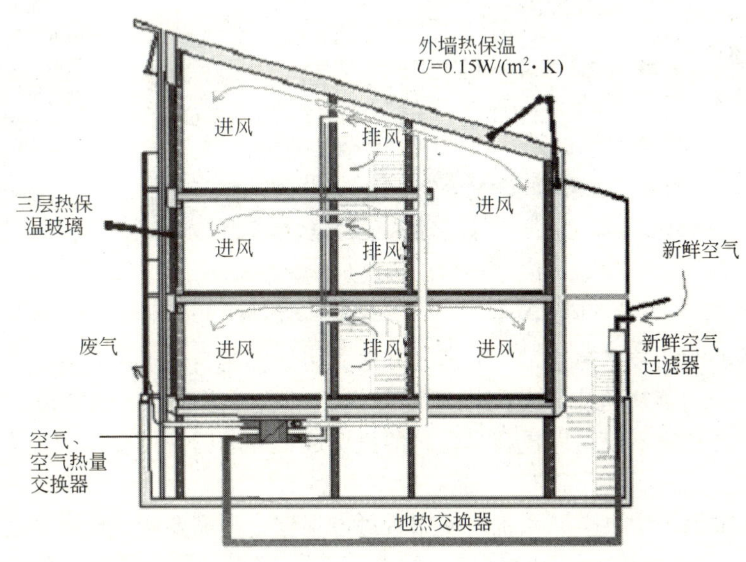

图 2 被动式建筑措施示意(来源：PASSIVHAUS INSTITUT，Darmstadt)

- 极好的外保温和建筑布局紧凑——所有的建筑实体外围护部分传热系数均要达到 0.15W/(m² · K)，尤其是要注意边、角以及内外连接部位，避免冷桥；
- 极好的外遮阳措施，夏季避免太阳直射，冬季尽量利用太阳光热；
- 性能极佳的窗，包括玻璃和窗框——窗体的传热系数均不能高于 0.80W/(m² · K)；
- 建筑要极为密封；
- 新鲜空气经过预热(冷)——在理想状态下可以通过地热交换器将要引入室内的新

鲜空气进行预热或预先制冷处理，不消耗电能；

• 回收排放废气的热量——在排放废气前将其通过一个热交换器，其75%的热量将转移给新风而留在室内；

• 部分热水装置利用可再生能源如太阳能热水器等。

1.3 混凝土核心激活技术

2002年8月建成的ARCOR大厦是一幢位于法兰克福市郊Eschborn中央办公区、设施一流、技术超前的高级办公楼。为保证经济的、同时也是生态的能源利用，法兰克福JO. Franzkf建筑事务所与Transsolar能源技术公司一起为这幢建筑面积45000m^{2}的高层建筑研发并使用了混凝土楼板核心激活技术，用于大楼的制冷和采暖(图3)。

图3 ARCOR高层办公楼，Eschborn

(建筑师：Jo. Franzke Architekten，法兰克福。来源：Jo. Franzke Architekten)

混凝土核心激活技术是通过充分考虑和最大可能实施自然通风、降低进风量到适度和利用建筑物自身的蓄热功能以及周围环境空气的自然降温能力达到节能、舒适环保的目的。使用者的舒适度不仅得以提高，建筑的投资成本和运营费用也得到最大程度降低。具体体现在：

• 由于取消了中央空调系统，使用者可以根据自身需要调节确定他的使用空间的温度，其基础温度(冷却最大量45W/m^2，供暖最大量37W/m^2)是通过混凝土核心激活技术得以保证的。

• 额外的需求量通过进风调节。这里每个小空间的使用者都可以自行控制。所必需的新风通过在架空地板内置的辐射式新风机经过一个加热或制冷的叶片装置被引入室内，并被调节到所需温度。各空间的出风口通过一个经过防噪声处理的构件，并有防止空气倒流功能，当室内气压增大时，空气将自动被排出到室外(见图4)。

与传统中央空调系统相比，这套体系的优点是非常突出的：

• 降低了投资成本；

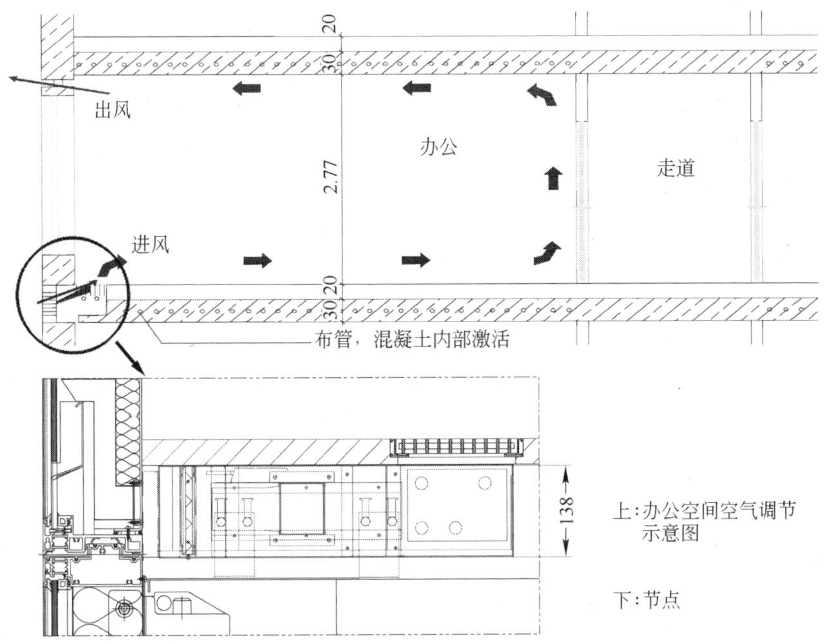

图 4　ARCOR 高层办公楼，办公空间空气调节示意图及节点大样
（建筑师：Jo. Franzke Architekten，法兰克福。来源：Jo. Franzke Architekten）

- 高灵活度，可符合单个用户不同需求；
- 极低的通风设备和管道面积；
- 大大降低了层高（办公空间内不需要吊顶）；
- 节省采暖和用电消耗；
- 噪声辐射降低到极限。

由于进风、出风口的精心设计，这些功能构件完美地融合于外立面当中，增强了立面的韵律从而提升了美学价值（图 5）。ARCOR 大厦的成功给西方建筑界做了极好的典范，

图 5　ARCOR 高层办公楼，立面局部，完全看不到进出风口
（建筑师：Jo. Franzke Architekten，法兰克福。来源：Jo. Franzke Architekten）

混凝土核心激活技术也在当今许多建筑中得以推广使用。

1.4 地热资源利用

大地是一个恒温体，地热是一个取之不尽用之不竭的能源。同其他可再生能源利用比较，地热不论冬夏、不论昼夜均可获取。夏季利用地下较低的温度制冷，冬季利用地下较高的温度取暖。德国近年来在地热利用方面进行的研究投入取得了明显的效果。大体上地热资源利用可以分为两大类：热量直接从土壤中获取——这种方式在全世界任何地方都可使用；热量从地下水中获取——要在当地地下水资源适宜情况下才可以使用。基本的原理相对简单：利用地热桩或地下铺埋的管道，通过循环水或循环空气不断进行热交换，将从地下获取的冷(热)量传至建筑物内部。其次将大地作为一个巨大的热存储器，将夏季从周围环境中收集到的热量存于地下，冬季将其取出使用，也在具体项目中得到验证和使用。

德国联邦环境部大楼(图 6)于 2005 年初建成投入使用，在其良好的通风系统的设计中使用了包括地热利用技术在内的多项节能措施。此项目通过地下长达 5km 的埋管，室外新鲜空气先通过地下管道进行热交换以及净化处理再进入室内，热交换器的热效率达到冬季约 86000kWh/a，夏季约 125000kWh/a，见图 7、图 8。

图 6 德国联邦环境部大楼(建筑师：Sauerbruch hutton，柏林。来源：自摄)

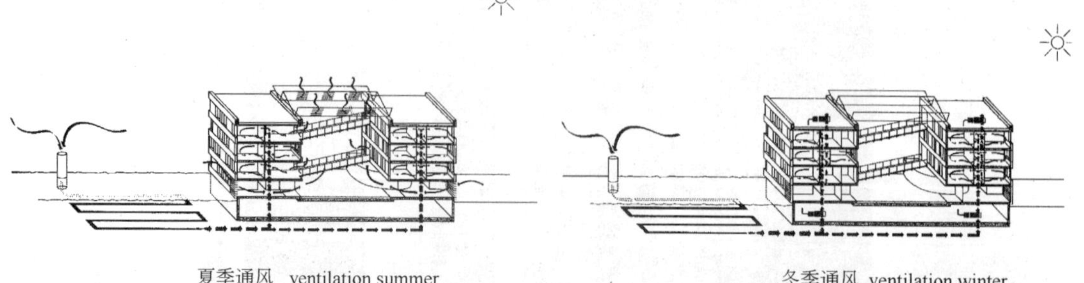

图 7 德国联邦环境部大楼，有地热资源利用的建筑通风设计

(建筑师：Sauerbruch hutton，柏林。来源：《Federal Environmental Agency Dessau》)

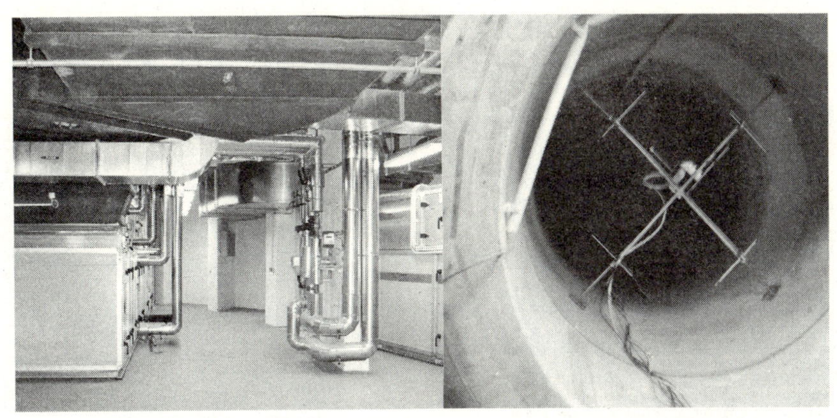

图8 德国联邦环境部大楼，地热交换器进口部位以及控制室(来源：自摄)

1.5 太阳能

太阳能利用是一个历史久远的话题。近年来太阳能在建筑中的运用得到极大的发展。遵循不同的原则和在不同层面上既有实践、实验型分析，也有理论、概念型研究。在降低所有建筑类型——包括小到单栋住宅大到大型展览场馆建筑——能量消耗的紧迫要求下，太阳能的利用势必越来越广，而且突破以往与建筑造型脱轨的局面，巧妙地与建筑整体结合，变成与建筑自身不可分割的组成元素的一部分。

建筑的外壳是体现建筑能量消耗的一个相当重要的系统。在与太阳能技术相结合的过程中，建筑外壳是介于建筑语言和太阳能技术的重要媒体。新近的太阳能技术的使用体现在外墙和屋面上，太阳能装置并非简单安装在上面，而是在建筑设计开始就作为建筑固定的一部分加以考虑，太阳能装置成为立面、屋面的一部分，有其经常的保护作用；或者反过来说，立面或屋面具有将太阳能转化为建筑能源的功能。这里太阳能装置元件的美观、与立面和建筑风格整体协调就成了要解决的关键问题。

现在几种常见的光电装置无论从功能上还是外观上都比以往任何时候得到了相当大的改进：单晶体硅光电池——生产难度较高但能量转化功能较好(14%～16%，见图9)；多晶体硅光电池——易于生产价格较单晶体硅光电池低，能量转化功能也略低(12%～14%，见图10)；非晶型硅光电池——价格更低，能量转化功能也更低(5%～7%，见图11)；

图9 单晶体硅光电池

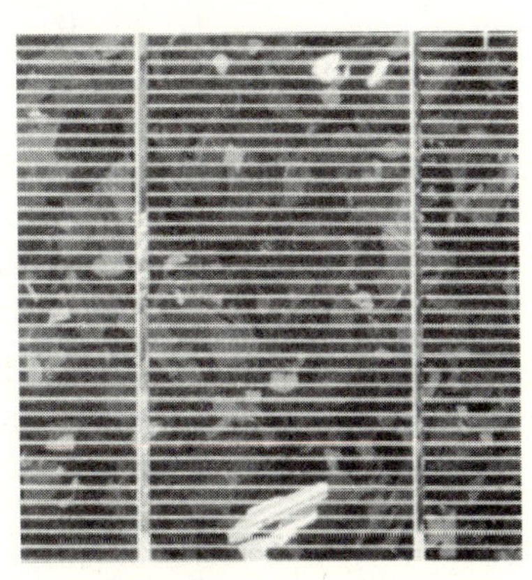

图10 多晶体硅光电池

CIS稀薄涂层硅光电池——新型电池技术，多用铜、铟、硒制成，金属用量极少，可作为大面积涂层用于任何表面，不拘形式，能量转化率可达到8%(见图12)。

图11 非晶型硅光电池

图12 CIS稀薄涂层硅光电池

太阳能技术元件作为融合到立面或屋面中的构件在已建成建筑中已经有大量的实例，如Hamm的Gründerzentrum(图13)，Thyssen Krupp Stahl厂房立面改造——欧洲目前最大的太阳能工程项目(图14)。但总体而言，基于建筑美学角度的考虑，材料的研发和改进仍是此领域的主要任务之一，另外建筑师和结构工程师以及其他工程师的紧密配合也是保证创新的太阳能技术的推广和应用的必然途径。

图13 Gründerzentrum(Hamm，德国。建筑师：Hegger Hegger Schleiff)

图14 Thyssen Krupp Stahl (德国。建筑师：Czerny-Gunia)

2 摸索如何在我国建筑实践中应用国外经验

不容忽视的一个事实，就是我国在节能绿色建筑方面的发展远远落后于发达国家的标准。建筑的能耗非常惊人，积极的一面是中央已经意识到了这个问题，正在加大力度改变这种局面。由于以往对建筑节能的忽视，我国在节能建筑设计和实践方面缺乏经验，尤其是在节能意识上面，必须尽可能学习发达国家的长处，借他山之石，弥补自身不足。笔者在德国学习工作多年，回国执教的同时，也在设计上与德国专家合作，在多个设计竞赛项目中都采用了合理的节能概念，并且配合到建筑整体造型与空间规划上。下面以一个例子介绍如何将德国先进的节能设计应用到我国的建筑设计中。

2.1 项目概况

上海申能集团是市政府下属大型国营企业，筹划在市区内环线以内靠近虹桥机场的一块基地上建设由集团母公司和其下属的申能股份公司和燃气集团公司共同使用的纯办公建筑，占地面积 14755m²，建筑面积地上与地下共约 3 万 m²。由于距离机场较近，建筑限高 40m。

2.2 概念性方案

2.2.1 建筑：与德国法兰克福 JO 建筑事务所和节能设计工程事务所 DS-Plan 一起，经过反复研究业主的使用要求、基地现状和出于建筑节能的考虑，我们提出了一个紧凑的“H”型总体布局——中央主要朝向为母公司，两翼分别为两个附属公司，空间上满足了三家公司既分又合、主公司位于中央占主导地位的要求。在“H”型平面前后两个围合空间处设计了中庭，正立面处为主要入口，两侧为展厅，兼备交通和展示作用；后面为大型会议和多功能厅、食堂和自用文娱设施配合室内屋顶花园(图 15)。

图 15 申能集团办公楼概念性方案设计，建筑表现

2.2.2 建筑与节能设计相结合：针对上海所处地域的特征，在设计建筑室内环境时要解决夏季炎热和冬季湿冷的问题。以往使用的中央空调系统虽然调节了温度，但其缺点是不仅能耗大，而且噪声大，风速使人易得空调病，舒适度极低。结构上由于风管高度限制，层高中浪费的空间较多，经济性差。空调机组设备的投资费用也很昂贵。但由于上海地区极端的气候条件，完全取消空调系统并不合适，所以设计中采用了综合方案，采取五大节能措施，以减小空调的负荷，使得设备的容量减小，降低设备投资和运营费用。具体体现在：

- 最佳的建筑保温措施——建筑总体设计紧凑，合理的窗墙比，极好的外墙保温，极佳的隔热玻璃和外遮阳措施；

• 通过混凝土楼板激活来调节室温：降低空调工作负荷，利用地热，使用效率极高的地源热泵，利用楼板极好的蓄热性夜间继续调节室温(图 16)；

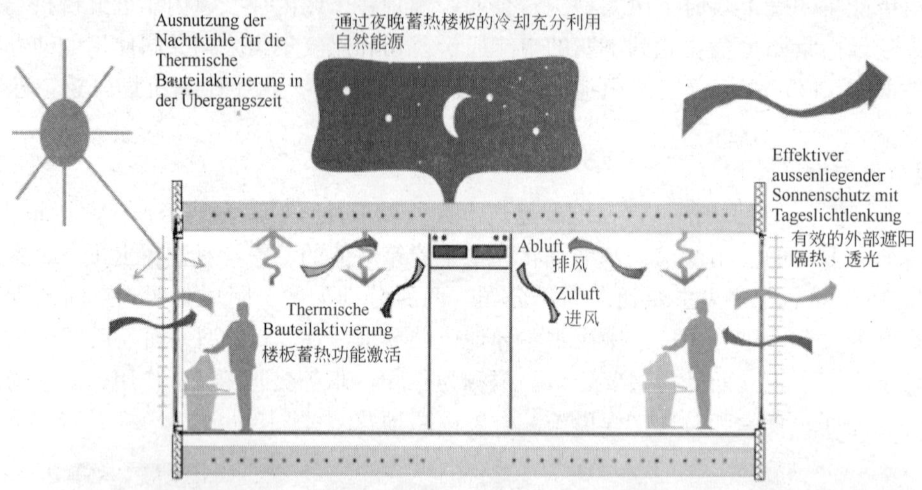

图 16 申能集团办公楼概念性方案设计，建筑热保温系统和通风设计

• 结构桩结合地热桩；

• 通过地热交换器预先调节送入中庭的空气——建筑设计与节能设计紧密结合，这个空间不仅有形象上的作用，同时也是一个大的恒温缓冲空间，其两侧的办公空间均可向中庭自由开窗；通过地热交换器送入中庭内的空气经过了预先净化处理，在一年大多数时间不需使用能源，仅在极端气温情况下经过地源热泵，消耗能量十分有限(图 17)；

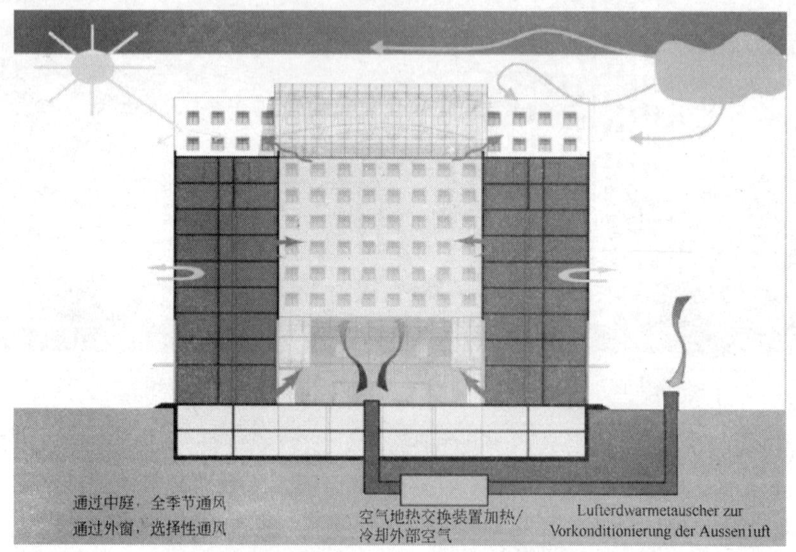

图 17 申能集团办公楼概念性方案设计，通过中庭和地热交换器的空气调节概念

• 屋顶和中庭顶部设置太阳能集电板，为太阳能辅助新风系统通过能源。

2.3 预期目标

通过这五项措施将达到经济、环保和舒适的目的。初步测算，初期投资仅增加 15%，而运营成本保守估计可节省 50%，更为重要的是由于尽量使用自然的可再生能源，减少了不可再生能源的消耗，二氧化碳的排放量也降低了 40%(图 18)。这是绿色节能建筑对于全世界以及未来的贡献。

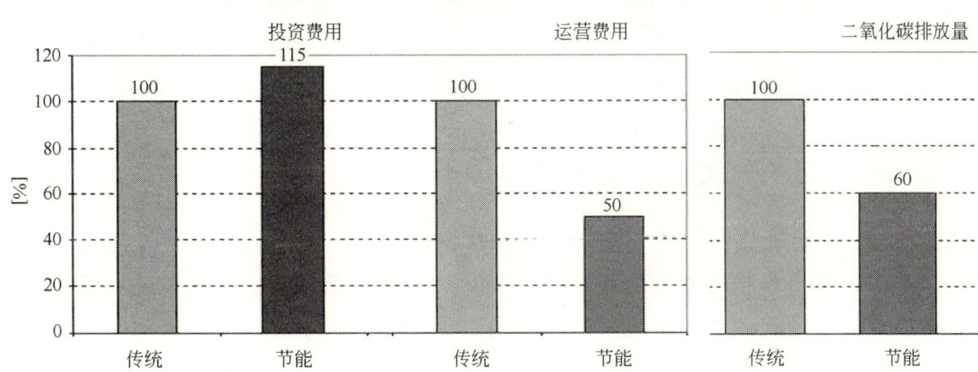

图 18　申能集团办公楼概念性方案设计，节能和不节能建筑在投资、运营费用和环保方面的比较

German Energy Saved Design Methods and Practice and How to be Combined in Our Design and Building Practice

Qu Cuisong

(Dr.-Ing. TU-D，Associate Prof. Faculty Architecture，College of Architecture and Urban Planning Tongji University，Shanghai　20092　China)

Abstract: Since middle of the 20. Century the industrialized countries came into median society. New building technique，new building materials turn the architecture to face a new situation. Thanks to this development many building types and design ideas in the past could be achieved now in many places. The new trends in Germany are more to pay attention to protect the environment，sustainable developing and the needs of humans. Centre activating technique of concrete，solar energy as energy recourse of building as well as new developed concrete building materials in the architecture are the main points of this article. Latest it will discuss how to us e those design methods in our design practice through a recent project.

参考文献

[1]　Arcor-Haus Jo. Frabzke Architekten，Christof Bodenbach(Hg.)

[2]　Gebäudehüllen，Konzepte Schichten Material，Christian Schittich(Hrsg.)

[3]　FederalEnvironmental Agency Dessau

[4]　Das Bauzentrum Baukultur，2001/2

[5]　Forschungsdokumente von Passivhaus Institut

[6]　TU-Darmstadt，Lehrstuhl Baugestaltung und Baukonstruktion，Lehrmaterial

构造设计学概念下建筑节能设计的探讨

刘启波
（长安大学建筑学院，陕西，西安　710061）

摘　要： 本文基于构造设计学的概念，力图在建筑设计中，尽量不依赖耗能设备，而在建筑形式、空间、布局和构造上采取措施，充分利用天然资源和地理条件，采取被动式构造设计手段，以改善建筑环境，实现微气候建构，这样既可满足生活舒适的要求，又可最大限度地减少能源用量。本文具体地从建筑设计中的自然通风、自然采光、遮阳以及围护结构的改善等几方面探讨了如何节约能源，从而保护生态环境。

关键词： 绿色建筑设计　建筑节能　构造设计学

1　前言

人类环境包括自然环境和人工环境，后者又包括实体(建造)和虚体(人文)环境。建筑是人工环境(实体与虚体)的重要载体，它对自然环境产生直接的冲击和影响，只有在建筑是可持续发展的前提下，人类环境才有可能成为可持续性的。[1]

建筑系统本身是巨大的消费者。它总会这样那样地给生态环境带来负面影响。传统的建筑活动在为人们提供生产和生活用房的同时，往往过度地消耗资源和能源。它所产生的建筑垃圾、城市废热等形成了严重的环境污染，造成全球范围内的人类生活质量下降和自然生态系统的破坏。中国经济在高速发展中造成了严重的环境污染和生态破坏，使中国的生态环境处于十分脆弱的状态，且至今生态环境的恶化仍在继续，建筑业对此负有重大责任。如果通过建筑设计降低建筑的耗能，就能减少全球的耗能，对整个生态系统的稳定有着不可小视的作用。

今天，我们要从绿色建筑设计的理念[2,3]出发再次认定建筑与环境关系的新内涵，那就是建筑的规划、设计要有利于促进人造建筑环境与自然环境的生态平衡，有利于形成自然生态的良性循环，有利于实现人类可持续发展的目标。因此，我们的设计意识就是环境设计意识，要求设计者具有更加生态化的责任感从事他的建筑设计工作，也意味着建筑设计要超越单一建筑建造的范围，走向设计整个环境，寻求获得最高的使用价值和对环境的最低影响。

那么如何体现实行低耗能、克服严重的环境污染，采取节约型建筑技术体系呢？有关专家提出了“构造设计学”的概念，即要求在建筑设计中，尽量不依赖耗能设备，而在建筑形式、空间、布局和构造上采取措施，充分利用天然资源和地理条件，采取被动式构造设计手段，以改善建筑环境，实现微气候建构，来满足生活舒适的要求。本文正是基于这样的概念，从建筑设计中的自然通风、自然采光、遮阳以及围护结构等几方面来阐述如何节约能源，从而保护生态环境。

2 关于自然通风的设计

建筑自然通风设计是建筑节能、创造室内舒适环境的有效手段，特别在2003年上半年突发的一场灾难——“非典型性肺炎(SARS)”袭击神州大地时，首要的防疫措施竟然是要保证自然通风！流行病学专家认为，无论居家、办公、行车，保持通风比消毒更重要。为此，在建筑设计中要求：建筑物留有适当的可开口位置，以充分利用自然通风，减少对机械通风的依赖，从而达到节能的效果；室内无通风死角，在自然状态下室内具有穿堂风；充分考虑不同地理气候条件下风向的影响。

通过流体力学原理分析，建筑通风的形成机理主要在于室内外温差造成的热压差(ΔP)，该热压差造成气流流动动力，并通过建筑洞口有效的高差设计，强化气流动力，而形成建筑通风[4]。有关ΔP形成的通风值的评价依据在于：洞口高差ΔH和空气密度差$\Delta\rho$，只要在设计中把握好二者，就能达到热压通风的目的。尤其是洞口高差ΔH，在设计中完全可以控制，其措施有：

(1) 建筑剖面设计的通风调整。在剖面设计中控制洞口高差，以尽量大的ΔH值来加强温差所产生的通风反应，尤其是在住宅建筑中，使出风口高度大于进风口高度，使室内气流呈上升趋势，这是符合室内外温差所造成热压之通风规律的，其上升力与通风走向相同，将有效强化通风质量。

(2) 建筑空间设计的通风调整。为了加强建筑通风，在建筑设计时，结合一些功能目的(楼梯间、中庭)，设计具备“烟囱效应”的空间体系，如日本建筑中的吹拔等，可起到拔风作用，以强化室内通风。

其次，还有压力体型系数的差C_1-C_2，建筑师虽无法改变室外风速V，但可以通过体型设计来控制C_1-C_2，如建筑迎风面的调整和建筑的导风调整。建筑形体控制方法的主要作用就在于通过对建筑物、构筑物的平面、剖面的形式，以及其形体间的空间组合关系的有效控制，并通过对所设计环境中植物的合理选择与配置来创造有利于居民生活的“再生风环境”。人工构筑物和绿化植物的遮挡作用的运用，是对建筑周围再生风环境进行调节的一种有效途径。

在延安枣园示范新窑的设计中，我们就充分利用了ΔP原则和导风的挡风板以形成负压区，通过楼梯间的拔风作用，使通风流道通畅，再加上地窖的送风(夏送凉风、冬送暖风)作用，改变了窑洞不能通风的弊端[5]，设计原理示意图如图1所示。又如马来西亚著名建筑师杨经文设计的槟榔屿州Mennara Umno大厦外墙中，外加了一种“捕风墙”的特殊构造设计(图2)，它在建筑两侧设阳台开口，开口两侧外墙上布置两片挡风墙，使两通风墙形成喇叭状的口袋，将风捕捉到阳台内，然后通过阳台门的开口大小控制过风量，形成“空气锁”，可以有效地控制室内的通风，这种做法值得借鉴。

另外，建筑自然通风设计还应考虑：南向是冬季太阳辐射量最多而夏季日照减少的方向，并且我国大部分地区夏季主导风向为东南向，所以从改善夏季自然通风房间热环境和减少冬季的房间采暖空调负荷来讲，南向是建筑物最好的选择；住宅楼的进深以10m至13m为宜，太浅或太深都不利于室内空气流动；建筑高度对自然通风也有很大的影响，

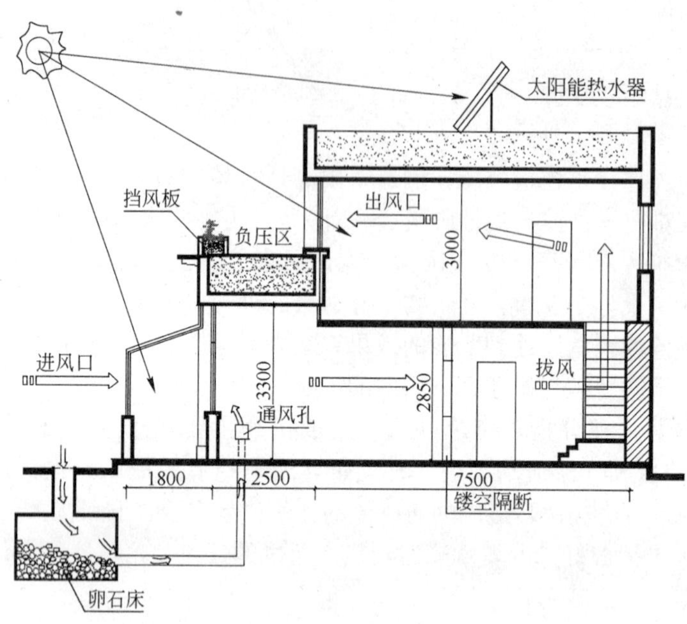

图 1 枣园示范新窑通风、采光示意图

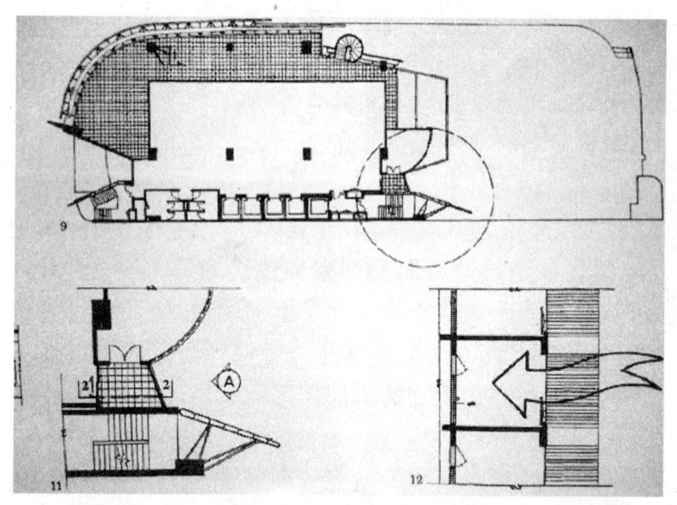

图 2 杨经文：槟榔屿州 Mennara Umno 大厦“捕风墙”设计示意图

一般高层建筑对其自身的室内自然通风有利，而在不同高度的房屋组合时，高低建筑错列布置有利于低层建筑的通风。

3 关于自然采光、遮阳的设计

自然采光能够非常显著地减少能耗和运行费用。建筑中照明的能耗约占建筑总能耗的40％～50％，而且由灯产生的废热所引起的冷负荷的增加占建筑总能耗的3％～5％，合理设计和采用自然采光能节省照明能耗的50％～80％，减少了电力需求和相关的因发电产生的污染和副产品。加之，国外的研究证明自然采光能形成比人工照明系统更为健康和更为兴奋的工作环境，长期不见日光或者长期在人工光环境下工作的人，容易发生季节性的情绪紊乱、慢性疲劳等病症。为保证一定的日照时间，建筑物在设计时还需考虑环境的遮挡问题，如住宅与阳光障碍物的距离一般不应小于障碍物高度的两倍。

自然采光需要建筑围护结构上的开口或洞口的位置正确，允许日光进入并充分分配和发散光线。设计良好的系统能避免因直射引起的会削弱视力和产生不舒适的多余的亮度和得热，为控制多余的亮度和反差，窗户上往往会设置一些附加件，如遮阳、百叶和格栅，如有可能，采用自动控制装置，当自然光足以维持理想的照明水平时，使灯光变暗或把灯关闭，在尽量节省照明能耗的同时维持所需的工作照明或环境照明，往往也是很理想的。另外，利用落叶树木在夏季可遮挡多余的过热的日光，冬季树叶枯落后，对日光的进入影响一般不大于20%。

自然采光环境除了居室日照外，还要考虑室外场地才能形成明媚愉人、有利身心健康的好环境。

4 关于围护结构及材料的设计

围护结构必须平衡通风和日光的需求，并提供适合于建筑地点的气候条件的热湿保护。围护结构的设计对于建筑在运行中的耗能是一个主要因素，因此，在设计时应考虑。

4.1 确定建筑物的位置和朝向，尽量减少气候因素对围护结构的影响

经调查，在其他条件相同情况下，东西向多层住宅建筑的传热耗热量要比南北向的高5%左右，尤其是在村镇住区，其建筑层数及容积率平均比城市为低，上述影响比城市更为显著。

4.2 确定合理的体形系数

$$\text{建筑物体形系数}=\frac{\text{建筑物与室外大气接触的围护结构面积 } F}{\text{上述外围结构所包围的体积 } V_0}$$

在其他条件相同情况下，建筑物耗热量指标，随体形系数的增长而增长，如果两幢建筑有相同的体积，设计得较紧凑的一幢热效率也较高。托马斯·赫尔佐格的“对角线立方体”就是应用了体形系数较小形式则冬季热损失较少的原理(图3、图4)。

图3 南方某高校遮阳百叶示意图

4.3 根据气候类型，考虑使用不同的围护结构材料，以减少热损失和达到保温隔热效果

这就要求：

（1）在干热气候中采用高热容量材料。

（2）在湿热气候中采用低热容量的材料。

（3）在温和的气候中，根据建筑位置和采用的供热/供冷策略选择材料。

（4）在寒冷的气候中采用密封和保温很好的围护结构。

图 4 “对角线立方体”示意图

例如，窑洞民居的被覆结构具有良好的保温隔热性能和热稳定性，但其正立面“窑脸”部分的热工性能极差，因此在延安枣园绿色示范村落，采用了毗联温室型窑居太阳房[5]，在冬季晴朗的白天，阳光间处于集热状态，当温室内气温高于窑洞内部气温时，与室内冷热空气自然循环，太阳能在被转换为热能后，又逐渐被输送扩散到窑洞内部，并有部分储存于室内储热墙等物体内，温室与窑洞形成连通的室内空间；夜间窑洞与温室间的门窗关闭，对于窑洞室内空间来说，阳光间成为其与外部空间之间的一“加厚”保温空气间层，室内热量经共用门窗传至温室，然后又传至门外，在周而复始的日周期循环中，两个阶段相继出现，达到了利用太阳能采暖的目的。

4.4 使用热效率高的材料

采用单一材料进行保温隔热已不合适，会大大增加建筑运营过程中的能耗，增加全寿命费用。目前，新型保温材料的物质种类很多，除考虑其物理性能外，还应了解它们的强度、耐久性、耐火及耐侵蚀性等是否满足要求，材料的选择要结合建筑物的使用性能、构造方案、施工方法、材料来源以及经济指标等因素。

4.5 确定合理的窗墙面积比

作为影响建筑能耗四大围护部件之一的门窗，一般是薄壁的轻质构件，是建筑保温、隔热、隔声的薄弱环节。它通过辐射传递、对流传递、传导传递和空气渗透等四种形式导致建筑物能量流失，绝热性能最差，普通单层玻璃窗的能量损失约占建筑冬季保温和夏季降温能耗的50%以上。因此增强门窗的保温隔热性能，减少门窗的能耗，是改善建筑热、光环境质量、实现建筑节能目标的重要步骤。由于玻璃的热传导系数大，无论就采暖还是空调来说，均应严格控制窗墙的面积比，表 1 列出其规定数值。

窗墙面积比 表 1

朝　向	窗墙面积比	朝　向	窗墙面积比
北	0.25	南	0.30
东、西	0.30		

4.6 外围护结构的气密性

在我国，目前仍有80%以上的门窗达不到节能要求，而通过门窗缝隙的空气渗透耗

热量，约占建筑物耗热总量的23%～27%，若加上门窗面积的传热耗热量，则约占全部耗热量的50%。由此可见，外门窗是耗热的重要渠道，是节能的重点部位。尤其是村镇住区，其钢、木门窗气密性较差，而只有达到每小时每米缝长的空气渗透量≤2.5m^3，才能达到节能优化设计。

4.7 利用传统的覆土屋顶、厚土坯围护结构及墙壁、屋顶的绿化隔热等

覆土屋顶之所以保温隔热性能好，其优点来自土壤的热工性质，厚重的土层所起的隔热作用使土中温升很低。干旱半干旱地区的突出特点是季节性和日温差大，但日温波动在土壤中仅有一定的深度，在此深度以外即无波动影响。最后，应大力提倡墙壁、屋顶的绿化，不但能够美化生活，增加绿色，而且对夏隔热、冬保温有节能的效益，在平屋顶、窑顶、墙面、阳台等处都可实施。

Discussion About Architectural Design of Economy on Energy Under the Concept of Constructional Theory

Liu Qi-Bo

(Architectural Institute of Changan University，Xi'an，Shanxi 710061 China)

Abstract: This paper based on the concept of Constructional Theory，with might and main strive to be independent of energy consumption equipment，and that take the measure in shape，space，distribution and construct of building，sufficiency utilize natural resource and geography factor，resort to passive form constructional artifice，so as to improve architectural environment and micro-climate construction. In this way as well can satisfy require of living coziness as can decrease energy dosage in high limit. This paper from reformation in ventilation，nature lighting，awning and enclosure construct to discuss how to economize energy resource，sequentially protect ecological environment.

参考文献

[1] 周若祁等. 绿色建筑. 北京：中国计划出版社，1999.6

[2] 刘启波. 绿色住区综合评价的研究. 国家自然科学基金委员会“九五”重点资助项目：黄土高原绿色建筑体系与基本聚居单位住区模式研究(报告09). 2001.4，西安

[3] 刘启波、周若祁. 绿色住区综合评价的研究. 新建筑. 2003(1)

[4] 宋德萱. 建筑自然通风设计导则初探. 21世纪建筑新技术论丛. 2000.7

[5] 刘启波、周若祁. 生态环境条件约束下的窑居住区居住模式更新. 环境保护. 2003(3)

绿色村落形态规划思想初探*

李立敏
（西安建筑科技大学建筑学院，西安　710055）
张　华
（河南工业大学土木建筑学院，郑州　450001）
吕红医
（郑州大学建筑工程学院建筑系，郑州　450001）

摘　要：本文立足于当前“城市化”进程影响下的广大农村，以一村为例，针对乡土建筑正在逐渐丧失其地方性特征，村落整体形态出现各种破缺现象，建筑体系产生强烈的单一化聚集及逐级拷贝等现象。力求探求其形成与发展的原因，试图指出绿色村落形态的动力机制和手段，调动其内在的潜力，实现村落的社会文化系统、自然生态系统、农工业经济系统以及建筑环境系统等的可持续发展。

关键词：绿色村落形态　动力机制　可持续发展

1　背景

1.1　绿色建筑与可持续发展

20 世纪 80 年代以来在全球兴起的“绿色运动”是人类在探索可持续发展道路上的重要方面。改善人类环境，实现可持续发展已成为当今世界各行业、各学科研究的行动纲领，绿色住区和绿色建筑的研究成为建筑行业可持续发展研究的主题。

1.2　中国农村问题

农村与农民是中国的发展和未来的坚强的后盾，中国农业和农村可持续发展问题得到中国政府、学术界和公众的高度重视，已列入《中国 21 世纪议程》。目前，我国农业和农村正处在转折关头，在新的形势下如何继续为农业服务是建筑学面对的重大课题。

1.3　“中国村庄的可持续发展研究”SUCCESS 项目

由奥地利 Oicodrom 申请、欧盟资助的项目，中国村庄的可持续发展研究 SUCCESS，纳入中国政府中欧合作第五框架之中。作为国内规划与建筑设计专业的主要负责人之一，我与国内外 15 个学科专业、17 个合作单位的 50 余名学者共同就中国不同地区 7 个不同类型典型村落的可持续发展进行了长达三年的研究工作。项目从可持续发展的角度勾勒出每个聚居村落的轮廓，并进而从生态、经济、社会文化和建筑等不同角度出发，阐述其未

*　本课题属中欧第五合作框架项目“中国村落可持续性前景研究”SUCCESS（EU Project：ICA-CT-2002-10007）。

来的可持续发展之路。

其中，河南焦作博爱县的下伏头村代表了自然衍生发展而形成，且面临城市化进程的典型村落。

2　下伏头村现状与症结

2.1　下伏头村现状

2.1.1　概况

下伏头村位于河南省焦作市博爱县与山西省沁阳县交界处，隶属于太行山脉。村子大约有将近370年的历史，四面环山，贯穿山西与河南两省的丹河支流村东绕村而过。村中自然及人文景观资源丰富，同时附近有北魏的摩崖石窟以及青天河水库等旅游资源。下伏头是一个自然兼行政村，共有160户540人，村中可使用耕地面积约170亩，户均2.1亩，年人均收入1000元。

2.1.2　村落形态分析(见图1)

(1) 自然生态系统

地处太行山脉，气候和自然条件较好，温度适宜，日照充沛，热量丰富，雨热同季，适合多种农作物生长和林果种植。山体矿山资源较为丰富，但开采过度，整个山体已造成了极大的破坏。能耗以煤为主，存在环境质量问题。

图1　下伏头总平面图

(2) 农工业经济系统

下伏头村是以基础农业耕种为主要产业，如小麦、玉米等。但土地面积极为有限，农业规模较小，以家庭为主，因此农业收入偏低。曾利用山地资源开发和种植林果业，但缺少技术和市场指导，造成产品滞销，现已换种杨树。

下伏头村的村镇工业，蚊香厂已关闭，耐火材料厂几近停产。村里还有些小型的粮食加工点以及一些小商业零售点等，对于村子基本不产生污染。在农民的收入中，其主要经济来源有三种，首先是种植和养殖业，其次是小型交通运输等，再次是外出务工的收入。其中，外出务工的收入在村民整个收入中占很大比例。[1]

(3) 社会文化系统

下伏头村最早的住户是董家，370年前由山西迁来至此。村中保留了一定的地方风土人情，村民热情好客，民风淳厚。村中的公共文化配套设施较为健全，村中设有小学、诊

[1] Dieter Wagner. Sustainable Regional Economy. Success. Deliverable 6：256～282。

所、商店、大队部等。形成了以大队部为政治中心，以及穿过照壁的树、土地牌位供奉点、水边石磨等多个公共聚会场所，但多集中于山下。村中年轻人口的流失比较严重，常住和留守的人口多为老人、妇女和儿童。青少年辍学问题严重。❶

(4) 村落整体形态与建筑生活系统

村落形态：下伏头村根据自然地势分为两个部分(图 2)：山上和山下。山上的旧村依山就势，形态比较灵活多变，与植被相互掩映。但道路随山体曲动，适宜步行，车行不畅，道路到院落多有过渡空间。山下的新村是 20 世纪 90 年代后陆续修建的，受城市化影响，呈兵营式排列，形态较为呆板。山下道路类城市道路呈网格状布局(图 3)，直线式道路主次分明。在山上旧村中，居住的相对分散，无集中的广场等，但是随着路径的曲折，每户门前都可以形成一个小的交往空间。而山下新村中公共场所如槐树广场、大队部等，空间比较大，人们交往比较集中。

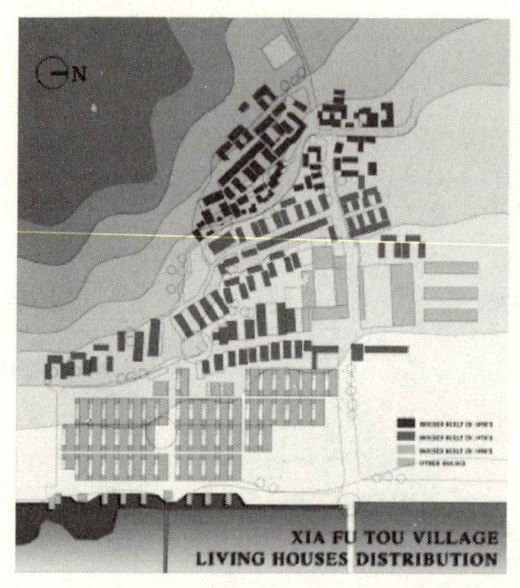

图 2 不同年代居住建筑

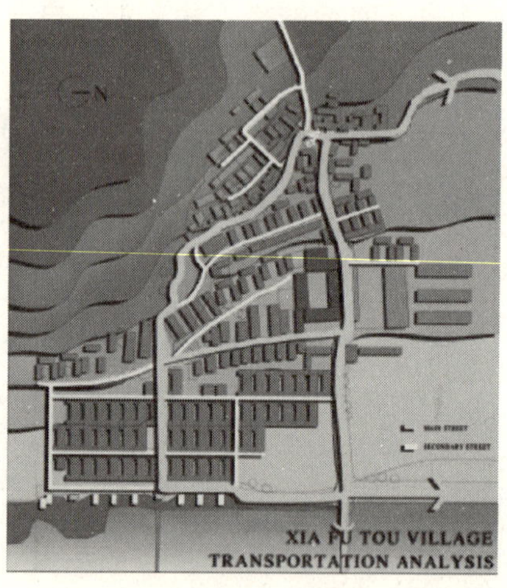

图 3 道路系统图

山上的旧宅一般为就地取材，多为青石砌成三合院形式，庭院较为开阔，院中大树为中心，有时院落对着街道打开，开放性较强。三合院形式能够营造较好的微气候，有效地组织院内的通风，且院中的大树有效的调节冬季和夏季阳光的辐射量，但其他的配套基础设施相对薄弱，譬如水、电的通讯设施等等。山下的新宅则打破了三合院的格局，由一栋二层的住宅及其附属的院子所组成。院子非常窄，院中植被较少，庭院内部的通风采光都不是很理想，但达成了较为便利的基础设施。

村中的整体卫生环境比较差，山上空间没有系统的排水和排污设施，亦不能完全适应现代生活的需要。

2.2 下伏头村症结

2.2.1 自然生态、社会文化、农工业经济和建筑生活各系统的严重脱节，无谓的耗散各系统能源，无法实现互动发展。

2.2.2 村中基础设施系统亟待加强完善，尤其是山上老村部分，使得人们有能动性

❶ Heidi Dumreicher & Bettina Kolb. Socio-Culture: Participation and Photo-interview. Success. Deliverable 6: 295～298。

地返回到传统建筑的人文景观当中生活。

2.2.3 村落发展无主导经济优势，导致大量年轻人口外流，形成“空心村”局面。

2.2.4 村民凝聚力不足，对村落建设缺乏参与意识。

2.2.5 与外界联系缺乏，信息闭塞，辍学的年轻人缺乏聚会交往和学习的场所。

2.2.6 村中山上与山下居住环境脱节，人流与能量流大量向下流动，同时目前村民们的主要活动区域和场所缺乏有机的联系性，同质性较强，缺乏一定的联动效应。

3 绿色村落体系

3.1 研究思路

3.1.1 更新传统观念

绿色村落的可持续发展研究，涉及因素庞多复杂，相互作用形成一个复杂精巧的网络，以传统的研究方法已无法面对这个复杂系统。因此无论从研究内容、思维方式、评价指标，还是研究方法等等方面都要加以更新(图4)。

观念 内容	传统研究观念	绿色村落研究观念
基础理论	纯自然科学	建筑学 规划学 生态学 经济学等多学科交叉
研究内容	物质运动规律	人与环境协调关系
思维方式	因果链	立体关系树、关系网
核心	力的作用	人为参与调控、符合自然生态规律
组分关系数	有限、关系简单	众多、关系复杂
参数	可统计处理	粗糙、不确定
目标	单目标	多维目标体系
评价指标	客观、公理，求标准	以符合自然生态发展规律为首要标准
调控方式	外部控制	自我调节，自我组织
研究方法	硬方法	软方法
研究途径	求解模型、优化过程	探索对策、实时调节
研究过程	结构问题→因果关系辩识→数学模型→解析优化→最优解→实施	功能问题→反馈关系辩识→生态模型→辩论探方→有效解(最优)→施治
研究结果	惟一或无解	多种途径但逐步趋向目标体系

图4 研究观念的更新

3.1.2 研究特点

(1) 强调综合性与协同性

项目遵循聚落的可持续性原则❶，将生态、经济、社会文化、农业发展以及建筑等的专业人员组织在一起共同开展研究工作，完成对于试点村落的综合评定，并协同地为村落的可持续性发展规划做出可行性方案。

（2）地方性与参与性 PRA

强调对于地方经验的搜集和村民的参与，项目研究的学者重视同农民交流，并和他们共同探讨自己村庄的可持续性发展道路。

（3）动态调整

强调将研究成果应用于实践，通过实践检验其可行性，并随时进行有针对性地调整（图 5）。

（4）局部实验性与“点穴法”❷

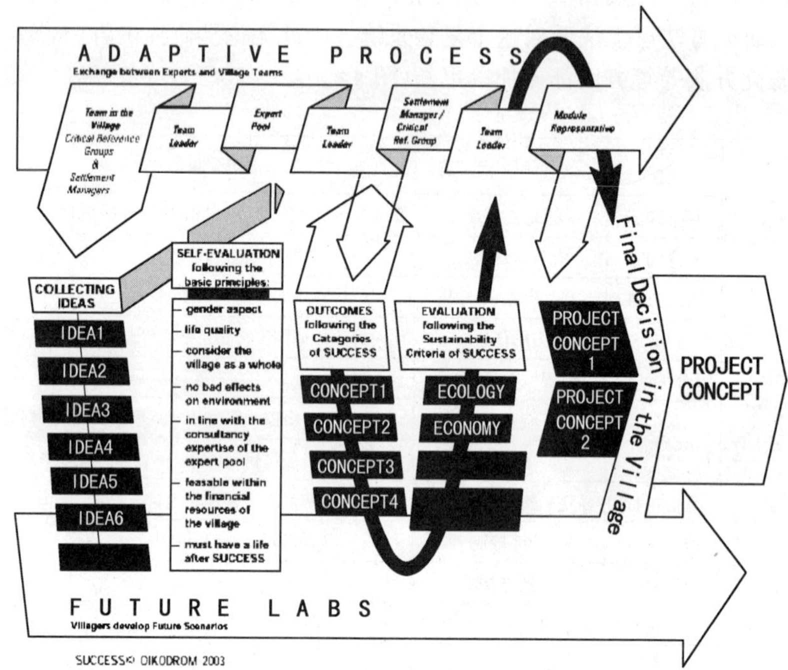

图 5　研究框架图❸

项目对于实验性项目的选择经过充分的探讨和论证，力求选择在村落整体形态中具有生态、经济、社会文化和建筑美学特点的关键性节点空间来建设小型的示范项目，并通过它的实施激活和带动整个村落的可持续性发展。

---

❶ 可持续发展是指在地方性，信息性，参与性，综合平衡与协调性等原则指导下，探索在可持续性资源总量之内的发展模式，以此，最大程度上地减少对于未来发展的消极影响，并进一步创造未来发展的空间和多种可能性。Heidi Dumericher & Richard S. Levine. SUCCESS. Deliverable 2. 2002。

❷ “中国村落形态的可持续性模式及试验性规划研究”博士论文 P5 吕红医。

❸ SUCCESS 项目报告 Deliverable 8. 2004. 10。

3.2 下伏头村绿色村落规划

3.2.1 规划设计的七项原则❶

(1) 村落发展的连续性；自组织规划思想将村落视为一个发展的过程，将其视为一个多种因素相互作用的综合体。

(2) 村落发展的地方性；自组织规划思想强调建立在普遍性基础之上的事物发展的特殊性，强调村落发展的地方性和个性。

(3) 村落发展的多义性；强调事物的多义性即强调事物发展的复杂性以及由此衍生的丰富的层次和多样性。

(4) 村落发展的自控性；自组织规划思想把村落看作是一个有机的“自然态”系统，是一个开放的具有自我调控能力、自我更新和修复能力的复合生态系统。

(5) 村落发展的开放性；自组织规划思想强调系统的开放性，并认为系统内部对于外界的任何输入都有非常敏感的反应。

(6) 村落发展的全益性；自组织规划思想追求整个村落形态系统的最优化，而不仅仅是某个因素的最优化，为了“权衡”各个不可兼顾的因素，仔细地折衷是必要的。

(7) 村落发展的全息性；村落发展的全息性原则源于生物学中的全息律，因而，自组织规划思想认为，村落形态系统与其组成的因子在性质上具有同一性的方面。

3.2.2 “点穴法”的应用

(1) 下伏头村“穴点”

通过对于下伏头村落系统的自然生态系统、乡土经济系统、社会文化系统以及建筑环境四个系统的互动关系的分析，我们不难看出，对于下伏头村的调整节点主要集中在这样几个场所：1)“树下”，2)“河边”，3)“山上”(村中的公共娱乐设施)，4)“路口”，5)“大队部”(图6)。

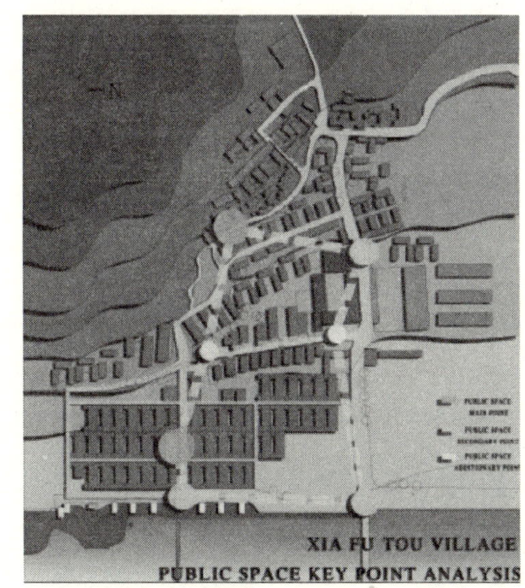

图6 下伏头村“穴点”分布图

(2)“穴点”的调控

由“河边”—“树下”—“队部”—“路口”—“山上”等建立起来的多重的复杂的联系，构成了不断与外界进行能量和信息交流的动态结构体系。

同时，每一个“穴点”又有机的调控着其周边的区域，与居住区，与自然生态建立起密切的联系。因此，在这个复杂的动态体系中，每一个“穴点”的活力及其能动性在很大程度上决定了整个村落形态的自组织模式的途径和发展方向。

随着“山上”大户董宅的衰落以及“乡村城市化”的加速和国家“退耕还林”政策的

❶ 村落形态的自组织演变及其规划思想初探 清华大学．首届全国博士生学术论坛论文集．吕红医 李立敏 382～389。

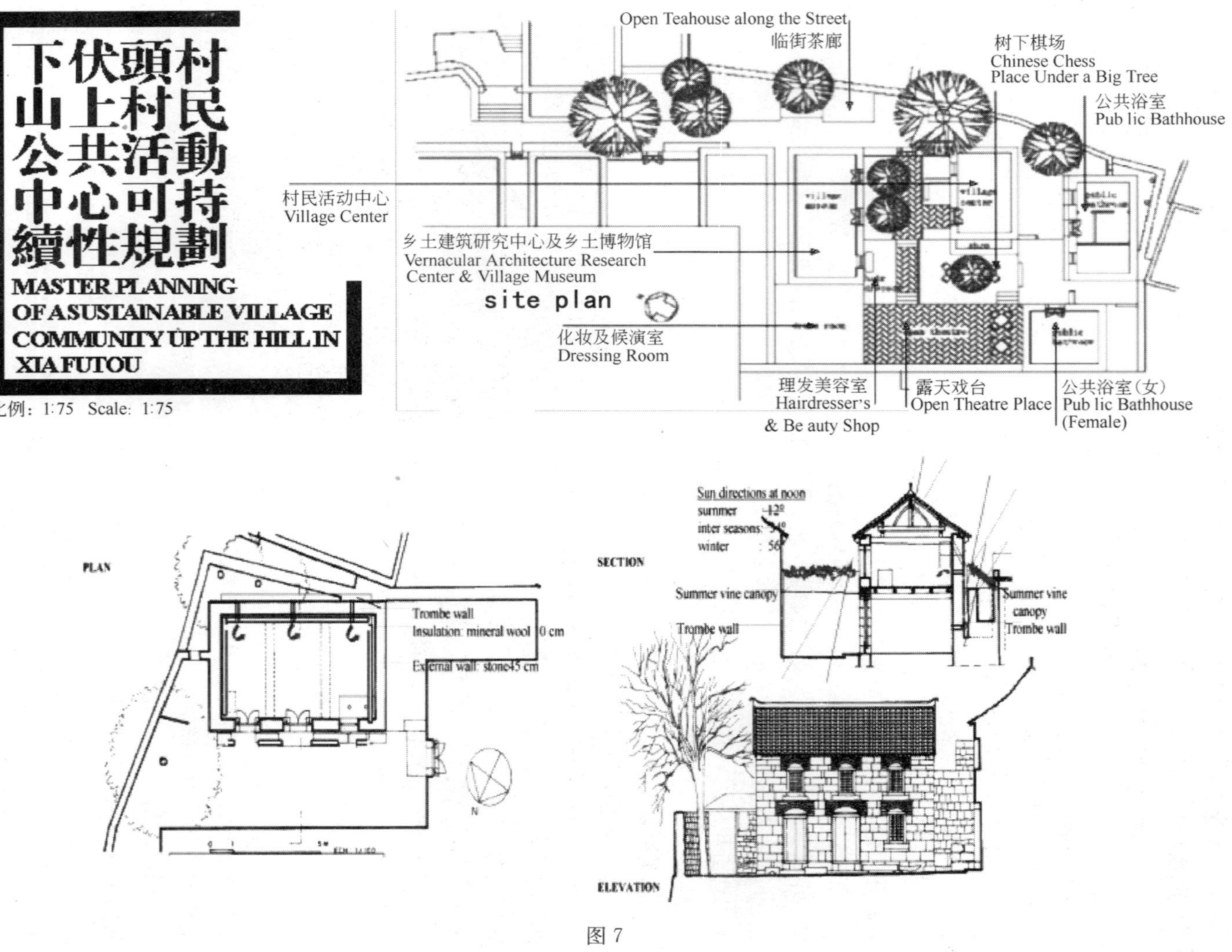

下伏頭村山上村民公共活動中心可持續性規劃
MASTER PLANNING OF A SUSTAINABLE VILLAGE COMMUNITY UP THE HILL IN XIA FUTOU
比例：1:75 Scale: 1:75
村民活动中心
Village Center
乡土建筑研究中心及乡土博物馆
Vernacular Architecture Research Center & Village Museum
site plan
化妆及候演室
Dressing Room
Open Teahouse along the Street
临街茶廊
理发美容室
Hairdresser's & Be auty Shop
露天戏台
Open Theatre Place
树下棋场
Chinese Chess Place Under a Big Tree
公共浴室
Pub lic Bathhouse
公共浴室(女)
Pub lic Bathhouse (Female)
PLAN
Trombe wall
Insulation: mineral wool 10 cm
External wall: stone45 cm
N
SECTION
Sun directions at noon
summer
inter seasons
winter
Summer vine canopy
Trombe wall
Summer vine canopy
Trombe wall
ELEVATION

图 7

导向，人们大规模向山下迁移。而这种迁移不但加速了“山上”的衰败，同时，大量占用了村中本已极为有限的耕地。这种趋向非常明显的导致了下伏头村村落结构体系的局部瘫痪和重心失衡的状态。而山上旧村的位置、尺度、规模，以及其多重意义和影响(生态价值、社会文化价值以及建筑美学价值等等)，使我们认识到对于“山上”穴点空间重振和综合开发利用对于整个村落结构体系具有举足轻重的影响。

在山上，利用废弃的董宅作为基础，建立“山上”绿色散点状小商业带，充分利用其建筑及景观优势，平衡和拉动“山上”经济杠杆的调节作用，主动吸纳外界的信息和能量，调节各“穴点”之间的有机互动并使之产生多重增值效应。功能上设计为公共浴室、露天戏台、树下棋场、村民活动中心、乡土建筑研究中心及乡土博物馆、美容美发等等(图7)。

3.3 穴点空间的实施建设与反馈

山上“穴点”空间作为首先调控的节点，先行开始公共浴室地建设，现已接近尾声。公共浴室结合地方气候特点，充分运用了太阳能、地能等能源资源，并且综合用水，制作小型中水设备，为山下提供部分生活用水(图8)。

图8　下伏头村公共浴室建设

通过公共浴室建设，已经有越来越多的人开始重新关注山上空间，在建设过程中已经开始了少量的经济活动。经过我们的努力，人们对山上空间重新认识，并频繁穿梭于山上老村与上下新村之间，村民有了更广阔的公共交往空间和交流场所，“山上”正在复苏。

4　结语

经过几年与项目组的中、外学者对于村落的发展进行了大量的论证和探讨，研究的内容涉及生态、社会文化、经济和村落形态等多个领域，对于绿色村落形态的可持续性规划前景的研究深感才刚刚拉开序幕。受“城市化”的巨大冲击，对于绿色村落系统如何保持其地域文化的内在活力和发展动力，保护其独特的形态特征和地方建筑特色等问题的研究，笔者仅以此文抛砖引玉，希望与更多的学者探讨。

The Theoretics Argumentation of the Green Village Patten

Li Limin[1] Zhang Hua[2] Lü Hongyi[3]

([1] College of Architecture, Xi'an University of A&T, Xi'an 710055 China;
[2] School of Civil Engineering and Architecture, He Nan University of Technology, Zhengzhou 450001 China; [3] Department Of Architechiture, Zhengzhou University, Zhengzhou 450001 China)

Abstract: By the influence of urbanization, affronting some confusing phenomena in the villages and local architectures, such as losing the local characteristic, monotonous and tendency to the city, we aim to discover the evolutionary reasons and find out the motivity and inner law of the village to make sure that the village, which includes social-culture system, rural economy system, rural ecology system and village pattern and rural building system, could develop in a green and sustainable way

参考文献

[1] 吴彤. 自组织方法论研究. 北京：清华大学出版社，2001. P3～4
[2] 郑春顺(法). 混沌与和谐. 北京：商务印书社，2002. P92～93
[3] 小拉尔夫・弗・迈尔斯(美)杨志信译. 系统思想. 四川：四川人民出版社 1986
[4] 费孝通. 乡土中国. 三联书店. 1985
[5] 凯文・林奇(项秉仁译). 城市的印象. 北京：中国建筑工业出版社，1999

绿色建筑中的控制原理*

夏 博 宋德萱 史 洁
（同济大学建筑与城市规划学院，上海 200092）

摘 要：本文从环境与建筑的关系出发，利用系统学、控制学的基本概念和理论，从一个全新的角度去研究绿色建筑。控制学在绿色建筑上的应用主要研究建筑的“可控性”，使绿色建筑适应环境的多变与差异，创造舒适健康的建筑环境。
关键词：绿色建筑 控制学 可控性

当前，绿色建筑与建筑的可持续发展已经成为全球共同关注的焦点。在人与自然的关系危机越来越突出的今天，对“环境—建筑”关系的理性探求就显得日益迫切。我国正在全面建设节约型社会，如何在我国特定的自然、社会、经济环境下建设可持续的、生态的、节约的建筑……面对这些问题，我们必须从新的角度，用新的研究方法去研究。

1 建筑与环境

环境作为建筑产生的前提，环境决定了建筑，环境是建筑设计的出发点。另一方面，建筑作为一种手段，创造并维持了舒适的建筑环境，是人对自然环境控制的一种手段。在建筑、人、环境组成的大系统中，环境是一切的“起因”，也是我们首先应该了解和认识的重要因素。

在建筑、人、环境组成的大系统中，环境包含了与主体相关的一切外围客体及其关联结构体系。“所有的环境都是多重要素整合的系统，建筑就是处于人、社会、自然这个整合系统中的主体事物。[1]”环境作为一个由多要素组成的功能复杂的系统，具有很多不确定性和差异性。

1.1 气候的差异性：由于地球公转的原因使其表面获得的太阳辐射存在很大的差异，和海陆间在物理属性上存在巨大差别，造成气候的差异性。

1.2 时段的多变性：即使在同一地区，外部环境也是在随着时间不断变化的。主要表现为四季和昼夜的周期性变化。环境的四季和昼夜变化强度随地区而不同。另外，除了周期性变化外，环境存在一些瞬时的偶然性的变化。

1.3 人体的差异性：不同年龄、性别、生活习惯、衣着、身体状况的人对环境的感觉是不同的，对建筑的要求也是不同的。因而建筑环境并没有一个绝对的标准，它只能是在一

* 国家自然科学基金(No. 50478101)资助项目。

个范围内，每个人都有一个适应自己周边群体的相对独立的标准。

2　绿色建筑中的控制原理

2.1　控制概念的引入

外部环境与建筑环境是一对互为因果的关系。一方面，外部环境是因，建筑环境是果，外部环境对建筑系统输入物质、能量和信息，物质流、能量流和信息流通过建筑输出到建筑环境，影响建筑环境；另一方面，建筑环境也对外部环境输出物质、能量和信息，反作用于环境。利用控制学理论以建筑环境为主要讨论对象，即从环境为因，建筑环境为果的角度去研究(图 1)。可以把这对复杂的因果关系简化成一般的函数关系公式(1)：

$$Y_t = F(X_t,\ M) \tag{1}$$

式中　Y_t——建筑环境子系统；

F——建筑子系统；

X_t——外部环境子系统；

M——不定因素(干扰量)。

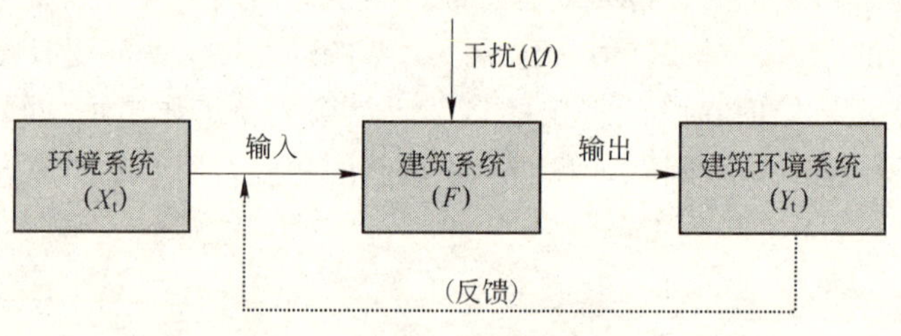

图 1　建筑内外环境的函数关系

函数关系表明了建筑环境与外部环境之间的因果关系，而控制的目的就是要根据外部环境的变化(X_t)以及使用者对建筑环境(Y_t)的不同要求，使建筑进行调整(即改变函数 F)，来获得舒适健康的建筑环境(Y_t)以适应变化了的环境条件、气候、个体等。

2.2　绿色建筑的控制目标

2.2.1　生态目标：保护自然生态环境，减少建筑对环境的影响。建筑的内、外环境作为一对因果关系，外部环境是我们控制的前提和基础。绿色建筑的最终目的是降低建筑环境的自然支持成本，即减轻建筑的环境负荷。而建筑与环境之间的关系归根结底是物质、能量与信息的交换，因而，绿色建筑就是对原有的建筑与环境之间的物质流、能量流和信息流模式的重建和整合的控制过程。

2.2.2　舒适目标：塑造一个舒适、健康、高效的建筑环境。舒适目标是控制的核心，是反馈控制的目标值和比较值，是整个系统调整的根据。现在的暖通空调技术完全可以绝对控制建筑环境，保证对每个人的“绝对”的舒适。但是它是以极高的环境成本为代价的；另一方面，这种稳态的舒适，造成了使用者对环境适应能力的下降，产生了很多健康问题。根据系统学原理，舒适的标准不应该是一个固定的值，它是根据外部环境动态变化的，是动态的舒适。英国剑桥大学的研究显示：对气候的适度变化适应机制的存在能减轻人的生理和心理压力，也就是说，室内气候的动态变化能对人体的舒适感觉起到增进作用[2]。因而当室内环境随外部环境呈现波向相近的变化时，不仅降低能耗，而且动态的环境对人体产生刺激也是生理和心理的需要(图 2)。

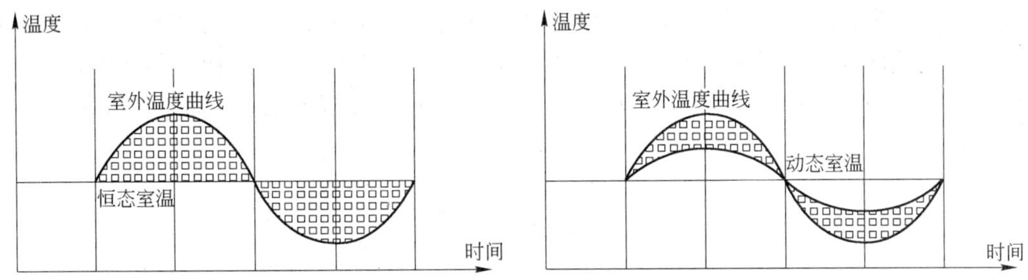

图 2　动态室温与恒态室温不同的能耗(网格部分面积可代表能耗)[1]

2.3　绿色建筑的控制原理

2.3.1　双极控制原理：由于人体的适应性，对频率出现最高的外界刺激敏感性减弱或消失，而对频率出现最低的极端刺激依然保持最敏感的应激性。也就是说，人体的适应能力最终总是使人对环境变化中出现频率最高的状态感觉最舒适。对于外部环境的周期性变化，控制技术利用负反馈原理，对外部环境周期变化中的双极端状态具有控制能力，以保持动态稳定的建筑环境和使用者的相对舒适。

2.3.2　整体性原理：系统整体性原理是系统论中的一个最基本的原理。它指系统诸要素集合起来的整体性能，也就是系统诸要素相互联系的统一性。对于绿色建筑的整体性原则就是：1)从大的生态环境去研究和设计建筑，以最小的环境成本取得舒适的建筑环境，保持“建筑—环境”系统的整体可持续发展。2)在研究建筑与环境的控制关系时，要整体关联地考虑局部与整体、形式与功能、能源与材料、设计与使用等等建筑与环境要素对建筑与环境关系的影响，把建筑与环境看作是一个要素相互关联的有机系统。

2.3.3　层次等级性原理：系统论的层次等级性原理指出层次等级结构是物质普遍的存在方式，另外不同层次等级的系统之间相互联系、相互制约，处于辩证的统一之中。绿色建筑的层次等级性原理包括两方面的内容：1)一栋建筑、一组建筑群对于整个生态环境来说是低层次的系统，但是它对整个环境和生态系统的影响是不可忽视的；2)对建筑与环境关系的研究要分类分层次的进行。室内外环境都是复杂的系统，有很多的子系统、子元素组成。当作研究、设计时，应该分类、分层次，由简单到复杂，做好绿色建筑设计的每一个方面。

2.3.4　系统有序性原理：系统学和控制学认为：任何系统必须保持开放性，才能使系统产生并且维持有序结构。热力学第二定律表明，对于一个没有热交换的封闭系统，系统的熵趋向增加，无序度增大，最后达到最大值而走向热平衡态，使系统成为“死结构”。建筑环境要保持稳定和有序，就必须保持其开放性，也就是必须要进行物质、能量和信息的交换，引入负熵。空调建筑是把建筑内部环境完全与外界隔绝，完全依靠设备利用大量的不可再生资源的引入，来保持室内环境的稳定，也就是以大量不可再生资源和生态成本为代价，引入负熵。而绿色建筑是要利用建筑系统把建筑环境与外部环境接通起来，通过控制有选择的使其进行物质、能量和信息的交换，把负熵引入室内环境，保持室内环境的稳定和有序。

2.3.5 目的性原理："所谓目的性，就是行为客体的一种反应或效应，这种反应或效应受行为结果的信息所控制，也就是说客体系统在和环境发生作用的时候，通过反馈不断调整自己的行为，使之逐渐趋达这一目标。[3]" 目的是反馈的依据，是控制的出发点。在进行绿色建筑设计时，无论是对整个建筑体系，还是建筑的子系统、子元素都应该有个明确的目标和期望值。根据目的才能确定具体的控制手段和控制方法，从定性的控制研究转变为定量的控制研究。

3 控制技术体系

绿色建筑控制技术根据施控装置的不同可以分为："零控制技术"和"反馈控制技术"。"零控制技术"是指在"建筑—环境"组成的大系统中，当外部环境变化时，建筑系统自适应的做出相应调节和无需不可再生能源的输入就可以保持建筑环境在舒适范围内的技术(无人的参与，无单独的施控装置，无反馈机制，无不可再生能源输入的建筑技术系统，属于开环控制系统)。"反馈控制技术"是指通过对受外界环境影响下的建筑环境信息的收集，再与建筑环境的目标值进行比较，对二者的差值进行分析评判，然后使建筑系统做出相应调整，改善建筑环境的技术系统(有控制装置和最少的不可再生能源的输入的技术系统，属于闭环控制系统)。

3.1 "零控制技术"

"零控制技术"不存在反馈机制，没有单独的控制装置，可以说建筑本身即是控制装置又是受控装置，在系统学里这种控制系统被称作"开环控制系统"(图 3)。绿色建筑控制技术里面，"零控制技术"相对来说结构较简单，投入成本较低，而且无需任何不可再生能源的输入，但"零控制技术"控制能力较弱，可控制范围较窄。从整体生态系统来看，"零控制技术"对环境的负面影响很小，因而在绿色建筑设计时，首先应该采用"零控制技术"来调整和控制建筑与环境的关系。

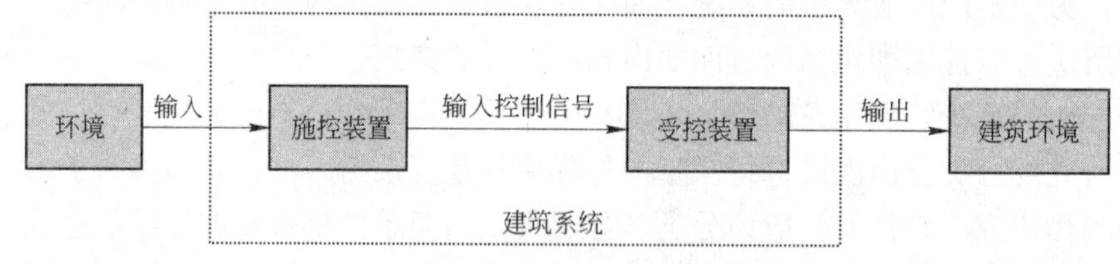

图 3 开环控制系统

在进行具体的绿色建筑研究和设计时，"零控制技术"主要包括以下三个方面：

3.1.1 总体环境布局的"零控制技术"：在绿色建筑总体环境布局中，根据当地气候条件合理规划建筑群体，确定建筑朝向、间距、绿化等，既要考虑利于夏季通风、遮阳、减少太阳辐射热，又需考虑冬季利用太阳能防止寒风侵袭，从而为创造冬夏兼顾的舒适的建筑环境提供可能。

3.1.2 建筑空间的"零控制技术"：根据外部环境的变化规律，通过对建筑的体型和内部空间的合理设计，最大限度的利用可再生能源，控制建筑环境的稳定。如 1987 年柏林住宅展览上，托马斯·赫尔佐格(Thomas Herzog)的"对角正方体"住宅的空间同时满足减少外界面总量和增加了南向外界面，很好的控制了建筑与环境的能量交换(图 4)。

3.1.3 建筑围护结构及构件“零控制技术”：建筑的围护结构及构件是绿色建筑设计和研究的重点，合理的建筑围护结构设计可以大大提高对建筑环境的控制能力。如控制围护结构外表面颜色，围护结构的能量储存技术以及借助围护结构的绿化控制建筑的环境。东京分国寺的蒲公英之家就是一个典型的融合绿化和建筑围护结构的例子(图 5)。

3.2 “反馈控制技术”

“反馈控制技术”相对于“零控制技术”是利用反馈机制，通过单独的控制装置对建筑系统进行控制。“反馈控制技术”不同于一般的暖通空调技术，他们主要的区别在于不可再生能源的利用和处理建筑与环境的关系上。“反馈控制技术”是调整控制建筑内外环境的之间的物质流、能量流和信息流，内外环境之间是开放的贯通的而不是完全隔绝的；而且“反馈控制技术”的控制过程追求最少的不可再生能源的输入。“反馈控制技术”在控制学里属于闭环控制系统(图 6)，它相对于开环控制系统结构较复杂，成本较高，但其效率高，有较大的控制能力。

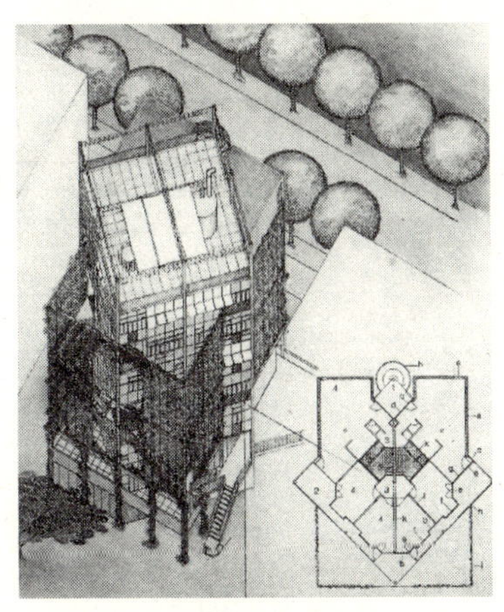

图 4 赫尔佐格的“对角正方体”住宅[4]

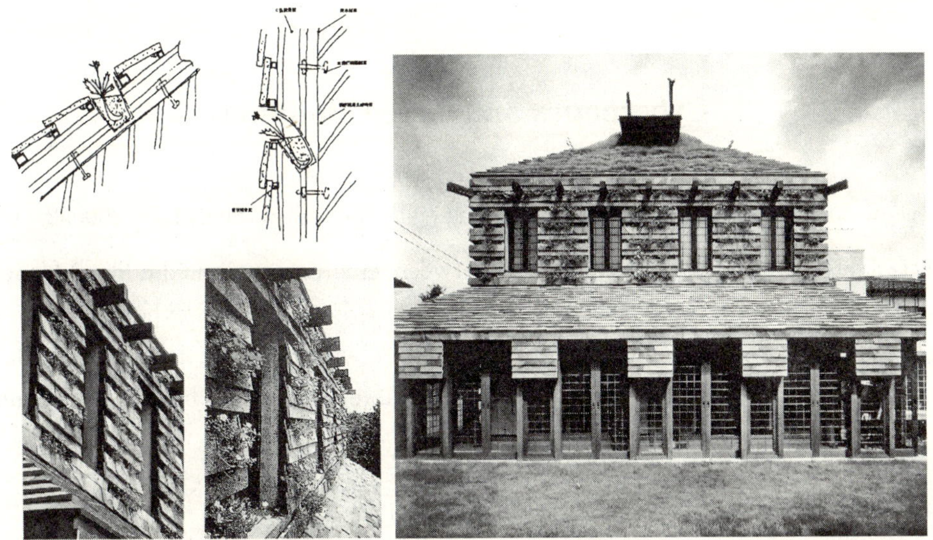

图 5 东京分国寺的蒲公英之家[5]

著名华裔学者傅京孙(K·S·Fu)根据控制器的不同提出的三种类型的智能控制系统，同样绿色建筑“反馈控制技术”根据控制装置的不同可以分为：人作为控制器的“反馈控制技术”，人机结合作为控制器的“反馈控制技术”和无人参与的“反馈控制技术”。三种技术系统是逐次由低级向高级发展进化的，代表了“反馈控制技术”的发展方向，也就是由完全的人智控制向建筑自主控制发展。

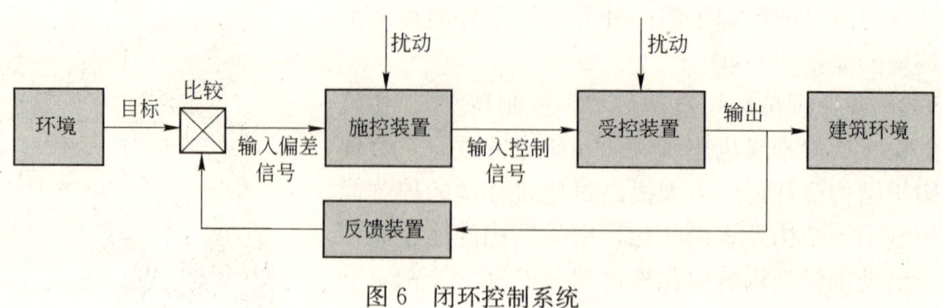

图 6 闭环控制系统

4 结语

系统学和控制学不是一种具体的科学方法，而是包括自然科学、社会科学和思维科学在内的共同的科学方法论，是一种更科学更规范化的方法。利用系统学和控制学的基本概念，可以从一个全新的角度去看待、去研究绿色建筑，为我国绿色建筑设计提供理论上的支持。

Cybernetics and Green Building

Xia Bo Song Dexuan Shi Jie
(College of Architecture and Urban Planning，Tongji University，Shanghai 200092 China)

Abstract: This paper studies the relation between environment and building from basic concepts and principles of systematology and cybernetics. The cybernetics in green building mainly researches building's "controllability" which causes the ecological building to adapt changing environment and creates comfortable and healthy building environment.

参考文献

[1] 吕爱民. 应变建筑——大陆性气候的生态策略. 上海：同济大学出版社，2003 年
[2] Sophia and Stefan Behling，Sol Power，New York：A publication for the READ Group，1996
[3] 曾广容，易可君等. 系统论·控制论·信息论概要. 长沙：中南工业大学出版社，1986 年
[4] 日华. 拖马斯·赫尔佐格的生态建筑. 世界建筑，1999-2
[5] 东京蒲公英之家. 日本. 世界建筑，2001-4

可持续节能建筑围护构造探讨

——以清华大学低能耗建筑为例*

李海英　马　欣

（北方工业大学建筑学院，北京　100041）

摘　要：本文以清华大学低能耗建筑为例，对可持续建筑的围护结构的构造设计进行分析和探讨，探讨重点在理论和详细构造方面，包括双层呼吸式双通道幕墙（宽通道和窄通道呼吸幕墙）、高性能玻璃技术、屋顶种植技术、相变蓄能楼面技术、太阳能利用、钢结构适合的高效轻质复合外墙等。

关键词：双层呼吸式双通道幕墙　相变材料　屋顶种植　太阳能利用

建筑的可持续发展，不仅是对建筑环境、建筑设备工程师的挑战，更重要的是对建筑师的挑战。节约能源，降低建筑能耗，在决定建筑能量性能的各种因素中，建筑的体型、方位及围护结构形式起着决定性作用，直接影响建筑物与外环境的换热量、自然通风状况和采光水平，对建筑的能量性能起着主导作用。

清华大学超低能耗节能楼紧邻建筑馆，总建筑面积超过 2900m²。超低能耗楼在围护结构、环境控制、能源系统、自控测量方面使用了多种国内外的先进技术，传统的建筑能量分析方法无法定性的分析这样较为复杂的建筑设计，特别是建筑物在全年的综合节能效应，因此在设计初期，暖通工程师和建筑师共同协作反复论证，通过计算机动态模拟风、光和热环境，建筑师及时准确的对围护结构方案进行调整和优化（图 1、图 2）。

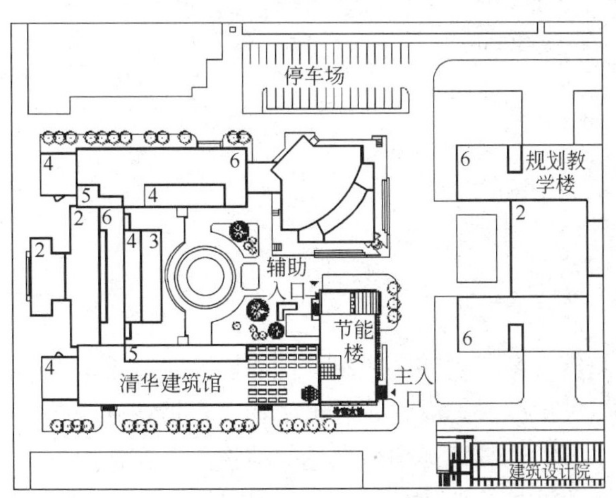

图 1　节能楼位置

*　清华超低能耗节能楼为北京市科委立项的重大科研项目，由建筑技术科学系江亿院士主持，设计负责人为栗德祥教授。作者在王余生和王丽娜两位导师指导下，主要进行外围护结构的构造设计和施工图绘制。

节能楼采用可循环利用的钢结构体系，外围护是以金属为饰面的多层复合轻质墙体和玻璃幕墙，钢构件支承外遮阳百叶，体现了钢结构建筑精密、细致的技术美感，尽可能选择可回收利用的材料，如石膏、加气混凝土、金属、玻璃等，在不同的方向，甚至相同朝向的不同开间，相同方向的不同层采用的围护做法也不尽相同，为今后开展各项与绿色建筑相关的实验数据的测量做准备(图 3、图 4)。

图 2 节能楼东、南立面

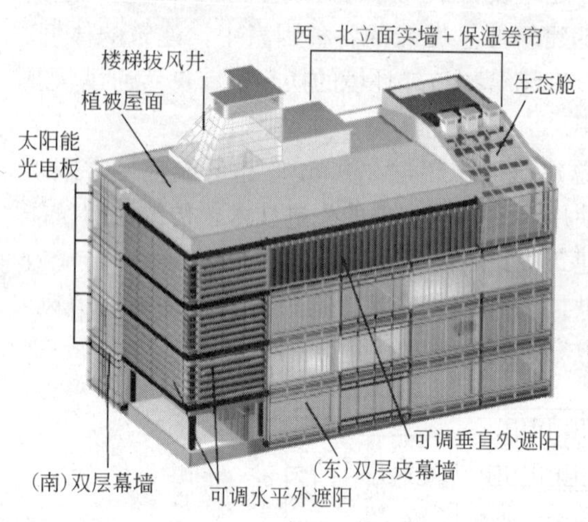

图 3 围护结构采用的方案

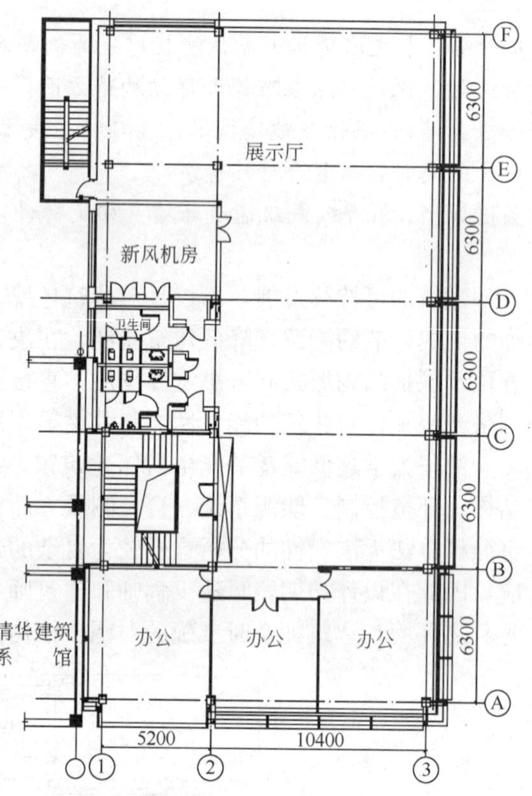

图 4 节能楼平面图

1 屋顶设计

屋顶种植能够有效的提高屋顶的隔热保温性能，同时改善生态与环境质量。在植物种类选择上，屋顶绿化以种植低矮的灌木、地被植物和宿根花卉、藤本植物等为主。为防止植物根系穿破建筑防水层，适宜选择须根发达的植物，不宜选用直根系植物或根系穿刺性较强的植物。节能楼屋顶种植土厚度为 250mm，构造依次向下为滤水层(无纺布)，排水层(陶粒 30～50)、防水保护层、防水层(EPDM)、找坡层、保温层(130 聚氨酯)、防水层(SBS)和结构层，综合传热系数达到 0.1W/m^2·K。在靠近女儿墙部位及屋顶中间纵横双向每 6m 间距设 600 宽走道，走道以两道砖墙架空，上铺活动盖板作为人行通道，走道下设屋面内排水口，砖墙最下一皮留空缝，以便滤水。楼梯顶部采用玻璃，强化热压通风效果(图 5、图 6)。

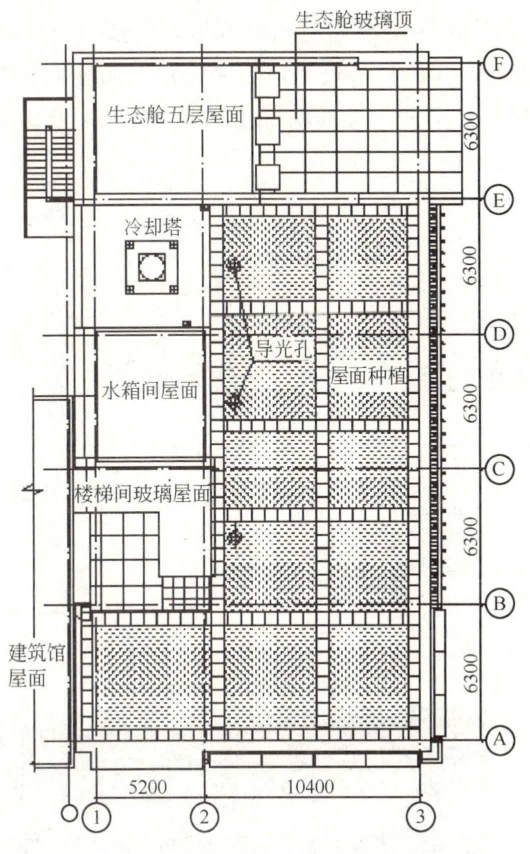

图 5　屋顶平面绿化布置

图 6　楼梯玻璃烟囱和蓄土种植池

2　幕墙设计和遮阳设计

双层呼吸幕墙是当今生态建筑中采用的一项先进技术，其实质是在双层幕墙之间留有一定宽度的空气间层作为气候缓冲空间，通常窄通道间层的宽度为 150～300mm，宽通道的间层宽度为 500mm 以上。其工作原理：在冬天，温室效应使间层空气温度升高，通过开口和室内空气进行热循环；夏季，通道内部温度很高，打开热通道上下两端的进、排风口，气流带走通道内部热量，降低了内侧幕墙的外表面温度，从而减少空调负荷。上下通风口的高度越大，热压通风的效果越好；当气流速度较小时，也可以辅助小型风机加速气流。从气流的组织和室内的关系来说，幕墙又可以分为内循环式和外循环式；从气流组织的高度来说，可以有多层串联式，和单层循环式。

由于空气间层的存在，能够在缓冲空间提供一个保护空间用以安置遮阳设施，比内置遮阳有更佳的遮阳效果；比外置遮阳易维护、免受直接的风雨和日晒的侵袭，同时还可兼具反射和折射的作用。

2.1 窄通道呼吸幕墙

节能楼在南立面使用窄通道幕墙，在1、2轴之间，一、二层采用内循环式窄通道双层幕墙，通道宽度为200mm，中间设宽50mm的电动百叶，为加速通道内风速，采用小型风机连通到室内通风系统；三、四层采用外循环式窄通道双层幕墙，夹层110mm宽，中间电动百叶宽度为25mm，在每层上下端分设排风口和进风口。无论是内循环还是外循环式内层幕墙均可开启，以便清洁；在凸出部分的两个侧面设置进风百叶，用以在过渡季节有更多的自然通风(图7、图8、图9左)。

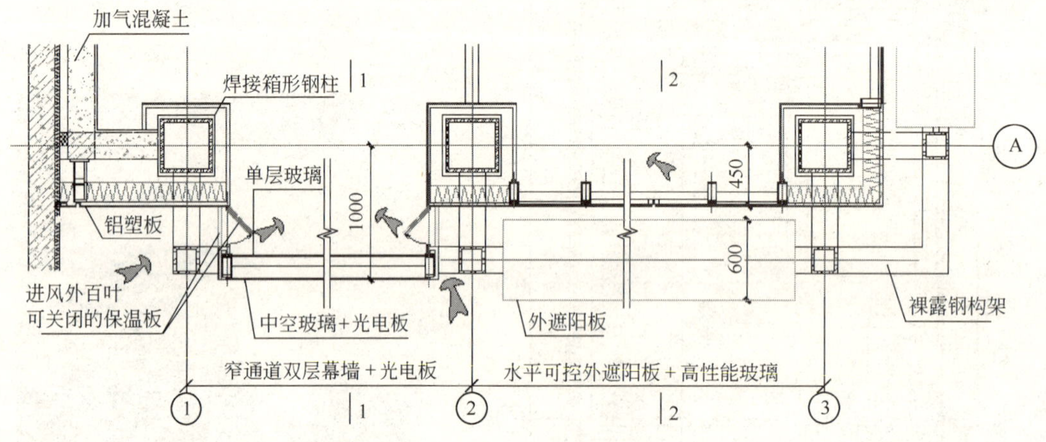

图7 南立面幕墙平面构造

2.2 太阳能光电转换和围护一体化

节能楼每层楼板和管道夹层的位置在立面安装光电池板，光电池板有一定的遮挡作用，因此在外侧看不到主体钢结构以及层间防火的处理，产生的电能还用于控制室外遮阳百叶的转动和幕墙窗扇的开启(图10)。

图8 南立面窄通道幕墙及内部百叶

2.3 宽通道呼吸幕墙

东立面1～3层为双层皮幕墙，外层为单层隐框幕墙，内层为双Low-E双中空玻璃，两层幕墙间隔约为600mm，人员可进入检修。以每层和每个开间划为一个独立单元，上下层以及左右开间不连通，噪声不会在通道内传播，在每层上部和下部分别设出气和进气的外旋窗，这样可在立面上形成不同的进风和出风口的位置关系，以便对比测量结果，但较好的方式为进气口和出气口错开开启，减小下层出风口对于上层进风口的污染，同时有利于双层通道内部风的循环。其中Ⓑ、Ⓒ轴间采用可拆卸盖板，能够改单层通风变为多层串联式通风，串联式通道热压通风效果优于单层循环式通风效果。在外侧幕墙的竖梃上设折光板，调节角度，可增加室内深部照度，紧靠内层幕墙玻璃以外100mm处设铝合金遮阳百叶，在每层的1.2m管道夹层位置，为遮挡层间防火，采用磨砂玻璃(图11～图16)。

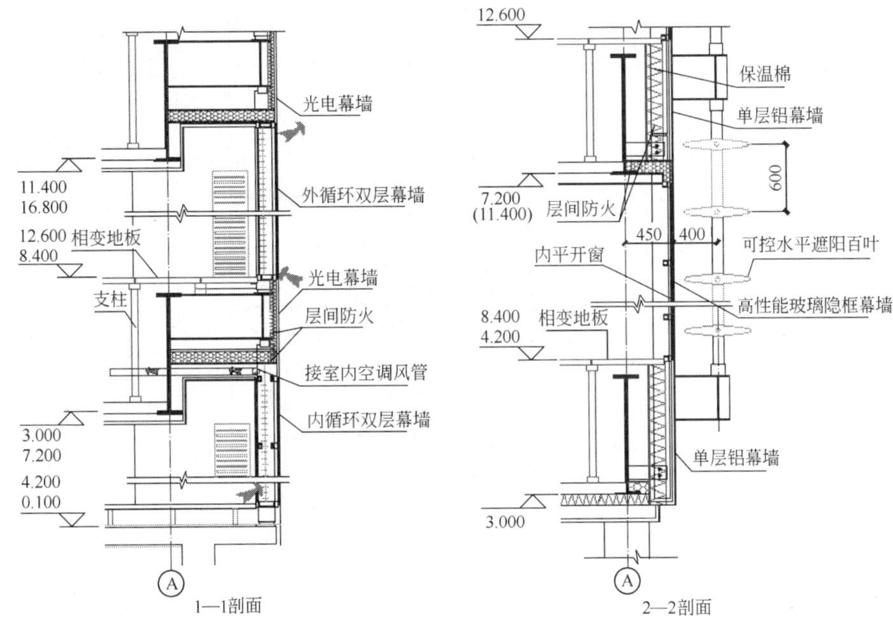

图 9　南立面窄通道幕墙剖面(左)和单层幕墙水平百叶剖面(右)

图 10　南立面光电幕墙和水平可控百叶

图 11　东立面宽通道双层幕墙和垂直外百叶

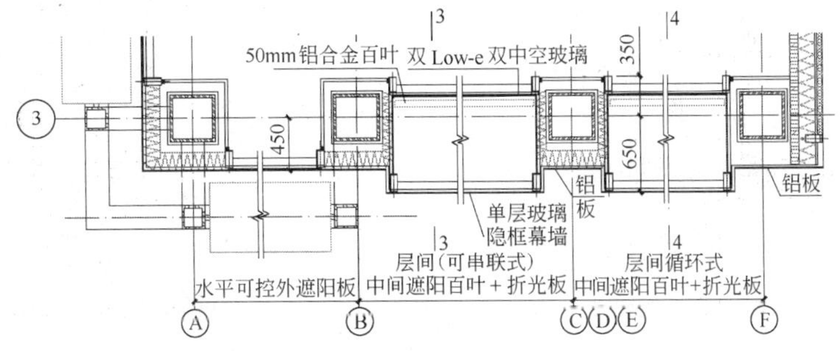

图 12　东立面宽通道幕墙 1～3 层平面构造

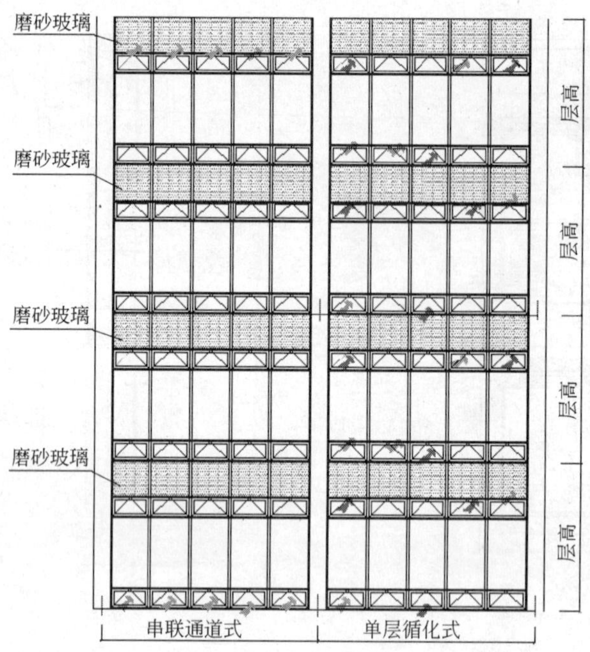

图 13 东立面串联式和单层循环式风口组织

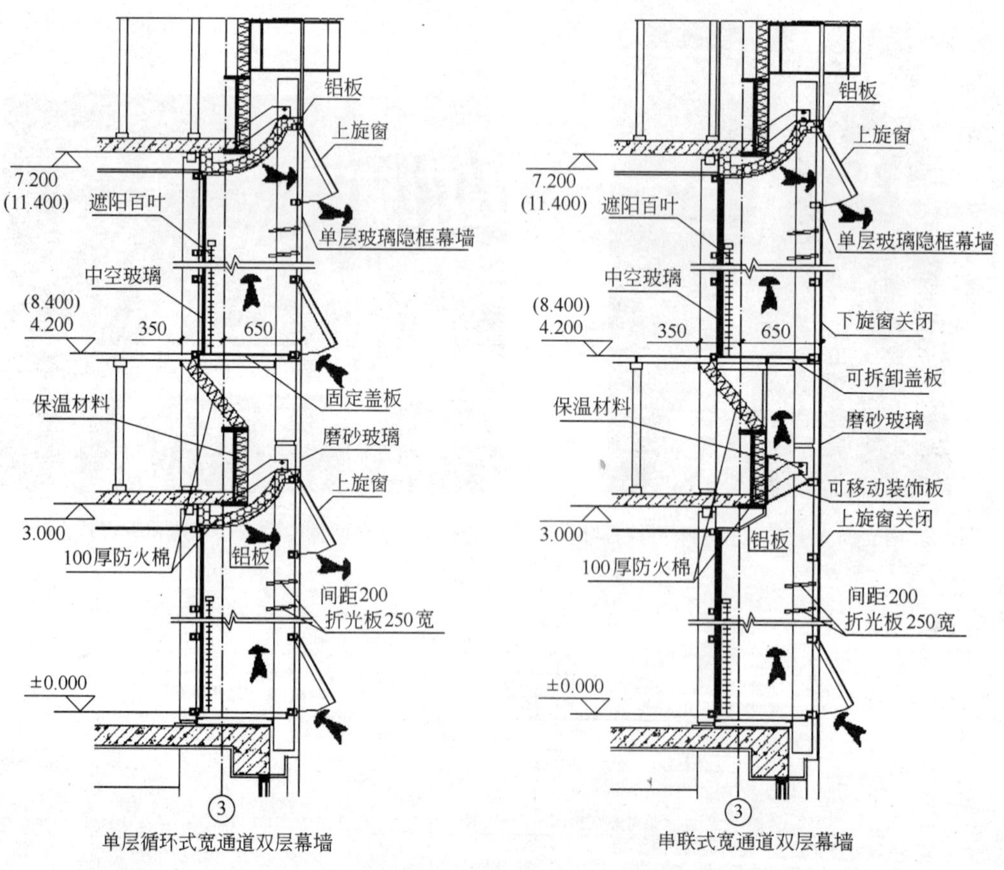

图 14 东立面宽通道双层幕墙单层通风(左)和串联通风(右)剖面构造

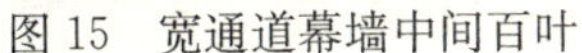

图 15　宽通道幕墙中间百叶

图 16　折光板

2.4　幕墙设计的防火处理

无窗间墙与窗槛墙的玻璃幕墙，应在每层楼板外沿设置耐火极限不低于 1.00h，高度不低于 0.8m 的不燃烧裙墙；玻璃幕墙与每层楼板、隔墙处的缝隙，应采用不燃烧材料严密填实。节能楼幕墙的层间防火，单层幕墙和窄通道幕墙的水平方向结合吊顶采用镀锌钢板和 100 厚岩棉封堵，层间夹层高度为 1200，垂直方向用埃特板和 200 厚岩棉，使上下楼层防火分区分开。双层幕墙是一种新颖的幕墙构造形式，因此在世界各国的消防规范中，尚没有针对双层幕墙建筑的特殊规定。宽通道双层幕墙的单层循环式幕墙，上下楼层防火分区被彻底分开，而串联式宽通道双层幕墙的防火要求在国内外来说还是个空白，目前也没有双层幕墙着火的案例，一些理论认为应在通道内设喷淋来保护内层幕墙，或者内层玻璃的耐火极限高于外层玻璃，外层玻璃先炸裂，这时即可视为建筑外面发生火灾，而不是烟囱效应使火焰和烟气在通道内蔓延。节能楼在设计时，即把中间通道火灾视为室外火灾，内层玻璃耐火极限高于外层玻璃(图 9、图 14)。

2.5　高性能玻璃和智能遮阳设计

玻璃幕墙的保温隔热的功能与很多因素有关，其中影响最大的是传热系数和遮阳系数。传热系数受材质和厚度的影响，而遮阳系数受到玻璃本身特性(太阳得热系数)的控制，又受到遮阳构件、窗帘等影响。清华节能楼玻璃采用高性能玻璃，如表 1 和表 2 所示。同时窗框为断桥隔热构造，采用了低热传导率的玻璃边部密封材料；窗外设外保温卷帘。

双层幕墙玻璃　　**表 1**

位　置	双层幕墙外层玻璃	双层幕墙内层玻璃
窄通道外循环双层幕墙(南)	T8	T8low-e+18A+4/PVB/T4
窄通道内循环双层幕墙(南)	T8low-e+12A+T10	T8
宽通道双层幕墙(东)	T6	T4low-e+9Ar+5+9Ar+T4low-e

注：表中 T 表示玻璃钢化；A 代表空气层；Ar 表示氩气层；V 代表真空玻璃。

高性能玻璃性能和普通玻璃对比 表 2

位　置	玻　璃	传热系数 W/m²·K	太阳得热系数 SHGC
单层幕墙(南)、西窗	T5+6A+4+V+4+6A+T5	0.93	0.48
单层幕墙(东)、北窗	T4low-e+9.Ar+4+9.5Ar+T4low-e	1.02	0.53
普通中空玻璃	6+12A+6	3.99	0.56

注：表中 T 表示玻璃钢化；A 代表空气层；Ar 表示氩气层；V 代表真空玻璃。

智能化遮阳是一套较为复杂的系统工程，是从功能要求到控制模式到信息采集到执行命令到传动机构的全过程控制系统。它涉及到气候测量、制冷继续运行状况的采集、电力系统配置、楼宇系统、计算机控制、外立面构造等多方面因素。节能楼单层幕墙外部设可控水平或垂直遮阳百叶，遮阳的支承钢构件必须从主体钢柱相连，在连接点处，虽然填塞了保温棉，但冷桥并没有完全避免。在冬天，叶片平行于入射光线，太阳入射光线进入室内，使室内升温。夜间，百叶叶片平行外幕墙，呈闭合状态，减少室内热量向室外的热量散失。水平百叶分为三个不同的功能区——采光、视野、遮阳/能量收集进行分别控制，每层较高位置的叶片控制角度，光线通过在百叶叶片和顶棚反射，提高室内深部照度；夏季较低位置的叶片转动到基本和入射光线垂直的位置，以阻挡直射光线的进入和遮挡室外的太阳辐射热，在冬季则平行入射光线。围护结构的测试包括各玻璃、窗框、遮阳百叶、保温墙体的表面温度、热流。控制系统可以采集室外各个测点的日照情况，来调节遮阳百叶的状态，减少建筑负荷(图 9 右、图 17、图 18)。

图 17　水平分组控制装置

图 18　水平和垂直外百叶

3　适应钢结构体系的轻质复合保温墙体

目前，较为适合钢结构建筑的外墙可以分为单一材料的砌块和板材，主要以加气混凝土砌块和板材为主；复合保温墙体可以分为工厂预制和现场复合两种。

低能耗楼西立面外墙基本由两个部分构成。一个部分是和建筑系馆墙体紧邻、设变形缝的部位，采用加气混凝土砌块，工人在节能楼内部施工比较容易，在钢柱和钢梁的位置，采用外砌加气保温块，阻止了冷桥。另一部分是西立面朝向内院的部分和北立面，采用现场复合的轻质保温外墙，300mm 厚，内侧板是 80mm "可呼吸" 脱硫石膏砌块，中间是 150mm 铝箔保温玻璃棉，外饰面板是 50mm 聚氨酯保温铝幕墙板，墙体传热系数达到 0.35W/m^2 · K。

现场复合避免了单一材料板材和工厂预制复合板材在安装中的通缝问题。多层材料交错安装，具有良好的保温隔热性，有效的避免了冷、热桥的产生，为隔绝雨水和室内潮气设置了多道屏障，空气间层不仅了减少内部冷凝水的产生，也使少量进入到内部的水分能够及时排走，便于设备管线的安装；同时墙体各层相对独立，易于维修更换，外幕墙板可以根据建筑师的要求，选择饰面层，内层板也可根据室内装饰的需要选材。

4 相变蓄能楼面和设备夹层设计

节能楼的外围护都是由热容量较小的玻璃幕墙、轻质复合保温外墙组成，较小的热惰性会使室内温度随室外的温度变化而波动剧烈，尤其是在昼夜温差大的季节，室内温度波动会很明显。为了使室内的环境更加稳定，采用了相变蓄热活动地板，把相变温度为 20～22℃的相变材料储存在 600mm×600mm 厚 50mm 的箱形地板块中。白天储存从幕墙进入的热量，夜晚缓慢释放白天接收到的太阳能热量，减少室内温度的波动(一般会小于 6℃)。现浇钢筋混凝土楼板浇筑在工字钢梁和桁架梁的下翼缘，相变活动地板以支柱支承在钢筋混凝土楼板上，活动地板架空层高度 1.2m，空调风道、各类水管、电缆、综合布线等均隐藏在架空层内(图 19、图 20、图 21)。

图 19　相变地板支撑方式

图 20　相变地板

图 21　结构和管道夹层

5 钢结构建筑防火构造设计

节能楼采用钢结构体系，具有自重轻、抗震好、工厂制作、施工快，环境污染小和可回收利用等优点，但钢结构建筑应注意防腐和防火。节能楼的钢柱涂抹厚型防火涂料，耐火极限能达到 3 小时以上；钢梁喷涂薄型防火涂料，耐火极限能达 2 小时以上；顶部生态舱玻璃屋顶钢结构支撑梁采用超薄型防火涂料，涂层≤3mm，装饰性较好，耐火极限能达 30 分钟以上；室外裸露钢结构只进行防腐处理，没有做防火处理。

6 结论

低能耗建筑是基于各项绿色技术而形成的，所以建筑形式也遵循了技术本身的规律，充分表达技术的特点。在设计中，没有刻意做外观形象设计，技术的特征及需要形成了围护的形象设计和构造设计，力求在建筑全生命周期(物料生产、建筑规划设计、施工、运营维护及拆除、回用过程等)中实现高效率利用各种资源(包括能源、土地、水系源、建筑材料等)；采用利于提高材料循环利用效率的新型结构体系；尽可能减少不可替代资源的消耗，减轻对环境的破坏和污染；推广和使用无害、无污染的绿色环保型建筑材料，保证室内品质。

The Envelop Constitution Design of Sustainable Buildings

——Take the low-energy consumption building in Tsinghua University for example

Li Haiying Ma Xin

(College of Architecture，North China University of Technology，Beijing 100041 China)

Abstract：The article discuss the envelop constitutions and theories of the sustainable Buildings taking the low-energy consumption building in Tsinghua University for example. The main points are double skin aspiratory curtain wall，high quality glasses，roof planting，phase change accumulating floor，energy using and light composite wall for steel structure.

参考文献

[1] 威纳尔·朗. 智能建筑围护结构. 建筑与可持续发展. 新建筑，2003，01：9～11

[2] 黄勇. 智能百叶与建筑遮阳. 华中建筑，2003，05：77～78

[3] 汪铮，李保峰，白雪. 可呼吸的表皮—积极适应气候的“双层皮幕墙”解析. 华中建筑，2002，01：22～27

建立“和谐城市”的技术支撑
——论城市生态技术学科的建构

吴志强　姬凌云
（同济大学建筑城规学院，上海　200092）

摘　要： 本文首先指出了建立城市生态技术学科的时代背景，其次分析了城市生态技术学科与和谐城市之间的关系，学科建构的目的和重要意义，最后以城市的非完全可控性为学科建设的理论基础，提出了城市生态技术学科的基本组成和结构框架。

关键词： 和谐城市　生态技术　城市生态技术学

1　背景

1.1　和谐城市

自20世纪中期可持续发展思想的广泛传播以来，近年来提出的“和谐城市”的理念更进一步，将可持续发展的思想落实到了城市层面。

从城市问题和城市发展趋势这两个基本经济社会背景出发，“和谐城市”的理念将城市作为行动的核心，解决问题的主要载体。它包括三个向度的基本理念：

（1）人与自然的和谐：既要有现代化的设施，又要和大自然融合，创造人类适宜的未来居住环境、创造21世纪的生态建筑，创造田园化的社区、创造具有地方特色的住宅。

（2）人与人的和谐：强调城市中不同的文化背景和不同的社会集团之间的社会和谐、重视区域中各城市之间居民生活的和谐，避免城市范围内的社会空间的强烈分割相对抗。

（3）历史与未来的和谐：保持城市发展过程中历史的延续性，保护文化遗产和传统生活方式。进新技术在城市发展中的运用并使之为大众服务，努力追求城市文化遗产保护与新的科学技术运用之间的协调。

“和谐城市”的思想不仅反映了经过20世纪的长期探索、检讨后的理性思考，更是面临21世纪的挑战对更为持续、公正和全面发展的理想人居环境的追求。

1.2　技术及技术学科的发展

科学技术水平是社会文明的标志。自工业革命以来，现代技术为人居环境品质的提高提供了强大的动力和武器，使人类初步具有了全方位、有深度的改造和利用自然的能力，创造了现代城市文明。

同时，技术的支撑作用和动力作用也具有两面性。在技术创造城市文明和财富的同时，也有技术不成熟导致的浪费，技术欠缺导致问题的无力解决，还有技术滥用导致的破坏，技术短视和功利主义带来的短期行为。技术本身的发展水平和人对待技术的主观态度决定技术发挥作用的强弱和后果的性质。

一个重要转变是人们对技术本体有了更深刻的认识，对技术与人和社会的关系有了更为全面和辩证的思考，技术观发生了重大的变化。

另一个重要转变是，技术的发展出现了一个新的源动力，那就是全球范围内开始普遍提倡的可持续发展的目标。在新目标的推动下，以环境的可持续为核心的生态技术迅猛发展，技术作为解决目前问题的重要工具被放到了显著的位置。

因此，当前更为紧迫的是需要认识到相对于理想目标当前技术的认识水平和实践水平的不足和差距。特别是，在可持续发展和“和谐城市”的目标下，技术如何为城市的和谐发展发挥更积极的作用，是尤其需要重视的关键课题之一。

2 城市生态技术学科与“和谐城市”

2.1 城市生态技术学科的概念

技术是由认知转化而来的指导行动的知识或方法。技术学科是这些知识、方法和手段构成的体系。城市生态技术学科是依据城市发展演变的规律以及人类活动对城市发展的能动性规律，形成的指导城市良性发展、和谐发展的行动知识或手段。它将城市作为整体来研究，考察城市各个组成部分结成不同体系时影响这些城市体系发展的各相关技术的特质，以此形成以城市为对象的技术知识、方法和手段综合体系。

2.2 城市生态技术学科与“和谐城市”的内在联系

城市发展的历史可以看成是人作为主体与客观世界的互动关系的变迁历程。从工业和技术史的角度看，前工业时期、工业时期和后工业时期三个发展阶段，反映了人类对自然的控制能力和方式的巨大转变。从前工业时期的对自然的依赖，到工业时期对自然的支配，人类对自然的控制力愈加强大，在自然面前已具备了相当的主动权。可以看到，技术在城市存在模式的演变历程中成为至关重要的因素之一。技术与城市的和谐性之间的内在联系，从环境的可控性的角度，可以建立以下的逻辑关系：

(1) 环境的可控性和不可控性。可控环境是人作为能动的主体可以决定其生成和发展的环境。不可控环境是独立存在有自身发展规律、人无法决定其生成和发展的环境。可控环境是主动性的(active)，人可以充分发挥主观能动性。不可控环境是被动的(passive)，人只能干预和影响。

(2) 城市是非完全可控环境。城市环境从某种程度上是可控的，是在遵循城市发展规律的条件下，可以由人的能动行为决定的。从另一个层面讲，它又是不可控的，它有自生发展的客观规律，只能干预、影响而不能决定。总体来说，越微观的环境可控性越强，越宏观的环境可控性越弱，物质性环境可控性强于非物质性环境，人工环境可控性强于自然环境。

(3) 城市发展不和谐的根源。进一步分析可看到，人与自然之间的对立是由于人类发展需求对自然资源的无限索取与自然资源的有限供应之间的矛盾，人与人之间的对立是在于生产力水平所决定的经济基础的对立，从而导致的上层建筑的对立和差异，历史与未来的不和谐是在于新兴文明为了占据有限的城市资源而产生的与传统遗存之间的对立。归根到底在于人居环境与非人居环境之间以及人居环境内部各类不同资源占有关系的对立和不平衡。

(4) 人居环境的可控性和人类对环境的控制力。要解决需求无限和供给有限这个矛盾，需要更强的创造、分配、保护、节省资源的能力。这个能力就是人类对人居环境的控制力。人居环境是非完全可控的环境，它的可控性在于它多大程度可以被人的主观能动性所决定，并被用来满足人类的需求。人居环境的可控性越高，超出人主观意识之外的不可知因素越少，妨碍控制力发挥的壁垒越低，越有利于主观能动性的发挥，人类实践活动的主动性越强。人居环境的可控性和人类对环境的控制力是相互促进，较强的环境控制力有利于创造可控性高的人居环境，反过来，可控性高的人居环境有利于人类的环境控制力。

(5) 城市技术学科的支撑作用。人类对环境的控制力，即创造、分配、保护、节省资源的能力，首先需要相应的城市技术子学科来提供技术手段，其次需要城市技术学科的关系框架来构筑各子学科的链接关系，与各控制力之间的链接关系相对应，使各子学科之间的互动发展能够满足各控制力之间的互动需求。

3 建构城市生态技术学科的目的和意义

建立城市生态技术学科的最终目的是为可持续的人居环境和“和谐城市”的发展目标提供技术学科的支撑。从理论和学科发展的角度，是为了提高对城市发展与技术之间相互关系的认识水平，完善现有技术学科体系。从实践的角度，是为了促进技术进步和技术创新提高人类作为能动主体的推动、协调、改善、控制城市发展的能力，提高以城市为主体的人居环境的可控性。建立城市生态技术学科，对于城市学科和技术学科的发展都有重要意义。从根本上看，城市技术学科研究的是城市与技术之间的关系，考察两者是如何相互作用的，技术通过什么样的途径、以何种程度来影响城市的生态发展，城市的运行规律又如何来影响技术的演变。城市生态技术学建构的意义主要表现在：

(1) 为城市学科提供系统化的技术运行的基本原理，对技术这个城市发展的重要影响因素，城市学科有更本质更全面的理解。

(2) 相比分散在自然科学领域的各类生态技术，城市学科可以直接利用城市生态技术学的成果来推进本学科的发展。

(3) 城市学科在实践方面运用能够获得城市生态技术的强大支持。

4 城市生态技术学科组成和结构

4.1 核心学科

核心学科包括横向学科体系和纵向学科体系。

4.1.1 横向学科体系。横向学科体系是以物质环境为出发点探索物质世界发展演变过程的技术体系。该体系包含了人类为达到发展目标处理人与物质空间环境关系所需的技术。城市作为物质空间环境，有其产生、发展、消亡的新陈代谢过程，并形成周而复始的循环过程。城市物质环境发展演变的各个阶段构成循环过程的节点。核心学科针对这些基本的节点，探索这些阶段物质环境的特性并发展相应的人类实践的手段和方法。由此可知，核心学科是横向学科框架的基本支点。根据现有技术水平，核心学科的主要类型包括：

(1) 能源技术：包括传统能源和新能源在内的各类能源开发和利用。

(2) 资源技术：与人居环境建设相关的各类资源的勘测、发掘、提取、加工、利用方面的技术。

(3) 建设技术：人居环境物质实体的各组成部分的建设技术。

(4) 环境技术：提高、改善、协调、控制人居环境质量的技术。

(5) 再生和循环技术：人居环境的更新、废物处理转化再利用、资源再造技术。

4.1.2 纵向学科体系。以人为出发点反映人的行动序列的技术体系。该学科体系包含了人类群体行为控制人居环境发展所需的处理人与人之间关系的技术。城市技术学科的纵向体系是要发挥技术的长处，来弱化群体行为的不可控性，旨在提高群体行为的效率。主要包括以下部分：

(1) 监测技术：获取信息的技术。

(2) 信息技术：处理、分配、传输信息的技术。

(3) 统筹与决策技术：对信息进行分析、判断，得到行动意向、计划、安排的技术。

(4) 组织技术：建立群体行动机制的技术。

(5) 管理技术：制定行为规范和执行行为规范的技术。

(6) 控制技术：干预行动以使之到达既定目标或要求的技术。

(7) 反馈技术：由行动产生信息的技术。

4.2 交叉学科

交叉学科是相邻核心学科的交叉部分。交叉学科将关联紧密的学科共同关注的领域作为对象，重点考察相邻学科之间的互动关系，形成将双方铆合为一体、促进双方联动的整合性技术。这些技术是单一学科的发展所达不到的，由此，交叉学科是有重要意义的独立学科。

4.3 城市技术学科体系的建构

横向学科轴与纵向学科轴建构成为一个水平学科网。轴上的学科是核心学科，网上的节点则是交叉学科。网状结构表达了两个维度的核心学科之间的交叉过程，力图在各个环节的技术中将人的因素和物的因素结合起来的过程。网络的各条轴线是每一个核心学科在另一个维度上进行跨越发展形成的学科系，网络上的节点是学科系中目前可以定位的分支学科。网络结构为学科的分类设定了逻辑框架，每一项技术都可以找到其大致的位置，同时每项技术都有明确的与其他技术的联系。

5 结论

面对21世纪，“和谐城市”将成为重要的城市发展理念。在应对20世纪的城市问题过程中，技术发挥了重要的作用，并还将在新世纪中发挥更重要的作用。但同时，现有技术水平距离实现城市生态发展的需求还有很大的差距，以城市为对象的生态技术学科也尚未形成，难以适应城市生态发展需求的问题也凸现出来。为了实现可持续发展和城市和谐发展的长远目标，发展能够支撑“和谐城市”理念的生态技术学科成为当前的重要课题。

城市生态技术学科的组成由核心学科和交叉学科组成，核心学科包括了以物的要素为核心的横向学科体系和以人的要素为核心的横向学科体系，两个维度的学科体系形成网状结构。

Establishing the Technological Support for Harmony City

——The Establishment of Science of Urban Ecological Technology

Wu Zhiqiang　Ji Lingyun

(College of Architecture&Urban Planning Tongji University, Shanghai　200092　China)

Abstract: First, the paper points out the background of establishing the science of urban ecological technology; then analyzes the relation between the science of urban ecological technology and harmony city; finally, proposes the basic composition and structure framework of the science of urban ecological technology, which is based on the theory of semi-controllability of urban environment.

参考文献

[1]　吴志强，陈秉钊，唐子来. 21世纪的城市建筑：走向三大和谐. 城市规划. 上海：同济大学出版社，1999(10)：20～22

[2]　常立农. 技术哲学. 长沙：湖南大学出版社，2003：1～17

[3]　张丽萍，吕乃基. 生态学视野下的技术. 科学技术与辩证法，2002(1)：55～57

建立绿色建筑科学体系的探索

饶 戎

（清华大学建筑学院，北京二十一世纪城市生态研究院，北京 100084）

摘 要：本文以近几年在生态技术、生态材料支持下的生态室内、绿色住区的研究与实践，及在生态规划指导下的生态村落、生态城市设计、生态景观设计的量化研究与规划编制为基础，初步进行绿色建筑科学体系的探索。包括绿色建筑目标、目的和价值标准，实践绿色建筑的方式、方法等。

关键词：绿色建筑 生态室内 绿色住区 生态村落 生态城市设计 科学体系

绿色建筑是一个古老的命题，也是一个全新的命题。说它古老，是因为绿色建筑源于人类先祖依赖自然、敬畏自然而选择的一种满足居住功能、保障生存的建造方式和方法。它以尊重自然规律、顺应自然为前提，以实现和谐、共生为原则的生存哲学和人类价值观为基础。说它是全新的命题，是因为当今人类面临全球生态恶化、环境破坏、资源危机、人口膨胀、物种灭绝等外部环境灾难带来的威胁自身存续的外部环境灾难带来的严重挑战，绿色建筑正在成为一种当今人类社会应对生态环境危机挑战和反省自身行为结果的重要修正和选择。

发展绿色建筑是中国的必由之路。中国作为一个资源有限、人口众多且经济增长与社会发展存在着深刻矛盾的发展中国家，土地资源的浪费、化石资源的匮乏、能源的低效率利用和森林资源的破坏，对社会整体发展的危害与制约日益突显。粗放片面地追求经济增长，低品质的城市化进程，城市规模与空间形态无序的畸形板结蔓延，直接导致城市生态安全、生态健康的问题。同时高能耗的城市运行，城市建设对环境的人为污染与扰动，使自然生态系统与社会生态系统遭受到难以弥合的割裂、破碎与损害。

21世纪的人类社会利用日益发展的科学技术创新能力，不断深入、客观地认识自身存在与自然存在变化发展的客观规律，以及它们矛盾与冲突的内在关系，寻找出人类社会对资源、环境的影响内涵，并进行有序、有度、有节制的控制与约束。人类的行为将决定人类生活的核心系统——自然生态系统的质量，它是社会生态系统与人类存续与发展的根本。绿色建筑本着和谐共生、健康安全、科学求是、永续发展的宗旨，将成为统合天人合一、和谐高效的人与自然完整的生态系统体系的基石。绿色建筑将实现由传统高消耗型建造模式向集约高效的绿色节约型发展模式的转变，是提高中国社会发展效率，节约城市建设、运行成本，顺应社会转变的必由之路，也是后工业时代当今世界建筑发展的必然趋势。

1 绿色建筑目标、目的和价值标准

绿色建筑作为现代社会的一个重要社会意志，具有明确的目标、目的和价值标准。绿色建筑的定义有很多，我们在从事绿色建筑的研究与实践过程中，主张绿色建筑的定义是：在建筑物的全生命周期中，最少程度地占有和消耗地球资源，用量最小且效率最高地使用能源，最少产生废弃物并最少排放有害物质，成为与自然和谐共生、有利于自然生态系统与人居系统共同安全、健康且满足人类功能需求、心理需求、生理需求及舒适需求的宜居的可持续建筑物。

绿色建筑的目标是要通过人类的建设行为，达到人与自然安全、健康、和谐共生，并满足人类追求适宜生存居所的需求、愿望与追求。目的则首先是要解决人类存续发展所必备的自然资源问题，为环境得以持续、稳定、均衡的发展提供保障；其二是控制和约束人类消耗自然资源的规模、水平与效率；其三是保持社会生态系统功能的完整性和丰富度，使历史文化的借鉴、传承与发展统合在建筑中得以表达，实现人类生存观的修正、优化与进步；其四是以科学的发展观实现人类社会可持续发展的诉求，通过科技水平的提高和高新技术的应用、推广，降低资源消耗，达到和谐、宜居的生态人居环境建设水平。发现新资源、再生资源、循环资源以及可替代资源，缓解并最终解决制约和威胁人类社会发展进步的自然资源与环境的瓶颈问题。

在明晰目标与目的的前提下，价值标准便成为固化和保证实现可持续科学发展的基础条件。自然创造了人，人依赖自然并向自然索取生存所必要的资源供给与环境条件，同时又用自己的行为结果不断的扰动、破坏甚至毁灭自然的规律结构与系统。自然的法则是不以人的意志而转移的客观存在，它可以从一种创造转化为另一种创造，继而毁灭她现有的创造结果。因此，价值标准的形成始终不是来自于人对人的强制和胁迫，而是来自于人类对生存要求与自然规律之间的科学认知、认同和认真的思辩。共同意志、社会意志是建构社会行政意志与法律意志的前提和基础。

2 实践绿色建筑的方式、方法与标准

基于绿色建筑的主张及其目标、目的和价值标准，实践绿色建筑的方式、方法是遵循科学发展观，通过具体的基础科学理论研究，落实科技成果与应用技术的转化。

绿色建筑应用方式的选择与确定具有功能性、地域性、社会性和特质性，须符合复杂的生态系统构成要求。以往的选择方式是基于人对建筑本体的依赖与需求，由现实条件、环境限制而决定，绿色建筑则强调建筑绿色功能因子的选择、绿色要素的关系组织、绿色功能系统集成的体系构成。

绿色建筑作为自然系统、建筑实体与人构成的运行使用体，它的标准是符合人与自然、人与建筑、建筑与环境的客观关系，达到并符合绿色建筑的定义和要求，实现绿色建筑的量化标准指标。

3 绿色建筑科学体系的实践与探索

对绿色建筑科学体系的具体研究与实践是推广应用的根本。在长期从事绿色建筑、生态景观、生态规划研究、教学与实践过程中，我们不但进行系统的基础理论研究，也通过多专业、跨学科专家团队交叉合作，以严谨创新的示范与实验工程，不断探索和验证应用绿色建筑科学体系的完善途径。

就绿色建筑研究与实践而言，通过建筑室内生态设计，绿色建筑功能、技术与材料的系统集成，绿色建筑适宜应用技术、新材料、循环材料、再生材料的研究与开发应用，生态景观、科学规划的研究与实践等，探索一条共同构成绿色建筑综合生态设计应用、推广的科学技术体系(表1)。

绿色建筑的科学体系示意 表1

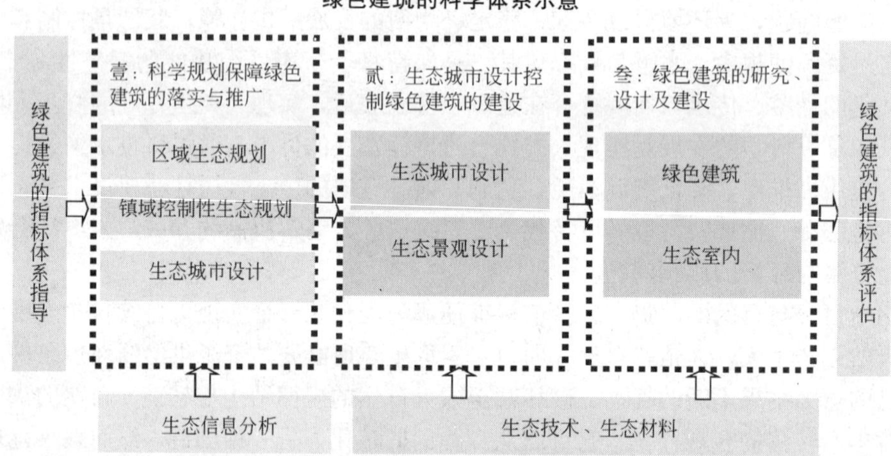

建构绿色建筑的技术系统主要涉及到：

(1) 建筑的生态室内空间、功能使用与环境品质的控制；

(2) 建筑设计与建造对能源、资源、风环境、光环境、水环境、生态景观、文化主张的系统组织；

(3) 实现绿色建筑节约与效率要求的新材料、新技术的选择与应用；

(4) 绿色建筑全寿命周期的组织、控制、使用与服务的系统管理；

(5) 绿色建筑对城市与村镇系统生态功能扰动、破损与阻断的控制、管理与修复。

4 绿色建筑科学体系探索的实践说明

实践一：绿色建筑技术策划、研究性设计施工及生态聚落与绿色民居模式。

示范工程——杭州北秀蓝湾绿色建筑技术咨询(建设部)。

研究项目——北京九渡河生态村落与绿色民居模式研究(中国—欧盟环境管理合作计划项目)。

背景：人类的建设对环境影响最大的不是城市的标志性建筑，而是城市中普通居民生活工作的居住与办公建筑，以及在城乡统筹一体化背景下的小城镇及村落建筑。前者数量占城市建筑总量的九成以上，对能源及资源的消耗量大；后者介于城市与乡村之间，与自然环境紧密融合，其能源、资源的消耗模式与大城市截然不同。所以，绿色建筑及其技术的实践除上届大会已介绍的海淀生态办公区（EOD）外，在经济适用房价格的建设部生态住宅示范工程杭州北秀蓝湾住区中，完成适宜生态住宅技术的集成，并在中国—欧盟环境管理合作计划项目怀柔生态规划的基础上探讨生态村落规划与山区绿色民居模式研究。

技术路线：建立经济适用的绿色住宅技术标准体系首先要制定绿色建筑产品的评定标准，结合绿色住区所在地特有的良渚文化，在原有建筑市场售价不变的基础上，在一定量的投资增加额度下，综合集成适宜生态技术体系，改造优化原建筑设计方案。

九渡河生态村落与绿色民居模式研究的技术路线为：在特有的现状自然基础环境及历史人文环境的基础上进行生态分析，并进行生态村落体系规划，研究生态建筑与系统周边资源、生态村落的生产生活、农业商业与环境的关系，完成生态村落人居系统模式设计、绿色系统模式设计及能源系统模式设计。

研究与设计实践说明：

依据北秀蓝湾绿色建筑评定的具体内容及技术体系为：

（1）建筑的适用性：考虑建筑的灵活使用性、热舒适性，选用绿色节能技术，设置控制环境噪声的绿色屏障，创造良好的室内环境品质。

（2）场地与环境：做好节地设计，降低环境负荷。绿地系统规划突出生态效益及生物多样性，达到杀菌、降噪、除尘、增氧、吸收有害气体、涵养水源、净化水体及调节风速和风压的作用。植物配置依据适地适树原则，同时选择乡土植物及种植屋面、垂直绿化技术（图 1a、图 1b）。尽量实现植物的免维护或很少维护。

（3）能源技术：针对不同区域对外墙、屋顶、外窗进行建筑热工优化，部分建筑节能综合指标比达到节能 65%，其余达到节能 75%。采取土壤源热泵、高舒适度平面式冷暖空调、天棚低温辐射采暖制冷系统、独立置换式送风及湿度控制等技术（下送上回置换式新风系统）及外遮阳系统。利用太阳能热水等可再生能源。

（4）水资源技术：实行雨、污、废水分流，收集优质杂排水，减少排入市政系统的污、废水和雨水量，不对生态环境产生负面影响；采用生物接触氧化和物化相结合的中水处理方法；以生物净化完成雨水收集利用及其回渗；雨水径流经过土壤渗透和生物过滤池根系净化后，进入景观水池，集满后，可自行溢流到小区的河里（图 1c）。防止因共用排水立管和排水系统渗漏所造成的交叉感染。

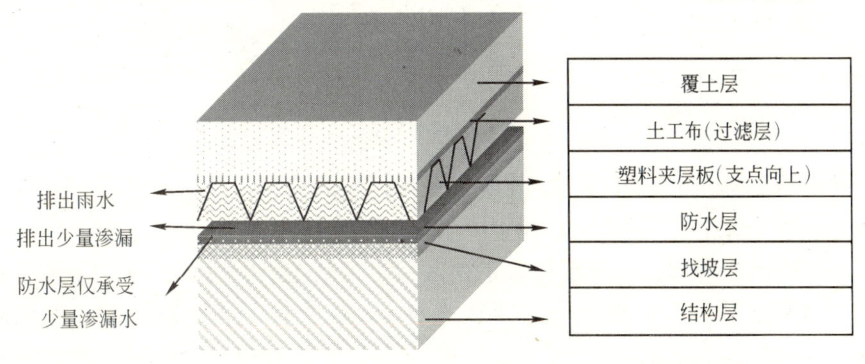

图 1a　杭州北秀蓝湾绿色住区屋顶绿化示意

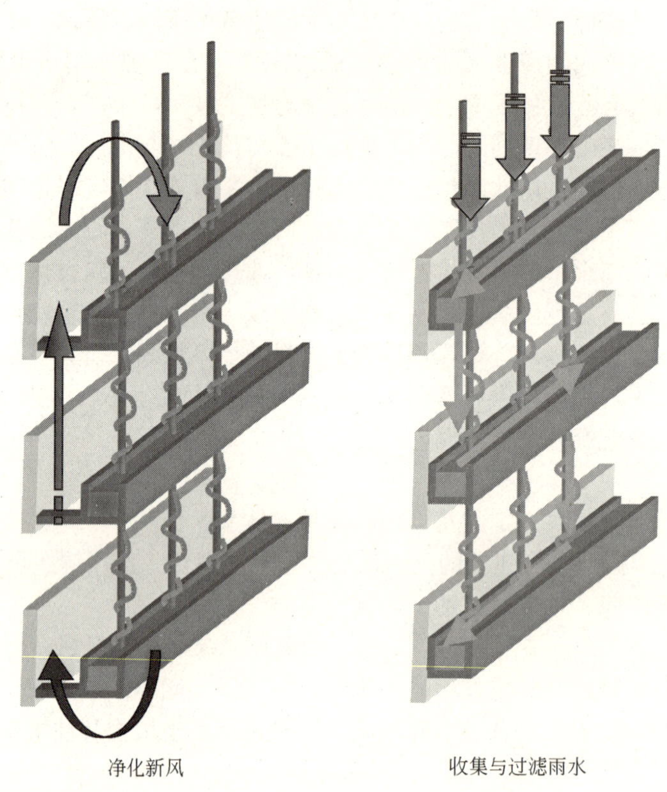

图 1b 杭州北秀蓝湾绿色住区垂直绿化示意

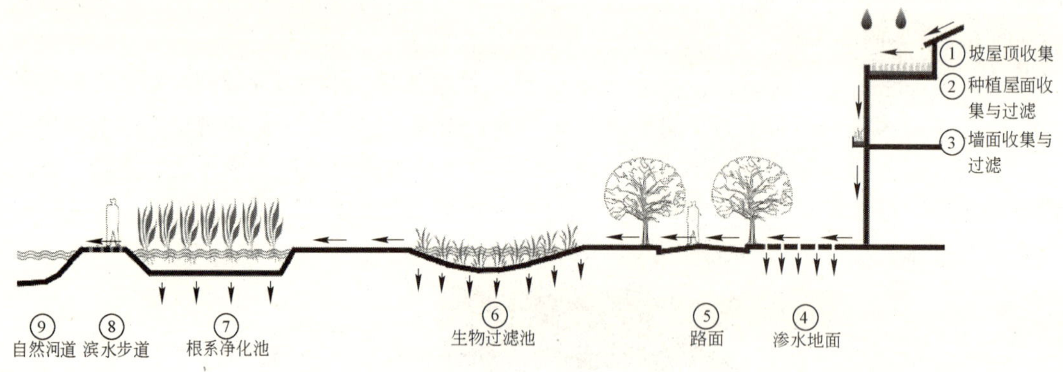

图 1c 雨水收集系统断面示意图

(5) 材料与资源：选用可回收、降解、循环再生的建材，绿色环境友好建材及本土建材，减少建材从开采、加工到运输过程中的能源消耗。

(6) 建筑的运营管理：以智能化数字社区系统控制建筑的运营管理。小区智能化系统集成，满足各系统功能、管理和信息共享的要求；优化控制各弱电系统，达到高效、经济、节能、协调的运行状态。

九渡河生态村落与绿色民居模式研究首先在生态规划的指导下，以科学分析为基础，对森林、居民地、景观点分布、交通组织、坡度、坡向、高程、太阳辐射、降雨量分布进行系统分析、甄别与量化判断；其次，进行村落体系分析与规划，明晰植物功能分区、控制人类生产区、组织村镇结构体系、针对长城景区每一个村落进行横剖面与纵剖面的解析，进行日照分析、风环境分析、研究植物生境、群落分布、生产区的构成；最终确定生态村落的规模与选址、体系规划，及生态村落建筑与周边环境的关系(图 2a、图 2b)。生态村落建筑的人居模式包括院落模式、人居交通体系与动植物扩散体系的关系、功能模式与围合模式；绿色系统模式包括生态安全框架对绿色建筑的渗透、立体化的由庭院到屋顶的立体植物空间、动物栖息空间等；资源系统集中包括被动式的能源利用，如阳光、生活余热等；主动式能源采用集中生物质气化利用技术，及自循环小型沼气池建立以沼气为纽带的农村能源生态模式(图 2c)。

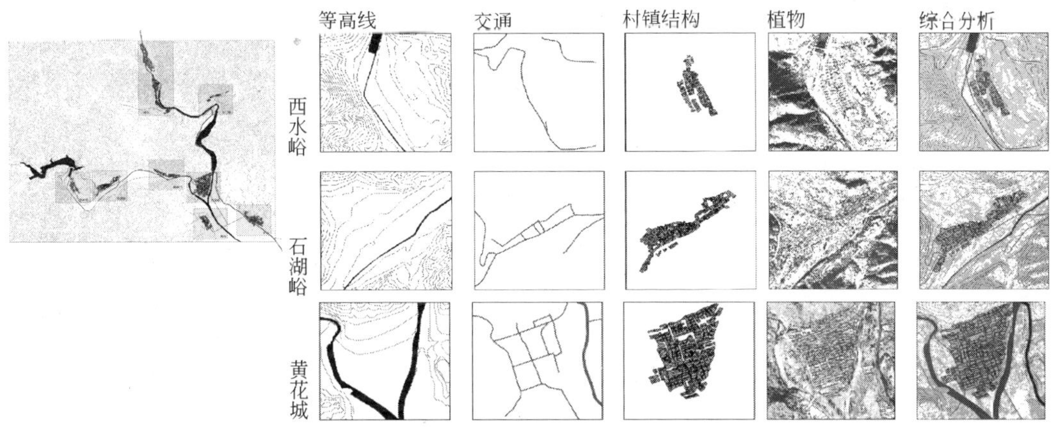

图 2a　九渡河生态村落的模式分析

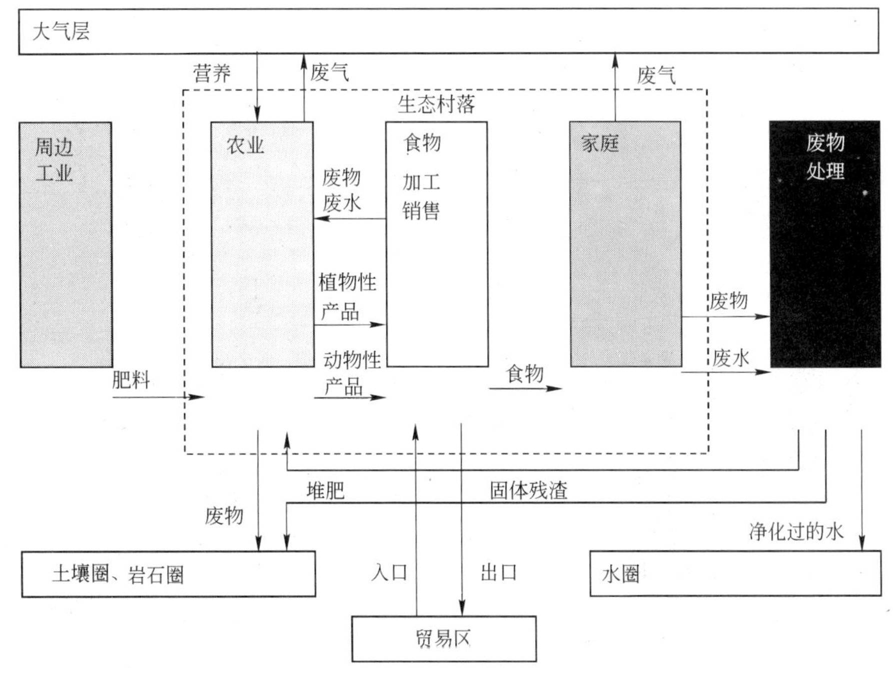

图 2b　生态村落与环境的关系

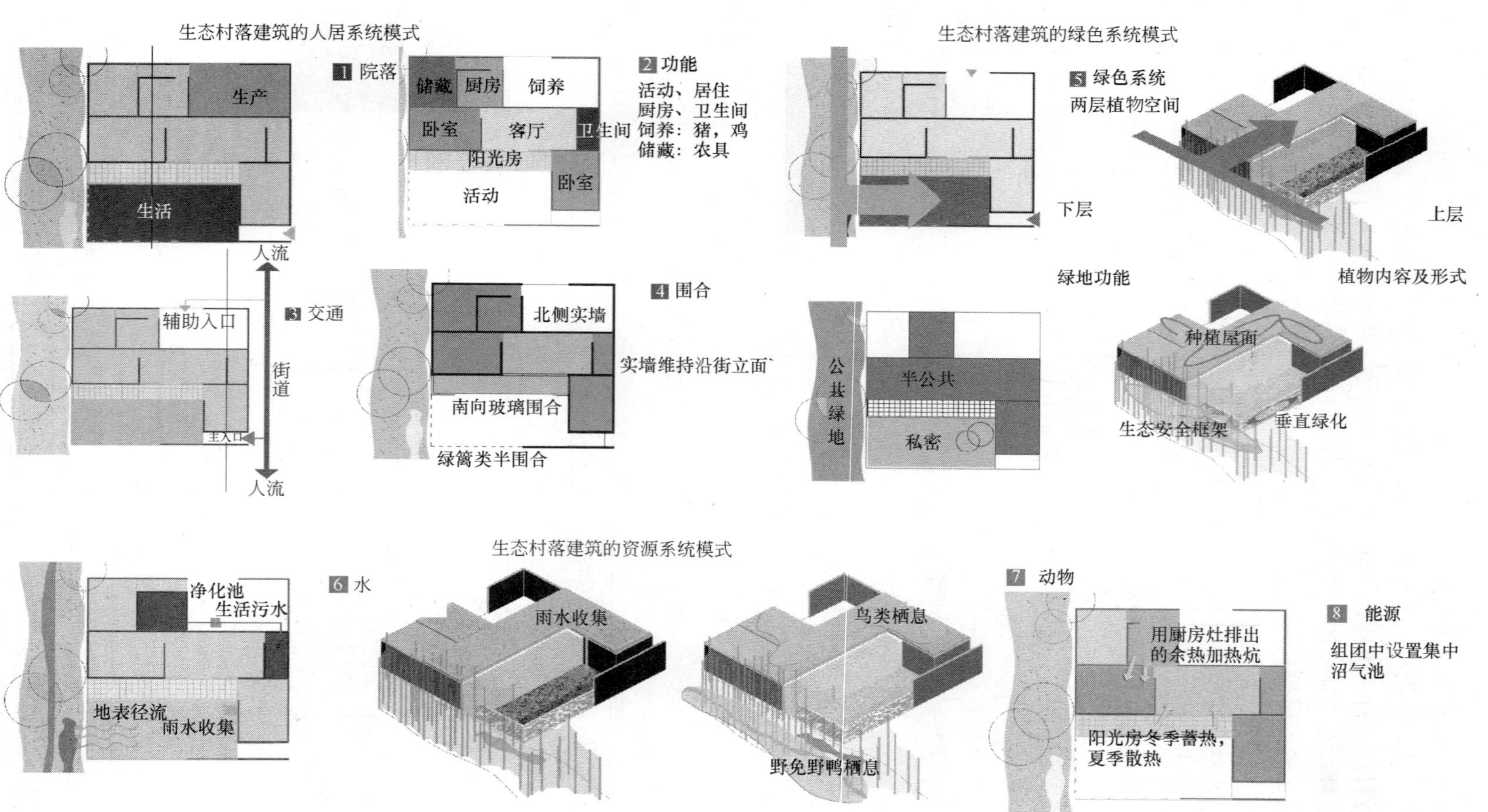

图 2c 生态村落建筑的模式设计

实践二：生态室内设计、研究与示范工程(北京市科学技术委员会)。

背景：室内作为与人关系最为密切的建筑空间，是绿色建筑的延伸，直接涉及到人对资源、能源及空间占有与使用的效率与规模问题。北京市科学技术委员会为了探索和示范可持续发展在建筑使用中的途径与方法，特设立本项目，在北京创业大厦15层进行室内科技创新成果集成研究与应用展示(图3a)。

技术路线：生态室内研究包括室内空间系统、能源系统、水系统、材料系统、墙体与新风系统、植物系统、光系统、智能系统和文化传承等，以及行室内系统功能集成组织与设计。通过与能源专家、水环境专家、材料专家、能源专家、植物专家及智能技术专家及施工机构密切的学科交叉与实施方案合作研究落实，将辅助、配合专业的提供模式转换为全程合作、评估与选择的优化团队协作模式。创新技术包括室内隔墙应用技术体系、室内通风系统组织、节能的冷热源辐射技术使用、室内功能植物系统的组织、循环材料与再生材料的选择与应用。

研究与设计实践说明：对原有的室内功能设施及空间进行淘汰和重组，优化室内存在的生态问题，建构全新的室内空间格局与功能系统组织。其中包括：废弃原有的新风系统、供热系统、厕卫系统、室内空间组织系统和消防组织系统。具体生态设计实施包括：

(1) 以素混凝土顶释放大面积的吊顶空间，加强空气流通与交换，减少空间对使用者的心理与视觉压力，提高使用空间的安全性、舒适度和健康指数。

(2) 能源系统：采取冷热源技术实现室内的采暖与降温，通过埋设的顶辐射、侧墙辐射达到室内均匀的调温与控制，并对外墙维护结构起到保温隔热作用。使室内最大的耗能大幅度降低，具有明显的节能作用(图3b)。太阳能技术的应用受已建成建筑的约束，无法使室内采暖与太阳能技术完整结合，只局部选用太阳能卫浴热水供给系统。由于太阳能光电技术转化效率低，存在成本核算与建造数量、铺设场地规模等限制问题，无法大规模采用，故仅用于厕卫照明的补偿(图3c)。

(3) 水系统的组织与应用：包括采用无水泡沫冲厕技术，节约大量日常冲厕用水；将浴室用水回送大楼中水回用收集系统；利用楼面雨水缓排技术实现雨水的收集与利用。降低了对市政雨污水排放资源的占有量，达到了降低雨污水处理成本和资源回用的生态目标。

(4) 材料系统：作为支撑生态室内的重要因素，材料系统具有特殊的价值与作用。本着新材料、再生材料和循环材料优先的原则，本次生态室内选择了：

麻秆再生材料：总后研发的70%废弃麻秆加30%废弃塑料的再生材料，用于门、地面等，具有杀菌、抗腐、隔声及整体性强的特点。

宝贵环保再生石材：70%的废弃石材或碎化建筑垃圾加30%水泥氧化还原为表现力丰富的再造石材，用于室内植物槽、外墙面装饰，并在内墙与通风结合(图3d)。

“绿盾”循环板材：在卫生间和沙龙展区采用软包装废弃材料再次加工利用制作的隔墙、活动展柜(图3e)和室内垃圾桶。

艺术型再生装饰板材：该特殊循环材料的原料来自于日常生活的牙膏皮、洗发和浴液瓶以及运动鞋鞋底材料等，通过艺术组织和热压塑性处理，形成特殊的肌理结构和唯美抽象艺术效果，具有室内装饰造型的价值和意义(图3f)。

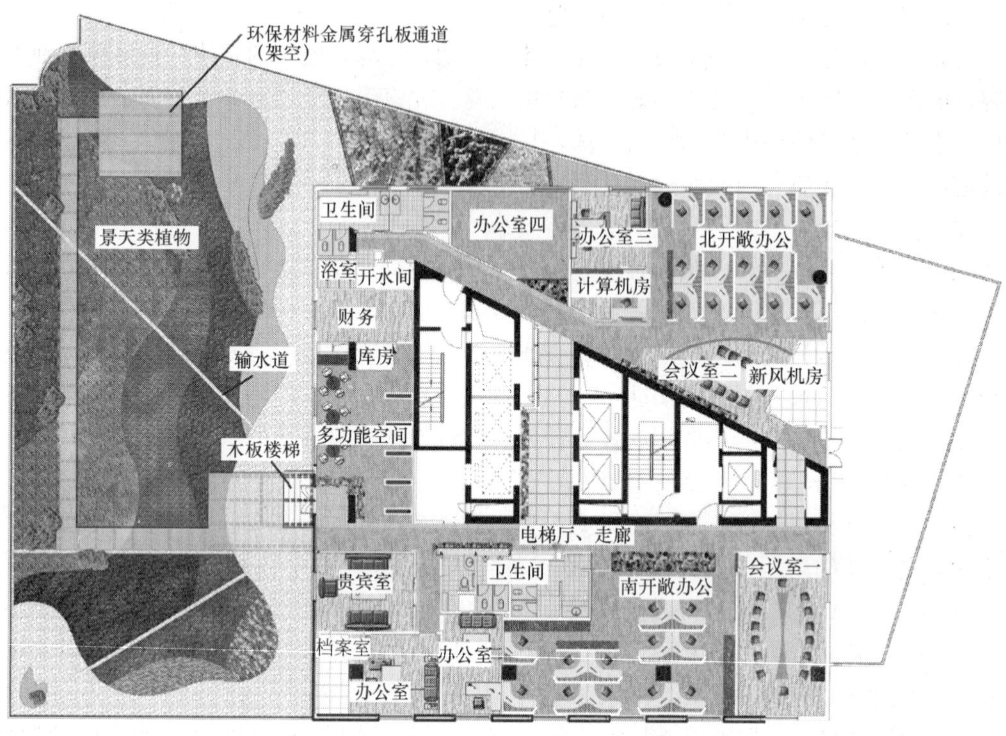

图 3a 北京科委室内平面布置图

图 3b 冷热源辐射技术

图 3c 太阳能技术

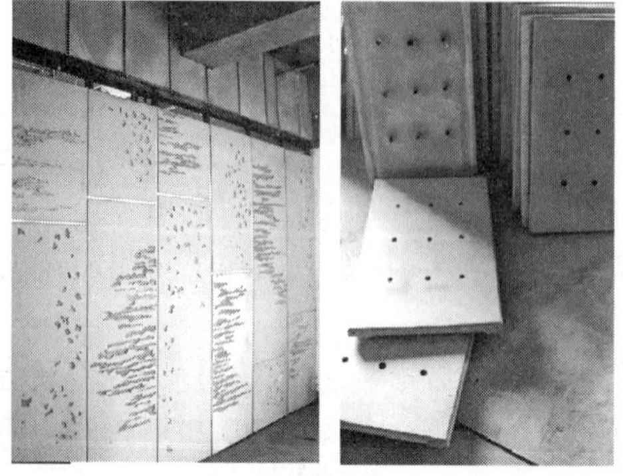

图 3d 环保再生石材

图 3e “绿盾”循环板材

图 3f　艺术型再生装饰板材

图 3g　钢弦墙体

图 3h　室内植物系统

图 3i　再生循环材料的传统表现

(5) 墙体与新风系统：管道集成是生态室内功能和技术研究设计的特殊之处。此次创造性地应用新型钢悬墙体石膏板墙体集成管道系统，具有良好的技术表现和功能发展。该墙体技术有柔性特点，曾在北京海淀区生态办公区试验项目中作为外维护墙体使用，具有防止温差变形、材料可回收利用、施工操作面小、稳定安全、构造空间可利用率高、可变性强、易于维护维修等特点。使用的材料涉及石膏板、再造石、植物、钢构玻璃、欧松板、钢网等，施工操作便宜、灵活，实现下送上回的墙体通风系统和各类管线入墙，不再需要吊顶管道空间(图 3g)。

(6) 室内植物系统：植物在室内多作为装饰材料或阶段性过滤材料，本次生态室内则根据植物的特殊物理、化学功能及形态特性，将其定义为“建筑的植物材料”。根据室内不同的使用功能分区，考虑植物生长的适宜室内生境、人对植物的感受及景观影响，组织具有不同降尘、降噪、杀菌、放氧、固碳的植物量组团，参与室内空气调节和优化空气质量。室内植物形式：植物槽、植物墙和植物围挡影壁(图 3h)。

(7) 智能系统：提高室内控制能力和降低能源与资源消耗的手段。本次选择性地使用了冷热源辐射的调控、室内植物滴灌智能控制、缓冲节能开关、无水厕卫自控泡末等技术。

(8) 文化传承在人的感知与行为密切关联的室内是一个重要的设计要素和使用题材。中国传统的建筑元素和室内装饰要素在门、隔墙、家具和室内雕塑中得到点题式的应用，特别是利用再生循环材料实现了艺术墙面造型(图 3i)。

实践三：北京屋顶绿化技术与材料的研究与试验工程。

背景：绿色建筑的建造、技术应用、使用方法、管理模式是实现人类对生态保护重要手段和研究方向。种植屋面技术是缓解、修复、连通以及建构生态人居系统与自然系统和谐共生的重要技术、产业和政策导向的依托内容。

从人文关怀的角度看，屋顶绿化构成的绿色环境系统及景观，包涵了丰富的文化及自然色彩，且具有愉悦人心理与精神的结构美、色彩美、形态美和韵律美。从城市空间效率及经济效益分析，屋顶绿化是开拓城市绿色空间的一种较为经济的途径。屋顶绿化几乎以等面积偿还建筑物所占的地面，有效缓解了建筑占地与城市绿地系统建设争地的矛盾，提高了城市空间生态效益承载。从城市视觉景观适宜性角度分析，屋顶绿化、垂直绿化修饰丰富了玻璃混凝土森林的冷漠视觉感受，加强了植物对建筑功能及建筑维护结构补充、优化的生态作用，降低日照耗能，避免日光反射对城市居民的视力损害。从城市建筑形态与美学价值上分析，屋顶绿化的柔化设计与刚性建筑材料的对比，具有丰富和充满生机的艺术效果，形成了多层次的城市空间景观格局。从降低城市建筑能耗、发展节约型城市、建立循环经济的角度分析，它通过提高建筑的效率和功能，加强局部生态小系统的服务功能，实现降低城市热岛和减少占用城市市政雨水排放资源的作用与价值。

屋顶绿化存在着各种形式，如：屋顶花园、空中花园、植被屋面(种植屋面)等，分别具有不同的价值取向。屋顶花园、空中花园注重美观价值，追求观赏性和园林造景效果，投资大，成本高，需人工养护，加大建筑荷载，不宜以生态功能为目的进行推广。植被屋面则注重生态效益，适宜的构造材料、植物和种植方法，使植被屋面从一开始就遵循节约、简约的原则，达到一次种植免维护、成本低、适宜推广的要求；同时构造上采取轻型材料，最大限度的降低对建筑结构的要求与破坏，提高了建筑的节能效率，符合发展中国家经济能力的客观现实。

技术路线：在气候条件好的南方，植被的成活率高，植被品种易于选择。而以北京为例的北方城市，冬天寒冷风大；夏季时而降雨量大，时而暴晒；春天风沙大，干旱，气温多变。因此北京种植屋面的实施具有很大的难度，其构造技术、种植基质和植物材料的选择是科研的关键所在，应遵循朴素实用、本地化、适宜性、效率优化及景观丰富多样化原则。为了植被屋面后期的免维护或少维护，种植屋面的创新为其构造体系、蓄水层与过滤层材料、乡土砾石的发掘与加工、边坡构造节点设计、种植基质配置和植物材料的选型。

研究与设计实践说明：北京种植屋面的研究通过选育理论上适宜的植物材料和构造材料、不同的构造系统和种植基质配方，在试验基地进行了近三年的北京环境适应性实验和筛选育种，取得了相应的科研数据及实验验证，完成了植物材料、构造系统和种植模式的筛选定型研究(图 4a、图 4b)。通过北京市可持续发展科技促进中心的支持，在近 800m^2 的北京创业大厦 15 层平台上实现了实用建筑屋面的中试施工实践，证明了针对北京条件下的屋顶绿化技术系统，具有适宜性、针对性、效益性和特殊性的特点，能做到稳定存活且免维护或低维护，所选出的植物具备抗寒、抗热、抗旱、耐涝；抗风、抗污染等特点。

(1) 植物材料的选择：选择防寒、抗旱、不怕涝、易成活的植物品种，同时要求四季有景，物种多样性类型达到＞8－10 种以上。中试植物材料第一阶段应用试验总共选用近 30 种，经过近 11 个月春夏秋冬的实验观察，基本适应种植生境，在种植初期的植物生长期内比较稳定，在植物生殖期内植物表现不很稳定，雨季以后开始逐步恢复和旺盛，入冬以后表现稳定(图 4c)。

图 4a　屋顶绿化实验

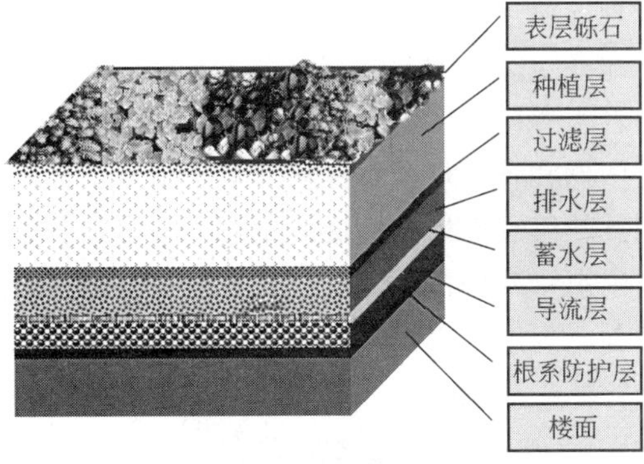

图 4b　种植屋面的构造示意

图 4c　种植屋面示范工程实景

（2）构造：比照欧州国家屋顶绿化构造，北京种植屋面的构造体系依照本地特殊的环境条件，增加了导流构造，强化蓄水构造，并提高了种植基质的涵养能力和透气能力(图4b)。为提高整体防风、防涝、防腐和构造的安全性，特殊设计了“边坡”——特殊构造节点。整体构造系统厚约8cm，种植系统厚度为3～15cm分区试验，总体厚度为11～23cm。

（3）构造材料：使用具有可降解、可再生、可循环材料和适宜的本地材料。针对性应用总后勤部士兵装备研究所合作研究的特有的构造材料，及相关学科专家发现并研制出的乡土砾石特殊添加材料，大大提高了构造效率和适应能力。

整个构造粗放维护(免维护)，不设灌溉系统，抗寒热(＋40℃～－25℃之间)，抗风(11级)，具顺畅的排水能力，轻体结构(自重不超过60kg/m²)，无土栽培，种植面雨水收集能力(＞40L/m²)，成本不得超出200元/m²，材料采购为本地范围内半径不超过500km。

实践四：生态城市设计、生态景观的研究与实践控制绿色建筑的建设。

实践研究：生态规划指导下的北京市海淀区国家湿地保护公园保护性生态景观设计与研究(北京海淀区规划分局)。

背景：生态景观设计具有与生态规划一脉相承的逻辑体系和科学体系，是落实总体规划、控制性规划、详细规划的科学设计方法。以往的城市设计多从使用功能和交通组织以及空间角度着眼，对科学地在空间结构及时空关系上把握生态系统的完整性、安全性和适宜性做得不够，直接影响着生态城市建设，无法为绿色建筑与绿色民居建设的选址、规模控制与建造结果提供科学依据和管理规范。在海淀北部地区，依照修正后《中关村科技园区海淀园发展区生态规划》的编制成果，完成北京翠湖(稻香湖)国家湿地保护公园生态景观研究设计。本案是根据生态资源分析与生态承载、生态规划建设用地、人口及水资源的规模、绿容率指标体系控制、生态基础设施和湿地保护范围与方法进行的(图5)。

图5　北京市海淀区北部国家湿地保护核心湿地总图

技术路线：结合景观生态学、植物学、地质学、水文水资源学，依照生态规划的成果，进行适合生态系统功能、资源结构以及环境条件的科学规划设计，明确限定湿地生态等级分区、环境构成、功能构成和系统组织。

研究与设计实践说明：确定湿地必要的规模与范围。依据现状存在的问题的性质和必要性，设定问题的政策调控机制，提出解决的要求和方法。对于建设选址、建设规模和建设标准依照生态规划的城市容度指标对总量进行科学的动态控制与释放，最大程度上控制、降低和缓解对生态系统，特别是湿地生态系统的扰动。明确确定水生态系统、植物生态系统的功能组织与规模范围。制定生态策略，从地质、水文、气候、土壤、植物、资源、能源、建筑和交通等方面建立完善的生态设施。在整体规范与控制的基础上，将湿地保护进行 1～3 级、缓冲区、过渡区的等级控制，制订湿地保护原则和保护措施，明晰湿地修复、恢复和重建的过程与方法，要求湿地保护区内交通组织的标准及设施要点及绿色建筑的适宜模式。

实践五：科学规划保障绿色建筑的落实与推广。

背景：随着可持续科学发展观成为我国的国策，中国“十一五”规划将注重生态保护与社会和谐发展定位发展战略，正在探索的科学规划被逐步落实。作为中国—欧盟环境管理合作计划项目，科学规划的创新体系研究与应用、生态规划的研究与编制成果，正在为城市规划的科学编制提供核心指导依据、前提条件和基础保障，并应用到总体规划、控制性规划、详细规划以及城市设计编制的全过程系统工作中。

中国—欧盟环境管理合作计划怀柔生态规划的落实是在北京市规划委员会主张下，与北京市城市规划设计院、清华安地建筑设计顾问有限公司、北京清华城市规划院共同合作完成。落实过程是首次多机构、多专业、交叉学科共同探索规划理论与技术路线的创新，被业内称为落实国家科学发展观的“科学规划的北京模式”、“怀柔模式”。

技术路线：以生态规划为科学基础，本着和谐共生、健康安全、科学求是、永续发展的宗旨，依照生态规划提供的核心指导依据，通过紧密探索与合作，共同完成各项规划编制与设计的理论研究、技术方法确定和编制内容的落实。

研究与设计实践说明：从总体规划编制工作中，主要体现在生态安全框架的法定落实与保障、土地利用的生态等级控制与空间分布；基于水资源、植物生物量及土地使用规模的人口规模控制；生态规划的生态承载指数控制下的资源使用与土地使用容量动态管理、评估与释放；针对性地在规划中明确要求建立生态保护、生态城市、宜居城市及城乡一体化统筹发展的具体要求。落实到具体细节包括：绿色建筑标准的制定、屋顶绿化技术的应用、节能节水技术的要求等。这是在中国规划编制技术体系中，首次将规划目标与落实规划的具体方法紧密结合的规划编制技术体系的创新(图 6a)。

在控制性规划编制中，依据生态规划编制成果进行深化，规划指标得以纵向深入。怀柔镇域体系与新城发展的控制性规划编制时，对局部资源分配与管理使用进行具体控制。利用整合、调节与配置等手段，实现保护与发展的最大、最佳及高效的选择与集成。在此基础上建立明确的节地、节水、节能、节材、产业结构和生态系统完整的法定管理与科学调控规划(图 6b)。

在详细规划到城市设计的编制中，主要是实现规划编制成果在行为与功能组织及空间上的落实，包括：在大型生态安全框架中斑块、廊道体系的内部结构与内涵的组织与应

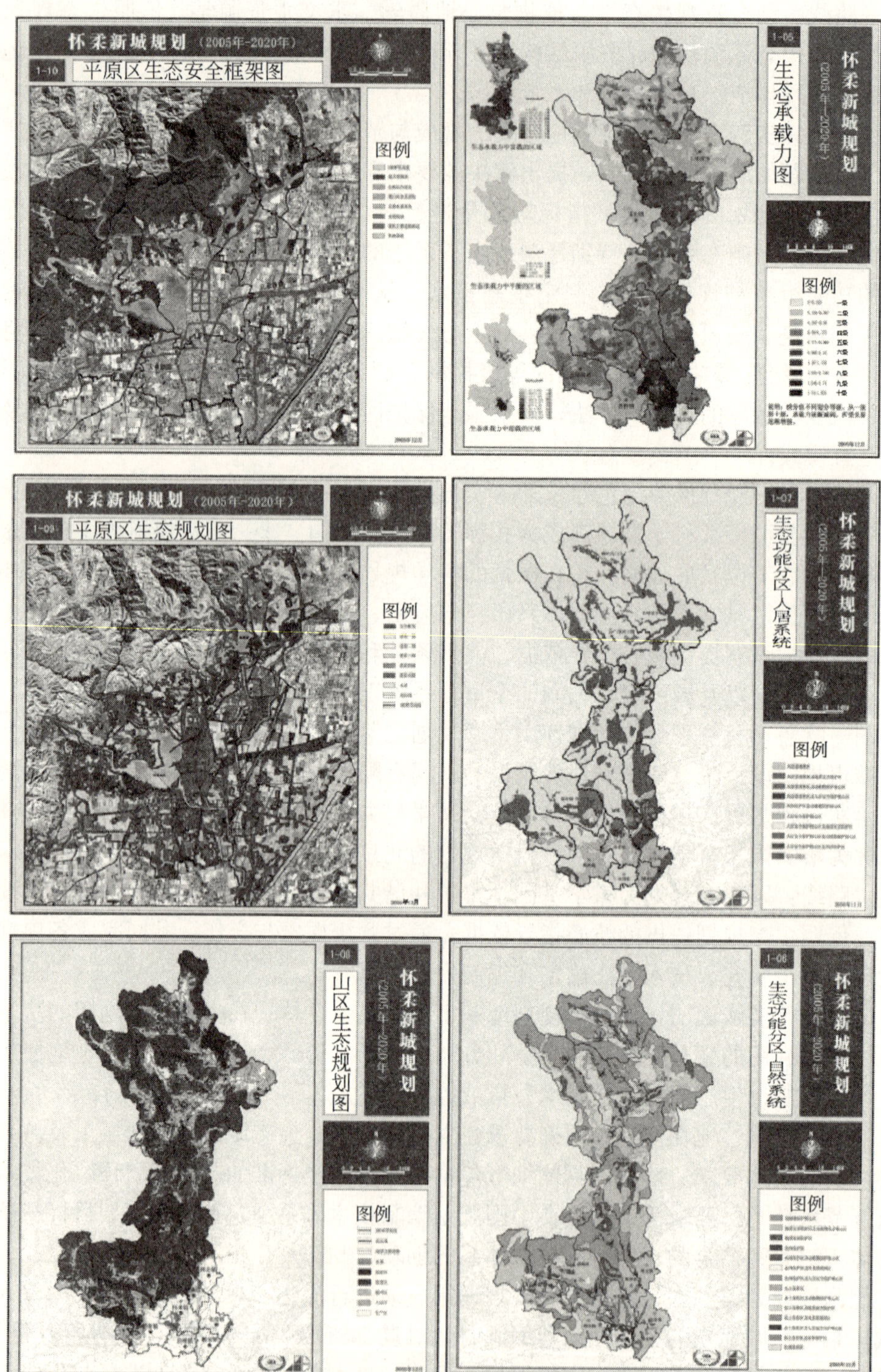

图 6a 怀柔总体规划落实生态规划

用；要求建立中型和微型斑块、廊道体系；适宜生境的植物群落、种群特点、景观功能的指导；生态设施的组织与建设；在人居系统规划设计中强调人的行为控制与规范；解决空

图 6b　生态安全框架的落实

间结构中人与自然交错存在的布局尺度、功能组织与分布等。在此基础上，研究提出城市设计的生态模式、设计要求与规范(图 7)。

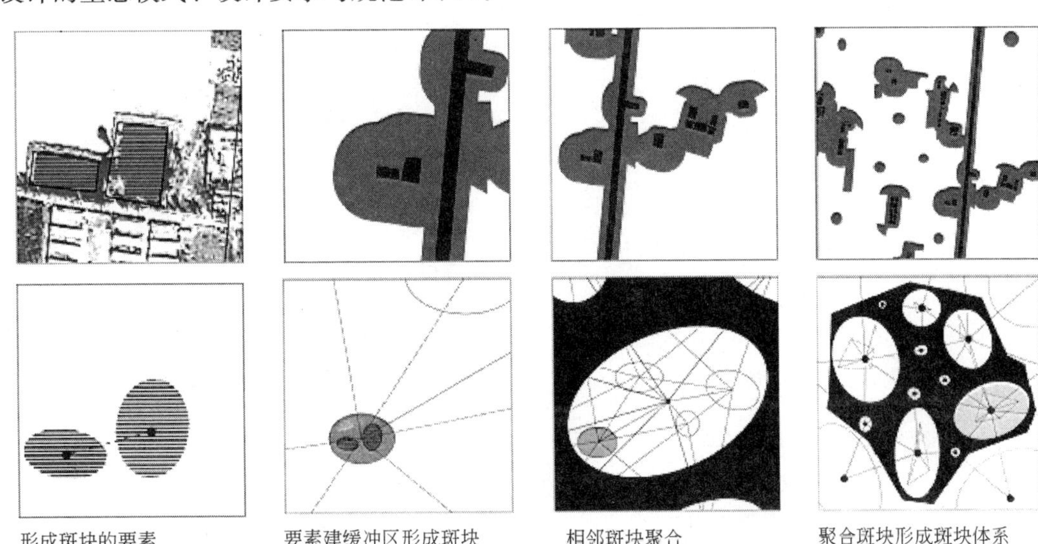

图 7　生态安全框架在不同尺度下的描述

从以上绿色建筑到生态规划应用的多项科研与实践中，充分证明了生态规划是绿色建筑推广应用的科学基础和法律保障。绿色建筑是实现生态规划并优化社会资源和自然资源分配与使用的重要科学途径。此外，通过对国内外有特点的绿色建筑评估标准的研究与国内标准的编制研究，尤其是《中关村科技园区生态人居环境标准体系》(2002 年)及《浙江省绿色建筑标准》(2005 年)的研究与编制，可以看到绿色建筑不但有利于人类建设节约型、生态型和宜居型社会，修复、恢复人类对自然生态系统长期的扰动与破损，更有利于人类主动减少对自身生存的生态环境的破坏，真正实现可持续发展的人类共同理念，人与自然和谐共生、天人合一的科学主张，让生态人居建设做到"山清水秀，鸟语花香"。

Explore the Scientific System of Green Building

Rao Rong

(Tsinghua University, 21 Century Urban Ecology
Academy of Beijing, Beijing 100084 China)

Abstract: Based on research and practical projects of ecological interior, green neighborhood, eco-village, ecological urban design, the article expatiates on establishing scientific system of green building. It discusses the goal, target and criterion of green building, and the mode, technique of green building construction.

参考文献

[1] Peter F. Smith, Architecture in a Climate of Change, Architecture Press, Oxford, 2001

[2] Rodney R White, Building the ecological city, CRC Press, Cambridge England, 2002

建筑的功耗综合指标与可持续发展

徐米甘
（北京维拓时代建筑设计有限公司，北京　100025）

摘　要：本文论述了对于人类的各种建设活动和古今中外各类建筑物进行可持续发展性能方面评价的重要意义；提出了加权评分和价值折算两种基本评价方法，计算建筑物在其全寿命期内或研究阶段内，已发生的或预期的，满足人们需求程度和消耗地球资源程度的功耗综合指标；并举例简述如何使用这一指标对各类建筑物在可持续发展意义方面作横向和纵向的比较。

关键词：可持续发展　功耗综合指标　绿色文明

1　现实与危机——绿色文明世纪之初的思考

近年来，“绿色、生态建筑”被老百姓所关注，也成为开发商炒作楼盘的卖点。问题是真正做起来，人们想到的只是如何加快发展速度，如何拉动和刺激消费；对于可持续发展与人类“生死存亡”的关系，往往只顾及自己眼前的享受，而忘掉了如何“不损害人类后代满足他们自身需求的能力”。

中国的经济发展曾因政治因素耽误了数十年时间，但目前正以罕见的速度向前猛追。我们追赶的目标有：国民生产总值、城市化率、人均居住面积、世界最高建筑、××产量世界第一等等，但是可持续发展方面我们要达到什么样的目标呢？

在世纪之交，关于21世纪时代特征有两种表述之争：信息时代，还是生态时代(可持续发展时代)。21世纪应该是绿色文明的世纪；从人类历史经过的三个文明的比较(见表1)可以看出，在绿色文明世纪，稀缺资本、资源、能源、人与自然的关系等方面均发生了极大变化，这是我们今天必须走可持续发展之路的根本原因。旧有文明的成果我们应该继承，但是当时代特征、生存条件发生了根本性的变化时，我们的思想和行动应该跟上变化，不能再局限在过去文明的框框内。

人类历史经过的三个文明的比较　　表1

时　代	色彩象征	稀缺资本	中心产业	资源	能源	技术	城乡关系	人与自然
农　业	黄色文明	劳动力	种、养殖业	生物	植物	农耕	相互依存	依赖自然
工　业	黑色文明	金融、物资	制造业	自然	矿物	工业	相互对立	掠夺自然
生　态	绿色文明	自然资源	生态、信息	智力	可再生	信息	缩小差别	和谐共生

历史上国家、民族间的竞争，首要的是经济实力的竞争，所谓“落后了就要挨打”。新世纪人类面临的各种天灾人祸，如911事件、非典事件、洪涝灾害等等提醒着人们，自己过得好，还要宽厚待人，还要善待自然；富而不义，先进了也会挨打；你逃过了，子孙后代也

会受到惩罚。绿色文明时代的"可持续发展"，谋求的是整个人类今天以至未来子孙万代的生存发展，人们应该有更加全面、完善的思想和行动；经济、政治、道德、法律应该有新的准则，来衡量人们行为的好坏对错。我们研究可持续发展，不仅要研究经济，还要研究政治、文化；不仅面向未来，还要面向历史。理论与实践均要求我们拿出尽量准确而又实用的定量计算方法，对人类的活动及其成果，包括建筑物，从可持续发展的角度进行分析、比较和研究。

2 人类活动的可持续发展评价

从宏观来说，对人类活动在可持续发展方面的意义进行评价，应考虑这一活动对于地球资源消耗的程度；但是，这一消耗程度随着人类文明的发展呈指数曲线增长；如何进行历史的、横向和纵向的评价？什么样的资源消耗程度是可接受的？

为回答这些问题我们引入功耗综合指标的概念。所谓"功"，是人类活动满足自身需求，包括物质和精神两方面的需求的程度；"耗"，是人类活动消耗地球资源的程度。

比如对历史上某一时期各个国家的政治、经济活动，在人类社会可持续发展意义上可这样进行评价：历史这一时期人类社会整体的发展客观上有一个平均文明程度，对地球资源的消耗客观上有一个人均消耗量；该时期人类社会可持续发展平均功耗综合指标为 1，表示此时期内某个个人按人均消耗量消耗地球资源，享受平均生活水平时的指标。某一国家的文明指数(该国文明程度与各国平均文明程度之比)除以资源消耗指数(该国人均资源消耗量与世界人均资源消耗量之比)，得到该国该时期的功耗综合指标数值；这一数值大于 1，说明该国在可持续发展方面领先于当时世界平均水平，反之亦然。使用这种方法，不仅可以对历史某一片断作横向比较研究，也可以对社会某一领域前后发展作纵向分析研究。通过这一研究，应能指明历史上某些民族、国家衰亡的根本原因。

3 建筑的可持续发展评价

3.1 建筑的功耗综合指标

建筑是人类进行生产、生活、娱乐各种活动的场所；是民族文化、文明发展、技术水平的标志；是人类用智慧和劳动，以各种建材，在大地上留下的立体史书。人类使用的建筑物对地球资源的消耗既包括使用过程中的消耗，也包括建造过程中的消耗。随着人类生活水平的提高，建筑对地球资源的消耗在人类各种活动总消耗中所占比例越来越高。因此，对古今中外各种建筑物在可持续发展性能方面的分析研究，有着重要的意义。

这里我们引入建筑物的功耗综合指标概念，它能够标示建筑物在其全寿命期内或研究阶段内，已发生的或预期的，满足人们需求程度和消耗地球资源程度的关系。这样的研究有什么意义呢？举例来说，某座建筑物可能在建造过程和将来使用中消耗较少的地球资源，但是在满足其他需求方面有重大缺陷，建成后长期空置，或建成不长时间就拆除重建，则其建造过程中消耗的大量资源并没能发挥实际效用，这座建筑物的功耗综合指标应该是很低的。当我们拆除一座旧建筑物，在原址上建起一座新建筑时，按已有评价方法，仅对新建筑本身进行评估，但功耗综合指标的评价方法，应该考虑这一建造活动的全部影响，计算其消耗的资源时要加上旧建筑建造时消耗的一部分资源(因为今后这部分资源本应继续发挥的功效中止了）。用这样的方法来进行可持续发展的研究，就要求我们在建设过程中，不要再以“拆除了多少旧建筑”作为一项成就而沾沾自喜。

3.2 建筑功耗综合指标的两种评价计算方法

3.2.1 加权评分法

加权评分法一般可在对历史较久远，或满足需求、消耗资源的价值不便计算的建筑物进行评价时使用。加权评分的计算项目，在消耗资源方面可以有建筑耗材、建造用地、建造耗工、维修耗材、环境影响、使用耗能、再生材料的利用、可再生能源的利用等；在满足使用需求方面可以有适用(舒适)性、安全性、寿命、环境价值、满足人们精神方面需求的性能(美观、旅游景观价值)等。

用加权评分法评价建筑物的功耗综合指标，一般可采用如下步骤：

(1) 根据评价对象的特点和评价目的，确定对评价有意义的所有项目。

(2) 收集评价项目的有关数据资料，进行必要的分析、取舍、初步计算等整理工作，并确定各项目的加权系数。

(3) 由具备资格的评审人员，根据预先确定的评价标准，对各评价项目进行评分；然后采用加权累积的方法，计算得出资源消耗和满足需求两方面的综合得分。

(4) 将资源消耗和满足需求两个综合得分相乘后开方，计算该建筑的功耗综合指标(表示资源消耗和满足需求两个综合得分的几何平均值)。

(5) 得出的建筑功耗综合指标，表明其可持续发展综合性能的优劣；而从资源消耗和满足需求两个综合得分和各评价项目得分的高低，可分析其主要问题所在。

举一个应用加权评分法的实例：假设有中世纪留存至今的两座建筑，一座是东方的木建筑，一座是西方的石建筑，用加权评分法来评价它们在可持续发展方面的优劣得失。表2列出了评价过程中的数据和最终结果。

从表2中数据可看出，在资源消耗评价中，木建筑的建筑耗材、建造耗工、使用能耗优于石建筑，其他三方面劣于石建筑；在满足需求评价中，木建筑仅在适用性能方面优于石建筑，其他三方面均劣于石建筑；计算结果，两者资源消耗的综合得分不相上下，而满足需求综合得分，木建筑低于石建筑12.5个百分点，功耗综合指标木建筑低于石建筑6.9个百分点。我国木造古建保存情况较差以及我国历史上伐木过度造成水土流失严重，就是以上评价结果的事实证据。

中世纪木、石建筑功耗综合指标　加权评分计算　　　**表2**

资源消耗评价(加权分/原评分)				满足需求评价(加权分/原评分)			
评价项目	加权系数	石建筑	木建筑	评价项目	加权系数	石建筑	木建筑
建筑耗材	0.3	21/70	24/80	适用性能	0.4	24/60	28/70
建造耗工	0.1	7/70	7.5/75	建筑寿命	0.3	25.5/85	15/50
维修耗材	0.1	7.5/75	6/60	景观价值	0.1	8/80	7/70
环境影响	0.2	16/80	13/65	安全性能	0.2	16/80	11/55
使用能耗	0.2	14/70	16/80				
材料再生	0.1	7.5/75	5.5/55				
资源消耗综合得分		73	72	满足需求综合得分		73.5	61
功耗综合指标：		石建筑　73.2		木建筑　66.3			

3.2.2 价值折算法

对于现存、在建和将建的建筑物，一般也可使用价值折算法来计算功耗综合指标，进行可持续发展性能的评价。价值折算的计算项目，与加权评分法相同，但计算中预测、预估的内容可能要多一些。价值折算法的评价步骤如下：

(1) 评价项目确定。同前节(1)。

(2) 数据资料收集整理。同前节(2)，但要着重注意经济数据的量化。

(3) 价值折算。根据预先确定的折算方法，对建筑物在其全寿命期内或研究阶段内，已发生的或预期的资源消耗或满足需求的具体价值进行折算。价值折算的单位形式有两种：货币单位形式或能源单位(如 kWh，标煤等)形式，可根据评价对象和目的选定。

(4) 计算功耗综合指标。将各个评价项目的折算结果分别相加，得到资源消耗和满足需求两个总价值；将满足需求总价值除以资源需求总价值，得到该建筑的功耗综合指标；其含义为“单位资源消耗价值所获得的需求价值”。

(5) 分析评价结果。根据各单项的折算结果、资源消耗和满足需求两个总价值和最后得出的建筑功耗综合指标，可以分析建筑物在可持续发展方面存在的问题、问题发生的原因，并得出总的评价结论。

应用价值折算法的实例：某公司要建造一座 5000m^2、寿命为 80 年的住宅楼，一个方案是在城区拆除一座旧住宅，建造一座 15 层塔楼；另一方案是在郊区一块较大的空地上另建一座 5 层板楼。表 3 列出了用价值折算法对两个方案进行评价计算的数据和结果。

某住宅拆建、另建方案功耗综合指标价值折算比较 **表 3**

资源消耗价值(万元)			满足需求价值(万元)		
评价项目	拆建方案	另建方案	评价项目	拆建方案	另建方案
建筑耗材	600	500	住宅自身价值	2300	2500
建造耗工	40	30	绿色生态价值	500	650
维修耗材	250	180	文化景观价值	80	100
环境影响	80	60	交通环境价值	200	150
建造用地	100	60			
使用能耗	800	740			
旧建拆除	400	0			
资源消耗总价值	2270	1570	满足需求总价值	3080	3400
功耗综合指标	拆建方案 1.357		另建方案 2.166		

从表中数据可看出，按 80 年远景预测，郊区住宅的全寿命价值应高于市区住宅，综合其他方面的比较，最终另建方案住宅的单位消耗价值所获得的需求价值，是拆建方案的 1.6 倍。

Composite Target for Architectural Uses-Expenses and Sustainable Development

Xu Migan

(Beijing Victer Star Architecture Co. , Ltd. , Beijing 100025 China)

Abstract: Discuss the importance of appraise on sustainable development for diachronic/existent/hade build/building architectures. Proposes two means of score value and cost convert for colligate guideline on effect and consume. This guideline is relation of satisfy demand degree with natural resources consume amount. Give two examples for the means.

参考文献

[1] 周凤起. 21世纪中国能源工业面临的挑战.《2002年新型建材发展论坛论文集》, 2002.6

[2] 聂梅生. 正本清源——中国绿色生态住宅.《中国住宅》, 2002.7

[3] 李明浩(韩). 可持续性主题下的绿色建筑战略,《建筑学报》, 2003.2

[4] 刘煜(澳). 国际绿色生态建筑评价方法介绍与分析.《建筑学报》, 2003.3

夏热冬冷地区大型建筑工程中双层皮玻璃幕墙的气候适应性设计策略研究*

王　振　李保峰

（华中科技大学建筑与城市规划学院，武汉　430074）

摘　要： 最近几年，双层皮玻璃幕墙建筑开始在中国出现。而且在当前相关话题如可持续发展、生态和节约能源当中，经常会推荐使用这种类型的玻璃幕墙。双层皮玻璃幕墙能够更好地提供夏季隔热和冬季保温，这已经成为欧洲双层皮玻璃幕墙建筑的一个普遍特征。然而，在中国夏热冬冷地区，关于双层皮玻璃幕墙气候适应性的经验和知识却一直是个空白。因此，双层皮玻璃幕墙在该地区的气候适应性研究具有现实意义。本论文基于不同条件下如朝向、遮阳、通风、能耗、淋水、蓄热和气流循环等实验的对比研究，探讨双层皮玻璃幕墙对室内热环境以及能耗的影响。此外，结合双层皮玻璃幕墙的技术特点，本文尝试总结双层皮玻璃幕墙的技术规则和气候适应性设计策略。本研究不仅为生态建筑的实现提供了可操作的技术支撑，而且为建筑设计的逻辑表达提供了最大的可能性。

关键词： 夏热冬冷地区　双层皮玻璃幕墙　气候适应性　设计策略

1　引言

最近几年，双层皮玻璃幕墙建筑开始在中国出现。而且在当前相关话题如可持续发展、生态和节约能源当中，经常会推荐使用这种类型的玻璃幕墙。双层皮玻璃幕墙能够更好地提供夏季隔热和冬季保温，这已经成为欧洲双层皮玻璃幕墙建筑的一个普遍特征。然而，在中国夏热冬冷地区，由于经济条件和国家能源政策的限制，长期以来该地区的建筑表皮设计，尤其是大型工程中的玻璃幕墙设计无法适应地域气候特点，身陷夏季无法充分防热，冬季无法充分御冷的“两难”局面。

建筑表皮即建筑的围护结构，通过控制外部状况对建筑的影响，为人们提供健康舒适的内部环境。“适应”是生命科学中的一个带有普遍性的概念，英文“Adaptation”一词来源于拉丁文“Adaptatus”，原意是调整、改变、特指对气候的适应。气候适应性在本文则意指为对气候适应的特性和能力。双层皮玻璃幕墙就是用内、外两层幕墙和中间的空气层共同组成的动态的、具有气候适应和表皮应变可能的建筑围护结构。它既有热屏蔽，引入太阳光，加速空气交换等功能，同时保证内外视线交流不受影响。如果说建筑的形态体现了建筑对于气候的静态性能，那么对于气候动态的适应，在很大程度上要通过对建筑表

* 得到国家自然科学基金项目资助，项目批准号为50578067。

皮的控制才能实现。建筑表皮实际上是人体适应气候能力的缺陷的补偿，表现出人区别于动物只能被动的适应环境的高明之处。夏热冬冷地区城市区建筑室内热环境受到微气候热环境、湿环境、风环境和大气环境等多个影响因素的共同作用，这些影响对建筑室内热环境产生直接或间接的作用。在夏热冬冷地区，对建筑的围护结构，尤其是玻璃幕墙进行具有生态意识的热工设计有其自身的特点。与我国其他四个热工气候分区不同，夏热冬冷地区需要同时考虑冬夏两季不同方向的热量传递以及在自然通风条件下建筑热湿过程的传递。对于夏热冬冷地区气候的极端性，作为建筑表皮之一的双层皮玻璃幕墙更能积极地体现这种补偿，也就更具备气候适应性的特征。双层皮玻璃幕墙实际上是建筑表皮设计策略与地域环境特点的适宜结合，是建筑作为气候过滤器或掩体的具体技术手段之一。

最初具有对双层皮玻璃幕墙研究性质的是 1930 年勒·柯布西耶在巴黎救世军旅馆(Cite de Refuge)中“mur neutralisant”的设想。后来由于缺乏资金，外层玻璃和制冷设备都被取消了，结果引发了建筑内部的恶劣环境。柯布西耶的多层玻璃幕墙或许是一次技术失败，但它指出了玻璃建筑将要面临的新挑战，同时也指明了未来建筑概念发展的方向。随后，双层皮玻璃幕墙技术在实践中逐步成熟，同时也暴露出一系列问题，继而引发了对双层皮玻璃幕墙的研究。Professor John Yellott 在美国亚利桑那州大学参照西方化学中心的双层皮玻璃幕墙建起同比大小的实体模型进行了遮阳系数、传热系数以及透光率等参数的测试，并证实了空腔缓冲层在太阳辐射得热时可以产生大规模的风塔效应/烟囱效应，同时加州伯克里大学利用 DOE-2 程序进行了相关能源分析。Werner Lang 在德国慕尼黑工业大学进行了双层皮玻璃幕墙的木遮阳和保温系统的实验研究。实验证实了木百叶在冬季的保温性能和夏季的遮阳效果。A. Zöllner 等在德国慕尼黑工业大学建造了足尺模型露天实验室进行了双层皮玻璃幕墙空腔内紊流状态下混和对流热传递方面的专业实验研究。National University Prof. K. W. Kim 等韩国学者运用实验的方法结合住宅阳台进行了双层皮玻璃窗的寒冷地区气候适应性研究，这是亚洲地区较早进行的此类研究。Pow Chew WONG 等在新南威尔士大学可持续建筑环境中心对双层皮玻璃幕墙在热带高层建筑中的运用做了节能效率方面的计算机模拟。针对双层皮玻璃幕墙技术的真正系统的研究则是最近十年的时间，其中又以德国最为突出。从 1999 年自今，德国出版了大量的关于适应气候的建筑表皮的专著。以上这些研究对气候适应性的建筑表皮和双层皮玻璃幕墙从概念到个案，从实验到实测以及计算机模拟等方面做了相关层次上的研究，且已经取得相当的成果。而国内较少从气候适应性角度对双层皮玻璃幕墙，以及对建筑表皮进行实验系统量化的研究以及设计策略方面的研究。到目前为止，双层皮玻璃幕墙在欧洲高纬度国家应用为多，并且大量的实测数据也证明了其冬季太阳能利用的优势，但较少见到在低纬度地区应用的报道。那么双层皮玻璃幕墙在中国夏热冬冷地区是否具有现实意义值得探讨。

影响建筑室内人的热舒适感的最终环境因素是室内热环境。建筑热环境不是一个孤立的系统，还要受到其他建筑物理环境需求和建筑功能、美观需求的影响和制约。因此，应当运用系统、整合的观念进行分析研究，从定性定量两个方面处理问题和解决问题。本文就是从定量的角度，运用足尺模型露天实验测试的方法，分析双层皮玻璃幕墙建筑室内热环境的影响因素和相互关系，重点探索使双层皮玻璃幕墙获得最佳隔热保温能力的设计原则和方法。

2 实验

2.1 实验条件及设备

由于双层皮玻璃幕墙空腔气流状态的边界条件不确定，许多计算机模拟试验的结论以及双层皮玻璃幕墙相关的理论计算的可靠性还存在问题。而且在欧洲相关研究报导中，大多数的实验使用的是人工辐射源以及小尺度的测试立面模型。因此，我们在华中科技大学建筑与城市规划学院建筑系馆楼顶设计了两个完全相同的实验房(图 1)以进行足尺模型露天实验的对比测试。

图 1 室外双层皮玻璃幕墙实验房

两个实验房在系馆顶层建造，初始状态均朝南，四周开阔无遮挡。实验房设计的尺寸为 3m(B)×2.4m(D)×4m(H)，双层皮玻璃幕墙有上下两层，左右两个立面单元(每个 1.5m 宽)，并且玻璃空腔间层顶部外加 600mm 高的可调节出风口，空腔间距可以调节(一般设置为 400mm)。结构为槽钢结构，玻璃幕墙的内、外两层皮通过连接杆件组成一个箱式整体。内层皮采用含有保温层的铝板幕墙，占一半面积的玻璃门和换气窗采用 6mm+8A+6mm 中空白色玻璃；外层皮采用点式玻璃幕墙，玻璃采用 5mm×5mm 普通夹胶白色玻璃。除玻璃面之外，其他 5 个面均用铝板幕墙，内壁为 1mm 厚薄钢板，内贴 80mm 厚聚苯保温板，外填 150mm 厚保温岩棉，外饰面为 3mm 银灰色浮碳喷涂实心铝板(图 2)。房背部有 6mm+8A+6mm 中空白色玻璃门，实验时门外侧满贴铝箔。

2.2 实验设备和实验测试点布置

实验中内、外层玻璃表面温度用热电偶数据采集器(DT 600)测量，室内外以及空腔空气温度则使用温湿度自记仪(RHLOG-T-H)。此外，热舒适分析仪(MICROTHEME IAQ)在空腔间层离室内地坪 0.9m 处放置，用来测试空腔风速以及温湿度。太阳辐射热监测记录仪(PC-2D)则记录室内以及空腔内太阳总辐射量。所有测试点如图 3 所示布置。同时，能源环境检测站(6160 Wireless Vantage Pro Plus)用来提供室外天气变化数据。

2.3 实验思路

双层皮玻璃幕墙并非单纯的技术装置或是建筑构造。它首先是作为一个要素，就像日光中庭、种植屋面等设计要素一样，融合于整个建筑设计的“生态圈”中，其次，它有其自身的子系统，其空腔及相关配件与内、外层皮共同组成一个可动态适应和积极应变的系统。双层皮玻璃幕墙的节能效益取决于整个系统和子系统的运行状态。本实验就是在第二个层次，即双层皮玻璃幕墙子系统上探索双层皮玻璃幕墙与传统单层皮玻璃幕墙在气候适应性上的差异，以及双层皮玻璃幕墙在夏季不同遮阳、通风、朝向、淋水和冬季不同蓄热、气循环等条件下对双层皮玻璃幕墙室内热环境和能耗的影响。

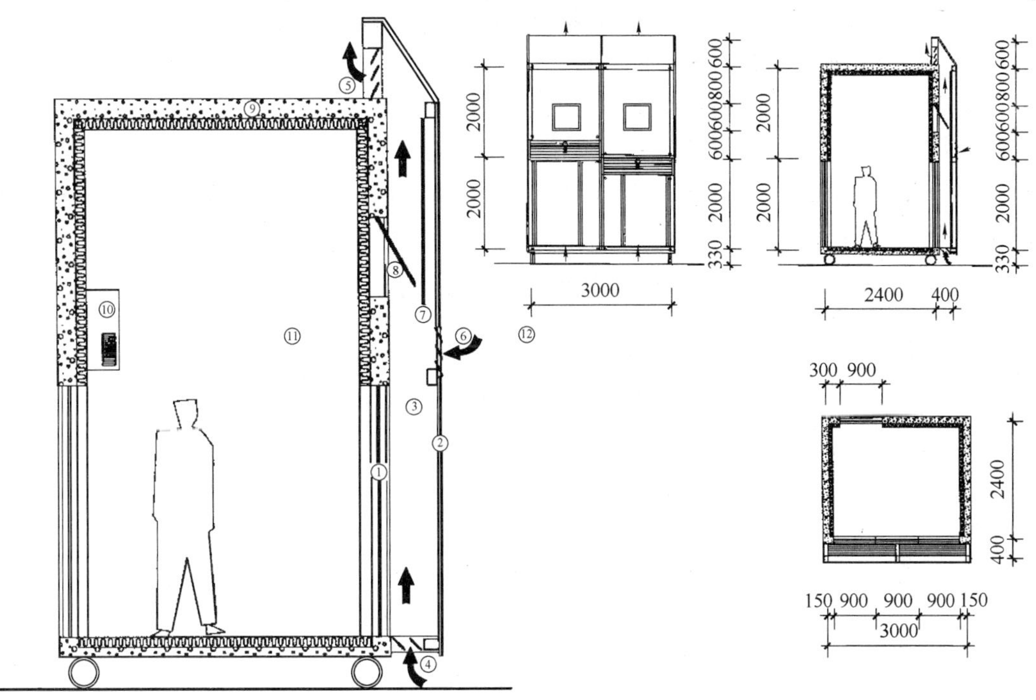

图 2　双层皮玻璃幕墙实验箱体结构和尺寸示意

①内层玻璃门(6mm+8A+6mm 中空白色玻璃，塑钢门框)；②外层玻璃幕墙(5mm+5mm 普通夹胶白色玻璃，点式联接)；③空腔间层(通常为 400mm 间距)；④底部通风百叶装置(可调节铝合金百叶)；⑤顶部通风百叶装置(可调节铝合金百叶)；⑥中部通风百叶装置(可调节铝合金百叶)；⑦遮阳百叶装置(通常为铝合金可调节薄百叶)；⑧内层上旋换气窗(6mm+8A+6mm 中空白色玻璃，塑钢窗框)；⑨墙体(1mm 厚钢板内壁，内贴 80 厚聚苯保温板，外填 150mm 厚岩棉；3mm 厚浮碳喷涂铝板外饰)；⑩空调系统(一匹功率)；⑪室内；⑫室外

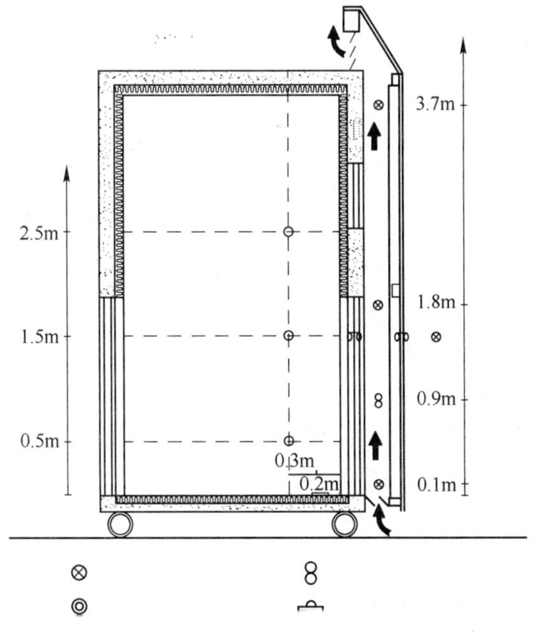

图 3　双层皮玻璃幕墙实验箱体测试点示意

如表 1 所示，本实验划分为夏、冬两大板块，夏季有朝向、遮阳、能耗、通风、淋水五个大系列，11 个小系列(CS×－×)，冬季有蓄热、气循环二个大系列，4 个小系列(CW×－×)。在相同的实验条件下，将双层皮玻璃幕墙与单层皮玻璃幕墙以及不同构造做法的双层皮玻璃幕墙进行对比。通过采集室内各点的温度、玻璃表皮温度、室内外太阳辐射值、室内外照度、空腔间层的空气流速以及能耗等数据，分析归纳各种条件下双层皮玻璃幕墙的技术策略。

双层皮玻璃幕墙实验结构说明 **表 1**

	双层皮玻璃幕墙 vs，单层皮玻璃幕墙		不同做法的双层皮玻璃幕墙对比	
夏季实验 Summer	· 遮阳模式	CS2-X 系列	· 遮阳模式	CS1-X 系列
			· 遮阳间距	
	· 不同朝向		· 遮阳材料	
			· 通风模式	
	· 制冷能耗		· 空腔间距	
			· 表面淋水	
冬季实验 Winter	· 蓄热材料	CW2-X 系列	· 蓄热模式	CW1-X 系列
	· 制热能耗		· 空腔间距	
			· 气循环模式	

2.4 实验综述

鉴于全部实验中的夏季实验的部分分析已在相关会议期刊❶上公开以及论文篇幅限制，夏季实验系列的分析过程就不在此论述。以下部分详细分析双层皮玻璃幕墙气候适应性研究中的冬季实验系列，并在此基础上整合和完成夏热冬冷地区大型建筑工程中双层皮玻璃幕墙的气候适应性的设计策略研究。

2.4.1 蓄热空腔能力的比较(冬季实验系列)

如图 4，CW1-6 实验数据显示：在冬季白天，无论双层皮玻璃幕墙空腔间距是多少，在不通风的情况下，空腔内及室内均积极得热，其温度远远高于室外温度(空腔内温度至晚上 10：00 左右才低于室外温度，室内保温则更是延长至 0：00 左右)。而且 A 实验房(200mm 空腔间距)的空腔内温度在上午 8：00 到下午 18：00 时段里比 B 实验房(400mm 空腔间距)都要要低，最大温差有 3℃。此时，室外峰值温度仅 12℃时。对于内层玻璃内外表面温度差，A 实验房则要高于 B 实验房，蓄热空腔间距的缩短意味着其对内层玻璃外表面热传递的相对减弱。大约下午 18：00 左右太阳辐射减弱直至无，双层皮玻璃幕墙空腔内温度开始低于室内温度，且 B 实验房室内和空腔内温度下降的斜率要大于 A 实验房室。这说明冬季白天较大的蓄热空腔有利于太阳辐射得热，夜间反而较小的保温空腔有利于延长室内及空腔向室外放热的时间。

❶ Whang Zhen. Research on thermal adaptability of Double-Skin Facades in Hot-summer and cold-winter zone. 1st International Tropical Architecture Conference (iNTA) - Architecture and Urban Design in the Tropical Rogions: Sustainability and Society. NUS, Singapore(第一届国际热带建筑研讨会——在热带地区的建筑和城市设计：可持续和社会. 新加坡国立大学. 2004.02.26～02.28)。

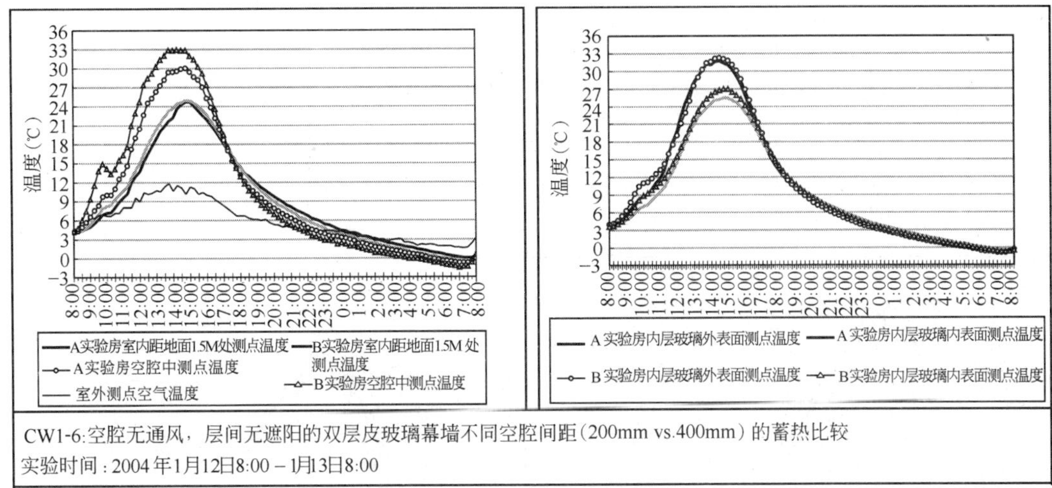

图4　实验房室内外、空腔内及内层玻璃表面温度曲线

CW1-7实验(图5)作为CW1-6的比较实验，其数据再次说明：蓄热空腔间距从400mm、300mm到200mm，双层皮玻璃幕墙空腔的冬季白天蓄热性能呈线性减弱的趋势。

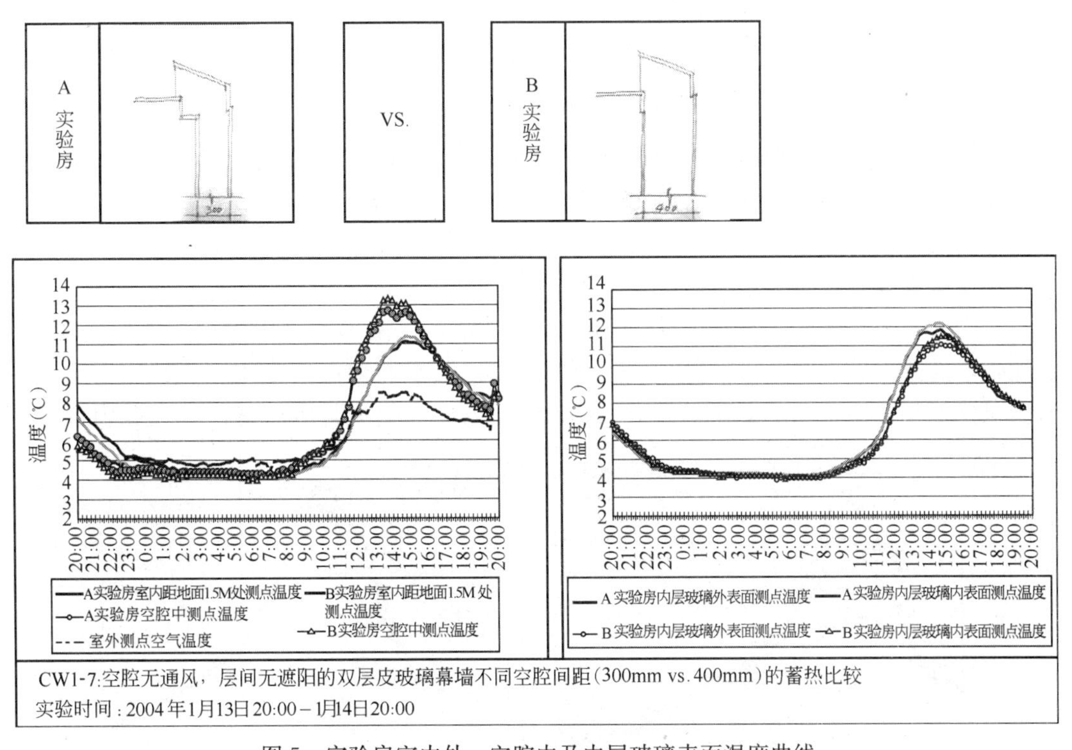

图5　实验房室内外、空腔内及内层玻璃表面温度曲线

2.4.2　气流循环的影响(冬季实验系列)

CW1-2实验(图6)比较了冬季双层皮玻璃幕墙无气循环和内循环对室内热环境的影响，其数据显示：在太阳辐射得热时期，尽管B实验房(内循环)的空腔内温度略低于A实验房(无气循环)，但B实验房的室内温度远高于对比房，最大温差竟达5.5℃。在下午

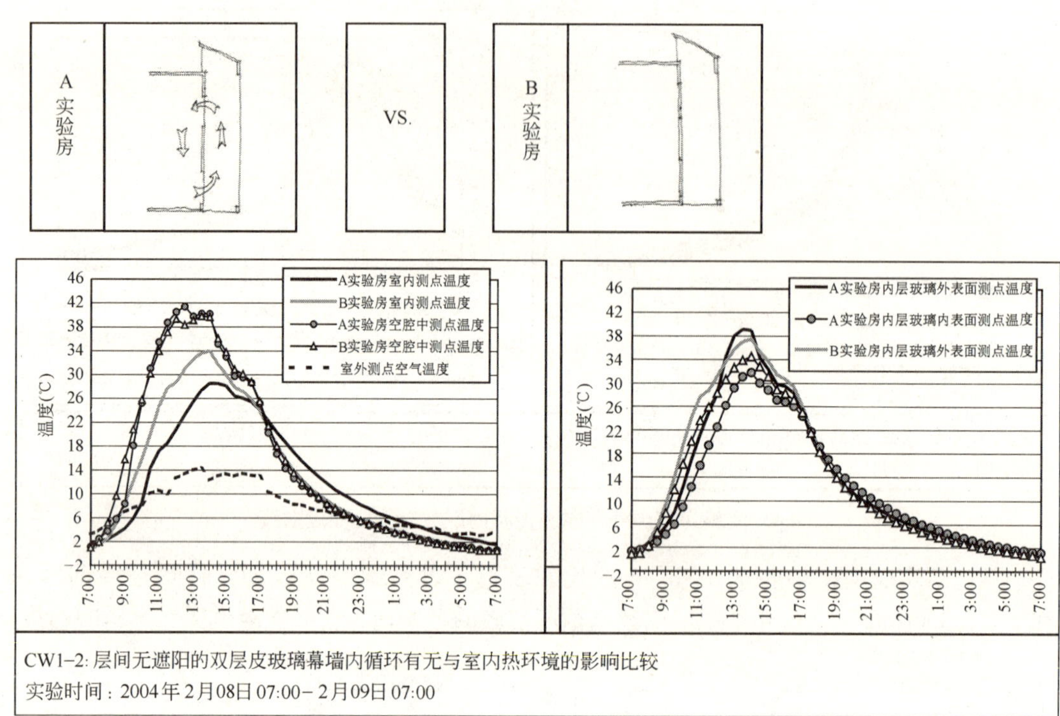

CW1-2: 层间无遮阳的双层皮玻璃幕墙内循环有无与室内热环境的影响比较
实验时间：2004年2月08日07:00- 2月09日07:00

图6 实验房室内外、空腔内及内层玻璃表面温度曲线

17：00左右，空腔不再从太阳辐射得热，B实验房的室内温度开始低于对比房，并且其温度曲线下降的斜率高于A实验房，这又说明较小的保温空腔有利于延长冬季夜间室内向室外的放热时间。总之，内循环在冬季白天更有利于提高室内空气温度；在冬季夜间，无气循环则有利于室内的保温延迟。

CW1-5实验(图7)对冬季双层皮玻璃幕墙内循环的气循环量对室内热环境的影响进行了比较，其数据显示：A实验房(一进一出)的室内及空腔内温度在太阳直接辐射时期略低于B实验房(二进二出)，同样在下午17：00以后，两实验房的室内及空腔内温度颇

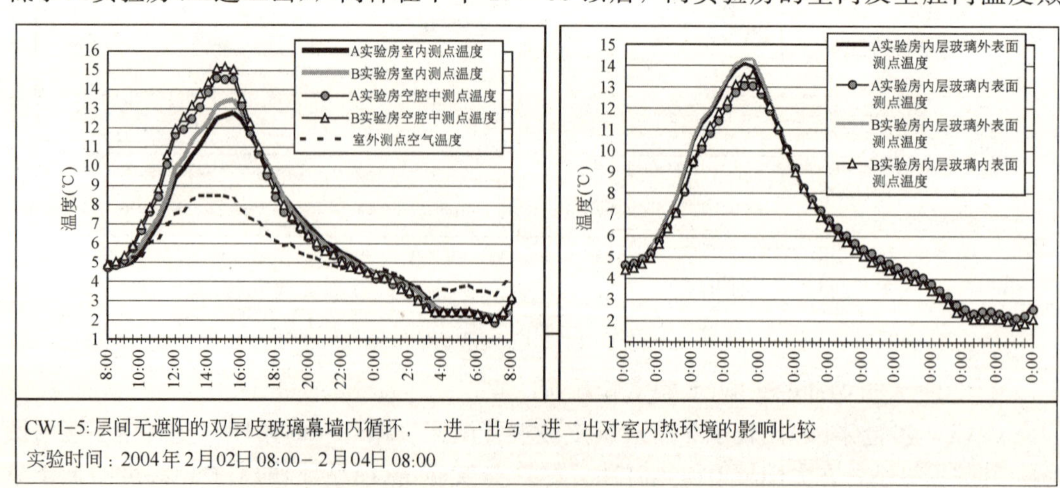

CW1-5: 层间无遮阳的双层皮玻璃幕墙内循环，一进一出与二进二出对室内热环境的影响比较
实验时间：2004年2月02日08:00- 2月04日08:00

图7 实验房室内外、空腔内及内层玻璃表面温度曲线

倒。这说明内循环的气循环量的增加在空腔表现为太阳辐射得热时期有利于提高室内空气温度，但在空腔太阳辐射得热量和空腔温差传热量达到平衡及其以后，则不利于空腔及室内保温。内层玻璃内外表面温度的交叉递变证实了室内和空腔内气循环的方向的变化。

3 实验成果

正是“归因于地球公转时地球表面获得的太阳辐射的差异与周期变化，能量分布的时空差异形成了自然气候梯度”。夏热冬冷地区的气候不同于其他地区，具有夏天热，冬天冷，且长年湿度很高的鲜明特点。作为围护结构的双层皮玻璃幕墙的气候设计在此地区必须同时考虑夏季隔热和冬季保温。无视环境而一味追求形式上的标新立异，是建筑设计的误区。实验证明双层皮玻璃幕墙在夏热冬冷地区仍具气候适应性的现实意义。笔者根据夏季朝向、遮阳、能耗、通风、淋水五个系列和冬季蓄热、气循环二个系列的实验成果，对夏热冬冷地区双层皮玻璃幕墙的技术规则归纳如下。

3.1 针对夏热的技术规则

双层皮玻璃幕墙的夏季防热首先是减少通过玻璃幕墙进入室内的太阳辐射；其次是要降低基于室内外温差的传热；最后还要控制因空气流通带入室内的热量。双层皮玻璃幕墙就其发挥隔热性能来说，通风是基础，遮阳是关键，淋水是补充。具体来说：

(1) 少使用玻璃幕墙。

- 朝向：

(2) 不同的朝向有不同的光热环境，注意玻璃幕墙的朝向。

(3) 双层皮玻璃幕墙在各个朝向均有较好的防热优势，尤其阳光直射的西向。

(4) 北向太阳散射以及环境反射不容忽视。

- 遮阳：

(5) 通过遮阳措施可以有效控制辐射得热。

(6) 使用层间遮阳的遮阳模式。

(7) 考虑使用部分活动式外遮阳。

(8) 利用外层玻璃的遮阳效率。

(9) 遮阳帘靠近外层玻璃有利于隔热，并保证遮阳位置距玻璃最少 15cm。

(10) 优先使用铝合金薄百叶遮阳帘，其次是木百叶遮阳帘或活动织物遮阳帘。

- 通风：

(11) 通过通风措施可以有效控制室外和空腔的温差热流。

(12) 在室外有合适自然风压的情况下，使用风口全开的通风模式；在室外静风的情况下，使用风口上下开的通风模式。若是箱-箱式或走廊式双层皮玻璃幕墙则使用错位通风模式，避免气流短路。

(13) 空腔通风间距≥300mm，有利于双层皮玻璃幕墙空腔与室外进行换热。

- 淋水：

(14) 通过表面淋水降低外层玻璃表面温度。

(15) 使用表面淋水时，关闭空腔通风百叶。

- 主动制冷：

(16) 在最不利的高温期，配合被动技术措施，使用主动制冷，降低能耗。

- 内外层玻璃材料：

(17) 合理选择玻璃隔热。

(18) 中空玻璃、太阳辐射得热最小，而可见光透过率最大的 Low-E 玻璃(太阳光谱的选择)、透明保温材料(高热阻)、可调玻璃(光敏、热敏、电敏)。

3.2　针对冬冷的技术规则

双层皮玻璃幕墙的冬季保温则要尽量争取太阳辐射得热，同时也要降低温差热流。就发挥保温性能来说，须重视气循环和蓄热材料的运用。具体来说：

- 蓄热保温空腔：

(1) 正确使用双层皮玻璃幕墙的空腔进行蓄热和保温。

(2) 双层皮玻璃幕墙的空腔间距可以调节。白天加大空腔(蓄热)以争取太阳辐射得热，夜间减小空腔(保温)延长室内放热时间。

(3) 若空腔间距不可调节，则设计较大的空腔间距以争取最大利用太阳辐射能。

(4) 空腔间距≤500mm。

- 气循环：

(5) 双层皮玻璃幕墙内层皮的上下有气循环换气窗孔。

(6) 避免外循环的气循环方式。

(7) 使用内循环的气循环方式实现温差对流。

(8) 气循环通气窗孔可以调节。白天开启换气窗进行室内与空腔的温差对流，并调节开窗大小以维持室内温度；夜间关闭换气窗实现保温。

- 蓄热体：

(9) 不影响太阳辐射得热的前提下，在空腔内侧使用蓄热墙(可考虑结合水墙或 Trombe/特朗勃墙)。

3.3　对上述“技术规则”的再思考

双层皮玻璃幕墙是一个具有动态调节能力的复合系统。该系统使得建筑表皮的“行为”可以随着外界气候条件的变化而动态调整，夏热冬冷地区季节气候差异大，尤其需要这种具有动态适应性的表皮。从这个意义上说，双层皮玻璃幕墙是符合夏热冬冷地区的建筑表皮类型。

在冬夏两季的双层皮玻璃幕墙实验中，我们发现：一方面不同隔热保温措施对室内热环境影响的差别不大，需要整合各种措施，最大限度发挥双层皮玻璃幕墙的综合隔热保温性能，如通风仅在有限范围内影响室内热环境；另一方面双层皮玻璃幕墙是一个系统(SYSTEM)而非简单的构件(ELEMENT)，某个部分的使用不当都可能影响系统的整体效益。如在夏季不通风、无遮阳的条件下，双层皮玻璃幕墙空腔形成强热源，其防热效果不如外遮阳的普通玻璃幕墙；而冬季若不关闭通风口，也会影响幕墙的蓄热效果。

此外，真正适宜的双层皮玻璃幕墙技术应尊重地方环境(使用者、气候、技术、经济、形式)，经济地使用当地资源，通过自然通风以及高质量的室内光、热环境，提高使用者的舒适感。如此，我们需要综合考虑：

- 利用太阳辐射得热，减少供热的能源需求。
- 依照太阳光条件，合理使用遮阳设施，减少制冷负荷。

- 提供有效的夜间换热。
- 考虑自然通风的换热作用。
- 提供良好的隔声性能。
- 延长立面构件的生命周期。

最后，双层皮玻璃幕墙空腔内不同设施、构造、细部、材料之间的不同组合还为建筑立面设计提供了选择的余地。在使用方式严格遵守当地气候“技术规则”的前提下，双层皮玻璃幕墙既可以满足建筑在不同季节、时段相互矛盾的要求，又可实现“光、薄、透”的立面效果。生态化、个性化及高技术、高情感的追求在此巧妙地得到了平衡。

4 结论

双层皮玻璃幕墙是由内、外层皮、空腔间层及相关配件等表皮技术手段共同组成一个可动态适应和积极应变的系统，其对气候要素的利用是建立在对表皮技术手段不同层次上运用的基础之上。对于双层皮玻璃幕墙，我们可以通过材料手段降低内外层玻璃表皮的传热系数以及选择恰当的遮阳系数，使用遮阳系统控制室内和空腔内太阳辐射得热量，运用通风系统控制室内外与空腔的温差热流，以及整合各种技术手段以适应气候的变化和达到节能的目的。

双层皮玻璃幕墙是基于气候适应性的建筑技术手段。不同类型的双层皮玻璃幕墙适用于不同气候特点的地区及不同的建筑。在欧洲高纬度国家侧重于冬季太阳能利用的双层皮玻璃幕墙的类型，不能随意强加到低纬度地区。实验数据证明双层皮玻璃幕墙在夏热冬冷地区仍具气候适应性的现实意义。通过实验研究，笔者分析了双层皮玻璃幕墙获得最佳隔热保温能力的措施和方法，并总结成若干技术规定，以便于对双层皮玻璃幕墙进行正确的设计和操作。这些技术规定是双层皮玻璃幕墙气候适应性设计策略的基础。最后从整体的观念出发，提出利用各种“亲环境”的能源，整合双层皮玻璃幕墙各种技术手段，以达到使用者、气候、技术、经济以及形式的平衡。

The Design Strategy of Double-Skin Glass Curtain Wall With Climatic Adaptability in Mega-Projects in Hot-Summer And Cold-Winter Zone

Wangzhen　Li Baofeng
(School of Architecture and Urban Planning HUST, Wuhan　430074 China)

Abstract: In recent years, the building with double-skin glass curtain wall has appeared in China. This type of glass curtain wall is recommended currently by arguments such as sustainability, ecology and energy-saving. Double-skin glass curtain wall are already a common feature of architectural competition in Europe for providing greater thermal insulation both in summer and winter. However, in Hot-summer and cold-winter zone in China, the experience and knowledge about climate environment for double-skin glass curtain wall are left as vacuum. So the study on double-skin glass curtain wall and its adaptability to the climate in this area has practical significance. Based on the comparing study of the experiments on double-skin glass curtain wall in various conditions such as orientation, sun-shading, ventilation, energy-saving, surface-watering, heat-storing and circulation of airflow, the thesis concerns their effects on indoor thermal environment and energy consumption. Integrated with technical characteristic of double-skin glass curtain wall, the paper tries to offer some technical rules for double-skin glass curtain wall and its design strategy for climate adaptability. The study not only affords the exercisable technical sustainment for the realization of ecological building, but may also provides the most possibility for logical expression of architecture design.

参考文献

[1] Michael Wigginton 编著，李冠钦译. 建筑玻璃. 北京：机械工业出版社，2002
[2] Michael Wigginton 著，高昇，王琳译. 智能建筑外层设计. 大连：大连理工大学出版社，2003
[3] 英格伯格·弗拉格等编，李保峰译，张凌云校. 托马斯·赫尔左格——建筑+技术. 北京：中国建筑工业出版社，2003
[4] 吕爱民著. 应变建筑——大陆性气候的生态策略. 上海：同济大学出版社，2003
[5] 涂逢祥主编. 节能窗技术. 北京：中国建筑工业出版社，2003
[6] Randall McMullan 著，张振南，李溯译. 建筑环境学. 北京：机械工业出版社，2003
[7] 颜宏亮编著. 建筑特种构造. 上海：同济大学出版社，2002
[8] 克里斯·亚伯著，张磊等译. 建筑与个性——对文化和技术变化的回应. 北京：中国建筑工业出版社，2003
[9] N. Grimshaw. Structure、Space and Skin：Nikula Grimshaw & Partners Buildings and Projects. Ernst & Sohn，1993
[10] Heinz W. Krewinkel. Glass Buildings—Material，Structure and Detail，Birkhäuser Publishers for Architecture，Switzerland，1998
[11] Andrea Compagno. Intelligent glass facades. Birkhauser Publishers，1998
[12] Martin Krampen，Glass Architects，Avedition，1999
[13] Schittich. Glass construction Manual. Birkhäuser，1999
[14] Österle. Double-Skin Facades-integrated Planning. Prestel，2001
[15] DAM Architektur Jahrbuch. Prestel，2001

可持续建筑与可持续建筑技术体系
——《建筑模网应用技术规程》编制

胡 英
（大连理工大学建筑与艺术学院，大连 116023）

摘 要： 文章首先论述了可持续建筑与可持续建筑技术体系；接着介绍《建筑模网应用技术规程》的内容，说明该可持续建筑技术体系的设计原则；在结语中提出：可持续建筑呼唤可持续建筑技术体系，可持续建筑技术体系支撑可持续建筑的平台，为此，要继续加大力度进行可持续建筑技术体系的科研工作。

关键词： 可持续建筑 可持续建筑技术体系 建筑模网

联合国世界环境与发展委员会于1987年在《我们共同的未来》报告中指出，可持续发展是“既满足当代人的需要，又不损害后代人满足需要的能力的发展”；可持续发展更是生态、经济、社会乃至技术、文化、美学等方面的协调发展。在建筑领域的可持续发展，是使建筑成为自然界物质和能量循环链条的一个环节，也称可持续建筑，可持续建筑的同义词还有生态建筑、绿色建筑等。自20世纪60年代以来，在全球范围内，有大批专家、学者进行理论研究和工程实践，同时，引起各国政府的高度重视。自20世纪90年代以来，许多国家研究制订了自己的绿色建筑标准和评估体系，如“美国LEED绿色建筑评估体系、德国的生态建筑导则LNB、英国的BREEM评估体系、澳大利亚的建筑环境评价体系NABERS、加拿大的GBTool、挪威的Eco Profile、法国的ESCALE、日本的CASBEE等”[1]。我国于2001年9月完成了“中国生态住宅技术评估体系”的制订，2003年8月又完成了“绿色奥运建筑评估体系”的制订。在该体系中，从规划、设计、施工和验收与运行管理四个阶段，分别从环境、能源、水资源、材料与资源、室内环境质量等方面对建筑进行评估，其目标是“为人类提供健康、舒适、高效的工作、居住、活动的空间，同时尽可能地节约能源和资源、减少对自然和生态环境的影响”[2]。由此可见，建筑师在创造各种人类活动空间时，须以可持续发展观为第一理念，这样才能使创造的空间成为可持续建筑的空间。可以说，可持续建筑的规划、设计、施工和管理的过程，“必将带来新的建筑运动，促进建筑科学的进步和建筑艺术的创造”。

新的建筑运动、建筑科学的进步及建筑艺术的创造体现在把各种可持续技术体系在建筑中进行科学和系统集成，集成后的技术体系称为可持续建筑技术体系。目前，在国内外，可持续建筑技术体系种类很多。可持续建筑技术体系的概念，从范围上讲，是一个整合的概念，应把它放到全球资源或整个区域内通盘考虑，许多技术问题，需要建筑师和各个专业的工程师乃至建筑行业和其他行业的共同合作来解决；从技术层面上讲，应使建筑物的规划选址合理，减少环境污染，资源高效循环利用，降低能源消耗，利用太阳能、风能等；从过程上讲，提高建筑物的保温隔热性能，做好建筑物防晒和自然采光的设计等；这些都是基本的可持续建筑技术体系。针对我国目前大量开发住宅楼盘的实际状况，建设部颁发了针对住宅的成套技术体系，该体系有 7 个方面内容，分别为：建筑与结构技术；节能及新能源利用技术；厨卫技术；管线技术；环境及其保障技术；智能化技术；施工建造技术。总之，可持续建筑技术体系是从建筑物生命周期全过程出发，是最能体现可持续发展能力的建筑技术体系。对建筑师而言，设计可持续的建筑，必须掌握可持续建筑技术体系；并且，运用何种可持续建筑技术体系直接决定可持续建筑的设计水平。所以，应对各种可持续建筑技术体系进行深入的科学研究，并及时应用，不断完善。

由《建筑模网应用技术规程》编制组研究、编制的建筑模网技术体系是可持续建筑技术体系之一。根据建设部建标［2002］第 84 号文件的要求，自 2002 年 8 月开始，前后历经三年多时间，编制组编写了《建筑模网应用技术规程》。目前，已经在征求意见稿的基础上完成了最后的送审稿，规程将于 2006 年在全国公开发行。建筑模网是由钢板网、竖向加劲肋及水平折钩拉筋共同组成的空间三维永久性模板，内浇混凝土，形成模网混凝土剪力墙结构。它集承重、围护、保温、防水等多种功能于一体，根据是否需要隔热保温分成保温型和普通型两种模网。《建筑模网应用技术规程》共分九章，二个附录，分别介绍了建筑模网的术语、材料、建筑与热工设计、结构与构造设计、施工与质量验收等方面的内容与要求。

本人负责该规程的建筑部分编写，通过在编写过程中参观、收集资料、整理资料、各专业讨论、反复征求意见和最终定稿的整个过程，对建筑模网的技术体系有了全面认识，该技术体系可以称为可持续的建筑技术体系之一，主要体现在两个方面：

1 建筑模网技术体系是建筑结构体系的变革

从以下 4 个方面论述：

1.1 建筑模网在工厂里加工、生产，把建筑业从传统的施工现场砌砖转到工厂批量生产，打破了几千年来秦砖汉瓦一统天下的局面，真正实现建筑产业化；

1.2 目前施工中大量用的空心粘土砖浪费农田，而往模网中浇灌的混凝土中确可以掺入粉煤灰等工业废料，利于环保；

1.3 在工厂生产的网片长度等于房屋层高的高度，所以，模网建筑除门窗洞口有异形模网拼接之外，其余全部为一层楼一片模网高度，施工工地几乎没有废弃物，完全符合无污染、无噪声的生态准则，是一种革命性的建筑结构体系；

1.4 建筑模网体系属于大开间剪力墙体系，比较适合建筑的灵活性、持续性的设计。作为变革建筑结构体系的新型建筑技术体系，其设计很容易掌握，以建筑专业为例：在规程中规定，模网混凝土内、外墙体的轴线定位均以模网混凝土墙中线定位；模网混凝土建筑的开间、进深尺寸，与网片的尺寸有关，建筑的开间、进深尺寸宜符合标准模网宽度任意组合的倍数加上一倍墙体厚度，模网混凝土剪力墙墙肢长度宜等于标准规格模网或其组合后的宽度，墙肢可由一片或若干片模网组成，见图 1。

同理，模网混凝土建筑的门、窗宽度也宜为标准模网宽度或为其倍数，门、窗高度可由设计者确定。模网混凝土建筑的墙身设计依次为：设在室外散水和室内标高之间的地梁

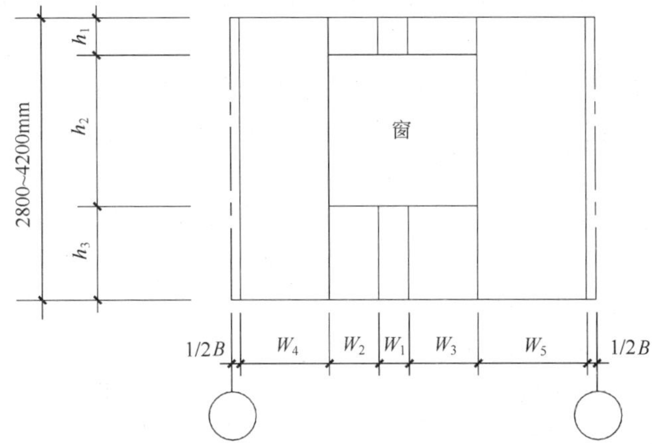

图 1　模网混凝土建筑开间、进深及开窗尺寸示意

注：(1) h_1、h_2、h_3 按工程设计；(2) B 为外墙厚度(不含外保温厚度)；(3) W_1、W_2、W_3、W_4、W_5 为五种标准模网的宽度，尺寸分别为 300mm、500mm、700mm、900mm、1100mm 可根据设计任意选择及拼接。

与模网构件相接处，不用设置防潮层；墙体的门窗洞口应预留，并应在洞口四周增设补网；阳台挑板与栏板的构造作法与其他结构形式的构造作法相同；模网墙体的女儿墙顶部应加构造筋，应做好顶部封口及周边保温的构造处理；模网建筑的变形缝处，设双层墙体。模网混凝土墙体的防火设计和埋线设计均同钢筋混凝土剪力墙设计。

2　建筑模网技术属于节能技术体系

目前，为推动全社会开展节能降耗，缓解能源瓶颈制约，建设节能型社会，国家发改委发布了《节能中长期专项规划》。在该规划中，提出了四个方面的目标：一是宏观节能量指标，二是主要产品单位能耗指标，三是主要耗能设备能效指标，四是宏观管理目标。作为耗能大户的建筑业(约占总能耗的 1/3)节约了能源，即在单位能耗指标、主要耗能设备能效指标下降的前提下，达到宏观节能指标。建筑模网技术为建筑外保温节能技术。它和内保温节能技术、中间夹芯墙节能保温技术共同组成了建筑外墙节能保温技术体系，但该外保温节能技术较之其他保温方式有更多的优点，主要体现在以下几个方面：

2.1　保护主体结构，保温型模网将聚苯乙烯保温板置于混凝土外侧，减少了空气中有害气体和紫外线对承重混凝土结构的影响，有效防止和减少墙体的变形，提高其耐久性，延长其寿命。

2.2　基本消除了在内、外墙交界部位、外墙圈梁、构造柱、框架梁、柱、门窗洞口以及顶层女儿墙与屋面板交界周边所产生的“热桥”的影响。

2.3　由于蒸汽渗透性高的主体结构混凝土处于保温层的内侧，墙体内部一般不会发生冷凝现象，无需设置隔汽层，墙体潮湿情况得到改善。

2.4　外保温墙体使蓄热能力较大的混凝土在墙体内侧，当室内温度上升或下降时，混凝土能够吸收或释放热量，有利于室温保持稳定。

2.5 外保温构造避免了内保温的外侧墙体砌筑灰缝和面砖粘贴不密实的问题，提高墙体的防水和气密性能。

2.6 外保温可以保持室内空气和墙体内表面有较高温度，从而有利于改善室内热环境。《建筑模网应用技术规程》根据该体系外保温的特点，在立、剖面及外饰面设计中规定，模网建筑的立面可以做任意线脚，建筑师可随意设计；外墙抹面同其他墙体一样，应做分缝处理；在抹灰之前，在门窗洞口的角部，沿 45°方向加铺 400mm×200mm 耐碱玻璃纤维网格布；外饰面宜采用涂料，但也可以采用其他饰面作法。当采用其他饰面作法时，应有可靠的技术保证措施。模网建筑的防水设计如同其他外保温墙体设计一样，有如下要求：在多雨水地区，应做双层抹灰，勒脚应采用水泥砂浆饰面；伸出墙外的凸出物，如雨篷、开敞式阳台、室外空调机搁板、遮阳板、窗套、外楼梯根部及水平装饰线脚等处，均应采用有效的防水措施；应采用 1∶2 水泥砂浆粉刷等有效的防潮措施，保护处于潮湿环境的墙体；宜做柔性防水设计，保护处于卫生间等处的特殊墙体。

结语：模网混凝土建筑已建和在建工程建筑面积已近百万平方米，基于这些工程实践，总结正、反两方面的经验与教训，在辽宁省地方标准《建筑模网混凝土结构技术规程》(DB21/T 1210—2004，J 10433—2004)的基础上，编制了全国的《建筑模网应用技术规程》，新增加了建筑设计内容。该技术规程可以在全国范围内指导建筑设计与施工。但是，有些相关技术，还需不断研究与完善。

可持续建筑既呼唤可持续建筑技术体系，又给各种可持续建筑技术体系提供平台，同时，也需要更多的可持续建筑技术体系支撑可持续建筑的平台；在该平台上，诸多的建筑技术体系都可以亮相、表演；并且，诸多的建筑技术体系有待于我们去发掘、研究和应用。

（注：文中论述的《建筑模网应用技术规程》的有关术语与将于 2006 年在全国公开发行的《建筑模网应用技术规程》的有关术语不一致之处，以公开发行的《建筑模网应用技术规程》的术语为准。）

Sustainable Architecture and the Technical System of Sustainable Architecture

——The Weave of《Technical Specification for Concrete Structures with Permanent Shuttering kit》

Hu Ying

(School of Architecture and Fine Art, Dalian University of Technology, Dalian 116023 China)

Summary: This paper comments on sustainable architecture and the technical system of sustainable architecture. Besides, it introduces the content of《Technical specification for concrete structures with permanent shuttering kit》and explains its design principle . In the epilogue, It brings forward that sustainable architecture calls the technical system of sustainable architecture and the technical system of sustainable architecture supports the platform of sustainable architecture. We must to do scientific research work of the technical system of Sustainable architecture.

参考文献

[1] 绿色奥运建筑研究课题组主编.《绿色奥运建筑评估体系》北京：中国建筑工业出版社，2003.8

[2] 吴良镛. 世纪之交的凝思：建筑学的未来. 北京：清华大学出版社，1999

国外绿色建筑发展概述与实例介绍

李百战[1]　姚润明[2]　丁　勇[1]　刘　猛[1]
（[1] 重庆大学三峡库区生态环境教育部重点实验室，重庆　400044；
[2] 英国剑桥大学马丁建筑研究中心）

摘　要： 本文论述了绿色建筑的发展背景、发展要点，针对绿色建筑在世界上的发展历程作了介绍，并结合绿色建筑实例介绍了目前国外绿色建筑设计的理念。

关键词： 绿色建筑　发展历程　实例介绍

1　引言

目前，人类面临两大问题：能源短缺和环境恶化。而这两者又是相互紧密联系，由此带来的气候变化已成为21世纪全球经济发展所面临的巨大挑战之一。据有关数据显示[1]，2002年是人类142年以来的第二个温度最高年，其地球表面平均温度大约比19世纪末上升了0.8℃，陆地平均温度比19世纪末上升了1.2℃。

众所周知，二氧化碳是导致气候变化的主要原因。在英国、美国等国家[2]，建筑业的能源消耗所产生的二氧化碳已占到全部能源消耗排量的40%。在中国[3]，目前建筑业的能源消耗也占到了全部能源消耗的27%，到2010年，该数据将上升到1/3。

在资源方面[4]，全球50%的土地、矿石、木材资源被用于建筑；45%的能源被用于建筑的采暖、照明、通风，5%的能源用于其设备的制造；40%的水资源被用于建筑的维护，16%的水资源用于建筑的建造；60%的良田被用于建筑开发；70%的木制品被用于建筑。建筑作为人类文明最重要的产物，耗费了地球约50%的资源，其已成为最不可持续发展的产业。

而在今后50年的建筑设计和建造中，还将有大量的建筑将维系着过去5000年的模式，因此，探索建筑的可持续发展模式已成为了建筑业发展的迫切所需。

2　绿色建筑的含义

毫无疑问，可持续建筑是目前建筑业发展的最重要的议题，大家对其的认识也多种多样，从最初的低能耗（low energy）、零能耗（zero energy）建筑到后来的能效建筑（energy efficient building）、环境友好建筑（environmentally friendly building），再到今天的绿色建筑（green building）和生态建筑（ecological building），谈论了各种各样的问题，以至于很多专业人士都提出：到底什么是可持续建筑？世界环境和发展委员会1987年提出[5]：可持续发展就是要既满足当代发展的需要而又不危及到下一代发展的需要。

绿色建筑是让我们应用环境回馈和资源效率的集成思维去设计和建造建筑。绿色建筑有利于资源节约(包括提高能源效率、利用可再生能源、水资源保护)；绿色建筑充分考虑其对环境的影响和废弃物最低化；绿色建筑致力于创建一个健康舒适的人居环境、降低建筑使用和维护费用。它从建筑及其构件的生命周期出发，考虑其性能和对经济、环境的影响[6]。

3 绿色建筑在世界上的发展及现状

20世纪60年代[7]，美籍意大利建筑师保罗·索勒瑞(Paola Soleri)把生态学(Ecology)和建筑学(Architecture)两词合并为“Arology”，提出了著名的“生态建筑”(绿色建筑)的新理念。20世纪70年代，石油危机的爆发，使人们清醒地意识到，以牺牲生态环境为代价的高速文明发展史是难以为继的。耗用自然资源最多的建筑产业必须改变发展模式，走可持续发展之路。太阳能、地热、风能、节能围护结构等各种建筑节能技术应运而生，节能建筑成为建筑发展的先导。

1992年[7]巴西的里约热内卢“联合国环境与发展大会”的召开，使“可持续发展”这一重要思想在世界范围达成共识。绿色建筑渐成体系，并在不少国家实践推广，成为世界建筑发展的方向。

2001年[8]国际可持续能源解决方案奖金在能效和可再生资源方面进行了资助，该奖金当年吸引了来自75个国家的共1000多个项目的竞争，最终给予奥地利的林茨市10万欧元的资助。

2001年7月[8]，联合国环境规划署的国际环境技术中心和建筑研究与创新国际委员会签署了合作框架书，两者将针对提高环境信息的预测能力展开大范围的合作，这与发展中国家的可持续建筑的发展和实施有着紧密关联。

30多年来[7]，绿色建筑的由理念到实践，在发达国家逐步完善，形成了较成体系的设计方法、评估方法，各种新技术、新材料层出不穷。在世界上的一些国家，绿色建筑的发展已初见成效，并向着深层次应用发展。

英国[9]已制定了一系列政策和制度来促进高能效在新建建筑和既有建筑的改造方面中的应用。在英国有很多来自于政府和其他组织的机制，这些机制和组织为推进新建建筑和既有建筑在能效和温室气体排放方面的不断进步作出了巨大贡献。英国在科技研究和革新方面也进行了大量投入，在很多可持续建筑领域都取得了显著的成果，例如太阳能光电系统、日光照明技术、低碳排量建筑、计算机模拟与设计、玻璃技术、地源热泵制冷、自然通风、燃料电池、热电联产等。

在奥地利[10]，目前有约24%的能源是由可再生能源提供。这在国际上都是发展较好的。在很多示范项目中，都大量应用了降低资源消耗和减少投资成本的技术。

澳大利亚[11]针对商业办公楼的绿色建筑评估工具近年来也发展很快，其绿色建筑委员会的评估系统——绿色之星(Green Star)，已被誉为新一代的国际绿色建筑评估工具。

在德国[12]，拥有公共绿地和具有环境友好性的建筑被大力发展，目前德国是欧洲太阳能利用最好的国家之一。在基础设施方面，德国非常注重种植物屋面、多孔渗水路面、各种排水设施、露天花园等低污染、低环境影响性的基础设施的利用。按照减少浪费、降低成本、创造健康社会的指导思想制定政策，正是德国在绿色建筑商成果发展的关键。

瑞典[13]充分利用太阳能、风能、水力作为能源生产的基础，其最大的太阳能应用项目就是将生物沼气和太阳能结合提供能量。为了保证环境和建筑的可持续发展，瑞典议会制定了14项用以描述环境、自然和文化资源可持续发展的目标。

在欧洲[8]，一个针对学校建筑能效设计集成的项目现在已经启动，其目标是要降低学校建筑50%～60%的热能消耗、降低30%～50%的电能消耗、减排CO_2达50%。该项目的示范在丹麦、挪威、瑞典、德国、西班牙、意大利、希腊7国进行。在示范项目中将会应用隔热玻璃、热回收、自然冷却、日光通道优化、高能效照明系统、被动式太阳能利用、热泵、高效通风系统技术。

加拿大[14]在推进设备能效标准和建筑能源法示范方面近年来做了很多工作。设备能效标准经过几年的努力已非常有效，其通过能效指导标签给出设备在一般情况下使用的能效情况。经济激励政策对于促进发展也是很有效的，目前已有300多个项目受到了相关奖励。

美国[15]联邦政府已经颁布了很多绿色建筑政策，并已取得了显著成效。事务管理处和预算审计处鼓励人们在进行新建筑设计以及建筑改造中结合能源之星(Energy Star)或LEED(the U.S. Green Building Council's Leadership in Energy and Environmental Design rating system)的方法开展工作。目前，已有9个部门在其新项目中使用了LEED或者是其他类似的方法。

目前，美国正在考虑成立一个更加权威的绿色建筑联合组织来引导绿色建筑的发展。这个组织要对绿色建筑的发展提供战略性指导，对于绿色建筑此相关的发展和实施政策进行识别，包括对行政法规的使用性进行识别。

在过去十年里，建筑环境评价系统得到了很大发展。很多评价系统都开始转向下一步工作，即建立标识系统，以便向使用者说明建筑的基本使用性能。目前在这方面做得最好的当属英国的建筑研究环境评价方法(Building Research Establishment Environmental Assessment Method，简称BREEAM)。目前英国有大约15%～20%的新建办公楼都使用了这种标识。作为该系统的子系统，加拿大BREEAM是BREEAM的北美版，目前已能满足加拿大的使用要求。与此同时，由在美国绿色建筑协会开发的LEED评价系统在美国政府和一些私营组织的支持下也得到大力发展，目前在美国已大范围使用。还有一些评价系统(这其中很多是基于BREEAM系统)在北欧的瑞典、挪威、丹麦、冰岛，以及中国香港等地也不同程度的发展。此外，一些专门针对生命周期评价(Life Cycle Assessment，简称LCA)的系统，如荷兰的ECO QUANTUM，德国的ECO-PRO，法国的EQUER和加拿大的Athena等也逐渐发展完善。

4 绿色建筑实例介绍

目前，绿色建筑在世界上的发展越来越成熟，下面将简要介绍两则已通过评估方法评估的绿色建筑，并以此为例介绍其设计理念。

美国：Interface Showroom and Offices[16]。

该建筑是位于美国亚特兰大市的一栋贸易大楼，建筑面积6500m^2，建筑完工于2004年4月。该建筑经美国LEED评估方法打分，获得44分，被授予铂金级。

大楼包括了8个商业空间、1个接待和陈列中心、1个产品介绍区、1个产品展览走廊以及一些样品和销售物质的仓库。大楼的使用功能包括零售店面(约占总面积的70%)、办公楼(约占总面积的15%)、会议中心(约占总面积的3%)、餐厅(约占总面积的3%)等。

该建筑的设计理念是亲近客户、创造良好的商业氛围，并同时为工作人员提供办公场所、为城市经济发展做出贡献、促进可持续发展、提供安全的人居环境。

该建筑是环境策略和建筑技术高度协作的结晶，也是美国第一个授予商业建筑的LEED评估铂金级。

该建筑位于亚特兰大中心，附近有便利的公共交通系统，同时还为自行车提供了充足的发展空间，甚至还为往返的人们提供了自行车租借服务。节水型的厨房水流开关、洗碗机、制冰机以及带感应器的节水型马桶等设备的采用，都为建筑节约了宝贵的水资源。

日光照明、高效率的照明灯具以及感应灯具等的应用，可为大楼节约大量照明电力。大楼里使用的设备、器具都是符合能源之星的要求。此外，大楼业主已购买了两年的风力发电用做建筑的能源供应。

设置在展厅的CO_2监控器可为室内提供可靠的通风保证。而建筑充分利用日光照明的思路同时也为建筑内提供了良好的视觉效果。

该建筑在建筑材料的选择上也同样是本着可回收利用、就地取材、低污染物排放以及经鉴定合格的原则，项目组在建造过程中就回收了约为总量85%的建筑废弃物。

英国：The Environmental Building[17]。

该大楼充分应用自然通风技术、最大限度的日光照明处理、优良的建筑保温隔热措施、最大限度的材料及废弃物的回收再利用，以求建筑在满足居住要求的情况下，实现最和谐的环境反馈。因此，该建筑荣获英国BREEAM评估方法的最高级别——优秀(Excellent)。

位于英国加斯顿港的该建筑是未来的能效办公楼性能规范的示范建筑，是由开发商、设计师、建造商以及能源部门共同作用的结果。作为该规范最重要的目标是要在现有基础上降低能源消耗和减少CO_2排放量达到30%。在这栋建筑中，将不再使用空调，从而可减少大量能耗；而大量利用日光照明，可减少照明能耗；此外，加强建筑隔热处理也是节能的一个方向。

该建筑独有的5个通风轴通道是为建筑内提供自然通风和自然制冷的关键部分。系统配置的冷凝燃气锅炉比一般的锅炉效率高出30%。

建筑采用了大面积玻璃以获得充足的日光照明，为了避免由于这种设计产生的热损失和太阳得热，以及冬季里的眩光现象，建筑对窗户、遮阳装置、控制系统等都进行了严格的设计和控制。

而在办公室里采用的最新的荧光节能灯，其1瓦电力可获得104流明。系统中的日光照明感应器，可根据日光照明的强弱，而自动调节照明系统的强度。

该建筑中太阳利用的另一个方面是光电系统的应用。建筑中的建筑一体化光电系统(building-integrated photovoltaic，简称BIPV)所产生的无污染电能足以提供给同样一栋大楼从 1997 年 5 月以来的所有照明负荷。

致谢：

本文的写作得到了英国外交部全球机会基金资助的中英合作项目“中国西部清洁能源发展与能效建筑设计”（UK-FCO GOF）的支持；得到了“重庆市小城镇建设科技重大项目——重庆市小城镇建设关键技术研究与示范”项目的支持。在此表示衷心的感谢。

The Introduction Of The Green Building Development And The Best Practices

Baizhan Li[1] Runming Yao[2] Yong Ding[1] Meng Liu[1]

（[1] Key Laboratory of the Three Gorges Reservoir Region's Eco-Environment, Ministry of Education, Chongqing University, Chongqing 400044 China;

[2] The Martin Centre for Architectural and Urban Studies, Cambridge University）

Abstract: This paper reviews the background, key issues and process of green building research and development in the world. The green building design concept and assessment methods have been demonstrated via some best practices.

参考文献

[1] 2002 Summer Hot, Dry Across Much of United States; Upper Mid-West Experiences Wetter-Than-AverAge Summer, http://www. noaanews. noaa. gov/stories/s999. htm

[2] Can we do Credible things to combat Global Warming, http://www2. env. uea. ac. uk/cred/Cafe%20Scientifique%20-%20Shanghai. doc

[3] 建设部建筑节能中心. 促进我国建筑节能工作的建议. http://www. efchina. org/documents/Ⅲ. 1-MOC _ Policy-CN. doc

[4] Brian Edwards and Paul Hyett, (2002) Rough guide to sustainability, London: RIBA Company.

[5] 中国科学院可持续发展研究组 2003 年中国可持续发展战略报告

[6] Green Buildings Introduction, http://www. sustainable. doe. gov/buildings/gbintro. shtml

[7] 国外发展, http://www. gbchina. org/lsjz4. htm

[8] European Green Building Forum, November 2001

[9] Green Buildings in the UK, UK: Climate Change and Renewable Energy CD, www. uksciencetech. com

[10] A Sustainable Future for Austria, Green Paper for Austria's Strategy on Sustainable Development

[11] The next generation Australia's green rating tool top in world ranking, Sustainable BUILDING, September/October 2003, www. re-focus. net

[12] Germany: model of environmental innovation, http://www. bizjournals. com/milwaukee/stories/2004/10/18/editorial4. html

[13] Area development Bo01-City of Tomorrow Malmö, Sweden, -2001, catalogue of best practice examples, European green building forum, 2001

[14] Nils Larsson, Canadian Green Building Strategies, The 18th International Conference on Passive and Low Energy Architecture, Florianópolis, Brazil, 7-9 November 2001

[15] Federal Environmental Executive, The Federal Commitment to Green Building: Experiences and Expectations

[16] http://leedcasestudies.usgbc.org/overview.cfm? ProjectID=409

[17] http://projects.bre.co.uk/envbuild/index.html

GREEN BUILDING DESIGN

Sarah Graham
(University College London)

Abstract: Climate change is a result of human activity and is largely attributed to the burning of fossil fuels to provide heat and power in buildings. This is the main driver for development of more sustainable buildings with a reduced impact on the environment. Although sustainable development of the built environment is concerned with social and economic as well as environmental impacts, there is an urgency attached to the latter. Environmental impact of buildings is largely associated with carbon emissions due to energy consumption, management of natural resource use is also important and includes water conservation measures within and around buildings; it also includes careful selection of building materials, choosing reclaimed or recycled products where appropriate. Early evaluation of different solutions is a very important aspect in the design of more sustainable buildings and there are limited tools to aid designers in carrying out these useful evaluations. The British Research Establishment Environmental Assessment Method (BREEAM) in the UK and Leadership in Energy and Environmental Design (LEED) in North America are two recognised means of benchmarking environmental performance of buildings against one another. The lessons learned thus far in developed nations provide a useful basis for developing countries with respect to design of more sustainable buildings.

Key words: sustainable building design, energy consumption, environmental impact, evaluation

1 Introduction

There are three themes within Sustainable Development; social, economic and environmental, the latter is the one most commonly thought of when considering design of sustainable buildings. The three themes must however be considered together.

In recent years the world-wide policy relative to the environment has shifted. Environmental issues have become increasingly important as one component of sustainable development. This is now a key element in determining environmental policy at an international, national, business and community level. Construction makes a significant contribution to sustainable development: it is a major sector of the economy and an essential contributor to our quality of life. Equally construction and maintenance form a massive consumer of resources and produce a large mass of waste and pollution of the environment.

2 Climate Change

Climate Change is a reality of modern life and a proven cause of increasing temperatures which lead to melting of the ice caps, rising water levels and extreme weather events.

Figure 1 shows the correlation between CO_2 concentration in the atmosphere and temperature and the graph clearly shows that the peaks coincide. This graph shows the measurements until 1950, post industrial revolution the CO_2 concentrations are higher than ever before and continuing to rise to dangerously high levels.

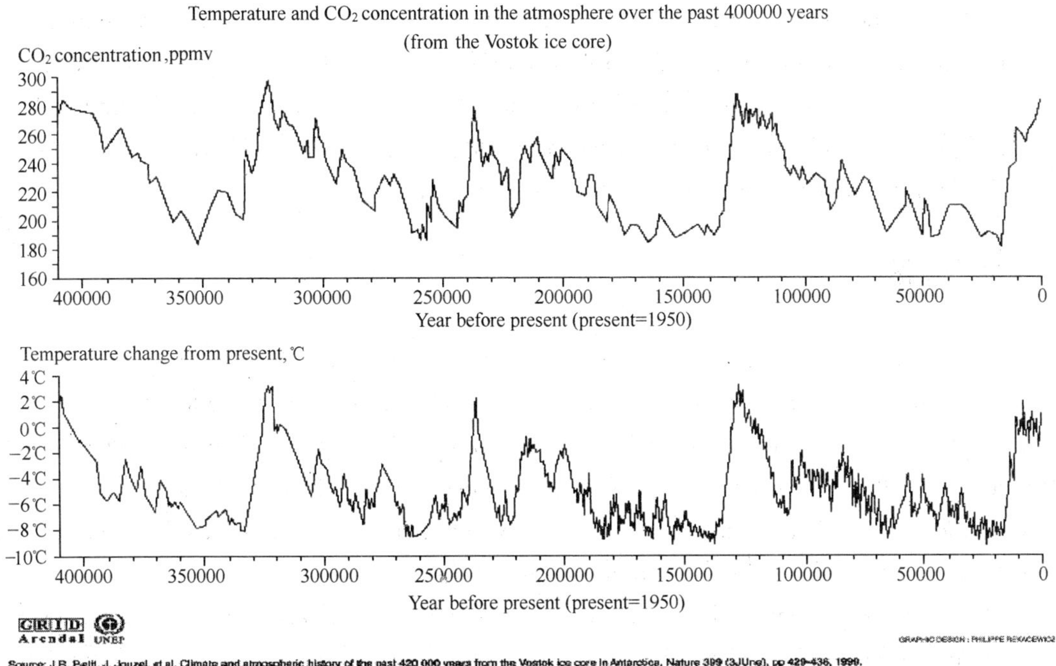

Figure 1 Climate Change over 400000 years to 1950

CO_2 is created by burning fossil fuels and since the beginning of the industrial revolution the amount of CO_2 in the atmosphere has so far risen by one third. As a result average temperatures are likely to rise by up to 6℃ over this century with potentially catastrophic consequences for life on the planet. Buildings in Western Europe are responsible for over 60% of all carbon emissions so we have a great responsibility to help limit the damage.

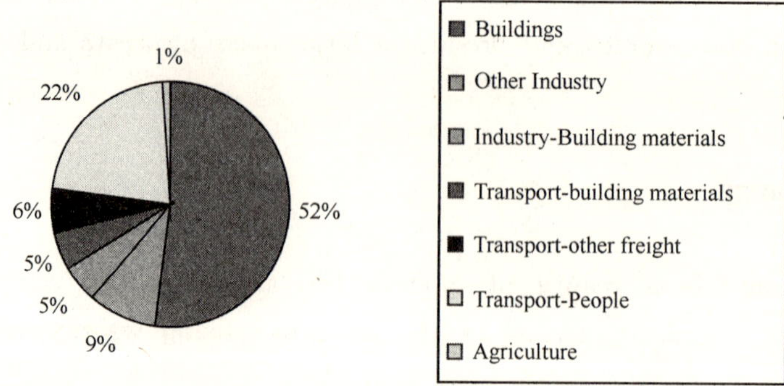

Figure 2 UK Energy Consumption(Source: BRE)

3 Energy Consumption

Figure 2 shows the breakdown of energy consumption in the UK, just over half of the total energy consumption is attributable to buildings; their construction and operation. A further 10% is arises due to the production and transport of materials for buildings. 22% arises from the transport of people between buildings. Energy Consumption is the primary area of improvement in relation to sustainable building.

4 Materials

Natural resource depletion is also a very important area related to construction and reducing environmental impact of buildings. BRE statistics conclude that the UK construction industry consumes 6 tonnes of building material per person per year and that between 2 and 2.5 tonnes per person is disposed of as waste. This indicates that management of waste and materials arising from construction needs to be addressed during design to find the most efficient and sustainable solution.

Materials environmental impact is as important as energy consumption during operation. In fact as buildings and systems become more energy efficient more emphasis will be placed on reducing embodied energy of materials. Materials environmental impact is more difficult to measure than energy consumption from operation of buildings. Examples of tools that have been developed to evaluate materials include Envest in the UK and Building for Environmental and Economic Sustainability BEES in North America.

5 Sustainable Development of the Built Environment

In simple terms sustainability is defined as follows;

'*To fulfill the needs of the present generation without compromising the needs of future generations'*

The UK Governments' most recent strategy for sustainable development at a national level is outlined in Securing the Future, published in March 2005.

The document states that 'Since 1999 scientific opinion has moved decisively to an almost universal consensus that climate change is happening and is the results of human activity.'

Sustainable development is presented in five principles with an explicit focus on environmental limits and four agreed priorities-sustainable consumption; climate change; natural resource protection and sustainable communities.

Furthermore the UK Government proposes consultation on a draft code for sustainable buildings to take place during 2005 with national roll out planned to start in 2006.

'Securing the Future' places sustainable development at the heart of land use planning and new planning guidance. Evidence of this is already appearing for example the London Plan launched 18 months ago by Mayor Ken Livingstone devised a ground breaking ruling, calling for 10% of the energy requirements of all major new developments to be met from renewable energy generated on site. Scores of councils nationwide have announced that they will follow suit, some with even more onerous stipulations.

Reducing emissions has become a major policy issue, the government's goal of a 20% reduction in carbon dioxide emissions by 2010(compared to their 1990 level)is much more ambitious than the UK's international legal obligation under the United Nations(UN) Kyoto Protocol of a 12.5% reduction in greenhouse gas emissions. More ambitious still is The Royal Commission for Environmental Pollution 60% reduction in greenhouse gas emissions by 2050.

In the design of sustainable buildings, evaluation of various aspects of the building at the earliest design stages is fundamental, informed decisions need to be made up front in the design process, the building needs to be considered holistically as the decision to increase glazing area will have a direct effect on energy consumption associated with heating and cooling. Various options need to be evaluated and feedback provided in terms of environmental impact and cost.

Clarke JA writes that 'better design of new buildings could result in a 50% to 75% reduction in energy consumption relative to 2000 levels and that intervention in existing stock could produce a 30% reduction these combined would significantly reduce a nation's energy bill, improve environmental impact and climate change mitigation and alleviate the stressful indoor conditions may citizens are exposed to. Energy efficiency could conceivably be viewed as an 'untapped, clean energy source of vast potential.' (Clarke JA 2001)

Clarke points out that the barrier to accessing this resource is less to do with technological constraints and more to do with ineffective decision support, especially true at the early design stage in the case of new build and at the remedial option selection stage in the case of existing buildings.

This is supported by the figures included in the mBIRD proposal submitted for DTI funding by a conglomerate of industrial and academic partners which claims that 'Construction impacts in terms of sustainability and carbon emissions depend on early stage decisions from conception. Bearing in mind the breakdown of total lifecycle costs into 1% design, 5% construction and 94% operation, a 5% reduction in operating costs from energy conservation over the facility's life is roughly equivalent to the initial capital cost of the building.

6 Buildings Environmental Impact

A common perception is that a project requires a green roof or a wind turbine in order to make it 'sustainable'. In fact, sustainability covers all aspects of a building from health and wellbeing of the users to enhancement of biodiversity in the local environment. The danger on many projects is that sustainability is thought of as a 'bolt on' which can be easily argued out of the design when it comes to reducing capital cost. Sustainable buildings need to be robust, adaptable, with reduced dependence on energy consuming plant and systems. Sustainability must be fundamental to the design and operation of the facility and as such requires an holistic approach and early evaluation of options to inform decisions.

Examples of tools that enable measurement of environmental impact of buildings during design include The British Research Establishment Environmental Assessment Method BREEAM and The North American Leadership in Energy and Environmental Design LEED. These assessment methods allow benchmarking of buildings against one another and are gaining popularity as the aim for more sustainable development of the built environment accelerates.

BREEAM provides an assessment of sustainability although it is largely focused on environmental performance of buildings it does consider the Health and Wellbeing of occupants. It does not provide any means of appraising the financial aspects of a building. BREEAM is currently under review in line with revisions to Part L of the Building Regulations. There are nine categories under which credits are awarded;

- Transport
- Energy
- Health and Wellbeing
- Pollution
- Water Conservation
- Ecology
- Land Use
- Management

- Materials

The assessments are carried out at various stages during design, construction and operation to ensure that the design intent is maintained through the life of the facility.

7 Conclusion

The buildings that we design, construct and operate have a significant adverse impact on the environment and this has manifest itself in recent years in the form of rising temperatures and associated extreme weather events including flooding, high winds, rising tides. In order to slow down this trend carbon emissions need to be cut quite drastically especially by developed countries around the world. Buildings designed to reduce impact on the environment need consider energy consumption and materials use at the earliest stages of design. More sustainable buildings require an holistic approach to design with targets set for carbon emissions and environmental performance measured throughout design, construction and operation so that it is available and easily understood by designers, builders and operators. A 'long life-loose fit' approach to building design means that buildings are flexible and robust enough to be used and adapted over the long term rather than demolished and replaced at 15 or 20 year intervals.

The developing nations can use the basis of lessons learned by developed counterparts to determine a more sustainable approach to development of their built environment.

绿色建筑设计

Sarah Graham
(University College London)

摘　要：气候变化是人类活动的结果。它在很大程度上归咎于为满足建筑用电和取暖而使用的化石燃料。减少对环境的影响是推动建筑向更加可持续方向发展的主要动力，尽管建造环境的可持续化发展同时受到社会、经济和环境因素的影响，但后者无疑更加紧迫。建筑对环境的影响主要体现在因能源消耗而产生的碳的排放，也体现在对自然资源的使用管理，包括建筑物内和周围设施的节约用水方案、建筑材料的精心选择、合理使用回收或再生产品。对这些不同的解决方案进行早期评估，是实现建筑物的更加可持续化设计的重要环节，而目前可辅助设计者使用的、进行有用评估的工具十分有限，英国的BREEAM和北美的LEED是已被认可的两个互相矛盾的建筑环境性能检测标准。发达国家的经验教训对发展中国家关于更加可持续发展建筑的设计提供了有用的依据。

References

[1] Securing the Future-the UK Government Sustainable Development Strategy. TSO March 2005
[2] Clarke JA., Energy Simulation in Building Design, Elsevier, 2001

［3］ Baldwin R，Yates A，Howard N，Rao S.，BREEAM 98 for offices -an environmental assessment method for office buildings，BRE 1998

［4］ US Green Building Council.，Green Building Rating System Version 2.0 Leadership in Energy and Environmental Design，March 2000

PRINCIPLES OF INTEGRATED PLANNING AND SUSTAINABLE SYSTEMS IN ECOLOGICAL SOUND BUILDING PROJECTS

Falk Kagelmacher
(Consultant & Advisor CIM/CAUPD/Ministry of Construction)
CAUPD, Rm. 537
(No. 9 Sanlihe Lu, Ganjiakou Beijing 100037 China)

Abstract: This article discusses the theory and the use of "Integrated Planning". The concepts of integrated planning will benefit for architecture as well as for urban planning. It describes a way, how to implement intelligent factors in the process, in order to achieve a sustainable, efficient and ecological sound living environment. There are different ways to create resource - efficient and sustainable living environments and there are many different factors for doing so. An integrated planning process, means, that the scope of the project needs to be evaluated in different aspects and from different point of views, such as: efficient planning, transparent communication, evaluation of use and life-span, flexibility, energy consumption, long-term calculations and finally social, economical and environmental impact. By carefully counter-weighting the different aspects, it is possible, to create the most beneficial and ecological sound project, with the least use of resources. I will discuss the background of Germany up to the present time, and I will discuss how these principles can be implemented in China.

Key words: integrated planning urbanization process energy efficiency intelligent

1 History and Development in Germany

1.1 Early Century-Weissenhofsiedlung Stuttgart

Stuttgart 1927. The capital of the Federal State of Baden-Wuerttemberg in South Germany initiated a markstone in architecture, which is still one of the most important examples of built modern ideas. The "*Deutscher Werkbund*", which can be translated with the "German Work Federation", together with the famous Architect Mies van der Rohe, and Dr. Theodor Heuss, who became German President afterwards, convinced the city of Stuttgart to appoint 17 architects to build visionary houses, following only one rule: a flat roof. Among them, some of the most famous architects of that time, people who have formed the image of German Architecture, and have laid the foundation of our architectural education.

17 architects who have become world-famous in the meantime such as Ludwig Mies van der Rohe and Le Corbusier planned a settlement here with 21 houses. 60 dwellings were created as a part of this for lower and middle income salaried employees and the elevated middle class in the wake of the dwelling building program of the city of Stuttgart this was realized in the form of the Weissenhofsiedlung.

The Weissenhofsiedlung was an experiment in all of this with the so-called "*Neues Bauen*" (New Construction), in which the largely young and progressive architects tried new construction matter and materials - all under the maxim of rationalization and typification. This was also intended to achieve a cost-reduction and time-related shortening of housing construction that had by no means yet been achieved at that time.

The result was an internationally recognized and admired modern settlement which was designed by the greatest variety of architects working together and which has been under Monument Protection since 1956. The Exhibition Manager, Mies van der Rohe, was able to achieve an admirable degree of overall harmony despite the large number of different architects who participated.

The Weissenhofsiedlung was intended to rationalize building construction by employing innovative materials, new construction methods, mass production and pre-fabrication techniques.

In order to give a complete overview of the achievements in the fields of technology, sanitation and art, the exhibition was complemented by an indoor exhibition, where building materials, technical installations, furnishings, furniture and simple household equipment were on display. Additionally, there was the "International Plan and Model Exhibition" featuring designs by national and foreign architects.

Based on today's views, the following criteria summarize the importance of the Weissenhofsiedlung:

1. Economic and social importance:

Typification, rationalization, utilization of industrial building methods, adherence to high technical and hygienic standards.

Optimization of floor plans, functionality during the planning stage

Building for mixed population groups, reflecting social diversity within a housing development.

2. Architectural and aesthetic importance:

No ornamentation, simple geometric cubes, undecorated facades(derived from interior func-tions), asymmetry, openness toward the outside, flowing spaces and transitions.

3. Architectural and historical importance:

Collaboration between the most important representatives of modern architecture at a single location-in a way that was unique in the whole world; this architectural monument and its creators have had a significant influence on 20th century architecture.

This happened in 1927 and it is still influencing our view on architecture.

1.2 WWII

The situation in Germany after the Second World War brought an end to the further development of the modern architecture. Many important architects, artists and creative people left the country. The Second World War left Germany in ruins.

Some German cities have been completely destroyed. Numbers show, that for example Cologne was destroyed by 98%. Berlin, Dresden, Leipzig, many German cities nearly disappeared.

But with remarkable effort and under a lot of difficulties, the people have rebuilt the remaining of the German history. At the same time, the people took the chance to improve the remaining. Sanitary facilities, which have not been common in dwellings before, are included inside the apartments. Water, gas and electricity have been installed. New regulations and laws have been established. In the following generation after the War, Germany was divided by the German Wall. Suddenly there was an East Germany and a West Germany with different governments. Berlin found itself in the 1960ies divided by a wall.

During these years, German cities have implemented different strategies in developing. Some cities, like Hannover and Mainz, tried to completely modernize their appearance. Some other cities, such as the western half of the divided Berlin, tried to establish a system of mixing old and new. New laws have been established to control the development and the land-use of the cities. New laws for social housing, meaning dwellings for the lower-income population, have been implemented. It became most important to create a big number of housing units for the people of Germany, whose cities have been destroyed. The need of efficient and fast buildable units with some basic ideas of using natural conditions arouse. For example, in the early 50ies, ideas of using central heating, and sunlight exposure suddenly became an issue.

Many modern German cities now have a unique mixture of historic buildings, new housing development from the 50ies and 60ies, with new and modern developments. The two parts of Germany, the East and the West part, developed in a different way with different strategies. The results are still visible today.

During this development, German planners and urban designers tried to implement the individual traffic as an important part of the modern development. The "*Autogerechte Stadt*", meaning "the city which is convenient for cars", was the idea for a better future. During the 70ies, new ideas, and the wish to look forward into a bright future, lead to the fact, that many old and historic buildings have been demolished. The people wanted to create new and clean architecture and forget the past. Nowadays, the thinking has changed. In many places, efforts are being made, to rebuild the historic remaining. Rebuilding in Germany means, we do not just rebuild a style. If a building was gone, we tried to find as much original material as possible to rebuild it. In cases, where this was not possible, people tried to rebuild as close as possible to the original.

1.3 Energy-Crisis from the early 70ies

In the early 70ies, Germany faced a very serious energy-shortage. The raw oil supply from the Middle East suddenly was limited. Energy became scarce and expensive. People had to think about resources. The fact, that it became more and more expensive to run and maintain a building, changed the way how technology was used in the building industry. Suddenly, the long-term running costs of a building became more and more important. Now we have the situation that the costs to run and maintain a building for just a few years, might easily exceed the building construction costs. "Facility Management" became an issue.

1.4 Current Development in Energy-and Resource Efficiency

This was leading to the 80ies, where the word "Umweltschutz" meaning "Environmental Protection" and the idea of a "Sustainable City-development" became more and more important. City-growth has to be controlled with land-use reforms. Clear distinctions between urban, rural and industrial regions have been made. To cope with the fast growth of the cities, a huge amount of housing had to be build, which, for example affected a dramatic change in the traffic policy. German cities had and still have a huge traffic problem. New initiatives have been launched to gain control about these issues. There are examples in Germany, where cities try to build back, meaning: making the roadways smaller! The idea is to limit the amount of traffic, as well as the speed. New and very restrictive car-parking policies are implemented. In certain cities, not even the citizens are permitted to park their vehicles close to the city centers. People are forced to use public transport. In bigger cities like Berlin, or Munich a private car driver might spend an hour looking for a parking place, driving around his home in bigger and bigger circles

2 EU-Standards and Implementation in the Building Sector

2.1 The "Six Priorities of the European Energy Policy"

In general, the six European Energy Priorities are as followed:

Increasing Energy Efficiency

Achieving a properly functioning internal market for gas and electricity for the benefit of all European citizens

Promoting renewable energy

Strengthening nuclear energy and security

Security of Europe's energy supplies and further developing external energy policy relations.

improving the links between energy policies and environmental and research policies

The "Umbrella Document" of the "Energy Performance of Buildings Directive" (CEN/BT WG 173 EPBD N 15 rev) describes the energy performance of both, new and existing buildings in terms of efficiently using energy resources and reduce impact on the natural environment. The document describes the European standards and defines the overall energy performance of a building. The newly established German Government, consisting of the left-wing social party SPD, and the conservative christian CDU, are talking about a law, to enforce renewable technologies into every new building. The various standards, described in the "Umbrella Document" of the European Union cover following areas:

Net energy for heating and cooling

Delivered energy use, for heating, cooling, ventilation systems, hot water and lighting, inclusive of system losses and auxiliary energy, and definition of energy ratings

Primary energy and CO_2 emissions

Ways of expressing energy performance (for the energy certificate) and ways of expressing requirements (for regulations)

Content and format of energy performance certificate

Boiler inspections

Air-conditioning inspections

(see Graph)

2.2 Implementation

A significant number of standards in Europe provide a solid legal framework for the planners. Research institutions are constantly developing new approaches to new technologies, with the backup from the governments. Also the industry is looking for new ways. A number of recent projects show this involvement. The GTZ (German Association for technological Cooperation), working on behalf of the German Federal Ministry for Economic Cooperation and Development (BMZ) for example, is researching in detail, about cities infrastructure, its use of resources, and its impact on the environment. Understanding the relations between these factors, helps to minimize losses, while increasing the living quality.

It shows, that in order to implement and to promote certain technologies, a strong legal framework is necessary. Without this backup, and without any motivation, new technologies will hardly be implemented. Right now, most of the implementations require some additional initial costs, which will in return on a mid-term base result in cost savings. Incentives have to be created to make it attractive for users and builders to implement those technologies.

3 Relevance for China

Like in many industrialized countries, the national use of energy in China shows, that

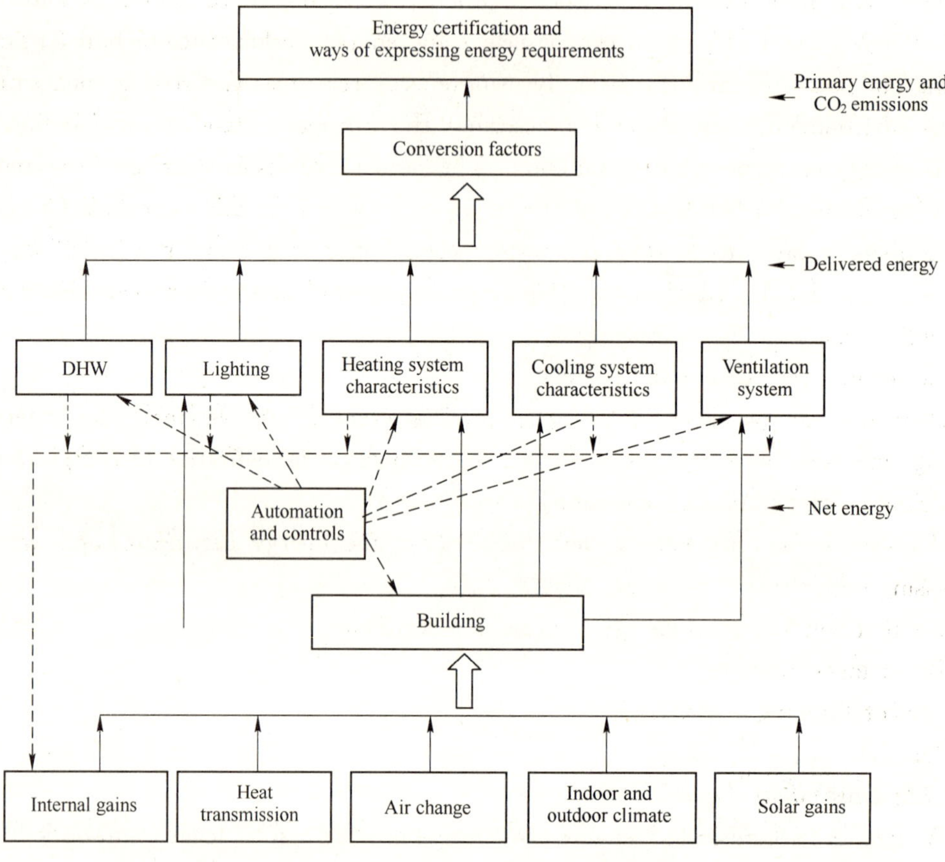

Graph "Umbrella Document" of the "Energy Performance of Buildings Directive"
(CEN/BT WG 173 EPBD N 15 rev)

roughly 50% of the total energy consumption is made up be the building sector. With simple measurements, 30% of that amount could be saved. Given the fact, that China wants to double the rate of urbanization, meaning that in the next 20 years, double the amount of people will live in cities, the area of resource-efficiency and ecological impact will become a major issue in the next few years. Also the cost of energy and natural resources will rise. An integrated planning approach and the use of green building technologies will gain more and more importance in a short future. From an economic viewpoint, it becomes absolute necessary to use resources efficiently and to design our living environment in an efficient way.

It is crucial to build a legal reinforcement in China fast. Without a legal framework and transparent standards, it is impossible for the builders, planners and users to oversee the process and implement the technologies. I want to reflect in short on the two major challenges in the development of China. First we will face tremendous challenges in the ongoing process of urbanization. Second, we will have a lot of work to do in the building sector itself. An important keyword for the future development of China is:

Sustainability-A system, which is able to work out of his own, without external forces: Conserving an ecological balance, by avoiding depletion of natural resources.

I will in short discuss about the main two areas, which are made up be the urban planning area and the building sector. In both cases exists a number of beliefs for the implementation of sustainability and resource efficiency. Basically there are two different approaches: an ecological approach, which uses passive means to save resources, and a high-tech approach, to save resources by implementing computerized control systems. Both approaches have its relevances, and can be used equally. For example, natural building materials and passive means of sun protection can be combined with computer-controlled heating and cooling devices.

3.1 Urbanization Process in China

According to current theories, China will double its urbanization by 2020. In order to create a sustainable city, it is necessary to consider some basic principles:

—Efficient use of existing inner-city areas.

—Mix-use. Avoid mono-functional areas such as sleeping cities. Combine living, working and recreation. Avoid commuting.

—Control urban crawl

—Define recreational urban areas.

Example of Urban Planning: The time and therefore the value of manpower, which people spend in cars, to go to work every day reaches a figure here in Beijing, which is influencing the GDP. Building housing projects in suburbs, is a short-term solution for a growing city. Inner city land use is a challenge, but a city has to offer a balance of several functions-a mix use.

Cities itself and the urbanization process as a major factor in the current development of China have a deep impact on the environment. They use natural resources such as energy, land, water, food and so on. On the other hand, they have a strong impact on the environment. They produce waste, waste water, pollution. We have to think about the responsible usage of natural resources, and we have to minimize the negative effects on the environment. At the same time, we have to cope with the fast growing demand of housing and living spaces.

3.2 Intelligent Building Technology

Using intelligent systems minimizes redundancies and maximizes results, while at the same time, can improve the inhabitants living quality.

High-Tech approaches, using the following technologies are a common parameter for intelligent buildings. With strategies of Facility Management, we can plan and layout the building in a way, that it needs less resources, (natural and human)to be maintained over a specified life-cycle.

Integrating all these approaches, may lead to an efficient and comfortable living and working environment, while using less resources. Donald A. Cogan Engineer, gives a simple definition of "Intelligent Houses" (Donald A. Cogan, Engineer, Montreal 2004)

3.2.1 Energy Efficiency

Computer controlled building systems can operate on many levels, to optimize processes. In terms of energy savings, a building can be controlled in a way, that it can react on the usage. Individual processes can be triggered and controlled. For example, individual heating costs can be observed easily.

Programmed start/stop of building processes

Optimal start/stop

Duty cycling

Set-point reset

Electric demand limiting

3.2.2 Life Safety systems

Intelligent systems in the observation, can minimize human dependencies and could automate a number of processes, such as security access control, health-check parameters for elderlies an so on.

Reduced manpower dependence

Closed-circuit television

Card access control

Smoke detection

Intrusion alarms

Emergency control of elevators, HVAC systems, doors

UPS

3.2.3 Telecommunication System

Telecommunication systems can facilitate a building in a flexible way. If the needs of the tenants might change, an intelligent system would be easy to change, in order to cope with the new requirements.

PBX telephone system

Cable-vision

Videotext

Electronic mail

3.2.4 Workplace Automation

Many tasks of a workplace can be automated, using a Centralized Data Processing Unit. The result may be a flexible working environment, which is easily able to adopt to different requirements.

Word Processing

Computer Aided Design

Information Services

3.2.5 Facility Management-Life Cycle Cost and Sustainability

In Europe, during the recent years, it became apparent that with the rising energy-

and manpower costs, the running and maintenance of a building can easily exceed it's building costs in a very short time. Right now, market structures and energy prices in China are different than in Germany, but in a short time, these factors will rise, and the importance of life-cycle calculations will become more and more important. Long-term cost saving may result in higher initial investment. But it is important to consider a projects performance for it's life-cycle.

4 Implementation of Resource Efficient Technologies and Sustainable Systems

There is no doubt, that is important to implement resource efficient strategies in urban development as well as in building projects. A major part of the building stock in China has very inefficient means of heating and cooling. (Retrofitting and Heating Reform) resulting in very uncomfortable living conditions and an excessive use of resources with a strong impact on the environment. We have to find ways how to deal with our existing situation and at the same time, we have to think about the future impacts on new buildings.

4.1 Legal Framework for Implementation of new Technologies in New and Existing Buildings

As long, as there is no legal framework, new technologies will have difficulties in the market. Example: With individual small-scale gas-turbines, with an efficiency of around 94%, it is possible to use clean natural gas to create energy for small and medium sized building projects. As long, as there is no legal foundation, which helps the users to define their implementations in detail, such as bringing excessive electricity to the power-lines, there will be no efficient use of these technologies. Germany is in the process of defining more and more legal frameworks, which gives benefits and therefore incentives for the users to implement these technologies.

5 Conclusion

There are a number of good technologies at hand in China. With some principles of Integrated Planning and Sustainable Systems these technologies could be used to create a better living environment with less impact on the nature and more efficient use of natural resources at the same time. Using an integrated planning approach in the design process allows to define specific tasks and needs in order to minimize redundancies and inefficiencies. The planner, the client and the different engineers have to set-up a common communication-platform. Integrated planning means, to take care for many factors, which may interfere each other and be aware of the life-span and the impact.

The technology is here at hand. Using it efficiently is the key. The principles of an integrated planning process is a way to create sustainable living environments.

整体规划原则和可持续性系统在健康生态建筑项目中的应用

Falk Kagelmacher
(Consultant & Advisor CIM/CAUPD/Ministry of Construction)
CAUPD, Rm. 537
(No. 9 Sanlihe Lu, Ganjiakou Beijing 100037 China)

摘 要： 本文主要就“整体规划”理论及其应用展开论述。“整体规划”的理念致力于创造可持续的、高效的、健康生态的居住环境，开拓了智能化元素在工程项目中应用的新途径，不但对建筑工程意义重大，同时对城市规划也十分有益。实现资源的高效利用和居住环境的可持续化的道路很多，影响其实现的因素也各有不同。以“整体规划”理论作指导，就要求对工程项目范围的进行全方位多角度的评估，包括：规划的高效性、信息的透明度、建筑物的使用寿命、适应性以及能耗等等，评估工作要做长期估算，并最终立足于对社会、经济和环境的影响。通过仔细对各方面的衡量，是能以最小的资源消耗代价，完成收益丰厚的健康生态建筑项目。在文中，将该理论到目前为止在德国的发展作为背景进行介绍，然后阐述如何在中国建筑项目中应用。

SUSTAINABLE URBAN DEVELOPMENT: THE ROLE OF TALL BUILDINGS

Ziona Strelitz
(ZZA Responsive Built Environments)

Abstract: The paper addresses holistic tall building design's contribution to urban habitability in the public and private realm, harnessing land saving through densification and promoting sustainability with energy-saving technologies. It draws on wider geographic application from London's experience as a polymorphic city of significant financial success, where a new generation of tall buildings has been supported by advanced modelling and analysis tools, generating an instructive model of integrated design theory, methods and practices. The focus is the strategic potential of building height in meeting multiple sustainability objectives, with distinctive opportunities and challenges that require multi-disciplinary processes to synthesise building systems and optimise design. The paper references the UK's advancing sustainability agenda, with urban policy and corporate governance requiring sustainable action in real estate development and procurement, meeting green criteria in embodied energy and operational performance. It describes the high internal environmental comfort standards of buildings like Fosters' tower for Swiss Re at 30 St Mary Axe, intelligent evaluation of external environmental and microclimate impacts, design innovation to limit energy use and use of renewables, and the creation of positive urban spaces around tall buildings. The paper calls for a cost and value management perspective to maintain sustainability through a building's life.
Key words: holistic tall building, land saving, energy saving, sustainability, cost and value

1 Introduction

At the 1st International Intelligent and Green Building Technologies and Products Conference & Expo convened by the Chinese Ministry of Construction in 2005, I presented a paper on '360° Sustainability', focusing on social and economic as well as environmental factors. On the environmental front, my focus was on urban habitability-making the city pleasant, healthy and efficient in the public realm, and designing comfortable buildings that energy efficient. A second paper targeted urban planning more focally. This was given by Rob Watson of the Natural Resources Defense Council, who appropriately proposed the need to re-address the urban grid of cities like Beijing, to facilitate pedestrian movement and more viable mass public transportation, thereby stemming the rising growth of private cars and their associated negative impacts through harmful emissions.

This paper builds on these vital themes, and looks to the contribution of carefully conceived and well designed tall buildings in meeting a holistic set of aims for green and intelligent development. It aligns with the full set of conference goals-for green building, energy efficiency, land saving, energy-saving technology and sustainable development, by referencing how skilful, integrated tall building design, informed by authoritative practitioners, can help to achieve intelligent, ecologically comfortable and energy-saving living.

2 New strategic guidance

The contribution of tall buildings in a programme of environmental design draws heavily from the experience of London, from which one can now derive an instructive case-based model of integrated design theory, methods and practices on sustainable building and urban design. China, of course, has many tall buildings of its own, and other cities in Asia and North America are well recognised both for their large numbers of tall structures, and for their height, including the tallest structures that have yet been built. However, this paper is not about a competition for height. Its purpose is to set out strategic design guidelines for planning, designing and evaluating tall buildings, that meet multiple objectives for sustainability. It draws from a recent book which I have edited and co-authored: 'Tall Buildings: A Strategic Design Guide' (ed. Strelitz, 2005), published by the Royal Institute of British Architects and the British Council for Offices.

The guide overviews the technical perspectives relevant to tall buildings. It is aimed at clients who are considering developing tall buildings, planning and building authorities, other public agencies and fellow practitioners. It sets out the particular opportunities for sustainable development that tall buildings offer, as well as the design challenges that they pose. Good development is always a multi-disciplinary process. In the case of tall buildings where the challenges are more extreme, collaboration and integrated thought are even more important. The key contributors who must work together to achieve a well integrated set of benefits include urban designers and town planners, architects, engineers, life safety experts, construction cost surveyors, real estate market economists, and specialists in building use and effective building interiors.

3 Looking back to inform the future: London's instructive story

All cities have a development narrative that is legible in their urban structure. London has evolved from a polymorphic series of discrete cities and villages that became 'knitted together' over time, covering an extensive surface area at relatively low vertical scale. London grew as a mercantile city, hosting many prosperous international businesses and wealthy patrons, and also as a capital city of a state in which the monarch is also head of the Church of England. In this context, the urban fabric was developed with many fine commercial, civic and cultural buildings, and also many churches, of which St Paul's Cathedral, designed by Sir Christopher Wren, was the most prominent for a long period. St Paul's has been highly cherished, to the extent that the requirements of London's planning regulations are considerably defined by the requirement to maintain open viewing corridors between key vantage points across the city and the Cathedral.

This status quo prevailed until the 1960s, when tall buildings began to emerge as part of London's post-war reconstruction, influenced by the vertical concrete dream of late modernism. Some of this new construction was commercial, but most of it was residential development, with almost all the tall residential development built as social housing. This programme coincided with a widespread initiative for high rise apartment blocks to replace low rise dwellings that were perceived as unsuitable because of their poor condition. The social dynamics of communities were not well recognised at the time, and in the process of demolishing long-standing neighbourhoods and moving people to tall new buildings, often in areas that were unfamiliar to them. Community groupings that had been bound together by established social bonds were fragmented. In the tall buildings to which people were moved, informal social control was absent, and the consequent anonymity and isolation that people experienced became associated with social decline and alienation. This was not helped by the fact that families with young children were moved into apartments high off the ground, leaving little sense of ownership or surveillance of the playgrounds at ground level that were then considered unsafe.

Two further factors condemned this first wave of tall residential buildings to lack of favour. One was their typical orthogonal architectural form and exposed concrete structures, which were unpopular, at least in combination. The next was a gas explosion in a residential tower called Ronan Point that was subsequently demolished, but not before the disaster had reinforced the view that these buildings were misconceived.

The first generation of tall office buildings faced a less negative reception, but these also had significant limitations. They were inefficient structures in terms of building core to usable floor plate area. In addition, they are now recognised as unsuccessful in the way in which they mediated with the public realm, particularly at ground level, creating difficult access both into and around them and other disruptive effects such as windy conditions.

4 The new sustainability agenda

That is the past. We are now in a different period, with new concerns-and fortunately with new capabilities. The international sustainability agenda has developed rapidly in the UK. It is apparent in corporate governance, with organisations choosing to report, or requiring their potential suppliers to state, their compliance with sustainable policies in both procurement and employment. This is given some expression in real estate design and selection, on which the agenda is becoming less elective, with planning authorities requiring evidence of programmes like 'green travel plans' to be associated with applications for new buildings, and planning approvals being influenced by the measures that schemes can demonstrate to meet green criteria in the energy that they embody in their design and construction, as well as their design provision for sustainable operational performance and management during the buildings' future life cycle. Environmental requirements are coming into even sharper focus with forthcoming regulations, for example, on energy labelling for buildings, and the ratio of opacity on building envelopes to limit solar gain and the associated energy therefore required to cool building interiors.

As in China, there is also concern in the UK with the take-up of greenfield land for urban development, especially where 'brown' land that accommodates outworn or outlived buildings or uses is available for re-use. Central Government's Urban Task Force, led by Lord Rodgers, has been championing the need for 'urban renaissance', by creating denser urban centres that not only make better use of land, but are amenable to greater reliance on mass public transportation and therefore a reduction in private car use (Urban Task Force, Report, 1999, 2005). London's government, the Greater London Authority, created by the Greater London Authorities Act of 1999, has created a technical planning framework to promote more sustainable development, through its Spatial Development Plan that targets an increase in density (The London Plan, 10 Feb 2004). The GLA has been approaching these objectives via a number of integrated planning and management measures, including mechanisms to limit private vehicle traffic and linked support to improve mass public transportation. Two notable policies in this connection are the introduction of the Congestion Charge for private cars entering the central city zone, and a significant increase in bus services.

5 How new generation tall buildings can contribute

At 20～60 storeys-the taller of these heights relating to recent proposals for tall buildings, the UK's tall buildings may be 'mid-high'. However, they inform a holistic understanding of and potential for sustainable design, with this theory and practice having a wider geographic and technical application than London. How can well planned and designed tall buildings promote sustainability? The argument presented will focus on the role that commercial buildings play and the potential they have to contribute to the built environment.

Signifying prosperity

Economic sustainability is vital. Cities are minded not to forgo economic opportunity in the short term. Tall buildings are widely regarded as a sign of economic success, which cities perceive as relevant to competitive advantage.

Promoting regeneration

Tall buildings can provide a new urban focus, signifying and strengthening the areas in which they are placed, through the 'visibility' they provide.

Occupier identity

These benefits of visibility have value as well in promoting economic sustainability for building occupiers. A tall building's landmark status and signature of height provide the advantage of easily identifiable address.

Supporting productivity

Functionally, tall buildings can contribute to economic development by extending the range of premises variety to accommodate businesses. The scope they offer to consolidate opportunity to consolidate the operations of large organisations in one venue supports productivity. Whilst large low buildings can provide a similar quota of space, they limit the scope for the pedestrian movement and mixed use that are necessary for sustainable modes of movement and good urban design. Where tall buildings attract numerous businesses in one sector, they form local economic clusters, promoting economic synergy in their localities. The scale of tall buildings also enables the inclusion of facilities for business support and staff amenity - from auditoriums to gyms - that are valuable to contemporary business in their operations, customer management, and staff recruitment and retention, as well as providing for the personal sustainability of staff who work in the building for prolonged periods on intensive, customer-focused activities such as negotiations.

Internal comfort

Building interiors are influenced by changing expectations and standards. Strong prevailing themes in the UK are space that is well suited to teams - including different depths of space, with a high environmental standard for individuals - including external aspect, natural light and preferably some degree of local control. Swiss Re's building at 30 St Mary Axe reflects this blend of corporate and individual objectives. Intelligent design can now model the performance of design conditions under different occupancy loads, and address the comfort challenges traditionally associated with tall buildings, in the performance of systems for vertical circulation, structural stability and wind - induced oscillation, internal environmental quality and life safety.

Urban design

Siting tall buildings at mass public transport nodes gives good, environmentally friendly access to large numbers of people, thereby promoting urban life and more sustainable travel. Where the buildings provide for a mix of uses - work, leisure, retail and/or residential functions, they can facilitate a further reduction in urban travel. Tall buildings can also provide for increased densities.

6 Contemporary business as a catalyst

Since the mid-1980s, London has replaced much of its office stock with new buildings (Duffy and Henney, 1989). This was fuelled by contemporary business forces and catalysed by the deregulation of the financial markets, bringing a large volume of new financial business to London, and in its wake also associated businesses, for example, legal services. These firms want good quality, strongly imaged accommodation, and they want 'big' space. Big in this context means large floor plates for trading, and also large amounts of space in one envelope(ZZA, 2002). Both of these requirements became more pronounced following a series of mergers and acquisitions, resulting in the emergence of very large businesses with commercial requirements for very large buildings.

7 Brave new world versus cityscape of memory and history

London grew out of many centres, and commerce is everywhere, but two areas are key in its critical mass of commercial life. One is the traditional financial centre, the City of London situated around the Bank of England. This is characterised not only by the planning constraints surrounding St Paul's Cathedral, but also by the complex lattice of medieval streets, resulting in a great diversity of land ownership, plot size and shape, rights of way and rights of light. In addition, there are many other heritage constraints, in the form of other important buildings, and sites such as the Tower of London and Environs World Heritage site located at the eastern boundary of the City of London financial district. The City had not been without any tall buildings, the notable structure being the 42-storey Nat West tower, now refurbished and rebranded as Tower 42. However, the planning stance towards tall buildings was protectionist.

This changed with the development and impetus of Canary Wharf. Built up on a disused docklands site, little by way of existing street pattern or viable structure was incorporated in its development. The area was fully masterplanned, given London's tallest building-1 Canada Square-to give it identity, and a decade later it is a buzzing district of commerce and retail activity. As more significant businesses have moved there-including Citigroup, HSBC, Barclays, Lehman Brothers and Clifford Chance-attracted by the big floor plates and the large volumes of space available there, the City has adopted a more enabling approach to approving tall buildings within its boundaries, rather than lose so much economic power to Canary Wharf. Nevertheless, the complex planning factors remain at play, and the character and benefits that they bring remain highly relevant to London's attractions and competitive advantage. Working together, the Government's statutory adviser on the historic environment-English Heritage, and the Government's non-statutory planning consultee-CABE, have formulated new guidance that sets out criteria for quality tall building design (CABE, English Heritage, 2003). Recent experience has shown that these can facilitate acceptable tall building design in a high land value area like the City of London, that continues to be managed with great planning care. It is into this setting that paradigm-shifting buildings like Foster's Swiss Re 'Gherkin' at 30 St Mary Axe have been inserted, becoming instant icons for contemporary London, to the extent that it was extensively used in promotional imagery to win support for London's successful bid to host the 2012 Olympic Games.

8 How intelligent design and evaluation tools can contribute

Tall buildings' environmental performance depends on numerous factors relating to location, design and specification. Recent tall proposals have harnessed the advanced modelling and analysis tools that are now available to develop energy-efficient and urbanistically complex designs, based on integration in performance of the building's orientation, envelope, services and management. These gains can work to balance tall buildings' inherent requirement for energy use on vertical travel, and through their relatively high ratio of external facade to floor area.

Integrated building design

The key is akin to two goals of this conference: the contribution of authority and leadership by experts, and their collaboration and exchange. This is necessary to optimise the design, based on appropriate development of the respective building systems and their successful synthesis and integration, through iterative design. The aims are to provide an efficient floor area and vertical transportation with a comfortable internal environment. Storey height, floor plate and core inter-react, whilst the vertical transportation, including the lift design for which options are now significantly expanded, have implications for the structure, design, core position and distribution of the mechanical systems. Each system impinges on the others and the overall outcome. "*With increasing height, the choice of component systems for structure, mechanical, cladding and life safety systems increases, as does the range of potential design solutions. For example, the designers of a building of* 15 *storeys would probably serve the building by a single group of conventional lifts, whereas at* 60 *storeys, zoned banks, with single-or double-deck cars, transfer floors and sky lobbies, and combinations of high speed shuttles and local lift groups become relevant, and permutations of these alternatives require evaluation*" (*ed. Strelitz*, 2005, *p.* 55).

Applying intelligence to the building exterior and environmental impacts

A tall building inevitably has visibility over a wide area. Its form and orientation will affect its visual impact on the urban prospect, calling for care to be taken in design development of distant, middle and local views, all of which can be intelligently modelled for design evaluation, using photomontages and other tools. Given the challenging and strict requirements that have been described in respect of urban views in the City of London, design intelligence is required on key aspects of form such as massing and core positioning. The fact that this can be successfully achieved, resulting in high standards of design, is evidenced by the planning approvals that have been given to tall building proposals such as KPF's design for Heron Tower and Richard Rogers Partnership's design for 122 Leadenhall Street.

There are also environmental impacts, external in the form of overshadowing and microclimate for which physical modelling tools like wind tunnels tests and computer simulations are available to test, inform and improve designs. Prior to identifying design solutions, intelligent evaluation is required in terms of the uses affected, as different tolerances are involved depending the specific impacts (for example, solar reflection or dispersion of emissions from buildings or vehicles), the affected uses (commercial versus residential activities being contrasting examples) and the time of day that the impact will occur. As part of a planning application, the timing and duration of any overshadowing will be considered in relation to established patterns of use. In terms of sustainability, the trend to a high level of night-time building illumination, notable in Asian cities like Hong Kong warrants reconsideration, not only for the energy usage this involves, but for the social impact of light pollution. The sunlight, daylight and overshadowing effects in relation to tall buildings are addressed in greater depth in "Tall Buildings in London: Guidance on Planning" (Linklaters, 2002).

Internal impacts are also contingent: the impact of windiness on potential oscillation or excessive infiltration to the building. As buildings play a large part in the generation of carbon dioxide emissions, the consequences of design for solar gain on active or passive energy management are pivotal, with intelligent design to reduce this important for sustainability, as will be developments in the generation of renewable energy in tall buildings, in which wind turbines may play a special role.

Creating positive urban spaces

Tall buildings' mediation with the ground plane carries not only the risks of negative microclimate effects, but also the positive potential to create attractive, useful urban spaces around buildings and public concourses on their lower floors. Fosters' building at 30 St Mary Axe is a good example of a building plan that tapers down at ground level, creating more public space. There is scope to do this in the design of tall buildings, as the reduction of area is compensated by the increase of floor area enabled by multiple floors at the upper levels.

When tall buildings are constructed at public transport nodes, they have special potential to meet sustainability objectives. In these cases, the mediation of the building with the transport terminal and related facilities such as retail uses need intelligent thought and modelling to provide routes and spaces for effective flows of people, goods and services, with the visual permeability provided by the vistas that are required for good, safe urban design. Examples of such designs in the UK are at the Canary Wharf underground station and in KPF's proposal for the new bus station beneath the station concourse at London Bridge.

Tall buildings can also play a distinctive role in making the urban fabric legible, by conferring identity on their locality. This has been well grasped in the GLA's approach to planning London, using the approach of 'tall building clusters' at certain agreed locations, rather than having them distributed as at random across the metropolitan area. This is helpful in modulating the experience of London's spaces, but when one is in them, and from afar.

Intelligent arrangement of interiors

Design for the interior should optimise the potentials of the floor plate and comfort for individuals, whilst promoting conditions that minimise reliance on artificial lighting and the impact of solar gain. These will in turn be influenced by the height of the section and the cladding on the external perimeter, underscoring the inter - reaction of design components and the importance of integrated design in securing outcomes that are sustainable for people and the environment. Dynamic facades should make a further contribution.

9 Cost

Tall buildings cost more to procure in spite of the economies that they offer through repetition of components and sequencing of work, the principal cost elements being structure, facades, vertical transportation and contractors' cost. Complex, more articulated buildings like those described for the City of London - where ingenuity with massing and form is required to fulfil the challenging footprints and planning requirements- are more expensive to build than regular orthogonal buildings. However, comparatively higher construction costs can be outweighed by the increase of space that they provide in areas of high land value, and the environmental benefits of travel reduction and land - saving, and the social benefits of refined, sensitive design also have a value.

Intelligence is also required for sustainability over the life cycle. Too short term a view on cost management can damage the building's durability in performance and or appearance. An intelligent perspective on value management is important to 'sustain sustainability' through the building's life.

持续城镇的发展：高层建筑在其中所起的作用

Ziona Strelitz
(ZZA Responsive Built Environments)

摘 要：本文主要讲述高层建筑整体设计对于城镇公共和私人区域宜居性的贡献，通过增强密集度实现节约土地，并利用节约能源技术推动可持续性发展。关注点是建筑高度在应对多重可持续性目标时的战略潜能，这意味着与众不同的机遇和挑战，即要求多重学科处理综合建筑系统，并优化设计。文中提倡在建筑的整个生命中，成本和价值管理对保持可持续性明显有效。

Bibliography

[1] Tall Buildings: A strategic Design Guide(ed. Strelitz Z, RIBA Publishing, BCO, 2005)
[2] Tall Buildings in London: Guidance on Planning(Linklaters, BCO, 2002)
[3] Urban Task Force, Report [1]; 29 June 1999
[4] Urban Task Force, Report [2]; 22 Nov 2005
[5] The London Plan(Published by The Greater London Authority, Feb 2004)
[6] Guidance on Tall Buildings(CABE, English Heritage, 2003)
[7] Tall Office Building in London: Giving Occupiers a voice(ZZA, BCO 2002)
[8] The Changing City(Duffy F & Henney A, 1989, Bulstrode Press, London.)

BUILDING ECOLOGY, BUILDING MATERIALS AND GREEN BUILDING

Prof. Rutuja Deshmukh-Jagtap
(B. Arch, M. E. Civil(Town and Country Planning), NICMAR PUNE CAMPUS, 25/1Balewadi, N. I. A. Post office, Pune 411045, India)

Abstract: A green building, also known as a sustainable building, is a structure that is designed, built, renovated, operated, or reused in an ecological and resource-efficient manner. Green buildings are designed to meet certain objectives such as protecting occupant health; improving employee productivity; using energy, water, and other resources more efficiently; and reducing the overall impact to the environment. Building "green" is an opportunity to use our resources efficiently while creating healthier buildings that improve human health, build a better environment, and provide cost savings. Green building measures cannot achieve their goals unless they work as intended. Building commissioning includes testing and adjusting the mechanical, electrical, and plumbing systems to ensure that all equipment meets design criteria. It also includes instructing the staff on the operation and maintenance of equipment. The study gives an overview of energy efficient design concepts, eco-friendly materials and a case study of major problems of India. Sustainable or 'green building' design and construction is the opportunity to use our resources more efficiently while creating healthier and more energy-efficient homes. Although there is no magic formula, success comes in the form of leaving a lighter footprint on the environment through conservation of resources, while at the same time balancing energy-efficient, cost-effective, low-maintenance products for our construction needs.

Key words: Passive Solar Techniques, Eco-Friendly material, Green Buildings, Sustainable Development

An energy-efficient building can be more comfortable, easier to maintain, and less costly to operate than an inefficient one. There are a variety of approaches to increasing the energy efficiency of buildings-whether the buildings are just now being designed, or are already built.

Once a building has been designed to minimise 'energy in use', designers must consider materials, specification to minimise the environmental impact and embodied energy of the construction materials.

Obviously once building energy use has been reduced, the energy and pollution embodied in materials increases from 10% of the total to a much larger proportion; at this point, therefore, it becomes important to consider the embodied energy and pollution of materials.

When we start to assess the impact of materials, we find ourselves returning to the consideration of all aspects of environmental impact including resource use and pollution, as well as energy use.

Energy Efficiency Design Concepts:

Buildings can be designed to meet the occupant's needs for thermal and visual comfort, at reduced levels of energy and resource consumption. Energy resources efficiency in new constructions can be effected by adopting an integrated approach to building design. The primary steps are listed below:

1 Incorporate Solar Passive Techniques in building design to minimize load on conventional systems. (Regulate building energy flows by natural means)

Passive Heating Strategies:

a) Direct Gain,

b) Indirect Gain,

c) Isolated Gain,

d) Solar Collectors.

Passive Cooling Strategies.

a) Cross and Stack Ventilation,

b) Evaporative Cooling,

c) Desiccant Cooling,

d) Solar Vapor Absorption Cooling.

Passive Heating/Cooling Strategies:

a) Earth Berming,

b) Earth Tubing,

c) Roof Ponds,

d) Cavity, Insulation,

e) Landscaping,

f) Water Bodies,

g) External Colour and Texture.

2 Design Energy Efficient Lighting, Heating, Ventilation and A/C Systems. (Use Energy Efficient Equipments, Controls and Operating Procedures).

3 Use renewable energy systems (Solar PV systems and water heating, wing powered systems) to meet a part of building load.

4 Use low energy materials and Methods of construction and reduce transportation energy. Discourage materials with high embodied energy.

In the process of Building design, Environmental forces namely Thermal, Luminous,

Acoustic and Aqueous environment, need to be addressed at the schematic design stage only, and using scientific methods.

Many clients and their architects are committed to tackling sustainability within their project briefs, increasing the need for building products to be supported by good quality information related to sustainable design issues.

The classic definition of sustainability comes from the Brundtland Report of the World Commission on Environment & Development of 1987, Our Common Future: "Sustainable development is development which meets the needs of the present generation without compromising the ability of future generations to meet their own needs."

More simply: "An activity is sustainable if, for all practical purposes, it can continue for ever." Sustainability involves the maintenance of all our environmental resources. It implies the need to conserve both energy and material resources, to maintain biodiversity and to reduce pollution to stabilise planetary conditions.

Once a building has been designed to minimise 'energy in use', designers must consider materials specification to minimise the environmental impact and embodied energy of the construction materials.

Obviously once building energy use has been reduced, the energy and pollution embodied in materials increases from 10% of the total to a much larger proportion; at this point, therefore, it becomes important to consider the embodied energy and pollution of materials.

When we start to assess the impact of materials, we find ourselves returning to the consideration of all aspects of environmental impact including resource use and pollution, as well as energy use.

What is Eco-friendly material?

Dictionary: describes a product that has been designed to do the least possible damage to the environment.

US EPA-EPP program defines as: "……product or services that have a lesser or reduced effect on human health and the environment when compared with competing products or services that serve the same purpose……"

Properties

ECO-FRIENDLY MATERIALS AND TECHNIQUES

- ➢ Biodegradable
- ➢ Reuse/Recycle
- ➢ Aids Energy Efficiency in building
- ➢ Durability and Life Span
- ➢ Reduction in air, land and water
- ➢ Pollution
- ➢ Local Availability
- ➢ Low Embodied Energy
- ➢ Reuse of waste product
- ➢ Renewable Source

Source of Material

a. Renewable Source

Rapid renewable sources e. g. wood from certified forests

b. Reuse of Waste

- Salvaged products-e. g. old plumbing door frames.
- Recycled contents-agriculture/industrial waste e. g. Bagasse Board

Performance

- Durable and life span

Material that are exceptionally durable, or require low maintenance e. g. PVC pipes.

- Reduce material use

These are energy efficient and also help reduce the dead load of a building e. g. Ferro cement.

Energy Conservation

- Materials that require less energy during construction-e. g. precast slab.
- Materials that help reduce the cooling loads-e. g. aerated concrete blocks.
- Products that conserve energy-e. g. CFL Lamps.
- Fixtures and equipment that help conserve water-e. g. Dual flush cisterns.

Recyclable

- Reuse or Recycle as different product-e. g. steel, aluminum.
- Biodegradable-that decompose easily-e. g. wood or earthen materials

EcoHousing Technical Criteria

- Efficient Building Materials has a total of 200 points.
- Use of certain materials is mandatory for a Eco-Housing Project.
- Increased use of other recommended alternatives will provide higher points.
- Verification is dependent on the Bill of Quantities and Specifications.

BASE MATERIALS FOR R. C. C AND STEEL SYSTEMS

Use of 25% Pozzolana material blended Portland cement(BPC)

- Pozzolana material-Flyash/Slag/Calcined Clay
- Increased of Pozzolana material in BPC up to 30%~50%
- Use sand and aggregate from pulverized debris and/or sintered fly ash for concrete and mortar.
- Use recycled steel forms and bars for reinforcement

Alternative structural system

- Ferro cement and/or precast components for beams, slabs, staircase, lofts, balconies etc.
- Read Mix Concrete.
- Use Resinous curing agents

Masonry

- Fly ash+sand+lime bricks/blocks,
- Pulverized debris+cement bricks/blocks,
- Industrial waste based bricks/blocks,
- Aerated lightweight BPC concrete blocks,
- Phospho-Gypsum based blocks
- Lato blocks

Mortar

- Use of 25% Pozzolana material blended Portland cement(BPC)
- Sand from Pulverized debris or sintered fly ash
- Increased use of Pozzolona materials content in BPC by direct addition of raw pozzolona materials.

Plastering

- Calcium silicate plaster/
- Cement plaster
- Fiber reinforced clay plaster/non erodible mud plaster/ phosphogypsum plaster
- Use resinous curing agents

Roofing and Ceiling

- Terrazzo floor for terraces and semi covered areas
- Fibre reinforced polymer
- Micro concrete roofing tiles
- Bamboo matt corrugated roofing sheets

Outdoor paving

- Fly ash/industrial waste/pulverized debris blocks in BPC
- Bedding sand to be from pulverized debris
- Terrozo flooring for terraces and semi open spaces

Flooring

- Use ceramic tiles(non-vitrified)
- Phospho-Gypsum tiles
- Bamboo board flooring
- Mosaic tiles
- Cement tiles

Windows, Doors and openings

- Ferro cement and precast R. C. C lintel, chajja and jails instead of R. C. C.
- Ferro cement and precast R. C. C Frames /Frameless Doors/ Bamboo reinforced concrete frames.
- Hollow recycled steel channels and recycled aluminum channels.

Electrical

- Use unplasticized Polyvinyl Chloride (PVC) or density polyethylene (HDPE) products.
- Use products with recycled aluminum and brass components

Water supply

- Use R. C. C. , unplasticized PVC, G. I, C. I. pipes instead of lead, A. C. pipes
- Use products with recycled aluminium and brass components
- Use polymer plastic(Random) for hot/cold water systems.
- Manholes and covers from precast cement concrete and high strength uPVC.

Wood work

- Timber used must be renewable timber or from salvaged wood.

Case study of Major problems of India

- Ever soaring fuel-bill
- Fertilizer subsidy
- Lack of Decentralized Infrastructure

Solution is Biomethanation of Biomass

- The solution is green
- The solution is sustainable
- The solution reverses the supply chain of energy
- It is a feasible solution

Biogas is green

- It is produced from agricultural crops which are renewable.

Biogas is clean

- Its use in combustion systems leaves minimal emissions.

Biogas gives sustainable development

- Biomethanation co-produces biocompost with biogas.
- Biocompost will go back to the soil to complete the bicycle and sustain the soil-humus high to sustain yields and give freedom from chemical fertilizers.

And above all

- This route reverses the supply chain of energy and fertilizer(organic carbon+minerals)

Products

- Power, steam and refrigeration
- Compressed natural gas
- Pipelines gas
- Cooking gas
- CDM benefits
- Biocompost

INDIAN STRENGTHS

- Plants(vegetation)are the best entrappers of solar energy and India enjoys the highest average solar flux in the world
- Climate conditions in most of India are conductive to biomethanation.
- CNG vehicles(Buses with 110 HP Tata Cummins engines, 3-wheelers of Bajaj Auto, Indica with petrol/CNG Version)Exist.
- Gobar gas programme is the most successful in the world.

FEASIBLE IDEAS FOR ORGANIZED AGRI-SECTOR

- Green CNG for cane transport
- Free all the bagasse for pulp of non-wood paper and newsprint(like TNPL, Tamil Nadu)
- Green power for more cane

INTANGIBE BENEFITS

- Clean cities-Emission free and biosanitation
- Local employment potential in rural areas, without need for upgradation of skills
- Decentralized development
- Permaculture

—Prevention of soil-erosion

—Co2-sewuestration

—Humidity control

—Improvement of landscape

Feedstocks

- MSW-least cost disposal
- Sweet sorghum/sudan grass
- Maize silage
- DOCs of Non-Edible oilseeds from biodiesel programmes and of castor

- Cane trash/cane tops
- Banana stalks/leaves

PROBLEMS OF POWER IN INDIA

- Power is the most critical infrastructure for development and it can not be imported. We have to develop our own power.
- Centralized power-plants have long gestation period
- T and D Losses are highest in the world
- Poor quality of rural electricity

PIPED GAS FOR COOKING

- Rural housing wife spends long hours for collection firewood for cooking
- Eye-sores and long cooking time are results of wood-stoves
- Blue-flame of biogas reduces the cooking to mere 3 hours(8hours) household income shoots up, children's education goes up, family health improves.
- Turn-the-knob convenience for quick modulation

BIOGAS AS FUEL

- It is clean-No particulate emissions, No Nox's Sox's, No sulphur.
- It is green-our growers can "Grow" energy
- High efficiencies of power-production through topping(combined)cycle.
- Can replace petrol/ diesel by 100%
- High energy density when converted
- To methane-can be put into national gas grid

SUSTAINABLE DEVELOPMENT

- Biocompost is co-enerated IT goes back to the soils to sustain humus and therefore the agricultural yields
- Reduces dependence on chemical fertilizers

Biomethanation Vs Gasification

- Biomethanation keeps the biocycle intact by recycling carbon in bioassimilable form, gasification and combustion return nothing
- Biogas has high energy density(5000～8600 KCAL/M3)Producer gas is very low at 1100 KCAL/M3.
- Plywood should be urea bonded and not phenol bonded.
- Use MDF board
- Use Mica Laminates and Veneers on composite boards instead of natural timber
- Bamboo Ply/Mat Board, Bamboo mat Veneer Composite
- Fibre Reinforced polymer board
- Bagasse Board
- Coir Composite Board
- Finger jointed plantation timber board
- Recycled laminated tube board

- Aluminium-foil+paper+plastic composite board

Water proofing chemicals

- Use of water based chemical instead of solvent based
- Use Epoxy resins instead of tar felt/pitch

Painting

- Use of cement paint/Epoxy Resin Paints for external surfaces
- Use of water based paints, enamels, primers and polishes.

We must inculcate the benefits of using low energy consumption and environmental protection building materials

建筑生态，建筑材料及绿色建筑

Prof. Rutuja Deshmukh-Jagtap
(B. Arch, M. E. Civil(Town and Country Planning), NICMAR PUNE CAMPUS, 25/1Balewadi, N. I. A. Post office, Pune 411045, India)

摘　要：绿色建筑，也被称为可持续建筑，是指以生态环保和资源节约的形式进行设计、建造、装修、运营或再利用的建筑物。绿色建筑的设计要达到某种目标，例如保护居住者的健康，提高雇员的工作效率，对能源、水和其他资源的利用更加有效，减小对环境的整体影响等。绿色建筑可以在有效利用资源的同时，创造更健康的建筑以提高人们的健康水平和改善环境，降低成本。绿色建筑只有通过工作中的刻意追求，才能获得。建筑需要对其机械、供配电和管道系统进行测试和调节以达到设计要求，以及指导工作人员如何对设备进行操作和维护。本研究回顾了节能的设计理念，生态材料的发展，并对印度存在的主要问题作了个案研究。“可持续建筑”或“绿色建筑”的设计和建造为我们提供了机遇，即更加高效的利用资源、创造更加健康和更加节能的住宅。尽管没有固定的模式，但仍可以通过保护资源，采取减轻对环境的影响的生活方式，同时平衡能源效率、成本效益、低维护费用之间的关系来达到我们的目标。

INTELLIGENT, GREEN & ENERGY EFFICIENT BUILDING & NEW TECHNOLOGIES

Subodh Shankar
(Chief Architect Planner, U. P. Housing & Development Board, India.
P. K. S. Naglakar & Ruchi Gupta
B. Arch, M. U. R. P, C. o. A, A. I. T. P)

1 Introduction

Owing to the fast depleting energy resources, energy scarcity and environmental pollution hitting high, the relevance of energy efficient buildings has assumed a greater significance. The building construction is one of the largest energy consuming sectors. There are numerous reasons why energy efficient design should form an integral part of building architecture, such as improved comfort conditions, long term economic gains and better environmental adaptation. Though an energy conscious design approach may at first appear to constrain the aesthetic features of a building, the architect can transform this drawback into an advantage by developing a better understanding of the building's interaction with the external environment and creating something which is not only visually appealing but also comfortable to live in and economical to operate.

Studies conducted by Intergovernmental Panel on Climate Change indicate that in 1990, the residential, commercial and institutional building sector consumed 31 per cent of global energy and emitted 1900 mega tonnes of carbon and by 2050, its share may rise to 38 per cent and 3800 mega tonnes, respectively. In India, it is estimated that about 25 per cent of the total energy demands are due to manufacturing materials required in building construction, while another 30 per cent go into the running needs of the building. The ministry of power approximation indicates that 20～25 per cent of the total electricity consumed in the government buildings are wasted because of inefficient design parameters of the building, thus resulting in a whopping energy financial loss of about Rs 150 crore a year. However, efficiency measures with paybacks in five years or less can reduce emissions by nearly 40% by 2050. Greater use of available, cost-effective technologies can radically reduce carbon emissions.

India's residential and commercial sectors consume 25% of its electricity, the major portion being utilized by buildings. The National Building Code of India recommends a minimum of 200 litres per capita per day of water supply for residences. However, the actual figure is less than half that. Inadequate collection and inefficient treatment of domestic wastewater degrades freshwater and contaminates groundwater. Wastewater treatment efficiency, even in Class I cities, is merely 30%.

The uncontrolled rise in population and rapid urbanization it is certain that the construction activities would increase, thus a larger share of energy in an already strained energy supply scenario, would be demanded. In this scenario creation of eco-sensitive building or 'green building' becomes important.

2 What is a "Green Building"?

A Green Building is the one which uses the resources judiciously, minimizes emissions and has efficient waste management system. It should be considered and designed to achieve the collective objectives of energy security and environmental protection and should impose minimal impact on the environment while enhancing user comfort and productivity. It is developed through an integrated approach to design. The aim is to achieve the desired comfort with the least input of conventional energy. Not only the new building stock can be targeted but also retrofitting of existing buildings with efficient technologies can result in substantial reduction of energy consumption.

2.1 Benefits of a "Green Building"

Green Building or an energy efficient building benefits the users in many ways.

- Reduced destruction of natural areas, habitats, biodiversity, and soil fertility
- Reduced air/water pollution(with direct health benefits as a result)
- Reduced water consumption
- Limited waste generation due to recycling/ reuse
- Increased user productivity
- Enhanced image and marketability

2.2 How can we achieve Energy Efficiency or make our Building Green?

In general, energy efficiency can be achieved in new buildings through:

- Using the passive solar techniques in building design to minimize the energy load;
- Minimizing energy losses by designing energy efficient lighting and HVAC (heating, ventilation & air-conditioning) systems;
- Effective and intelligent use of renewable energy resources such that it can take care of a part of the building load; and
- Use of low energy materials and methods of construction.
- Efficient Waste and Water management.

- **Using the passive solar techniques in building design to minimize the energy load**

The first stage of eco-friendly and green building design is to incorporate passive solar design interventions and reduce the loads on conventional systems. Passive systems provide the thermal and visual comfort by using natural energy sources and sinks such as solar radiation, ambient air, internal gains, sky, wet surfaces, and vegetation. The solar passive design varies according to the climatic condition prevalent at the site. Buildings designed for passive solar and day lighting incorporate design features such as large south-facing windows and building materials that absorb and slowly release the sun's heat.

Basic energy sources in an eco-friendly building complex **Table a**

Sky	Sun	Air	Earth	Water
-day lighting -heat sink	-heating -electricity generation -day lighting -greenhouse effect -solar chimneys	-ventilation -heat sink	-roof gardens -earth berms for insulation	-roof ponds -fountains for humidification -rainwater harvesting

Source: TERI

In cold climates, the building needs to adopt measures to maximize its solar heat gains by embracing strategies such as maximum exposure to south, windows to capture heat, dark colored surfaces, high thermal mass and insulation to retain the captured warmth of the sun or use of design elements such as trombe wall and sun spaces, among others.

In hot and moderate climates, the strategy is to admit light while rejecting heat. A building designed for a hot climate should use measures to reduce solar gain such as smaller window sizes, shaded walls, minimum exposure to west and east or use of design elements such as solar chimneys and wind towers, among others, to maximize ventilation. Use of building materials like energy efficient glass could reduce heating or cooling demands by 8～10 per cent. However, one has to resort to advanced techniques of passive conditioning such as roof ponds, trombe walls, wind towers, evaporative cooling, etc. in extreme climatic conditions.

Further, hard paved parking lots, pathways, and courts could be minimized as they generate a heat island effect that causes a sharp temperature rise. Lighting accounts for 20～50 per cent of the electricity requirements of a building and several measures could reduce the consumption of energy. For instance, motion sensors could be installed for security purposes instead of constant lighting.

- **Minimizing energy losses by designing energy efficient lighting and HVAC systems;**

Energy efficiency of HVAC and other systems plays an important role in determining the energy consumption pattern of a building. Energy conservation is possible by judicious design of lighting and HVAC (heating, ventilation and air conditioning) systems, controls and operation strategies. Increasing insulation levels in conditioned buildings is regarded as the most cost-effective investment in energy efficiency. Thermal insulation of external walls, roofs and floors, and double-pane windows can reduce energy consumption for space heating by lowering heat losses through the envelope of the building. Energy consumption for cooling is also reduced because of lesser heat gains from outside through the envelope. Energy efficient windows with their high thermal insulating values and spectral selectivity can make air conditioning systems work more effectively. This can lead to reduction in AC loads, lower consumption of electrical energy and reduction in peak load demand [Bandyopadhyay 2001].

The use of energy efficient glazing helps in minimizing unwanted solar gains in summer and heat losses in winter, while maximizing the amount of useful daylight in buildings. Lighting load constitutes about 10 to 15 percent of the total electrical load of a building, and so energy efficient lighting systems, such as compact fluorescent lamps and fluorescent tubes with electronic ballast, are generally recommended instead of conventional lighting fixtures to reduce the lighting load. Although the initial cost of such installation is high, they last longer and the running cost is also less. As electricity is efficiently converted to light in energy efficient lamps, the amount of heat generated is also less.

- **Effective & intelligent use of renewable energy resources**

Renewable energy systems are installed for meeting a partial load of the building, thus considerably reducing the overall electrical and thermal load. Solar energy could be utilized for a variety of purposes and in a number of ways: generating electricity, providing hot water, and heating, cooling, and lighting buildings. Solar photo voltaic (PV) can provide electricity for lighting. Solar thermal systems may be used for heating water or space heating and transpired solar collectors can preheat air for the building's ventilation system. Solar water heaters generate hot water at 60 to 85℃ and can conveniently be used for domestic use. The usefulness of a solar water heater can be ascertained from the fact that a standard 100 liter capacity system can be installed as an alternative to electric geyser3 for residential use that can save 1500 units of electricity annually and could prevent emissions of 15 tons of carbon dioxide into the atmosphere every year [MNES 2001]. It is estimated in India that the use of 1000 solar water heater of a 100 liter capacity each can contribute to a peak load saving of 1 MW. In cold and sunny climates, solar air heaters can be used extensively to reduce demand for electricity and firewood for space heating.

Solar electricity can be generated by integrating solar cells in buildings, on roofs and facades, or by installing grid connected photovoltaic plants. It is an expensive proposition where grid connected electricity is available, but for locations where conventional electricity is not available or the grid is far away, it makes an ideal alternative. Further, 'building integrated PV'(BIPV) systems may be installed for the generation of electricity, and surplus electricity generated during non-working hours could be fed to the grid. Inclined roofs, if oriented in the right direction, are an ideal support structure for PV modules.

In moderate to high wind speed locations, roof top aero generators may be installed for the generation of electricity. A sunspace or solarium, which comprises of a combination of direct and indirect gain systems, may be used to heat up the living space by convection and conduction through the mass wall. Technologies have been commercialized to convert the solid and liquid waste generated in a building to productive use in the form of bio-fertilizer, gas for power generation, cooking etc. The long-term objective is thus to reduce building energy loads in a cost-effective manner such that renewable sources of energy can meet and exceed energy demand in the building.

- **Use of low energy materials and methods of construction.**

The choice of building materials also substantially contributes towards reducing the energy load of buildings. The use of conventional energy can be minimized by use of low energy materials, efficient structural design and reduction in transportation energy. Thus when building an energy efficient structure, it is necessary to closely examine the issues of building materials and to make appropriate decisions according to local conditions. Bamboo which is called the poor man's timber in India is also getting recognized globally as a suitable building material because it is eco-friendly and highly suitable for energy efficient buildings.

3 Energy Efficient Building in India

In India, the concern for energy scarcity and the environmental problems associated with the consumption of energy in the building sector has resulted in continued interest in energy efficient building design. The Ministry of Non-Conventional Energy Sources (MNES), of the Government of India has undertaken several initiatives to promote solar efficient building design in the country.

The RETREAT (Resources Efficient TERI Retreat for Environment Awareness and Training) complex of TERI (The Energy and Resources Institute) at Gurgaon in Haryana has amply demonstrated that 'zero energy in-zero waste out' building could be created and run efficiently. The residential training complex for executives, according to TERI officials-harnesses both traditional and modern means of tapping renewable sources of energy to offer modern amenities like lights, air-conditioning, cooking and laundry at substantially reduced costs.

The complex apparently saves 50 per cent of energy costs over conventionally designed buildings. The building has been designed as south facing for the winter heat gains. The summer gains are minimized using deciduous trees and shading. The south side of the structure is partially sunk into the ground to reduce heat gains and losses. Instead of the conventional grid electricity, the RETREAT has a bio-gasifier plant that generates 50kW power by burning bio-refuse such as twigs, branches, straws and crop residue. By avoiding conventional air conditioning methods, Earth Tunnel system has been constructed that collects atmospheric air, cools it under the earth and then pumps it to all residential and conference halls to maintain steady temperature at 20℃ in winter, 28℃ in summer and 30°C in monsoon. The lighting is provided by a combination of innovative day-lighting by means of skylights, use of compact fluorescent lamps, high efficiency fluorescent lamps and lighting controls to reduce consumption. The efficient lighting system has resulted in a lighting load of only 9kW compared to minimum requirement of 28kW in a similar conventional building.

Solar hot water systems integrated with the parapet wall meets the entire hot water requirement of the building. The 'Root Zone' facility recycles the waste water with the help of phragmites and the recycled water is used for gardening purposes. The annual energy savings by use of the passive design and efficient devices in the building is about 250000 units of electricity and 2100 LPG cylinders and in rupee terms is about Rs 10.5 lakh. Not only the RETREAT complex, the TERI office complexes in Bangalore and Mukteshwar (Uttaranchal) have also been designed and constructed following the principles of 'Green Building'.

4 Efforts in the Field of Achieving Green Building

While going in for area-specific, need-based local action, it is imperative to be a part of the global alliance network working for environment protection and conservation of nature.

India figures in two such projects. Pune is the only city participating **in Global Green Mapping System** (Young World, September 9,2002). The endeavor asks everyone to come together and create a local environment map that identifies links and maintains natural and cultural heritage sites in their area.

Another smart solution is the concept of the green building. TERI (The Energy and Resources Institute) is now trying to popularize the 'Green Building' concept in the country and is in the process of drafting a rating system for Indian buildings called **TERI Green Building Rating System (TGBRS)** on the lines of the internationally accepted rating systems such as the US-based LEED (Leadership in Energy and Environment Design) and the UK-based BREAM (Building Research Establishment's Environmental Assessment Method) and also current building practices in India. The TGBRS, through its qualitative and quantitative assessment criteria rates a building on the degree of greenness and is applicable to both new and existing commercial, institutional and residential buildings.

The building performance guidelines and standards are based on established national and international building design codes. Data for building energy analysis is will be gathered from the client through a questionnaire or checklist. The energy analysis incorporates either an energy audit (for existing buildings) or a computer modeling and simulation (for new buildings, at the design stage). Apart from quantifying a building's energy performance, the TGBRS recognizes efforts to adopt eco-friendly solutions. This is particularly important in India where awareness of building-related environmental issues is rather limited. Rewarding good intentions goes a long way in disseminating information and knowledge and encourages the community to adopt sustainable practices. The TGBRS awards credits for the building's green performance based on the following attributes: Site planning, Building envelope, Building systems, Water and Waste management, and Green design and practices.

The TGBRS is, however, a voluntary scheme to start with, but the effort will provide useful inputs to the upcoming mandatory building codes/guidelines being developed by the Bureau of Energy Efficiency (BEE) and the Ministry of Non-Conventional Energy Sources, Government of India. The BEE has been entrusted with the task of implementing and enforcing the Energy Conservation Act 2001, which has many provisions to enable energy conservation across all sectors. In the building sector, the Act prescribes the development of suitable guidelines for the development of energy conserving building codes and practices.

5 After all, How can we Achieve the Results?

After having discussed most of the relevant issues on energy efficiency, in brief, we finally want to put forward the basic idea of integrating the efforts being taken up in the field. Apart from integrating these efforts it is all the more essential to implement these efforts as per the ground realties. A joint or rather integrated effort is the only solution to achieve energy efficiency in all terms and hereby save the depleting environment.

For this the first step would be to prepare a comprehensive program on the guidelines of the efforts being taken up in the field of energy conservation, in all parts of the world. Energy conservation should be considered as a mission and taken up as seriously as Disaster Management Efforts or Health Improvement Efforts. The program would involve the following steps:

- Awareness Program for developing willingness amongst the individuals to take up the task.
- Training Programs for Policy makers, Architects, Planners, Engineers, Construction Professionals, Builders, Material Manufacturing Industries etc. should be conducted to give proper knowledge in the field and also transfer Best Practices.
- The trainees should be given Certificates that empower them to perform in the field. These professionals can also be empanelled by the Governments to ensure dissemination of knowledge at various levels. These professionals would be expected to train others.
- The architects and planners share the responsibility of energy efficient deigns at different level of planning/ design. For example, the layout of a township in itself ensures energy efficiency in many ways-like traditional city of Jaipur, Lucknow etc. whereby the street patterns ensured comfortable living conditions without use of any electric fixtures. The courtyard planning concept followed by traditional Indian societies have always

ensured best living conditions, ever identified, in India. Basically one has to start understanding the importance of Vernacular Architecture and identify measures of mainstreaming the same.

- Enforcement of existing acts and measures to ensure implementation of techniques with participation of private-public sectors, NGOs etc.
- TGBRS has emerged as a good concept and has to be widely promoted and legally enforced.

To conclude, we would mention that achieving energy efficiency should be taken up as a responsibility rather than a task. A responsibility that would bear fruits in future, may be for our future generations. It has been widely identified that inspite of our efforts in the field we still are far behind when it comes to eco-technology. Efforts have always been made to promote use of non polluting and non conventional energy resources, but the results are still awaited. It is a promising sector and is expected to develop further in near future with the active cooperation of architects, engineers, consultants and policy-makers.

Energy efficiency has to be taken up as a Revolution.

Bibliography

[1] 'Green' buildings can reduce energy bill by almost half: TERI by Deepak Goel, Article in Business Line

[2] "Green retreat" by Soma Basu, Article in The Hindu

[3] Govt buildings waste Rs 150 cr of power each year: Shahi Ramesh Menon, Feature Article at TERI website

[4] Green building Debajit Palit Feature Article at TERI website

[5] Website of TERI

[6] Website of Center of Excellence

[7] Website of World Sustainable Development Forum

[8] Website of World Energy Efficiency Association

[9] www. makingindiagreen. org

[10] www. ci. austin. tx. us

[11] www. worldchanging. org

GREEN BUILDING

Sachin Jain
(Lecturer, NICMAR, Pune, India)

As now a days due to urbanization and migrations populations are concentrated in small area and all countries has mega cities. Hence due to over population we need more houses and these houses are build without considering environment, which will cause global warming and more pollution. Hence lately a new concept of green building has come where we consider building as a part of our environment and it is so design and constructed so that it will cause minimum disturbance to surrounding. This is a relatively new concept in India, its promotion requires greater focus.

Benefits of green buildings

- Energy efficiency
- Resource conservation
- Environment friendly
- Greater comfort and lower utility bills
- Covers community design issues
- Good indoor air quality
- Better performance & longer durability
- Low maintenance cost
- Beneficial to health.
- Truly cost-effective

We have to consider following steps for green building

- **Natural Planning**
- **Natural Building**

Natural Planning

As "consumers" we are frequently confronted with life style decisions that can impact our environment. There are a few choices in this life that can make a big difference in what the quality of life will be for those who follow us. Going with the flow of our culture is hard to avoid, and unfortunately the flow is not in the right direction for evolving a sustainable future. One of the most momentous choices that any of us will make is the kind of house we live in. Following a list of twelve principles of basic architecture that can guide us in for designing green housing and to have good planning.

❖ **Design as per required and think small** means Small can be beautiful and cozy. The trend lately has been toward huge mansion-style houses. Large houses generally use a tremendous amount of energy to heat and cool. This energy usually comes from the combustion of fossil fuels, depleting these resources and emitting greenhouse gases and pollutants into the air. Also, the larger the house, the more materials go into its construction materials which may have their own environmental consequences. A home should be just the right size for its occupants and their activities. The key to this is efficient use of space, good organization, and keeping possessions to a manageable level.

❖ **Use maximum of natural light and heat with the Sun** means nothing can be more comfortable for body and mind than living in a good solar-heated house. Good passive solar design will provide just enough sunlight into the rooms to be absorbed by the surrounding thermal mass (usually masonry materials), so that the heat will be given back into the room when the sun goes down. The thermal mass is a kind of "heat battery" that stores the warmth, absorbing it to keep the room from getting too hot during the day. Equally important to thermal mass is insulation (such as straw bales or crushed volcanic rock) that will keep that heat inside. Thermal mass materials need to be insulated from the outside, or else they will just bleed that warmth right back out. A rock house might have tons of mass, but be uncomfortably cold because of this energy bleed. So a good solar design will utilize materials of the right type in the right places, blending thermal dynamics with utilitarian design.

❖ **Use nature to keep your house cool** means a well designed solar house is both warm when you want it and cool when you want it; that is to say, the temperature tends to stay fairly even. A good way to keep temperature cool is to dig into the earth. About six feet under the earth, you will find that the temperature varies by only a few degrees year round. While this temperature (about 50～55 degrees F.) might be too cool for general living comfort, we can use the stability of the earth's temperature to moderate the thermal fluctuations of the house. If we dig into a south-facing hillside to build, or berm the north part of the house with soil, we can take advantage of this. The part of the house that is underground needs to be well insulated, or the earth will continually suck warmth out of the house.

❖ **Use renewable energy** means there are many ways to conserve the use of fossil fuel; using the sun, wind, or water to produce electricity are among them. If we choose to do this, we will be forced to be careful in the way we use our electricity because it is limited. Whether we get our electricity from alternative sources or from the grid, it pays to choose energy efficient appliances. Front-loading clothes washers, for instance, use much less electricity, water and soap than the top-loaders. Compact florescent lights use about a third of the electricity of standard bulbs. Many appliances use electricity by just being plugged in (known as a phantom loads); be sure to avoid this.

❖ **Use methods to conserve water** means the use of low water capacity toilets, flow restrictors at showerheads and faucet aerators are fairly common now. More radical conservation approaches include diverting gray water from bathing, clothes washing and bathroom sinks to watering plants; catching rainwater from roofs and paved areas for domestic use and switching to composting toilets. These can be very effective and safe means of water conservation if done carefully to avoid bacterial infestation; be sure to comply with all local laws that regulate these strategies. Landscaping with drought tolerant, indigenous plants can also save an enormous amount of water.

❖ **Use local available materials** means in almost all localities, nature has provided us with some wonderful materials to build with. Because these materials require little processing or transporting, the environmental and economic costs are low. Some are renewable resources (like trees and straw), and some may be so abundant that their supply seems almost inexhaustible (like rocks and sand). One of the beauties of building with local materials is that they seem to fit well with the feeling of the place, naturally.

❖ **Use natural materials and environment** means our home is an extension of our self. It reflects how we feel about our environment and it nurtures us. Our home is very much a part of us, just as the shell of a hermit crab becomes part of the crab. The best barometer for whether our home is healthy or not is how we feel when we spend time there. If we feel vibrantly healthy and truly enjoy spending time at home, then it's got to be good. How we feel about our home can be influenced by many factors. Air quality, temperature, noise, light, ambience, vibes, etc. can all affect our relationship with our home. Some of these affect us physically, while others are more psychological but of course these two aspects affect each other. Among the natural materials that are both aesthetically pleasing and benign it would include: stone, glass, lime or mud plasters, adobe or rammed earth, bricks, tiles, untreated wood (simple oil finishes can be used), cork, paper, reeds, bamboo, canes and grasses, all natural fibers (linen, cotton, wool, etc.). Linoleum is more natural than it realized before. It is made from powdered cork, linseed oil, wood resin, wood flour mixed with chalk, and pressed onto a backing of burlap or jute. Light and color are other elements that affect how we feel. Both harsh, overly bright lighting, and dark, dingy spaces can be oppressive. We want our homes to elevate our moods, not depress them. Choose colors that we find soothing, or stimulating, whichever is appropriate for the function of the room. Try to arrange for natural day lighting as much as possible.

❖ **Conserve nature and save forests** means use wood as decoration. Cull dead trees for structural supports. Use masonry, straw bales, paper crete, cob, adobe, rocks, bags of volcanic rock, etc., instead of wood. Unfortunately it is difficult to get away from lumber in making a roof, so consider making a dome from materials that can be stacked. Domes are also more energy efficient and use fewer materials for the same space as a box. A conventional straw bale house only diminishes the amount of wood used by about 15%. Homes can be made with certified sustainably harvested trees.

❖ **Use old and recycle materials** means it is satisfying, fun and economical to reuse old building materials. In our society there has been a stigma attached to "used" stuff. We value "new" above all else, or at least the advertising forces would have us believe that. The value of recycling building materials, or anything else for that matter, is that the cost is likely to be a fraction of the same thing in the "new" category. The savings can be substantial.

❖ **Planning must be done for long life and build to last** means a prime ingredient of sustainable architecture is durability. If a building doesn't last for a long time, it would be a waste of a lot of embodied energy, from both a human and a resource perspective not to mention the economic loss.

❖ **Try to have more plants and grow our food** means few things give us more pleasure than living amongst green plants, especially when they are edible. It is best to design an attached greenhouse into the original house plan, but it is often quite feasible to retrofit one onto an existing house. We need some space on the south side of the house wide enough to accommodate a growing bed and space for access. Trees or other things that would limit the available sunlight should not obstruct the area. It could have an outside entrance, but it doesn't have to.

❖ **Use natural energy to store our food** means let nature keep your food cool. In the old days people relied on pantries and root cellars to help keep produce and other provisions fresh. Ice boxes made way for refrigerators, which are obviously much more convenient, but somehow the use of cool pantries and root cellars also fell by the wayside. This is too bad because these spaces have functions that a refrigerator simply can't replace. Root cellars can store large quantities of produce from the time of harvest until the next summer. Cool pantries can store some produce, but also all manner of other foodstuffs and kitchen supplies can be kept there. Cool, dry storage is the best way to preserve most food. The cool of the earth can keep a totally beamed pantry or root cellar cool; the night air can also be used to cool a storage room.

Natural Building

"Natural building" is an umbrella term than connotes any sort of building that is accomplished with the use of natural materials primarily, as opposed to the use of man-made or industrial materials. There is, of course, a blurring of this distinction when any specific material or building technique is examined, because the influence of technology is all pervasive in today's world. Still, it is worthwhile to focus on those ways of building that minimize the use of products that require considerable embodied energy for their manufacture or transportation. The objective is to build with simple techniques that don't further pollute the environment, consume more fossil fuel, or unnecessarily extract the resources of Mother Earth. Such techniques, by their very nature, have an aesthetic value that tends to blend in with the environment and "feel" natural. Hence it has two main parts

- Material
- Systems

Materials

❖ **Earth bag** Building with earth bags (sometimes called sandbags) is both old and new. Sandbags have long been used, particularly by the military for creating strong, protective barriers, or for flood control. The same reasons that make them useful for these applications carry over to creating housing: the walls are massive and substantial, they resist all kinds of severe weather (or even bullets and bombs), and they can be erected simply and quickly with readily available components. Burlap bags were traditionally used for this purpose, and they work fine until they eventually rot. Newer polypropylene bags have superior strength and durability, as long as they are kept away from too much sunlight. For permanent housing the bags should be covered with some kind of plaster for protection There has been a resurgence of interest in earth bag building since architect Nader Khalili, of the Cal-Earth Institute, began experimenting with bags of adobe soil as building blocks for creating domes, vaults and arches. Khalili was familiar with Middle Eastern architecture and the use of adobe bricks in building these forms, so it was natural for him to imagine building in this way. The Cal-Earth Institute has been training people with his particular techniques, and now the whole field has expanded considerably with further experimentation by his students and others.

Khalili's ideas of building with earth bags that are laid in courses with barbed wire between them, and come up with some hybrid concepts that have proven to make viable housing. Instead of filling the bags with adobe soil, it can be filled with crushed volcanic rock. This creates a very well insulated wall (about as good as straw bale) that will never rot or be damaged by moisture. As a covering for the earth bags papercrete can be used. This seems to be a very good solution to the need to seal the bags from the sun and the weather, without necessarily creating a vapor barrier the walls remains breathable. Papercrete may not be a good choice in warm and humid climates, because mold could form on it.

❖ **Strawbale** Building with bales of straw has become almost mainstream in some parts of the countries, especially in the Southwestern United States. Many localities have specific codes for strawbale construction, and some banks are willing to lend on this technique. Straw is a renewable resource that acts as excellent insulation and is fairly easy to build with. Care must be taken to assure that the straw is kept dry, or it will eventually rot. For this reason it is generally best to allow a strawbale wall to remain breathable; any moisture barrier will invite condensation to collect and undermine the structure. Other possible concerns with strawbale walls are infestation of rodents or insects, so the skin on the straw should resist these critters.

There are two major categories of building with strawbales: load bearing and non-load bearing. A post and beam framework that supports the basic structure of the building, with the bales of straw used as infill, is the most common non-load bearing approach. This is also the only way that many building authorities will allow. While there are many load-bearing strawbale buildings that are standing just fine, care must be taken to consider the possible settling of the strawbales as the weight of the roof, etc. compresses them.

❖ **Cordwood** construction utilizes short, round pieces of wood, similar to what would normally be considered firewood. For this reason this method of building can be very resource efficient, since it makes use of wood that might not have much other value. Cordwood building can also create a wall that has both properties of insulation and thermal mass. The mass comes from the masonry mortar that is used to cement the logs together, and the insulation comes from the wood itself and the central cavity between the inside and outside mortars. Like strawbale walls, many building authorities require a post and beam or similar supporting structure and then using cordwood as an infill, even though the cordwood method creates a very strong wall that could support a considerable load. This method produces a look that is both rustic and beautiful. The process of building is similar to laying rocks in mortar, where the logs are aligned with their ends sticking out to create the surface of the wall and mortar is applied adjacent to each end of the log. Typically the logs are not coated with a moisture barrier, but are allowed to breath naturally. It is possible to include other materials into the matrix, such as bottle ends that would provide light to enter the wall. Recent experiments with the use of cob instead of cement mortar to join the logs have been encouraging and this method may provide a somewhat more ecological approach to cordwood building. In this case special care should be taken to have large eaves to keep water away from the wall.

❖ **Papercrete** Papercrete is a fairly new ingredient in the natural building world. It is basically re-pulped paper fiber with portland cement or clay and/or other dirt added. When cement is added, this material is not as "green" as would be ideal, but the relatively small amount of cement is perhaps a reasonable tradeoff for what papercrete can offer. Care must be taken to utilize it properly, or you could be courting disaster. Eric Patterson and Mike McCain, who independently "invented" papercrete (they called it "padobe" and "fibrous cement") and they have both contributed considerably to the machinery to make it and the ways of using it for building. The paper to be used can come from a variety of sources and is usually free. We can use newspaper, junk mail, magazines, books, etc. , which we can get from our local dump or from the waste bin at our post office. Depending on the type of mixer that is used to make pulp out of it, the paper might be soaked in water beforehand or not. It can be a mixer used a small electric motor mounted directly to a shaft with a couple of four inch square blades on it, rather like a milk shake maker. This shaft was suspended in a plastic 55 gallon drum where the mixing took place. . Fill the tank nearly full of water; add about one wheelbarrow full of dry paper, one sack of portland cement, and some sand, depending on how we plan to use the mix. Then drive slowly around the block, back over a drain box with 1/8 inch mesh on the bottom, and dump the slurry into the box via a drain hole in the bottom of the tank. After about a half hour of draining the excess water from the slurry, the papercrete is like soft, workable clay, but not nearly as messy. This is the material that we can use to plaster both the inside and outside of my earth bag house. The slurry can just as easily be pumped or dumped into forms to set up that way. Cured papercrete acts like a sponge unless it is coated with something to stop the entry of water. In one earthbag /papercrete house one have allowed the papercrete to breath fully, so that it absorbs an enormous amount of water when it rains. It is a whole new concept for a roof: a sponge that welcomes the moisture, and then simply gives it back to the atmosphere through evaporation. It can have large cracks (up to about 1/2 inch wide) in the initial layer of papercrete on the earthbags, and still have not any water getting through into the house.

Other properties of papercrete are:

❖ It is dimensionally very stable both through the process of taking in moisture and drying out and in a wide range of temperatures.

❖ It will hold fasteners to some extent, especially screws, without cracking.

❖ It is highly insulating (about R-2 1/2 per inch).

❖ It does not support flames, but will smolder for days if it does catch fire. The more cement and mineral material that is added to the mix, the more fire proof it becomes.

❖ It will support molds if it remains warm and moist for too long.

❖ It will wick moisture from the ground into the wall if it buried in dirt.

❖ It becomes soft and will deteriorate if kept damp (especially underground) for too long.

❖ It resists rodent and insect infestation.

Paper adobe is similar to papercrete, but instead of cement used to bind the paper fiber into a solid, clay is used as the binder. This can work well if the material is kept absolutely dry; otherwise it will become soft and could deform.

❖ **Lightweight concrete** Lightweight concrete, weighing from 35 to 115 pound per cubic foot, has been used in the United States for more than 50 years. The compressive strength is not as great as ordinary concrete, but it weathers just as well. Among its advantages are less need for structural steel reinforcement, smaller foundation requirements, better fire resistance and most importantly, the fact that it can serve as an insulation material. It can cost more that sand and gravel concrete, and it may shrink more upon drying. Lightweight concrete may be made by using lightweight aggregates, or by the use of foaming agents, such as aluminum powder, which generates gas while the concrete is still plastic. Natural lightweight aggregates include pumice, scoria, volcanic cinders, tuff, and diatomite. Lightweight aggregate can also be produced by heating clay, shale, slate, diatomaceous shale, perlite, obsidian, and vermiculite. Industrial cinders and blast-furnace slag that has been specially cooled can also be used. Pumice and scoria are the most widely used of the natural lightweight aggregates. They are porous, froth-like volcanic glass, which come in various colors and are found in the Western United States. Concrete made with pumice and scoria aggregate weighs from 90 to 100 pounds per cubic foot. The rock from which perlite is manufactured has a structure resembling tiny pearls and when it is heated it expands and breaks into small expanded particles the size of sand. Concrete made with expanded perlite weighs between 50 to 80 pounds per cubic foot and is a very good insulating material.

❖ **Pumicecrete** Pumicecrete has been used in constructing buildings for many years. It is simply concrete that uses crushed volcanic rock as an aggregate rather than conventional sand and gravel. Both pumice and scoria, when used in this fashion, render a product that is much lighter than concrete. It also transforms what is usually considered a thermal mass material into something that is much more of an insulator (about R-1. 5 per inch), because of all the trapped air. This is very useful, because it makes it possible to actually build a load-bearing structure with an insulating material, as with earthbags filled with the same crushed volcanic rock. When mixing pumicecrete, the idea is to use just enough wet cement to coat the aggregate so it will adhere to the surrounding particles. Too much cement will defeat the purpose of maintaining all of that trapped air; about three bags of portland cement per cubic yard of aggregate is recommended. Once the material has set up a bit, the surface can be washed to expose the natural color of the stone. The rough texture of pumicecrete is ideal for adhering to further plasters that might be used. Pumicecrete is best placed on an ordinary concrete foundation, and most applications require a cement bond beam at the top of the wall, for structural strength and to tie the roof structure to. Entire domes of pumicecrete have been successfully constructed. A wall thickness of at least 14 inches is advised, with thicker walls providing more stability and insulation.

❖ **Cellular Lightweight Concrete** (CLC) Extensive research has been undertaken in the use of industrial waste consisting of fly ash from power plants as a raw material for manufacturing building materials. The large volume of waste has become one of the most significant problems of environmental protection, as its disposal is expensive and non-productive. Experiments show that this waste material can be used for the production of high quality bricks, blocks and other building elements which are less energy intensive than their conventional counterparts. This research has yielded patented technology for the production of concrete-like blocks based on oil-shale and coal fly-ash, cured under normal atmospheric conditions.

A particularly interesting material that has been developed is ash-based cellular concrete, which in addition to being based on industrial waste, is also manufactured through a low-energy process. The manufacture of conventional cellular concrete of comparable properties requires very high energy input.

This material has been used in over 40 countries over the past 25 years to build residential and commercial buildings. It is an air-cured lightweight concrete that can be produced at the project site, using standard concrete equipment and molds. A typical mix for making blocks is:

Portland Cement.......... 190kg
Sand..................... 430kg
Fly Ash.................. 309kg
Water.................... 250kg
plus a foaming agent

❖ **Perlite and vermiculite concrete** This type of lightweight concrete has a long history of industrial and construction uses; it can be highly insulating, and is especially useful where its light weight is an advantage, such as on roof structures.

❖ **PSP** PSP refers to a method of building developed by Mike Oehler, who wrote *The Fifty Dollar and Up Underground House Book*. Mike definitely has some ideas worth relating, although they won't appeal to everyone. He did indeed build a fifty dollar house that he is still living in. Much of the savings that Mike has been able to attain is through a combination of using recycled materials, logs harvested from his own land, and a simple method of building underground.

PSP stands for Post/Shoring/Polyethylene. The framework of the building is created with posts that are preserved in various ways and planted in the earth. These posts serve to support both the walls and the ceiling. The space between the posts is planked with used dimensional lumber, such as from wood pallets. This is what he calls the "shoring". Then the whole thing is wrapped in polyethylene plastic before it is backfilled with earth, making a truly underground home. Instead of conventional flooring, Mike advocates using the existing earth, finely raked and smoothed, and then carpeting thrown over it. He has come up with a system for designing rooms that can provide daylight and proper drainage for wide variety of arrangements. His thinking goes way beyond what most underground architects have accomplished.

Manufactured Systems

There are quite a few systems of building that are based on special manufactured products. These systems could be considered "alternative" since they do not represent typical, mainstream building techniques. Green building systems must interested in promoting ways of building that utilize natural materials that require little industrial processing and transportation, it is recognized that not everybody wants a house that is made of dirt, straw, field stones, curved tree parts, etc. It is a fact that many of the manufactured systems, once the house has been created, provide an excellent, durable, energy-efficient domicile. Even though the house may be constructed of considerable steel, cement, synthetic insulation, or other materials that are high in "embodied energy," over the lifetime of the house the energy saved in comparison to more conventional construction may be considerable. Hence following are some manufactured systems. From any of these manufacturers for including them here can be used to get good green buildings.

❖ Autoclaved Aerated Concrete(ACC)

In a way it is like man-made pumice, in that it is masonry, but is still rather lightweight. It is highly insulating and completely fireproof. If one held a one-inch thick piece of it in his or her hand while a blow torch heated the opposite side, and they could barely feel it get warm! It can be cut with any carbide-tipped tool into very precise shapes, then the blocks can be "glued" together with ordinary masonry tile products. Precut blocks are available that have channels in them for standard rebar bond beam applications. This material has been used for buildings in Europe for a couple of decades, but is just now being tried in the United States. It is fairly expensive, but I would expect it to be extremely durable and create well-insulated and attractive structures. Interior and exterior finishes can be spayed on to provide a uniform, durable and colored surface. It has occurred to me that this stuff could be precut into "kit houses" that were architecturally intricate and beautiful, and would just need to be assembled by numbers.

❖ Insulated Concrete Forms(ICFs)

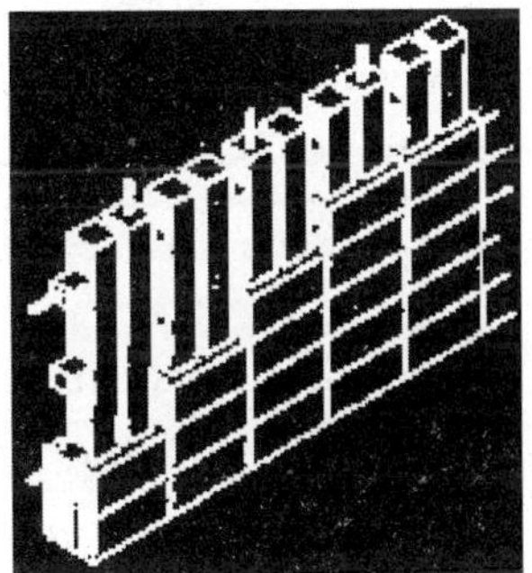

Fig. 1 Farmwork and concreting by ICF method

This describes a whole class of systems that employ modular insulated forms into which rebar is threaded and concrete is poured to create an insulated monolithic wall structure. The forms are generally very fast and easy to assemble, the concrete can be pumped into the forms virtually any time of the year because the cement is insulated as is cures, and the result is a very strong, durable, insect and fire proof, and fairly well-insulated wall.

The various modules are assembled more or less like Lego toys, where one units locks into the others around it in some way. Mostly they are made out of Expanded Polystyrene (EPS) reinforced with steel or plastic ties. Some are made of EPS that has been recycled and then glued back together with cement. Some of them are made out of mineralized sawdust or wood chips bound with cement. Most of these are in block form, but some are available as panels or planks. Many of them are billed as "green" or environmentally friendly, either because of the recycled content or because of the thermal properties of the wall. They definitely reduce the use of milled lumber in building a house, which is good.

Main objection to these approaches to building, besides the fact that they utilize massive amounts of manufactured insulation, steel and cement, is that the structure of the wall itself does not provide optimal placement of insulation and thermal mass. Ideally we want most of the thermal mass of the building to be on the inside where it can absorb the heat provided, and stabilize the interior temperature. This mass should then be well insulated from the outside air. With most ICF systems the mass (concrete and steel) is embedded within the insulation in the finished wall, where it helps stabilize the fluctuations of temperature somewhat, but also does bleed energy to the outside. A couple of inches of Styrofoam is not sufficient insulation for the thermal mass from the outside, and the insulation of the inside prevents the mass from doing its job efficiently. Reports of the the overall insulation of these ICF systems vary widely, but typical R values are right around R-20. When compared to the R-40 attained with straw bale or earthbag/volcanic rock walls, this is not so great.

❖ **Thermal wall forming system** This system for forming concrete is a hybrid of standard forms and ICF's. It offers the tremendous advantage of placing the concrete mass on the inside of the building shell, where it can do the work of moderating temperatures effectively. The outside of the form is composed of 4 inches of EPS insulation, with ties that remain to attach to the exterior siding. The company that manufactures this system is in Canada. For more information.

❖ **Concrete Form with Insulation**(CFI)

This unique wall system, manufactured by allwallsystem. CFIs are forms for poured concrete walls that stay in place as a permanent part of the wall assembly. The forms consist of concrete board on the inside and out held together by galvanized steel studs every 7.25 inches with 4 " of rigid insulating material held by 1" spacers in the center. The forms have window and door openings pre-cut as well as electrical conduit and boxes installed. Red iron steel (rebar) runs vertically from every slab every 4' and horizontally in the tie beam area, then the form is braced and poured full of concrete. The poured concrete creates a "post and beam system" including a continuous tie beam along the top, as well as walls 1 " thick solid concrete next to each concrete board surface.

Once the forms are poured, the walls are ready for conventional drywall finishing on the inside, and any conventional finish such as stucco or siding. Electrical conduit boxes are already in place for the electrician as well. This unique combination of steel studs, cement board, rebar and poured concrete creates a wall of great strength, reasonable insulation (R-20), low air infiltration and good thermal mass.

❖ **Shotcrete systems**

Shotcrete, sometimes called ferrocement, is sprayed concrete. There are both wet and dry methods of doing this. The aggregate must be quite fine, more like mortar to be able to do this, and usually the percentage of Portland cement to sand is quite high to create a very strong product. A distinct advantage of systems that employ shotcrete is that the concrete shell of the building can be exposed to the inside, where its thermal mass can be

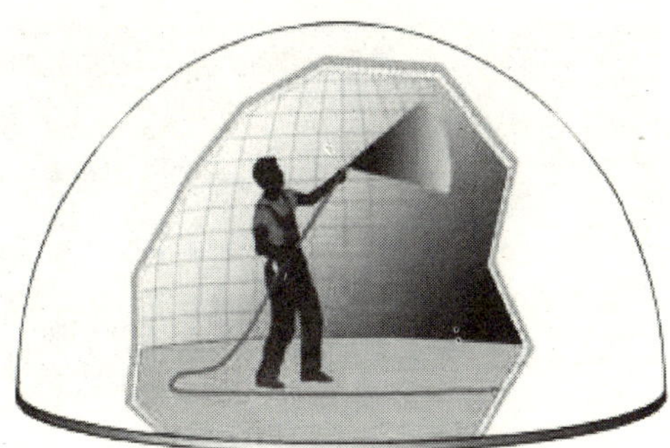

Fig. 2 Dome construction

fully utilized. Usually some form of steel mesh is provided to give tensile strength to the wall, which is then insulated in various ways from the atmosphere. Monolithic Domes One of the more imaginative systems for using shotcrete is Monolithic Domes' approach. They have pioneered the use of an inflated balloon-like structure which acts as a form for the eventual building. They actually have two different ways of doing this: one retains the fabric of the form as a part of the exterior of the building, and the other allows the form to be reused several times for other buildings. In the first instance, polyurethane foam insulation is spayed on the inside of the form and then allowed to set up. Then rebar is assembled beneath this to conform with the shape established, and the shotcrete to sprayed onto the steel reinforcing from the inside. Once this has set up, the resulting building is about as tough as anything made by man. Domes of this sort have withstood hurricanes and earthquakes; even cannonshot would not knock one of these down!

The second of Monolthic's methods of building uses a similar inflated balloon, but the rebar is tied outside of it and the shotcrete is applied from the outside. Once this has set up, the fabric can be deflated and removed for reuse. In this case the polyurethane insulation would also be sprayed on from the outside, with some sort of protective covering for durability. It relies on a great deal of machinery working optimally, and is not particularly pleasant to be around. There is the blower that keeps the balloon inflated, the urethane foam spraying equipment, the compressor to run the shotcrete machine, the cement mixer and equipment to keep it fed with materials, lights to be able to see what is going on, masks to keep from breathing all of the fumes and particulates in the air, etc. This requires a well organized crew of workers who are trained in their special tasks.

❖ **Structural Insulated Panels**(SIPs)

There many kinds of structural insulated panel systems, including the above mentioned vertical shotcrete ones. Instead of cement used as a skin over the insulation, other systems mostly use oriented strand board (OSB) onto which the finished surfaces can be attached. The insulating core is typically EPS or polyurethane foam, but SIP's manufactured with MEPS is preferable because the process does not emit ozone-depleting gas. The advantages of these systems are that they are air tight, well insulating

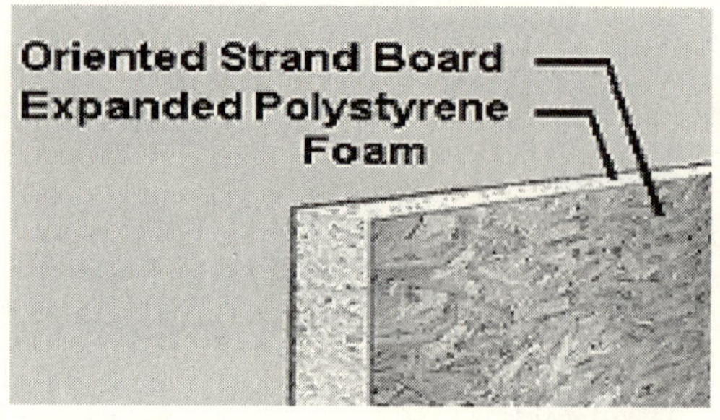

Fig. 3 Panel detail in SIP

(depending on the type and thickness of the insulation) and easy to install. Also the panels can be used as insulated roof components. It is estimated that a SIP building might use 30% less wood than a conventionally framed one.

One disadvantage of these wood-clad panels is that they don't provide any thermal mass as part of their composition. While SIPs with EPS cores will burn, it has been demonstrated that they remain structurally sound for a lengthy period during a fire and do not emit fumes any more hazardous than those of wood products. Another possible concern is with insects or rodents nesting in the insulation since this can be an ideal habitat for them.

❖ **Light-guage Steel** Light-gauge steel is being used more often these days to replace wood studs for framing a house. The cost is comparable to wood, and much of the steel is recycled. Steel is actually stronger and lighter than wood, and is termite and fire proof. Also the use of steel conserves our forest resources.

Fig. 4 Light gauge steel structure

❖ **CONICS** is a radical roof system design that reaches for the ideal of a minimal architectural form that is both functional and inspirational. Made entirely of a 1/2 inch plywood skin, CONICS mimic the natural landforms of ancient mountain ranges which approximate an interlocking network of regular and inverted cone segments. Built to last, these mountain ranges have evolved into this cone-based geometry because it maximizes resistance to earthquakes, wind, water, ice, and other forces of nature. CONICS borrows this geometry to create a minimal system where the plywood acts as structure, sheathing and roofing. CONICS are made entirely of tiled 4'x4' plywood sheets that are simply attached (no pre-bending required), The resulting architectural form is both concave and convex, creating interior spaces that undulate in ways no other architecture has achieved. CONICS can be constructed with unskilled labor very quickly and the cost for CONICS is as low as $ 2.00 (US) per square foot (materials only).

❖ **Cement** The use of slag cement and fly ash can enhance the quality and reduce the environmental impact of the use of cement.

❖ **Renewable Energy** solar electric panels, hydroelectric systems, wind turbines, potable power systems, backup, generators, invertors, power panels and charge controllers etc. shall be used in green building.

PLANS

PLAN 1 Sun Earth House Paul Shippee, Designer

This earth-covered, passive solar, 1863 sq ft house achieved the very best performance in a HUD - sponsored energy survey when it was first built in the late 1970's. For three years after the house was built, it was computer - monitored, and according to the National Solar Data Network, the SunEarth house out - performed hundreds of passive solar houses in the country. The house was designed, built, and marketed by Colorado Sunworks. The furnace was put to rest during its first winter because the pilot light was wasting natural gas. All of the space heating demand is supplied by the passive solar system.

Fig. 5 Sun earth house

The heating system is a direct gain, passive solar system and drum wall. The south side exposes 300 square feet of glass windows. The windows are two panes of insulated glass that run floor to roof. Behind the windows are 54 large, vertically stacked barrels. Sunlight directly warms these 55-gallon oil drums that are painted with a flat black finish and filled with water. During the day, the water and interior concrete walls inside the house absorb the sun's heat. The heat is released slowly after the sun sets.

At night, when temperatures begin to drop, a blower pushes polystyrene beads between the two panes of glass, providing insulation to keep the daytime heat inside the home. On winter nights, this moveable Beadwall insulation converts the large window areas to R20 heat loss barrier. Six vertical skylights are arranged on the north side of the earth roof. Maximum solar energy takes place during the winter, and minimum solar energy occurs in the summer.

Winter Mode

Stored solar heat is released from the water containers as needed. Heat flows naturally by low temperature radiation and by warm air convection to the north side of the house, thus balancing comfort zones throughout the living space.

Summer Mode

Interior thermal mass is cooled down on summer nights by providing a natural air now path. Daily heat gains aree rejected by this method through turbine roof ventilators. The cooled massive house is then closed up on hot summer days for comfortable living.

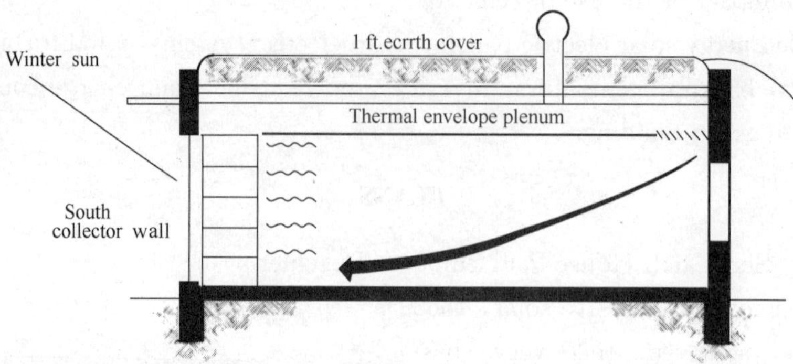

Fig. 6 Winter mode

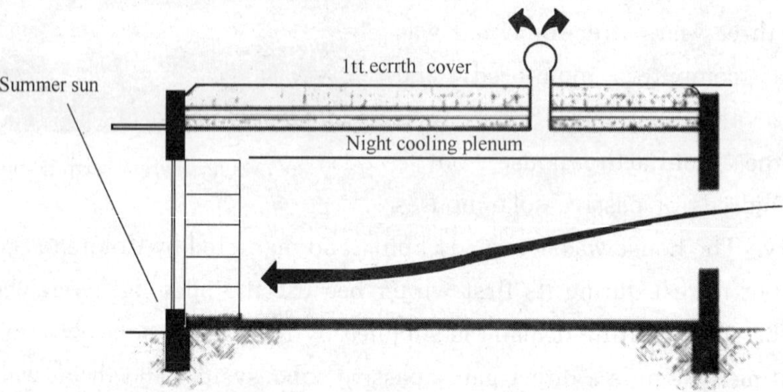

Fig. 7 Summer mode

PLAN 2 Desert Barn Joe Prudden, Architect

Fig. 8 Energy efficient house

Desert Barn uses energy efficient design principles and construction. The building plan allows for the maximization of floor area, with minimized wasted space. The design incorporates sustainable materials for their recycled content, recyclability, and their manufacturing process. Chemically stable materials and construction techniques ensure a healthy indoor environment. Passive design for heating and cooling, and the careful placement of living spaces, assure comfort through all seasons. Building material and massing integrates the building into the landscape and terrain. **The roof is designed to catch**

rain water to store in a cistern for later use. The building functions as a holistic working unit.

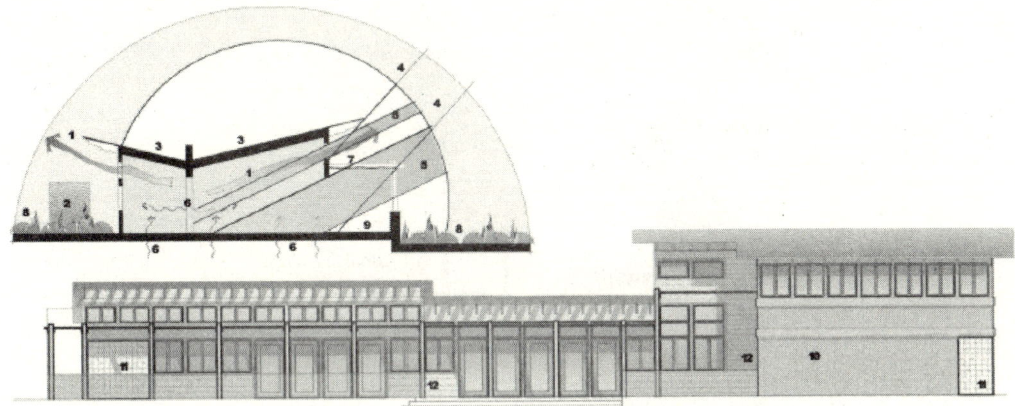

Fig. 9 Detail of house

1. natural ventilation (cooling through convection) 2. rain water storage 3. rain water collection 4. summer sun shading 5. winter access (sun is allowed to heat the home) 6. concrete walls and floors 7. retractable awning 8. natural landscape 9. outdoor living 10. corrugated metal 11. opaque wall system 12. recycled natural wood siding

PLAN 3 Sunworks Greenhouse Paul Shippee, Designer

Fig. 10 Green house

This attached solar greenhouse is designed for the Rocky Mountain (8000 Degree Day) climate. It provides about 100000 BTUs per day to help heat the house, as well as the 300 sq. ft. of floor space for growing plants. It is a 2×6 wood frame structure with a large glass facade (210 sq. ft.) aimed at the winter sun. The glass is insulated at night by a Beadwall system to help maintain minimum air temperatures of 55F degrees. A small fan moves hot air out of the greenhouse and into the main house delivering heat and humidity on clear winter days. The fan is necessary to prevent temperatures in the greenhouse from climbing over 90F at mid-day. Barrels of water stacked two high along the north wall for thermal storage rise about 15F degrees on a sunny day. They store heat for night time use, preventing daytime overheating.

Glass is tilted 58 degrees to the horizontal to gain maximum solar aperture at this site. It has to be shaded by exterior curtains is the summer. The 350 cfm fan brings heat in winter to the house, dropping 15F before it returns to the greenhouse through an existing door. Summer ventilation is by roof ventilator and by cross-breeze through doors.

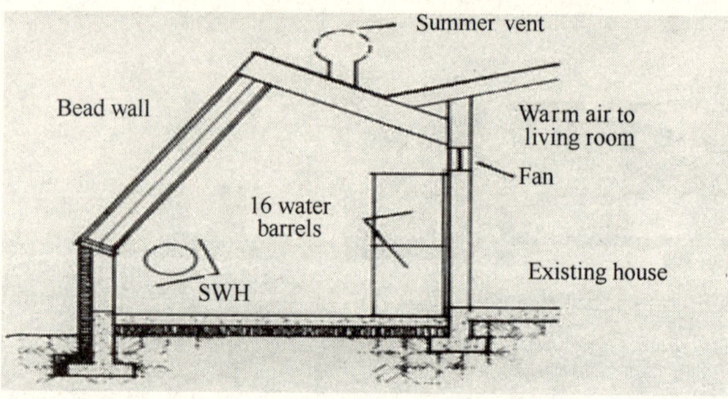

Fig. 11 Section of house

RELIABILITY DESIGN FOR BUILDING SERVICES

Shaomin Wu Derek Clements-Croome
(School of Construction Management and Engineering,
The University of Reading, RG6 6AW, UK.)

Abstract: The need to place reliability and cost at the forefront of design for building services systems is becoming increasingly important as failures, inadequate maintenance policies and support issues directly and indirectly affect their performance. This paper presents approaches to integrating reliability design into a through life business model (TLBM) for building services systems. After investigating each phase in the whole life time span of systems, we introduce a method on how reliability design can be applied in those phases. We then analyze the current approaches adopted in UK construction industry based on questionnaires designed to investigate the usage of reliability in the industry. Suggestions are made based on TLBM and the analysis of the questionnaire.

Keywords: *Building services systems*, *through life business model*, *maintenance policy*

1 Introduction

Building services encompass mechanical, electrical, security, information and communication systems. The services are installed to support the functions required so it is essential that they can be carried out without interruption, meeting the user needs, providing reliability and cost effectiveness throughout the whole life cycle. Since building services systems are complex and encompass many different kinds of components, the ability of the systems to continually perform is of vitally importance.

Reliability is an important aspect of a building services system. Reliability is the ability of an item to perform a required function under given conditions for a given time interval. If an unreliable system does not perform a required function, the business loss can be huge. For example, based on research in office building services systems, Wu and Derek (1998) have identified the relationships among the principal costs incurred within a system's whole lifecycle: initial capital costs; maintenance and building operating costs; and business operating costs. They found that the operation and maintenance of the building will cost eight times the capital costs; and the business operating costs will be two hundred times the capital costs over the life of the building. In other words, business losses are forty times as large as maintenance costs. Rigorous reliability design and maintainability planning can ensure system's performance and therefore increase business value.

From a long-run time viewpoint, building services systems have several different phases(John et al., 2003): client requirements, design, installation, commissioning, operations, maintenance and disposal phases. At each phase, there are elements related to reliability. Suitable approaches to integrating reliability design into each phase can achieve optimal and durable life cycle cost (LCC) that represents the summation of the costs at each stage of the progress. Searching for a point in which combines optimal lifecycle cost and reliability is very important. For example, reliability, capital costs and maintenance costs vary with the different choices of components or subsystems. There is a trade-off among these factors. Components with high reliability are often more costly to buy but cheaper to maintain, whereas cheaper components are often less reliable but more expensive to maintain. There is also the influence of system configuration on overall reliability.

There is little research on the integration of reliability design and lifecycle costing. After analyzing each phase in building service systems' whole life time span, a methodology is proposed here to integrate reliability design into each phase of systems' lifecycle. In order to investigate the status of the application of reliability deign in the construction industry in UK, a questionnaire has been designed to increase management awareness of the status of reliability design in the whole life cycle costing. We analyze the responses to the questionnaire and make recommendations.

The second section first review the state-of-the art of research on reliability in building service systems, and then introduces approaches to integrate reliability design with lifecycle costing. The third section analyzes the questionnaire. The last section summarizes the achievements from this paper.

2 Reliability design for building service systems

2.1 *Research on Reliability of Building Services*

Knowledge about reliability as related to building services systems is mainly based on mechanical systems, electrical systems, and fire and hydraulic systems.

In mechanical systems, Staahl and Wallace (1995) compared life-cycle cost between traditional cooling systems and natural convection based systems. Neumann and Fenton (1993) selected compression equipment from industrial refrigeration systems based on optimisation of life cycle costs; Doi (1996) and Yamada et al. (2001) analysed reliability for cooling systems. Mohamed et al. (1998), Molodezhnikova et al. (2001) and Ponta et al. (1999) introduced some approaches to reliability analysis for heating systems but only relevant to manufacturers. Taylor et al. (1993) presented some research on reliability, maintainability and availability based on their impact on the electrical power, heating, ventilation, and air conditioning of the facility using functional availabilities. Software tools for evaluating ventilation reliability were introduced by Kronvall (1998) and Tombuyses and Arien (1996). Data modelling for ventilating systems and infusion pump systems have been conducted by Baker(2001).

In electrical systems, Kay (1996) and Li and He (2000) conducted research on reliability of electrical systems Baker (2001) developed models based on data from electrical monitors. Little has been studied in relation to building services electrical systems.

In fire and hydraulic systems, Jones and Bukowski (2001) provide a review of operational reliability and performance estimates for fire detection, automatic suppression and, to a limited extent, construction compartments. Risk analysis on fire systems has been researched by many authors, typically Magnusson et al. (Magnusson et al., 1997) and Frantzich (1998).

All the above-mentioned research has in some way dealt with reliability of services systems but in isolation from the business functions that they are supporting. There is even less research investigating the way to integrate reliability design into each phase of the whole life time span. From inception to disposal, a building services system may go through several cost-associated stages of LCC. Purchase cost is widely adopted as the principal (and sometimes only) criterion for equipment selection. This single criterion is simple to use but often results in poor performance resulting in a waste of money. The capital cost of some equipment is easy to afford, however, owing to costs with respect to operation, maintenance and disposal; the extremely complex capital cost of a system is normally less than its LCC. Depending on the equipment type, the ownership cost over the life cycle span may vary from ten to one hundred times the procurement cost (Dhillon, 1989). To make better financial decisions, therefore, LCC should be taken into consideration and carefully analysed at each stage of the equipment's life from cradle to grave.

2.2 *Reliability design and maintainability design in the whole lifecycle*

According to John et al (2003), future building services systems need to identify and address the cost drivers proactively. A Through Life Business Model (TLBM) for building services is therefore introduced. The TLBM is a proposed framework under which synthesis and integration of integrated logistic support concepts can be achieved. The TLBM is divided into five phases and each phase is further divided into sub-levels. The first level with five phases is as follows: client requirements; briefing; design; installation and commissioning; operations and maintenance; and disposal/recycle/reuse.

Reliability impacts on all of the above six phases and should take into account human, environmental and economic issue. We analyze the impact of reliability on the phases as follows.

(1) System reliability requirements should be introduced into the client requirements and hence specified in the brief.

(2) Reliability design and maintainability design need to be performed as part of design process.

(3) The installation and commission phase involves initial logistic support analysis covering spare parts, test and support equipment. Designers, contractors, commissioning agents, facilities managers and manufacturers are all involved at this stage.

(4) During the operations and maintenance phase, the maintenance policy, spares analysis, logistic support, software maintenance are influenced by reliability. Information such as mean time between failure (MTBF), mean time between maintenance (MTBM), mean downtime (MDT) and related costs should be collected.

(5) Disposal, reuse and recycle are influenced by reliability. Residual life of the system may be analyzed in order to reuse.

Although reliability may be analyzed at all of the phases in TLBM, reliability design and maintainability design are the most important aspects that need considering. These two aspects may interact each other, and they should be conducted from three levels: environmental factors, system structure and item's reliability properties.

(1) Reliability Design: In literature, reliability design also is referred as reliability optimization(Kuo and Prasad, 2000).

a. Environmental factors: Health and Safety Analysis (HSA), and Environmental Effects Analysis (EEA) should be carried out. HSA reviews any aspects from a health and safety perspective, and EEA focuses on the possible effect of the system from and /or on the environment. Since reliability of the system may be impacted by the environmental factors, stress-strength analysis may therefore need taking into consideration. The major categories of stress are electrical, thermal, mechanical and chemical. Two design approaches can be applied: selecting components with sufficient strength against maximum load and protecting part against excessive stresses. Second, because unreliable systems/subsystems may influence on environment at disposal time, system reliability should hence be considered during the design time in order to reduce negative environmental effects;

b. System structure: Reliability Block Diagrams Analysis (RBDA) and Reliability Allocation should be conducted in the system structure level. The purpose of RBDA aims to review the system structure, to calculate system reliability and to identify weak points using importance measures and improve the system. Reliability allocation should be conducted based on the RBDA in order to increase system's reliability by adding suitable redundancy. One must note that redundancy has its advantages and disadvantages. Its main advantage is the quickest way to improve reliability. Its disadvantage may involves that sensors and switching may increase cost and reduce reliability, it may exceed size, weight or power constraints, it may also increase maintainability requirements;

c. Item Reliability Properties. Based on RBDA and item's intrinsic reliability properties,

Failure Modes and Effects Critical Analysis (FMECA) and Fault Tree Analysis should be performed here. The purpose of FMEA is to identify possible failure modes of each critical subsystems/components, possibly relative frequency of faults occurrence, suitable preventive maintenance and corrective actions concerning accident. FTA analyzes the effects of the sub-systems/components on the performance of the overall system.

(2) Maintainability design. Maintainability is the probability that the system that is DOWN will be returned to the UP mode within a given time and according to given procedures. Based on RBDA, FMECA, FTA and information provided by equipment's suppliers, maintainability evaluation, corrective/preventive maintenance tasks analysis and support elements definition should be conducted. Detailedly, the following aspects are important.

a. Maintenance Policy Development: Maintenance for building services systems costs money but preserves value. Down time is a major obstruction in the improvement of productivity. The basic maintenance methods are broken down into two different policies, corrective maintenance policy and preventive maintenance policy. Preventive maintenance includes periodic maintenance, planned maintenance and predictive maintenance. The goal of maintenance is to reduce building services systems' down time by providing a "heads up" on pending equipment failures and improve the recording of equipment maintenance data to provide a quantitative basis to improve maintenance strategies, more cost effective spares inventories, and a better understanding of maintenance costs and opportunities to reduce those costs.

b. Spare Analysis: Maintaining an adequate spares inventory, while crucial for minimising equipment down time, does involve costs for storing idle stock. There is the potential to develop algorithms to optimise spares inventories, which may result in the greatest reduction in equipment down time for the lowest cost. Optimal spares inventory analysis identifies the optimal set of components to hold in inventory to minimise inventory costs while maximising system equipment availability. Optimal spares analysis helps to identify the right parts (and number of parts) to hold for repair, based on the data on system and component failures. The spares analysis, in conjunction with maintenance cost analysis, helps identify the best maintenance strategies for availability improvement.

c. Labour Needs Analysis. Maintenance personnel are responsible for any maintenance for the building services systems.

In practice, Standards (Standard, 1992, Standard, 1994, Standard, 1993) have presented helpful information on reliability design and maintainability design.

3 Analysis of questionnaire

In order to understand the state of the art of the application of reliability and maintainability design in building services systems, we design a questionnaire that consists of 17 main questions. These questions are categorized three classes: company information, reliability design and maintenance policy selection.

Based on company information for our industrial partner dataset, 216 questionnaires were sent by surface mail. Out of the questionnaires, the total number of respondents was 26. This means that the response rate is only 13%. As Mitchell and Jolley[25] warned, even if one starts with an unbiased sample, by the end of the study the sample may become biased because people often fail or refuse to respond to a questionnaire. They showed that a typical mail survey response rate might reach only 10 percent. We therefore regard the response rates is acceptable.

We analyze some of questions from the questionnaires as follows.

3.1 Company information

In the first two questions, we asked the age and size of the company. Based on the survey results, one of the 28 companies has been in business between 11 and 20 years, the rest companies have been in business more than 21 years. Of all the companies, seventeen companies employ more than 500 staffs.

Question 3 asks company's main business. Some of companies may involve more than two areas. As shown in Figure 1, of the total responses, 17% of them engage "building services design". However, all of them should consider reliability important in their business.

As shown in Figure 2 based on question 4, of the total responses, 24% employ whole life support staffs for maintenance. 19% employ staff in charge of safety problems.

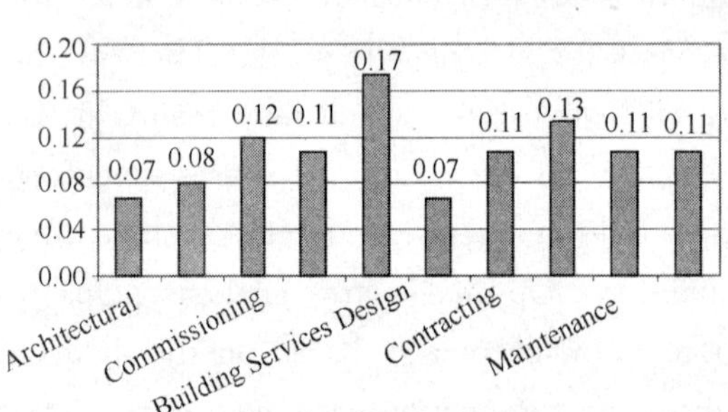

Figure 1 What kind of services does your organisation provide?

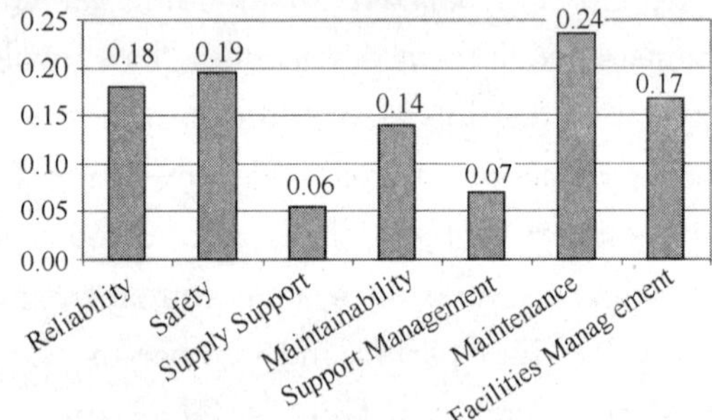

Figure 2 which types of whole life support staff are employed?

3.2 Reliability/Maintainability Design

In the question "*Do you take Life Cycle Costing into account when designing reliability for systems?*", only 7% of the responses answer that they never do that, whereas 48% always or frequently take life cycle costing into account when design reliability for systems.

When being asked '*Does your organisation use any standard for design for reliability/maintainability?*', 70% of responses design their systems using standards (see

Figure 4). Standards such as British Standards[26], USA Military Standard[27] are guides for reliability and/or maintainability design.

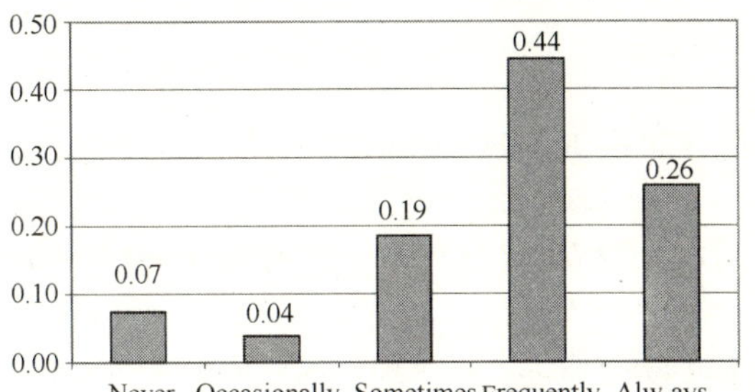

Figure 3 Do you take Life Cycle Costing into account when designing reliability/maintainability for systems?

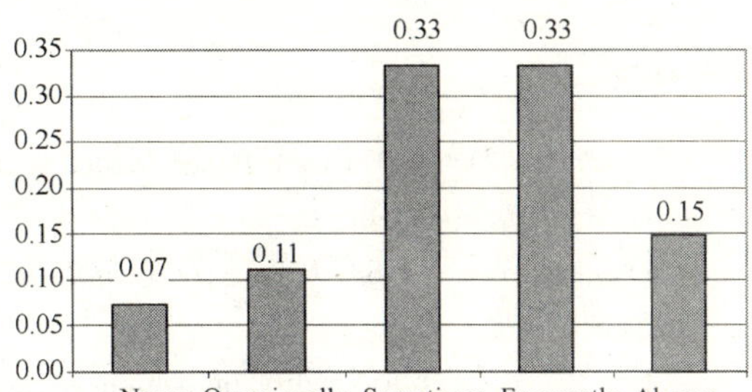

Figure 4 Does your organisation use any standard for Design for Reliability/maintainability?

As shown in Section 2, reliability should be considered from the '*Client Brief*' and '*Design*' phase of the TLBM. Maintenance personnel should therefore be involved in the very early phases. According the question '*Which phase is maintenance personnel involved in for Design for Reliability*? 8 of the 28 companies never or occasionally have their maintenance personnel involved during design review (see Table 1).

Which phase is maintenance personnel involved in for Design for Reliability/Maintainability? Table 1

	Never	Occasionally	Sometimes	Frequently	Always
Design Review	4	4	4	10	4
Purchase Review	4	2	9	7	2
In Service Support	3	0	3	9	8
Others (please list)					

4 Conclusions and Further Work

This paper firstly summarised the latest progress in reliability-related topics of building service systems and then presented a methodology on integrating reliability design and maintainability into the Through Life Business Model, and it finally analyzed questionnaires designed for investigating the state of the art application of reliability design and maintainability design in UK companies.

建筑物设施的可靠性设计

Shaomin Wu Derek Clements-Croome
(School of Construction Management and Engineering,
The University of Reading, RG6 6AW, UK.)

摘 要：建筑物设施的可靠性设计正变得日益重要。本文全面寿命商业模型考虑可靠性设计，并介绍了每个阶段的可靠性设计方法，同时分析英国建筑业的可靠性设计情况。

References

[1] Baker, R. D. (2001) Data-Based Modeling of the Failure Rate of Repairable Equipment, *Lifetime Data Analysis*, 7, 65～84

[2] Dhillon, B. S. (1989) *Life cycle costing: technique, models and applications*, Gordon and Breach, New York

[3] Doi, Y. (1996) Reliability of helium refrigeration systems for the TRISTAN detector magnets, *International cryogenic engineering conference*, 165～168

[4] Wu, S, Clements-Croome, D (2005) Critical reliability issues in building services systems, The fourth international conference on quality and reliability, Beijing

[5] Frantzich, H. (1998) Risk Analysis and Fire Safety Engineering, *Fire Safety Journal*, 31, 313～330

[6] John, G., Loy, H., Clements-Croome, D., Fairey, V. and Neale, K. (2003) Through Life Support for Building Services Systems, *Worldwide CIBSE/ASHRAE Gathering of the Building Services Industry*: 24～26 *September* 2003

[7] Jones, W. and Bukowski, R. (2001) Using High Reliability Detection for Fire Service Response in Buildings, *Automatic fire detection*, 574～591

[8] Kay, G. N. (1996) In *Preservation and Rehabilitation*, Vol. 1258 (Ed, Kelley, S. J.) American Technical Publishers Ltd, pp. 248～263

[9] Kronvall, J. (1998) In *Ventilation Technologies in Urban Areas*, Vol. Issue 19 Oscar Faber Plc, pp. 168～170

[10] Kuo, W. and Prasad, V. (2000) An annotated overview of system reliability optimization, *IEEE Transactions on Reliability*, 49, 176～187

[11] Li, J. and He, J. (2000) Seismic reliability analysis of electric power network system, *Earthquake engineering; Earthquake engineering frontiers in the New Millennium*, 239～242

[12] Magnusson, S. E., Frantzich, H. and Lundin, J. (1997) Risk Analysis-An Introduction, *COST Action E5 workshop on fire safety of medium-rise timber frame residential buildings*, *Espoo, Finland*, *June* 2～3, 1997, 44～58

[13] Mitchell, M. L. and Jolley, J. M. (1996) *Research design explained*, Harcourt Brace College Publishers, Fort Worth

[14] Mohamed, A., Lemaire, M., Mitteau, J. C. and Meister, E. (1998) Finite element and reliability: a method for compound variables-application on a cracked heating system, *Nuclear Engineering and Design*, 185, 185～202

[15] Molodezhnikova, L. I., Zhilkin, E. A. and Bairakova, O. Y. (2001) Effectiveness and Reliability Increase of Heating System, *International scientific and practical conference of students, post-graduates and young scientists; Modern techniques and technology MTT'* 2001, 32

[16] Neumann, S. B. and Fenton, D. L. (1993) Life-Cycle Cost Analysis Applied to Selection of Compression Equipment for Industrial Refrigeration Systems, *Transactions- American Society of Heating Refrigerating and Air Conditioning Engineers*, 98, 148

[17] Ponta, A., Tripodi, C. and Bertocci, S. (1999) Reliability analysis of Torino Sud district heating

system, *Proceedings of the Annual Reliability and Maintainability Symposium*, 336～342

[18] Staahl, L. and Wallace, M. H. (1995) Life cycle cost comparison-traditional cooling systems vs. natural convection based systems, *Intelec*, 11～3

[19] Standard, B. (1992), Reliability of Systems, equipment and components- Part 14: Guide to formal design review, BS 5760～14

[20] Standard, B. (1994), Reliability of systems, equipment and components-Part 11: Collection of reliability, availability, maintainability and maintenance support data from the field, UK

[21] Standard, U. D. (1993), Reliability and Maintainability MOD Guide to Practices and Procedures, 00-41/Issue 3

[22] Taylor, R. A., Dayton, R. M., Lam, H. S. and Sweetra, C. M. (1993) Reliability, Maintainability, Availability Study on Impacts of Facility Electrical Power and Heating, Ventilation, and Air Conditioning on NAS Functional Availabilities, *Meeting the challenges of change*, 543～565

[23] Tombuyses, B. and Arien, B. (1996) Reliability Analysis of a Ventilation System with the CAMERA Software, *ESREL: Probabilistic safety assessment and management* '96; *ESREL*'96-*PSAM-III*; *June* 24-28, *Crete*, *Greece*, 1235～1240

[24] Yamada, O., Sawada, Y., Harada, M., Yokozuka, T., Yasukawa, A., Moriya, H., Saito, N., Kasai, K., Uda, T. and Netsu, T. (2001) Improvement of the Reliability of the C4 for Ultra-High Thermal Conduction Module with the Direct Solder-Attached Cooling System (DiSAC), *Electronic components and technology conference*, 1144～1148

LEED FOR HOMES: AN INTRODUCTION TO THE PILOT PROGRAM OF THE U. S. GREEN BUILDING COUNCIL'S NEWEST RATING SYSTEM

Ann V. Edminster M. Arch.
(owner, Design AVEnues)
and Jay Hall Ph. D.
(Building Knowledge, Inc.)

Abstract: Currently in "pilot" operation across the U. S., LEED® (Leadership in Energy and Environmental Design) for Homes is being developed to reward new, high-performing homes, which offer several benefits: healthier indoor air, energy and water efficiency, improved durability, environmentally preferable materials, reduced stormwater run-off, and climate-appropriate landscaping-at reduced net monthly cost and higher market value.

The LEED for Homes rating system and delivery infrastructure were designed to target mainstream U. S. builders of single-family detached homes, duplexes, row houses, townhomes, and multifamily structures up to three or four stories. Further analysis will determine whether it may also apply to high-rise housing. The program had to be affordable, easy to use, and provide high value to home builders and buyers while distinguishing green building leaders. Rating criteria were formulated to represent "best practices" in each LEED topic area.

LEED for Homes is similar in its basic architecture to other LEED systems, but in response to needs of home builders, it was built around existing successful programs. It incorporates innovative measures addressing durability planning, home size, and resource efficiencies inherent in dense, urban, multifamily housing.
Keywords: Home, Residential, High-performance, Resource-efficient

OVERVIEW

LEED® (Leadership in Energy and Environmental Design) for Homes is a new green building rating system from the U. S. Green Building Council that is being designed to help home buyers identify newly constructed homes that are both environmentally friendly and more affordable to own. A certified LEED Home-which, like all buildings certified under LEED rating systems, is third-party verified-will offer the following valuable benefits as compared to conventional US homes:

√ Healthier indoor air;

√ At least 30 percent greater energy efficiency;

√ At least 30 percent greater water efficiency;

√ Improved durability and more environmentally friendly building materials;

√ At least 30 percent less stormwater run-off ;

√ Climate-appropriate landscaping; and

√ Higher market value.

Currently in its "pilot" phase, the LEED for Homes Rating System is being created through USGBC's consensus processes by a national team of green home-building experts. The pilot program began in July 2005, operating in markets across the United States. As of early December 2005, 50+ builders in eleven test markets had committed to building and rating homes using the pilot version of LEED for Homes. The rating system is expected to be released for public comment, approved, and nationally launched in early 2007. Green building leaders from several other countries have also expressed interest in adapting LEED for Homes for use outside the U. S. , in keeping with precedent set with LEED for New Construction.

WHAT IS A GREEN HOME? A LEED HOME?

LEED for Homes is being developed to distinguish high performance homes, which are designed and built to perform substantially better than conventional U. S. homes. There are currently more than 50 local green home-building programs across the U. S. , each with a somewhat different approach. LEED for Homes will be the first national leadership standard for green home-building. Considerable modification is likely to be necessary to make LEED for Homes an effective tool for "greening" homes outside the U. S. , including recalibrating the rating criteria appropriately in reference to design and construction norms in other countries.

"Green homes" have often been perceived and portrayed as offbeat structures, an image acquired during the era of early passive solar design and reinforced, to some extent, by the "natural building" movement, which includes cob (mass adobe) cottages and straw-bale homes. However, this image is a narrow stereotype that doesn't begin to capture the scope or potential of green home-building. Most homes in the LEED for Homes pilot, in fact, are visually indistinguishable as "green homes," per se. Instead they are distinguished by intelligent design and quality construction that greatly reduce their environmental impacts as compared to their conventional counterparts.

WHAT ARE THE BENEFITS?

The goal of a green home is to step more lightly on Mother Earth. The LEED for Homes pilot rating system therefore rewards a wide range of features designed to reduce the "environmental footprint" of the home-to better steward the land on which it is built, save energy and water, make efficient and responsible use of materials, and provide a healthy indoor environment. It also rewards builders who are constructing homes on "location-efficient" sites-that is, in locations that are near a variety of everyday needs and amenities and thus foster more energy- and resource-efficient lifestyles by encouraging walking and bicycling instead of the use of motor vehicles.

The tangible benefits to occupants of a LEED Home include reduced utility bills, cleaner indoor air, improved fitness (from increased walking and biking), and reduced likelihood of durability and other performance problems (e. g. , moisture intrusion, mold, uneven heating and cooling, etc.). With major increases in energy prices expected in the winter of 2005-6, most home buyers will be drawn to the highly energy-efficient home designs. For many others, the health benefits offer the greatest appeal. Asthmatics and others who suffer from environmental sensitivities may find that homes designed with improved air quality translate to better health and reduced medical expenses. And few Americans would not benefit from a more active lifestyle.

Builders of green homes enjoy market distinction, improved risk management, and often premium prices. LEED Homes have their performance verified, which means that home buyers are assured of a superior product with tangible benefits including 30 to 70 percent energy savings, increased value, and healthier living.

WHAT'S THE COST?

Although conventional wisdom has it that a green home carries a significant price tag, that is not necessarily the case. As shown in Table 1, a LEED Home can deliver sufficient improvement in energy and water efficiency to more than offset the nominal increase in sticker price.

Cost of a LEED Home vs. a Conventional Home **Table 1**

	Conventional Home	LEED Home	Difference
Sticker Price	$ 300000	$ 308500	+ $ 8500
Monthly Mortgage	$ 1439	$ 1480	+ $ 41
Monthly Energy Bill	$ 180	$ 126	− $ 54
Monthly Water Bill	$ 30	$ 20	− $ 10
Net Cost $ / Month	$ 1649	$ 1626	− $ 23
$ / Day			− $ 0.76

BUT IS IT *REALLY* GREEN?

The LEED for Homes pilot rating system has been crafted as a "leadership standard," intended to recognize roughly the 25 percent of builders who employ the greenest building practices.

The U. S. green building industry, though diverse and rapidly evolving, is pretty well in agreement about the basics. Where performance thresholds are set, however, and which are the *right* green features are more subject to debate. Thus when USGBC set out to create LEED for Homes, the Council assembled a committee of home-building and other technical experts from around the U. S to take on the task of reaching consensus decisions about the issues.

LEED for Homes also is built on experience of many innovative and successful local, regional, and national programs. These include the Austin Green Builder Program (the first in the U. S.), Built Green Colorado, Vermont Builds Greener, EarthCraft House in Atlanta, EPA's Energy Star Homes and DOE's Building America program, among others. The pilot rating system has undergone numerous drafts and review cycles with builders, the LEED Homes Committee, and additional technical experts.

The resulting LEED for Homes pilot rating system is the product of a consensus among numerous contributing organizations and individuals who collectively represent a number of large and technically savvy constituencies. This substantive strength is reinforced by the delivery mechanism, which employs third-party, on-site verifiers to ensure that LEED Homes are built to perform as intended, and an auditing procedure by USGBC to further back up the credibility of the builders' claims regarding the "greenness" of the homes.

WHO IS BUILDING LEED HOMES?

During the LEED for Homes pilot demonstration phase, eleven "Provider" organizations in different parts of the U. S. are working closely with selected builders of pilot LEED Homes. The USGBC may consider expanding service into new areas midway through 2006. In early 2007, following approval of the rating system by the USGBC membership, the fully chartered rating system is expected to become available nationally.

HOW DID LEED FOR HOMES COME ABOUT?

The USGBC is self-described as, "a coalition of the country's foremost leaders from across the building industry that promotes buildings that are environmentally responsible, economically profitable, and healthy places to live and work." Its LEED green building rating systems[1] are the USGBC's principal tool in its efforts to transform mainstream building practices to more sustainable models, by rewarding and recognizing those who are developing the greenest buildings in the country.

Before tackling the development of the LEED for Homes pilot rating system, the USGBC focused exclusively within the commercial construction arena. Residential construction alone, though, represents roughly half of all construction activity, as measured in dollars spent. And there are nearly two million new housing units built in the U. S. each year, about 80 percent of which are single-family homes. So home building is an important market for the USGBC to address in terms of meeting its mission and objectives.

The LEED for Homes pilot program was formulated under the same premise that has driven development of the other products in the LEED suite - that is, by offering an opportunity for builders and developers to distinguish their products from competitors' through the achievement of a credible, tiered, third-party label, USGBC is creating a market driver for higher levels of "green" performance. The program is intended to spur builders and developers to incorporate greater efficiencies in their projects.

WHAT TYPES OF HOMES ARE INCLUDED?

The scope of LEED for Homes includes individual residential (or primarily residential) structures on individual sites. These may range from single-family detached homes to duplexes, row houses, townhomes, and multifamily structures up to three or perhaps four stories. Although the rating criteria were initially drafted with single-family homes in mind, the Council - supported by a generous grant from the Home Depot Foundation-has since expanded the scope of the Rating System to include multifamily and affordable housing. Work on tailoring the rating criteria for these types of projects is occurring during the pilot, with the goal of incorporating rating criteria appropriate for multifamily and affordable housing in the version of LEED for Homes that will be presented to the USGBC membership for balloting.

For the initial phases of program development, the rating system will only address new construction; remodeling projects will be addressed in a later version of LEED for Homes or in an entirely new product, based on market needs.

The focus on new construction arose due to strategic reasons: the new construction market offers access to mainstream production home builders, who are responsible for an ever-growing segment of the home construction marketplace. Thus LEED for Homes targets a builder

---

[1] The LEED "family" includes rating systems for new commercial buildings (LEED-NC), commercial interiors projects (LEED-CI), existing building operations and maintenance efforts (LEED-EB), and speculative commercial "core and shell" projects (LEED-CS).

audience that includes a range from leading-edge quality production homebuilders, through to innovative green custom builders and affordable housing developers.

THE DEVELOPMENT PROCESS

LEED for Homes differs from its antecedents-the home-building programs upon which it was based - in a number of ways. First, of course, it is intended to be both comprehensive of the full range of green building issues and national in scope, meeting a market need for builders in areas of the country that are not served by local and regional programs (the vast majority of the U. S.).

A fundamental issue the LEED Homes Committee faced was making the rating system and its delivery infrastructure accessible, relevant, and appealing to the target audience. This meant that it had to provide compelling value to both the builders and the home buyers and it had to be user-friendly. And from USGBC's perspective, the system had to have sufficient rigor for the Council to uphold its mandate of recognizing leadership in the marketplace.

The development of the rating criteria started with a consideration of the suitability of the overall LEED structure for the home-building arena. Were the LEED categories appropriate and workable? The committee found that they were. The six existing LEED categories were used, and two new ones were added:

- Location and Linkages (new)
- Sustainable Sites
- Water Efficiency
- Indoor Environmental Quality
- Energy and Atmosphere
- Materials and Resources
- Homeowner Awareness (new)
- Innovation and Design Process

Another important early step was to investigate the range of existing local and regional green building programs and catalog the assessment measures used in those programs. This process yielded a master list of around 1400 different measures, many of which were aimed at the same basic outcomes, but approached those outcomes differently. They also represented a range in terms of rigor-some set thresholds higher or lower than others. The investigation of existing programs also examined the programs' structure-such issues as number of points, number of different evaluation measures, distribution across issues and categories, etc. The programs exhibited a very wide range across these parameters; the most dramatic range was number of points, which ranged from 65 to nearly 1800. The LEED Homes Committee felt strongly that widespread acceptance meant that our program needed to be as lean as possible, which drove a very strategic focus.

The next step was to identify the primary strategies to be addressed in each LEED category and assign weights to them. These weights were then balanced across categories, targeting a 100-point total. The pilot version of LEED for Homes includes 108 points, which may be pared down to the targeted 100 post-pilot. Technical subcommittees then evaluated all the relevant measures in the master list of 1400 and narrowed that list to include only those few that represented the appropriate level of rigor and most best environmental performance for the effort invested by the builder.

Rating criteria were then drafted for each of these primary strategic areas, formulated to represent "best practices" in each area. Mandatory requirements were established based on the committee's sense about what elements a home must have to be considered green, and what is economically feasible for builders to do. The rest of the system's rating measures are optional, i. e. , the builders may choose among the optional measures to aggregate sufficient points to earn a rating. The pilot will be the final proving ground for the rating system.

The LEED for Homes pilot will continue throughout 2006. Once the data-gathering phase of the pilot test is complete (mid-2006), the LEED Homes Committee will incorporate refinements based on the pilot experience, and present it to the USGBC membership for review. Following two rounds of public comment, the rating system will be balloted by the USGBC membership. Following approval of the rating system, LEED for Homes will be made available across the U. S. The official launch is expected late in 2006 or early in 2007.

THE STRUCTURE OF LEED FOR HOMES

Like the other LEED products, LEED for Homes has a four-tiered structure, with the different award levels based on points earned for incorporating various green design and construction measures in the home (along with the completion of the mandatory measures). Of the total of 108 points available, 30 are required at the lowest rating level-"certified," then 50 for silver, 70 for gold, and 90 for platinum.

The basic unit of the rating system is called a "credit." A credit is a strategy or set of strategies with a common intent-that is, a particular environmental benefit those strategies are designed to achieve.

Beyond those fundamental structural similarities to its sister LEED rating systems, LEED for Homes bears little resemblance to them. It is distinguished by a number of key features, described in the following paragraphs.

One intent, one credit. Within LEED for Homes, there is a one-to-one relationship between credits and intents, such that each action related to the improvement of lighting efficiency, for instance, resides in the same credit. A credit may include mandatory and/or optional elements. The mandatory ones must be completed in every LEED Home, and with one notable exception (home size-more about that later), do not earn points. The builder may select from among the optional elements to earn the number of points needed for the desired rating level.

Integration with LEED for Neighborhood Developments (LEED-ND). LEED-ND is being developed in collaboration with the Natural Resources Defense Council and Congress for the New Urbanism to establish a definition of green neighborhood development that will be implemented nationally to foster the design of exemplary green developments. LEED for Homes foreshadows LEED-ND in the credits in the Location and Linkages category, whose focus is decisions made before construction about the location and type of site chosen for development.

Integration with successful Energy Star Home programs. Home builders already face a wide array of marketing programs for their homes. Among the most successful of these is Energy Star Homes, developed by the U. S. Environmental Protection Agency (EPA). Because USGBC anticipates that the same builders who have embraced Energy Star Homes will be the early adopters of LEED for Homes, it was important that these programs be integrated in order to ensure the highest level of acceptance among Energy Star builders and ease their adoption of a new program. Thus Energy Star Homes, along with its complementary sub-programs-the Indoor Air Package and the Advanced Lighting Package-are embraced within LEED for Homes. In practical terms, this means a builder of an Energy Star Home automatically meets the mandatory measures in the Energy and Atmosphere category (and may earn up to 16 more points, depending on the home's HERS[1] score); using the Advanced Lighting Package earns an additional 3 points; and using the Indoor Air Package automatically meets the mandatory measures plus earns 10 points in the Indoor Environmental Quality category. (Energy Star appliances can earn up to 3 more points.)

Regional variability: climate & precipitation zones. Throughout the development of LEED for Homes, stakeholders emphasized the importance of making the system responsive to local conditions. This was a significant challenge for a national program, but respect for the inherently local nature of green building won out. LEED for Homes responds to local conditions in two ways: with climate zones and precipitation zones. The climate zones are in fact an integral part of the Energy Star Homes program and are reflected in the credits based on that program. There are three precipitation zones: dry (0～20 inches annually), normal (20～40 inches), and wet (more than 40 inches). These affect the number of points available for credits related to landscape design, irrigation, and durability (because of the prevalence of moisture damage as a durability problem in U. S. housing).

Durability planning. Durability is another aspect of LEED for Homes that has been viewed as essential by all parties involved in its creation-and yet it has proven one of the most challenging to implement. This is because the set of appropriate design,

[1] Home Energy Rating System.

specification, and construction measures for any individual home to foster durability depends on a large number of variables-site, climate, microclimate, exposure, natural disaster risks, orientation, soil type, building systems, etc. -which can occur in virtually infinite combination. Thus the durability planning credit is structured to guide the builder through a decision-making process of identifying his/her risks and developing appropriate responses to those risks. The mandatory component of the credit simply requires that the builder enumerate the durability measures incorporated in the home. The optional component of the credit includes third-party verification that the builder-identified durability measures were in fact incorporated.

Home size incentive. The most controversial credit in the rating system is the home size credit. The exception to the mandatory/no points rule, this credit awards points to those who build homes smaller than the U. S. national average, and deducts points from those who build larger homes, creating an incentive for larger homes to make up for their greater consumption of materials and energy by working harder at greening their homes in other ways. Although this approach violates LEED's generally benign "reward-only" philosophy, ultimately it was the only mechanism that the Homes Committee felt was both equitable and environmentally defensible. This credit carries a significant weight because-although it resides in the Materials and Resources category-it also reflects energy impacts.

Affordable/multifamily basis for many credits. Affordable homes-and in particular multifamily homes-typically represent a higher level of resource efficiency than their market-rate counterparts. Of course this is for economic reasons; nevertheless, from the perspective of equity, this inherent efficiency could not be ignored. Credits that are based on these resource efficiencies include home size, location credits for infill and compact development, site credits exempting compact projects from landscaping and irrigation requirements, and an air quality point for no attached garage (having no attached garage reduces the likelihood of vehicle emissions infiltrating into living space).

National third-party verification and rating service network. If USGBC is successful in attracting even a very small portion of the 2-million-unit-per-year homes market, USGBC's existing rating infrastructure easily could be overwhelmed. Thus it was imperative that not only a new rating system be developed, but also a new infrastructure for serving participating builders. USGBC, consequently, has formed a strategic alliance with the Residential Energy Services Network (RESNET, www. natresnet. org), tapping into the existing community of HERS raters throughout the U. S. to provide verification services to builders of LEED Homes.

MAKING THE CASE FOR LEED FOR HOMES

One of the largest hurdles yet remains before USGBC-marketing LEED for Homes successfully to mainstream builders. The marketing effort will rely substantially on effective communications strategies; but those strategies will need to be founded on compelling information. Perhaps the most compelling angle to the LEED for Homes story is the one based on cost and return. The table on page 2 tells this story: a LEED Home delivers sufficient reductions in utility costs to more than offset the increased sticker price, while delivering improved indoor environmental quality, reduced likelihood of durability problems, verified performance, and a smaller environmental footprint.

There is ample evidence in market studies that there is already considerable consumer interest in homes that fulfill this promise. That interest will only increase in years to come, as energy and water prices continue to escalate and concerns about national security and climate change come ever more to the fore. LEED for Homes is designed to be an agent for change, a starting point for a "course correction" for U. S. home building and a model that may be adapted to respond to different national contexts based on their own unique housing types and market circumstances.

* * *

Ann Edminster, M. Arch., LEED AP v2.0, has designed more than 100 home remodels and has spent the last decade as a green building consultant. She is a member and past co-chair of the LEED Materials & Resources Technical Advisory Group. She has worked with the USGBC's LEED Homes Committee since its inception and served as its chair and co-chair for the past four years. She lives in Pacifica, CA.

Jay Hall, Ph. D., has been consulting with USGBC on the development of LEED for Homes for the past two years. His background includes 20 years developing voluntary market transformation programs including several of the US EPA's ENERGY STAR initiatives and Fannie Mae's Emissions Trading Initiative. Since 1995, he has managed the consulting team that developed and supported the EPA's ENERGY STAR for Homes program. He resides in Annapolis MD.

LEED FOR HOMES：美国绿色建筑协会最新评估体系示范项目介绍

Ann V. Edminster M. Arch.
(Owner, Design AVEnues)
and Jay Hall Ph. D.
(Building Knowledge, Inc.)

摘　要：目前，LEED® (Leadership in Energy and Environmental Design) for Homes评估体系的试点操作正在美国进行，用于奖励那些新建的、高绩效的住宅。这些住宅需带来以下几项益处：提供更健康的室内空气，有效利用能源及水资源，耐久性得到提高，使用环保型材料，减少降水流失和适应当地气候的景观设计——在降低每月净成本和提升市场价值的前提下。

LEED for Homes 评估体系和其外围内容着眼于美国单户式独立住宅、双排住宅、联排住宅、连栋别墅和多户 3～4 层住宅的主流建筑商。进一步的分析将确定其是否应用于高层住宅。有别于绿色建筑的示范项目，这一项目对于营造商和买主来说，必须是可负担的、简单易行的并可以提供较高的价值。评价级别的阐述要代表每一个 LEED 主题领域的“最好实践”。

LEED for Homes 在其基本建筑方面与其他 LEED 体系类似，但鉴于住宅建筑商的需求，它又是基于现有成功项目基础之上的。它在选址、耐久性规划、住宅面积、固定密度内的资源效率、城镇和多户住宅方面都体现了其创新举措。

HK-BEAM-ENHANCING SUSTAINABLE BUILDINGS IN MAINLAND CHINA

Peter C K Chan Daniel Chan
(HK-BEAM Society, Hongkong)

Abstract: HK-BEAM (Hong Kong Building Environmental Assessment Method) was introduced to Hong Kong in 1996 and is an environmental performance based assessment scheme purpose designed for high rise buildings. With similar climatic conditions and same relevance on the range of environmental issues, lesson learned from the adoption of HK-BEAM in its ten years of operation can be applied to buildings in mainland China.

HK-BEAM scheme is completely voluntary and is owned by the HK-BEAM Society, a non-profit making organisation consisting of industry professionals. The latest HK-BEAM standards (Version 4/04 and Version 5/04) covers all "New" and "Existing" building types, including office, residential, mall, hotel, school, hospital, institutional and mixed complexes-centrally air-conditioned, naturally ventilated or mixed mode.

HK-BEAM assessment embraces a range of good practices into a pool of criteria using a life cycle approach. The comprehensive assessment framework encompasses exemplary environmental practices in planning, design, construction, commissioning, operation, maintenance, and management. However, more emphasis has now been placed on the wider attributes of a sustainable building, particularly safety, hygiene, security, amenity and overall quality. This approach encourages buildings that fulfil their intended functions whilst minimizing the resulting impacts on the environment. Finally, the scheme provides a benchmark for sustainable buildings and third party recognition to building performance excellence.

As of December 2005, over 100 private and public sector premises covering 6.2 million square meters and including 52000 residential units have been submitted for HK-BEAM certification. This level of adoption makes HK-BEAM one of the most widely used voluntary green building labelling schemes in the world, particularly in terms of number of buildings, percentage of total stock assessed, and buildings per capita.

This paper outlines the latest framework of HK-BEAM, including rationale on the assessment approach and criteria, and explanations of its achievements and benefits. All of which provides valuable experience and unique insight for buildings in China intending to achieve "green", energy efficiency and sustainability objectives of HK-BEAM.

Keywords: Environmental Sustainability, Building Performance Assessment, Green Building Labelling for China

1 INTRODUCTION

The Hong Kong Building Environmental Assessment Method (HK-BEAM) was launched in 1996 to measure, improve, certify and label the whole-life environmental sustainability of buildings. More than 100 premises have so far adopted independent and voluntary certification within the scheme's first decade, and steady updates to the assessment framework and criteria have heightened its relevance in addressing sustainable-design issues.

From its inception, HK-BEAM has been the only scheme of its kind to originate locally and be widely applied across the region. HK-BEAM assesses projects in terms of their whole-life site, materials, energy, water, indoor environment and innovative aspects. Improvements are identified during assessment, and buildings are labelled as *Platinum*, *Gold*, *Silver*, *Bronze* or *Unclassified* accordingly.

2 THE FUNDAMENTALS AND PRINCIPLES OF HK-BEAM

2.1 A holistic perspective

Although much has been written about the social, economic and environmental aspects of sustainable development, what constitutes a sustainable, green, healthy or ecology-friendly building remains a matter of viewpoint.

HK-BEAM views sustainable building as embracing (in order of priority) safety, health, comfort, function and efficiency while protecting local, regional and global ecosystems during construction and operational lifetimes. HK-BEAM therefore integrates aspects including:

- Hygiene, health, comfort and amenity;
- Land use, site impacts and transport;
- Use of materials, recycling and waste;
- Water quality, conservation, recycling; and
- Energy efficiency and conservation.

2.2 An emphasis on the indoor environment

Urban environment poses numerous challenges in providing good indoor environmental quality. Indoor air quality (IAQ) determines both health and comfort in terms of thermal aspects, lighting, noise, etc, while hygiene is also a major issue, particularly after the SARS outbreak in Hong Kong during 2003. The needs of building users, however, must be balanced against the wider carrying capacity of Mother Earth. HK-BEAM emphasises indoor environmental quality (IEQ) and amenities while addressing the local, regional and global impacts with which they are linked.

2.3 A label of building quality

HK-BEAM provides a single performance label for overall building quality in respect of safety, health and comfort (important to users, occupants, buyers and tenants), and the environment and society (important to the community as a whole).

Clients ultimately decide on the merits of certification, but assessments assure quality-not as a subjective promise but as a measured reality. HK-BEAM is notable in that new buildings are not fully certified until they are complete. This came about by popular demand from clients to ensure that design and construction practices are actually implemented and reflected in the final certificate for the built product.

2.4 Whole-life cycle for all building types

HK-BEAM assesses performance across the whole life-cycle using two standards:

- HK-BEAM version 4/04 for new buildings, which covers planning, design, construction, commissioning and refurbishment; and
- HK-BEAM version 5/04 for existing buildings, focusing on management, operation and maintenance.

Assessments encourage detailed design analysis, using computer-aided tools including computational fluid dynamics to predict project aspects such as the anticipated microclimate. Design is just one stage of the development process, however. HK-BEAM also helps to focus attention during construction and operation so that the design intent can be fulfilled, and also provides feedback so that plans meet construction and operational needs.

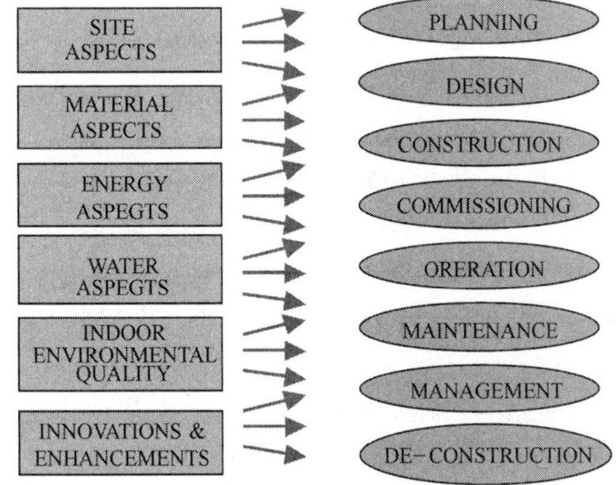

Fig. 1 The HK-BEAM Whole-Life Assessment Framework

All buildings under single ownership can be assessed using HK-BEAM, including commercial, residential, retail, catering, industrial, educational, institutional and hotel buildings. Any mix of central cooling, mechanical or natural ventilation is catered for, whether to the shell or fitted out. In all cases, HK-BEAM focuses on what the designer, builder and commissioning agent can achieve.

2.5 Dynamic, evolving and consensus-based

Collective knowledge and opinion as to what constitutes a sustainable building continues to evolve. Industry standards and regulations are constantly on the move. HK-BEAM is thus updated periodically to keep pace with feedback from an increasing number of stakeholders.

Hundreds of individuals and organisations have reviewed and advised on HK-BEAM. The criteria, methods and credit allocations cannot always satisfy all. However, the real value of HK-BEAM lies not in its scientific rigour but in the actual improvements and increased awareness that it helps to create.

2.6 Facilitating continual improvement

HK-BEAM strives to progressively improve performance and innovation. As a voluntary scheme without financial or other incentives, its success is determined by market acceptance of the criteria, the cost of assessment, the relative weightings that determine the overall grades and the perceived benefits to the client.

It is not expected that every building will meet all requirements. By meeting some, however, a building will have less impact than one in which the requirements have not been met. Experience shows that improvements, of one kind or another, are implemented in every project assessed.

3 HK-BEAM CRITERIA AND COVERAGE

Despite its name, the Hong Kong Building Environmental Assessment Method (HK-BEAM) extends beyond the environment and covers an increasing number of building-related socio-economic issues such as land utilisation, site amenity and acces sibility, pedestrian comfort, environmental hygiene and occupant health and well-being. HK-BEAM continues to evolve as an environmental sustainability assessment and, as parameters become better defined, ultimately towards a building-sustainability assessment to improve performance. The following factors determine the framework of criteria.

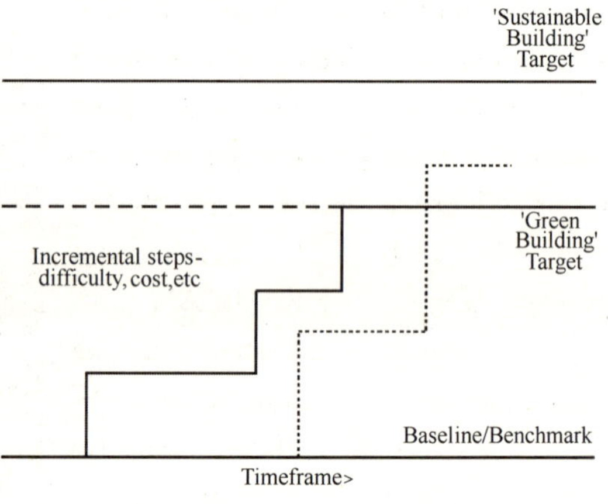

Fig. 2 Incremental Performance Improvements

3.1 Credit allocations and weightings

Credits are allocated and weighted to take into account consensus both locally, from surveys and informed opinions of those involved in the development of HK-BEAM, and internationally, by analysis of similar methods operating elsewhere.

As with other schemes, HK-BEAM does not assign credits or weightings strictly in accordance with the significance of given impacts. It is not currently possible to assign economic costs to such diverse aspects as health, global warming and resource depletion, for example.

3.2 Validity and practicality

HK-BEAM embraces only those issues for which there is good evidence of the problems they cause, and for which effective performance criteria can be defined. Where there is consensus among stakeholders that an issue is important, and where a reasonably objective assessment can be made, the issue is included.

Certain aspects attributable to buildings and their use have yet to be included, either because their impacts are not well defined, or because performance criteria have not been established. These will be included in future updates when information becomes available to enable their objective assessment.

3.3 Transparency and clarity

All criteria and assessment methods need to be transparent in their benchmarks, baselines, data and assumptions, as well as in how these are accounted for in the assessment and consequent rating. This is particularly the case when taking a holistic view of building performance, with an emphasis on life-cycle impacts, with priorities established and weighted for global, local and indoor aspects.

3.4 Setting the baselines

Legal requirements and industry norms constitute the zero-credit baseline in HK-BEAM, and in certain cases, legal compliance is a prerequisite. First-level credits are achieved for good practice in line with local and overseas guidelines, and further credits achieved are commensurate with building size, complexity and age. Criteria are updated periodically as legal requirements evolve. Issues becoming subject to legislation no longer count for credits, and these are deleted or amended in subsequent revisions of the scheme.

3.5 Adopting recognised standards

Whenever available, HK-BEAM embraces codes of practice and technical circulars issued by Hong Kong authorities and institutions to build on and increase familiarity with these. In the absence of local standards, international standards are tailored to the Hong Kong context. The HK-BEAM Society develops new criteria as a last resort.

It is often found that performance criteria differ between authorities in aspects such as thermal comfort. In these cases, HK-BEAM allows clients to specify what they consider to be appropriate for their buildings, with assessments seeking to establish that the specified performance is acceptable and has been achieved.

3.6 Prescriptive versus performance-based

Prescriptive assessments can preclude buildings without prescribed features from obtaining good results, regardless of actual performance, and tend to encourage feature-based design and rigid construction and operating practices. In line with emerging international consensus, HK-BEAM is therefore performance-based as far as possible.

Prescriptive measures are only adopted with requirements for their effective implementation to improve performance. Where issues are rather subjective, HK-BEAM uses checklists to facilitate equitable and consistent assessments.

3.7 Self assessment

Where appropriate, assessments to demonstrate compliance with the HK-BEAM criteria are undertaken by qualified persons acting on behalf of the client, submitting documents, data and reports as evidence. This promotes self-learning in the adoption of the practices and processes recommended by HK-BEAM for clients, designers, contractors and managers. In other cases, issues are examined based on evidence collected by the assessor during site inspections and building surveys, in the interests of efficiency.

3.8 Flexibility and objectivity

Given that HK-BEAM embraces a wide range of building types, variable in scale, location and mix of uses, its criteria and methods need to be flexible yet reasonably objective to avoid undue controversy. HK-BEAM does not seek to be overly prescriptive-it instead encourages owners and operators to consider alternative approaches that meet the same objectives.

Client representatives can submit alternative method statements stating their own proposed criteria and assessment method for review by the HK-BEAM Society. If endorsed, these are absorbed into assessment guidance notes and future revisions to the HK-BEAM standards as a means of continual improvement.

3.9 Coverage

HK-BEAM defines more than 100 best practice criteria used across the planning, design, construction, commissioning, operation, management, and maintenance process:

- *Site aspects*: land use and location, site layout optimisation, transportation, accessibility, ecology, amenity, site and neighbourhood interfaces, site emissions and management, etc

- *Materials aspects*: optimisation in design and operation, innovative construction methods, building flexibility and durability, avoidance of environment-damaging materials, waste minimisation, etc

- *Energy aspects*: passive/low-energy design, microclimate, plant/equipment efficiency, renewable energy, annual energy consumption reduction, etc

- *Water aspects*: potable water quality, water economy and recycling, effluent management, etc

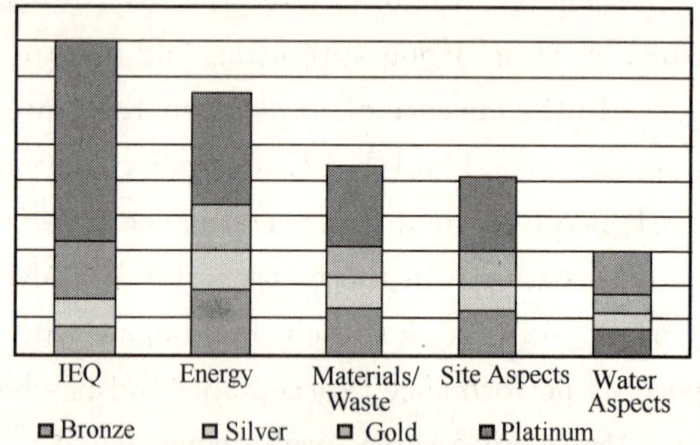

Fig. 3 HK-BEAM: Emphasis and Performance Requirements

- *Indoor environmental quality*: safety, security, hygiene, amenities, thermal comfort, ventilation effectiveness, indoor air quality (internal and external pollutants), natural/artificial lighting, acoustics and vibration, etc

Innovative techniques: innovative techniques and enhancements beyond those stipulated in the HK-BEAM criteria indicated above

4 ASSESSMENT AND CERTIFICATION PROCESS

HK-BEAM aims to reduce the environmental impact of buildings while improving quality and user satisfaction by adopting the best techniques, practices and standards compatible with prevailing economic constraints. Going beyond its environmental label, the success of HK-BEAM should be measured by its impact in actually improving building performance.

4.1 The two-stage assessment process

At the start of the assessment process, the client receives questionnaires and checklists which detail the required information. For new building projects, the assessor then meets the project team of the client, the consultants and the contractors to discuss assessment details. For existing buildings, the assessor meets with the management team of the client, the property or facility managers and their contractors.

Information gathered is used as the basis for a provisional assessment report that identifies exactly which credits have been achieved and what actions may result in improved performance. Clients can evaluate potential changes to the design or management processes at this stage.

For new developments, refinements to the building are submitted to the assessor on an ongoing basis for real-time review and approval. Site visits continue throughout the construction process to ensure that the agreed measures are properly implemented, with credits withdrawn when, for whatever reason, this is not the case. The final assessment report and certificate are issued on completion of the building and after provision of all necessary information.

For existing developments, the emphasis is on how to operate and maintain a building. Recommendations therefore address energy efficiency, indoor quality, health and hygiene issues. The assessor provides information and advice on issues such as how to do energy audits, where to buy materials and how to train staff, so the client receives considerable guidance throughout the assessment process. On completion, the assessors undertake a final review to determine the extent to which recommended improvement measures have been taken on board and the client receives a certificate with a final grade for the building.

HK-BEAM assessors have noted that the assessment process has a partnering effect which encourages dialogue during the progression of a project. Feedback received from HK-BEAM clients frequently reveals that assessment meetings offers the only opportunity for the whole design team and contractors to sit around a table and discuss specifically about environmental matters.

4.2 Innovations and additions

HK-BEAM does not presume to be comprehensive in its coverage of aspects and hence encourages clients to submit proposals for others not covered. While innovative solutions are encouraged, they do not necessarily justify credits. The HK-BEAM Society examines proposals that demonstrate performance gains and, after weighing the environmental gains, awards an appropriate number of credits.

4.3 Assigning the HK-BEAM score

The overall grade is based on the level of applicable credits gained, with a minimum percentage of IEQ credits needed to qualify for an overall grade (*see* Table 1 *below*).

Bonus and innovation credits do not count towards the total number available, but do count towards the total number achieved, which qualifies for an award classification.

To cater for the overall assessment of a complex with different building types, HK-BEAM weighs up the available energy and IEQ credits based on the normally occupied floor area of each type.

Credits needed for HK-BEAM ratings **Table 1**

	Overall	IEQ
Platinum	75%	65%
Gold	65%	55%
Silver	55%	50%
Bronze	40%	45%

4.4 A process of improvement

In addition to measuring actual performance, HK-BEAM also gives guidance on improvement - the best performance that can be obtained from a building given its occupancy and use. This is cited by most clients as a key benefit of the process, with information and guidance originating from ongoing research and previous assessments.

Measuring actual performance determines operating costs, use of resources and by-products, and provides the basis for determining improvement measures. While new buildings can be assessed at any stage of the production process, the most benefit is derived if this begins during planning so that designers can integrate improvements without retrospective and more costly changes. Existing buildings can be assessed at any time throughout their operational life.

5 CONCLUSION: LESSON TO BE LEARNED FOR CHINA

The design, construction and use of buildings consume enormous quantities of resources, creates untold environmental pollution and generates much of our waste. Building design and management shape our urban environment and control our overall quality of life. In short, as a large sector of the economy, the construction industry can make a significant contribution to a more sustainable environment. These issues are of paramount significance for China as a developing nation. It is not then surprising that for example, China's Ministry of Construction has a commitment to reduce energy consumption of its buildings over the coming decades.

HK - BEAM has provided a unique stepping - stone towards embracing such sustainability issues, through partnership of stakeholders in the building industry. The inherent comprehensive nature of the life cycle approach adopted in HK-BEAM has lead to the adoption of many building designs and management measures, which are environmentally beneficial, and often financially rewarding, which might otherwise have been overlooked. For example, with an additional capital investment of HK $1 million, an annual cost savings of HK $ 0. 7 million was realised for a new office premises (Lincoln House) assessed under HK-BEAM. Another existing office building adopted a series of energy improvement measures identified and recommended under HK-BEAM and resulted in an annual energy costs reduction HK $ 1 million, with a pay back period of less than 3 years.

In conclusion, through the adoption of HK-BEAM to over 100 high rise buildings, many lessons and valuable experience can be learned. China as a developing nation with similar climatic conditions and facing virtually identical environmental sustainability issues is ideally positioned to take full advantages of this, and embrace and apply HK-BEAM principles and physiology for their buildings.

香港建筑环境评估法——提升内地楼宇的可持续发展性

陈志强　陈炳祥
（香港环保建筑协会，香港）

摘　要：香港建筑环境评估法(HK-BEAM)于1996年被推出，是一套适用于多层大厦的环保表现评估法。基于两地天气及环境因素的相近，这十年间从推行HK-BEAM中吸取的经验相信可应用于内地建筑。参与此HK-BEAM认证是自发的，乃香港环保建筑协会(HK-BEAM Society)所有。该协会为非盈利性组织而会员包括专业人士。新版本的HK-BEAM(4/04版及5/04版)可应用于任何“新建”或“现存”的大厦，包括办公楼、住宅、商场、酒店、学校、医院、公共机构及综合用途建筑——使用中央空调、自然通风或混合设计。

HK-BEAM网络所有合适的环保标准于每一评判项目内以及采用生命周期作评估。此标准提供全面的评估框架，涵盖规划、设计、施工、启用、作业、保养及管理等方面的环保实务典范。然而，新标准更加侧重可持续发展楼宇的全面特性，尤其是安全、卫生、保安、设施及整体质素，鼓励建筑设计在符合原定用途之余，更同时尽量减少对环境造成的影响。HK-BEAM可提供一套可特续发展的基准及获得外界认可。

时至2005年12月，过百座约6200万m^2的私营机构及公共部门建筑物包括52000余住宅单元已接受了HK-BEAM认证。如此应用规模使得HK-BEAM成为全球应用最为广泛的自愿环保建筑标签计划之一，尤其以建筑物数量、市场物业百分比及大厦面积计算。

本文大致介绍了新版本的 HK-BEAM 架构包括认证流程，评估准则及简介其对香港环保的成绩与得益。本文亦提供一些例子给内地楼宇作参考，以达致 HK-BEAM 内提及的“绿化”、能源高效及可持续发展性的目的。

B 建筑智能化与绿色建筑

绿色建筑智能系统工程的探索

程大章　刘　钢
（同济大学，上海　200092）

摘　要：本文对绿色建筑的理念与我国的相关政策进行了分析，指出智能化技术与信息技术是绿色建筑的重要技术组成部分。文章阐述了绿色建筑智能系统工程的信息、控制、管理与决策的内容，揭示其大系统多目标共存、多学科交叉，多技术结合的特点，并提出绿色建筑中智能系统工程面临的6个新课题。

关键词：绿色建筑　智能系统工程　能源　资源　信息　控制　管理　决策

进入21世纪，全球城市化进程加快、经济快速增长，但是人口急骤增加、资源萎缩、环境恶化，使人类面临着严峻挑战。在建筑物的建造和使用过程中不仅消耗大量的能源与资源，而且破坏了环境。据研究数据反映人类的建设行为及其成果——建筑物在生命周期内消耗了全球资源的40％、全球能源总量的40％，建筑垃圾也占全球垃圾总量的40％。因此，对于节约能源与资源，提高使用效率，缓解能源资源短缺的矛盾，保护和改善环境，各国政府与工程界都进行了不懈的研究与实践，提出了绿色建筑的理念。绿色建筑强调人与自然的和谐，避免建筑物对生态环境和历史文化环境的破坏，资源可达到循环利用，室内环境舒适，符合中国政府实行的可持续发展的基本国策，因而受到了各级政府、工程界与房地产业的高度重视。

1　绿色建筑的实践

1.1　绿色建筑的理念

人类的建筑经历了掩蔽、舒适、健康三个阶段。第一阶段是低能耗甚至无能耗的阶段，第二和第三阶段是高能耗的阶段。随着人们对全球生态环境的关注和可持续发展思想的深入，建筑物走向第四阶段——绿色建筑。美籍意大利建筑师保罗·索勒端(Paola Soleri)把生态学(Ecology)和建筑学(Architecture)相结合，在20世纪60年代首次提出“绿色建筑”(Arology)的概念。第四阶段建筑的主要特征是大量利用可再生能源(Renewable Energy)和未利用能源(Unused Energy)，强调能源节约和建筑材料资源的循环使用，尽量减少建筑建设与使用过程中对自然生态环境的损害。综合国内外专家的研究，绿色建筑可理解为在建筑的“全生命”周期内，最大限度地保护环境、节约资源(节能、节水、节地、节材)和减少污染，为人们提供健康、适用和高效的使用空间，最终实现与自然共生的建筑物。

绿色建筑是一个有机的整体概念。它贯穿于建筑物的规划、设计、建筑、使用以及维护的全过程，覆盖建筑物的整个生命周期。绿色建筑注重与资源的和谐，包括日光利用，空气流通，景观环境等的综合考虑，为使用居住者提供健康舒适的空间，并对周边环境形成积极影响。绿色建筑关注建筑材料与能源的合理利用与节约，因而建筑物建造与使用过程的每个环节都应最大程度地节约能源与材料。绿色建筑将环保技术、节能技术、信息技术、网络技术渗透到居民生活的各个方面，即用最新的理念，最先进的技术去解决生态节能与居住舒适度问题。总之，绿色建筑是自然环境的一部分，与之共同构成和谐的有机系统。

1.2 政府积极推进，建设行为从生态意识走向生态尽责

中国正处于工业化、城市化加速发展时期。我们不仅要注重单体建筑的效果，更需要从区域来考虑降低能源资源消耗、保护环境的总体效果。中国现有建筑总面积 400 多亿 m^2，预计到2020年还将新增建筑面积约300亿 m^2。中国政府积极调整经济结构、转变经济增长方式，提出鼓励发展节能省地型住宅与公共建筑，制定并推行更严格的节地节能节水节材(简称“四节”)标准，以促进城镇建设发展质量和效益的提高。

发展节能省地型住宅与公共建筑，必须用城乡统筹、循环经济的理念，挖掘建筑“四节”的潜力。“四节”都有各自的要求，必须统筹考虑，综合研究。中国政府正在研究制定促进绿色建筑产业现代化的技术经济政策，采取有效措施，区域统筹，分类指导，推进建设节能省地型的绿色住宅和公共建筑。这些工作在城镇体系规划、城市总体规划、近期建设规划、控制性详细规划等不同层次的规划中得以体现，充分研究论证能源、资源对城镇布局、功能分区、基础设施配置及交通组织等方面的影响，确定适宜的城镇规模、运行模式，加强土地、能源、水资源等利用方面的引导与调控，实现能源资源的合理节约利用，促进人与自然的和谐。这需要以科技创新为支撑，组织科技攻关；尤其是新型能源、可再生能源、绿色新材料及新产品的开发与推广应用更需要引进、消化、吸收国际先进理念和技术，增强自主创新能力，发展适合国情具有自主知识产权的适用技术。在工程领域，正在加大标准规范的编制力度，形成较完善的建筑“四节”标准规范体系，并加强标准执行的实施和监管。在此同时，政府充分重视存量建筑的改造，把治理污染、降低能耗作为日常的基本工作，并推进供水、供热、污水处理等市政公用事业改革，探索创新的体制机制。

综上所述，中国政府近年来在绿色建筑领域所做的这些工作，反映了政府、工程界和房产业的建设行为正从生态意识走向生态尽责。这是中国可持续发展的重大转折点。

2 绿色建筑对智能化与信息化的需求

为了实现绿色建筑的建设目标，在工程中涉及大量的技术与政策问题，详见表1。

绿色建筑涉及的技术、工程与政策 **表1**

类别	内容
区域规划	城镇体系规划、城市总体规划、近期建设规划、控制性详规
建筑设计	自然采光、自然通风、室内设计、结构设计

续表

类　别	内　容
建筑材料	墙体保温材料、门窗、墙面材料、涂料、结构、隔墙材料、遮阳百页
建筑设备	照明节能控制、空调节能控制、节水型设备控制，变频调速应用，热能回收
能源系统	太阳能/风能/地热利用、热电联产、区域供冷热、吸收式制冷、冰蓄冷、燃料电池
资源利用	雨污水再生回收、生活垃圾再生利用(沼气等)、建筑垃圾再生利用
管理信息	环境监测、生态监测、能源与资源综合管理信息、社区信息共享、建筑智能化系统、社区通信网络系统
生　态	绿化设计、生态系统设计、环境设计
技术标准	建筑节能标准、建筑节能评估指标体系、绿色建材技术标准、绿色建筑评估标准、节能与环保设备技术标准
政策法规	建筑节能标准的执行条例，供电、供水、供热、污水处理等市政公用事业体制改革，新能源与可再生能源推广应用奖励制度

由表中1所列项目可见，在建筑设备、资源利用、管理信息、生态等领域，有大量需要解决的智能控制与信息管理的课题。如果不能有效地实现各类设备系统的智能控制，不能完备地进行建筑物建设、运行与更新过程的信息管理，绿色建筑的主要目标是不可能达到的。对此，中国的建设工程界给予了高度的重视，2005年3月建设部召开了首届国际智能、绿色建筑与建筑节能大会并发表了《北京宣言》，2006年3月建设部、科技部、国家发改委和环保总局将联合举办第二届国际智能、绿色建筑会议；浙江省正在编制的《绿色建筑标准》同样把建筑智能化系统列为绿色建筑的主要组成部分。

3　绿色建筑的智能系统工程

3.1　绿色建筑智能系统的内容概要

(1) 信息

信息是绿色建筑智能系统的基础，在绿色建筑工程中，需要广泛采集环境(大气、水体、日照、气象等)、生态(植物、动物等)、建筑物(结构、地基、建材等)、设备(能源、空调、水处理、供电、固体废弃物处理等)、社会(安全、资讯、服务、管理、行政等)等领域的信息，为人类社会活动与生活提供准确可靠的信息，为绿色建筑的控制、管理与决策提供良好的工作基础。

(2) 控制

绿色建筑的控制包罗万象，有绿色能源——从太阳能、风能、地热应用到区域热电冷三联供系统等的发配电系统控制；有利用峰谷电价差的冰蓄冷系统的控制；有采用最优控制技术进行室内照明控制与通风空调控制，以充分利用自然能量来采光、通风，实现低能耗建筑，若辅以可再生能源则成为零能耗建筑；有可以随环境温度、湿度、照度而自动调节的智能呼吸墙；有应用变频调速装置对所有泵类设备的最佳能量控制；有自动收集雨水、处理污废水，提供循环使用的水处理设备控制系统……

(3) 管理

给出一堆新建材与新设备并不是绿色建筑，我们需要的是有机集成后的绿色建筑功能，因此要能有效实现绿色建筑功能，必须具有相应的管理。绿色建筑的管理涉及环境、生态、能源、资源、建筑物、设备、社会、安全、通信网络等。这些子项之间相互关联，但有些子项之间的目标相互矛盾，因此管理不能仅限于单一子项，而要信息共享充分协调相互关系，于是绿色建筑的集成信息管理系统应运而生，在一个统一的平台上，实现绿色目标的综合管理。

（4）决策

绿色建筑的生命周期中为了保障生态目标的实现，在建设、运行管理与更新改造有大量的事物需要决策，并付诸执行。这是因为技术不断进步，人们认识逐步深入，而随之形成的新产品与商用新能源类型的增加，以及环境、资源及政策的变化，都需要在今天能够承担的前提下，进行技术方案、经济、生态责任等方面的决策。

3.2 绿色建筑智能系统工程的特点

（1）大系统，多目标共存

为实现绿色建筑的建设目标，在大型建筑与建筑群中必须设置环境、能源、资源、生态、安全、信息等监控管理系统。虽然众多的子系统构成一个有机整体，但是各子系统的目标是不相同的。例如，在有多种能源可供选用时，如果全部采用可再生能源，可能由于其不稳定性影响设备系统工作，于是安全与资源保护之间出现了矛盾。又如，全部采用可再生能源的运行成本在目前的技术水平下较高，如何在经济承担能力的范围内进行生态尽责，又是一个问题。因而，通常需采用多目标模糊优化控制来权衡利弊，寻求一个可行的实施方案。

（2）多学科交叉，多技术结合，协同支撑

绿色建筑中智能系统的监控管理对象分属土木工程、环境工程、生态工程、能源工程、信息工程、化学工程、设备工程，在一个子系统中，往往需要采用多种技术手段来实现目标。如模仿“人体表皮组织—呼吸系统—神经系统”应激性能的气候调节设备(呼吸墙)，就要综合运用新型墙体材料、过滤器、气候调节设备、室内外空气参数检测设备、智能控制装置，以室内外空气品质、建筑能耗及居住人的舒适度为综合目标进行自动调节。

因而，在绿色建筑的工程建设中，不存在以何类专业为主以何阶段为主的说法与理由，因为每一专业都是绿色建筑建设链的一部分，任何阶段都在绿色建筑的生命周期之中。按有效实施的要求，绿色建筑的生命周期必须依靠一个协同支撑的模式，如图1所示。

图1中的政府代表、业主代表、物业管理代表、建筑师、智能系统工程师、设备工程师、建筑公司代表、安装公司代表、绿色产品供应商代表等构成绿色建筑建设与运行工作的主体群，他们需要交流意见协同工作。他们的工作并不独立，是在一个以分布式计算机网络系统为基础，由绿色建筑政策法规库、绿色建筑设计数据库、绿色建筑评估系统、绿色建筑知识表达模型库、绿色产品数据库、生态环境数据库等组成的实施平台支撑下进行的。

3.3 绿色建筑向智能系统工程提出的挑战

由于智能系统与绿色建筑共生同存，智能系统的功能、性能及运行成本是绿色建筑不可分割的一部分。绿色建筑要实现的大部分目标及实现方案都离不开智能系统。但是建设

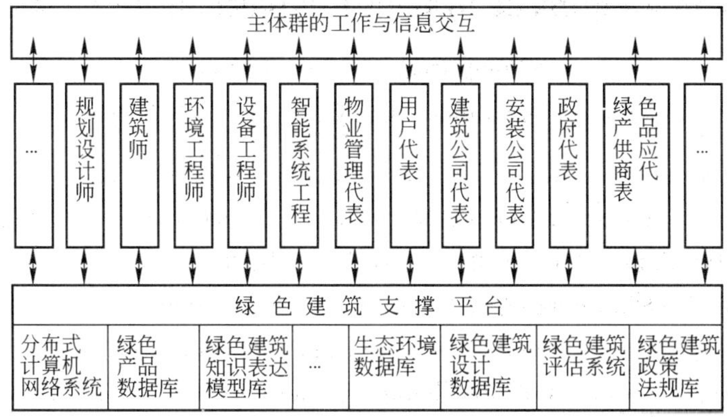

图1　绿色建筑实施的协同支撑模式

业中的智能系统与传统工业系统的工作环境和要求有很大差别，因而面临着许多新的课题。

(1) 绿色建筑的智能系统虽然其结构为多子系统组合的大系统，但建设业对于监控管理系统则要求低成本，高可靠性。

(2) 绿色建筑的智能系统大多是由没有专业技能的人群来直接操控的，因此，系统的人机界面不仅需个性化、可视化、更要人性化、简易化。

(3) 绿色建筑中大量使用的检测器件，涉及到各类物理量、化学量、生物量。这些器件不仅要求长期(几年，十几年)准确可靠工作，而且成本要低。如冷热源能耗自动计量等至今尚无理想技术方案。

(4) 为能增强安全系统功能，生物特征信息处理与数字图像分析应用需要有进一步的突破。

(5) 为实现低能耗与零能耗建筑，在能源与资源管理上，在区域能源供应方面，结合设备系统功能实行负荷预报控制与能源优化控制，尚需积累实践经验。

(6) 对绿色建筑众多设备系统进行在线故障诊断，实行容错控制，以增强设备系统的安全性，保证居住使用者的便利，这需要进行实践与探索。

4　结束语

绿色建筑不同于传统建筑，其建设理念跨越了建筑物本体而追求人类生存目标的优化，是一个大系统多目标优化的典型案例。同时，绿色建筑必须采用大量的智能系统来保证建设目标的实现，这一过程需要信息、控制、管理与决策，智能化、信息化是不可缺少的技术手段。建设部仇保兴副部长在《中国的能源战略与绿色建筑前景》一文中提出：“以智能化推进绿色建筑，节约能源，降低资源消耗和浪费，减少污染，是建筑智能化发展的方向和目的，也是绿色建筑发展的必由之路。”由于绿色建筑在我国刚刚起步，其中大量的课题有待人们去探索与解决。本文仅是笔者在学习研究过程中的一些认识，以期引起更多的同行对绿色建筑的智能系统工程给予注意。

The Study of Intelligent System Engineering in Green Building

Cheng Dazhang　Liu Gang
(Tongji University, Shanghai　200092　China)

Abstract: The paper analyses the idea of Green Building and the policy of Green Building in China and points out that intelligent technology and information technology are important parts of Green Building. This author described the content of information, control, management and decision-making of Green Building engineering, and revealed the characters of Green Building, such as multipurpose, and multisubjects etc. At last, the author pointed out 6 new items of intelligent system engineering in Green Building.

参考文献

[1] 建设部科技司,《智能与绿色建筑文集》编委会. 智能与绿色建筑文集. 北京：中国建筑工业出版社，2005
[2] 中国建筑科学研究院. 绿色建筑技术文集. 北京：中国建筑工业出版社，2005
[3] 程大章. 智能建筑楼宇自动控制系统. 北京：中国建筑工业出版社，2005
[4] 刘飞. 绿色制造的理论与技术. 北京：科学出版社，2005
[5] 大林芳久. 建设业のグリ-ン战略. 东京：海象社，2002

智能建筑节能的调研与分析

程大章
（同济大学高等技术学院电子信息系，上海　200092）
王长庆
（同济大学高等技术学院楼宇系，上海　200092）
王坐中
（同济大学电子信息学院，上海　200092）
黄　洁
（同济大学高等技术学院，上海　200092）

摘　要：本文通过对上海市部分智能建筑调研发现，由于历史、管理等原因建筑能源消费中存在着较大的浪费现象，建筑物的节能空间有待进一步提升。针对目前的节能形势、智能建筑能耗情况，就智能建筑可采取的节能手段以及如何进一步提升智能建筑节能水平的空间进行探讨和分析。

关键词：智能建筑　建筑能耗　节能技术

前言

2005 年 8 月 14 日，“全民节约，共同行动”大型主题活动在北京人民大会堂拉开序幕，中共中央政治局委员、国务院副总理曾培炎出席并讲话。曾培炎说，人口多、人均资源不足是中国的基本国情，这就决定了我们必须建设节约型社会。2005 年 8 月 15 日，“上海百万家庭节约活动”在浦东陆家嘴中心绿地拉开序幕，上海市委副书记、市长韩正出席活动并与市民进行交流。

在发展中的中国，能源的合理节约使用不仅是一个经济问题，而且已成为重要的战略问题，从政府到普通民众对此达成了广泛的共识。在建筑业，从建设部到各级建设主管部门，从建设方到设计施工与物业管理企业都对节能高度重视，制定了大量的规范、措施来节约建筑物的能源消耗。近期，我们对上海市部分智能建筑进行调研时发现，由于历史、管理等原因建筑能源消费中存在着较大的浪费现象，建筑物的节能空间有待进一步提升。

1　节能形势

当前，经济的不断发展与能源供应的相对紧缺成为制约我国社会与经济发展的一大矛盾。近几年，电力、原油、水资源等能源都出现了供应紧缺局面。

(1) 电力：2004年全国电力缺口在4000万kW，共有24个省级电网出现拉闸限电。2005年上半年全国电力供需形势总体呈现“前紧后松”状态。一季度全国共有26个省级电网出现拉闸限电现象；进入二季度，全国电力供需形势虽整体有所缓和，但部分地区仍较紧张，二季度共有18个省级电网出现拉闸限电现象。

(2) 石油：原油进口首次超过亿t大关，达到1.2亿t，比上年增长了34.8%。

(3) 水资源：全国城市缺水总量达60亿m^3。全国有11个省、市的100多座县级以上城市供水短缺，其中，严重缺水城市占56%。

建筑物对于水、电、油等能源的消耗是巨大的，据统计目前建筑能耗已占到社会总能耗的30%。21世纪头20年，将是我国建筑业的鼎盛时期，2020年全国房屋建筑面积将接近2000年数量的2倍。目前我国建设高潮持续不断，每年建成的房屋面积高达16～20亿m^2，超过所有发达国家年建成建筑面积的总和，可见建设规模极其巨大。但遗憾的是，不仅近400亿m^2的存量建筑中99%为高耗能建筑，新建建筑中90%以上仍可能属于高耗能建筑，单位建筑面积采暖能耗为发达国家新建建筑的3倍以上。按照目前建筑能耗水平发展，到2020年，我国建筑能耗将达到10.89亿t标准煤，超过2000年的3倍；仅空调高峰负荷将相当于10个三峡电站满负荷运行。问题相当严重，情况十分紧急，建筑节能已迫在眉睫。

2 智能建筑能耗情况

近期，我们对上海市12幢智能建筑进行了调研，其中一项重要内容就是了解目前智能建筑的能源使用情况。这12幢智能建筑按其使用功能可以分为三类：

(1) 酒店类：包括A酒店、B酒店两幢大厦；

(2) 政府、企业专用办公类：包括C政府办公楼、D政府办公楼、E企业办公楼、F企业办公楼四幢大厦；

(3) 商用、半商用出租型办公类：包括G、H两幢商用写字楼，I、J、K、L四幢半商用写字楼。

下面将就上述12幢大楼能源消耗情况作一分析。

2.1 各大厦使用能源情况(表1)

各大厦所使用能源情况 **表1**

楼宇名称	电	天然气	柴 油	备 注
A酒店	✸	✸		
B酒店	✸	✸		
C政府办公楼	✸	✸		
D政府办公楼	✸			VRV系统
E企业办公楼	✸	✸		
F企业办公楼	✸		✸	
G商用写字楼	✸		✸	
H商用写字楼	✸		✸	
I半商用写字楼	✸	✸		燃油锅炉改成燃气锅炉

续表

楼宇名称	电	天然气	柴　油	备　注
J 半商用写字楼	✹		✹	
K 半商用写字楼	✹		✹	
L 半商用写字楼	✹			

说明：用天然气或者用油主要根据建设时能源系统的设计而选择，有的大厦原配置油气两用锅炉，在当前油价高涨的情况下就全部使用天然气，节约费用；有的大厦原先配置的是燃油锅炉，后来改造成燃气锅炉，也得到了很大的实惠；对于那些没有条件进行油改气的大厦来说，现在燃油的费用很高，运行成本的压力很大。

2.2 各大厦年用电量、电费统计(表 2、表 3)

各大厦年用电量统计(04 年数据)　　**表 2**

楼宇名称	面积 m^2	全年总量(度)	最高月份用量(度)	全年单位面积用量(度/m^2)
A 酒店	65,000	10,639,400.00	1,220,708.00	177.32
B 酒店	53,000	6,864,360.00	886,660.00	127.12
C 政府办公楼	28,000	2,779,800.00	382,400.00	99.28
D 政府办公楼	7,000	776,805.00	84,000.00	110.97
E 企业办公楼	28,000	5,290,966.00	572,439.00	170.68
F 企业办公楼	64,000	7,356,144.00	964,242.00	118.65
G 商用写字楼	68,000	7,885,590.00	780,840.00	140.81
H 商用写字楼	49,000	4,789,206.00	510,000.00	122.8
I 半商用写字楼	77,000	9,664,900.00	1,144,100.00	125.52
J 半商用写字楼	100,000	16,520,000.00	1,672,000.00	165.2
K 半商用写字楼	50,000	4,470,870.00	616,800.00	89.42
L 半商用写字楼	60,000	15,057,822.00	1,626,597.00	250

各大厦年电费统计(04 年数据)　　**表 3**

楼宇名称	面积 m^2	全年总费用(元)	最高月份费用(元)	全年单位面积费用(元/m^2)
A 酒店	65,000	7,556,812.39	875,213.73	125.95
B 酒店	53,000	5,299,796.19	800,279.83	98.14
E 企业办公楼	28,000	4,232,772.80	457,951.41	151.17
C 政府办公楼	28,000	2,223,840.00	305,920.00	79.42
D 政府办公楼	7,000	621,444.00	67,200.00	88.76
G 商用写字楼	68,000	6,219,276.04	643,431.21	111.06
H 商用写字楼	49,000	3,763,152.38	419,542.73	96.49
I 半商用写字楼	77,000	8,045,300.00	942,200.00	104.48
J 半商用写字楼	100,000	12,636,732.00	1,313,496.00	126.37
K 半商用写字楼	50,000	3,576,696.00	493,440.00	71.53
F 企业办公楼	64,000	5,884,915.20	771,393.60	93.41
L 半商用写字楼	60,000	12,046,258.00	1,301,278.73	200.77

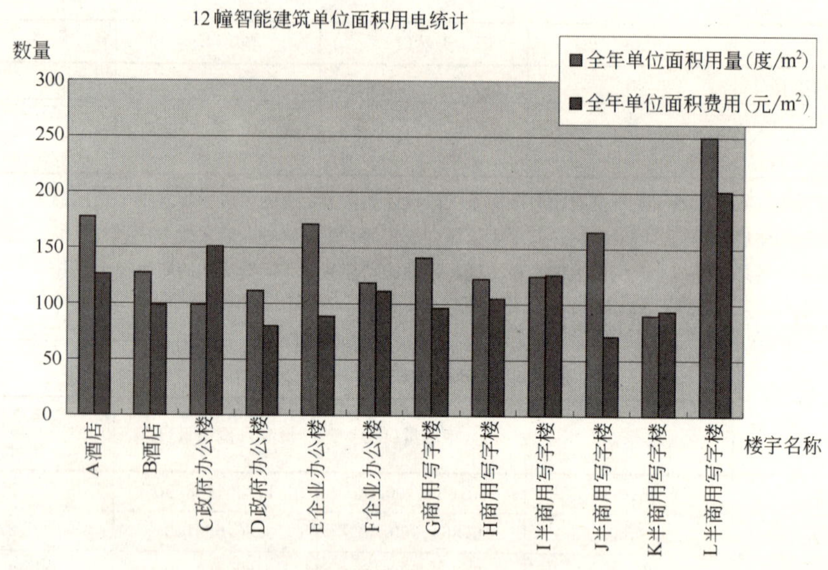

图 1 12 幢智能建筑单位面积用电统计

图 1 显示出 04 年全年单位面积用电量居于前三位的大厦分别是 I 半商用写字楼、E 企业办公楼和 A 酒店，以此可以作如下分析：

(1) I 半商用写字楼是一幢包含 3 万多 m^2 大型商场的综合性大厦，由于商场对空调、照明的特殊要求，造成大厦电能消耗巨大。如果扣除商场的生产性用电，大厦的全年单位面积用电量和纯商业办公楼相当，趋于合理。可见，商场是耗电大户，特别需要采取有效措施节约用电。

(2) E 企业办公楼是一幢自用型机关办公楼，与同类企业办公楼相比其用电明显较大。据了解大厦物业管理公司是根据自身办公和保证领导机关的需要，楼宇内舒适度要求较高，因而能耗较大。然而，实际上原 BA 系统设计的功能不完整，加上大厦物业管理不够完善是造成能耗较大的重要原因，因此提高物业管理公司管理水平也是发掘节能空间中的重要一环。

(3) 酒店类建筑与其他类型的建筑相比能耗一般较高，并且根据不同星级的酒店，具有较高星级的酒店其能耗也相应较高。A 酒店是一幢年营业额过亿的五星级酒店，酒店对于能耗节约非常重视，投运后采取了一系列的节能措施并且增加了很多先进设备来降低能耗，保障服务质量。

2.3 各大厦主要燃料年用量统计(表 4、图 2)

各大厦主要燃料年用量统计(04 年数据) **表 4**

楼宇名称	面积 m^2	燃料类型	全年燃量	全年燃油(气)费(元)	全年单位面积费用(元/m^2)
A 酒店	65,000	天然气	768336m^3	1,997,674.00	30.73
B 酒店	53,000	天然气	400000m^3	1,040,000.00	19.62
C 政府办公楼	28,000	天然气	139800m^3	363,480.00	12.98
D 政府办公楼	7,000	(用电)			

续表

楼宇名称	面积 m²	燃料类型	全年燃量	全年燃油(气)费(元)	全年单位面积费用(元/m²)
E 企业办公楼	28,000	天然气	662413m³	1,722,276.00	61.51
F 企业办公楼	64,000	柴　油	200 T	760,000.00	11.88
G 商用写字楼	68,000	柴　油	450 T	1,710,000.00	25.15
H 商用写字楼	49,000	柴　油	320 T	1,216,000.00	31.18
I 半商用写字楼	77,000	天然气	115384m³	300,000.00	3.9
J 半商用写字楼	100,000	柴　油	465 T	1,767,000.00	17.67
K 半商用写字楼	50,000	柴　油	180 T	684,000.00	13.68
L 半商用写字楼	60,000	(用电)			

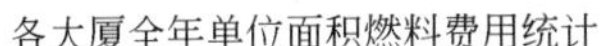

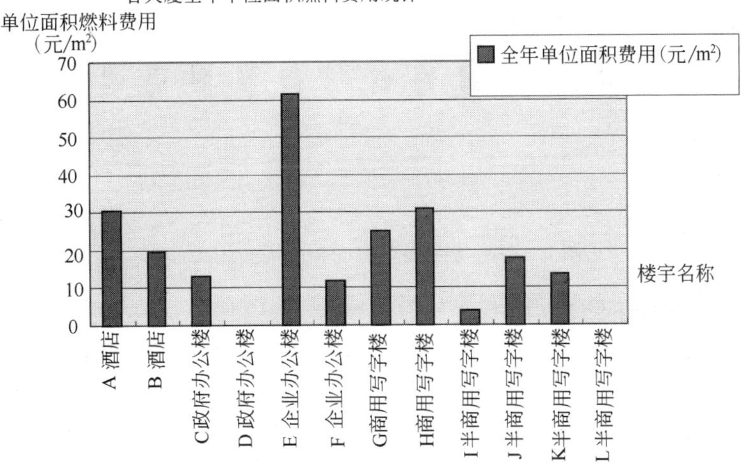

图 2　各大厦全年单位面积燃料费用统计

在燃料费上，受油价上涨关系，用柴油作为燃料的楼宇在每平米燃料费上要远远高于使用天然气的楼宇。

2.4　各大厦年单位面积能耗费用统计(表 5)

各大厦年单位面积能耗费用统计(04 年数据)　　**表 5**

楼　宇　名　称	面积(m²)	全年单位面积费用(元/m²)
A 酒店	65,000	156.68
B 酒店	53,000	117.76
C 政府办公楼	28,000	92.4
D 政府办公楼	7,000	88.76
E 企业办公楼	28,000	212.68
F 企业办公楼	64,000	105.29
G 商用写字楼	68,000	136.12
H 商用写字楼	49,000	121.67

续表

楼　字　名　称	面积(m²)	全年单位面积费用(元/m²)
I半商用写字楼	77,000	108.38
J半商用写字楼	100,000	144.04
K半商用写字楼	50,000	85.21
L半商用写字楼	60,000	200.77

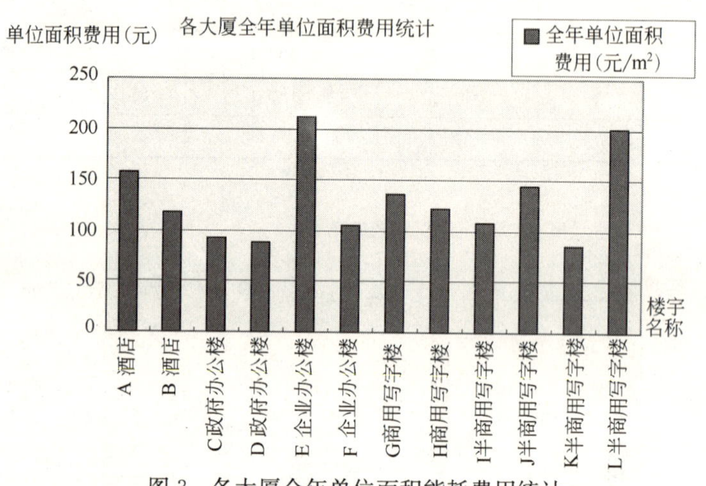

图 3　各大厦全年单位面积能耗费用统计

由图 3 可以得出如下结论：

(1) 每幢大厦的能耗状况依各大厦的设备使用状况和节能手段不同而不同，可以看到一些节能措施和节能政策贯彻较好的大厦其能耗相对较低。

(2) 在一些政府办公大楼，因为大楼是内部使用，所以对设备可以做到集中控制，而一些商务出租的办公大楼就不能做到物业部门的完全控制。比如说现在市政府提倡大楼内的空调温度要在 26℃以上，在调研中发现办公大楼一般只能做到在公共区域控制到 26℃，而租用单元是由租户自己控制，达不到市政府的这个规定。就这一点来说，政府办公大楼可以做到更好的节能效果。

(3) 从调研的数据来看，同样功能、规模相近的建筑，他们的能耗可以相差一倍以上。可见采用必要的方法与措施之后，建筑能耗是能够降下来的，是我们可以去发掘的节能空间。

3　智能建筑节能措施

智能建筑由于配置了大量的建筑设备，采用了先进的技术手段，建成了功能强大的智能化系统，因而在节能方面更加灵活、便捷，可以具有比传统建筑更多的节能手段。例如在空调系统中可以采用的节能措施有：

(1) 冷冻水温度设定：提高冷冻水温度，可以达到节能效果。在保证舒适的前提下，现场操作人员根据每个季节及每天室外温度的变化情况，设定冷冻水的出水温度。

（2）对于大堂、走廊、办公室主要能保证舒适为前提，适当放宽控制要求；提高设定温度，如进门的前厅在夏季将温度设定值设在28～30℃，比室外低4～5℃，人们已感觉舒适；廊道设定值定在27～28℃已满足要求；办公区定在26℃左右，同样可以令人感觉舒适。

（3）新风量控制：根据季节变化，进行合理的新风量有效调节是节能的另一个措施。在设计工况（夏季室温26℃，相对湿度60%；冬季室温22℃，相对湿度55%）下，处理一公斤室外新风量需要冷量6.5kW，热量12.7kW，故在满足室内卫生的前提下，适当减少新风量，有显著的节能效果。可以实现新风量控制的措施有以下几种方法：

• 在回风位置设置CO_2或空气品质检测器，根据回风CO_2气体浓度或空气品质自动调节新风风门的开启度，改变新风量。

• 根据室内人员变动规律，并采用统计学的方法，建立新风风阀控制模型，以相应的时间而确定的运行程式进行程序控制新风阀，以达到对新风的控制。

• 对新风机组、空调机组及通风设备编制相应的时间程序、假日时间程序及事故程序，控制器除了能对机组实现PID控制及TEP（Time/Event Program）控制，还包含如下的节能条件：间隔运行、最佳启动、最佳关机、设定值再设定、夜间净化，并且当机组滤网发生压差报警时，能及时通知清洗滤网，保持空调风的清净程度。

（4）夜间设定：夏季通过大量引入夜间室外的低温新鲜空气，置换室内污浊的空气，以提高室内空气的品质，减少能量消耗。

（5）控制优化：以最佳组合方式控制设备的启停。通过楼宇自动化系统（BAS）对空调设备进行预冷、预热的最佳启停时间的计算和控制，以缩短不必要的预冷、预热宽容时间，达到节能的目的，同时在大厦预冷、预热时，关闭室外新风阀，不仅可以减少设备容量，而且可以减少获取新风带来的冷却或加热的能量损耗。

（6）风机水泵变频调速：一幢建筑物中有数百台风机水泵，由于空调的水系统与风系统运行状态是变化的，而且都小于设计的最大工况，因此广泛应用变频技术不仅可以改变流量使设备处于最佳工作状态，而且降低10%的流量可以节省27%的电力消耗。

另外，调研的12幢智能建筑都积极响应政府号召，根据中央关于“全民节约、共同行动”的有关精神，以及上海市的“百万家庭节约活动”的有关精神和《上海市建筑节能管理法》、《上海市能源标准汇编》、《上海市节约能源条例》等文件的指导，各大厦都积极贯彻中央精神，把节能工作做到实处。

在调研的12幢智能建筑中，各物业管理单位都把上海市关于节能方面的文件张贴到了醒目的位置，要求从领导到普通员工切实贯彻市委市政府精神。同时，出租型建筑物的物业管理单位还将文件下发到大厦内的各个公司，传达市委市政府精神，宣传节能的重要性。据各楼宇物业人员介绍，在开展“全民节约、共同行动”的活动以来，下班后办公室内照明忘记关闭的现象少了，大楼的室内公共区域的温度也普遍达到26℃以上。在商用写字搂的私用区域，各单位也普遍提升了自已区域的平均温度，使温度达到或接近26℃的标准。据不完全统计，建筑运行过程中大楼空调平均室内温度设定每提高1℃，整个大楼就可以平均节电8%。又如D政府办公楼物业人员根据平时大厅进出的人流不大的实际情况，关闭了大厅的空调，减少了不必要的能耗。

4 智能建筑节能水平的提升

智能建筑节能，首先需要考虑的是人的因素。因为再好的节能建筑、节能设备与节能技术，如果物业管理、维护人员的节能意识不强、技术水平不高而不利用这些设备、技术，建筑节能也只能是一纸空谈。只有充分重视节能，并且在日常管理、维护过程中主动发掘、创造与运用这些节能措施，才能真正实现建筑节能。在对12幢智能建筑调研过程中我们发现，物业管理与建筑节能具有强烈的相关性，因此，提升建筑节能水平首先需要进一步增强物业管理机构的节能意识，其次需要在建筑设备管理的技术与评价指标上创新来引导建筑节能，第三是通过引进节能技术与设备进行工程改造以降低建筑的能耗水平。

4.1 增强物业管理公司节能意识

此次调研的12幢甲级智能建筑都具备有智能化系统。这几年的实际运行的事实证明当初的系统设计比较合理，采用的技术也比较先进，满足了整个大厦的功能需求。在大厦的使用与运行中，智能化系统为物业管理人员简化操作、方便维护发挥了很大的作用，物业管理人员都认为建设智能化系统非常必要，投入很大的人力、物力去了解、学习、掌握智能化系统的使用，使其真正发挥作用。

但是，也有个别物业管理公司对智能化系统认识不足，以为建设初期选择了先进的技术设备、安装了超前的软件系统就足够了，忽视了对智能化系统后期的维护保养。例如，有的维护人员对空调机组温度传感器出现故障不去及时更换，存在检测误差也不去校正，导致BA系统因为得不到准确的现场数据而运行异常。传感器对于BA系统所起的作用犹如人的眼、耳、鼻对于大脑所起的作用，没有前端的正确反馈，大脑就不可能下达正确的指令，因而整个系统就不可能正常工作。久而久之，有的物业人员干脆把部分设备长期置于手动档，使BA系统失去了应有的作用，造成能源浪费。

智能化系统功能的体现与物业管理息息相关。在良好的使用、维护过程中，物业人员能够不断地对智能化系统进行改进，从而提升系统功能，更好地为大厦管理服务。物业管理公司是大厦节能的执行者，只有物业公司充分认识到了节能的必要性和紧迫性，节能的措施才能真正落到实处、真正被使用。节约能耗不仅为国家节约了能源，更减少了运营成本，对物业管理公司而言是一件两全其美的事情。然而，调研中也发现个别物业公司的节能意识淡薄，凭借自己财力雄厚不把建筑节能当回事。对此需要加强教育提高认识，并采取一些必要的手段。

能源相对紧缺与经济的快速增长是我国发展中的一对重要矛盾。作为建筑节能的最终执行者，物业管理机构更应当积极发挥自己的聪明才智、综合运用各种技术手段在建筑节能方面发挥更大的作用。

4.2 通过技术创新节能

物业管理公司主动想办法、积极实践推行节能。技术创新的节能方式不仅使大楼物业从技术中尝到了节能节电的甜头，也是推动智能建筑技术发展、提高企业自主创新能力的最好体现。在调研的12幢智能建筑中，有一些虽然只是很小的改进，却对节能节电有着很大的作用。

例如，某大厦为了解决高温天气下热泵机组风冷散热不足、冷量不够导致的效率低能耗高问题，在热泵机组外自建了冷水喷淋设施，把水喷淋到风冷散热片上，用水冷辅助风冷进行散热，大大提高了原有热泵的工作效率。原来热泵机组运行的电流为400A，采用水冷风冷相结合的技术措施后，热泵机组的电流下降到320A，下降了80A。仅这一措施就可以节电近20%。

又如，某大厦针对冷水管积垢与沉淀物多，管道流量不够的情况，采用了管道水倒流反冲管道内的堵塞沉淀物。通过长时间摸索，该大厦物业管理人员发现单纯提高空调水系统的送水温度并不能很好的节能。为了解决大厦内各区域温度不均和送水流量不够的问题，增加一台冷冻水泵，加大流量，取得了很好的效果。截止到今年7月份，该大厦用电比去年同期节省了4万，用油节省了20多t。由于今年上海高温天气多于往年，该大厦在气温升高的情况下还能使能源用量大幅度下降实属不易。

再如，某大厦的物业管理公司针对20台自动扶梯在无人使用时长时间空载运行的情况，自主引进了红外线检测装置，在检测到扶梯一定时间没有人使用后，就立即把扶梯切换到低速运行或停止运行状态，当有人进入扶梯区域准备使用扶梯时，则立即启动扶梯。该大厦在扶梯上仅此就节电75%。

总之，这类看上去简单的节能改进，往往也能取得良好的节能效果。正是这些点点滴滴的改进，使建筑物的节能水平得到不断的提高。

4.3 通过设备改造进行节能

这种方式已经成为现在节能的主要手段。在调研的智能建筑中，节能设备的引进是一种常见而有效的方式。常见的节能手段是变频技术的使用。一些楼宇对冷冻水泵，冷却水泵的控制增加变频器，还有一些大楼对风机加入变频器，这些变频器均由BA系统进行状态控制。对于风机和水泵设备，其风量与转速的一次方成正比，风压H与转速的平方成正比，轴功率N与转速的三次方成正比。采用变频器进行调速，当风量下降到80%时，转速也下降到80%，而轴功率N将下降到额定功率的51.2%；如果风量下降到60%，轴功率N可下降到额定功率的21.6%，因此采用转速控制方式来调节风量，在节能方面具有显著的效果。如果对一个额定75kW、50Hz的风机进行变频控制，当它的频率降低到40Hz的时候，所消耗的功率仅为额定功率的53.3%。也就是说，采用变频技术把风量降低到原来的80%的时候，所消耗的风机功率仅占原来的53.3%。物业管理机构了解了变频技术的这一优点，纷纷在楼宇中采用变频技术对风机水泵的控制进行改造。

有不少建筑物的能源系统在建设时存在缺陷或不合理的问题，在能耗费用的压力下，物业管理机构不惜化大代价进行改造。除了油改气外，还有一些大厦新置一台小容量的冷冻机来配合大容量冷冻机工作，使得冷冻机组群可以处于满负荷状态从而使运行效率最高，在低负荷运行时减少了大型机组效率低下的现象。

小结

建筑物建成投运后，其能源的消耗量是巨大的。目前，在大多数存量建筑缺乏完善的节能技术措施，节能的空间也是巨大的。我们的调查数据反映，完备、规范的物业管理可以有效节省能源消耗，提高服务质量。尤其需要指出的是，通过加强管理，充分发挥BA系统的作用，增加建筑设备的优化控制措施，能够快速地把一部分浪费的能源降下去。

The survey and analysis of Intelligent Building energy-saving

Cheng Dazhang
(Department of Electronics & IT Tongji University, Shanghai 200092 China)
Wang Changqin
(Department of Building Facility Management Tongji University, Shanghai 200092 China)
Wang Zuozhong
(Department of Electronics & IT Tongji University, Shanghai 200092 China)
Huang Jie
(Department of Building Facility Management Tongji University, Shanghai 200092 China)

Abstract: Through investigating of some intelligent building construction in Shanghai area, we find out a lot of cases of serious construction energy waste due to history and poor management reasons, we think there is more room for increasing the energy efficiency. According to recent trend of energy saving and consuming, we did our new study on the possible methodology in order to increase the level of energy consumption for the intelligent building.

参考文献

[1] 龙惟定，程大章主编. 智能化大楼的建筑设备. 北京：中国建筑工业出版社，1996
[2] 程大章主编. 智能建筑楼宇自控系统. 北京：中国建筑工业出版社，2005

现代建筑能源环境智能化管理与平台构建研究

王东伟
（上海现代建筑设计(集团)有限公司，上海 200041）

摘　要：文章提出了建筑能源环境管理平台的概念。在现代建筑特别是大型公共建筑中，如何建立面向能源环境管理的平台，在系统间数据共享、协同工作的基础上，实现建筑能源环境监测，达到节能和环保的目的。本文就其功能、系统结构、及其与BAS、IBMS和FMS的关系进行了探讨。

关键词：建筑　节能　能源　环境　智能化　管理　平台

1　前言

随着我国经济的飞速发展，能源与发展之间的矛盾已经成为了社会各界普遍关注的问题。作为世界上最大的建筑市场，我国每年新建房屋建筑面积近20亿m^2，占全球的一半，仅北京或上海一个城市的新开工建筑面积，就相当于整个欧盟的总量。但是我国的单位建筑面积能耗是发达国家的2～3倍以上。根据近30年来能源界的研究和实践，目前普遍认为建筑节能是各种节能途径中潜力最大、最为直接有效的方式，是缓解能源紧张、解决社会经济发展与能源供应不足矛盾的最有效的措施之一。利用智能和IT技术，改变传统的建设方式和管理手段，以实现建筑运行的低能耗和管理手段的高科技化为目标，本文提出了建筑能源环境管理平台的概念。

本文所定义的建筑能源环境包括建筑能源和室内环境两部分。在现代建筑特别是大型公共建筑中，已包含了建筑设备管理系统(BAS)和物业管理系统(FMS)及相关的智能化系统，部分建筑物还在此基础上设置了一体化集成平台，如何建立面向能源和室内环境的管理平台，在系统间数据共享、协同工作的基础上，实现建筑能源环境监测，达到节能和环保的目，本文就管理平台的功能、系统架构及其与BAS、IBMS和FMS的关系进行了探讨。

2　建立建筑能源环境分析指标体系

通过对建筑能源环境要素的分析，建立建筑能源消耗的指标体系，以及室内舒适化环境与能源消耗的关系，以便精确把握建筑能源消耗的实际情况，通过管理方案，并结合智能化手段，实现设备运行和能源使用的合理化，达到以最小的能耗和污染排放，满足舒适的工作、生活和居住要求，实现人与自然的和谐共处。

建筑能源包括能源的供给和消耗。按照国际通行的分类，建筑能耗是指民用建筑(包括居住建筑和公共建筑以及服务业）使用过程中的能耗，主要包括采暖、空调、通风、热水供应、照明、炊事、家用电器、电梯等方面的能耗。我国城镇建筑能耗按性质可分为五大类：

- 北方地区采暖能耗；
- 除采暖外的住宅能耗（照明、炊事、生活热水、家电、空调）；
- 除采暖外的一般性非住宅民用建筑能耗（办公室、中小型商店、学校等）；
- 大型公共建筑能耗（写字楼、星级酒店、大型购物中心等）；
- 工业建筑能耗。

据统计，我国建筑能耗除采暖外，住宅能耗中的用电量为10～30度/m^2每年，大型公共建筑的用电量达100～300度/m^2每年，为住宅建筑用电量的10倍以上。目前这类建筑仅占城镇总建筑面积的5%～7%，但已有扩大趋势。与发达国家相比，我国这类建筑的平均能耗值高于日本水平，与美国的平均值大体接近。

如何建立建筑能耗和环境指标的统计和管理目标体系，是建筑能源环境分析的目的，下面提出一种建筑能源和室内环境关键要素分析，见图1，表1、表2。

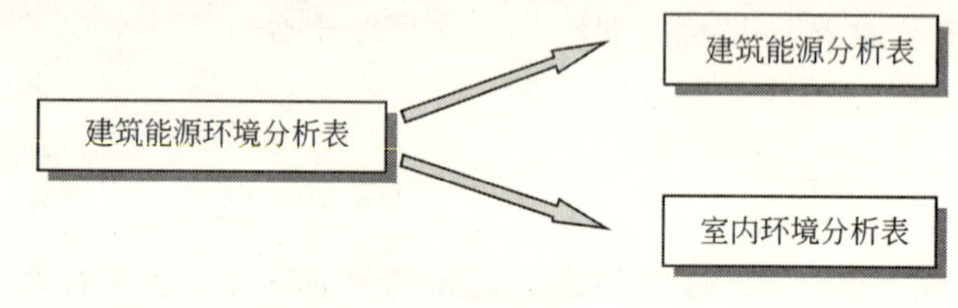

图1

建筑能源管理要素分析表 **表1**

建筑能源要素分析					
1	建筑物能源消耗量	每m^2耗量	5	按设备分类消耗量	冷冻机、锅炉等
	• 月、年统计 • 运行时间包括：建筑物和空调	人均耗量		• 月、年统计	泵（一次、二次）
		运行时间·m^2			空调机
2	按能源种类消耗量	电			送排风机
	• 月、年统计	煤气	6	按楼层消耗量	照明
		油		• 月、年按楼层统计	插座
		水			动力
3	按用途分类消耗量	空调	7	能源利用效率	冷热源设备
	• 月、年统计	照明		• 月、年统计分别进行两次比较	冷、热生成
		插座			水输送
		卫生			空气输送
		电梯			综合（水＋空气）
		其他	8	其他	VAV效果
4	按空调分类消耗量	冷热源		• 实际值与定值之比乘运行时间，按月、年统计 • 利用量与消费量之比，按月、年统计	VWV效果
	• 月、年统计	泵			根据CO、CO_2浓度控制换气量
		空调机			太阳能使用率
		送排风机			蓄热利用

室内环境要素分析表　　表 2

序号	室内环境要素分析		
1	室内温湿度	(1) 室内温度实现率 在合理范围内实际值与设定值之差绝对值/空调运行时间	夏季
			冬季
			平均(冷热平均)
		(2) 室内湿度实现率 在合理范围内实际值与设定值之差绝对值/空调运行时间	夏季
			冬季
			平均(冷热平均)
2	空气清洁度	(1) 室内浮游粉尘浓度	通过实际值与标准值的比较
		(2) 室内 CO 浓度	
		(3) 室内 CO_2 浓度	
		(4) 其他环境指标	
3	其他	室内照度	通过实际值与设定值的比较，进行月和年平均统计

通过对建筑能源环境各要素值的分析和比较，首先，可以指导建筑在设计、施工、运行和维护等各阶段的能源环境目标，其次，跟踪建筑能源消耗和环境参数的全过程，各设备之间的能耗关系和规律，从而能够对症下药，促进各阶段之间能源环境要素的实现，优化和改善建筑能耗和室内环境目标。

3 建筑能源环境智能化管理平台构建

建立了建筑能源环境分析的指标体系，如何用智能化方法采集分析要素的实际值、优化和改善目标值，这就是智能化管理平台的目标：通过对建筑能源运行数据的自动采集、精确统计和综合分析，进行节能和室内环境效果评估，并利用这个平台实现 BAS 和 FMS 进行数据交换和共享、相关数据的协调和有针对性的处理，以达到指导、改善和优化建筑能源环境的管理。

3.1 管理平台功能

建筑能源环境智能化管理平台主要具备以下功能：

1. 建筑能源环境数据的采集、分析和处理功能；

2. 建筑能源环境数据共享、交换和管理的平台功能；

3. 系统根据能源环境要素和管理指标建立节能和环境优化数学模型、设定控制参数，实现智能化管理，利用管理平台可以进一步实现：

- 能源环境状况实时化、可视化；
- 利用各项指标，具有与其他建筑物的可比性；
- 数据连续采集和分析，对能源环境具有可预测性；
- 可分析设计值和实际值的差异原因；
- 可评价改善和优化后的效果。

图 2 描述了管理平台的功能结构。

3.2 管理平台系统架构(图 3)

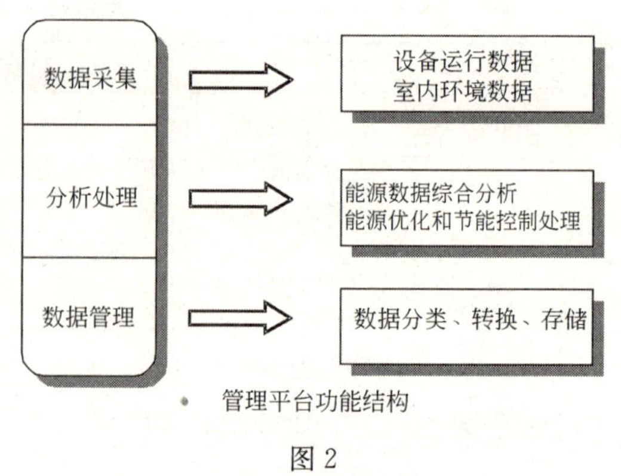

图 2

建筑能源环境智能化管理平台的系统架构描述了管理平台与 BAS 和 FMS 的数据流关系，以及各系统的功能之间的关系。管理平台的数据采集通过 BAS 获取，经过节能和环境优化数学模型分析和处理后的数据，又可作为 BAS 的设定值和控制参数；FMS 作为设施和运行管理系统，同样接受管理平台处理后的数据，以指导其设施管理、维护计划，能源用量、收费和设施运行调整，以及制定科学、合理的政策指导和制约用户对能源使用；三个系统通过管理平台实现数据交互、功能协调和分别处理，实现建筑节能和室内环境改善的目标。

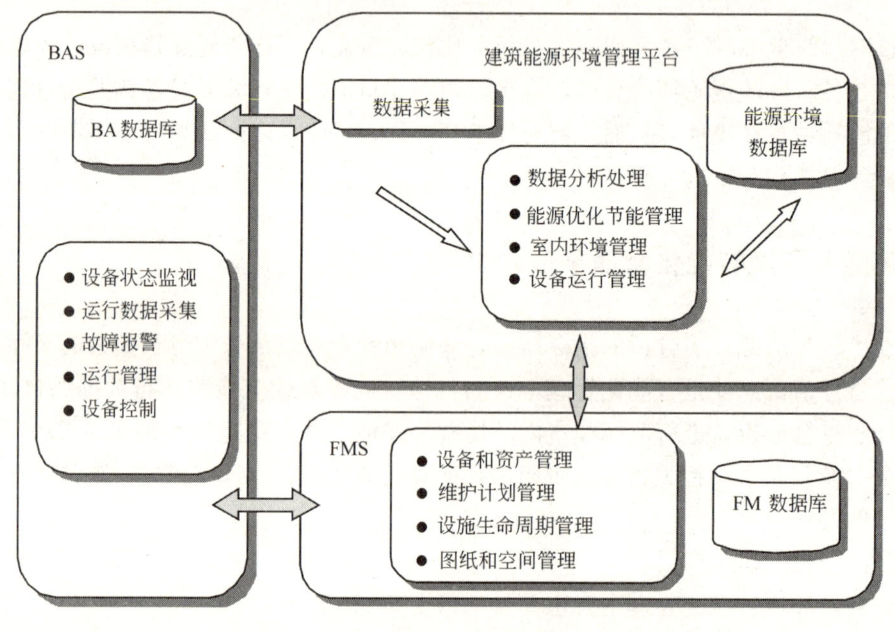

图 3

3.3 管理平台与 BA 等系统的关系

目前，大部分 BA 系统都宣布具备能源管理模块，但管理对象和节能目标、参数设定依据以及系统的开放性等各项因素，对 BA 系统设计者和使用者来说了解并不透彻，而建筑设施管理系统(FM)，在现有的 BA 系统或 IBMS 中也没有突出其面向能源环境的功能或数据关联。

建筑能源环境管理平台的价值在于它描述了三个系统间的功能分工和数据关系，同时，也为系统集成提出了新的目标，使 IBMS 从设备全局监控和跨系统联动，向以能源环境为驱动目标进行集成的方向演进。建筑能源环境管理平台是一个独立系统或嵌入在其他系统之内，将视其使用目的和已设系统的情况而定。

4 结束语

本文探讨在智能建筑已有相关管理系统的基础上建立建筑能源环境管理平台，以期实现建筑能源和室内环境的智能化管理，并对现有相关系统的概念和功能进行了扩展，具有实用价值和现实意义。

建筑节能是一项潜力极大的绿色发展机制。它对促进能源的节约和有效地保护环境，无疑起着积极的作用。我们应当抓住这一全球发展机遇，使我们的建筑更具时代特色，做到人、建筑、自然的更加和谐，为子孙万代留下一片洁净而广阔的蓝天，为经济的可持续性发展创造更大的空间。

Research on Intelligent Management & its Platform Technology for Energy Environment in Modern Buildings

Wang Dongwei

(Shanghai Xian Dai Architectural Design (Group) Co. , Ltd Intelligent Building Design & Research Institute, Shanhai 200041 China)

Abstract: The paper proposes the concept of buildings energy & environment management platform. In modern large scale buildings, how to build a energy & environment oriented management platform, where data from relevant system can be shared and cooperative work can be done, based on it, buildings energy & environment can be monitored, the energy saving and environment protecting. can be realized. The article studied the platform function, system architecture, and its relations with BAS、IBMS and FMS.

参考文献

[1] 建筑节能，中国建筑工业出版社，2005 年 7 月

[2] 佐藤孝輔モニタリガ/BEMS の 現狀 からみた 理想的な姿と推奨ァプローチ，2004 年 1 月

[3] 岡恒 晃 改正省ェネ法とBEMS 導入支援事業概要，2003 年 10 月

智能建筑节能分析及优化方法

李明海[1] 鲁 娟[1] 武六元[2]

（[1] 西安建筑科技大学 建筑设计研究院；

[2] 西安建筑科技大学 建筑学院绿色建筑研究所，西安 710055）

摘 要：本文从智能建筑可持续发展战略、智能建筑节能规划、工程技术经济性分析等方面分析了智能建筑节能问题，提出了系统优化的原则，并对智能建筑BAS系统控制方案优化方法进行了探讨。

关键词：智能建筑 可持续发展 节能分析 系统优化 BAS系统

智能建筑节能是世界性的大潮流和大趋势，同时也是中国改革和发展的迫切要求，这是不以人的主观意志为转移的客观必然性，是21世纪中国建筑事业发展的一个重点和热点。节能和环保是实现可持续发展的关键。可持续建筑应遵循：节约化、生态化、人性化、无害化、集约化等基本原则。这些原则服务于可持续发展的最终目标。

从可持续发展理论出发，建筑节能的关键又在于提高能量效率。因此无论制订建筑节能标准还是从事具体工程项目的设计，都应把提高能量效率作为建筑节能的着眼点。智能建筑也不例外，业主建设智能化大楼直接动因就是在高度现代化、高度舒适的同时能实现能源消耗大幅度降低，以达到节省大楼营运成本的目的。依据我国可持续建筑原则和现阶段国情特点，能耗低且运行费用最低的可持续建筑设计包含了以下技术措施：(1)节能；(2)减少有限资源的利用，开发、利用可再生资源；(3)室内环境的人道主义；(4)场地影响最小化；(5)艺术与空间的新主张；(6)智能化。

20世纪70年代爆发能源危机以来，发达国家单位面积的建筑能耗已有大幅度的降低。与我国北京地区采暖度日数相近的一些发达国家，新建建筑每年采暖能耗已从能源危机前的300kWh/m^{2}降低至现在的150kWh/m^{2}左右。预计在今后不会很长的时间内，建筑能耗还将进一步降低至30～50kWh/m^2。

创造健康、舒适、方便的生活环境是人类的共同愿望，也是建筑节能的基础和目标。为此，21世纪的智能型节能建筑应该是：*a*. 冬暖夏凉。*b*. 通风良好。*c*. 光照充足。尽量采用自然光，天然采光与人工照明相结合。*d*. 智能控制。采暖、通风、空调、照明、家电等均可由计算机自动控制，既可按预定程序集中管理，又可局部手工控制。既满足不同场合下人们不同的需要，又可少用资源。

1 智能建筑节能的现状及发展方向

1.1 智能建筑节能现状

虽然建设智能化大楼直接原因就是在高度现代化、高度舒适的同时能实现能源消耗大幅度降低，以达到节省大楼营运成本的目的。然而实际上在智能化目标定位中明确提出节能要求的不多，建成的确有节能功效的智能建筑更是罕见。主要原因有以下几点：房地产投资商认为能源建设是政府业绩和投资环境的标志之一，建筑节能看不见摸不着，所以宁可在建筑的豪华和设施的先进上花巨资，却不愿为节能多花一分钱；系统与设备、设备与设备之间的控制集成成功率不高，相关系统结合“接口”界面如通信协议、网络构架的标准化、统一性不够；物业管理水平跟不上，运行优化无从谈起；没有从通过可再生资源利用技术来提高建筑的可持续性战略高度重视智能建筑的节能；缺少正确有效的建筑设备能耗评估方法，限制了智能建筑节能研究更好地开展。

1.2 智能建筑节能的发展方向

中国的建筑节能工作已经走过了20多年，在市场经济的推动下，随着住房体制改革的前进，房屋用能费用理所当然地要由住户承担，节约建筑用能势必逐渐成为广大居民的自觉要求，加上改善大气环境愈来愈迫切要求减轻建筑用能带来的污染，建筑节能将是大势所趋，人心所向，是国家民族利益的需要。

智能建筑节能是一门新兴科学，与原有专业分工不同，它包含有建筑、施工、采暖、通风、空调、电器、家电、建材、热工、能源、环境、检测、计算机软件等许多专业内容，是许多专业学科边缘交叉并结合形成的，在社会需要的推动下，许多高等院校、科研院所和生产单位都在围绕不同方面进行研究开发，正在出现蓬勃发展的新局面。

2 智能建筑的节能分析和评价原则

2.1 做好智能建筑的节能规划

节能规划要从可持续发展的战略高度出发，采用新方法、新思路。节能要从原先的拾遗补缺，变为在技术经济分析可行后优先考虑的方案，要以提高能源利用效率和利用效益为中心。总的节能目标要根据经济发展、能源平衡、能源消费弹性系数和节能率来编制。建筑节能方面，建筑物的设计和建造应当依照有关法律、行政法规的规定，采用节能型的建筑结构、材料、器具和产品，提高保温隔热性能，减少采暖、制冷、照明的能耗。逐步开展建筑物的节能认证。目前建筑节能标准可以分为两大类，即规定性标准和效益型的标准。规定性的节能标准给出一定的节能指标要求，如外墙的最大传热系数和最大窗墙比等；效益型的节能标准对一些指标并不规定过死，只要所设计的建筑物总能耗满足标准要求即可，所以设计人员有更大的设计灵活性。

2.2 智能建筑的经济性分析应坚持节能原则

对智能建筑进行技术经济性分析是评价智能建筑节能效益的依据。

技术经济就是用尽可能少的劳动消耗和劳动占用为社会提供更多、更好的使用价值，即要求以最少的劳动消耗和劳动占用，取得最大的经济效益。一个建设项目的设计，从资源利用、工艺流程、总体布置、设备选型、能源消耗及工程效益等等的考虑，无不关系到技术经济，所以技术经济在工程设计中所占的地位自然是十分重要的。不重视技术经济的分析研究和评价，或者不始终做好技术经济工作，就会导致建成投产的项目效益不高，花了不少投资而没有达到预期的效益。

2.3 采用层次分析法对智能建筑的综合评价也离不开节能原则

层次分析法常常被用在对智能建筑的综合评价中。

层次分析法是系统工程学中一种能处理具有复杂因素在内的技术、经济和社会问题。这些问题往往很难用定量的模型或模拟来分析，因为其中所含定性因素很多，而且需要考虑决策者的心理因素、知识经验和决策水平等。而层次分析法则能通过建立所谓判断矩阵的过程，逐步分层地将众多的复杂因素和决策者的个人因素综合起来，进行逻辑思维，然后用定量的形式表示出来，从而使复杂问题从定性的分析向定量结果转化。

利用层次分析法对智能建筑评价采用以下评价准则：满足用户需要的程度；建筑物的节能程度；建筑物使用者舒适性；建筑物管理者管理的方便程度；在现有基础上进行扩充的可能性；整个智能建筑的性能价格比。

3 智能建筑 BAS 控制方案的优化

3.1 智能建筑能量控制与管理系统的优化

智能建筑楼宇自控系统将建筑内所有设备集成一个系统，实现信息共享，进行综合管理，其作用和效益是巨大的。要实现这些作用和效益，就必须实施优化。建筑智能化工程的最优化设计与常规设计相比，有以下特点：(1)可以从系统的各种可能结构和参数中找到最佳匹配，使整体效能最佳，从而提高系统的效率，降低投资和运行费用；(2)可以对系统及其过程进行定量化的状态模拟，减少控制环节，提高可靠性与稳定性，发生故障概率降到最低可能限度，系统响应输出最优化；(3)它有不变条件下的优化控制及优化控制条件的两种型式，通过优化控制方案达到节能目的的是一种“主动节能”，它有别于墙体结构、门窗的形式和设置的改造的“被动节能”。智能建筑能量管理与控制系统优化的基本出发点、优化原则及技术措施见表1。

智能建筑能量管理与控制系统优化的基本出发点、优化原则及技术措施　　表1

基本出发点	优化原则	技术措施
VAV系统与BA联网	优化VAV送风静压和新风量，节约能耗，改善室内空气品质	在温度控制、新风量控制、送风压力控制及变风量末端装置温度控制基础上，优化送风静压(基于VAV末端装置的风量调节风门开度)和新风量的设定(基于室内人数)
VAV系统智能控制	运用智能对系统进行全局控制，不需系统建模，解决控制回路耦合带来的诸多控制问题	神经网络控制、模糊控制、专家系统
空调水系统变流量控制	冷热媒水供回水温度保持不变，根据冷热负荷调节冷热媒水的流量，提高系统的运行效率，避免无功损耗，达到充分节能	根据冷热媒水回水温度参数，及时调节水泵工作特性曲线(变流量)使冷热媒水的回水温度趋于恒定

续表

基本出发点	优化原则	技术措施
智能变配电	克服“只监不控”，自动调峰，调节负荷率，平峰填谷，三相平衡，降低运行费用	利用BA系统与冰球蓄冷设备、应急发电机等配合，用电高峰卸除不重要机电设备，投入应急发电机，释放存储冷量。配电网自动化系统
智能照明控制系统	定时及实时灯光自控及故障指示，自动实现合理的能源管理，节省能源，降低成本，保护灯具	模块化结构，分布式网络计算机控制
电梯	防止交通阻塞，节约能源，扩大楼宇有效使用空间。远程监控与智能故障报警诊断，提高运行与维修效率	综合控制计算机预测交通需求、阻塞形式；在群控系统中采用模糊理论、神经网络、专家系统。基于时变隶属度计算的模糊推理
蓄水池进水	利用数据处理、网络功能推进新的工程目标	用电动阀门与水池内的液位装置取代浮球阀，实现程序控制
蓄水池出水		利用程序控制来管理，使池内水位降至某一定值时生活泵自动停止，防止消防水量平时被动用，抬高生活泵吸水管所造成的死水区水质变坏，低水位时水泵工作失稳
变频调速供水	新技术实现传统的工程目标，使节能效果向理论值靠近	压力传感器安装在本给水系统的最不利点，对水泵变压变量供水

3.2 基于节能策略的智能建筑BA系统优化方法(主要针对空调系统)

3.2.1 控制策略的优化

PID控制：空气处理机的DDC通常采用PID控制，选择合适的PID参数对空调系统的稳定运行是非常关键的。PID系数高，空调对室内温度波动的反应特性曲线陡，达到设定温度的过渡过程较短；相反，PID系数低，达到设定温度的过渡过程较长。但并不是PID系数越高越好，否则易引起DDC控制系统失稳，表现为室内温度的振荡和水侧的电动调节阀周期性的来回运动无法在固定开度上运行。PID能解决大部分场合的空调控制，但对于影剧院等大热惯性空调场合，靠高的PID系数来提高空调机组对负荷变化的响应速度是不足以解决问题的。这时可以采用双级控制，即分别在空调的送风道和室内安装温度传感器，室内的温度设定由主DDC控制器完成，水阀的驱动由副DDC根据风道温度传感器和主DDC的指令完成，由于风道温度变化速度快于房间温度的变化，这一控制方式加速了系统对温度波动的响应。必须注意的是，为防止水阀被人为关死或水系统供水不足等异常情况对控制系统的影响，副DDC通常只需采用比例控制算法(P)，不可加入积分分量(I)。在实际的工程设计中，BA系统对空调的节能控制有多种手段可以采用，例如室内外焓值比较法、二氧化碳等污染物浓度检测法确定新风量，基于日程表的定时操作等等。工程设计中可以视需要灵活运用，以达到最优的效果。例如，办公、商场等场合，夏秋季在清晨时通过程序启动空气处理机(或新风机)，利用室外凉爽空气对室内全面换气预冷，既节约新风能耗又提高了室内空气品质。

3.2.2 控制权的优化

通常BA遵从的是中央控制站集中管理的原则。有时也有其不便的一面。在某些场合

（如会议室）将空调、通风系统的参数的设定功能放置在现场可能更符合使用者的需要。DDC本身并不提供这样的功能，需要专门部件来实现。例如：Honeywell T7780A 数字显示墙挂式 Lon 分站，可以在液晶面板上显示房间的温度，通过 4 个按钮完成设定房间温度，风机速度、启动/停止风机等功能，并能通过 LonTalk 总线驱动空调箱的 DDC 控制器执行相应的动作。这类功能接近 VRV 控制面板的设定器给房间的使用者带来极大的便利和舒适性，必要时应积极采用。

3.2.3 直接数字控制器（DDC）的优化

主流 BA 系统供货商都能提供大中小不同处理能力的 DDC，冷冻机房、热力站监控点是密集场合应优先采用大型控制器，以减少故障率和控制器间的通讯。这种控制器的典型特征是有强大的处理器（如：摩托罗拉的 68302）和 1MB 以上的 RAM，尤其是能够和I/O扩展模块连接达到输入输出功能的扩展，例如 SIEMENS MBC 可以挂接 40 个I/O模块，Trend IQ251 控制器允许有 8×16 点的接口扩展能力。对空气处理机、新风机、通风机一般采用中型或小型的控制器即可。近年来，可编程逻辑控制器件（PLC）进步很快，其应用不再局限于工业场合，在空调通风的现场设备控制工程中不应将其排斥在外。

3.2.4 关于 BACnet 和 LonWorks 的支持

BACnet 和 LonWorks 的提出是为了在不同层面解决控制系统的互联互操作问题。LonWorks 采用现场总线控制技术（FCS），突破了以往的集散型控制技术（DCS）。它的提出是 BA 技术的一次飞跃，是今后 BAS 发展的技术趋势。但目前受到各种条件限制 LonWorks 技术优势还不能完全发挥出来，工程实现并不完善。若自控系统规模不是很庞大，最好不用全面采用 LonWorks 技术。如果仔细分析目前主流的 BA 产品，会发现其实它们在不同程度上部分采用或部分支持现场总线技术。这种取长补短的混合模式在当前是切合工程实际的，可以实现技术的平稳过渡。BACnet 协议由美国 ASHRAE 制定并颁布，是现行美国国家标准及欧共体预备标准。运用 BACnet 协议，可以使空调冷热源主机自带的控制器直接进入 BAS 控制网络。但实施中应注意空调主机和 BA 供货商对 BACnet 的支持程度和技术协调，目前非标准的数据通讯格式仍大量存在于主机设备中。总之，在设计 BA 系统的过程中切不可一味追求技术的先进性。

3.2.5 控制网络优化

在满足扩展性和灵活性的前提下，控制网络的拓扑结构应尽可能简化、清晰，无论基于 RS485 总线或基于 LonTalk 总线的控制网络都是如此。分支、分级多的网络管理复杂、可靠性低。LonTalk 总线在理论上可以组成任意拓扑结构的网络，这种布线设计的随意性如果运用不当，在工程实践中仍然是有技术风险的，并可能增加系统的投资。小型工程尽可能运用基于 RS485 总线的控制网络，采用“手拉手”的布线方式，大型工程可以考虑楼层网络分级。

3.2.6 BAS 监控中心

BAS 监控中心负责监控整个空调、通风、动力系统，一般与消防控制、安保监控等合用一室。由于该机房通常远离冷冻机房、锅炉房，在这里远程操作这些关键设备是不合适的。推荐的做法是在冻机房和锅炉房现场控制室另设置一台监控分站，由该分站负责冷冻机、锅炉监控功能，并且该分站功能受权局限为冷热源设备。

4 结论

智能建筑节能是一门新兴科学，与原有专业分工不同，它包含有建筑、施工、采暖、通风、空调、电器、家电、建材、热工、能源、环境、检测、计算机软件等许多专业内容，是许多专业学科边缘交叉并结合形成的。对智能建筑的分析和评价应坚持节能的原则。确立智能建筑能量管理与控制系统优化的基本出发点、优化原则及技术措施对于智能建筑节能实现具有重要的现实意义。

BAS 系统是实现智能建筑节能的有效途径之一。智能建筑 BAS 控制方案的优化是整个智能建筑节能优化方案实施的具体体现，它包含了建筑物内部主要耗能单元的节能优化。通过对 BA 控制系统(主要是暖通空调系统)的传感器、执行器、控制器、网络等若干环节的探讨，力图使 BA 系统更好地服务于受控的空调通风系统，最大限度地节约建筑物能源。

Study on Control of Heating Plant Based on Online Adaptive Temperatures Predictor

Li Minghai　Lu Juan　Wu Liuyuan
(Institute of Architecture Design, Xian University of Architecture and Technology, Xi'an 710055　China)

Abstract: Aiming after analyzing the strategy of the developing intelligent building, the plan of energy efficiency in intelligent buildings and the engine economy system, this article indicate the strategy of optimal rules to solve the problems of energy efficiency in intelligent buildings mentioned above. Using a compare method, the control means of BAS system in intelligent buildings were detailed. It pointed that it can be used for energy efficiency in building.

参考文献

[1]　涂逢祥. 建筑节能 [M]. 北京：中国建筑工业出版社，2001
[2]　吴建兵. 楼宇自动化系统(BAS)的优化设计 [D]. 上海：上海众慧科技实业公司，2003
[3]　蒋澄. 楼宇自控系统设计浅析 [D]. 江苏：常州亚细亚电子技术有限公司，2004

基于突发事件的智能建筑设备虚拟运行系统的设计

余　庄　高　威
（华中科技大学 建筑与城市规划学院 智能建筑研究所，武汉　430074）

摘　要： 本文介绍了虚拟现实技术，通过对现有应用于智能建筑设备管理与研究系统的分析，提出将虚拟现实技术应用于智能建筑相关研究中的优势与意义。主要讨论了基于突发事件的智能建筑设备运行虚拟现实系统的设计与架构。通过对系统仿真过程的分析、研究后得出系统的设计方法及流程，介绍了基于MFC/VEGA的虚拟现实系统开发技术，描述系统的系统结构以及系统实现的关键技术。

关键词： 智能建筑　虚拟现实　突发事件　VEGA

1　引言

随着信息化、网络化的高速发展，人们对建筑物的高效性、多功能性、安全性等方面的要求也随之提高，智能建筑应运而生。在智能建筑中，一旦发生诸如火灾、非法入侵等突发事件，都可能造成人员伤亡和经济上的巨大损失，如何将各种先进的信息技术应用于智能建筑中，使之更为安全、有效地运作已成为智能建筑发展所面临的一个问题。

虚拟现实(Virtual Reality)是把客观上存在的或并不存在的东西，运用计算机技术，在用户眼前生成一个虚拟的环境使人感到沉浸在虚拟环境中的一种技术。虚拟现实技术是综合了计算机图形学、计算机仿真、传感技术、显示技术等许多学科的基础上发展起来的一种计算机应用新领域[2]。

1993年，Burdea G在Electro 93国际会议上发表的“Virtual Reality System and Application”一文中，提出了虚拟现实技术三角形。即三个“I”特征：Immersion（沉浸）、Interaction（交互）、Imagination(构想)，如图1所示。它是虚拟现实系统的三个基本特征，表示虚拟现实系统能够使人沉浸到计算机系统所创建的系统中，能够利用多种传感器与多维化信息的环境发生交互作用，能够从定性和定量综合集成的环境中得到感性和理性的认识。

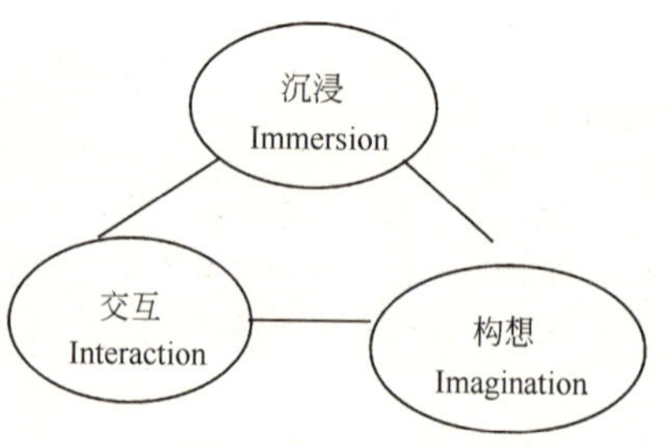

图1　“三角形”虚拟现实技术体系结构

突发事件(Emergency)是指一切与智能建筑的结构、系统、服务和运营有关的，影响建筑物安全、节能、高效率和高舒适性等目标实现的活动和过程。这是一个广义和抽象的概念。本文只讨论与建筑智能化系统有关的突发事件，主要包括建筑智能化系统在运行过程中出现的过程故障、自适应和优化过程中非例行操作、非法入侵以及火灾等。

目前应用于智能建筑设备管理与研究的系统大多数还停留在二维平面上，仅仅提供数字或图形化的结果来供研究。将虚拟现实技术运用于处理突发事件的智能建筑设备运行研究中，不但可以提供一个逼真的三维实时虚拟环境，模拟出当突发事件发生时建筑内部设备的运行状态，弥补许多在现实研究中的缺陷，更重要的意义在于它还可以方便的获得大量的实验数据供我们选择、处理、分析，以此减少突发事件带来的人员及财产损失。

2　系统的设计及描述

基于突发事件的智能建筑设备运行虚拟现实系统的设计是一项复杂的系统工程，技术层次涉及到各个独立的实体及虚拟场景模型建立、仿真环境生成、通道设置、图形状态设置、观察者设置、光源设置、实体运动方式设置，以及碰撞检测、环境效果和特效等内容，知识层次涉及到决策支持、计算机仿真、人工智能、计算机图形图像学、工程数学和安全疏散理论等知识，无论从技术支持和理论知识来讲，都是值得深入研究的。特别是对于不同的应用研究领域，还要根据其特点来进行具体分析，然后建立系统。

本文所描述的设备运行虚拟现实系统是依托于决策支持系统的，它们的逻辑关系如图 2 所示。智能建筑的数据采集系统采集数据源，经由数据集成系统形成格式统一的数据存储于数据库中。在数据库的基础上，形成 CBR 的知识库，形成决策支持系统。当有突发事件发生时，决策支持系统提供决策，并通过决策支持接口将决策信息提供给设备运行虚拟演示系统。关于本系统中基于处理突发事件的智能建筑系统集成相关部分内容参阅文献［1］。

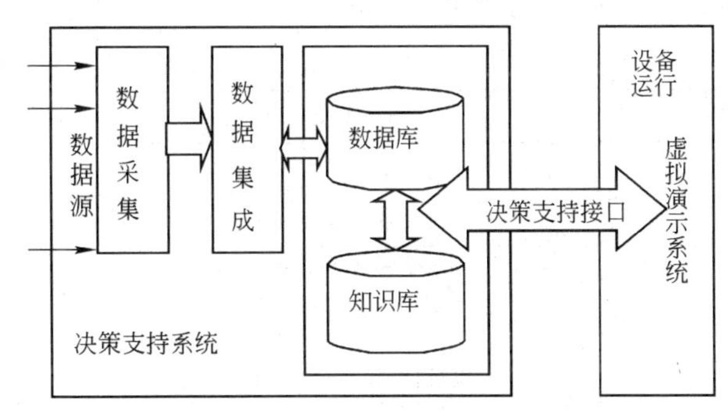

图 2　决策支持系统与虚拟演示系统关系图

设备运行虚拟演示系统的仿真过程为：当突发事件发生时，决策支持系统对突发事件发生状态下建筑设备应采取的运行策略提供决策，并向虚拟演示系统提供决策支持接口。场景内的各设备通过决策支持接口依照决策指令进行运行。用户依照所观察到的场景内智能建筑设备运行效果，经由主计算机系统控制外围设备，如计算机鼠标、游戏控制杆、数据手套等操作设备来模拟参与者的运动。参与者依照场景内建筑设备的运行情况，来选择自己的运动状态，系统对参与者的运动效果实时的反映输出到外围设备，如计算机显示器、立体眼镜等上。参与者可以根据这一个循环过程所产生的结果决定下一个循环过程所应进行的操作，由此实现真实的模拟和控制效果。在整个过程中，系统将参与者有效的运动信息写入后台数据库文件，供研究人员处理、分析。

2.1　系统架构选择及设计流程

基于以上对系统整体架构及仿真模拟过程的讨论，可参考选用 MutilGen Paradigm 公司 Vega 作为主要的虚拟场景开发平台，辅助使用 3D MAX、PolyTrans、Multigen Creator、OpenGL 来进行开发。Vega 是一个为建立实时仿真和虚拟现实应用程序而开发的高性能软件环境和工具库。它由图形用户界面程序 Lynx、一组与 Lynx 一起工作的工具(如预览工具、定位工具等)、Vega 库和可调用函数的头文件组成。采用 Visual C++中的 MFC 来调用 Vega 进行仿真程序设计可以获得功能更完善、界面更友好的实时仿真程序。但值得注意的是，由于 Vega 提供的是 C 语言应用接口程序 API，所以决策知识系统所提供给虚拟现实系统的后台数据库及知识库必须选用能够用 C/C++驱动的数据库，可参考选用如 MS SQL Server2000、ORACAL 等来进行架构，这样系统程序可方便的调用数据库数据来进行设备的决策演示。

根据系统设计、仿真过程、Vega 设计流程以及 Visual C++开发的一般过程，设备运行虚拟现实系统的设计过程总体上可分为五个部分：首先，分析准备层次的设计，如设定虚拟演示的策略和内容，确定建筑设备运行规则等。其次，进行建模层次设计，运用多种工具建立能被 Vega 调用的实体模型。再次，仿真环境层次设计，使用 Vega 的 Lynx 应用程序将所建的实体模型调入仿真环境，并设置相应参数，生成程序定义 ADF 文件。然后，进行开发层次的设计，在 MFC 开发环境中，通过对 Vega 系统函数和窗口函数的访问，对 Vega 应用程序提供实时控制。最后，对试验数据进行分析处理，完善知识库，给出建筑物设备在突发事件下的危险性等相关的评价与结论，给出人员安全疏散的建议。系统的详细设计流程如表 1 所示。

系统详细设计流程 表 1

序号	层　次	主要内容或参数	参考工具	数据库	相关层
1	分析准备层	设定演示策略和内容			
		建筑设备运行规则等	人工智能工具	知识库	
2	建模层	建筑实体及建筑设备模型	3D MAX PolyTrans MultigenCreator	实体模型库	1
		建筑周边地形地貌模型		地形地貌模型库	1
		建筑物内部物品及装饰模型		装饰物品模型库	1
3	仿真环境层	通道、观察者、物体、运动体、环境、场景等设置	Vega Lynx OpenGL	程序定义 ADF 文件	2
		火、烟雾、爆炸等特效		特效模型库	1
		相交矢量检测设置			
4	开发层	基于 Vega 的 Visual C++程序框架设计	Visual C++6.0 Vega API	程序文件	3
		特效及碰撞检测程序开发			2
5	分析处理层	分析实验数据		实验所得数据库	
		完善知识库及分析处理		知识库	1

2.2 系统结构描述

从设计流程的角度(如表 1 所示)，建筑设备运行虚拟现实系统的总体结构如图 3 所示。

3 系统的关键性技术描述

虚拟现实系统的实现主要可以分为虚拟环境制作和虚拟仿真驱动两部分。仿真环境的

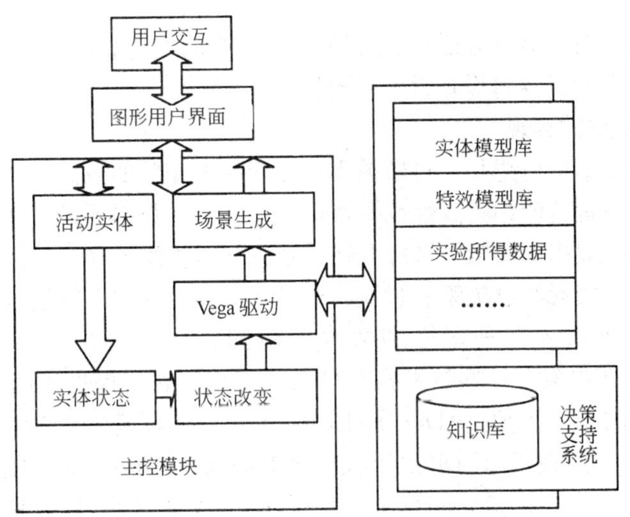

图3　建筑设备运行虚拟现实系统结构图

制作主要包括表1中所描述的建模层所要完成的任务，它要求构造出逼真的三维模型和制作逼真的纹理和特效。这一个步骤生成的仿真效果与真实自然环境中的物体外观越接近，最终的效果便越能够让人产生真实感。虚拟仿真驱动主要包括：场景驱动、模型调动处理、分布交互、算法及决策驱动等，它要求高速逼真地再现仿真环境、实时响应交互操作等，这一步骤需要开发者对人们在自然环境中的观察习惯非常熟悉，以及对各种设备在与观测者位置产生相对变化时的运动情况作出适当的判断，以使得最终的仿真效果不仅拥有接近真实的外观，更让用户有身临其境的沉浸感。

3.1　基于MFC/VEGA的虚拟现实技术

将表1中建模层次生成的模型在仿真环境层进行封装得到ADF文件后，利用Vega程序进行场景驱动。基于Vega的MFC应用程序分三个主要步骤：初始化、定义和系统配置。初始化就是调用vgInitSys()函数初始化系统，并创建共享的内存区和信号区。定义就是创建需要的事件和需要的类来定义系统。系统配置就是使ADF文件的定义与函数调用结合起来，由vgConfigSys()函数完成。

为了便于开发者能较容易的开发出基于MFC的Vega应用程序，Vega通过继承MFC中的CView类而派生出一个子类zsVegaView。这个zsVegaView类提供了一个虚函数，在此函数中启动一个Vega线程，还以虚函数的形式定义了特定的应用要进行操作的通用接口，因此用户的应用程序只需从zsVegaView派生出新类并根据需要重载必要的虚函数即可。

3.2　观测模式及特效处理

在仿真过程中，观察员必须观测到设备的运行状态，所以将观测模式设置为两种模式，模式一是对单个设备做观测，模式二是建筑剖面总体观测。在观测模式一下，需要将观测者的观测点进行移动，满足对特定的建筑设备运行模式观测的需求，这通过vgPosVec()和vgPos()这两个函数来实现；在观测模式二下，将观察员的观测位置设置为(0，0，0)，截取建筑剖面图，使得观测者能对智能建筑体内的所有需要模拟的设备有一个总体的观测。

当进行设备模拟时，会产生爆炸、火焰、碎片等特效，这些特殊效果的显示通过函数vgProp()来完成，同时需要将设备模型移出场景。

3.3 数据库访问机制的实现

系统对知识库搜索的思想是：将数据库和知识库中相关的数据项与系统场景中的设备进行绑定，各种设备运行的相关属性都和数据库中相关的数据项进行映射。当突发事件发生时，按照决策系统提出的决策，选择合适的数据项提供给场景中的设备，然后触发场景中的设备进行运动，实现经过决策后智能建筑设备在突发事件下的运行状态模拟。

Visual C++提供了多种多样的数据库访问技术——ODBC API、MFC ODBC、DAO、OLE DB、ADO等。这些技术各有自己的特点，它们提供了简单、灵活、访问速度快、可扩展性好的开发技术。ADO是目前在Windows环境中比较流行的客户端数据库编程技术。ADO是建立在OLE DB底层技术之上的高级编程接口，因而它兼具有强大的数据处理功能(处理各种不同类型的数据源、分布式的数据处理等等)和极其简单、易用的编程接口。ADO中最重要的对象有三个：Connection、Command和Recordset，它们分别表示连接对象、命令对象和记录集对象。

4 结论与展望

本文主要介绍如何将虚拟现实技术应用于基于突发事件的智能建筑设备运行研究中，提出了详细的系统结构和仿真过程，分析了设计流程，描述了相关关键性技术的实现思想。

虽然虚拟现实的技术很早就被提出，但是近年才得到大的发展，所以它还被认为是一项新的技术，每天都在不断的变化发展。现在虚拟现实技术在各个领域的应用价值已经逐渐被人们所认识，将虚拟现实技术运用到智能建筑的相关研究中是一种趋势，相信由于虚拟现实技术的加入，必将推动建筑智能化系统的发展，使之进入一个更精彩的领域。

The VR System Design for Equipments' Motion of Emergency in Intelligent Building

Yu Zhuang Gao Wei

(Intelligent Building Institute, School of Architecture and Urban Planning, Huazhong University of Science and Technology, Wuhan 430074 China)

Abstract: This paper introduces the technology of the virtual reality. Through the analysis of the existing equipments' management system for the intelligent building, we get the advantage and the significance of using the VR technology in the study of the intelligent building. Chiefly discusses the design and accomplishment of the virtual reality system of equipments' motion for accident occurring in intelligent building. Through analysis and investigate, we get this system's design method and the flow path. Also we introduce the virtual reality development base on MFC/VEGA, describing the key technology in this system's accomplishment.

参考文献

[1] 余庄．基于处理突发事件的智能建筑系统集成［J］．第一届国际智能与绿色建筑技术研讨会智能与绿色建筑文集，622～627．北京：中国建筑工业出版社，2004

[2] 王国庆，等．虚拟现实（VR）技术及其应用［J］．航空计算技术，1994，（2）

[3] 龚卓蓉．Vega 程序设计［M］．北京：国防工业出版社，2002

[4] 张培红，陈宝智，刘丽珍．虚拟现实技术与火灾时人员应急疏散行为研究［J］．中国安全科学学报，2002，12（1）：46～50

[5] Wolfe,R.，and C. Liu，1988，"interactive Visualization of 3D seismic Data：A Volumetric Method" IEEE Computer Graphics and applications，1988（July）

[6] 龚卓蓉．Lynx 图形界面［M］．北京：国防工业出版社，2002

[7] 汪箭，聂小林，季辉，等．虚拟现实技术在火灾领域中的应用［J］．计算机仿真，2002，19（2）：28～31

[8] R. Brooks. Intelligent room project. in：Proceedings of the 2nd International Cognitive Technology Conference（CT _ 97），Japan，1997

智能建筑的后续发展

程大章　冯　威　沈　晔　陆明浩　陈　欢
（同济大学社区信息化与智能建筑研究中心，上海　200092）

摘　要： 本文根据对12幢甲级智能建筑运行、管理、功能提升与节能改造的调研，分析了建筑智能化系统的维护费用数据与建筑物寿命周期成本的关系，提出了智能建筑后续发展的意义与实现目标。

关键词： 智能建筑　后续发展　建筑物寿命周期成本

随着文明的进步，人们对工作与生活的环境要求不断地提高，建筑物的功能与建设标准也逐步提升。在中国大规模的城市化进程中，智能建筑作为现代建筑技术的核心，获得了巨大的发展。数以千计的智能大楼，如雨后春笋般建立在中国大地，成为建设成就的重要标志。虽然建筑被称为凝固的音乐，但是智能建筑的建成并不意味工作终结，还需要不断地改善与创新。

1　智能建筑后续发展是现代城市建设的主题

如今我们已经认识到只要是按现代方式运作的行业如银行、证券、期货、保险、商场、贸易商社、政府机构、科研机构、医院、学校、图书馆、体育场馆、机场、住宅等，它的建筑物都具有智能化的倾向。智能建筑作为现代城市的活动中心，已充分表现了它在经济、文化、科技领域中的重要作用。在信息技术智能化、信息网络全球化和国民经济与社会信息化的信息革命浪潮冲击下，中国信息化进程在大踏步地前进。全国国民经济与社会信息化工程的启动运作，为智能建筑的发展提供了优越的外部环境。

城市的规模越来越大，建筑群的功能特征日趋明显，出现了中央商务区 CBD(Center Business District)、休闲商务区 RBD (Recreation Business District)、工业园区、行政中心区、经济开发区、住宅小区等特定功能的区域。现代城市管理必须采用信息化手段对这些区域的建筑群与建筑设备进行综合管理，这对智能建筑提出了新课题。美国“911”事件之后，建筑物的防灾、减灾及反恐等安全问题凸现，建筑物中消防、安防、防灾及应急供电等已不可缺少，唯有借助智能化的应用系统，才能使之精准、有效、稳定与可靠地工作。当我们把视角从一栋建筑物扩展到一组建筑群乃至一个城市时，显然智能建筑的运行与管理对于现代城市的作用与价值是巨大的。

建筑物的寿命一般为60～80年，建筑智能化系统设备的寿命则多为6～10年，智能化系统的技术更新周期较短，通常仅4～6年。建筑物投入运行后的功能与需求往往与当初的设计目标有一定的偏离，因此，智能建筑需要在系统功能、设备配置及技术等多方面后续发展，才能适应社会的需求。2005 年我们受上海市智能建筑试点工作领导小组办公室委托，对上海从 1997 至 2004 年评定的 12 幢甲级智能建筑的智能化系统运行情况进行调研。从总体来看这些工程的智能化系统运行状态良好，为发挥建筑物的功能、提高建筑物的管理水平提供了有力的技术支持，而且在智能建筑建成后还进行了大量的设备改造、更新、功能提升等工作。但是，上海未参加智能建筑等级评定的建筑物约还有数百幢，运行与管理情况则差异很大。显然，智能建筑需要后续发展，这将是现代城市建设的重要主题，也是一个值得研究的问题。

2 建筑物寿命周期成本 BLCC 揭示了后续发展的内容

建筑物除了建设的投入外，运行、管理及设备更新等费用是不可缺少的。图 1 所示的建筑物寿命周期成本(BLCC-Building Life Cycle Cost)是典型办公建筑各项费用的比例图。

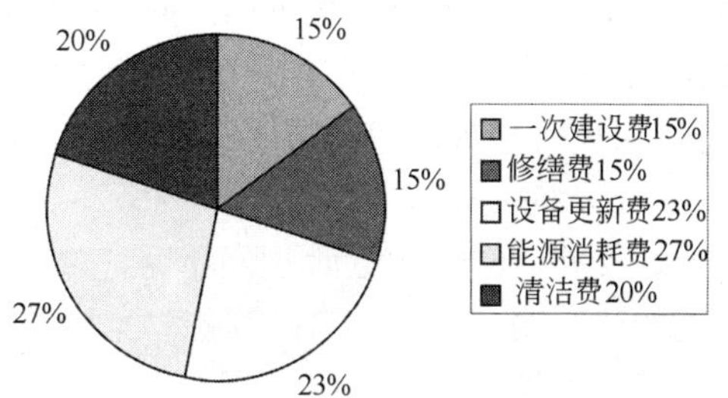

图 1 建筑物寿命周期成本(BLCC)费用比例

从 BLCC 图中可知，在建筑物的寿命周期内，设备更新费用大于一次建设投资，高达 23%。采用智能化技术对建筑物各类设施设备进行监视与故障诊断，能够及时发现渐发性故障与突发性故障，实现预防性维修。此项工作能大大地延长设备的使用寿命，减少建筑物寿命周期内更新设备的次数，有效地缩小设备更新费的总量。这在可靠性理论与工程实践中都已得到了证实。

BLCC 图中的能源费达 27%，几乎是建筑物一次建设费的 2 倍。建筑物的能耗虽然是由多项消耗构成，但是采用智能化系统对各类设备进行精确优化的控制与合理有效的管理，可以降低无谓的能源消耗，提高物业服务的质量与效率。这已由建筑节能理论与工程实践得到了证实。而且，如采用 BA 系统进行照明系统的节能控制(按照度、时间、夜间最低照度、分区等)，可以有效降低电耗与照明设备的运行时间，而照明设备因此而延长寿命减少了照明设备的维护更新费用。这不仅可使 BLCC 图中的能耗费与设备更新费用减少，更重要的是对于广义的环境保护具有巨大价值。

当然，智能化系统后续的设备更新与维保要产生费用，但是由于日常的维保费用要远远低于设备一次性投资，而能得到设备使用年限明显的增加。尤其是对于在实际运行尚未发挥作用的智能化系统设备或对不能满足需要的设备，那么我们的功能提升和设备改造就是必须的，而且改造带来的实际功效提升要远大于改造费用。从调研的情况可以看到，很多大厦采用自主创新方式进行功能提升所需的费用并不大，但是却带来了很好的效果。所以设备改造和功能提升也是降低智能建筑生命周期成本的重要手段。

3 智能化系统的运行概况

被调研建筑物的智能化系统在其业务中起了重要的作用，总体运行情况如下：

3.1 建筑设备监控系统(BA 系统)

BA 系统的对象是各类机电设备和设施，如冷水机组、空调机组、通排风机、热交换器、锅炉、水泵、配电柜、照明、电梯等，大多采用进口设备如 Johnson(美)、Siemens(德)、Honeywell(美)、Invensys(英)等。BA 系统实现了设备工作状态设定、起停、调节的自动化，能根据设定温度和室内空气状态自动对室内温度、送风量等参数进行调节，并控制相应的执行机构进行动作。近年来变频技术引入设备系统，使电动机转速可调，在满足设备工况的同时大大降低了设备的能耗。管理员在工作站上可以及时了解各设备的运行情况和重要参数。对于偏离运行状态的设备，BA 系统发出警报通知管理人员及时维修排错。BA 系统根据办公与工作时段，自动关闭不必要的照明与空调设备，尽量节省能源消耗。

因电梯设备供应商强调自身系统的安全性与独立性，大多数 BA 系统不能对电梯运行状态进行监视；冷热源系统均只能进行监视与单机连锁控制，未实现按负荷自动进行台数控制。

3.2 火灾自动报警和联动控制系统(FA 系统)

FA 系统通过智能探测器和网络实现对火灾报警与消防联动控制，为建筑内用户提供安全、便利、舒适的生活办公环境，对保障人员生命与财产安全起着举足轻重的作用。被调查的建筑物多采用进口的智能型集中报警系统，可靠性高，误报率低。有些大厦可实现预报警、报警和联动三级报警与控制；显示、记录运行与报警的各种信息；管理数据库(地址、地区名称、连动控制方案等)；并具备各种自控功能和定期自动试验功能。

为防止 FA 系统误报带来误动作的损失，大多数大厦都将 FA 系统的联动控制开关放在手动位置。发生火灾报警时，需值班人员核实后手动消防操作，往往难以快速响应。瑞吉红塔大酒店则把消防联动放在自动档，能自动确认并动作。在一次小规模的火灾事故中，系统自动完成了灭火，获得了消防局的高度评价。总体上 FA 系统均工作状态正常，也都出现过不同程度的误报，主要是由吸烟、粉尘及装修所致。

3.3 公共安全防范系统(SA 系统)

SA 系统为金融业务楼、商业办公楼、宾馆酒店等建筑内办公和居住的人们提供便利及安全的保障。一般包括：出入口控制、防盗报警、闭路电视监控、巡更、内部对讲和停车场管理等系统。久事复兴大厦采用了 CT2 移动电话，瑞吉红塔大酒店和远洋大厦采用了 DECT 移动电话，取代了传统的对讲机。

由于多数大楼建设较早，监控系统仅少数几家建设时采用了数字录像机，其余的大楼均在后期维护和系统升级中将模拟录像机改换成硬盘数字录像机，并对监视屏进行了升级和更新，二中院的部分显示屏还改为大屏幕等离子屏。少数行政办公楼的防盗报警系统接入了 110 或区域报警中心。

3.4 综合布线系统(GCS)

所有大楼均采用了光纤到层作为垂直子系统，5 类或者 6 类线作为水平子系统的布线方式。酒店类建筑物都设有无线网络，方便客人无线上网。

3.5 智能化集成管理系统(BMS/IBMS)

BMS/IBNS 将 BA、FA、SA 等相对独立的系统、功能、信息作有机的结合，进行信息综合管理，实现统一协调联动。但实际运行情况表明 BMS/IBMS 并没有发挥预期的作用。在被调查的智能建筑中，为数不多的几家大楼建立了BMS(IBMS)系统，尚在运行的仅有二中院等。具体原因有以下几方面：

- 工作目标不明确。BMS(IBMS)工作站和BA、SA、FA等系统的终端均置于中控机房，管理人员通过BMS(IBMS)来监控各个系统显然并非必要。因BMS(IBMS)不能替代专业系统的监控，它将各个系统的信息与功能集成，使之服务于建筑物的综合决策和管理。所以其作用是辅助大楼的管理，主要服务对象是高层管理人员。
- BMS(IBMS)没有与物业管理业务相结合，仅是集成监视没有实用价值。
- 某些BMS(IBMS)系统的技术与功能不成熟，系统运行不稳定。

3.6 防雷接地系统

各建筑都进行一年一次的防雷检测，定期测试接地电阻阻值。测试防雷接地系统的工作形式一般有两种：气象局签订防雷接地测试合同或大楼物业机构自行测试。

4 智能化系统的维护情况

建筑设备投入使用后不可避免地会发生故障，需要物业管理人员利用智能化系统及时发现并排除故障，保证建筑物的正常运行。但是智能化系统自身也需要维护。

4.1 智能化系统的维护方式与费用

在被调研的建筑物中，智能化系统的一些子系统(如：防雷接地系统，消防报警与联动控制系统)需由专业资质的公司进行维护。其他智能化子系统除极个别大厦由物业公司全部自行维护外，其余大楼均采用外包专业维护＋自行专业维护的形式。外包维护部分由大厦和专业公司签订维保合同，免费维保期后按合同和实际故障情况进行维保，支付维保费用。2004年度12幢大楼智能化系统的维保费用见表1。

2004年度智能化系统的维保费用 **表1**

智能化大楼	维护费用(RMB/年)单位：万	智能化大楼	维护费用(RMB/年)单位：万
G	16	K	(自己维保)
H	20	L	25
B	30	D	(尚处维保期内)
I	120	C	150
J	43	F	(尚处维保期内)
A	150	E	2

4.2 常见故障情况

除了近年投入使用的建筑外，调研的情况反映，常见的故障主要出现在以下几方面：

- 执行机构(电动阀)出现问题最多。进口的执行机构的故障率较低，国产的执行机构故障较多。较早投运的L大厦采用进口电动调节阀，在近10年的运行中，仅有6个出现故障。而像K大厦采用国产电动调节阀经常出现阀门限位失灵、阀门锈死或开度不足等现象。
- 火灾探测器：温感、烟感探测器暴露在装修环境中及处于工作环境比较差的地方常会出现老化、损坏或由于粉尘等污染物产生消防误报。出现故障的部件，大都采用清洗与更换的方法来解决。

- 电视监控设备：电视监控设备由于多为 24 小时通电工作，摄像机与监视器老化到影响监视效果时，只能进行更换。

4.3 自行维护保养的情况

各部门的管理人员分别对自己负责的设备进行定期维护。为了缩短发现设备故障的时间，期货大厦等楼宇还采用了巡检和用户报修相结合的手段，以巡检为主报修为辅，变被动报修发现故障为主动发现故障，缩短了系统故障的时间。管理人员在巡检同时对设备进行一定程度的保养，巡检人员随身携带诸如润滑油、维修工具等物品，对常发生锈死的执行机构等部件添加润滑，对易松动部件进行加固，以事前维护的方式来减少故障，延长了设备的寿命。

5 智能化系统功能提升

由于建设时的缺陷、建筑物功能的变化、长期运行后智能化系统与设备的性能下降等情况，提升功能与系统改造是必然的。被调研的建筑物均进行过不同程度的智能化系统的改进和功能提升。各大厦或依靠自身工程人员的力量，或依靠设备提供商和维保公司，在满足用户的舒适度要求的基础上对现有设备进行改进，在功能欠缺之处引入新的设备，进一步完善了楼宇智能化系统功能。

5.1 酒店类智能建筑

酒店类智能建筑功能提升的目的是提高酒店入住的舒适度，方便客人生活、休息，降低运行能耗。在原有酒店网络的基础上，引入无线网络，使大堂，会议室等公共区域有无线网覆盖，使入住的客人进入 Internet 进行商务办公。B 酒店的能源系统把原来的燃油锅炉改造成燃气锅炉，既提高了环保性能又在油价高涨的情况下节省了开支。A 酒店的冷却水与冷冻水的水泵增加变频控制；高档套房可按客人要求进行送风量控制；为避免过渡季节启动大型制冷机组增加两组 VRV 空调；并进行了一卡通系统与电视系统的改造。

5.2 政府、机关、企业专用办公大楼

该类建筑的智能化系统目标是保证机关、企业工作的顺利进行，为办公人员和领导提供良好的环境。C 大楼增设对风冷散热片进行喷水散热的供回水管路，显著提高热泵的工作效率；增加两台 UPS 作为备用电源；增加程控交换机的冷备主机；局部冷量不足的区域增加 VRV 设备，避免加班时间开启大型冷水机组；BMS 数据库的服务器更新。E 大厦设备多采用自行维保，由于原有工程质量存在一定问题，物业管理人员在近期对停车场系统和中控室的布线进行改进。D 大楼系统投运时间不长，但 VRV 系统室外机组因间距小散热困难冷量不足，物业部门准备对此重新布置。

5.3 商用办公楼

该类建筑主要是保证出租用户的办公需要，同时根据建筑自身特点进行系统改建，使之更经济更高效的运行。被调研的建筑主要改进项目：将模拟录像机改为数字录像机，加大图像存储容量；改善空调系统的控制效果；BA 系统的操作软件升级。

5.4 功能提升情况汇总

各大楼具体的功能提升情况如表 2 所示。

各大楼功能提升情况汇总表　　表 2

大楼名称	功能提升项目
G	(1)空调二通阀的改造，使风机盘管电动阀在中央控制站上可控；(2)与通信运营商合作对公共区域进行通讯信号的覆盖
H	消防系统和闭路监控系统的点位调整
B	(1)模拟录像机改为数字录像机；(2)上网方式增加(天翼通+FTTB)；(3)一卡通改造；(4)有线电视的增加以及卫星电视的改造；(5)增加两组 VRV 空调
I	(1)电动扶梯采用红外技术控制；(2)VAV 送风控制改进；(3)模拟录像机改为数字录像机
J	(1)取消电话控制照明服务，节省的 I/O 点用于冷却塔的监控；(2)BA 系统的控制器采用专用 UPS 供电；(3)模拟录像机改为数字录像机；(4)增加 50 个摄像点
A	(1)冷却泵，冷冻泵的变频控制；(2)增加无线上网方式；(3)高级套房根据客户需求进行送风量变频控制
K	采用空调冷水管路倒冲技术，用反冲水流除去空调管路中的污垢，提高流量与换热率
L	(1)EBI 软件升级；(2)商场区域将增加空调管路，提高商场冷量，解决商场温度过高的问题
D	拟进行 VRV 室外机组的位置改建
C	(1)弱电机房的扩建；(2)对热泵机组的风冷散热片用冷水喷淋降温；(3)增加两台 UPS 备用电源；(4)增加程控交换机的冷备主机；(5)BMS 数据库的服务器更新
F	(1)将锅炉的控制状态连入 BA；(2)VAV 经试验采用送风控制
E	(1)停车场系统的改造；(2)中控机房布线改进

6　必须注重智能建筑后续发展的三大问题

智能建筑的后续发展涉及到多方面，智能化的物业管理是核心；生命周期成本是智能建筑运行经济性的考察指标；节能措施是缓解我国能源紧张，创建节约型社会的重心；绿色环保是智能建筑应为社会可持续发展承担的义务。

6.1　智能建筑的后续发展理念的实现必须有体制与机制的保障

智能建筑后续发展的理念已经逐渐被建筑物的拥有者与管理者所接受，但是要得以实现必须有体制与机制的保障，即在人员、制度、经费与技术上得到保证。

由图 2 中表示了智能建筑后续发展的概念及内在的四者相互关系。在智能建筑后续发展中，物业管理直接决定了其他三个因素能否发挥作用。人的因素第一，所以，智能建筑的功能发挥需要优秀的物业管理团队。经验丰富的物业管理企业是智能化大厦良好运行的保障，是采取各种先进技术进行功能提升、降低 BLCC，以经济高效地管理大厦的工作主体。

建筑物智能化系统需要专业的维护保养，从发达国家的大楼设施管理的经验及国际惯例来看，大型设备与复杂系统的维护保养都采用专业化、社会化的方式进行，由系统集成商或设备供应商完成，平时定期派员前来测试维护，以及早发现隐患，设备故障时能及时(2～4 小时内)响应排除。这样可降低对物业管理人员的专业技术能力要求，减少高技能的人力费用和备件的贮存量。最主要的是提高了智能化系统安全、正常运行的保障能力。因此，在体制上，物业管理机构与专业公司应形成两个不同的智能化系统维保的工作层

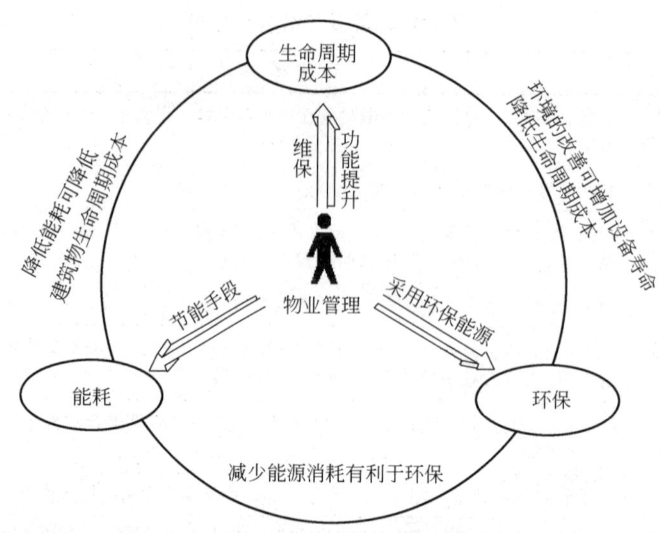

图 2　智能建筑后续发展的概念

面，并且以合同方式确定两者的工作界面。系统集成商不能仅限于完成工程，还须承担长期的维护保养责任，否则应取消其系统集成的工程资质。

建筑物智能化系统的建设需要投资，管理和发挥系统功能同样需要投入，这就是图 1 的 BLCC 揭示的意义。因此，重建设轻管理的传统观念影响应该更新，在建筑智能化系统的运行维护与设备更新上给予足够的预算。

除综合布线系统外，各类智能化系统的日常专业维护保养费用视其系统的特点与工作责任(一般维护：器件损坏另收费；全责维护：包换损坏器件等)的大小，每年收费在设备总价的 2%～5%左右。随着接近使用寿命年限，维护保养费用还将升高。各类智能化系统的运行还需要一些易损易耗的器材，据统计这笔费用约为设备总价的 5‰左右。

智能化系统都是由计算机与电子设备组成，这些设备的运行虽然多数不存在机械摩耗，但由于器件老化和环境污染，一些设备在投运 4～6 年后性能明显下降；信息系统则经 4～5 年的运行后，由于设备技术进步和软件功能更新，使用者往往已不能再忍受使用旧系统而要求升级改造……这些设备与系统的更新费用是昂贵的，如何评价他们的老化状态，科学地确定更新周期，以保证充分发挥投资的效益，还有待积累经验。

由于中国目前的物业管理是一个低利润的行业，智能化系统的维护保养与设备更新费用如何支出，尚未形成规范的运作机制，尤其是智能住宅小区更存在严重的问题，这将是我们今后面临的重要课题。

6.2　节能改造是智能建筑的后续发展的重要内容

智能建筑单位面积的能耗是评价大厦智能化系统与运营管理水平的重要指标，在目前我国经济高速发展的时期，能源高度紧张，建筑物节能改造更是智能建筑后续发展的重要内容。

虽然建筑节能的途径有多种，对于已投入使用的建筑物而言，主要是依靠加强管理、精确与优化的控制和引进节能设备进行系统改造来实现。节能的政策导向是中央政府的“全民节约、共同行动”精神及市政府的规定，物业管理机构通过要求用户改变空调设定温度、减少不必要的照明等措施来节能节电。在调研中我们了解到，大多数建筑物的 BA

系统主要是对建筑设备进行基本的工作状态监视控制，尚未实现节能的优化控制。如果广泛采用变频器按实际所需流量调节风机与水泵的工作状态、按照最大限度使用室外空气能量控制空调系统运行、根据实际负荷量自动控制冷热源设备投运台数、按工作面的实际照度需求控制照明设备——，那么建筑能耗再降低 20%是可以达到的。无论是理论分析还是国外的工程实践，这一观点都已得到证实。

能耗费用占据了智能建筑运行费用的很大部分，所以节能是降低智能建筑运行费用的重要手段，也是降低 BLCC 的主要途径。因为能耗的降低也意味着设备运行量的降低与使用年限的延长，减少能耗还进一步减少因发热形成的设备损耗。可见节能是智能建筑后续发展的重要工作，也是我们创建节约型社会，走可持续发展道路的重要途径。

6.3 智能建筑的后续发展要加强管理信息化

现代城市的建设需要大量的智能建筑，现代城市的管理更需要智能化与信息化系统的支持，来提高城市的管理水平与突发事件应急能力。为此，智能建筑的后续发展要加强管理信息化系统的建设。

智能建筑的系统集成 BMS/IBMS 是应用信息化技术实现建筑物智能化管理的重要工作平台，以往许多智能建筑的建设方与工程承包方都认为 BMS/IBMS 是智能建筑技术的最高境界，投入了大量的人力财力来实施，但却由于诸多原因国内能达到开发目标并投入正常运行的系统集成项目为数甚少。从调研的情况分析，系统集成工程的健康发展需重视以下问题。

- 认识系统集成的意义。

对于投资者(建筑物业主或管理机构)来说系统集成 BMS/IBMS 可归纳两个管理和两个服务，用四个“面向”来描述：面向设备的管理，以提高工作的效率和质量；面向客户的管理，以提高工作的效率和质量；面向客户的服务，以提供完善的服务信息和便捷的服务方式；面向领导的服务，以提供完整的信息供分析，作为辅助决策与应急指挥的依据。

对于政府行政及各类机构专用的建筑，其中的最后一个面向尤为重要，这是由于虽然物业管理事务目前大多已经采用服务购买的方式外包，但是行政机构的内部管理部门需要对物业管理公司的日常工作进行监督，而集成化智能建筑信息管理系统则是最有效的技术工具。

- 确认 BMS/IBMS 的功能目标。

智能建筑的后续发展对 BMS/IBMS 提出了以下要求：建立统一应用平台、统一操作界面；实现各设备和系统的联动，优化系统控制；建立统一数据库，实现信息资源共享；为提高综合管理效率设置功能，以节省能耗降低运行成本；建立应急处理与决策的辅助系统，增强建筑物的安全性。

计算机网络与数据库技术的进步，使信息系统的信息传输、信息处理的能力与性能发生了巨大的变化，智能建筑各类设备与系统中无论是 TCP/IP、ASHRAE 的 BACNET 通信协议还是 LONWORKS 技术的应用，都是追求开放的通信接口和高速的信息传输，及时地获得更多的信息，以对建筑内的人流、物流及信息流进行全面的监控与管理。如目前国内 BA 系统的设计目标大多定位在运行控制与管理，对建筑设备与系统状态检测点的确定仅是运行控制所必需的信息。这是个严重的问题，如果不能全面及时地掌握各设备系统的信息，就不能进行综合分析，尤其是对于渐发性故障不能提出预警。设备运行信息的综合分析有利于物业设施管理的设备故障诊断、设备运行状态优化、设备维护保养、降低设备能耗、提高服务质量等诸多工作项目。但是如果必要的基础信息不能自动采样取得，那么现代物业管理目标的实现是很困难的。

智能建筑从初期的单一设备的自动控制，发展到今天的综合优化控制、在线设备故障诊断、全局信息管理和总体运行状态协调等高层次的集中管理分散控制方式，面向建筑物设备的智能化系统已经将信息、控制、管理、决策有机地融合在一起了。随着大型建筑物、建筑群的建设，特别地，随着 2008 年北京奥运会和 2010 年上海世博会场馆建设，公共建筑中发生造成人员伤亡、财产损失和重大社会影响的突发事件的机率越来越高，如何预防发生以及突发事件的应急处理，控制事故所产生的后果是我们在建设与管理大型建筑物、建筑群应重点关注的。利用 BMS/IBMS 为技术平台，可以建立应急处理与决策的辅助系统，在突发事件时有效地对事件产生的后果进行控制，努力的降低危害程度。

- 以科学方法建设 BMS/IBMS。

智能建筑的信息系统日趋大型化、复杂化，子系统众多，关联程度复杂，涉及多学科和技术，已不能用单纯的软件工程方法或控制工程方法分析，必须采用系统工程的方法和思想，站在系统的高度应用综合集成方法对复杂系统进行分析、组织、设计和实施。

系统集成不是简单的一套系统或软件，而是一种思想，一种哲理，是一种指导信息系统建设的总体规划、分步实施的方法和策略。它需要按照用户的需求，根据建筑物智能化系统的实际情况，恰当合理地选择相应的技术和策略，最佳选择和配置各种软硬件产品和资源，组合成完整的、能够实现用户具体应用需求的集成方案。因此，所谓系统集成是针对某个应用目标而提出的全面解决方案(Total Solution)的实施过程，系统集成包括人员的集成、组织机构的集成、管理与技术的集成、综合信息的集成等。从狭义上讲，系统集成特指计算机信息系统集成，包括硬件平台、软件平台、开发工具和应用系统的集成，以及相应的咨询、服务和技术支持。

BMS/IBMS 不仅要为用户的需求和系统建设的总目标提供全方位的解决方案，而且还要得到工程实施，并提供相应的服务与咨询。作为一个系统工程，其内容涉及项目开发和实施的整个过程，从系统规划、系统生成到系统维护；从系统的功能模型设计到技术的实现；从系统的立项、建设中的管理到系统的验收，所有的工程行为共同组成了系统的工程空间。准确把握系统集成的时期与重心。系统集成时期的选择是十分重要的，进行系统集成应是各组织机构及其工作流程相对稳定的时期，否则此时因最基层的工作流程与数据流不稳定而可能导致工作失败。对于智能建筑来说，系统集成工程的有效性体现在长期的建筑物物业管理工作中，一切的技术方法设备都是为需求服务的，尤其是系统的数据库建立与系统维护没有实际用户(物业管理机构与人员)的介入，都是注定要失败的。因此，有效的 BMS/IBMS 工程是智能建筑后续发展的工作重心，需要智能建筑工程建设者与物业管理机构及其人员的紧密合作。

7 结束语

我们历时 2 个月对上海 12 幢甲级智能建筑进行了调研，在此工作基础上思考了智能建筑的后续发展问题，并提出了智能建筑后续发展的内容与工作思路。文章可能不完善不全面，但是我们认为智能建筑后续发展是现代城市建设的重要主题，而且其重要性不亚于智能建筑的建设。希望业界人士能对我们的观点进行批评，共同推进智能建筑事业的健康发展。

The Continuing Development of Intelligent Building

Cheng dazhang　Feng wei　Shen ye　Lu minghao　Chen huan
(Research Center of Information & Intelligent Building of
Tongji University, Shanghai　200092　China)

Abstract: This article contains the running state, management, function upgrade and energy saving for 12 first class intelligent building with analyzing its relationship between the maintenance expenses and BLCC. In the end, the article put forward the significance and target of the continuing development of intelligent building.

智能温控器控制效果实验测试技术研究

王　健　王亚慧　郝学军　曹国涛
（北京建筑工程学院电气工程与自动化系，北京　100044）

摘　要：为了测试某智能温控器在风机盘管空调系统中的控制精度与节能效果，利用某宾馆中两个户型完全相同，且上下紧邻的空调房间，进行了该智能温控器与普通双位控制的温控器的对比实验。在典型夏季工况下，近一个月的连续实测表明了该智能温控器控制精度高、冷量节省明显。本文总结了关于该类实验的大量经验，包括如何进行实验前的准备、实验的一般指导原则、操作顺序、影响实验误差的因素以及出现意外情况时所应采取的措施等。

关键词：空调　温控器　控制精度测试　节能

1　引言

能源是城市生存和发展的必要条件，是城市功能正常运转的基本保证。在《北京市“十一五”能源发展及节能规划》（初稿）（以下简称《规划》）中建筑节能作为“十一五”期间能源节约重点措施，已在法律地位上获得了社会各界的普遍重视。为大力推进建筑节能，在《规划》中明确规定，新建居民建筑严格执行节能 65％的设计标准；对现有建筑逐步实施节能技术改造；实现公共建筑节能 50％的设计标准，建立和完善建筑节能验收和监理机制[1]。在建筑能耗的各项组成部分中，暖通空调系统消耗能量最大，占建筑物中总电力用电的 50％左右，有些建筑物的空调耗能量甚至还要高，这充分说明了空调节能的必要性，也从另一侧面说明了目前我国的空调节能有很大的潜力可挖。

风机盘管式空气调节器（FCU）以其布置灵活，调节方便，节约建筑空间和投资费用低等特点，在全国各地的宾馆、饭店和一些大型民用建筑中得到了广泛应用，但是在风机盘管系统的运行与维护中，也存在诸如凝结水问题、噪声问题、故障率高等许多问题，同时 FCU 与变风量空调系统（VAV）相比，其节能效果较差。而在现行建筑中完全用变风量空调系统取代 FCU，在经济和技术上都是不现实的，因而对现有 FCU 进行局部改造已成为可能的节能发展方向之一。在 FCU 的诸多改造方向中，FCU 末端温控器的改造是很有前景的，已有文献表明[2]，不采用简单的双位控制而采用先进控制策略的温控器，不仅能提高房间温度的控制精度，而且能取得极好的节省空调房间耗冷量的效果。

通过为风机盘管系统温控器改造作节能效果测试的过程，结合笔者的工作经历，简述做该类测试研究时所应注意的问题。

2 测试工作的整体思路

测试工作的总体思路为：在夏季工况下，在两个结构完全相同的密闭空调房间内，分别装设两种风机盘管温控器(一种为普通双位控制温控器，一种为智能温控器)，在房间设定温度相同、房间热舒适度相近的情况下，比较其温度控制的精度及在相同时间内消耗的冷量。

3 测试房间的选择和约束条件

应选择建筑面积相等，户型完全一样，朝向一致的两个空调房间做对比测试，同时应尽量使这两个房间所处的环境温度一致。因此，在本次测试中选择了宾馆的上下两相邻客房做测试房间。客房里的电器、灯具等散热装置及其摆放位置具有一致性，同时应使散热装置的工作时间，居住人员的起居规律尽量一致，这些都为测试的准确性提供了条件。

4 温控器的设置位置

在控制系统中，由于是取温控器所在位置的那一点的温度作为房间内的温度，因而温控器的设置位置必须具有代表性。如果房间内的温控器出于操作方便的目的而放置在了不能代表整个房间温度场分布平均值的位置，则应加长温控器连接线，使温控器尽量靠近送风回流区。温控器也有显示设定值与实际设定值不一致的情况，应提前做好校正工作。

5 温度的精确测量

作为热工测量的常用感温元件，热电阻的测量范围、灵敏度、精度、稳定性等性能指标使其亦可应用于温控器节能效果测试中房间温度的精确测量。在笔者的工作中，选用了分度号为 Pt100 的铂热电阻传感器。使用热电阻测温时要特别注意线路电阻的影响，因为线路电阻的任何变化最终都将折合为被测温度的误差。为了克服环境温度变化对导线电阻的影响，应尽可能采用三线制接线方式。本测试中，数据采集设备使用的是安捷伦公司生产的多功能数据采集单元(Agilent 34970A Data Acquisition/Switch Unit)，为了达到精确测量温度的目的，在传感器与数据采集卡(20 - Channel Armature Multiplexer)之间采用四线接法，如图 1 所示。

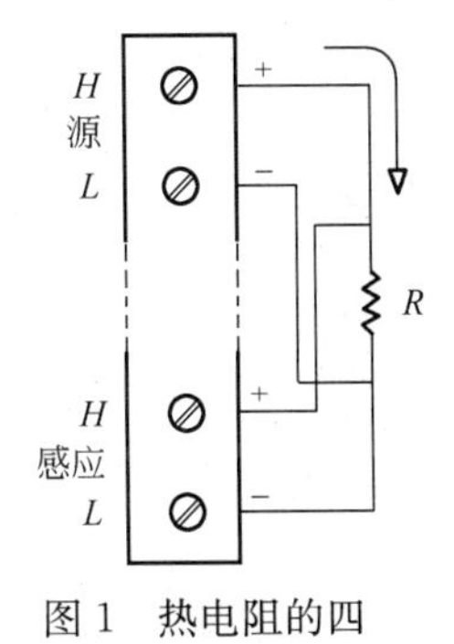

图 1 热电阻的四线接法示意图

在 HP 34901A、34902A 中通道 n(源)与通道 n+10(感应)、n+8(感应)是自动匹配的。热电阻范围：100、1k、10k、100k、1M、10M、100MΩ；热电阻类型：0.00385、0.0391。

此处应注意的两个问题：

5.1 封装好的热电阻传感器的引出线一般为颜色相同、长短一样的三根线(如图 2、图 3 所示)，因此应首先设法区分开这三根线：用万用表随机地测量两根引出线间的电阻，若电阻值为零，则这两根线为 1、2 端；若电阻值为 100 多欧姆，则这两根线为 1、3 端或 2、3 端，再用万用表随机测量另外一个两根引出线组合的阻值情况，以达到区分引出线的目的。

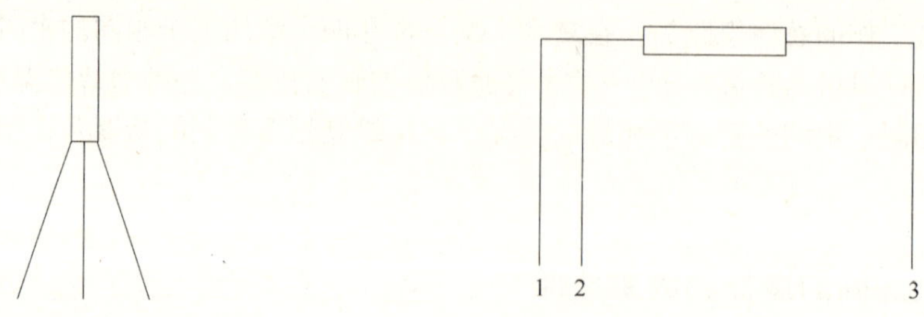

图 2 热电阻传感器外观示意图　　　图 3 热电阻传感器结构示意图

5.2 由于数据是在数据采集仪处集中采集的，而测温点却分布在每个房间的各个位置，因此在现场实测的过程中，需要加长热电阻传感器的引出线。但是本次测试所用的双绞线在常温下有大约 0.15Ω/m 的电阻率，因而需要在裁剪好延长线并与热电阻传感器焊接好之后对其进行校验。实验室和工业用的铂热电阻温度计校验方法有两种：比较法和两点法。结合实测现场的条件和所用软件(Agilent Benchlink Data Logger，其设置界面如图 4 所示)的特点，本实测选用两点法做校验。两点即在当地大气压下纯净的冰水混合物的温度(0℃)与纯净水的沸点温度(约 100℃)，纯净水可选用矿物质较少的品牌。修正当地大气压对水沸点的影响的修正公式为：

Setup 12/04/05 21:23:42 - Data 12/04/05 21:23:45 - Agilent BenchLink Data Logger

File Edit View Scan Channel Graphics Instrument Help

None Off

Channel			Measurement				Scaling (Mx+B)				Alarm Limits					Comment
ID	Sc	Name	Function	Range	Resolu	Advan	Scal	Gain(	Offset	Label	Test	High	Low	H/W	Advan	
101	☑		DC volts	Auto	5.5 digits	Default	☐	1.0	0.0	VDC	Off	0.0	0.0	Alarm1		
102	☐		DC volts	Auto	5.5 digits	Default	☐	1.0	0.0	VDC	Off	0.0	0.0	Alarm1		
103	☐		DC volts	Auto	5.5 digits	Default	☐	1.0	0.0	VDC	Off	0.0	0.0	Alarm1		
104	☐		DC volts	Auto	5.5 digits	Default	☐	1.0	0.0	VDC	Off	0.0	0.0	Alarm1		
105	☐		DC volts	Auto	5.5 digits	Default	☐	1.0	0.0	VDC	Off	0.0	0.0	Alarm1		
106	☐		DC volts	Auto	5.5 digits	Default	☐	1.0	0.0	VDC	Off	0.0	0.0	Alarm1		
107	☐		DC volts	Auto	5.5 digits	Default	☐	1.0	0.0	VDC	Off	0.0	0.0	Alarm1		
108	☐		DC volts	Auto	5.5 digits	Default	☐	1.0	0.0	VDC	Off	0.0	0.0	Alarm1		
109	☐		DC volts	Auto	5.5 digits	Default	☐	1.0	0.0	VDC	Off	0.0	0.0	Alarm1		
110	☐		DC volts	Auto	5.5 digits	Default	☐	1.0	0.0	VDC	Off	0.0	0.0	Alarm1		
111	☐		DC volts	Auto	5.5 digits	Default	☐	1.0	0.0	VDC	Off	0.0	0.0	Alarm1		
112	☐		DC volts	Auto	5.5 digits	Default	☐	1.0	0.0	VDC	Off	0.0	0.0	Alarm1		
113	☐		DC volts	Auto	5.5 digits	Default	☐	1.0	0.0	VDC	Off	0.0	0.0	Alarm1		
114	☐		DC volts	Auto	5.5 digits	Default	☐	1.0	0.0	VDC	Off	0.0	0.0	Alarm1		
115	☐		DC volts	Auto	5.5 digits	Default	☐	1.0	0.0	VDC	Off	0.0	0.0	Alarm1		
116	☐		DC volts	Auto	5.5 digits	Default	☐	1.0	0.0	VDC	Off	0.0	0.0	Alarm1		
117	☐		DC volts	Auto	5.5 digits	Default	☐	1.0	0.0	VDC	Off	0.0	0.0	Alarm1		
118	☐		DC volts	Auto	5.5 digits	Default	☐	1.0	0.0	VDC	Off	0.0	0.0	Alarm1		
119	☐		DC volts	Auto	5.5 digits	Default	☐	1.0	0.0	VDC	Off	0.0	0.0	Alarm1		
120	☐		DC volts	Auto	5.5 digits	Default	☐	1.0	0.0	VDC	Off	0.0	0.0	Alarm1		
121	☐		DC current	Auto	5.5 digits	Default	☐	1.0	0.0	ADC	Off	0.0	0.0	Alarm1		
122	☐		DC current	Auto	5.5 digits	Default	☐	1.0	0.0	ADC	Off	0.0	0.0	Alarm1		

Scan Setup / Graphics Setup 1

Activate/Deactivate channel　HP34901A　(None)　(None)　SIMULATED

图 4 Agilent Benchlink Data Logger 的设置界面

$$t_{沸}=[100+28.0216(p/p_0-1)-11.642(p/p_0-1)^2+7.1(p/p_0-1)^3] \tag{1}$$

式中 p——当地大气压；

p_0——标准大气压[3]。

在 Agilent Benchlink Data Logger 中有数据校正公式 $Mx+B$，其中的系数 M、B 的求取方法就用热电阻温度计在冰水混合物中的温度 t_0 与在沸水中的温度 t_{100}，解二元一次方程组

$$\begin{cases}0=M\times t_0+B\\ t_{沸}=M\times t_{100}+B\end{cases} \tag{2}$$

即可解得系数 M、B，该方程组的简化解法为：

$$\begin{cases}B=-t_0\\ M=\dfrac{t_{沸}+t_0}{t_{100}}\end{cases} \tag{3}$$

将 M、B 值输入软件即可在线校正输出值。

6 空调房间耗冷量的计算

空调房间耗冷量的基本计算公式为：$A=B+C$。其中 A 代表风机盘管系统提供给空调房间的冷量；B 代表抵消围护结构传热量所消耗的冷量；C 代表空调房间内发热体(包括发热装置和人)实际所消耗的冷量。

1. 风机盘管系统提供给空调房间的冷量由供回水温差 t_1-t_2、冷水流量 Q、水的比热容 C，三者相乘再对时间取积分得到。供回水温度作减法之后得供回水温差 t_1-t_2，供回水温差，冷水流量，水的比热容三者之间作乘法得风机盘管系统提供给空调房间的冷量的瞬时值，这两步简单计算也可由热量计算仪来实现自动计算

$$P=Q\times(t_1-t_2)\times C \tag{4}$$

供回水温度亦由 Pt100 热电阻测量，在测量过程中应将传感器紧贴水管铜质对接处(铜的导热系数大)的外表面并紧固外面的保温材料(同样地，如果需要测量房间壁面温度，则应考虑相应措施来尽量防止光照所带来的测量误差)。若使用涡轮流量计测量水流量，则应注意使用磁电法时易受电网工频干扰的问题。例如流量计的仪表系数为766.050 1/L，则电网所叠加的 50Hz 电信号刚好能引起$\dfrac{50\times10^{-3}\times3600}{766.050}\approx0.235(\mathrm{m^3/h})$的误差。这个误差是很大的，必须消除。可以用 RC 滤波电路并联在流量信号的输出端来滤除高频干扰。同时，注意必须接地，地线可用冷水管中的立管及其金属连接部位；同时也可利用热量计算仪、数据采集仪的瞬时流量上下限报警功能，手动排除受干扰的数据。

2. 定量测量出抵消围护结构传热量所消耗的冷量是一项极其复杂的工作，在本次测试中采用多点热流仪进行定性大致测量，本实测中所用的是日本 KEM 生产的型号为 MV100 的多点式热流仪，其测量精度高，可接传感器的数量较多。

7 结论

本次测试的数据表明，在不影响房间使用人员的休息及特殊使用的情况下，在夜间自

动提高房间温度的设定值，可以节省大量的空调冷量。这是在运行管理中节能的重要途径之一，因为在《规划》中已有明确的指示，要完善管理节能。因此，建议在温控器内设置系统时钟芯片与光敏器件，并根据实际情况编写相应的程序写入单片机，以自动控制房间温度的夜间设定值。

综上所述，对温控器进行技术改造，提高其控制效果，增强其智能化水平是空调系统节能，乃至建筑节能的发展道路之一。但是温控器节能效果测试实验所涉及的技术繁杂、仪器设备的数量大、种类也比较繁多，且易受干扰，因此，对温控器节能效果测试的技术研究是很有必要继续进行的。笔者只是结合自己的工作经历简单总结了做该类测试的经验、办法，其实也可以选择不同的测试思路、方案，同样也可以选择各异的测试设备，为空调系统节能、为建筑节能作好测试工作。

A Study on the Experiment Measuring Technology of the Control Effect by Intelligent Thermostat

Jian Wang　Yahui Wang　Xuejun Hao　Guotao Cao
(Department of Electrical Engineering and Automation Beijing Institute of Civil Engineering & Architecture, Beijing 100044 China)

Abstract: In order to measure the control accuracy and energy - saving effect of an intelligent thermostat used in a FCU(Fan-Coil air-conditioning Unit), a comparison test experiment was made in two identical air - conditioning rooms which were adjacent in vertical in a commercial hotel, with the intelligent thermostat used in one room and a common off - on thermostat in the other. The experiment lasted almost one month continuously under classical summer conditions. The results showed the high precision and obvious cooling-saving effect. This paper summarized a lot of experiences about this kind of experiments, such as preparations for the experiments, common guiding principle and operating sequences, factors affecting the error of the experiment and emergency measures if something unpredicted happened.

参考文献

[1] 北京市“十一五”能源发展及节能规划(初稿). 北京市发展和改革委员会 . 2005 年 8 月
[2] 郝学军等 . FHK-1 型智能温控器控制精度与节能效果实验研究 . 家电科技 . 2004 年第 5 期
[3] 吕崇德主编 . 热工参数测量与处理(第二版). 北京：清华大学出版社，2001 年

建筑智能化系统的安全问题及对策

唐　宏　王　波

（重庆大学计算机学院建筑智能化研究室，重庆　400044）

摘　要： 本文从系统角度提出了建筑智能化系统的安全问题。阐述了建筑智能化系统安全问题的重要性，以及建筑智能化系统安全与安全防范自动化系统的区别与联系。详细分析了建筑智能化系统中存在的安全问题，包括控制网络、信息网络和无线网络等方面的安全问题，并提出了相应的对策。这对于保证建筑智能化系统的安全运行，具有重要的参考价值。

关键词： 建筑智能化　智能建筑　网络安全　系统安全

1　引言

在智能建筑中，包含了各个自动化子系统，如BAS、FAS、SAS、CAS、OAS。它们共同组成一个建筑物的智能化系统。这些子系统既可以独立完成其领域内的相关工作，也可以和其他子系统协同，完成某些较为复杂的功能。建筑物要具有更高的智能，它的各个子系统协同工作的能力就应当越强。目前已有许多技术，不但能够将各个子系统集成起来，还能够利用INTERNET将远距离的不同建筑智能化系统集成为一个更大的建筑智能化系统。这样的建筑智能化系统在赋予建筑更高智能的同时，也给自身带来一定的安全隐患。目前对信息领域的安全性研究已经相当广泛，而对于建筑智能化系统这样一个集信息、控制网络于一体的系统的安全性也应当足够重视。可以想象，如果一个精通信息网络和控制网络的破坏者蓄意对建筑智能化系统进行攻击，造成的后果可能是十分严重的。因此对建筑智能化系统的安全性进行研究是非常重要而且必要的。

2　建筑智能化系统的安全问题

2.1　建筑智能化系统的安全与安全防范子系统的区别与联系

提到建筑智能化系统安全时，很容易与安全防范子系统混淆。它们是两个不同的概念，建筑智能化系统安全是指整个建筑的智能化系统的安全，即IBMS的安全。而安全防范子系统用于对建筑内的人和财物的安全进行保护，防止偷窃、破坏、暴力等的发生。安全防范子系统可包括：闭路电视监控系统、出入控制系统、入侵报警系统、巡更管理系统、停车库管理系统。同时建筑智能化系统安全和安全防范子系统又有很紧密的联系，建筑智能化系统安全包含了安全防范子系统的安全；安全防范子系统对建筑智能系统的安全是一个有力的保证，它可以防止非法人员接近建筑智能化系统，窃取建筑智能化系统的重要信息数据、对建筑智能化系统进行物理破坏。

2.2 物理安全威胁

在考虑任何系统安全时，首先应当考虑到物理安全。物理安全威胁包括自然威胁和人为威胁。自然威胁主要是指由于自然原因造成的对设备硬件的损坏，如自然灾害（雷击、火灾、水灾、地震等），正常使用情况下的设备损坏（指设备使用寿命到期或设备质量不好导致的使用损坏），非正常使用情况下的设备损坏（如突然停电或工作环境恶化）。人为威胁包括人为故意威胁和人为无意威胁。人为故意威胁是指人为故意通过物理接近的方式窃取网络信息（如搭线监听、窃取密码），或破坏网络的物理设施（如盗割通信电缆、破坏通信设备）。人为无意威胁是指人为由于操作失误和意外疏忽造成的设备损坏（如删除文件、格式化硬盘、带电拨插硬件等）。可见物理安全是其他安全的基础，攻击者可以通过物理接近窃取相关信息资料，为进一步攻击做准备，或者直接对系统硬件进行物理破坏，因此物理安全在整个系统安全中至关重要。

2.3 控制网络的安全威胁

在建筑智能化系统中，控制网络仍然处于相当重要的位置，特别是在 BAS、FAS、SAS 中，控制网络完成检测、控制、执行等功能。由于控制网络最初都是用于相对独立、封闭的系统，并且使用专用的技术，它所面临的安全威胁主要来自物理威胁，因而并没有充分考虑安全问题。从发展角度看，建筑智能化系统中的控制网络技术主要有 BACnet 和 LonWorks。

BACnet 已经提供了较好的安全机制，但是还不够完善。如它只支持 DES 加密；它的认证协议容易遭到中间人（man-in-the-middle）、重放（replay）等方法的攻击；它的密钥管理也没有很好定义。此外如果在设计 BACnet 网络时，如果对安全问题不够重视，攻击者可以利用 BACnet 提供的一些应用层服务和网络层的报文对 BACnet 网络进行信息窃取、应用 DoS 攻击、网络层 DoS 攻击等。

LonWorks 的安全性要比 BACnet 差。LonWorks 的安全机制只采用了简单的质询回应（challenge-response）服务器认证方式，服务器并不能鉴别客户端，攻击者可以容易地假冒客户端；LonWorks 的通讯不提供具有认证、完整性、保密性的包。

BACnet 和 Lonwoks 都提供了相应的在 IP 网上通信的能力，在提供更大网络互联能力的同时，也带来了更多的安全问题。

2.4 信息网络的安全威胁

在建筑智能化系统中，除了对建筑设备、安防设备、消防设备等的自动控制外，建筑中的人们各种信息的需求越来越大。CAS 和 OAS 中均包含了大量的信息，并使用高速计算机网络来传递信息，构成了建筑智能化系统的信息网络。建筑智能化系统能够将信息网络和控制网络集成起来，使得控制网络的一些状态、控制、管理等信息也能够在信息网络中被查询、传递、处理。信息网络广泛使用 IP 技术，因此信息网络面临的安全威胁与普通计算机网络的安全威胁相同。这些安全威胁包括：口令攻击、端口扫描、网络监听、IP 欺骗、拒绝服务、缓冲区溢出、计算机病毒、计算机蠕虫、特洛伊木马、逻辑炸弹等。随着网络技术的发展，这些安全威胁的种类及其实现技术还在不断的发展变化。

2.5 无线网络的安全威胁

在建筑智能化系统中，可能用到许多无线技术，如 802.11、Bluetooth、Zigbee 等。目前应用最多的是由基于 802.11 的无线局域网，其他技术应用较少，其安全问题也相对较小。由于无线技术的特殊性，它比有线网络更容易受到攻击。无线网络易受到的攻击有：无线信号窃听；插入攻击；相同频率大功率无线信号的干扰；拒绝服务攻击；基站伪装等。802.11 采用有线等价保密(Wired Equivalent Privacy，WEP)协议和服务集标识符(Service Set Identifier，SSID)提供一定的安全保护，但这些措施存在着不少安全问题。SSID 是一个可选项，许多实际产品默认允许具有任何 SSID 的工作站都可以登录接入点，并且 802.11 允许接入点通过信标帧以明文形式广播自己的 SSID，这使得 SSID 很容易被窃取。同时 WEP 协议也被证明具有许多漏洞，其采用的开放系统认证及共享密钥认证均存在一定的安全问题。

3 建筑智能化系统的安全对策

3.1 物理安全对策

对于物理安全威胁，在建筑智能化系统建设时，应从设备，设备环境、线路及管理等方面考虑网络的安全。设备选型时，应当选择可靠性高的设备。建立良好设备运行环境，包括适宜的温、湿度，良好的接地以预防雷击，良好的设备防震处理，以及自动的火灾、水淹报警机制。通信线路应尽量封闭于管道中，采用安全性较高的光纤作为传输媒体。加强安全管理，定期对设备运行状况进行巡检，发现安全隐患；通过安全巡逻，交接班制度等预防物理接近和对网络设施的破坏；加强密码管理；加强安全教育，尽量避免人为疏忽、失误影响网络安全。

3.2 控制网络安全对策

对控制网络的安全威胁，一方面控制网络的协议制定者已经意识到安全问题的重要性，正在从协议体系结构方面引入更多、更强的安全机制。同时我们在使用现有网络的时候，应当增强安全意识，不仅要从物理安全方面重视，还要充分利用协议目前所提供的安全功能。

3.3 信息网络安全对策

对于信息网络安全威胁，目前有许多安全技术及解决方案，最主要的技术包括防火墙、入侵检测系统、认证技术、加密技术、访问控制技术、安全审计、VPN 技术、病毒防治、数据保护与恢复等。

3.4 无线网络安全对策

对于无线网安全威胁，首先在选择接入点的位置时，应当尽量使合法用户的信号加强，合法用户范围之外的信号减弱；其次 SSID 不要使用默认值，关闭接入点的 SSID 广播；启动 WEP 的 128 位密钥模式，定期修改 WEP 密钥；关闭 DHCP 服务，并在接入点上进行 MAC 地址过滤。同时在整个网络设计时，应将无线网络视为非信任设备，放置在非信任网段中，采取一定措施将非信任网段与内部安全网络隔离。

4 结束语

综上所述，我们可以看到，在建筑智能化系统中存在着诸多的安全问题。而在实际的应用过程中，由于各个建筑智能化系统的规模不同，其包含子系统类型，子系统的集成程度及集成技术也千差万别，因而对安全的要求也不尽相同。我们应当高度重视建筑智能化系统的安全问题，但并不是要对所有的建筑智能化系统都采用最强的安全措施，而是要制定出满足实际安全需求的、合理的安全方案。目前，我国大力提倡绿色建筑，它要求建筑物更加环保、节能。建筑智能化系统作为绿色建筑的一部分，促进绿色建筑的发展，而建筑智能化系统安全则是绿色建筑不可缺少的保障。

The Security Problems and Countermeasures of Building Intelligentization System

Tang Hong Wang Bo
(Building Intelligentization lab, College of Computer Science,
Chongqing University, Chongqing 400044 China)

Abstract: This article puts forward the security problem of building intelligentization system at the system view. It explains the importance of the security problem of building intelligentization system and then describes the difference and relationship between building intelligentization system security and Safety Automation System. After analyzing the security problems existing in building intelligentization system in detail, including control network, information network and wireless network, the corresponding countermeasures are pointed out. It has important referential value for ensuring the safe running of the building intelligentization system.

参考文献

[1] 邓亚平. 计算机网络安全［M］. 北京：人民邮电出版社，2004.9. 2～3

[2] Nobuo Okabe Issues of Control Networks When Introducing IP［EB/OL］http://www.taca.jp/docs/saint2005/ presentation-workshop-nobuo _ okabe.pdf

[3] David G. Holmberg. BACnet Wide Area Network Security Threat Assessment［EB/OL］http://fire.nist.gov/bfrlpubs/build03/PDF/b03034.pdf

[4] 姜楠，王健. 移动网络安全技术与应用［M］. 北京：电子工业出版社，2004.11. 198～199

模糊控制在变风量空调中的应用*

刘维福　叶安丽
（北京建筑工程学院电气工程及其自动化系，北京　100044）

摘　要：常规 PID 控制器不能在线整定参数，当对象存在结构非线性、参数时变性或模型不确定时，其参数更加难以整定，且最佳参数容易漂移。本文以变风量空调房间为研究对象，采用模糊控制与 PID 控制相结合，构成模糊自适应 PID 控制器，利用模糊控制器在线调整 PID 控制器的参数进行大量的仿真研究，然后在仿真实验的基础上将其应用于实际控制。

关键词：模糊控制　模糊自整定 PID　模拟仿真

1　引言

变风量空调系统由冷热水机组、空气处理机组、风阀及风管、送/回风机以及空调房间构成。控制系统一般可以分为五部分：

1. 室内温度控制，即变风量末端装置控制；
2. 室内正压控制，即送、回风机匹配控制；
3. 送风静压控制；
4. 送风温度控制；
5. 新风量控制。

五个部分相互独立，又相互关联，各回路之间存在着很强的耦合性。当某个房间的温度下降时，该房间的末端装置的风阀就会关小，从而导致总风管内的送风静压升高，其他房间的送风量增加。这时，这些房间末端装置的风阀就会关小以恒定各自的送风量。这又将导致系统静压的进一步升高。当达到一定程度时，静压控制器就将降低送风机的转速以减小风量，回风机风量也随之减少。系统静压又回到原来的水平，这样各末端装置的风阀又开始开大。由于系统压力的变化必将影响到新风量的变化，从而导致送风温度的变化。送风量变化使得控制器调节新风、排风、回风 3 个风阀的开度，这 3 个风阀阀位的变化将致使整个系统的静压和流量发生变化。这时，系统处在一种频繁的调节当中。风阀时而开大时而关小，送进室内的风量也是忽大忽小[1]。

因此变风量系统对控制的要求比定风量系统要高，要建立一个适合那些工程控制的数学模型比较困难，无法运行和运行情况不佳的系统，常常都是因为不恰当的控制系统所造

* 基金项目：北京市重点实验室项目(KF200501)。

成的。VAV空调系统是一个干扰大、高度非线性的复杂系统，因此，传统PID控制的效果往往并不理想。在过去的几年里，模糊逻辑控制逐渐被应用到暖通空调领域，并取得了比传统PID控制更好的控制效果。本文针对VAV末端装置采用模糊控制进行了探讨。

2 模糊控制理论

模糊控制是以模糊集合论、模糊语言变量及模糊逻辑推理为基础的计算机智能控制。基于规则的智能控制，以模糊数学为基础。模糊控制系统的基本原理框图如图1所示[2]。它的核心为模糊控制器，如图中划线框中所示，模糊控制器的控制规律由计算机的程序来实现。微机经中断采样获得被控量的精确值，然后将此量与给定值比较得到误差信号 e，一般选误差信号 e 作为模糊控制器的一个输入量。把误差信号 e 的精确量进行模糊化变成模糊量。误差 e 的模糊量可以用相应的模糊语言来表示，得到误差E的模糊语言集合的一个子集 e(e 是一个模糊矢量)，再由 e 和模糊控制规则 R(模糊算子)根据推理的合成规则进行模糊决策，得到模糊控制量 u：[2]

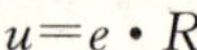

$$u=e\cdot R$$

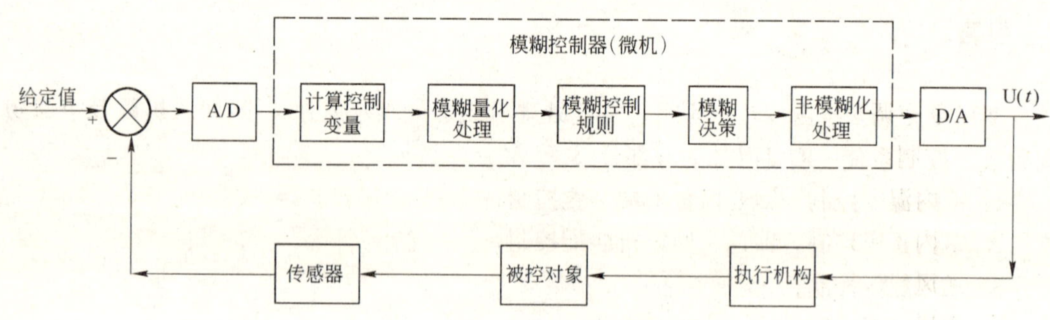

图1 模糊控制器的基本原理框图

3 模糊控制的设计

3.1 设计思路

随着计算机技术的发展，人们利用人工智能的方法将操作人员的调整经验作为知识存入计算机中，根据现场实际情况，计算机能自动调整PID参数，这就是古典的PID和先进的专家控制相结合。模糊自适应PID控制器以误差 e 和误差变化 ec 作为输入，可以满足不同时刻的 e 和 ec 对PID参数自整定的要求。利用模糊规则在线对PID参数进行修改，便构成了自适应模糊控制器。模糊自整定PID控制器由参数可

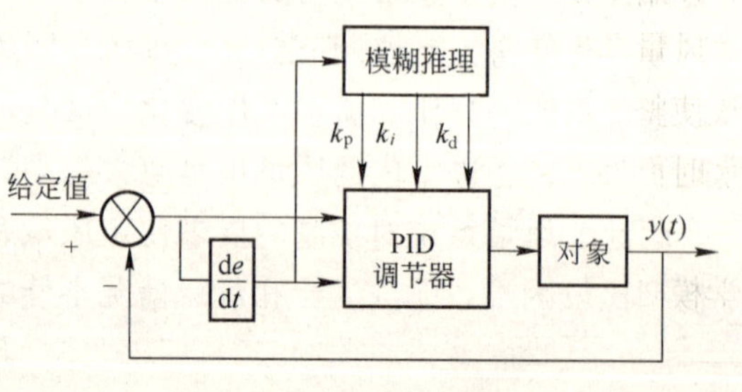

图2 自适应模糊控制器结构

调整 PID 控制器和模糊控制器两部分组成，其控制原理框图如图 2 所示。其设计思想是：先建立 PID 控制器的三个参数与偏差 e 和偏差变化率 ec 的模糊关系即模糊规则，然后以偏差 e 和偏差变化率 ec 作为输入量，通过模糊规则对 PID 参数进行在线修改以满足不同时刻偏差 e 和偏差变化率 ec 对 PID 参数自调整的要求。本系统中，模糊控制器的设计将是设计的核心，因为模糊控制器的好坏直接影响到 k_p、k_i 和 k_d 的选取，从而影响到控制系统的精度。[3]

模糊控制器的设计主要包括以下几项内容：

（1）确定模糊控制器的输入变量和输出变量；

（2）选择模糊控制器的输入量及输出量的论域并确定模糊控制器的参数（如量化因子、比例因子）；

（3）设计模糊控制器的模糊规则；

（4）确定模糊和反模糊化的方法，即模糊推理及模糊运算的方法；

（5）编制模糊控制算法的应用程序。

3.2 模糊控制表的生成

1. 确定模糊变量的赋值表

模糊变量误差 e、误差变化 ec 及控制量 u 的模糊集和论域确定后，须对模糊语言变量确定隶属函数，即所谓对模糊变量赋值，就是确定论域内元素对模糊语言变量的隶属度。

2. 建立模糊控制表

模糊自整定 PID 是在 PID 算法的基础上，通过计算当前系统误差 e 和误差变化率 ec，利用模糊规则进行模糊推理，查询矩阵表进行参数调整。模糊控制设计的核心就是总结工程设计人员的知识和实际操作经验，建立合适的模糊控制规则表，得到针对 k_p、k_i，k_d 分别整定的模糊控制表。

（1）k_p 的模糊规则（表 1）

表 1

Δk_p \ ec / e	NB	NM	NS	ZO	PS	PM	PB
NB	PB	PB	PM	PM	PS	ZO	ZO
NM	PB	PB	PM	PS	PS	ZO	ZO
NS	PM	PM	PM	PS	ZO	NS	NS
ZO	PM	PM	PS	ZO	NS	NM	NM
PS	PS	PS	ZO	NS	NS	NM	NM
PM	PS	ZO	NS	NM	NM	NM	NB
PB	ZO	ZO	NM	NM	NM	NB	NB

（2）k_i 的模糊规则表（表 2）

表 2

Δk_i \ ec / e	NB	NM	NS	ZO	PS	PM	PB
NB	NB	NB	NM	NM	NS	ZO	ZO
NM	NB	NB	NM	NS	NS	ZO	ZO
NS	NB	NM	PM	NS	ZO	PS	PS
ZO	NM	NM	NS	ZO	PS	PM	PM
PS	NM	NS	ZO	PS	PS	PM	PB
PM	ZO	ZO	PS	NM	PM	PB	PB
PB	ZO	ZO	PS	PM	PM	PB	PB

(3) k_d 的模糊规则表(表 3)

表 3

Δk_d \ ec / e	NB	NM	NS	ZO	PS	PM	PB
NB	PS	NS	NB	NB	NB	NM	ZO
NM	PS	NS	NB	NM	NM	NS	ZO
NS	ZO	NS	NM	NM	NS	NS	ZO
ZO	ZO	NS	NS	NS	NS	NS	ZO
PS	ZO	ZO	ZO	ZO	ZO	ZO	ZO
PM	PB	NS	PS	PS	PS	PS	PB
PB	PB	PM	PM	PM	PS	PS	PB

3.3 去模糊化

k_p、k_i 和 k_d 的模糊控制规则表建立以后，可根据以下方法进行 k_p、k_i 和 k_d 的自适应校正。将系统误差 e 误差变化率 ec 变化范围定义为模糊集上的论域 e，ec={−6，−5，−4，−3，−2，−1，0，1，2，3，4，5，6}。其模糊子集 e，ec={NB，NM，NS，ZO，PS，PM，PB}，子集中元素分别代表 {负大，负中，负小，零，正小，正中，正大}。设 e，ec 和 k_p、k_i，k_d 均服从正态分布，因此可以得出各模糊子集的隶属度，根据各模糊子集的隶属度赋值表和各参数模糊控制模型，应用模糊合成推理设计 PID 参数的模糊矩阵表查出修正参数代入下式计算：

$$k_p = k_p' + \{e, ec\}_p$$

$$k_i = k_i' + \{e, ec\}_i$$

$$k_d = k_d' + \{e, ec\}_d$$

在线运行过程中，控制系统通过对模糊逻辑规则的结果处理，查表和运算，完成对 PID 参数的在线自校正。其工作流程图如图 3 所示。

3.4 仿真结果分析

利用我校的北京市重点实验室变风量空调系统，该系统有 5 个变风量末端。各个末端通过 DDC 进行控制检测，由 BACnet 现场总线负责各末端的通信。DDC 中包含有 PID 控制算法，可直接控制阀门的开度。末端阀门采用的是 6 片对开式叶片组成节流式调节风门，调节范围为 100%～10%；模拟的空调房间面积 $25m^2$，室内设定温度 26℃，VAV 末端送风温度 18℃，室内负荷 5kW，使用 2 个 1kW 的电热炉和 30 个 100W 灯泡来模拟。

在某一时刻加开一部分灯，来模拟房间内负荷变化作为系统的扰动量，得到 VAV 末端风阀开度的变化过程如图 4 所示。同时，采用 Matlab 仿真软件中的模糊控制工具箱来对 VAV 末端的开度调节进行仿真研究。在相同条件下的模糊 PID 控制下的风阀开度变化曲线。

由图 4 可以看出，采用模糊控制后的风阀开度变化幅度较 PID 控制更小，能实时跟踪温度的变化，对模型的不确定性具有较好的适应能力。在实际运行时，控制表可根据运行结果进行修正。

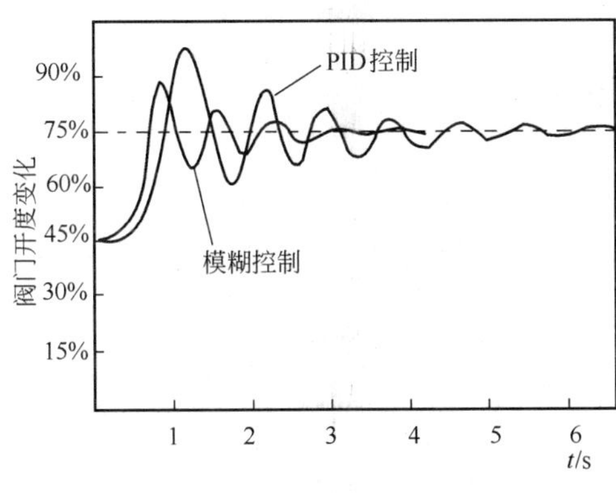

图 4　VAV 末端风阀控制输出

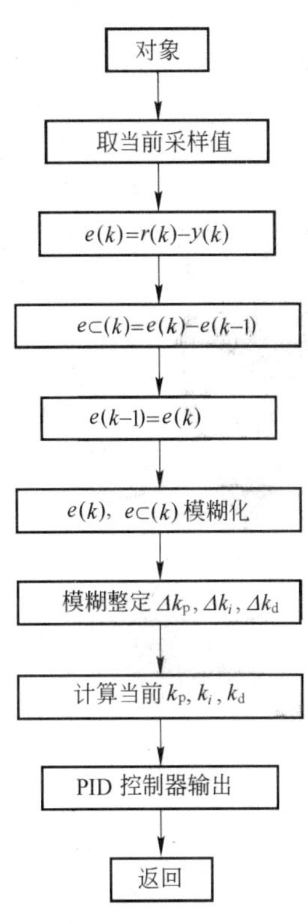

图 3　在线自校正工作流程图

4　结论

本文提出的模糊参数自整定 PID 控制器结合了 PID 控制和模糊控制的优点，实现了对 PID 参数的实时在线整定。与常规 PID 温控系统相比，该控制器不仅控制精度更高，对受控对象模型的适应能力更强，并且结构较为简单，计算量小，易于工程实现。仿真实验克服了工程实践中的盲目性，为实际温控系统的设计与调试提供了理论参考依据。

Fuzzy Control Used in VAV System

Liu Weifu　Ye Anli

(Department of Electrical Engineering and Automation, Beijing Institute of Civil Engineering and Architecture, Beijing　100044　China)

Abstract: The PID controller can't adjust parameters online and its optimum parameters are inclined to drift with nonlinear, time variable or uncertain model system. Combining PID with fuzzy theory with researching on the VAV air conditioned room, the Fuzzy self-tuning PID controller is proposed. Before application, a lot of simulation experiments upon fuzzy controller to identify the parameters of PID controller is performed.

参考文献

［1］ 李萍，李峰，赵虎，居滋培. 模糊自整定 PID 控制器的设计和仿真. 仪器仪表学报 2002

［2］ 先进 PID 控制 MATLBA 仿真. 北京电子工业出版社. 2004

［3］ 胡益雄，袁 锋，张丽萍. VAV 末端装置的模糊设计探讨. 长沙铁道学院学报 2002 年 12 月第 20 卷第 4 期

［4］ 吴爱国，高德云. VAV 空调系统的智能控制. 制冷学报. 1999

［5］ 李慧，段晨旭. 变风量空调系统末端装置的性能分析. 山东建筑工程学院学报. 2005

人工神经网络在空气处理单元换热器建模中的应用

胡钦华　　　　　　　　　　　　李奎山
（东莞理工学院教务处，东莞　523106）（东莞理工学院国际交流处，东莞　523808）
苏廷弼　　　　　　　　　　　　董阿妮
（香港城市大学建筑系，香港）（东莞理工学院城市学院，东莞　523106）

摘　要： 本文应用人工神经网络技术对安装在空气处理单元（Air Handler Unit，AHU）中的换热器建立了静态和动态神经网络模型。该模型的所有预测值基本都在实际测量值的95％～105％之间。实际测量值和神经网络模型预测值之间的平均相对误差（Mean Relative Error，MRE）小于2.5％，充分说明模型逼近实际对象的精确性。本文中同时给出了该神经网络模型的权值和阈值，它可以作为工程上一个很好的参考。本文的实验结果显示，神经网络，尤其是BP神经网络，可以作为一个很容易的建模工具来获得空气处理单元中换热器的神经网络模型，同时也很容易应用在类似的热系统中。

关键词： 换热器　换热率　人工神经网络　BP算法　模型

1　简介

人工神经网络目前的研究热点很大程度上源于它能模拟人脑智能，具有自学习能力。它可以通过学习输入-输出关系对来自学习，而不需要精确的差分方程模型。所以它们广泛应用于模式识别、函数估计、优化、预测、自动控制、建模和快速信息处理等很多方面。从这点可以看出，人工神经网络可以学习和预测一个系统的性能。因此神经网络建模已经成为经典建模方法的一个替代选择，而且经常应用于经典建模方法难于处理的方面。

换热器是广泛应用于制冷系统和空气处理系统的一种重要设备。它的机理复杂以至于很难用热力学第一原理来预测它的静态和动态性能。在过去的几年里，Yasar Islamogly采用神经网络技术建立了换热器的差分模型[1]。Yasar Islamogly采用神经网络技术建立了换热器的稳态模型来预测换热器的换热率，Arturo Pacheco-Vega et al.[2]研究了利用换热器的神经网络模型来估计换热率的精确性问题，Bechtler et al.[3]运用神经网络建立了蒸汽压缩液态热泵的稳态性能模型，Gerardo et al.[4]提出了使用神经网络对换热器进行动态预测和控制的方案。许多公开刊物上也列出了神经网络在HVAC系统中应用的大量实例，并且神经网络模型的精确性也得到证实[3]。从上面提到的参考文献中我们可以得出结论，人工神经网络在工程应用中更有利于热力学分析。几乎所有的神经网络在换热器中的应用都与制冷有关。神经网络从未应用在空气处理单元换热器的建模中。因此本文重点研究人工神经网络方法在空气处理单元(AHU)中换热器建模上的应用。

由于物理模型需要系统的几何参数，而这些参数在工程上通常都是很难获得的。其他的建模方法希望在不使用物理或者热力学模型的前提下得到换热器的性能预测。在本文中提出了一种新的建模方法来预测换热器的性能，它把系统当作黑箱处理而不考虑系统内部的信息。该模型是神经网络的应用，特别是多层神经网络的实际应用。该研究领域已经成为一个很有前景的研究领域并且在近几年越来越流行，因为神经网络比其他方法能够更快的解决问题。神经网络在制冷系统和空气调节中的应用以多层神经网络应用最多，这些方法在制冷机的建模中也同样适用[5~9]。

2 神经网络

人工神经网络是由权值向量和学习算法组成的系统，其学习算法采用由一组线性模式输入来产生一组数字模式输出表示实际的输出。在各种不同用途的人工神经网络的多种结构中，多层前馈神经网络是目前最流行的。在 1980 年到 1990 年，许多作者给出了结论性的证明，即人工神经网络可以在任意精度逼近任一非线性函数。因此人工神经网络在许多领域得到广泛应用，包括函数逼近和模式识别。人工神经网络模型代表了系统辨识和预测中的一种新的方法。神经网络类似“黑箱”运行但不需要系统的详细信息。相反，他们通过学习先前记录的数据来学习输入参数和可控或者不可控变量之间的关系，同非线性回归方法类似。神经网络的另外一个优点是他们具有通过相关参数来处理大而复杂的系统的能力。

在本文中选择多层前馈神经网络是因为它具有以下几个优点[11]：

(1) 自由的线性假设；

(2) 具有很大程度的自主；

(3) 更有效处理非线性函数问题。

前馈网络一般由三层神经元组成，即输入层从外部接受输入，隐含层接受输入层信息并将信息输出到输出层，输出层最后将处理后的信息输出，而且很多情况下，输出层把输出误差方向传播。网络的连接强度依赖于神经元之间的权值，这些值在训练过程中更改。训练通过给网络施加特定的数据信息集并通过训练算法来使得网络得到期望的输出。尽管有很多其他的方法，但误差反传算法(BP 算法)[10]是使用最普遍的训练人工神经网络的算法。我们使用它是因为它的优良特性有助于我们得到精确的模型。本文采用批训练模式，也就是所有的输入和目标值都完成后才更新权值和阈值。本文中多层前馈网络使用的性能函数是 MSE—即网络输出和目标值之间的均方误差。

3 系统和实验数据的描述

本实验是在香港城市大学江森自控实验室的 HVAC 系统上完成的。所研究的换热器(表面积 0.3068m^2(0.82m×0.44m))是安装在该系统 AHU 中的。该实验系统由两部分组成。一部分是空气处理单元 AHU，带有送风管道和回风管道；另一部分是制冷系统，由蒸发器、压缩机、冷凝器和热力膨胀阀组成。风道、换热器、变频器和风机是 AHU 的主要组成部分。通过改变由变频器控制的送风风机的转速来改变通过换热器的空气流量。同时改变冷冻水管上的三通水阀的控制电压来改变通过换热器的冷冻水流量。在实验期间，可以通过两种方式来改变：一种是通过控制面板来手动调节，另外一种是通过计算机由程序直接控制。

实验在 2003 年 3 月至 6 月进行，数据采集是通过与计算机相连的 ADLINK 的数据采集板来得到的，采样间隔 1 分钟。冷冻水进出换热器的温度和空气温度都是采用美国江森自控有限公司提供的电子温度传感器测得的，精度为＋/－0.2K。冷冻水的流量采用精度为全刻度＋/－1％的流量表测得，同时我们采用精度为量程＋/－5％的风速传感器来测量通过换热器的风速。空气流量则通过下面的公式计算得到：

$$m_a = V_a \times S_h \times p_a \tag{1}$$

式中，a 代表空气，h 代表换热器，m_a 代表空气流量，S_h 代表换热器的表面积，p_a 代表空气密度。

换热率 Q 采用下面的公式计算得到：

$$Q = m_{chw} \times c_{chw} \times (T_{chw}^{out} - T_{chw}^{in}) = m_a \times c_a \times (T_a^{in} - T_a^{out}) \tag{2}$$

式中，chw 代表冷冻水，m_{chw}代表冷冻水流量，c_{chw}和 c_a 分别是水和空气的比热，T_{chw}^{out}和 T_a^{out}分别表示从换热器出来的冷冻水和空气的温度，T_{chw}^{in}和 T_a^{in}分别表示进入换热器的冷冻水和空气的温度。由于空气侧和冷冻水侧的换热率有轻微差别，所以使用二者的平均值。此解决方案由计算机通过 VB 编制而成。

4　换热器的人工神经网络模型

BP 模型的建立是在基于实验数据的基础上，在 MATLAB 下实现的，因为MATLAB 提供了矩阵运算和易于编程的特点。我们通过 VB 开发的计算机实时监控程序使得我们的 BP 模型得以应用。三通冷冻水阀的控制电压由四阶的逆 M 序列给出，驱动送风机的变频器的控制电压由两个不同频率的正弦信号叠加而成。当系统启动并进入稳定状态时，我们通过计算机给系统施加控制信号。从而使得冷冻水流量和空气温度有规律的变化，并且冷冻水和空气温度的变化处于同一工况下。

4.1　换热率的静态预测

在稳定状态下，神经网络可以在给定工况下预测换热率。神经网络模型的目标是只要求那些对工程师来说易于获得的输入参数。本实验中是指冷冻水的进出换热器的温度，进入换热器的空气温度，以及冷冻水和空气的流量(五维的输入空间)。模型输入和输出参数的值限制在表 1 给出的训练值的范围内。使用 BP 网络训练时输入值超出范围则无法达到训练精度。所以神经网络的输入输出被规范化到区间［0，1］内。该模型用来预测换热器的换热率(一维输出空间)。

神经网络的结构如图 1 所示。输入层(5 个)为 m_a，m_{chw}，T_a^{in}，T_{chw}^{in}，T_{chw}^{out}，输出层(1 个)为 Q，隐含层神经元个数选择为 10 个。为了决定神经网络的结构并利用实验数据来预测换热率，通过调节隐含层神经元的个数和学习速率及动量项来检查误差的收敛情况，最终隐含层选择了 10 个神经元，学习速率和动量因子分别选择为 0.01 和 0.9。利用 445 对实验数据中的 345 对数据来对网络进行训练，经过 137187 步网络训练结束，并且误差性能函数 MSE 收敛到 0.0000999998(目标设置为 0.0001)。剩下的实验数据仅用作最后的模型验证。

余下的100对数据用来对神经网络模型进行验证，由于这些数据未参与神经网络的训练，所以是相对独立的。

模型参数范围表 **表1**

参数	最大值	最小值
进入换热器的空气温度(℃)	29	22
换热器出口空气温度(℃)	19	14
冷冻水进入换热器温度(℃)	15	8
换热器出口冷冻水温度(℃)	19	12
空气流量(kg/s)	1.5	0.2
冷冻水的流量(kg/s)	1.0	0.4

图2对实验测量值和神经网络模型的预测值进行了比较。同时通过实验数据和预测值进行了相对误差计算。图3显示了神经网络模型的相对误差。从图中我们可以看到所有的预测值都在实验测量值的95%～105%之内。平均相对误差(MRE)为1.38%，最大相对误差在第68个样本数据处，为4.87%。

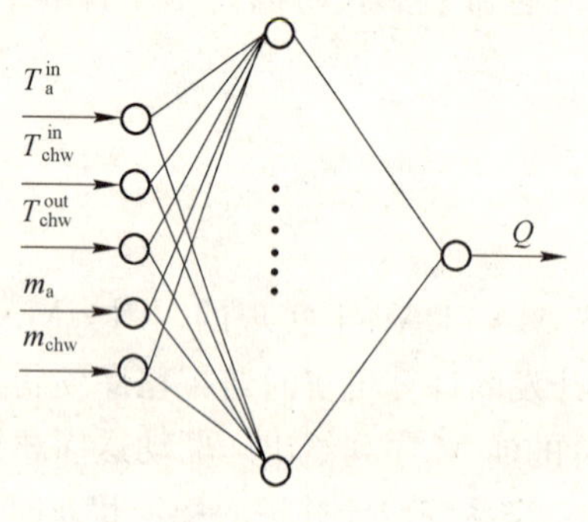

图1 换热率神经网络预测模型结构

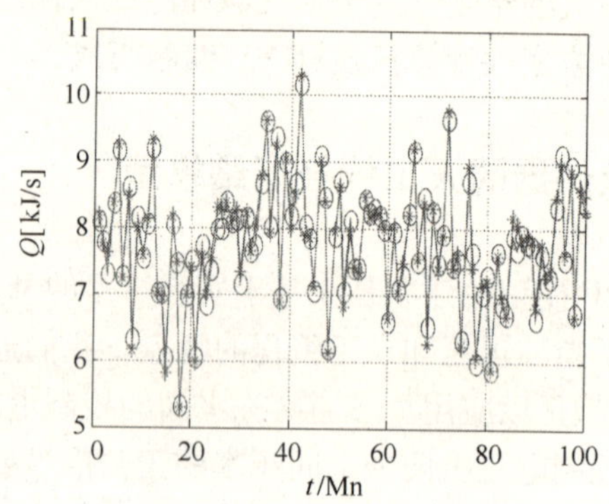

图2 预测值和测量值比较

注：*标识测量值；o标识预测值。

作为参考，该换热器的静态神经网络模型的权值在附录Appendix I中给出。

4.2 换热器动态性能的预测

上节讨论的静态神经网络模型仅对该换热器的性能进行了精确建模，而对控制目的而言，远远不能满足于静态预测，因为该系统是时间相关的系统。我们可以利用冷冻水通过换热器的出水温度 T_{chw}^{out} 来控制压缩机的旁路运行。由于本实验系统中冷机规模小并且负荷相对稳定，所以这种能量控制方法能够用在本实验系统中。送风风机的速度和冷冻水的温度是由送风温度 T_a^{out} 来决定的。我们可以通过实验来得到三者之间的最佳匹配关系。目的是为了使得风机和压缩机的输入功率最小。而安装在空气处理单元中的换热器则是将空气调节系统和制冷系统两个部分连接为一个整体的关键部件。所以对换热器的动态行为进行仿真显得尤其重要。本文中对于控制方案未进行讨论。

通过神经网络模型来对该换热器进行动态预测。网络输入选择为 m_a、m_{chw}、、T_a^{in}、T_{chw}^{in}，输出选择为 T_a^{out} 和 T_{chw}^{out}，图5给出了测量数据 m_a 和 m_{chw}，而 T_a^{in} 和 T_{chw}^{in} 则在图6中

给出。在对一个系统进行动态建模的时候一个很重要的问题是要考虑系统的阶次。我们在前面已经给出了相关变量，而神经网络在对一个未知阶次的系统进行仿真时，阶次越高，就越是需要更多过去时刻的变量的值来作为神经网络的输入参数。Gerardo 指出，如果要选择一个整数作为系统的阶次，一般可以选择 2 阶而不需要选择更高的阶次[4]。在本文的实验中，系统有两个回路—空气回路和冷冻水回路，通过阶跃响应我们可以选择两个回路的阶次均为 1 阶。所以只需要一个过去时刻的变量值来进行预测。该动态神经网络模型在图 4 中给出。

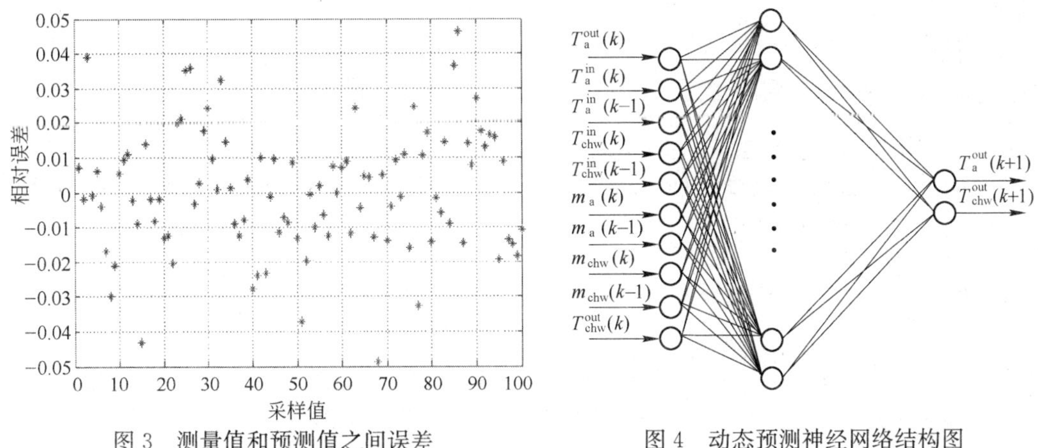

图 3 测量值和预测值之间误差

图 4 动态预测神经网络结构图

用 502 组实验数据中的 410 组数据对该神经网络进行训练，经过 484240 步之后网络训练结束，网络的性能函数 MSE 收敛到 0.001(目标值为 0.001)。剩下的实验数据仅用作最后的模型验证。

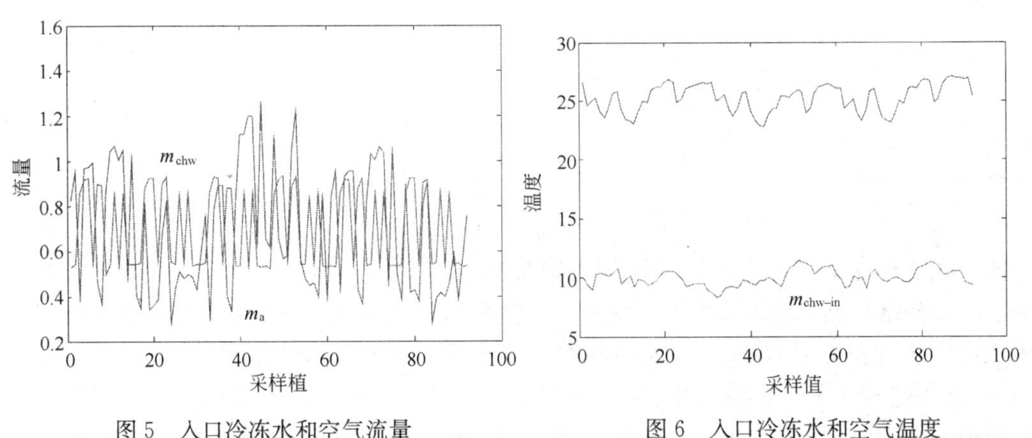

图 5 入口冷冻水和空气流量

图 6 入口冷冻水和空气温度

用余下的 92 组实验数据对该神经网络的模型进行验证。图 7 和图 9 分别比较了冷冻水通过换热器的出水温度、换热器出来的空气温度的实验值和预测值。图 8 和图 10 则给出了二者实验值和预测值之间的相对误差。从图中我们可以看出几乎所有的(冷冻水出水温度数据的 90%，从换热器出来的空气温度数据的 98.9%)预测值都在实验值的 95%～105%范围之内。对于冷冻水出水温度的预测中，平均相对误差为 2.45%，最大相对误差在第 66 个采样点处为 11.42%。在换热器出口空气温度的预测中，最大平均相对误差和

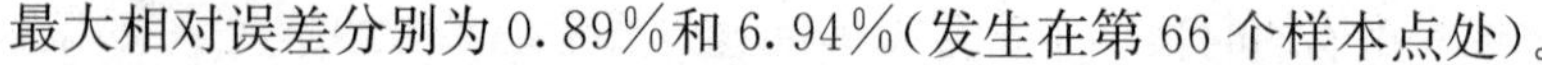

最大相对误差分别为 0.89%和 6.94%(发生在第 66 个样本点处)。

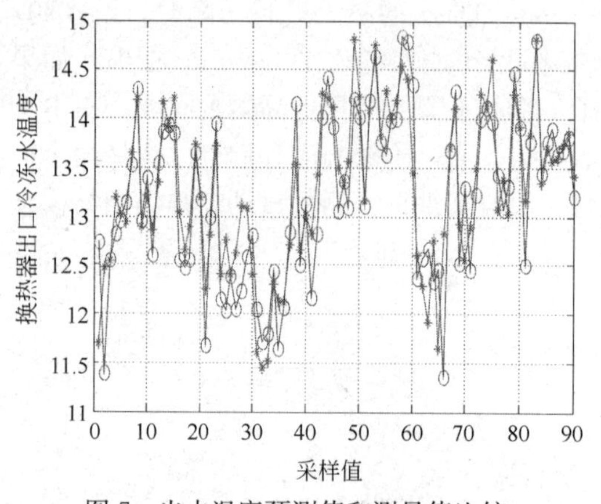

图 7 出水温度预测值和测量值比较

注：*表示测量值，o 表示预测值。

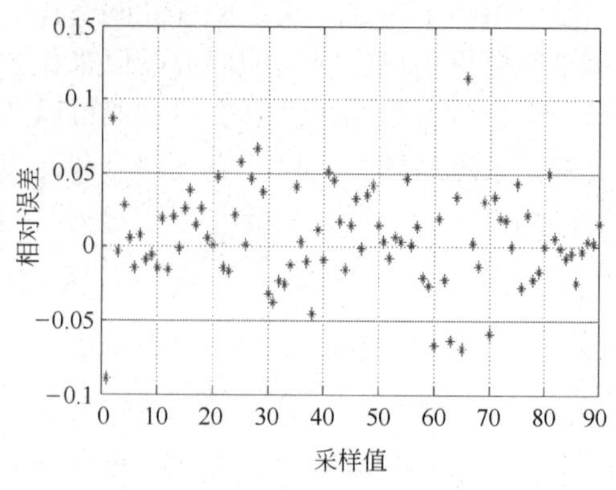

图 8 出水温度测量值和预测值之间的相对误差

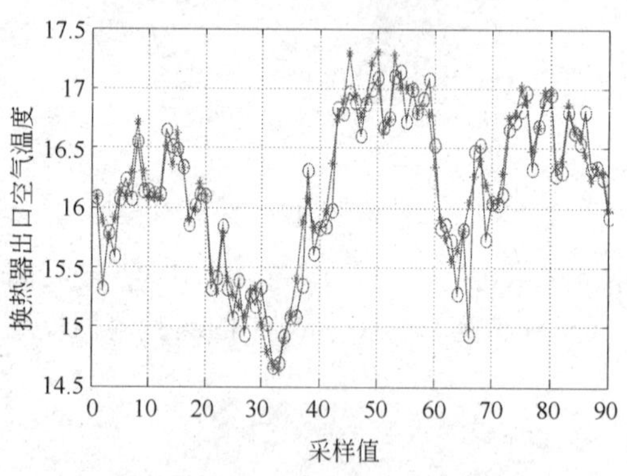

图 9 出口空气温度预测值和测量值比较

注：*表示测量值，o 表示预测值。

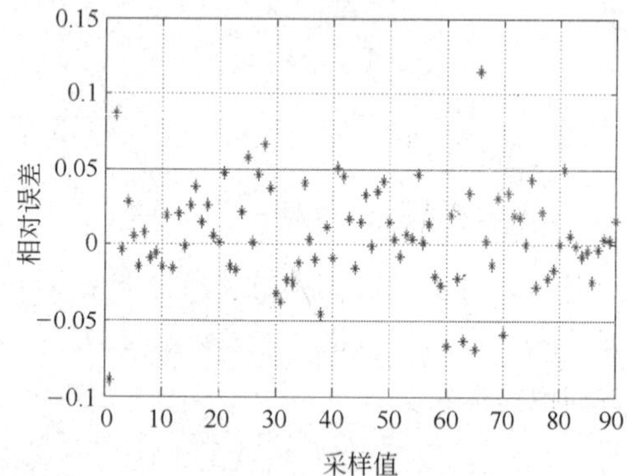

图 10 出口空气温度预测值和测量值之间的误差

5 结论

文中成功建立了空气处理单元中的换热器的静态和动态神经网络模型。换热器换热率的静态神经网络模型预测值很精确，集中在实测数据的 95%～105%范围之内。并把这种技术扩展应用到换热器动态行为的预测中。几乎所有的动态预测值都位于实测数据的 95%～105%范围之内。这些结果显示，神经网络模型可以在一定范围内作为替代热力学第一定理的合理选择。进一步的研究重点放在利用神经网络对制冷系统和空气调节系统的每个部件进行神经网络建模，并引入神经网络控制技术到系统中。

6 致谢

作者在此感谢香港城市大学提供支持(City University of Hong Kong for the financial support of the Strategic Research Grant #7001535)。

Building Up Neural Networks Models for a Heat Exchanger Mounted in Ahu

Hu Qinhua[1] Li Kuishan[2] Su Tingbi[3] Dong Ani[4]

([1] Dean's office of DongGuan University of Technology, DongGuan, 523106; [2] International Communication Office of DongGuan University of Technology, DongGuan, 523808; [3] Building and Construction department of City University of Hong Kong, Hong Kong; [4] City College of DongGuan University of Technology, DongGuan, 523106)

Abstract: This study uses the neural network technique to obtain the static and the dynamic models of the heat exchanger mounted in the air handler unit (AHU). Almost all of the predicted values are within 95%～105% of the actual measured values. And the consistence between measured and ANNs predicted results was achieved by a mean relative error(MRE)<2.5%. The weights and biases of the neural network models were given in the paper. It is the better reference for engineers. The results from this study show that neural networks, especially the BP network, can be an easily modeling tool for obtaining the models of heat exchangers and similar thermal system.

参考文献

[1] Yasar Islamoglu, A new approach for the prediction of the heat transfer rate of the wire-on-tube type heat exchanger-use of an artificial neural network model, Applied Thermal Engineering 23 (2003)243～249

[2] Arturo Pacheco-Vega, Mihir Sen, K. T. Yang, Rodney L. McClain, Neural network analysis of fin-tube refrigerating heat exchanger with limited experimental data, Int. J. Heat and Mass Transfer 44 (2001)763～770

[3] H. Bechtler, M. W. Browne, P. K. Bansal, V. Kecman, Neural networks-a new approach to model vapour-compression heat pumps, Int. J. Energy Res. 2001: 591～599

[4] Gerardo Diaz, Mihir Sen, K. T. Yang, Rodney L. McClain, Dynamic prediction and control of heat exchangers using neural networks, Int. J. Heat and Mass Transfer 44(2001)1671～1679

[5] D. J. Swider, M. W. Browne, P. K. bansal, V. Kecman, Modeling of vapour-compression liquid chillers with neural networks, Applied Thermal Engineering 21(2001)311～329

[6] A. T. P. So, T. T. Chow, W. L. Tse, A neural-network-based identifier/controller for modern HVAC control, ASHRAE Transactions 101(2)(1995)14～31

[7] A. Palau, E. Velo, L. Puigjaner, Use of neural networks and expert system to control a gas/solid sorption chilling machine, Int. J. Refrig 22(1)(1999)59～66

[8] F. J. Ferrano, K. V. Wong, Prediction of thermal storage loads using a neural network, ASHRAE Transactions 96(2)(1990)723～726

[9] M. Kawashima, C. E. Dorgan, J. W. Mitchell, Hourly thermal load prediction for the next 24 hours

by Arima, Ewma, LR and an artificial neural network, ASHRAE Transactions 101(1)(1995) 186～200

[10] D. E. Rumelhart, G. E. Hinton, R. J. Williams, Learning internal representations by error propagation, in: Parallel Distributed Proceeding: Explorations in the Microstructures of Cognition, vol. 1, MIT Press, Cambridge, MA, 1986, pp. 318～362

[11] F. F. Farshad, J. D. Garber, J. N. Lorde, Predicting temperature profiles in producing oil wells using artificial neural networks, Engineering Computations 17(2000)735～754

中国智能、绿色建筑一体化发展*

尚春静　　　　　　　　　　刘长滨
(北京交通大学经管学院，北京　100044)(北京建筑工程学院，北京　100044)

摘　要： 本文针对未来10年到30年巨大的国际智能建筑市场和国际上智能化系统在节能、绿色建筑发展中的举足轻重作用，分析了我国现阶段的智能与绿色建筑的发展特点，指出我国智能与绿色建筑一体化发展是建筑业产品实现循环经济和全面协调科学发展观的战略途径。本文构建了我国智能绿色建筑的体系结构，给出了智能绿色建筑一体化发展模型，阐述了智能绿色建筑一体化发展的政府行为。

关键词： 智能建筑　绿色建筑　节能　协调发展

1　引言

据美国有关专家预测：2014～2040年这一时期的建筑物大多是智能型的，即在建筑物中安装有多种智能传感器，用于检测应力、沉降、裂缝、腐蚀，以及可能出现的其他问题。借助于环境传感器，可检测空气的温度、湿度和污染程度等情况，并通过墙壁和房顶对空间进行过滤。所有这些信息都能很快反馈到一个中心处理机构，以保证建筑物的可靠及居民的安全与健康。

能源成为21世纪可持续发展的关键问题之一。欧美等发达国家近年在智能建筑设计中常融合绿色建筑设计的理念，将智能建筑设计和高技术绿色设计方法相结合。通过建筑、设备和智能化系统来提供节能、环保的生活空间，防治大气和水的污染、防治电磁污染等，突出智能建筑的“智能”功能，“以人为本”，节约资源，实现人类聚居环境的可持续发展。

随着智能系统的不断发展，智能系统将与绿色建筑共生同存，智能系统的功能、性能及运行将是绿色建筑不可分割的一部分，绿色建筑要实现的大部分目标及实现方案都离不开智能系统。

因此，如何使智能化系统与绿色建筑一体化发展实现建筑过程、建筑产品与环境、资源协调发展是我国建筑业对循环经济和全面协调科学发展观的一个战略途径。

2　智能、绿色建筑一体化发展的含义

虽然智能科学与技术已经有了不少的进步和发展，但是，究竟什么是智能？至今却还

* 国家“十五”科技攻关重点项目(2004BA209B01)。

没有一个公认的统一的答案。普遍认为：智能的本质是“对感知到的信息的恰当响应”。这个恰当的响应，其内容可能是一个结论、一个答案、一个决策或者一个规划等等，其形式可以是信息的输出、信息的存储或者由信息驱动的行为。

绿色，普遍是指有利于保护生态环境，最低限度地影响环境，同时高效率地利用资源。

智能绿色建筑一体化发展即是建筑物以“可持续”(既满足当代人的需要，又不损害后代人满足需求的能力)为核心，通过智能化手段与绿色理念的融合来实现人、资源、环境三者的最优化发展。

3 国际智能系统融入绿色建筑的发展

在 2000 年 USGBC 美国绿色建筑委员会 U. S. Green Building Council 发布的第二版《绿色建筑评估体系》7 项评估体系总指标总计 69 分中，建筑物设备监控系统 BAS 的影响约有 14 分，占总分 20%，占绿色建筑入围(26)分 54%，由此可见，智能化系统对绿色建筑的影响。

3.1 智能系统在建筑节能中举足轻重

美国建筑界正在研制通过信息和智能技术实现建筑节能。近年在美国得克萨斯州开发的 500 栋新型建筑物能够利用微型计算机管理包括照明、加热、取暖、制冷和空调设备等在内的全部耗能系统，控制自己能源的消耗，与普通建筑物相比，可以节约能源 20%左右，节能效果十分显著，而且可以减少管理人员。美国能源管理部门计划在全国推广这项新技术。

最近美国研制出一种由计算机控制的旋转房屋。它里面安装了一种如同雷达的红外线跟踪器，天一亮屋内的马达就会启动；接着整座房子便像向日葵似地迎着太阳缓缓转动，一直与太阳保持着最佳的角度，从而确保屋内能够射入最多的阳光；在太阳下山后，房屋又慢慢地回复到原位。这种新式房屋一方面充分利用太阳能驱动房屋运动，确保了室内的日常供热和用电；另一方面能够储存宝贵的光能，以便阴雨天和晚间使用。

2002 年 9 月 1 日，英国政府建筑研究部门宣布，为了促进可持续发展战略的实施，英国政府将实施百万“绿色住宅”建筑计划，今后 10 年内建设 100 万栋“绿色住宅”。据悉，在英国现有的约 2500 万个家庭中，绝大多数家庭的能源使用和节能效率都不高，平均每年要耗费英国能源产量的 1/3，而新建“绿色住宅”将大大节省能源。

3.2 智能系统促进建筑以人为本，可持续发展

以生物方式感知建筑内部的状态和外部环境，并及时做出判断和反应，一旦灾害发生，能进行自我保护的生命建筑已有科学家在研究试验。如美国的彼得·弗尔教授把光纤直接埋在建筑材料中作为建筑物的“神经”。光纤是光纤传感器的一部分，通过感知光信号的相对变化特征，从而反映出建筑物变形和振动的情况。美国南加州大学的罗杰斯研究小组，则在建筑物的合成梁中埋植记忆合金(SMA)纤维，由电热控制的 SMA 纤维能像人们的肌肉纤维一样产生形状和张力变化，从而根据建筑物受到的振动改变梁的刚性和自动振动频率，减少振幅，使框架结构的寿命大大延长。此外，生命建筑还应具有“大脑”，称为人工神经细胞，能自动调节和控制建筑物内的无数光纤维传感器，能在突发的建筑事故中，具有判断能力，驱动执行器有条不紊地工作，或是由神经网络处理，或是送往远端

的中央处理器处理。同时，生命建筑还设置了自动适应系统，以便在必要时自动接换各自的传感器等。

2004 年英国建筑革新中心展示了建筑业信息化项目“从 3D 到 Nd 模型”从而撬动了 3 维模型整合计划、成本、进度、风险、可持续、易维护、声学和节能仿真到 n 维模型的建筑生命周期管理，实现建筑可持续发展。英国一家在建筑标准和建筑隔离技术方面负责向政府提供咨询的机构说，采用新技术建造的住宅所耗费的能源将是传统建筑方法的十分之一。

日本的智能建筑突出“以人为本”，大量采用新材料、新技术；充分利用信息、网络、控制与人工智能技术，注重功能，兼顾未来发展与环境保护。

4　我国智能与绿色建筑的发展特点

4.1　智能、绿色建筑以政府主导为特点

1994 年我国发表了“中国 21 世纪议程”，议程提出建筑节能，城市规划、建设与管理信息系统的开发与建设和城市建筑与生命线工程的防灾灭灾综合发展等项目是中国人类居住区持续发展的优选项目。

1996 年又发表了“中华人民共和国人类住区发展报告”，对进一步改善和提高居住环境质量提出了更高要求和保证措施。如：合宜的住宅，方便的出行及服务；安全的智能化的管理；节地节能的建筑；洁净的空气；污水资源化；垃圾减量，无害处理；噪声控制等。

2004 年 2 月，建设部制定《建筑节能试点示范工程(小区)管理办法》，同时，作为科技奥运十大项之一的“绿色奥运建筑评估体系”项目顺利通过验收，作为我国第一套绿色建筑项目标准，首先应用于奥运建设。

2004 年 8 月，建设部颁布施行《全国绿色建筑创新奖管理办法》，并于 2005 年 2 月公示了 53 个拟推荐授奖的“全国绿色建筑创新奖”项目，其中包括 28 项技术与产品类的技术与产品，工程类的 14 项综合工程、4 项节能工程、7 项智能工程。

4.2　智能化系统与绿色建筑逐渐融合

在我国已经有多个城市的公共建筑建立了基于计算机控制网络技术的开放式智能监测控制系统，并实现计算机控制网络与以 Internet 为代表的数据网络的一体化集成，以实现对建筑能耗和室内热湿环境及空气质量进行远程监测控制，并优化系统运行，最大限度地减少建筑对能源的需求，减少环境污染，以实现城市经济建设可持续发展的目标。如北京某建筑建立了基于 LonWorks 计算机控制网络技术以及 Web 浏览方式的开放式智能监测控制系统，实现对建筑能耗和室内热湿环境及空气质量进行监测控制，优化系统运行，并可以对其进行全年能耗分析评价。

在建设部干部学院对《2004 年楼宇自控市场研究报告》中，针对良好的楼宇自控系统与节省智能建筑的运营费用关系的调查统计显示，人们普遍认为良好的楼宇自控系统将节省智能建筑运营费用的 20%左右。

由于智能化系统的监控和信息管理作用，使得其在室内环境控制(如空调、设备)中占据了至关重要的地位。我国于 2004 年推出的每两年评选一次的绿色建筑奖项中就设置了智能创新专项奖和包括了智能化系统的绿色建筑创新综合奖，这体现了我国智能建筑与绿

色建筑的相互促进发展。

总而言之，我国建筑产品正在逐步向智能绿色建筑一体化战略阶段发展。

5 我国智能与绿色建筑一体化实施的战略问题

5.1 构建智能绿色建筑的体系结构

由于智能与绿色建筑一体化发展是近年来认识上的突破，如何促进两者共同协调发展还处于探索阶段，本文对于智能绿色建筑体系的架构也仅是一种初步的探索，见图 1。智能绿色建筑体系结构应包括其目标、过程和内容等多方面的集合，应涵盖智能与绿色建筑的生命周期全过程与时代发展相符的方方面面。

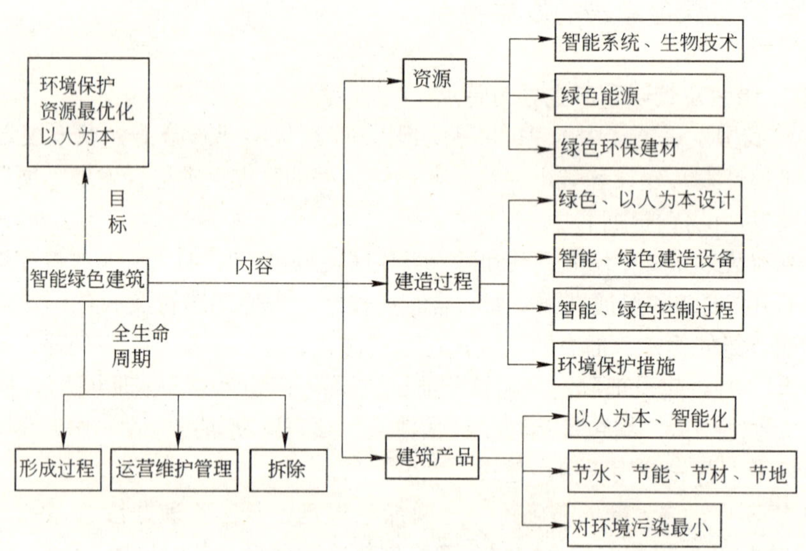

图 1 智能绿色建筑的体系结构

5.2 智能绿色建筑一体化实现的战略途径

智能绿色建筑一体化实现的战略途径是资源优化利用。我国目前对建筑物资源的优化利用研究非常活跃，但大都集中在某项单项技术上，如废弃物的利用技术，人、机、时等资源优化利用方面，而从建筑物整个系统生命周期内动态控制资源利用方面研究的较少。

本文从系统的角度，从智能绿色建筑系统全过程和整体的角度提出了一种动态智能绿色建筑系统的资源流和信息流发展模型，见图 2。

5.3 智能绿色建筑一体化发展的政府行为

5.3.1 促进智能绿色建筑市场规范组织完善

智能绿色建筑市场规范组织的建立与完善，对智能绿色建筑的生产、销售过程的抽查和检验，有着强有力的组织保障。

(1) 充分发挥行业协会中介组织的市场作用

发展和完善市场行业协会中介组织，着力解决中介组织结构分布不合理以及管理体制、组织形式、运作方式行政色彩浓厚的问题。

(2) 行政管理部门逐步有序退出市场主体地位，转向对市场参与者资质和竞争行为的

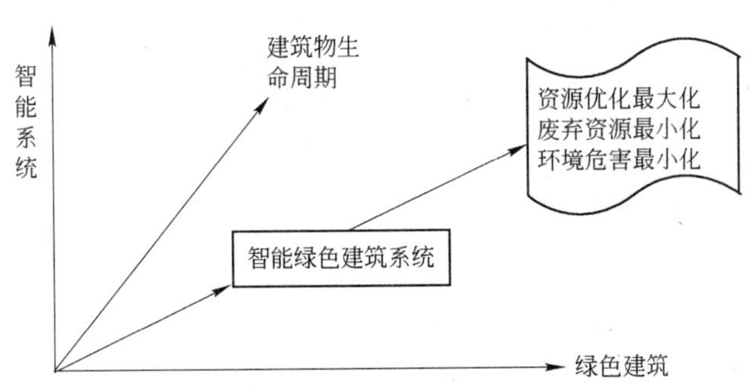

图 2 智能绿色建筑一体化发展模型

管制。

首先，政府要主动充当智能绿色建筑市场开放的第一推动，实行政企分开，组建数家竞争性公司，并坚决而稳妥地完成从行政控制价格转向市场竞争定价的体制，依靠市场价格机制来配置资源、引导企业行为。

其次，政府必须从直接的市场准入审批和价格管制，转向管制市场参与者的资质和竞争行为。政府的管制机构，要逐步扩大管理范围，缩小管制重点。由于在各种产品和服务之间存在广泛的替代竞争，政府管制部门的设置，要逐步超越按照计划管理生产时代的局限，形成对涉及整个智能绿色建筑市场的综合性管制，关照各类产品和服务之间的联系，消除妨碍替代性竞争的因素。

最后，智能绿色建筑市场的竞争局面一旦形成，政府部门要因势利导，让市场机制在市场结构的重组中发挥更大的作用。

5.3.2 调整产业政策，建立促进技术应用和发展的政策及相应的激励机制

智能绿色建筑体系是建筑业、房地产业和国家可持续发展的追求的战略目标，智能、绿色环保技术发展异常迅速，各行各业中的高新科技都会直接或间接地反应到智能绿色建筑中。因此，政府应通过调控投资、价格和税收等政策予以引导和支持企业紧跟先进技术，吸收、消化和创新，并逐步建立起对智能绿色建筑灵活有效的激励机制，如对建筑中使用优质材料的减税、建筑泄漏能量多的建筑加收使用税等。

5.3.3 改革行政管理方式

（1）充分运用间接管理、动态管理和事后监督管理等手段，减少对智能绿色建筑项目的直接的、微观的和事前的干预；

（2）充分运用行政指导、行政奖励和行政给付等方式的作用，减少行政审批、行政处罚、行政强制等强制性行政手段的使用。

Integrating Chinese Intelligent and Green Building

SHANG Chun jing[1] LIU Chang Bin[2]

（[1] Beijing Jiaotong University，Beijing 100044 China；[2] Beijing Institute of livil Engineering and Architec ture，Beijing 100044 China）

Abstract: This paper analyzed the characteristics of Chinese intelligent building and green building, and clarifies the development relation of international intelligent systems and green building. The paper points out it is a favorable approach to integrate Chinese intelligent building and green building for the development of both, and structures the system and model of the intelligent and green building; last, this paper expatiates the government behavior in the development of the intelligent and green building.

参考文献

[1] 程大章，刘钢. 绿色建筑智能系统工程的探索. 智能建筑与城市信息，2005. 第10期

[2] 杨鲁宁，王剑，马向阳. 我国绿色建筑的特点分析及其评价方法初探. 山东环境. 2003. 第6期

[3] 李德英，寿大云. 绿色建筑能源综合利用与智能控制技术研究. 安徽建筑工业学院学报(自然科学版)第12卷，第4期. 2004

[4] 丁艳. 智能建筑科技节能总动员. 智能建筑与城市信息. 2005. 第5期

[5] 尚春静，刘长滨. 21世纪的中国智能建筑. 回归绿色与自然. 首届国际智能与绿色建筑技术研讨会. 北京：中国建筑工业出版社，2005

[6] 白胜芳. 国外绿色建筑对中国的启示. 绿色中国. 2005.10

电梯节能的最佳响应时间初探

刘润宗　王　波
（重庆大学计算机学院建筑智能化研究室，重庆　400044）

摘　要：针对传统电梯调度算法未考虑适当设置电梯响应时间，只是简单地使电梯对任意请求都作即时反应，从而导致电梯平均满载率偏低、电梯有效利用率不高及电能浪费的状况，本文指出应综合人们心理容忍度、作息时间、建筑高度和性质、楼层、电梯数量以及用电季节等各种因素，设置恰当的电梯响应时间和选择相应的电梯及时调度算法，以求得最佳的节能及经济效益。

关键词：电梯平均满载率　电梯响应时间　及时调度算法　建筑节能

1　问题背景

传统电梯调度算法性能指标偏好于平均等待时间和最长等待时间，而对资源利用率并不重视。然而目前随着经济的快速增长和人口的不断增加，我国淡水、土地、能源、矿产等资源不足的矛盾更加突出，环境压力日益增大，国内城市电力资源普遍紧张，为此国家已明确提出“十一五”是我国全面建设小康社会、加快推进社会主义现代化的关键时期，必须统筹协调经济社会发展与人口、资源、环境的关系，加快建设节约型社会，坚持资源开发与节约并重，把节约放在首位的方针，紧紧围绕实现经济增长方式的根本性转变，以提高资源利用效率为核心，在生产、建设、流通、消费各领域节约资源，提高资源利用效率，减少损失浪费，以尽可能少的资源消耗，创造尽可能大的经济社会效益。因此，将电梯调度算法的能耗指标摆在重要位置加以考虑是非常必要的。

2　电梯响应时间的提出及节能效果的理论分析

传统的电梯调度算法只是简单地将电梯响应时间设为 0，使电梯对任意请求都作即时反应，从而导致电梯平均满载率偏低，电梯利用率不高。从下文的分析可知，不超过人们心理容忍度的等梯时间最长可达 70 秒左右。而当前调度算法的良好性能可使人们平均等梯时间为 20 秒左右[1]，由此可见电梯响应时间的设置范围可在 0～50 秒之间。于是我们可通过设置适当的电梯响应时间和电梯及时调度算法，把电梯平均等待时间和电梯最长等待时间控制在用户基本满意的范围之内，而同时提高电梯平均满载率，减少电梯启停次数和运行里程，从而减少电梯损耗，实现建筑节能。

为了说明这点，下面将讨论电梯平均满载率、电梯单向运行次数、运行里程和电梯启停次数与电梯响应时间的关系：这里电梯满载率是指电梯轿厢内实际人数与额定人数之比，电梯单向运行是指电梯由出发层向上(下)运行直至最高(低)层，包括其间电梯的启停。电梯单向运行时，若为空载则电梯亦须启停一次，若非空载则期间每服务不同楼层呼叫一次，电梯须启停一次，所以电梯启停次数可由电梯单向运行空载次数与异层呼叫次数之和来度量。

某时段 T 内乘客呼叫的随机序列 Q，设其呼叫总数为 m，电梯单向运行的期望时间为 t，电梯满载时额定人数为 P，则电梯单向运行期望次数 $n=T/t$，电梯平均满载率 $k=(m/n)/P=(m\times t/T)/P$，电梯期望运行里程 $s=n\times(a\times H)=(T/t)\times a\times H$($a$ 为一小于 1 的正常数，H 为建筑高度)。记同一次电梯单向运行中，目的层相同的乘客期望比率为 $v=v(k)$，电梯单向运行空载次数与总次数的期望比率为 j，则电梯启停次数 $y=m-n\times k\times P\times v(k)+n\times j=m-m\times v(k)+n\times j=m\times(1-v(k)+n\times j$，显然平均满载率 k 越大，则 v 越大，j 越小，电梯启停次数 y 亦越少. 可见对于同一建筑同一时段 T 而言，T、m、P 与 H 均相等，k 和 s 的大小取决于 t，且 k 与 t 成正比而 s 与 t 成反比。当采用下文介绍的及时调度算法，并设响应时间为 w 后，电梯单向运行期望时间 t 则变为 $t+f(w)$，由以上分析可知电梯平均满载率 k 提高了，电梯期望运行里程 s 和电梯期望启停次数 y 则减少了。而电梯能耗主要取决于电梯期望运行里程和电梯期望启停次数，由此可见，设置恰当的电梯响应时间，的确可以减少电梯损耗，实现建筑节能。

3 及时调度算法的设计

这里先讨论单部电梯的情形：

电梯调度算法常采用 4 个队列来存放来自各楼层的先后呼叫：上优先队列，上次优先队列，下优先队列，下次优先队列。电梯及时调度算法也采用这样的结构，并在传统算法的基础上增加电梯响应时间。设 t_0 时刻电梯静止在第 k_0 层，此时第 k_1 层传来呼叫 C_1，电梯并不立即响应，而是等待响应时间 $W(k_0)$，在 t_0 至 $t_0+W(k_0)$这段时间，电梯可能接收到 C_2、C_3、…、C_n 的呼叫序列。将这些呼叫按传统算法分配到相应的队列：上优先队列和下优先队列。并按第一个呼叫 C_1 所在层 K_1 与 K_0 关系决定电梯运行方向：若 $K_0>K_1$ 则电梯下行，若 $K_0<K_1$ 则电梯上行。电梯运行过程中所接收到的呼叫也按传统算法分配到四个队列中相应的一个。此运行过程中，电梯到达呼叫层 C_i 时，一般不再额外等待响应时间 $W(C_i)$，以避免用户乘梯和候梯时间过长，除非 C_i 为第一层 C_L 或最高层 C_H，因为此时轿厢内所有乘客都会走出电梯。当电梯到达第一层 C_L 或最高层 C_H 时，电梯必须等待响应时间 $W(C_L)$或 $W(C_H)$后才继续运行，此段时间内仍会有来自其他层的呼叫到达，将它们按传统算法加入到已有的相应队列中。等待响应时间 $W(C_L)$或 $W(C_H)$后，按照队列情况，电梯继续上行或下行或静止。

多部电梯的情形也类似处理，只是传统群控算法在决定哪部电梯执行调度时，要考虑电梯响应时间带来的影响。

4 影响电梯响应时间设置的因素

4.1 约束条件

4.1.1 人们心理容忍度

平均等待时间不能超过人们的心理容忍度。

一般而言，人们是否愿意乘电梯，是看乘电梯能否给他们带来好处，这包括省力，省时这些因素。但倘若乘电梯要花更大的代价，比如等梯乘梯时间比人们走楼梯时间多得多或乘梯要付费时，人们就不一定乘电梯而宁可自己走楼梯。对于上楼梯，尤其是要上六层楼以上时，由于比较花力气，除非有急事，人们一般愿意等待电梯的到来。但对于往下走楼梯时，若楼层不高，如六层以下，人们往往不愿等电梯而自己走楼梯下去。然而，从节能需要出发，六层以下的人的确不必要坐电梯下楼。那么当我们考虑乘梯等待时间的上限时，我们可让之接近成年人以一般速度上六层楼梯所需的时间。而这个时间大概在 70 秒左右。当然，在电力资源特别紧张时，该时间上限还可适当延长。

4.1.2　建筑性质

根据建筑物的不同用途，这里主要考虑办公楼、商场或住宅。

办公楼是人们工作的地方，生活节奏较快，时间紧张，遇到需紧急处理的事情概率较大，因而应将其电梯响应时间设置为较小值。商场是人们消费娱乐场所，住宅是人们居住的地方，这些都是人们放松休闲的地方，时间较充裕，遇到需紧急处理的事情概率较小，因而可将其电梯响应时间设置为较大值。

4.2　客流

4.2.1　人们的作息时间

对于星期一至星期五，人们因工作学习原因，一般有着较规律的作息时间，因而电梯客流量有着明显的上高峰和下高峰。在客流高峰时，电梯平均满载率高，候梯人数多，电梯响应越快越好，这时电梯响应时间可设为 0。而在空闲时段，则可根据该建筑客流特征的统计分析，设置适当的电梯响应时间，以提高电梯平均满载率，减少电梯损耗，实现建筑节能。

4.2.2　建筑高度

建筑高度影响用户乘梯时间，及电梯客流分布，进而影响电梯响应时间的设置。

4.2.3　楼层

按第一层、最高层、特殊需要的楼层、一般楼层进行分类考虑。

第一层和最高层是电梯服务的终端，在这两层轿厢内所有乘客都要出电梯。而绝大部分要上楼的乘客都须从第一层进，可见此层是呼叫比较密集的地方，将此层电梯响应时间设为较大值，则可显著提高电梯平均满载率。同样，把最高层电梯响应时间设为较大值，则在响应时间内收到来自最高层以下各层下行呼叫的概率较大，也可明显提高电梯平均满载率。对于一般楼层电梯响应时间的设置，则可明显小于第一层和最高层，但应按楼层高度递减的设置响应时间。而对于有特殊需要的楼层，如政府机关所在楼层，则可将电梯响应时间设为 0。

4.3　电梯载客能力

4.3.1　电梯数量

一般而言，增加电梯数量，可缩短用户平均等待时间和用户最长等待时间，避免单部电梯过于频繁启停而过快折旧。但从节能角度，在满足用户基本需要的情况下，我们应尽可能减少电梯数量，这可取得与设置响应时间相同的效果。当然，在电梯载客能力仍有过剩的情况下，我们可在减少电梯数量的基础上，设置响应时间，以求得最大的节能和经济效益。

4.3.2　用电峰谷

每天用电高峰时段，电力资源较紧张，电费较高，而此时正是人们工作时段，电梯处于空闲时段，候梯人数不多[2]，可以将电梯响应时间设置为较大值，以减少建筑电力负荷和节省物业管理成本。每天用电低谷时段，电力供应较充足，电价较低，此时则可将电梯响应时间设为较小值，以提高乘梯舒适度及满足客流高峰的要求。

4.3.3 用电季节

夏季用电高峰期，城市电力资源普遍紧张，常出现拉闸限电的情况，此时则可通过将电梯响应时间设为较大值，以减少电梯能耗，缓解电力资源紧张的局面。冬、春、秋用电低谷期，供电较充足，则可将电梯响应时间设为较小值，以提高乘梯舒适度。

5 结语

本文着重改善电梯调度算法的能耗指标，设计了电梯及时调度算法，提出了电梯响应时间，电梯平均满载率、电梯单向运行次数等概念，分析了它们与电梯运行里程、电梯启停次数的函数关系，从而论证了设置恰当的电梯最佳响应时间的确可减少电梯损耗，实现建筑节能。最后，分析了人们心理容忍度等多种因素对电梯响应时间的影响，实现了对电梯节能最佳响应时间的初步探讨。

A Primary Research on the Best Elevator Response Time for Saving Energy

Liu Runzong Wang Bo

(Building Intelligentization lab，College of Computer Science，Chongqing University，Chongqing 400044 China)

Abstract：Traditional elevator dispatching algorithms haven't considered how long the best elevator response time should be. Whereas，they just make elevators give the fastest response to any call. Thus，the elevator system only reaches a low average elevator load ratio and can't perform efficiently enough. This article points out that in order to gain the biggest economic profit with lowest energy，a proper elevator response time and a suitable elevator timely dispatching algorithm should be chosen，considering factors as follow，people's patience，people's schedule，building's height，building's character，floor，the amount of elevators，seasons，etc.

参考文献

[1] 宗群，蔡昱. 基于电梯交通流概率仿真模型的空闲交通模式电梯调度方法. 控制与决策 2002.5：339～342

[2] 郑延军，张惠侨，叶庆泰，朱昌明. 电梯群控系统客流分析与仿真. 计算机工程与应用 2004.22：140

LonWorks 技术在学生公寓能源管理系统中的应用

孙启利[1]　侯晓鹏[2]　韩　宁[2]　燕　飞[2]

([1] 中国新兴建设开发总公司　北京　100036；[2] 北京林业大学工学院　北京　100083)

摘　要： 考虑到当前高校物业管理改革的需求以及学生公寓安全、高效地使用能源的迫切需要，实施学生公寓能源的综合信息管理，可以提高公寓的物业管理效率、降低运行成本，减轻管理维护人员的劳动强度。依据上述思想，介绍了基于 LonWorks 技术的学生公寓能源管理系统的结构及系统特点、系统功能以及监控软件的设计思想。

关键词： 公寓管理　LonWorks　能源管理

1　引言

随着高校招生规模的不断扩大及在校学生数目的增多，对学生公寓的机电设施采用传统的人工监督和管理的状况显然已跟不上现代化物业管理的要求，与当前节约型社会的发展趋势明显不符。同时随着高校后勤改革的进一步深入，对学生公寓的水、电的使用提出了智能化供电、水电定量免费等更多、更细的要求。此外用电设备的增多增加了学生公寓发生火灾的可能性，基于先进测控网络技术的火灾预警系统成为学生公寓迫切的需求。为了提高校园内学生公寓及公寓群的管理效率和服务质量，本文开发了针对学生公寓能源使用特点的能源运行管理系统，实行对水电消耗的实时监控与计量。该系统同时可以扩展到学生公寓的其他机电应用设备的智能化和网络化管理[1]。

2　LonWorks 技术简介[2、3]

LON(Local Operating Networks—局部操作网络)是美国 Echelon 公司于 1992 年推出的一种智能型分布式现场总线控制网络，它使用具有全分布控制与通信联网功能的大规模集成电路的神经元芯片构成的智能网络节点，通过网络收发器及通信媒体，将各个节点构成全分布式局部操作网络，节点之间通过 LON 总线互相传递信息。LonWorks 技术的通信协议—LonTalk 协议在各种现场总线技术中惟一实现 ISO/OSI 网络参考模型的七层协议，是直接面向对象的协议。LonTalk 协议提出了一种重要的信息传递机制—系统定义的网络变量，是可以被网络其他节点通过网络环境操作的专用变量。

LON 总线及其系统具有以下优点：

(1) 与其他现场总线相比，LON 总线最大的优势是具有网络处理能力。它通过网关很容易与异型网互连，并支持多种网络拓扑结构，可以与多种网络系统沟通，以其独特的技术优势把计算机技术、网络技术、控制技术和通信技术结合起来，实现了工业测控及管理的统一。

(2) 采用超大规模集成电路的神经元芯片，使每个节点的应用变得简单可靠。固化的LonTalk协议具有检测应答、自动重发、请求/响应等功能，易于实现冗余功能，保证了通信的可靠，实现了控制功能的全分布，使风险分散，提高了系统的可靠性。

(3) 使用神经元芯片、LonWorks收发器及基于LNS网络操作系统的网络管理工具，可方便灵活地构建新一代的控制应用系统。

(4) LON总线是惟一实现ISO/OSI网络参考模型7层协议的现场控制网络，它在充分考虑控制系统特殊要求的基础上，建立了每一层的直接面向对象网络协议。

(5) Lon总线支持无线、电源线、红外线、双绞线、射频、光纤、同轴电缆等多种传输介质，满足多种特殊的应用场合，数据传输速率可达1.25Mbit/s。

(6) 提供完善的系统开发工具，为用户开发、组态网络提供了方便。

3 学生公寓管理系统的组成

根据校园中学生公寓的分布特点及管理系统中对能源管理的具体要求，整个管理系统由下而上可以分成三个层次：数据采集与控制层、数据传输层和数据管理层，如图1所示。

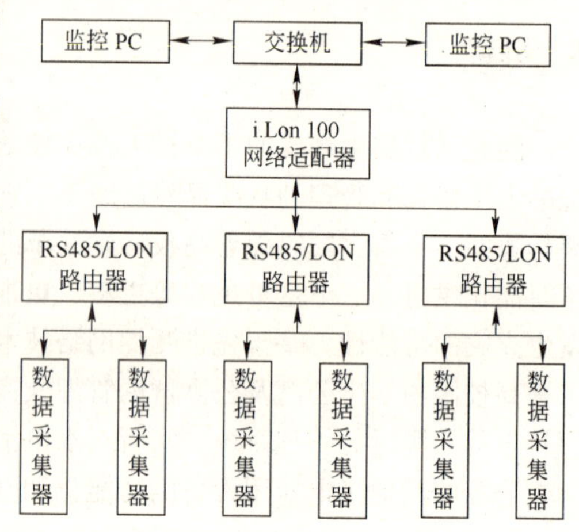

图1 学生公寓能源管理系统框图

数据采集与控制层由各采集与控制模块(WitMeter)组成，完成公寓房间内水、电及火灾信息的采集和水电供应及报警装置的控制。每个采集与控制模块(WitMeter)是一个以单片微处理器为核心的测控单元，通过其RS485通讯接口实现与上层网络的信息交互。

数据传输层由RS485/LON路由器、i.Lon100网络适配器组成。RS485/LON路由器完成RS485协议与LonTalk协议间的相互转换；i.Lon100网络适配器可以将底层的测控网络无缝地接入校园以太网中，易于利用已有的网络设施实现信息的采集与管理。

数据管理层由以太网网络设备、管理PC及管理软件组成。

该系统的特点之一是：将基于RS485的测控模块与LonWorks控制网络、基于TCP/IP的校园Intranet相结合，构建了能源管理的综合信息管理系统，在实现系统需求基础上有效地降低了系统的建设成本。目前由于基于LonWorks控制网络技术的专用芯片及其他辅助工具价格偏高及没有被广泛被使用，导致纯LON系统的用户成本比其他基于单片机的系统成本偏高。为了在保证系统功能和技术先进性的基础上尽可能降低用户成本，系统设计中采用了LON+RS485的系统结构。RS485测控单元主要侧重于信号的采集、预处理与设备控制；而LON网络主要侧重于分布式测控信息的网络化传输。

该系统的特点之二是：LON网络的分布式测控特性适合学生公寓的分布特点。通常学校的学生公寓可能分布在校园不同的地方甚至不同的校区，这些公寓的能源信息的集中综合管理成为管理人员非常关心的问题。近几年 Internet 及校园 Intranet 的建设已经非常成熟，易于实现远程信息的集中管理，因此上述问题的核心内容是如何实现基于LonWorks技术的测控网络与 Internet 的集成。i. Lon 系列网络设备提供了很好的解决方案。

该系统具备如下所述的功能：

(1) 水、电计量数据和火灾信息的实时检测与上传；

(2) 水、电控制信号和警报信号的实时下传与控制；

(3) 与管理中心的管理软件配合，实现各房间能源消耗的计量与统计；

(4) 与管理中心的管理软件配合，实现超负荷断电、超设定用量断电断水、火灾报警。

4 管理软件设计

将上述能源信息的集中管理需求与公寓学生、宿舍、卫生及维修等信息的管理功能相结合，采用 VB 程序设计语言和 SQL Server 数据库设计了学生公寓综合信息管理软件系统，其结构框图如图 2 所示。

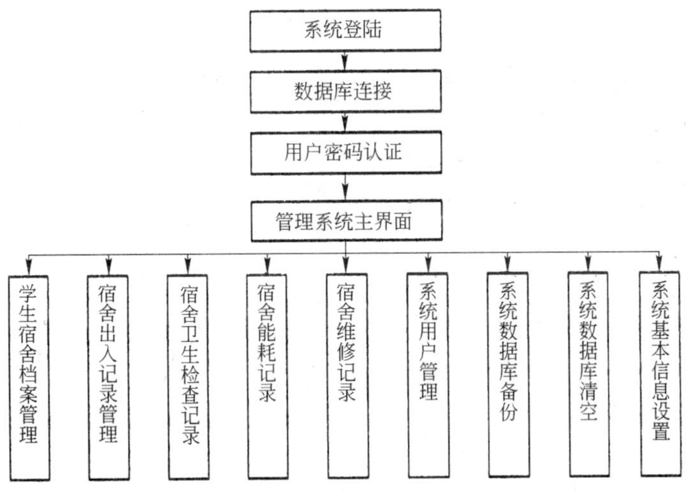

图 2　管理软件结构图

VB 管理软件以 DDE 方式与 LonWorks 控制网络完成数据交换，由 Echelon 公司提供的 LNS DDE Server 作为数据服务器。DDE(Dynamic Data Exchange，动态数据交换)是 Microsoft Windows 提供的一种数据共享技术，其实质是一种 Windows 应用程序之间的数据交换机制，应用程序之间可以进行各种类型的数据交换。LNS DDE Server 是一个允许任何与 DDE 兼容的 Microsoft Windows 应用程序，在不需要复杂编程的情况下就可以监视和控制 LON 网络的软件包。LNS DDE Server 在 Lonworks 设备和 VB 管理软件之间交换表示测控信息的网络变量和应用报文。支持 DDE 的 VB 管理软件可作为 DDE 客户，监视和控制整个 LonWorks 网络。

管理软件的部分运行界面[4,5](宿舍管理界面、学生管理界面、学生管理报表界面、宿舍能源用量查询界面、能源使用交费界面和账单打印界面)如图 3 所示。

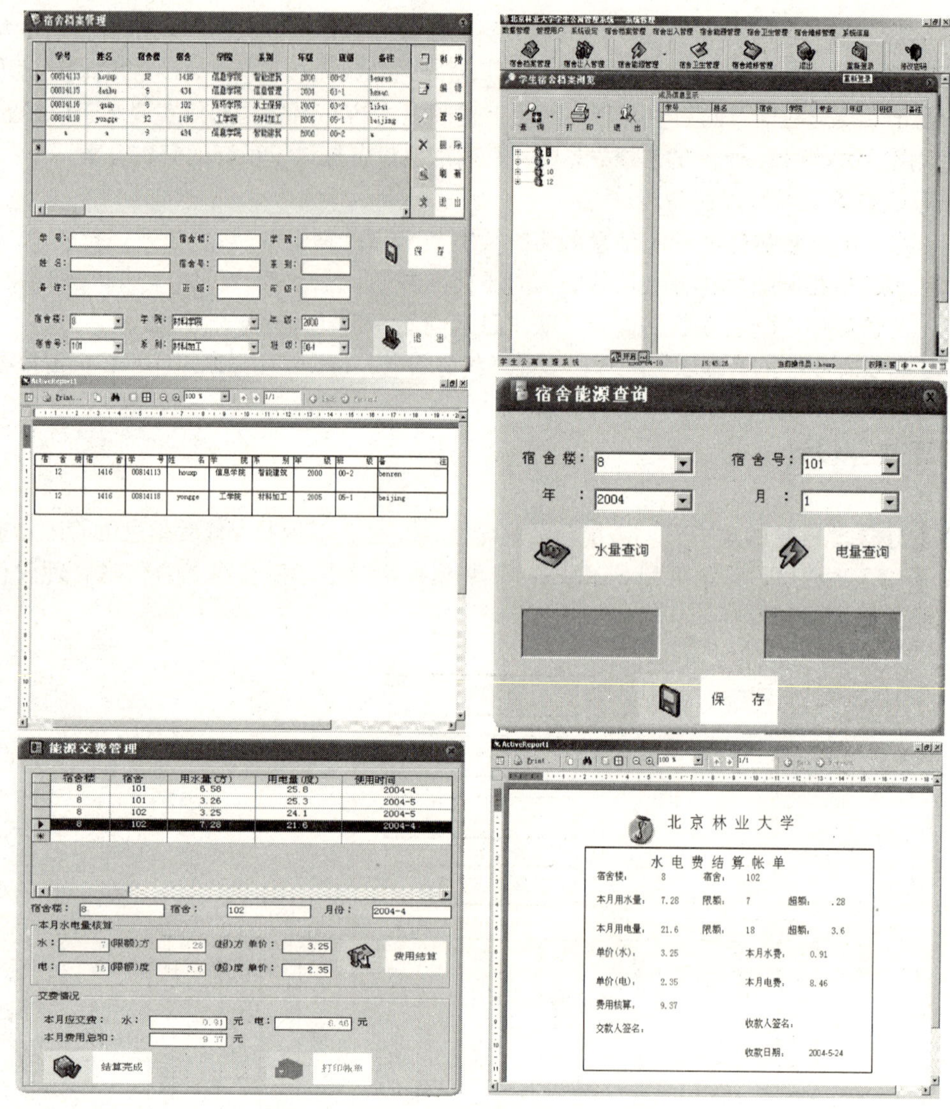

图 3 管理软件界面

5 结束语

本系统主要针对大学生学生公寓这一应用对象进行了物业管理软件和能耗管理系统开发，希望可以将学生宿舍管理人员从繁重的工作中解放出来，实现高校学生公寓管理的信息化和网络化，将学生公寓的能源使用更高效和更安全。

Application of Lonworks Technology in Energy Management System for Student Apartments

Qili Sun[1]　Xiaopeng Hou[2]　Ning Han[2]　Fei Yan[2]

([1] China Xinxing Construction & Development Corporation　Beijing　100036;
[2] School of Technology, Beijing Forestry University　Beijing　100083　China)

Abstract: Under demand for the property management reformation in universities and the urgent requirement for safely and efficiently using energy implementing the integrated energy information management in university apartments could improve management efficiency, reduce operation cost and lighten intensity of all staff. According to this above idea, the system frame, the system trait, the system function and the monitoring software are in detail introduced.

参考文献

[1]　李慧丽，武高洁. 学生公寓管理信息系统的建设与研究. 中国林业教育. 2004.6

[2]　沈晓琴 李奇等. 基于 LonWorks 和 RS485 的智能小区网络设计. IB 智能建筑与城市信息. 2005.2

[3]　Dietmar Dietrich. LonWorks 技术：开放的控制网络与应用中的分布系统

[4]　曾琦. Visual Basic 6.0 编程实例与技巧. 中国水利出版社. 1998

[5]　于英杰. Visual Basic 6.0 数据库开发与专业应用. 国防工业出版社. 2002.1

天然采光技术对室内环境的优化控制之探讨

夏　林　陆明浩

（同济大学建筑设计研究院，上海　200092）

摘　要：本文通过对控制百叶旋转角度的天然采光技术的探讨，提出了室内光环境和热环境的控制优化方案和舒适节能指标，并对构建的控制模型进行了分析。

关键词：温度　照度　太阳能增益　节能

前言

20 世纪 70 年代能源危机后，能源、环境问题举世瞩目。建筑物如何充分利用天然光，节省照明用电，已引起了国际建筑和照明界的高度重视。

天然光是取之不尽、用之不竭的绿色能源。在进行天然采光时，我们通过关闭或调节部分照明设备来节约照明用电，同时还可降低照明设备向室内的散热，减少空调负荷。

但任何事都有其两重性，天然采光也不例外。例如：在夏季，我们在使室内获得天然光、增加室内照度、减少照明用电的同时，也大大增加了室内太阳辐射得热，从而增大了空调的负荷。而工程实践表明，采取一定的外遮阳措施是比较有效的。它可以降低室内太阳辐射得热，减少夏季空调负荷。

百叶作为一种结构比较简单的外遮阳设施，其应用已相当普遍。因此，本文将探索如何控制百叶旋转角度，在保证室内光环境的前提下，对室内光环境和热环境进行最优控制，以达到节能的目的。

1　理论分析

1.1　基本符号

$\mu_{COMf_\Phi}(\Phi_i)$和$\mu_{COMf_T}(T_i)$——舒适度性能指标；

Q_{heat}——加热功率，即空调向室内送暖或送冷时的功率；

Q_{vent}——通风而引起的热损；

Q_{trans}——通过墙的传导而引起的热损；

Q_{int}——由机器或人产生的内部热量增益；

Q_{sol}——太阳能增益，即太阳能引起的室内热量增益：

其中 $Q_{heat}=(Q_{vent}+Q_{trans})-(Q_{sol}+Q_{int})$

$q_{sol,o}$——外部太阳能辐射，即室内向外流失的热量；

Φ_i——室内工作面照度；

T_i——室内温度；

$T_{i,\mathrm{nom}}$——最优室内温度设定值；

B_{pos}——百叶旋转角度；

Λ——模糊符号，代表模糊与；

θ——太阳高度角(图 1，太阳高度角示意图。)

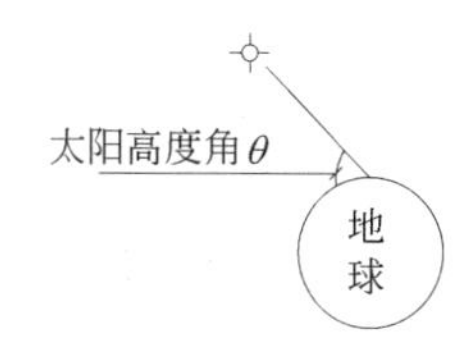

图1　太阳高度角示意图

1.2　理论分析

1.2.1　舒适节能指标的制定

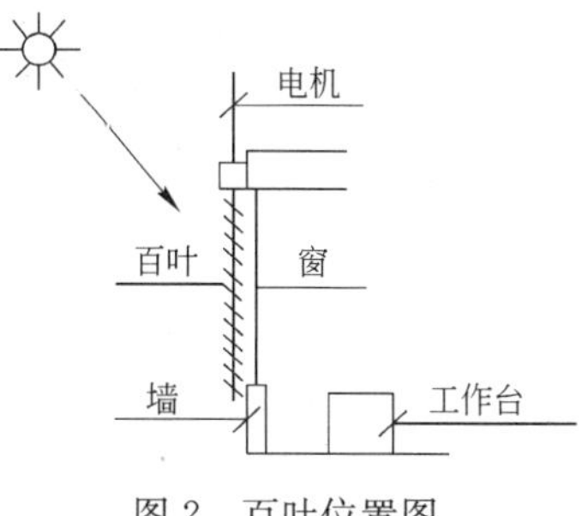

图 2　百叶位置图

在图 2 百叶位置图中，利用百叶控制室内工作面照度 Φ_i 和室内太阳能增益 Q_{sol} 时，我们应区分冬夏季工况。在夏季，Q_{sol} 对室内温度 T_i 具有很强的影响，即 Q_{sol} 越大，T_i 越高。而在冬季，Q_{sol} 的贡献则是降低加热功率。因此，在夏天由照度和室内温度所定义的性能指标 $\mu_{\mathrm{COMf\_\Phi}}(\Phi_i)$ 和 $\mu_{\mathrm{COMf\_T}}(T_i)$ 就显得相对重要。而在冬天，由照度和空调加热功率所定义的舒适度指标 $\mu_{\mathrm{COMf\_\Phi}}(\Phi_i)$ 和 $\mu_{\mathrm{eco\_B}}(Q_{\mathrm{heat}})$ 则为主要的性能指标。因此，我们由下式定义模糊确定。

$$\mu_{\mathrm{D}}(B_{\mathrm{pos}})=\begin{cases}\Lambda[\lambda\mu_{\mathrm{COMf\_\Phi}}(\Phi_i),\ (1-\lambda)\mu_{\mathrm{eco\_B}}(Q_{\mathrm{heat}})] & \text{冬季}\\ \Lambda[\lambda\mu_{\mathrm{COMf\_\Phi}}(\Phi_i),\ (1-\lambda)\mu_{\mathrm{COMf\_T}}(T_i)] & \text{夏季}\end{cases} \tag{1}$$

式中，Λ 为模糊符号，代表模糊与；$\lambda\in[0,1]$ 为加权系数。冬季工况下，可用 λ 来加权舒适和节能的指标，在夏季工况，用户可通过 λ 来调整在两种舒适指标中偏向哪一种(即按实际需求进行调节)。如：在夏季，偏重于照度要求，则 λ 可偏向 1，若忽略室内照度要求而偏向于对室内温度的要求，则 λ 可偏向 0。

1.2.2　算法方程的建立

为计算方程式(1)，我们须建立预测百叶角度对工作面照度 Φ_i、室内温度 T_i 以及加热功率 Q_{heat} 影响的模型，即

$$\begin{cases}T_i=f(B_{\mathrm{pos}})\\ \Phi_i=g(B_{\mathrm{pos}})\\ Q_{\mathrm{heat}}=h(B_{\mathrm{pos}})\end{cases} \tag{2}$$

(1) 温度模型

下面假设内部的温度值 $T_{i,\mathrm{nom}}$ 为已知，由 $T_{i,\mathrm{nom}}$ 可定义误差 $e(k)$ 为：

$$e(k)=T_{i,\mathrm{nom}}(k)-T_i(k) \tag{3}$$

式中的 k 为时标，用于 B_{pos} 的 PI 算法为：

$$B_{\mathrm{pos}}(k)=B_{\mathrm{pos}}(k-1)+K_{\mathrm{p}}\left[\left(1+\frac{\Delta t}{l_{\mathrm{I}}}\right)e(k)-e(k-1)\right] \tag{4}$$

式中，K_{p} 为比例增益；Δt 为采样时间；τ_{I} 为积分时间常数。对 $T_{i,\mathrm{nom}}$ 求解方程(3)、(4)，可得到简单的预测模型为：

$$T_{i,\mathrm{nom}}(k)=T_i(k)+\frac{1}{1+\frac{\Delta t}{l_{\mathrm{I}}}}\left\{e(k-1)+\frac{1}{K_{\mathrm{p}}}[B_{\mathrm{pos}}(k)-B_{\mathrm{pos}}(k-1)]\right\}\equiv f[B_{\mathrm{pos}}(k)] \tag{5}$$

该式相当于 $T_i=f(B_{pos})$，利用上式求解最优化结果，并将其代入方程式(1)，计算舒适节能指标 μ_D。

(2) 亮度模型

由于太阳高度角 θ 每小时变化 15°，工作面照度 Φ_i、百叶位置 B_{pos} 的变化均随太阳高度角 θ 的变化而近似线性变化。因此，假设在此 0°～15°的微小范围[1]内，工作面照度 Φ_i 变化和百叶旋转角度 B_{pos} 成正比。

$$\frac{\Phi_i(k)-\Phi_i(k-1)}{B_{pos}(k)-B_{pos}(k-1)}=\alpha(k) \tag{6}$$

比例常数 $\alpha(k)$ 取决于房间的结构特性、百叶以及外部太阳能辐射 $q_{sol,o}$。假设整个百叶关闭时($B_{pos}=0$)$\Phi_i=0$，由下式估算 $\alpha(k)$ 的值：

$$\alpha(k)=\frac{\Phi_i(k)}{B_{pos}(k)} \tag{7}$$

(3) 太阳能增益模型

为了估算太阳能增益 Q_{sol}，假设 Q_{heat} 的实际值是受到正比于 B_{pos} 和太阳能辐射 $q_{sol,o}$ 的变化影响而变化。即

$$Q_{heat}(k+1)=Q_{heat}(k)-\gamma[B_{pos}(k)-B_{pos}(k-1)]q_{sol,o}(k) \tag{8}$$

根据窗户的表面积 S_W 和传输系数可估算参数 γ 值，即

$$\gamma=S_W g_W \tag{9}$$

2 建模与分析

2.1 模型建立

上海位于北纬 31 度 14 分，东经 121 度 29 分，四季变化明显。冬至日正午太阳高度角 35 度 24 分，夏至日正午太阳高度角为 82 度。

根据办公室布局特点分析建立一个 6m×4m×3m(长×宽×高)的办公室办公单元模型(图 3，办公单元尺寸示意图)。办公单元有一面面南的外墙，墙上外窗长为 3.2m 居中，窗高 1.5m，框下延距地 0.9m，上延距梁 0.6m，室内工作面高度为 0.8m，工作面照度要求为 400lx(CIE 照度要求)。地面反光率等于 0.2，办公室内表面天花板/地板/内墙的反光率分别为 0.75/0.35/0.75。

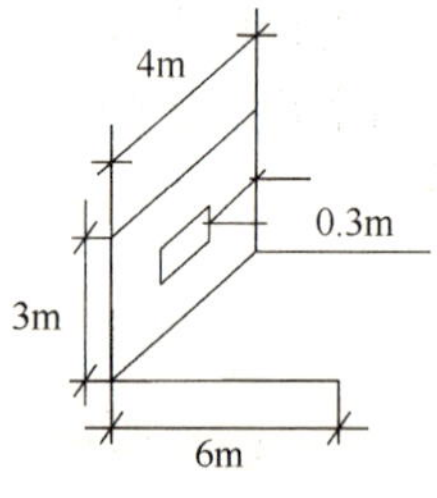

图 3 办公单元尺寸示意图

根据《建筑电气专业设计技术措施》中规定：

【第 6.2.1 条】 室内一般照度的照度均匀度(参考平面的最低照度与平均照度之比)应大于 0.7；

【第 6.2.2 条】 在采用混合照明方式时，一般照明在工作面上产生的照度不宜低于混合照明所产生的总照度的 1/3～1/5，且不宜低于 50lx。

当采用反射光照明时，其照明计算可以参照以下式子进行。

---

[1] 只有 B_{pos} 的小方位变化这种线性模型才有效，必须考虑在每个采样周期内 B_{pos} 的变化。

$$E=\frac{\phi*\eta}{A_n}*\frac{\rho_m}{1-\rho_m}$$

式中 E——照度(lx)；

ϕ——灯具总光通量(lm)；

A_n——房间面积总和(m^2)；

ρ_m——房间平均反射系数，$\rho_m=\frac{\rho_1 A_1+\rho_2 A_2+\cdots\cdots\rho_n+A_n}{A_n}$。

因此，在不考虑天然采光情况下，按规范和工程实践经验得：工作面照度为400lx，照明功率应满足13W/m^2。则该办公单元总照明功率为13×24=312W。当采用功率为18W的节能灯和3W的整流器，安装高度为2.8m，需要均匀安装该型节能灯15盏。

安装方式：每一个灯座中安装三盏节能灯，共计5个灯座。其中，1个灯座居中嵌顶安装，其余4个灯座对称均匀嵌顶安装。照明回路分为3个回路，回路1控制居中灯座的节能灯3盏，其余2个回路分别控制两边的2个灯座的节能灯(图4)。

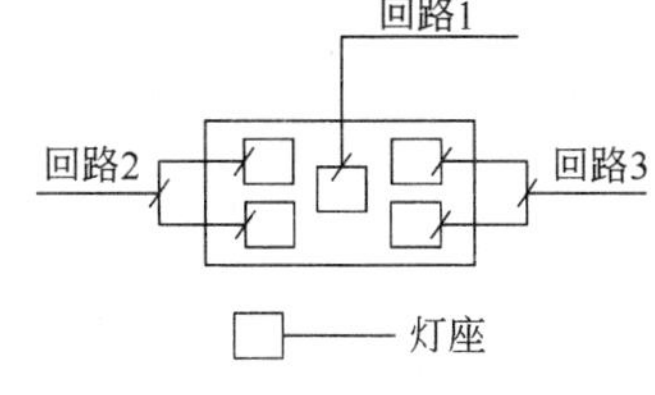

图4 天花灯座布置图

用电量分析(一个24m^2的办公单元一年的用电量计算如下)：

每小时用电量(灯+整流器)：

(18+3)W×15÷1000=0.315kW

如果一天以8小时工作时间计，则该单元一天的总耗电量为2.52kW·h，一年12个月每月工作22天，则一年用电为665.28kW·h。显见，一个几千平方米的商务办公楼，单每年在照明上的耗电量就是巨大的。因此，如果在办公建筑中引入天然光，减少人工照明时间，其资源节约的效果将是显而易见的。

2.2 控制策略与算法实现

可以结合空调系统，根据室内外空气焓值的差值，即$\Delta h=h$室外$-h$室内。将工作单元的百叶工况全年分为3个区：夏季工况区、过渡季工况区和冬季工况区。如图5，百叶控制系统全年工况分区所示。

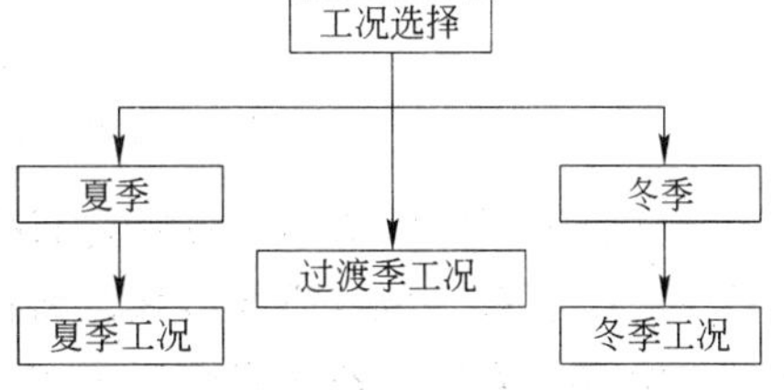

图5 百叶控制系统全年工况分区

当焓差$\Delta h=h$室外$-h$室内>0时，系统运行在夏季工况区。此时太阳能增益Q_{sol}对温度T_i具有很强的影响，Q_{sol}越大，T_i越高。因此，我们应在满足工作面照度的情况下，尽可能减小百叶的旋转角度以减少太阳辐射对工作单元的影响。当焓差$\Delta h=h$室外$-h$室内<0时，系统运行在冬季工况区。此时，应尽可能大的打开百叶，以最大限度的引入天然光和太阳能，以增加工作面照度且减小空调负荷。当室内外空气焓值相差不多时，则系统处于过渡季工况，在满足正常工作照度的情况下，用户可以按需对百叶系统进行手动调节。

本文即针对夏季工况工作面照度小于400lx，在保证夏季舒适性能指标的情况下，探索对百叶旋转角度B_{pos}进行最优化调节控制，以达到光、热环境的统一，最终实现节能的目的。图6为夏季工况下Φ_i<400lx时百叶控制流程图。

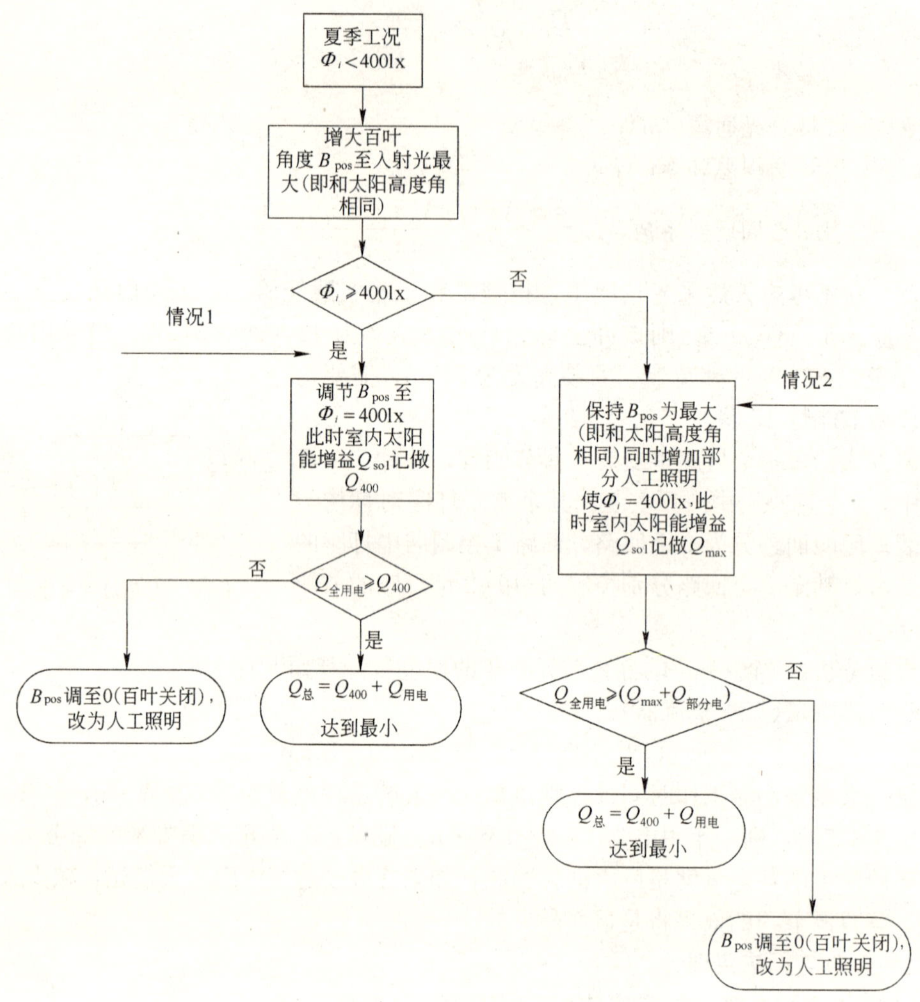

图6 夏季工况下$\Phi_i<400$lx时百叶控制流程图

靠墙安装照度传感器，其安装高度接近工作面高度。若工作面照度为400lx时，考虑外界散射等因素，则该照度传感器所得值应为500lx左右。令该照度传感器每隔10分钟进行一次数据采集。若其值小于500lx，则意味着工作面照度小于400lx。

以情况1为例。此时，增大百叶角度至Ω_0，使得工作面照度为400lx，记此时刻为K_0。由前文太阳能增益模型算法可得，其房间每小时得热为$Q_{400}=Q_{heat}(K_0)$。Q全用电表示全部采用人工照明，满足工作面照度为400lx时，每小时的耗电量。该值为0.315kW（前文已得）。

若$Q_{400}\geqslant Q_{全用电}=0.315$kW，则关闭百叶，全部改为人工照明。

若$Q_{400}<Q_{全用电}=0.315$kW，则自动打开回路1，同时计算此时的耗电量为$Q_{回路1}=0.063$kW，调节百叶角度至Ω_1，使得工作面照度为400lx，记此时刻为K_1。由前文太阳能增益模型算法可得，其房间每小时得热为$Q_{heat}(K_1)$。此时的总能耗$Q_{总能耗1路}=Q_{heat}(K_1)+Q_{回路1}=Q_{heat}(K_1)+0.063$kW。

同理，关闭回路1，仅打开回路2。调节百叶角度，使工作面照度为400lx，记此时刻

为 K_2、百叶角度 Ω_2，房间每小时得热为 $Q_{heat}(K_2)$，总耗能为 $Q_{总能耗2路}=Q_{heat}(K_2)+Q_{2路}=Q_{heat}(K_2)+0.126\text{kW}$。

接着，同时打开回路 1 和回路 2。调节百叶角度，使工作面照度为 400lx，记此时刻 K_3，百叶角度为 Ω_3，房间每小时得热分为 $Q_{heat}(K_3)$。总耗能为 $Q_{总能耗1+2路}=Q_{heat}(K_3)+Q_{1+2路}=Q_{heat}(K_3)+0.189\text{kW}$。

最后，关闭回路 1 且同时打开回路 2 和回路 3，得到百叶角度 Ω_4，记此时刻为 K_4，房间每小时得热分为 $Q_{heat}(K_4)$。总耗能为 $Q_{总能耗2+3路}=Q_{heat}(K_4)+Q_{2+3路}=Q_{heat}(K_4)+0.252\text{kW}$。

分别比较 $Q_{总能1路}$、$Q_{总能2路}$、$Q_{总能耗1+2路}$ 和 $Q_{总能耗2+3路}$，其最小值时的百叶角度 $\Omega_{最小能耗}$ 即为总能耗最小的百叶角度。再通过此角度和前文所述算法方程，得出最优节能舒适指标 $\mu_D(B_{pos})$。

3 总结

有关数据资料表明：照明用电占整个商业建筑能耗的 25%～40%，而天然采光可以节省 52%的照明用电。如何更有效的提高天然光的利用率，实现天然采光和节能，应引起设计者的关注。

上述方案模型是对照明节能中天然采光技术的一种探讨，实践中还需要包括建筑师在内的各方面专家的配合和指导。

总之，充分利用天然光，节约照明用电，对我国实现可持续发展战略具有重要的意义，同时具有巨大的生态效益、环境效益和社会效益。但是，由于天然采光也会使夏季室内热环境恶化，所以，在做天然采光设计的同时，我们还应从多方位进行节能设计，不应顾此失彼。

The Discussion of the Daylighting Technique for the Optimization Control of the Room Environment

Xia lin　Lu minghao
(The Architectureal Design&Research Institute of Tongji Univeriesity,
Shanghai　200092　China)

Abstract: The article discusses the daylighting technique of the blinds control, and then puts forward the optimization project and energy saving standards. Finally, the article analyzes the model which is established in the study.

参考文献

[1]　戴瑜兴，汪鲁才. 智能大厦照明系统的节能技术措施. 电工技术杂志，2000(11)
[2]　蒋美其. 绿色照明节能技术研究. 建筑技术开发，2003，11(11)
[3]　吴硕贤，夏清. 室内环境与设备. 北京：中国建筑工业出版社，1996
[4]　彭小云，邰惠鑫. 天然采光的生态方法. 建筑设计研究，2002

[5] 窦振中. 模糊逻辑控制技术及其应用. 北京：北京航空航天大学出版社，1995

[6] 刘曙光，魏俊民，竺志超. 模糊控制技术. 北京：中国纺织出版社，2001

[7] ZIMMERMANN，H.-J.，Fuzzy Sets，Decision Making，and Expert Systems；Boston：Kluwer，1987

[8] ROMMELFANGER，H.，Fuzzy Sets，Decision Support System；Heidelberg：Springer，1994

[9] 全国民用建筑工程设计技术措施电气. 建设部工程质量安全监督与行业发展司. 中国建筑标准设计研究所，2003

研究智能建筑弱电理论 促进智能与绿色建筑发展

张言荣
（山东农业大学机械与电子工程学院，泰安 271018）
张宏庆
（山东省交通厅公路局，山东 250002）

摘 要：文章报告了作者开展智能建筑弱电体系研究的情况，简述了 IBWP 体系几个方面的研究：IBWP 体系的内涵和三层结构；IBWP 体系的特点和应用发展前景。

关键词：智能建筑弱电 理论 体系 三层结构

1 引言

智能建筑这一建筑业的高新技术自 1984 年在美国诞生后，开始在世界发展。由于智能建筑技术发展历史短，在理论，技术和系统工程等方面经验都十分匮乏，急需研究总结出一套完整的系统体系，推动智能建筑的可持续发展。

我们根据智能建筑的这一发展形势需求，1999 年以来开展了“智能建筑弱电理论及系统体系”研究工作，创立了“智能建筑弱电理论及系统体系”（The weak power theory of intelligent building and its systemic economy）简称 IBWP 体系或 IBWPSy.，获 2004 年山东省科技进步三等奖（证书号：K2004-3-206-1）；2003 年通过省级项目鉴定，专家鉴定为“国际领先水平”，“形成了新的学科体系”。下面对 IBWP 体系及其内容做简要介绍。

本研究项目揭示了智能建筑弱电的理论、技术和系统以及它们之间的紧密关系等方面的规律。这套完整体系的建立对统一认识、应用智能建筑弱电规律和工程技术、促进智能建筑发展起到重要作用。下面介绍 IBWP 体系的内涵、特点、应用前景等。

2 研究技术方案

1. 以智能建筑环境为背景，运用辩证唯物哲学思想研究 IB 中信息科学以及信息论、控制论、系统论（以下简称“三论”）的衍生应用与发展。

2. 采用理论密切联系实际的方法，围绕国内外相应标准及其在 IBWP 工程中的应用与发展开展研究。

3. 以信息科学以及信息论、控制论和系统论的理论概念和主要内涵，研究它们在智能建筑中的运用、衍生情况，由它们与建筑科学、建筑学的有机融合研究并揭示 IBWP 理论的产生及其理论基础，确定 IBWP 体系的理论——智能建筑弱电理论 IBWPTh（Intelligent Building Weak Power theories）。

4. 把项目研究过程作为对 IB 进行广泛调查、深入探讨和总结的过程。突出理论研究、工程设计和工程实施三方面有机结合的方法。

3 建立了 IBWP 体系

本研究建立了以 IBWP 理论、技术以及技术工程所构成的一个完整的 IBWP 新科学体系。IBWP 体系是信息科学、“三论”等现代科学技术与建筑科学、建筑学、建筑技术等交叉融合而形成的边缘科学。

3.1 IBWP 体系是多学科的交叉、综合运用的边缘学科

IBWP 体系是以唯物辩证哲学原理和理论联系实际的方法作为基本方法，以信息科学、“三论”为理论基础，揭示它们(信息科学等理论)和现代弱电科学技术同现代建筑科学技术相互交叉融合的规律。包括：信息科学等理论与建筑科学理论的交叉融合；现代弱电技术与建筑设计技术和艺术的交叉融合；信息科学等理论与弱电技术和弱电系统工程在建筑环境中的交叉融合，IBWP 工程技术与招投标及监理等工程管理技术的交叉融合。从而形成了 IBWP 这一新的边缘学科，形成了具有 IBWP 边缘学科理论、技术、系统三层结构的 IBWP 体系。

3.2 揭示了 IBWP 体系的理论基础

研究信息科学和“三论”的产生、发展、内涵以及它们之间的关系，使我们得到启发。这些理论原理随着 21 世纪信息时代高新科学技术的发展，它们的传统属性和内容也必然得到发展，这些普遍科学理论和研究方法，会渗透到当代各行各业中去，它们也必定广泛应用到现代建筑中，从而形成了我们的研究基础，即以信息科学、“三论”和建筑科学为理论基础，把它们的基本原理和研究方法，运用到分析研究 IBWP 体系中去，揭示出智能建筑弱电一整套科学技术规律。

研究可知：IBWP 体系就是在智能建筑领域进一步认识信息(例如计算机信息、语音、图像信息等)在智能建筑中的传输和交换、处理等应用规律；就是研究并揭示利用这些当代信息和技术如何对智能建筑中的各种设备、系统进行控制的原理和方法；形成通信网络系统、建筑设备自动化系统、办公自动化系统、安全防范自动化系统、消防自动化系统等最优的智能建筑系统。

3.3 明确了 IBWP 与 IB 的关系

IBWP 这一概念，人们习惯上主要指的是应用于智能建筑中的语言、数据、图像等信息的产生、传输、控制、处理等信息类弱电及技术，它属于信息类弱电。

建筑智能化主要靠建筑弱电来实现，通过 IBWP 才能实现“向人们提供一个安全、高效、舒适、便利的建筑环境”。新时代赋予智能建筑的信息化、绿色环境、以人为本的文明文化等新内容也主要依靠 IBWP 直接或间接地实现。IB 为 IBWP 的现代化运用提供了一个建筑平台、环境，而正是 IBWP 的运用才能实现建筑智能化，形成智能建筑。IBWP 是智能建筑的命脉和关键，可以说离开 IBWP 就无法实现建筑的智能化，就没有智能建筑。

4 IBWP 体系内涵及三层结构

4.1 IBWP 体系内涵及三层结构示意图

IBWP 体系内涵及三层结构可用图 1 来概括。IBWP 体系主要由 IBWP 理论、IBWP

技术、IBWP 系统三部分构成，并分为三个层次：IBWP 理论(上层，理论层)、IBWPT(中间层，方法层)、IBWPS 层(下层，实现层)。

4.2 IBWP 体系的内涵

如图 1 所示，IBWP 体系主要内涵如下：

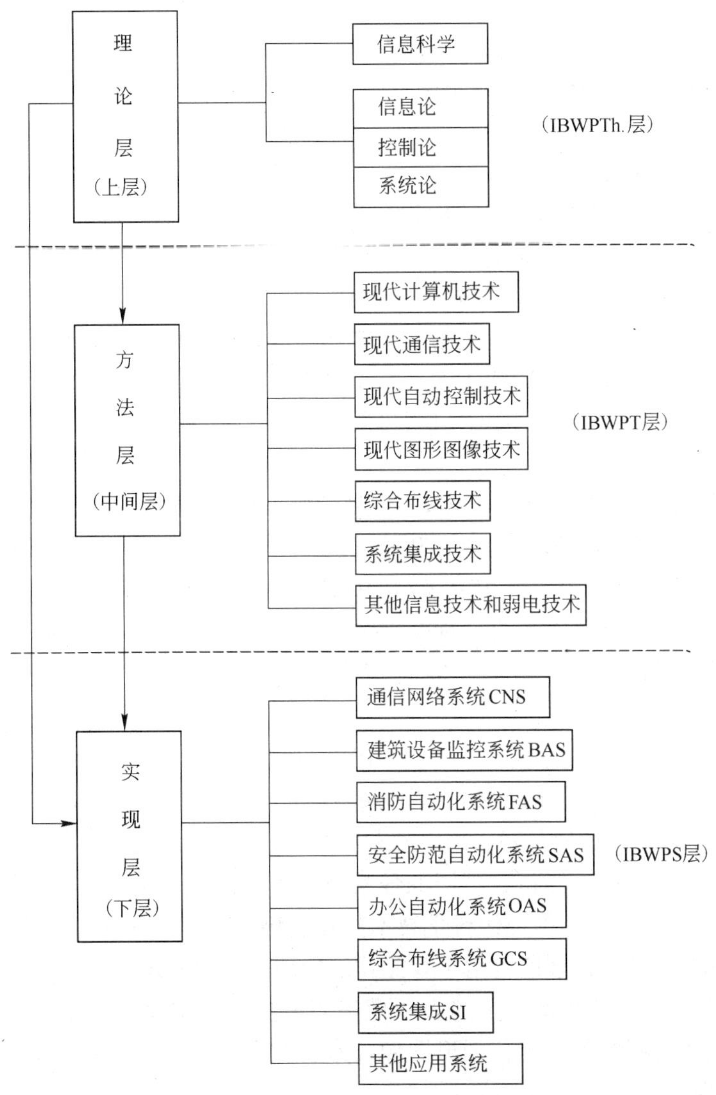

图 1　IBWP 体系内涵及三层结构示意图

(1) IBWP 理论是信息科学、信息论、控制论、系统论应用于智能建筑的派生理论和应用理论。信息科学、信息论、控制论、系统论和建筑科学，是 IBWP 体系的理论基础。它是 IBWP 体系的理论层。

(2) 信息科学、信息论、控制论、系统论与建筑科学、建筑学的交叉融合，即它们在智能建筑中的运用形成了 IBWP 理论。

(3) IBWPT 是信息科学、信息论、控制论、系统论在与建筑科学、建筑学、建筑技术交叉融合应用过程中形成的具体方法、途径和手段。IBWPT 是运用于智能建筑的现代技术的体现。它是 IBWP 体系的方法层。

(4) IBWPT 的主要内涵是信息技术等多项现代弱电技术，主要有现代通信技术、现代计算机技术、现代控制技术、现代图形图像技术、综合布线技术、系统集成等。

(5) 智能弱电系统是以 IBWP 理论为理论依据，运用 IBWPT 方法实现智能建筑功能环境的最终体现和成果。它是 IBWP 体系的实现层。它是 IBWP 体系实现价值和效益最主要的部分。它包括硬件(环境)和软件(环境)，体现出各主要 IBWPS 的关系，归纳出目前常用的 55 个子系统，并画出了子系统图。

4.3 三层结构

4.3.1 三层结构

图 2 示出了 IBWP 体系的三层结构为：

(1) 理论层：IBWP 理论是信息科学“三论”应用于智能建筑的派生理论和应用理论。

(2) 方法层：IBWPT 是信息科学、“三论”在与建筑科学、技术交叉融合应用过程中形成的具体方法、途径和手段，是运用于智能建筑的现代弱电技术方法。

(3) 实现层：IBWPS 是以 IBWP 理论为理论依据，运用 IBWPT 方法实现智能建筑功能环境的最终体现和成果。我们归纳出了目前常用的 55 个子系统。

4.3.2 理论层

IBWP 体系理论层，是体系的上层，它是体系的理论依据和灵魂，它决定了体系的方向和实质。它是方法层和实现层的源泉和基础。

IBWP 理论主要内涵有：(1)IBWP 体系的理论基础；(2)IBWP 体系是多学科交叉融合的边缘学科体系；(3)IBWP 理论是信息科学、“三论”的派生理论和应用理论；(4)IBWP 理论及体系与建筑科学的关系。

4.3.3 方法层

方法层是 IBWP 体系的中间层，是贯彻 IBWP 理论，指导实现具体系统工程的方法途径和手段。这些方法、途径和手段是在 IBWP 理论的应用中形成的，所以本层的基础是理论层，具有逻辑上承上启下的地位和作用。没有中间层，IBWP 理论和方向就无法实现，成为空头理论，没有先进技术作支持，下层的各系统工程就无法优化。方法层的主要内涵是：

(1) IBWPT 是信息科学等现代科学技术同建筑科学技术交叉融合应用而产生的方法。

(2) 这些方法是 IBWP 理论运用于智能建筑的现代弱电技术的体现。

(3) IBWPT 是以信息技术为主的多项现代弱电技术，主要有：现代通信技术、现代计算机技术、现代控制技术、现代图形图像技术、综合布线技术、系统集成等。

4.3.4 实现层

IBWPS 及工程是 IBWP 体系的实现层，它处于三层结构的最下层，是 IBWP 体系的应用层。IBWPS 及工程发展、实现的理论根据是 IBWP 理论，丰富多彩、功能各异的 IBWPS和工程又是运用先进的 IBWPT 多项技术方法、途径和手段而得以实现的。IBWPS 及其工程直接向人们提供安全、舒适、节能、绿色等若干功能智能建筑环境。它是 IBWP 体系服务于人类和社会的具体体现，本层具有十分重要的经济价值和社会价值，是 IBWP 体系最重要的组成部分。

IBWPS 与 IBWP 理论和 IBWPT 有十分紧密的关系，它们相辅相成，共同构成了一个完整体系，理论——方法——实现结果之间的这种唯物辩证关系，完全符合事物构成发展的普遍规律——唯物辩证理论。

IBWPS是对当前智能建筑功能的综合和体现。按智能建筑弱电的功能，我们总结出了55个当前世界上流行的功能子系统，画出了它们的组成图，涵盖了中国标准GB 50314—2000中规范的智能建筑各主要子系统。

5 IBWP体系应用及发展前景

IBWP边缘科学体系的特点和体系应用情况等都表明它具有广阔的应用与发展情景。

1. IBWP体系创立了一个促进IB发展的新的边缘学科，IBWP体系的产生，创立了一个现代高科技与建筑科学相融合的新边缘学科。它丰富了建筑科学技术宝库，适应了世界IB业可持续发展的前景。

2. IBWP体系是一个完整的应用型科学体系，IBWP体系集IBWP理论、技术、系统工程等方面的新理论观点、新方法、新内容于一体，已广泛应用于科研教学、系统工程设计、施工、招投标监理等整个智能建筑领域。

3. IBWP体系以应用为主要目标，严格把IBWP理论和技术通过贯彻执行国内、外标准落实到智能建筑各种系统工程的设计、施工和管理中去。总结出的工程实践经验，是可以直接应用于工作中的理论、方法和技术，广泛运用于科研教学、系统工程设计、施工、招投标监理等整个智能建筑领域。

4. 已经应用的情况，证实了IBWT体系具有广阔的前景。IBWP体系在工程设计、施工、检测验收、招投标、监理等方面已得到了广泛的应用。来自全国6个省市29个单位的应用情况如下：近三年新增产值33691万元人民币，平均每个单位1162万元人民币；新增利税4552万元人民币，平均每个单位157万元人民币。29个单位类型包括IB工程设计研究院所、系统集成公司、工程施工公司和招投标监理公司。

IBWP体系已应于大学生的培养和IB专业人员的专业培训。来自4个应用单位的情况表明，在有的大学以IBWP体系为依据确定“电气工程自动化”、“建筑环境与设备工程”等专业的专业方向和课程设置。2003年，4个单位的培训人数为1300余人。

据来自6位高级工程师的情况，他们在近2年应用IBWP体系作指导，完成了120多项如2008年北京奥运会部分比赛场馆等国内外大中型IBWP系统工程项目的设计和施工。上述应用单位的情况表明，他们运用IBMP体系后本公司和员工的技术素质、工作质量普遍提高，对外信誉大大提高，取得了巨大的社会效益和经济效益。

由上所述，IBWP体系应用已获得一定的经济效益和社会效益，并且具有广泛的发展前景。注意的是，智能建筑是21世纪建筑业的高新技术，必须融合信息时代多项弱电高新技术，才能满足人类工作和居住、生活建筑环境的需求，因而IBWP体系的研究和应用将会越来越普及和深入。可以预见IBWP体系具有广泛的应用发展前景。

6 结论

我们根据IB可持续发展需求，开展了IBWP体系研究，创立了IBWP体系这一边缘学科——IBWPSy.。它的理论基础是信息科学、“三论”和建筑科学以及它们的融合；

IBWPSy. 的内涵主要有 IBWPTh.、IBWPT 和 IBWPS，并分为三层结构，分别是理论层、方法层和实现层。IBWPSy. 是一个应用型学科体系。它的研究过程也是它初步应用的过程，它的新学科特点及生命力和已应用情况证明了 IBWPSy. 具有广阔的应用发展前景，它丰富了 IB 知识宝库，对促进智能与绿色建筑发展会起到巨大的推动作用。

Study the Weak Power Theory of Intelligent Building Promote the Development of Intelligent and Green Building

Zhang Yanrong(College of Mechanic & Electronic Engineering Shandong Agriculture University, Taian 271018 China)

Zhang Hongqing(Transportation Hall Highway Bureau of Shandong, Shandong 250002 China)

Abstract: It is reported that the author develops the Intelligent Building Weak Power system research, briefly stated the research of some aspects about the IBWP System: the connotation and three layers construction of the IBWP System; the characteristics and the prospects of development and application of IBWP System.

参考文献

[1] 张言荣等编著. 智能建筑安全防范自动化技术(P7～20 丛书绪论). 北京：中国建筑工业出版社，2002 年 10 月第一版

[2] 李佩珊、许良英主编. 20 世纪科学技术简史. 北京：科学出版社，1999 年 9 月第二版

[3] 鲍世顾、孟潮主编. 杰出科学家钱学森论山水城市与建筑科学. 北京：中国建筑工业出版社 1999 年 6 月

英国曼彻斯特智能体育场案例分析*

洪　桔[1]　陈　震[2]　李　恒[3]　徐　骞[3]

([1] 北京建筑工程学院城市研究所，北京　100044；[2] 英国瑞丁大学建设管理与工程学院智能建筑研究所，英国 RW66AW；[3] 香港理工大学建筑及房地产学系，香港)

摘　要：本文在总结国内外相关文献的基础上，通过对英国曼彻斯特新体育场的案例分析，较完整地介绍了当代国际智能体育建筑对于多学科多专业知识与合作的具体要求，为建筑学、建筑结构、建筑设备、建筑施工和建筑管理等专业领域的在体育场馆智能化建设中的有效合作提供了若干富有建设性经验总结。本文为我国智能体育建筑的发展提供了颇具价值的参考资料。

关键词：智能建筑　体育场　体育馆　曼彻斯特

1　引言

英国曼彻斯特市新建大型露天体育场(以下简称曼城新体育场)位于英国曼彻斯特市缅因路，同时也是曼城足球俱乐部总部的所在地。曼城足球俱乐部实际上就是曼城新体育场的经营者。曼城新体育场拥有醒目的卵形整体结构造型，周边簇拥着八座环形塔楼(塔楼具有疏导观众、设备服务和支撑屋面结构的多重作用)，波浪起伏的屋顶，加之桅杆与钢缆支撑体系共同塑造了当代体育建筑的全新风貌(参见图 1 和图 2)。除了足球比赛和竞技比赛，曼城新体育场设计的灵活性还能够让它成为大型流行音乐会以及其他单项大型商业活动的主办场地。不仅如此，曼城新体育场还以其创新的智能体育建筑理念，先进的智能建筑系统和成功的俱乐部经营方式，成为英国乃至世界最引人入胜的体育中心之一(TSG，2005)。

本文以曼城新体育场为案例，在归纳总结国内外相关参考文献的基础上，较完整地介绍了当代智能体育建筑的若干特点，包括以人为本的智能建筑设计，以高效率为目标的智能建筑设备系统，和以高回报为前提的建筑经营理念。

2　建筑设计

曼城新体育场最初是为 2002 年第 17 届英联邦运动会的竞技比赛而建造。在英联邦运动会之后，曼城新体育场经历了为期 10 个月的改造，包括去除跑道和增设看台，开挖

* 得到中国电子技术标准化研究所中国智能建筑评估系统研究项目的资助。

图 1 曼城新体育场建筑效果图(PanStadia，2001)

9 万 m^3 土方以降低 6m 赛场地平从而增设第三层座席等，最终拥有座席 47500 个。曼城新体育场改建的目的是让它成为欧洲最现代化的体育场，并以其完美的创新设计把观众带到观赏足球比赛的全新境界。根据英国足球赛场的传统，曼城新体育场在观众视觉和听觉上被定调为提供一个强烈且动感的竞技场地。场内每一个座席都在距离赛场中心 100m 的圆周之内，而且保证每一位观众拥有完美通畅的视野。曼城新体育场的建筑设计承包商 Arup 联合体在建筑设计上的别具匠心，使得它能够在球赛中提供最高水准的观赏舒适性和视觉及听觉等的可及性的同时让观众群情激奋。与智能建筑的设计相一致，体育场馆智能化设计所包含的范围除了建筑设备系统的智能化之外，还有建筑设计中设计者们对人与人造环境的最佳和谐的深刻考虑。

建筑设计承包商 Arup 联合体在曼城新体育场的设计过程中制定了一套针对性的设计准则，包括：

- 在观众层面上，体育场要全面提升建筑的内涵，包括观众到场后的愉悦感，观众在场内的无障碍通行，场内的喧嚣与宁静区间划分，以及比赛中观众的舒适但激动人心的空间感等；
- 在建筑设计层面上，体育场要有视觉上的层次感和动态感，能够清晰地反映和突出曼城新体育场的建筑功能；
- 在城市规划层面上，体育场要能够吸引公众，成为曼城能够满足公众需求的好去处，和成为曼城旧城改造的标志；
- 在城市风景层面上，体育场要以其屋面和看台所具有的超几何学的雕刻造型成为一种尝试。

以下介绍曼城新体育场核心部分的设计，包括看台、屋面、通行、电气与照明、通风和安全等内容。

2.1 看台

看台设计的目的之一是让观众就座后立即感受到体育场的主要空间，即赛场和看台。从看台放眼望去，观众席之上的超环面屋面轮廓配合几何学造型的径向放射状檩条结构形成了升降有序的屋面内周长，颇具动感地把观众的视线引向赛场。每当高朋满座之时，整个体育场内部就有如波动的海洋。曼城新体育场为球迷设计了 47500 个座席。考虑到体育场多种经营的需要，例如音乐会，观众的实际容量可以增加到 6 万人。看台的设计充分考

图 2　曼城新体育场看台及场地
(Dennis Gilbert，architecture. com)

虑到观众的多种需求。首先，尽可能多地保证了任何一个观众的视线与赛场中心点的最远距离都在 100m 的范围内。就球赛而言，靠近赛场中线(东西向)两侧的看台是最佳的位置，看台的设计因此依照既定的原则，包括赛场东西侧各设三层座席，南北侧各设两层座席(参见图 2)。其次，考虑到伤残人员的需要，整个体育场还设置了 226 条通往看台的坡道，并且在每一层为他们设置了专用座席。能够行走的伤残人员也可以使用这些坡道进入临近的看台。此外，体育场为视觉受损观众提供耳机用来收听现场评论；听觉受损的观众也可以使用耳机得到公告广播。这些充满人性化的无微不至的建筑设计为观众提供了舒适宜人的观赏空间。

曼城新体育场建筑设计上的独具匠心，为其提供了大量安全实用的建筑空间。看台主体采用钻孔灌注桩和现浇钢筋混凝土结构体系，而阶梯看台体系则由延伸的楼面板和预制钢筋混凝土阶梯板组合而成。支撑看台板的主斜梁采用 7.6m 间距，是走廊长度的一半，其目的是为了在看台下创造更多的使用空间；次斜梁的分布则根据整个体育场看台的座席几何分布确定。与看台结构相毗邻，在体育场外侧周边分布着八座圆环形螺旋坡道塔楼，除了作为进入体育场的通道使用之外，它们为体育场提供一些辅助功能，包括洗手间和设备间等，此外，它们还支撑着屋面结构桅杆(参见图 1)。巨大看台下面为观众创造了三个室内大厅，成为入口和通往看台的坡道或阶梯甬道之间的缓冲空间，大厅的设计保证从体育场入口到大厅内任何一处服务设施之间的距离小于 40m，观众可以在那里舒适宜人的环境中从容不迫地得到令人满意的多项餐饮服务或者交流对球赛的观感等。

2.2　屋面

曼城新体育场的突出特点在于它与众不同的动感造型，而与之相应的就是其屋面结构体系。屋面结构体系采用金属结构，由若干结构子系统组成，包括屋面结构的桅杆支撑体系，从钢筋混凝土看台后部支撑立柱顶向赛场悬伸的(纵向)屋面檩条体系，纵横连接屋面檩条体系与支撑体系的缆索网以及屋面板等。屋面结构体系的 个显著特点就是采用长达 40m 的钢桅杆支撑体系。这种屋面结构体系不但提高了建筑结构体系的工作效率，而且也成为建筑学设计的出发点，整个体育场的建筑设计实际上是围绕上述屋面结构的构思展开的。

曼城新体育场的屋面围护结构非常典型地体现了建筑学、结构工程和设备工程在体育场项目规划设计中的综合与统一。以屋面板下的铝制衬板盘为例，它们固定在檩条体系上，从结构功能上考虑，它们支持着直缝排列的铝板屋面板，并且在横向加强檩条体系的刚性；从使用功能上考虑，它们为结构布线提供了隐蔽空间，提高了看台的声学效果，并且非常经济地解决了天花板在视觉上的清洁问题。此外，从缆索网向赛场内侧方向延伸有10m 宽的透明聚碳酸酯纤维板屋面板，这一方面增加了看台的采光，同时又为赛场草坪的生长创造了条件。不仅如此，采用这部分透明的聚合物材质屋面板还能够巧妙地利用阳光在赛场上创造光线的和谐过渡，这对于电视直播尤其重要。屋面的超几何曲面造型实际上创造了从 1.5°到大约 17°的倾斜度，保证屋面平面始终向场外倾斜。屋面的排水系统则因势利导地使用铝质天沟板把雨水径流疏导去屋面南北两侧特制的排水管。这些设计上的综合，为体育场创造了和谐的氛围。

2.3 通行

曼城新体育场的设计者们通过对既有体育场研究，总结和汲取了若干经验与教训，认为场内大堂需要保留观众服务空间，而且在从大堂到看台座席的全程为观众创造类似于商业步行街似的舒适和礼节空间。这种充满人性化的设计让观众在半场休息期间到大堂享用茶点的时候，再也不会经历艰巨的爬行，取而代之的是从容不迫的经历。

为了向观众提供舒适，安全和方便交流的通道，体育场采用了螺旋甬道方便观众进出高层大堂和看台，而放弃使用传统的楼梯。这种设计构思最终又与体育场的建筑、结构和设备等项设计巧妙地结合起来，形成体育场曲面墙壁外侧引人注目的八座圆筒式塔楼。每座塔楼的直径为 10m，它既是体育场主体结构的一部分，支撑桅杆及其传递的屋面荷载；同时又是核心设备用房。为了减少在塔楼的核心筒上开洞，所有的空调终端机和阀门都安置到塔楼的上部，而主要的输送管线都经由地面以下出塔楼并与体育场内的管线网连接。这些安排最终为观众创造了完美的场内通行。

2.4 电气与照明

在费用与风险分析的基础上，曼城新体育场采用外环形综合电气管线布局，通过塔楼与体育场主体之间的两层连桥，连通塔楼的设备间与体育场内与电、水、气、暖以及通讯等项服务相关的设备。在体育场内部，电缆在体育场上层形成水平环绕，通过指定的管道井形成垂直连接。这种综合布线的格局为整个体育场包括看台之后的大堂服务提供了有效支持。价值工程的研究实践证明，这种集成化的设计非常经济和有效。

曼城新体育场严格按照国际足联的有关规定设置体育场的照明系统。为了预防由于电力系统中断而造成事故，体育场安装了备用发电机，在紧急情况下通过一部分带热触发控制装置的泛光灯提供 800lx 的照明，而这种照度水平也是电视摄像的最低亮度要求。体育场的临时电力供应来自两个 2MW 变电站，分别位于体育场东西两侧看台外的塔楼内。为了预防一台发电机发生故障，影响对于热触发泛光灯的临时电力供应，系统采用发电机 A—电力网—发电机 B 的自动切换设置，其目的就是在紧急情况下最大限度保障体育场的照明。

2.5 通风

曼城新体育场的通风系统由设置在东西看台外侧各层的中央空调机组组成。室外空气进入系统时要经过过滤后才能与系统内再循环的空气混合，室外的冷空气还要经过热水升温后才通过热绝缘的通风管道向场内输送。室内回风管一般没有热绝缘，但如果回风管经过未配备暖气设备的空间，为了防止冷凝和减少散热，回风管就要作热绝缘。回风管里的空气分别被送回系统再循环，或经高空或地面的排气口排出。

2.6 安全

曼城新体育场的安全系统在幕后支持着体育场的正常运作。除了下文将要介绍的智能体育场系统之外，分布在整个体育场的 15 部固定彩色摄像机和 57 部全功能旋转摄像机为体育场的中央监控系统提供高品质的画面，实时监控观众座席，看台和通道的状况。中央控制室通过这些画面对整个体育场实现全天候的安全监督与控制。

体育场的防火设计采用了防火工程的概念。其目的是在建立高标准的防火体系的同时，促进设计创新和控制造价。火灾的发生会导致观众的整体逃生，所以体育场的建筑设计不得不在艺术创新和防火减灾之间求得平衡。以大堂中出租营业区的防火设计为例，这些地方不但是火灾易发区，而且也是火灾发生后威胁观众逃生的危险区。为了解决保护逃生通道和维持建筑设计意图之间的矛盾，设计者们提出“吸尘器系统”的概念。每一个出租营业区安装双层齐地面的防火卷帘门，之间利用通往卫生间排气系统的排烟管道创造空气负压带，从而保证出租营业区的烟不会扩散到大堂观众集散地。出租营业区的烟与温度较低的空气混合，从而还降低了排烟管的温度。计算表明，排烟管内的烟由于被来自大堂的自由流通的空气所稀释，温度大大降低，所以排烟管不必采用防火处理。通过使用计算机模拟，防火工程的方法向法定主管部门演示了体育场的安全性；同时，不论是初始成本还是长期成本都得到很好地控制，从而在防火工程指导思想下，让以“吸尘器系统”为代表的防火设计顺利通过了方案评审。

3 智能体育场

曼城足球俱乐部与若干世界著名的系统供应商包括 Hewlett-Packard 公司，Cisco 系统公司以及 Fortress GB 公司等通力合作，引进智能体育场馆的设计理念并且建造了一整套智能化体育场系统；这也是全球第一套智能体育场系统，得到来自这些国际著名公司的顶级技术与创新的大力支持。概括来说，曼城新体育场的智能系统在 Cisco Aironet 无线局域网(WLAN)技术和 Fortress 智能体育场软件和射频验证(RFID)技术的基础上，集成了一套 802.11b(WiFi)无线网络系统，用来实施俱乐部的智能卡系统。这套系统能够支持曼城新体育场在无人监控状态下观众的快速入场，支持曼城新体育场和其他任何球迷俱乐部内客户关系管理数据的自动采集和建立，而且支持遍及曼城新体育场的安全可靠的无线实时数据的查询与传输，以及电子交易等多项任务。曼城新体育场智能系统的设计和实施实际上是曼城足球俱乐部的经营需要，其面临的挑战包括：

- 建立曼城足球俱乐部的品牌价值和得到最广泛的信赖；
- 加强曼城足球俱乐部与球迷和赞助厂商联系；
- 获得更多的客户信息；
- 从体育场及其可能的周边环境获得收益；
- 在多功能的曼城新体育场得到投资回报等。

在智能体育场系统的设计和施工过程中，Hewlett-Packard 公司向曼城足球俱乐部提供了完整的服务，包括智能体育场无线网络系统规划、现场勘察、网络系统设计、网络系统安装、HP iPAQ 的集成以及系统测试等。此外，Cisco 公司，Fortress GB 公司和 Computer Software 公司等著名企业为智能体育场系统提供了相关的硬件和软件系统支持。具体而言，智能体育场系统的硬件系统包括 Cisco 公司的 Aironet 1200 系列无线接入点(支持 802.11b)，Hewlett-Packard 公司的 HP iPAQ 5550 掌上电脑(配备无线接受器)以及 Fortress GB 公司的旋转闸门控制箱等；软件系统包括 Fortress GB 公司智能足球俱乐部软件，用以支持无线驱动智能卡和售货机等，以及 Computer Software 公司的客户关系管理数据库等。

曼城新体育场智能系统的建成与投入使用达到了设计方案的预期效果，取得了非常显著的社会和经济效益。其中颇为突出的效果分析如下：

- 智能卡系统创造了让观众以每分钟 1200 人的速度涌入体育场的历史纪录。
- 在第一个足球赛季，85000 名球迷购买了智能卡并且开始累积球迷分，而且智能卡回购服务项目的收入高达 50 万英镑。
- 新客户关系管理数据库帮助俱乐部开发更多的创新服务和发掘新的赞助厂商。
- 用智能卡联合赞助厂商的做法提升了曼城足球俱乐部的品牌价值。

以下详细介绍曼城新体育场智能系统的核心技术，包括智能卡和无线网络。由于商业秘密的限制，本文不涉及与这些核心技术设计方案相关的细节。

3.1 智能卡

由 Fortress GB 公司为曼城足球俱乐部开发的智能卡系统集成了多种使用功能，包括入场券的销售和购买，运动场内的通行控制，现场人员管理，俱乐部会员管理，和电子钱包使用等。这些功能满足了足球俱乐部和球迷观众的日常使用需要。具体而言，曼城足球俱乐部智能卡能够满足以下各项需求：

- 基于现场的入场券购买和销售；
- 基于互联网的入场券购买和销售；
- 电子入场券；
- 电子通行证；
- 电子钱包；
- 电子交易；
- 考勤制度和观众到场记录；
- 准确而完整的数据库；
- 俱乐部的电子商务；
- 俱乐部的信息共享；
- 会员计划；
- 家庭参与计划；
- 其他。

观众入座率是衡量体育场收入的重要指标之一。如果不能通过传统的人工检票方法让 48000 多名观众快速入场，或者说最大化收益，那么只有采用新的技术手段。在这种指导思想之下，曼城新体育场摒弃了过去对门票销售状况的静态的记录，采用动态的客户关系管理数据库对门票销售进行管理。为了完善基于智能卡的客户关系管理，曼城足球俱乐部制定了“一人一卡一记录”的原则，即每一张智能卡都有一位持卡人，而且与持卡人的会员身份保持一致，如此一来，持卡人的每一个购买行为都能够被记录到俱乐部的客户关系管理数据库里。俱乐部的管理者们就能够根据这些历史数据对持卡人的购买行为进行分析，从而制定更有针对性的销售策略和完善俱乐部的经营。为了刺激持卡人进行消费，俱乐部推出了智能卡积分制，持卡人可以年复一年地积累得分，然后利用自己的积分换取好座位或者享受折扣等项服务。智能卡的设计同时能够满足曼城之外，乃至世界各地球迷们的需求。此外，俱乐部的旅行者智能卡服务能够同时提供购票和住宿服务。

与传统的纸票不同，数字式智能卡在其功能和使用上的灵活性能够让俱乐部为球迷和赞助厂商提供多种多样的个性化服务。俱乐部向球迷推出多种智能卡的选择，包括曼城赛季卡、曼城会员卡以及曼城球赛卡等。通过对客户关系管理数据库的数据挖掘，俱乐部发现多数持有曼城赛季卡的球迷错过一些球赛。基于以上发现，俱乐部与 Fortress GB 公司合作，开发并推出了一项新的回购计划。在该项计划下，俱乐部以信用积分的形式回购曼城赛季卡持卡人的座位，然后再销售这些回购得到的座位。回购计划的实施能够在每个赛季为曼城足球俱乐部带来大约 50 万英镑的收益。此外，俱乐部还在每场球赛之前的一周通过电邮向数据库中智能卡的持卡人发布座席状况报告，从而避免了昂贵的广告费用。

采用智能卡之后，球迷们不必再在售票处索然无趣地排队而消磨更多时间。对于俱乐部而言，希望球迷们到体育场内消磨更多时间，从而意味着会增加体育场内的商品零售额。通过客户关系管理数据库，俱乐部实际上能够立即知道哪一位球迷在哪一场球赛何时到场。统计表明，48000 名观众的绝大多数能够在球赛开始前 30 分钟内入场，而且入场越早的球迷在体育场内消费越多。为了显著增加球迷们在体育场的消费，俱乐部通过电子邮件，向球迷提供额外的积分和折扣活动信息，其目的就是在球赛开始前的 45 分钟内，多吸引 50%的球迷入场。

智能卡系统投入使用后，足球俱乐部的管理变成了一种商务活动，在不断为球迷提供花样翻新的优质服务以满足观众各种需求的同时，谋求俱乐部的最大化盈利。以曼城新体育场现场商品零售为例，商业分析认为曼城足球俱乐部通过利用智能卡销售赛季门票，简化现场观众购票过程，并且方便观众入场等一系列服务内容的改进会让观众在体育场内外有更多的闲暇时间，从而进一步增加曼城新体育场在足球赛季里各项零售商品的销售。智能卡在曼城新体育场的应用有力地增强了曼城足球俱乐部经营上的主动性和对赞助厂商的广告安排，增加了观众积极参与这座智能体育场各项娱乐活动的积极性，加强了观众与体育场之间紧密的情感关联，促使赞助商源源不断地进行投资，从而为提升曼城足球俱乐部的竞争力创造了大量的机会。这种利用智能卡支持与驱动的紧密关系使得观众，赞助商和俱乐部/体育场处于多方共同受益的良性循环，最终促使曼城新体育场成为当今整个欧洲的最佳体育场之一。

3.2 无线网络系统

Hewlett-Packard 公司的专家们为曼城足球俱乐部规划、设计和安装了一套基于无线网络的智能体育场系统。这套系统采用 Cisco Aironet 1200 系列无线接入点，Fortress GB 公司的用于自动售货机和旋转闸门的无线驱动控制箱，以及集成在 HP iPAQ 掌上电脑里的无线访问与接收装置等。在这套无线网络的基础平台之上，各种无线网络支持设备，包括自动售货机，旋转闸门和 HP iPAQ 掌上电脑等能够实时访问俱乐部客户关系管理数据库和进行数据交换。

智能体育场的核心是 Fortress GB 公司开发的一套智能足球俱乐部软件。它集成了由 Computer Software 公司开发的客户关系管理数据库，有效地支持曼城足球俱乐部的日常工作，具体包括：俱乐部会员管理，电子交易，在线博采，雇员考勤和设备管理等。系统的实际运行表明，曼城足球俱乐部的管理者对 HP iPAQ 掌上电脑的功能，对同时使用有线和无线网络系统，以及整个系统的兼容性，稳定性和可靠性感到非常满意。

取代传统的纸质票据，球迷们现在使用的是集成了射频芯片的智能卡。智能卡储存了多种有关持卡人交易细节的信息，包括持卡人的永久客户编号，指定的入场大门牌号和座席。Fortress GB 公司为曼城新体育场的各个出入口设置了集成有射频驱动装置的入口系统，地点包括 84 处闸门，高级会员出入口、企业出入口以及路障等。球迷们只要在他们指定的大门向访问控制器出示通过智能卡，就能够立即过闸入场。与此同时，无线智能卡系统立即实时更新智能卡和俱乐部的客户关系管理数据库，记录持卡人的入场时间。

智能体育场的自由应答无线智能卡系统每分钟能够在体育场范围内确认和记录 1200 张智能卡，同时让票房前的长龙彻底消失。与以往球赛前用几小时排队购票的历史相比，球迷们现在可以简单地打电话或者通过俱乐部的网页买票。Fortress 智能足球俱乐部软件与俱乐部的票务系统相连，在客户关系管理数据库中记录每一笔交易，而这些记录又同时能够在无线状态下被所有集成有射频芯片的设备访问，包括体育场内的管理人员使用的 iPAQ 掌上电脑。为了防止智能卡的滥用，系统只允许持卡人在每场比赛到场一次。当有人使用打折卡的时候，系统会立即发出信号，现场管理人员就能立即识别是否有成年人使用儿童卡。在读卡出现障碍的时候，系统也会立即发出通知，然后由现场管理人员根据 iPAQ 掌上电脑的记录进行处理或者对数据库进行更新。俱乐部的统计表明，这种灵活变通的设置，在每场球赛能够增加 3000～4000 英镑的收入。

曼城新体育场的智能体育场系统同样兼顾了体育场的安全需要。在监控室里，保安人员能够通过智能体育场系统输出的实时数据实现对现场状况的监控，包括监督场内人流，以及控制场内球迷密集区。当场内某一部分出现拥挤的时候，保安监控室就会对现场管理人员发出到场指令；等到球赛结束之后，监控室会对当场情况进行分析，并调整现场管理人员的分布，为随后的球赛做好准备。此外，现场管理人员通过 Fortress 智能足球俱乐部软件的考勤功能界面报到之后，他们的考勤信息就自动进入俱乐部的工资管理系统。

Hewlett-Packard 公司的专家们在曼城新体育场智能系统的设计与安装过程中，充分考虑了系统的日常管理与维护以及未来的系统升级。曼城足球俱乐部在 Fortress GB 公司和 Cisco 公司的协助下，采用了 Fortress 智能足球俱乐部软件；这套软件能够支持电子交易数据传输和在线博采，同时记录体育场内部各项设施和座席的使用情况。在体育场内部，无线网络系统能够向球迷们提供各项方便，包括电子化交易和无处不在各种数字化访问设备等。此外，球迷们也不必再在中场休息的时候排队下注，取而代之的是去场内就近的 Ladbrokes 公司的代理商那里找一部 iPAQ 袖珍电脑。HP 掌上电脑还能够实时接收球赛照片。这就大大方便了球迷们和记者们的需求。此外，曼城新体育场的新闻记者席为那些争分夺秒的记者们提供集成在智能体育场系统中的安全可靠的互联网设施，保证他们能在第一时间把报告和图片发送出去。与此同时，曼城足球俱乐部的高级会员还能在他们座席附近的售货机订购商品并享受专递服务。

智能体育场系统能够同时为曼城足球俱乐部的五千多企业账户提供个性化服务。智能体育场系统支持下的客户关系管理实现了曼城足球俱乐部寻求与球迷和赞助厂商紧密联系的经营策略。此外，智能体育场系统与曼城新体育场的完美结合，使曼城足球俱乐部真正成为市场为导向的营销组织。事实上，客户关系管理数据库帮助曼城足球俱乐部从更深层次上有效地对球迷们的需求进行理解和分类，从而可能为赞助厂商们提供各种量体裁衣式的赞助项目。这些对赞助厂商的服务，诸如金融业和旅游业的服务项目，已经通过无线智能卡系统向球迷们推广。对于曼城足球俱乐部而言，这套灵活和可升级的智能体育场系统有效地推动各项经营活动。

4 结论

本文通过对曼城新体育场在建筑、结构、设备和管理上突出特点的总结，较全面介绍了以曼城新体育场为代表的当代智能体育建筑的发展现状，包括设计上的创新与综合和管理上的高效与务实。以人为本的指导思想是当代智能体育建筑的出发点，综合全面的设计技术和有效的多方合作是实现设计创意的保证，先进的建筑技术和信息与通讯技术是实现创新设计方案的基础。限于资料有限和对商业秘密的尊重，本文没有提供更多更具体的内容，也没有对曼城新体育场的各项技术经济指标作量化的分析。

结合正在进行的中国智能建筑评估系统研究，本文作者认为，在智能建筑的设计阶段按照建筑的类型对其进行评估的做法能够为智能建筑设计提供一套完整的工作程序，协调不同专业的工作内容，促进各专业设计的合作，从而有效提高设计质量。以曼城新体育场项目为例，极富创造力的设计者们提出了若干创新的设计思路，包括建筑可持续发展的前提、建筑防火工程的概念、以及价值工程的应用等；然而，本文作者相信智能建筑评估系统的应用会从整体上综合上述创新思路，并提升建筑设计的水平。

A Case Study on Intelligent Buildings: the City of Manchester Stadium

Ju Hong
(Beijing Institute of Building Engineering, 1 Zhanlanguan Road, Beijing 100044 China)
Zhen Chen
(The University of Reading, Reading, RW6 6AW, UK)
Heng Li Qian Xu
(The Hong Kong Polytechnic University, Hong Kong, China)

Abstract: This paper presents a case study on the City of Manchester Stadium, which is the first intelligent stadium in the world, based on relevant literatures. The authors aim to provide a comprehensive introduction about how collaboration and creativeness can work effectively in the design of modern intelligent sports buildings. This paper historically describes building and services technologies adopted in the City of Manchester Stadium, and provide valuable references to the design of intelligent stadiums in China.

参考文献

[1] Hewlett-Packard，Customer success：Manchester City Football Club. Hewlett-Packard Development Company，L. P. 2004

[2] MCFC，Stadium Info：Introduction. Manchester City Football Club. UK. 2004

[3] PanStadia，Fireproof Methods. *PanStadia International*：*Quarterly Report*，8(1)，81～050. PanStadia International，UK. 2001

[4] Plasmor，Countdown to the Games：A superb stadium for Manchester. Points New，Issue 31，Plasmor Limited，UK. 2001

[5] J. Harvey，City of Manchester Stadium completes transformation from athletics to football. Arup News. 23/08/2003

[6] K. Fletcher，Revitalised Victorian. Regional Report-Manchester. Building Services Journal：the magazine of CIBSE. July，2003

[7] D. Patel，et al.，Designing the City of Manchester Stadium. The Arup Journal. Issue 1，25～35，2003

木材加工车间环境监控网络的设计

吕兵伟　韩　宁　张军国　燕　飞
（北京林业大学工学院，北京　100083）

摘　要： 考虑到 LonWorks 控制网络技术的优越性，将其应用到木材加工车间环境监控系统中。在分析该系统的构成、功能以及工作原理基础上详细介绍了系统的设计方法。通过对木材加工车间现场各种环境污染源监测和相关处理设备的网络控制，实现了对车间噪声、粉尘、温度、湿度以及火灾的综合监测监控，有效改善了木材加工车间的环境状况，保证了车间环境指标符合相关环保标准。
关键词： LonWorks　OPC　木材加工　环境监控

1　引言

随着我国木材加工业的蓬勃发展，木材加工企业的污染问题日趋严峻，尤其是加工车间污染问题更为严重。如：高噪声、木质粉尘、有机挥发物、电磁辐射以及火灾等。虽然现阶段各企业也采取了相应解决措施，应用了各种污染监测和处理设备，但其应用效果很不明显。例如：木材加工车间作业位置噪声值远远超过我国《工业企业噪声卫生标准》规定的 85 分贝(A)噪声级；车间内产生大量的木屑粉尘，其浓度远超过威胁作业工人安全的最高允许浓度 10mg/m^3；木材加工车间内的温湿度难以平衡稳定在一定的范围内；木材加工车间火灾事件频繁发生，造成极大的财产损失。究其原因，一方面是我国木材加工车间采用的各种污染监测和处理设备技术水平相对落后；另一方面这些设备的运行都是单独分散的，缺乏科学有效的网络组织及智能管理。针对上述原因，本文应用先进的 LonWorks控制网络技术，把车间内各种污染监测和处理设备合理地组织起来，加之科学的管理监控，形成一个合理监控、有效管理的环境监控网络系统。

2　系统总体结构

木材加工车间环境监控网络系统正是本着对加工车间内不同地理位置上的噪声、粉尘、温度、湿度以及火灾的综合监测监控的目的，结合车间现有的相关设备，增加必需的设备，协调各现场设备的联动运行，保证整个系统能够正常工作。该系统按照功能层次划分可分为三个部分：现场控制级、数据交换级和过程管理级，其结构图如图 1 所示。

(1) 现场控制级：现场控制级是监控系统的基础，负责完成现场监控任务，它由连接到控制网络上的智能节点构成现场智能网络。智能节点主要用于接收和处理来自测量单元中传感器的现场数据，控制执行器操作和执行网络通信等任务。现场控制级中的测量单元和执行单元主要包括噪声传感器、粉尘浓度传感器、温度传感器、湿度传感器、火灾探测器等现场信号输入设备及变频器、电机、报警器等现场运行设备。这些输入输出设备的输入或输出作为智能节点的I/O点接入智能节点中。现场控制级又可细分为车间噪声监测模块，车间粉尘监控模块，车间温湿度监控模块和车间火灾监控模块。

(2) 数据交换级：数据交换级完成过程管理级与现场控制级之间的数据交换功能。

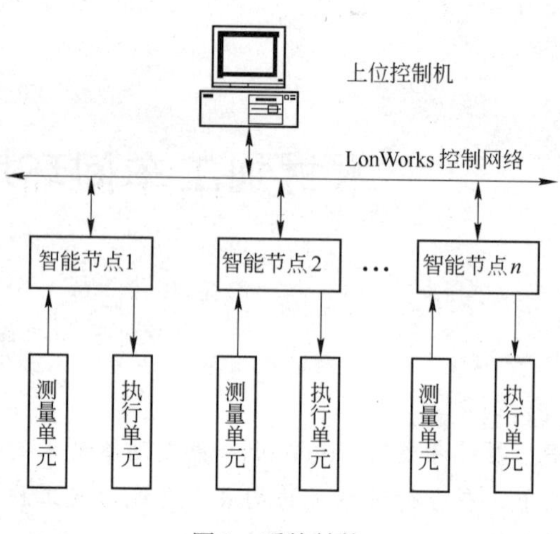

图1 系统结构

(3) 过程管理级：过程管理级作为系统的人机交互接口主要完成系统监控任务，主要功能为系统参数设置、智能节点状态检测、监控数据动态显示、数据分析及报警提示、网络通信管理、实时数据采集、数据打印等。

3 系统实现

系统通过对木材加工车间各种现场环境污染源监测和处理设备的网络控制，有效实现了对车间噪声、粉尘、温度、湿度以及火灾的综合监测监控，保证了车间环境指标符合相关环保标准。其系统实现的主要关键技术包括基于LonWorks控制网络技术的智能节点设计和基于OPC技术的数据交换技术。

3.1 基于LonWorks控制网络技术的智能节点设计

LonWorks(LON，Local Operating System，局部操作系统)是由美国Echelon公司20世纪90年代初推出的一种控制网络技术，是目前国际上控制领域中最热门的通用控制总线技术之一。它具有全数字化、分散、双向传输和多分支的特点。基于LonWorks控制网络技术的系统具有开放性和互操作性、分布式无中心控制、系统组态灵活、开发周期短、应用介质范围广泛等诸多优点。现在LonWorks技术应用范围非常广泛，如灌溉管理、分散和过程控制、电梯控制、能源管理、环境监视、火灾报警、智能建筑等领域，可以说，LonWorks技术已经走进社会生活的各个领域。

LonWorks技术的核心是神经元芯片(Neuron Chip)。神经元芯片主要有3120和3150两大系列。本系统中智能节点的设计采用了3150神经元芯片，结构框图如图2所示。

节点设计过程中，应用Echelon公司提供的NodeBuilder 3节点开发工具进行节点程序的设计与调试。各个智能节点所承担的任务、所需共享的数据及其相互作用用通过神经元芯片(Neuron Chip)专用编程语言Neuron C来编写。

本系统中由LonWorks智能节点负责现场数据的采集和现场设备的控制，应用Echelon提供的网络集成工具LonMaker对系统的LonWorks网络进行集成和维护。系统节点具体

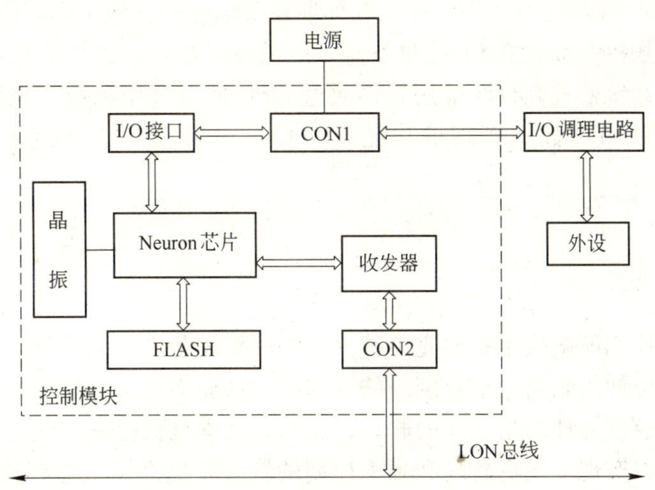

图2　基于3150控制模块的节点结构

功能实现如下：

(1) 车间噪声监测：由现场噪声传感器采集到车间现场噪声信号传到智能节点中，经过数值转化，以噪声值的形式供主控机调用显示。做到现场噪声的实时监测和显示，这些噪声数据可以一直保留作为车间降噪的参考。

(2) 车间粉尘监控：由现场粉尘传感器采集到车间现场木粉尘浓度信号传送到智能节点中，智能节点作出判断，发出控制信号，控制现场执行器—除尘器电机启停除尘器运行，以达到车间除尘效果。

(3) 车间温湿度监控：由现场温度和湿度传感器采集到现场温度和湿度信号传送到智能节点，智能节点发出控制信号控制执行器启停车间通风设备、加湿设备，以达到车间调温、调湿的效果，保证车间的温湿度稳定。

(4) 车间火灾监控：车间中固定堆积木屑的位置安装温度传感器专门测量木屑堆的温度，以防止木屑自燃。该温度传感器将采集到的木屑堆温度传送到智能节点，智能节点作出判断，发出控制信号启动防木屑自燃报警信号，提醒车间作业人员及时清理木屑；车间内的点式差温火灾探测器、红外式火焰探测器和CO气体火灾探测器将采集到的相关现场信号传送到智能节点，智能节点发出控制信号，控制现场火灾声光报警设备和车间水灭火设备的运行，以实现车间防火功能。

3.2　基于OPC数据交换技术实现的LonWorks控制网络与上位机之间的通信

OPC(OLE For Process Control，即用于过程控制的OLE)，是一种基于OLE的通信标准。OPC重点解决应用软件与过程控制设备之间的数据的读取和写入的标准化及数据传输等功能。OPC规范采用客户/服务器模型(如图3所示)，将数据采集端视为OPC服务器，其他的访问端视为OPC客户。OPC规范提供了服务器与客户之间进行数据通信和交换的通用标准，它是解决应用程序和设备间的通信的理想手段。它作为一个现场总线与计算机之间的信息传输通道，使得现场设备与控制系统的连接更为简

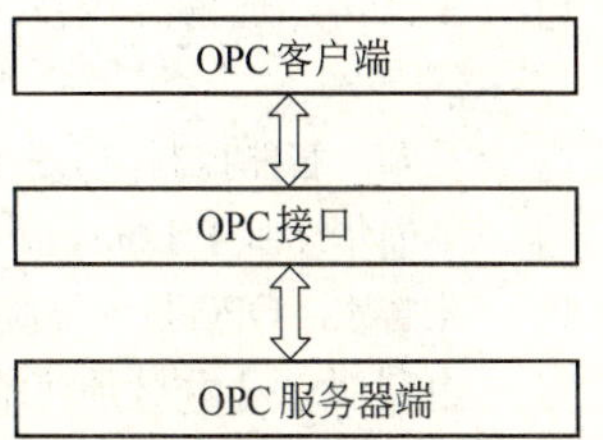

图3　OPC客户、服务器模型

单、方便和灵活，OPC 技术作为一项工业标准在我国已经得到了广泛的关注。

本系统在现场控制级建立 OPC 服务器，提供现场数据；过程管理级即为 OPC 客户，过程管理级中的系统监控软件等应用软件通过 OPC 接口调用和修改现场数据，完成数据存取、系统报警、历史数据管理等功能。OPC 接口采用 OPC 规范提供的自动化标准接口。

4 结束语

本系统采用控制网络技术将原先单独分散的测控设备转变成为网络节点，把它们连接成网络系统后便可加入到企业信息网络的行列中，从而沟通了生产过程的现场设备之间及它们与更高层的控制管理网络之间的联系；还使其基本过程控制、报警、计算功能等完全分布到现场完成，实现了现场管理和集中控制的统一，提高了信号传输的可靠性和精度，也大大的简化了系统结构。系统中 OPC 技术的采用使得系统管理级与现场设备间的数据交换更加透明、快捷、高效。该系统的实施运用可以有效改善木材加工车间的环境状况，有着不可低估的经济效益和社会效益。

Design of the Environmental Monitoring Network in the Wood Processing Workshop

Lü Bingwei Han Ning Zhang Junguo Yan Fei

(School of Technology，Beijing Forestry University，Beijing 100083 China)

Abstract：According to the advantage of the LonWorks technology，its first application in wood processing workshop environmental monitoring system is given. After the structure，the functions and the operating principle of this system are introduced，its detailed design way is proposed. Through monitoring noise，dust，temperature，humidity and fire in workshop and controlling the related equipments a synthetical monitoring and controlling system is implemented. It greatly improves environmental condition of the wood processing workshop.

参考文献

[1] 王江涛. 现场总线技术在立体停车库控制系统中的应用研究. 中国仪器仪表，2004，(2)：24～26

[2] 马莉. 智能控制与 Lon 网络开发技术. 北京航空航天大学出版社，2003

[3] 英向华. 基于 Lonworks 总线的监控系统设计及应用. 微处理机，2001，11(4)：31～33

[4] 张智杰. OPC 技术与应用. 信息技术，2002，(3)：29～31

[5] 申利明. 对木材加工企业劳动安全卫生状况的实证研究. 南京林业大学学报，1999，(4)：37～42

[6] 林秀兰. 林产工业污染及防治. 厦门大学出版社，2001

GREEN INTELLIGENT BUILDINGS: AN ENERGY FOCUS

Derek Clements-Croome
(School of Construction Management & Engineering, The University of Reading UK.)

Abstract: Buildings and their occupants generate large energy and water consumptions with consequential pollution including waste. There is a global awareness that CO_2 levels need to be drastically reduced. In order to do this, the design and management processes need an integrated approach which emphasize value in terms of the environment, social needs and whole life performance. Changes in society as well as climate patterns offer many challenges. Advances in technology will affect systems, products and processes offering opportunities to succeed in responding to these challenges but always remembering that vernacular architecture demonstrates many effective yet simple methods of environmental control in various climates of the world. (Clements-Croome, 2004)

Times between innovations have become shorter. New thoughts, ideas, technology and progressive business models change society. Barriers to innovation are reducing; global outlooks are commonplace; rapid rates of change demand sharp perception; the need for depth and breath of knowledge stimulates cross-disciplinary collaboration. All these factors affect the way we design, manage and care for buildings.

There is the challenge of making the world a more sustainable place whilst also meeting rising expectations of the processes, products and systems we use. Sustainability challenges our thinking and acceptance of materialism, quality of life, lifestyle and individual responsibility.

Key words: renewable energy resources, sustainable energy design, energy efficiency and effectiveness

Introduction

In a recent strategic planning initiative by the Airconditioning and Refrigeration Technology Institute in US the principal issues believed to be important were:

- Advancing technologies in many related fields;
- Changes in energy availability and cost;
- Embedded industry practices in architecture, construction, and refrigeration;
- Expanding government mandates and requirements;
- Lifestyle changes;

- Impact of international initiatives;
- Increased awareness of Indoor Environment Quality(IEQ) and its link to health and productivity;
- Public pressures to improve the environment;
- Shortage of skilled installation and service people.

Some of the most important future trends and drivers were given as:

- Environment concerns and regulations;
- Expanding global food supply chains;
- Growing demand for energy efficiency and source diversity;
- Impact of globalization;
- Increasing customer expectations;
- Increasing integration of building technologies;
- Increasing requirements for indoor environmental control;
- Shortage of qualified labour;
- Urbanization.

Buildings are composed of several layers each with sets of components. The *Shell* is the structure which has a lifetime of about 50 years bounded by the *Skin* with a shorter lifetime; the *Services* are all systems covering information technology systems, power, air, water and transportation having a lifetime ranging from 3 to 20 years; *Scenery* is the layout of partitions including false ceilings which change about every 5～7 years; *Set* is the furniture and contents which can be changed every few weeks or months even. Brand (1995) interprets these factors and recasts them in terms of six S's but here a seventh S representing the *Senses* of the occupant has been added:

- *Site*-the geographical setting;
- *Structure*-the foundation loadbearing elements;
- *Skin*-exterior surfaces;
- *Services*-operational systems of a building, including information technology and communications systems;
- *Space plan*-the interior layout including the *Scenery*;
- *Stuff* or *Set*-the contents including furniture;
- *Senses*-the human being.

There are two other S's that one could debate and that is *Spirit* and *Soul*. Certainly buildings should be a sensory experience and the quality of that experience will depend on the interaction between the individual and the building together with the environment that it creates. Building, systems and people should be interactive. Building services within the architecture are the root of this experience.

Buildings have evolved throughout history and many lessons can be learnt by studying vernacular architecture (Oliver 2003). See for example the book by Lim Jee Yuan *The Malay House* (Pulau Pinang; Institute Masyarakat 1987) who describes how it creates a near perfect solution of control of climate, multi-functional use of space, flexibility in design and a sophisticated pre-fabricated system which can extend the house with growing needs of the family. There are many examples world-wide of how people have ingeniously designed buildings to respond to climate using natural passive environmental design. Some new buildings imitate the best of mature buildings; they increase in sophistication while retaining simplicity.

Lessons from Nature

Termites, tardigrades, beavers, spittle bugs, sharks, spiders and many other insects demonstrate how Nature applies certain underlying principles in a spectacular variety of imaginative forms which:

- Economise the use of materials;
- Maximise structural strength;
- Maximise the enclosed volume;
- Produces extremely high strength-to-weight ratios;
- Utilise stress and strain as a basis for structural efficiency;
- Create energy efficiency through form without external power;
- Create form that enhances air circulation;
- Use local materials for building;
- Use curvilinear forms that disperse and dissipate multi-directional forces;
- Integrate aerodynamic efficiency with structural form;
- Produce nothing that is toxic to the environment;
- Design structures that can be built by a single organism.

There are now active research programmes in biomimetics which aims to see how we can learn from Nature.

Integrated Services Structures

Today new buildings or the refurbishment of existing buildings explore the attributes of microzoning and personal control systems. The idea is to use central distribution systems with individual user responsive nodes. The quality of the environment can affect work performance, and there is a need to take this into account when bringing in new technology for the purpose of improving the performance of business organisations. Loftness(1997) lists seven basic infrastructures to achieve workplace needs. They are:

- Fresh air and temperature control;
- Lighting control;
- Daylight and view, including reduced direct solar gain;
- Privacy and working in quiet conditions;

- Network access;
- Multiple data, power and voice connected systems;
- Ergonomic furniture with environmentally sensitive finishes.

Grids provide a distribution system to handle the flow of supply air(or water or power)and provide connecting points to the appropriate equipment, whilst the end points, or nodes, provide flexibility in location, density and type of service.

Building services systems are designed to provide environmental control, mobility, communications, facilities, fire protection and security. It is the balance between building contents, the organisation and services that determines if the value objectives are achieved. Ultimately people, not machines, are the vital ingredient of the system, and this applies to management and the organisation, as well as to the interaction between people and their total environment.

The different climates throughout the world have affected our thinking in terms of designing the building shell. Clothing is the first protective layer for the human being, and usually there is a large volume of air between that and the building fabric. On the other hand, it is the building fabric that receives the first impact from the rigours of weather, so it is sensible to investigate the possibility of using the building structures as a form of climatic moderator. The ultimate intelligent building has an environmentally responsive envelope that can control the transmission of heat, light and sound according to the needs of the occupants. There are examples of this already in Sweden, where the air vent window and the hollow block floor have been dominant over the last thirty years, and in Germany, where the Gartners ventilated form of construction has been used for several years.

A family of dynamic building facades is emerging in which materials and building services are uniting in smart or intelligent building skins, and thereby energy-intelligent, self-aware buildings. This opens up opportunities for the use of alternative energy sources such as the sun and wind. Photo and electro-chemical variability of surface properties also offers a form of energy control and, together with modern information sensing and control devices, makes the envelope of the building self-regulating.

These horizons present an exciting future for architecture and building services. We should not be apologetic for exposing buildings services where that is appropriate. But one can also see a vista where services as we know them, like radiators, convectors and other heating devices or airconditioning units, will not be necessary in the working space.

There is a basic need to admit daylight but reduce the heat and glare from solar radiation. Laar and Grimme (2002) give a review of daylight guidance systems including various Venetian blind developments; controllable louver systems; holographic optical elements; prismatic systems; micro-grates; plexiglass daylight for redirecting low and high altitude sunlight; transparent insulation; the light sculptor as used in the cupola of the Reichstag in Berlin which admits reflected daylight into the chamber whilst having a moveable shield to prevent solar radiation transmission.

Impacts of Global Warming

In 1972 the fuel crisis started off an ever increasing interest in energy and its consumption in buildings. New office buildings were sealed tightly with tinted windows or were double and triple glazed; tighter control was enforced by building management over lighting and ventilation and air temperature. Money was saved, public credit was taken for energy conservation and people became subject to building sickness syndrome due to being sealed in with chemicals outgassing from the carpets and building materials which is still common. Now this is not enough, water, waste management and pollution are equally important considerations as energy in terms of sustainability.

Over the 20th Century the energy needs of a person taken as a global average increased by a factor of four. As life expectancies increase, populations increase, material wealth increases; all of these things put a very high demand on the eco-system of the earth. Most people live and work in buildings and these already use about half of the world's energy. Edwards(2002) and others point out the following factors:

- *materials*: 50% of all resources (sand, gravel, clay, iron ore, wood) globally go into construction;
- *energy*: 45% of energy is used to heat, light and ventilate buildings and 5% to construct them;
- *water*: 40% of water used globally is for sanitation and other uses in buildings; 16% of water is consumed in construction;
- *land*: 60% of prime agricultural land lost to farming is used for building purposes;
- *timber*: 70% of timber products end up in building construction.

Globally it has been estimated that as many buildings will be designed and constructed over the next 50 years as have been over the last 5000 years so the need for a sustainable approach to architecture is urgent and has to be robust.

The world projected carbon dioxide emissions are shown in the following table and broadly follow expected increases in energy consumption.

World Carbon Dioxide Emissions 2001～2025 **Table 1**

Region	Carbon Dioxide emissions (million tonnes)		
	2001	2010	2025
Industrialised Countries	3179	3572	4346
East Europe/Russian Federation	856	1038	1267
Developing Countries	2487	3075	4749
Total	6522	7685	10361

(Energy Information Administration, US at www. eia. doe. gov/iea)

The biggest rate of growth is in developing countries particularly Asia.

The greenhouse gases in order of importance are shown in the following table.

Greenhouse Gases(DETR 2000a; Fawcett 2002) **Table 2**

Gas	Greenhouse Gas Emissions (%)	Key sources
Carbon dioxide (CO_2)	84	Fossil fuel energy used (households, commerce industry, transport, power stations), land use change
Methane (CH_4)	8	Agriculture, waste, coal mining, Natural gas distribution
Nitrous oxide (N_2O)	7	Agriculture, industrial processes, fuel combustion
Hydrofluorocarbons (HFCs)	1	Refrigerants, general aerosols, solvent cleaning, firefighting
Perfluorocarbons (PFCs)	0.1	Electronics, refrigeration/air conditioning
Sulphur hexafluoride (SF_6)	0.2	Electrical insulation, magnesium smelting, electronics, training shoes

The Intergovernmental Panel on Climate Change (IPCC 2001) states that human influences(anthropogenic effects) will continue to change atmospheric conditions throughout the twenty first century. Increases in global mean temperatures are expected in the range 1.4~5.8℃ by the year 2100. Water levels are anticipated to rise by a further 0.09~0.88m by the year 2100; half of this is due to thermal expansion and the remainder due to melting of glaciers and the Greenland ice sheet. There are a number of other impacts such as increases in heat waves, flooding, coastal erosion, and malarial disease; increases in agricultural outputs will be affected in some countries. Colls(2002) discusses the impact of these effects in the UK and also in developing countries.

Designing for Sustainability

How can green buildings be assessed? There are a number of systems being developed in various countries. The real difficulty is data availability and also there are not enough benchmarks against which to judge whether a particular building is successful in achieving a good green building rating. Life cycle assessment(LCA) methods attempt to assess the environmental performance of systems and constructions as well all the elements of which they comprise. The difficulty here is that often the performance of an element in its own right often changes when it is linked to a system as a whole. Understanding system behaviour is paramount. The basic methodology for LCA involves defining the goal and scope, inventory analysis, impact assessment and interpretation as set out in BS EN ISO 14040: 1997. Another difficulty is tracking all the flows in and out of a particular system but in general the objective function is to *minimise life cycle flows from and to Nature*(Trusty and Horst 2003).

There are now several assessment tools which can be used to aid the achievement of good design. One example is the SPeAR Investment Scheme devised by Arups. The *environmental* factors which are important to consider include low energy and water consumption, production and utilisation of waste, and a reduction in environmental pollution. All of this can be achieved by using a *whole life value* approach which results in reduced life-cycle costs for the owners due to greater systems reliability and reduced use of resources. The *social* gains are a better quality and healthier environment inside and outside of buildings, and the contribution made towards helping the global problem of sustainability by accommodating green issues into ones lifestyle.

The Construction Industry Council (www. dqi. org. uk) with Gann et al (2003) have developed a *Design Quality Indicator*, not dissimilar to the Building Quality Assessment described by Williams (2000), which attempts to assess and measure the value of a completed building using a generic questionnaire comprising over 90 statements covering:

- *Build Quality*: engineering performance; durability; integration of systems; construction.
- *Functionality*: design of spaces for working in; use; access.
- *Impact*: effect of building on environs; does building create a sense of place, form and materials; internal environment; innovation.

The design quality indicator(DQI)is a toolkit that has been developed to improve the design of buildings by providing feedback and capturing perceptions of design quality embodied in buildings(Gann et al., 2003). It is composed of three main elements, the *conceptual framework* focuses on function, build quality and impact; a *gathering tool* which uses a questionnaire to collect general data about function, impact and build quality; and a *weighting mechanism* which was developed through the use of a simple multi-criteria assessment algorithm. Gann et al., (2003)is continuing to develop the DQI by a series of case study evaluations with the aim of seeing the DQI maybe used in a more extensive way as a tool for thinking during the design.

The Through Life Environmental Business Model (TLEBM) is a representation of sustainable(green) issues and whole life performance of a building (Clements-Croome, Jones, John and Loy 2003). The TLEBM is the process whereby the technological issues and economic issues of whole life performance about a building are planned. The TLEBM framework addresses the processes and sustainable issues using an integrated logistic analysis approach. There are three layers to this model. The first is the connectivity of the supply chain processes and the interrelated working conditions of the practitioners' of the industry; the second addresses the sustainable aspect that embraces the economic, social and environmental concerns within each phase of the building life cycle and the third is the value associated to the project on the minimisation of the impact or risk assessment to these strands.

Edwards(2002)summarises the aspirations of sustainable development by emphasising the three 'E's *energy*, *environment and ecology* and the four 'R's *reduce*, *reuse*, *recycle and recover*. These should feature in any green policy for companies and building developers and must embrace operational and embodied energy; waste; water; transport energy and a respect for biodiversity. They represent principles for society as a whole.

Carbon Emissions

CIBSE(2002)compares the carbon emissions of naturally ventilated and airconditioned offices based on the Energy Efficient Best Practice Programme ECON 19 of 1998 issued by BRECSU. The results in the following table show that a good natural ventilation scheme can save over 40% in carbon emissions compared with good airconditioning but there is a big difference between good and typical emissions. This means typically designs can be much improved. Note that 1kg CO_2 emission is equivalent to 0.273kg carbon.

Relative Carbon Emissions(CIBSE 2002)**Life Cycle Energy** **Table 3**

System Basis		Annual Carbon Emission(kg/m^2)
CIBSE(2002)		13
Natural Ventilation	—good	12
	—typical	20
Airconditioning	—good	20
	—typical	37

In February 2003 a White Paper entitled *Our Energy Future: Creating a Low Carbon Economy*(*Cm*5761) announced that the UK is committed to cutting greenhouse gases by 60% by 2050. It is hoped that the carbon dioxide emissions will be reduced by 25million tonnes by 2020, by improving the energy efficiency in homes and industry, by improvements in transport, and from carbon emissions trading between businesses. It is intended to open up the market for hydrogen fuel cells by 2010.

Energy Consumption

A major aspect of sustainability is *energy consumption*. Carbon dioxide emissions from fossil-fuel power stations are high, whereas those from renewable resources such as solar power, wind power and biomass are very low. Nuclear power also offers low emissions but public perception about waste disposal remains cautious. The economics of solar photovoltaic systems are becoming increasingly attractive; by 2010 they will be a viable alternative in many situations. The total windpower capacity in the UK is 6.5GW onshore and 11.4GW offshore: at present we are using under 10% of this. With the advent of the hydrogen fuel cell there is the opportunity for electricity purchasers to become producers. A fuel cell relies on hydrogen and oxygen to be fed into it in order to produce electricity and pure water; there is no emission of carbon dioxide. The hydrogen can be provided from natural gas which is already provided to many homes. In the future it is expected that hydrogen will be generated from renewables such as solar energy. This form of combined heat and power offers the possibility of 2 million homes being connected by 2010, effectively making them mini-powerstations(DEFRA 2001). It is likely that fuel cells energised by centralised hydrogen production plants will become common after 2020.

Globally the average energy consumption is 2kW per head but there is a wide variation between users. The US is the highest consumer with an average of 10kW per head then Western Europe with 5kW per head whereas India is in the order of 0.25 kW per head. Colls(2002)gives the following figures for global energy consumption in 1998.

Global Consumption of Energy in 1998(Colls 2002) **Table 4**

Energy Source	Consumption(Mtoe *)	Consumption(%)
Oil	3500	41.1
Natural gas	1900	22.4
Coal	2100	24.7
Nuclear	800	9.4
Hydro-electric	200	2.4
Total	8500	100

Note: * Mtoe = Million tonnes oil equivalent. All the different fuel consumptions are translated into a common currency by expressing them as the weight of oil that would contain the same energy. 1 Mtoe is 42 PJ.

The scale of the problem can be appreciated when one realises that in 1860 the world consumption was 500 million toe and this has risen in over 140 years to nearly 9 billion toe.

Energy

The Energy Conservation in Buildings and Community Systems Programme (www.ecbcs.org)is operated by the International Energy Agency(www.iea.org). There are some 46 projects described as Annexes. Low Exergy Systems for Heating and Cooling of Buildings is Annex 37(www.lowex.net). Exergy is energy which is entirely convertible into other types of energy. It enables the potential, the quality as well as the quantity of energy to be calculated. Thus the exergy is high in power plants, moderate in heat pumps but low when dealing with air near room temperature(see ECBCS News June 2005, Issue 41, pages 4~9)

Renewable Energy

The control of the internal environment relies mainly on fossil fuels. This is not sustainable and more of the energy required for heating and ventilation will have to be derived from renewable sources.

Kirkwood(1998)discusses the impact of climatic change and the action that is required by Government and clients in understanding the significance for building design, construction and management. There will be a gradual reduction in the use of fossil fuels and an increase in the use of renewable energy resources. The following table shows the global potential for the various renewable energy resources.

Global Renewable Energy Potentials(Kirkwood 1998) **Table 5**

EJ thermal equivalent annual contributions(1 Exa Joule = 10^{18}J)			
Energy Source	1990	2025	Long term
Hydro-electricity *	21	35~55	>130
Geothermal	<1	4	>20
Wind	—	7~10	>130
Ocean	—	2	>20
Solar	—	16~22	>2.600
Biomass	55	72~137	>1.300
Total	76	130~230	>4.200

* Hydropower accounts for about 19% of the world electricity supply; largest producers are Canada, US and Brazil.

The sun provides 173000TW and at present we use 12.6TW compared with 2.5TW for other renewables. Today solar sources provide about 10% of the energy worldwide. Global turnover for *solar photovoltaics* is expected to grow from $1.2m per year to more than $8~$10m in 2012. The UK Government has pledged £23m over three years to step up PV activity and this should treble UK capacity to 6MW by 2005 although this is still lowercompared to levels of 114MW in Germany, 280MW in Japan(aim is 5000MW by 2010)and 12MW in Netherlands.

Worldwide *biomass* offers considerable scope for sustainable development. Current use is 55EJ. The energy potential is about 4500EJ(1EJ=10^{18}J); the annual bio-energy potential is 2900EJ but at present only 270EJ is considered available on a sustainable basis and at competitive prices. Yet progress with renewable resource has been slow. Low combustion efficiency and some health issues remain problems. There are a range of technologies from landfill gas to municipal waste combustion, sewage gas, agricultural residues and from specially grown willow and miscanthus grass gasified in new turbines to liquid bio-fuels. In Germany several thousand MW of capacity are about to benefit from an increasingly dynamic biomass market. In Finland it is common to see the co-firing of coal and woodchips, and biomass provides around 20% of overall energy needs. In Sweden the figure is 15% and is set to grow further. In Austria there are more than 500 district heating biomass systems and 1 million efficient domestic woodchip burners are used for heating. Wood fuels are estimated to cover nearly 6% of the world energy requirement. Denmark has supported tax and investment policies for biomass as well as electricity market rules which means that utilities have to take the power generated on a priority basis. This is in contrast to the UK where the market actually penalises more intermittent renewables such as wind. Biomass in the UK only serves under 1% of energy needs; in the US this figure is 1.6%

The total world *wind* capacity was about 17500 MW at the end of 2000. It is anticipated that wind will provide 15%~20% of Europe's electricity within 20 years. Germany with over 6000MW has the highest capacity but Denmark with over 2000MW has the highest level per capita. Spain and the US also have significant energy production from wind sources. Typical installation costs for onshore wind farms are about $ 1000 per kW and for offshore around $ 1600 per kW. The UK is located favourable for wind energy but only has about 10% of the wind capacity of Germany at the present time but in July 2003 the Government announced a substantial increase in off shore wind farm finance. This is partly due to the difficulties that developers have in gaining planning approval. The UK industry anticipates construction of about 11000MW wind power capacity. Noise, bird life, visual aspects and integration into supply networks are issues that need to be taken into account when planning wind farms.

The following table shows the predicted expansion in renewable energy in the EU.

A predicted expansion in renewable energy use in EU(Edwards 2002) **Table 6**

Type of Energy	1995	2010
Biomass	45Mtoe *	135Mtoe
Photovoltaics	0.03GW	3GW
Solar Collectors	6.5Mm^2	100Mm^2
Wind	2.5GW	40GW
Geothermal(Heatpumps)	1.3GW	5GW

* 1Mtoe=42GJ.

Energy Efficient Buildings

The following features are important:

- High levels of thermal insulation for the fabric; use values lower than those recommended in national Building Regulations, which are subject to change anyway, say U=0.1 to 0.2W/m^2℃.
- Natural ventilation should be used wherever possible but excessive infiltration avoided. Connected spaces and conservatories can help to promote air movement. In houses natural infiltration needs to be about 0.4 air changes per hour, or alternatively make the home as tight as possible but use trickle ventilators to ensure sufficient fresh air in Winter. Lack of fresh air can give rise to sick building syndrome.
- Utilise daylight and passive or active solar design but exclude excessive solar gain and glare.
- Use high efficiency lighting with dimming circuits.
- Use individual control measures which are effective.

- Use systems, equipment and appliances which are effective as well as efficient.
- Employ effective facilities management procedures, which includes helping building owners to save money and hence save resources.

Consider the use of bio-mass boilers, condensing boilers, combined heat and power (CHP) systems. Waste can be fed to energy plants operating CHP systems with a potential to generate electricity and provide community heating. CHP is flexible in terms of size and fuel source. It can be as low as 30kW for house upgrades and as high as 2000MW in industry.

Lighting

Daylight is very important for people inside building from a physiological and a psychological point of view. An overview of recent developments including daylight guiding Venetian blinds, mirror louvres, optical mirror elements, holographic optical elements, prismatic systems, transparent insulation, moveable glass louvres and 'light sculptor' systems is given by Laar and Grimme(2002). However artificial lighting will be necessary.

Select efficient lamps as shown put into the following table. There is an ECO labelling scheme for fluorescent lamps.

Typical Electric Lamp Efficacies(Rawlings 1999). **Table 7**

	Lumens per watt
Incandescent lamps	
Tungsten halogen	20
GLS(household lamps)	12
Reflector lamps	10
Fluorescent lamps	
T8 triphosphor	88
T8 halophosphate	80
Compact fluorescent	50
Discharge lamps	
Low pressure sodium	160
High pressure sodium	100
Metal halide	70

Remember that longer service life reduces consumption of resources. The colour, shape and material of the luminaire affect the distribution of light so white reflectors are 60% reflective but foil is 95% reflective. Highly reflective wall coverings indirectly make the space feel lighter. It is important to use appropriate control fittings for dimming or switching off lights.

Recent technological advances are in Light Emitting Diodes(LED). There are many benefits of using LED lighting systems especially as they have lower running costs than incandescent bulbs and they last around 100000 hours.

Solar Energy

To take advantage of solar radiation one needs to consider the form, fabric and orientation of the homes. Some guidance is given in the following table.

Rules of Thumb for Solar Design(Rawlings 1999). **Table 8**

Form	• create sun spaces, lighting ducts, light shelves
Orientation:	• main glazing to face 30 degrees either side of due south • reduce north glazing • minimise tree over-shadowing • on housing estates build to a density of≤40 properties/ha • design atriums/roof lighting in accordance with the position of the sun in both summer and winter
Fabric:	• fabric transmission losses may be reduced by improving insulation or by reducing the mean inside air temperature

External blinds are more efficient than internal blinds where it is possible to fit them.

- When selecting glazing considering the energy embodied in the frame as well as that which is saved by the glazing choice.
- Net curtains reduce solar gain by up to 20%.
- Effective solar water heating provides almost free hot water in terms of running costs during daylight hours; in some situations pumping will be required.

Photovoltaics

There are three main types of photovoltaic systems as shown in the table.

Properties of Common PV Cell Types(CIBSE 2000) **Table 9**

Property	Mono-crystalline Silicon	Poly-crystalline Silicon	Thin film amorphous silicon*
Cell efficiency(%)	13~17	12~15	5~10
Module efficiency(%)	12~15	11~14	4~7.5
Appearance	Blue/black, homogeneous†	Blue, multi-faceted	Grey, brown or black, matt

* For amorphous silicon lower values are for single junction(stablilised efficiencies), higher values are for multiple junction.

† The usual colour is blue/black but a range of other colours including bronze, magenta and light blue are available.

Photovoltaics arc still costly but the costs are reducing, currently it is likely that the installation of solar panels for domestic hot water are a more likely solution with the gradual introduction of PV. Solar panels collect the sun's energy as hot water which can be used directly in the heating system or as a heat source for a heat pump heating system. As insulation levels of dwellings have increased the use of solar energy has become more attractive.

Fuel Cells

Fuel cells work using an electro-chemical process whereby hydrogen combines with oxygen from the air to generate DC electricity, water and heat. Currently fuel cells cost about £2660 per kWe compared to about £700～1000 per kWe for conventional combined heat and power(CHP) equipment(BCO 2003). It is expected that eventually their production and consequent use will increase and will reduce the price to a more competitive level. Fuel cells can be used as a reliable form of on-site generation to replace the more conventional diesel generator or CHP unit. American Society of Heating, Refrigeration and Airconditioning Engineers(ASHRAE) Research Project 1058, explains that fuel cells systems have the potential to revolutionise the way power is generated and for building applications they offer modularity, high efficiency across a wide range of low conditions, minimal environmental impact and opportunities for integration into co-generation systems. *Proton exchange membrane* fuel cells are most commonly described but there are other types such as *alkaline* fuel cells but these require pure hydrogen and oxygen and are unlikely to be commercialised for buildings; *phosphoric acid* fuel cells have some potential for small power generation systems; *solid oxide* fuel cells are best suited for large scale power generators that could provide electricity for factories or towns; *molten carbonate* fuel cells are again more suitable for large generators. There are other types of fuel cells but these are unlikely to be used in buildings.

In the future it should be possible to buy a fuel cell for something approaching $1000 per kW supplied from the natural gas supply, to provide both heat(and hence replacing the boiler), and electricity(hence replacing the grid connection).

A useful summary of the characteristics of fuel cells is given by Galliers(2003) also Mott, Green and Wall(2003), together with Davis Langdon & Everest, of the characteristics of fuel cells to help designers.

Biomass

3 billion people worldwide rely on biomass fuels and coal for cooking and heating; biomass accounts for 80% of household energy consumption in developing countries. Environmental, economic and social benefits from the utilisation of modern biomass for energy purposes is rapidly becoming appreciated. By using biomass to displace fossil fuels then carbon dioxide emissions are avoided, biomass being carbon neutral due to the

recycling of the carbon via photo-synthesis. Use of woodchips derived from short rotation coppice and various waste materials save fossil use but can also act as a carbon sink. Other benefits from modern biomass use relate to improved quality of life, lower emissions of harmful substances compared to fossil fuel use, local employment opportunities and other benefits of sustainability resulting from land use change.

Water

Water will become a critical resource issue in many parts of the world by 2025. It is essential for public health, agriculture and industry. The world's water crisis is described in the *World Water Development Report* (WWDR 2003) produced by the United Nations. Water covers about 70% of the earth's surface but only 2.53% is freshwater and some two thirds of this is locked in glacial formations and permanent snow cover. The majority of water is salt water and it maybe that over time much more attention needs to be given to de-salination processes. The problem is also the distribution of water which is highly available in South America which holds 26% of the world's supply for 6% of the world's population; in contrast Asia has 60% of the world's population but only 36% of the water. The critical continents are Asia, Africa and Europe. Water related deaths are over 2 million per year. Irrigation at present accounts for 70% of all water withdrawals and this will increase by about 40% over the next thirty years. Likewise industry calls on 24% which will also increase by some 40%. The worst scenario is that by 2050 7 billion people in 60 countries will face water shortage; the best scenario envisaged is that 2 billion people in 48 countries will be affected.

Baths typically use 80 litres and showers 30 litres, although power showers can use as much water as baths. It is advisable to try and persuade people to take a shower rather than a bath on most occasions. Meters can encourage householders to use water more efficiently although in 2003 only about ten per cent of UK domestic consumers have a water meter. 25% of all domestic water consumed becomes waste water, 2% is used externally and 2% is lost within the household. Savings of up to 70% can be achieved by water efficient appliances. Rainwater and greywater use are described by Rawlings(1999) and Brewer et al(2001).

Waste and Recycling

In the UK construction site and demolition waste amount to 72 million tonnes per year; a similar amount of waste is generated by commerce and industry. If we include production construction waste rises to 122 million tonnes. The 1995 UK Government White Paper *Making Waste Work* included targets for increasing the use of waste and recycled materials as aggregates from about 30 to 55 million tonnes *per* year by 2006.

In waste management the priorities should be given to reducing the amount of waste, then reusing it, followed by recovery and finally disposal. At present it is increasing by 3% per year, 60% of domestic refuse might be recoverable and this is nationally worth £2bn per year(1997 figures). The UK Government has been committed to recycling or composting over 25% of household waste by the present time. In 1999 only 5%~7% of UK household waste was recycled compared with 30%~50% across Europe and the USA(Girardet 1999). Composting schemes can remove 200%~400kg per household per annum from the waste stream. Recyclates and organic composts have commercial value(Rawlings 1999).

The National Waste Management Plan for the Netherlands aims to ensure that 83% of waste substances must be applied to useful purposes by 2012. Recycling and the use of non-recyclables as fuel will be the main methods of achieving this target. In Japan the aim is to reduce landfill waste from 56 millions tonnes to 28 million tonnes by 2010 using recycling and better conservation. In Germany the average recycling composting rate is 45% but with a wide variation of 24%~65% across its 16 regions.

The Way Ahead

In his CIBSE Presidential Address entitled *Adapt or Die*? given on May 8, 2003 Terry Wyatt highlighted four major trends which were *globalisation*; *standardisation and the integration of systems*; *climate change and the emergence of the carbon management market*; *and rising concerns about health and wellbeing within buildings*. This paper has emphasised the importance of reducing the carbon emissions and all this depends on having efficient and effective design and management processes besides ensuring there is efficient feedback between buildings operating in practice and new and refurbished ones that are being planned. Off-site construction, development of leaner processes, as well as new aspects in the processes such as continuous commissioning and post-occupancy evaluation help to ensure that targets can be achieved and improved. Sustainable buildings which are well managed are also healthier buildings for people to work in, this results in higher productivity. The life cost ratios 1 : 10 : 200 show the relative values of design, construction, operation and the impact of these processes on people and hence business. Heerwagen(1998), quotes studies which show that sustainable design, construction and building operation can have far reaching impact on an organisation.

A further cutting edge project was reviewed in the Energy and Environmental Magazine (DEFRA May/June, 2003). The Beddington Zero Energy Development Housing Scheme in Sutton, (www.bedzed.org.uk) also Dunster(2003) features housing, community facilities, workshops as well as green space. The building materials are selected from natural renewable or recycled sources; a combined heat and power unit(135kW electric output)produces all the heat and electricity from tree waste(850 tonnes of woodchips per year); the houses are highly insulated($U = 0.1\text{W/m}^2$℃)and use argon filled triple glazed windows. Each house has an unheated passive solar sunspace which provides up to 30% of the typical homes space heating needs. There is a water strategy to reduce consumption by at least 30% using rain and recycled water. Every home uses recycling bins. As part of BedZED's zero carbon total energy strategy there is a green transport plan.

The much famed Eden project in Cornwall demonstrates how an enclosed botanical environment can be achieved to grow plants from many different regions of the world. There are two biomes, one covering humid tropics(over 330000m^3)and a warm temperate one(over 85500m^3) made from a transparent film material ethylene tetrafluroethylene co-polymer(ETFE)(Barrett et al., 2002). There were triple layers of ETFE giving a combined U value of 1.4W/m^2℃. Barrett et al., (2002)described how biomes work to create the environment favourable for plant growth by letting the sun shine on the enclosure which is positioned close to a south rock face which acts as a storage system. The light transmission of a single ETFE layer is about 95%. The surface of the biomes is made up of a series of triple layered ETFE air cushions, giving a light transmission of about 80% and a solar gain of about 90%. The fabric is self cleaning. The water conservation was a major design consideration and rain water is collected, stored and reused.

The intelligent building should be sustainable, healthy and responsive to the changing needs of individuals, business organisations and society. Globally there is a need to agree how we manage renewable energy exploration and innovation for short, medium and long-term use. Architecture, whilst using emerging technologies appropriately, needs to respect what history has taught us about passive environmental design examples which abound in vernacular architecture throughout the world.

绿色智能建筑——聚焦能源

Derek Clements-Croome

(School of Construction Management & Engineering, The University of Reading UK.)

摘　要: 建筑物和其中的居住者都消耗大量的能源和水，也因此产生了包括垃圾在内的污染。彻底地控制二氧化碳在大气中的含量现在已成为全球性的共识。为达到这一目的，设计和管理需要一个着重对环境影响、社会需求和整个生活性能进行衡量的整体步骤。社会的变化像气候一样带给我们很多挑战。技术上的进步将通过影响我们的体制、产品和生产工艺，来提供成功应对这些挑战的机遇；同时也在不断提醒我们，在全球多样的气候中，许多看似简单却有效的环境控制理论往往在本国的建筑项目中获得证实(Clements-Croome，2004)。

目前，创新步伐越来越快。新的思想、观念、科技以及创新型的商业范例改变着我们的社会。创新的阻碍正在减少；世界前景趋于正常；高速率的变化需要敏锐的理解力；对知识的深度和新鲜度的需求也加速了各学科的交叉融合。这些都影响着我们对建筑的设计、管理和维护的方法。

当我们在提升对所使用的生产工艺、产品和体制的期望值时，也面临着创造一个更加可持续的世界的挑战。可持续发展对我们的思维方式、对唯物主义的认知、生活质量、生活方式和个人责任等方面都提出了挑战。

References

[1] Arup, 2003, http://www.arup.com/communications/knowledge/intelligent.htm ASHRAE, 2003, *ASHRAE Handbook: HVAC Applications*, 39.14～39.17(ISBN 1-931 862-23-0).

[2] Barrett, C., et al., 2002, *Creating the Eden Environment*, The Arup Journal 1/2002, 3～12.

[3] BCO, 2003, *Fuel Cells for Offices?* A BCO Guide to the New Technology by Roberts and Partners (British Council for Offices).

[4] Brand, S., 1995, *How Buildings Learn*, (Penguin Books).

[5] Brewer, D., Brown, R., Stanfield, G., 2001, *Rainwater and Greywater in Buildings*, Technical Note TN7/2001(Building Services Research and Information Association).

[6] CIBSE, 2000, *Understanding Building Integrated Photovoltaics*, TM25(ISBN 1903287065)

[7] CIBSE, 2002, *HVAC Strategies for Well-Insulated Airtight Buildings*, TM29(ISBN 1 903287 189).

[8] Clements-Croome, D. J., 2004, *Intelligent Buildings: Design Management & Operation*, Thomas Telford, London(Available in Chinese from 2006)

[9] Clements-Croome, D. J., Jones, K. G., John, G., et al., 2003, *Through Life Environmental Business Modelling for Sustainable Architecture*, CIBSE Proceedings of Conference on Building Sustainability, Value and Profit, Edinburgh, September 24～26.

[10] Colls, J., 2002, *Air Pollution*(Spon Press).

[11] DEFRA, 2001, *Electricity's Super-Highway*, Energy and Environmental Management, September/October, Page 9(Department of Environment, Food and Rural Affairs).

[12] DETR, 2000a, *Climate Change: Draft UK Programme*, Department of the Environment Transport and Regions, the Stationery Office, London.

[13] Dunster, W., 2003, *BedZED: Beddington Zero-Fossil Energy Development* in Sustainable Urban Design by Randall Thomas(Spon).

[14] Edwards, B., 2002, *Rough Guide to Sustainability*(RIBA Publications).

[15] Fawcett, T., Hurst, A., Boardman, B., 2002, *Carbon UK*, Industrial Sustainable Development Group, Environmental Change Institute, University of Oxford. (ISBN 1-874 370-311).

[16] Galliers, S., 2003, *Fuel Cell Technology*, Building Services Research and Information Association, BG9/2003.

[17] Gann, D. M., et al., 2003, *Design Quality Indicator As a Tool for Thinking*, Building Research and Information, 31, 5, Sept～Oct, 318～333.

[18] Giradet, H., 1999, *Creating Sustainable Cities*(Green Books: Bristol).

[19] Heerwagen, J. H., 1998, *Productivity and Well-Being: What are the Links?* American Institute of Architects Conference on Highly Effective Facilities, Cincinnati, March 12～14.

[20] IEA, 2002, International Energy Agency Factsheet *Renewables in Global Energy Supply* November.

[21] IPCC, 2001, *Climate Change- Third Assessment Report*(Cambridge University Press).

[22] Kirkwood, R., 1998, *Solar Dawn: Challenging the Horizon*, CIBSE Presidential Address May, London.

[23] Laar, M., Grimme, F. W., 2002, *German Developments in Daylight Guidance Systems* Building Research and Information, 30, 4, 282~301.

[24] Loftness, V., et al., 1997, *The Intelligent Workplace Retrovif: Investing in the Hidden Infrastructures for Workplace Flexibility and Performance*, Proceedings of the Intelligent Buildings Congress, Tel-Aviv, Israel in New Methods and Technologies in Planning the Construction of Intelligent Buildings II, Edited by A. Lustig (IB/IC Intelligent Buildings Congress Steer Group Ltd., Israel). ISSN 0793-2448.

[25] Mott, Green & Wall et. al., 2003, Building Services Journal, September 16~18

[26] Oliver, P. D., 2003, *Dwellings: The Vernacular House World Wide* (Phaidon Press) also see *Encyclopaedia of the World Vernacular Architecture*.

[27] Rawlings, R., 1999, *Environmental Rules of Thumb*, Technical Note TN12/99 (Building Services Research and Information Association).

[28] Trusty, W. B., Horst, S., 2003, *Interpreting LCA Tools in Green Building Rating Systems*, wbtrusty@fox.nstn.ca and horst@early.com., ATHENA Sustainable Materials Institute in Ontario and Pasadena, also see The Austin Papers: Best of the 2002 International Green Building Conference, page 53 (ISBN 0-929884-10-9).

[29] Williams, B., 2000, *An Introduction to Benchmarking Facilities* (Building Economics Bureau Ltd.)

WWDR, 2003, World Water Development Report, *Water for People: Water for Life*, ISBN 92-3-103881-8 (Berghan Books UK), www.unesco.org/water/wwap

[30] Wyatt, T., 2003, *Adapt or Die?* Presidential Address for CIBSE, May 8th.

7 December 2005

INTELLIGENT BUILDING MANAGEMENT SOLUTION: MEETING ASPIRATIONS OF CLIENTS IN THE CHANGING WORLD

Lau Po Chi

(*Chairman of Activities & Events Committee, Asian Institute of Intelligent Buildings*)

Bonnie Lam

(*Member of Activities & Events Committee, Asian Institute of Intelligent Buildings*)

Abstract: This paper gives an overview of the paradigm changes faced by property management in recent years and the progress made in addressing these changes. The main focus will be on advances in technology, and life cycle property management. Modern approaches will be dealt with, such as analysis for sustainable development, and ways of incorporating scenario intelligent dimensions in design and management approaches. In the context of sustainable development, intelligent buildings and management, architectural and engineering ambience, multiple use of intelligent concepts in the areas of design, operations and management will also be discussed.

Key Words: intelligent building, property management, architectural, engineering, life cycle

Introduction

Property management has a definite life cycle that generally falls into three distinct but linked stages. They are design and development stage, hand-over stage for on-going management, and the operations and management stage. The operations and management stage is the most **important** and **longest** stage in the management cycle. It begins with the first business as resident/tenant taking up occupation. A great of activity take place during this important and longest stage. The focus of activity has changed over time as both occupation and building became matured.

The authors believe that property management, like any other business management of any business, is a management behaviour focusing on technology, people and environment. By incorporating advanced technology and environment, intelligent property management maintains the balance between people and environment in which they are living and/or working more comfortable, safe and efficient.

Changing Face

Traditionally, property manager once considers caretaker who made sure that the electricity was on, that roof did not leak, the area was clean and the rent was collected, today faces new responsibility and challenges. While still accountable for the built-asset, property manager has now to perform different tasks from multiple uses of advanced technology and "smart" building materials, legal to financial. Their lines of responsibility have been substantially changed. The scope of responsibility include: arranging insurances; designing and processing renovation/refurbishment bids; drawing up annual plan/budgets, hiring, managing, and motivating staff; coordinating repairs and maintenance services; improving landscaping; overseeing property security; compliance with government regulations including Building Management Ordinance and Deed of Mutual Covenants, reporting income and expenses; attending meetings of various committees and liaising with various government departments and other related bodies. To accomplish these tasks, property manager of today must evaluate and respond to increased expectations, changing market conditions and world events, be aware of new developments in technologies, professional practices(best practices), and find additional ways to enhance the value of built-assets. At all times, they must maintain client relationship and provide excellent customer services that are so integral to their business management. In recent years, and especially in the past few years, numerous external environments and factors have affected how property manager works. As government regulations/law, insurance requirements, environmental concerns, and outbreaks of diseases, health and safety, competitions, political and social issues, property management is required to specialize in order to cope with the complexities and challenges.

The concept of life cycle management

The concept of life cycle management underlines the importance of early involvement and participation by property manager from day one of the inception of the development from stage one throughout to the final stage of the "end-product" for on-going management. The work includes feasibility study, design and constructions, safety requirements and commissioning of procedure manual for best practices and bench marking for on-going management of the development. It plays a key role in prolonging the life of all facilities of the development and lowering the life cycle cost management, as well as enhancing its market value. The most challenging jobs of the on-going management are to maintain the property at top notch conditions at all times(as if it was just completed yesterday)and sustain its life span to the full extent of its designed life. In Hong Kong many buildings, like Two International Finance Centre, the designed life is up to a life span of 120 years. The new paradigm of management has to meet the challenges of the sustainability in the development in tiptop conditions at all times throughout its life.

Technological Advances

Numerous advances in technology such as building automation system and control are helping property managers to their jobs more effectively. As in most industries, computerization has provided some relief regarding data collection, manipulation, analysis, and storage as well as new and better means of communications. Many organizations including Hong Kong Housing Department, Hong Kong Housing Society, MTRC, KCRC of the HKSAR have developed terminal server platforms that allow their personnel to connect company networks over the internet and run applications in web browsers.

For security, policing patrols, and maintenance, repairs inspections, electronic patrol managers and palms are provided to facilitate the management staff to carry out their day-to-day operations and management more efficiently and effectively.

The following sections continue with the identifications and discussion of intelligent building dimensions combined with innovative and technological advances, operations and management.

Management of Change to the Buildings of Increasing Intelligence

The term "intelligent building" (IB), originated in early 1980s in the United States, is used to denote buildings with sophisticated telecommunications, property management and data networking services that provided shared tenant services(STS) to the occupants. [1]

In mid-80s, a developer in North America even described IB as "a building that is fully leased". According to this interpretation, any feature that helped the building to be fully leased out can be considered as intelligent. The property managers of the intelligent buildings at this time might have become a semi-marketer so as to help the building to be more "intelligent".

However, IB has finally described as "an intelligent building combines innovations, technological or not, with skillful management, to maximize return on investment." In an International Symposium on the Intelligent Building in Toronto in May 1985. The Intelligent Building Institute at that period has proposed the IB should be defined as "one

[1] Harrison, A., Loe E. & Read, J. (1998). *Intelligent Building in South East Asia*. E & FN Spon(London) Publisher.

that provides a productive and cost-effective environment through optimization of its four basic elements-structure, systems, services-and management and the interrelationships between them. IBs help business owners, property managers and occupants to realize their goals in the areas of cost, comfort, convenience, safety, long-term flexibility and marketability." [1]The role of the property managers in such developments has therefore gradually been changed from the marketing oriented to more user focus and long-term sustainability.

High Technology image in the IBs

Regardless of the different packaged economic trends in these years, the name IB is still holding a high technology image to most of the public. In the United States, IB is first addressed in four main areas:

- Energy efficiency;
- Life-safety systems;
- Telecommunications systems;
- Workplace automation.

In fact, the ultimate dream in the design of an intelligent building at the time it came to the market is to integrate the above four operation areas into one single computerized system. Over. time, these four areas have been gradually merged into two broader categories: property management(energy and life-safety)and information systems(telecommunications and workplace automation). Property managers also have to update themselves in these computing systems from time to time.

In 1991～1992, DEGW and IT consultants Teknibank undertook a major research project to assess the status of intelligent buildings in Europe. The project, Intelligent Buildings in Europe (IBE), defined IB as any building that "provides a responsive, effective and supportive intelligent environment within which the organization can achieve its business objectives." The model states that the three main goals of an organization occupied a building are property management, space management and business management. From the time on, the knowledge of interactive management has begun to be involved into the management of IB.

Besides, effective space management has become crucial in this "management of change" and the minimization of operating costs. As well, business management is another management of the organization's core business activities. In most cases, this can be characterized as a combination of the processing, storage, presentation and communication

[1] Donald A. Coggan, PE(1996). *Smart Buildings*. Ingenieur-Engineer.

of information with the assistance of the property managers. [1]

Management of Intelligent Facilities

As intelligent building is one of the very marketable products in the property field today, different kinds of "intelligent packages" have been introduced to the market in these years. Although some developments with broadband provided might have also once claimed themselves as intelligent buildings, there are in fact no commonly accepted facilities for the IBs. Anyhow, it would still be more convincing to the users if some basic facilities would not be neglected in the IBs.

Intelligent Building Index(IBI)

In 2001, Asian Institute of Intelligent Buildings (AIIB) introduced the Intelligent Building Index (IBI) to the industry. The IBI (10 modules) provides a more holistic approach to building design and management. It covers a much broader different views of issues in buildings. These indexes were developed from extensive surveys and feedbacks from different groups of people. The weighting of each index will vary with different types of building under the IB ten modules as below:

1. Environmental friendliness-Green Building;
2. Space utilizations and flexibility;
3. Home comfort;
4. Working efficiency;
5. Culture;
6. Image;
7. Earthquakes, disaster and structure damages;
8. Construction process and structure;
9. Life Cycle Cost-Operations and maintenance with emphasis on effectiveness;
10. Health and sanitations.

Since the inception of the IBI in early 2001, it is seen that there is an increasing demands for the introduction of a reorganized intelligent home standard. AIIB has conducted a research with a view to introducing recognized standard for intelligent home to meet the need of the industry. AIIB believes a home must provide a safe, convenient, healthy and efficient environment for users to live, relax enjoy and work. The recommended standard for intelligent homes entitled "Intelligent Home Index" (IHI) will have 10 performances.

---

[1] Harrison, A., Loe E. & Read, J. (1998). *Intelligent Building in South East Asia*. E & FN Spon(London) Publisher.

1. Safety and security.
2. Environmental friendliness.
3. Efficiency.
4. Cost effectiveness.
5. Convenience.
6. House keeping.
7. Comfort.
8. Culture.
9. Availability of innovativeness facilities.
10. Activities recommendations.

With the rapid advancement in technology coupled with AIIB concepts of intelligence increasingly more "intelligence" can be incorporated in buildings both commercial and residential nowadays. The final decision of whether to adopt intelligent building concepts and the extent of the "intelligence" to be incorporated, would be dependent on the cost and benefit analysis and many other factors, such as the case to operate and maintain the system, the availability of spare parts due to fast changing technology, the type of system in use and the amount of investment involves.

Generally Accepted Features in an IB

1. Communication facilities:

a. Video recording:

Besides the traditional Closed Circuit Television(CCTV)that is solely monitored by the management staff, the tenant can also have the access to get into certain part of the system(say swimming pool and entrance area)and allow them to monitor or identify their kids and visitors with the help of their home computers.

b. Telecommunication system:

For quick and easy security management, the tenant can have their telephone lines directly linked-up to the management office, the alarm company, the hospital and clinics, or even the police station and fire station with some prior settings.

2. Security and alarm system:

- Infrared sensor:

  It is used to help or to alarm the tenants and the management companies if any strangers have invaded into the restricted area or to get into the office without permission after the normal office hour.

- Monitoring alarm system:

  These systems usually included temperature sensor, smoke detector, gas detector, washroom floor water level sensor etc.

- Emergency cut-off system:

Examples of such systems included alarms, automatic gas and water supply cut off settings etc.

- Management monitoring control system:

  This system indicates or displays the exact location or facilities that required attention.

- Alarm Zoning System:

  The tenants or the property managers can activate their alarm system zone by zone according to the individual or situational requirements.

- System self-monitoring alarm:

  To minimize the system run-down-rate, it can automatically do the self-monitoring or self-testing regularly.

3. Fire Services

Common examples for the management are:

- Automatic Fire Alarm and Detection System:
  - ➢ all fire alarm would sound and signal in the property management office in the event fire happens
  - ➢ the system is also directly linked to the Fire Services Department
  - ➢ a micro-processor is based to pinpoint the source of fire
- Fire Protection System that has:
  - ➢ a wet hydrant system at each escape staircase with a hose connection on each floor
  - ➢ direct links to the Fire Services Department
  - ➢ hose reels on every floor level
  - ➢ break-glass alarm points linked to the automatic fire alarm and detection system
- Automatic Sprinkler System
- Automatic Drencher System

Besides receiving alarm and event information from the full spectrum of building systems, fire fighters can also assess to override commands to smoke control systems, elevator systems, electrically locked doors, and auxiliary power and pumping systems during emergency.

4. Energy Saving

Although some people still query about the total energy consumption of the more advanced systems, the Department of Energy of the United States estimated energy savings can result from the schedule of maintenance of about 5% to 15% of the total energy bill.

5. Building Automation System(BAS)

BAS is the services available with office automation, and the communication capability available with a multi-dimensional telecommunications system that provided not only voice but also data teleconferencing, telex, and facsimile communications.

It also allows users of big open offices or assembly halls to activate the power and ventilation system by simply dialing a few access numbers. Users of a too big open office often find that too remote and complicate to activate the lighting and air-conditioning systems of a partial area in the office. Double to triple energy often wastes when a small number of staff is working over time after the normal office hours. BAS system in this way can save the users from the problems of getting to the correct power control and to save energy by simply dialing a number code and access into the management control system. For example, open office of over 20,000sq. ft. in Two IFC of the Hong Kong is divided into 72 zones. The control centre can assist their tenants to activate and adjust the systems in the designated area by request.

Property managers got to have an overall knowledge of all these systems and have to train the operational staff to operate as well as to educate the tenants to operate such systems. Property managers can indeed help the tenants.

Green Concerned Elements

Environmental green concept has been receiving increasingly concerns from the property owners today. Besides the traditional double-glazed curtain wall, many new commercial buildings have used the triple-glazed windows for helping the buildings to retain more cool air inside the office. In some specially designed buildings, the photovoltaic panels are even established for energy production inside the buildings. Like One Peking in Hong Kong, the spare power produced by the photovoltaic panels is used for the operation of electronic blinds in the tenants' offices. Under normal situation, this energy can be stored inside the battery for about one week's time. Due to absence of moving parts, the system does not require extensive maintenance and quietly pumping renewable energy to the building system.

To maintain the system to be operated in a stabilized performance, the extra absorbed power is stored into the battery as a buffer to condition unstable solar power generation period, e. g. cloudy sky.

Although not every property manager can have a strong technical background, the property managers of the buildings with such new systems should at least got a basic understand of the new facilities in the market so as to educate both the tenants and the management staff to make good use of these new facilities.

Using Robots to clean the curtain wall

Besides the environmental green concept mentioned above, there is also something look fancy but still useful come to the market from time to time. For example, to use robots in cleaning the curtain walls of the IBs may become another fancy product in these years.

In order to minimize the tenants from the disturbances from the "human gazing" outside their office windows during curtain wall cleanings, "robots" (automatic curtain wall cleaning system) are used in some IBs. In fact, the propeller type of wall-climbing robot has been used in the building industry for decades in both Japan and the United States. In Hong Kong, The Center has adopted the similar type of automatic cleaning system in 1998. Of course, the management also has to obtain the knowledge about the maintenance of the new management systems whenever requires.

However, for tall buildings with large window size, property managers may have to take other considerations sometimes. For example, like 2IFC in Hong Kong, it takes a team of window cleaners operating three gondolas from 7:00am~12:00 midnight seven days a week to complete one round of window cleaning. As the building is slightly narrowing up on the upper floors, the flexible extension arms gondolas are also adopted for this cleaning purpose. In the other words, property managers today should be able to identify and apply the right management to the right building.

Improve Indoor Air Quality

The introduction of workplace computers, printers, photocopiers, and fax machines has indeed increased indoor pollution. Electrical and telecommunications facilities in office buildings are under pressure to satisfy the demands created by the rapid growth of computer and networking technologies. These factors have a definite impact on worker productivity. New technology can be used to create Intelligent Buildings that address these problems by providing a healthier, more productive, and less energy-intensive work environment. As these are critical factors for business success, owners of Intelligent Buildings have a clear advantage in attracting and retaining 21st century tenants.

Different kinds of minor office sickness (such as headache, throat ache, red eyes, rhinitis etc.) are commonly existed in the tiny enclosed work places. To avoid "sick building syndrome", good air-circulation is important to the office building. With the good air-circulation system, minor office illness can also be minimized.

A variable air volume (VAV) system is able to conserve energy by reducing the volume of air delivered to a space based on its thermal requirements so as to reduce the power that is required for the fans. Of course, clean and purify air would also be important to the users of the building. Nowadays, NCCO (Nano-Confined Catalytic Oxidation Air Cleaning/Deodorizing System) technology has been becoming popular in many offices. Again, property managers got to update their knowledge about the new products in the market from time to time.

Intelligent Building Reduces Operational Costs

Different forms of battery backups and standby generators are used to secure an uninterruptible power during emergency. Such emergency power devices can activate the emergency lights, fire services systems and life safety equipment. It provides power to the tenant's computing system, as well as gives air conditioning to the pre-determined areas. The installations of the electric wirings and cables are also flexible in the intelligent buildings.

Air-conditioning

Air-conditioning usually dominates most of the electricity usage. If less air has to be transported, costs are bound to come down. Conventional air-conditioning equipment uses 50% of its energy just to move air around. Water as the means of transport reduces running costs by more than 90%.

To improve the quality of the indoor air can increase motivation and keep people healthier and happier at their work, because it creates exactly the right "working atmosphere".

For the best result, the management has to search for the system with less additional investment necessary for a separate heating system as it is "thrown in free" with the air-conditioning system. The spaces that have been taken up by radiators can then be used for something more productive. Some systems can also be operated as hybrid system i. e. heating on the north side, cooling in the rooms along the south side, all with one and the same system.

Lighting Systems

"Smart" light resources under good lighting system can provide more then lighting. Suitable choice of decorative lights in use can change the mood of the house and can also be a help to improve health. All lighting can be controlled through the energy management system. The three basic approaches to automated lighting control are programming, dimming, and occupying sensors, which can present savings of 30% to 50% of the energy used for lights. Computer-controlled lighting systems use a twisted-pair network to connect the lighting CPU to the transceiver panels that are scattered about the building. Good intelligent management should be able to provide training to the operational level on regular basis.

Standby Power System

Besides the high power daily operating and back-up generators, many buildings like 2IFC in Hong Kong has adopted the dual power system. For example, like Two IFC, the energy back-up generator can provide half of the usual lift services and air-conditioning services for about 10 hours during emergency. As well, a standby oil tank, which is also claimed to be the largest oil tank in Hong Kong, is prepared for emergency fuel storage during urgent situation. Of course, the property managers in this case should also have a thorough knowledge about risk management.

Security Management

In nearly all buildings Hong Kong, CCTV cameras are installed. Many buildings have also adopted smart card or Octopus card access control system. To smooth the handover procedure between the contractors and the management company, many new buildings today have invited the input of the management company during the development stage. This allowed the management company's requirements in some technical tenders and offer opinions on the kind of materials to be used for floors, carpeting and wall coverings.

Real Cost Saving for the Building

The true cost of a building is not simply its cost of construction. During budgeting, the management must include operating and maintenance costs over the structure's life span. IB yields cost reductions in all of these areas by optimizing automated control, communication, and management systems. They also guard against repair costs, employee time/productivity loss, revenue loss, and the loss of customers to competitors.

Automatic Refuse Collection System

Although many kinds of sophisticated new systems are being introduced to the IBs nowadays, the property managers also have to select the right product to the right building. The automatic refuse collection system (ARCS) would be one of the recent examples in Hong Kong. ARCS seems to be working pretty well in The Science and Technology Parks, but it failed to gain the acceptance of management in the government residential estates.

ARCS is a system that has been designed for the collection of refuse disposed by domestic flat residents in a more efficient and effective, but less disturbing procedure. Under this system, refuse can be collected and sent to dispose with less chances of exposing the trash into the air. Thus, better environment and less inconvenience can be provided to the residents. During garbage collection, the refuse can also be compressed before transportation. As a result, less moving frequencies are required and transportation cost can also be minimized. Because the conventional refuse chutes or refuse collection systems are cheaper in set-up, management and maintenance. There are only a few government estates have such an investment in Hong Kong.

However, it gained the recognition of the non-residential development like Science Park, but its "seeped smelt" is highly rejected by the residential tenants of numbers of government estates. Besides, maintaining the updated market knowledge of the new products in the market, the management also has the wisdom to select the right systems for the right building.

In fact, to most of the resident, the rating order of importance and the willingness on money spent for a healthier building are:

1. better property management;
2. better air quality;
3. less noise;
4. more cross ventilation;
5. better lighting.

In the order words, to enable the occupiers to stay in a more satisfactory atmosphere, the property managers should indeed communicate with the architects and developers well before the opening of the new buildings.

Occupant Safety Management

This includes smokes detector, fire alarm system, airflow control system(building pressurization) and indoor air quality control system. A computer based automatic fire detection and alarm system should be installed. When a detector senses a fire, the property management system can immediately identify which detector has been activated and correctly pinpoint the fire application for fire fighting personnel.

The management also has to fully understand the application of different emergency management detectors. For example, the fire detection system is an entire computer based automatic fire detecting and alarming system. It contains the smoke and heat detection systems, break-glass units, sprinkler flow detectors and a control panel. This system monitors the condition of the detectors and the break-glass units that are located inside the plant room, switch rooms and offices. It automatically gives visual and audible fire alarm signals to alert the occupants, the central fire control centre and the property management centre in the event of fire. When a detector alarmed by a fire, the property management system can immediately identify which detector has been alarmed. The fire location can therefore be pinpointed and communicated to the Fire Services Department immediately.

Lift and Escalator Systems

Complex control computers are used for scheduling the vertical transportations. The lift and escalator systems can optimize the service by preventing cab over-crowding and minimizing energy consuming.

Such computer softwarc(c. g. Micronic 10-popular software used in the market) allows the system to provide more frequent services to the highly used floors. It enables different patterns of services to be given to the users during the morning up-peak and evening down-peak hours. As well, the strategic locations in the parking area can be operated when necessary.

In large installations, each lift has a microprocessor that communicates directly with the mechanical room controller. The sensors on the cab are home-wired with twisted pair to the lift controller.

For security reasons, the lifts can also be programmed to service some designated floors with the command of a smart card or access code. The management has to understand the characteristics of the new systems in the market so as to explain to the buildings' tenants and visitors when required.

With the growing population, tall buildings are becoming more and more common today. To let the tall building users to reach the designated floor promptly, fast traveling life system are also required by the tall buildings. For example, Taipei 101(the world's tallest building today) owns the world's fastest elevators. Such systems can help the management in minimizing complaints caused from long waiting time when using the lifts.

Understanding the helpful products in the market

Different kinds of intelligent packaged computing property management systems are available in the markets today. With proper application of these systems, the intelligent property management today can indeed save a lot of works and manpower. For examples, these systems can help in:

1. Recording data like property, owner, tenant, and tenancy information.
2. Billing processing, like:

—Automatic generation of monthly debit notes.

—User-definable charge types and charges codes.

—Full audit trail of all the transactions generated and posted.

—Fully integrated with accounts receivable and general ledger.

3. User-definable Database Extension, like:

—Diary records can be attached to the whole estate or any particular building or any individual unit for keeping the time-related matters such as seasonal events, reminders or maintenance schedule.

—Associate records can be grouped by category and can be attached to any estate, building or individual unit. Unlimited user-definable entities can be added to the database to provide enough flexibility for property management information.

4. Drill-down inquiry, like:

—Billing and outstanding aging information by unit, tenant, and user-definable analysis.

—Full audit trial for tracing back to the original document or voucher.

—On-line aging inquiry.

5. Management reports with user-definable selections:

—Aging Analysis.

—Billing History.

—Auto-pay List.

—Deposit List.

—Income & Revenue Forecast.

—What If Analysis on Projected Income.

—Performance Report by Property Manager.

—Tenant Contract Expiration Report.

6. Payment and inquiry through Internet or phone.

—Tenants can browse their account status through Internet.

—Tenants can make payment through Payment Gateway on the Internet.

Benefits of the computerized recording new systems:

To the property Managers:

- ➢ Verification of alarm from outside the premises by owner, monitoring center or police.
- ➢ Recording human presence in the premises.
- ➢ Scheduled automation tasks-turn on or off household appliances, etc.
- ➢ Surveillance of domestic staff, cleaners.
- ➢ Vacation home monitoring.
- ➢ Campus security monitoring-man guarded installations.
- ➢ Total security and video systems for retail premises.

To the tenants:

- ➢ Event notification-security or other individual requirement.
- ➢ Remote verification of property status.
- ➢ Multiple communication mechanisms.
- ➢ Remote control, from anywhere.
- ➢ Off-site data storage.
- ➢ Automation and control.
- ➢ High reliability.
- ➢ Critical data storage e. g. video images.

With the Support of such systems, both the management and the tenants can be benefited from:

- ➢ More speedy and accurate data input.
- ➢ Quicker repairing process.
- ➢ Better focusing the job schedule, e. g. completion date of new building or a repair job.
- ➢ Better managing re-inspection process.

➢ Easier analysis of contractor performances and defect reasons.

Using eBusiness Applications in Property Management Industry

Besides helping the management in most of the analyzing and recording works, the eBusiness Applications today also helps the management in most of the management and clubhouse facilities payment.

Using Smart Card in Management

To enable the property managers to integrate the controlling system such as lift control, door access control, and carpark control system. Smart card system, such as the Octopus card in Hong Kong is widely used:

Lift Control System

Lift traveling cables should be reserved for the connection between the readers and the corresponding control computers. Each control computer in lift computers should be equipped with 2 serial devices on the configuration diagram. One would be responsible for collecting data packs sent by those card readers while that other for communicating with the corresponding lift control processors.

Access Control System

Each smart card/Octopus reader for door access control system can be connected with two LCD indicating lights(one for accept; one for reject)and one buzzer while one dry contact has been provided by those readers. The dry contact will be connected with the solenoid of door lock to execute the access control while indicating lights and buzzer will show the processing result to the users.

Carpark Control System

When the loop detector of a card reader machine is energized, the dry contact will be closed so that the control computer can be notified with the arrival of vehicle. The smart card/ Octopus reader will then be ready for the registration of card ID. Once when a complete ID is received by the control computer, the data processing will then be executed. If the request is accepted, the drop gate machine will operate according to the control computer's instruction. The carpark control system shall work independently for the basic control logic such as safety operation.

IB-Tourists Visit Points

Because of its high-tech image and special building designs, intelligent buildings always become the visiting points of the tourists. In China, for example, Jin Mao Building in Shanghai, Taipei 101 in Taiwan, Two IFC in Hong Kong, Macau Tower in Macau…

Property managers of today not only have to understand the building technical management of the daily operation, they also have to have the knowledge of dealing with different visitors from different institutes, interest groups or even tourists from different places of the world.

Sustainable of the IB's High-tech Image

In 1970s when personal computer is invited into the offices, it is predicted that less paper would be used in the following years. However, the reality is turned out to be an increase of 100 lb per person paper used in the past 25 years in the United States of America. Many people prefer to print out the emails on the paper instead of reading them on the computer. The invention of email has simply increased 40% of the paper consumption. However, it is believed that the new generations have been using to read the emails directly on the computer. The concept of paperless office would still be a trend in the future office development. With the similar concept, intelligent building should also be able to provide the users with a more efficient and effective working and living environment. Of course, it still requires the contribution of the landlord, the architects, construction managers, property managers' effort to building an environment of real intelligent.

Concluding Remarks

This paper has shown that in recent years significant progress has been made in the field of property management from rudimentary basis to advanced technology and knowledge based management, especially in the context of intelligent buildings and management. It covers a wide spectrum of architectural engineering, modern technology and methodologies. In the modern and competitive world, sustainable development and management must have engineering, managerial, financial and intelligent features and capability to respond to the rapid changing pace of technology and to meet the increased demand for a safe, healthy and comfortable home with quality-value for money services. These are the main waive of thedriving forces of the future for quality home with intelligent dimensions at heart. With the explosion of technology coupled with the emergence of intelligent concepts increasingly more "intelligence" can be incorporated in buildings both commercial and residential management nowadays.

The final decision of whether to adopt intelligent building concepts and the extent of "intelligence" to be incorporated, would be dependent on the life-cycle costs and benefits analysis and many other factors, such as the cost with which to operate and maintain the facilities and the systems, the availability of spare parts due to fast changing technology, the type of systems in use and the amount of investment involved.

智能大厦管理的解决方案
持续转变中的世界保持达到业主的要求

刘宝志
（亚洲智能建筑学会活动及事务执委会主席）
林婉文
（亚洲智能建筑学会活动及事务执委会会员）

摘　要：本文主要环绕物业管理经常面对因时代进步，先进科技而产生的变化及物业周期性转变等等的管理问题。着手处理持续性建筑物的分释，如何融入在未来想象中物业设计及管理的尺度，建筑物的产生应考虑到的智能大厦及管理，建筑学的概念，工程学的实际性，在设计范围内应是具有多元化的智能概念，如操作上，管理上也需同时讨论。

Bibliographies

[1] 龙惟定，程大章（1997）. 智能化大楼的建筑设备. 北京：中国建筑工业出版社

[2] Bernaden, J. A. & Neubauer, R. E. (1998). The Intelligent Building Sourcebook. Fairmont Press. Construction & Contract News. Issue no. 2, 2004

[3] One Peking, Tsim Sha Tsui, Landmark Constracts. *Building Journal Hongkong China*. May 2003

[4] One Peking, Tsim Sha Tsui. *Building Journal Hongkong China*. May 2004

[5] Research carried out by CII-HK on Environmental Healthy Buildings in 2004

[6] The Centre sales brochure, (1998). Cheung Kong Holdings Limited. Vertical Challenges. *Building Services Professional*. April 2004

[7] Watson, G. H., Tucker, R. L. & Walters, J. K. (1993). *Automation and Robotics in Construction X*. Elsevier Science Publishers B. V

THE INTELLIGENT BIOCLIMATIC BUILDING

Serge Salat Jean-Christophe Visier Bruno Mesureur
(Centre Scientifique et Technique du Batiment)

Abstract: The term "intelligent building" is used to describe the responsive automated function of a building. The intelligent building has certain intelligence-like capabilities (such as responding to stimuli) to optimise its mechanical, electrical and enclosure systems to serve the users and managers of the building.

The building's intelligent systems have software programs that imitate to a limited extent the intelligent behaviour of humans. The intelligent bioclimatic building of the 21st century requires control centres analogous to those of the human nervous system, interrelated control and synthesis of all servicing, total environmental integration and balance, and systems that are monitored and harmonious.

The main challenge in the design of intelligent control systems for energy performance in buildings is to find the balance between implementation costs, operating costs, energy consumption, indoor climate quality, user satisfaction and the contribution to a sustainable building.

The development of a control strategy for a specific building will depend not only on technical parameters such as building type and design, ventilation system, external noise and pollution, solar shading and internal loads, but also such parameters as user attitudes and user expectations.

Key Words: Intelligent Building, Energy Efficiency, Comfort, Sustainability, Bio Climatic Building

1 CONTEXT AND OBJECTIVES

To increase efficiency, the "intelligent" building is equipped with sensors, active components and communications networks. Electronics are now to be found in all building components: windows, ventilation, heating and air conditioning systems, lighting, floors and ceilings. It is clear that this use of electronics is destined to increase dramatically.

The term "intelligent building" is used to describe the responsive automated function of a building. The intelligent building has certain intelligence-like capabilities (such as responding to stimuli) to optimise its mechanical, electrical and enclosure systems to serve the users and managers of the building.

The building's intelligent systems have software programs that imitate to a limited extent the intelligent behaviour of humans. The intelligent bioclimatic building of the 21st century requires control centres analogous to those of the human nervous system, interrelated control and synthesis of all servicing, total environmental integration and balance, and systems that are monitored and harmonious.

The main challenge in the design of intelligent control systems for energy performance in buildings is to find the balance between implementation costs, operating costs, energy consumption, indoor climate quality, user satisfaction and the contribution to a sustainable building.

The development of a control strategy for a specific building will depend not only on technical parameters such as building type and design, ventilation system, external noise and pollution, solar shading and internal loads, but also such parameters as user attitudes and user expectations.

Objectives of a control system

—to maintain thermal, visual and indoor air quality comfort;
—to give priority to passive techniques in order to reach comfort requirements;
—to be flexible i.e. to be able to satisfy the different users' preferences (e.g. by using smart cards containing the users' preferences, inserted in a specially designed reader unit that supports the proposed application);
—to achieve a response of the controlled variables without over shootings and oscillations that can cause energy waste.

2 STRATEGIES

Control strategies will improve energy efficiency and indoor climate in spaces through smart management of switching facilities between passive techniques and active systems.

—These optimized strategies will contribute to increasing energy efficiency and a growing use of renewable energies in buildings.
—Indoor climate will be improved (thermal and visual comfort, indoor air quality, noise), contributing to a more healthy indoor environment.
—Pollutant emissions will also be reduced by means of smart control of hybrid technologies.

Composition of an automatic control system

—one or more sensors: necessary to measure the parameters required for the implementation of any control strategy. For natural ventilation control, classical parameters to be measured may include indoor and outdoor temperatures, CO_2 (or air quality) level, wind velocity and direction, and rain detectors. The selection of the right parameters depends on the control strategy i.e. control overheating, promote comfort or improve indoor air quality;
—one or more actuators;
—a controller: the heart of the control system. It is comparable to a personal computer receiving and treating data from the sensors and sending orders to the actuators. The control strategy is stored in the memory of the controller, and continuously implemented at fixed time intervals. Modern controllers are designed with the aim of handling all types of sensors, actuators and control strategies.

S- a supervisor: implemented in a BEMS(Building Energy Management System), a computer is necessary in order to supervise control decisions e. g. 8 to decide which windows to open according to wind direction or to close all windows in case of fire.

Sensors required

—presence or motion sensors
—temperature sensors: basic components of a control system, used for measuring indoor and outdoor temperatures. For example, they indicate whether fresh air should be introduced as a cooling means;
—carbon dioxide(CO_2) sensors: used in order to evaluate indoor pollution caused by occupants only. Most of these sensors are based on infrared absorption spectrometry.
—mixed gas sensors(air quality sensors): used to estimate indoor air quality based on multi-gas sensing
—wind speed and wind direction sensors: essential when natural ventilation is implemented. Wind direction is measured in order to select which openings on which facades should be opened, while wind speed is mostly used in order to adjust opening positions and to close them if the wind is too strong;
—rain detectors: necessary to prevent water ingress into buildings and should be placed on top of the roof.

Function and control tasks

The control strategy should determine both control times and rates, and thus different control modes in relation to different weather conditions. The actual control strategy should reflect the demands of the building's owner, users' needs and requirements in terms of standards and regulations:

- indoor air quality during occupied and unoccupied hours;
- room temperature during occupied hours in summer;
- solar shading;
- night ventilation during summer;
- preheating of ventilation air;
- room heating and cooling;
- fan assistance and alternating passive and active systems.

Control strategies

- *Control based on indoor pollution*: this strategy consists in monitoring an air quality index(i. e. CO_2 or air quality sensor) and opening or closing windows and vents as follows:

—if measured CO_2 concentration is above a given level, open vents;

—if measured CO_2 concentration is below a given level, close vents.

- *control based on indoor and outdoor temperatures*: this strategy is a "free cooling strategy". The aim of such a strategy is to determine whether outdoor air can be used for cooling purposes or not. Free cooling is switched on if outdoor temperature is lower than indoor temperature.

- *integral control strategy*: in this case, both indoor air quality indices and temperatures are taken into account in the control strategy. Vents, windows or dampers are operated depending on the requirements of these two criteria.
- *mixed-mode control strategies*: in bioclimatic buildings, passive systems will not be sufficient to achieve thermal comfort throughout the year. Mechanical cooling systems are therefore necessary. Specific control strategies need to be developed in order to integrate control of passive and active systems. These control strategies are defined as "mixed-mode control strategies".

■ Technological developments based on artificial intelligence and fuzzy logic

Conventional control strategies proposed to date for indoor comfort are limited to On-Off and conventional PID methods(Proportional- Integral- Derivative). The On-Off control in buildings has shown that the controlled variable swings continuously. Moreover, the On-Off control in thermal comfort means control of indoor temperature only, without taking into account other critical thermal comfort variables.

Technological developments based on artificial intelligence techniques (neural networks, fuzzy logic, genetic algorithms, etc.) offer significant advantages over classical control systems by incorporating users' demands, reduced energy use and improvement of the indoor environment.

Artificial intelligence has attracted the growing interest of researchers in various scientific and engineering fields. The number and variety of applications of fuzzy logic and neural networks ranges from consumer products and industrial process control to medical instrumentation, information systems and decision support.

Fuzzy logic is based on the way in which the human brain deals with inexact information. Fuzzy systems are structured numerical estimators. They start from formalized insights about the structure of categories that exist in the real world and then formulate fuzzy IF-THEN rules that represent expert knowledge. They combine fuzzy sets with fuzzy rules and they produce complex non-linear behaviour.

Comparisons of fuzzy systems(FS), neural networks (NN), genetic algorithms (GA) and conventional control theory *

	FS	NN	GA	Control theory
Mathematical model	SG	B	B	G
Learning ability	B	G	SG	B
Knowledge representation	G	B	SB	SB
Expert knowledge	G	B	B	SB
Nonlinearity	G	G	G	B
Optimization ability	B	SG	G	SB
Fault tolerance	G	G	G	B
Uncertainty tolerance	G	G	G	B
Real time operation	G	SG	SB	G

* *The fuzzy terms used for grading are good(G), slightly good(SG), slightly bad(SB) and bad(B).*

■ Smart buildings and Internet-based energy services

The supply of Internet - based energy services will enable orchestrated energy management of individual buildings or blocks of buildings and potentially decrease building energy consumption and increase indoor comfort conditions.

The energy - related services in buildings require the integration of information technology, intercommunication and telecommunication. A solution to Internet - based energy services is the gateway approach. According to this approach, the general structure of the system requires the combination of the following elements:

- a Local Network installed in the buildings. This network supports the intercommunication between the various connected devices(mainly sensors and/or actuators). The controlled appliances are end-point devices located on the local network.
- a Gateway that is used to connect the local network with the external/access network;
- an External/Access Network that enables the establishment of two-way communication;
- one or More Service Providers to provide and support various energy services.

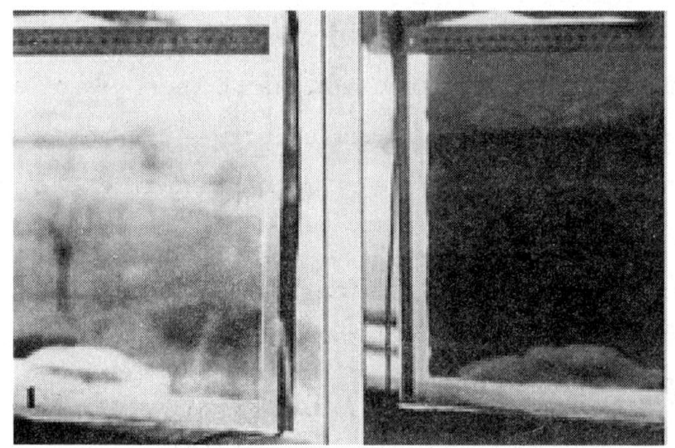

1. *Electro chromic glass is the result of adding dynamic properties to a stable material. Electro chromic glass can turn opaque within a few seconds of an electrical current being applied.*

■ Smart materials

As Philip Ball puts it: "Smart materials represent the epitome of the new paradigm of materials science, whereby structural materials are being superseded by functional ones". Smart materials carry out their tasks as a result of their intrinsic properties. In many situations they will replace mechanical operations. We will see "smart devices in which the materials themselves do the job of levers, gears and even electronic circuitry". There is even the prospect of "A house built of bricks that change their thermal insulating properties depending on the outside temperature so as to maximise energy efficiency".

Smart materials are already on the market, like thermo chromic or electro chromic glass. At present electro chromic glass is a sandwich construction with a gel that changes its light emission properties in response to an electric current. This will dispense with the need for mechanical blinds and solar shades and will give individuals much greater control over their immediate environment.

The really exciting advances are in active smart materials. An active system is controlled not only by external forces but also by some internal signal. In smart systems an active response usually involves a feedback loop that enables the system to "tune" its response and thus adapt to a changing environment rather than be passively driven by external forces.

■ Smart fluids

By introducing a strong electrical field, we can change certain fluids to a near solid state. They are called "electro rheological fluids" (rheology is the study of the viscosity and flow capacity of fluids). They can be made intelligent by coupling them to sensor devices which detect sudden movement. They have the potential to replace a range of mechanical devices.

There is yet another dimension to the characteristics of smart materials-materials that learn, that get smarter as they get older. They have an inbuilt degree of intelligence and are capable of optimising their performance in response to feedback information. What we will see in the near future are smart structures equipped with an array of fibre optic "nerves" that will indicate what a structure is "feeling" at any given moment and give instant information of any impending catastrophic failure.

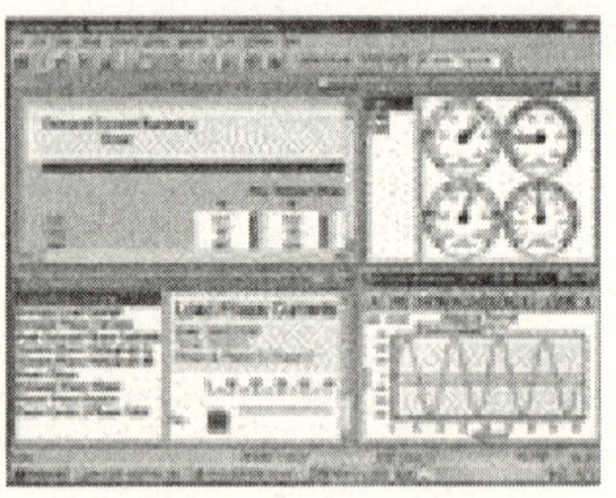

Energy Management System & Building Control Command Products

Power Logic power monitoring and control system by Schneider Electric

Environmental performance

By installing PowerLogic system, a 2% ~ 4% reduction in energy costs can be realized, but that's just the "tip of the iceberg". Typically, an additional 2% ~ 5% can be saved through better equipment utilization and avoiding unnecessary capital purchases. As much as 10% of potential savings can be found by improving power system reliability.

3 ADVANTAGES

■ Energy performance of automatic control

The combined control of active and passive systems, e.g. night ventilation cooling and mechanical cooling or hybrid ventilation, generally requires the use of so-called "logic

control" implemented by various rules in order to determine whether the passive or active systems should be operated.

Many digital controllers offer this possibility to implement logic control rules.

The use of automatic control for the control of passive elements and its integration into a BEMS system, when compared with classical manual control as usually implemented, could result in:

- energy savings for the building considered;
- improved and satisfactory indoor comfort levels even without auxiliary cooling;
- no or limited additional cost within certain conditions.

One of the great advantage of automatic control systems is their flexibility. Once the right actuators and sensors have been installed, the control strategy can be modified at any time.

Weather data acquisition

Weather data acquisition is closely related to energy services because the energy performance of buildings depends on the local climate. The data logging of weather parameters can be conducted at the site where the building is located or by using measurements from a nearby meteorological station. Another convenient approach is to measure some of the required parameters in situation(such as outdoor temperature and relative humidity)and use the data supplied by a neighbouring meteorological station to acquire the rest of the necessary data.

智能化生物气候建筑

Serge Salat　Jean-Christophe Visier　Bruno Mesureur

(Centre Scientifique et Technique du Batiment)

摘　要:"智能建筑"一词常用于描述建筑的自动反应功能。智能建筑拥有某些类智能的性能(如对刺激做出反射),可以优化建筑的机械、电力和维护系统,为用户和管理者提供更好的服务。

建筑智能系统具有在某种程度上模仿人类行为的软件程序。21世纪的智能化生物气候建筑要求控制中心要类似于人类的神经系统,具有相关控制和全面维护的集成,达到环境的融合与平衡,并实现系统的监控与协调。

对建筑在能源性能方面的智能控制系统的设计,其主要挑战是找到实施成本、运行成本、能源消耗、室内气候质量、用户满意度和对可持续建筑的贡献等方面的平衡点。

某一特定建筑控制系统的发展,不仅取决于如建筑类型及设计、通风系统、室外噪声及污染、遮阳和室内负荷等技术参数,还取决于用户的态度和期望值等因素。

Bibliography

[1] The Sustainable Design Guide, High Environmental Quality Cities and Buildings for China, CSTB.

STANDARDISATION & MODULARISATION OF INTELLIGENT BUILDING GONTROLS AS A BASIS FOR REDUCING CAPITAL COST, LIFECYCLE COST AND RISK IN HEALTHCARE CONSTRUCTION PROJECTS

Roger Moncur

(Technology Manager TAC Satchwell)

Abstract: The paper will address the following key issues in large healthcare construction projects: Cost both Capital and Operational / Lifecycle; How to ensure operational effectiveness; Minimisation of energy consumption; Project delivery; Patient comfort and the Minimisation of Risk through all phases of the project.

The focus for the paper is in the application of intelligent building controls over open protocols as a means of addressing key issues in healthcare construction projects. Issues discussed will include the management of the design phase to reduce capital construction cost, the use of off-site fabrication and testing as a means to de-risk construction programmes, cost reduction through standardised distributed control of ventilation and other electrical systems. In particular we will address the significant advantages of delivering intelligent building systems on a single data backbone as a means of both reducing cost and of ensuring maximum availability and maximum operational effectiveness.

A case study (Manchester Royal Infirmary) will be used to illustrate how these advantages can be delivered. Specifically we will demonstrate how it is possible to generate savings of 10% on a Facilities Management cost package of £5.8m. The paper will also demonstrate capital cost savings of in excess of 15% on a £15m intelligent integration solution for a large PFI hospital. In addition it will explain how the benefits of our approach are available to projects of any size-not just large PFI's.

Operational effectiveness issues will be addressed in two ways. The paper will show how an information-based approach to maintenance and repair delivers significant advantages in terms of availability for mission critical building elements such as Operating Theatres. It will also show how an open standards based systems approach makes information-based intelligent maintenance a reality making it possible to move from planned maintenance cycles to condition and reliability based maintenance whilst improving facility availability. Additional information will be provided on improving patient comfort and safety - hence enhancing throughput rates through an effective standardised integrated control solution.

Energy consumption in healthcare facilities is a significant cost. By connecting metering and other systems via the same standards based protocols as other intelligent building systems it is possible to present operational staff with a single viewfinder of key building parameters. When energy consumption is correlated with other factors such as, for example, temperature set points, occupancy rates and lift movement it becomes possible to manage the facility holistically ensuring minimum energy consumption without compromising comfort or safety. The paper will illustrate how this is achieved and how it affects operational facilities staff.

The conclusion will examine the overall benefits of the proposed approach, illustrated by a real example, and will show that for the full benefits to be achieved, a team approach to design is essential. The paper will argue that the entire professional team has a role in the requirements analysis and will demonstrate how a focus on capital cost, lifecycle cost, operational risk and on construction risk leads to better buildings that are cheaper to build and to operate.

Key Words: Cost, energy consumption, intelligent building control, design management

1 Introduction

1.1 This paper

This paper and its associated presentation is intended to show how standardisation and modularisation of the application of intelligent building controls based upon 'Open Systems' can reduce capital cost, provide a platform for reduction in lifecycle cost and reduce risk in healthcare construction projects. This paper also shows how attention to design and project management are important aspects of cost and risk reduction.

1.2 The project

The project is a redevelopment of the existing Manchester Royal Infirmary(MRI) for the Central Manchester and Manchester Children's University NHS Trust. The Shareholders in the project, members of the Private Finance Initiative(PFI) Partnership Catalyst are:

Bovis Lend Lease Holdings Limited

Halifax Bank of Scotland and

Sodexho Investment Services Limited

The £370 million scheme to redevelop a cluster of 5 out-of-date teaching hospitals in inner Manchester rates as one of the largest PFI healthcare projects unveiled to date. It is also the first of the next generation of PFI hospitals. The complex is to provide a total of 1660 hospital beds and spreads over an area equivalent to four city blocks around the existing Manchester Royal Infirmary on Oxford Road.

The project is being completed in phases. The early phases include a new energy centre and one of the largest multi-storey car parks in the country. We are interested in particular in the building control and monitoring systems, essentially all those systems that manage the physical comfort and safety of the staff and patients throughout their time within the hospital and its immediate surroundings.

2 Where do the costs lie?

In order to understand where cost savings can be made on a project such as this we need to see how the cost of a building over its lifetime can be broken down.

For the majority of buildings it has been established that the cost of construction only accounts for 25% of the total cost of a building over its lifespan. At least 50% of this total cost is accounted for in running the building with the remaining 25% going on modifications and additions.

Savings made during construction are obviously very valuable but they are typically one-off costs. By paying proper attention to how the building is to be controlled, monitored and managed once it occupied there is far greater scope for savings during this operational phase. In PFI projects the system availability requirements add an additional layer of cost management where high visibility of the status and running cost of the building is critical.

2.1 Cost Reduction Stages

At MRI, the PFI partnership, Catalyst, realised that the earlier the content of the building control and monitoring systems could be established, with the supplier on-board to be able to participate in the design, the more effective would be the delivery of these systems with a corresponding reduction in the level of risk. Catalyst also realised that integration of systems and system providers could provide an effective means of achieving an efficient build and a series of buildings that could be efficiently controlled and managed.

It was soon established that by concentrating the supply of the BMS, Fire System, Access/Security and CCTV under a single contractor the physical cost of project management, preliminaries and the like, could be reduced and the flow of information and full design buy-in could provide construction cost savings.

The value of this process was that it allowed a constructive dialogue to be set up with the mechanical and electrical consultants. One of the results of this was a change to a distributed control system with advantages that will be referred to later. Further value engineering exercises involved the fire system and the main communications infrastructure and numerous other details that will give operational benefits.

Much of this work has been done to enable the system to be more energy efficient and have a higher level of availability, reducing the effects of system downtime and enhancing the view of the system for the operating company.

2.2 Integrated Approach

Distilling this down we have a common systems delivery approach and a common communications infrastructure-two key elements to the effective delivery of this project.

The common systems delivery route has optimised the design of the project and generated savings that would not have been realised if systems such as the BMS and fire detection system had been supplied by separate contractors.

2.3 Open Standard Control Systems

A wide range of systems is being deployed throughout the hospital. Systems concerned with the environment, power, safety, security and management. Combining these systems and the information they provide into an integrated solution that is effective in its operation and manageable by those required to use it is a challenge.

Key to the implementation of a cost effective building control system has been the introduction of systems employing open standard communications protocols.

The adoption of a site-wide IT infrastructure serving the building control systems has allowed these different systems to share a common network, employing TCP/IP to carry BMS, CCTV, access and security data and reducing the need for duplicated stand-alone networks. Standard IT methodology and equipment has been deployed reducing the need for bespoke products from specialist manufacturers.

The phased construction of the project means that as the buildings are completed the various systems within them are brought onto the complete network via series of IT nodes, where rack-mounted IT switches allow the distributed control and monitoring systems to be visible to users throughout the network.

Other open standards becoming more common in the building controls industry, such as Lonworks, Modbus and OPC have been employed across a range of control and monitoring systems including variable-speed drives, meters, electrical monitoring and discrete control systems. This has enabled more efficient integration between systems supplied from a range of manufacturers. This type of integration also allows modular construction methods to be applied that benefit the construction and operation phases, reducing the need for large centralised motor control panels.

An example of the way these systems are being installed in a modular way is in the creation of standardised controls packages, incorporating a variable speed drive and associated measurement and monitoring devices and local cabling, which are being installed off-site into packaged plant. These can be functionally tested and commissioned before being delivered to site and connected into the communications network.

There are a number of advantages in using this approach including reducing peaks in commissioning labour requirements and the inevitable problems this causes for management, access, power and the like. This method also reduces the potential for bottlenecks on site caused by large motor control panels. It reduces the amount of controls wiring and during operation it reduces the amount of equipment affected when work has to be carried out on a control panel, reducing the effect on the operational areas of the hospital.

2.4 Where does this approach reduce cost?

Looking at this example in more detail, a variable speed drive supporting an open standard communications protocol reduces the need for signal cabling, supports local I/O reducing the overall point count in the building management system (BMS), reduces energy by matching the fan speed to the actual volume requirements, allows performance to be fully monitored and incorporates its own energy metering.

When such a drive is incorporated into a distributed controls solution with off-site fabrication, testing and commissioning, construction savings can be realised from reduction in site cabling and commissioning costs. Studies by independent consultants have examined this distributed controls model using integrated open standards based control systems, and compared it with traditional non-integrated, centralised closed protocol solutions and have demonstrated potential savings of up to 15% or more.

2.5 Operational Costs

As we have seen earlier the most significant cost of a building is represented by the operational period of the building's life. 50% of the total cost is identified during this phase and the savings to be made here are mainly the result of good information and good management.

This cost includes energy and maintenance, as for all buildings, but in addition there is cost to the operating company associated with penalties for non-availability of critical systems, a particular aspect of running a PFI building.

2.6 Different management approaches can affect the energy costs of a building.

The first element of the operational cost to consider is the use of energy.

A building management system that is not supported properly following hand-over will tend to drift into a situation where energy consumption rises over time. A well-managed system will keep energy costs relatively stable over time, but a system that has an energy strategy applied to it will demonstrate a reduction in energy costs over time. This process is easier to achieve when good information can be presented to the operator from an effectively integrated system.

2.7 Energy Metering

The completed MRI site will have over 250 integrated electricity meters monitoring LV panels together with energy data from nearly 600 variable speed drives. Energy profiles will be created for each LV panel based upon an agreed operational load.

Metering devices using open standard protocols are the most reliable way of providing energy data to an integrated system, when compared with pulse output meters. A greater amount of monitoring data can be more easily and reliably accessed and collected.

2.8 Energy Logging and Monitoring

It will be important to log and monitor the ongoing electrical loads using tools based on-site and off-site. The employment of such tools can allow operators to be quickly alerted to significant excess consumption and the reasons for this departure from the agreed profile can be determined.

Other management tools can continuously monitor the performance of the BMS control routines, comparing set-points with outputs to look for changes in system efficiency that may lead to excessive energy use or point to system failures.

2.9 Facility Management

The second element of the operational cost relates to facility management.

An integrated solution enables more up-to-date maintenance techniques to be applied that take more account of the failure rate of individual system components in relation to their importance to the system operation. In some cases this may mean running an item of plant to destruction or it may mean using data from a meter or a drive to help identify when system performance has degraded to a point at which failure is likely and remedial action is required to pre-empt a fault and its resulting costs.

Such techniques can help significant savings to be realised over more conventional scheduled maintenance regimes where plant is visited on a fixed time schedule, little or no maintenance is required but the cost of the visit has still been incurred.

2.10 Operational Penalties

The third operational cost in this example relates to availability.

In a PFI project such as this the operating company has a duty to provide maximum system availability. Failure to do so may result in penalty costs to the operator depending on the nature of the system and how long it has been off-line.

Part of the design process has included a risk analysis to take into consideration the effects of system failure or downtime. Where necessary, redundancy has been built in to critical systems and this is backed up by the monitoring techniques described earlier to reduce the risk of system failure and the resulting penalties.

2.11 Construction Cost Revisited

Returning to the areas of cost saving during construction it is worth summarising the important components:

It is important to include the controls specialists early in the design phase. This helps savings to be identified that can influence changes in design and implementation that can influence construction and operational phases.

By employing off-site modularised fabrication and testing there are savings in on-site wiring, commissioning and testing costs together with improvements in resource management. This is made easier by employing control systems based on "Open Systems".

Distributed control centres reduce commissioning bottlenecks and reduce downtime risk during operation.

Having the integrated controls solution supplied from a single source, with on-site design and project management reduces construction overheads.

The result of the effort put into construction and operational cost saving can result in savings of up to 15% capital cost and 10% or more life-cycle cost.

2.12 Reducing Risk

A project such as Manchester Royal Infirmary has a lifespan of 30 years and is intended to operate cost effectively and efficiently while being profitable for the operating company.

As a result the risk to the operating company needs to be reduced where possible by approaching the whole project with risk in mind. This means designing out as much risk as possible by considering how plant and systems are required to operate, where they are to be located, how is access to be provided, what are the implications of system failure and a host of other factors. The solutions may include incorporating redundancy in networks and systems where necessary, repositioning equipment and access and providing additional monitoring and management.

During operation the maintenance regimes must manage systems correctly thereby avoiding costly failures and excessive energy costs resulting in downtime and penalties. The plant must operate in accordance with the agreed regimes and profiles and any variation to these must be easily identified and attributed. The net result is that the required financial returns are achieved and the financial investment is secure.

3 Conclusion

To conclude, the Manchester Royal Infirmary project is an excellent example of where a team approach has been employed to encompass the construction and operation of the hospital as a single solution.

The requirements of the operator and the need to control construction costs have been understood and solutions found to achieve a cost effective design that will provide the operator with the necessary data for good management.

Integration of systems and networks is an important aspect of providing a manageable environment that will allow life-cycle costs and benefits to co-exist with the need to provide a safe and comfortable environment for patients and staff.

智能建筑控制的标准化，来减少医疗卫生建筑项目的成本，生命周期的费用和风险

Roger Moncur

(Technology Manager TAC Satchwell)

摘　要：本文讨论大型医疗卫生建筑项目中的关键问题。

资金和运作成本/生命周期；如何确保运行效率；使能耗降至最小；项目交付使用；

病患者的舒适度和整个项目阶段的风险最小化。

本文的中心是通过开放协议，应用智能建筑控制，阐述医疗卫生建筑项目中的关键问题。问题的讨论将包括加强设计阶段的管理，来减少基本建设成本，对离站点的使用和测试来减少编程的风险。通过标准分布的通风控制和其他电气系统来减少费用支出。我们将特别提到通过单一数据链接智能建筑系统的显著优势，既减少费用支出，又确保了最大的运作效率。

案例分析(曼切斯特皇家医院)将提到我们的优势何在。我们将特别说明如何在£5.8m 设备管理成本上节省10%。本文还将说明在一家大型的£15m PFI 医院的智能综合解决方案中，如何节省超过15%的成本。另外还将阐述运用我们的方法如何受益于各种规模的建筑项目，而不仅仅是大型 PFI 建筑项目。

运行效率问题将从两方面阐述。一方面以信息途径来维护系统，根据建筑物的特殊要求，例如手术室，来达到实用性的要求。另外标准的开放系统途径来维护，并时时增进设备的实用性。附加的信息将提高病患者的舒适度和安全性——通过有效的标准的集成控制解决方案来提高效率。

在医疗卫生设施中的能耗是十分显著的。基于标准协议，通过计量法和连接其他智能建筑系统，使运营人员观测单一主楼参数成为可能。当能耗和其他因素联系起来，例如，温度设定，占有率和电梯运动，设备的全盘管理成为可能，以确保能耗的最小化，而不影响舒适度和安全性。本文将说明如何做到这些，以及对操作人员会产生什么的影响。

结论将检验被提议的方法，用实例说明，演示所有的优势，和一整套方法。本文还将探讨专业团队在需求分析中的作用，以及通过关注成本控制，生命周期，运作风险和建筑风险，倡导更经济，更易于操作的建筑。

A REVIEW OF INTELLIGENT BUILDINGS ASSESSMENT

Zhen Chen[1] Derek J. Clements-Croome [2]Ju Hong[3] Heng Li [4] Qian Xu[5]
([1] School of Construction Management &Engineering, The University of Reading, Whiteknights, PO Box 219, Reading, UK; [2] School of Construction Management &Engineering, The University of Reading, Whiteknights, PO Box 219, Reading, UK; [3] Associate Professor, Beijing Institute of Civil Engineering and Architecture, Beijing 100044; [4] Department of BRE, Hong Kong Polytechnic University, Hong Kong; [5] Department of BRE, Hong Kong Polytechnic University, Hong Kong)

Abstract: This paper provides a review of the assessment methods of intelligent buildings (IBs). Based on a review of rating method currently used for building assessment, 6 rating systems for IB assessment are compared according to assessment clusters such as Architecture, Engineering, Environment, Economics, Management, and Sociology. The 6 IB rating systems include AIIB method developed by the Asian Institute of Intelligent Buildings(AIIB), Hong Kong, China; BRE method developed by the Building Research Establishment Ltd., UK; CABA method developed by the Continental Automated Building Association(CABA), Canada & USA; IBSK method developed by the Intelligent Building Society of Korea (IBSK), Korea; SCC method developed by the Shanghai Construction Council(SCC), China; and TIBA method developed by the Architecture and Building Research Institute, Ministry of the Interior, Taiwan, China. Although the AIIB method is identified as the most comprehensive assessment system, its four weaknesses are explained. The paper concludes that an innovative building approach using analytic network process will bring advantages to IB assessment.

Key words: Intelligent building; Assessment; Methodology; Review

1 Introduction

Sustainable building design, construction and operation require innovations in both engineering and management areas at all stages of a building's life. The lifespan of buildings is composed of a series of interlocking processes, starting from initial architectural and structural design, through to actual construction, and then to maintenance and control as well as to eventual demolition or renovation of buildings.

Inside this lifespan, essential requirements are generated from considerations of social, environmental, and economic issues for high-efficient energy-saving building systems in compliance with building codes and regulations. In this regard, building assessment is becoming popular in order to have a standard method to evaluate new and existing building design. For example, the U. S. Green Building Council [1] developed the LEED (Leadership in Energy and Environmental Design) Green Building Rating System as a voluntary, consensus-based national standard for developing high-performance, sustainable buildings. The Japan Sustainable Building Consortium [2] developed the CASBEE(comprehensive assessment system for building environmental efficiency) system as a new environmental assessment system to meet both the political requirements and market needs for achiving a sustainable society. The Building Research Establishment Ltd. [3] from UK developed the BREEAM (Building Research Establishment Environmental Assessment Method) to assess the environmental performance of both new and existing buildings. Meanwhile, Intelligent Buildings(IBs) are also under assessment according to their IB related characteristics and actual circumstances. For example, the Asian Institute of Intelligent Buildings(AIIB) [4] from Hong Kong developed an IB Index system to specifically assess the performance of IBs; and the BRE developed a matrix tool called MATOOL for assessing the performance of intelligent buildings [5]. Although a new international benchmark of IB assessment is under developing by the Continental Automated Building Association [6] in Canada, there is not a standard sustainable or green IB assessment tool, and this leads to the research being presented in this paper.

2 Building assessment methods

According to the latest literature [10], an IB is one that provides a productive and cost-effective environment through optimizations based on its three basic elements-people (owners; occupants; visitors etc.); products(materials; fabric; structure; facilities; equipments; services); and processes(automation; control; systems; maintenance; performance evaluation)-and the interrelationships between them. IBs use integrated and intelligent systems to provide a rewarding experience for the building owners, property managers, occupants and visitors to achieve their goals. These goals include the lifespan high energy efficiency, the environmental-friendly built environment with substantial safety, security, well-being and convenience, a lower life-cycle cost, and long-term flexibility and marketability, which lead to achieve a high-level of buildings that have the highest social, environmental and economic values. Meanwhile, IBs use advanced information and communication technologies to develop embedded data collection and information networks through which its services systems are automatically controlled to respond using an approach similar to the sensor system of human beings, guided by predictions based upon knowledge of the past situations of the building and usage, maintained in an integrated data base. Thus, IBs should be sustainable, healthy and technologically aware, meet the needs of occupants and business, and should be flexible and adaptable to deal with change.

Practitioners use assessment methods to evaluate the design or the performance of IBs. There are three main kinds of assessment methods including building rating, computer simulation and facilities management [10]. The rating method relies on a series of factors/indicators related to the design and the performance issues together with their defined scales to rate an IB. The simulation method uses artificially settings based on real-world data from the operation of IBs. The facilities management method use experts' knowledge to achieve goals in practical IB design, construction and operation. The applications of the first two kinds of assessment methods can be at either design or operation stage of any IB under evaluation, while the third method can be applied at all stages of the IB life cycle.

The authors are conducting an extensive literature review on conventional building assessment systems in order to grasp the progress of building rating method and develop innovative approach to IB assessment. Current building assessment systems under review include:

—Assessment Standards for Certifying Intelligent Buildings(ASCIB, by Intelligent Building Society of Korea(IBSK), Seoul, Korea)[3],

—Building Quality Assessment(BQA, by Building Economics Bureau, UK),

—Building Research Establishment Environmental Assessment Method(BREEAM, by Building Research Establishment Ltd. (BRE),UK) [3],

—Building Sustainability Assessment Tool(BSAT, by the Department of Trade and Industry, UK) [11],

—Building IQ Rating Criteria(BIQRC, by Task Force 1-Intelligent Building Ranking System, Continental Automated Building Association(CABA), Ottawa, Canada) [6],

—Comprehensive Assessment System for Building Environmental Efficiency (CASBEE, by-Japan Sustainable Building Consortium(JSBC), Japan) [2],

—Design Quality Indicator(DQI, by Construction Industry Council, UK),

—Environmental Performance Express of Buildings(Eco-Quantum, by IVAM, The Netherlands),

—Assessment Framework & Green Building Tool (GBTool, by the International Initiative for a Sustainable Built Environment(IISBE), Canada) [12],

—Green Mark for Buildings(GMB, by Building and Construction Authority, Singapore) [13],

—Hong Kong Building Environmental Assessment Method(HK-BEAM, by HK-BEAM Society, Hong Kong) [14],

—Intelligent Building Assessment(by Architecture and Building Research Institute, Ministry of the Interior, Taiwan, China) [7],

—IB Index(by Asian Institute of Intelligent Buildings(AIIB), Hong Kong, China) [4],
—IB Rating(by Shanghai Construction Council(SCC), Shanghai, China) [8],
—Leadership in Energy and Environmental Design/Green Building Rating System (LEED, by U. S. Green Building Council, USA) [1],
—A matrix tool for assessing the performance of intelligent buildings(MATOOL, by Building Research Establishment Ltd. (BRE), UK) [5],
—National Australian Built Environment Rating System(NABERS, by Department of the Environment and Heritage, Australia) [15],
—Office Scorer(Sustainable Refurbishment/Redevelopment Decision Support Tool for office buildings, Building Research Establishment Ltd. (BRE), UK) [18],
—Sustainable Project Appraisal Routine(SPeAR, by Arup, UK) [16], and
—Sustainability Checklist (Assessment of the social, environmental and economic impact of a proposed development, by the South East England Development Agency(SEEDA), UK) [17].

According to the literature review focusing on the building assessment systems, the authors noticed that there are several successful applications of rating methods for building performance assessment. For example, the LEED *Green Building Rating System®* is a voluntary, consensus-based national standard for developing high-performance, sustainable buildings in the United States [1]. The *Environmental Assessment Method* by the Building Research Establishment Ltd. (BREEAM) is adaptable to assess the environmental performances of both new and existing buildings in the United Kingdom [3, 19]. The *Standard Assessment Procedure* (SAP) of the National Home Energy Rating (NHER) is the UK's premier energy labelling scheme recommends by the UK Government for home energy rating [20]. On the other hand, although simulation methods can provide more reliable results than rating methods using various conditions in the building lifespan based on objective and subjective settings in computer programmes; there is not a comprehensive simulation tool for practitioners to conduct IB assessment at present. On the contrary, popular simulation approaches mainly focus on only one part of building performance such as thermal environment or acoustic environment, and it is a difficult task to develop a tool for complete performance simulations of the total environment in buildings. In this regard, rating systems have been widely adopted in building performance assessments, and the simulation method is often adopted in building design.

Among these building assessment systems, there have been several rating methods designed for IB assessment, and there are some new rating systems under development as well. Table 1 gives a summary of representative methods based on current practice in IB assessment. According to the literature review, the authors identified six assessment clusters of indicators centring on Architecture, Engineering, Environment, Economics, Management, or Sociology. Among the five IB assessment systems listed in Table 1, the AIIB method, i. e. IB Index method [4] is the most comprehensive one that covers all of the seven assessment clusters, and the SCC method [8] is mostly focused on the one assessment cluster, i. e. Engineering. The CABA method [6] aims to benchmark the IB assessment in a more general way but is still under construction. And the BRE method, i. e. MATOOL [5] and the IBSK method [7] have less coverage of assessment clusters than the IB Index. Therefore, the AIIB method is currently the most comprehensive method for IB assessment.

3 The limitations of building rating method

One problem of currently used building rating methods is that they actually pay less attention to functional variation in different types of buildings, which influence not only the emotional as well as the physical well-being of human beings, but also the design and the management of buildings. In other words, each assessment procedure conducted under each rating method actually uses a generic platform of indicators applied to all kinds of buildings therefore do not differentiate one building from another regarding their various features. As a consequence, assessment results of different kinds of buildings actually lack the power of comparability regarding the features of IBs. For example, AIIB method adopts 29 sub-indicators to assess the performance of lift and escalators[4]; however, there is not a practical guide regarding how to compare two designs for one IB project if one uses'a lift but another does not. In fact, it is not sensible to say buildings with a lift are more intelligent than buildings without them but a common generic platform will ensure all buildings have consideration given to aesthetics, function, convenience, flexibility, adaptability, reliability and health. In addition, the IBSK method [7] uses occupation density(occupation area for one person)as one indicator to assess *Architectural Design* of IBs, and the building with larger occupation area(a low occupation density)will get a higher score; however, one cannot say easily that a supermarket is much more intelligent than an office building because occupation area in supermarket is larger than that in an office building. In fact, buildings are classed according to their patterns of use at the design stage or management stage. For example, *The Town and Country Planning (Use Classes)Order* [21]regulates building class into four main categories with 16 classes depending on the purposes of building utilization in town and country planning and has been widely adopted in building design in the UK [4]. *The NYC Building Classification Codes* [22], on the other hand, provides a complete, comprehensive list of each Building Classification Code, and has been officially used to classify all properties and parcels from private homes to amusement parks by the City of New York. The lack of flexibility in current rating methods for IB assessment and the preference of classification in building design, construction and management indicate that innovations are required to develop flexible techniques for more objective assessment results of IBs.

Another problem of current building rating methods for IB assessment is that their calculation processes are not convincing enough to provide a reasonable assessment result.

For example, the AIIB method, i. e. IB Index [4]aims to provide a quantitative composite approach to IB assessment using 10 indicator clusters based on the Cobb-Douglas utility function [23]. However, the recommended method for the IB Index calculation (see Equation 1 and 2 in Equations in Appendix) is not actually reliable due to the following four reasons:

—the criteria of the AIIB method lead to non-determinism;

—the calculation method of the AIIB method is a non-sequitur;

—the calculation results from the AIIB method are non-unique, and

—the assessment procedure is based on non-organization principle/judgment.

Brief explanations to these reasons are given below:

The non-determinism led by the criteria of the AIIB method means that the assessment scores for each IB result from the evaluation criteria has questionable validity. As assessment results from each rating method depend upon a set of criteria denoted with a group of IB indicators, it is important to select the most appropriate group of indicators that are able to stand the test, and indicators adopted in a rating method that have less relevance to the IB will reduce the accuracy in assessments. For example, *Special feature (s) recommended by the auditor* is adopted in the AIIB method as an IB indicator in all most every category including *Green*, *Space*, *Comfort*, *Working efficiency*, *High-tech image*, *Safety and Structure*, and *Practice & Security*. It is clear that different auditors will give different scores to these indicators even though all auditors deal with the same building because of their knowledge and their various understanding of the fuzzy definition during assessment. Based on this consideration, evaluations of IB indicators are of necessity required.

On the other hand, the AIIB method adopts, from the field of economics, the celebrated Cobb-Douglas utility function as its calculation method in the process of assessment [4]. The Cobb-Douglas utility function is a standard utility function applied to describe matching output to input in a production processes and it is used commonly in both macro and micro economics [24, 25]. However, there is no clear information to support concerns about the application of the Cobb-Douglas utility function to the rating procedure according to personal discussions between the authors and other researchers in either the Cobb-Douglas utility function or rating procedure fields. In fact, the AIIB did not provide a reasonable explanation of reasons to adopt the Cobb-Douglas utility function in the calculation of a 10-module IB Index algorithm. Although the Cobb-Douglas utility function is one of the most widely applied utility functions in microeconomics, its major drawbacks such as the limited scope of effective regions and the harsh constraint terms to parameters definitely affect its utility in applications [26~30]. It is actually hard to define a physical model to describe this 10-module IB Index algorithm beyond the Cobb-Douglas utility function. Moreover, according to the second law of thermodynamics, which requires that any process that takes place at non-zero speeds must consume a minimum finite amount of

The main categories of criteria adopted in rating methods for IB assessment **Table 1**

Assessment clusters	Main modules by each assessment system					
	AIIB method [4] (Hong Kong, China)	BRE method [5] (UK)	CABA method [6] (Canada/USA)	IBSK method[7] (Korea)	SCC method[8] (Shanghai, China)	TIBA method [9] (Taiwan, China)
Architecture	Comfort	Built Environment	—	Archit. Design	—	Health & Sanitation
	Health & Sanitation	—	—	—	—	—
	Space	—	—	—	—	—
Engineering	High-tech Image	Functionality	Automation	Electrical System	Communication	Info & Comms
	Safety & Structure	Responsiveness	Comms	Info & Comms	Earthing	Safety & Security
	Working Efficiency	Suitability	Security	Mechanical System	Facility Control	Structured Cabling
	—	—	Structure	System Integration	Fire Accident Control	System Integration
	—	—	Systems	—	Int. Integration	—
	—	—	—	—	Office Automation	—
	—	—	—	—	Power Supply	—
	—	—	—	—	Security	—
	—	—	—	—	Structured Cabling	—
Environment	Green	—	—	Environment	Environment	Energy Consumption
Economics	Cost Effectiveness	Economic Issues	—	—	—	—
Management	Practice & Security	—	Property	Facility	Property	Facilities
Sociology	Culture	—	—	—	—	—

exergy(the quality of energy), so production isoquants(combinations of inputs that yield the same output) [31]cannot be of the Cobb-Douglas type [32]. In these cases, the necessary and the sufficient conditions of applying the Cobb-Douglas utility functionto the 10-module IB Index algorithm therefore require more study.

In addition, the AIIB method allows subjective weights of different building modules but this can lead to confusion about the interpretation of the assessment results. Table 2 recalls an example by the AIIB [4], in which the rate of weight comparison between two building modules are set as $w_x : w_y = 2 : 1$, and the results of IB Index for each kind of building and the rank of their intelligence are in accordance with common intuition as to which kind of building is more intelligent. However, the function adopted in the IB Index calculation(refer to Equation 3)does not always lead to a sensible result. For example, let $w_x : w_y = 3 : 1$, the IB Index values for each building are then different from the ones under $w_x : w_y = 2 : 1$, and the sequence of building intelligence also changes(see Table 2). The AIIB method cannot provide a unique result, as different auditors may make different conclusions, which definitely cause complexity and variance in IB assessment.

An experimental verification of the AIIB IB Index method **Table 2**

Buildings	Scores		IB Index			
	Module x	Module y	$w_x : w_y = 2 : 1$	Rank of Int.	$w_x : w_y = 3 : 1$	Rank of Int.
A. Smart Tower	70	50	63	1	64	2
B. Balanced Building	60	60	60	2	60	3
C. Mechanical Plant	100	20	59	3	69	1
D. Tree House	20	100	34	4	30	4

Regarding the non-organization assessment procedure adopted in current building rating systems, the authors find that it is difficult to recognize *Organization* factors from current systems besides the *Management* cluster, in which only property management issues are concerned. Based on the summary in Table 1, the non-organization principle/judgment existed in current building rating systems can definitely lead to partial assessments in which evaluators will miss their chance to study the culture, the structure and the occupants of all factors, which influence the performance of the building.

Theoretically speaking, logical defects in the currently used building rating methods, such as the IB Index method, may lead to an invalid IB assessment. It is thus required to provide an alternative method to evaluate the characteristics of IB, under objective and real life conditions, in which all indicators are taken into account, not only their values but also their interrelationships.

4 Further development

Based on this review, the authors propose an alternative approach to IB assessment by means of analytic network process(ANP) [33]. In a test drive using the ANP, the authors also noticed that 43 indicators can be extracted from 378 elements of IB Index by using a quantitative indicator selection approach. In fact, this extraction also indicates that most elements adopted in the IB Index are repeated and need simplification. As mentioned above, the IB Index has a comprehensive classification of indicators; therefore a most appropriate group of indicators can be selected from it for the ANP based assessment. To overcome the shortcomings that exist in the current IB Index method, the proposed ANP approach will provide an innovative way for IB assessment, in which both the value of indicators and their interrelations are taken into account.

5 Conclusions

This paper provides a review of various systems for building assessment focusing on intelligent buildings. Based on comparison among currently used IB assessment systems, it has been identified that rating method has its limitations in the assessment of IBs. In this regards, the authors propose an innovative approach for IB assessment using analytic network process, in which both scores and interrelations of indicators can be included.

智能建筑评价方法综述

陈震[1] Derek J. Clements-Croome[2] 洪枯[3] 李恒[4] 徐骞[5]

([1] School of Construction Management & Engineering, The University of Reading, Whiteknights, PO Box 219, Reading, UK; [2] School of Construction Management & Engineering, The University of Reading, Whiteknights, PO Box 219, Reading, UK; [3] Associate Professor, Beijing Institute of Civil Engineering and Architecture, Beijing 100044; [4] Department of BRE, Hong Kong Polytechnic University, Hong Kong; [5] Department of BRE, Hong Kong Polytechnic University, Hong Kong)

摘 要: 本文在总结国内外相关文献的基础上，对现有的智能建筑评价方法进行了总结和评价。通过对20余个建筑评价体系的总结，本文对六个针对智能建筑的建筑评价体系进行了比较。比较的方法是对各个智能建筑评价体系采用的指标体系在统一的评价域(包括建筑、工程、环境、经济、管理以及社会学)范围内的覆盖程度进行配比，从中发现最优的智能建筑评价体系。进行比较的六个智能建筑评价体系分别来自亚洲智能建筑协会、英国BRE、北美CABA、韩国智能建筑协会、上海建委和台湾建筑研究院。比较发现，尽管亚洲智能建筑协会的评价体系在指标的覆盖上最全面，但是它自身的缺陷也不容忽视。本文最后指出以网络分析法为核心的创新方法将为智能建筑评价提供有力的支持。

References

[1] USGBC, LEED: Leadership in Energy and Environmental Design, the U. S. Green Building Council

(USGBC), Washington, USA, 2005

[2] JSBC, Comprehensive Assessment System for Building Environmental Efficiency(CASBEE), Japan Sustainable Building Consortium(JSBC), Institute for Built Environment and Energy Conservation (IBEC), Tokyo, Japan, 2004

[3] J. Anderson, D. E. Shiers, and M. Sinclair, The Green Guide to Specification: An Environmental Profiling System for Building Materials and Components(3rd edition), Blackwell Science Publishing, Oxford, UK, 2002

[4] AIIB, IB Index(3rd edition), Asian Institute of Intelligent Buildings(AIIB), Hong Kong, 2005

[5] R. Bassi, MATOOL: a matrix tool for assessing the performance of intelligent buildings, in: Proceedings of the BRE Seminar on Intelligent Buildings, Building Research Establishment Ltd., UK, 2005

[6] CABA, Building IQ Rating Criteria. Task Force 1-Intelligent Building Ranking System, Continental Automated Building Association(CABA), Ottawa, Canada, 2004

[7] IBSK, Assessment Standards for Certifying Intelligent Buildings 2002, Intelligent Building Society of Korea(IBSK), Seoul, Korea, 2002

[8] SCC, Shanghai Intelligent Building Appraisal Specification, Shanghai Construction Council(SCC), Shanghai, China, 2002

[9] S. L. Wen, Manual of Intelligent Building Assessment. Architecture and Building Research Institute, Ministry of the Interior, Taiwan, China. ISBN: 957016137X. 2003

[10] D. J. Clements-Croome, Intelligent Buildings: Design, Management and Operation. Thomas Telford, London, 2004

[11] S. Sayce, A. Walker, A. McIntosh, Building Sustainability in the Balance: Promoting Stakeholder Dialogue, Estates Gazette, UK, 2004

[12] R. J. Cole, Review of GBTool and Analysis of GBC 2002 Case-Study Projects, International Initiative for a Sustainable Built Environment(IISBE), Canada, 2002

[13] GMB, Green Mark for Builfings, Building and Construction Authority, Singapore

[14] HK-BEAM Society, Hong Kong Building Environmental Assessment Method, HK-BEAM Society, Hong Kong, 2004

[15] DEH, National Australian Built Environment Rating System, Department of the Environment and Heritage(DEH), Australia, 2004

[16] Arup, SPeAR®: Product overview, Arup, London, 2002

[17] A. Amato, A Comparative Environmental Appraisal of Alternative Framing Systems for Offices, PhD thesis, Oxford Brookes University, UK, 1996

[18] J. Anderson, K. Mills, Refurbishment or Redevelopment of Office Buildings Sustainability Comparisons, Building Research Establishment Ltd. (BRE), Garston, UK, 2002

[19] I. Dickie, N. Howard, Assessing Environmental Impacts of Construction - Industry Consensus, BREEAM and UK Ecopoints, BRE, UK, 2000

[20] NHER, Home Energy Rating: Introduction, National Home Energy Ratings (NHER), The National Energy Centre, Milton Keynes, UK, 2004

[21] ODPM(1987) *The Town and Country Planning (Use Classes) Order* 1987. Office of the Deputy Prime Minister(ODPM), UK

[22] Profiles, New York City Building Classification Codes, Profiles Publications, Inc., New York, USA, 2004

[23] A. R. Schotter, Microeconomics: A Modern Approach, 3/E, Addison-Wesley, Boston, USA, 2001

[24] P. B. Meyer, Econterms, EconPort(A microeconomics digital library), USA

[25] E. E. Vargas, D. F. Schreiner, G. Tembo, D. W. Marcouiller, Computable General Equilibrium Modeling for Regional Analysis, Regional Research Institute, West Virginia University, USA

[26] K. J. Arrow, H. B. Chenery, B. S. Minhas, R. M. Solow, Capital-labor substitution and economic efficiency, The Review of Economics and Statistics 43(3)(1961)225～250

[27] C. W. Cobb, P. H. Doughlas, A theory of production, American Economic Review, 28 (1928) 139～165

[28] D. F. Heathfield, S. Wibe, An Introduction to Cost and Production Functions, Macmillan Education, London, 1986

[29] X. Qi, Effective utility function and its criterion, in: Proceedings of the 2nd China Economics Annual Conference, Northwest University, China, 2002

[30] X. Yin, A tractable alternative to Cobb-Douglas utility for imperfect competition, Australian Economic Papers, 40(1)(2001)14～21

[31] M. J. Osborne, Tutorial on the theory of the firm and industry equilibrium, on-line tutorials, Department of Economics of the University of Toronto, Toronto, Canada, 1997

[32] S. Islam, Effect of an essential input on isoquants and substitution elasticities, Energy Economics 7 (3)(1985)194～196

[33] T. L. Saaty, Decision Making with Dependence and Feedback: the Analytic Network Process, RWS Publications, Pittsburgh, USA, 1996

INTELLIGENT OPEN CONTROL SYSTEMS FOR ENERGY SAVINGS & COMFORT

Stephen Woodnutt
(Managing Director, Delmatic UK-stephen)

Abstract: This paper provides a review of the assessment methods of intelligent buildings (IBs). Based on a review of rating method currently used for building assessment, 6 rating systems for IB assessment are compared according to assessment clusters such as Architecture, Engineering, Environment, Economics, Management, and Sociology. The 6 IB rating systems include AIIB method developed by the Asian Institute of Intelligent Buildings(AIIB), Hong Kong, China; BRE method developed by the Building Research Establishment Ltd., UK; CABA method developed by the Continental Automated Building Association(CABA), Canada & USA; IBSK method developed by the Intelligent Building Society of Korea (IBSK), Korea; SCC method developed by the Shanghai Construction Council(SCC), China; and TIBA method developed by the Architecture and Building Research Institute, Ministry of the Interior, Taiwan, China. Although the AIIB method is identified as the most comprehensive assessment system, its four weaknesses are explained. The paper concludes that an innovative building approach using analytic network process will bring advantages to IB assessment.

Key words: Energy savings, Intelligent building, Interoperable systems

Overview

It is an undisputed fact that buildings consume energy, burning precious and scarce resources and impacting adversely on the environment through the generation of greenhouse gases. Yet there are many ways in which the environmental impact of buildings can be reduced and this presentation shows how the latest open building control systems can achieve energy savings of 70% while also enhancing occupant comfort and convenience.

Open Control Systems

The two main energy consuming services in a building are lighting and air-conditioning/heating. When someone mentions lighting or heating controls we typically think of time clocks, presence or motion detectors and time delay switches-all intended to provide energy savings. But these traditional approaches are inefficient and cannot compare with the savings and convenience of latest addressable, open protocol systems.

Latest control systems use communication protocols that are standard around the globe. Standard protocols provide many advantages including the ability to connect hardware from various suppliers on a single bus as well as for systems to be supported and maintained without reliance on a single vendor.

One of the leading internationally-adopted communication protocols is LonWorks. Lon is used by many building services systems so different types of systems-lighting, access and security, fire, air-conditioning, etc-all communicate and work seamlessly together using a common language and standard message formats. This enables systems to share information and avoids unnecessary duplication of hardware. For example a single presence detector in an office can monitor occupation of the area and, using a signal broadcast in standard Lon format, can communicate with the lighting control, air-conditioning control, and access control systems. When someone enters the office, lighting switches on, the air-conditioning increases in speed and the security system checks whether the area should be occupied at that time-if not, a security guard can be sent to investigate. When the person leaves lighting and air-conditioning is promptly turned off achieving valuable savings.

Open protocol systems share network cabling and equipment with other building services(so avoiding unnecessary duplication of hardware and infrastructure) and pool information to enhance the overall operation of the building services. Integrated internet and web technology allows remote monitoring of multiple sites over the internet while system reports and alerts can be relayed to maintenance staff via mobile phones or PDA.

While Lon excels at enabling building services to talk together, the Dali dimming protocol has been developed for the lighting industry. Dali is the latest advance in digital dimming technology, replacing analogue 1～10V & digital DSI dimming ballasts. Dali enables ballasts from any manufacturer to respond to standard, published commands and offers accurate digital dimming between 1% and 100% while providing lamp failure data in real time. The unique benefit of Dali is that ballasts may be individually addressed, switched, dimmed & monitored along a single buswire and the ease of installation enables flexibility and energy efficiency to be achieved as a retrofit in existing buildings without changes to the existing mains wiring.

Maximum building energy efficiency is achieved by combining the dimming efficiencies of Dali with the interoperability efficiencies of Lon and I thought it would be interesting to look at some major international applications of Dali and Lon systems.

The first project to consider is the unique new headquarters of Swiss Re in London.

The futuristic Swiss Re Tower is a 41storey, 180m high, 76,000 m^2 commercial office building and is London's first environmentally progressive tall building. Its unique curved shape is designed to create ventilation up through the building and allow natural cooling for 40% of the year. The spiralling lightwells bring daylight deep into the building to minimise the need for artificial lighting and further reduce energy consumption.

The Swiss Re Tower incorporates the latest Dali addressable dimming technology which enables light fittings to be individually addressed, switched, dimmed and monitored along a single buswire. The lighting management system includes ceiling - mounted multisensors which relate lighting to occupation of areas as well as the amount of daylight. The multisensors energise lighting when an area becomes occupied and automatically set the lighting level to take account of the amount of daylight in the area-so lighting is only on were staff are working and the energy consumed is minimised when free daylight from the sun is available. Reducing lighting levels also reduces cooling loads thereby further enhancing the greenness of the building.

If we look at the system architecture we see the network buswires throughout the building. The Dali luminaires connect to the DELi control modules and the Dali sub-systems connect via the Lon bus. Multi-sensors provide daylight-linked and presence related control of lighting with local dimming control from hand-held transmitters in cellular offices. Multi-circuit switching modules control lighting in core areas and a floor control router on each floor links to a head-end PC which manages and monitors the complete building through graphical software. Dali emergency test and monitoring units within emergency lamps monitor lamp and battery performance and pass information to the head-end PC for proactive maintenance.

But the energy saving benefits of Dali dimming apply not only to commercial office buildings but may also be enjoyed on housing developments. The next case study relates to the application of energy efficient lighting management to public housing blocks in Singapore where a pioneering project has achieved 70% energy savings while enhancing lighting control and maintenance.

A typical residential block lighting installation (before enhancement) has a standard and uniform level of lighting throughout the night and uses a very labour intensive way of inspecting and replacing failed lamps with maintenance engineers having to visit and tour blocks on a regular basis.

The new system has been installed without changes to the mains wiring and enables the Town Council to:

- optimise the use of energy and achieve considerable savings for public housing electricity bills;
- adjust the lighting levels to suit residents' needs and the various designs of the housing buildings;
- control and set various lighting levels according to times of the day for individual or group of lamps;
- detect in real time lamps which have reached the end of their life or have failed;

- generate data such as hours-run, required lamp replacement schedule for proactive maintenance;
- manage and monitor multiple sites from central location;
- adopt a pro-active maintenance approach and achieve faster response time to replace defective lamps.

The new system has achieved savings of 70% in energy consumption as shown in the following two tables.

Table 1 shows how the previous arrangement of lighting being fully lit between 1900 hours and 700 hours has changed to a phased regime where lighting is dimmed by different amounts at programmed times.

Comparison of the Annual Power Consumption between the existing and new lighting system. Table 1

Comparison of the Annual Pawer Consumption between the Existing and New Lighting System

Kwh(two blocks)

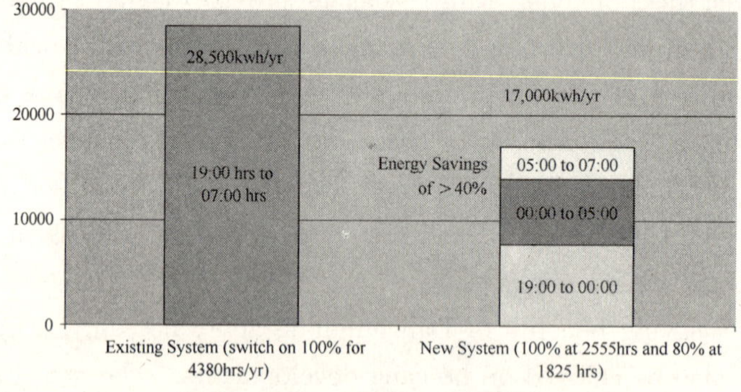

Table 2 shows the results of the energy monitoring of the blocks and the 30% savings achieved by changing the luminaire ballasts from magnetic type to high frequency digital type and the additional 40% savings achieved through dimming controls reducing light levels during the night.

Blocks Energy Monitoring Chart Table 2

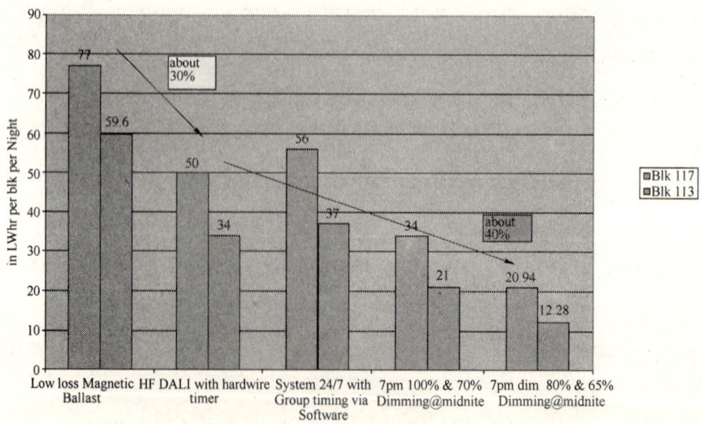

The Town Council can remotely manage and monitor the lighting installation in all the blocks and the graphic screens on the remote PC display the active status of lighting as well as logging lamp run time and reporting on lamp failures.

So far we have focussed mainly on the energy savings achieved through digital controls and Dali dimming. Let us turn our attention now to Lon and the advantages of the open building services protocol. One advantage of an open protocol for building services is that hardware from multiple suppliers can be integrated together and configured through software supplied by a third party.

Such is the case at Victoria Prison in Hong Kong. For this project energy-efficient presence detectors and switching/dimming lighting control modules were purchased by the Hong Kong network integrator who configured the system using third-party software and created an interoperable system for the benefit of the client.

Another prime example of Lon integration(and world-class energy-efficiency) is the new headquarters of ADIA in Abu Dhabi.

ADIA is the leading investment company within the UAE and the 80, 000 sq. m building(the tallest in Abu Dhabi) comprises a totally glazed facade and is a prominent landmark on the Abu Dhabi waterfront. The building is the first in the region to include fully integrated systems for lighting, HVAC and window blinds to maximise energy savings and optimise user control and comfort.

Lighting, air - conditioning and window blind controls are supplied by three individual companies and yet are fully integrated and share a common buswire network. Infra-red desktop multi-controllers provide individuals with the ability to switch and dim lighting, adjust the room temperature and raise, lower and tilt the window blinds, so providing individuals with total control of their local environment: the desktop controller also contains a thermostat which measures the room temperature. To optimise energy efficiency, local control is permitted within certain parameters(for example which limit the permitted temperature range) The desktop-controller beams infra-red messages to a ceiling-mounted receiver which transmits lighting, HVAC and blind Lon messages onto the network: the receiver which also contains a presence detector to switch off lighting and reduce the air-conditioning speed when the office is vacated to maximise energy efficiency.

So how does the integrated system work at ADIA? The system operation & communication is shown in figure 1.

The diagram shows the lighting controller, blind controller and VAV controller(each from a different supplier)as well as the infra-red desktop multicontroller and the infra-red multisensor receiver. The multisensor includes a presence detector which sends occupancy messages to the lighting controller and, via the Lon coupler, to the VAV controller. Infra-red signals are transmitted from the desktop controller to the multisensor which puts the various commands onto the shared Lon bus. Lighting on, off and dimming commands

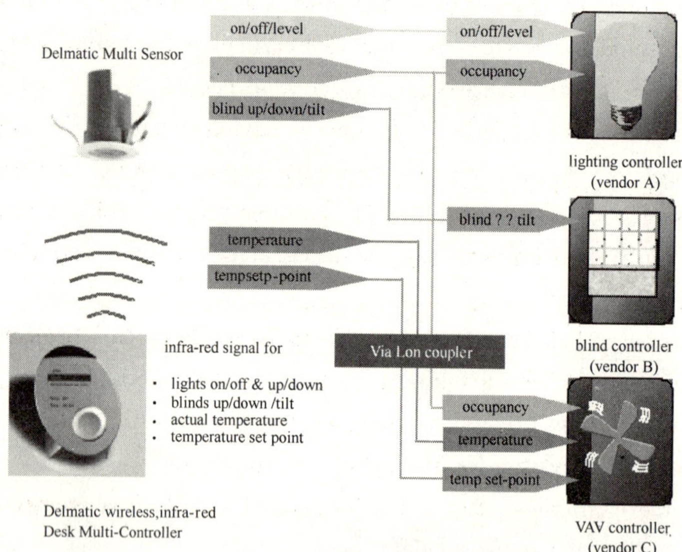

Figure 1

are received by the lighting controllers, while blind up, down and tilt commands are received by the blind controllers which are connected to the same Lon bus. Current room temperature and user temperature set-point data are sent via the Lon Couple to the BMS which adjusts the speed and temperature in the room.

So how does this work in practice?

When an office is unoccupied, the air conditioning runs in background mode: this operation is purely a function of the BMS system. When somebody enters the office the multisensor broadcasts a Lon occupancy signal onto the network buswire: lighting switches on to a programmed level and the BMS increases air-conditioning to occupied speed.

While the office is occupied, the presence detector within the multisensor keeps lighting on and air-conditioning remains at occupied speed. At intervals, the desktop controller transmits temperature signals to the multisensor which broadcasts the actual temperature and set-point data to the BMS which adjusts room temperature.

If an individual wants to adjust lighting levels within his office he can do so from his desktop multi-controller. Lighting levels can be raised or lowered and commands are sent via the multisensor to the lighting controller. The individual can also adjust the temperature within his office from the desktop multi-controller and the new set-point is transmitted to the multi-sensor and(via the Lon coupler) to the BMS which adjusts the temperature within the room

Blinds can be raised, lowered and tilted from the desk controller and commands are sent directly to the blind controllers on the shared Lon bus.

Once the office has been vacated, the detector broadcasts an unoccupied message onto the Lon network: lights dim to off and air conditioning reverts to background speed so that energy usage is very closely linked to occupation of office areas.

We have looked at the system hardware and architecture. Now let us look briefly at software and the features that are available to the client and facilities manager. The graphical software is a window into the lighting installation and represents a graphical configuration and monitoring tool using building layouts as the basis for binding and system management.

Calling up a floor graphic displays real-time active status of lighting and shows every lamp against background floor layouts. Each time a lamp is switched on or off the operation is reflected in the graphics. Virtual wiring flexibility enables lamps to be programmed into soft groups which can be controlled together from switches, sensors, desktop controllers, telephones and desktop web browsers. Logged data showing the number of hours each luminaire has been in use is compared with the manufacturers' recommended lamp life to generate optimum relamping schedules, while the latest Dali systems enable lamp failures to be detected for prompt and efficient maintenance.

At the Hong Kong and Shanghai Banking Corporation's world headquarters in London software trend logging records the number of hours each lamp has been in use and calculates the energy consumption based on the lamp's wattage. The system than totals the energy consumption per department or cost-centre so that individual sections of the bank are charged for their energy consumption-a good way to focus the attention of heads of department onto the energy costs (and environmental impact) of their section of the operation.

We have come to the end of our international tour of Dali and Lon projects. Delmatic have been supplying control and management systems for more than 45 years and we are often asked what we think the future holds. We believe the future will follow three trends:

- LonWorks and LonMark integration of building services.
- Dali digital dimming becoming the norm for fluorescent lighting.
- Increasing convergence of building services and IT via IP.

And the purpose of these trends is to enhance the efficiency of our buildings and thereby benefit our planet and our children.

智能型开放控制系统实现既舒适又节能的效果

Stephen Woodnutt

(Managing Director, Delmatic UK-stephen)

摘　要：建筑物不仅消耗大量宝贵的能源，而且造成环境污染，加剧温室效应，这已是不争的事实。

本文介绍了智能型开放控制系统如何降低建筑物对环境所造成的不良影响。通过对商用建筑和住宅建筑的调查显示，无论是新建的还是改造的建筑物采用这一系统后，不仅为使用者提了供舒适方便的环境，而且节能效果能达到70%。

本文还介绍了两个在世界上广泛采用的开放式通信协议，即LonWorks总线技术和Dali总线技术。前者应用于建筑物自动化系统，后者应用于数字式节能照明系统。

在介绍的典型案例中，照明与空调系统集成到一起实现了控制与节能管理。开放系统可以共享通用的硬件和电缆，简化了系统，降低建设成本，而且用户可以从多个渠道选择供货商和服务商。采用无线通信和浏览器技术能对多座建筑物进行远距离监控和物业管理，实现最佳控制，达到节能和降低运行成本的目的。

ENERGY SAVING DESIGN-LIGHTING CONTROLS IN BUILDINGS

Rajan Venkateswaran, B. Arch
(Chief Architect, L&T-ECC Division, Chennai- 89, India)
R. Nagalingam, B. Arch
(Senior Architect, L&T-ECC Division, Chennai- 89, India)
Larsen &Toubro, ECC-Division
(Mount Poonamallee road, P. B. No: 979, Manapakkam, Chennai-89, India)

Abstract: The general objective of this paper is to explain the various types of lighting controls and their benefits with a case study.. This paper elaborates on "Energy efficient Lighting" dealing with specialized Lighting controls like Occupancy sensor, Time scheduling, Bi-level switching, Daylight dimming with a case study.

Key words: Occupancy sensor, Time scheduling, Bi-level switching, Daylight dimming

1 Introduction:

Good lighting design enhances architecture. Good energy efficient lighting design enhances both the design and the performance of buildings. Knowledge of the opportunities and constraints of available and emerging lighting technologies is key to success with this crucial design element. Lighting systems offer extraordinary opportunities for cost-effective energy savings. In addition, many strategies for reducing lighting energy used often can improve the visual environment.

The energy efficiency can be achieved by:

- Delivering the appropriate amount of light for the tasks that will be performed in the space, and distribute that light in a way that prevents glare.
- Take advantage of natural daylight whenever possible, while avoiding direct sunlight.
- Install appropriate lighting controls.
- Use of Building Automation Systems, high-efficiency fluorescent systems, high-intensity-discharge(HID)systems, electronic ballasts etc.

The main objective of this paper to explain the various types of lighting controls and their benefits with a case study.

2 Lighting controls:

Lighting controls can increase the value of a commercial building by making it more comfortable, productive, and energy efficient. But before any discussion about the latest and greatest control gadgetry or technique, it's important to consider what's going on inside that building: Who's working in all those different spaces? What are they doing? What configuration and tools will help them? "Architecturally speaking, the objective is to design environments for people."

The benefits that can be achieved for using lighting controls are not widely recognized. While reducing Lighting energy is a major reason to install lighting controls, employee satisfaction should also be given equal importance. By incorporating controls that vary the time of operation and the intensity of electric lighting systems, designers and owners can significantly improve the value of commercial buildings.

These controls may be used to make buildings more:

Comfortable, by allowing users to tailor lighting levels to their personal tastes.

Productive, by optimizing the work environment for whatever tasks are at hand. Eyestrain and fatigue may be reduced, which allows workers to concentrate better and for longer periods of time.

Energy efficient, by keeping lights off when they are not needed or by reducing lighting levels to match the task at hand.

Certainly, the advanced lighting controls contributed to that improvement. The lighting controls that make such benefits possible can be quite sophisticated, but in general, they perform two basic functions:

they turn lights off when not needed, and they modulate light output so that no more light than necessary is produced. The equipment required to achieve these functions varies in complexity from simple timers to intricate electronic dimming circuits. Each of these technologies can be applied individually with much benefit, but by creatively combining them, designers can deliver even greater value to their clients.

3 Types of Lighting controls are:

- Time based.
- Occupancy based.
- Ambient / Daylight based.

4 Occupancy Sensors:

Occupancy sensors are the most common lighting control used in buildings today.

Two technologies dominate: infrared and ultrasonic. Infrared sensors detect temperature changes in a room, and work well where the entire room is within the sensor's field of view. Ultrasonic sensors use high frequency sound, much like bats do, to detect motion(even around corners). Dual-technology sensors (Fig. 1) use both methods, increasing accuracy and flexibility, but at a higher price. Even though lamp running life may be somewhat shortened by increased switching due to occupancy sensors, the overall chronological life of lamps is usually extended by the reduced daily burn hours.

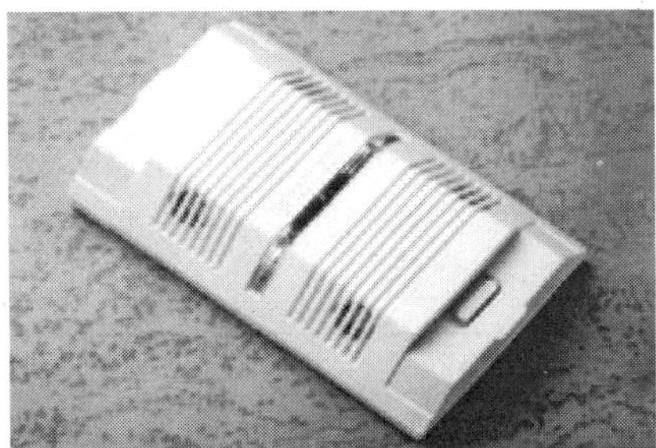

Fig. 1 Occupancy sensor-dual technology model

5 Mounting:

Occupancy sensors can be mounted on walls at the wall switch, on the ceiling or at the workstation. Wall-mounted sensors (Fig. 2) are suitable in smaller rooms-offices, bathrooms, and equipment rooms that are only intermittently occupied. In larger spaces or wherever the lighting load is higher, it is better to mount the sensor in the ceiling. Some units can be mounted in the corner or on the wall near the ceiling.

Fig. 2 Ceiling mounted occupancy sensor

6 Sensor coverage:

Ultrasonic sensors can detect motion at any point within the contour lines(Fig. 3). Infrared sensors "see" only in the wedge shaped zones, and they don't generally see as far as the ultrasonic units. Some sensors see further ahead than to the side. The ranges

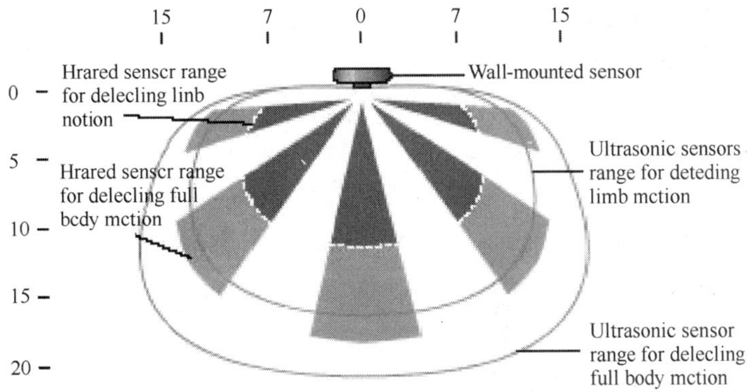

Fig. 3 Sensor coverage diagram

shown here are representative some sensors may be more or less sensitive.

7 Savings:

Typical range of Savings from occupancy sensor **Table 1**

Type of room	Energy savings %
Private office	13%~50%
Open plan office	20%~28%
Classroom	40%~46%
Conference room	22%~65%
Rest room	30%~90%
Corridors	30%~80%
Storage area / closet	45%~80%

Source: Advanced Lighting Guidelines, U. S EPA

Savings range by a factor of two or three in most applications, actual savings may differ.

8 Switching Options:

For wall - switch - mounted and ceiling - mounted applications, these options are available:

- Automatic-on. The occupancy sensor switches the lights on automatically when occupancy is detected and off automatically, after a preset time delay, when a lack of occupancy is detected.
- Manual-on(switch-mounted). The occupant must manually flip the switch to active the lights, while the sensor switches the lights off automatically in case the occupant forgets when exiting.
- Bi-level switching(switch-mounted). The two switches in a double-gang wall box each control one ballast in each two-ballast fixture, enabling dual switching between 50% and 100% of light output based on occupancy. The switches can also be connected to bi-level electronic ballasts-one that provides for 100% light output for all lamps and the other that provides a lower light output.

9 Time Scheduling:

Large open office areas work well with simple time scheduling-automatic switching at fixed hours of the day. Overrides allow users to turn on the lights after hours(using wall switches or telephone dial-up codes). Time scheduling can be accomplished with simple time clocks or more sophisticated computer controls(Fig. 4). To save more energy, time scheduling systems can be designed so that lights are turned on manually rather than automatically at the beginning of the day, but are turned off automatically at 1- or 2-hour intervals after close of business.

Fig. 4 BUS system-NETWORKABLE MICRO-MODULES

10 Bi-Level Switching:

Some people prefer lower overhead lighting levels(especially if daylight is available). Lower light levels are often preferred for computer use, meetings or tasks that are not visually demanding. Bi-level switching can provide simple manual control. For example, in a typical 3-lamp fluorescent fixture, the outer lamps are switched separately from the middle lamp, allowing the user to switch on one, two, or all three lamps. This low-cost measure is a minimum control requirement in some state energy codes, and can provide a simple means of load shedding during peak hours if the bi-level lighting circuits are remotely controllable.

11 Manual Dimming:

In rooms where different light levels are needed at different times, such as conference rooms and some private offices, the use of manually-operated dimming controls is a common solution. These controls can be either wall-mounted or, for convenience, use wireless remote controls(like the controls for a TV or VCR).

12 Automatic Daylight Dimming:

Automatic daylight dimming, or "Daylighting," uses a light sensor(Fig. 5) to measure the amount of illumination in a space. Then, light output from dimming ballast is adjusted to maintain the desired level of illumination. The combination of daylight dimming with appropriate task lighting is often very effective.

Corridors and open cubicles near windows, particularly those with task lighting, are good candidates for Daylighting controls. Private offices with windows can also be equipped with individual daylight sensors. Initial commissioning and calibration of light sensors and controls is critical for effective Daylighting, however; poorly calibrated daylight sensors can result in little or no savings, and may annoy occupants.

Fig. 5 Photocells

13 Demand Limiting:

During peak demand period's utilities often charge significantly higher prices for electricity. Remote operation of dimming ballasts or bi-level switching helps operators to respond to price signals or utility requests to shed load to help avoid power outages.

14 Typical Lighting Control applications:

shows the Typical Lighting Control Applications. (For office space) Table 2

Type of Control	Private Office	Open Office - Daylit	Open Office - Interior
Occupancy Sensors	++	++	++
Time Scheduling	+	++	++
Daylight Dimming	++	++	0
Bi-Level Switching	++	+	+
Demand Lighting	+	++	++

++ = Good savings potential

+ = some savings potential

0 = not applicable

15 Lighting Controls: Where and Why:

Lighting controls can save energy and reduce peak demand in offices and other facilities. Controls save money while providing user convenience and an improved lighting environment. There are several different kinds of controls. The choice of control type should be based on lighting usage patterns and the type of space served.

Areas with intermittent occupancy are well suited to occupancy sensors. In large, open office areas with many occupants, scheduled switching("time scheduling")is often an effective energy-saving strategy. In daylit offices, properly adjusted daylight sensors with dimming ballasts make sense. Because some workers prefer lower lighting levels, bi-level manual switching is another option. Advanced lighting controls can be used for demand limiting to allow building managers to reduce lighting loads when electricity demand costs are high.

Some types of lighting are not well suited to certain controls. For example, daylight dimming and occupancy sensing are not generally appropriate for high intensity discharge (HID) lighting (which requires a delayed re-start), whereas time scheduling is usually a good match for HIDs.

16 Appropriate Illumination Levels:

Proper illumination levels depend on the type of work being performed, and on occupant preference. Recommended illuminance levels for offices range from 30 to 60 footcandles*, but the quality of the visual environment can have a substantial impact on the "appropriate" amount of illumination. In well-designed office spaces, with light-colored surfaces, appropriate task lighting, and careful placement of lights and furniture to avoid glare and shadows, much lower illuminance levels are acceptable, and usually even preferred.

17 Installation and Maintenance:

Proper placement and orientation of both daylight and occupancy sensors is essential. Placement of controls should take into account furniture placement as much as possible. Occupancy sensors must be able to sense all occupants to avoid turning off lights while the space is occupied. At the same time, "false-on" incidents can be triggered by an automatic on/off sensor * that is exposed to passersby in an adjoining hallway.

Set time scheduling controls so that the switching times and intervals make sense for the occupants and usage pattern of the space. Occupants need to know how to override the schedule easily when needed.

Commissioning and calibration of lighting controls are essential if energy savings are to be achieved and maintained. Occupancy sensors with sensitivity set too high can fail to save energy, but occupancy sensors with too low a sensitivity or too short a delay time can be annoying to occupants. Similarly, improperly adjusted daylighting controls can dim the lights too low, causing occupants to override them (e. g. by taping over the sensor), or can fail to dim the lights at all.

* Automatic on/off sensors switch on when they sense an occupant and off when they don't. Better energy savings are usually attained from sensors that are manual-on/automatic-off. These avoid unneeded turn-ons.

18 Control ideas:

Table 3, below provides control ideas for several different room types and usage patterns.

Table 3

Type of room	Use pattern	If.	Then…
Private office	Primarily one person coming and going	Daylighted	Consider manual dimming, automatic daylight driven dimming, or automatic on/off
		Occupants are likely to leave lights on and occupants would be in direct view of a wall mounted sensor	Consider a wall mounted occupancy sensor
Open plan office	Usually occupied	Daylighted	Consider automatic daylight driven dimming
		Varied task from computer usage to reading	Consider manual dimming
		Lights left on after hours	Consider centralized controls and/or occupancy sensors
Classroom	Usually occupied & Occasionally occupied	Occupied by different students and teachers	Consider ceiling or wall mounted occupancy sensor and manual dimming. Make sure that minor motion will be detected
		Lights left on after hours	Consider centralized controls and/or occupancy sensors
Cafeteria/ Lunchrooms	Occasionally occupied	Occasionally occupied	Consider ceiling or wall mounted occupancy sensor Make sure that minor motion will be detected
Retail Store	Usually occupied	Different lighting needs for retail sales, stocking and cleaning	Consider centralized controls or preset scene-dimming controls

Source: Lighting design lab

19 Case study: EDRC Building, Chennai.

The above concepts of "energy efficiency" are implemented in the State of art EDRC building(Engineering, Design & Research Center)of L&T Chennai.

Besides lighting control, this building has, Automation / Control for:

HVAC

Sanitary

Water system.

Fig. 6

Fig. 7 The "TIME" based lighting control in side the Main hall is shown here
Soft timers achieve this. Also seen manual over ride being used to have selective areas

Fig. 8&9 The "OCCUPANCY" based lighting control inside the cabin is hown here . This is achieved by PIR(Passive Infra Red)motion detectors

20 Added advantages due to intelligent lighting control:

- Reduction of man-hours in maintenance operation.

Fig. 10 PIR(Passive Infra Red)motion

Fig. 11 Interior View of office area

- Bi-stable relay was used in lighting control resulting in energy savings.
- Lighting control over 2500 fixtures.
- Lighting scheduling for busy hours / lean hours.
- The total building cost: 2100 lakhs apprx.
- The electrical system cost: 700 lakhs apprx.
- The building automation system cost: 67 lakhs. (inclusive of software, attendance management system, surveillance system etc.)
- This is< 3% of the over all building cost.
- Approx. 8640 units/annum is saved. (This is only for main building lighting. The saving accrued on control in cabins, ambient based lighting is not included.)
- Reduced impact on power bills.
- Due to the reduction in burning hours lamp life getting enhanced.

21 Conclusion:

Lighting controls have traditionally been used to create moods. Today, they are also used as part of a high quality energy efficient lighting system that integrates daylight and electric light sources to provide a comfortable and visually interesting environment for the occupants of a space. The "Energy crisis" prevailing in the country due to unsustainable and inefficient use of most Energy resources, has brought home the urgent need for developing a strategy for the conservation of energy and its resources. Improvement of Energy use efficiency will have to be paid more attention from now on as otherwise the demand will outstrip the availability in a few decades from now.

建筑物中的节能照明控制设计设备

Rajan Venkateswaran, B. Arch
(Chief Architect, L&T-ECC Division, Chennai-89, India)

R. Nagalingam, B. Arch
(Senior Architect, L&T-ECC Division, Chennai-89, India)
Larsen &Toubro, ECC-Division
(Mount Poonamallee road, P. B. No: 979, Manapakkam, Chennai-89, India)

摘　要：本文的主要目的是说明建筑物中各种的照明控制以及他们的效益。论文通过个案研究，详细介绍了“节能照明”在专业照明控制中的综合运用，如“占用状态传感器”，“定时调度控制”，“双档开关”以及“日光调光装置”等。

References

[1] www. energydesignresources. com.
[2] www. 1rc. rpi. edu
[3] www. eere. energy. goy
[4] Interiors today

ENERGY-SAVING STRATEGIES USING LONMARK OPEN SYSTEMS

Barry Haaser
(Executive Director LonMark International)

Abstract: Although open systems are relatively new in China, they are commonly used to create intelligent, energy-saving buildings throughout the world. This global phenomenon involves taking a common networking standard and embedding it in a wide variety of products and subsystems deployed in buildings. By logically linking products from different vendors together in a control network, it is possible to create very intelligent, energy-efficient buildings. Common interaction between individual devices-rather than stand-alone subsystems, such as lighting, access control, energy management, energy meters, and life safety systems-can provide direct economic savings on the initial cost of deployment as well as year-to-year energy savings. This paper describes the network standard commonly used and provides examples of buildings benefiting from this approach.

1 Introduction

There is a natural evolution in technology that is completely transforming the building industry worldwide. This evolution involves taking simple computing and networking concepts and applying them to the devices installed in buildings. Perhaps the best way to understand this concept is to look at mainframe computers: It wasn't that long ago when mainframes ruled the computing world. There used to be a common phrase used in the computer industry, stating, "Nobody was ever fired for buying from IBM." This phrase emerged because only a few short years ago, everything was purchased from a single company: computers, data storage, terminals, printers, paper, and service contracts. These systems used a variety of proprietary technologies, such as microprocessors, communication protocols, and operating systems.

In the 1980s, everything changed with the introduction of the personal computer. We rapidly witnessed the demise of mainframe and minicomputers as the computer industry adopted open systems. Soon we were able to purchase products from multiple suppliers to either produce our own computers or create networks of computers. This was made possible through the adoption of standards for microprocessors, communication protocols, and operating systems.

There is a similar transformation taking place in the building-automation industry. Old mainframes or legacy building-automation systems are being replaced with less expensive, open systems. Rather than purchasing everything from a single vendor, facility professionals are purchasing the best products, from the best vendors, at the best price through open competition. For the first time, facility professionals are able to reduce the cost of deploying and maintaining systems while lowering energy costs.

Since approximately 80% of the energy consumed in commercial office buildings comes from heating, ventilating, air conditioning, and lighting, it is only natural to develop ways to optimize building efficiency. By incorporating open, interoperable solutions in buildings, it is possible to create dynamic interactions between individual devices, subsystems, and systems-resulting in dramatic improvements in operational efficiency and energy savings.

2 Introducing LonMark® Open Systems

Just as we witnessed the adoption of open standards in the computer industry, the same evolution has taken place in the controls industry. For over a decade, LonMark International has been creating and publishing guidelines for open, interoperable controls. The leading manufacturers around the world use the LonMark open systems standard as the foundation for products installed in buildings; including lighting, security, access control, building-management systems, energy meters, emergency generators, and more.

Each manufacturer uses the standardized LonWorks® network platform to create intelligent, interoperable products. The products all use a common communication protocol that makes it possible to connect them together in a single building network. By installing twisted-pair wiring, it is possible for independent integrators to purchase products from different vendors that have been certified by LonMark International and install them into a single network. The myriad of workstations that typically occupy valuable space in the control room can be replaced with a single, Web-based computer interface. Multiple buildings can be easily linked together-providing real-time access to essential data.

3 Boeing Abandons Legacy Systems For Energy-Saving Open System

Boeing was very frustrated with the limitations of their legacy Energy Management System(EMS). This system locked facilities into one system, which required expensive gateways to connect to other manufacturers' equipment. This built-in obsolescence provided very limited access to the user interface, which required a software-license fee for each computer tied to the system. Tired of being stuck with obsolete protocols-combined with increased pressure to control costs and optimize facility performance-Boeing decided that they needed to adopt an open system for their facilities.

Boeing was attracted to the LonMark open system because it is supported by hundreds of manufacturers, and it would allow them to marry the best equipment and software from numerous suppliers-delivering "the best of the best." They recognized the power of competitive bidding on projects so that multiple integrators could work on jobs. Best of all, maintenance could either be performed in-house or competitively bid by independent integrators. This new, flexible system would eliminate expensive gateways to connect vendor equipment together-delivering a more flexible solution with lower implementation and support costs.

Boeing began the process of updating the facilities systems by replacing its copper cabling with a fiber-optic network. Systems integrators connected the existing fire panels, smoke detectors, and sprinklers in several buildings to a new, central-monitoring system. The previous system required extensive "point-to-point" wiring and several very large and expensive gateways. The entire system was dramatically streamlined and condensed into a very small, self-contained box.

Fig. 1 Boeing Fire Panel System Before

Fig. 2 Boeing Fire Panel System After

In the next phase of the project, Boeing replaced their old, proprietary EMS system by installing the "best of breed" products from leading suppliers-such as controllers for the central plant, variable air-volume controllers, programmable controllers, variable-frequency drives, and a variety of I/O.

In the final phase of the project, Boeing set out to solve a common problem involving the constant use of the building's lights, which were on 24-hours-a-day, seven-days-a-week. The lighting-control system was ineffective, so Boeing retrofitted the building with a new, open lighting-control system. The new, open system was tied together with a Web-based Graphical User Interface (GUI) that allows building occupants to view their space temperatures and override their lighting from their desktop computers. Maintenance and engineering have access to the Web-based GUI that also provides links to all maintenance manuals and drawings.

Boeing was able to reduce their energy costs by 37% over the first 18 months following project completion. This equated to a reduction of USD 215,000 per aircraft.

4 Singapore Management University Reduces Campus Energy Costs

Located in the heart of the city, the Singapore Management University's(SMU)City Campus is the first of its kind in the historic Bras Basah district. The SMU campus is part of an ambitious city-planning exercise to revitalize community life in the heart of the city. When the university moved from its old Bukit Timah campus to the new city campus they wanted to utilize a state-of-the-art Intelligent Building Management System(IBMS).

Unlike commercial office buildings that operate with a more or less constant load through fixed office hours, universities are dynamic and constantly changing. The student population moves within the campus and energy loads(especially for cooling)are shifted from one location to another. The number of classes and lecture halls in use varies from one day to the next. The university's timetable is nowhere near as predictable as a commercial office building's. Efficient management and control of air-conditioning equipment is especially important since energy used for cooling traditionally takes up to 60% or more of the energy-consumption bill. Proficient management of varied and changing load patterns presented here can only be handled by intelligent, automated controls that are built upon the latest technological innovations and that can react to these changes.

To leverage the benefits of the LonMark Open System, the air-conditioning and mechanical-ventilation(ACMV)system was designed to use devices from different vendors to gain maximum effectiveness. The system is not locked into a single vendor, making future expansion easier. The ACMV system is made up of eight chillers in two different plants with a combined maximum capacity of 5550 tons; over 110 air-handling units; 1500 variable air-volume(VAV)boxes, mechanical-ventilation fans, and dampers. The system is capable of managing the ACMV system to optimize cooling from the chiller plants in order to maximize energy efficiency. The distributed-control system provides essential coordination and feedback from a user-friendly front-end to optimize energy efficiency and real-time load control. The distributed-control system, installed in the chillers in each plant, supports local control to shifting load patterns, thus eliminating the need for expensive central controllers.

The system has seamless connectivity to an Internet Protocol(IP)backbone in order to maximize the investment in the IT infrastructure, thereby eliminating the need for homerun wiring across the huge campus. A Web-based Human Machine Interface(HMI) was developed for the IBMS system so that facility personnel can use a single, user interface to view and interact with the system. While fully integrated, the Web-based HMI can also view other subsystems using different protocols.

Expanding on their control-network infrastructure, SMU decided to achieve additional

Fig. 3 Singapore Management University Control Room

energy savings through lighting control. Because the lighting-control system installed in halls, theaters, group study rooms, and other parts of the campus use the same protocol, it was easily integrated into the Electronic Facility Management System(EFMS). Lecturers and speakers now have complete control over the lights in their rooms-ensuring that facilities will be well-lighted, safe, and secure. Further integration is supported by linking lighting controls with access controls. Unauthorized access to facilities can trigger alarms and activate an "all lights on" event to help locate intruders. When authorized personnel access facilities, lights are automatically switched on. When people leave rooms and lecture halls, lights are automatically turned off after ten minutes to help reduce energy bills.

Further expanding on their investment in network infrastructure, SMU connected other elements to the system, including electric power metering(main incoming), sub-metering, fire-alarm monitoring, lift monitoring, and synchronization of the campus-wide digital master-clock system to ensure students get to their classes on time.

The City Campus was completed in July 2005, so it is too early to determine the true impact of the various energy-saving solutions. Based on previous experience with open systems, SMU is confident that they can achieve at least 20% savings in electrical consumption and are confident that even more savings can be achieved in the new City Campus, where more devices and systems are linked together; providing even more choices and flexibility in implementing energy-saving strategies.

5 Echelon Achieves Significant Savings Through Integrated Facility

As the inventor of the LonWorks platform, the foundation of a new generation of open, interoperable products for industry, Echelon Corporation's new headquarters in San Jose, California is a showcase for the company's technology. Completed in 2001, this

22,860 m^2 facility consists of private offices for the company's 170 headquartered employees. This created quite a challenge for the designers to optimize energy efficiency for so many rooms.

The open-systems specification required the facility to integrate the key building sub-systems-security, lighting, lift, and HVAC systems - together using LonMark Guidelines-certified devices from multiple manufacturers into a single, intelligent building. This open-systems approach, for example, allowed VAV controllers from two different suppliers to be installed in the building. Similar examples exist for other forms of automation deployed throughout the building.

For the facility manager, Echelon's new office provides the unprecedented ability to monitor and control lighting, security, HVAC, lifts, and many other functions in the building. The operator can be at the building site or at a remote location and access the devices via the Internet using a PC. This remote control is enabled by using Internet Servers, which use an IP backbone to route the building's sub-systems data to the remote operator.

As soon as the building became operational, facility engineers started "fine tuning" the building in order to reduce energy consumption. One of the first projects involved installing intelligent temperature sensors throughout the building. This was a relatively easy task since the sensors were simply connected to the existing twisted-pair wire network. Next, engineers started implementing schedules for the lobby lights and light-harvesting strategies in hallways to automatically adjust to outdoor ambient-light levels.

In addition to a full Web-enabled interface for facilities managers to monitor the building's control systems, Echelon's office provides several benefits to Echelon's employees: By creating a unique Web interface for each office, Echelon has made it simple for employees to setup their individual workspace preferences. Employees can call-up the building's interface to select individual lighting levels and temperature settings. An advanced occupancy-sensing strategy coordinates lighting and temperature to desired levels when the office is in use. Energy usage is minimized when the space is unoccupied. This eliminates waste, conserves energy, and cuts costs.

Recognizing the impact of natural gas on energy bills, engineers adjusted boiler temperatures based on time of day and usage patterns. Occupancy audits were completed to monitor employee work habits to identify additional energy savings, such as not lighting unoccupied offices. Coordinating with the local energy utility, building engineers implemented load-shedding programs to reduce energy consumption during peak periods. Once a notice arrives from the utility, the building is automatically able to turn-off non-essential lights, raise building temperatures, strategically shut-off VAV valves, and increase remote-terminal-unit setback temperatures.

Due to Echelon's energy-saving strategies, the company was able to significantly reduce its energy bills; despite increases in electricity and natural-gas consumption, as shown in Fig. 4 and 5.

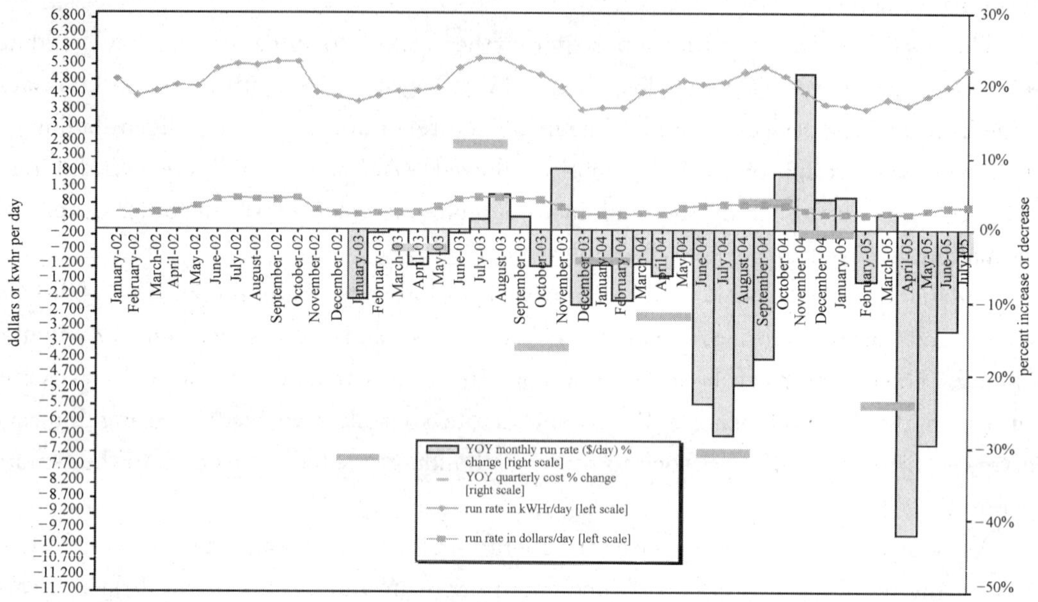

Fig. 4 Gas Savings Over 1 Year

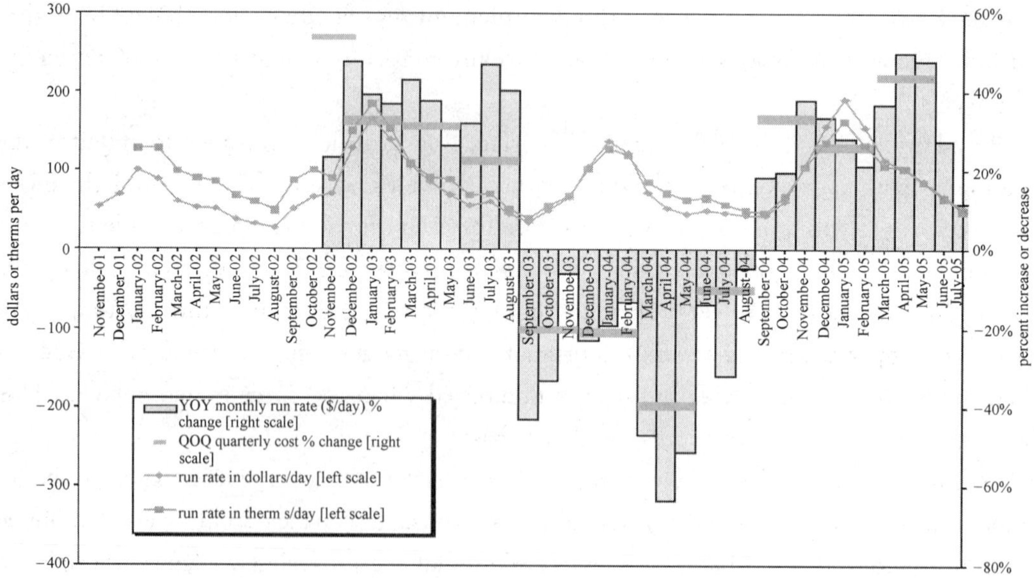

Fig. 5 Electricity Savings Over 1 Year

Conclusion

Adoption of open, interoperable solutions in buildings can have a dramatic impact on energy consumption without causing any discomfort or inconvenience to its tenants. The ability to shop for "best of breed" products gives facility professionals the ability to shop around for the best products at the best price with the best features. Not being locked into a single vendor provides more granular control of building systems and better access to important operational data. Creating a cohesive network connecting all essential devices and subsystems together gives facility professionals the tools required to create better energy-reduction strategies. By simply creating interoperability between HVAC and lighting, it is possible to achieve savings of 30% or more.

When planning new buildings, it is extremely important to balance requirements for "green" and intelligence. By investing in the right open-system architecture, it is possible to minimize the building's environmental impact while improving building efficiency.

利用 LonMark 开放系统的节能策略

Barry Haaser

(Executive Director LonMark International)

摘　要：开放式系统虽然在中国还相对较新，但在世界上却普遍用于智能和节能建筑。这一全球性现象涉及到采用通用的网络标准，并使其广泛应用于建筑所涉及的产品及其子系统中。通过逻辑上联接，将不同供应商的产品整合到同一个控制网络中，是可以实现真正智能和节能建筑的。个体设备之间的交互作用，而非独立的子系统，如照明设备、门禁控制、能源管理和计量、安全防范系统可直接带来前期经济上成本的降低，并与年度同比的能源节约效果同样好。本文描述了通用的网络标准及从中受益的建筑实例。

References

[1] LonMark magazine, 5th Edition(2005). Singapore Management University Implements Open, Interoperable Solution for Their New City Campus. San Jose, California: LonMark International

[2] Jeffrey Habermann(2005). Boeing Long Beach Building Automation System. Paris: LonWorld Expo and Conference 2005

[3] Mark Kendall(2005). Energy Reduction on the Echelon Campus. Paris: LonWorld Expo and Conference 2005

AN INFORMATION VISUALIZATION APPROACH TO INTELLIGENT BUILDINGS ASSESSMENT

Ju Hong[1]; Zhen Chen[2]; Heng Li[3]; Qian Xu[3]
([1] Beijing Institute of Civil Engineering and Architecture;
[2] University of Reading; [3] Hong Kong Polytechnic University)

Abstract: This paper presents a Knowledge-oriented Information Visualization (KIV) approach to facilitating the implementation of building rating systems such as the Asian Intelligent Building Index (AIIB) for the post-assessment of Intelligent Buildings (IBs). The KIV approach is introduced by using a prototype of KIV for IB assessment. The KIV model consists of several toolkits, including an IB Casebase for the Applications of Sustainable Technology (called IB-CAST), a web-oriented information visualization toolkit for IB assessment using Microsoft Office (called IB-Radar and IB-Compass), and a Geographical Information System (GIS) toolkit for bridging knowledge (called IB-GIS) in IB assessment etc. A case study is used to demonstrate how the KIV approach can be applied to support IB assessment.

Key words: Intelligent building; Assessment; Methodology; Model; Information visualization

1 Introduction

For building assessment, the currently most popular way is to use rating method. For example, the U. S. Green Building Council (USGBC, 2005) has developed the LEED Green Building Rating System for developing high-performance, sustainable buildings. Regarding the assessment of Intelligent Buildings (IBs), the authors have reviewed six IB assessment systems adopted or to be adopted all over the world, including AIIB method developed by the Asian Institute of Intelligent Buildings (AIIB, 2005), BRE method developed by the Building Research Establishment Ltd. in UK (Bassi, 2005), CABA method developed by the Continental Automated Building Association in North America (CABA, 2004), TIBA method developed by the Architecture and Building Research Institute in Taiwan, China (Wen, 2003), IBSK method developed by the Intelligent Building Society of Korea in Korea (IBSK, 2002), and SCC method developed by the Shanghai Construction Council in China (SCC, 2002). Among these IB assessment systems, the IB Index is identified as the most comprehensive one with more potential utilizations in IB assessment.

However, it is also noticed that all building assessment systems stop at achieving a score for each building under assessment. Regarding how to improve the building based on assessment results, there is still lack of necessary replenishments. One possible explanation to this query is to further review audit paper for detailed information regarding which part of the building need to be revised to get a higher score. However, this is usually time consuming and not accurate. In this regard, toolkits to support the post-assessment of IBs are in demand.

Information visualization is an inherent means in the building professions no matter whether computer is used or not. It can be realized by means of various diagrams, maps or photos etc. For example, Bouchlaghem, Sher, and Beacham(2000)made an application of digital imagery and visualization materials to aid building technology related modules in higher education. In order to facilitate the post-assessment of IBs, including active gap identification and reliable decision support, this paper will introduce a knowledge-oriented information visualization(KIV) approach is introduced in this paper. The KIV approach comprises a group toolkits including a casebase for the knowledge management of IBs, called IB-CAST, a radar chart made in Microsoft Excel to illustrate current and gap scores of IBs based on their rating, called IB-Radar, an internet-enabled portal for IB information navigation based on IB-Radar chart, called IB-Compass, and a geographical information system for spatial information management of IBs, called IB-GIS. A case study will be conducted to demonstrate how to use the KIV approach in the post-assessment of IBs.

2 Methodology

The methodology adopted in this research comprises several methods focusing on detailed objectives mentioned above. First of all, an extensive literature review is required to explore achievements in the areas of information visualization, knowledge management, decision support, and building assessment etc. After that, system analysis will be used to develop a prototype for applying the KIV approach to IB assessment. For each toolkit of the KIV approach, system analysis and development techniques will be adopted to realize their various functions. In addition, software tools such Microsoft Office will be used to make toolkits easy to be used in the building professions.

3 The KIV model

The KIV model is a prototype to describe the theory of the KIV approach for IB assessment. Figure 1 illustrates a conceptual KIV model.

In order to realize the function of IB assessment for the KIV approach, several toolkits are integrated to an IB Data & Knowledge Management Platform(IBIMP), which

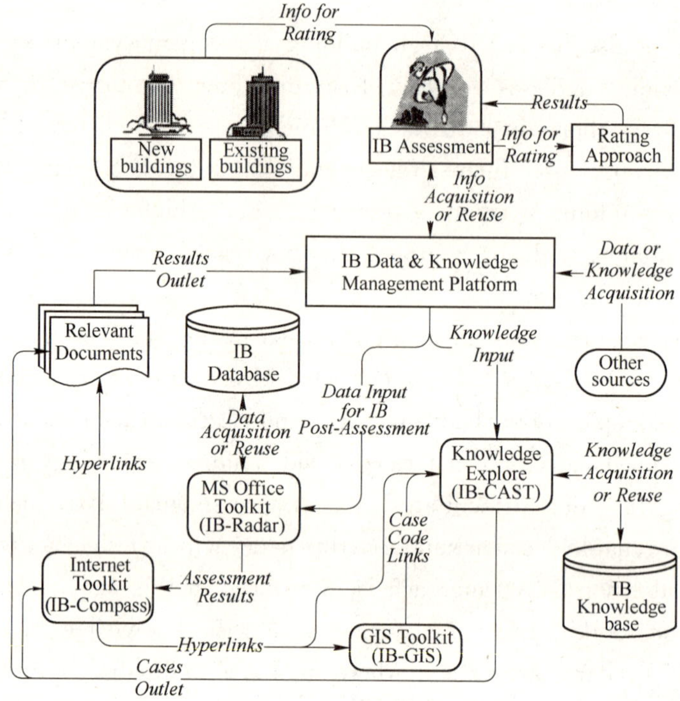

Figure 1 the prototype of KIV for IB assessment

is an open information platform by system users including Building Designers, Construction Contractors, Property Managers, Recyclers or Manufacturers, and Educators, etc. to share their data or knowledge with the community or their peers (Hong, et al., 2006). There are three main toolkits, including an IB Casebase for the Applications of Sustainable Technology (called IB-CAST), a web-oriented information visualization toolkit for IB assessment using Microsoft Office (called IB-Radar and IB-Compass), and a Geographical Information System (GIS) toolkit for bridging IB knowledge between IB-Compass and IB-CAST in IB assessment (called IB-GIS). For each toolkit, its functions and relations to other components of the KIV model are explained below:

3.1 The IB-CAST Toolkit

The IB-CAST is a casebase for the knowledge management of IBs. It is a result of a series of research focusing on system prototype, knowledge classification, case assessment, and knowledgebase. The system prototype is a product model for IB-CAST system development. The knowledge classification is a category of buildings for sorting IB-CAST case materials. The case assessment is a case selection process for evaluating case materials for the IB-CAST based on a knowledge-driven multi-criteria decision-making model. The knowledgebase is a consummative knowledge system with various e-book formats. To facilitate the utilization of IB-CAST, both online and offline versions are developed. For example, the IB-CAST portal is to be used for online education, training and consultancies in the area of IB, and the IB-CAST DVD is an e-book detached from IB-

CAST portal and to be used for offline education, training and consultancies in the area of IB. In order to facilitate knowledge management, IB-CAST shell as a KM tool is adopted to accumulate knowledge materials for the IB-CAST portal. Figure 2 illustrates one of the main interfaces of the IB-CAST(Baltsoft, 2004), in which a case study on the Hong Kong International Finance Centre(IFC)complex is shown.

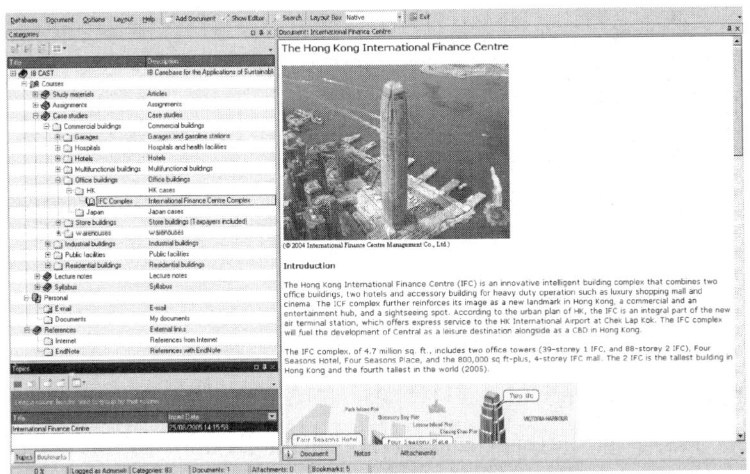

Figure 2 A screenshot of the IB-CAST

3.2 The IB-Radar Toolkit

The IB-Radar is a radar chart made in Microsoft Excel to illustrate current and gap scores of IBs based on their rating. It uses an n-dimension diagram to illustrate the status of an IB in accordance with its scores corresponding to each dimension. The number of dimensions of the IB-Radar equal to the number of clusters, which are used to classify indicators for IB rating. For example, the Asian Intelligent Building Index(AIIB, 2005) adopts a multi-dimension IB rating system covered ten models including Green Index module, Space Index module, Comfort Index module, Working Efficiency Index module, Culture Index module, High-tech Image Index module, Safety and Structure Index module, Management Practice and Security Index module, Cost Effectiveness Index module, and Health & Sanitation Index module. The IB-Radar is useful in describing the status of an IB regarding its score, and indicating the gap between current status and further efforts. It is a visual measurement for IB assessment, and can facilitate decision-making in the design or management of IBs. Figure 3 gives an example to show how IB-Radar can be applied in IB assessment. The example has been conducted by using an existing IB case, previously conducted by Chow and Choi(2005)using the IB Index(AIIB, 2005). There are two main radar charts adopted by the IB-Radar to illustrate current status of the building under rating from the IB Index(AIIB, 2005), and further effort required for a better design or utilization with respect to expected targets for the building. Based on visual information provided by the IB Radar(see Figure 3), building auditors can use the IB-Radar as an effective visualization approach to showing the results of assessment and measuring the gap from current status to ideal status of the IB under assessment. It is called IB post-assessment. Therefore, the IB-Radar is a necessary complement of building rating method such as the IB Index(AIIB, 2005)in IB post-assessment.

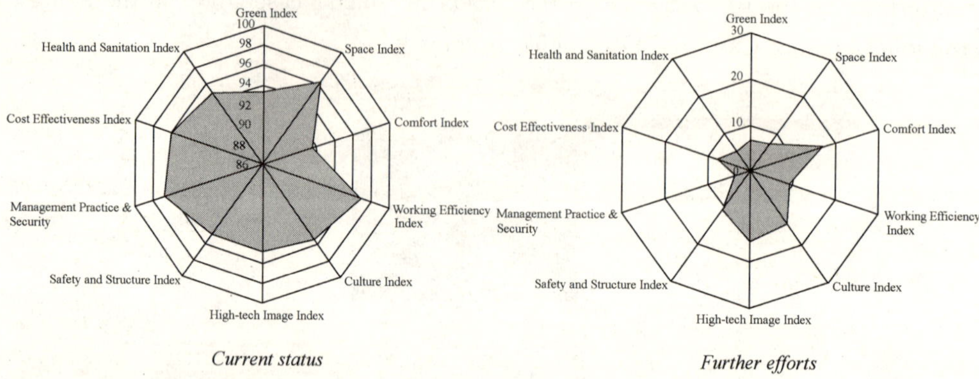

Figure 3 A screenshot of the IB-Radar using MS Excel

3.3 The IB-Compass Toolkit

The IB-Compass is an internet-enabled portal for IB information navigation based on IB-Radar chart. Generally speaking, the IBIMP (see Figure 1) can lead all potential participants from each life-cycle stage of IBs to release or to acquire relevant data and knowledge via Internet. As a matter of fact, all participants including Building Designers, Construction Contractors, Property Managers, and Academic Educators, etc. can benefit from the IB-CAST and the IB-Radar regarding the IB post-assessment and the content management of IB projects. However, they are actually demanding a more efficient delivery approach to facilitating IB post-assessment or others relevant to content management of IBs. As mentioned by JC&A(2004), of all the applications of the Internet in the design professions, none has more wide-ranging significance than web-based project management. In this regard, the internet-enabled portal called IB-Compass is developed to further support the post-assessment of IBs. Figure 4 gives an example of the IB-Compass in Microsoft Internet Explore. As shown on the IB-Compass(see Figure 4), different IBs are located at different places on relevant axes based on their IB scores to each dimension of assessment. The ten-dimension radar chart for the IB Index(AIIB, 2005) is adopted here. Although this example uses the chart of IB-Radar for the HK IFC complex(see Figure 2)as a compass, the compass can generally use an empty radar chart too. By using the IB-Compass, IBIMP users can go through each IB project for relevant data and knowledge previous collected by the IB-CAST or the IB-Radar by clicking on embedded hyperlinks to each building. Hyperlinks embedded to IBs will lead to other webpage or knowledge management toolkits such as the IB-CAST or the IB-GIS(see Section 3.4 below).

3.4 The IB-GIS Toolkit

The IB-GIS is a Geographical Information System (GIS) for spatial information management of IBs. The IB-GIS is another information visualization toolkit to support the

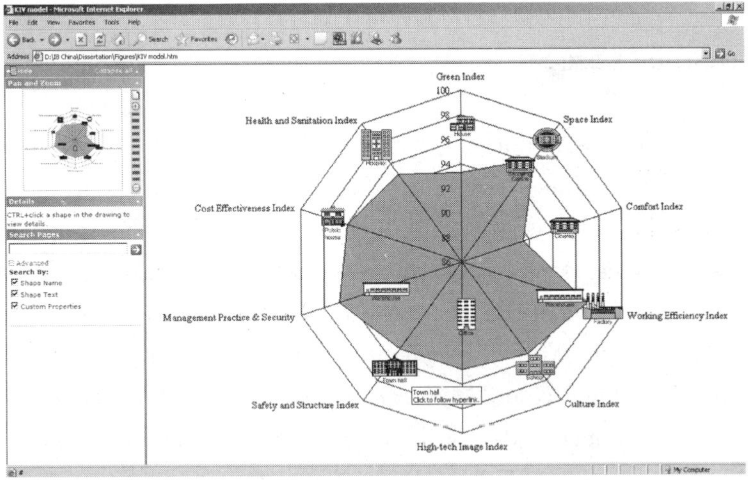

Figure 4 A screenshot of the IB-Compass in MS Internet Explore

post-assessment of IBs. The main reason why the GIS is adopted is that a comprehensive IB assessment actually requires an extensive information support. In other words, although the IB-CAST and IB-Radar can provide some relevant information for IB post-assessment, they are not able to support decision-making regarding suggestions to single building development or local area regeneration. In this regards, the GIS is introduced to the IBIMP(see Figure 1) to integrate a decision-making function. As illustrated in Figure 1, the IB-GIS is linked from the IB-Compass, and linked to the IB-CAST. To further support decision-making within the IB-GIS, integration with the IB-CAST is required. Figure 5 gives an example to apply GIS for IB information management. The HK IFC Complex is also used to demonstrate its usability. Google Earth is adopted as a novel GIS shell for the IB-GIS because it uses aerial colour photography as mapping interface which involves much information than hand-drawn maps.

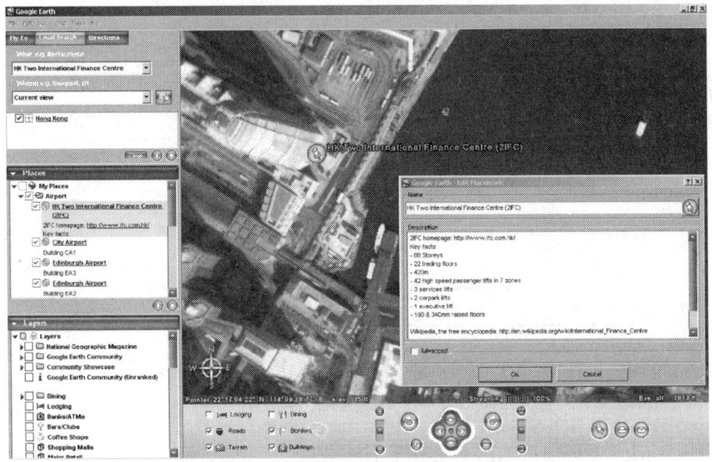

Figure 5 A screenshot of the IB-GIS using Google Earth

4 An Experimental Case Study

The Hong Kong International Finance Centre (IFC) is an innovative intelligent building complex that combines two office buildings(IFC I(39 storeys) and IFC II(90 storeys)), two hotels and accessory building for heavy duty operation such as luxury shopping mall and cinema. It is located adjacent to the narrowest crossing of Hong Kong (HK) Victoria Harbour, one of the most beautiful urban sites in the world, marking a new gateway to the HK Island. After the completion of the IFC II in 2003, it became the new headquarters for the Hong Kong Monetary Authority, and the ICF complex further reinforces its image as a new landmark in Hong Kong, a commercial and an entertainment hub, and a sightseeing spot. The project reflects the importance of HK as a world financial centre. According to the urban plan of HK, the IFC is an integral part of the new air terminal station, which offers express service to the HK International Airport at Chek Lap Kok. The IFC complex is playing an important role to accelerate the development of Central as a leisure destination alongside as a CBD in Hong Kong.

According to the characteristics of IFC II, an IB Index based assessment was conducted to the tallest building in Hong Kong by the AIIB in 2004(Chow and Choi, 2005), and it has been identified that the IFC II is the most intelligent building in Hong Kong compared with all other buildings their had audited. Table 1 gives some of their assessment results for the IFC II in accordance to the ten modules adopted by the IB Index(AIIB, 2005).

The assessment results of the IFC II(Chow & Choi, 2005) **Table 1**

	Modules	Scores		
		Current	Target	Gap
1	Green Index	93	100	7
2	Space Index	96	100	4
3	Comfort Index	91	100	9
4	Working Efficiency Index	97	100	3
5	Culture Index	95	100	5
6	High-tech Image Index	95	100	5
7	Safety and Structure Index	95	100	5
8	Management Practice and Security Index	97	100	3
9	Cost Effectiveness Index	96	100	4
10	Health & Sanitation Index	95	100	5

For the post-assessment of IFC II using IB-KIV approach, the following steps are adopted:

- Step 1: make a case study using IB-CAST(see Figure 2);
- Step 2: make radar charts using IB-Radar(see Figure 3);
- Step 3: make compass pages using Microsoft Office(see Figure 4);

- Step 4: make place-mark with description using Google Earth(or other GIS shell) (see Figure 5);
- Step 5: study the gap information indicated in IB-Radar;
- Step 6: click on relevant hyperlinks on IB-Compass for improvement;
- Step 7: review relevant information from IB-CAST and IB-GIS;
- Step 8: make decisions to fill in the gaps.

In this case study, target scores to each module are set to 100 (see Table 1), accordingly gap scores to each module are calculated using Table 1 in Microsoft Excel. Based on this result, it is identified that Comfort and Green issues are the most important ones for improvement in IFC II. In order to find appropriate solutions to improve the levels of Comfort and Green, suggestions are searchable from other existing buildings marked on the IB-Compass(see Figure 4)(The compass background can be replaced by a radar chart of Gaps), which links to IB-CAST and IB-GIS to provide detailed suggestions. Regarding how to improve the status of Comfort Index and Green Index of IIFC II, this paper doesn't provide more details.

5 Conclusions

Although building rating method such as the IB Index is currently adopted in IB assessment, it is noticed that toolkits to support post-assessment of IBs are in demand. In this regards, a knowledge-oriented information visualization(KIV) approach is introduced in this paper. The KIV approach comprises a group toolkits including IB-CAST(a casebase for the knowledge management of IBs), IB-Radar(a radar chart made in Microsoft Excel to illustrate current and gap scores of IBs based on their rating), IB-Compass(an internet-enabled portal for IB information navigation based on IB-Radar chart), and IB-GIS(a geographical information system for spatial information management of IBs). A case study on the IFC II is conducted to demonstrate how to use KIV approach in the post-assessment of IBs.

智能建筑评价的信息可视化方法

Ju Hong[1]; Zhen Chen[2]; Heng Li[3]; Qian Xu[3]

([1] Beijing Institute of Civil Engineering and Architecture;

[2] University of Reading; [3] Hong Kong Polytechnic University)

摘要：本文在总结国内外相关文献的基础上，提出了一个面向知识的信息可视化方法用于支持对智能建筑的后评估。本文详细介绍了面向知识的信息可视化方法的模型原理，并对支撑其在智能建筑后评估过程中发挥作用的若干软件工具及其相互关联作了概要介绍。面向知识的信息可视化方法的软件工具包括：智能建筑知识库，智能建筑风向标，智能建筑罗盘和智能建筑地理信息系统等。本文最后通过案例分析，对如何在智能建筑后评估中系统使用面向知识的信息可视化方法及其工具软件作了说明。

References

[1] AIIB(2005)*IB Index*(3rd edition). Asian Institute of Intelligent Buildings(AIIB), Hong Kong, China

[2] Baltsoft(2004)General Knowledge Base - Overview. Baltsoft, Co., Ltd. <http://www.baltsoft.com/product_gkb.htm>(26/11/2005)

[3] Bassi, R. (2005)MATOOL: a matrix tool for assessing the performance of intelligent buildings, *Proceedings of the BRE Seminar on Intelligent Buildings*, Building Research Establishment Ltd. (BRE), UK

[4] Bouchlaghem, N., Sher, W., and Beacham, N. (2000)Computer Imagery and Visualization in Civil Engineering Education. *Journal of Computing in Civil Engineering*, 14(2), 134～140

[5] CABA(2004)*Building IQ Rating Criteria*. Task Force 1 - Intelligent Building Ranking System, Continental Automated Building Association(CABA), Ottawa, Canada

[6] Chow, L. K. H., and Choi, T. W. K. (2003)The new Intelligent Building Index(IBI) for buildings around the world - a quantitative approach in building assessment and audit. *Proceedings of the 1st International Conference on Intelligent and Green Building Technologies*. China Architecture and Building Press, Beijing, China. 793～807

[7] Hong, J., Chen, Z., Li, H., and Xu, Q. (2006)An integrative methodology of intelligent buildings assessment. The 2nd International Conference on Intelligent Green and Energy Efficient Buildings (IGEBC2006), 28-30/03/2006, Beijing, China

[8] IBSK(2002)*Assessment Standards for Certifying Intelligent Buildings* 2002, Intelligent Building Society of Korea(IBSK), Seoul, Korea

[9] JC&A(2004) Internet-enabled project management. Jonathan Cohen and Associates, Berkeley, USA. <http://www.jcarchitects.com/project_manage.html>(26/11/2005)

[10] Marshall, B et al. (2003)Convergence of knowledge management and e-learning: the GetSmart experience. In *Proceedings of ACM/IEEE 2003 Joint Conference on Digital Libraries*(JCDL 2003), 27～31 May 2003, IEEE Computer Society, Houston, 135～146

[11] SCC(2002)*Shanghai Intelligent Building Appraisal Specification*. Shanghai Construction Council (SCC), Shanghai, China

[12] USGBC(2005)LEED: Leadership in Energy and Environmental Design, the U.S. Green Building Council(USGBC), Washington, USA

[13] Wen, S.L. (2003)*Manual of Intelligent Building Assessment*. Architecture and Building Research Institute, Ministry of the Interior, Taiwan, China

DELIVERING BUILDING SUSTAINABILITY WITH INTELLIGENCE-THE ROLE OF THE CENTRE OF EXCELLENCE FOR INTELLIGENNCE IN BUILDINGS

Alan Kell, Alison Nicholl
(IBEXCELLENCE, England)

1 Introduction

The Centre of Excellence for Intelligence in Buildings(ibexcellence)is an open network promoting the development and delivery of intelligent building solutions as an integral part of achieving a sustainable built environment. A sustainable built environment meets social, economic and environmental needs at local and global levels.

The Centre of Excellence for Intelligence in Buildings provides a focal point for disseminating best practice and supporting knowledge exchange in the successful application of intelligence in buildings. ibexcellence is funded and directed by sponsors from leading commercial, public sector and government organisations.

2 Why is ibexcellence needed?

Over the past 20 years many different buildings have been labelled as "intelligent". However, the application of intelligence in buildings has yet to deliver its true potential. Relatively few clients, or their professional advisers, have sufficient experience and understanding of contemporary intelligent building solutions, and what they have to offer in terms of improved performance and value.

At the same time, industry now has a well-established intelligent building solutions capability but it is finding it difficult to demonstrate and prove the benefits of these solutions. Many on the demand-side remain unsure, and indeed sceptical, of the benefits and the value that investing in improved building intelligence can actually provide.

The Centre of Excellence will redress this balance by bringing together demand side, supply side, and other stakeholders to independently evaluate and communicate the added-value that intelligent building solutions can provide, both to building projects and to building performance over time.

The effectiveness of open networks like ibexcellence in promoting innovation is well demonstrated by the success of the INTEGER Intelligent & Green Partnership, which over the past ten years has succeeded in demonstrating the value and opportunities of integrating intelligence in buildings into environmental performance. The Centre of Excellence for Intelligence in Buildings is applying equivalent principles in order to integrate building intelligence into all aspects of building performance including profitability, energy, health, business and community development.

3 Aims of ibexcellence

3.1 To provide an independent forum for knowledge sharing, networking, research and learning.

3.2 To facilitate the communication and transfer of experience and knowledge between key stakeholders including suppliers, consultants, clients, policy makers and established business networks, with the aim of raising awareness and understanding of the wide-ranging benefits of intelligence in buildings.

3.3 To support the development of professional and technical skills to improve the quality, performance and value of intelligent building solutions.

3.4 To promote collaboration, research and funding to support the application and evaluation of intelligence in buildings.

3.5 To organise and support regular workshops, seminars, and international conferences which facilitate the development and dissemination of global best practice.

3.6 To produce publications supporting the successful application of intelligent building solutions including case studies, market reviews, management guides, and business directories.

4 Structure of ibexcellence

Companies and organisations seeking to directly influence and drive the development of intelligent buildings solutions are invited to join as sponsoring partners. Other organisations may participate in the activities of ibexcellence as Associate Members. Associate membership is free and available to any organisation sharing the aims of the Centre, and willing to be pro-active and participative in achieving these aims in cooperation with other members.

More than 50 organisations have already joined ibexcellence, drawn from across the full spectrum of interest in intelligent building solutions and representing a wide variety of interests. Members include many of the major international businesses committed to providing intelligent building solutions. By definition, the Centre of Excellence is open to organisations sharing its aim, so this membership will certainly grow. A list of current members is available on www. ibexcellence. org

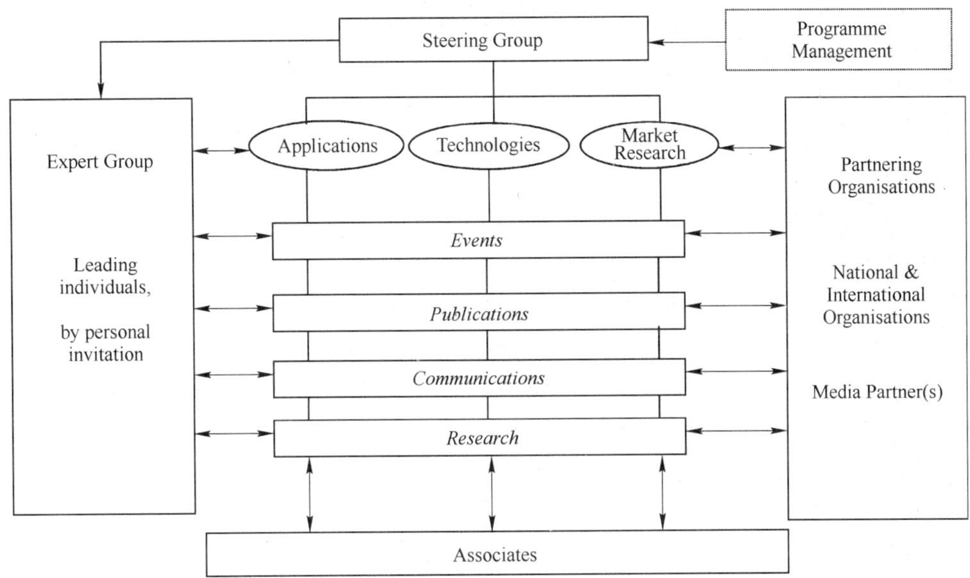

Fig. 1 Structure of ibexcellence

- The steering committee is made up of sponsors and supporters of ibexcellence from commercial, public sector and government organisations;
- Associate membership of ibexcellence is currently free of charge and is open to any organisation who shares the centre′s aims of developing and delivering intelligent building solutions as an integral part of a sustainable built environment;
- Membership of the expert group is by invitation and is made up of leading experts in specific intelligent buildings fields including academics, industry representatives and end-user representatives;
- Partnering organisations are drawn from national and international bodies whose members are committed to developing and delivering intelligent building solutions as an integral part of a sustainable built environment. ibexcellence will work in partnership with these organisations to achieve this aim.

5 Principal Activities

5.1 Events

ibexcellence arranges seminars, workshops and conferences throughout the year, which focus on areas of specific importance to its members. These include subjects such as the value of building intelligence in particular types of buildings (health, education, housing etc.), specific international markets (China, USA, Europe etc), and significant technology developments. This programme culminates in a high profile conference on best practice in intelligence in buildings which takes place each autumn.

ibexcellence also participates in major intelligent buildings events around the world, including Realcomm in the USA, BuilConn in the USA and Europe, and the International Intelligent & Green Buildings Conference & Expo in Beijing.

5.2 Publications

The Centre of Excellence publishes good practice guides, directories, case studies and market reviews highlighting the added value that intelligence in buildings can offer. These publications are disseminated widely through the members of the ibexcellence network.

5.3 Research

ibexcellence works with its members to identify, develop and fund research opportunities to enhance and develop the application of intelligence in buildings.

ibexcellence will also work with research partners to ensure that intelligent buildings research is effectively aligned with the requirements of its members and that results are successfully disseminated to key stakeholders.

5.4 Communications

The ibexcellence website, www.ibexcellence.org is being developed as a high profile web portal for disseminating information about intelligent buildings, including details of other events, case studies of intelligent building solutions and news about current technologies.

ibexcellence is also developing relations with media partners in order to include editorial on intelligence in buildings to ensure that the message is spread to a wider audience.

6 ibexcellence & China

ibexcellence is partnering with the Chinese Ministry of Construction China Intelligent Green Association (CIGA) and the Digital Community Control Network Industrial Alliance. Agreements formalising these arrangements are being signed at IGEBC in March 2006. These agreements will enable best practice in intelligence in buildings in China and Europe to be developed and shared, creating mutual benefits and opportunities for UK, European and Chinese companies. This cooperation is taking place within the framework of the China-UK Construction Working Group, which meets in Beijing for the first time on 31st March 2006.

Under this agreement CIGA will launch an ibexcellence column in its journal and ibexcellence will provide case studies and related information on the latest technologies, standards and successful projects of relevance to China and assist CIGA with its international journal publication. CIGA and ibexcellence will look at energy saving, environmental protection, ecology, renewable energy, digitalisation and sustainability. Cooperation will take place at a commercial and academic level and will include: technology exchange, technology consulting, skills and technology cooperation.

ibexcellence will also be supporting Digital Control Network Industry Alliance members, providing assistance and advice on entering the European market. Areas of support will include product standards, product examination and product promotion.

7 Involvement with ibexcellence

The Centre of Excellence for Intelligence in Buildings is an open network. Companies and organisations seeking to directly influence and drive the development of intelligent buildings solutions are invited to join as sponsoring partners. Other organisations may participate in the activities of ibexcellence as Associate Members. Associate membership is free and available to any organisation sharing the aims of the Centre, and willing to be pro-active and participative in achieving these aims in cooperation with other members.

Please visit ibexcellence on stand number 187 at IGEBC or the centre at the address below.

建筑智能化与可持续发展
在智能建筑领域中卓越智能建筑中心的作用

Alan kell, Alison Nicholl
(IBEXCELLENCE, England)

摘要：在过去的20年里，很多不同的建筑都被戴上了“智能建筑”的帽子。但是智能建筑的真正能力与本质远远没有达到它的潜力。只有不多的投资商或者是他们的顾问，在智能建筑方面有足够的经验与理解。这使得智能建筑的深层潜力无法有效的展示于众。

与此同时，业内已经有了优秀的建筑智能技术，但是发现展示证明这些技术产品的机会却很困难。在需求商方面，大部分人还对智能建筑的印象是不稳定，充满怀疑。对智能建筑是否能够将其价值与功能显示出来，对是否真可以提供给建筑智能充满了疑问。

卓越智能建筑中心填补了这个空缺，对供应商，需求商，业内其他人士对智能建筑不同的理解下功夫，把几方联系到起来。通过密切的交流，信息的互换，树立智能建筑在业内的信心，通过真正的实例来证明而外的投资会给建筑带了更多回报。

C　建筑节能与绿色建筑

热湿独立控制空调系统在生态办公楼的研究和应用

朱伟峰　邓　良　叶　倩　陈建萍
（上海市建筑科学研究院有限公司，上海　200032）
江　亿　刘拴强　李　震
（清华大学，北京　100084）

摘　要：对目前传统空调方式存在的问题进行了分析，以此为背景，上海建科院和清华大学在上海兴建的生态办公楼中研制了一种热泵驱动的热湿独立控制的空调系统，对此系统进行了分析和测试。

关键词：热湿独立控制　空调系统　生态建筑

1　研究背景

在上海市建筑科学研究院位于莘庄基地的建筑面积约 2000m^2 的生态建筑示范楼中，研究和应用了一种热湿独立控制的新型空调系统。研制和应用该空调系统是在人们对室内空气品质 IAQ(Indoor Air Quality)日益关注、全社会对节约能源和保护环境高度重视的背景下进行的。

在室内空气品质方面：近年来，随着人们对健康问题的日益重视，室内空气品质得到了人们的广泛关注，并提出了病态建筑综合症 SBS(Sick Building Symptom)的问题。室内空气品质可以简化成人们普遍关心的两个问题：即人体舒适和人体健康。一般来说，温度、相对湿度、空气流动及噪声水平主要影响人体的舒适，而室内污染物主要影响人体的健康。室内污染可以有多种分类方式，例如，可以把室内空气的污染物分为四类，即微生物粒子，可吸入颗粒，气态物质污染及汽态物质污染[1]。室内空气污染已对人们带来很大的损失，例如因空气微生物的污染造成的呼吸道疾病给人民健康造成极大的灾难和损失。据统计，我国每年发生的呼吸道感染病例 500 万，损失达 10 亿元[2]。在美国，IAQ 问题已变成一个主要的环境问题并且已有立法要求业主对该问题负责。在目前建筑中，空调系统普遍使用表冷器换热，通过降温将空气中的水分冷凝下来，以达到降温除湿的目的。潮湿的表冷器表面，是滋生大量霉菌的最好环境，同时也能造成其他病菌和病毒的聚集和滋生[3]。特别是在空调箱停机期间(如下班或放假)，在空调箱的密闭空间内，细菌浓度能骤然上升几十万倍[4]。这些微生物通过送风系统大量进入室内，会对室内的空气品质造成严重的影响。当前，用表冷器同时处理热、湿负荷的空调方式仍然是国内外应用最普遍的空调方式，空调系统本身就可能是一个很大的污染源。在这样的空调系统中，要解决好霉菌等问题已成为一个难题。

在节约能源方面：随着中国的改革开放和经济发展，商业建筑的面积日趋增大，建设规模和建设速度逐渐加大。我国自 1991 年起新建建筑平均每年 10 亿 m^2，城镇约 3 亿 m^2。随着经济建设的发展，商用建筑(包括写字楼、宾馆饭店、大中型商场等)大量兴建，致使用电量增长迅猛、环境恶化、全球变暖、用电峰谷差增大。保护资源和环境的可持续发展已成为中国的基本国策，建筑节能上升到前所未有的地位，商业建筑的能耗问题已引起了政府、社会的普遍关注。在商业建筑能耗中，空调系统的能耗是最主要的部分，通常达到商业建筑全年能耗的 50%左右。所以研究节能、高效的空调系统是社会发展的必然需求。

在环保方面：地球臭氧层的破坏会导致地球表面许多物种的消亡、粮食减产、人类皮肤癌患者的增加。因此，臭氧层的破坏，也是人类当前面临的主要环境问题之一。而目前空调系统广泛采用的制冷剂 HCFC 和 CFC 类物质，对臭氧层有明显的破坏作用。所以空调系统大量减少甚至完全替代 HCFC 和 CFC 类物质的使用，也是空调行业的发展趋势。

寻求一种健康、节能、环保的空调方式是社会发展的必然需求，可以说是空调系统今后发展的重要方向。

目前，国内外也有采用固体吸附材料除湿的系统，可以消除换热器的冷凝水表面，实现热、湿分开处理。其主要有固定床式和转轮式两种。固定床式固体吸附除湿装置是通过改变空气侧流向实现间歇式的吸湿再生；转轮式除湿得到了更广泛的应用，它可实现连续的除湿和再生。这两种除湿方式有着致命的弱点就是运行过程的混合损失大，影响效率，另外，很难实现等温的除湿过程，而除湿过程释放出的潜热使除湿剂的温度升高，吸湿能力大打折扣，整个过程传热传质的不可逆损失大，效率太低。相反，其造价较高。所以目前在国内外都没有进行大规模的商业应用。

在上海市建筑科学研究院生态建筑示范楼中，由上海建科院和清华大学共同研制的热湿独立型空调系统，正是为研究一种能够实现热、湿分开处理，消除换热器冷凝水表面，又能实现较大幅度的节能，造价合理，可实现产业化和大规模商业应用的空调系统的一种尝试。

2 热湿独立控制型空调系统的主要形式研究

热湿独立控制型空调系统包括新风除湿机、再生器、热泵、储液灌、干工况盘管等主要设备，如图 1 所示。

各设备的主要功能及系统的主要原理如下。

新风除湿机向室内输送新风，其包括全热回收段和冷却除湿段。首先通过全热回收段使室外高温高湿的空气和室内较低温度和湿度的排风进行一次全热交换，经过全热回收段的新风再进入除湿段与高浓度溶液(即溴化锂与氯化锂混合溶液)接触进行热湿交换，并利用一定量的冷水冷却除湿过程。这样，通过新风除湿机，新风被除湿降温，同时浓溶液吸湿后浓度降低，成为稀溶液，最终将被送入再生器浓缩再生。

再生器的主要功能是浓缩除湿后的稀溶液，在此过程中，需要有较低品位的热源驱动。

在新风除湿机和再生器之间还分别设置了浓溶液和稀溶液两个储液罐。其主要用来储

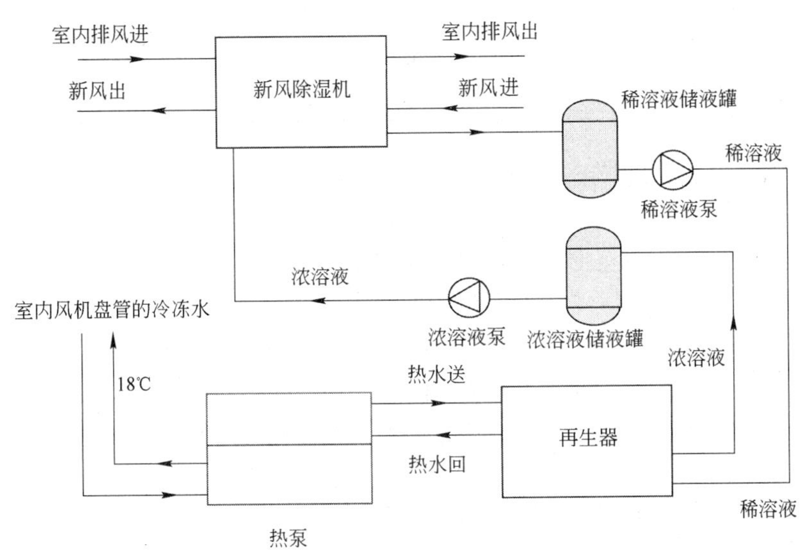

图 1　热湿独立控制型空调系统原理图

存溶液，相当于一个蓄能的装置。湿负荷小的时候储存浓溶液，负荷大的时候用储存的浓溶液作为补充来除湿，从而减小了系统的容量和相应的投资。

系统中还设置有一个高温热泵，其主要作用是一方面产生约 18℃的高温冷水为风机盘管提供冷源，处理空调的显热负荷，另一方面，由热泵排热产生热水，驱动再生器，为处理空调的湿负荷提供能量。这样，通过同时利用热泵的制冷和排热，以及对空调排风进行全热回收，提高了整个空调系统的能效比。

干工况的风机盘管，作为房间的空调末端设备，处理房间的显热负荷。

这样，该空调系统使用新风除湿机处理空调的湿负荷(即潜热负荷)；使用干工况风机盘管处理空调的显热负荷；同时利用热泵提供的高温冷水给风机盘管供冷，利用热泵的排热驱动再生器。

3　生态示范楼热湿独立控制型空调系统主要设计参数

3.1　室、内外空调设计参数(见表 1)

表 1

	干球温度(℃)	相对湿度(%)	湿球温度(℃)
室外空气设计参数	34.2	64	28.2
室内空气设计参数	26	55	19.5

3.2　建筑的空调总负荷及新风负荷(见表 2)

生态楼建筑面积约 1900m^2，但除去有特殊环境要求的实验室(其空调系统另配)外，约有空调面积 1200m^2。建筑内人员总数设计 30 人，新风量设计指标为每人每小时 30m^3。经计算，空调系统的总负荷及新风负荷如表 2 所示。

表 2

	总负荷(kW)	新风负荷(kW)
空调设计负荷	110	50

3.3 空调房间室内换热盘管干工况能否得到保证的分析

室内干球温度 26℃，相对湿度 55%，绝对湿度 11.6g/kg 干空气，新风送风的绝对湿度最低可以处理到 7～8g/kg 干，取新风送风绝对湿度 8g/kg 干，则每处理 1g 湿量，需要的新风送风量为 0.231m³，即 0.231m³/g。人体产湿按照 100g/(人·h)计算，去除一个人的产湿量，需要新风量为：100×0.231 =23.1m³/h。而目前新风量设计标准 30m³/(人·h)，于是，可以得出，在设计的新风量下，新风能够将房间人员的产湿量全部带走，不存在房间内空气结露的危险。房间内存在的风机盘管，能够在干工况状态下运行。

4 室内风机盘管是否保证干工况的测试结果

该新型空调系统的一个显著特点就是实现热湿负荷的独立控制，室内风机盘管实现干工况运行，消除室内带冷凝水的换热表面。为验证此效果，特以 202 和 208 两个不同类型的房间为代表，对典型工况下冷冻水的供水温度和这两个房间的室内空气露点温度进行了测试。202 房间位于生态办公楼二层北面楼梯西侧 5 间办公室的正中间一间，长 6.6m，宽 4.2m，为生态办公楼内的标准办公室。208 房间位于为生态办公楼二层北面楼梯东侧，相当于 4 间标准办公室的大小，目前为大楼内惟一与大厅连通的一个大开间办公室。测试结果如图 2。

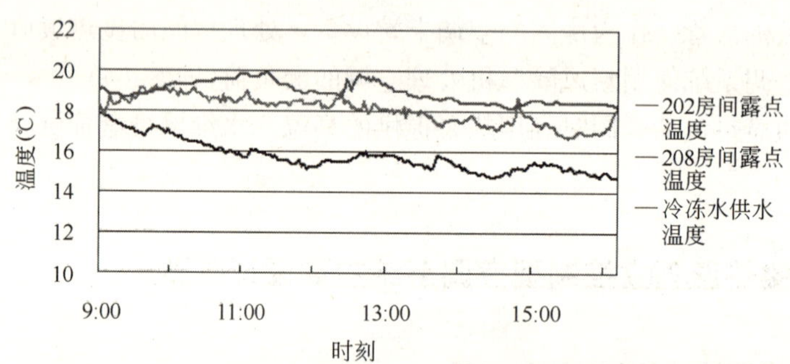

图 2 空调房间露点温度与供水温度关系曲线

如上图所示，冷冻水供水温度在 18～20℃左右，202 房间的露点温度基本在14～16℃之间，可以保持在冷冻水供水温度以下，所以风机盘管不会出现凝水，可以运行于干工况。该房间新风承担湿负荷，末端设备承担显热负荷且运行于无凝水的干工况，这是热湿独立控制空调系统的典型特征。生态办公楼其余房间的特点和测试结果和该房间基本一致。

208 房间的风机盘管基本上可以运行在干工况，但个别少数时间点存在结露的危险。经查，主要原因是 208 房间与大厅之间的空气有通风百叶连通，而大厅由于植物等因素，空气湿度较高，导致 208 房间湿度偏高。两个房间干球温度相同的条件下，208 房间的空气相对湿度比 202 房间高近 10%，导致 208 房间结露危险性有所增加。只要能够有效控制该房间外湿负荷的进入，将能够保证室内风机盘管运行在干工况。

通过测试表明，在正常情况下新型空调系统能够保证室内的风机盘管运行在干工况。

5 结论

这种热泵驱动的热、湿负荷独立控制的空调系统通过避免使用有凝结水的盘管，解决了目前空调系统中存在的霉菌滋生问题。同时通过除湿机内盐溶液的喷洒除去空气中的尘埃、细菌、霉菌及其他有害物，有效改善室内空气品质。并通过新排风之间的全热回收、同时利用热泵的制冷量和制热量等措施提高系统的能效比。其在解决室内空气品质、节能、以及环保等方面都比目前的传统空调方式具有非常突出的优势，具有很好的应用前景。

目前在上海市建筑科学研究院生态示范楼中研究和应用该热湿独立控制型空调系统仍然仅仅是对这种新的空调方式进行探索和研究的开始。其仍有不少的设计、运行、控制以及参数优化等问题有待进一步的分析和研究。我们将以建成的生态示范楼及其空调系统为平台进行进一步的实测和研究。

Independent Humidity Control Air Conditioning Systems in Green Building

Zhu Weifeng　Deng Liang　Ye Qian　Chen Jianping
(Shanghai Research Institute of Building Sciences，Shanghai　200032　China)

Jiang Yi　Liu Shuanqiang　Li zhen
(Tsinghua University，Beijing 100084　China)

Abstract: The problems of the traditional air conditioning systems are analyzed. On the basis，one independent humidity control air conditioning systems is studied and used in one green building. Studies the characteristic of the new air conditioning systems.

参考文献

[1] Liu Rea-tiing，Mark A. Huza. Filtration and indoor air quality：a practical approach. ASHRAE journal February 1995，18～23

[2] 主编(澳) Chu Cordia，Rod Simpon. 主译李立明，王临虹．生态大众健康：公共卫生从理想到实践．北京：北京医科大学．中国协和医科大学联合出版社，1997 年

[3] 江亿．“非典”问题引起的对今后空调系统方式的思考．暖通空调，2003 暖通空调与 SARS 特集，p4～7

[4] 沈晋明，邓伟鹏．正确认识空调系统防范非典型肺炎．暖通空调，2003 暖通空调与 SARS 特集，p16～19

[5] 江亿，李震．除湿法空调及系统．2002 年全国暖通空调制冷学术年会论文

[6] Andrew Lowenstein, Davor Novosel. THE SEASONAL PERFORMANCE OF A LIQUID DESICCANT AIR CONDITIONER. ASHERA Transactions 1995, p679～685
[7] Yu. I. Aristov, G. Restuccia, G. Cacciola, V. N. Parmon. A family of new working materials for solid sorption air conditioning systems. Applied Thermal Engineering 22 (2002)
[8] 赵荣义．空气调节(第三版)．北京：中国建筑工业出版社，1994

绿色建筑的运行管理策略
——以上海生态办公示范楼为例

韩继红　刘景立　杨建荣

（上海市建筑科学研究院有限公司，上海　200032）

摘　要：建筑的运行阶段占整个建筑全生命时限的95%以上，因此运行管理模式和策略关系到绿色建筑建设的成败，是真正实现绿色建筑内涵的关键之一。本文围绕节约资源、环境友好、以人为本、运行高效的绿色建筑目标，以上海生态办公示范楼为例，研究探索绿色建筑运行管理策略，旨在引起业界对绿色建筑运行管理的重视，提升我国绿色建筑的建设运营水平，促进我国绿色建筑的推广应用。

关键词：绿色建筑　运行管理

1　前言

当今时代，绿色建筑已经成为建筑领域的国际潮流。

当前，我国已经启动了绿色建筑的推广政策，提倡建设节能、节水、节地、节材的环保的建筑（“四节一环保”），这就要求改变以往对艺术形式的过分重视，改变或改良传统的建造方式，并提升建筑的运行管理水平。这势必为我国的建筑设计、建造和运行管理的理念带来一次深刻的变革。

绿色建筑最大特点是将可持续性和全生命周期综合考虑，从建筑的全生命周期的角度考虑和运用“四节一环保”的目标和策略，才能实现建筑的绿色内涵，而建筑的运行阶段占整个建筑全生命时限的95%以上。可见，要实现“四节一环保”的目标，不仅要使这种理念体现在规划、设计和建造阶段，更需要提升和优化运行阶段的管理技术水平和模式，并在建筑的运行阶段得到落实。

早在上个世纪80年代就有国家研究建筑的整体运行管理策略，并将实现此类策略的建筑称之为智能建筑或绿色建筑，目标是在为居住和使用者提供人性化使用空间和舒适的室内环境的前提下总体控制和协调建筑的运行，实现节约资源、减少运行能耗及延长建筑寿命的目的。如1984年在美国诞生了世界上第一栋采用智能技术参与运行管理的建筑即智能建筑，随后的10多年内，这种类似的建筑运行管理模式不断涌现，如美国得克萨斯州开发的500栋新型建筑能够根据建筑的室内环境状况控制建筑设备和构件的运行、开启状态，达到自动控制能源消耗的目的，与普通建筑相比，采取这样的智能运行管理模式可以节约能耗20%左右。日本东京的麻布地区修建的一幢房屋，有一个半露天式庭院，室内的感应装置能够随时测量出天气的温度、湿度和风力等，然后将各种数据及时输送给地下的管理系统，后者以此为根据，控制门窗的开闭和空调器的开关，使房间保持住户感到最舒适的状态；在刮风下雨的不良天气情况下，门窗会立即自动关闭，室内空调则随之开始运转。芬兰赫尔辛基市 VIIKKI 生态住宅小区为节约能耗、提高能源利用效率协调建筑运行中的采暖通风和热水供应系统，集中供热系统低温回水用于地板供暖；机械通风系统的废热被独立回收并供给具有特殊构造的窗或玻璃阳台以加热进入的室外空气，促进房间的自然通风。与此同时，国外在绿色建筑实践中由于不重视运行管理策略而使技术运用成为摆设的例子也屡见不鲜，如为了有效改善室内热环境和自然采光效果，英国某绿色办公建筑投入大量资金使用了大面积可自动控制的外遮阳百叶，但业主入住后还是按照当地的习惯安装使用了内百叶；而另一栋建筑则具有良好的自然采光设计，但在白天的建筑运行中员工们仍然保持拉上窗帘打开人工照明的做法，这些例子说明仅重视绿色建筑的设计手段和建设技术的运用，而没有合理的关注其运行管理策略，就会使得前期的投入难以发挥应有的功效。

在我国，建筑领域长期以来过多的注重建筑的形体设计，而运行过程的技术管理则有待提高；“建筑是凝固的音乐”，只有运用合理、先进的运行管理策略才能赋之以生命；低水平的运行管理模式会缩短建筑的寿命，更会造成建筑功能和资源的极大浪费，因此，针对我国的实际情况，要实现绿色建筑的目标，建立科学合理的运行管理策略，提高绿色建筑的运行管理技术水平是重中之重，随着我国从政策上对绿色建筑的推动和各种相关的建筑技术进步，我国的建筑运行整体管理技术和水平已经呈现更新和进步的态势。相关绿色建筑运行管理策略和技术的研究已经起步。本文围绕节约资源、环境友好、以人为本、运行高效的绿色建筑目标，以上海生态办公示范楼为例，研究探索绿色建筑运行管理策略，旨在引起业界对绿色建筑运行管理的重视，提升我国绿色建筑的建设运营水平，促进我国绿色建筑的推广应用。

2 上海生态办公示范楼运行管理策略

作为可持续发展理念的具体实践和探索，上海生态办公示范楼基于上海的地域特征和经济发展水平，提出了“节约能源、节省资源、保护环境、以人为本”的绿色建筑理念和建设目标。为此，由上海市建筑科学研究院总体负责并会同上海建筑相关领域 12 个交叉学科团队协同攻关，不仅在建筑设计和建设过程中重点研究和应用了建筑节能、自然通风、自然采光、太阳能利用、健康空调、绿色建材、智能控制、生态绿化、水资源利用和舒适环境等十大类绿色建筑关键技术体系，同时，还积极开展了运行管理技术和策略研究，在建筑的运行过程中总体协调和控制关键技术的各个分支，重点针对建筑运行的节能高效和室内舒适环境，研究不同的运行工况下相应的智能集成运行监控模式，确保实现了总体建设目标。

2.1 开敞式大办公室的 PSALI 混合照明的运行管理

上海生态办公示范楼二楼的大开间办公室位于整幢建筑的阴面，大面积侧窗为室内采光创造了良好的条件。为了充分并合理的利用自然采光，该办公室采用自然采光和人工照明相结合的混合照明方式，不仅可节约大量的人工照明用电，而且对提高室内采光均匀度、改善室内光环境的质量都具有重要的经济、生态效益。由于长期的生理适应性，人们更喜欢在自然光环境下工作和生活，但实际的建筑采光设计中，很难在室内的所有部分都提供足够的自然光照明。通常在靠近采光口的位置获得的照度高，距其越远照度越低，使得室内的照度不够均匀甚至某些区域不能获得满足工作可视度所要求的标准照度（见

图 1）。所以，在充分利用自然采光的基础上就需要将人工照明做为自然采光的辅助，两者结合使用，使得采光口的近处和远端的照度都能达到要求的照度标准并使之尽量保持均匀。这样的人工辅助照明系统即室内恒定辅助人工照明(PSALI，Permanent Supplementary Artificial Lighting in Interiors)。这种混合照明方案的设计和运行目标就是白天室内的自然光和人工光能够舒适合理的协调起来，节约照明能耗并形成舒适的光环境。大开间办公室的 PSALI 混合照明系统依靠特有的整流器装置和感光元件(图 2)，采用智能调光系统，实现随自然光的变化及时调节室内工作平面高度的照度水平，使室内天然光和人工照明在考察位置形成的总体照度保持大致的稳定和均衡。为此，照明灯具沿进深方向分为三列相对独立控制，每一列灯具的控制感光元件为照度传感器，当考察位置的照度传感器感知的照度值低于系统设置的室内照度控制目标时，智能控制系统开启并调节对应一列灯具的亮度，以使传感器感应位置的照度达到设定的目标值。一般情况下，离侧窗最远的一列灯具开灯时间越早、持续时间越长；而靠近侧窗的一列灯具由于自然采光一般天气情况下在该处工作面贡献的照度可以达到 400lx 以上，因此很少开启，达到了节约照明能耗的目的。

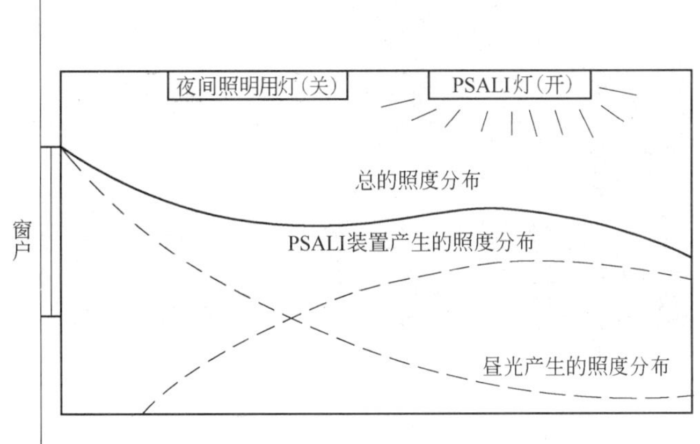

图 1　侧窗采光和 PSALI 混合照明效果

图 2　PSALI 混合照明智能调光系统及其照度传感器

PSALI 混合照明系统在设计和施工安装过程中严格控制灯具的选型和光源的选用，采用的蝠翼型灯具不但高效节能而且可以降低灯具表面的亮度、增大发光面积、减少眩光；而所选用的三基色 T5 荧光灯管色温中等、显色性好，与自然光可以很好的调和共同营造了舒适的室内光环境。

2.2　中庭遮阳篷运行管理

中庭是整幢大楼的共享空间，底层中部的室内花园对阳面的交流区和阴面的展示区形成视觉分隔和空间过度；二层周边设“跑马廊”，直通数个屋顶花园，打破了一般办公楼类似旅馆客房的单调格局，使整个实验办公楼能成为一个让人心情开朗的场所(见图 3)。中庭的存在为大楼内部贡献了充分的自然采光和明媚的阳光，提示着人们时刻变换的信息、拉近了室内员工与外界自然的距离；同时在过渡季节和冬季，中庭引入的阳光也成为室内“热源”，增加室内热舒适水平并有利于降低室内空调能耗。

图 3　绿色建筑办公示范楼的中庭

中庭在贡献室内自然采光和节能、热舒适性的同时，在夏季的时候却会成为室内过多热量的不利因素，不但使室内过热而且大大增加了室内空调能耗，因此对中庭遮阳根据季节、日照、气温的变化而实现灵活控制显得非常重要。

本大楼采用自动卷帘遮阳篷(图 4、图 5)。在夏季完全伸展时，可遮挡大部分太阳辐

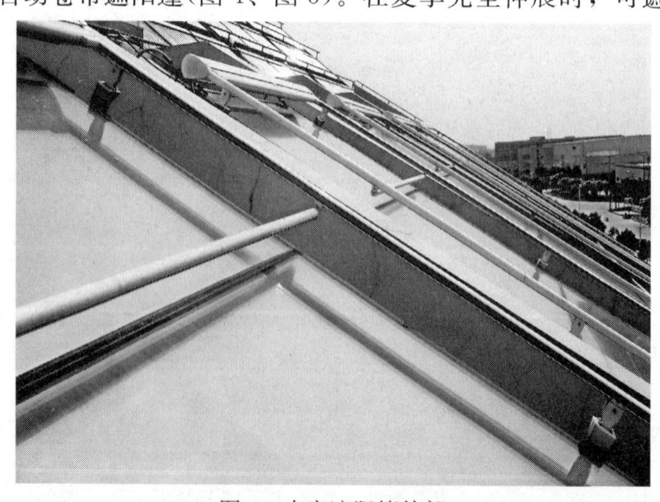

图 4　中庭遮阳篷外部

射和光线，减少眩光的同时能够引入足够的内部光线；冬季时可以完全打开，使阳光进入建筑空间，提高内部温度的同时也提高了照明水平；在过渡季节，则根据室外日照变化自动控制中庭遮阳篷的运行模式，具体见图 6。

图 5　中庭遮阳篷内部

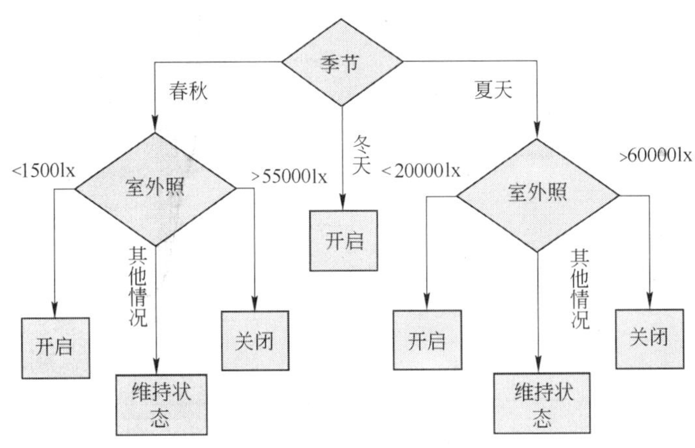

图 6　中庭遮阳篷运行管理模式

其中，夏季室外照度大于 20000lx 时定义为晴天、低于 20000lx 时定义为阴天；春秋季节室外照度高于 55000lx 时定义为晴天、低于 15000lx 时定义为为阴天。因此，夏季室外照度低于 20000lx 时即阴天时遮阳篷打开，大于 60000lx 即晴天时关闭；春秋季节室外照度低于 15000lx 时遮阳篷打开，高于 55000lx 时关闭。

2.3　室内热舒适运行管理技术

上海生态办公示范楼除利用太阳能空调等手段控制室内热环境的舒适性之外，最大的亮点是对自然通风的挖掘和利用。

自然通风不消耗任何机械能而是依赖室内外的一些自然条件，使室内产生气流流动的通风方式。在建筑的设计(见图 7、图 8)和运行管理模式中利用建筑内外的有利条件，合理充分的发挥自然通风的贡献，是实现本大楼夏季和过渡季节室内热舒适的重要技术手段。

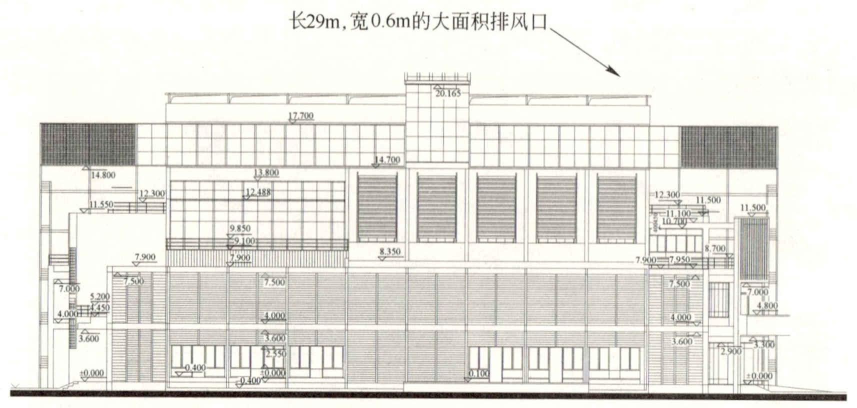

图7 上海生态办公示范楼自然通风烟囱设计

上海属于夏季炎热地区，因此示范楼应用自然通风的主要方式是在过渡季节主要依靠自然通风模式保持室内热舒适和健康的室内空气品质，并推迟太阳能空调开启时间；在夏季夜间利用自然通风带走室内余热，为室内降温，从而减少白天工作人员办公时空调系统的运行时间并有助于白天空调系统为室内迅速降温，不仅可以帮助达到建筑节能目的而且是保持室内空气健康品质的绿色手段。图 9 为示范楼自然通风模型示意图。

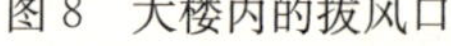
图8 大楼内的拔风口

图9 生态办公示范楼自然通风模型

2.3.1 过渡季节自然通风运行模式

自然通风条件下，室内可接受的热舒适温度可以达到 28℃，而空调系统产生的室内环境属于中性热舒适的稳态环境，室内可接受的热舒适温度一般较低，如 24℃。因此，在过渡季节温度较高的上海可以通过良好的自然通风运行模式达到节能并明显提升室内的热舒适水平的目的。

此外，良好的自然通风也是保持室内空气清新的最有效的途径。上海生态办公建筑示范楼实施自然通风策略的关键在于开窗关窗的运行模式，结合上海市地域气候特点和在过渡季节大楼内温湿度测试分析，确定本大楼的自然通风开关窗户的运行模式如图 10 所示。其中：

- T_o 为室外温度，$T_{o\text{-}dp}$ 为室外露点温度；

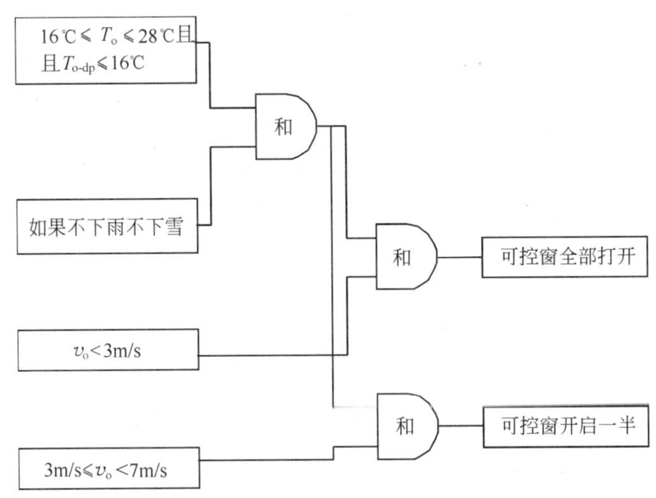

图 10　过渡季节自然通风运行模式

- v_o 为室外平均风速，监测时，按 20 分钟的平均值计算，测试数据可从同地址的节能楼气象站获得；
- 在不满足条件的情况下，窗户处于关闭状态。

2.3.2　夏季夜间自然通风运行模式

上海生态办公示范楼夏季室内制冷主要由太阳能空调系统完成，主要制冷区域为大楼东侧的二楼办公室。工作时间(8：30～17：00)空调系统开启，而下班后“人去楼空”，室外气温却开始下降，这时通过采取自然通风的运行管理模式将室内余热散去，可以为第二天的早晨提供一个清凉的办公室室内环境，不仅有利于空调节能，更有利于让有限的太阳能空调负荷发挥最佳的降温效果，使办公室在日间经历高温的时段室内温度控制在舒适范围。

夏季夜间自然通风智能管理模式见图 11。17：00(下班时间)以后，如果室内温度超过 24℃时，出现以下情况之一则按照各自情况的时段将侧窗打开，同时促进自然通风的

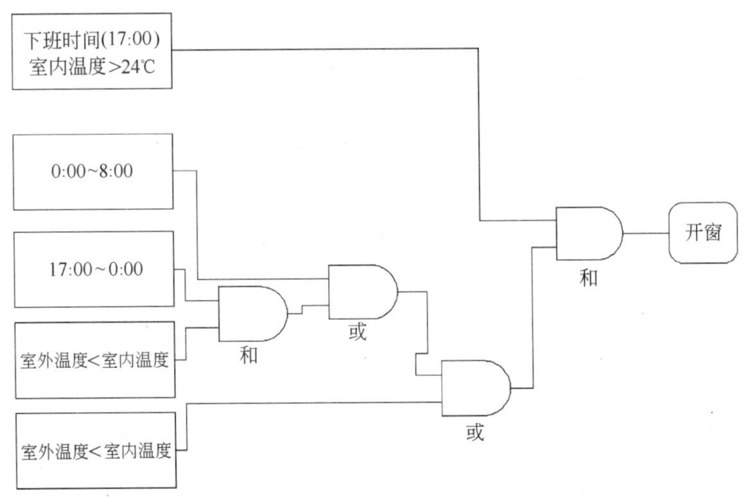

图 11　夏季夜间自然通风智能管理模式

通风风道(见图 8)开启：

- 早晨 0：00～8：00 时段内；
- 17：00～0：00 时段内室外温度低于室内温度；
- 17：00～8：00 时段内室外温度低于室内温度；

通过 2005 年整个夏季的跟踪测试、对比和分析发现，当“日间开启太阳能空调系统，而夜间按如上模式控制自然通风模式”时，在开窗到关窗期间大楼一层的展示厅与室外温差达到 0.5℃左右，二层主管办公室与室外温差可达到 0.5℃左右；而当采取“日间开启太阳能空调系统，而夜间不采取自然通风模式”时，在对应的时段内大楼一层展示厅与室外温差为 1℃左右，二层主管办公室与室外温差为 0.8℃左右。这表明绿色建筑办公示范楼采用如此夜间自然通风运行模式对于降低空调能耗、提高室内热舒适具较好的作用。

3 小结与展望

本文研究探索了上海生态办公示范楼的节能高效运行和室内热舒适环境管理模式和策略，对一些工况下运行模式的实施做了介绍，无疑对促进我国绿色建筑的建设水平和运行管理策略的研究是一次极其宝贵的实践积累。

为提高绿色建筑运行管理水平，进一步的研究将包括建筑系统节能运行管理模式、建筑资源节约的关键管理技术以及建筑环境智能调控与安全防灾技术等，从能耗同济、系统优化、可再生能源、废弃物管理、环境综合调控等方面进行开展。上海生态办公示范楼不仅是绿色建筑关键技术的展示平台，而且是相关技术的研究和实验平台，在这新一轮的研究中将发挥更大的作用。

Operation management strategies of green buildings

——With the analysis on SRIBS eco-building

Han Jihong　Liu Jingli　Yang Jianrong

(Shanghai Research Institute of Building Sciences，Shanghai　200032　China)

Abstract: For a building，the operation period takes more than 95% of the whole life. Therefore，the operation management becomes the most important stage for green buildings to be sustainable. By taking the example of SRIBS eco-building，this paper focuses on resource saving，environmental protection，human orientation and efficient management of green buildings. The management technique and strategies are analyzed to emphasize its importance，to promote the management level and improve the application of green buildings in the future.

经济视角下绿色建筑节能问题的解读与思考

赵 辉 袁 镔

（清华大学建筑学院，北京 100084）

摘 要： 基于可持续发展观念的绿色建筑，目的是减少资源和能源的消耗，实现资源和能源的最优化利用。建筑节能是绿色建筑的核心问题之一。作为一种经济活动，建筑节能是人们基于效益最大化的选择行为，是个人、集体、城市等多方面利益主体的博弈。本文从三个经济学的概念——成本与收益、贴现率、外部效应，解读建筑节能实践，思考建筑节能对策。

关键词： 绿色建筑 建筑节能 成本与收益 时间效应 外部效应

经济与环境具有辩证统一的关系，环境是经济发展的基础，经济发展对环境产生反作用力。作为人类经济活动重要内容之一的建筑活动自古以来就对环境产生了巨大的影响，晚唐诗人杜牧在《阿房宫赋》中的诗句“蜀山兀，阿房出”描绘了这种景象。近几年，中国建筑规模十分巨大，每年建成房屋达 16 亿 m^2～20 亿 m^2（中国统计年鉴），相当于欧洲和美国每年新建建筑的总和，带来巨大的资源、能源消耗和环境影响。建筑能耗已成为中国目前影响经济发展与环境健康的重要问题之一。建筑节能是中国可持续发展战略的必然选择。

基于可持续发展观念的绿色建筑是以现代生态环境观念全面审视以往建筑理念后，逐渐提炼出来的。绿色建筑在空间上关注建筑活动对全球生态环境、周边生态环境和居住者生活环境的影响；在时间上关注建筑“全寿命周期”的影响，包括原材料的开采、运输与加工、建造、使用、维修、改造和拆除等环节。绿色建筑的目的是减少资源和能源的消耗，实现资源和能源的最优化利用。建筑节能是绿色建筑的核心问题之一。

建筑节能是一种生产和消费行为，建筑节能不仅受技术因素与环境因素的影响，作为经济活动的产物，建筑节能和经济因素紧密相关。经济活动是人们的一种选择的行为，人们的日常生活、行动，每一步都在选择。从经济视角来看，建筑节能是人们基于效益最大化的选择行为。因此，建筑节能的实践工作非常复杂，矛盾多种多样，并不简单的是一个技术问题，建筑节能是个人、集体、城市等多方面利益主体的博弈。

1 绿色建筑节能中的成本与收益问题

成本与收益是绿色建筑节能经济选择行为的基础。建筑节能需要采取一定的技术措施，需要一定的成本。建筑节能的成本由谁来承担或者收益由谁来享受，这是建筑节能能否顺利开展的关键，而在实际工作中，却存在成本与收益不符的现象。例如居住建筑外墙保温所广泛采用的聚苯板保温墙体，按照我国原有的节能 50%的标准，一般保温层做到 60mm 左右即可达到，造价在 40 元/m^2 左右，提高节能标准，则需要加大保温层厚度，投入也相应增加，按照实际情况统计，大体上每增加 10mm，造价增加 4～8 元/m^2，也就是整个外墙保温的 1/10～1/5。考虑到外墙成本在总造价的比重，投入增加还是较多的，这部分成本主要由业主来承担，但是，现行的采暖收费制度是按建筑面积收费，而不是实际的能耗，使得节能建筑和普通建筑同样缴费，业主享受不到建筑使用过程中节能带来的好处，严重的阻碍了业主的建筑节能积极性。

从发达国家的经验来看，建筑节能并不能靠开发商和业主完全承担，需要政府的投入。例如德国 1994 年开始实施的太阳能供暖建设项目——位于汉堡伯拉姆费尔德的生态村，是当时欧洲最大的项目。整个生态村花费了 600 万马克，其中 400 万用于建造收集器、蓄水池和热管网，平均每个住户单元花费 33000 马克，联邦政府承担了 50%的费用，汉堡市政府承担了 100 万马克。2002 年由英国著名的生态建筑师比尔·邓斯特(Bill Dunster)设计的贝丁顿零能耗发展项目，被誉为英国最具创新性的住宅项目，位于伦敦附近的萨顿(Sutton)市，在这个项目的设计和运作中，世界自然基金会为其提供了资助，并且萨顿市政府也以低于正常价格的地价作为鼓励。

2 绿色建筑节能中的时间效应问题

绿色建筑需要从建筑的全寿命周期来考虑，建筑能耗是“从摇篮到坟墓”的全过程，包括建筑材料的采集、加工制造、运输过程，施工过程，使用过程和分解过程，其中建材和建造过程中所消耗的能源仅相当于建筑物使用 3～5 年中所消耗的能源。建筑使用过程中的能耗主要包括采暖、空调、通风、热水供应、照明、炊事、家用电器、电梯等方面的能耗，其中采暖空调通风能耗约占建筑总能耗的 2/3 左右。因此，提高建筑热工性能将会产生长期经济效益，虽然这也会增加一部分建筑材料和能源消耗，但从建筑全寿命周期能耗考虑还是很划算的，但现实中为什么一些业主不愿增加建筑的早期节能成本投入，换取长期的收益呢？为什么一些新的建筑节能技术不能很好的在新建筑中得到应用呢？这里有一个时间效应——贴现率问题。

时间效应是指将来发生的费用与效益比现在发生的同等数额的费用与效益的价值低。相隔的时间越长，其现值就将越小。折现是将未来费用与效益调整为现值的过程。人们使用折现有两个主要原因——时间偏好和资金的机会成本。时间偏好是指每个人都喜欢先收益后支出，资金的机会成本是指现在的一笔资金比将来同等数额的资金价值更高。因此，高贴现率使遥远以后发生的环境破坏和节约的成本通过折现可以减少到很小。建筑全寿命周期中占主要部分的使用能耗是在一个长期过程(50 年到 70 年)中发生的，高贴现率使这个长期费用在现在看来显得并不重要，使得业主对建筑节能缺乏动力。

3 绿色建筑节能中的外部效应问题

建筑节能受环境因素与经济因素的影响，而谈到环境与经济就会涉及到一个重要理论——外部效应理论。外部效应理论是本世纪初(1910年)由著名经济学家马歇尔提出的，简单的来说，所谓外部效应就是实际经济活动中，生产者或消费者的活动对其他消费者和生产者产生的超越活动主体范围的影响，它是一种成本或效益的外溢现象，根据影响的效果可以分为外部经济和外部不经济。

建筑节能实践中存在外部效应，例如，在供热体制改革的分户热计量试点工作中，由于分户墙和楼板的传热，使得各户之间存在热传导现象，可以从相邻的用户无偿得热或者白白失热，造成一些用户不开暖气室内温度也很高，或者开足暖气室温仍很低的现象，得热与失热取决于相邻用户之间的室内温度，这是一个典型的具有外部效应的例子。业主在开暖气(采暖消费)的过程中，会对相邻的用户带来有益的收益(得热)，而用户不开暖气或少开暖气的情况下，会对相邻的用户带来额外的负担(失热)。因此，一个业主的采暖消费行为会对其他人产生影响，使得分户热计量变得不公平，供热改革遇到挫折。可以说，我们的一些节能工作往往从技术上是可行的，环境效益是明显的，但往往没有顾及到人的经济行为选择，使得建筑节能工作出现偏差。

外部效应还表现在建筑能源市场价格上。经济学家认为，稀缺性赋予商品或劳动以价值。当物品和服务具备市场时，其稀缺性通过其价格来反映，但是，由于存在外部效应问题，即没有考虑对环境的影响，造成市场失效，使一些环境资产的市场价格并不反映其真实的稀缺性，例如，我国建筑广泛使用的能源——煤炭的价格很低，并没有充分考虑煤炭对环境造成破坏的经济成本，造成了对煤炭资源的过量开采和使用，而建筑中其他一些可利用的清洁能源和可再生能源(例如太阳能、地热、天然气等)同煤炭在经济成本竞争中处于绝对劣势。当然，建筑能耗问题不仅是经济、环境问题，还是个社会问题，并不能仅仅靠简单的提高能源价格来解决，但是，降低建筑能耗应是绿色建筑努力的方向。

4　绿色建筑节能的思考

建筑节能需要经济措施来解决。由于建筑节能存在成本与收益、时间效应、外部效应等经济问题，使得建筑节能成为“市场机制部分失效”的领域。从国际上建筑节能实践的经验来看，建筑节能需要采取控制和市场两种方法，控制的方法主要是强制性的建筑节能法律法规、条文规范，市场的方法主要是政府对建筑节能的经济激励措施。我国由于相应的法律基础薄弱，而且没有财政税收的经济激励政策，不仅“市场机制部分失效”，政府的节能政策干预也因缺乏力度而失效，在这种情况下，再好的节能技术推广起来也是困难重重。因此，在建筑节能发展中，法律法规和经济激励机制是必不可少的。

我们任何一种行为带来的后果分成两部分，内部性的和外部性的，我们每个人的决策只考虑对自己的，给自己带有什么，这就是说个人的最优行为和社会的最优行为是不一样的。从外部效应理论来理解，建筑节能措施就是如何使每个人都能够对自己所有的行为负责任，如果他对所有的行为负责任，个人最好的东西也是对社会最好的东西。使一个行为的外部后果内部化，这就是法律法规和激励机制的一个核心问题，因此，建筑节能问题还在于机制的创新。

总之，绿色建筑节能问题涉及到社会、经济、技术、环境等多方面的因素，需要根据各地的自身条件、发展状况，因地制宜的采取措施，才能取得预期的效果，否则就会事与愿违。

Explication and Consideration on Energy Efficient Green Building From the View of Economy

ZHAO HUI YUAN BIN

(School of Architecture Tsinghua University, Beijing 100084 China)

Abstract: According to sustainable development's idea, green building is to reduce the depletion of the resources and energy, and make optimal use of resources and energy, thus, one of the core problems in green building is energy efficient building. As a kind of economic activities, energy efficient building is the individual choice behavior according to oneself income maximize, and the game of various benefits corpus such as individual, collective, city, and so on. From the three economic concept——costs and income, time effect, externalities, I clarify the practice and consider the measures of energy efficient building.

参考文献

[1] 李克国，魏国印，张宝安主编. 环境经济学. 北京：中国环境科学出版社，2003

[2] 夏菁，黄作栋. 英国贝丁顿零能耗发展项目. 世界建筑 2004/08

[3] 汉堡伯拉姆费尔德生态村 德国. 世界建筑 2002/12

地表水水源热泵技术的发展现状及系统模拟

孙宗宇　杜国付　冯晓梅　邹　瑜
（中国建筑科学研究院空调所，北京　100013）

摘　要： 本文介绍了地表水水源热泵系统国内外研究和应用的现状，以及存在的问题，分析了地表水水体的换热过程。在 TRNSYS 环境下开发了地表水水体及地表水换热器模块，对地表水温度、换热器出口温度、及地表水体的各项得热量等参数进行了全年动态逐时模拟。提供了一套地表水水源热泵系统换热计算和能耗计算的手段，并为地表水水源热泵系统的设计和经济分析提供了依据。

关键词： 地表水水源热泵系统　地表水换热器　TRNSYS　全年动态模拟

1　地表水水源热泵系统应用现状

1.1　地表水水源热泵系统简介

地表水水源热泵系统（SWHPs，Surface Water Heat Pumps system）是地源热泵系统（GSHPs，Ground-Source Heat Pump system）的一种系统方式。它是利用地球表面的水源，如河流、湖泊或水池中的低温低位热能资源，并采用热泵原理，通过少量的高位电能输入实现低位热能向高位热能转移的一种技术。如果空调建筑附近有可利用的海、河流、湖泊或水池等水体，SWHP 系统有可能是最具有节能优点而又最经济的系统。同土壤源或地下水水源热泵系统相比，地表水系统节省了挖掘所需的费用，因而很多情况下可以节省一部分的初投资费用。

1.2　地表水系统分类

SWHP 系统可以分为闭式系统和开式系统。

闭式系统与土壤源热泵系统相类似，由潜在地表水水体中的地表水热交换器，以及与之相连的水—水或水—空气热泵组成。地表水热交换器由多重并联的换热管组成，其管材多为高密度聚乙烯。水泵驱动水或防冻溶液，在热泵的水—制冷剂换热器（蒸发器或冷凝器）和地表水换热器之间进行循环。闭式系统将地表水与管路内的循环水相隔离，保证了地表水的水质不影响管路系统，防止了管路系统的阻塞，也省掉了额外的地表水水处理过程。但换热管外表面有可能会因地表水水质状况产生不同程度的污垢沉积，而影响换热效率。

开式地表水水源热泵系统中，地表水经水处理装置后直接抽取至热泵机组，或通过一个中间换热器与闭式系统相连接。开式直连系统适用于地表水水质较好的工程，并需要进行除沙、除藻、除悬浮物等处理。开式间连系统也应进行一定的水处理。并且由于中间换热器一般比闭式系统的地表水换热器效率低，会损失一部分热量，因此较多的应用于一些深层地表水实例当中。

当地表水（湖水）较深时，底层水温受外部气候影响非常小，全年温度波动不大，这时有可能由湖底的水直接供冷或预冷而不经制冷机（Deep Lake Water Cooling，DLWC）。

1.3 地表水水源热泵技术的国内外发展现状

地表水热泵技术在国外出现较早，应用较为成熟。20 世纪三四十年代开始就出现了成功的工程应用。1939 年，瑞士就出现了以河水做为低位热源的 175kW 的热泵系统。目前，在欧美和日本，地表水系统的应用较为普遍。其中开式系统（包括直连和间连）多用于大型和超大型的项目中，如区域供冷供热系统（District Heating and Cooling，DHC）。这类系统从数量上较少，但规模较大，并且都取得了非常可观的节能和减排效果。如瑞典的斯德哥尔摩市利用波罗的海建成的约 246,000 kW 的 DHC 系统，集中向市区进行供冷和供热。加拿大的多伦多市投资 1.7 亿美元兴建 DLWC 系统，利用安大略湖深层湖水进行集中供冷。美国康奈尔大学利用深层湖水建成的集中供冷系统，为校区提供约 63,000kW 的冷量，取代了原有的制冷机。

闭式 SWHP 系统对地表水水质基本无特殊要求，所以在国外的应用也更为广泛。如图 1、图 2 所示的位于美国印第安纳州的一个 150kW 的地表水换热器的安装现场照片，该工程使用浅塘水作为低位热源，采用高密度聚乙烯换热管。图 3 是位于美国俄克拉何马州大学的湖水换热器的安装现场照片，采用了铜管。

图 1

图 2

在我国，地源热泵技术近年来发展较快，其中地下水和土壤源热系统得到了大量的工程应用。而地表水系统也逐渐得到了专业人员的重视，近年来在部分地区也得到了工程应用，系统型式大部分为直连或间连的开式系统。其中的一些工程中，只在夏季抽取地表水作为空调系统的冷却水。闭式系统由于需要在水体内设置换热器，而换热管的布置和安装方法在国内还没有被广泛的掌握，因此目前的应用还比较少，但也在少数的工

图 3

程项目中进行了尝试。另外，近两年海水源热泵系统在沿海地区受到了广泛的关注，一些海水源热泵系统已经开始策划和实施。我国地表水资源丰富，很多需要供暖和空调的建筑物周围存在着可利用的水体，因此地表水水源热泵系统具有较为广泛的应用前景。

地表水的水温受气候影响较大，尤其是浅表水，全年处于波动状态。掌握地表水的水温动态变化规律是实施 SWHP 系统的前提。但设计人员往往对某地区的地表水温情况不了解，得到的实测数据往往不够全面。现有的设计资料只是给出在水温恒定的情况下(或范围较大的区域内，如南方、北方)，单位表面积水体换热量的推荐值及单位换管长度的换热量经验值。这使得设计人员无法准确估算特定地表水资源所能提供的吸热量或放热量，也无法准确评估 SWHP 系统对地表水体热环境的影响。另外，对于闭式 SWHP 系统，在特定水体条件下，如何计算合适的换热面积也是需要解决的问题。

2 地表水水体的换热过程和变化规律

2.1 地表水换热的组成部分及对水温的影响

如果不考虑流入流出带走的热量，地表水与外界的热交换主要包括：太阳辐射、天空辐射、与空气的对流换热、蒸发和与大地的热传导。如图 4 为地表水传热过程示意图。

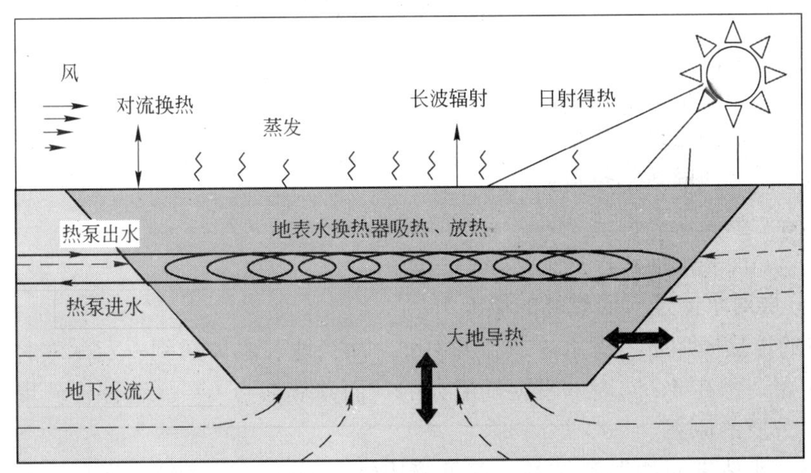

图 4 地表水传热过程示意图

其中，太阳辐射得热占有很大比重，可以达到 950W/m^2。它主要被水体上部分吸收，除非湖水很清澈的情况。大约 40%的太阳辐射在表面被吸收，其余能量的大约 93%在人眼可见的深度内被吸收。

通过对流传热的得热非常小，通常只相当于太阳辐射得热的 10%～20%。当湖面温度较低时，热量通过对流换热传递到湖水中，风速的增加会使湖水的对流换热量增加。湖水的冷却主要通过表面蒸发完成，风速对蒸发换热量的大小也有很大的影响。对流得热或散热在夏季只占一小部分，因为温差相对较小。

在晴朗的夜晚，温度相对较高水体表面将向温度较低的天空进行辐射传热。这部分热量在水体冷却中占有不小的比例。比如当温差达到 14K 时将有 160W/m^2 的热量通过辐射传递出去。

最后一项热传递是通过大地导热的热量，它所占的比例并不大。

2.2 浅表水体数学模型的建立

当水深不超4～5m时，可以假设水体充分混合，即整个水体温度一致，忽略内部的温度梯度，建立单节点传热模型。由热力学第一定律，有：

$$q_{in}-q_{out}=V\cdot\rho\cdot C_p\frac{dT}{dt} \tag{1}$$

式中 q_{in}——水体吸热量；

q_{out}——水体放热量；

V——水体的体积；

ρ——水的密度；

C_p——水的比热；

$\frac{dT}{dt}$——水体温度对时间的导数。

根据2.1所述，浅表水水体的吸热放热主要包括太阳辐射、长波辐射、对流换热、大地导热、蒸发换热，这里增加浅表水换热器的换热量。则水体温度随时间的变化可以表示为微分方程：

$$\frac{dT}{dt}=\frac{q_{solar}+q_{longvave}+q_{convection}+q_{ground}+q_{evaporation}+q_{fluid}}{V\cdot\rho\cdot C_p} \tag{2}$$

式中 q_{solar}——太阳辐射得热量；

$q_{longvave}$——长波辐射得热量；

$q_{convection}$——对流换热得热量；

q_{ground}——大地导热得热量；

$q_{evaporation}$——蒸发换热得热量；

q_{fluid}——浅表水换热器的换热量。

2.3 温度分层对水体竖向温度场的影响

2.2中的传热模型不考虑水体的温度梯度，而当地表水体达到一定深度时，会产生明显的分层现象。水在4.0℃时的密度是最大的，而并非冰点0℃。这种物理现象是地表水体产生温度分层的原因。

在冬季，水体表面的温度较低，并可能冻结，而底层温度较表面可能高出3～5℃。水体的竖向温度场趋于稳定，进入所谓“冬季停滞区”。春季，当水体表面温度升高并超过4℃时，表面和底层产生循环，这使得整个水体的温度变得相对平均。随着温度的上升，循环只在水体的上部分进行，这种过程一直持续到整个夏季，水体上部受气候的影响较大，不断进行着吸热放热的过程。这些传热过程和水体循环一般对底部的影响不大，而大地导热量相对较小，所以水体底部维持相对较低的温度。这样水体上部可能达到21～32℃，底部则可能维持在4～13℃，而在中部较小的深度范围内形成一个斜温层。这时水体达到了比较稳定的竖向温度分布状态，称为“夏季停滞区”。随着秋季和冬季的来临，水体上部温度逐渐降低，整个水体不断进行着因密度差产生的循环，一直到进入“冬季停滞区”。很多湖水的温度分层情况与前述的理想状况非常接近，而造成与理想分层状态差异的因素主要包括：

a. 流入流出水体的影响；

b. 深度较小；

c. 水体内的波动；

d. 风力的扰动；

e. 冬季气温较高而达不到“冬季停滞区”。

值得注意的是，温度分层对提高换热量一般是有利的。因为在夏天需要向地表水体排热时，水体底部温度低于平均值；而冬季需要从水体中吸热时，水体底部温度一般高于平均值。

3 TRNSYS 环境下的地表水水源热泵系统动态模拟

3.1 TRNSYS 系统简介及浅表水体组件模型的建立

TRNSYS(Transient System Simulation Program，即瞬时系统模拟程序)是一套建筑热环境模拟和分析的软件。TRNSYS 软件最早开始于 1975 年，是由美国威斯康新大学太阳能实验室的研究人员开发，并在欧洲一些研究所的共同努力下逐步完善，目前版本已达到 16.0。TRNSYS 软件最大的特点在于其模块化的分析方式。在进行空调系统模拟时，利用的是设置—计算—反馈的模拟方法，计算更为精确。TRNSYS 所有标准模块的源代码是开放的，用户可以任意修改和添加模块。

在 TRNSYS 环境中来建立地表水换热模型的好处是：

a. 可以利用 TRNSYS 中已有的标准组件，提供瞬时输入量，如利用全年气象参数计算水平面的太阳辐射量、天空温度、环境温度、风速等；

b. 可以利用 TRNSYS 中已有的数学函数库和瞬时计算功能；

c. 建立好的地表水换热模块可以与其他模块相连接，从而建立地表水热泵系统的模型，或与建筑模型相结合进行计算。

利用 2.2 所描述的地表水水体温度的数学模型，在 TRNSYS 中建立浅表水换热模型。其功能是：根据一定的参数设置，按一定的时间步长，接收气象文件数据和必要的输入数据进行计算，输出地表水水体的瞬时温度，换热器出口的瞬时温度，地表水体各项得热量的值。在地表水换热模型应用时，应充分收集水体的各项信息，并采用与实测数据相结合的方法，以提高计算的准确程度。

3.2 SWHP 系统模拟

在 TRNSYS 环境下，建立了地表水热泵系统模型，其中包括的主要组件模型为：气象数据、天空温度、辐射、地表水体换热、热泵、建筑物负荷计算、水泵、数据输出等。冷机组件采用部分负荷下恒定出水温度的控制方式。通过这个 SWHP 系统模型，主要可以实现下列计算功能：

a. 水体的逐时温度、各部分换热量的逐时值；

b. 对闭式系统，可以计算水体中换热器的进、出口温度及换热量逐时值；

c. 热泵机组蒸发器和冷凝器侧表逐时的进、出口温度、制冷(制热)量、能效比等；

d. 整个系统的逐时能耗。

图 5 为计算的某浅表水体各项日得热量的变化曲线。可以看出，日射得热、蒸发换热占有较大比重，长波辐射部分对水体冷却也有比较稳定的影响。对流换热量有较大的瞬时值，但累计到一定时间后的总换热量较小。

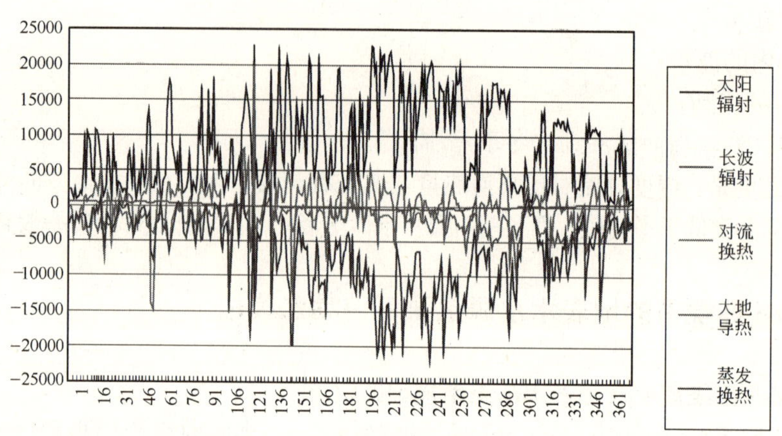

图 5 某浅表水体各项日得热量的全年变化曲线

图 6 为北京市 5 月 15 日至 10 月 1 日某地表水体，无流入流出的计算结果。可以看出，地表水的日均温度与室外日平均温度有着类似的趋势，但明显平缓一些，其差值−6～6K 之间。如果存在水量的流入流出，地表水的惰性还要更大。图 7 是得到的地表水换热器进、出水温度及水体温度在 7 月 8 日～10 日的模拟计算结果。

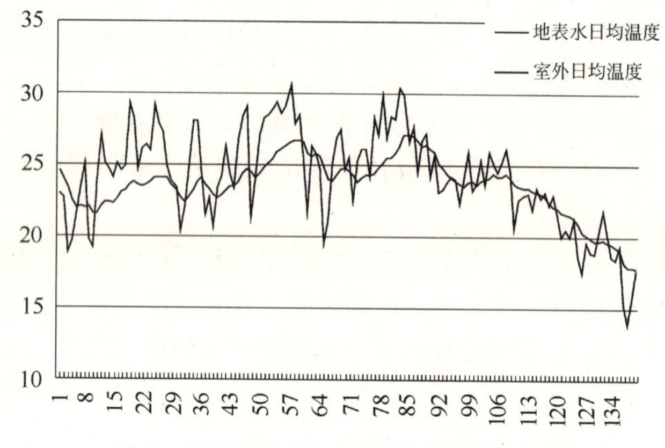

图 6 夏季地表水与室外日平均温度对比

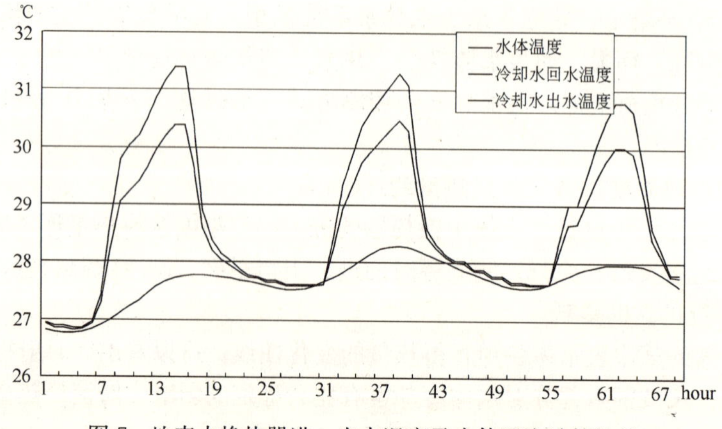

图 7 地表水换热器进、出水温度及水体温度计算结果

4 结论

本文分析了地表水水源热泵系统的应用和研究现状，分析了浅表水水体的热过程并建立了数学模型。在TRNSYS环境下开发了地表水水体及地表水换热器模块，对地表水温度、换热器出口温度、及地表水体的各项得热量等参数进行了全年动态逐时模拟。并在此基础上建立了典型SWHP系统的模拟程序，在此模型下可以对SWHP系统进行设计计算，也可以对整个系统进行全年逐时能耗计算和经济分析。

An Analysis of Current Situation and System Simulation of Surface Water Heat Pumps System

Zongyu Sun Guofu Du Xiaomei Feng Yu Zou
(Institute of Air Conditioning，China Academy of Building Research，Beijing 100013 China)

Abstract: In this Article，the research and applications of surface water heat pumps system (SWHPs) are introduced，and some problems of SWHPs are also discussed. On the basis of analysis of heat exchange of surface water，a simulation module in TRNSYS is developed，which can be used in dynamic simulation of SWHPs. With this module，the temperature of surface water，inlet and outlet water of exchanger，and the quantity of every heat exchange can also be calculated. And in the process of design and economy analysis of SWHPs，calculations of heat exchanges and Energy consumption can be made.

参考文献

[1] ASHRAE HANDBOOK. 2001 FUNDAMENTALS，2001
[2] 徐伟等译. 地源热泵工程技术指南. 2001. 11
[3] 何涛. 地表水水源热泵系统应用实例分析. 中国建筑科学研究院空调所，全国暖通空调年会论文集. 2004
[4] 湖南大学土木工程学院，刘婷婷，彭建国，张国强，陈晓. 水源热泵在湖南地区推广的可行性探讨. 全国暖通空调年会论文集. 2004
[5] 湖南大学，陈晓等. 利用湖水的水源热泵系统应用分析. 全国暖通空调年会论文集. 2004
[6] GROUND SOURCE HEAT PUMPS: MEETING GLOBAL CHALLENGES THROUGH LOAD NETWORKING，James E. Bose，Ph. D. 2005 INTERNATIONAL MEETING OF HEATINGPUMP
[7] Lake-Source Cooling，Tim Peer，P. E.；W. S. (Lanny) Joyce，P. E.，ASHRAE Journal，vol. 44，no. 4，p. 37-39
[8] Lakewater Applications of Water-to-Air Heat Pumps，Kavanaugh，S. P.，Pezent，M. C.，ASHRAE Trans，1990，vol. 96. Part 1，paper AT-90-4-2，813-820，11 figs，2 tabs，refs. 1990
[9] Optimal Design of a Cold-Sea-Water Chiller，Bhargava，A.，Trabia，M.，Moujaes，S. F.，ASHRAE Trans. 1994，Vol. 100，Part 2，Paper number 3795，122-130，4 figs.，1 tab.，refs
[10] Precooling and Direct-Cooling Coils for Groundwater，Lake Water，and Water-Loop Heat Pump

Systems, Kavanaugh, S. P. , ASHRAE Trans. 1992, vol. 98, Part 1, Paper number AN-92-11-1, 1003-1008, 4 figs. , 3 tabs. , refs

[11] Thermal Storage Case Study: Combined Building Mass and Cooling Pond, David Arnold, Ph. D. , F. Eng. Member, ASHRAE 2000 Winter Meeting

[12] A Model for Simulating the Performance of a Shallow Pond as a Supplemental Heat Rejecter with Closed-Loop Ground-Source Heat Pump Systems, Andrew D. Chiasson, 2001

[13] TRNSYS 15 Reference Manual, 2000

建筑分布式能源系统应用探讨

张蓓红
（上海市建筑科学研究院有限公司，上海　200032）

摘　要：本文在介绍分布式能源系统概念、组成和系统特点的基础上，对分布式能源系统在建筑中应用的关键技术作了分析，指出动态负荷模拟和准确预测、系统型式和排热回收装置的选择、电力负荷的合理取值和优化运行方式等是建筑分布式能源系统应用成功与否的关键所在。分布式能源系统应用推广与技术、政策和市场因素密切相关。

关键词：分布式能源系统　负荷模拟　优化运行

引言

我国能源消费已经进入世界前列，但人均资源占有量相对匮乏，且能源利用率低，浪费严重。同时，我国温室气体 CO_2 的排放总量仅次于美国，已经超过 30 亿 t[1]。我国政府把保护环境、节约能源、实现可持续发展作为工作的重中之重，于 1994 年通过和确立了《中国 21 世纪议程》，1997 年 11 月 1 日颁布实施了《中华人民共和国节约能源法》。2000 年由国家发展计划委员会、国家经济贸易委员会、建设部和国家环保总局联合下发了《关于发展热电联产的规定》，再次明确了国家鼓励发展热电联产的政策，强调国家支持发展以天然气为燃料的燃气轮机热电冷联产项目，特别强调了国家积极支持发展燃气轮机联合循环热电联产和小型燃气冷热电联产。

本文通过介绍分布式能源系统概念、系统组成和特点，对分布式能源系统在建筑中应用的关键技术进行重点分析，并提出笔者的观点。

1　建筑分布式能源系统概述

1.1　分布式能源系统的概念

分布式能源系统(distributed energy system，DES)是小型的、模块式的、以高效的热电冷联产或可再生能源发电技术为主、靠近用户(负荷)端的能源系统。分布式能源系统以其规模小、灵活性强等特点，通过不同循环的有机整合可以在满足用户需求的同时实现能量的综合梯级利用，从而避免了冷能和热能远距离传输的困难，同时大大提高了供电的可靠性。

1.2　系统型式及特点

建筑分布式能源系统由以下几部分组成：发电设备(小型或微型燃气轮机、内燃机或燃料电池等)、供热或制冷设备(溴化锂吸收式冷热水机组、电制冷机组)、锅炉或蓄热系统、汽—水换热器、调节装置(使蒸汽参数符合用户要求)以及控制系统等。与新能源相关的分布式能源系统，目前尚处在研究阶段，还没有实现商业化。如燃气轮机—太阳能辅助循环、太阳能(风能)—燃料电池联合循环、燃料电池—小型燃气轮机联合循环、燃气轮机—热泵联合循环等。这些系统的开发利用为我国未来的能源结构的调整，建设新一代 DES 指明了方向。

分布式能源系统特别适合于组成分布式的热电联供或冷热电联供系统，正在形成所谓“第二代能源系统”，即：在一次能源上，以气体燃料为主(天然气、煤层气、地下气化气、沼气等)，可再生能源(太阳能、风能以及潮汐能等)、核能以及城市固体垃圾废物等为辅，利用一切可以利用的资源；在二次能源上，以分布在用户端的冷热电联产为主，其他分布能源技术为辅，实现能源梯级利用。同时依靠现有电力系统和大电网设施进行补充和支持；在环境保护上，将污染资源化，争取实现零排放的目标。

从上世纪 70 年代开始，分布式能源系统在西方国家，特别在美国兴起。随着信息技术革命和天然气的广泛利用，天然气供应管网的发达与电力网的充分发展，以及新能源发电的不断推广，使得以效益规模为法则的分布式能源系统为代表的第二代能源系统得到进一步发展，其所涵盖的技术也从热电冷联产机组向可再生能源发电技术(如太阳能光伏发电、风力发电、生物质能发电、太阳能热发电等)、以及燃料电池及氢能等方面拓展。

分布式能源系统可以满足特殊场合的需求，如：不适宜铺设电网的偏远地区或散布的用户；对供电安全稳定性要求较高的特殊用户如医院、银行等；能源需求较为多样化的用户，需要电力的同时还需要热或冷能的供应，如医院、旅馆等。国内外的实践表明，分布式能源系统在温室气体排放、环境保护、能源的高效利用等方面表现出了极大的优势，从而使其拥有巨大的市场潜力和商业价值，DES 将成为能源动力产业发展的主攻方向。

2 建筑分布式能源系统应用的关键技术

2.1 采用动态模拟技术进行建筑热、电、冷负荷的详细计算；充分采用各种节能措施降低建筑能耗

我国公共建筑的特点是功能复杂，往往将办公、公寓、商场、餐饮、娱乐等多种设施集中在一幢高层建筑之中。在设计分布式能源系统时，必须考虑不同使用功能建筑的同时使用系数和参差系数对冷(热)负荷的影响。建议采用详细的和先进的负荷计算方法。在预可行性设计阶段，当设计条件不具备时，也可采用指标估算方法，但应有一定的科学依据，越是大型的建筑越是需要考虑负荷参差率。

方案设计阶段必须采用建筑负荷动态模拟技术来预测建筑全年的冷热负荷情况，预测成果不仅要包括各不同季节(冬季、夏季、过渡季)典型日的负荷小时变化情况，还应有建筑全年的热、电、冷负荷的延时曲线。图 1、图 2 示出了某医院建筑在夏季典型日的负荷变化和全年负荷变化情况。获得这些数据是进行分布式能源系统方案设计的重要前提。

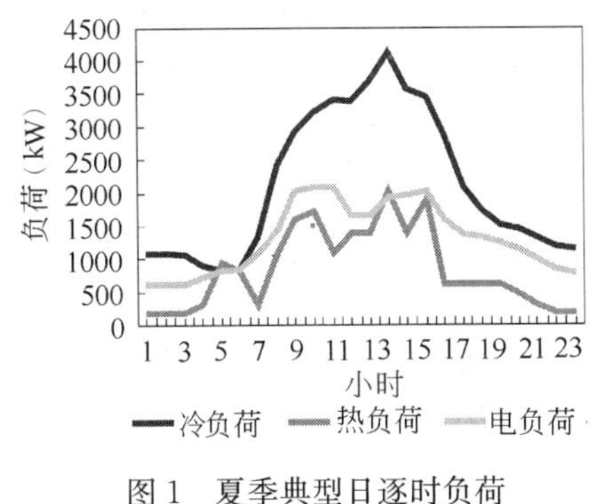

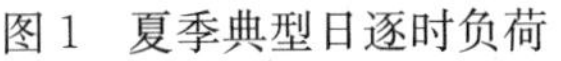

图1 夏季典型日逐时负荷

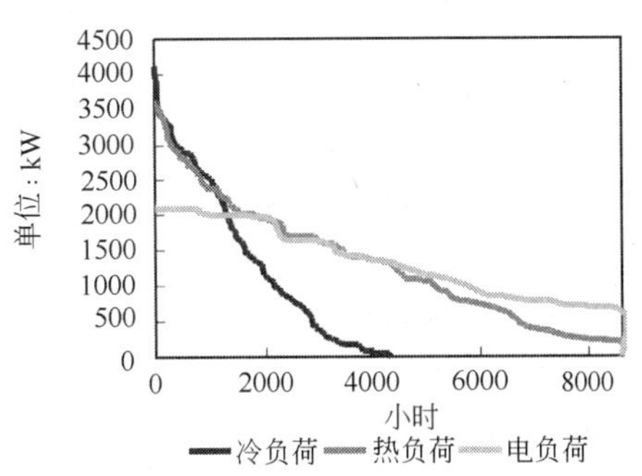

图2 全年延时热电冷负荷曲线

2.2 不同用途的建筑物应采用不同的分布式能源系统和不同的排热回收装置

不同用途建筑物冷、暖、热水、电的峰值负荷不同，全年负荷量也不同，故应根据负荷的性质选择分布式能源的燃气机类型以及相应的排热回收装置。选择的原则是做到两个平衡，即：分布式能源系统发电量和建筑用电量之间的平衡；分布式能源系统排热量和建筑采暖、制冷、生活热水用量之间的平衡[2]。两者之间平衡得越好，分布式能源系统的经济效益和节能效益越好；反之，多余排热只能排出，增大了系统能耗，降低了能源利用率。

分布式能源的燃气机类型常见的是燃气轮机和燃气内燃机。燃气轮机分为大中型（>20MW）、小型（1～20MW）和微型（<1MW），分布式能源系统中常使用的是小型和微型燃气轮机，微型燃气轮机的压比约为4.0，透平进口温度约为1000℃，出口烟气温度约为650℃，回热器排温（出口烟温）小于320℃。燃气轮机和燃气内燃机的主要区别见表1。从表1可知，燃气轮机的排热量全部为450℃～550℃烟气的排热量，排热量/发电能力较大，烟气温度较高，当建筑冷、热负荷较大或需要蒸汽负荷量较大时，可采用这种方式。当建筑要求生活热水负荷较多时，由于生活热水所需温度较低，故可采用燃气内燃机方式。

燃气轮机和燃气内燃机的主要区别 **表1**

	燃气轮机	燃气内燃机
发电效率（%）	27～39（小型、微型）	30～45
排热量/发电能力（Mcal/h/kW）	1.8～1.96	1.51
排烟温度和排气排热量（℃）	450～550（全部排热量）	400～450，约为全部排热量的35%
水套出口水温和水套排热量（℃）		55～90，约为全部排热量的65%

排热回收是分布式能源特有的技术，利用排热量的多少和排热装置的效率直接决定了分布式能源系统的经济性、节能性，同时还对系统的初投资、占地和系统的可靠性、耐久性和维护保养性有很大的影响。选择排热回收装置的一般原则是使分布式能源排热量和建筑采暖、制冷、生活热水用量之间达到平衡，以便充分地利用排热量，提高分布式能源系统的能源利用率。

2.3 电力负荷的合理取值和优化运行方式等是分布式能源系统应用成功与否的关键

电力负荷取值是确定发电机装机容量的主要依据。发电机装机容量的取值是一项优化设计的问题，应以经济效益和节能性最佳为目标，综合考虑建筑热电冷负荷波动、设备的性能特性以及系统的运行策略等约束条件后，对发电机装机容量和其他主要设备容量进行最优的取值匹配。从已实施的分布式能源系统工程可知，发电机装机容量约为公共建筑电力峰值负荷的 60%时[2]，经济性较好。

在考虑系统设备容量选择的时候必须兼顾运行策略的优化。单一的“以电定热”或“以热定电”都是不足取的。前者会造成大量能量的浪费，而后者虽然能够保证发电设备余热的利用，经济性却不佳。运行策略的优化必须在保证系统获得最佳的经济性的前提下灵活选择，它对设备容量的配置又有着重要的影响，这是相互紧密联系的两个方面，是决定热电(冷)联产系统设计成败的关键。

分布式能源系统的运行方式一般有以下几种：(1)满足电力负荷的运行方式(以电定热)。(2)满足热负荷的运行方式(以热定电)。(3)优化运行方式，即根据电、热负荷需求进行台数控制，当夜间电力负荷大量减少时，发电机停止运行。

下面引用图 3 来说明系统不同的运行策略对经济性的影响[3]。图 3 为三种不同的运行策略即以电定热、以热定电和兼顾电、热负荷需求的优化运行方式随天然气价格变化的不同运行效果(系统图式略)。图中横坐标 α 为气电价格比，即天然气价格相对于电价的价格：

$$\alpha=\frac{\text{天然气价格(元/MJ)}}{\text{平均电力价格(元/MJ)}} \tag{1}$$

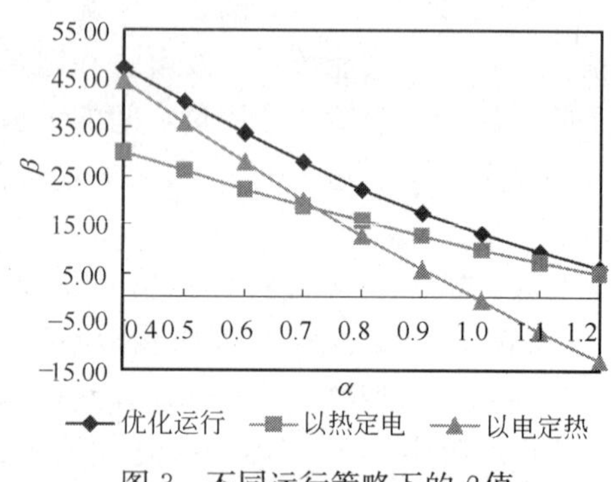

图 3 不同运行策略下的 β 值

纵坐标 β 为代表分布式能源系统经济性的指标，定义为分布式能源系统年度能耗费用相对于常规能源系统[3]的缩减比例，即：

$$\beta=(C_1^0-C_1)/C_1^0\times 100 \tag{2}$$

式中 C_1——分布式能源系统全年能耗费用，万元；

C_1^0——常规能源系统全年能耗费用，万元。

由图 3 可见，位于最上方的曲线为优化运行的曲线，无论 α 如何变化，其在三个方案中始终保持最优的经济性。在 $\alpha=0.7$ 时，以热定电和以电定热两根曲线交于一点，说明此时这两种不同的运行策略造成的结果是一致的。当 α 小于 0.7 时，以电定热的策略较之以热定电更占优势，而 α 大于 0.7 以后则刚好相反。这说明了以电定热或以热定电都不是绝对的，分布式能源系统的运行策略要受到负荷水平、能源价格等多种因素的制约。

3 建筑分布式能源系统应用的影响因素

分布式能源技术的应用与地域条件、各种技术的可行性、资源的可获得性和环境要求、社会的认可度、乃至监管和市场条件等密切相关[4]。影响分布式能源系统发展的主要因素可归结为技术、政策、市场等几方面。

3.1 技术因素

技术因素涵盖了与分布式能源系统相关的各个技术问题。具体地说，就是分布式能源设备的技术性能，如发电机组、并网设施、输配电网的运行，以及电网安全、可靠及稳定运行的标准等。电网公司所要求的技术指标对于分布式能源开发商来说常常是不确定的，容易引起争议的技术问题包括电网的容量、电网改造、并网的接入点、并网电压、线路保护技术等。

3.2 政策因素

政策因素包括直接和间接地影响分布式能源项目的政策，以及制定、实施相关政策的体制结构等。支持政策的空缺或不确定是分布式能源发展的重要障碍。我国现有的电力监管体制未能充分认识分布式能源系统的价值，尤其是环保效益、提高电网服务质量、保障电力供应等方面的益处。另外，分布式能源能否纳入区域规划也是关系到分布式能源项目建设和运行的重要因素。

3.3 市场因素

市场因素包括分布式能源技术自身的经济特性和市场结构。前者指的是设备及运行成本，以及融资、燃料和产品合同等。后者指的是市场参与者的数量和规模大小、定价和交易机制、竞争的力度、进入市场条件等。许多分布式能源系统的开发商、运营商往往不是实力雄厚的电力公司，多为新成立的中小型企业，因此在技术、经验和能力方面不及有实力的大公司。电力公司的市场操纵力可能直接影响分布式能源项目的发展。对于中小规模的独立发电商来说，电力公司可能对分布式能源开发商制造各种并网的障碍，并压低分布式能源的上网电价，对非电力部门的分布式能源开发商采取抵制的策略；燃气公司也可能会抬高热电机组燃料供应的容量费等。对于天然气热电联产设备而言，气电价格比的迅速攀升也极大地影响了分布式热电联产技术的推广。

4 结语

城市化水平的发展，商业和民用对能源需求的急剧增加，能源需求的主体呈现出多样化的态势，这种需求的特点突破了传统意义上的热电冷联产主要服务于工业工艺用热用冷的局面，从而为系统的发展提供了更广阔的市场空间。技术的发展，特别是小型、微型燃气轮机和燃料电池以及其他新能源技术的进步为热电冷联产赋予了新的内涵。可以预见，分布式能源系统将是未来楼宇空调发展的首选。

本文重点介绍了分布式能源系统应用中的关键技术问题，提出了一些参考意见。建设分布式能源系统时要根据实际情况，因地制宜，实现多种能源综合利用，做到经济效益和社会效益相统一。

Discussion on Application of Distributed Energy System in Buildings

Zhang Beihong

(Shanghai Research Institute of Building Sciences, Shanghai 200032 China)

Abstract: With the introduction of composition and system characteristics of distributed energy system(DES), this paper lay emphasis on the key technologies of DES applications in buildings. Several important issues are put forward, namely dynamic load simulation, selection of generator type and heat recovery, system sizing and optimal operation, etc. The promotion of DES is closely related to technical, policy and market aspects.

参考文献

[1] 张洪伟等. 分布式能源系统的方案选择及性能分析. 暖通空调. 2004, 34(5)

[2] 李先瑞. 分布式能源与建筑的融合. 节能与环保. 2004.9

[3] 张蓓红, 龙惟定. 热电(冷)联产系统运行调节的双目标规划. 暖通空调. 2005, 35(10)

[4] 黄东风. 分布式能源发展的影响因素分析. 能源工程. 2005.1

居住建筑能耗与各建筑因素关系分析

王志刚　鄢　涛
（深圳市建筑科学研究院，深圳　518031）

摘　要：不同地区各建筑因素以及气象参数对建筑能耗有着不同程度的影响，通过三水平线性回归得到能耗与建筑因素的回归方程。由建筑各因素与建筑能耗的相关性分析，明确了不同地区在建筑节能中的侧重点。

关键词：建筑能耗　建筑因素　相关性

1　建筑类型的选择

根据对城市中居住建筑类型调查结果，选择三种建筑类型，分别为别墅型建筑、小高层建筑和高层建筑。模拟用建筑模型见图 1～3。建筑南北朝向。三类建筑体形特征见表 1。

图 1　6 层小楼

图 2　4～5 层

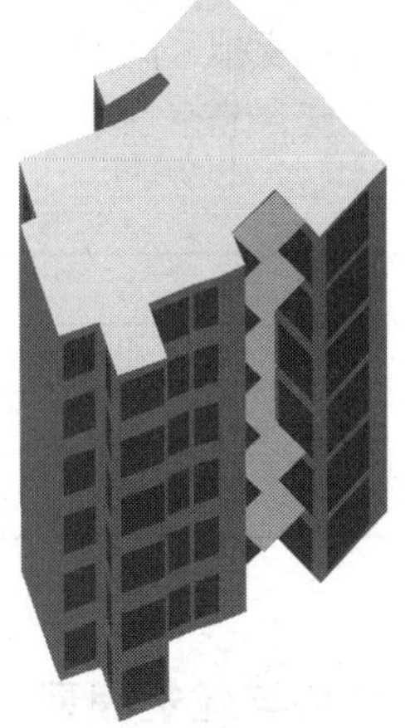

图 3　20 层高层建筑
（模型建 6 层）

三类建筑特征　　表 1

建筑类型	平均窗墙面积比 CM	体形系数 S
6 层小楼	0.19	0.38
4～5 层	0.3	0.41
20 层高层	0.4	0.35

2 模拟条件

根据我国建筑气候区域划分[1]，分别选择广州、重庆、北京代表夏热冬暖地区、夏热冬冷地区、寒冷地区。模拟条件采用三水平条件，所有因素皆为三水平。选用这个三个城市的气象参数对不同组合的建筑模拟采暖能耗和空调能耗。模拟设空调温度 26℃，采暖温度 18℃，一次换气次数。围护结构热工性能参数三水平值如下设定如下：

外墙 K 值：1.5，1.0，0.5W/m^2·K；

屋面 K 值：1.0，0.7，0.5W/m^2·K；

外窗 K 值：4.7，3.35，2.0W/m^2·K；

外窗 SC 值：0.7，0.5，0.3。

3 气候参数统计

度日是指每日平均温度与规定的标准参考温度(或称温度基准)的离差。因此，某日的度日数，就是该日平均温度与标准参考温度的实际离差。T_B 为基础温度，T 为日平均温度，对于采暖度日数(HDD)，如果该日的平均温度小于基础温度则该日 HDD=T_B-T，如果该日温度大于基础温度则 HDD=0，全年度日数就是逐日度日数的总和；空调度日数(CDD)，如果该日的平均温度小于基础温度则该日 HDD=0，如果该日温度大于基础温度则 HDD=$T-T_B$，全年空调度日数就是逐日空调度日数的总和。

本次模拟采用 DOE-2 气象参数，由逐时气象参数计算得到日平均气象参数，分别对广州、重庆、北京三城市的日平均气象参数进行采暖度日数和空调度日数统计，结果见表 2。

三城市采暖度日数和空调度日数 **表 2**

城 市	采暖度日数 HDD	空调度日数 CDD
广 州	388.4	328.4
重 庆	1073.4	241.38
北 京	2605	85.26

4 能耗与建筑因素关系分析

分别对前述三栋建筑各 81 个组合方案用广州、重庆、北京的气象参数进行采暖能耗和空调能耗模拟。对模拟结果进行多元线性回归分析，显著性水平 α=0.05。

采暖、空调及总能耗与各因素线性回归系数 **表 3**

项目	采暖能耗(kWh/m^2)			空调能耗(kWh/m^2)			总能耗(kWh/m^2)		
	北京	重庆	广州	北京	重庆	广州	北京	重庆	广州
C	−21.85	15.19	6.33	−22.91	−16.52	−38.21	−44.76	−1.33	−31.88
CM	−31.51	27.26	6.82	22.29	23.25	36.54	−9.22	50.51	43.36

续表

项目	采暖能耗(kWh/m²)			空调能耗(kWh/m²)			总能耗(kWh/m²)		
	北京	重庆	广州	北京	重庆	广州	北京	重庆	广州
S	183.93	−34.18	−16.59	39.60	32.63	89.77	223.54	−1.55	73.19
QK	17.34	6.36	1.83	1.72	2.87	4.23	19.06	9.23	6.05
FK	5.29	2.19	0.65	0.67	0.76	1.34	5.96	2.95	1.99
CK	8.00	2.37	0.89	−0.28	−0.04	−0.14	7.73	2.32	0.74
SC	−36.38	−5.52	−4.43	16.41	15.66	28.80	−19.97	10.15	24.37
R	0.963	0.975	0.952	0.952	0.967	0.959	0.966	0.987	0.963

采暖、空调及总能耗与各因素皮尔逊相关系数　　表 4

项目	采暖能耗(kWh/m²)			空调能耗(kWh/m²)			总能耗(kWh/m²)		
	北京	重庆	广州	北京	重庆	广州	北京	重庆	广州
CM	−0.354	0.598	0.475	0.425	0.471	0.348	−0.266	0.680	0.481
S	0.423	−0.426	−0.418	0.018	−0.044	0.118	0.458	−0.328	0.004
QK	0.507	0.556	0.447	0.201	0.336	0.282	0.598	0.577	0.407
FK	0.074	0.092	0.077	0.038	0.043	0.043	0.090	0.089	0.064
CK	0.518	0.459	0.481	−0.072	−0.012	−0.021	0.537	0.322	0.111
SC	−0.433	−0.197	−0.442	0.785	0.747	0.783	−0.255	0.259	0.668

注：表中符号 *C*——常数；*CM*——平均窗墙面积比；*S*——体形系数；*QK*——墙传热系数；*FK*——屋面传热系数；*CK*——窗传热系数；*SC*——综合遮阳系数；*R*——复相关系数。

从表 3 复相关系数 *R* 可以看出，采暖能耗、空调能耗以及总能耗与建筑因素有着很强的线性相关性，其复相关系数都大于 0.95。

采暖能耗、空调能耗、总能耗与各因素的皮尔逊相关系数的大小以及正负表明了该变量对能耗的影响相关程度。除寒冷地区建筑 *CM* 值与建筑采暖能耗及总能耗呈一定的负相关性以外，其他地区 *CM* 值与各种能耗的影响呈正相关性，即能耗随着 *CM* 的增大而增大。寒冷地区大的 *CM* 值有利于日照进入房间减少采暖负荷。体形系数 *S* 对寒冷地区采暖负荷有较大的关系，应严格控制体形系数。由于体形系数 *S* 与 *CM* 值存在一定的相关性，体形系数 *S* 对夏热冬冷和夏热冬暖地区的能耗影响还应进一步分析。

外墙、屋顶的传热系数 *K* 值与能耗呈正相关性，且与采暖能耗的相关性大于空调能耗的相关性。外墙对总能耗的影响程度由北向南递减，但对整个由北到南区域，外墙与总能耗的相关性较大，其值为 0.4～0.6。屋顶的 *K* 值与整栋建筑能耗的相关性很小，但对顶层的影响大，有必要采取屋顶保温。

外窗的传热系数 *K* 值与采暖能耗呈正相关性，其相关度与外墙传热系数相近；外窗的传热系数与空调能耗呈负相关性，但相关程度绝对值很小，可忽略外窗的传热系数对空调能耗的影响。但对于总能耗，外窗的传热系数与能耗呈正相关性且由北往南递减。

采暖能耗与外窗遮阳系数 *SC* 值呈负相关性，与空调能耗呈正相关性，且在所有对空调能耗影响因素中相关性最大的，其相关程度接近 0.8。对于寒冷地区，采暖能耗为主，遮阳系数与总能耗呈负相关性。夏热冬暖地区遮阳系数与总能耗相关系数是所有因素中相关性最大的因素。

从总能耗角度分析建筑的各因素与总能耗的相关性得到，寒冷地区外墙 K 值与外窗的 K 对能耗有正的相关性，应侧重保温，外窗的遮阳系数要求低；夏热冬冷地区建筑能耗与建筑平均窗墙面积比、外墙的传热系数有较大的正相关性，外窗的传热系数与遮阳系数对总能耗的影响次之，建筑节能中应侧重保温，兼顾隔热；夏热冬暖地区建筑能耗与外窗的遮阳系数有最大的相关性，平均窗墙面积比和外墙的传热系数次之，建筑节能应侧重建筑外遮阳，同时应注意隔热。

5 能耗与度日数关系

为消除体形系数、平均窗墙面积比等因素的影响，只考虑度日数对能耗的作用，分别将别墅、小高层、高层建筑在广州、重庆、北京不同组合下所模拟得的能耗平均后再进行回归分析。

平均处理后的能耗结果 表 5

城市	建筑	采暖度日数	空调度日数	总能耗 kWh/m²	采暖能耗 kWh/m²	空调能耗 kWh/m²
广州	6 层小楼	388.4	328.4	26.1	4.1	22.1
广州	4～5 层	388.4	328.4	33.1	4.3	28.8
广州	20 层高层	388.4	328.4	33.1	6.0	27.0
重庆	6 层小楼	1073.4	241.4	31.1	19.7	11.4
重庆	4～5 层	1073.4	241.4	36.6	21.7	15.0
重庆	20 层高层	1073.4	241.4	41.8	26.4	15.4
北京	6 层小楼	2605	85.3	75	69	6
北京	4～5 层	2605	85.3	66.4	56.9	9.5
北京	20 层高层	2605	85.3	80.7	71.1	9.6

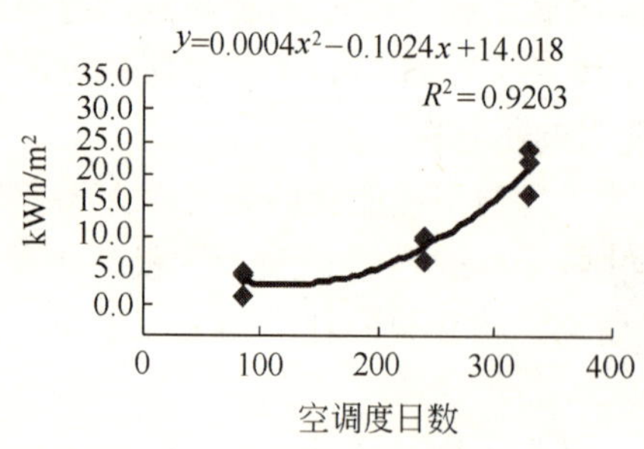

图 4 空调耗电量与空调度日数关系

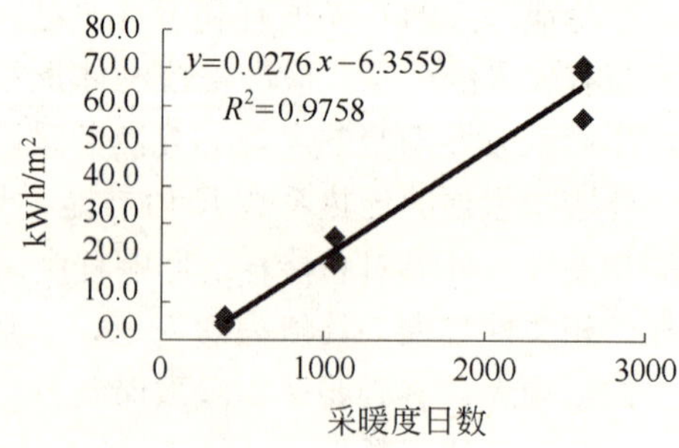

图 5 采暖耗电量与采暖度日数关系

将总能耗与采暖度日数和空调度日数进行线性回归，显著性水平 $\alpha=0.05$，结果如下：

$$总能耗=-214.49+0.09\times 采暖度日数+0.64\times 空调度日数$$

复相关系数平方：$R^2=0.945$。

总能耗与采暖度日数相关系数为 0.956，与空调度日数相关系数为－0.945，表明总能耗与气象参数有很大的相关性，寒冷地区能耗要大于夏热冬冷和夏热冬暖地区总能耗。

6 总结

通过三个区域的建筑采暖能耗、空调能耗以及总能耗与建筑各因素的相关性分析以及能耗与采暖度日数和空调度日数分析可以得到以下结论：

（1）能耗与建筑因素的线性相关度≥0.95，可以通过相关的回归公式估算建筑能耗。

（2）外墙的传热系数与总能耗呈正相关性，其相关程度由北向南减小，整体相关度较高。寒冷地区应注重外墙保温。

（3）屋顶传热系数对整栋建筑能耗影响很小，但考虑对顶层能耗的影响应适当考虑。

（4）外窗传热系数与采暖能耗呈正相关性，与空调能耗呈负相关性且相关系数很小，可见外窗的保温对采暖有利，寒冷地区应注意窗户的保温，而外窗的传热系数对降低空调能耗作用不显著。

（5）窗的传热系数与空调能耗呈负相关性，在空调季节传热系数低的窗户应能有足够的开启面积，以便利用自然通风降温。

（6）外遮阳系数对寒冷地区能耗有负作用，寒冷地区可不考虑遮阳；夏热冬冷地区应适当考虑遮阳，而夏热冬暖地区应重视外窗遮阳。

（7）建筑能耗与气象参数有很大的相关性，空调能耗与度日数呈平方关系，采暖能耗与采暖度日数为线性关系。寒冷地区能耗要大于夏热冬冷和夏热冬暖地区能耗。

Analysis of Building Energy Consumption Based on Energy Plus

Wang zhigang　Yan Tao

(Shenzhen Institute of Building Research, Shenzhen　518031　China)

Abatract: Different Building factors and meteorological parameters have different extent impacts to Building energy consumption. This paper gets a regression equation of energy consumption and Building by tri-level linear regression method based on three typical residential buildings and shows that we should lay emphasis particularly on special building factors in different climate areas.

参考文献

[1]　刘加平. 建筑物理[M]. 北京：中国建筑工业出版社，2000. p19～20

绿色生态建筑与暖通空调

张惊涛　　　　　　　　　　　　胡孟娣
（上海华东建设发展设计有限公司，上海　210050）（浙江科技学院，浙江　310010）

摘　要： 简要介绍了当今世界各国有关绿色生态建筑的可持续发展探索与实践，以及国内外已建成的有代表性的绿色生态建筑。概括叙述了它们一些全新的设计理念与技术特点，分析了暖通空调专业在这些绿色生态建筑中的作用，以及与其他专业的配合问题。
关键词： 绿色　生态　建筑　技术特点　暖通空调

1　资源危机呼唤绿色生态建筑

自从发明房子以来，人们将自然界里获得的50％以上的资源和48％的耕地用来建造各类建筑及其附属设施。这些建筑在建造与使用过程中又消耗了全球能源的50％、水资源的42％。而全球环境总体污染中，50％的温室效应、50％的臭氧层破坏、40％的水污染、24％的空气污染与建筑有关。建筑垃圾则占人类活动产生垃圾总量的40％。早在2000多年前，我国古代著名思想家荀子就提出："强本而节用，则天不能贫"。然而，全世界建筑能耗发展趋势却与荀子的教诲背道而驰。

与全球建筑的严峻资源消耗形势相比，我国建筑的资源消耗尤其是能源消耗更显危机重重。据统计，我国建筑能耗总量在能源总消费量中所占的比例，已从上世纪70年代末的10％，上升到近年的27.5％。其中建筑采暖、空调、照明占15％左右，建筑建造能耗为12％左右，采暖空调的能耗已占建筑总能耗的55％。目前，我国的房屋建设规模堪称世界第一，每年新建成建筑面积近20亿m^2。但95％以上都是高能耗建筑，合格的节能建筑不足4％。

和同纬度许多发达国家相比，我国冬天气候更冷，夏天气候更热，南方空气湿度还很高。在这种湿热环境下，我国房屋的保温隔热性能却要比发达国家差得多，相当于发达国家在同等条件下采暖耗能的2～3倍。一般而言，GDP增长8％，能源消耗增长约为5％，而我国2002年和2003年的能源相应增长为14.5％和13.5％，相当于2年时间耗尽了5年的能源供应。中国的能源供应正面临深刻的危机，传统的经济增长方式已走到尽头。

并且，我国的能源利用效率很低，仅为33％，比发达国家落后20年，相差10个百分点。能源消费强度大大高于发达国家及世界平均水平，约为美国的3倍，日本的7.2倍。目前中国正处于工业化和城镇化快速发展阶段，剧增的城市扩张侵占土地、破坏生态环境等现象日益严重。要在未来15年保持GDP年均增长7％以上，每年新建房屋达16～20亿m^2，所以必将面临巨大的资源约束瓶颈和环境恶化压力。严峻的事实告诉我们，中国要走可持续发展道路，发展节能与环保的绿色生态建筑刻不容缓。

目前，生态、节能和可持续发展是21世纪全球面临的最迫切的课题，落实到房屋建筑领域，便派生出绿色生态建筑的概念。学术界有关绿色生态的定义有许许多多，各有侧重。但是比较一致些的观点认为，所谓绿色不仅仅指绿化，而是一种对建筑环境进行全面系统的环保设计的理念。所谓生态是指在不影响人们健康和舒适感的前提下，对各种资源实施节约与再利用，尽可能地减少废弃物，保持和周围生态环境协调融合。绿色生态建筑强调建筑与人文、环境、资源利用及社会发展的和谐统一。

2 国内外绿色生态建筑的实践

目前，绿色生态建筑已成为当前建筑领域研究探索的热点。国内外设计界，尤其是欧洲设计界近20年来，已进行了大量卓有成效的研究探索。特别是近十年来，已取得了很多开拓性的成果，期间全世界已建成了上百座具有示范性意义的绿色生态建筑。

在绿色生态建筑方面德国走在世界的前列，所以建成的绿色生态建筑最多，也最有启发意义。由于地理原因，德国能源匮乏，石油几乎100%进口，天然气80%进口，所以开源节流和保护环境是德国政府的一贯政策。2000年，德国颁布了促进新能源技术开发的《可再生能源促进法》、《未来投资计划》、《家庭使用可再生能源补助计划》、《生物能发展规划》等。2002年颁布了《能源节约法》，2003年颁布了《10万屋顶太阳能发展计划》。按照一系列新法规，德国的建筑允许能耗要比2002年前的能耗水平下降30%左右。而早在1978年，德国就修改过一次建筑节能标准，使得其后的建筑能耗比1978年前的老建筑减少了60%以上。

欧洲各国生态住宅区对选用的建筑材料也都有一定的要求，德国慕尼黑市规定住宅建设中不准用铝材窗。原因是铝材在制造过程中能耗大，污染大。欧洲许多国家近年来绿化率高，森林处在良性循环中，每年都要伐掉一定量的木材，所以国家鼓励使用木制品。尤其木材比金属、钢筋混凝土传热系数小得多，故欧洲许多国家都将木材作为生态性建材，在住宅建设中得到普遍使用。许多住宅采用木制品作骨架，墙体和建筑配件。北美的加拿大和美国也因为同样原因将木材作为生态性建材，在住宅建设中得到普遍使用。

英国政府也非常重视建筑节能，从1986年开始就制订了国家节能计划，并将建筑节能由低到高分为10个等级。目前英国的新建住宅基本上都达到了最高节能等级的要求，并且住宅的内部舒适程度也因节能构造的增强得到了明显的提高。按新标准设计的节能型住宅比传统住宅的能耗要减少75%。

瑞典在1975年通过的新住宅规范当中关于节能的指标，比我们国家刚刚出台的节能标准里面的指标还要高。三层玻璃现在在瑞典非常常见，三层固定玻璃窗扇中间带百叶，可在关闭的情况下通风。除了节能以外，瑞典还广泛利用可再生能源，采用代用能源。比如在集中供暖方面，瑞典城市集中供暖已经覆盖80%，而在这80%的供暖当中只有10%采用的是化石燃料，其他的燃料来源主要为垃圾焚化和工业废料等代用能源。

瑞士自20世纪80年代以来大力推广高舒适度低能耗的绿色生态建筑，这些建筑物让采暖、制冷设备的开机时间大大缩短。一年中的大部分时间的室温都可保持在20～26℃，提高了居住者的舒适度，降低了2/3能源消耗。如平常夏天最热的时候有51天时间需用空调，采用绿色生态技术之后只需 7 天。

作为美国最主要的建筑行业协会之一的美国绿色建筑委员会(USGBC)在美国能源部的大力支持下，于 2000 年在 1998 版基础上，推出 LEED(Leadership in Energy and Environmental Design)的绿色建筑评估体系第二版。作为世界权威的认证机构，迄今为止，除了为美国 50 个州提供绿色建筑认证外，USGBC 还接受了全球 20 来个国家的近 1700 个建设项目绿色建筑的认证申请。认证程序按五项评估标准为每个申报工程打分，最高值为 69 分。这五项评估标准为：项目选址的可持续性、改善与提高用水效率、材料和资源利用、能耗减少与大气污染控制、室内空气质量。其中：69～52 分为白金奖、51～39 分为金奖、38～33 分为银奖、32～26 分为认证奖，另外还有一项奖励整体设计具有示范性创新思路的特别大奖。作为 LEED for Core & Shell 全球第一批试验性项目之一的我国杭州“西湖天地二期”工程，恰巧是 USGBC 的第 1000 个申报 LEED 绿色建筑认证的工程，目前正在申报 LEED 白金奖。

我国政府近些年来围绕节能、环保和可持续发展，颁布了一系列新的建筑规范及标准。包括《夏热冬冷地区建筑节能标准》、《住宅建筑节能设计标准》、《公共建筑节能设计标准》等。建设部在今年 2 月提出：到 2010 年我国新建建筑争取 1/3 以上能够达到绿色建筑和节能建筑的标准。同时，全国城镇建筑的总耗能要实现节能 50%的目标。

2004 年 9 月，我国第一座生态建筑示范楼在上海落成。其综合能耗为普通建筑的 1/4，再生能源利用率占建筑使用能耗的 20%，再生资源利用率达到 60%，室内综合环境满足舒适健康要求，见图 1、图 2。

图 1　上海生态建筑示范楼外景

2005 年 3 月，国内又一幢超低能耗示范楼在清华大学落成，楼内集中展示了近百项国内外最先进的建筑节能技术产品。据测定，建筑围护结构导致负荷仅为常规建筑物的 10%，冬季可基本实现零采暖能耗。包括照明、办公设备、空调通风系统在内，示范楼单位面积全年总用电量约为每平方米 40kW，而北京市高档办公建筑则为每平方米 100～300kW，见图 3。

说起绿色生态建筑，大家都会关心它的造价会比普通建筑增加多少。按目前统计，绿

图 2　上海生态建筑示范楼内景

色生态建筑要增加 8%～10%的造价。由于绿色生态建筑采用的新技术措施与设备名目繁多，各自相应增加的投资回收年限从置换通风的 2.5 年到太阳能光电板的 90 年，相差很大。所以我们只能大致估计整体增加的投资回收年限约为 5～13 年左右，视具体工程到底采用了哪些技术而定。

图 3　清华大学超低能耗示范楼落成典礼

3　绿色生态建筑的技术特点

为达到节能与环保目标，绿色生态建筑在建筑选址、建筑造型、绿色建材、生物多样化、绿化、保水、室内环境控制、综合节能、智能化系统、减少 CO_2 排放、水资源的节约与回收利用、减少废弃物、污水与垃圾处理、防潮调湿防霉、降噪减震等等方面采用了大量新颖实用的技术手段。其实，绿色生态建筑涉及的技术手段还远不止这些，打算尝试运用的还有很多。对于上述每项技术，都包含许多值得探讨和尝试的内容，不同技术之间可能还存在密切的相互关联。

就拿综合节能来说，涉及到建筑造型、开窗率、保温与密封、低透热高透光玻璃、外遮阳、建筑自然导风构造、暖通空调技术、节水技术、能源系统管理、太阳能利用、自然导光等等。

当然，暖通空调技术在绿色生态建筑中责无旁贷地扮演着举足轻重的角色。目前，已经被国内外绿色生态建筑广泛采用的暖通空调技术有：空调合理分区、通风空调合用与切换系统、大空间分层空调、VAV、VWV、全热交换系统、新风预处理、蒸发制冷、地板与吊顶辐射供暖供冷、置换通风、自然通风利用、太阳能制冷制热、天然冷热源利用、电热冷联供、水环热泵系统、智能化控制、余热型吸收式制冷机、蓄冷蓄热系统等等。

下面着重介绍绿色生态建筑的几种与暖通空调专业关系密切的绿色生态技术：外遮阳、自然通风、天然冷热源、智能化系统。当然其他很多暖通空调技术也很重要，但一般来说我们相对熟悉一些，设计中也不时会运用。但下面这几种技术却是近些年来受到绿色生态建筑思潮的特别青睐，而我们又不太熟悉。

3.1 外遮阳技术

我们知道夏季如能减少进入室内的太阳辐射，就能有效地减少空调冷负荷、减少空调运行能耗、降低空调设备投资与占地空间。冬季尽可能地让太阳辐射进入室内，可以减少采暖供热负荷及其投资与运行费用。目前，成功的绿色生态建筑主要采用两种外遮阳技术：主动遮阳与被动遮阳。

主动遮阳方法主要借助建筑技术，使得整幢建筑物的外表构造刚好巧妙地挡住夏季强烈的太阳直射光线。2002 年建成的英国伦敦市政厅是目前世界上采用主动遮阳法巧妙遮挡太阳直射光线的一个杰出范例，见图 4。日本东京天然气公司总部大楼的朝北向中庭也是一个很出色的主动遮阳实例。

图 4　巧妙遮挡夏季太阳直射光线的英国伦敦市政厅

所谓被动遮阳主要指外遮阳百叶具有自动可调特性。其原理是：利用太阳跟踪装置提供的太阳位置，自动调节电动遮阳板的角度，使之在夏季有效阻挡太阳直射光线进入室内，冬季使更多的太阳光进入室内。自从上世纪80年代后期英国建筑师Richard Rogers在伦敦Loyds Headquarters设计中，首次采用中间设有自动可调遮阳百叶的双层玻璃幕墙后，这种巧妙的方法很快受到广泛地注意和模仿。所以Richard Rogers也成了“高技派”建筑师的代表人物。

上世纪90年代中期以来，可调外遮阳技术在世界各地建成的绿色生态建筑中，使用的非常普遍。具有代表性的有：德国国会、Geisenkirchen Science Park；英国伦敦Helicon商店、卡斯廷BRE Office of Future办公楼、伦敦劳埃德船务注册公司大楼；IBM马来西亚公司总部大楼等。国内新建成的上海生态建筑示范楼、清华大学超低能耗示范楼也都采用了自动可调外遮阳百叶的技术。

上述建筑的自动可调电动遮阳板的安装方式有两种，一种将可调节遮阳板设于双层玻璃幕墙中间，见图5。另一种方法是将可调节遮阳板设于双层玻璃幕墙外侧，见图6。当然绿色生态建筑的玻璃幕墙采用的都是目前最好的中空低辐射玻璃、热反射镀膜玻璃、光谱选择透过性玻璃、透过率可调玻璃等等。

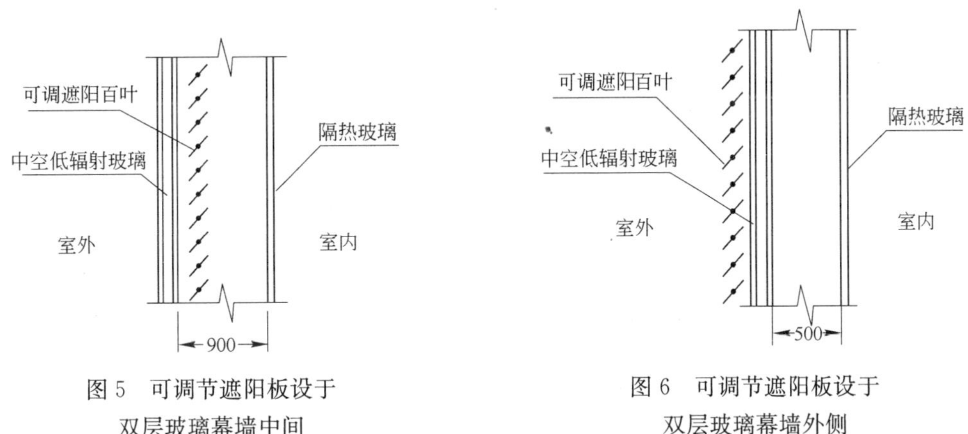

图5　可调节遮阳板设于双层玻璃幕墙中间

图6　可调节遮阳板设于双层玻璃幕墙外侧

巧妙有效的外遮阳技术为大大减少建筑空调冷热负荷、减少空调运行能耗、降低空调设备投资与占地空间，提供了必不可少的有力保障。有分析资料显示，这种双层玻璃幕墙与传统的单层玻璃幕墙相比，采暖时可以节约能源42%～52%，制冷时可以节约能源38%～60%。

3.2　自然通风技术

自然通风是绿色生态建筑中的核心技术之一，在降低建筑物能耗、改善室内空气质量、提高室内人员舒适感和健康水平起了很大作用。尽管世界各地建成的绿色生态建筑所采用的自然通风方式不尽相同，但是其基本原理都是相同的。即利用室外的风压作用和室内的热压作用，使室外新鲜空气进入室内后流经人员工作区，再排向室外。经笔者分类，目前绿色生态建筑所采用的自然通风方式主要有：中庭方式、通风塔方式及分散方式等。

3.2.1　中庭方式

中庭方式是指建筑物各房间利用外窗自然进风、或有组织的新风输入，室内的污染空气流入中庭空间，再经中庭顶部天窗或排风机排向室外大气。采用中庭自然通风方式的著名绿色生态建筑有德国国会大厦（见图 7）、法兰克福商业银行、柏林 DAIMLER BENZ OFFICE；英国剑桥大学法学院、北安普顿 BARCLAYACARD 公司大楼；加拿大多伦多约克大学计算机学院、上海生态建筑示范楼；日本东京天然气公司总部大楼、东京的日本 MATSUSHITA ELECTRONIC 公司信息传播中心（见图 8）等等。

图 7　德国国会大厦中庭的半椭球形玻璃穹顶

3.2.2　通风塔方式

通风塔方式是指建筑物内空气流动的出口不是中庭顶部，而是一个或数个矗立于屋顶之上的通风塔。室外空气从外窗进入每个房间后流向走道，经由排风竖井进入屋顶通风塔，再由塔内排风扇排向大气。也有不少通风塔不设风机，而由通风塔烟囱效应将室内混浊空气排向室外。采用通风塔方式的著名绿色生态建筑有德国法兰克福 J. WATTER THOMPSON 广告公司总部大楼、RWEAG 大厦；英国剑桥 LONICA 电讯总部大楼、卡

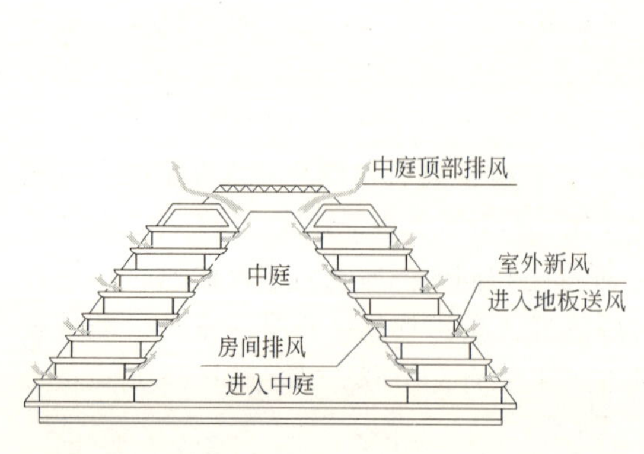

图 8　东京—日本 MATSUSHITA ELECTRONIC 公司信息传播中心

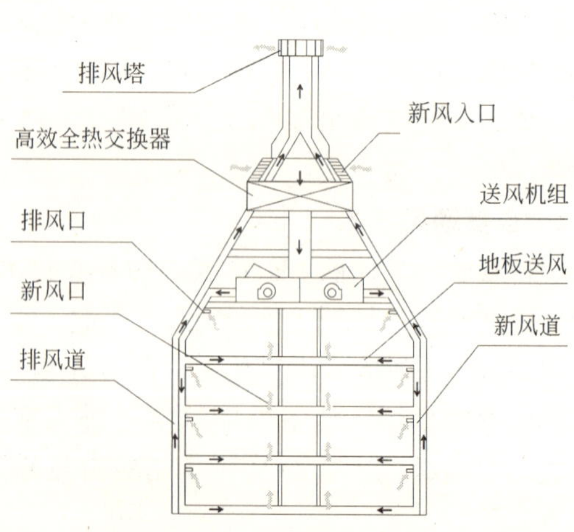

图 9　伦敦 PORTCULLIS HOUSE(2～4 层为英国国会办公楼)自然通风塔

斯廷 BRE 未来建筑研究组织办公楼、伦敦 PORTCULLIS HOUSE(见图 9)、考文垂大学图书馆；清华大学超低能耗示范楼等等。加拿大多伦多约克大学计算机学院除了采用中庭方式外，其建筑物很大部分还采用通风塔方式。

3.2.3　分散方式

分散方式是指绿色生态建筑自然通风的排风不采用集中方式，既不利用中庭顶部天窗，也没采用屋顶通风塔，而是采用分层进、排风方式。具体气流组织为：室外新风从每个房间地板下或贴近地面处被引进，流经房间工作区后由窗户上部百叶排向室外。采用分散方式自然通风的绿色生态建筑较少。典型的有：英国伦敦 HELICON 大厦；法国国家电力公司大楼；柏林 DAIMLER BENZ OFFICE 与中庭不相邻的办公室也全部采用分散方式的自然通风，见图 10。如果采用双层玻璃幕墙的话，分散式自然通风就会复杂起来，必须采取冬夏季切换的方式。总的原则是：夏季底部进入双层玻璃幕墙的空内排风经加热后由上部排至室外；冬季底部进入双层玻璃幕墙的室外新风经加热后由上部进入室内。

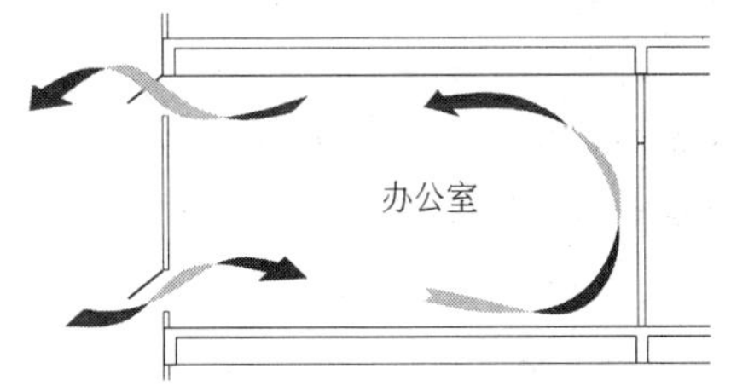

图 10　办公室夏季采用分散式自然通风

3.3　天然冷热源

绿色生态建筑一大技术特点，就是尽可能地利用自然界的免费能源。有不少绿色生态建筑，其空调系统就采用了天然的冷热源。比如：太阳能制冷制热、地源热泵、湖水热泵、湖水冷却、新风地下预处理等等。例如：德国国会就利用距地表 300m 深处的地下水层作为大楼的补充冷热源，新风还用埋地管道进行冷热预处理；英国伦敦 Portcullis House 的新风在全热交换器能力不足时，将使用地下水作预冷预热处理；上海生态建筑示范楼使用了太阳能制冷制热技术；清华大学超低能耗示范楼利用太阳能集热再生新风除湿系统的溶液、太阳能光伏玻璃发电用于调节电动遮阳板和双层玻璃幕墙上的风口等等。

3.4　智能化系统

智能化控制系统在绿色生态建筑中是必不可少的核心技术之一。智能化控制系统能自动采集室外的日照参数，测定玻璃、窗框、遮阳百叶、墙体表面的温度与热流；测定室内空气温度、相对湿度、风速、CO_2 浓度、冷热源设备运行参数、空调末端设备的运行参数、工作区各点的照度等。依以上采集测定到的实时数据，自动地调节遮阳百叶的角度；调节外窗、热压通风道及双层玻璃幕墙空气进出风口的开启程度或方向；调节控制各种暖通空调冷热源设备、空调末端设备以及各种辅助设备运行方式以及运行参数、空调系统新风量大小等等。

当然，智能化控制系统在绿色生态建筑中的作用远不止这些。除了用于暖通空调系统、给排水系统与照明系统的调节控制之外，还用于：计算机网络、通信、数字电视、卫星电视、公共广播、火灾报警、消防控制、电子监控、安防、门禁、对讲、物业管理、抄表、电子通告、车库引导与收费、家庭智能管理等等。

4　暖通专业在绿色生态建筑设计中与其他专业的配合协调

绿色生态建筑是全球可持续发展战略与当代高新技术在建筑领域的综合体现与完美结

合。任何一幢成功的绿色生态建筑决不是某几个人拍拍脑袋、随便凑合一下便可以设计建造出来的。而必须由一支每个成员都能深刻理解可持续发展战略、熟悉全球绿色生态建筑发展动向、设计理念超前、精通各自专业理论、熟练掌握各自专业技术、并对其他相关专业也有着充分了解、起码也能比较了解的多学科设计团队协同攻关，才能完成的一项艰巨复杂的任务。

这支多学科团队除了建筑师、结构工程师、建筑设备工程师、材料工程师、建造师之外，很可能还需要生物学家、医学家、植物学家、物理学家、化学家、气象学家甚至心理学家等等的加盟。在绿色生态建筑设计与建造过程中，一些政府相关部门的政策支持与行政协调常常也是必不可少的。

比如，在绿色生态建筑方面做了很多成功探索的德国著名 LOG ID 事务所的确就是上述各学科专家学者的优秀组合。而建于上海的我国第一幢生态建筑示范楼，也是由 12 个学科的 60 多家产学联合体组成的综合团队协同攻关的结晶。

对于暖通空调专业来说，由于本专业在建筑领域是出了名的能耗大户、用电大户、建筑内部空间占用大户。暖通专业的设计好坏，将直接影响建成后的建筑物能否达到绿色生态标准。所以在设计过程中，暖通专业人员与其他专业人员，特别是建筑专业和电气控制专业人员的协调配合工作就显得格外重要，当然也会显得格外繁复。从以上的文字叙述和图例中，我们可以从中大致领略到许多新技术手段之妙和将之成功实施所面临的复杂性和艰巨性。

设计中，暖通工程师关心的主要问题有：对整个建筑物能耗种类和数量的确定、建筑物生命周期(Life Cycle Assessmengt，LCA)的各种能耗的分布特性、建筑物布置、构造、材料选用是否有利于暖通专业健康舒适与节能、建筑有否自然通风构造和这些构造是否有效、建筑有否提供自然通风出入口的冬夏季节转换措施、室内外绿化对通风气流有否影响、能否和如何采用天然冷热源、可否采用太阳能制冷和供热、是否可以采用置换通风、能否采用辐射采暖和制冷、是否需要和怎样蓄热储冷、如何避免家具和室内物品损害室内空气质量、是否拥有室内空气质量定期检测手段、如何采用经济高效的智能调节控制系统、如何采用能耗计量收费系统等等。实际上，暖通工程师关心的问题一定还可以有更多。

以上问题提醒我们暖通空调工程师，除了不断努力提高自己的本专业理论水平和设计能力之外，很有必要大大加深对可持续发展战略的领会、对绿色生态建筑其他相关学科与专业的掌握程度以及提高设计过程中与其他各专业的配合协调能力。

Green Ecological Building And Hvacr

Zhang Jingtao[1] Hu Mengdi[2]

([1] East China Construction Development Design Co., Ltd. Shanghai 210050 China;

[2] Zhejiang University of Science & Technology, Zhejiang 310010 China)

Abstract: Reviews the researches and practices of green ecological buildings to sustainable development around world in today. Presents a mumble of typical green ecological building build in domestic and overseas. Statements some new design ideals and technical characteristics of them. Analyses the functions of HVAC in green ecological buildings, and the coordinated problems with anther specialties.

参考文献

[1] 李华东主编．高技术生态建筑．天津．天津大学出版社，2002.9
[2] 建设部科技司．首届国际智能与绿色建筑技术研讨会文集．北京．中国建筑工业出版社，2005.3
[3] 美国绿色建筑委员会著．彭梦月译．建设部建筑节能办公室中文版策划．绿色建筑评估体系．北京．中国建筑工业出版社，2000.7

建筑智能分布复合式能源系统的发展前景

连志敏
（新西兰硕普国际科技咨询有限公司
深圳市硕普智能科技有限公司，深圳 518033）

摘　要： 本文通过对我国建筑及小区节能的探讨，较全面阐述了通过智能化手段，建筑分布复合式能源系统的结构组成、运行方法。从一个新的方向说明：以可再生能源为基础，以网络化、智能化为条件、对多种能源进行优化组合，对建筑系统进行综合提升，在建筑机电设备系统实现大幅度节能是可以实现的。

关键词： 智能生态建筑社区　智能分布复合式发电　智能分布复合式冷热联供　能源梯级利用

1　前言

能源是人类生存和社会发展的原动力和基本要素。随着社会经济的飞速发展，能源日益成为影响经济继续健康增长的瓶颈。目前全世界都在推动新一代的能源系统，它主要具备这样几个特点：一能源多样化；二可再生能源优先；三冷热电系统联产；四智能化控制和信息化管理；五网络化；六绿色、环保。具体在建筑上，我把它称为：建筑智能分布复合式能源系统。

本文想从一个新方向，高起点来分析我国的建筑节能之路。从宏观的角度，归纳综合节能技术及产品。利用智能技术，以可再生能源为基础，实现节能最大化，系统最优化，投资最少化。创造低能耗智能生态社区。

2　建筑分布复合式能源系统的一般描述

民用建筑、特别是住宅小区耗能有以下几个特点：一耗能位置分散；二时间不集中；三可以采用可再生能源满足冷热能源需求及部分电力供应。民用建筑易于进行能源结构改革推广应用新能源系统。

根据能源产能位置和能源负荷位置，合理均衡的安排分布式能源系统。如居住小区，尽可能将能源机房安排靠近各自居住楼，以减少能量输送衰减。

根据该建筑或小区的的气象、地理、水文等自然条件，合理的布置规划复合式能源结构，并优先使用可再生能源，将传统能源作为后备和补充。主要有以下几个方面：

- 建筑小区规划的生态型配置

根据甲方要求及设计规划意图，根据小区自然地形及当地气象、水文、地质条件，根据智能控制系统的技术条件，合理配置能源结构（如太阳能、风能、潮汐发电、地源热泵、空气源热泵、家庭天然气热水器、天然气锅炉、煤油锅炉、热力管网等）及能源中心的产能方式、比例关系，以解决小区生活热水、采暖热水、空调冷水、部分绿色发电的能源供应。反过来对小区的能源设备的布局、景观水面、水深、绿化等进行规划。

- 能源结构的生态型配置

根据当地条件及建筑结构条件，经过周密计算选出最优化的能源配置顺序，一般 3～4 项，并合理配置其比例关系(如太阳能、水源热泵、热力管网或太阳能、土壤源热泵、天然气锅炉)，采用楼宇智能控制系统对其进行闭环开环自动控制，以求达到节能最大化。

- 智能系统与能源系统的合理配置

采用楼宇智能系统与智能家庭交互式控制系统的无缝结合实现智能控制的最优化组合及最经济配置。楼宇智能控制系统对于小区的生活热水、采暖热水、空调冷水设备、变配电、电梯、三级安防等进行监控。家庭智能控制系统对于家庭的采暖、空调的环境温度、湿度、空气质量、安全防范、电器进行智能控制。

- 建筑系统的最优化设计

尽可能保护原生态环境，利用自然资源，在区域的建筑系统和单体建筑系统合理的优化组合，将产能负荷与耗能负荷合理配置，减少投入浪费及运行耗损。在低成本、高产出的原则下、优化设备配置，尽可能做到系统综合，管网综合，设备综合。

- 管网系统的合理组合及设计

在小区及家庭智能系统的合理有序监控下，将小区管网及单体管网综合配置，如将生活热水与采暖热水合并，太阳能、热泵、备用能源系统管网合并，采用冷热联供热泵，建筑设施的综合利用等。这将减少初期投入成本，提高运行效率。

3　以可再生能源为基础的智能分布复合式的建筑发电系统

它主要是由建筑风力发电系统(靠近海岸地区可增加海潮发电系统，小城镇可增加生物质发电系统)、太阳能光伏发电系统及建筑小水利发电系统所组成。建筑风力发电系统是由一种水平旋转的扇叶组成的发电机多台联网而成，它可以像积木一样灵活匹配。它有体积小，噪音小，利用建筑小气候，小风可发电，大风不用阻尼等特点。它可根据建筑规划将风机设置在特定位置，改变风力的密度及气动路线，将风能利用率提高到 80%以上，输出功率是旋浆式风电机的几倍。太阳能光伏发电系统主要是分布在建筑屋顶及外墙。建筑小水利发电系统是利用建筑的高位水箱，小区的标高位差，利用峰谷电价进行发电。

它主要有以下优势：

- 弥补大电网在安全稳定方面的不足，大大提高供电可靠性。为重要部位及边远地区提供价廉物美的补充电力能源。
- 贴近负荷用户，减少高低压配电投资，减少配电及远距输送损耗。
- 复合式发电为可再生能源发电提供了新的方向。它可有效地解决可再生能源发电的不稳定性、分散性、利用率低等问题。

由于它的一次投资长期受益的特点，我们要把它与楼宇里的后备发电系统区别开来。

该系统可无停息运转，发出的电力三级预案处理：(1)直接入网使用；(2)进入蓄电池；(3)多余电力进行电解氢，把氢气储存起来做燃料。可再生发电与市电网由智能系统控制自动切换，以发电优先，主要供应小区设备系统。

4 以可再生能源为基础的智能分布复合式建筑冷热能源联供系统

可再生能源具有运行费用低、环保性能好等突出优势。智能系统是建筑复合式能源系统的神经，可再生能源是它的基础，以达到节能最大化是它的目标。我们知道：人们在建筑内需要的舒适温度只是在10～35℃(大部分地区只是10～20℃)的范围内进行调节。使用低品位的可再生能源完全可以胜任。长期以来，我们却使用着高品位能源：煤、油、电等大量消耗在民用建筑领域。加剧了大气变暖，增加了环境污染，加重了居民负担。

热水太阳能、热冷水热泵可以产生50～60℃的热水和10℃左右的冷水，而这些都来自阳光或浅地层的能源，我们统称太阳能。但它们每种也具备着可再生能源的弱点：不稳定性、分散性。如热水太阳能受天气影响较大，同时制冷效果不佳。冷热水热泵相对传统空调夏季节能效果不明显，且运行费用尚有下降空间。在工程中，系统各自独立，资源不能共享，往往容易一次投入设备负荷偏大，致使单位造价偏高，造成市场推广困难。

智能分布复合式能源系统的理念是：以楼宇智能系统串联各自独立的能源系统使其网络化连接、智能化管理，集中控制，分散运行，达到资源最大化，节能最大化。如按平均四口之家的生活热水计算，如果独立投资太阳能热水器需3～4m^2、4000～6000元，而系统使用只需2～3m^2、2000～3000元。太阳能产热水性价比指标较好，冷热型热泵在京津地区夏季需求的冷负荷与冬季需求的热负荷基本相等的条件下，效率是最佳。所以需要把两者优化组合成一个复合能源系统，根据能量需求确定合适的比例关系，实现节能的最优化组合。

同时，为了减低一次设备投入，应充分发挥智能系统的调节作用，利用蓄冷蓄热技术及高低谷电价差，实现二次节能和设备投资最小。此外应将传统能源作为复合能源系统中的后备能源，如燃气锅炉、热力管网等，以减少为应对冬季用热高峰而投资过多的设备。

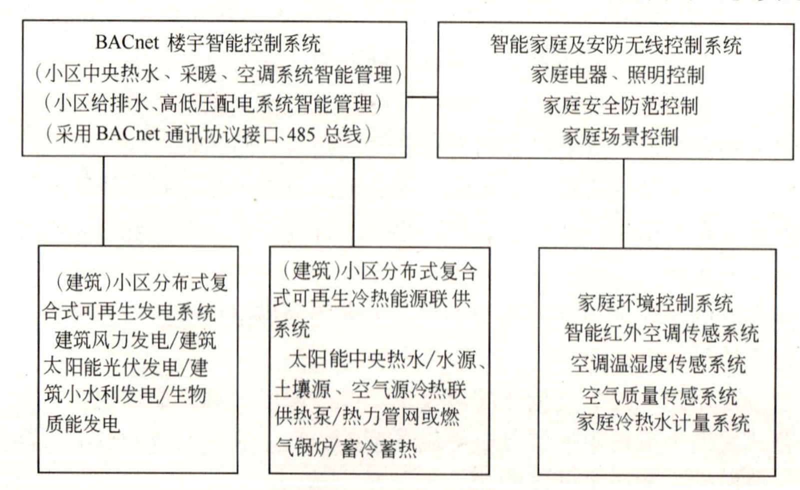

图1 建筑智能复合式能源系统

在极端不利的气候条件下，适当使用传统能源可能是最好的选择，我们要细心地寻找这个最佳平衡点。如在广东地区，当冬季气温低于6℃时，关闭空气源热泵，开启燃气锅炉是较为经济的方法。但在有的地区，可能选择使用煤油锅炉。这不能千篇一律，要根据各地的条件因地制宜。

建筑智能复合式能源系统见图1。

5　在系统上的综合节能措施

建筑节能是一个系统工程，需要站在统筹全局的高度，广泛吸纳各个方面的节能技术、措施，将节能系统不断完善；如将分散的机组组成智能化网络，使其资源共享、根据需求自动调节设备开启数量及状态。采用智能型的可变量的终端，根据人数的多寡自动调节风机速率及能量，如采用节能变压器，设备动力负荷上安装节电器，智能系统对于高低压配电进行调控，安装家庭节电器，使用家庭智能系统减少家用电器待机耗能等许多成熟的技术、产品。如使用合理，实现在能源系统对比节能50％～80％是完全可以作到的。

能源的梯级利用也是节能的重要组成部分。家庭生活废水中余热的提取，是水源热泵产热的一个捷径。与中水系统合并提能，在气候适度的南方对于地表水的利用，使景观水池、瀑布可产能、可发电，节省打井费用，在气候适度的海滨使用海水作为水源热泵的能源，将大大提高热泵产能效率。

6　对于老建筑的节能改造

对于老建筑、特别是耗能大户，公共建筑的节能改造也是迫在眉睫的事情。在技术上可以实现的一个好的办法是，对于公共建筑中央空调系统进行节能运行诊断与优化这一节能节电技术。该技术采用先进的中央空调系统在线监测技术、人工智能诊断技术、空调系统节能优化运行仿真技术和计算机控制技术，实现对现有公共建筑空调系统节能运行诊断，分析全年节电潜力，提出可行性运行方略，通过对空调系统运行模型的优化调整和改造，达到节能节电的效果。一般地，对大型公共建筑中央空调系统，年节电量可以达到20％～40％以上，技术改造投资在2年内就可收回。照明系统加装节电器，可节能20％～40％。

7　结论

我国经济正在从粗放型向节约型转轨，硕普智能愿意与各界同仁共同努力，研究和探讨在建材、设备、系统及规范中如何改进以达到更好节能的问题，使我国建筑系统能够与时俱进，赶上先进国家的能耗水平，共同打造智能、环保、节约型的社会。

本文通过对建筑智能分布复合式能源系统特点、系统组成及发展趋势的阐述，说明可再生能源是建筑及社区能源系统的基础和主力军，传统能源是补充。智能分布复合式发电系统是城市电网的补充，建筑或社区动力系统的首选。智能分布复合式冷热能源联供系统是建筑生活热水、采暖用热、空调用冷的最佳供源及优化组合。它以多种可再生能源系统优先及主供，传统能源为后备和补充。以智能化手段将发电及产能系统和谐、有机组成一个智能化控制、网络化管理、高效节能、公平结算的综合能源系统，实现能源设备系统节能50%～80%，建设具有中国特色的低能耗智能生态建筑社区。

The Development Prospect of Building Intelligence Distributes the Compound Types Energy System

Jimmy Lian
(SNOOP INTERNATIONAL CONSULTANT (NZ) LIMITD Auckland, New Zealand)
(SHENZHEN SNOOP INTELTECH CO. LTD. Shenzhen 518033 China)

Summary: This paper expresses a study on building and residence area of China to energy efficiency, The expound to explanation the new energy system structure and system carry through of the distribute and the compound type energy go through intelligence means. From a new direction elucidation: with renewable energy as foundation, with the network and intelligence as qualification, It is a the combination to various energies, proceeding to synthesize to promote to the building system, aimed at driving forward energy efficiency in building HVAC and electricity the equipments system.. It can achieve to a economy energy society.

分布式绿色能源生态小区建设

万海滨
((英国)中国工程师协会　理事长，英国)

摘　要： 针对国家十一五期间将大力推进建设节约型社会，提出根据当地可再生能源资源的情况，采用系统优化集成技术，综合实现可再生能源利用最大化，建设局域微型电网，实现分布式能源供应，解决集中能源供应方式的诸多问题，使能源利用结构和方式更合理和科学。建议国家在十一五期间，开展示范项目建设和关键技术的研究。

关键词： 可再生能源　绿色能源　分布式能源供应　局域微型电网

最近(2005年7月)国务院向全国各省、自治区、直辖市人民政府，国务院各部委、各直属机构下发了《国务院关于做好建设节约型社会近期重点工作通知》，进一步表明了国家在十一五期间将大力开展建设节约型社区的决心。

改革开放以来，特别是中央提出加快两个根本性转变以来，我国推进经济增长方式转变取得了积极进展，资源节约与综合利用取得一定成效。但总体上看，粗放型的经济增长方式尚未得到根本转变，与国际先进水平相比，仍存在资源消耗高、浪费大、环境污染严重等问题。随着经济的快速增长和人口的不断增加，我国淡水、土地、能源、矿产等资源不足的矛盾更加突出，环境压力日益增大。“十一五”是我国全面建设小康社会、加快推进社会主义现代化的关键时期，必须统筹协调经济社会发展与人口、资源、环境的关系，进一步转变经济增长方式，加快建设节约型社会，在生产、建设、流通、消费各领域节约资源，提高资源利用效率，减少损失浪费，以尽可能少的资源消耗，创造尽可能大的经济社会效益。

建设节约型社会就要树立和落实以人为本、全面协调可持续的科学发展观，坚持资源开发与节约并重，把节约放在首位的方针，紧紧围绕实现经济增长方式的根本性转变，以提高资源利用效率为核心，以节能、节水、节材、节地、资源综合利用和发展循环经济为重点，加快结构调整，推进技术进步，加强法制建设，完善政策措施，强化节约意识，尽快建立健全促进节约型社会建设的体制和机制，逐步形成节约型的增长方式和消费模式，以资源的高效和循环利用，促进经济社会可持续发展。

1　可再生绿色能源

自上世纪90年代，我国经济进入高速持续增长期，工业化、城镇化已成为社会经济发展的主流。同时，人们也注意到经济的高速增长、工业化和城镇化的快速推进对自然资源和生态环境的不良影响。对自然资源，尤其是对化石能源的过度开发，必然导致资源枯竭期提前来临，这不可避免地将成为我国经济发展的瓶颈。在另一方面，过度依赖化石能源，会造成大气污染、温室效应，破坏人类赖以生存的生态系统。随着社会的发展，能源和环境压力会越来越大，发展可再生能源将成为必然的选择，充分利用可再生能源将为建立“资源节约型”和“环境友好型”社会，走可持续发展道路奠定物质基础。

近两年来，国际石油市场强势震荡，石油价格始终在高价位盘旋。减少对石油、煤等不可再生资源的开采与利用，将是我国能源发展战略的必然选择。因此，充分利用太阳能、风能和沼气等生物新能源和可再生能源对实现国家经济社会可持续发展和促进生态环境改善有着重要的意义。

我国每年城乡新建房屋建筑面积近 20 亿 m^2，其中 80%以上为高能耗建筑；既有建筑 400 亿 m^2，95%以上是高能耗建筑。目前我国单位建筑面积能耗是发达国家的 2～3 倍以上。这也是使得我国能源短缺问题更加严重。因此，在建设现代城镇的过程中，必须充分提高能源的利用率，扩大绿色能源所占份额，控制环境污染。建设真正的生态园区和绿色建筑。

可再生能源是指在自然界中可以不断再生、永续利用、取之不尽、用之不竭的资源。它对环境无害或危害极小，而且资源分布广泛，适宜就地开发利用。可再生能源主要包括太阳能、风能、水能、生物质能、地热能和海洋能等。由于对这些能源的开发利用对环境无害或危害极小，因此它们又被称为“绿色能源”。我国地域辽阔，在某些地区，太阳能、生物质能、风能资源非常丰富。而我国在利用可再生能源方面，与欧美发达国家相比处于落后状态。

目前世界各国的电力能源供应基本是以集中方式为主。即无论是采用煤炭、石油，还是水力发电，以及垃圾发电，大都集中建立电厂(电站)发电，通过高压输变电网络远距离输送供应送到用户终端。这样一来，高压输变电及保护设备投资和运行及维护费用都相当昂贵，并且损耗十分严重。此外集中供电的风险问题也十分突出，比如近年美国、英国和俄罗斯都发生了大面积停电事故，造成极大经济损失。集中供电方式，近年还要面对恐怖袭击的威胁。

2 分布式能源供应新概念

分布式能源供应是一个刚刚出现的新概念，所谓分布式能源供应，就是结合当地的可再生绿色能源资源情况，在一个局部范围(例如一个城镇)内综合利用可再生能源，建立分布式绿色能源供应系统，部分解决该地区的能源供应问题，不足部分由社会公共能源供应补充解决，不仅包括对电力能源，也包括对燃气、水、热能的供应的考虑。这将是对传统能源供应方式的重大革命。因此，我们认为推进“分布式可再生绿色能源系统”技术和相关设备产品的研究和建立相关示范工程，对中国建设节约型社会具有特别重要的意义。

可再生能源的大规模独立使用，从技术角度，特别是在能源控制和管理方面还存在一定的难度。这是由于在目前的实际情况下，可再生能源转化的电能还难以并网发电。以风电为例，为了给风电机组提供备用，电网调度不得不预留超出常规水平的燃煤机组，以保证供电连续性。从整体上讲，风电并没有根本上替代常规电能，而且发电费用高。其他可再生能源也面临同样的问题。

可再生能源的开发利用有它的特点，例如：容量小(大型风场除外)、间歇性强，适宜就地开发利用，用于满足本地负荷。而从能源管理角度分析，能源的生产地应尽量靠近终端用户，以降低成本，提高能源供给系统的可靠性。因此，可再生能源的开发利用应该本地化。在此，在一个局部区域，根据具体可再生能源资源情况，建设“局域微型电网”，利用“局域微型电网”技术将各类可再生能源转化所得的电能在一个独立的“局域微型电网”中统一管理，以提供连续可靠的绿色电能供应。“局域微型电网”还可以与区域超高压电网相连，只取电、不送电，以便在“局域微型电网”电量不足时，由区域高压电网提供补充。

3 “局域微型电网”建设

如前所述，任何一种可再生能源的单独使用，必然有它的局限性。这也是在大范围内推广使用可再生能源的一个障碍。从能(电)源供应连续性角度加以分析，应该主张在“局域微型电网”建设中同时使用多种可再生能源，使它们彼此之间互为补充，以达到能源供应的安全性和连续性。因此，生态园区的能源输送和管理将变得尤为重要。本项目中，电力传输将采用先进的微型电网概念，能源管理将以互联网为平台，实现智能控制和能源综合管理，使得控制和管理过程更为有效。

我们为此策划了一个“局域微型电网”建设的示范项目，其目的是：利用真实的负荷状况和现有的可再生能源生产转化装置，将各类可再生能源，根据当地的资源特点，形成互补的规模组合，在一个统一平台上提供绿色能源综合服务，建设绿色生态园区，并且展示“局域微型电网”技术在控制管理可再生能源方面的优势，以及降低可再生能源生产成本的潜力。

“局域微型电网”建设示范项目可以小到一个居民小区，也可以大到一个城镇。微型电网的能源供应绝大多数来自于本地生产的绿色能源，做到能源的就地消化，做到最大程度的无污染、无浪费，资源可再生循环利用。考虑到上面提到的可再生能源容量小，具有间歇性的特点。“局域微型电网”将考虑间或从区域电网中购买一部分电能，以作补充。

我们还考虑了在示范项目园区内的各种能源的综合安排和互为补充，比如：生物质能转化为沼气，既可以用于发电，也可以用作园区公共燃气使用；当园区沼气不足时，也可以由公共燃气给予补充。

在项目规划上，我们还将充分考虑节能和园区的生态环境建设。例如热泵的使用、热电联产、中水处理和雨水收集回用、从建筑结构和建材选择设计上考虑节能问题等。

示范项目的亮点是：将不同的可再生能源技术和生态技术集成为一体，在一个局部区域，用最少的投资建设最科学合理的生态环保绿色能源生态环境，形成示范效应。

示范项目采用积木式的开放型结构，以利于分阶段实施，便于推广，这对中国这一具有广阔地域和各地自然条件千差万别的国度，在城市化进程中全面推行绿色能源建设方针具有巨大的意义。在生态园区拟采用以下可再生能源技术：

光伏(太阳能电池)发电系统：将太阳能直接转换为电能。

太阳能聚热系统：包括太阳能热水系统和太阳能取暖、制冷系统。

太阳光照明：利用透射式采光机，采集阳光，再通过光纤传导，将太阳光引进建筑内。

微型燃气轮机热电联产系统：该系统将以沼气为燃料。

风力发电系统。

热泵系统：利用地温地热能制冷。

吸收式制冷和加热系统：本系统将利用热电联产或太阳能聚热系统生产的热力。

图1是能提供绿色能源服务的生态园区的一个示意图，示范项目的主要思想已体现在图中。例如：各种绿色能源设备连网运营，互为补充；基于互连网的能源管理；积木式开放结构；微型电网等。

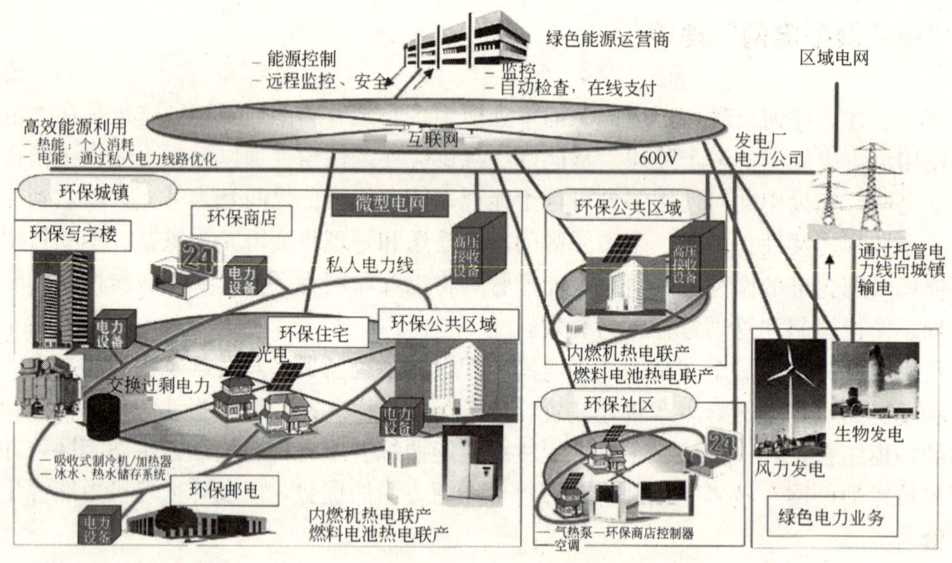

图1 提供绿色能源服务的生态园区示意图

4 项目示范与关键技术研究

目前，已有十余家中外企业和研究单位表示愿意参与示范项目的建设。其中，多数企业来自欧洲，这些企业在可再生能源利用和生态园区建设方面，在世界上处于领先地位。例如，（英国）i&i 智能建筑设计咨询公司在伦敦建立了世界上第一个综合利用可再生能源的示范生态建筑。这些企业将带来各自领域的最先进技术，这必将为我国消化吸收世界上相关的先进技术提供一个良好的契机。

可再生绿色能源分布式供应示范项目，其中有两大关键技术需要解决：一个是：根据当地的可再生能源资源情况，所制定一个协调合理、综合投资性价比高的系统集成方案技术。另一个就是："局域微型电网"建设管理技术。我们相信：系统集成技术将随着该项技术的不断推广应用，各种可再生能源技术和设备的发展，不断趋于完善。而"局域微型电网"建设以及相应控制设备和管理技术，也将作为一个崭新的行业领域而快速得到发展。

也许，这种分布式绿色能源供应方式的推广和发展，将改变整个世界的能源供应方式和政策，也未可知。

Construction of Eco-Communities Using Distributed Green Energy

Wan Hai Bin

(The Association of Chinese Engineers in the UK, UK)

Abstract: Energy problems are attracting more and more attention in both developed and developing countries. At present, fossil fuels are still a dominating energy source, the reserve of which is limited and will be exhausted ultimately. Furthermore, the excessive use of fossil fuels is causing serious pollution problems. Therefore, a great deal of effort is being made to seek new fuels and renewable energy, or called green energy, as an alternative to fossil fuels in order to resolve energy problems and protect environment. As energy consumption in construction accounts for a considerable portion in total energy being used in the country, it is envisaged that distributed energy supply systems will be implemented during the period of 11th five year plan, which are based on local resources, and green energy and various energy conservation techniques will be used for small communities which consist of a number of commercial and residential buildings. This paper proposes a demonstration to show how green energy, including solar, wind, micro CHP, fuel cell and bio-mass, can be effectively used in a small community. The demonstration uses a micro electric grid for electricity transmission and an intranet for energy management in the community. The purpose of using the micro grid system is to increase the overall efficiency of the operation of green energy included, reduce the total production cost, and make the community self-sufficient in terms of energy. Some new energy saving techniques, such as CHP, Heat pump and absorption chiller, are also included in the demonstration. The concept proposed in this paper can also be used for eco-cities and eco-towns.

基于神经网络的地源热泵传热模型研究

陈建萍

（上海市建筑科学研究院，上海　200032）

摘　要：地源热泵因其显著的节能和环保特点，受到越来越多的重视。本文介绍了地源水—水热泵实验系统设置。本文引入地下埋管传热系统的神经网络，并建立了神经网络预测模型，对埋管换热进行模拟计算和预测，从计算结果可以看出计算精度较高，可以推广和指导工程实践。

关键词：地源热泵　神经网络预测模型

地源热泵因其显著的节能和环保特点，受到越来越多的重视，美国、德国及瑞典等西方国家的地源热泵技术尤其是供暖方面的应用日趋成熟。由于受初投资大、本地地理气候影响大等限制，我国地源热泵尚处于初级开发阶段。这就需要结合我国的具体地理气候特征，进行模拟研究，积累大量设计、施工及运行的数据资料，以制定出设计技术规范和应用软件。

在我国长江流域等冬冷夏热地区，大地初始温度为15～20℃，地源热泵既可实现冷暖联供，又可降低能耗，具有重要的应用意义。在长江流域城市建成的浅埋竖管地源水—水热泵实验装置上，进行了连续一年暖冷联供的观察和测试，获得大量运行和技术数据，并据此建立地下埋管换热器的神经网络预测模型。

1　实验装置简介

热泵的实验系统由以下三部分组成(平面布置原理见图1)：

(1) 室外地下埋管换热器部分：地下埋管多管群换热器的总设计换热量为10kW。地下管群实验装置共5组，组间采用并联同程式，考虑到管间换热影响，组间是错排布置。套管管径分别为ϕ50、ϕ75、ϕ90三种，内管为ϕ15，埋深10m，管材采用普通PVC管。

(2) 室内热泵机组部分：由风冷热泵空调器自行改装成的水—水热泵套管换热器连接地下埋管侧的进出水(冬季作为蒸发器，夏季则为冷凝器)，另一端套管换热器连接用户侧的进出水。水—水热泵系统采用较简便的制冷剂不换向、水侧换向方法进行，即夏季e、f、g、h四个旁通阀关闭，a、b、c、d阀门开启，冬季阀门的关、开与夏季相反。如图1，虚线表示的管路为冬季供热工况，实线表示的管路为夏季制冷工况，而埋管侧和用户侧的水路循环则冬夏不变。

(3) 冬供暖夏供冷用户系统：热泵系统分别供给空调房间A和空调房间B。A间

$15m^2$，地板下埋设了冷暖地板蛇行盘管；相邻的 B 间亦为 $15m^2$，内设立式风机盘管。

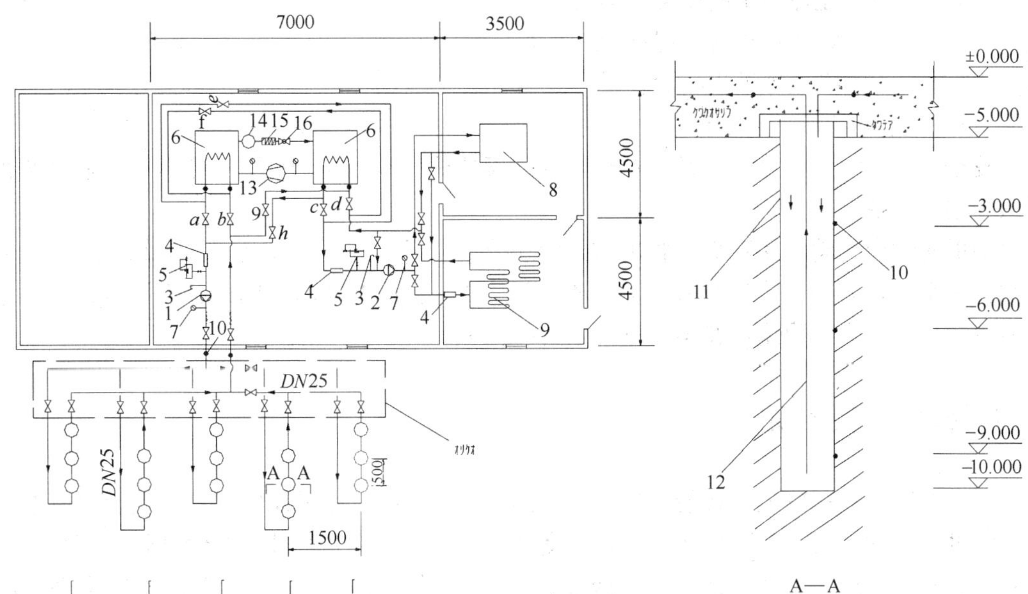

图 1　地源水—水热泵实验装置平面布置原理图

1. 埋管侧水泵；2. 用户侧水泵；3. 放气阀；4. 转子流量计；5. 膨胀水箱；
6. 套管式换器；7. 压力表；8. 风机盘管；9. 冷暖地板；10. 热电偶测点；
11. 套管；12. 内管；13. 压缩机；14. 贮液器；15. 干燥器；16. 膨胀阀

2　测试方法和数据处理

温度测试对象为地下埋管壁温、埋管换热器总进出水温、室内外温度及用户管路系统的水温，测点布置参见图 1。温度测试采用铜—康铜热电偶温度传感器，适用温度范围 −200～400℃。

流量测试采用 LZJ-40 转子流量计，精度 1.0 等级。实验的水流量范围为冬季 960～1060 L/h，夏季 760～1345 L/h。为研究地下埋管的换热特性和热泵运行性能，建立埋管侧和用户侧的热平衡，需要测试埋管侧管路系统和用户侧管路系统中各自的流量，即地下埋管进出水流量 G 和进出地板及风机盘管的总流量 G_0。

记录累计耗电量的精密电表与记录时间的电控电子钟采用连锁控制，当间歇运行的水—水热泵一启动，电表和时钟瞬间同时开始通电记录，两表的对时差即为一天的热泵机组运行时间。热泵耗电量包含了压缩机和埋管测水泵的能耗。

实验测试分四个阶段：冬季供暖和夏季供冷(机组昼夜间歇运行)；夏—冬过渡季和冬—夏过渡季(机组停机不停测)，全天 24 小时不间断测试，以了解大地温度恢复状态。除因停电、机器维修故障停机近 10 天，全年机组间歇运转 2808h，测试 6900h。

地下换热器日换热量 Qh(kWh)采用下式计算：

$$Qh=\frac{Q\times h}{1000}=G\rho c_{\mathrm{p}}(T_{进}-T_{出})h$$

式中 G——循环水流量，m^3/s；

ρ——管内流体即水的密度，kg/m^3；

c_p——水的定压比热容，kJ/(kg·℃)；

$T_{进}-T_{出}$——埋管换热器进出水温差，℃；

h——热泵日运行时间，h。

夏季 $T_{进}>T_{出}$，Q 为正，埋管换热器对土壤放热；冬季 $T_{进}<T_{出}$，Q 为负，埋管换热器从土壤中吸热。

为更科学合理地使用数据，找出数据的规律，从最小二乘意义上，所测数据经过在数学软件中编程预处理，再进行最佳拟合和误差分析。且对在实验中因设备损坏检修、停电等造成的少数不完整数据，在拟合基础上进行插值。

3 神经网络模型介绍

神经网络模型是以人工智能为其理论基础，模仿人的大脑的思维方式：大量的简单神经元经过复杂的网络结构而形成的一种高度复杂、非线性、并行处理的信息处理系统。它将非线性系统看成一个“黑箱”，采用系统辨识的方法，不去分析其内部的反应机理，而只是研究对象主要控制变量和输出变量之间的相互关系，并且用一个超曲面来描述从 p 维欧氏空间(输入变量)到 q 维欧氏空间(输出变量)之间的非线性映射关系。

多层网络可以解决非线性问题这一结论早已有了。但由于有隐层后学习比较困难，限制了多层网络的发展，BP 算法促使多层网络的研究重新得到重视，BP 算法中有两种信号在流通：

(1) 工作信号(用实线表示)，它是施加输入信号后向前传播直到在输出端产生实际输出的信号，是输入和权值的函数。

(2) 误差信号(用虚线表示)，网络实际输出与应有输出间的差值即为误差，它由输出端开始逐层向后传播。

关于多层前馈网络学习的反向传播(或称 BP)算法详见参考文献 [3]，限于篇幅，本文从略。

4 地下埋管换热器夏季实验的神经网络预测模型

本次夏季实验有两组套管运行，选其中一组套管，共有 54 组数据(8 月 10 号、11 号数据因故障原因未能测得)，采用其中的 39 组数据对神经网络模型进行训练，用其中的 15 组数据对神经网络模型的预测值进行检验。

用于训练神经网络的 39 组数据如表 1。

用于对神经网络预测模型进行检验的 15 组数据如表 2。以下为预测埋管壁温的神经网络模型。

对于已知输入而求壁温的模型，本模型采用对壁温影响较大的 6 个参数作为输入层，分别为：室外温度 Tw，埋管出水温度 $C3$，埋管进水温度 $J3$，热泵日运行时间 H，地下埋管循环水流量 G，地下换热器日换热量 Qh。

用于训练神经网络的夏季实验第二组套管数据表 **表 1**

日期 (Date)	天数序号 (Num)	−3m 管壁温度 $L9-3$ (℃)	−6m 管壁温度 $L9-6$ (℃)	−9m 管壁温度 $L9-9$ (℃)	室外温度 Tw (℃)	埋管出水温度 $C3$ (℃)	埋管进水温度 $J3$ (℃)	热泵日运行时间 H (h)	地下埋管循环水流量 G (L/h)	地下换热器日换热量 Qh (kWh)
706	1	30	25.9	25.2	28	27.03	29.98	10.4	1345	47.91
707	2	29.5	25.3	24.6	27.5	24.23	28	10.3	1345	60.64
708	3	29.3	27.18	25.18	28	27.2	30.33	10.38	1345	50.74
709	4	30.4	25.68	24.94	27	25.44	27.87	10.57	1345	40.11
710	5	29.32	25.45	24.69	27.5	27.05	29.36	10.62	1345	38.31
711	6	30.35	26.05	25.5	27.5	26.53	28.86	10.52	1345	38.28
712	7	30.4	26.33	25.43	27.5	26.41	30.7	10.47	1345	70.15
713	8	31.5	26.78	26.04	26.5	26.26	29.58	10.6	1345	54.96
714	9	28.6	25.6	25.05	28	24.83	28.66	10.55	1345	63.1
715	10	28.2	25.28	24.46	28	24.7	28.28	10.22	1000	42.48
721	16	31.64	27.84	27.24	30	27.46	30.69	10.4	1125	43.88
722	17	32.3	27.3	26.85	32	27.45	31.12	10.45	1125	50.1
723	18	33.31	28.41	28	31.5	28.05	31.3	10.43	1125	44.28
724	19	34.36	29.68	29.3	32	28.71	32.14	10.45	1125	46.82
725	20	33.54	29.93	29.58	32.5	28.91	32.48	9.65	1125	45
726	21	32.03	29.11	28.08	33	26.8	31.17	10	1125	57.08
727	22	31.81	28.03	27.56	33	27.3	30.92	9.62	1245	50.34
728	23	31.7	27.64	27.33	32.5	27.1	31.33	9.8	1245	59.93
729	24	31.18	28.95	28.35	31.5	28	31.2	9.62	1245	44.5
730	25	32.63	28.83	28.53	27.4	28.22	31.25	10.32	1245	45.2
805	31	35.6	31.25	31.15	27.5	30.7	33.7	10.37	850	30.7
806	32	34.4	30.7	30.6	27.5	30.2	33.2	10.5	850	31.09
807	33	35.05	30.81	30.73	29	30.01	33.8	10.5	850	39.28
808	34	34.27	30.12	30.1	28.5	29.4	32.7	10.51	850	34.23
809	35	35.6	30.7	30.5	28	30.4	33	10.3	850	26.43
812	38	34.32	30.69	30.48	30	30.12	32.88	10.52	850	28.66
813	39	37.09	32.54	31.65	29	31.94	35.15	10.68	850	33.84
814	40	36.71	32.04	31.6	29	31.41	34.54	10.88	850	33.61
815	41	36.33	32.15	31.8	28.5	30.9	33.8	10.33	850	29.57
816	42	36.1	31.6	31.44	29.5	31.15	34.19	10	850	30
817	43	34.94	31.9	31.1	29	29.83	33.01	8	850	25.11
818	44	35.85	32.95	31.9	29	31.42	35.9	8.67	850	38.33

续表

日期 (Date)	天数序号 (Num)	−3m管壁温度 L9−3 (℃)	−6m管壁温度 L9−6 (℃)	−9m管壁温度 L9−9 (℃)	室外温度 Tw (℃)	埋管出水温度 C3 (℃)	埋管进水温度 J3 (℃)	热泵日运行时间 H (h)	地下埋管循环水流量 G (L/h)	地下换热器日换热量 Qh (kWh)
819	45	37.21	32.8	32.6	27	31.51	36.19	9.82	850	45.36
820	46	37.1	32.48	32.1	28	32.35	35.55	8.13	850	25.68
821	47	37.42	33.1	32.39	28.5	32.86	35.65	9.82	760	24.18
827	53	36.43	33.17	32.6	26.5	32.19	35.91	9.43	760	30.96
828	54	37.2	33.34	33.16	28	32.16	35.63	10.12	760	30.99
829	55	35.8	33.3	32.7	24.5	31.9	34.7	9.7	760	23.97
830	56	35.3	31.4	30	24.5	29.6	33.3	9.5	860	35.1

用于对神经网络预测模型进行检验的夏季实验第二组套管数据　　表2

日期 (Date)	天数序号 (Num)	−3m管壁温度 L9−3 (℃)	−6m管壁温度 L9−6 (℃)	−9m管壁温度 L9−9 (℃)	室外温度 Tw (℃)	埋管出水温度 C3 (℃)	埋管进水温度 J3 (℃)	热泵日运行时间 H (h)	地下埋管循环水流量 G (L/h)	地下换热器日换热量 Qh (kWh)
716	11	30.02	25.82	24.92	28	23.39	26.57	10.28	1000	37.96
717	12	30.7	25.9	25.6	29.5	25.3	30	10.3	1000	56.21
718	13	30.8	27.02	25.74	30.5	26.6	30.99	10.33	1000	52.65
719	14	32.13	28.55	27.61	31	27.97	31.74	10.33	1000	45.22
720	15	31.43	27.77	27.32	31	27.38	30.5	10.38	1000	37.6
731	26	33.88	30	29.85	28	29.51	32.83	9.9	1245	47.51
801	27	35.48	31.18	31.06	27	29.33	32.55	10.18	1245	47.39
802	28	35.29	31.56	31.35	25	31.53	34.44	10.4	1245	43.75
803	29	34.89	30.78	30.77	24.5	30.41	34.33	10.35	850	40.04
804	30	36.34	32.44	32.24	25.5	32.06	35.04	10.38	850	30.53
822	48	37.07	32.78	32.65	29	32.59	35.01	10.5	760	22.42
823	49	35.51	31.27	31.14	28	31.25	33.83	9.5	760	21.63
824	50	35.3	30.7	30.6	26.5	31.8	33.83	9.77	760	17.5
825	51	36.35	31.78	31.7	28	32.32	35.18	9.42	760	23.77
826	52	36.93	32.41	31.73	27.5	32.45	35.48	9.97	760	26.66

隐层为5个神经元，激励函数为tansig函数；输出层为三个神经元，采用purelin线性阈值单元，分别对应−3m、−6m、−9m处的壁温。

预测模型的神经网络结构图如图2。

采用表1中的数据对神经网络进行训练，在最大训练次数为100的情况下，神经网络训练均方误差、训练均方权重、有效参数与训练次数的关系如图3、4，可以看见网络的训练误差迅速减小，结果非常理想。

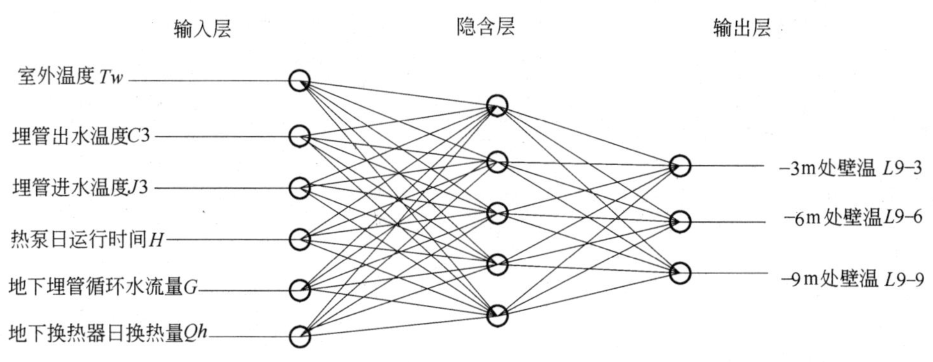

图 2　预测埋管壁温的神经网络模型

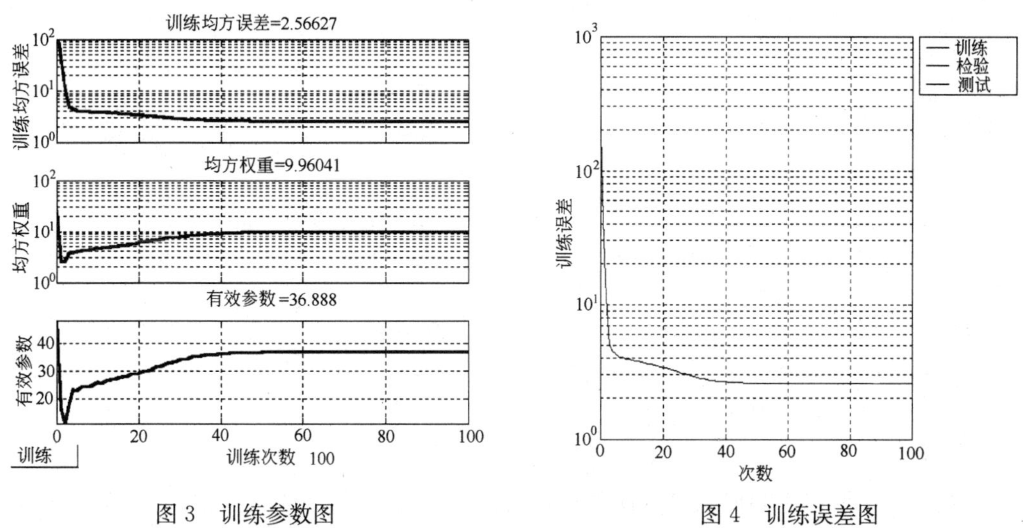

图 3　训练参数图　　　　图 4　训练误差图

训练后将表 2 中的数据输入神经网络进行预测和检验。地下埋管换热器第 2 组套管 −3m、−6m、−9m 壁温的预测结果与实测结果的对比图与相关系数图分别如图 5～7，图中虚线为实测值，实线为预测值。

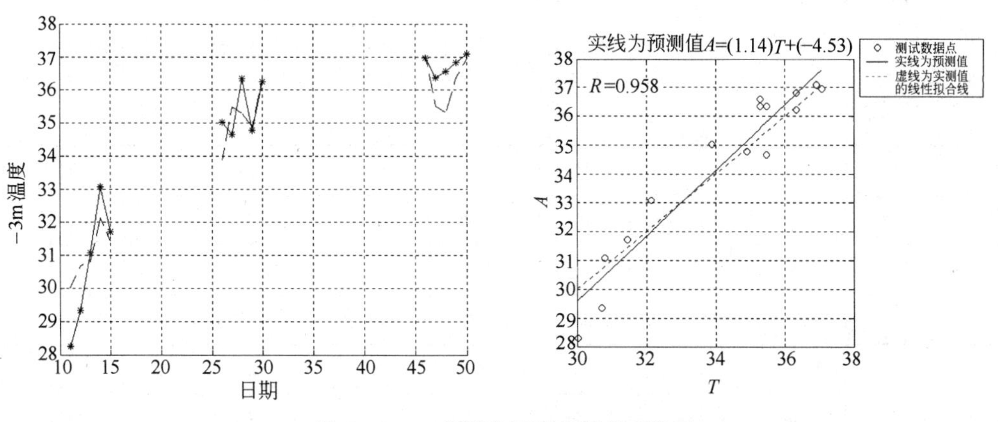

图 5　−3m 套管壁温图和相关系数图

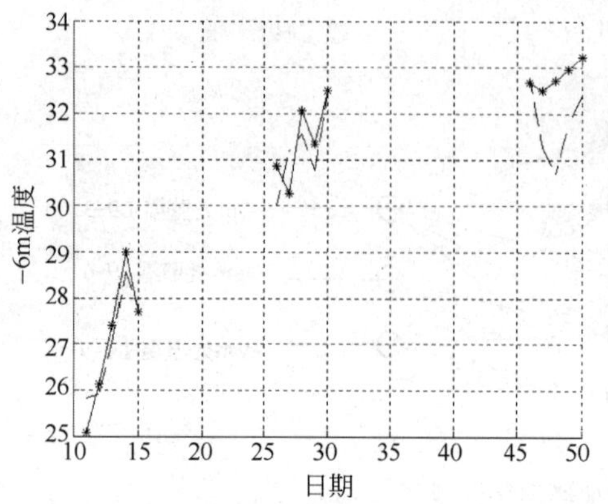

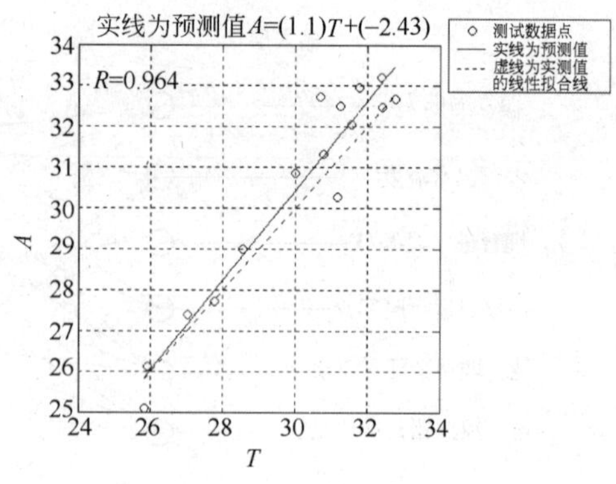

图 6 −6m 套管壁温图和相关系数图

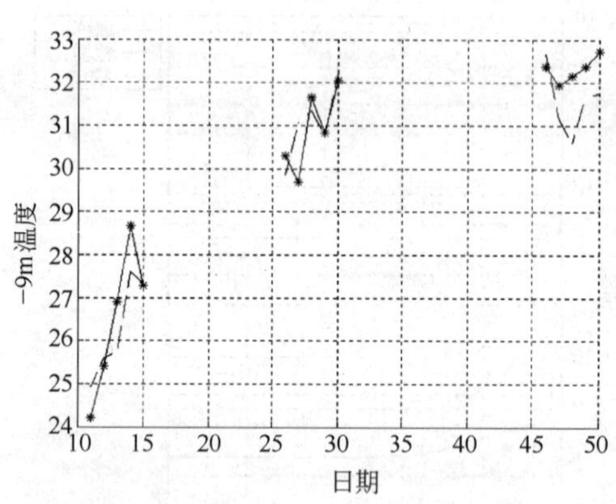

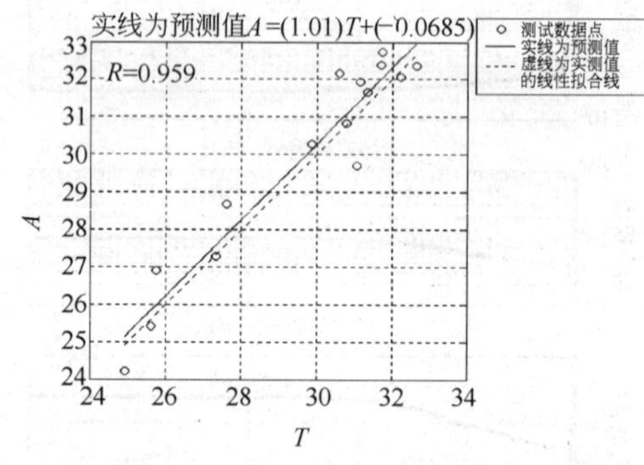

图 7 −9m 套管壁温图和相关系数图

壁温的预测值和实测值的绝对误差和相对误差如表 3。从表中可以看出预测的精度相当高，相对误差都在 7％以下，$L9-3$、$L9-6$、$L9-9$ 的平均相对误差分别为 2.07％(最大误差为−4.70％～＋3.51％)，2.22％(最大误差为−2.9％～＋6.58％)、2.24％(最大误差为−4.41％～＋5.13％)。

壁温预测值和实测值 **表 3**

Date	Num	L9−3				L9−6				L9−9			
日期	天数序号	实测值(℃)	预测值(℃)	绝对误差(℃)	相对误差(％)	实测值(℃)	预测值(℃)	绝对误差(℃)	相对误差(％)	实测值(℃)	预测值(℃)	绝对误差(℃)	相对误差(％)
716	11	30.02	28.61	−1.41	4.70	25.82	25.07	−0.75	2.90	24.92	24.21	−0.71	2.85
717	12	30.7	29.35	−1.35	4.40	25.9	26.13	0.23	0.89	25.6	25.42	−0.18	0.70
718	13	30.8	31.09	0.29	0.94	27.02	27.4	0.38	1.41	25.74	26.9	1.16	4.51
719	14	32.13	33.09	0.96	2.99	28.55	28.99	0.44	1.54	27.61	28.67	1.06	3.84
720	15	31.43	31.71	0.28	0.89	27.77	27.71	−0.06	0.22	27.32	27.27	−0.05	0.18
731	26	33.88	35.03	1.15	3.39	30	30.85	0.85	2.83	29.85	30.29	0.44	1.47

续表

Date	Num	L9—3				L9—6				L9—9			
日期	天数序号	实测值(℃)	预测值(℃)	绝对误差(℃)	相对误差(%)	实测值(℃)	预测值(℃)	绝对误差(℃)	相对误差(%)	实测值(℃)	预测值(℃)	绝对误差(℃)	相对误差(%)
801	27	35.48	34.67	−0.81	2.28	31.18	30.27	−0.91	2.92	31.06	29.69	−1.37	4.41
802	28	35.29	36.34	1.05	2.98	31.56	32.06	0.5	1.58	31.35	31.66	0.31	0.99
803	29	34.89	34.78	−0.11	0.32	30.78	31.34	0.56	1.82	30.77	30.83	0.06	0.19
804	30	36.34	36.24	−0.1	0.28	32.44	32.49	0.05	0.15	32.24	32.05	−0.19	0.59
822	48	37.07	36.97	−0.1	0.27	32.78	32.67	−0.11	0.34	32.65	32.37	−0.28	0.86
823	49	35.51	36.36	0.85	2.39	31.27	32.51	1.24	3.97	31.14	31.93	0.79	2.54
824	50	35.3	36.54	1.24	3.51	30.7	32.72	2.02	6.58	30.6	32.17	1.57	5.13
825	51	36.35	36.83	0.48	1.32	31.78	32.96	1.18	3.71	31.7	32.38	0.68	2.15
826	52	36.93	37.09	0.16	0.43	32.41	33.22	0.81	2.5	31.73	32.74	1.01	3.18
平均值					2.07				2.22				2.24

5 结论

本文是在浅埋套管换热器地源水—水热泵实验装置上，进行了冬夏暖冷联供全年测试，获得大量数据，对地源热泵的应用和设计很有参考价值。

地下埋管换热器的换热性能与其影响因素不是简单的线性关系，而是高度的非线性关系，传统数学方法的多元回归模型，精度较低。在无法通过统计回归分析方法用显示数学模型来揭示地下埋管换热器与其影响因素之间的关系的情况下，神经网络模型是建立地下埋管换热器与其影响因素的非线性关系的有力工具。本文成功地采用神经网络方法建立了隐式数学模型，从计算结果可以看出神经网络的模拟值与实验值相当一致，计算精度高。可加以推广和指导工程实践。

Modeling Research on Ground-Source Heat Pump By Neural Network

JianPing Chen

(Shanghai Research Institute of Building Science，ShangHai　200032　China)

Abstract: Ground source Heat Pump(GSHP) is becoming the more and more focus of the world's attention as a HVAC technique of energy saving and environment protection. This paper first introduced the experiment for Ground-Source water/water Heat Pump.

The heat transfer model of underground heat exchanger was introduced, and established predication modeling using neural network. From the computer simulation results, it was concluded that with NN modeling the precision was very high. It is worth to develop and apply for engineering practice and different situation.

参考文献

[1] 蒋照能等．空调用热泵技术及应用．机械工业出版社，1997

[2] thihault J. and Grandjean B. Nueral network methodology for data ananlysis. Int，J. Heat mass transfer，1991(34-8)，p2063～2070

[3] D. Aball，R. D. Fischer，D. L. Hodgett. Design Methods for Ground-Source Heat Pumps. ASHRAE，DC-83-08，No. 4

深圳地区居住建筑外窗遮阳性能模拟分析

田智华[1,2]　刘俊跃[1]　康侍民[2]
（[1] 深圳市建筑科学研究院，深圳市建筑节能重点实验室，深圳　518031；
[2] 重庆大学城市建设与环境工程学院，重庆　400044）

摘　要：本论文首先对深圳地区的居住建筑外窗及其遮阳现状进行调研分析，并利用WINDOWS5.2模拟软件对深圳地区建筑外窗的常用形式及其节能外窗的性能参数（包括传热系数*U*-值，太阳辐射得热系数SHGC，以及可见光透过率VT等）进行模拟分析。以一栋实际建筑为基础，利用Resfen5.0对不同外窗及其遮阳组合方式的采暖空调总能耗、各朝向的建筑外窗能耗以及峰值能耗等进行比较分析。最后根据模拟分析结果，得出了几点深圳地区居住建筑外窗遮阳性能的结论。

关键词：居住建筑　外窗　遮阳性能　模拟分析

引言

深圳市的气候特点是冬季温和短暂，夏季炎热漫长且相对湿度大，一般从4月份开始进入夏季，直到10月份才转入秋季，这种高温高湿气候的持续时间可达7个月[1]。因此决定了夏季防热是该地区建筑节能的重点，而建筑遮阳是该地区最有效的建筑节能措施之一。但是目前深圳市乃至整个南方地区建筑设计的潮流却是开窗面积越来越大，大面积飘窗、凸窗和全玻璃幕墙建筑比比皆是，建筑设计中几乎没有考虑建筑遮阳的节能设计。因此在夏季通过建筑外窗进入室内的太阳辐射热越来越多。据调查2004年深圳市夏季最大空调耗电量中有50%是用来消除建筑外窗的辐射得热和传导得热。本文将在调研的基础上，对深圳地区建筑外窗及其遮阳组合方式的遮阳性能进行模拟分析。

1　深圳地区建筑外窗及遮阳现状分析[2]

2004年12月至2005年4月我们对深圳地区的18个具有代表性的住宅小区（其中包括高层住宅、小高层住宅、普通多层住宅、联排别墅等）进行了详细调研和分析，共计109栋建筑，其中对窗户调研分析的参数包括窗墙比、外窗可开启面积与所在房间地面面积的比值、窗户构造形式和外窗遮阳设施等。调研结果分析如下：

（1）窗户构造形式：除两个小区采用中空玻璃和双层透明玻璃外，调研建筑中绝大部分采用普通透明玻璃铝合金门窗和塑钢门窗，均没有达到《深圳市居住建筑节能设计规范》（以下简称《节能设计规范》）中对外窗传热系数$K \leqslant 4.7\mathrm{W/(m^2 \cdot K)}$的要求。

(2) 外窗遮阳设施：所有调研建筑都没有采用活动式外遮阳形式；除建筑构造遮阳(包括阳台遮阳)外，所有调研建筑都没有采用水平遮阳或者垂直遮阳措施，但是几乎所有的建筑在建成后都采用窗帘等内遮阳措施。这与《节能设计规范》中的遮阳设计要求不符，规范要求建筑外窗应设置活动式遮阳设施，建筑外窗的太阳辐射透过率不应大于0.3。

(3) 窗墙比：所有调研建筑的平均窗墙面积比较大，并有不断增大的趋势。另外为了追求住宅的外视景观，许多建筑都采用近乎密闭式的大玻璃窗、落地玻璃(阳台)门等设计手法，但极少考虑遮阳和自然通风措施。

(4) 外窗可开启面积与所在房间地面面积的比值：客厅、主卧和次卧的平均可开启外窗面积与地面面积的比值都大于 10%，该指标达到了《夏热冬暖地区居住建筑节能设计标准》8%的要求，其中联排别墅的平均窗地面积比较大。

2 深圳地区建筑外窗性能参数的模拟分析

2.1 建筑外窗构造形式

调研建筑中除两个小区采用中空玻璃和采用双层透明玻璃外，其他的建筑采用无色透明玻璃的铝合金门窗和塑钢门窗(其中绝大部分为铝合金普通玻璃窗)。表 1 为深圳地区常用建筑外窗构造形式，表 2 为深圳地区节能外窗构造形式。

深圳地区常用外窗构造形式 **表 1**

序 号	建筑外窗构造形式	备 注
1	3mm 普通透明玻璃+铝合金窗框(不断热)	不考虑遮阳
2	3mm 普通透明玻璃+塑钢窗框	不考虑遮阳
3	3mm 普通透明玻璃+12mm 空气间层+3mm 普通透明玻璃+铝合金窗框	不考虑遮阳
4	3mm 普通透明玻璃+12mm 空气间层+3mm 普通透明玻璃+塑料窗框	不考虑遮阳

深圳地区节能外窗构造形式 **表 2**

序 号	建筑外窗构造形式	备 注
1	6mmLow-E 玻璃+铝合金窗框(不断热)	不考虑遮阳
2	6mmLow-E 玻璃+铝合金窗框(断热)	不考虑遮阳
3	6mm 低透射率 Low-E 玻璃+铝合金窗框(不断热)	不考虑遮阳
4	6mm 低透射率 Low-E 玻璃+铝合金窗框(断热)	不考虑遮阳
5	6mm 低透射率 Low-E 玻璃+塑料窗框	不考虑遮阳
6	6mmLow-E 玻璃+塑料窗框	不考虑遮阳

2.2 建筑外窗性能参数的模拟分析及结果

采用 WINDOW5.2 对深圳地区的几种常用建筑外窗以及几种节能外窗的性能参数进行了模拟分析。WINDOW5.2 可以对窗户的传热系数(*U*-值)、太阳辐射得热系数(SHGC)、可见光透过率(VT)、遮阳系数(SC)、冷凝热阻(CR)等进行模拟分析。WINDOW5.2 的模拟计算方法采用美国国家门窗协会(NFRC)标准、ISO—15099 和

ASHRAE—SPC442 标准，并且可以动态显示窗户的温度场分布，以及可以链接 DOE—2E 和 EnergyPlus 等能耗计算软件，是一个专用的窗户性能模拟分析软件[3]。计算模拟的环境及边界条件采用美国国家门窗评级协会的 NFRC 100—2001(summer)标准。室外温度：32℃；室外太阳辐射强度：783.0W/m²；室外平均风速：2.8m/s；室外对流换热系数：采用 ASHRAE/NFRC 模型，取 15.00W/(m²·K)；室内温度：24℃；室内对流换热系数：采用 ASHRAE/NFRC 模型。模拟的结果见表 3 和表 4。

深圳地区常用建筑外窗性能参数的模拟结果 **表 3**

序　　号	U 值(W/m²·K)	SHGC①	VT②	使用情况
3mm 普通透明玻璃＋铝合金窗框(不断热)	6.54	0.795	0.751	最常用
3mm 普通透明玻璃＋塑钢窗框	4.93	0.703	0.721	有所使用
3mm 普通透明玻璃＋12mm 空气间层＋3mm 普通透明玻璃＋铝合金窗框	4.13	0.713	0.681	很少采用
3mm 普通透明玻璃＋12mm 空气间层＋3mm 普通透明玻璃＋塑料窗框	3.6	0.624	0.652	很少采用

深圳地区节能外窗性能参数的模拟结果 **表 4**

序　　号	U 值(W/m²·K)	SHGC①	VT②	备　　注
6mmLow-E 玻璃＋铝合金窗框(不断热)	4.27	0.509	0.652	不考虑遮阳
6mmLow-E 玻璃＋铝合金窗框(断热)	3.432	0.472	0.652	不考虑遮阳
6mm 低透射率 Low-E 玻璃＋铝合金窗框(不断热)	4.43	0.391	0.418	不考虑遮阳
6mm 低透射率 Low-E 玻璃＋铝合金窗框(断热)	3.595	0.355	0.418	不考虑遮阳
6mm 低透射率 Low-E 玻璃＋塑料窗框	2.89	0.316	0.401	不考虑遮阳
6mmLow-E 玻璃＋塑料窗框	2.73	0.429	0.652	不考虑遮阳

注：①SHGC(Solar Heat Gain Coefficient)，太阳辐射得热系数，一般的取值范围是 0.1～0.9 之间，SHGC 值越小表示进入室内的太阳辐射越小，SHGC 与遮阳系数 SC 的换算关系，SC≈1.19SHGC；②VT(Visible Transmittance)，可见光透过率。

2.3 与节能设计标准的比较分析

深圳地区的建筑节能工作尽管起步比较晚，相对南方地区其他城市来说还是开展的比较好，是夏热冬暖地区第一个制定居住建筑节能设计标准的城市。表 5 列出了国内外几种节能设计标准和规范对深圳地区居住建筑外窗性能参数的规定，尽管要求有所差别，但都对建筑外窗的太阳辐射得热系数有比较严格的规定。根据以上的模拟结果并对比居住建筑节能设计标准，可以得出以下的几点结论：

(1) 深圳地区居住建筑最常采用的单玻铝合金外窗和单玻塑钢窗均不符合现有的建筑节能规范，但是单玻塑钢外窗的传热系数已经接近《深圳市居住建筑节能设计规范》中规定的建筑外窗传热系数小于 4.7W/(m^2 · K)的要求（单玻塑钢窗的传热系数 U-值为 4.93W/m^2 · K），其他的两种外窗构造形式的传热系数符合节能设计规范对传热系数 U-值的要求。

(2) 深圳地区居住建筑目前采用的几种外窗构造形式的太阳辐射得热系数 SHGC 值均不符合现有节能设计规范的要求，由于均采用普通透明玻璃，所以对太阳辐射的遮挡效果相当差，而最常用的单玻铝合金外窗的 SHGC 值几乎是节能设计规范要求的近两倍左右。

(3) 深圳地区居住建筑的外窗几乎全部没有考虑外遮阳形式，而窗户外遮阳是深圳地区居住建筑最主要的节能技术措施之一。

几种节能规范对深圳地区居住建筑外窗性能参数的规定[4~6]　　**表 5**

规范名称	U 值(传热系数)	太阳辐射得热系数(SHGC)
《深圳市居住建筑节能设计规范》(SJG 10—2003)	≤4.7	≤0.3
《夏热冬暖地区居住建筑节能设计标准》(JGJ 75—2003)	无要求	0.3～0.6①
IEEC 国际节能规范②	≤4.3	≤0.4
ASHRAE 节能规范③	无要求	≤0.4

注：①根据外墙传热系数的不同和窗墙面积比的不同取相应的综合遮阳系数值；②适合窗墙比≤30%的 3 层及低于 3 层的居住建筑的外窗；③适合窗墙比≤40%的 3 层以上居住建筑的外窗。

3　深圳地区居住建筑外窗遮阳性能的模拟分析

RESFEN5.0 计算模拟软件是居住建筑外窗的能耗及其经济性模拟分析软件，该软件能够计算新建和既有居住建筑不同朝向外窗的全年能耗及能耗费用，以及建筑外窗的峰值能耗。用户只需要输入当地的气象数据（RESFEN5.0 中已经有我国大部分城市地区气象数据资料，而以前的 RESFEN 版本只有美国和加拿大等城市的气象数据资料），建筑外窗的朝向，电力和燃气费用，建筑的详细信息（包括建筑外墙构造、地板方式、HVAC 系统类型等），以及窗户的大小、形状、遮阳方式（目前的 RESFEN 有 8 种典型的遮阳组合方式供选择），需要输入建筑外窗的性能参数（包括传热系数 U-值，太阳辐射得热系数 SHGC，空气渗透率 VT），由于 RESFEN5.0 可以动态链接 WINDOWS5.0 软件，可以利用 WINDOWS5.0 的模拟计算结果来分析建筑外窗的节能特性[7]。由于考虑到 RESFEN5.0 版本目前并不能定量的改变建筑外窗的遮阳方式。因此只分析在具体遮阳方式下不同建筑外窗的节能性能。

3.1　建筑模型和边界条件

利用 Resfen5.0 模拟软件，我们对调研的某一别墅建筑采用不同建筑外窗及其遮阳组合形式的总体能耗、各朝向的建筑外窗能耗和峰值能耗等进行比较和分析。该建筑共有两层，总建筑面积为 370m² 左右，外墙为 180mm 厚的实心粘土砖，内墙为 180mm 厚的加气混凝土。表 6 为模拟建筑外窗的类型和参数，表 7 为模拟建筑外窗的遮阳组合类型。

模拟建筑外窗类型和参数 **表 6**

序号	建筑外窗类型	朝向	面积 (m^2)	SHGC	VT	U 值 ($W/m^2 \cdot ℃$)	空气渗透率($1/s \cdot m^2$)
1	3mm 透明玻璃＋铝合金窗框(不断热)	北向	20.85	0.795	0.751	6.54	1.02
		东向	13.66				
		南向	13.36				
		西向	31.36				
2	3mm 透明玻璃＋12mm 空气间层＋3mm 透明玻璃＋铝合金窗框(不断热)	北向	20.85	0.713	0.681	4.13	1.02
		东向	13.66				
		南向	13.36				
		西向	31.36				
3	6mm 低透射率 Low-E 玻璃＋铝合金窗框(不断热)	北向	20.85	0.391	0.418	4.43	1.02
		东向	13.66				
		南向	13.36				
		西向	31.36				

模拟建筑外窗遮阳组合类型 **表 7**

序 号	遮 阳 组 合	备 注
1	不采取任何遮阳方式(None)	None
2	采用内窗帘遮阳(Interior)，夏季 SHGC=0.8，冬季 SHGC=0.9	Int
3	外挑 0.6m 遮阳板(Overhang)	Ovh
4	同时采用内窗帘遮阳和外挑遮阳板	Int +Ovh

3.2 建筑外窗遮阳性能的模拟分析及结果

利用 Resfen5.0 模拟软件，对以上模拟建筑分别采用不同的建筑遮阳形式后的总体能耗、各朝向的建筑外窗能耗和峰值能耗等进行模拟分析。图 1 和图 2 分别为外窗采用 3mm 透明玻璃铝窗时模拟建筑单位建筑面积的采暖空调能耗和峰值负荷，图 3 和图 4 分别为模拟建筑各朝向单位建筑外窗面积的空调能耗和采暖能耗。图 5 和图 6 分别为外窗采用 3mm 双层透明中空玻璃铝窗时模拟建筑单位建筑面积的采暖空调能耗和峰值负荷，图 7 和图 8 分别为模拟建筑各朝向单位建筑外窗面积的空调能耗和采暖能耗。图 9 和图 10 分别为外窗采用 6mm 低透射率 Low-E 玻璃铝窗时模拟建筑单位建筑面积的采暖空调能耗和峰值负荷，图 11 和图 12 分别为模拟建筑各朝向单位建筑外窗面积的空调能耗和采暖能耗。

从图 1～图 12 可以得出以下几点结论：

(1) 在深圳地区的建筑能耗以制冷空调能耗为主，采暖能耗只有制冷空调能耗的 1/5 左右。

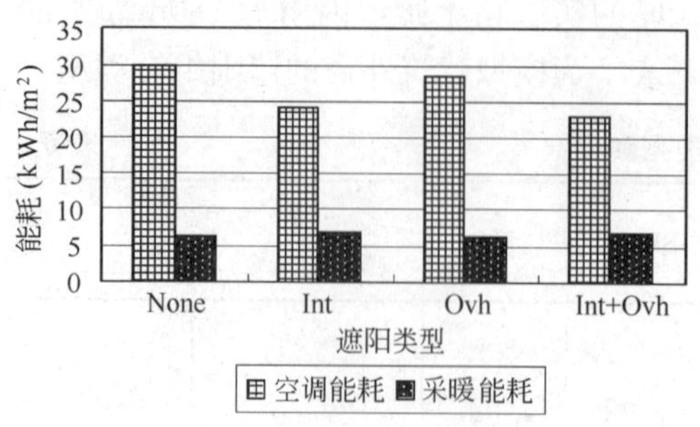

图 1 模拟建筑全年单位建筑面积采暖空调能耗（3mm 透明玻璃铝窗）

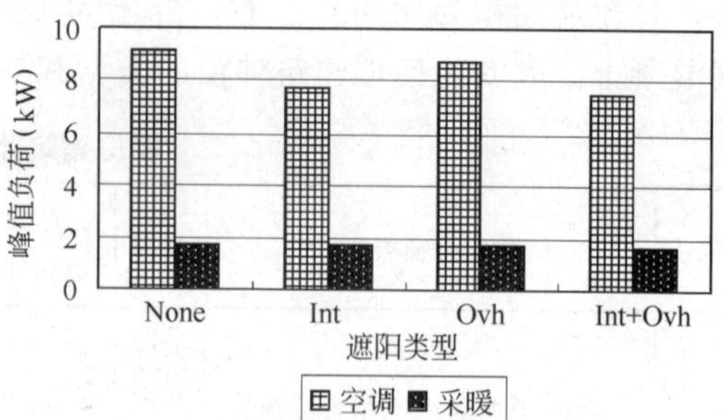

图 2 模拟建筑全年单位建筑面积采暖空调峰值负荷（3mm 透明玻璃铝窗）

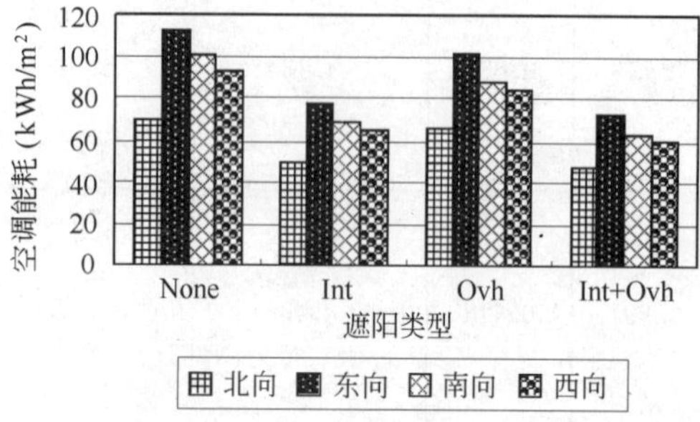

图 3 模拟建筑各朝向单位外窗面积空调能耗（3mm 透明玻璃铝窗）

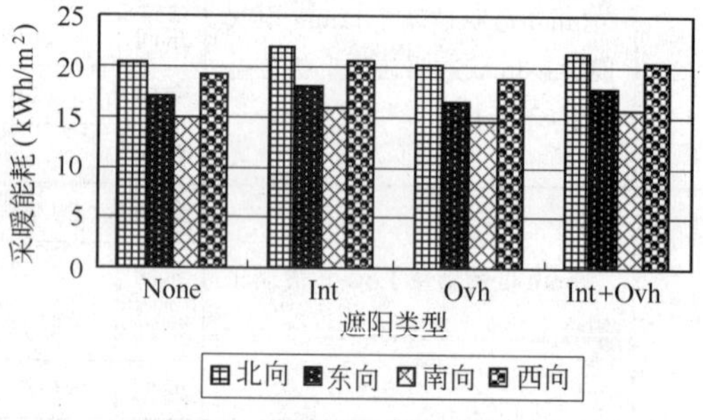

图 4 模拟建筑各朝向外窗面积采暖能耗（3mm 透明玻璃铝窗）

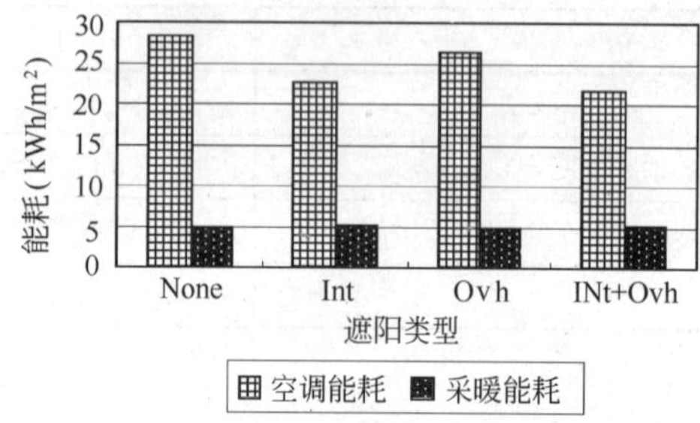

图 5 模拟建筑全年单位建筑面积采暖空调能耗（3mm 透明玻璃＋12A＋3mm 透明玻璃铝窗）

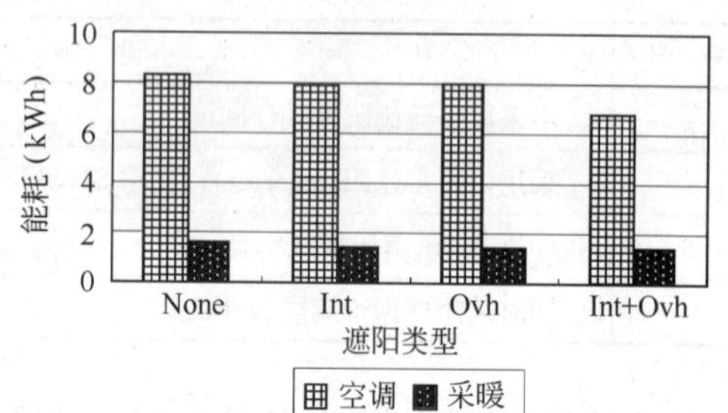

图 6 模拟建筑全年单位建筑面积采暖空调峰值负荷（3mm 透明玻璃＋12A＋3mm 透明玻璃铝窗）

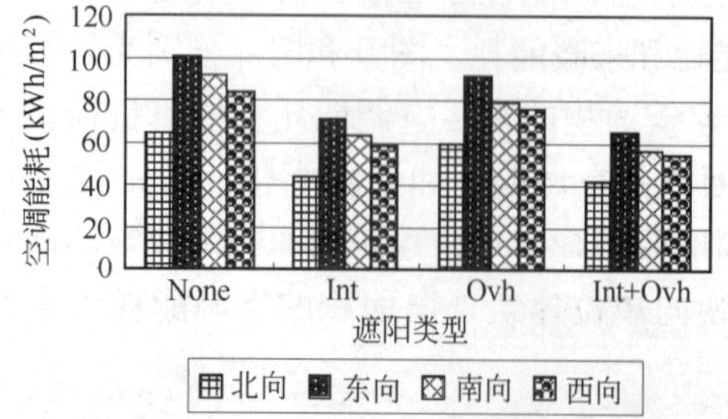

图 7 模拟建筑各朝向单位外窗面积空调能耗（3mm 透明玻璃＋12A＋3mm 透明玻璃铝窗）

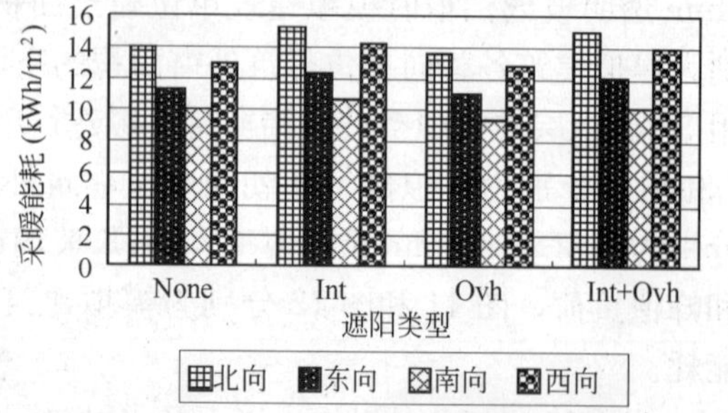

图 8 模拟建筑各朝向外窗面积采暖能耗（3mm 透明玻璃＋12A＋3mm 透明玻璃铝窗）

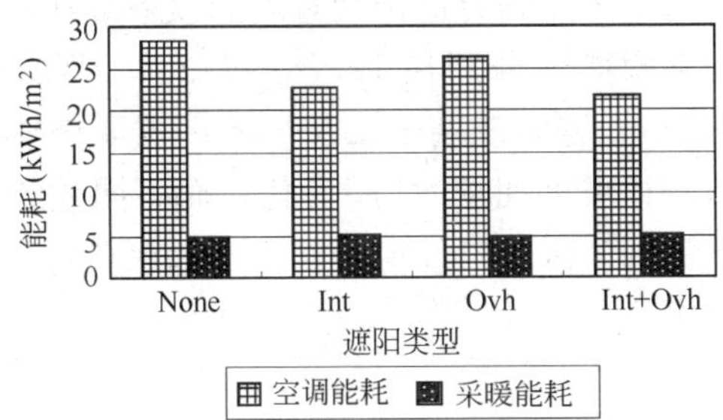

图 9　模拟建筑全年单位建筑面积采暖空调能耗（6mm 低透射率 Low-E 玻璃铝窗）

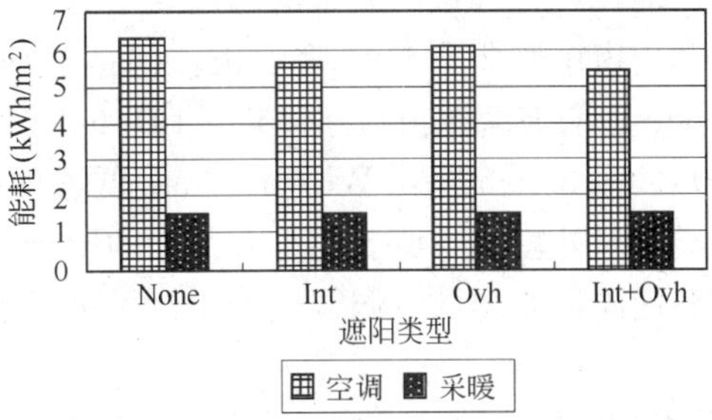

图 10　模拟建筑全年单位建筑面积采暖空调峰值负荷（6mm 低透射率 Low-E 玻璃铝窗）

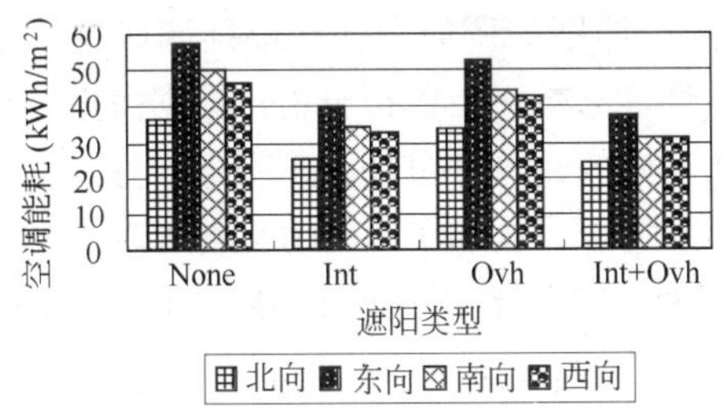

图 11　模拟建筑各朝向单位外窗面积空调能耗（6mm 低透射率 Low-E 玻璃铝窗）

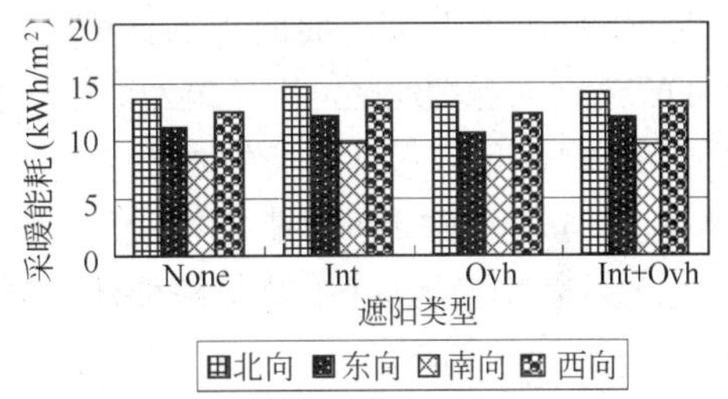

图 12　模拟建筑各朝向外窗面积采暖能耗（6mm 低透射率 Low-E 玻璃铝窗）

(2) 遮阳装置对采暖能耗影响的很小，一般在 6.4～6.8kWh/m²。遮阳装置对采暖峰值负荷几乎没有影响，固定在 1.6kW～1.7kW 之间。而建筑外窗为中空玻璃铝窗或者低透射率 Low-E 玻璃铝窗时，可以减少采暖能耗和采暖峰值负荷，一般为 5.1～5.6kWh/m² 和 1.5kW～1.6kW 之间。

(3) 遮阳装置对模型建筑全年单位建筑面积制冷空调能耗和制冷空调峰值负荷影响比较大，分别在 16.59～30.14kWh/m² 和 5.43～9.22kW 之间，其中当建筑外窗为 6mm 低透射率 Low-E 玻璃铝窗并采用内窗帘遮阳＋外挑遮阳板后空调能耗和峰值负荷最小，而当建筑外窗为 3mm 透明玻璃铝窗且没有采取任何遮阳措施时最大，其他组合方式次之。从以上的分析可知建筑外窗空调能耗的最大节能贡献率为 45%，峰值负荷的减少率为 41.1%。也就是说采用合理建筑外窗及其遮阳组合方式可以显著的减小建筑空调能耗。

(4) 从以上可以分析得出各朝向单位外窗面积的采暖能耗很小且变化波动幅度不大，其中北向外窗最大，其次为西向和东向外窗，南向外窗最小。采用内遮阳装置会相应增加建筑外窗的采暖能耗，但是增加的幅度很小，采用外挑遮阳板会减少建筑外窗的采暖能耗，但是减少的幅度也相当有限。模拟建筑单位外窗的采暖能耗分别在 20.01～21.86kWh/m²（3mm 透明玻璃铝窗）、13.63～15.22kWh/m²（3mm 透明玻璃＋12A＋3mm 透明玻璃铝窗）和 13.22～14.54kWh/m²（6mm 低透射率 Low-E 玻璃铝窗）之间变化。因此可以分析得出在深圳地区遮阳装置对采暖能耗的影响很小，相对空调能耗来说可以忽略不计，而不同建筑外窗类型对外窗的采暖能耗有很大的影响。

(5) 各朝向单位外窗面积的制冷空调能耗很大且变化波动幅度很大，其中东向外窗最大，其次为南向和西向外窗，北向外窗最小。采用遮阳装置后会相应减少建筑外窗的制冷空调能耗，各种遮阳装置的遮阳效果有所不同，其中采用综合遮阳装置后的效果最好，其次为内窗帘遮阳和外挑遮阳板。模拟建筑单位空调能耗分别在 72.78～112.42kWh/m^2(3mm 透明玻璃铝窗)、65.47～101.64kWh/m^2(3mm 透明玻璃＋12A＋3mm 透明玻璃铝窗)和 37.33～56.74kWh/m^2(6mm 低透射率 Low-E 玻璃铝窗)之间变化，而不同建筑外窗类型对外窗的空调能耗也有很大的影响。因此深圳地区建筑外窗节能主要考虑东向、南向和西向，北向的空调能耗很小几乎可以不予考虑。

4 结论

本文首先对深圳地区的居住建筑外窗的参数(包括构造形式、外遮阳设施、窗墙比、外窗可开启面积与所在房间地面面积的比值等)进行深入和详细的调研分析。根据调研结论采用 WINDWS5.2 对深圳地区常用建筑外窗和节能外窗的性能参数进行模拟分析，根据模拟分析的结果，利用 Resfen5.0 模拟软件对模型建筑分别采用不同的建筑外窗及其遮阳组合方式后的总体能耗、各朝向的建筑外窗能耗和峰值能耗等进行模拟分析。得出了以下的几点结论：

(1) 目前华南地区(主要是深圳地区)居住建筑外窗面积有增大的趋势，因此外窗的能耗有显著增大的趋势。

(2) 目前华南地区(主要是深圳地区)居住建筑所普遍使用的单玻铝窗是不节能的，不符合建筑节能设计规范的要求。深圳地区的建筑能耗主要是夏季空调能耗，采暖能耗只有制冷空调能耗的 1/5 左右。

(3) 遮阳装置对建筑采暖能耗和采暖峰值负荷影响很少，内窗帘遮阳会增加采暖能耗而外挑遮阳板会减少采暖能耗，但是其增加或者减少的程度很少。不同建筑外窗类型对采暖能耗和采暖峰值负荷有比较大的影响。

(4) 遮阳装置和建筑外窗类型对模型建筑全年单位建筑面积制冷空调能耗和制冷空调峰值负荷影响很大。从以上的分析可知建筑外窗及其遮阳组合方式对空调能耗的最大节能贡献率为 45%，峰值负荷的减少率为 41.1%。也就是说采用合理建筑外窗及其遮阳组合方式可以显著的减小建筑空调能耗和空调峰值负荷。

(5) 各朝向单位外窗面积的采暖能耗很小且变化波动幅度不大，其中北向外窗最大，其次为西向和东向外窗，南向外窗最小。各朝向单位外窗面积的制冷空调能耗很大且变化波动幅度很大，其中东向外窗最大，其次为南向和西向外窗，北向外窗最小。

Simulation Analysis an Shading Performance of Windows for Residential Buildings in Shenzhen

Tian Zhihua[1,2] Liu Junyue[1] and Kang Shimin[2]

([1] Shen Zhen Institute of Building Research, Shen Zhen Key Laboratory for Building Energy Efficiency, Shenzhen 518031 China;

[2] Faculty of Urban Construction & Environmental Engineering, Chongqing University, Chongqing 400044 China)

Abstract: The status of exterior windows and its shading devices for residential buildings are investigated and analyzed. Performances parameters (including U-factor, solar heat gain coefficient and visible transmittance etc.) of general and energy efficiency windows are simulated with WINDOWS 5.2 based on results of investigation. The total HVAC energy consumption, windows energy consumption, and peak energy consumption of different windows and its shading combinations are compared and analyzed based on one investigated building in Resfen5.0 software. At last, several conclusions on shading performance of exterior windows for residential buildings in Shenzhen based on hereinbefore simulation results are obtained.

参考文献

[1] 马晓雯、范圆圆、侯余波、付祥钊、刘俊跃. 深圳居住建筑夏季自然通风降温实验研究. 暖通空调，2003，第5期

[2] 田智华，康侍民，刘俊跃等. 深圳地区居住建筑外窗性能参数调研及其能耗模拟分析. 2005年西南暖通空调年会，2005.9

[3] WINDOWS5.2 Use Manual

[4] 深圳市居住建筑节能设计规范(SJG 10—2003)

[5] 夏热冬暖地区居住建筑节能设计标准(JGJ 75—2003)

[6] www.nfrc.org

[7] RESFEN 5.0 Use Manual

基于全寿命周期评价的住宅节能技术分析

王卫仑[1] 谢静超[2] 刘 伟[1] 邢 锋[1]

（[1] 深圳大学，深圳 518060；[2] 日本东北大学，仙台 980-8579）

摘 要：本研究，在考虑住宅全寿命周期内家庭生活方式及成员构成变化的基础上，对住宅全寿命周期的能耗及成本进行数值分析。结果表明，随着地域的不同，各种节能技术对住宅全寿命周期能耗及全寿命周期成本所带来的节约效果是不同的；特别是对于夏热冬冷地区，仅采用围护结构的高断热高气密技术，在降低能耗的同时也会增加住宅全寿命周期的能耗成本；只有同时结合太阳能技术，才能获得住宅全寿命周期能耗最佳的节能性与经济性。

关键词：全寿命周期评估 节能技术 全寿命周期能耗与成本 成员构成变化

1 前言

节能性建筑物外围护体系的高断热高气密性以及光伏发电等建筑节能技术正在逐步地普及推广中。目前对建筑物节能的评价逐步转向全寿命分析，即不单考虑正常使用期间建筑物的能耗，而是全面计算建筑物建设期间、维修保养期间和拆除期间的能耗，以建筑物全寿命能耗 LCE 及全寿命能耗成本 LCC 来分析评价各种节能措施的适用性。本研究从住宅建筑正常使用期间，居住者的生活方式(Life-Style)以及此期间内居住者数量变化对能耗的影响入手，研究各种节能技术措施对不同地区住宅的 LCE 及 LCC 的有效性，提出更优化的低减环境负荷与建筑能耗成本的住宅节能手法。

2 正常使用期间住宅能耗评价

2.1 能耗预测模型

图 1 给出了正常使用期间住宅能耗计算模型的总体流程。此计算流程考虑了建筑物使用者的人员构成、生活方式等个性条件。同时为了比较本计算模型的有效性，对某一住宅进行了为期一年的能耗实测，以期与计算结果比较，来验证此计算模型的有效性。

本文所选用的实测建筑物为二层独立住宅，夫妇两人使用，总面积 $176.3m^2$，热损失系数为 $2.33W/m^2 \cdot K$，同时在建筑物南侧屋面设置了 3.0kW 的太阳能光伏发电系统，其倾斜角度为 26.6°。从 2002 年 11 月到 2003 年 10 月对该建筑物进行了为期一年的能耗实测，按所使用设备类别分类的能耗实测结果见图 2。同时，图 2 也记录了计算结果，实测值/计算值的平均值为 1.03，变异系数为 9.44％。

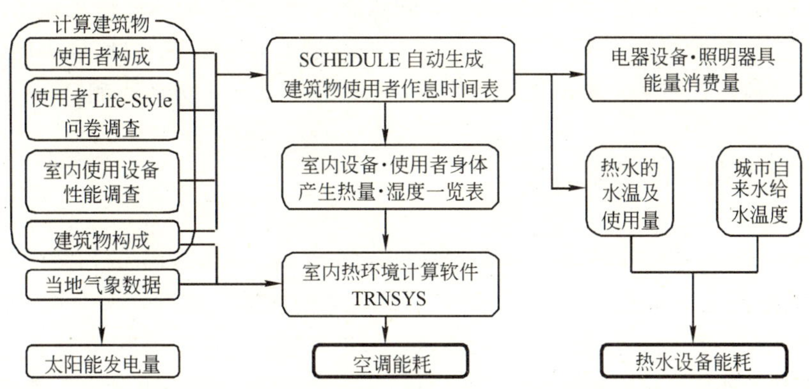

图 1　正常使用时能耗预测模型的计算流程

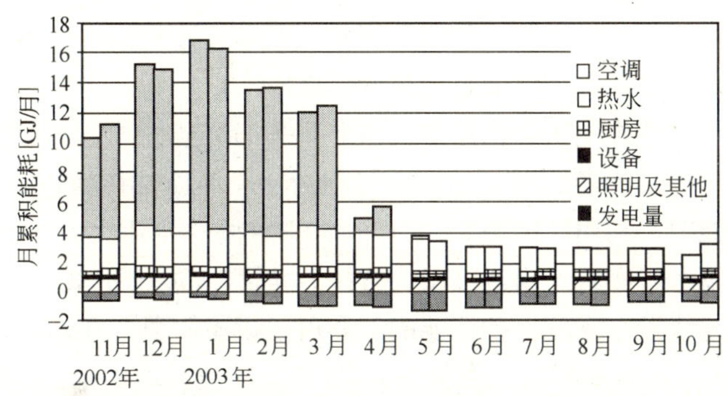

图 2　能耗·发电量的实测值与计算值

2.2　家庭人口构成变化对能耗的影响

考察同一住宅在其使用期间，不同的家庭人口构成对建筑能耗的影响以及在此使用期间内家庭人口变化对能耗的影响，将此影响因素在 2.1 的计算模型中体现出来，会得到更加符合住宅使用全过程实际的能耗评价。基于这种考虑，针对某一标准住宅，设定不同的家庭人口构成、就业状况等人口构成变化条件，按照 30 年使用期来计算家庭人口构成变化对能耗的影响，计算结果见图 3。

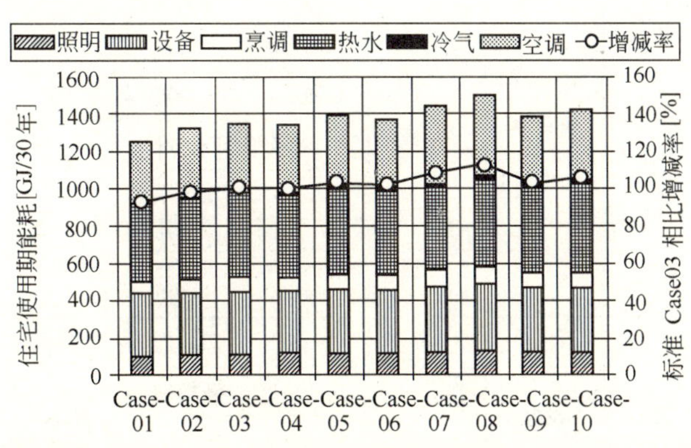

图 3　不同家庭构成的能耗比较

由图 3 可以清楚地知道，家庭成员的不同状态对建筑使用期总能耗是有比较大的影响的，最大值(4 人核心家庭，其中丈夫为住宅勤务，妻子为家庭主妇)比最小值(2 人家庭)高出 25%。因此，在住宅全寿命能耗分析中，必须要考虑这部分的影响。

3 建筑的全寿命周期节能评价

通过上述分析，本文在住宅建筑全寿命能耗评价中，采用 2.1 的预测模型并且考虑 2.2 所提出的家庭人口构成影响因素。计算对象住宅同前述，空调效率由外气温与设备效率的关系式导出。[3]同时，计算时以 2.2 中 Case03 为基准，考虑此 Case03 的家庭成员构成以及在使用期间成员构成的变化，计算时设定的地域范围从严寒地区一直到夏热冬暖南区。

3.1 各种节能措施的设置

结合目前国内外建筑节能的技术与研究现状，在本文建筑能耗全寿命评估计算中，对所研究住宅设定下面 4 种有代表性的节能技术措施：

(1) 高断热高气密性外围护结构体系，参照日本旧节能标准、新节能标准、次世代节能标准以及加拿大 R2000 节能标准设定 4 个节能水平，确定相应的外围护结构及门窗等开口部位的断热性能值，利用 TRNSYS 软件来计算空调的能耗。

(2) 光伏发电技术，建筑物南侧屋面设置倾斜角度为 26.6°的 3.0kW 太阳能光伏发电系统，通过对常用光伏发电系统设备性能的调查以及设备手册，在本次计算中考虑其发电量逐年递减 1%，其动力调节器的耐用年数为 15 年，发电模块的耐用年数为 30 年。

(3) 高气密性蓄热地面，采用这种方法可以提高建筑物整体的热容量，也就是提高了建筑物的热稳定性。此类地面参照日本建筑学会的钢筋混凝土地面构造规定，依次为防水层、隔热层、混凝土地面层以及面层。与普通地面相比有较高的热容量和气密性。

计 算 Case **表 1**

<table>
<tr><th>case</th><th>高断热高气密</th><th>光伏发电</th><th>高气密蓄热地面</th><th>附设温室</th><th>case</th><th>高断热高气密</th><th>光伏发电</th><th>高气密蓄热地面</th><th>附设温室</th></tr>
<tr><td>Case1-1</td><td>旧标准</td><td rowspan="4">不设置</td><td rowspan="8">不设置</td><td rowspan="12">不设置</td><td>Case4-1</td><td>旧标准</td><td rowspan="4">设置</td><td rowspan="12">设置</td><td rowspan="4">不设置</td></tr>
<tr><td>Case1-2</td><td>新标准</td><td>Case4-2</td><td>新标准</td></tr>
<tr><td>Case1-3</td><td>次世代</td><td>Case4-3</td><td>次世代</td></tr>
<tr><td>Case1-4</td><td>R2000</td><td>Case4-4</td><td>R2000</td></tr>
<tr><td>Case2-1</td><td>旧标准</td><td rowspan="4">设置</td><td>Case5-1</td><td>旧标准</td><td rowspan="4">不设置</td><td rowspan="8">设置</td></tr>
<tr><td>Case2-2</td><td>新标准</td><td>Case5-2</td><td>新标准</td></tr>
<tr><td>Case2-3</td><td>次世代</td><td>Case5-3</td><td>次世代</td></tr>
<tr><td>Case2-4</td><td>R2000</td><td>Case5-4</td><td>R2000</td></tr>
<tr><td>Case3-1</td><td>旧标准</td><td rowspan="4">不设置</td><td rowspan="4">设置</td><td>Case6-1</td><td>旧标准</td><td rowspan="4">设置</td></tr>
<tr><td>Case3-2</td><td>新标准</td><td>Case6-2</td><td>新标准</td></tr>
<tr><td>Case3-3</td><td>次世代</td><td>Case6-3</td><td>次世代</td></tr>
<tr><td>Case3-4</td><td>R2000</td><td>Case6-4</td><td>R2000</td></tr>
</table>

(4) 附设温室，在建筑物的南侧设置 2m 进深的玻璃温室，其地面采用(3)中提到的高气密性蓄热地面。温室和主体建筑的门窗开闭条件，按照 Life-Style 问卷调查的结果，设定为冬季夜间 18：00～7：00 及夏季白天 7：00～18：00 关闭使用空调设备，冬季白天 7：00～18：00 及夏季夜间 18：00～7：00 开启。

按照这 4 种节能技术措施的组合，可以得到不同的计算方案，见表 1。

3.2 建设期、维修期及拆除期的能耗计算条件

使用全寿命能耗计算工具 Leaf 来计算建筑物建设、维修、拆除时产生的环境负荷以及费用。这里要注意的是，在计算建设期间各种能耗时，采用了如下的设定值：①材料生产能耗=Σ建筑物各构成材料的重量×生产单位材料环境负荷指标；材料运输能耗，假设运输距离为 100km；②现场施工以及拆除时能耗，以建筑物的面积为准，按比列套用单位指标；③建筑维修能耗，依照结构及材料的耐久年限来计算维修所需材料量，同时使用上述①、②的方法，计算出维修时的能耗，此处由于正常维修的工作量比较小，所以此部分施工时的能耗忽略。

3.3 可持续住宅能耗最优设计

设置光伏发电设施，虽说在建设期间多使用大约 80GJ 的能耗，然而在使用过程中，能得到很好的节能效果，经过 6 年住宅使用期，建设时多消耗的能量就可以完全回收，因此在住宅的全寿命过程中，使用光伏发电设施对于住宅节能是非常有效的，计算结果见图 4。

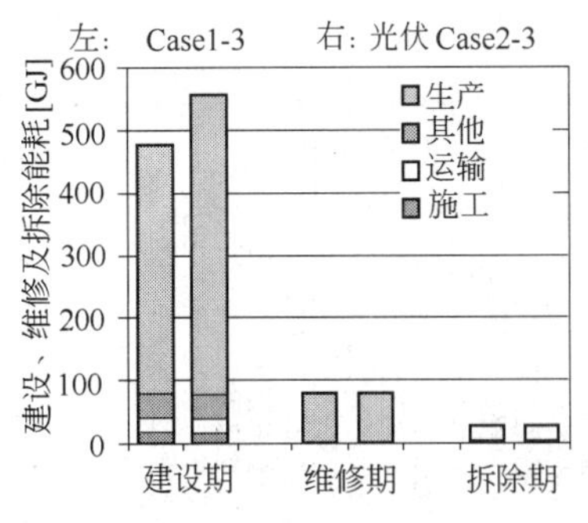

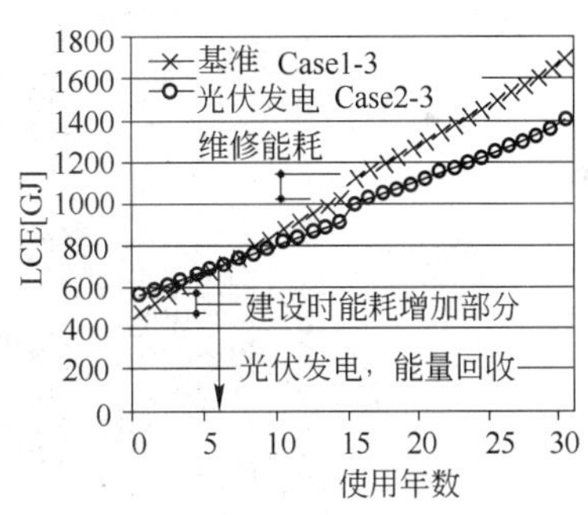

图 4 设置光伏发电 LCE 比较

各地区(分别为严寒地区、寒冷地区、温和地区、夏热冬冷地区和夏热冬暖地区)不同计算 Case 的计算结果见图 5，此计算结果包括了建筑物全寿命周期(建设、使用、维修和拆除)的总能耗。从图 5 中，我们可以清楚地看出，采取节能措施虽说会使得建设时能耗增加，但是在比较长的使用期间能耗有递减的趋势。同时，节能水平从旧节能标准 Case1-1 提高到 R2000 水准 Case1-4，如果只采用高断热高气密性外围护体系节能措施，对于严寒地区，建筑物全寿命能耗可以削减 30%；夏热冬冷地区，能耗削减 10%；但是对于夏热冬暖地区的建筑，其全寿命能耗几乎没有变化，换句话说，在这样的地区过高地采用高断热高气密性外围护体系节能措施来提高节能水平是没有效果的。

同时，如图 6 所示，比较了采取各种节能措施时住宅建设期能耗的增加量以及此部分增加量在以后的使用期内由于节能措施的采用其回收的期限。对于高断热高气密节能措施

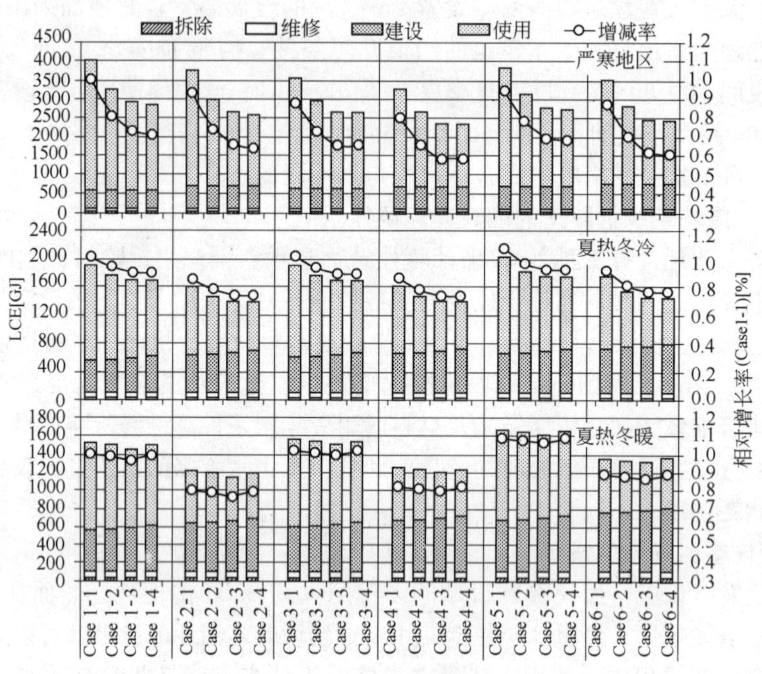

图 5 能耗·发电量的实测值与计算值

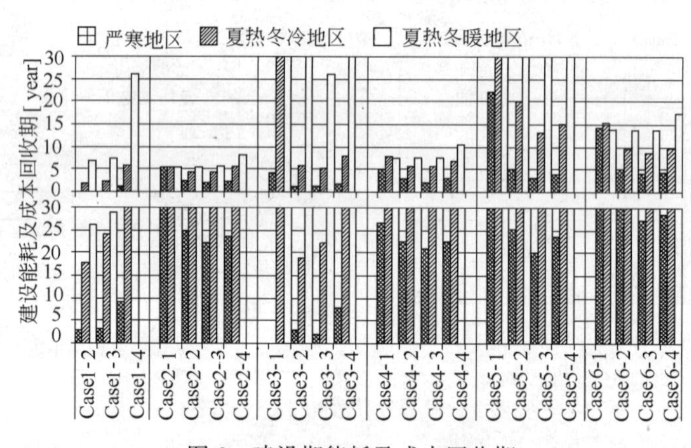

图 6 建设期能耗及成本回收期

来说，在夏热冬暖地区节能效果不太明显，其成本回收时间随着节能水准的提高而大幅度增长，一直到 R2000 水准的 26 年；光伏发电措施，有很好的节能效果，但是成本回收时间是个长期的过程，对于夏热冬冷地区和夏热冬暖地区基本上是无法完全回收成本的；至于蓄热地面与附设温室，同样表现出了在严寒地区成本回收很快，而温暖地区成本回收很慢的趋势。

图 7 是采用不同节能技术措施的 LCE 与 LCC 的降低效果比较。LCE 在高断热高气密性措施下可以削减 10％～30％，然而 LCC 几乎是不能降低的，特别是采用高水准的 R2000 标准，LCC 会增加 2％～4％；光伏发电措施，各地区能耗大体上都有 20％的削减，但是除了在严寒地区，其 LCC 都有不同程度的增加，同时也应该知道如果不联合使

用高断热高气密性措施，LCE 的降低也是不能达成的。

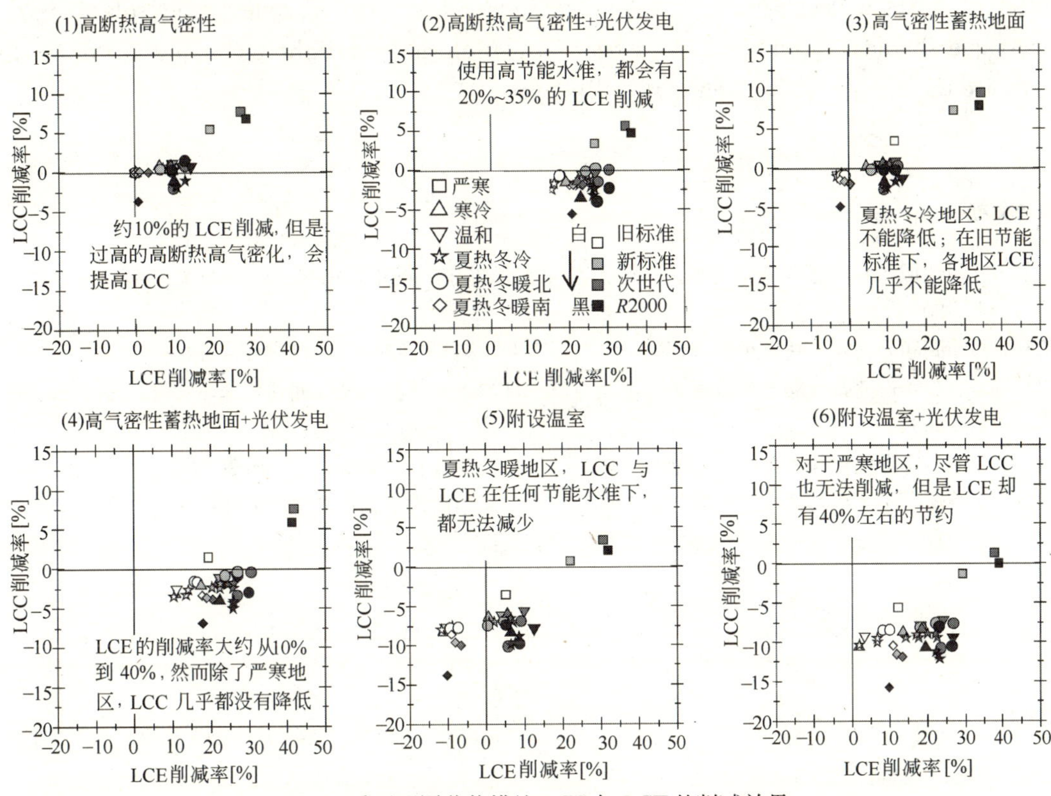

图 7 采用不同节能措施 LCC 与 LCE 的削减效果

4 结论

对于所有的节能技术措施来说，不同地域有不同的 LCE 及 LCC 效果，在节能计划中，了解这一点是非常重要的。高断热高气密性节能技术如果被过度地提高水准，也是会增加 LCC 的。总的来说，在建筑物的全寿命能耗评价过程中，使用者的生活方式以及使用者的成员构成和变迁，也是必须要考虑的因素。

Study on Energy-Saving Techniques in Residential Building based on Life Cycle Assessment

Weilun Wang[1] Jingchao Xie[2] Wei Liu[1] Feng Xing[1]

([1] ShenZhen Univ., ShenZhen 518060 China;

[2] Tohoku Univ., 980-8579 Japan)

Abstract: In this paper, a number analysis is carried out to evaluate the energy consumption and cost of residence building based on Life Cycle Assessment by taking the variation of member over life cycle of the used residence building into account. The results is shown the efficiency of energy-saving is various by using each energy-saving technique in different areas. Especially, in hot summer and warm winter zones, the solar energy technique should be combined with heat insulation and air tight techniques in order to optimize the energy-saving and cost in life cycle.

参考文献

[1] 九州地域における住宅用 PVシステムの発電量推定. 酒井孝司等. 日本太陽エネルギー学会志. 2004

[2] 標準問題の提案 住宅用標準問題. 宇田川光弘. 第 15 回熱シンポジウム. 1985

[3] エネルギー消費実態と実效性ある削減對策. 日本建築学会. 第 3 回住宅エネルギーシンポジウム. 2004

[4] 住宅の次世代省エネルギー基準と指针. (財)建築環境・省エネルギー機構. 2000

[5] ライフサイクルエネルギー計算システム— Leaf. 生活價值創造住宅開発技術研究組合. 2001

绿色节能照明及昼光照明的引入

钱颖初

（美国 EMSI 环境管理咨询有限公司，上海　200031）

摘　要： 本文从舒适、节能两个角度出发对传统照明系统设计进行分析，指出其在当前能源危机大前提下所面临的挑战，从而阐明实施绿色节能照明设计的重要性。在简要综合分析当前可用绿色照明系统设计方法、技术、产品之后，提出昼光照明是实现绿色照明的重要途径，也是实现绿色(可持续发展)建筑的重要实施策略之一。并进一步结合案例深入探讨其设计方法，同时通过计算机模拟对其技术、经济可行性及舒适性进行确认，从而验证昼光照明设计的实践价值。

关键词： 照明能耗　热负荷　绿色节能照明　昼光照明　建筑外形设计　空调设备减容

概述

照明系统从自身构成来讲，相对简单，由光源、灯具、线路、控制设备等组成。但要在节能的前提下，同时满足视觉舒适性和热舒适性，设计就趋于复杂，牵涉到空调系统、建筑设计等各个方面，因此需要全盘考虑，整体设计。

为求得一个合理的照明设计，以下先对其涉及的两个舒适性：视觉舒适性和热舒适性(相关照明部分)进行讨论和定义。

1　照明舒适度探讨

1.1　视觉舒适性

人眼睛之所以能够看到物体，是因为物体发射或反射人体所能感知范围波长的光线。拿太阳光来讲，其波长范围从零趋于无限大，人眼能感知到的只是其中的一部分，我们称其为可见光，波长范围在 380nm～780nm 之间。这部分光占到了太阳光总能量的 50%左右。我们去评价光线属性的时候，通常采用以下几个参数：照度、色温、显色性等，涉及到舒适性，同时必须考虑物体表面亮度、眩光、背景光等因素。

照度即为单位面积内光通量的数目，是用光中较为复杂的环节。我们很多人认为照度高就好，其实不然，和暗一样，过亮的环境也会使眼睛疲劳。

色温即指光的冷暖，和人的心理健康关系密切。餐厅用暖色调，以更好衬托食物的色彩，增进食欲。卫生间用光偏冷，更清爽和干净。甚至对于同一个空间，随着季节的变换，要求也不一样，冬天卧室适用暖色调 3000K，夏天卧室适用偏冷色调 4000～4500K。

显色性评价光是否能真实反映物体的颜色(即在日光下的颜色)，对显色性要求有一定的绝对性，即 $R_a>85$，当然某些空间的要求相对可以低一些，如仓库，车库等。

表面亮度不仅和照度有关，还和物体表面的物理性质有关，同样的照度，浅色物体的表面亮度就高，表面亮度其实可以理解为照度在具体物体上产生的最终结果。

眩光是指较强的光直接射入人眼，引起强烈的不舒适，甚至暂时的失明，具有很大的危害性，如迎面而来汽车前灯所产生的光线，清晨或傍晚东西向窗户射入的直射日光。

背景光是一种辅助性照明手段，但直接影响到视觉舒适度，适当的背景光能够突出主题，营造气氛，但如应用不当，则往往造成喧宾夺主、人眼调节过大等问题。

1.2 热舒适性(照明部分)

照明设备在发光的同时，大量产生热量。这种热量主要以两种方式进行扩散，主要是对流方式，即设备产生的热量由其周边的气流带走，其次为辐射方式，即以热辐射的形式，直接使照射物体温度升高。这两种方式中，前者我们可以通过合理的空调设计，在照明设备附近设置回风口，使室内的热负荷及时排除。但对于热辐射而言，因为其具有定向强，而且是通过照射物体表面而发热，往往引起局部热舒适性失调。如在很多电梯厅和门厅处，处于照度和定向照明的考虑，多数采用碘钨灯(即石英灯)，即使在冬天，人在下面的时候都不舒服，感到头顶有灼热感，同时也过于耀眼。可以通过合理分布光源和光源选择来解决。

2 当前照明设计现状简析

考察目前的照明设计，专业性普遍不高，特别是照明配电负荷设计和最终的场地照明设计

制冷负荷

	Space Sens.+Lat. kW	Plenum Sens.+Lat kW	Net Total kW	Percent Of Total (%)
外部热负荷				
天窗日照	0.00	0.00	0.00	0.00
天窗热传导	0.00	0.00	0.00	0.00
屋顶热传导	0.00	105.69	105.69	0.96
玻璃日照	4,031.09	0.00	4,031.09	36.56
玻璃热传导	−66.70	0.00	−66.70	−0.60
墙体热传导	61.31	28.66	89.97	0.82
隔墙	0.00		0.00	0.00
暴露的楼板	0.00		0.00	0.00
空气渗透	0.00		0.00	0.00
小结	4,025.70	134.35	4,160.05	37.73
内部热负荷				
照明	1,175.97	293.99	1,469.96	13.33
人员	2,506.01		2,506.01	22.73
小电器	1,469.96	0.00	1,469.96	13.33
小结	5,151.94	293.99	5,445.93	49.39
吊顶负荷	−84.22	84.22	0.00	0.00
通风负荷	0.00	0.00	1,326.76	12.03
余量调整	0.00		0.00	0.00
排风热量		93.95	93.95	0.85
送风风机热量			0.00	0.00
回风风机热量		0.00	0.00	0.00
管道得热		0.00	0.00	0.00
再热系统			0.00	0.00
总热负荷	9,093.42	606.51	11,026.69	100.00

图1 照明负荷

往往由不同单位完成，相互之间缺乏沟通，往往造成配电容量过大，照明线路更无优化可言。

据资料显示，目前建成商业建筑中20%～30%左右的能耗是用于各种各样的照明电器的，可见人工照明对于能源的需求之大。而另一方面，照明电器在发光同时又附带产生了大量热量，拿白炽灯来讲，其输入电能的75%直接转变成了热能。这些热量增加了室内的热负荷，引起空调系统的装机容量增大，运行费用增高，另外照明电器产生的局部热传导和热辐射现象，也容易引起人的不舒适感。

以哈尔滨市中心的一个商业中心为例[1]，夏天大部分热负荷来自于室内，其中照明占到13%(图1所示，下划线部分为照明引起的热负荷)。

因此，在满足照明舒适度的前提下，降低照明配电负荷密度，不仅能使照明系统本身节能，而且能有效的降低空调系统的装机容量和今后的运行费用，可谓一举两得。

3 绿色室内照明策略

绿色照明强调对室内空间实施有效的照明，并尽可能利用天然光。以下对涉及绿色照明的技术和设计方法做一个简单的分析。

3.1 优化照明设计

3.1.1 合理分配照度，使用局部照明

照度的分布应尽量根据使用功能制定，不同场所，配合不同照度。如：中餐厅的平均照度为500lx，而西餐厅的平均照度在200lx即可。

对空间实施环境/工作/重点分层次照明，从而提供灵活的照明控制，降低环境照明水平，有利于节能。控制原则为——环境照明为作业照明的20%～50%，可以使人眼更为舒适，并提高对局部的注意力和敏感性。如：办公室的平均照度200lx即可，但局部工作区域可配备较高的照度500lx(如图2所示)。

图2 环境和局部照明

3.1.2 眩光的控制

照明设计时，应避免眩光的产生，如加设光源遮蔽装置，并控制一些镜面反射的产生在高视觉舒适性要求的空间可采用反射或反射直射混合的光源。

3.2 选用节能照明光源

应在满足实际功能及美观需求的前提下，选择适合的节能光源。举例来说，白炽灯(除却一些特殊的功用，比如频繁的开关，之外)完全可以用应用紧凑型荧光灯来替代，能源利用效率可以提高到原来的5～6倍。线性荧光灯也应从T8升级到T5，等等。

3.3 选用高效照明灯具

优先选用配光合理、效率高和保持率高的灯具，在保证照明质量的前提下，优先选用

---

[1] 该商业中心实施LEED™国际绿色建筑认证，并于2005年获得银级证书，以上照明负荷分析，为其实施绿色照明策略之前的模拟结果。

开启式灯具，少采用带有格栅、保护罩的灯具，并保证光线向有效部位集中。开启式灯具的效率不宜低于75%；带有灯罩的灯具效率不宜低于55%；带有格栅灯具的效率不宜低于60%。在具有高视觉舒适性的空间，可以采用反射类的灯具，但必须降低其环境光照度要求，实施重点区域照射的策略。

3.4 照明配电系统的优化

对照明的回路和布线应进行优化设计，其中分散补偿和优化配电方式易于实现，是减少线路损耗的必要条件。电子镇流器的应用可使系统光效比提高15%，节电20%以上，功率因数可达0.9以上，线路损耗会相应减少，另可进行调光，可配合昼光照明一起使用。

3.5 照明控制系统的优化

使用控制器是降低能耗的有效途径，可以利用被动红外线、超声技术、光感应、时间控制，或其中两者结合的感应器进行信号输入。它们可以安装在吊顶上或室外，控制较大面积，或与墙壁上的开关进行整合，控制较小区域。

3.6 避免光损失

光损失的方式有两种，主要是指人工照明系统的光损失，其一、从建筑开口部位向外泄漏，如窗户；其二、被深色的建筑表面吸收。对于光泄漏而言，不但无谓造成了室内照明能耗的浪费，有时也对室外环境造成了光污染。对于光损失，我们可以通过对开口部分加装光发射性好的物体，以及在室内物体表面采用浅色调的方式建议解决。当然这里有一个室内色彩设计的问题，需要协调解决。

3.7 昼光照明策略的引入

“昼光照明”在国内很少提，一般都误称为“日光照明”，其实两个概念很不一样，英文中，日光是Sunlight，而昼光是Daylight。为什么要区分，简而言之，因为日光非常强烈(直射阳光的照度往往在100000lx以上)，不能直接用于视觉照明。而昼光是经过弱化的阳光，在人视觉舒适的宽容度内。对于一些白天需要照明的空间，如办公室、教室而言，昼光照明具有重要的意义(如图3所示)。以下，我们将对其进行主题分析。

图3 Michigan大学法学馆
(建设设计：Gunnar Birkerts)

4 合理使用昼光照明

4.1 昼光照明特点

合理的引入昼光照明，可谓是加强和自然沟通的一种手段，不仅节能，而且在视觉舒适性上也是一种提高，甚至能够培养健康的心理。其特点分列如下：

- 合理引入昼光照明后，照明系统电耗将明显下降，最大潜力可降低到原来的50%；
- 制冷机组容量可降低到原来的93%～98%，运行费用也会明显下降；
- 昼光是全色系的光线，对人的眼睛最为友好，可以提高商品的吸引度，激发顾客潜意识的购买欲望；

- 有利于加强室内外的沟通，改善建筑的整体形象；
- 树立绿色建筑形象，有利于塑造发展商的品牌效应。

4.2 技术和经济分析

以下我们将结合哈尔滨市某商场的实际设计情况，从技术和经济上论证昼光照明实施的可行性。

结合《建筑采光设计标准》GB/T 50033—2001 及客户的要求，商场照度最终定为500lx，在模拟过程中，我们对建筑模型做了部分简化，输入了哈尔滨当地的实际气候情况。

4.2.1 确定昼光采光面积

立面采光面积确定的情况下，分别将顶光的采集面积定为 1%、2%、4%、6%、10%，通过软件对其投入产出效益比进行分析(如图 4 所示❶)：

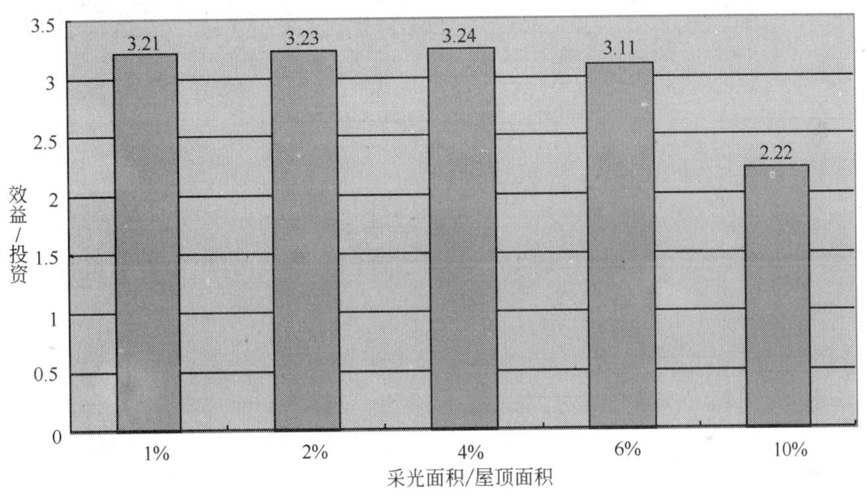

图 4 投资效益分析表

通过上面的经济分析，我们可以看到该建筑在顶光采集面积为屋顶面积的 4%～5%时，其效益是最优化的。这里需要说明的是，因为我们在具体采集昼光的时候，不宜采用直接引入的方式，而是采用反射的间接方式，由于一般采光板的发射率在 60%～70%，因此采光面积控制在 5.7%～8.3%，是一个比较合适的面积。下面是对原有设计和 4%面积采光的方案的经济比较。

4.2.2 昼光照明能耗分析

柱状图中(如图 5 所示)可以看到在采用顶部昼光照明的情况下，照明年能耗得以大大下降，同时用于制冷的能耗也有所下降，具体指标比较见表 1。

昼光照明节能经济分析 **表 1**

	照明系统(kWh/m²)	原/昼光设计	整个建筑系统(kWh/m²)	原/昼光设计
原有设计	171.2	43%	1248	92%
昼光照明设计	72.7		1145.3	

❶ 该分析相对量有效，绝对量仅具参考价值。

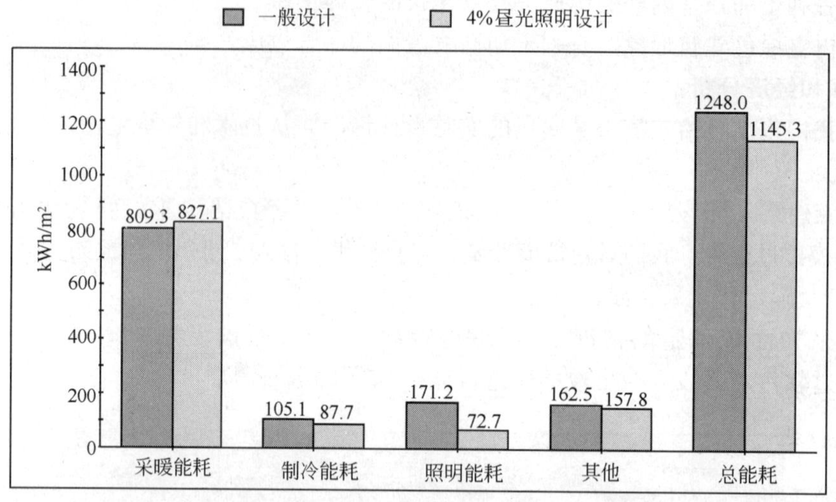

图 5 年能源消耗比较表

4.2.3 昼光照明对用电峰值的影响

如图 6 所示，采用昼光照明之后，每月的用电峰值也有所下降，特别是在夏季，这说明了昼光照明并未如我们原来预计的给室内带来更多的热负荷(以下章节将详细证明)。

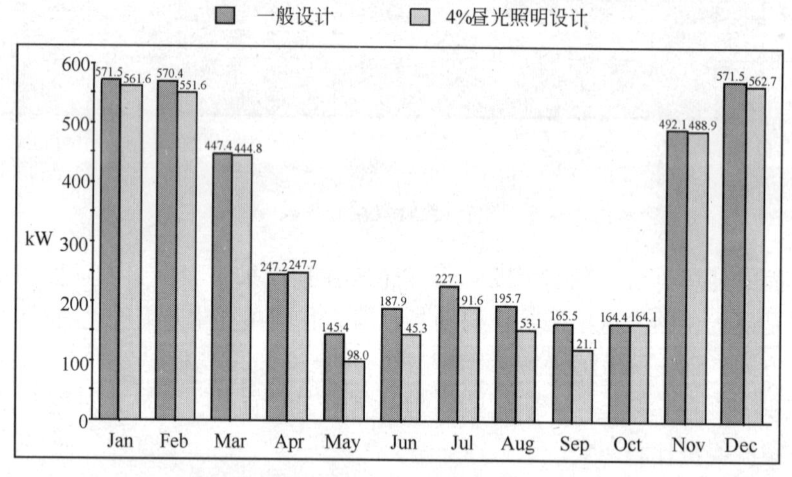

图 6 各月用电负荷峰值情况

4.2.4 昼光照明冷负荷分析

一般认为昼光照明会导致室内冷负荷增加，空调配置容量增大，其实未然。为阐明这个问题，现将模型扩展到项目的实际面积(58,000m^2)，在调整了相关设计参数后，建立以下四个模型：

(1) 930m^2 天窗，普通玻璃 U=3W/(m^2·K)，Sc=0.93；

(2) 无天窗；

(3) 930m^2 天窗，普通玻璃 U=3W/(m^2·K)，Sc=0.93，采用昼光照明；

(4) 930m^2 天窗，low-e 玻璃 U=1.36W/(m^2·K)，Sc=0.39，采用昼光照明。

灯光的开启时间 **表 2**

工作日			周末		
午夜 0 点	8 a. m.	0	午夜 0 点	8 a. m.	0
8 a. m.	9 a. m.	50	8 a. m.	9 a. m.	50
9 a. m.	9 p. m.	100	9 a. m.	9 p. m.	100
9 p. m.	10 p. m.	10	9 p. m.	10 p. m.	20
10 p. m.	午夜 0 点	0	10 p. m.	午夜 0 点	0

冷负荷分析比较表 **表 3**

	采光天窗面积(m^2)	天窗 W/(m^2·K)	天窗遮阳系数	天窗日光负荷(kW)	天窗传热负荷(kW)	人工照明冷负荷(kW)	总体冷负荷(kW)	年度能源费用(RMB)
方案一	930	3	0.93	589.54	24.8	1209.97	7,422	6,573,124
方案二		/	/	0	0	1209.97	6,848	6,404,072
方案三		3	0.93	589.54	24.8	732.44	6,983	5,895,896
方案四		1.36	0.39	211.18	12.29	738.24	6,636	5,779,968

注：电费取 0.71RMB/kWh，天然气取 16.07RMB/m^3。

从表 3 可以得出：当天窗安装普通玻璃时，其峰值冷负荷会高于没有天窗，但其年能耗还是优于无天窗方案。而当安装高性能玻璃时，无论年能源消耗，还是总体冷负荷，天窗设计的表现都优于无天窗方案。就冷负荷而言，两者相差 470kW 的负荷，按 RMB500 元/kW 冷量来计算，即可节省制冷机投资 RMB235,000 元。使用 Low-e 玻璃后，效果更好，由日光产生的冷负荷只有 211kW。

4.2.5 小结

从以上数据分析可以看到，在引入昼光设计后，虽然在采暖能耗方面有所增加，但综观全局：

- 照明系统的能源消耗；
- 整个建筑系统的能源消耗(如图 7 所示)；

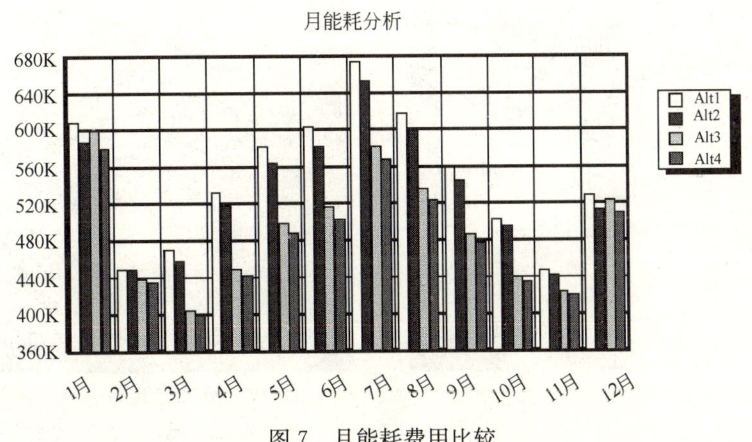

图 7 月能耗费用比较

- 整个空调系统(特别是投资较大的制冷机可减容 5%)的配置容量;
- 用电峰值。

整体效益是正面的，但要注意：在引入昼光照明时，采用的天窗的密闭性能一定要好，其 U 值也要控制在 $2W/(m^2 \cdot K)$ 以下，建议采用 Low-e 高性能玻璃。

4.3 昼光照明的设计方式

限于篇幅，这里只对昼光照明的设计作一个简单说明。其一般可考虑的措施包括调整建筑物朝向，使用浅色板、增加建筑周边、室内外永久遮阳装置、高性能玻璃和光感应器等。为了最大限度实现光线透过率，可将窗户安装在墙的较高位置，或使用高侧窗或反光板。将储藏室、厕所和其他使用率不高的房间安排在建筑的中心区，而经常使用的房间安排在建筑周围区域。天窗和采光屋顶可使用折光板漫射光线，减少眩光。如图 8 所示，某商场项目的昼光照明设计。

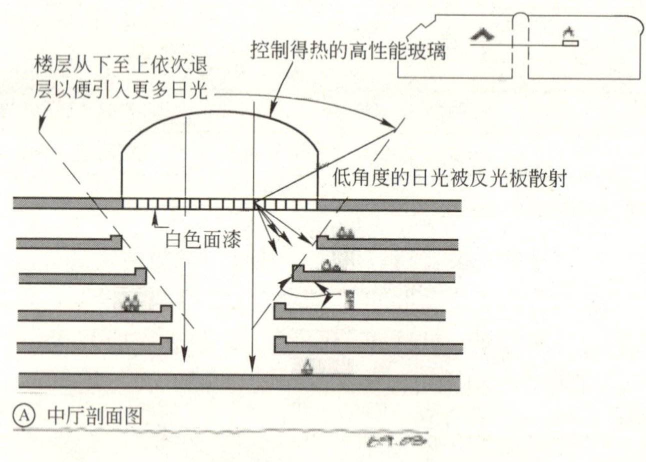

图 8 昼光照明设计图

反光板还可以作为遮阳板，同时结合遮阳来考虑，这样一来，不仅达到昼光照明的效果，使建筑深处得到较为充足的光线，而且能够避免近窗户处光线过量，引起炫光(如图 9、10 所示)。在设计过程中，最好能够事先通过计算机模拟确定反光板(遮阳板)的最佳安装位置、角度和宽度。

图 9 不考虑昼光照明——局部过亮

图 10 考虑昼光照明——设置反光板和遮阳板，室内光线柔和明亮

根据日照强度情况控制人工照明是昼光照明的重要措施之一，在节能的前提下，提供极好的视觉舒适。当日光充足时，光感应器发出的信号会关闭人工照明或调暗光线。此外，控制器也可以和建筑整体能耗系统整合，在能耗高峰时段降低照明负荷。感应器应该根据生产商的建议进行安置并进行调整，这样日光的短时间变化(如偶然经过的云朵的影响)不会启动感应器。

Daylighting，Lighting For Green Building

Qian Yingchu

(Environmental Market Solutions，Inc. Shanghai 200031 China)

Abstract: Based on the analysis of current general lighting design，this paper makes it clear we are facing the big challenge from both Energy Saving and Comfort. The Green Lighting Design could be the solution. To make the article more complete，some green lighting design methods are referred first，then Daylighting，which is one of the most important ways to be energy saving and make the building more green. Some case studies are given to help readers to understand how to design this system. Furthermore，with detailed information and computer simulation，the paper proves daylighting is the effective way to add values for the building.

参考文献

[1] ANSI/ASHRAE/IESNA Standard 90. 1-2004，American Society of Heating，Refrigerating and Air-Conditioning Engineers，Inc. 2004

[2] R. Sullivan，L. Beltran，E. S. Lee，M. Rubin，and S. E. Selkowitz. Energy and Daylight Performance of Angular Selective Glazings，1998

[3] Jennifer O'Connor. TIPS FOR DAYLIGHTING

[4] 张绍纲主编. 绿色照明工程实施手册. 北京：中国建筑工业出版社，2003

[5] 杜异主编. 照明系统设计. 北京：中国建筑工业出版社，1999

[6] 房志勇主编. 建筑节能技术教程. 北京：中国建材工业出版社，1997

住宅建筑节能设计的经济性方案分析

曹毅然　范宏武　刘明明　李德荣　张蓓红

（上海市建筑科学研究院，上海　200032）

摘　要： 为了分析住宅建筑节能设计在满足节能标准要求的同时，如何花费最少的成本，能达到同样的节能设计效果，以上海地区的典型住宅为建筑物理模型，采用建筑物能耗分析软件 DeST 进行动态模拟。并对结果进行综合分析，根据计算结果，得出上海地区住宅建筑节能设计的经济性优化方案。

关键词： 节能　设计　经济性　优化

上海第一个节能评估机构——上海建科建筑节能评估事务所自 2003 年 12 月成立至今，已完成了 100 多万 m^2 的节能评估案例。在这些项目的评估过程中，笔者发现这样一个具有共性的现象：许多开发商在做节能项目时，可以选择多种节能设计方案，这些方案都能达到国家节能标准[1]的要求。但是作为开发商来讲，在达到同样节能效果的前提下，他们需要一个成本最低的设计方案，所以从住宅建筑节能设计的经济性角度出发，来探讨建筑节能的问题，是非常必要的。

1　基本概念

在《夏热冬冷地区居住建筑节能设计标准》JGJ 134—2001 中明确提出夏热冬冷地区的节能设计的强制性要求，围护结构各部分的传热系数｛K［W/（m^2·K）］｝和热惰性指标如表 1、表 2 所示。

围护结构各部分的传热系数　　表 1

屋　顶	外　墙	外　窗	分户墙和楼板	底部自然通风的架空楼板	户　门
K≤1.0 D≥3.0	K≤1.5 D≥3.0	按表 2 的规定	K≤2.0	K≤1.5	K≤3.0
K≤0.8 D≥2.5	K≤1.0 D≥2.5				

当设计的居住建筑不符合上述表中的各项规定时，则必须计算和判定建筑物节能综合指标。建筑节能综合指标应采用动态方法计算。

从基本概念可以看出，上海地区判断建筑物节能设计是否达标可以有两种方式，一种是各项完全满足规定性指标要求，一种是可以允许若干项不满足规定性指标要求，但是通过动态计算的综合性指标必须达到要求。

不同朝向、不同窗墙面积比的外窗传热系数 **表 2**

朝 向	窗外环境条件	外窗的传热系数 K[W/(m^2·K)]				
		窗墙面积比≤0.25	窗墙面积比>0.25且≤0.30	窗墙面积比>0.30且≤0.35	窗墙面积比>0.35且≤0.45	窗墙面积比>0.45且≤0.50
北(偏东60°到偏西60°范围)	冬季最冷月室外平均气温>5℃	4.7	4.7	3.2	2.5	—
	冬季最冷月室外平均气温≤5℃	4.7	3.2	3.2	2.5	—
东、西(东或西偏北30°到偏南60°范围)	无外遮阳措施	4.7	3.2	—	—	—
	有外遮阳(其太阳辐射透过率≤20%)	4.7	3.2	3.2	2.5	2.5
南(偏东30°到偏西30°范围)		4.7	4.7	3.2	2.5	2.5

2 影响因素

根据项目评估的情况可知影响建筑物节能效果的主要为屋顶、外墙和外窗这三块。所以在影响因素中主要对这三块作主要分析。

2.1 屋顶

上海地区的坡屋顶节能设计方案一般分为结构层、找平层、保温层、防水层、保护层；平屋顶设计方案一般为结构层、找坡层、找平层、保温层、防水层、保护层。在这两种设计方案中，保温层设置的厚薄对屋顶的 K、D 值影响很大。屋面的保温层材料一般为 XPS(挤塑聚苯板)或聚氨酯。

2.2 外墙

上海地区的外墙节能设计方案一般分为结构层、找平层、粘结层、保温层、防护层、饰面层。外墙保温的应用体系一般分为以下几种：膨胀聚苯板(挤塑聚苯板、矿棉板)薄抹灰系统、胶粉聚苯颗粒保温浆料系统、砂加气块系统、聚氨酯外墙保温板系统。在设计中，保温系统的材料选择和厚薄设置都对外墙的 K、D 值影响很大。

2.3 外窗

上海地区的外窗节能设计方案一般采用铝合金中空玻璃窗、断热铝合金中空玻璃窗、塑钢中空玻璃窗、断热铝合金 Low-E 中空玻璃窗这四种形式。其中采用比较普遍的为铝合金中空玻璃窗和断热铝合金中空玻璃窗这两种形式。

3 综合分析

在实际的项目评估过程中，实际建筑节能设计方案中围护结构各部分单项指标总会存在一项或几项不能达到国家标准要求，所以需要通过计算建筑节能综合指标来进行综合分析。这时候，对于节能设计方案优化来讲，如何组合好屋顶、外墙、外窗的节能设计方案就显得尤为重要了。下面通过一个实例来进行具体分析。

3.1 项目简介

具体分析的实例是同济大学建筑设计研究院设计的位于上海市区的一个高层楼盘。由于建筑师为了追求建筑的立面效果，所以该楼盘南向和东向的窗墙比设计的都突破了规范的要求，所以开发商特地委托上海建科建筑节能评估事务所为该楼盘设计一个性价比较高的节能设计方案。通过对建筑设计方案的技术解读，得出该楼盘的基本信息如下：该楼盘为剪力墙结构，33 层高的条式建筑。外墙准备采用膨胀聚苯板作为外保温材料。屋面准备采用挤塑聚苯板作为保温隔热材料。而外窗的选取则满足综合评估达标即可。四个朝向的窗墙比分别如下所示：南向：0.50、北向：0.32、东向：0.35、西向：0.11。该楼盘的建筑节能计算分析的模型如图 1 所示。

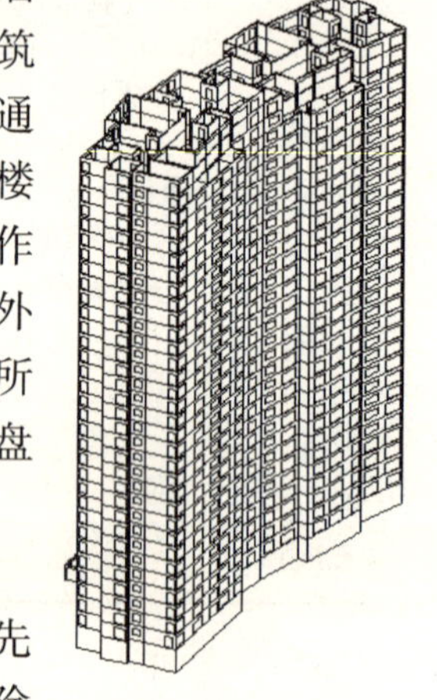

图 1

3.2 优化思路

根据项目的基本信息可以得出如下的节能设计优化思路。首先可根据外墙、屋面、外窗在建筑面积中所占的比例采用排除法排除一项对能耗影响较小的因素。由于该楼盘是高层建筑，屋面在整个建筑面积中所占的比例很小，所以可以首先确定屋面的节能设计方案，通过热工参数的计算分析，使得屋面的热工设计指标满足标准要求即可，不用超出很多。确定完屋面的节能设计方案后，接下来就可以对外窗和外墙作综合分析了。根据开发商的要求，窗框必须用铝型材，所以外窗就有三种选择：普通铝合金中空玻璃窗、断热铝合金中空玻璃窗、断热铝合金 Low—E 中空玻璃窗。而外墙的保温材料厚度可以根据不同的外窗选项来选择。最后可以通过综合比较来确定哪种组合为经济性最优组合方案。

3.3 优化方案

根据标准要求和初步热工计算可得到如下几组窗墙组合方案：*A*. 南、北、东向外窗采用断热铝合金 Low-E 中空玻璃窗、西向外窗采用普通铝合金中空玻璃窗、外墙采用 20mm EPS 外保温。*B*. 南、北、东、西向外窗采用断热铝合金中空玻璃窗、外墙采用 25mm EPS 外保温。*C*. 南、北、东、西向外窗采用普通铝合金中空玻璃窗、外墙采用 35mm EPS 外保温(实际方案还可根据委托方要求进行个性化调整，如厨卫外墙不做保温等)。

A 方案是窗墙指标都符合标准中的规定性指标要求，所以该方案不用再进行综合计算了。B 方案通过建筑物能耗分析软件 DeST 进行动态模拟后得到整栋建筑的能耗值为 44.5、而标准参照物的整栋建筑的能耗值为 45.3。根据对比评定法[2]可知 *B* 方案综合能耗指标符合标准要求。*C* 方案通过建筑物能耗分析软件 DeST 进行动态模拟后得到整栋建筑的能耗值为 44.1，而标准参照物的整栋建筑的能耗值为 45.3。根据对比评定法可知 *C* 方案综合能耗指标符合标准要求。三个方案的经济性比较如表 3～5 所示。

方　案　一　　**表 3**

类　　别	市场价格	施工面积	总　　价
断热铝合金 Low-E 中空玻璃窗	550 元/m^2	5326.53m^2	2929591.5 元
普通铝合金中空玻璃窗	300 元/m^2	176.88m^2	53064 元
20mm EPS	450 元/m^3	9392.79m^2	84535.11 元
			3067190.61 元

方　案　二　　**表 4**

类　　别	市场价格	施工面积	总　　价
断热铝合金中空玻璃窗	400 元/m^2	5503.41m^2	2201364 元
25mm EPS	450 元/m^3	9392.79m^2	105668.89 元
			2307032.89 元

方　案　三　　**表 5**

类　　别	市场价格	施工面积	总　　价
普通铝合金中空玻璃窗	300 元/m^2	5503.41m^2	1651023 元
35mm EPS	450 元/m^3	9392.79m^2	147936.44 元
			1798959.44 元

由以上三个方案可以看出在同样达到国家节能标准的前提下，方案三比方案一和方案二的经济性成本更低。

4　结论

根据上文所述，可得出如下结论：

（1）一个建设项目在满足国家建筑节能设计标准的前提下可以有好多种节能设计组合方案。

（2）通过专业软件的模拟计算，专业节能评估机构可以为开发商设计出性价比更优的建筑节能设计方案。

Analyse on Economical Design Approach of Residence Energy Efficiency

Cao Yiran Fan Hongwu Liu Mingming Li Derong Zhang Beihong
(Shanghai Research Institute of Building Sciences, Shanghai 200032 China)

Abstract: this article analyzes how to optimize the design and invest the lowest cost about the design of residence energy efficiency. In this article, we take typical Shanghai residence as calculational model, run dynamical simulation with DeST software and analyze the simulation results. In the end, we get the optimization about the design of residence energy efficiency of Shanghai region.

参考文献

[1] 《夏热冬冷地区居住建筑节能设计标准》(JGJ 134—2001)
[2] 卜震，曹毅然，范宏武，陆善后，李德荣 . DeST-h 住宅节能评估软件在节能建筑计算与评估中应用的二个工程实例 . 上海建材发展与应用，2003，(3)

新型太阳墙内自然对流换热分析

杨 昭 徐晓丽

（天津大学热能研究所，天津 300072）

摘 要：兼有被动冷却和加热的复合太阳墙从全年运行看：具有节能降耗、改善空气品质、提高舒适性的优点符合亲和自然的绿色生态建筑理念，是一极具发展潜力的太阳能墙体，适宜深入研究和大力推广。

本文基于层流和RNG、RELIABLE的K-ε紊流模型不同工况（Gr数）下空气夹层中的对流换热进行数值模拟。通过与文献中的Nu实验关联式比较，表明所选计算模型合理，适宜对太阳墙内的对流换热进行分析和预测。

关键词：太阳墙 CFD 准则方程 生态建筑

前言

如今，节能已经成为世界性问题。同时由于现代建筑中“病态建筑”的出现，越来越多的人们崇尚自然通风。太阳能建筑以其节能、亲和自然的特点成为研究热点。传统特朗勃集热墙体的主要缺点是：厚重集热墙表面温升较慢，进而集热时滞性大；冬季夜间，温度较高的集热墙在向室内散热的同时向室外传热而导致热损失较大。夏季由于集热墙表面温度较高，室内易产生过热现象。新型特朗勃集热墙通过对结构和建筑材料的改进，有效的克服了传统特朗勃墙的缺点，作为被动式太阳能建筑的主要形式，极具发展潜力。随着建筑主体材料的更新，需要对其热过程进行较详细的分析，以进行优化设计。近年来，CFD技术在太阳能建筑应用的可靠性已得到验证[1]。本文通过建立三维模型，研究了新型集热墙内的自然对流换热。

1 模型结构

夏季，复合特朗勃墙可以看作是一个太阳能烟囱(图1)。在浮力的作用下，室内的空气从风口6吸入金属板和厚重墙体间的空气间层内，而房屋北侧较冷的空气从风口7进入室内。这样，室内由于自然通风而被冷却。在室外室温较高不适宜开启风口7时，室外热空气直接从空气间层下部进入加热后由上部排出而不进入室内以防止过热。冬季，金属板与厚重墙体间的空气吸收太阳能被加热后在浮力的作用下向上流动从上风口流入室内(1、2关，3开)，室内冷空气从风口6流入空气间层，形成一个循环。此外，在过渡季或室外温度不是很低的情况下可将风口4、5打开，6关闭，室外的新鲜空气向上流动通过风口3进入室内，在一定程度上，复合特朗勃墙可以提高室内空气的 IAQ 且有利于节能。

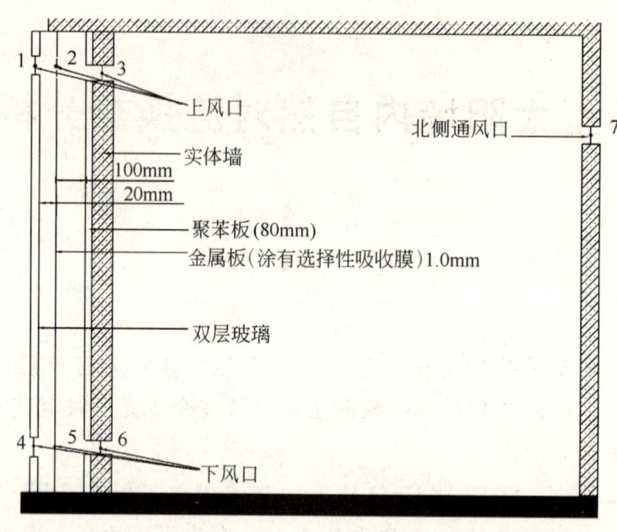

图 1 带特朗勃壁太阳房结构图

2 数学模型

关于新型太阳墙的动态热响应和保温、隔热、预热效果已在另一文中介绍。本文主要研究空气夹层中的对流换热，同时以实验准则方程进行验证和校核。

针对新型太阳墙的不同功效，通道形式可分为：封闭空间通道(用于冬季作为热缓冲层)和耦合壁面开口形式。后者又可分为同侧开口(冬季房间加热和夏季通道隔热)和异侧开口(过渡季空气预热和夏季通风冷却)两种形式。

2.1 建立模型为了准确地分析特朗勃壁内的温度场分布，建立了三维结构模型。在处理中，特将一部分室内空间做为计算区域，以便较真实地反映通风口的热状况

在模型中做了如下假定：

(1) 由于室内的湿度变化不是很大，忽略湿度变化对热平衡的影响。

(2) 忽略各层间的接触热阻。

(3) 材料热物性视为常量。

(4) 忽略空气渗透换热作用。

(5) 应用 Boussinesq 模型。

2.2 控制方程

基于流固耦合求解

(1) 流体区域两方程模型为：

$$\frac{\partial(\overline{\rho\phi})}{\partial t}+\frac{\partial(\rho\overline{u_i\phi})}{\partial x_i}=\frac{\partial}{\partial x_i}\left(\Gamma\frac{\partial\overline{\phi}}{\partial x_i}-\rho\overline{u_i'\phi'}\right)+S \tag{1}$$

流体区域源项：在方程中添加 $C_3\rho\beta g\frac{\mu_t}{\sigma_t}\frac{\varepsilon}{k}\left(\frac{\partial T}{\partial x_j}\right)$ 和 $\rho\beta g\frac{\mu_t}{\sigma_k}\left(\frac{\partial T}{\partial x_j}\right)$ 为源项。即考虑动量

方程中浮力的影响。

（2）固体区域传热方程：

$$\rho c\frac{\partial T}{\partial t}=\lambda\left(\frac{\partial^2 T}{\partial x^2}+\frac{\partial^2 T}{\partial y^2}+\frac{\partial^2 T}{\partial z^2}\right)+S_E \tag{2}$$

固体区域源项：S_E 为能量体积源项。在本模型中为各固体介质层所吸收的太阳辐射热量和辐射换热量。

吸收太阳能辐射量为：

$$Q_r=\alpha A(\tau_b I_b+\tau_d I_d) \tag{3}$$

基于能量平衡原理，计算各介质层的太阳能辐射吸收率[3]。

各表面间的辐射换热量为：

$$Q_k=\varepsilon_k A_k\sum_{j=1}^{N}G_{ik}\sigma(T_i^4-T_K^4) \tag{4}$$

在本文中，由于主要研究对流换热，所以，假定壁面相对辐射系数很小，忽略辐射换热的影响。

（3）对于耦合外壁面开口的通道形式在进口处需考虑风压的作用，满足伯努力能量方程，即：

$$p_0+\frac{a\rho u_0^2}{2}=p_1+\frac{a\rho u_1^2}{2}+\Delta p_{loss} \tag{5}$$

近壁面处理：考虑到低 Re 数流动，在模拟中选用几种不同分模型分别进行模拟，其中，以增强壁面函数做近壁面处理，精度较高。

3 结果分析

3.1 封闭空间通道

对于封闭空间通道形式中的换热，参考文献关于流场的分析可以看作是等壁温有限空间对流换热。本文为了选取合适的计算模型，分别选用可 k-ε 模型及改进的适宜于低 Re 数的 Reliable k-ε、RNG k-ε 方程进行模拟；壁面处理分别选用标准壁面函数、不平衡壁面函数和增强的壁面函数法。计算结果显示，选用不同的模型和方法对计算结果有很大的影响（见图 2）。为验证所选模型的正确性，选用有限空间内的自然对流准则方程进行校核。[3][8]

$$Nu=0.197(Gr_\delta \mathrm{Pr})^{1/4}\left(\frac{H}{\delta}\right)^{-1/9} \tag{6}$$

$$Gr_\delta=8.6\times10^3 : 2.9\times10^5$$

$$Nu=0.073(Gr_\delta \mathrm{Pr})^{1/3}\left(\frac{H}{\delta}\right)^{-1/9} \tag{7}$$

$$Gr_\delta=2.9\times10^5 : 1.6\times10^7$$

$$Nu=0.23A^{-0.131}Ra^{0.269}$$

$$10^3\leqslant Ra\leqslant5\times10^6$$

$$4.9\leqslant A\leqslant78.7 \tag{8}$$

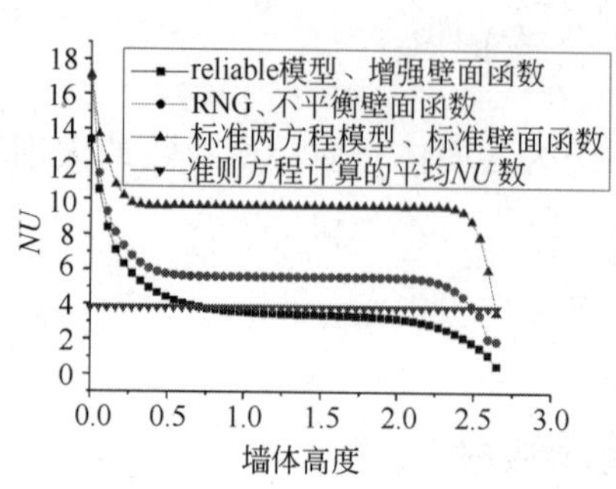

图 2 不同模型和方法下的壁面

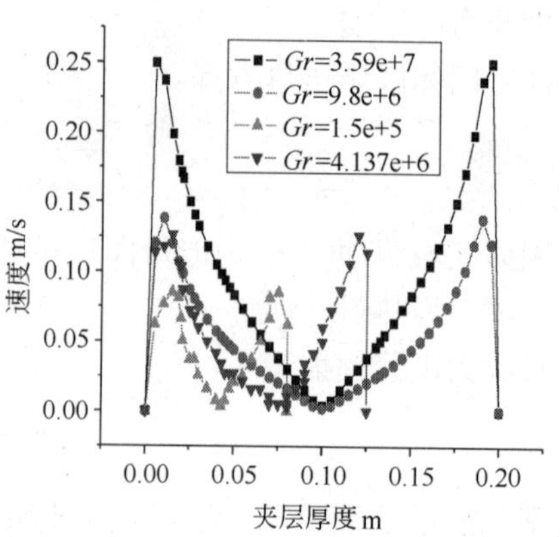

图 3 不同 Gr 下通道横断面速度分布

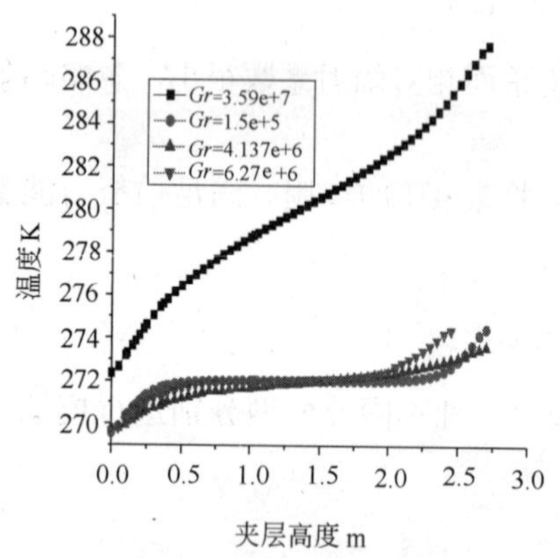

图 4 不同 Gr 下通道纵断面温度分布

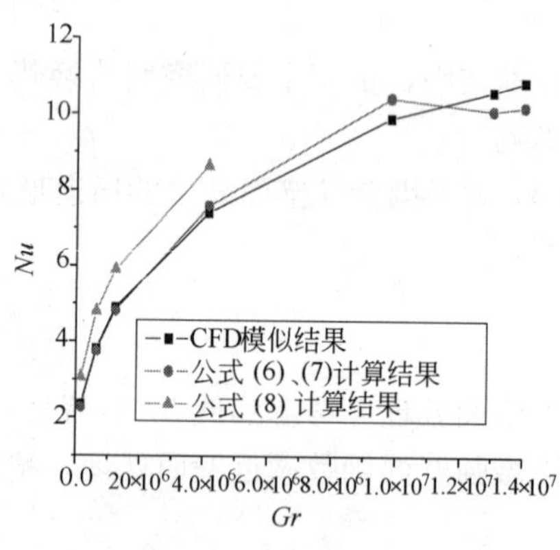

图 5 模拟计算与关联式计算

由图 2 可知，对于有限空间内的自然对流换热采用标准的两方程模型，其计算结果误差较大。壁面 Nu 数是准则方向计算结果的 2.63 倍；而对于近壁面区域的处理也尤为重要，选用不同的壁面函数，Nu 数值差别约 1.7 倍。Reliable k-ε 方程及增强的壁面函数进行模拟，计算结果与准则方程的计算结果最大偏差为 17%（见图 5），且在低 Gr 数时，精度更高。故其模拟结果可信。对有限空间自然对流横截面速度场分析可以看到：换热主要发生在壁面处，中间部分的流体基本处于静止状态。随着 Gr 值的增加，速度随着 Gr 增加而增加且 Nu 数也增加，热交换效果增强。图 4 所示的纵截面温度分布图也说明了这个问题。Gr 较小时，截面温度分布曲线变化平缓，只在靠近壁面处由于黏性作用，壁面温度梯度较大，而在中间部分温度几乎保持恒定。而 Gr 较高时，纵截面温度近似呈指数分布。

3.2 耦合壁面开口

对于耦合壁面开口的太阳墙有限空间内的对流换热分析，可以认为是等热流边界条件作用下的自然对流换热。

1. 异侧开口：以过渡季太阳墙的预热作用进行分析。

由图 6、图 7 看到：入口处速度分布很均匀，随着热气流向上运动，速度随热流密度和空气夹层高度的增加而增加，由于粘性作用，壁面处速度为 0。基于质量守恒定律，在固定热流密度下，不同截面上速度积分为一定值。表面温度变化与速度变化趋势相同。在金属板附近由于层流底层影响，温度达到最大值。在固体区域内，温度分布呈现线性分布，这主要由于墙体的高度是厚度的5倍多，传热以一维导热为主。不同的温度线斜率表示了不同的热传导量。夹层温度随热流密度的增加而增加。

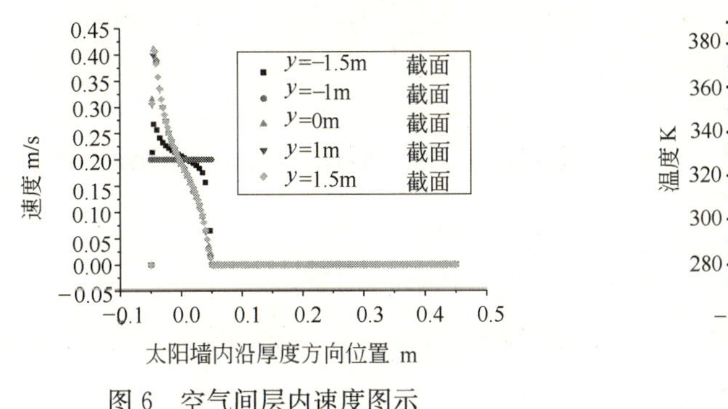

图6 空气间层内速度图示

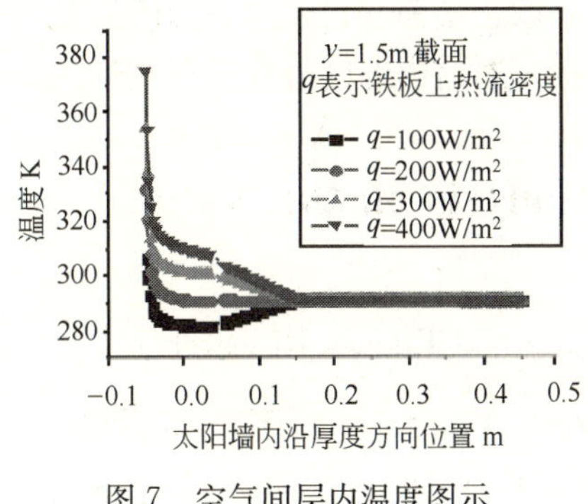

图7 空气间层内温度图示

2. 同侧开口(以冬季太阳墙的被动集热功能进行分析)

参考文献[9]对流换热系数关联式为(图9)：

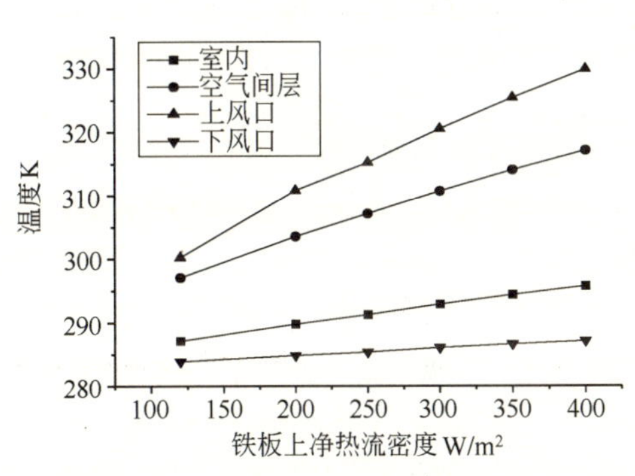

图8 冬季新型太阳墙及室内温度

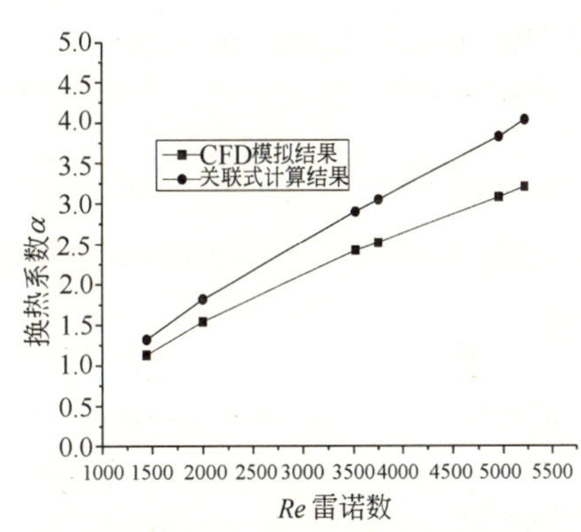

图9 模拟计算与关联式计算

$Re<2000$ 时：

$$\alpha_{c1}=\left[4.9+\frac{0.0606\left(\dfrac{RePrD_h}{h_v}\right)^{1.2}}{1+0.0856\left(\dfrac{RePrD_h}{h_v}\right)^{0.7}}\right]\frac{\lambda_g}{l_g}$$

当 $Re>2000$ 时：

$$\alpha_{c1}=0.0158Re^{0.8}\frac{\lambda_g}{l_g}$$

可以看到，模拟值与关联式计算值的偏差最大为25%，考虑模型假设带来一定的误差，认为模拟结果可信。

4 结论

(1) 对于新型太阳墙夹层内流动属于低 Re 数流动，选用标准两方程计算时会带来较大的误差，宜采用 reliable 型两方程模型，壁面宜采用增强壁面函数法。

（2）对新型太阳墙的换热分析可得知：新型太阳墙内通过不同的风门组合，可实现集热及通风功效。

（3）通道内的自然对流换热主要集中在壁面处。封闭式通道中间空气层几乎处于静止状态；耦合壁面开口通道内速度分布与壁面的热流密度有关，满足能量守恒原理。

（4）通道内温度分布呈指数形式。

Analysis on the Natural Convection in the New Style Solar Wall

Yang Zhao Xu XiaoLi

(Thermal Energy Research Institute, Tianjin University, Tianjin 300072 China)

Abstract: From the yearly operation, the composite solar wall with passive heating and cooling thermal performance has the advantages of comfortable、high AIQ、energy saving. It is such a ecological、potential architecture that suitable for popularization and thoroughly study. Based on the laminar flow model and the RNG、RELIABLE K-ε turbulent model, numerical simulations were made under different conditions (the value of Gr). The simulations were verified reasonable for analysis on the natural convection and the thermal performance of solar wall by comparing with the Nu correlations from the literature which are based mainly on experimental data.

参考文献

[1] R. Letan, V. dubovsky, G. ziskind. Passive ventilation and heating by natural convection in a multi-story building. Building and Environment 2003; 38: 197～208

[2] Heinrich Manz. Total solar energy transmittance of glass double facades with free convection. Energy and building 2004(36): 27～136

[3] S. H. Yin, T. Y. Wung, K. Chen, Natural convection in a enclosed within rectangular cavities, Internatioanal Journ Mass Transfer 1978(21): 307～315

[4] 严启森、赵庆珠．建筑热过程[M]．北京：中国建筑工业出版社．1986 年：P1～P99

[5] 李元哲、狄洪波．被动式太阳房原理及设计[M]．北京：能源出版社．1989 年：P1～P122

[6] 涂逢祥．建筑节能[M]．北京：中国建筑工业出版社，2004，(42)

[7] 陈启高．建筑热物理基础[M]．西安

[8] 陶文铨．传热学[M]．北京：高等教育出版社，1998

[9] 王德芳．集热墙不稳定传热问题的反应系数解法．太阳能学报．1990 年，第四期：P363～P36

EnergyPlus 建筑能耗模拟与实例分析

鄢　涛　卜增文　刘俊跃　马晓雯
（深圳市建筑科学研究院，深圳　518057）

摘　要： 介绍了能耗模拟程序 EnergyPlus 的主要功能和特点。以深圳地区为例，分析了 IWEC 气象数据中温度和太阳辐射值偏差对能耗模拟的影响。通过对比能耗模拟值与实测值，对采用该软件进行建筑能耗动态模拟的适用性做出初步分析，并提出了部分建议。

关键词： EnergyPlus　能耗　动态模拟

1　概述

上世纪最后 20 多年里，美国能源部资助的两个建筑能耗逐时模拟程序—BLAST 和 DOE-2 在世界范围内得到了广泛应用，其中 DOE 软件据估计已经为人们节约了 200 亿美元[1]。但是随着时间的推移，人们发现进一步扩展它们的功能变得异常困难。基于上述原因，美国能源部联合了美国陆军建筑工程研究实验室、伊利诺斯大学、劳伦斯伯克利国家实验室等机构，从 1996 年开始开发一个新的建筑能耗软件 EnergyPlus。这个软件集合了 BLAST 和 DOE-2 大部分特性和功能，于 2001 年 4 月发布使用。

EnergyPlus 被认为是新一代能耗模拟软件，但与其他大型能耗模拟软件一样，依然存在一些需要完善的地方，比如：界面不太灵活，系统节点输入不可视，导致用户输入时容易出错；定义每面墙时需要输入所有角点的坐标，且对于分隔不同热区的内墙或楼板等需输入两次，而 DOE-2 围护结构的输入就比 EnergyPlus 简洁；不能模拟系统环路压力变化；复杂的系统模拟比较耗时，且易发散；提供的系统模板尚不全面；设备性能曲线定义较复杂等。

2　软件气象数据简析

EnergyPlus 中国的气象数据采用的是国际能耗计算气象数据 IWEC（International Weather for Energy Calculation）。IWEC 包括了众多国际城市的数据，主要适合于逐时能耗模拟软件（DOE-2，BLAST 和 EnergyPlus），可用来模拟建筑典型年能耗量。IWEC 不适合于模拟风能等可再生能耗系统，因为 IWEC 在选取典型月时着重强调的是干球温度和太阳辐射，风速等参数则无法保证其典型性。

下面以深圳为例，分析 EnergyPlus 模拟深圳建筑能耗的适用性。由于 IWEC 中没有深圳的数据，要模拟深圳的建筑能耗，必须解决气象数据的问题。深圳毗邻广州，而国际气象数据库中有广州的数据，因此可以考虑利用广州的气象资料代替深圳地区气象资料。下面通过对比IWEC广州的部分气象资料与深圳地区气象台气象参数，判断这种代替引起的误差有多大。图1为深圳地区近30年来的干湿球温度月平均值与IWEC气象数据的对比情况。4～10月干球温度二者平均每月相差0.01℃；其余月份干球温度平均每月相差1.55℃。从对比的结果来看，4～10月的温度二者相差非常小；而1、2、3、11、12月5个月二者差别相对较大，但深圳市空调季节的室内外温差只有5～6℃，温差传热对能耗的贡献并不大，且这一段时间恰好为深圳的非空调季节。在深圳非空调期居住建筑和大部分公用建筑不需要供暖，全年能耗构成中与气候有关的只有空调通风系统能耗，照明、办公设备等基本与室外气候无关。故非空调期的温度偏差对全年的能耗模拟而言影响很小。因此对于深圳地区的大部分建筑，利用IWEC广州数据中温度数据可以满足该地区全年能耗模拟的要求。

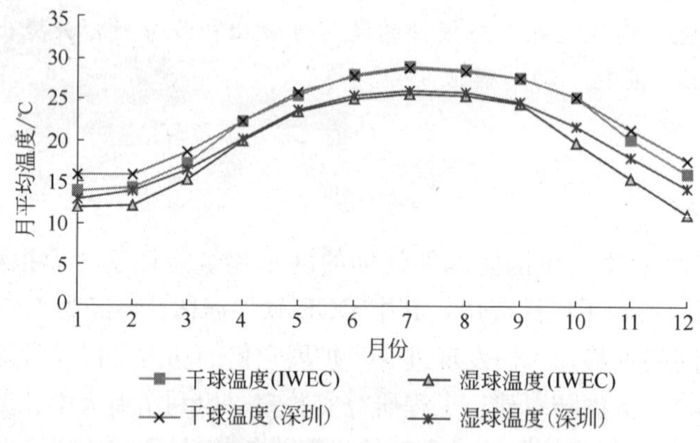

图1 IWEC广州月平均温度与深圳地区累年月平均温度对比

图2为IWEC广州数据的太阳直射与太阳散射图，全年散射分布较均匀，而夏季的几个月其太阳辐射值特别是直射值则明显偏低。对比广州累年实际辐射数据，IWEC广州数据与实际偏差较大(DOE-2广州气象数据类似)。

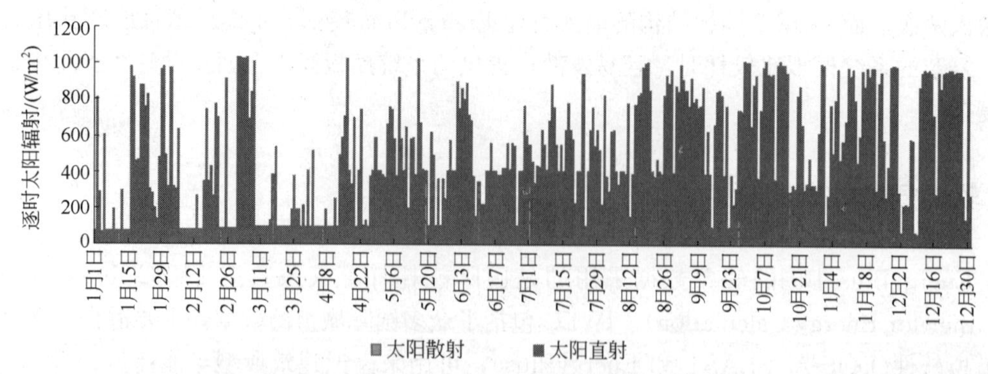

图2 IWEC广州辐射逐时变化曲线

如图3，对比深圳与IWEC广州辐射数据，全年平均每月相差12%，4～10月7个月平均每月相差15%。如果采用IWEC广州数据，模拟结果与理想值必存在差偏差，可通过增加遮挡物使建筑物完全接受不到太阳辐射，对比加遮挡前后模拟的能耗结果，初步考

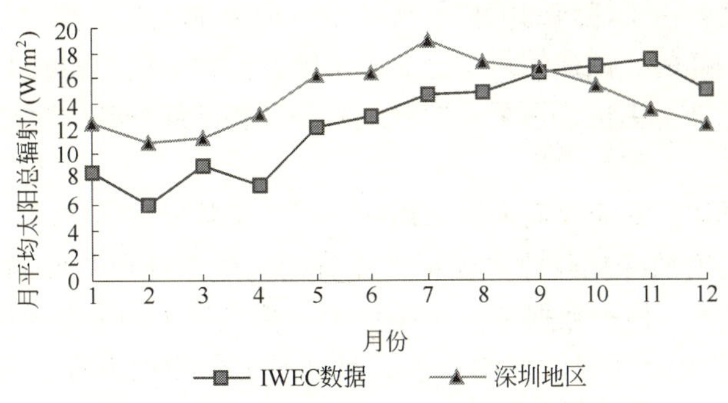

图 3　IWEC 广州总辐射值与深圳地区总辐射对比

察太阳辐射对能耗的影响。

表 1 为某政府办公楼加遮挡前后 8 月能耗模拟对比结果。完全消除太阳辐射的影响后空调系统能耗整体减少 17%。因为模拟采用的辐射数据比深圳实际值偏小 15%，由此引起空调能耗偏低约 3%左右。对于窗墙比较大、遮阳不好，而墙体 *K* 值又比较小的建筑，太阳辐射在建筑冷负荷构成中会占据更大的比例。对于这类建筑，因 EnergyPlus 气象数据的偏差，引起空调系统耗电量减少量会大于 3%，但在 10%范围以内。通过更改气象文件中的太阳辐射数据后重新模拟计算，结果类似。在使用 EnergyPlus 进行节能设计时，该偏差引起的误差基本可以忽略，因为设计建筑与参考建筑使用了同一气象数据，绝对值的意义不大。当需要进行某些精确的模拟，对结果的绝对值要求较高时，太阳辐射的数据必须进行修正。EnergyPlus 提供了相关的气象数据转换器，当然对选用的其他气象数据如 TMY 气象数据依然存在相同问题，也需要进行适当修正[3]。

办公楼加遮挡前后 8 月能耗模拟结果对比　　　　**表 1**

项　　目	冷水机	水泵	末端及冷却塔
未加遮挡能耗(万 kWh)	1.6	0.7	0.35
加遮挡能耗(万 kWh)	1.2	0.7	0.3

3　A 建筑模拟与实测对比实例分析

3.1　建筑与模型概况

2003 年深圳建筑科学研究院实测了深圳某政府办公楼 A 大楼。大楼建筑面积 8625m^2，空调面积 7300m^2，常年作为办公楼使用，采用新风加风机盘管系统。建筑外形如图 4。室内人员活动规律、设备使用规律比较稳定，大部分房间使用功能相同，且房门全天基本为开启状态，通过实测发现，走道与大部分房间温差很小，为系统分区简化提供了可能。对于使用功能单一、系统简单的建筑，分区时可以进行一些简化，特别是对于强调建筑总体能耗的模拟，甚至可以只分一个区。尽量少分

图 4　A 建筑模型图

区可降低模拟的难度，由于建筑的围护结构不变，内部的人员、照明、设备等设定不变，故总体能耗水平是一样的。该建筑每层分为六个区，围护结构的输入中同时考虑了窗户的遮阳、周围建筑的遮挡。主机、水泵、照明及办公设备等部分基础数据采用实测值。

3.2 模拟结果

由于实测的是具体的几天，而模拟选用的全年气象数据是固定的，为了对比模拟的室内温度与实测室内温度的变化规律，模拟对比日的天气情况应与实测日天气情况比较相似，实测日为8月6号，模拟数据选择8月10号，室内室外温度对比情况见图5。房间设定温度为24℃，模拟值为理想的控制状态；测试日上午室内温度在设定温度上下波动，下午室外温度下降，室内人员减少，但空调依然开启，导致房间温度下降较大。模拟值与实测值存在一些差异，但变化情况相似，在下午17时空调停机后室内温度上升，并在一段时间后超过室外温度。

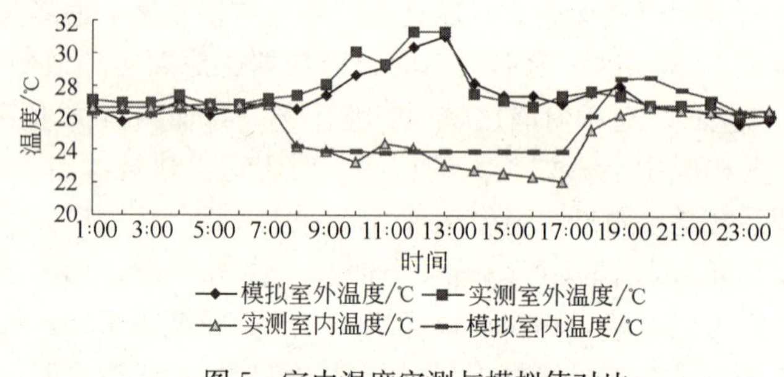

图5 室内温度实测与模拟值对比

实测日空调运行时段大楼平均总电能消耗测试值为254kW，模拟结果中与实测日天气情况相似的8月10日大楼平均总电能消耗模拟值为234kW，相差8%。图6和图7是大楼能耗分配实测与模拟的对比情况。由于每个模型都包含了很多可变量，用户可以适当调整某些变量使模拟结果与长期的实际能耗水平趋于一致，但这需要大楼有足够的能耗及气象基础数据，基础数据越多，则模拟结果的准确性及可预见性就越强，价值也越大。

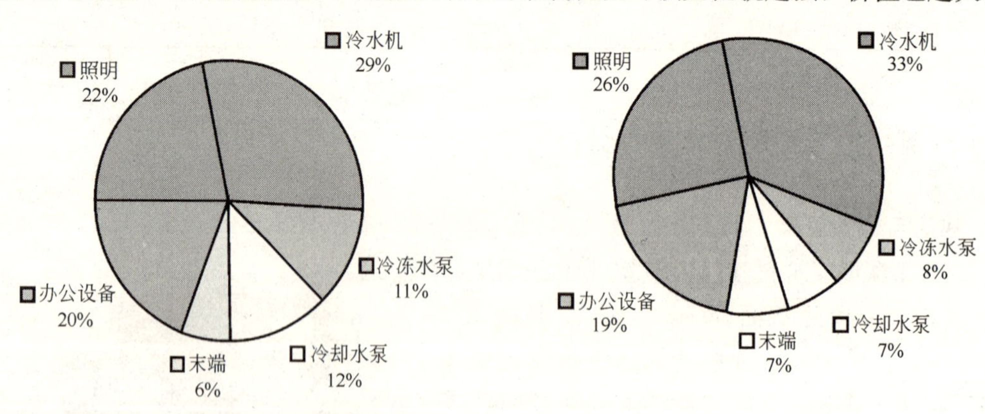

图6 A大楼能耗分配模拟

图7 A大楼能耗分配实测

4 某办公建筑B模拟实例分析

B建筑为某大型综合性办公大楼，建筑面积约3.5万m^2。外观如图8。通过对该大楼建模并进行专项的节能分析，可以初步考察 EnergyPlus 对复杂建筑能耗模拟的能力。下文以对主机开启、冷冻水出水温度以及新风控制为例进行说明。

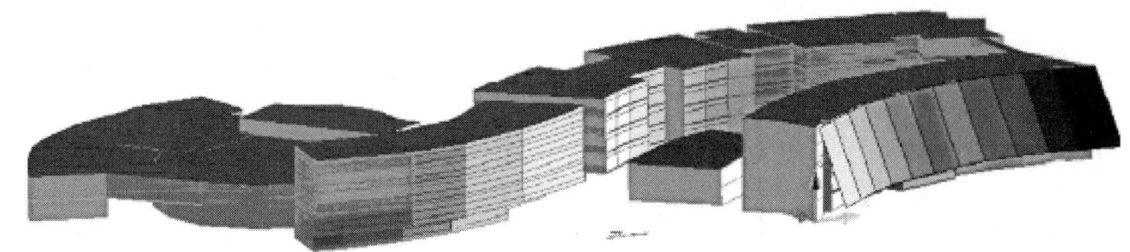

图 8　B 大楼模型

（1）主机控制的节能效应。在办公大楼中，全年的人员活动规律长期来看比较稳定，每日早上 8：30 上班，下午 18：00 下班。图 9 显示的是模型建筑在主机提前一小时停机后房间温度变化。房间 1 和房间 3 为外区南向房间，其余房间位于北向或内区。主机在 17：00 停止供冷后，所有房间因为人员活动与室内蓄热的释放，温度上升。房间 1 和房间 3 温度上升较快，大约 45 分钟后超过 27℃。因此在下班前 45 分钟停止供冷可基本满足后面时段的舒适度要求。事实上，由于条件限制，模拟的情况是风机停止向房间供冷。在实际的运行控制中，往往可以提前更久时间如一个小时关冷水机，但水泵和末端风机依然运行，环路储存的冷冻水冷量释放到房间，使房间温度不致于上升太快，在工作人员离开前依然保持房间温度可以接受。从图 5 也可看出类似变化。

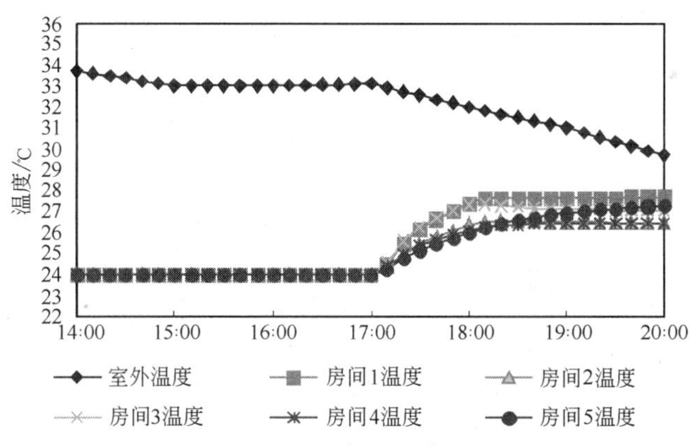

图 9　主机停机后室内外温度变化

（2）控制冷冻水出水温度的节能效应。提高冷冻水温度有利于提高主机的 COP 值，但却不利于末端盘管的换热，降低主机能耗的代价是末端能耗的增加。为了考察对于特定情况下，如何使总的能耗达到最低的冷冻水出水温度，可以通过软件模拟全天空调系统能耗。如图 10，冷冻水出水温度提高有利于主机能效比的提高；主机能耗下降趋势变缓，末端能耗升高幅度变大。随着冷冻水出水温度的提高，总的能耗先下降，在冷冻水出水温度为 9℃～9.5℃时达到最小，随后能耗增加。而全年能耗模拟表明，冷水机出水温度设定为 9℃时全年冷水机和末端运行能耗之和相对于冷水机出水温度设定为 7℃时下降 5%。

不同规模或不同类型的系统其合适的出水温度可能不同，但不管对那种情况，总可以

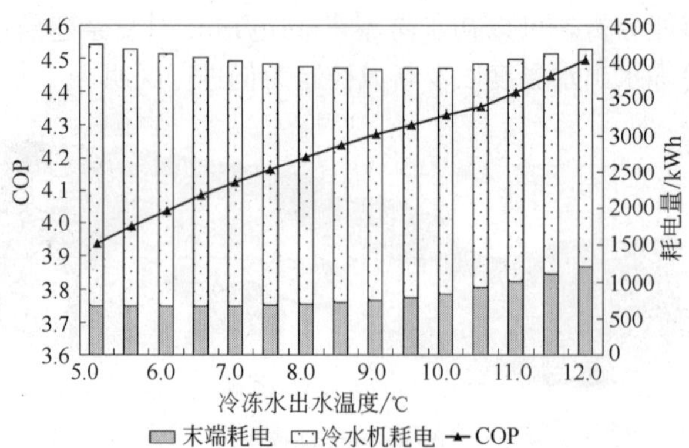

图 10 某日主机冷冻水出水温度与 COP 及系统耗电模拟

找到一个最佳的冷冻水出水温度，而不应该只是固定的选择 7℃。在实际运行中可以通过调整出水温度对运行能耗进行对比，选取合适的出水温度。

(3) 控制新风量的节能效应

从卫生角度出发，室内人员必须有一定量的新风量，但新风量取的过多，新风能耗就会增加。在满足室内卫生要求的前提下，减少新风量有比较显著的节能效果。

对新风的控制有几种方式。下面比较三种方案。方案一为固定新风量控制方式；方案二为基于室内 CO_2 控制方式；方案三为焓控制与主机连锁控制方式。

从模拟结果图 11 来看，基于焓的新风控制即方案三的主机耗电最低。但不同的月份新风控制方式对主机能耗的影响存在量的差异。在过渡季节，由于室外气象条件对全新风运行有利，采用方案三的节能效果最为显著。

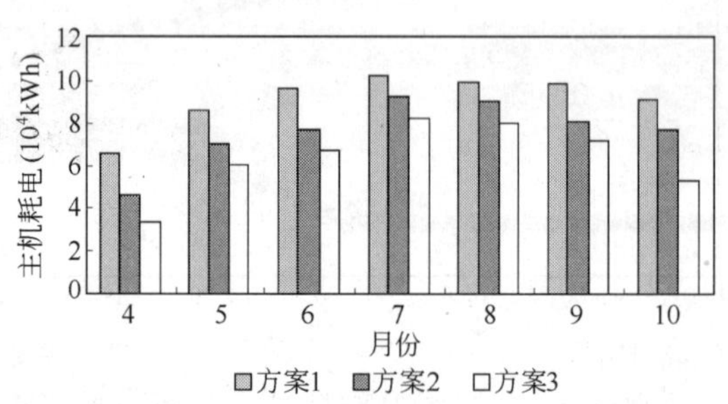

图 11 不同新风控制方案主机模拟结果

5 总结

通过对深圳和广州的气象数据分析，认为 EnergyPlus 气象数据库 IWEC 中广州资料的温度数据用于深圳地区能耗模拟是可行的；辐射数据虽存在偏差，但对于进行全年节能评估的影响并不大。进一步完善深圳能耗模拟用气象数据，能耗模拟结果将更准确。

通过能耗模拟的应用实例以及定性分析认为，EnergyPlus 模拟结果可以接受。以此为基础可根据用户需要和软件提供的丰富的系统模板进行各类模拟，以辅助各类试验或节能设计的深入进行。

Dynamic Simulation and Analysis of Building Energy Consumption Based on EnergyPlus

Yan Tao　Bu Zengwen　Liu Junyue　Ma xiaowen
(Shen Zhen Institute of Building Reseach ShenZhen　518057　China)

Abatract: Introduces the main features and capabilities of a new building energy simulation software-EnergyPlus. Estimate the impact of temperature and radiation data's deviation in IWEC on energy simulation in EnergyPlus. Analyses the applicability of this software in Shenzhen and presents some suggestions.

参考文献

[1] D. B. Crawley, L. K. Lawrie, F. C. Winkelmann, etc. Software Tools for Building Envelopes: EnergyPlus: New, Capable and Linked. In: Proceedings of the 2002 International Green Building Conference, Austin, 2001

[2] D. B. Crawley, L. K. Lawrie, Curtis O. Pedersen, etc, EnergyPlus: Creating a New-Generation Building Energy Simulation Program. In: Energy & Buildings, 2001, 33(4): 319~331

[3] 苏华，田胜元，王靖等. TMY与随机气象模型的准确性. 暖通空调. 2004. 34(1): 5~7

[4] Richard K. Strand, An Example of the Integration of Thermal Simulations into an Existing HVAC Course, In: Proceedings of the ACSA Technology Conference, Austin, 2001

中央空调制冷机组废热回收

郑少义[1]，郑爱玲[2]　张伟廷[2]　马廷全[3]

（[1] 新加坡南洋理工大学，新加坡　639798；[2] 广州市立能中央空调节能顾问有限公司，广州　510080；[3] 新加坡经济能源私人有限公司，新加坡　199566）

摘　要： 本文介绍了中央空调制冷机组废热回收的技术原理，特点及好处。以具体实例对中央空调制冷机组废热回收用于制备生活热水的项目进行了详细的节能经济性分析。该项技术具有技术成熟、可靠，项目改造成本低，投资回收期短的特点，在中国南方地区具有广阔的应用前景。

关键词： 中央空调　废热回收　节能环保　合同能源管理

1　引言

改革开放后，中国用 20 年多点的时间实现了经济总量翻两番的目标。面对能源效率较低的情况，政府也一直关注节能降耗，能源利用效率逐步提高。但是，回顾 20 多年来的发展，中国总体上还没有走出高增长、高消耗、高污染、低效率的粗放型扩张和外延为主的经济增长方式。据测算，1980～2000 年，中国单位产值能耗下降 64%，现在的单位产值能耗仍比世界平均水平高 2.4 倍，是德国的 4.97 倍，日本的 4.43 倍，美国的 2.1 倍，印度的 1.65 倍[1]。传统高速度、高消耗、高污染、低效率的发展模式已经无以为继，中国急需将传统的发展模式转变为经济、社会和环境相协调的发展模式。

现代社会的能源消费可以概括地划分为工业、建筑和运输三大群体。随着社会的进步和中国城镇化速度的加快，建筑用能已经成为中国能源消费的主体之一。21 世纪头 20 年，将是中国建筑业的鼎盛期，2020 年全国建筑面积将接近 2000 年的 2 倍。目前中国每年建成的房屋达 16～20 亿 m^2，超过所有发达国家年建成建筑面积的总和，可是不仅既有的近 400 亿 m^2 建筑中 99%为高能耗建筑，新建筑中 95%以上仍属于高耗能建筑，单位建筑面积能耗比同等气候条件下发达国家单位建筑面积能耗高出两到三倍[2]。

中国“十一五规划建议”提出要努力建设资源节约型和环境友好型社会，而推进建筑节能则是其中的重要一环。中国“2020 年建筑节能远景规划目标”提出：2010～2020 年间，在全国范围内有步骤地实施节能率为 65%的建筑节能标准；2015 年后，部分城市率先实施节能率为 75%的建筑节能标准；大中城市基本完成既有的高耗能建筑和热环境差建筑的节能改造。专家指出，中国虽然发布了居住建筑和公共建筑节能标准，但只有极少数城市执行得较好。实现节能目标任重道远。

面对中国既有建筑的巨大保有量且绝大多数都是高能耗建筑的现实，如何实现既有建筑的节能改造，大规模提高能源使用效率和环境保护，促使既有建筑逐步向绿色建筑转变是极其重要和紧迫的。

建筑物在使用过程中的能耗主要包括空调及通风、采暖、照明、热水供应、办公设备、动力等方面的能耗。其中暖通空调系统能耗约占建筑物总能耗的 1/2～2/3 左右。提高暖通空调系统的能源使用效率对整个建筑的节能具有决定性作用。

随着科技的发展和社会的进步，原有的空调制冷、供热水的方法已逐渐不适应现代化的发展。以现有的宾馆为例，空调只能制冷，而将热量排放到大气中，浪费了大量热能。另一方面，生活用热水又要采用燃油、燃气锅炉或电热锅炉加热，不但消耗大量的能源，且造成不同程度的环境污染。对既需要热水供给又需中央空调制冷的现有建筑实施中央空调制冷机组废热回收措施，将大量被排放到建筑物外的热量重新回收利用到对生活热水的加热中，充分挖掘其最大能耗设备(中央空调)的节能潜力，必将会有非常可观的节能效益和良好的环保效果。

2 中央空调制冷机组废热回收技术

2.1 废热回收技术原理

中央空调系统往往是既有建筑的最大的单一能耗系统。中央空调系统流程如图 1 所示。它包括四个循环：制冷剂循环、冷却水循环(针对于水冷机组)、冷冻水循环和空气调节循环。中央空调制冷机组在运行过程中会排放大量的废热，热量等于空调系统从室内环

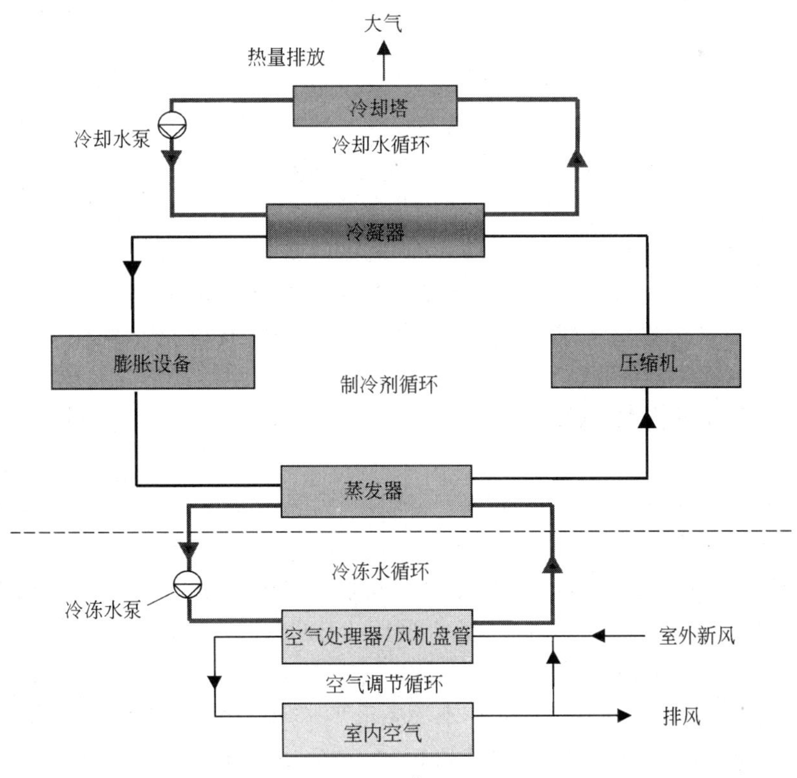

图 1 中央空调系统流程示意图

境吸收的总热量加压缩机电机的发热量。水冷机组通过冷却水塔，风冷机组通过冷凝器风扇将全部热量排放到大气中去。如果能够把这部分热量利用起来，则可以实现单向能耗，双向输出，大大提高制冷机组的能源效率。废热回收技术就是要用中央空调制冷机组排出的部分或全部热能来加热生活用水，可少用或停用现有的生活热水制备系统，节省燃料费用，同时对于水冷机组可部分或全部地减少冷却塔的负荷，对于风冷机组可部分或完全地停用冷凝器。

2.2 废热回收技术的特点

废热回收技术主要有两个特点，其一是废热利用，获得免费生活用热水；其二是提高机组工作效率、延长机组可靠运行寿命。废热回收技术利用这部分热量来获取生活热水，实现废热回收利用并部分或完全地替代生活热水制备系统的目的。废热回收技术应用于水冷机组，还可降低原冷凝器的工作温度，使其制冷效率更高；应用于风冷机组，使其部分实现水冷化，兼具有水冷机组高效率的特性；所以无论是水冷、风冷机组，经过废热回收改造后，其工作效率都会显著提高，同时机组故障减少，运行寿命延长。

废热回收技术的核心是废热回收器。对于水冷机组的废热回收，废热回收器将中央空调排放的废热连同自身所做的功一起加热生活热水到所需的温度；对于风冷机组，废热回收器其实就是安装在冷凝器与压缩机之间的一个高效制冷工质—水热交换器，其主要功能就是将制冷工质吸收的热量经过热交换加热生活热水，起到废热回收的作用。

2.3 废热回收技术的好处

热回收量大——在一般空调使用工况下，中央空调制冷时排掉的热量，远远可以满足制备生活热水所需的热量，通过中央空调废热回收改造工程，它可以替代锅炉的工作，节省大量的燃料。

保护环境——由于利用废热提供所需的生活热水，大大减少了供热锅炉向大气排放 CO_2 气体，从而减少了地球大气变暖的温室效应，同时直接减少了向大气的废热排放量，降低城市的热岛效应。完全或部分地代替冷却塔，及减少冷却塔运行过程所消耗的大量补充自来水，对临近生活区的冷却塔，可降低噪音污染。

提高空调机组效率，节省机组能耗，空调机组压缩机的一部分热量经过废热回收器吸收后，原冷凝器的热负荷减少(风冷机)或降低工作温度(水冷机)，热交换效率提高，空调机组的效率提高，耗电量也将减少，同时，由于采用废热回收技术，机组的负荷减少。

3 系统运行经济可行性分析

在此，以中国南方地区某四星级酒店为例对该废热回收器进行经济可行性分析。该知名四星级酒店建筑面积 45800m^2，空调面积 26000m^2，客房 430 间，并设有中式、西式大小餐厅、咖啡屋及会议室等。酒店采用中央空调系统和全年性 24 小时生活热水供应系统(60℃)。该中央空调系统主机选用两台 CTE425FVIG3JI 水冷机组，单台机组制冷量为 401t 和一台制冷量为 156t 的 CTEHID5D5DBF 水冷机组，冷却水塔共有三台：两台大的每台冷却量 500t 和一台冷却量为 250t；酒店生活热水采用燃油锅炉加热提供。

在空调运行时间较长的季节(每年 5～10 月)里，酒店生活热水平均每天的耗热量为 1.55×10^6 kcal/d，可完全停用燃油热水锅炉；在空调间断运行的季节(每年 4 月、11 月和

12 月)，酒店生活热水平均每天的耗热量为 3.34×10^{6}kcal/d，新旧热水系统同时提供生活热水，燃油热水锅炉仅在废热回收系统提供热水不足时才启动。只有在冬季(每年 1～3 月)完全不用中央空调的 3 个月时间，才全部由燃油热水锅炉提供生活热水。

虽然，生活热水是全天 24 小时提供，根据客房实际的使用情况，估计燃油热水锅炉每天实际的运行时间为 20 个小时。对于该水冷机组的废热回收，废热回收器在回收空调废热后经热泵进一步的加热将生活热水加热到所需的温度，在此取废热回收器热泵的热效率 COP=5.0；夏季里每天的运行时间为 20 个小时，而过渡季节里每天的运行时间为 5 个小时；同时，假设夏季里自来水供水温度为 20℃，经空调系统冷凝热加热后温度达到 35℃，而过渡季节里自来水供水温度为 18℃，经空调系统冷凝热加热后温度达到 30℃。

3.1 夏季废热回收节能经济可行性分析

根据该酒店燃油(重油)热水锅炉的油料消费情况可知，在夏季(每年 5～10 月)里，酒店生活热水平均每天的耗热量为 1.55×10^{6}kcal/d，只要制冷量为 401t 的一台机组运行，所回收的热量已足够保证夏季生活热水使用，可完全停用燃油热水锅炉。

在每年的夏季(每年 5～10 月)里，采用废热回收器制备生活热水过程中废热回收器热泵的能耗为：

$$\frac{1.55\times10^{6}\text{kcal/d}}{20\text{h/d}\times860\text{kcal/kWh}\times5.0}\times20\text{h/d}\times30\text{d}\times6\times\frac{60-35}{60-20}=40552\text{kWh}$$

取电单价为 1.08 元/kWh，则整个夏季(每年 5～10 月)里，废热回收器的能耗成本为：

$$40552\times1.08=43796\text{ 元}$$

酒店最近一年度(2004 年 9 月～2005 年 8 月)按月生活热水耗油费用见表 1。

酒店最近一年度按月生活热水耗油费用 **表 1**

时　　间	用油量(公升)	价格(元/t)	成本(元)
2004 年 9 月	5100	2280	11162.9
2004 年 10 月	7100	2320	15813.1
2004 年 11 月	12000	2590	29836.8
2004 年 12 月	12000	2490	28684.8
2005 年 1 月	20000	2250	43200.0
2005 年 2 月	15000	2290	32976.0
2005 年 3 月	12000	2390	27532.8
2005 年 4 月	11000	2820	29779.2
2005 年 5 月	8400	2950	23788.8
2005 年 6 月	7300	2940	20603.5
2005 年 7 月	5400	2980	15448.3
2005 年 8 月	5000	3040	14592.0

整个夏季(每年 5～10 月)里所需生活热水由燃油(重油)锅炉来取得，需要购买重油的费用为：101408 元。其省下来的费用：101408－43796=57612 元，其实就是废热回收器在夏季里的直接节能效益。

3.2 过渡季节热回收节能经济性分析

根据该酒店燃油(重油)热水锅炉的油料消费情况可知，在过渡季(每年4月、11月和12月)里，酒店生活热水平均每天的耗热量为3.34×10^6kcal/d，只要制冷量为401t的一台机组运行，所回收的热量已足够保证夏季生活热水使用；可是，在这一过渡季节里，中央空调系统只是间断运行，平均每天运行5个小时左右。估计在这一过渡季节里，废热回收器回收的中央空调系统排热量可满足生活热水的一半需求。

在每年的过渡季节(每年4月、11月和12月)里，采用废热回收器制备生活热水过程中废热回收器的能耗为：

$$\frac{3.34\times10^6\text{kcal/d}}{20\text{h/d}\times860\text{kcal/kWh}\times5.0}\times5\text{h/d}\times30\text{d}\times3\times\frac{60-30}{60-18}=12483\text{kWh}$$

取电单价为1.08元/kWh，则整个过渡季节(每年4月、11月和12月)里，废热回收器的能耗成本为：

$$12483\times1.08=13482\text{元}$$

表1所示，整个过渡季节(每年4月、11月和12月)里所需生活热水由燃油(重油)锅炉来取得，需要购买重油的费用为：88300元，这其中的一半热量是可以由废热回收器来提供的。其省下来的费用：88300/2－13482＝30668元，其实就是废热回收器在过渡季节的直接节能效益。

3.3 节能改造项目总投资及其投资回收期

由3.1和3.2的计算可知，要实现中央空调系统废热回收并满足酒店生活热水所需，该废热回收器的最大设计功率为：

$$\frac{3.34\times10^6\text{kcal/d}}{20\text{h/d}\times860\text{kcal/kWh}}=194.2\text{kW}$$

安装一个功率为200kW的废热回收系统的总投资约为185000元(设备成本及安装费用)；结合3.1和3.2的经济性分析，每年的夏季和过渡季节里，废热回收器共节省的燃油费用为57612＋30668＝88280元；所以整个节能改造项目的投资回收期为：

$$\frac{185000}{88280}=2.10\text{年}$$

有一点需要强调的是，这里算的投资回收期只计算了废热回收器在运行季节里所节省下来的燃油费用，还未考虑由于空调排热被吸收后，冷凝效果改善所带来的节电量以及冷却塔负荷降低或完全停用所带来的节电量。

4 商业运作模式探讨

根据有关资料介绍，我国仅北纬32度以南的14个省市，就有约宾馆、酒店等5万家以上，由于这部分地区空调运行期较长，每天运行时间也较长，实施中央空调系统废热回收项目投资效益非常明显，根据原有系统的改造难度，投资回收期一般在15～25个月之间。

合同能源管理服务是一种以减少的能源费用来支付节能项目全部成本的节能投资模式。这种节能投资方式允许用户使用未来的节能收益为现有设备进行改造或升级，以降低目前的运行成本和获得持续的节能效益。实施节能方案的服务和费用由节能服务公司提供，节能服务公司和业主共同享有节能效益，合同期满后，业主将继续享有全部的节能效益。合同能源管理服务在中央空调系统废热回收节能项目上的应用将大有前途。

5 结论

对于宾馆、酒店、医院、公共浴室和一些食品加工厂等需要中央空调供冷同时还要供生活热水的场所，利用废热回收技术把中央空调系统冷凝放热过程中排放的热量用来制备生活热水，不仅减少了排到大气中的热量，还提高了能源利用效率，节省了能源费用，具有非常明显的节能经济效益。在长江以南地区，此项技术应用于宾馆、酒店和医院等场所的现有空调系统上，不影响现有空调系统的运行，技术改造成本低，投资回收期短，具有广阔的应用前景。

Waste Heat Recovery on Central Air-conditioning System

Tay Seow Ngie[1]　Zheng Ailing[2]　Zhang Weiting[2]　Ma Tingquan[3]

([1] Nanyang Technological University　639798　Singapore;

[2] Guangzhou LiNeng Energy Conservation Consulting Co., Ltd,

Guangzhou　510080　China; [3] Economic Energy Pte Ltd, 199566　Singapore)

Abstract: The technical principle, characteristics and advantages of waste heat recovery on central air-conditioning system have been introduced in this paper. A detailed economic analysis with a project example of waste heat recovery on central air-conditioning system for producing living hot water has been done. With mature and reliable technique, low cost and short payback, this technique has wide application potential in southern China.

参考文献

[1]　谢振华. 大力发展循环经济实现经济社会与环境协调发展. 在 2004 年中国科协年会上的报告

[2]　涂逢祥，王庆一. 建筑节能研究报告. 建筑节能. 中国建筑工业出版社，2004

既有住宅节能改造技术研究
——谈常德路德怡苑高层住宅楼节能改造

孙生根　赵　鸿　杨永巍　周　俊　杨　靖
（上海市房地产科学研究院，上海　200031）

摘　要：上海市首个住宅楼节能改造工作已经完成。在改造前由于墙体马赛克起壳严重，利用红外热像检测技术对外墙损坏情况进行检测，然后提出节能改造技术方案，并通过节能计算软件进行计算，最后确定改造方案。通过实际应用，并对改造前后房屋的热工性能进行检测，得到该房屋的节能改造技术是成功的，该项技术具有较好的应用前景，对既有建筑节能改造具有示范作用。

关键词：既有建筑　节能　改造　红外热像　检测

在构建节约型国民经济和促进整个社会可持续发展成为时代重任的时候，在建筑行业积极推广使用节能建材、发展利用可再生资源，成为相关部门及企业的使命，也成为现代化城市发展的必然选择。在城市，发展节能建筑已经成为城市文明和可持续发展的象征。有关专家说，我国城市节能建筑的发展，还处于起步阶段，除了大力发展新的节能建筑，还需对既有的建筑进行技术改造，在既有的约 400 亿 m^2 城乡建筑中，99％为高耗能建筑；它们的单位建筑面积采暖能耗，相当于气候条件相近的发达国家节能建筑的 2～3 倍。但对这些建筑的节能改造工作迟迟难以启动，仍然在无节制地消耗着大量能源。面对如此危机的能源情况，今年上海市开始了对既有建筑进行节能改造的工作，选择常德路的几幢老住宅楼作为市房地局综合整治计划项目、市建委既有建筑节能改造试点工程以及市和普陀区两级政府实事工程。在该项工程中首先利用红外热像检测技术对外墙坏损情况进行检测，然后提出节能改造技术方案，并通过节能计算软件进行评估，最后确定改造方案。通过实际应用，并对改造前后房屋的热工性能进行跟踪检测评估后，可以看到该房屋的节能改造技术是成功的。该项技术具有较好的应用前景，对既有建筑节能改造具有示范作用。

1　工程简介

常德路的几幢地处原本市危房重点改造的“药水弄”区域。1989 年建设，1991 年竣工，层高为 21～26 层。根据设计要求，大楼为钢筋混凝土剪力墙结构，外墙表面采用水泥砂浆粘贴马赛克，总建筑面积为 4.5 万 m^2，现有住户 616 户。由于 10 多年来日晒雨淋、风化等自然影响，这三幢大楼的外墙面出现多处损坏，并有局部马赛克砖剥落现象，2003 年曾发生 $3m^2$ 左右的外墙马赛克连同粉刷层突然坠落，差点造成人员伤亡。大楼墙外的空调大多都进入了“高龄”期，长期太阳暴晒、酸雨侵蚀，许多空调支架锈蚀严重，残缺不全，且空调安装杂乱无章，有碍观瞻。因长期受风化锈蚀，加上缺少保养，许多住户阳台上的雨篷及晾衣架破旧不堪，摇摇欲坠，情况比较危险。屋面因放水材料年久失修，有渗水现象，保温效果较差。

2 红外热像检测技术检测建筑物外墙

2.1 原理

红外热像检测是测量通过物体的热量和热流来鉴定该物体质量的一种方法。当物体内部裂缝和缺陷时，它将改变物体的热传导，使物体表面温度分布产生差别，利用红外热像仪测量它的不同热辐射，可以查出物体的缺陷位置。由于空气的热导远低于瓷砖、砖、混凝土等建筑材料，因此当热流从表面进入建筑物饰面层时，即会在“空鼓”等缺陷部位受到空气阻挡发生“热堆积”，使该处的红外热像呈“热斑”等特征。由红外热像“热斑”出现的部位、持续时间等特征推知存在饰面层粘结质量的区域范围。

2.2 检测过程

2.2.1 通过目测检查，发现大楼外墙存在图 1～图 4 所示的破坏情况。

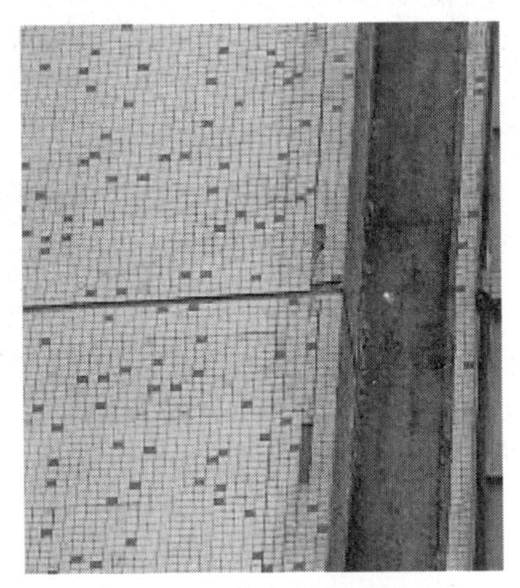

图 1 马赛克局部脱落

图 2 阳角处竖向严重空鼓开裂

图 3 窗洞右侧墙面马赛克大面积脱落

图 4 阳台围墙东南棱角竖向裂缝

2.2.2 红外热像技术检测结果如图 5～图 8 所示。

图 5 外观检测无明显异常

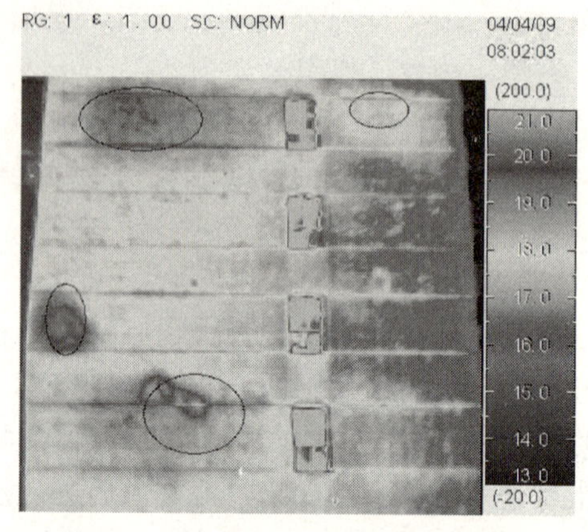

图 6 红外检测发现墙面多处空鼓

图 7 外观检测无明显异常

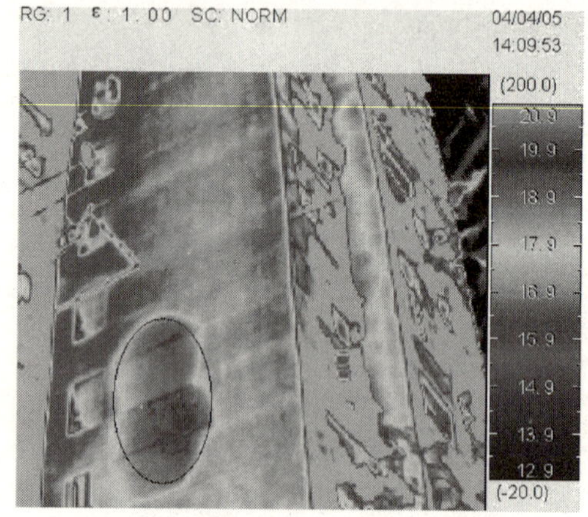

图 8 红外检测发现墙面多处空鼓

2.3 检测结果

通过红外热像检测技术检测，并应用自行编制的红外分析软件进行分析，得到这几幢楼的外墙空鼓率均在 30%左右，对人民的生命及财产形成极大的威胁。同时该房屋没有进行节能设计，热工性能较差。

3 节能改造方案

根据对常德路几幢高层建筑的房屋安全及节能检测，其外立面存在着温差裂缝引起的渗漏，影响了其使用功能，还因雨、雪、冻、融、干、湿等自然力因素循环侵袭，加剧对结构的破坏，引起了饰面层及粉刷层空鼓，存在着严重的安全隐患。建筑能耗综合指标为 62.38kWh/m^2，属非节能住宅，通过采用节能改造来降低建筑能耗，防止围护渗漏、外饰面坠落等危险发生，保护主体结构，延长建筑物的使用寿命。常德路高层节能改造主要为墙体和屋面节能改造，其中外墙节能改造难度较大。从施工不影响居民日常生活，也不减少室内居住面积及有利于结构稳定性方面考虑，本次节能改造采用外墙外保温技术。可应用于常德路几幢高层建筑节能改造的外保温技术有聚氨酯外墙外保温系统、聚苯板薄抹灰外墙外保温系统、胶粉聚苯颗粒外墙外保温系统。因外保温体系被置于马赛克旧饰面外测，直接承受来自自然界的各种因素影响，因此对用于既有建筑节能改造的外保温系统提出了更高的要求。在外保温构造设计时充分考虑了与旧饰面的粘结、热应力、风、火、水及地震等因素对系统的影响。采用了既有建筑 193 聚氨酯外墙保温体系涂料外饰面技术，该技术包含既有建筑旧饰面界面处理技术、系统组成材料性能兼容技术、系统抗裂技术、系统耐久性技术。具有其他各种外墙外保温体系相同的优越性的同时，还具有如下优势：①该系统与老建筑外饰面粘结牢固，系统安全性高。②现场喷涂、发泡，无拼缝产生的热桥负面影响。③可塑性强，适合于所有形状和类型既有建筑围外立面饰面。④该系统集防水保温十一体，在既有建筑节能改造同时解决了老建筑外墙面渗水问题。⑤该系统整体性强，柔性变形量呈梯形释放，变形适应能力强，抗裂性能优异。

原屋面为珍珠岩保温材料和放水卷材，因放水材料破损，水渗至珍珠岩保温层内，保温性能大大下降，若将原屋面防水保温层全部铲去，工程量太大，且影响居民日常生活。故在原保温层上直接重置 2cm 聚氨酯防水保温层，解决屋面渗漏问题，提高隔热性能。

由于该大楼的外窗已进行了部分改造，同时该项工程没有专项资金，且高层住宅的窗墙比较小，经过评估，外墙和屋面节能改造后，建筑能耗综合指标达 37.20kWh/m^2，已达到节能 50%的目标，故未进行外窗改造。

4 节能改造

4.1 外墙节能改造

常德路几栋高层住宅楼 193 聚氨酯外保温系统节能改造施工工艺主要包括：旧饰面基础处理、旧饰面界面处理、喷涂 193 硬质聚氨酯泡沫塑料、193 硬质聚氨酯泡沫塑料平整处理、滚涂 193 彩色防水涂膜、批嵌 193 纤维增强抗裂腻子(内置耐碱网格布)、批刮 193 柔性外墙腻子、滚涂 193 弹性涂料。

因考虑到常德路几栋高层住宅楼在完成节能改造后，其形成的复合墙体与原重质墙体基层不同，外保温系统由于保温层内置密度小，强度低，复合墙体呈现软基底的特性，如在复合墙体上打洞，安装空调或晾衣架，使保温系统的复合墙体局部损坏，造成外墙渗漏，甚至由于空调架或晾衣架与复合墙体连接不够牢固，造成安全隐患。因此，在本次常德路围护结构节能改造试点工程前，完成了建筑物外立面空调板、滴水管、落水管、晾衣架等方面的改造。通过采用“植筋法”在统一位置增设空调板，在外墙外保温系统施工完成后，将空调外机统一移到空调板上加以固定。对于拆除空调架后，遗留下的膨胀螺栓孔，用密封胶进行防水处理，以免留下渗水隐患。对于需安装的空调滴水管及需移位的落水管、晾衣架，在基础墙体上预留螺栓或预埋件，在完成外墙外保温系统施工后进行统一安装(图 9、图 10)。

4.2 屋面节能改造

原屋面为珍珠岩保温材料和放水卷材，因放水材料破损，水渗至珍珠岩保温层内，保温性能大大下降。若将原屋面防水保温层全部铲去，工程量太大，且影响居民口常生活。故在原保温层上直接重置 2cm 聚氨酯防水保温层，解决屋面渗漏问题，提高隔热性能。

图 9 基层处理

图 10 现场喷涂

5 节能改造评估

主要从以下几个方面对房屋的节能情况进行评估：

5.1 软件评估

使用 CHEC 夏热冬冷地区居住建筑节能设计分析软件对其中一幢大楼进行节能改造前后的评估计算，得到表 1、表 2 所示数据。

节能改造前大楼节能计算表 **表 1**

层数(26)，体型系数(0.29)，朝向(南)		
项 目	构 造	热工参数
外 墙	20mm 水泥砂浆＋240mm 钢筋混凝土＋20mm 水泥砂浆	K_p＝2.95(W/m²·K) D＝2.84
外 窗	6mm 普通双层玻璃，6mm 普通单层玻璃 窗墙比(东 0.17 南 0.30 西 0.12 北 0.18)	设计传热系数 K＝4.20W/(m²·K) K＝6.40W/(m²·K)
屋 面	25mm 水泥砂浆＋15mm 水泥膨胀珍珠岩＋20mm 水泥砂浆＋120mm 钢筋混凝土	K＝2.70W/(m²·K)
分户墙	20mm 水泥砂浆＋240mm 钢筋混凝土＋20mm 水泥砂浆	K＝2.43W/(m²·K)
户 门	普通钢框实体门	K＝4.52W/(m²·K)
楼 板	15mm 水泥砂浆＋150mm 钢筋混凝土＋15mm 水泥砂浆	K＝2.87W/(m²·K)
能耗综合指标：62.38kWh/m²		
不符合规范 JGJ 134—2001 第 5.0.5 条的要求		

节能改造后大楼节能计算表 **表 2**

层数(26)，体型系数(0.29)，朝向(南)		
项 目	构 造	热工参数
外 墙	15mm 硬脂聚氨酯＋20mm 水泥砂浆＋240mm 钢筋混凝土＋20mm 水泥砂浆	K_p＝1.13(W/m²·K) D＝3.00

续表

层数(26)，体型系数(0.29)，朝向(南)		
项　目	构　　造	热工参数
外　窗	6mm 普通双层玻璃，6mm 普通单层玻璃，窗墙比(东 0.17　南 0.30　西 0.12　北 0.18)	设计传热系数 $K=4.20W/(m^2 \cdot K)$ $K=6.40W/(m^2 \cdot K)$
屋　面	20mm 硬质聚氨酯＋25mm 水泥砂浆＋15mm 水泥膨胀珍珠岩＋20mm 水泥砂浆＋120mm 钢筋混凝土	$K=0.91W/(m^2 \cdot K)$ $D=2.09$
分户墙	20mm 水泥砂浆＋240mm 钢筋混凝土＋20mm 水泥砂浆	$K=2.43W/(m^2 \cdot K)$
户　门	普通钢框实体门	$K=4.52W/(m^2 \cdot K)$
楼　板	15mm 水泥砂浆＋150mm 钢筋混凝土＋15mm 水泥砂浆	$K=2.87W/(m^2 \cdot K)$
能耗综合指标：37.20kWh/m²；参考建筑节能综合指标：38.53kWh/m²		
比改造前的节能综合指标降低了 40%		

5.2　热工测试

使用热流温度巡检仪、红外热像仪等对其中一幢大楼的外墙进行热工测试，得到图 11 和表 3 所示结果。

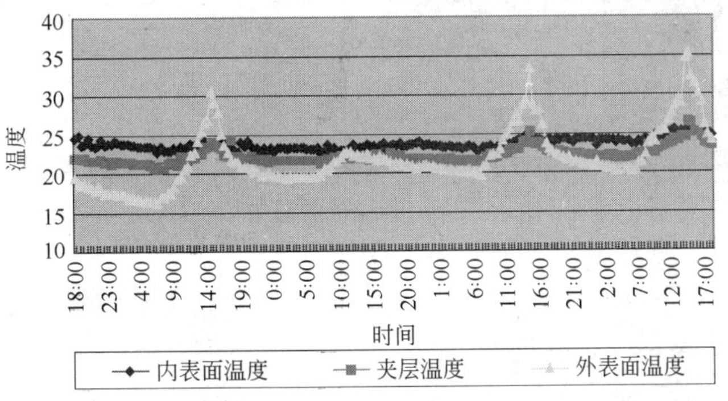

图 11　各层测点温度变化图

各时间段传热系数实测值 $K[W/(m^2 \cdot K)]$　　**表 3**

位　置	东立面	南立面	西立面	北立面
传热系数实测平均值 K	1.21	1.31	1.34	1.09

5.3　红外热工缺陷检测(见图 12、图 13)

5.4　检测结果(图 14、图 15)

(1) 改造后墙体传热系数满足上海市节能 50% 的要求。

(2) 外墙面没有明显的热工缺陷。

(3) 使用 CHEC 夏热冬冷地区居住建筑节能分析软件计算，符合规范 JGJ 134—2001 第 5.0.5 条的要求。

(4) 改造前外墙内外温差与改造后外墙内外温差相差 4℃左右。

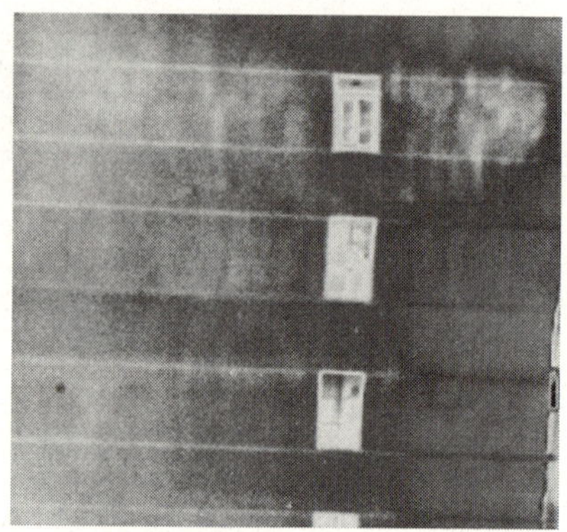

图 12 未发现墙面有异常现象

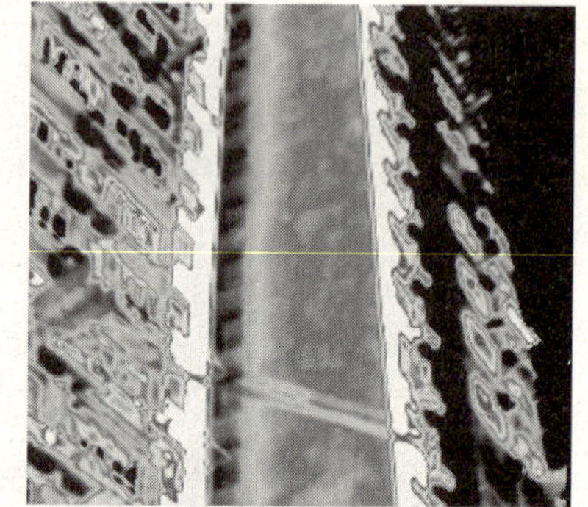

图 13 未发现墙面有异常现象

图 14 改造前

图 15 改造后

6 结论

通过对常德路几幢住宅楼的节能改造并进行测试，可以看到在该项改造工程中应用的各项技术都取得了满意的结果，解决了老建筑几大问题：①解决了外墙马赛克跌落而引起的安全问题；②解决了外墙外保温问题；③解决了外墙的渗漏问题；④解决了屋面的渗漏问题，提高了屋面隔热性能；⑤解决了室外空调机架的整治，特别是把“解危”和“节能”有效结合起来，是一次重大的突破和创新，对既有建筑节能改造具有示范作用。

The Investigation of Energy Efficiency Technique for Existing Buildings

——Discussion On Energy Efficiency At Changde Road Yide Garden High Dwelling Building

Sun Shenggen　Zhao Hong　Yang Yongwei　Zhou Jun　Yang Jing

(Shanghai Real Estate Science Research Institute, Shanghai　200031　China)

Abstract: The first building energy-efficiency transformation was accomplished. Before transformation, the mosaic of wall was terribly exfoliated. Infrared thermography inspection technique was used in inspecting the exterior wall. And the energy-efficiency evaluate software was used to calculate the energy-consumption. After inspecting and evaluating, the project of transformation was conformed. By testing the building thermal quality, we can conclude that the technique of energy-efficiency transformation of the building is successful, and this technique has fine foreground.

参考文献

[1]　夏热冬冷地区居住建筑节能设计标准 JGJ 134—2001

[2]　民用建筑热工设计规范 GB 50176—93

[3]　住宅建筑围护结构节能应用技术规程 DG/TJ 08—206—2002

[4]　住宅建筑节能检测评估标准 DG/TJ 08—801

[5]　采暖居住建筑节能检验标准 JGJ 132

[6]　聚氨酯彩色防水保温系统技术规程 DGJ/CT 022 193

寒地城市哈尔滨的供热及清洁能源研究

董　洋　李桂文
（哈尔滨工业大学建筑学院，哈尔滨　150006）

摘　要：本文围绕哈尔滨这个寒地城市的供热问题出发，对比了国内外寒地城市的供热方式区别，在介绍我国寒地城市供热中能源利用基本情况的基础上，分析了哈尔滨市供热方式与能源利用的主要问题，提出了哈尔滨市供热中能源利用及清洁能源的发展趋势，最后，对哈尔滨市的供热发展进行了展望。

关键词：寒地城市　城市供热　清洁能源

1　寒地城市供热基本概况

1.1　国内城市供热基本概况

解放前，除上海，南京等城市局部有一些自备锅炉热水采暖系统外，从未形成联网的城市供热设施，城市供热设施事实上等于零。解放后随着第一个五年计划的开始，哈尔滨、吉林、北京、保定、太原、兰州、西安、郑州等城市为援建项目配套建设的热电厂除了供工业用热外，同时也向自备的生活区住宅供热，城市供热设施的建设局限于北方地区。1986年后不仅北方地区的城市建设注重城市供热设施的发展，其他地区也着手城市集中供热配套建设规划，上海市、常州市等城市除继续考虑工业用热外，开始注意住宅区公建的热水供应和冬季的热水采暖，空调北移，采暖南移的趋势已露端倪，全国有70多座城市已设有不同水平的城市区域集中供热设施。2000年全国三北地区集中采暖供热普及率为23.6%，热电厂和区域锅炉房是典型的集中热源，其他小型锅炉，电能和燃气为热源的供热方式作为不同层次的集中供热结构作为补充。

1.2　国际寒地城市供热方式基本概况

美国是最早创建区域锅炉房集中供热的国家，由于其最早发现本土拥有丰富石油和天然气储量，所以发展以煤炭为燃料的热电厂和区域锅炉房的同时，以石油和天然气为燃料的热源也得到发展。据统计，天然气已开始作为城市供热设施热源中重要的能源，比例达到60%以上，其余补充能源按比例分别为电能、燃油锅炉，部分区域锅炉房。

日本占30%以上的家庭选用天然气作为住宅供热热源，荷兰、加拿大等国家用天然气作供热热源的用户已占天然气用户的95%以上。

俄罗斯（前苏联）在50年代由于主体能源是煤炭，所以从提高综合能源利用效率考虑，着重发展燃煤的热电厂，作为城市供热设施热源。50年代后期前苏联发现大储量气田，但其能源分配政策规定天然气重点供发电和工业用气，所以采暖用热仍以热电厂和区域锅炉房为热源。80年代以后，开始发展天然气作为部分建筑采暖系统的热源，其所占比例也开始有所上升。

2 供热方式中能源利用情况

2.1 供热中能源利用状况

全国现集中供热的城市中由于多采用热电厂和集中供热锅炉房的供热方式，所以一次能源即化石能源燃煤是供热中基本能源利用方式。2000年全国生产用热的能耗量为7000多万吨，标准煤、化石燃料的大量消费产生大量二氧化碳进而导致地球变暖，氧化硫和氮化硫的氯氧化物则形成酸雨，这些都严重破坏了地球环境。

2.2 国家的能源政策及发展战略

国家计委、科委、经贸委制定的《2010年新能源和可再生能源发展纲要》中明确指出，按照社会主义市场经济的要求，加快新能源和可再生能源的发展和产生建设步伐。

从长远来看，大力发展新能源和可再生能源可以逐步改善以煤炭为主的能源结构，促进常规能源更加合理有效地利用，缓解与能源相关的环境污染问题，使我国能源与环境的发展相互协调，实现可持续发展目标。

新能源和可再生能源产业发展目标是：加速技术和产品的推广应用，增强我国设备制造和生产能力；建立产业化配套服务体系、健全法规和机制，实现新的能源和可再生能源开发利用的商业化发展，具体任务是：

（1）促进大型高效太阳能热利用产业发展。

（2）推动并网风电的商业化发展，加快国产化进程。

（3）积极推广地热采暖和地热发电技术。

（4）推进大中型沼气工程建设，开发生物质能高效利用设备。

（5）推进新技术产业化。

2.3 第二代能源系统—清洁能源

为了解决地球环境危机和能源危机的双重压力，符合中国能源发展政策战略要求，清洁能源的利用及推广及产业化成为必然。清洁能源种类很多，它既有化石能源中的天然气、煤气或由化石能源转化来的电等，又包括核能、太阳能、地热能、水能、海洋能及新型再生生物能等，据统计太阳能年辐照总量大于502kJ/m^2、年日照时数在2200小时以上的地区约占我国国土面积的三分之二以上，具有良好的开发条件和应用价值，风能资源理论储量为32.26亿kW，而可开发的风能资源储量为2.53亿kW，地热资源的储存条件也较好，其远景储量相当于2000亿万t标准煤以上，已勘探的40多个地热田可供中低温直接利用的热储量相当于31.6亿t标准煤，生物质能源也十分丰富，农作物秸杆产量每年约7亿t，可作能源的资源约为2.8～3.5亿t。太阳能热水器、地热采暖已较成熟并具有相应市场，形成初步的产业基础。低温地热的开发利用已经进行了城镇供热和综合利用等多面试点，技术基本成熟，1998年供暖面积已达800多万m^2，太阳能建筑一体化技术的发展使得太阳能热水供应、空调、采暖工程成本逐渐降低，也将是太阳热水器潜在的巨大市场。

3 哈尔滨市供热方式及存在的主要问题

3.1 概述

哈尔滨市位于我国东北北部，是我国纬度最高的省会城市，冬季平均气温为－14.2℃，最低温度－33℃。取暖期为179天/年，供暖热指标在64～75W/m^{2}之间，全国最高。供暖能源消耗标准煤500万吨/年，为全国比重最大。

3.2 供热现状

哈尔滨市是我国最早实行集中供热的城市之一。20世纪60年代随着“三大动力”的发展，建成了以哈热电厂为热源的动力热源。进入80年代，相继建设了哈尔滨发电厂，哈尔滨化工热电厂及一批企业自备热电站和区域锅炉房供热系统。哈尔滨热电厂装机容量300MW，年发电量12.4亿度，生产供热能力24GJ/h，采暖供热能力838GJ/h。哈尔滨发电厂装机容量75MW，年发电3.5亿度，生产供热能力119GJ/h，采暖供热能力894.7GJ/h，哈尔滨化工热电厂装机容量36MW，年发电1.9亿度，生产供热能力134.75GJ/h，还有哈制药厂、哈一工具厂、东安厂热电站等共14个企业自备电站，总装机容量75MW，生产供热能力576.63GJ/h，采暖供热能力832GJ/h。集中式燃煤锅炉房100余座，总容量2764MW，生产供热能力1159.GJ/h，采暖供热能力5533GJ/h，分散式小锅炉房(7MW以下)3200座，生产供热能力78037GJ/h，采暖供热能力12512GJ/h。

哈尔滨市城区供热目前有五种方式，一是以热电联产为主的集中供热；二是以大型区域锅炉房为主的集中供热；三是以小型锅炉房为主的分散供热；四是以煤气、电力和燃油能源为主的单体分散供热；五是以小煤炉为主的分散供热。

截至2004年，哈市房屋建筑面积为12220万m^2。

热电联产实现集中供热面积1798万m^2，热电联产集中供热普及率14.7%；企业自备电站实现集中供热面积385万m^2，企业自备电站供热普及率3.2%；大型区域锅炉房实现集中供热面积3829万m^2，大型区域锅炉房集中供热普及率31.3%。

截至2004年止，哈市集中供热总面积为6012万m^2，集中供热普及率为49.2%。分散小锅炉房(含燃煤、燃油、电力)实现分散供热面积4408万m^2，分散供热普及率34.4%；分散小煤炉供热实现供热面积1800万m^{2}时，分散供热普及率15.0%。分散供热总面积为6208万m^2，占全市供热总面积50.8%。

3.3 存在主要问题

(1) 环境质量问题

哈尔滨市目前尚未达到国家二类环境功能区标准，空气污染指数停留在70～90之间，采暖期间甚至超过100，大气环境质量急需改善，市区总悬浮微粒物年均值为0.368mg/m^3，日均值标准率为70%，最大日均值出现在第一季度，为0.763mg/m^3，超过标准1.54倍。二氧化硫全年日均值为0.026mg/m^3，最大日均值在第一季度为0.094mg，降尘全年均值为39.03t/km^2，月超标2.59倍，最大日均值在第一季度，为55.88t/km^2。通过对综合指数统计得出：道外区大气污染指数程度最高，然后依次是道里、太平、南岗、香坊、动力、最后是平房，我们知道，影响大气质量的主要因素是烟尘，而烟尘污染主要来自锅炉。

(2) 效率问题

目前由于哈尔滨市小型燃煤锅炉供热仍然大量存在，锅炉容量大多为蒸发量4～6t/h，也有历史上留下的0.5t/h蒸发量的锅炉。这些锅炉房基本上是个工矿企业及房屋开发公司自筹自烧，热效率基本在60%以下，有相当一部分实际运行热效率连50%也不到，既造成环境污染又使能源浪费严重，由于采用间歇供暖方式，供暖质量不能保证，严重影响了供暖费的收缴，由于收费困难，设备更新改造资金不足，造成70%的锅炉设备陈旧，带病运行，锅炉过滤低，能耗大，对大气环境污染严重，这些因素相互制约，恶性循环，影响了正常的供暖工作。

4 哈尔滨市供热中能源利用及清洁能源发展趋势

4.1 一次能源的利用及发展

由于我国能源结构决定，在供暖方式日趋多样化的今天，一次能源即化石类能源燃煤锅炉的主导地位仍不可取代，热电联产是对低品位能源和高品位能源的有机利用，同时由于热电联产锅炉都为大型锅炉，热效率85%以上，再加上采用先进的燃烧技术，加大清净煤技术的开发与利用力度，侧重优化煤炭的燃烧方式，提高其燃烧效率，使热电联产成为高效节能而又适合我国国情的集中供热方式，并同时应有步骤，有层次的淘汰小型的燃煤供热锅炉，按《哈尔滨市原煤散烧污染防治办法》等法规文件，划定了原煤散烧控制区和无燃煤区，加快淘汰在用小型分散燃煤锅炉的步伐。相对而言，用非热电联产集中式燃煤锅炉供热，锅炉热效率设计除尘效率比小锅炉强，却无法同热电联产相比，它可作为热电厂供热的调峰锅炉，对热电联产供热时给予必要的调节和补充，远期也将逐步渡入热电联产。

4.2 开发利用清洁能源及发展

按国家新能源发展战略的要求，选择适合哈尔滨市又利于推广的节能、环保、高效的供热方式，大力开发利用清洁能源是我市增强城市载体功能，直接关系到经济可持续发展和社会安定。

(1) 大力推广地热系统

大地因特有的构造，储存了太阳能。哈尔滨市的地下水温终年保持在8～14℃，目前在哈尔滨市环保局采用的新型地热供热方式就是打两眼井直通地下水层，用提升泵从一个水井中抽取地下水，通过热交换将热量提取出来，地下水的温度降至6℃左右，再将降温的地下水灌回地下，地下水只提供热量，并不参与循环，所以不会受到污染，提取的热量再经过一次热交换传递给供热循环水，使循环水达到设计温度送热给用户，其特点是环保、节能、高效、安全、运行维护都非常简单，且不会对周围环境造成任何污染，运行费用仅为燃煤锅炉供热的三分之二。另外对于那些工业企业中废热水量较大的企业既解决了供热问题，也避免了热水排放造成的浪费。

哈尔滨市地热资源的勘察工作已进行了6～7年，经地表地质工作和其他地质科学研究资料对比，发现江北有地热异常区，面积近400km^2。往地下每增加100m，温度上升5～6℃。地下2000m左右能达到90℃，丰富的地热资源将对建设哈尔滨松北绿色新城区建设起到极大的推动作用。

(2) 促进太阳能利用及产品化进程

太阳能数量巨大，且取之不尽用之不竭，全国有 2/3 以上的地区年辐射总量大于 502 千焦/平方米。按照北京绿色奥运规划，2008 年太阳能将成为北京奥运会上主要能源，太阳能采暖技术，已列入建设部节能技术政策范畴和建筑节能“十五”计划。随着对导致全球气候变暖的温室气体排放量的进一步限制，太阳能产业将与计算机和电信产业一样，成为 21 世纪发展最为迅猛的工业领域之一。

哈尔滨日照数在 2400 小时以上，太阳年辐射总量为每平方米 111.6 万大卡，居全国中等水平。近年来，太阳能热水器市场年增长 20%以上，热水供应，洗浴用水等太阳能热水系统所产生的源源不断的热水由于优越的经济性，良好的实用性，无与伦比的环境效益成为太阳能热水供应大力发展的前景，且由于太阳能热利用与建筑节能一体化技术的发展使太阳能供热工程也在逐步发展。

(3) 低温核供热

核能是一种有广泛应用前景的新型清洁能源，过去多用于发电，近几年已逐步应用于供热，它具有热值高(比煤高 270 万倍)，污染可控制，节省空间等优点。清华大学核能所开发的 SMW 试验性供热堆已于 1989 年投入运行。哈尔滨市在低温核供热的研究上已有初步规划设想，站址拟选建于道里区西侧，该站包括两座反应堆及相应的控制系统，核常规系统和常规辅助系统，核反应堆为低温、低压、轻水堆，采用双壳，自然循环一体化布置，压水运行，技术上是可行的，可在借鉴其他国内城市经验的基础上在城市供热远景发展上加以利用。

(4) 加快燃气化发展

哈尔滨市现有燃气供应主要为哈依煤气供应，天然气供应仅在局部地段。哈依煤气系引自依兰县达连河镇，输气管线总长度 270km，设计日最高产气量 240 万 m^3/日左右，家庭锅炉燃气普及率达 92%。但是不仅要发展燃气锅灶，引水洗浴设备，更要大力发展家庭单体燃气锅炉才能达到全面燃气化。据初步估算，仅按道里区最终供热面积 1555 万 m^2 估算，按设计热指标计算，采暖日需气量设计值为 520 万 m^3，煤气厂日最高产气量仅为 242 万 m^3，所以仅从气量上看，目前哈尔滨市的煤气产量远远不能满足道里区民用采暖的需求。而且如用煤气供热，必须考虑冬夏季气量不平衡的调峰问题。目前尚不具备大容量的调峰手段，在煤气输配上，整个输气干线和市区各网均需做大的改建和调整。

在气源问题上，按国家计委的工作重点已基本结束西气东输项目，正着手俄气南下的规划工作，看俄罗斯远东天然气输送工程已提到日程上来，据国家计委提供资料来看，俄罗斯乌拉尔以东的西伯利亚和远东地区，是世界上最大的能源宝库，拥有全球 1/3 的天然气资源，引进那里的天然气，是解决我国能源紧缺的一个重大战略。1999 年中俄双方签订了引进天然气可行性研究协议，2002 年 7 月，双方已在供气线路方面达成共识，即俄罗斯天然气通过东线主要供应我国东北地区，每年向我国出口天然气 200 亿 m^3，稳定供气时间 30 年，规划 2007 年送气，这使哈尔滨市供热工程向燃气化发展有了气源上的保证，将会从根本上改变哈尔滨市的供暖能源结构，化石能源中的燃煤将由化石能源中的清洁能源天然气所取代，到时，燃气的消费价格将远远低于电费甚至于低于燃煤的费用。

(5) 根据具体情况推广采用电锅炉

电能是清洁能源，电锅炉供热是清洁、高效、环保的供热方式。由于电力市场供过于求，哈尔滨市出现了电力过剩现象，哈尔滨市第三发电厂等电厂机组用量，为避免机组闲置带来的损失，有效利用电网峰谷差价，电力部门制定了一些优惠政策，如享受优惠峰谷电价，免征全部供电工程贴费和增容费等，这使电能成为解决中心区，太阳岛风景区等区域的最佳供能方式之一。

（6）分户供热管理方式也是节能，增效的有效途径

分户供热增强人们商品意识，通过经济杠杆调节，提高居民节能意识，并导致企业经营理念的改变和规模化产业结构的调整，产生很大的经济效益，提高热效率，节约了能源，实际上也就是减少了燃烧能源产生的大气污染物量，用户节能25%和相应热网改造，可使同一锅炉容量增加1/3的热面积，减少30% 燃料，除尘效率进一步提高，环境效益更好。

5 哈尔滨市供热发展趋势

城市供热设施是现代城市地方性基础供能设施，是现代高效利用热能，提高城市能源利用效率，降低大气污染的公用供热系统，它的发展水平反映一个城市的社会和经济的发展程度，城市供热设施的发展依赖于城市经济发展水平，城市供热设施发展水平应与城市经济发展水平同步，且应与城市社会基本消费承受能力，工商企业水平的提高相适应。

哈尔滨市是我国北方特大城市，采暖期长，其地理位置使发展集中供热有着广阔的前景，大力发展集中供热事业不仅改善城市生态环境，也将改善城市投资环境，促进城市可持续发展。供热事业的发展既要尊重我市能源结构特点，与我国节约能源发展政策相吻合，又必须符合环保要求。所以，发展以热电联产集中供热为主，以区域锅炉房为辅，地热、太阳能、天然气、核能、电能等多种清洁能源并存的供热格局就是哈尔滨市供热发展的大趋势。

Study on the Heating and Clean Energy in the cold city of Harbin

Dong Yang　Li Guiwen

（College of Architecture，Harbin Institute of Technology，Harbin　150006　China）

Abstract: This article has based on the heating question of coldly city in Harbin, contrasted the different heating ways of domestic and foreign coldly cities, introduced the situation of coldly city heating and energy using in our country, analyzed the main question of heating ways and energy using in Harbin, then proposed the development tendency of energy using and clean energy. At last, carried on the forecast to the Harbin heating development.

参考文献

[1] 沈贞珉等. 燃气供热锅炉技术培训教材. 北京：航空工业出版社

超低能耗住宅示范楼节能设计与效果分析

范宏武　陆善后　李德荣　曹毅然　张蓓红
（上海市建筑科学研究院有限公司，上海　200032）

摘　要：上海超低能耗住宅示范楼是为探索适合上海地域特征低能耗建筑发展模式而建造的。为尽可能降低该楼在使用过程中的采暖空调能耗，采用计算机仿真技术进行了节能设计。并参照《夏热冬冷地区居住建筑节能设计标准》，采用DeST软件对其全年采暖空调能耗进行了分析与评价。结果表明，该建筑全年采暖空调能耗值为11.3kWh/m^2或13.9kWh/m^2，远远低于现行国家节能设计标准，实现超低能耗的节能设计目标。

关键词：超低能耗建筑　节能设计　节能效果分析

前言

树立全面、协调和可持续的科学发展观已成为我国的基本国策，而对于城市建筑来说亦必须由传统高消耗型发展模式转向高效生态型发展模式。低能耗建筑正是实施这一转变的必由之路，因此成为当今世界建筑发展的必然趋势。

影响建筑能耗的因素有很多，如气候条件、室内热环境舒适要求、建筑热工性能、节能技术、采暖空调设备和经济性等。本文从上海地区气候特征出发，借鉴国内外最先进建筑技术的科研成果，从设计规划着手，对建筑围护结构、材料、设备和可再生能源利用等多个方面进行全面考虑，主要采用：(1)高效的外墙外保温系统；(2)先进的节能门窗系统；(3)高效遮阳技术；(4)倒置式保温与种植屋面技术，以期实现超低能耗的设计目标。

本文主要通过计算机仿真模拟技术对建筑物围护结构进行了节能设计，并根据《夏热冬冷地区居住建筑节能设计标准》，对超低能耗示范楼的能耗水平进行分析评价。

1　建筑概况

超低能耗住宅示范楼位于上海建科院工业园区，建筑面积总计277.5m^2，体形系数0.55。各朝向窗墙比分别为：南向0.45，北向0.22，东向0.01，西向0.06。

2　围护结构节能设计

建筑外墙是室内外空间的一道屏蔽，其设计是建筑节能设计的一个重要组成部分。该示范楼外墙若不采取节能措施，传热系数将超过2.0，在冬夏两季使用空调的情况下，将会有较多的热量通过外墙流失，因此有必要进行节能设计。

图1为不同朝向外墙的平均传热系数对建筑全年采暖空调能耗的影响情况。从图1可知，随着外墙平均传热系数的降低，建筑全年采暖空调能耗随之明显改善。考虑到施工工艺及保温材料使用的性价比，该建筑外墙平均传热系数统一设计为0.30W/(m^2·K)。具体构造为砂加气200mm填充墙加EPS100外墙外保温体系。

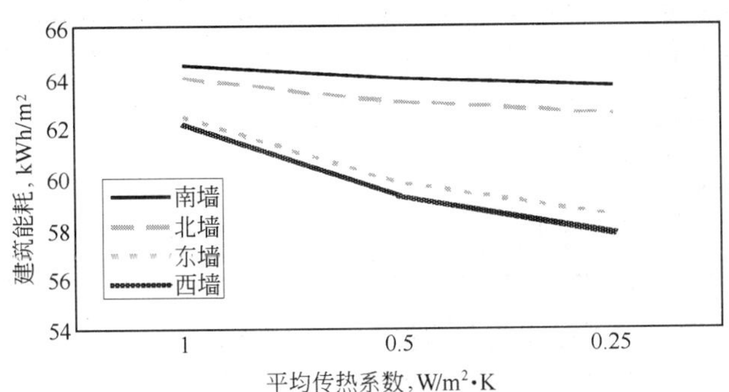

图1　墙体平均传热系数对建筑能耗的影响

屋面由于直接受太阳辐射面积大、时间长而成为节能设计的关键部位。节能屋面设计考虑因素较多，从施工难易角度出发，屋面更适合采用复合保温隔热构造、通风屋面与种植屋面形式。种植屋面是一种良好的隔热保温屋面，它对于降低建筑室内空调采暖负荷，改善室内热环境，调节室外气候，节约建筑能耗具有重要的作用，在实际工程中已广泛应用。因此，该示范楼设计采用两种节能屋面技术：种植屋面屋面和复合保温隔热屋面。

图2为屋面平均传热系数对建筑能耗的影响情况。从图2可以看出，当屋面平均传热系数低于0.5以后，其对建筑能耗的影响程度明显减弱。因此该示范楼屋面平均传热系数分别设计为0.23W/(m^2·K)和0.30W/(m^2·K)，具体构造型式如表1所示。

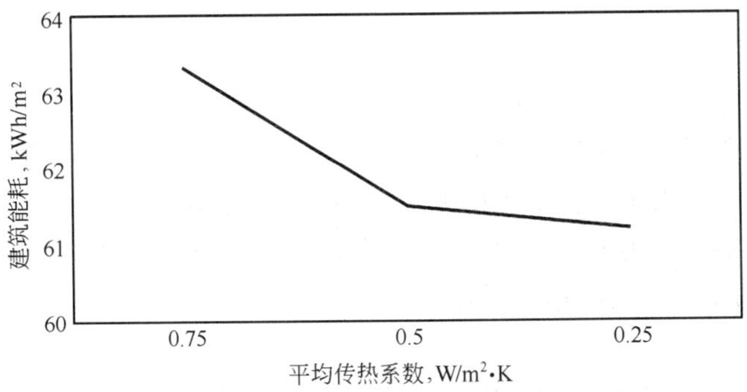

图2　屋面平均传热系数对建筑能耗的影响

超低能耗住宅示范楼屋面构造型式及材料组成　　**表1**

应用部位	保温体系主要构成(单位：mm)
平屋面	混合砂浆20+混凝土结构层120+陶粒混凝土找坡层(平均厚度100)+水泥砂浆找平20+防水层3+XPS100+隔离布+抗渗细石混凝土40(内设Φ4@200双向钢筋网片)+防水层3+疏水板20+轻质种植土600
坡屋面	混合砂浆20+混凝土结构层120+水泥砂浆找平20+防水层3+XPS100+隔离布+抗渗细石混凝土40(内设Φ4@200双向钢筋网片)+浅色外饰面涂层

由于外窗对建筑能耗的影响很大，因此在此示范楼中，外窗选择采用了真空低辐射中空塑钢窗，玻璃传热系数约 1.2W/(m² · K)，外窗平均传热系数为 1.5W/(m² · K)，玻璃遮阳系数 0.72；天窗采用夹胶钢化低辐射中空塑钢窗，玻璃传热系数 1.8W/(m² · K)，外窗平均传热系数取值为 2.5W/(m² · K)，遮阳系数 0.69。具体玻璃参数如表 2 所示。

超低能耗住宅示范楼窗户玻璃使用参数　　**表 2**

部　位	玻 璃 类 型	可见光透过率,%	遮阳系数	玻璃传热系数,(W/m² · K)	品　牌
外　窗	N4+K4+A9+N5Low-E 真空	67	0.72	1.2	新力基
天　窗	6+6 夹胶+12A+6Low-E 钢化	71	0.67	1.8	上海耀皮玻璃有限公司

由于太阳辐射对建筑空调能耗的影响很大，因此该建筑考虑采用外窗遮阳技术。但由于太阳辐射的方向性，考虑到冬季的采暖要求，不同朝向的外窗应采用不同的外遮阳形式。为实现最大可能的节能，本建筑采用了活动遮阳型式，具体如表 3 所示。

超低能耗住宅示范楼遮阳方式　　**表 3**

部　　位	玻 璃 类 型	品　　牌
南向(一层)	电动软遮阳+遮阳铝合金百叶	太阳能集热器遮阳由山东力诺瑞特提供 电动遮阳、遮阳百叶和内遮阳百叶由德国 WAREMA 遮阳公司提供
南向(二层)	太阳能集热器遮阳+遮阳铝合金百叶	
西向	遮阳铝合金百叶	
北向	内遮阳百叶	
天窗	电动软遮阳	

3　节能效果分析

该建筑由于设置了一个中庭，增加了建筑节能分析的复杂程度。为全面了解该建筑实际能耗水平，本文采用建筑能耗动态模拟软件 DeST 对中庭不用空调和中庭用空调两种工况进行分析评价，建筑分析模型如图 3～5 所示。计算条件参照《夏热冬冷地区居住建筑节能设计标准》中的 5.0.4 条规定设置。

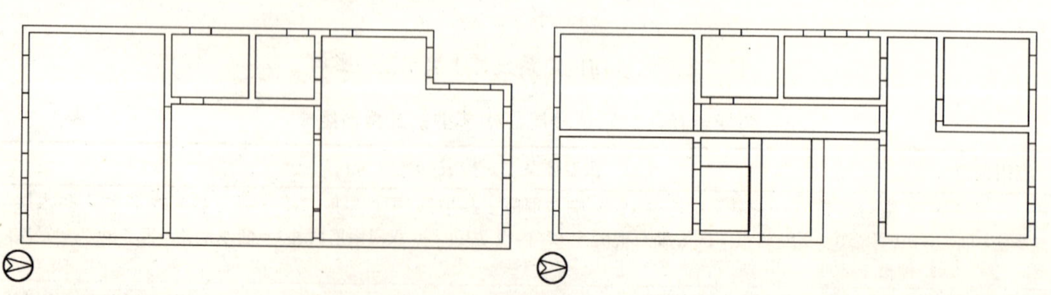

图 3　超低能耗建筑 1 楼模拟分析模型　　图 4　超低能耗建筑 2 楼模拟分析模型

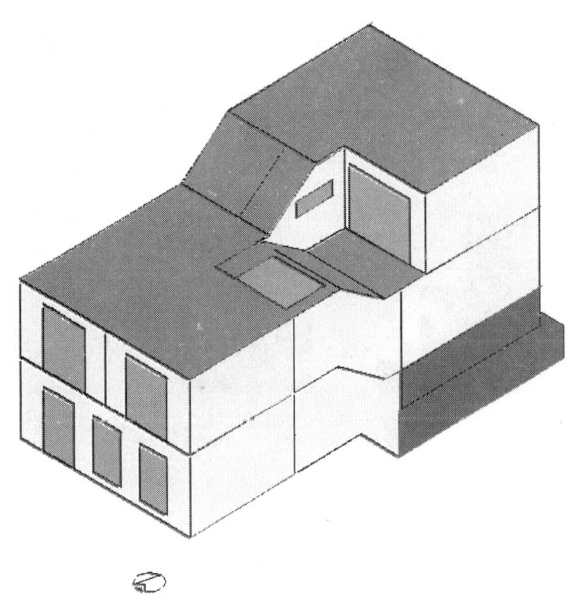

图 5　超低能耗建筑模拟分析模型示意图

（1）围护结构节能效果分析

通过与普通建筑、标准建筑（节能 50%）比较，分析评价该示范楼建筑围护结构的节能情况，其中，普通建筑外墙、屋面、外窗的平均传热系数分别为 2.0W/(m^2·K)、1.5 W/(m^2·K)和 6.4W/(m^2·K)；标准建筑的外墙、屋面平均传热系数分别为 1.5W/(m^2·K)和 1.0W/(m^2·K)，外窗传热系数则根据《夏热冬冷地区居住建筑节能设计标准》确定为南向 2.5W/(m^2·K)和北、东、西向 4.7 W/(m^2·K)。两种不同工况的建筑能耗计算结果如图 6 和图 7 所示。

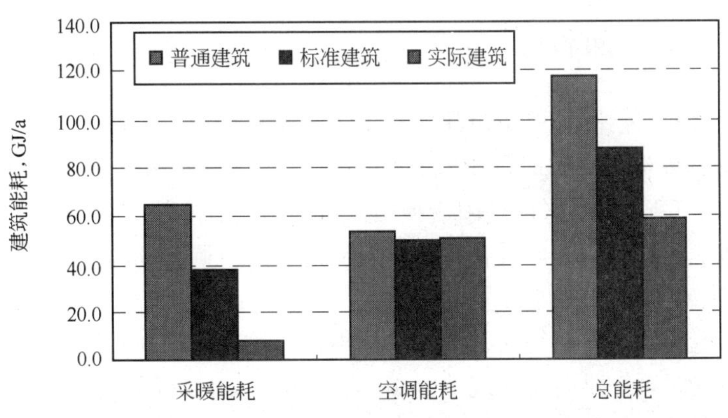

图 6　中庭不空调时围护结构节能情况

根据计算结果，在不采取遮阳的情况下，该示范楼中庭不用空调时全年采暖空调能耗为 59.0GJ，与普通建筑和标准建筑相比，围护结构分别实现节能 50%和 32%；中庭空调时全年采暖空调能耗为 68.4GJ，与普通建筑和标准建筑相比，分别节能 53.5%和 36%。

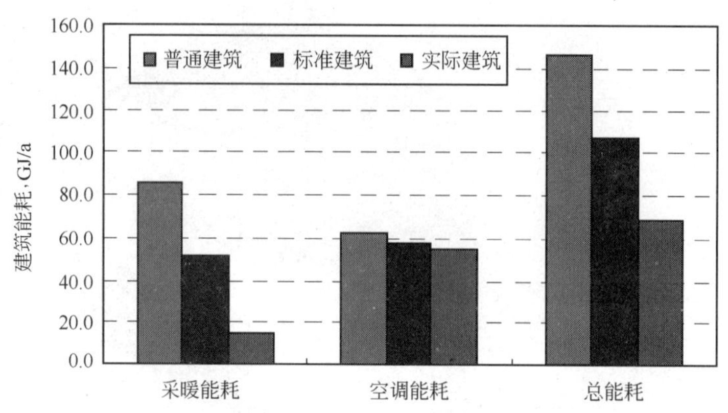

图 7 中庭空调时围护结构节能情况

(2) 遮阳系统节能效果分析

在其他围护结构热工性能已经确定的情况下，采用 DeST 软件对建筑不遮阳和遮阳两种工况的能耗进行计算。计算过程中，遮阳系统的遮阳系数取值为 0.5，从 4 月中旬～10 月中旬，每天 8：00～18：00 进行遮阳，计算结果如图 8 和图 9 所示。

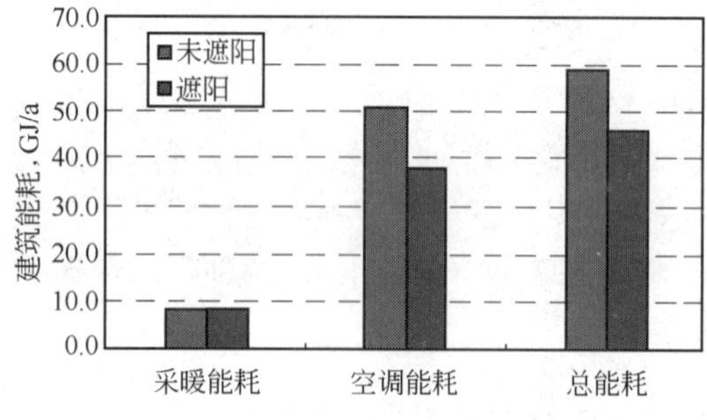

图 8 中庭不空调时遮阳系统节能情况

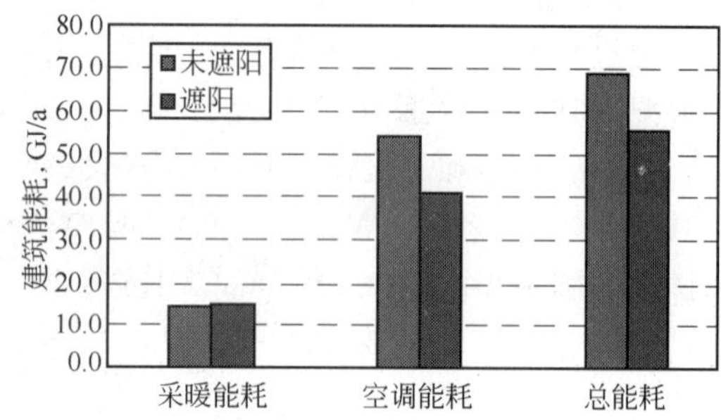

图 9 中庭空调时遮阳系统节能情况

从计算结果可知，活动外遮阳对建筑夏季空调能耗具有明显作用。对于该建筑，采用遮阳系数为 0.5 的活动外遮阳时，中庭不用空调时建筑全年采暖空调总能耗为 46.0GJ，中庭空调时为 55.4GJ。与不采取遮阳措施相比，其综合节能效果分别为 22%和 19%。这说明，对于上海地区，采用遮阳设备可明显降低建筑的空调能耗需求。

(3) 超低能耗建筑实际耗电量分析

为尽可能降低建筑空调系统对常规能源的消耗，该建筑采用地源热泵和辐射吊顶空调系统。根据地源热泵系统供暖可实现节能 40%～60%，制冷可节能 50%～ 70%，可估算出地源热泵和辐射吊顶系统采暖 COP 可达 3.0 以上，制冷 COP 可达 4.5 以上。因此，本文系统采暖 COP 取值为 3.0，制冷 COP 为 4.5。结果如图 10 所示。

根据计算结果，当中庭不用空调时，建筑全年采暖空调耗电量约 3100kWh(11.3 kWh/m^2)，中庭空调时约 3860kWh(13.9 kWh/m^2)，远远低于《夏热冬冷地区居住建筑节能设计标准》中 55.1kWh/m^2 的节能指标。

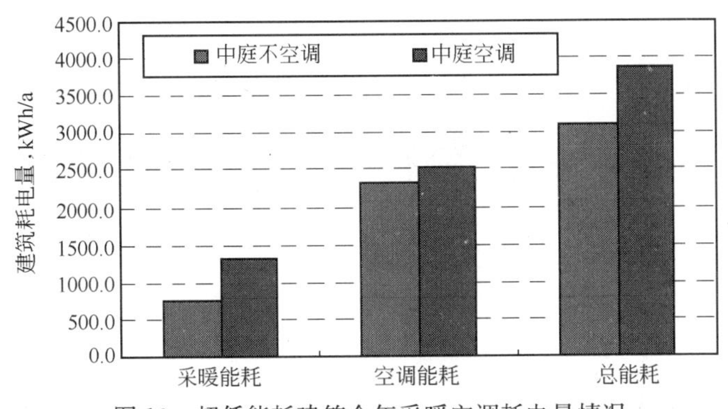

图 10 超低能耗建筑全年采暖空调耗电量情况

4 结论

为实现住宅示范楼的超低能耗设计目标，本文根据计算机仿真模拟分析结果，对建筑外墙、屋面和外窗遮阳进行了相关节能设计与分析。结果显示，该建筑全年采暖空调能耗值为 11.3kWh/m^2 或 13.9 kWh/m^2，远远低于《夏热冬冷地区居住建筑节能设计标准》的节能指标，实现超低能耗目标。

Simulation Analysis on the Energy Efficiency for Building with Low Energy Consumption in Shanghai

Fan Hongwu Lu Shanhou Li Derong Cao Yiran Zhang Beihong
(Shanghai Research Institute of Building Sciences, Shanghai 200032 China)

Abstract: To explore the developing mode of the building with Low energy consumption in Shanghai, the energy efficient design of the demo-house is carried out by the simulation analysis. The energy performance of this building is also studied by the simulation tool DeST and the result shows that, contrast with the energy efficient standard of China, the energy consumed for heating and cooling of this building is very low, only about 11.3kWh/m^2 or 13.9kWh/m^2, that means the demo-building is a lower energy consumption building.

夏热冬暖地区居住建筑节能设计实践
——以深圳市振业城为例*

叶　青[1]　卜增文[1]　马晓雯[1]　刘俊跃[1]　陆培康[2]　徐　丹[3]
([1] 深圳市建筑科学研究院，深圳　518031；[2] 深圳市振业(集团)股份有限公司，深圳　518001；[3] 华森建筑与工程设计顾问有限公司，深圳　518069)

摘　要：本文以深圳振业城为例，对夏热冬暖地区大型居住小区的建筑节能设计进行了详细分析，涵盖了围护结构节能设计、自然通风的利用以及太阳能热水的广泛利用。

关键词：夏热冬暖地区　建筑节能　性能化设计　自然通风

1　前言

1.1　工程简介

振业城位于深圳市龙岗区横岗镇六约(图1)，南临深惠路，北接环城北路，东为梧桐路，西为深峰路，地块距深圳市区18km，用地面积约41.7万m^2，总建筑面积约63.5万m^2，建筑容积率为1.3，建筑覆盖率为25%，绿地率为40%。分三期建设，一期用地面积19.25万m^2，建筑面积14.7万m^2，住宅建筑面积12.13万m^2，以联排别墅为主，少量多层单元式住宅；二期、三期以高层为主。

图1　振业城

1.2　设计依据

本项目从规划设计阶段开始，建筑设计单位华森建筑与工程设计顾问有限公司即进行了部分节能设计。后来甲方邀请深圳市建筑科学研究院建筑节能研究中心，按照建设部建筑节能示范项目要求进行了自然通风和建筑围护结构的专项节能设计。设计的依据为《夏热冬暖地区居住建筑节能设计标准》JGJ 75—2003(以下简称《标准》)中的1条推荐性条文和5条强制性条文，同时也参考了《深圳市居住建筑节能设计规范》的相关条文。

* 美国EF基金资助项目。

条文 4.0.1　居住区的总体规划和居住建筑的平面、立面设计应有利于自然通风。

条文 4.0.4(强制)　外窗的面积不应过大，各朝向窗墙面积比应符合下列规定：北向不大于 0.45；东、西向不应大于 0.30；南向不应大于 0.50。

条文 4.0.5(强制)　居住建筑的天窗面积不应大于屋面总面积的 4%，传热系数不应大于 4.0W/(m^2·K)，本身的遮阳系数不应大于 0.5。

条文 4.0.6(强制)　居住建筑屋顶和外墙的传热系数应符合表 1 的规定。

屋面和外墙的传热系数 K[W/(m^2·K)]　　**表 1**

屋　顶	外　墙
$K \leqslant 1.0$，$D \geqslant 2.5$	$K \leqslant 2.0$，$D \geqslant 3.0$　或　$K \leqslant 1.5$，$D \geqslant 3.0$　或　$K \leqslant 1.0$，$D \geqslant 2.5$
$K \leqslant 0.5$	$K \leqslant 0.7$

注：$D<2.5$ 的轻质屋顶和外墙，还应满足国家标准《民用建筑热工设计规范》GB 50176—93 所规定的隔热要求。

条文 4.0.7(强制)　居住建筑采用不同平均窗墙面积比时，其外窗的传热系数和综合遮阳系数应符合表 2 的规定。

南区居住建筑外窗的综合遮阳系数限值　　**表 2**

外　墙 ($\rho \leqslant 0.8$)	外窗的综合遮阳系数 S_W				
	平均窗墙面积比 $C_M \leqslant 0.25$	平均窗墙面积比 $0.25<C_M \leqslant 0.3$	平均窗墙面积比 $0.3<C_M \leqslant 0.35$	平均窗墙面积比 $0.35<C_M \leqslant 0.4$	平均窗墙面积比 $0.4<C_M \leqslant 0.45$
$K \leqslant 2.0$，$D \geqslant 3.0$	≤0.6	≤0.5	≤0.4	≤0.4	≤0.3
$K \leqslant 1.5$，$D \geqslant 3.0$	≤0.8	≤0.7	≤0.6	≤0.5	≤0.4
$K \leqslant 1.0$，$D \geqslant 2.5$ 或 $K \leqslant 0.7$	≤0.9	≤0.8	≤0.7	≤0.6	≤0.5

条文 4.0.10(强制)　外窗(包括阳台门)的可开启面积不应小于外窗所在房间地面面积的 8%或外窗面积的 45%。

1.3　设计原则

针对深圳市的气候特点，并结合深圳市的经济发展现状和本项目的特点，确定了振业城一期的节能设计原则。

(1) 采取整体设计策略：从总图规划、建筑单体设计、材料选择与经济技术比较的全过程考虑节能设计，而不仅仅是在单体建筑阶段进行，可以更加有效实现建筑节能设计，降低修改返工成本。

(2) 从建筑节能的经济性、地方性、阶段性出发选择适宜的节能技术路线；我们要立足深圳现有的建筑材料资源，用简单成熟的技术和较低造价实现建筑节能，而不是高科技、新材料、高造价来实现高节能率；如没有采用外保温技术，也没有采用铝合金百叶活动遮阳技术；否则造价将会成倍增加。

(3) 适应气候的节能设计原则：根据深圳市的气候特点，采取以“自然通风+间歇式空调”为主的被动式设计策略，避免出现“节能建筑不节能的情况”。如：有的建筑保温隔热性能非常优越，但是不能够自然通风，即使在过渡季节也要空调，全年能耗远高于一般自然通风较好的建筑，因为这类建筑夏季空调能耗大，但是在过渡季节可以完全不用空调而达到热舒适。

2 振业城小区建筑节能设计

节能设计中我们将重点放在了利用自然条件和本地资源上，首先考虑自然通风，其次利用本地的建筑材料如墙体材料和门窗材料。

2.1 自然通风设计

2.1.1 深圳市的自然通风条件(表3)

深圳市室外气象数据统计表 表3

月 份	5月	6月	7月	8月	9月
室外温度低于28℃天数	24	14	11	11	11
室外温度低于28℃的天数占月总天数比例	77%	47%	36%	36%	37%
主导风向平均风速 m/s	3.20	2.95	2.90	2.80	2.95

根据对深圳市气象统计分析，5～9月，每月有10天以上室外气温低于28℃，其中5月份室外气温低于28℃的天数超过70%，对应的室外风速在2.8～3.2m/s之间，具备采用自然通风降温的条件。

建筑的总体布局决定了建筑的朝向，并与基地的通风和热环境状况密切相关，因此适应气候的总体建筑布局是建筑节能设计的起点。该工程采取适应气候的总体建筑布局。

2.1.2 疏导基地和建筑群体的通风

在规划设计中，合理、高效的利用自然通风是节能的一个重要手段，在深圳市尤其重要。深圳市在可利用自然通风的季节，其主导风向为东偏南22.5度或东南风，为减少夏季空调运行时间，并保证春、秋季不使用空调时的室内热舒适性，在群体空间布局上，采用斜列、错列等方式以疏导气流。同时依据地势使建筑在空间上错位。图2和图3是振业城一期的自然通风模拟结果。

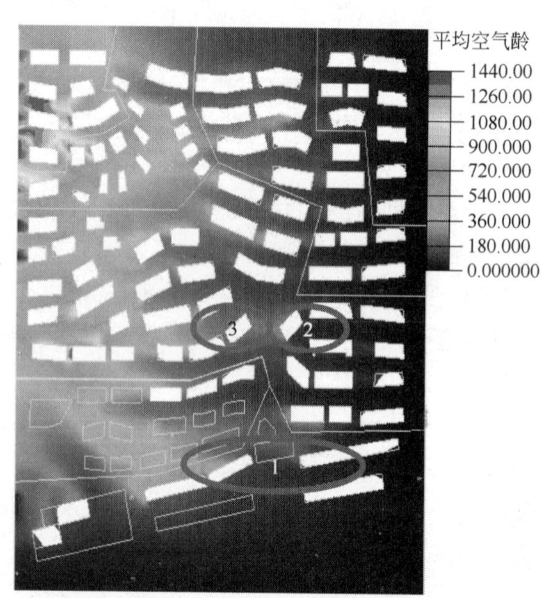

图2 小区改前空气龄图

按照《标准》条文4.0.10的要求进行外窗可开启面积的设计外，我们对该项目一期进行了小区、组团和户型的自然通风的计算机模拟设计，以期通过小区换气、建筑换气、室内换气三个层次充分利用深圳市良好的自然通风条件。

根据小区和组团的自然通风计算机模拟结果，我们发现在小区西北角出现一个空气滞留区（模拟图中区域为1，2，3，如图2），空气龄达26分钟，这里的通风状况将会非常差，这里的住户即使户型很好，南北通透，但是也将无法实现自然通风。因此，我们建议对该小区的原规划设计进行了调整，如图3。

修改1：剪掉部分商住楼，增大室外风入口宽度；修改2：取消该栋楼，消除对主导风的阻挡；修改3：将该联排别墅去掉一户，将室外风引入小区内部。

通过对规划设计的调整，小区西北角的建筑周围室外空气龄明显降低，特别是去掉了几个气流死角，改善了建筑周围的自然通风环境，如图3。

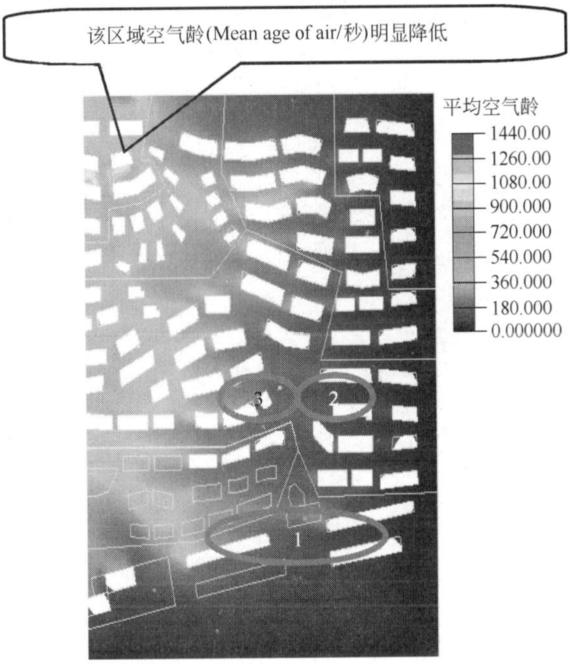

图3　小区改后空气龄图

2.1.3　建筑单体自然通风设计

夏热冬暖地区的南区，在充分考虑建筑遮阳的前提下，怎样高效利用自然通风成为建筑节能设计的主要内容之一，这一点在深圳尤为重要。因为深圳地处海边，具有良好自然通风的气候条件，夏季昼夜海陆风交替，在空调季节，利用好自然通风可采用间歇式通风技术降温，减少开启空调的时间，节约能源。

要强化自然通风，就需要外窗的可开启面积较大，如何在较小的外窗面积上最大限度的利用自然通风是本次节能设计的重点之一。

建筑设计时作了如下考虑：单体建筑室内分隔简洁，建筑进深适中，建筑户型前后没有阻挡；对单个居住空间，在有两面外墙的情况下，均在不同的外墙上开设了可开启的外窗，且可开启的面积之和大于外窗所在房间地面面积的8%，为居住空间夜间利用自然通风降温创造了条件，如图4和图5。

根据对户型的自然通风模拟结果，我们对原设计中自然通风不利户型的室内分隔、外窗的位置、外窗的可开启尺寸等进行了调整。

结论：通过移动外窗位置，增大外窗可开启面积，该户型室内的气流更加畅通，室内空气龄明显降低，特别是餐厅和客厅两个居室的室内空气新鲜程度提高。

自然通风设计基本不需要或很少增加建造成本，而带给住户的是舒适健康的居住环境和巨大的节能贡献率，如前文所述，即使夏季，也可以有46%的时间利用自然通风解决热舒适，而不需要空调。

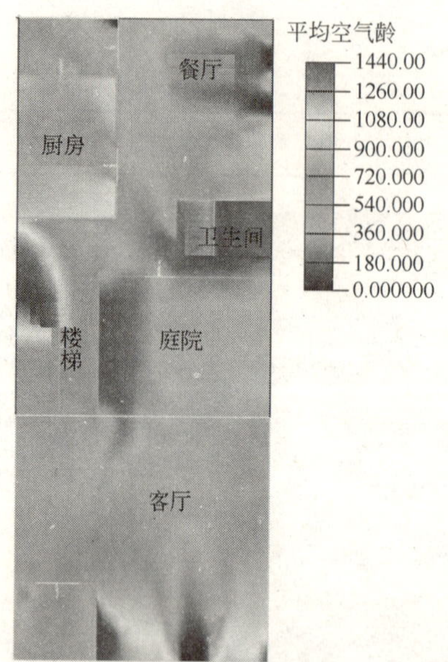

图4 改前空气龄图

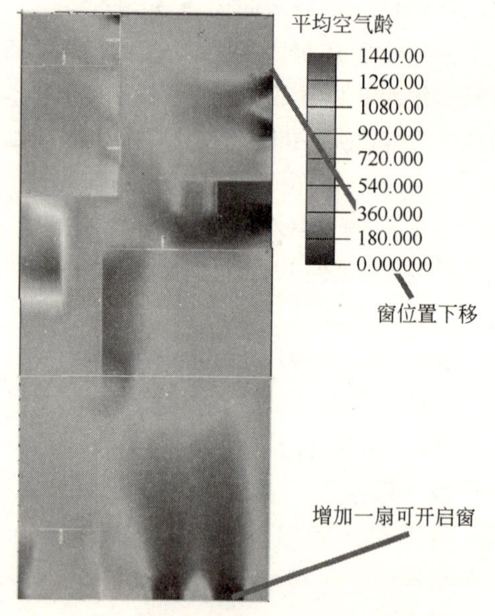

图5 改后空气龄图

2.2 建筑热工设计

围护结构的节能设计必须要增加成本，增加多少与采用什么样的材料，在什么地方使用什么样的节能产品或节能材料非常有关系。例如在深圳，有的节能设计照搬北方做法，采用断热型材的铝合金双层白玻璃窗，造价是普通窗的3～4倍，但是还没有达到节能设计的要求。有的节能设计认为西晒严重，不利于节能，那么在西面外墙采用外墙外保温，而其他朝向采用普通的灰砂砖；屋顶保温层做得非常厚，结果也是不满足节能设计标准，开发商花了大量的投资，却没有获得应有的结果。

围护结构节能设计如何设计呢？

2.2.1 节能率50％的达标途径选择

《标准》中提供了两条达标途径，规定性指标和性能性指标。本项目由于是示范项目，我们对建筑围护结构节能设计采用性能和规定性两个方法，通过技术、经济比较，得出最佳的围护结构节能设计方案，并实施(图6)。

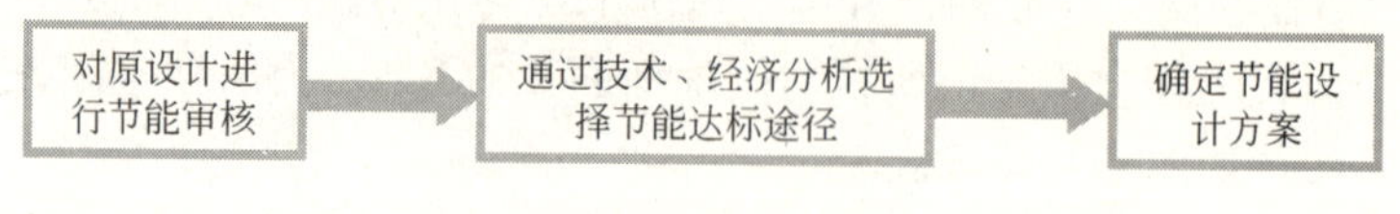

图6 节能设计步骤

2.2.2 围护结构节能计算步骤

(1) 按规定性指标进行节能设计。

逐条对照上述审核结论中不符合《标准》要求的设计参数，修改这些设计参数，使之达到《标准》规定性指标的要求。通过详细的技术经济分析，确定的节能设计方案为：

(一) 窗墙面积比：减小东向外窗的面积，使窗墙面积比低于0.30。

(二) 天窗：取消现有建筑施工图中的全部天窗。

（三）外墙：改外墙是为了减轻外窗节能的压力，建筑的外墙全部采用加气混凝土砌块墙体，其平均传热系数 K=1.33W/(m^2·K)，热惰性指标 D>3.59。

（四）屋顶：屋顶的保温层 40mm 不变，传热系数 K=0.641W/(m^2·K)，热惰性指标 D=3.45>2.5。传热系数优于节能设计标准 1.0W/(m^2·K)的要求。

（五）外窗（含阳台门透明部分）：减小各朝向外窗的面积，使建筑的平均窗墙面积比 C_M≤0.30。采用遮阳构造，外窗的综合遮阳系数限值为 S_W≤0.7。

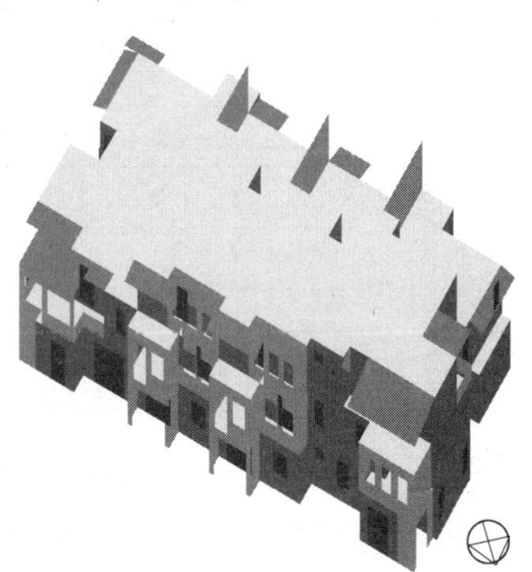

图 7　节能设计的 DOE 模型

（2）按性能指标进行节能设计。

为了保持原设计风格，即：1)不减小东向外窗的面积；2)不减小天窗面积。因此，需要通过性能化的能耗分析来修改原设计，使之达到《标准》的节能要求。

（一）选择计算工具

采用美国劳伦斯伯克力国家实验室开发的 DOE-2 动态能耗计算软件作为计算工具(图 7)，对所需要的建筑进行建模计算，以其中 1G4 栋建筑为例。

（二）计算参照建筑的能耗指标：通过计算可知，参照建筑空调年耗电量为：40.62kWh/(m^{2}a)。

（三）计算设计建筑的节能综合指标

输入的外围护结构保温构造如表 4～6。

外　　墙　　**表 4**

代　码	墙体材料或保温材料	传热系数 W/(m^2·K)
A	190mm 加气混凝土砌块墙体	0.912
B	25mm XPS 板外墙外保温体系	0.776

外　　窗　　**表 5**

代　码	传热系数 W/(m^2·K)	遮 阳 系 数
A	4.5	0.6
B	4.5	0.5

屋　　顶　　**表 6**

代　码	屋顶保温材料	传热系数 W/(m^2·K)
A	25mm XPS 板屋顶保温体系	0.886
B	40mm XPS 板屋顶保温体系	0.641

设计建筑的节能综合指标计算结果如表 7。

表 7

方　　案				设计建筑节能综合指标 kWh/(m^2a)	是否达到标准的要求
序　号	外　墙	屋　顶	外　窗		
1	A	A	A	41.32	否
2	A	A	B	39.01	是
3	A	B	A	40.24	是
4	A	B	B	37.97	是
5	B	A	A	40.13	是
6	B	A	B	37.89	是
7	B	B	A	39.03	是
8	B	B	B	36.81	是

从计算结果可知，如果想保持原设计建筑的外观不变，则只要选择序号为 2～8 的外围护结构节能方案，该建筑就能达到《标准》的要求，是节能建筑。

（四）确定节能设计方案

(1) 采用上述规定性指标的节能方案，减小了大量的外窗面积，去掉了天窗，因此节省了大量的初投资。

(2) 通过性能化设计，可以：不用减小东向外窗和天窗的面积、外窗综合遮阳系数可以提升到 0.6。

通过与建设方和设计单位的共同技术经济论证，本项目采用以下设计方案：

◆ 减小各朝向外窗面积，使各朝向窗墙面积比满足《标准》的要求。即北向≤0.45；东、西向≤0.30；南向≤0.50；

◆ 外窗：由于深圳地区的外窗主要是控制太阳辐射得热，而内院的窗四周有遮挡，北向的窗相对于其他朝向来说受太阳辐射的照射也较弱。同时，为了尽量减小东、西向和南向外窗的太阳辐射得热，优化节能设计。因此，选择外窗构造：北向外窗和内院外窗采用普通玻璃窗($S_C=0.8$)，其他朝向外窗采用单层 Low-E 玻璃窗($S_C=0.5$)。并且在确定外窗面积时，满足下式：

$$\frac{0.8\times(\text{北向外窗面积}+\text{内院外窗面积})+0.5\times\text{其他朝向外窗面积总和}}{\text{整个建筑的总外窗面积}}\leqslant 0.7$$

从整个振业城一期来看，外窗综合遮阳系数的选择优于《标准》的规定值($S_C=0.8$)，是《标准》规定值的 0.875 倍。北向外窗和内院外窗采用普通玻璃窗($S_C=0.8$)，其他朝向外窗采用 Low-E 中空玻璃窗($S_C=0.5$)。

◆ 取消全部的天窗；原有天窗的面积过大，达到 6.3%，超过标准的 4%。

◆ 建筑的非承重外墙：全部采用加气混凝土砌块墙体；既符合国家及深圳市墙体革新政策，同时在热工性能方面也优于原来使用的粘土砖。外墙构造为：20mm 石灰水泥砂浆＋190mm 加气混凝土砌块＋20mm 水泥砂浆。外墙平均传热系数为 1.26W/(m^2·K)，热惰性指标 $D=3.59$。优于节能设计标准 1.5W/(m^2·K)的要求，是标准要求的

0.84倍。

◆ 屋顶：采用30mm的挤塑板屋面保温体系，K=0.786W/(m^2·K)；屋面构造做法为：20mm水泥砂浆+40mm细石混凝土+30mm挤塑板+2mm防水涂膜+15mm水泥砂浆+30mm陶粒混凝土+180mm钢筋混凝土。屋顶传热系数K=0.805W/(m^2·K)，热惰性指标D=3.41。优于节能设计标准1.0W/(m^2·K)的要求，是标准要求的0.8倍。

(3) 节能率验算

按规定性指标进行完节能设计修改之后，我们又按性能性指标的要求对最终的施工图进行了节能验算，结果如表8。

表8

代表性建筑	单位建筑面积空调年耗电量 kWh/(m^{2}yr)		是否符合标准要求	设计建筑能耗是参照建筑能耗的倍数	节能率
	参照建筑	设计建筑			
C9栋	34.07	32.25	是	0.95	52.67%
D9栋	38.14	37.90	是	0.99	50.31%
E5栋	35.51	32.19	是	0.91	54.67%
F8栋	45.15	41.71	是	0.92	53.81%
G9栋	45.03	42.23	是	0.94	53.11%
H16栋	42.00	41.50	是	0.99	50.60%

结论：各代表性住宅都超过了节能50%的要求，是节能建筑。

3 总结

通过甲方的初步估算，振业城一期的住宅节能设计增加的总投资为：2500万元左右。该投资包括：屋顶保温材料、LOW-E中空玻璃窗、太阳能热水系统。振业城一期住宅建筑面积约12万m^2，折算到单位建筑面积，约206元/m^2。如果不包括太阳能热水器的投资，则节能设计增加的造价约177元/m^{2}当然这一数据也是粗略计算，如果详细计算，造价应当还会下降。

通过对建筑的自然通风、建筑热工的节能专项设计，振业城一期达到了《标准》节能50%的要求，并已通过了国家3A级住宅性能预审。

总的来说，作为夏热冬暖地区居住建筑节能示范工程，深圳振业城一期节能示范工程的节能设计采用了较为全面的整体设计策略和性能性指标设计手段，在适应地方气候、结合地方经济和地方材料方面都进行了大胆的尝试，采取以传统技术更新为主的技术路线，为夏热冬暖地区和深圳市建筑节能工作的开展拓宽了视野，进行了有益的尝试。

但是还是存在一些问题，一是成本偏高，是否可以达到实现建筑节能50%，增加的造价不要超过50元/m^2。二是建筑墙体材料单一，现在惟一的选择就是加气混凝土砌块。如果是高层住宅，剪力墙面积过大，超过40%，即使使用加气混凝土砌块也无法满足传热系数小于1.5的要求，势必增加窗户的节能压力，增加造价。

本项目的顺利完成要感谢：广东省建设厅、深圳市建设局、龙岗区建设局、EF基金会专家的大力支持。

Energy Efficiency Design of Residential Building in Hot Summer & Warm Winter Zone

Ye qing[1] Bu Zengwen[1] Ma Xiaowen[1] Liu Junyue[1] Lu Peikang[2] Xu Dan[3]

(［1］ Shenzhen Institute of Building Reseach，Shenzhen 518031 China；

［2］ Shenzhen ZhenYe Limited Liability Company，Shenzhen 518001 China；

［3］ Hua Sen Architecture and Engineering Design Consultant Company Limited，Shenzhen 518069 China)

Abstract：This paper analyses the building energy efficiency design method for the Zhenye Town，a large community in Shenzhen，a typical city in hot summer & warm winter zone. The comprehensive design includes building energy efficiency analysis，natural ventilation and solar water heater use.

高层住宅太阳能一体化设计体系研究*

史　洁　宋德萱
（上海同济大学建筑与城市规划学院，上海　200092）

摘　要： 本文结合中国建筑太阳能利用现状，针对上海高层住宅的节能发展，提出了高层住宅太阳能一体化设计的可操作系统：太阳能技术集合策略、系统的设计方法和政策引导的运作机制三方面的有机结合；强调了未来高层住宅的可持续性设计，不是单一技术体系的运作，而是多技术体系与建筑系统的整合过程。

关键词： 高层住宅　一体化设计　建筑节能　太阳能利用

在中国建筑建设中，高层住宅发展的速度和规模令世人瞩目，同时建筑能耗的问题也日益凸现。据最新统计，2000 年中国建筑能耗已占当年全社会终端能源消耗量的 27.8%，接近发达国家建筑用能占全社会能源消费量 1/3 左右的水平 。目前利用可再生能源，提高建筑能效的研究与实践已逐步展开，如何将太阳能利用技术与中国国情下的高层建筑完美的结合，创造低能耗高舒适度的健康居住建筑，已成为住宅节能建设中的一个最新亮点。

1　中国的“阳光事业”

太阳能的推广利用工作被誉为“阳光事业”。中国地域辽阔，有着丰富的太阳能资源。据估算，我国陆地表面每年接收的太阳辐射能约为 50×10^{18}kJ，全国各地太阳年辐射总量达 335～837kJ/cm^2·a，平均值为 586kJ/cm^2·a，属太阳能资源丰富的国家之一。

我国自 1977 年始，从甘肃民勤县建造了第一座被动式太阳房，至今已推广约 1000 万 m^2（建筑面积）。目前我国被动太阳房已进入规模普及阶段，并开始由群体建筑向住宅小区、太阳村、太阳城发展。我国被动太阳房采暖节能达 60%～70%，平均每 m^2 建筑面积每年可节约 20～40kg 标准煤，发挥着良好的经济和社会效益。经过数年的研究和开发，太阳能热水器的推广应用范围在不断扩大，其发展在我国最为迅速，已形成为一个产业。新型太阳能集热器的试制成功，使太阳能热水器可作为建筑构件应用在建筑的外围护结构上。

图 1　辛庄生态办公样板楼

*　国家自然科学基金资助项目（批准号：50478101）。

当前我国正尝试建设各种生态节能的示范工程，如上海辛庄工业园区生态办公样板楼（图 1）。该工程中尝试应用了真空管太阳能热水器、PV 光电板、太阳能空调和地板采暖系统等太阳能复合技术。总体而言，国内这些年在太阳能利用的过程中，从理论研究、模拟试验、材料构件的开发、示范房屋及设计工程等方面，已逐步积累了经验，为我们在建筑中进一步开发利用太阳能奠定了良好的基础。

2 上海高层住宅节能与太阳能利用潜力

上海是中国最大的综合性城市，人口普查数据显示，2000 年的人口密度达到 2588 人/km^2，远远高于全国人口密度每平方公里 132 人的数值，居于全国首位。因人口聚集且受地价影响，在城市聚居区建筑密度日趋提高，高层尤其是小高层住宅建设发展迅猛。高层住宅对于建设用地紧张、寸土寸金的上海，无疑已成为最适宜的建筑方式。

作为非采暖地区的上海，由于气候的原因，冬夏两季多通过消耗电能来换取室内热舒适度，据统计，百户空调安装率已高达 80%左右，其能耗对城市环境造成了极大的压力。2001 年 10 月，国家建设部颁发了我国《夏热冬冷地区居住建筑节能设计标准》，同时上海市建委会同有关部门组织编制了《上海市“十五”期间建筑节能实施纲要》，提出了推进建筑节能的发展思路和工作目标，计划到 2010 年，全部新建住宅建筑由执行节能标准的 50%提高到 60%。目前为使建筑节能工作有序推进，上海市依据国家标准，已逐步建立和完善了有关地方标准，基本形成适合上海特点的围护结构建筑节能技术体系；实施了近百万平方米节能住宅项目工程试点。

太阳能与建筑一体化的发展，不仅能有效的节能和环保，而且也为建筑外观设计拓展了新的创作空间。太阳能资源因受气候条件、地理纬度等环境状况的影响，其分布具有明显的地域性。我国共划分了四个太阳辐射资源区，上海属夏热冬冷地区，处在太阳能辐射资源Ⅲ区[❶]。判断太阳能资源充足与否，主要依据太阳日照时数与日照百分率两个因素。依据对它们的比较分析，在夏季，夏热冬冷地区的太阳能辐射资源与我国的北方及西部地区基本持平；在冬季虽不及我国北方，却明显超过欧洲，且与北美基本持平[❷]。足见在我国夏热冬冷地区，对太阳能的开发利用有着巨大的潜力。目前上海已制定 2005 年～2007 年开发利用太阳能《行动计划》，力争 3～5 年内在光伏技术和产业方面走在全国前列，在太阳能光热利用上取得显著成效。应用目标是到 2007 年，安装与建筑结合的太阳能热水系统 10 万 m^2(集热面积)。

3 太阳能建筑一体化设计体系的可操作途径

3.1 建筑中太阳能技术的集成策略

对太阳能的利用主要有光热转换、光电转换、光化学转换三种方式。在建筑中主要利用的是光热与光电转换。建筑中太阳能利用技术体系主要可归纳为三个层次：被动式太阳

❶ [我国的太阳能资源/Html]. 西部开发网，http：//www.west-develop.com/resouce/ny/ny-08-11-01.Html。

❷ 李保峰. 适应夏热冬冷地区气候的建筑表皮之变化设计策略研究. 北京：北京清华大学，博士论文，2004。

能系统、太阳能热利用技术系统、太阳能光伏发电系统。太阳能在建筑中的利用应依据可持续性设计的原则，不应是以往的单一类型体系的技术利用，建筑作为一个复杂而统一的整体，对太阳能的利用应采用适用技术的集成策略，首先是建筑本身的基本设计，其次是在合理的前提下，尽可能的利用被动技术与自然能源，最后用主动技术来弥补前两个层面所不能满足的要求。

3.2 高层住宅太阳能利用的系统设计方法

将太阳能技术体系纳入建筑设计的整个过程，建立一个有效的系统的设计程序。

3.2.1 小区规划中的群体组合与太阳能的利用

在规划中建筑群体的不同方位、体形、间距、高低及道路网的布置，广场绿地的分布等都会影响规划区的微气候，影响建筑的日照和通风，影响到建筑的能耗。为合理的规划小区，确保每栋建筑的有效日照和最大的接收太阳能，可利用“太阳围合体”(Solar Envelope)对建筑形态进行控制。“太阳围合体”方法是针对特定的区域空间，通过调整围合建筑各立面的法线方向，使建筑在不遮挡临近建筑物日照的情况下达到最大的体积容量(图 2)[1]。目前“太阳围合体”设计可以方便的在计算机中进行操作。

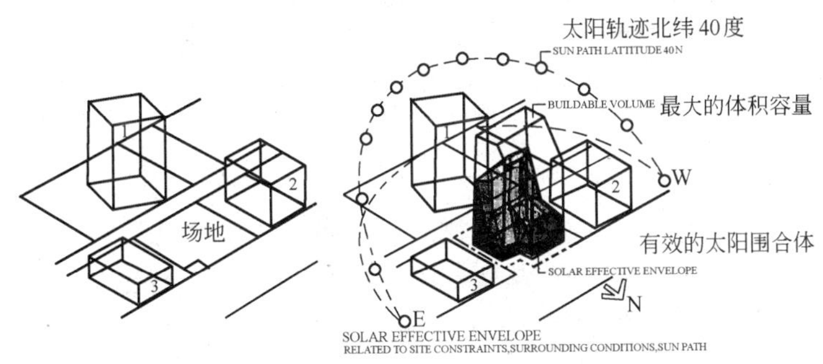

图 2 太阳围合体

3.2.2 建筑单体方案与太阳能的利用

1. 体型：对高层而言，其体型主要取决于建筑的进深与高度的相对比值。按它的外部体形可分为塔状、板状两大类。在这方面，美国南 California 大学以洛杉矶区域为参照，进行了一个 10 年的住宅研究，通过对城市住宅的 V/S(建筑体积与外表面积比)[2]与密度(每英亩的住宅单元数)的分析研究，发现它们的大小影响着建筑设计表现形式与用能的方式。研究结果显示：对于 3～7 层的建筑，在城市建设发展与节能方面有着较为宽泛的选择优势，也是利用被动式太阳能与低能耗策略的最佳层数；当 V/S 值超过 10 时，建筑使用高能耗是无法避免的；这也从太阳能的利用角度，揭示了住宅建筑不宜追求过高或过密，否则设计中主要应对的是内部能耗问题，而很少能结合自然来进行能效设计。

2. 平面空间：高层住宅平面类型大多为点式或板式。据调查，上海居民喜欢房间在有通风的前提下主要卧室一定要有直射阳光，即卧室和客厅要求朝南，它不但可以增加

---

[1] Ekkachai Mahaek. Solar effective envelope design advision(SEEDA). Dissertation Abstracts International, Volume：64-04，Section：A，page：1108.

[2] Ralph L. Knowles. [The solar envelope：its meaning for energy and buildings]. School of Architecture，Energy and Buildings 35(2003)15～25.

太阳照射的时间，也可以为家庭晒衣被提供场所。据此在将阳台封闭处理成太阳房时，要与通风系统相结合，合理组织气流，给居室带来温暖的同时，也改变其湿冷的状况(图3)。

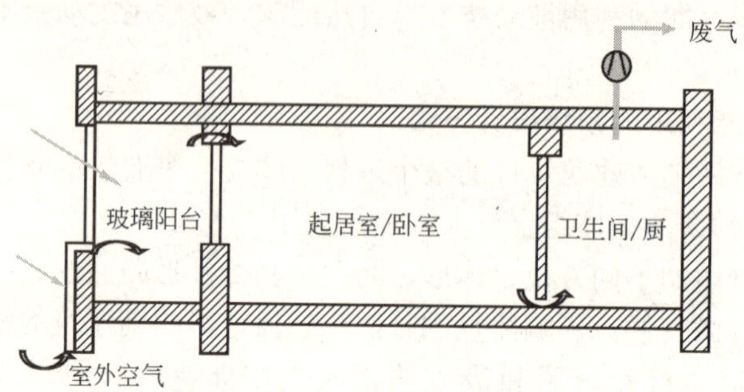

图3 太阳房与通风组织的结合

3.2.3 建筑构件与太阳能元件的一体化设计：

太阳能建筑一体化意味着把传统的建筑围护结构从能量散失的部分转换成能量吸收部分，是将太阳能技术元件与建筑构建的一体化，但这并非简单地在建筑上安装一些太阳能元件，而是将他们与建筑物本身一体化成建筑的组分。这可以从如下的高层住宅外界面进行系统的结合。

墙体：外墙体是高层住宅接收太阳最多的表面。太阳能的应用将进一步优化墙体，现已革新出多种墙体构造和材料，如集热蓄热墙、透明绝热材料以及附加于墙体的集热器等。现代的集热墙颜色不再拘泥于传统的黑色，如深绿、深红、深蓝都是目前建筑中常选用的颜色，这些颜色对太阳的吸收效果只稍微逊色于黑色，但却活跃和丰富了立面。除此之外，太阳能集热器、光伏电池还可与外墙组成“复合型”墙体，如太阳能保温墙板、太阳能集热器墙体(图4)、光伏墙体等，这些板兼装饰、能量与美观于一体。近来发展的太阳墙，是一种将太阳能集热与通风相结合的技术。

屋面：因不受遮挡，其对与太阳能部件具备良好的结合条件可以选择集热器或光电板的最佳接收辐射的角度，也可以根据造型的需要设计出优美的屋面外形(图5)。

图4 皇明太阳能利用综合示范楼

图5 青岛千禧龙花园集热器屋顶

阳台：高层建筑楼顶面积相对紧张，可以利用向阳的阳台。阳台围栏外壁较适合安置分户式集热器系统，也可代替栏杆使用，真空管横向布置还可增添建筑物横向线条。另外还可以将阳台封闭，处理成集热太阳房。

遮阳装置：阳光可以增加人们的舒适程度，但有时阳光也会带给人不舒服。如在夏季，人们并不需要太强的阳光，这意味要把自然光引向它所需要的地方，也包括排除强而热的直射太阳光。所以除了聚集热量以外，还要能选择性的处理透过玻璃的阳光与能量。在夏热冬冷地区，利用窗户的外遮阳“可调节附加构件”可以很好的改善窗户的保温、隔热性能。目前新型 PV 的遮阳板遮阳的同时，还可获取能量(图 6)。

图 6　德国某建筑光伏遮阳板

3.3　完善的评价与引导机制

太阳能利用技术体系的日益成熟和种类的多样，让我们在设计中对太阳能技术有了更多地选择余地，但同时如何有效与最大限度的利用宝贵的太阳能资源，需要建立一个以太阳能为主的能源综合优化评价体系，以便建筑师在方案初始阶段，针对不同类型不同地域的建筑，利用计算机模拟软件，对各种能源利用体系的选用进行有效的评价。

对太阳能利用应给予示范导向和税收等激励政策，尤其对于高层住宅则应实行税收激励政策、能源投资机制及业主有偿使用相结合的策略。首先要制定激励政策，拓宽资金渠道，强化科技投入。同时编制设计规范、标准及其相关图集，建立产品(系统)检测中心和认证机构，完善施工验收及维护技术规程等工作；这些是将太阳能利用列入建筑工程设计的环节，也是作为一个新型“专业”纳入建筑体系的前提。

综上所述，对太阳能的利用要具有地域性，建筑不是被生硬的强加于一个特定的气候区域，它意味着要针对我国的社会发展、技术进步、经济能力、区域气候、生活需求和地域文化等多种因素，通过集合的技术策略、系统的设计方法和政策导向机制三方面共同建立有效的设计体系，从而推进太阳能在高层住宅中利用的快速发展。

Integrated Desgin Systems Reaserch Of Solar Energy And High-Rising Dwellings

Shi Jie　Song Dexuan

(College of Architecture and Urban planning,

Shanghai TongJi University, Shanghai　200092　China)

Abstract: This article unifies the present situation of China buildings of solar energy use, in view of the Shanghai high-rise dwellings energy conservation development, proposed the operating system of the integrated design of high-rising dwellings and solar energy to be possible: Solar energy technology system strategy, whole design method and policy guidance operation mechanism which are organic synthesis; It's Emphasized that the future high-rising housing sustainable design, will not be the sole technical system operation, but will be a integrated process of the multi-technical system and the building system.

参考文献

[1] 王长贵．郑瑞澄．新能源在建筑中的应用．北京：中国电力出版社，2003 年．18

[2] 李保峰．适应夏热冬冷地区气候的建筑表皮之变化设计策略研究．北京：北京清华大学，博士论文，2004

[3] 夏云，夏葵．生态可持续建筑．北京：中国建筑工业出版社，2001

[4] (德)英格伯格・弗拉格．托马斯・赫尔佐格 建筑＋技术．北京：中国建筑工业出版社，2003

[5] (日)森保洋力．高层、超高层集合住宅．北京：中国建筑工业出版社，2001

[6] Ralph L. Knowles. [The solar envelope: its meaning for energy and buildings]. School of Architecture, Energy and Buildings 35(2003)15～25

[7] 顾斌．沪港两地高层住宅比较研究．上海：同济大学硕士论文，P23～25

[8] Volker Wittwer. [Building Integration Of Solar Energy Systems]. Solar Energy Vol. 67, Nos. 46, pp. 181～187, 1999

[9] 高辉，刘泉．太阳能利用与建筑一体化设计．华中建筑，第 22 卷，2004-01

某大剧院观众厅空调气流组织数值分析

毛洪伟[1]　周孝清[2]　罗志焱[3]

（[1] 深圳市建筑科学研究院，深圳　518055；[2] 广州大学，广州　510405；
[3] 广州珠江外资设计院，广州　510405）

摘　要：采用CFD技术对某剧院座椅送风空调方式的初步设计方案及其各种改进方案进行了模拟分析，最终确定了合理的设计方案。

关键词：CFD技术　气流组织　座椅送风

引言

空调气流组织设计合理与否直接决定能否实现良好的空调舒适性。但随着现代建筑的发展，各种造型新颖、内部复杂、功能多样的建筑层出不穷，同时各种新型的空调方式技术也不断涌现，使得以往只用简单公式或经验公式进行气流组织计算已经不能满足现有要求。而伴随着计算机技术的快速发展，CFD(Computational Fluid Dynamics)技术将是新型建筑空调气流组织设计分析的利器。本文介绍了CFD技术在某大剧院观众厅空调气流组织设计上的应用。

1　工程概况

某大剧院观众厅为三层(图1)，一层池座面积约为880m^2，座位1200个(包括乐池)；

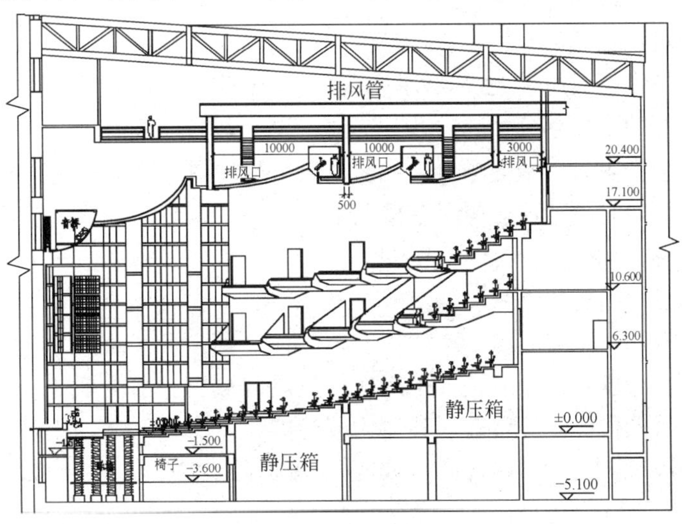

图1　观众厅剖面图

二、三层楼座面积各约为 250m²，座位各 300 个。二层楼座上各有一道挑台灯，功率分别为 50kW；屋顶有二道面光灯林，功率分别为 80kW；观众厅前端二侧都有耳光灯，功率也为 80kW。

观众厅空调系统为集中式全空气系统，观众厅的室内热负荷为 268kW，初步设计采用的空调送风温度为 21℃，送风的风量池座和楼座分别为 80000m²/h、44000m²/h。气流组织方式初步采用座椅送风和屋顶回风的方式。

2 模拟计算

2.1 物理模拟

根据琴台大剧院观众厅的建筑图纸，建立如图 2 的物理模型。在建模过程中做了以下几点的简化：

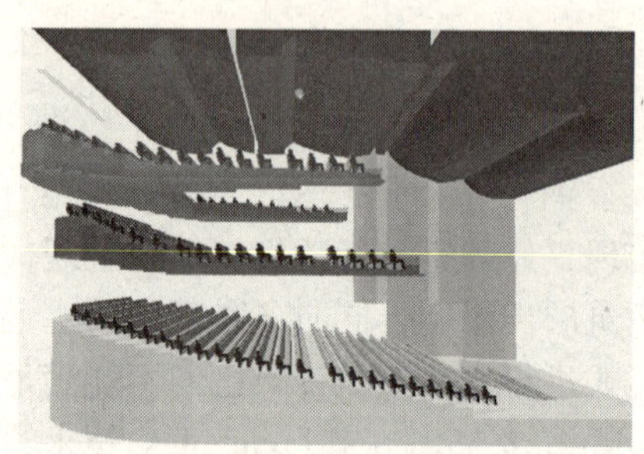

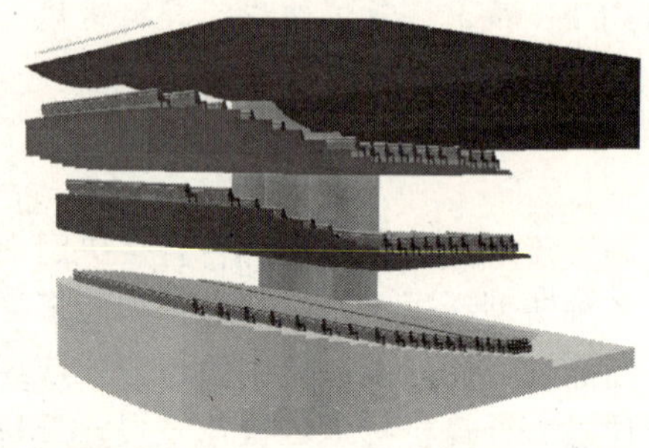

图 2 观众厅模拟模型

1. 工程中座椅是成一定的弧度排列的，在建模过程中简化成直线排列，忽略座椅弧度对观众厅的温度和流速场不会造成太大的影响。

2. 忽略观众席中间的二条宽 1m 的过道。这种简化同样不会对观众厅的温度和流速场造成过多的影响。

2.2 模型热负荷设置

2.2.1 人员负荷设置

影剧院是人员密集场所，尤以观众厅为甚，人体散热和散湿形成的冷负荷和湿负荷成为观众厅空调负荷的主要部分。

对影剧院、会堂等人员密集场所，辐射散热引起的围护结构蓄热量相应减少，可近似将散热量全部作为即时冷负荷。所以在模型中，观众厅人员的全热散热量为 108W/人(静坐状态)，集群系数 0.89。[1]

2.2.2 灯光负荷设置

(1) 挑台灯热负荷

已知二道挑台灯设在楼座前端，正好处在空调区域内。由文献 [1] 可知：挑台灯除 35%的热量随有效光投射到舞台外，其他的最终扩散到观众厅内。因此每一道挑台灯的散热量为：32.5kW，其中：50×40%=20kW(对流扩散)，50×25%=12.5kW(辐射扩散)。

(2) 耳光灯热负荷

观众厅前端二侧的耳光灯有独立的灯光小室，所以对观众厅的负荷影响较小，可以采用文献［1］的公式进行计算，得散热量为10kW。

(3) 面光灯热负荷

而屋顶的二道面光，有独立的面光小室，又在观众厅空调区之外，设独立排风系统后可认为对观众厅空调负荷没有影响。

2.2.3　围护负荷设置

由于观众厅没有外墙，屋顶也有格楼，所以这里忽略围护负荷。

综上所述，则模拟模型内的室内热负荷为：

$$Q_{模型室内}=人员负荷+照明负荷$$
$$=108\times1800\times0.89+10\times1400+32500\times2+20000$$
$$=272.2\text{kW}$$

这与初步设计中室内热负荷268kW基本吻合。

2.3　模拟计算与分析

2.3.1　方案1

主要参数设置：

(1) 气流组织方式：座椅送风、屋顶回风；

(2) 送风量：池座送风量80000m²/h，楼座送风量44000m²/h；

(3) 送风温度：21℃(考虑送风温升后，即送风口温度)；

(4) 假设观众厅预冷温度为26℃。

模拟结果：

如图3所示，当空调稳定运行后，观众厅内中间断面平均温度是26.7℃，说明送风温度合适，但整个温度场并不均匀，存在很多局部的相对“高温”区。这些“高温”区形成的原因不尽相同，对人员舒适性的影响也不相同。

图3　观众厅中间纵断面温度分布

图3中A、B、C三个“高温”区是由于层高较低以及挑台底部的凹槽结构使得上升的热气流无法有效地扩散出去，从而大量滞留在这个区域形成“高温”区。这将使得这些区域无法达到空调要求。

图3中D高温区是由于二股上升气流相遇，形成一个涡流区，使得这个区域的上升

热气流不容易扩散出去从而形成“高温”区。这个“高温”区位于人员活动区域的上空(如图3)，而且温差也不会太高(图中所示约为2℃)所以它对人的舒适性影响不会很大。

图3中E“高温”区是由于上升的热气流在天花的凹槽处淤积，而此处又离回风口较远，从而形成“高温”区。它在天花板下面，远离人员活动区域，温度一般在30℃，对人员活动区域的辐射影响也很小，所以对人员的舒适性影响可以忽略不计。

图4中的F、G“高温”区是由于观众厅天花前宽后窄，考虑回风口成排排列，在观众厅的中段三层楼座上空没有正对的回风口，热空气不能及时排出而形成“高温”区。

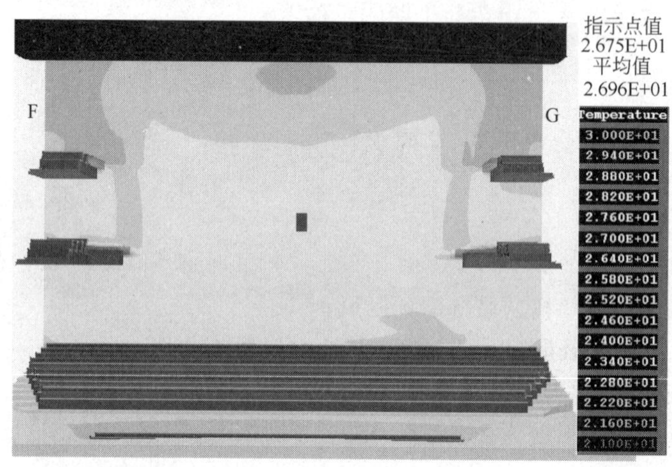

图4 观众厅中间横断面温度分布

所以要使得观众厅有较满意的空调效果，则A、B、C、F、G三个“高温”区要设法消除。

A“高温”区可以考虑把屋顶的回风口安装的离后墙更近一些来消除。

B、C“高温”区可以考虑在后墙上各安装二个回风口，以此来及时的排走热气流。

F、G“高温”区可以在正上空加设回风口来消除。

因此为克服方案1的不足，采用以上三个措施得到方案2。

2.3.2 方案2

主要参数设置：

(1) 气流组织方式：座椅送风、屋顶回风，一层和二层后墙上的四个入口大门上方各安装一个回风口(如图5)。一层的二个回风口尺寸为1000cm×800cm(净面积)，每个风口风量为8600m^3/h(相当于风口风速为3m/s)；二层二个回风口尺寸为1000cm×500cm(净面积)，每个风口风量为3600m^3/h(相当于风口风速为2m/s)。同时要把二个回风口移到三层楼座二侧上空，把最后排的风口移

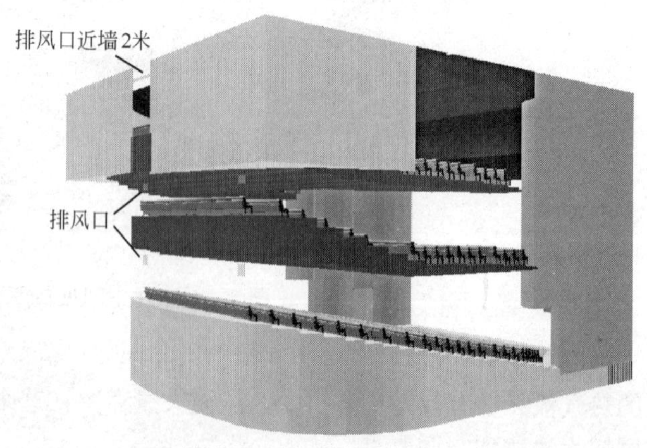

图5 方案2模型

到离墙 2.0m 左右地方。

(2) 送风量：池座送风量 80000m³/h，楼座送风量 44000m³/h。

(3) 送风温度：21℃(考虑送风温升后，即送风口温度)。

(4) 假设观众厅预冷温度为 26℃。

模拟结果：

结果如图 6 所示，方案 2 与方案 1 相比，C 区的改善比较明显，C 中区中心温度降到 28℃以下，在人员活动区上部(人员的头部以上位置)的温度则在 27℃左右，人员活动区下部温度由于座椅送风装置的效果温度在 21～26℃之间。A 区的高温区收缩成墙面附近，而沿墙附近是走道，没有布置座位。座椅区人员活动上部温度在 27℃左右，又由于上排风口的影响 A 区的风速也较大在 0.2～0.4m/s。因为剧院等夏季观众衣着单薄，考虑空气流速影响，当轻薄衣着、静坐时，干球温度 27℃，流速 0.2m/s，或干球温度 28℃，流速 0.5m/s，均为舒适条件[1]。所以无论 C 区还是 A 区如果综合考虑温度和风速的影响都可以认为是舒适的。

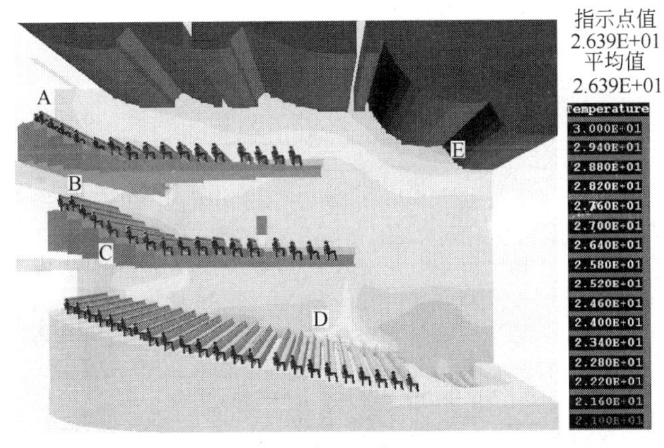

图 6 观众厅中间纵断面温度分布

F、G 二个区温度(图 8)与改进方案 1 相比(图 4)有很大的改善，温度降到在 27℃左右，同样考虑这二个高温区也在人员活动区的上部，风速也相对较高，所以也可以认为基本能达到人员舒适性要求。

从图 6～图 9 中可以知道，除几处温度相对高点的地方外，观众厅大部分人员活动区域内的温度在 21～26℃，流速在 0.2m/s 以下。而温度达到 27℃左右的地方流速也在 0.2～0.4m/s 范围之内，而且只存在人员活动范围的上部，在人员活动范围中下部由于座椅送风装置的影响温度比较低。

所以综合考虑观众厅人员活动区温度和流速的影响，可以认为方案 2 的空调方式是可以基本满足要求的。

同时方案 2 存在几点不足：*a*. 一层二个排风口总风量 17300m³/h，回风量比较大。*b*. A 区(三层楼座后部)温度还是偏高，说明调整屋顶回风口位置对消除 A 区“高温”区的效果没有直接在后墙上安装回风口的效果好(与 B 区相比)，这主要是因为层高过高的影响。

对于第一个不足从模拟结果来看主要是由于风口的布置间距较大，所以只能加大排风量才能排走中间位置的热气流。因此可以考虑在一层后墙上拉出排风管，并在风管上均匀布置四个排风口的办法来排除热空气，这样可以大大减小此处的回风口。对于第二个不足可以考虑在三层楼座后墙上也各安装二个回风口，于是得到方案 3。

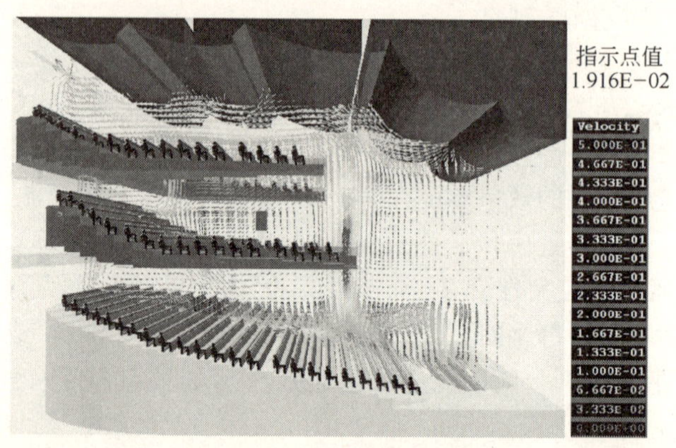

图 7 观众厅中间纵断面流速分布

图 8 观众厅中间横断面温度分布

图 9 观众厅中间横断面流速分布

2.3.3　方案 3

主要参数设置：

(1) 气流组织方式：座椅送风、屋顶回风，二、三层后墙上的四个入口大门上方各安装一个回风口(如图 10)。一层则拉出风管安装四个回风口，尺寸为 500cm×500cm(净面积)，每个风量 2700m³/h(相当于 3m 风速，四个风口总风量 10800m³/h)；二、三层的回风口尺寸为 1000cm×500cm(净面积)，每个风口风量为 3600m³/h(相当于 2m 风速，各层的总风量 7200m³/h)。

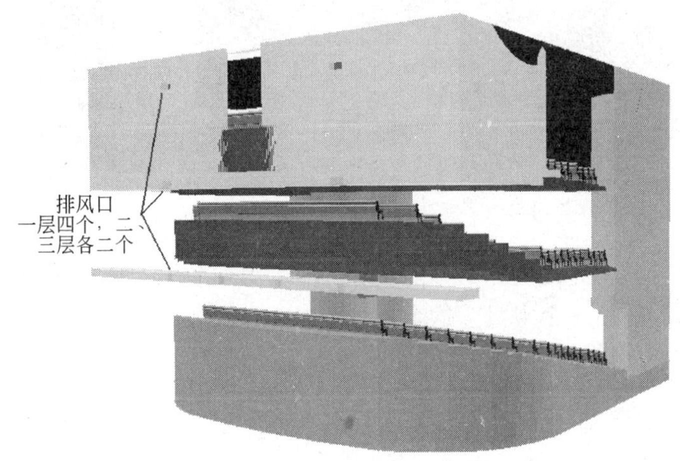

图 10　方案 3 模型

(2) 送风量：池座送风量 80000m³/h，楼座送风量 44000m³/h。

(3) 送风温度：21℃(考虑送风温升后，即送风口温度)。

模拟结果：

如图 11 所示，方案 3 比方案 2 的效果好的多，A、B、C 都基本消失，特别是 A、C 的效果最明显，C 区中心温度降到了 27℃左右；A 区最高温度则降到了 27℃以下。从图 11、12 中可以知道，观众厅大部分人员活动区域内的温度在 21～26℃，流速在 0.2m/s 以下。所以可以认为改进方案 3 的空调方式比改进方案 2 的效果好，可以满足空调要求的。

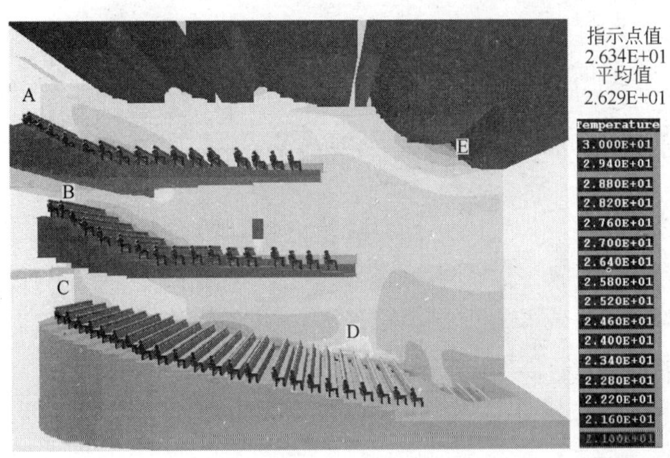

图 11　观众厅中间纵断面温度分布

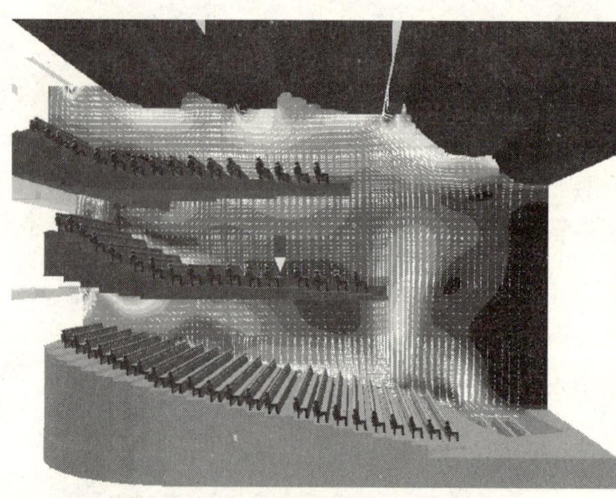

图 12 观众厅中间纵断面流速分布

但是如果装修等其他方面的原因不能在一层后墙上安装横向风管，则此方案将不可能实施。所以这儿再分析另一个方案，这个方案不但要克服方案 2 的二个不足，同时也限定不能在后墙上安装横向风管，即方案 4。

2.3.4 方案 4

主要参数设置：

(1) 气流组织方式：座椅送风、屋顶回风，一层、二层和三层后墙上的六个入口大门上方各安装一个回风口(如图 13)。一层的二个回风口尺寸为 1000cm×800cm，每个风口风量为 6000m³/h(相当于 2.1m 风速，各层的总风量 12000m³/h)；二、三层的回风口尺寸为 1000cm×500cm(净面积)，每个风口风量为 3600m³/h(相当于 2m 风速，各层的总风量 7200m³/h)。同时在二层楼座静压箱下面设一长为 24000cm、宽为 10cm 的条缝送风口，送风量设为 3500m³/h，送风温度 21℃。因为二层楼座的温度相对比较低，所以条缝送风口的风量主要来自二层楼座的送风量。

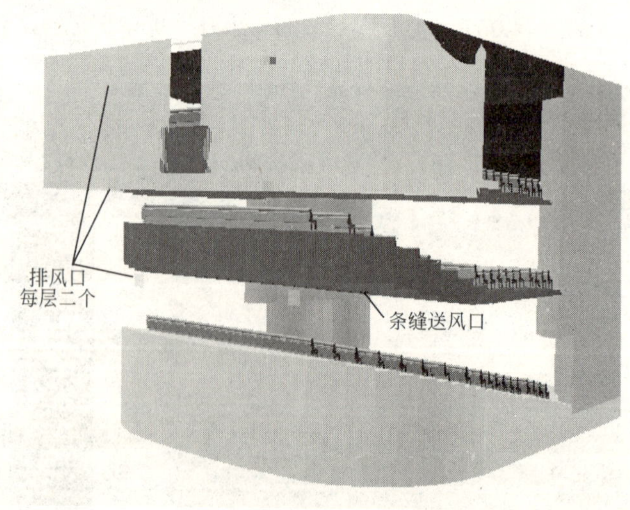

图 13 方案 4 模型

(2) 送风量：池座送风量 84000m^3/h，楼座送风量 40500m^3/h。

(3) 送风温度：21℃(考虑送风温升后，即送风口温度)。

模拟结果：

如图 14 所示，方案 4 与方案 3 相比空调不相上下。A、B 区基本一致。而 C 区改进方案 4 的空气温度明显要低一些，由于条缝送风口本身出风速度小，所以条缝送风口附近的人员区内的风速都在 0.2m/s 以下，不会产生太大的吹风感。所以改进方案 4 是比较理想的空调气流组织方案。

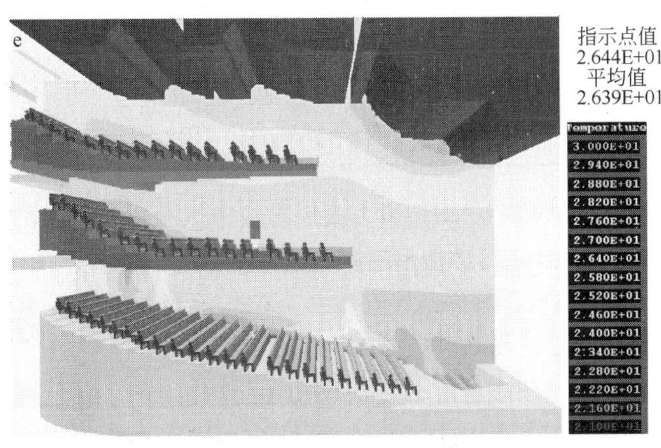

图 14　观众厅中间纵断面温度分布

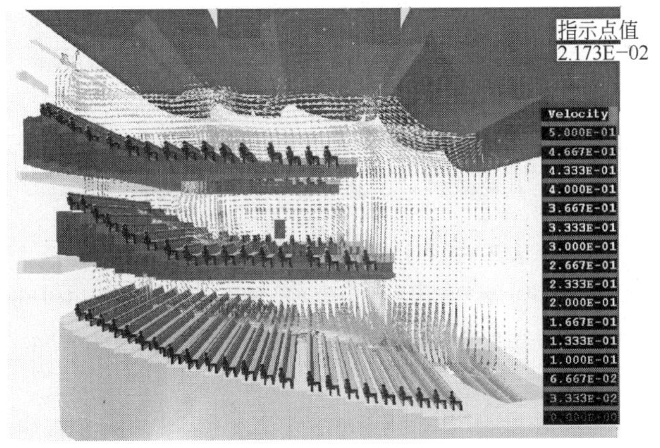

图 15　观众厅中间纵断面流速分布

3　结论

由上面的模型分析可以得出以下几点结论：

(1) 方案 1 的回风口设置不合理，因此需要改进。

(2) 如果设计采用模拟分析改方案 2 的话，应该再在三层楼座后墙二个入口上方各安装一个回风口，风量与二层相同，并且条件可能的话应该再增加一层回风口的回风量。

各方案比较 **表1**

<table>
<tr><th rowspan="3">方案</th><th rowspan="3">气流组织方式</th><th colspan="6">后墙侧排风口设置</th><th rowspan="3">条缝送风量</th><th rowspan="3">空调效果评价</th><th rowspan="3">备注</th></tr>
<tr><th colspan="2">一层池座</th><th colspan="2">二层池座</th><th colspan="2">三层楼座</th></tr>
<tr><th>个数</th><th>风量/个</th><th>个数</th><th>风量/个</th><th>个数</th><th>风量/个</th></tr>
<tr><td>方案1</td><td>座椅送风顶回风</td><td colspan="6"></td><td rowspan="3">无</td><td>差</td><td>局部高温区多</td></tr>
<tr><td>方案2</td><td rowspan="2">座椅送风顶、后侧回风</td><td>2</td><td>8600m³/h</td><td>2</td><td>3600m³/h</td><td></td><td></td><td>一般</td><td>一层侧排风量大，高温区改善有限</td></tr>
<tr><td>方案3</td><td>4</td><td>2700m³/h</td><td>2</td><td>3600m³/h</td><td>2</td><td>3600m³/h</td><td>良</td><td>一层侧排风要拉出横向风管</td></tr>
<tr><td>方案4</td><td>座椅送风顶、后侧回风，一层条缝送风</td><td>2</td><td>6000m³/h</td><td>2</td><td>3600m³/h</td><td>2</td><td>3600m³/h</td><td>3500m³/h</td><td>好</td><td>高温区消失，侧排风量适中</td></tr>
</table>

（3）由表1可以看出：方案3、4是该观众厅比较理想的空调方式，推荐采用。

（4）随着现代建筑的发展，各种造型新颖、内部复杂、功能多样的建筑层出不穷，同时各种新型的空调方式技术也不断涌现，这将对空调气流组织设计提出越来越多的挑战，而日益成熟的CFD软件会是应对这些挑战的有效工具。

Numerical Analysis Air Distribution of Air Conditioning in the Auditorium of Grand Theatre

Mao Hongwei[1] Zhou Xiaoqing[2] Luo Zhiyan[3]

（[1] Shenzhen Institute of Building Research，Shenzhen 518055 China；

[2] Guangzhou University，Guangzhou 510405 China；

[3] Guangzhou Pearl Piver Foreign Investment Architectural Desianing Institute，Guangzhou 510405 China）

Abstract: Simulation analysis about the original planning scheme as well as the revised schemes of air distribution of air-conditioning mode to a theatre are completed by the use of CFD technology，accordingly leading to a reasonable planning scheme.

参考文献

[1] 王鸿章，李惠风等编著. 影院空调设计. 中国建筑工业出版社，P43～47

[2] 电子工业部第十设计研究院主编. 空气调节设计手册(第二版). 中国建筑工业出版社

低辐射镀层塑料薄膜与建筑节能窗

戴义生　PETER KATER　JOHN MEADE
（美国韶华科技公司，美国　CA94303）

摘　要： 本文通过现代建筑对玻璃的性能要求，在探讨了太阳辐射和玻璃传热的一些基本参数之后，介绍了低辐射镀膜的塑料薄膜在建筑节能玻璃中的应用。薄膜型双中空玻璃和透明热反射薄膜夹层玻璃是分别用于寒带保温和热带遮阳的高性能玻璃产品。而透明热反射自贴窗膜则是应用于现有建筑玻璃节能改造的解决方案。通过与现有产品的数据比较说明了这些产品的性能特点。

关键词： 中空玻璃　夹层玻璃　低辐射镀膜　透明阳光控制

1　建筑节能与舒适对玻璃的要求

现代建筑对玻璃提出了越来越高的要求。建筑玻璃是构成窗户和外围护的重要材料，对建筑的综合能量效率和舒适程度有重要影响。玻璃在建筑中的功能可以归纳为：

- 作为室内与室外的视觉界面，提供居住者与环境的视觉联系；
- 作为阳光能量的采集界面，提供日间自然光源和热源；
- 作为室内空间与外界的热流界面，形成独立的热系统。

玻璃对建筑能耗的影响有两个方面：太阳能的有效利用和热阻隔性能。太阳辐射是建筑采光和控温的基本条件。所以，有效利用太阳能的有利因素和防止太阳能的不利因素是建筑玻璃技术的一个主要组成部分。热阻隔包括对辐射、对流和传导三种热交换方式的阻隔。其中对热辐射的阻隔与主要辐射源阳光热辐射直接相关，因此也属于有效利用太阳能的范畴。而对传导和对流热的阻隔则是对玻璃或玻璃制品本身的热传导性能提出的要求。

2　节能玻璃产品的主要性能参数

在建筑玻璃中经常使用的一个与太阳辐射有关的参数是遮阳系数 S_C。在传统玻璃产品中着色玻璃和热反射玻璃具有较低的遮阳系数，但它们共同的特点是可见光透过率比较低。在通过降低阳光得热节能的同时，也降低了窗户的自然采光效率。为了更精确地定义同样遮阳系数下的通透程度，可以用光效系数描述玻璃选择性透过，或透明遮阳特性：$S=T_{vis}/S_C$，其中 T_{vis} 为可见光透过率。该参数的物理意义为在单位总阳光能量透过中可见光的比例，当 $S>1$ 时玻璃被认为有良好的光谱选择性。

建筑玻璃的隔热保温效率由 U 值描述。这一参数的计算涉及玻璃或玻璃系统的内外表面的辐射传热、材料的传导传热和内外表面空气的对流传热的综合效果。由于各种玻璃的材料热传导率大致相同，总传热就与其他两个因素密切相关。它们都取决于玻璃对远红外辐射能量的吸收能力。低辐射镀膜可以显著降低该吸收及再次辐射，从而降低由此产生的热量流失。

3 LOW-E 塑料薄膜及其复合玻璃产品

经过多年的开发，玻璃上的多层低辐射镀膜已经在技术上成熟并得到广泛应用。与此同时，在塑料薄膜上的多层低辐射镀膜技术也得到了长足的进步[1]。先进的卷绕式大型溅射镀膜生产技术已经可以稳定地提供幅宽 2m、长度达几千米的优质低辐射薄膜。它的出现为玻璃制造商提供了采用玻璃—塑料薄膜复合技术的可能，从而实现生产、运输和使用的高效率和灵活性。塑料薄膜为低辐射膜系提供了新的载体，它与传统玻璃加工技术的融合催生了一系列具有新功能和新应用的复合产品。

3.1 薄膜型双中空玻璃[2]

这是一种将低辐射塑料薄膜张悬在两片玻璃之间形成的双中空玻璃结构。可以采用单面和双面镀膜的低辐射塑料薄膜，也可以采用不同光学性能的薄膜及内外片玻璃的搭配以适应不同需要。使用低辐射塑料薄膜可以显著降低系统的 U 值，使之成为同样重量和厚度下隔热保温性能最优异的玻璃产品。

3.2 热反射型透明遮阳夹层玻璃

具有透明热反射功能的低辐射塑料薄膜可以夹层在两片玻璃之间成为兼有阳光隔热和安全功能的夹层玻璃产品。由于薄膜在运输保存和使用上的便利性，这一产品使没有大型玻璃离线镀膜设备的玻璃加工商得以生产高性能的透明阳光隔热玻璃，并可以利用玻璃的组合实现灵活多样的设计。

3.3 热反射型透明遮阳窗用自贴薄膜

上述低辐射塑料薄膜经过涂胶和防划伤等后处理工序成为窗用贴膜。与市场上常见的吸收型遮阳膜或金属热反射膜不同的是，这种产品具有更高的可见光通透感，适应于升级改造所有由于采用透明玻璃而造成日照过热的建筑。由于贴膜的可见光透明性，改造后的建筑可以显著改善能量效率和舒适程度，同时又保持了原有的外观和设计构想。

4 寒带节能玻璃—薄膜双中空玻璃的特点

4.1 独特的双中空结构[3]

中空玻璃以两层玻璃之间的隔热层为防止热传导的主要方式。普通中空玻璃提高保温隔热性能的途径是增加隔热层厚度或数量，但这些途径都会大大增加窗体的厚度和重量，进而对建筑的整体设计和成本造成负面影响。张悬低辐射薄膜可以在基本不增加厚度和重量的情况下增加隔热层数量和效率，使原来普通中空玻璃难以达到的隔热保温性能成为可能。图 1 列

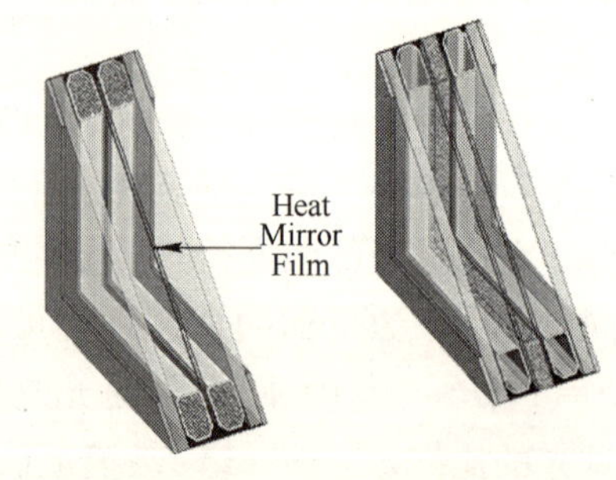

图 1 低辐射塑料薄膜中空玻璃结构

出了两种典型的低辐射塑料薄膜中空玻璃结构。当使用三腔结构时，中空组件的厚度仅比普通中空玻璃稍有增加，却可以达到实际使用玻璃系统的最高隔热保温效果。

4.2 优异的节能遮蔽功能

低辐射塑料薄膜双中空玻璃的独特结构使之具有远高于其他所有中空玻璃产品的隔热性能。同样是双中空玻璃，低辐射塑料薄膜双中空玻璃比 3 层玻璃组成的双中空玻璃有更高的隔热性能。独特的多层光学干涉膜系设计具有对阳光热辐射的选择性红外反射功能，提供了一种透明阳光控制的解决方案，可以在不增加室内照明能耗的同时降低空调系统的制冷负荷。表 1 列出了各种玻璃产品 U 值，遮阳系数和光效系数的比较。由表中数据可见在同一厚度下，低辐射塑料薄膜双中空玻璃的 U 值远低于透明玻璃双中空玻璃并低于大多数低辐射镀膜玻璃。其中典型的透明遮阳组合同最高性能的离线低辐射镀膜中空玻璃有相似的遮阳系数和光效系数。

各种中空和双中空玻璃的隔热性能比较 表 1

玻璃种类	厚度	U 值	遮阳系数	透过率	光效系数
6 透明+12 空气+6 透明	23.4	2.7	0.81	0.79	0.98
6 在线 LE[a]+12 空气+6 透明	24.26	2.1	0.59	0.60	1.02
6 离线 LE[b]+12 空气+6 透明	24.03	1.64	0.41	0.63	1.54
6 透明+6 空气+3 透明+6 空气+6 透明	26.29	2.16	0.75	0.73	0.97
6 透明+6 空气+PET 薄膜+6 空气+6 透明	23.4	2.17	0.74	0.70	0.95
6 透明+6 空气+TC88[c]LE 薄膜+6 空气+6 透明	23.4	1.58	0.56	0.64	1.14
6 透明+6 空气+SC75[d]LE 薄膜+6 空气+6 透明	23.4	1.76	0.42	0.61	1.45

注：所有数据取自美国罗伦斯伯克利国家实验室（LBNL）管理的国际玻璃数据库（IGDB）[4]，玻璃和玻璃组合的性能参数由该实验室开发提供的模拟软件 OPTICS5[5] 以及 WINDOWS5[6] 计算得出。

a——IGDB 910，AFG Industrial，Comfort E2；b——IGDB 968，AFG Industrial，Comfort TiAC36；c——IGDB 1511，Southwall Technologies TC88；d——IGDB 1510，Southwall Technologies SC75。

传统的中空玻璃是透明或着色玻璃制造的，其紫外透过率约为 50%。低辐射镀膜玻璃的紫外透过为略有降低，也达到 10%～42%。低辐射塑料薄膜中空玻璃中的薄膜使用了高性能长效紫外遮断成分，使紫外线透过率小于 0.5%。防紫外功能使该产品在博物馆，美术馆等需要高度自然光照明而又要防止紫外老化和破坏的工程得到广泛应用，也经常应用于高级私人住宅。

4.3 斜面和采光顶应用优势

中空隔热层内气体分子在重力方向的对流运动会增加水平截面上的热传导，普通中空玻璃在倾斜使用时保温效率会大大降低。例如在 27 度倾角度斜面使用时普通中空玻璃的隔热系数相对与竖直状态降低了 33%。当使用低辐射塑料薄膜中空玻璃时由于双隔热层的作用气体分子的对流被有效抑制，同样的倾角使用状态下隔热系数只降低 1%[7]。由于在斜面和天顶使用时对重量厚度的要求更加苛刻，在不增加的重量和厚度下只有低辐射塑料薄膜中空玻璃该产品可以达到最佳的隔热保温效果。

4.4 方向配置灵活性和多种选择

低辐射塑料薄膜中空玻璃中，作为功能性关键材料的镀膜薄膜具有舒适的中性色调，而外观颜色完全可以通过外片玻璃进行调整。在一个典型的办公型建筑案例中，北东方向的窗户可采用高可见光透过率的产品以达到最大限度利用自然光和加强保温性能。南西方向因为有强烈日照，为了减少阳光得热造成的过热现象，通常需要降低遮蔽系数至0.35以下并保持高自然光透过率，可采用透明阳光控制型薄膜。由于水平或斜面安装造成的隔热能力降低，采光顶中空玻璃侧重最佳保温性能及尽可能高的自然光透过。这部分的薄膜可选择高度热反射的品种并兼顾可见光透过率。

5 热带节能玻璃—热反射型透明遮阳夹层玻璃和自贴薄膜的特点

5.1 膜系的光谱选择性

用于热反射型透明遮阳产品的膜系具有高可见光透过和高红外线反射的特点。其反射波段的起始边从可见光波段尾端急速上升，正好有效反射阳光中不具视觉效应的近红外辐射。近红外部分的阻隔使进入室内的阳光总通量大大降低，从而在不牺牲可见光通透性的前提下减少阳光得热进而降低镀膜的遮阳系数。膜系具有很高的光效系数，表现出优异的光谱选择透过和反射性。图2示出该膜系典型的透过和反射光谱。

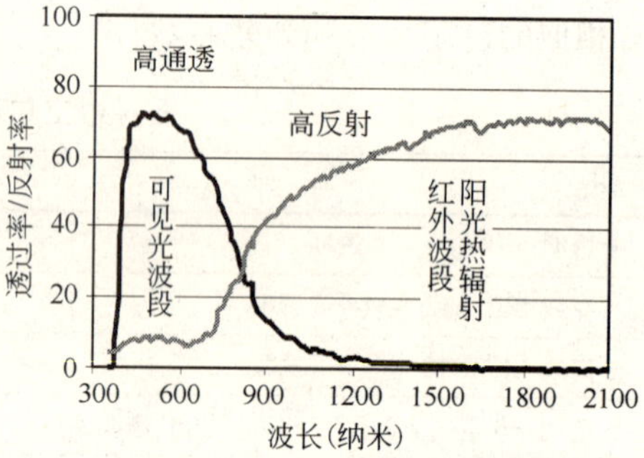

图2 热反射型透明遮阳膜系的光谱特征

5.2 透明遮阳性能

上述透明热反射镀膜的塑料薄膜与玻璃集成的复合产品具有高红外反射的主要特点。以自贴窗膜为例，表2列出染色，金属反射和选择性遮阳三类产品中代表性的光学和阳光控制参数进行比较。染色和金属反射薄膜具有相似的阳光能量透过率，但染色膜可见光透过在阳光能量总量中的比例远小于金属反射膜。这类薄膜的热辐射透过大于可见光，低遮阳系数主要由低可见光透过率实现。相比之下金属反射膜具有全光谱均匀的反射和透过特征，其阳光控制效率要高于染色膜。而透明热反射遮阳薄膜在保持近70%可见光透过的条件下阳光能量透过仅为35%，说明热辐射部分阳光能量几乎全部被有效阻隔，其低遮阳系数主要是由降低热辐射透过来实现。

各种薄膜的隔热性能比较 表2

产品编号	IGDB2801	IGBD232	IGBD1562
类别	染色薄膜	金属反射	透明遮阳
阳光能量透过	45.4%	42.6%	35.0%
阳光能量反射	9.3%	12.3%	36.4%
可见光透过率	26.9%	42.9%	68.5%
可见光反射率	8.9%	12.6%	9.3%
阳光能量吸收	45.3%	45.1%	28.6%
遮阳系数	0.67	0.64	0.50
光效系数	0.40	0.67	1.38

5.3 在建筑夹层玻璃中的应用

图3为透明热反射薄膜夹层玻璃的结构。与普通夹层玻璃不同的是它需要两层PVB中间膜将热反射薄膜夹住。为了充分利用薄膜的红外反射功能，外片应使用透明玻璃而用内片调整色调和透过率。

如图4所示，当同时需要冬季保温和夏季遮阳时，复合的夹层玻璃可以作为中空玻璃的外片使用。在以遮阳为主的热带地区，复合的夹层玻璃可以直接用于幕墙和采光顶。在采光顶应用中，透明遮阳的夹层玻璃对于降低采光顶荷载和提高制冷的能量效率具有突出优势。由于薄膜以卷型包装运输和保存，薄膜夹层玻璃具有很大的灵活性和时效性，特别适用于中小玻璃加工企业生产多品种、小批量、高性能的夹层玻璃产品。膜系的高导电性也为屏蔽电磁辐射干扰及信息安全应用提供了可能性。

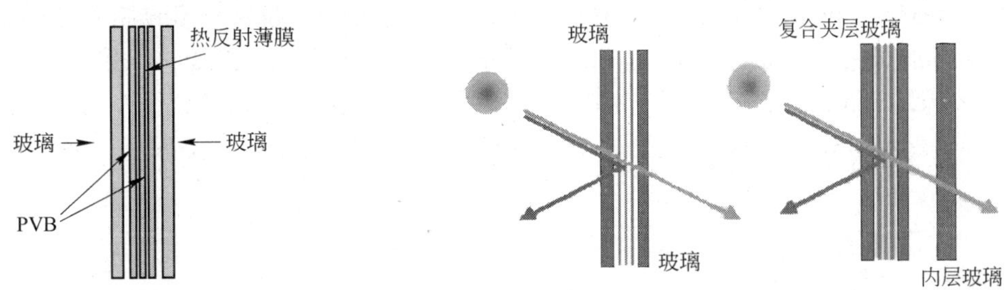

图 3　透明热反射薄膜夹层玻璃的构造　　图 4　透明热反射薄膜夹层玻璃的单层和中空结构

5.4　在自贴窗膜中的应用

透明热反射镀膜的塑料薄膜经后加工工序，与保护薄膜夹层、涂胶和表面强化后就成为自贴式窗用膜。虽然车用贴膜已经受到广泛采用。建筑贴膜的使用才刚刚起步。在现有建筑的节能改造以及舒适性升级方面，贴膜是一种经济有效的解决方案。

和夹层玻璃一样，以透明热反射镀膜的薄膜为功能性基材制造的窗用贴膜具有高通透和反射阳光热辐射的基本特征。出于阳光控制节能和舒适考虑，改造项目最关注的性能参数是遮阳系数和可见光透过率。表 3 列出了比较典型的透明玻璃贴膜前后的性能参数。由表可见，透明玻璃加贴透明热反射膜后显著降低了阳光总能量透过。由于可见光透过率下降不多，该阳光能量的阻隔主要针对热辐射部分。贴膜后的遮阳系数由 0.92 下降为 0.52。由此可见，透明热反射膜在节能改造中最适用于装有透明玻璃的项目。加贴薄膜不仅可以显著降低阳光得热和空调能耗，薄膜的高透明性还可以保持原有建筑的设计风格。

玻璃加贴透明热反射膜前后的隔热性能比较　　表 3

	IGDB890	
	透明玻璃	透明玻璃贴膜
阳光能量透过	80.10%	33.30%
阳光能量反射	6.92%	21.50%
可见光透过	88.30%	67.50%
可见光反射	7.65%	9.15%
阳光能量吸收	12.98%	45.20%
遮阳系数	0.96	0.52
光效系数	0.92	1.30

6　结束语

低辐射镀膜的塑料薄膜为中小玻璃加工商制造高性能加工玻璃提供了新的可能性。在寒带的保温应用中，薄膜型中空玻璃具有低热导系数，高效采光顶，遮蔽紫外和朝向优化等方面的独特功能。在热带的阳光控制应用中，透明热反射夹层玻璃及贴膜具有高通透，高隔热，加工改造便利等方面的显著优点。随着中国建筑节能水平的不断提高，这些新技术新产品必将得到越来越多的认识和采用。

Low-E Coated Plastic Film and Energy Efficient Windows

戴义生 PETER KATER JOHN MEADE
(Southwall Technologies. Inc. , Palo Alto，CA94303 United States)

Abstract: Solar radiation and heat transfer parameters in glazing systems are discussed before introducing new products integrating low-E coated plastic film into architectural glass, including insulating glass with suspended Low-E PET films for cold climate, and Low-E PET laminated glass/applied films for hot climates. The features of the products are elaborated through performance data comparison with related existing products.

参考文献

[1] E. O. Dietrich，R. Ludwig，Vacuum Web Coating，An Old Technology with a High Potential for the Future，40th Annual Technical Conference Proceedings，Society of Vacuum Coaters，1997，P354

[2] Joseph S. Amstock 主编，王铁华/李勇译，建筑玻璃实用手册，清华大学出版社，2004 年 4 月，第 17 章，250～266

[3] Timothy E. Johnson，Low - E Glazing Design Guide，Butterworth Architecture，USA，1991，P13，P34

[4] Lawrence Berkley National Laboratories，International Glazing Database V14. 2，http：//windows. lbl. gov/materials/IGDB/

[5] Lawrence Berkley National Laboratories，Window 5. 2 V17a，http：//windows. lbl. gov/software/window/window. html

[6] Lawrence Berkley National Laboratories，Optics 5 V1. 02，http：//windows. lbl. gov/materials/optics5/default. htm

[7] Ing. A. J. Piers，U-value Calculations of 3 IGU Systems，TNO Report 033-10201，TNO Science and Industry，The Netherlands，May 2005

浅析节能建筑的空调系统应注意的几个问题

刘 寒[1] 申 剑[1] 戴 藤[2]

（[1] 北京航空工业规划设计研究院，北京 100011；[2] 枣庄市房产管理局，枣庄 277100）

摘 要： 空调节能已经日益成为人们所关注的大事。空调专业的设计师为了实现建筑物的整体节能，应该认真贯彻国家 2005 年 4 月 4 日颁布的《公共建筑节能设计标准》(GB 50189—2005)；制冷、供热系统宜根据不同的情况采用节能的热泵系统；设计空调风系统、水系统时，从多方面入手实现节能的目的。

关键词： 节能建筑 保温材料 热泵技术 热回收

能源是人类赖以生存和发展的基本条件。进入 21 世纪以来，能源供应的紧张对各国的经济发展和社会生活产生了巨大冲击，全世界都充分认识到节约能源的重要性。由于建筑能耗在社会总能耗中占有重大比例，建筑节能已成为世界节能浪潮的主流之一。在我国，2004 年以来，全国大面积的电荒让普通家庭意识到了能源的重要性，国家权威部门指出，我国目前已经面临能源短缺的危机。而随着我国近年来经济的快速发展，各地旅游宾馆、大商场、办公及商用建筑、住宅越来越多，建筑能耗日益成为人们所关心的大事。

建筑能耗是指建筑使用过程中的能耗，包括采暖、空调、照明、热水、家用电器和其他动力能耗。其中，以采暖和空调能耗为主，占建筑总能耗的 50%～70%。因此，各种公共、家庭用空调数量的激增，使空调能耗激增，夏季空调耗电、冬季采暖耗能日渐成为能耗大户，空调节能、建造节能建筑已成为刻不容缓的大事。我国抓建筑节能工作必须从现在做起，重点解决的是北方城镇采暖和大型公共建筑的能耗问题，在具体节能途径上应该在优化建筑结构、降低采暖系统能耗、减少空调照明能耗上下功夫，从建筑节能的角度构建节约型社会。

1 建筑热工设计

设计节能建筑的空调系统，首先应该重视的问题是为什么空调耗能如此巨大？除了空调数量庞大外，房屋围护结构热工性能差是一个重要原因。由于很多原来建造的房屋没有采取保温隔热措施，冬季室内的采暖热量通过门窗、墙体、屋面和地面散发出去很多，夏天室外的太阳辐射热和热空气通过门窗、墙体和屋面传到室内，造成夏季室内外温度接近甚至高于室外，居民不得不开空调制冷降温。空调向室外排放大量热气，使室外气温越发升高，室外温度的升高，空调不停地运行，形成城市热岛效应，能耗增大。

所以，暖通专业在设计空调系统时，为了要降低建筑的空调耗能，第一步就是要配合建筑专业确定建筑内、外墙、屋面、地面及其他围护结构的建筑构成方式、使用保温材料的情况。根据国家2005年4月4日颁布的《公共建筑节能设计标准》(GB 50189—2005),应根据我国的地域情况,按照严寒地区A区、严寒地区B区、寒冷地区、夏热冬冷地区和夏热冬暖地区来区分建筑围护结构的热工设计标准。其中,首先特别要注意的是控制好建筑的体形系数和窗墙比,体形系数、窗墙比越大,单位建筑面积对应的外表面面积就越大,单位建筑面积对应的传热系数就越大,传热损失就越大。而体形系数和窗墙比又是牵涉到建筑造型、平面布局、采光通风等各方面的因素,需要既能达到建筑师的创作意图、达到建筑功能,又能权衡利弊,兼顾不同类型的建筑造型,尽可能的减少房间的外围护面积,使体形不要太复杂,凹凸面不要过多,以达到节能的目的。

过去主要是通过暖通空调设备的使用来满足温度要求的,而现在要通过加强建筑物的保温和隔热能力,来尽量满足这一要求,减少通过外围护结构的能量浪费,使室内环境接近人体舒适度的要求和降低建筑的能耗。因此,在校正建筑专业选用各种围护结构的材料时,应特别注意其保温效果。现在,我国的建筑都广泛的使用了大面积的玻璃幕墙和各种窗户,在外墙、屋顶、地面上使用的保温材料类型、热工性质都比较成熟的情况下,窗户就成为了节能的最薄弱环节,其保温和隔热的效果在整体节能效果中显得尤为重要。如果窗户的保温性能不过关,当室内外温差过大时,窗子结露,继而产生的自流水的问题给用户会带来很大的麻烦。尤其在选择使用低辐射玻璃时,应特别注意其材料性能,否则就容易出现冬季效果尚可,但夏季反而耗能会增大的现象。在对窗户节能设计的各种措施中,外遮阳措施是建筑物在夏季或东西向的房子可采取的非常好的节能方法,只是需要解决外遮阳的建筑美观问题。目前,可以参考的国外的遮阳节能窗的做法也多了起来,如铝合金遮阳卷帘、木条遮阳,还有在双层玻璃中间加遮阳百叶的做法。

不可否认,使用了节能材料的建筑的初投资额会比以前增加一些,根据有关方面资料,在欧洲,一些西欧国家以法律形式明文规定,建筑必须达到一定的节能要求,如法国与瑞典等国,对建筑物的隔热型玻璃都有强制的规定,因而使得节能技术与材料都比较成熟,使用比较普遍,大批量生产也使成本不断降低。在西欧地区建设节能型建筑,成本一般增加5%~10%左右。但关键问题是节能型建筑的运营费用将会大大降低,其运营费用大约只相当于普通建筑的70%,而且总体费用将逐年递减。

2 热泵技术的应用

设计节能建筑的空调系统,如何减少一次高品位能的利用是一个很关键的技术问题。如果能利用各种在土壤、太阳能、水、空气、工业废热中蕴藏着无穷无尽的低品位热能无疑是一种成功的节能措施,热泵技术正是现阶段实现这个目标的最佳选择。其可以通过输入较少的高品位能源把广泛存在于大自然的绿色低品位的热能提高到可以在建筑用能的温度(如采暖、生活热水)。

根据热泵系统的热力循环型式,通常将热泵分为蒸汽压缩式热泵、气体压缩式热泵、蒸汽喷射式热泵、吸收式热泵、热电式热泵。其中,蒸汽压缩式热泵是在目前研究和使用最为普遍的方式,按照其使用的低温热源的种类,基本都属于空气源热泵、地源热泵、水源热泵和太阳能热泵四种类型:

1. 空气源热泵：将室外的空气作为低位热能，获取方便，设备基本上都是使用一个气—液换热器与热泵机组耦合，几乎不会对环境造成影响。因此该系统具有系统组成简单、年运行时间长、初始投资较低、技术比较成熟的优点，在地方气候条件适宜，特别是冬季气候较温和的地区，是一种性价比出色的节能方法。但该系统的缺点也是很突出的，室外空气随季节的变化不断变化，温度、湿度对热源的影响明显，热泵的年效率不稳定。在湿度较大或者冬季天气寒冷的地区，其制热量的变化与建筑热负荷的需求趋势正好相反，温度低、湿度大会使热泵效率会大大降低，甚至无法工作。由于除霜技术尚不完善，虽然很多的产品都宣传可以正常工作在零下 20℃甚至更低的温度，但在实际工程应用中，常常是耗费大量的电热能源的情况下才能达到理想效果，这与节能的目的是相违背的。因此在寒冷及高湿度地区，热泵蒸发器的结霜问题已成为节能中很大的技术障碍。

2. 地源热泵：利用地表浅层中蓄存的低品位热能(土壤、地层、地下水)作为热源，冬季热泵从浅层的土壤中取热，用于建筑供暖，同时蓄存冷量以备夏用；夏季热泵逆向运行，将建筑物内的热量转移到地下对建筑进行降温，同时蓄存热量以备冬用，因此这是一种典型的可以再生的能源。现在，对这种地表浅层中蓄存的低品位热能的利用技术已经比较成熟。地表 5m 以下温度一年四季相对稳定，夏季比环境空气低，冬季比环境空气高，热容量大，变化幅度比较小，既能保持热泵高效、稳定运行，又可利用岩土的天然蓄能能力，且对周围环境影响较小，维护费用低。但是，大地土壤的温度较低，地下埋管内的载能流体与管外的土壤之间的换热系数小，能流密度很低，为了获取足够的热量，就需要的地下换热器的表面积足够大，就需要有足够的地下埋管换热器，因此将占据较大的地下和地上空间，初始投资是比较大的。

3. 水源热泵：水源热泵与地源热泵相似，但是以低温水作为低温热源，可高效地利用量大面广的地下水、地表水、电厂冷却循环水及工业废水、污水等作为低位热源，热泵 COP 一般可达到 4～5，节能效果十分明显。根据其热源的温度的情况可以实现一机多用，满足供暖、空调及生活热水的需要，波动范围也要远小于空气温度的变化，因此其系统的全年运行性能稳定、工作可靠、运行费用低。但是，水源热泵也有其致命的弱点，它要受到可利用的水源条件、水层的地质结构、水资源使用政策以及能源结构和价格等因素的限制。使用电厂冷却循环水及工业废水、污水等作为低位热源，应用的局限性较大；使用地下水，由于水资源已经是一种日益紧缺的资源，因此在应用之前必须做详细的水文地质调查和技术经济性能分析。而且由于会造成水资源的浪费，为避免地下水的严重流失，通常还要求采用地下水回灌技术，同时保证地下水不受污染，这在目前仍然是一个难题；利用地表水，可以在靠近江、河、湖、海等大体积自然水体的地方比较经济的实现，但其缺点和空气源热泵类似，受到自然条件的限制，热泵的换热可能会对水体中生态环境产生无法预计的影响，而且环境温度越低时热泵的供热量越小，而且热泵的性能系数也会降低。此外，不同面积、深度和温度的地表水单位体积能够承担的冷、热负荷需要根据具体情况进行计算。

4. 太阳能热泵：太阳能热泵不同于普通的太阳能直接供热系统，也不同于以太阳能光电或热能发电驱动的热泵机组，而是利用太阳能集热器作为蒸发器热源的耦合热泵系统。能源清洁、安全，几乎不会对环境产生环境污染。传热设备可以采用廉价的低温集热器，集热成本非常低。虽然太阳能对于整个地球来说全年是非常稳定的，但我们可利用的部分本身并不具有稳定的特点，其能流密度低、供能不连续，这就需要采用较大的集热和蓄能装置，并且需要配备相应的辅助热源，其较大面积的集热器占据了较大的空间，造成系统初投资较高。

综上所述，可见这些类型的热泵技术可以有效地提高一次能源利用率，减少空调用能中高品位能源的消耗，但是各有其利弊，设计者应根据所设计建筑的地域特点和具体的设计情况及甲方要求进行灵活的选用。在合适的地方，也可以采用多种热泵耦合的方式提高效率和性价比。

3 空调设计中的相关问题

在设计节能建筑的空调系统过程中，设计者在空调系统各个环节都应该从细处入手，切实落实好节能的观念，规避一些不好的设计习惯。

1. 在选择冷热水机组主机配备容量时，出于保护性参数的考虑，常常会将各空调系统的峰值叠加，然后再加上安全系数，从而导致最后冷热水机组配备容量偏大，这样会导致大部分时间在部分负荷下低效率运行，造成了很大的投资和能源浪费。主机余量过大也会造成水泵等其他输送动力设备的容量过大，整个管路特性远离最佳工作点，以至于总体能耗过大。

2. 仅仅考虑到减小冬、夏两季的新风负荷，而将新风口及空调机组的新风入口按照冬、夏两季的风量设计，致使过渡季节仍需要开启冷水机组，也将使空调能耗显著增大。

3. 室内温度的设定上，由于盲目追求舒适性或者忽略了普通空调系统的自控系统，将导致系统在实际运行中没有方法可以根据人员的变化和实际负荷进行必要的调整，致使供冷量、供水量和送风量都大于实际的需求，由此就会造成大量的能量浪费。

4. 在水系统设计中应增加必要的控制手段，可在运行时根据流量的变化与功率的变化比成三次方关系，增大送回水温差，减小水流量，以降低水系统设备的能耗。同时也可以采用变流量技术，改善水泵运行工况，使其保持在最高效率点运行，以达到节能的目的。

5. 重视冷热源的回收。空调系统能源利用率最大限度的提高也是实现空调节能的主要途径。现在，我国空调冷热回收利用主要是通过系统中安装能量回收装置，用排风中的能量来处理新风，就可以减少处理新风所需的能量，降低机组负荷，达到节能的目的。在选择热回收装置时，应当结合当地气候条件、经济状况、工程的实际状况、排风中有害气体的情况等多种因素综合考虑，以确定选用合适的热回收装置，从而达到花较少的投资，回收较多热(冷)量的目的。换热器的布置形式和气流方式对换热性能也有影响，热回收系统设计要充分考虑其安装尺寸，运行的安全可靠性以及设备配置的合理性，同时还要保证热回收系统的清洁度。热回收设备可以与不同的系统结合起来使用，利用冷凝热，以节约能源。

综上所述，可见设计节能建筑是一个在设计理念上的改变，涉及到设计的方方面面，在这里谈到的只是其中的几个问题，也借鉴了不少的资料，希望能给同行们在设计相关建筑时提供一些有益的参考。相信随着设备工艺技术的不断更新和设计理念的进步，我国的节能建筑会稳步迈向国际先进水平。

Brief Analysis of Several Questions of Energy Conservation Building's Air-Conditoning System

Liu Han[1] Shen Jian[1] Dai Teng[2]

([1] China Aeronautical Project and Design Institute, Beijing 100011 China;
[2] Rear Estate Adminisliration Bureau, Zaozhuang 277100, China)

Abstract: The air-conditioning energy conservation has became more and more important recently. To achieve the goal of energy conservation of the whole buildings, the air-conditioning designer should earnestly implement the "Public building Energy conservation Design Standard" (GB 50198—2005) which was promulgated on April 4th, 2005. The energy conservation heat pump system should be used in the refrigerating and heating systems based on the different conditions. In order to save energy, various aspects should be considered when designing the ventilation system and water system.

参考文献

[1] 涂逢祥主编. 建筑节能. 北京：中国建筑工业出版社，2005.1

[2] 蒋能照主编. 空调用热泵技术及应用. 北京：机械工业出版社，1997.6

[3] 徐伟等译. 地源热泵工程技术指南. 北京：中国建筑工业出版社，2005.5

[4] 公共建筑节能设计标准(GB 50189—2005)

[5] 尉迟斌主编. 实用制冷与空调工程手册. 北京：机械工业出版社，2001.9

[6] 赵键主编. 建筑节能工程设计手册. 北京：经济科学出版社，2005.7

提高城市住宅节能水平，建设节约型社会

朱玉梅　刘加平

（西安建筑科技大学建筑学院，西安　710055）

摘　要: 根据国务院部署大力发展节能省地型建筑，建设节约型社会的要求，全面实施建筑节能50%设计标准，进而推动城市节能65%标准的实施。我国大部分城市目前正在制订或者已经执行居住建筑节能65%标准。如何将建筑的性能化节能指标转化为围护结构的热工性能指标，需要对节能50%设计标准的住宅的热工性能现状进行分析总结，才有可能在技术上找到合理地提高住宅节能水平的设计方法。本文以西安市为例，归类总结近年来城市住宅围护结构的构造做法和热工性能，其节能指标与节能50%、节能65%标准的比较，找出目前住宅的热工性能指标与新标准的差距。据此提出应优先采用高效节能保温窗；墙体保温材料的选择要综合考虑保温层的特点、造价等因素；同时，对于因抗震设计而产生的热桥的影响应引起高度重视，以寻求新的构造方式加强热桥部位的保温。

关键词: 城市住宅　节能　围护结构　热工性能

1　引言

近年来，随着我国建筑规模的迅速扩大，预计到2020年底，全国房屋建筑面积将达686亿m^2，其中城市为261亿m^2。如此巨大的建筑规模，在世界上是空前的[1]。一方面要新建大量的住宅，另一方面又要使住宅有一个较舒适的室内热环境，这就需要消费大量的能源。因此建筑能耗持续迅速增加是不可避免的趋势，建筑能耗占全国总能耗的比例，也将从现在的27.6%快速上升到33%以上。尽管我国人均能源占有量约为世界平均水平的40%，能源消费总量已达世界第二，我国面临严重的能源短缺[1]。节能是国家的一项基本政策，节能法里明确提出建筑要节能，建设部等相关部门陆续制定了建筑节能管理办法，编制了一批建筑节能设计标准，北京、天津、辽宁、吉林、内蒙古、陕西、甘肃、宁夏等省、市、自治区还制订了地方性的实施细则。

中国是一个城市密集，人口众多，城市化快速发展的国家。但建筑技术水平、建筑物能耗、建筑居住的舒适度却一直与发达国家之间存在很大的差距。只有不断提高建筑节能技术水平，才能推动建筑业走资源节约型的发展道路。当前，中国正按照全面落实科学发展观的要求，加快推进科技进步，大力发展节能省地型住宅，建设节约型社会。最近在国务院下发的《国务院关于做好建设节约型社会近期重点工作的通知》中明确提出建筑要节能，以及新建建筑必须严格实施节能50%的设计标准，推动北京、天津、西安等大城市实施节能65%的标准等具体措施。

西安市位于北纬 33°39′～34°45′，东经 107°40′～109°49′之间关中平原中部，属暖温带温暖半湿润大陆性季风气候[2]。西安地区在我国建筑热工设计分区图上位于寒冷地区，属采暖地区。西安的居住环境以多层砖混为主，并有部分中高层和高层建筑。多层住宅的平面和立面比较规整，一般由多个单元组合而成，体形系数基本上都保持在 0.35 以下。本文以西安市住宅为特例说明城市住宅节能水平与节能标准的差距。

2 住宅的节能指标

住宅的节能指标包括建筑物耗热量指标和耗煤量指标。其中建筑物总传热耗热量的 80％是由外围护结构传热损失的。

我国目前对采暖地区执行的是 1996 年实施的节能 50％的《民用建筑节能设计标准(采暖居住建筑部分)》JGJ 26—95。根据标准要求，在满足体形系数、窗墙面积比的前提下，对围护结构的传热系数限值作了规定，对于西安地区，它们分别是：屋面❶ 0.80W/m^2·K，外墙❶ 1.00W/m^2·K，外窗 4.70W/m^2·K；建筑物耗热量指标和耗煤量指标分别是 20.2W/m^{2}和 9.7kg/m^2。按照建设部规划的建筑节能下一阶段目标——节能 65％的要求，在节能 50％标准的基础上，综合各方面因素，西安地区围护结构传热系数限值应分别是：屋面❶ 0.50W/m^2·K，外墙❶ 0.80W/m^2·K，外窗 4.00W/m^2·K；建筑物耗热量指标和耗煤量指标应分别是 14.57W/m^2 和 7.02kg/m^2·(标准煤)。

2.1 屋面的节能指标

表 1 是不同屋面的构造做法及其热工性能，其中屋面 1 是西安市 1980 年以前住宅的构造形式，屋面 2、3、4、5、6、7 是近十年来采用西安市常用保温材料的构造做法。从表 1 可以看出，除屋面 1 外，其余几种屋面均能满足标准中规定的传热系数限值 0.80W/m^2·K，且传热系数值相差很小，但就保温层自重和保温层造价来看，却相差甚远。其中 50mm 厚聚苯乙烯泡沫塑料板的重量是 1kg/m^2，而 160mm 厚水泥膨胀珍珠岩板是 72kg/m^2，同时它们的价格也相差很多。所以在选用时要综合考虑这些因素。

不同屋面的构造做法及其热工性能 **表 1**

编号	构造做法(由上到下)	传热系数 K(W/(m^2·K))	保温层自重 (kg/m^2)	保温层造价 (元/m^2)
屋面 1	两毡三油防水层；②；150mm 水泥珍珠岩板保温层；40mm 1∶6 水泥焦渣找坡层；④；⑤	1.01	40	22.95
屋面 2	①；②；80mm 憎水膨胀珍珠岩板；③；④；⑤。	0.74	20	43.17
屋面 3	①；②；160mm 水泥膨胀珍珠岩板；③；④；⑤。	0.79	72	63.62
屋面 4	①；②；120mm 水泥聚苯板保温板；③；④；⑤。	0.79	54	23.25
屋面 5	①；②；50mm 聚苯乙烯泡沫塑料板；③；④；⑤。	0.78	1	14.83
屋面 6	①；②；50mm 硬质岩棉板；③；④；⑤。	0.78	6	10.19
屋面 7	①；②；30mm 挤塑泡沫板；③；④；⑤。	0.75	1.14	18

注：1. ① 4mm 防水层；② 20mm 水泥砂浆找平层；③ 30mm 1∶6 水泥焦渣找坡层；④ 120mm 预制钢筋混凝土空心板；⑤ 10mm 石灰砂浆抹灰。

2. 保温层造价参考《陕西省建筑工程预算定额》(99)或西安市目前市场平均报价。

❶ 表示屋面和外墙的传热系数限值适用于体型系数≤0.30 的建筑，以下均同。

2.2 外墙的节能指标

表2是不同外墙的构造做法及其热工性能，其中外墙1是西安市1980年以前多层砖混住宅的构造形式，外墙2是近十年来西安市普遍采用的做法，也是目前绝大多数新建住宅的构造形式，外墙3、4、5、6、7是根据西安市目前常用保温材料设计的节能墙体。

不同外墙的构造做法及其传热系数　　表2

编号	构造做法(由内到外)	主体传热系数 K(W/m^2·K)	平均传热系数 K_m(W/m^2·K)
外墙1	20mm石灰砂浆；240mm粘土实心砖；20mm水泥石灰砂浆	2.04	2.24
外墙2	20mm石灰砂浆；240mm承重多孔砖；20mm水泥石灰砂浆	1.62	2.18
外墙3	50mm ASA保温板；10mm空气层；240mm承重多孔砖；20mm水泥石灰砂浆	0.77	1.09
外墙4	20mm石灰砂浆；40mm憎水膨胀珍珠岩板；240mm承重多孔砖；20mm水泥石灰砂浆	0.88	1.25
外墙5	20mm石灰砂浆；100mm蒸压粉煤灰加气混凝土块；240mm承重多孔砖；20mm水泥石灰砂浆	0.82	1.12
外墙6	20mm石灰砂浆；25mm聚苯乙烯泡沫塑料板；240mm承重多孔砖；20mm水泥石灰砂浆	0.83	1.22
外墙7	20mm石灰砂浆；40mm硬质岩棉板；240mm承重多孔砖；20mm水泥石灰砂浆	0.88	1.27

从表2可以看出，外墙1、2的平均传热系数值均远远大于标准中规定的限值1.00W/m^2·K，说明目前绝大多数住宅外墙不符合节能要求。按节能要求设计的外墙3、4、5、6、7其平均传热系数值略大于限值，这是因为西安市属于八度抗震区，对抗震要求较高，所以构造柱、圈梁多，加之外墙的内保温构造做法，使得热桥影响显著。同时从表2还可以看出，内保温外墙平均传热系数值比主体传热系数值高许多，总体保温效果较差。

2.3 外窗的节能指标

表3是不同窗户类型的热工性能及其价格对比情况。其中单层钢窗是西安市1980年以前住宅所采用的外窗类型，其余三种是目前西安市常见的窗户类型，尤其单层塑钢窗是近十年来大多数住宅的外窗选择。从表3可以看出，单层钢窗已远远不能达到现行节能50%标准，单层塑钢窗不能满足节能65%的要求，单框双玻塑钢窗和中空玻璃塑钢窗均能满足节能65%要求，尤其是中空玻璃塑钢窗，保温效果较单层塑钢窗要好得多，价格增加幅度却不大。

不同窗户类型的热工性能及其价格对比　　表3

窗户类型	空气间层(mm)	传热系数 K(W/(m^2·K))	平均价格(元/m^2)
单层钢窗	—	6.4	160
单层塑钢窗	—	4.7	200
单框双玻塑钢窗	12	2.7	250
中空玻璃塑钢窗	12	2.6	240

注：平均价格参考西安市目前市场报价。

2.4 住宅的节能指标

建筑物的耗热量、耗冷量指标综合反映了建筑设计和围护结构热工性能的优劣，因此是节能建筑的重要控制指标[3]。

本文以西安市近十年来的典型住宅为依据说明城市住宅与节能标准的差距。

住宅1—1基本情况：

砖混结构，7层，一字形平面，4个单元。长75.78m，宽11.64m，一般层高2.8m，顶层层高2.9m，总高19.7m。1～5层南、北向外墙为370mm厚，其余墙体为240mm厚。

南北向布置，楼梯间不采暖。阳台用塑钢窗封闭。

屋面：从下到上依此为：10mm水泥石灰砂浆抹灰；120mm预制钢筋混凝土空心板；20mm厚1∶3水泥砂浆找平层；焦油聚氨酯防水涂料隔气层；40mm厚1∶6水泥焦渣找坡层；30mm挤塑泡沫板保温层；20mm厚1∶2.5水泥砂浆找平层；两毡三油防水层；砌115×115×300砖磴上卧铺35mm厚490×490预制钢筋混凝土架空板。

外墙：由内到外：20mm石灰砂浆粉刷；370mm KP1型承重多孔砖(240mm KP1型承重多孔砖)；20mm水泥砂浆粉刷。

外窗：采用塑钢窗。

楼梯间隔墙：240厚KP1型多孔砖，两边石灰砂浆粉刷。

户门：三防门。

图1为住宅1—1西头单元的平面简图。

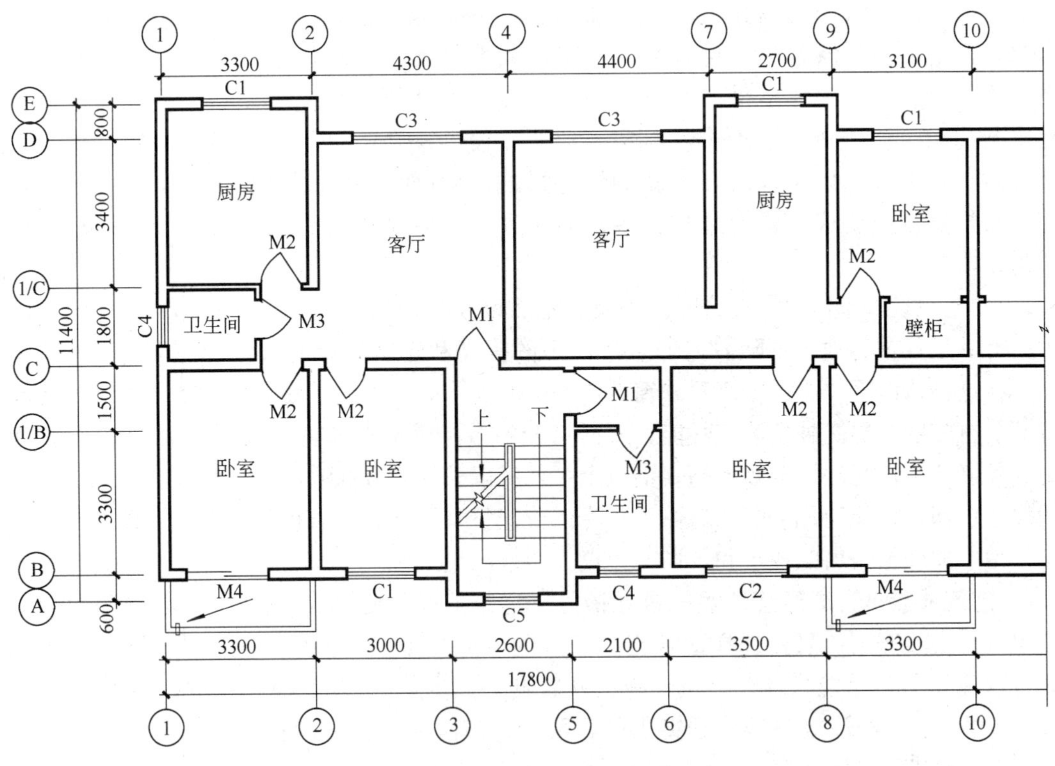

图1 住宅1—1西头单元平面简图

经计算，住宅1—1的耗热量指标和耗煤量指标分别是21.41W/m² 和10.31kg/m²，它与节能标准要求的限值的比较见表4。

住宅1—1节能指标与节能标准限值的比较 **表4**

项目 \ 节能指标	建筑物耗热量指标(W/m²)	采暖耗煤量指标(kg/m²)
住宅1—1	21.41	10.31
节能50%	20.2	9.7
节能65%	14.57	7.02

3 结论和建议

3.1 同全国绝大多数城市一样，西安市1980年以前建造的住宅属非节能住宅。采暖居住建筑没有采取保温措施：围护结构单薄，门窗密闭性差，传热系数大，热损失严重。导致建筑物耗热量和耗煤量巨大，浪费惊人；室内热环境恶劣，冬天寒冷、夏天炎热，室内潮湿，结露，严重影响居民的健康。建议按照有关规定进行节能改造。同时，从近十年来绝大多数住宅的节能情况看，存在同样问题。由于对节能工作执行不力，导致目前很多“节能”住宅与现行节能标准相差甚远，建议对这些住宅加强保温措施。

3.2 中国是一个多地震的国家，尤其是北方一些城市。所以，建筑设计中对抗震要求较高，住宅的热桥问题显得比较突出。屋面大多是外保温结构形式，基本上不存在问题。而外墙目前国内大多采用的还是内保温做法，使热桥状况雪上加霜，外墙平均传热系数高出主体传热系数近半就是实证。所以说，这是一个不容忽视的问题。建议加强热桥部位的保温措施以及寻求新的结构形式。同样，住宅楼梯间也是抗震的薄弱环节，按抗震规范要求设置构造柱较多，热损失大，建议楼梯间采用与外墙相同的保温措施。

3.3 住宅建筑的节能设计，应根据外墙、屋面以及窗户选型综合考虑。窗户在建筑节能中占有举足轻重的地位。随着人们生活水平的提高和建筑设计的需要，外窗面积不断增加。在这种情况下，必须提高对外窗热工性能的要求，才能真正做到住宅的节能。技术经济分析也表明，提高外窗热工性能，所需资金不多，每平方米建筑面积约10～20元，比提高外墙热工性能的资金效益高3倍以上[4]。所以采用高效保温窗不仅可以有效减轻外墙的保温负担，还能简化节能构造体系，降低节能投资。我国有着成熟的单框双玻塑钢保温窗和中空玻璃保温窗的技术和市场，建议大力推广使用。

3.4 建筑节能是提高经济效益的重要措施。建筑节能需要投入一定的资金，但投入少、产出多。实践证明，只要选择适合当地条件的节能技术，若使用4%～7%的建筑造价，即可达到30%的节能指标。建筑节能的回收期一般为3～6年，与建筑物使用周期60～100年相比，其经济效益是非常突出的，可见，节能建筑在一次投资后，可在短期内回收，并能长期受益[3]。

Improve the Energy Efficiency on Urban Buildings and Build Up a Saving Society

ZHU Yumei　LIU Jiaping

(Xian University of Architecture and Technology, Xi'an　710055　China)

Abstract: According to the deployment of State Department, the demand of greatly developing the energy efficiency and saving soil of buildings and building up a saving society, it should implement the standard of building energy efficiency 50% in all over the country, moreover impel the execution of energy efficiency 65% standard of urban building. In our country, many cities are establishing or have carried out the dwelling building energy efficiency 65% standard. How could we change building energy efficiency index into envelope thermal property index? Only though analyzing and summarizing the situation of existing building thermal property under energy efficiency 50% standard could we find a reasonable design method which improves energy efficiency level of urban building used technology. This paper took Xi'an as an example to found out the gap between thermal property index of existing building and new design standard though classifying and summing up the structure forms and thermal properties of building envelopes in resent years and comparing the energy efficiency index to energy efficiency 50% and 65% standard. In conclusion paper put forward to energy-saving windows should be firstly adopted and electing thermal insulation materials of wall should take into account some factors such as structure form and cost. Moreover the effect that is generated by thermal bridge resulting from earthquake design must be greatly pay attention to and seek new structure forms to strengthen thermal insulation of thermal bridge position.

参考文献

[1]　涂逢详. 建筑节能(43). 北京：中国建筑工业出版社，2004

[2]　西安市地图集编纂委员会. 西安市地图集. 西安地图出版社，1989 年 10 月

[3]　赵键. 建筑节能工程设计手册. 经济科学出版社，2005 年 7 月

[4]　中国建筑业协会建筑节能专业委员会. 北京市建筑节能与墙体材料革新办公室编著. 建筑节能：怎么办. 北京：中国计划出版社，1997.10

[5]　杨善勤. 民用建筑节能设计手册. 北京：中国建筑工业出版社，1997

[6]　中国建筑西北设计研究院编.《民用建筑节能设计标准陕西省实施细则》(陕 DBJ 24—8—97). 西安，1997

天津市首座超低能耗示范楼系统分析

王一平[1]　任建波[1]　朱　丽[2]　田　玮[1]
（[1] 天津大学化工学院，天津　300072；[2] 天津大学建筑学院，天津　300072）

摘　要： 建筑能耗占我国社会总能耗的1/4强，随着国家经济的持续发展和城乡建设的加快，这一比重还将逐步上升。天津大学化工学院王一平教授等提出了将太阳能等可再生能源技术和三步节能改造相结合的设计理念，并应用该理念对天津市一座1248m^{2}的民用住宅进行改造。本文对该系统从结构原理、工作方式、经济效益和社会效益等方面进行了全面系统的分析，由此阐明太阳能等可再生能源和建筑相结合对于降低建筑能耗、改善城市环境的意义。

关键词： 三步节能　太阳能集/散热系统　建筑能耗

1　引言

随着常规能源的枯竭和环境的恶化，人们意识到节能问题刻不容缓。建筑行业作为全社会的耗能大户更应该充分认识的节能工作的重要性。我国现有建筑的95%达不到节能标准，单位面积的建筑能耗为发达国家的3～5倍[1]，2000年我国建筑能耗在全社会能源消费量的比重占到了27.8%[2]。另外，由于人们生活水平的提高，对室内舒适性的要求也日益提高，空调、暖气大规模使用，这对社会造成了沉重的能源负担和严重的环境污染，已成为制约中国可持续发展的突出问题。

天津大学化工学院王一平教授等提出了将太阳能等可再生能源技术和三步节能改造相结合的设计理念，并应用该理念对天津市一座1248m^{2}的民用住宅进行改造。

2　系统介绍

该系统为坐落于天津市蓟县翠屏湖科学园内的一栋三层民用住宅，建筑物周围没有建筑物遮挡。天津市的相关地理参数为，地理纬度：北纬39.08°，地理经度：东经117.07°，海拔高度：2.5m。

天津市属于寒冷地区，该系统的改造设计分两步进行：首先，加强围护结构的保温隔热能力，在现有建筑外墙的基础上，加置保温材料，降低外围结构的传热系数；在减小窗户尺寸和结构形式的基础上，改推拉窗为平开窗，同时将单层玻璃改为双层中空玻璃窗，降低外窗散热引起的热量损失。其次，在建筑屋顶安装新型瓦式太阳能集/散热器，为建筑物提供制冷和供暖能力，取消原有的集中供暖系统，结合水源热泵满足三步节能后的供暖或制冷负荷指标，降低系统初次投资费用和系统的运行费用。

2.1 外围护结构保温改造

2.1.1 外墙保温改造

外墙的改造过程中，在现有建筑外墙的基础上粘结 80mm 厚聚苯板，聚苯板外采用抗裂砂浆复合网格布，外面粘结饰面层。改造后外墙的平均传热系数下降为 0.38W/(m^2·K)，小于三步节能后的 0.6/(m^2·K)[3]。

2.1.2 窗户改造

采用两扇整体式玻璃窗体，一扇固定不开，一扇用于室内通风。原有框架的去除可在原有窗户尺寸 1.6m×1.8m 的基础上，将窗体面积减小。缩小方法为四角分别减少 0.05m，减小 11.4%的窗户面积。同时将窗户采用双层中空玻璃窗体，中空玻璃结构尺寸为 5+5+5。改造后外窗的平均传热系数降为 2.6W/(m^2·K)，小于三步节能后的2.8W/(m^2·K)[3]。

2.1.3 屋顶改造

在建筑物 267m^2 顶部外围面积安装新型太阳能集/散热器。该集散热器制成瓦状结构，与建筑物实现完美结合。太阳能集热器热水系统采用定温放水，以提高热泵机组的工作能效。通过测量的水体温度切换电磁阀，获得不高于 45℃的热水。冬季满足住宅供暖需求并为建筑物提供生活用热水，不足的热量由水源热泵提供；夏季，白天利用太阳能集散热器为建筑物提供生活用热水，夜晚利用集散热器与天空的长波辐射产生冷水为建筑物供冷。图 1 为建筑物改造前后的屋顶构造图。

(*a*) 改造前　　(*b*) 改造后

图 1　屋顶改造前后

2.2 水源热泵及冷热存储系统

2.2.1 水源热泵系统

采用户用型水源热泵机组，采用太阳能集散热器提供的热工质或冷工质作为热泵工作的热源或冷源。通过智能控制系统，在不同冷热源间，即热水水箱、蓄热蓄冷箱、集热器之间相互切换工作状态，达到满足居住舒适度、降低建筑能耗和经济最优化的目的。

2.2.2 冷热存储系统

水源热泵系统通过水环路实现热量的空间转移，然而每时每刻需要转移的热量或冷量与所需供热量或供冷量之间很难平衡，同时为解决太阳能的间歇性和不稳定性，在每个楼层设置了一个 2m^2 的水箱，三个水箱并联连接。夏季，利用谷电通过热泵将冷量储存起来用于白天的补充供冷；冬季，将白天的热量储存于水箱，夜晚同样利用谷电通过热泵补充供热。利用水体密度与温度间的关系，实现在三个并联水箱中同时储蓄冷水和热水。密度大的冷水和密度小的热水在三个水箱中分层，密度大的冷水储存在一楼水箱中，密度小的热水储存在三楼水箱中。

2.3 系统运行方式

该系统冬、夏两季运行方式各不相同，可以通过改变自动控制方案在不同季节之间自由转换。

2.3.1 夏季工作方式

夏季，对于气温高、太阳辐照度较好的天气，把水箱中的水用泵输送到屋顶热散热器中，当集散热器出口水温达到45℃时，水泵启动向集散热器中输送低温水，同时将集散热器中的45℃热水输送并存储到三楼水箱，为建筑物提供生活用热水；集散热器出口水温低于45℃时，水泵停止运行。夜晚，启动水泵向集散热器中输送水箱中的水，利用集散热器与天空的长波辐射，降低集散热器中水的温度，当集散热器出口水温低于17℃时，水泵启动向热散热器送水，将低温水输送到一楼水箱，用于室内制冷；当集散热器出口水温高于17℃时，水泵停止运行。

若遇极端天气情况，启动热泵向三楼水箱输送45℃的热水，为建筑物提供生活热水；向一楼水箱输送低于17℃的冷水，为建筑提供制冷。

2.3.2 冬季工作方式

冬季，对于晴朗的天气通过测量集热板表面温度和一楼水箱底部温度的温差，来控制向屋顶集散热器输送低温水的水泵，当集热板表面温度高于一楼水箱底部温度时，启动水泵向集散热器送水；当集散热板表面温度低于一楼水箱底部温度时，关闭水泵停止向集散热器送水。为了防止冬季夜晚以及极端天气下集散热器冻裂，通过测量集散热器入口水温来控制电磁阀，为集散热器放水。为了提高太阳能利用的效率，冬季系统的运行不采用定温放水系统，从屋顶集散热器出来的热水直接进入热泵，加热到45℃输送到三楼水箱，为建筑提供生活用热水和采暖。

2.3.3 春、秋季工作方式

在春、秋季，系统可通过控制系统方便地在制冷与供暖之间切换。更多的是通过太阳能集热器为建筑提供生活用热水。

3 系统经济效益分析

热泵错峰的使用使制冷能耗仅占三步节能数据的8%，占三步节能后分户单冷+集中供暖系统制冷、供暖总能耗的5%。与三步节能后的分户制冷+集中供暖系统对比，本示范项目的制冷供暖系统的年均运行费用仅为128元，占对比系统的0.5%，投资回收期为32年，40年使用期内的投资收益为16万元，总计节约电力为1.83GWh，相当于节约368t标准煤。

与三步节能后的单冷+集中供暖系统进行对比，分析系统的初始投资费用和年运行费用，结果见表1所示。

太阳能集散热器系统经济性对比分析 **表1**

系统	本项目	单冷+集中供暖	差别
初始投资(万元)	69.3	26.9	42.4
运行费用(元/年)	128	20442	20314

续表

系　　统	本　项　目	单冷＋集中供暖	差　　别
投资回收期(年)	21	—	—
使用期经济收益(万)	40	—	—
节能效益(t标准煤)	368	—	—
节能效益(GWh电力)	1.83	—	—

从对比结果可以看出，三步节能改造＋太阳能一空气集散热系统的安装，可以大大降低建筑制冷供暖的负荷和对传统能源的消耗量。在产生可观的经济效益和节能效果的同时，具有很好的社会和环境效益。

4　社会效益分析

目前，天津地区人均建筑面积为21.5m^2，人口按1000万计，天津地区的节能建筑面积所占百分比仅为5%。将所有建筑依照二步节能标准计，则每年的住宅供暖指标超过266万t标准煤。若采用三步节能标准将原有建筑进行改造，并利用本项目提出的可再生能源系统，则每年可节约255万t标准煤，相当于减排二氧化硫5万多t、二氧化碳110万t，对于消除温室效应，提高城市居住环境质量有着不可低估的作用。如果97%的高能耗建筑能在现有的基础上实现零能耗，则会为国家节约大量的能源，同时为保护生态环境、实现可持续发展做出巨大的贡献。天津地区，乃至我们整个国家是一个能源消耗大国，建筑业又首当其冲，有限能源正在逐渐减少。利用取之不尽的自然能源，既可节省国家有限的能源，又可减少对大气和环境的污染，更符合国家建筑节能与环保的方向。

5　超低能耗建筑发展展望

随着太阳能等可再生能源利用技术的发展，将太阳能、风能、水力、生物质能等技术与建筑相结合，实现零能耗甚至负能耗、低污染甚至零污染的目标是超低能耗建筑的发展方向。零能耗建筑最大限度的控制和节约不可再生能源的使用，具有节能效果。与此同时，还能够产生环保的效果，降低温室气体排放，维持生态平衡，保护环境。

在天津地区乃至全国大面积推广类似于本系统的超低能耗建筑具有重要的现实意义和指导意义。

The Technical and Economic Analysis of the First Energy-Saving Building in Tianjin

Wang Yiping[1]　Ren Jianbo[1]　Zhu Li[2]　Tian Wei[1]

([1] School of Chemical Engineering and Technology Tianjin University, Tianjin　300072, China; [2] School of Architecture Tianjin University, Tianjin　300072, China)

Abstract: In China, building energy consumption accounts for approximately 25% of the total energy use and is increasing with the economy development. The combination of solar energy technology and three step energy efficiency is developed by Prof. Wang Yiping in Tianjin University and based on this method we have rebuilt the residence about 1250 square meter of Tianjin, China. The paper fully analyzed the characteristic of the proposed new system, which includes the design method, operation mode, economic benefit, and social efficiency. The results indicate that it is important to combine the new energy and building to reduce the building energy consumption and improve the urban environment.

参考文献

[1] 陈景玺，吕志强，田波．建筑节能的现状与对策［J］．辽宁过程技术大学学报，2004，23(3)：P330～332

[2] 国家建设部，建设部能源“十五”规划纲要［J］．施工技术，2002，31(3)：P1～6

[3] 《天津市居住建筑节能设计标准》(DB 29—1—2004)．中华人民共和国国家标准

辐射地板动态模拟软件的研究

高殿策　马晓雯
（深圳市建筑科学研究院，深圳　518031）
卢　军　付祥钊
（重庆大学，重庆　400044）

摘　要：以 VB6.0 为开发平台，以有限差分法、反应系数法为理论基础，开发出辐射地板动态模拟软件。该软件可以逐时模拟地板表面温度、室内气温以及各壁面温度随供水温以及室外气象参数的变化值，同时还可以逐时计算地板表面向上、向下的传热量以及地板结构内能的变化量。

关键词：辐射地板　数学模型　模拟软件　空调

1　问题的提出

地板辐射采暖系统在我国北方地区已得到日益广泛应用，将辐射地板作为夏季空调降温方式的研究也在国内有条不紊地进行[1]。在辐射地板的工程应用中，普遍受关注的问题是：如何确定地板表面的温度，如何确定地板表面向室内的散热量，如何确定地板结构等。由于地板表面的传热受众多因素的影响，各个因素间又相互干扰，一个因素的改变将直接或间接影响其他因素，因此对于不同的地板结构、不同的供水温度、不同的房间构造、不同的气象参数，上述问题的答案是不一样的，不能用统一的指标来标定不同结构的辐射地板系统。

目前常采用的实验方法能够反映地板传热的一般规律，但存在较多限制，除了供水温度和水流量可以改变外，其他影响地板传热的各扰量：如地板结构的各层材料的厚度、物性参数，埋管的管径、间距，室内热源，室外气象参数等很难一一进行实验，各因素间的正交实验就更加困难。实验方法的缺憾是：无法全面了解冷却地板的传热特性，对于不同的地板结构、不同的盘管铺设方式，很难确定地板表面的温度和换热量。

辐射地板模拟软件可以解决上述问题，为实验做必要的补充。本文将通过改进的地板传热数学模型，致力于实用辐射地板动态模拟软件的开发，以期利用开发出的软件对辐射地板系统进行全面的动态模拟，达到优化辐射地板系统的目的。

2　改进的辐射地板传热数学模型简介

对辐射地板传热的数学模型已基本建立起来，即传热过程大致可以分为如下三个阶段[2]：1)工质水通过与管壁的对流换热把热量传给管外壁；2)管外壁通过热传导与其周围的地板结构层、覆盖层进行热交换；3)地板表面通过对流与辐射与室内壁面、空气、人体进行热交换。目前，许多辐射地板模拟程序则对上述数学模型进行了多种假设和简化：在第二阶段将地板内部传热按稳态传热进行计算；在第三阶段，将室内气温及壁面温度简化为一个定值。这些简化方法在工程应用中能够体现出地板传热的规律，具有简洁、快速的优点，但若应用于研究领域，则稍显不够精确。若将室外气象参数的影响引入模拟程序，并且将地板的内部换热按非稳态传热考虑，这样形成的模拟程序将综合影响地板传热的各种因素：盘管铺设方式、地板结构方式、水系统参数、房间围护结构参数、室内扰量以及室外气象参数等。

辐射地板的铺管形式常采用“双回型”布置方式，如图1所示。为了简化计算，只取一组供、回水管之间的平板部分作为模型计算单元，如图2所示。地板的热传递是一个典型的三维传热过程，由于沿管长方向温度变化相对缓慢，温度梯度很小，因此可以忽略 z 方向的导热，整个平板的传热过程简化为二维传热过程，且平板内部无内热源。

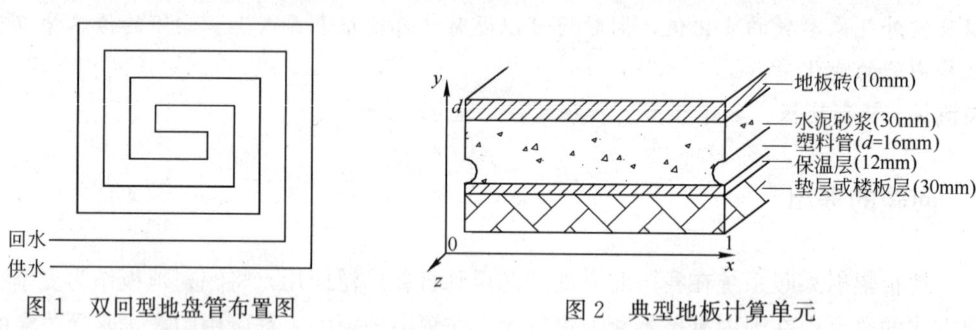

图1 双回型地盘管布置图　　　图2 典型地板计算单元

2.1 盘管内的水与管壁的换热

盘管内水与管壁的换热属于管内强迫对流换热。由于盘管的长度一般比管径大得多，忽略管内进口段的影响，为简化计算，按常壁温处理。该阶段数学模型详文献［2］，此处不再赘述。

2.2 管壁与地板表面的换热

当把地板内部的传热过程当作二维非稳态传热过程时，整个地板内部温度场满足如下微分方程：

$$\frac{\partial t}{\partial \tau}=\alpha\left(\frac{\partial^2 t}{\partial x^2}+\frac{\partial^2 t}{\partial y^2}\right) \tag{1}$$

式中，α 为材料的导温系数，m^2/s。

地板边缘围有轻型保温材料，可以认为绝热；对每一个计算单元而言，由于温度场的对称性，沿着 x 轴的两面可以近似认为是绝热面；地板表面同时受对流换热与辐射换热的作用，采用复合换热系数来考虑。

2.3 地板表面与房间的换热[3]

地板表面与房间的换热包括对流换热与辐射换热两部分。并且，地板与房间的热交换受到室外气象参数、建筑围护结构物性参数、房间内扰的影响。图3是一个典型的房间热传递示意图。从图中可以看出，房间的热平衡是多种因素共同作用的结果。

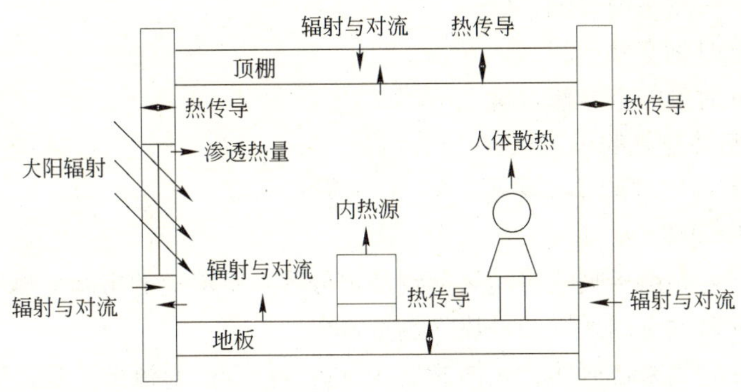

图 3　房间热平衡示意图

对于房间内的每个围护结构内表面，都可以建立一个热平衡方程式。用文字表示为：导热量＋与室内空气的对流换热量＋各表面之间互辐射热量＋直接承受的辐射热量＝0；

若用方程式表示，对于 n 时刻第 i 表面而言，应满足如下方程式：

$$q_i(n)+\alpha_i^{\mathrm{c}}[t_{\mathrm{a}}(n)-t_i(n)]+\sum_{j=1}^{N_i}C_{\mathrm{b}}\varepsilon_{ik}\varphi_{ik}\left[\left(\frac{T_{\mathrm{k}}(n)}{100}\right)^4-\left(\frac{T_i(n)}{100}\right)^4\right]+q_i^{\mathrm{r}}(n)=0 \tag{2}$$

式中　$q_i(n)$——由于两侧温差第 i 围护结构内表面所获得的传热得热量，$\mathrm{W/m^2}$；

α_i^{c}——第 i 围护结构内表面的对流换热系数，$\mathrm{W/m^2\cdot ℃}$；

$t_{\mathrm{a}}(n)$——室内气温，℃；

$t_i(n)$、$t_{\mathrm{k}}(\mathrm{n})$——第 i 和第 k 围护结构内表面温度，℃；

C_{b}——黑体辐射常数，等于 $5.67\mathrm{W/m^2\cdot ℃}$；

ε_{ik}——第 i 围护结构内表面与第 k 围护结构内表面之间的系统黑度，约等于第 i、k 表面自身黑度的乘积，即 $\varepsilon_{ik}\approx\varepsilon_i\cdot\varepsilon_{\mathrm{k}}$；

φ_{ik}——第 i 围护结构内表面对第 k 围护结构内表面的辐射角系数；

N_i——房间内不同围护结构内表面总数；

$q_i^{\mathrm{r}}(n)$——第 i 围护结构内表面直接获得的太阳太阳辐射得热量和各种房间内扰的辐射热量，$\mathrm{W/m^2}$。

由于各围护结构具有蓄热能力，扰量又在不断变化，因此，上述围护结构内表面热平衡方程式并非是简单的代数方程，而是一组微分方程，可以运用传热和内外表面吸热反应系数法求解，可参照文献［3］。

2.4　室内空气的热平衡

房间空气的热平衡可以用文字叙述为：

与各壁面的对流散热量＋各种对流得热量＋空气渗透得热量＋空调系统显热除热量＝单位时间内室内空气显热量的增值

用数学表达式为：

$$\sum_{k=1}^{Ni}F_{\mathrm{k}}\alpha_{\mathrm{k}}^{\mathrm{c}}[t_{\mathrm{k}}(n)-t_i(n)]+[q_1^{\mathrm{c}}(n)-q_2^{\mathrm{c}}(n)]+\frac{L_{\mathrm{out}}(n)(c\rho)_{\mathrm{out}}[t_{\mathrm{out}}(n)-t_{\mathrm{a}}(n)]}{3.6}$$

$$-HE_{\mathrm{s}}(n)=V(c\rho)_{\mathrm{a}}\frac{t_{\mathrm{a}}(n)-t_{\mathrm{a}}(n-1)}{3.6\Delta\tau} \tag{3}$$

式中　$t_a(n)$——n时刻室内气温，℃；

$t_{out}(n)$——n时刻室外气温，℃；

$L_{out}(n)$——n时刻的空气渗透量，m^3/h；

$(c\rho)_{out}$——室外空气的单位热容，$kJ/m^3 \cdot ℃$；

$(c\rho)_a$——室内空气的单位热容，$kJ/m^3 \cdot ℃$；

V——房间体积，m^3；

$q_2^c(n)$——n时刻由于吸收房间热量致使水分蒸发所消耗的房间显热量，W；

$HE_s(n)$——n时刻空调系统的显热除热量，W；

$q_1^c(n)$——n时刻来自照明、人体显热和设备显热等的对流散热量，W。

3　模型求解

3.1　地板内部导热微分方程求解

本文欲建立一个非稳态地板传热模型，在求解地板内部导热微分方程时采用了有限差分法。在差分格式上选择了完全隐式差分格式。

3.2　房间热平衡方程组求解

在已知地板表面温度的前提下，可以通过求解房间热平衡方程式(2)和房间空气热平衡方程式(3)所组成的矩阵，得出逐时的房间各壁面的温度、室内气温的逐时变化值。

3.3　地板传热与室内热环境的耦合求解

将地板传热方程与室内热平衡方程进行联合求解，二者的结合点为地板表面温度，这样就形成了一个动态模拟过程。在已知供水温度、水流量、盘管的长度、地板的结构参数和物性参数、围护结构的尺寸和物性参数、室外气象参数以及室内各壁面温度和气温的初始值的情况下，就可以动态的模拟出室内各壁面温度和室温的逐时变化值。

上述求解过程编成程序，即可在计算机上分析管径、管间距、覆盖层厚度等地板构造参数，以及水温、水量等运行参数，还有室外气象参数等对地板表面温度和地板换热量的影响，帮助辐射地板的设计和运行的优化，详细的计算框图见图4。

4　模型结果验证

以重庆市某住宅单元套房内卧室夏季工况为验证模型，住宅外墙体结构为200mm加气混凝土两侧抹灰，内墙为150mm厚陶土空心砖，单层塑钢窗。盘管采用PEX交联聚乙烯管，管径为$DN16\times2mm$，“双回型”布置，供水温度16℃，流量400kg/h。房间各壁面初始温度以及室温初值统一采用计算时刻的室外气温。室外气象参数采用重庆典型室外逐时气温。

对上述所建立的数学模型以及边界条件进行联合求解，即可得出自上午9：00开始的地板表面温度的逐时变化值。将模型计算值与相似条件下的实验值进行对比，如图5所示，可以发现模型曲线的值基本上要稍低于实验曲线，在运行开始阶段这种差异要明显的多，随着时间推移差异渐渐变小，最大偏差为7.76%，平均偏差为1.8%。造成偏差的原因是：①模型中计算中没有考虑太阳辐射，实际实验中太阳散射辐射会有少量进入室内；

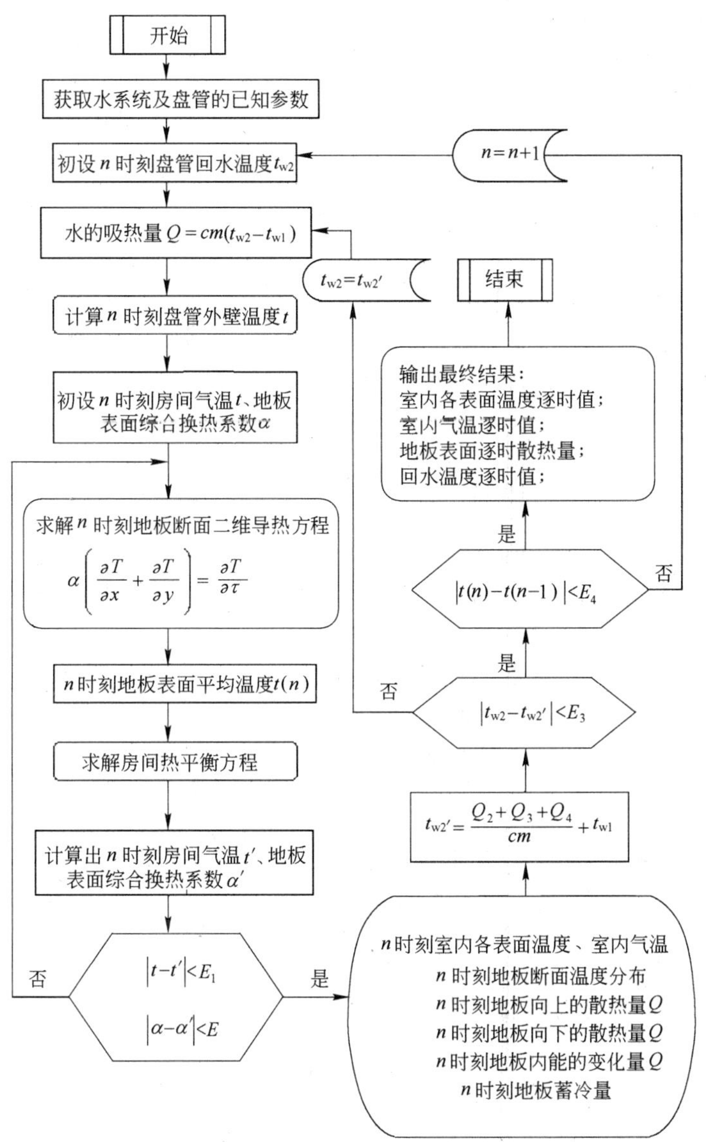

图 4　辐射地板计算框图

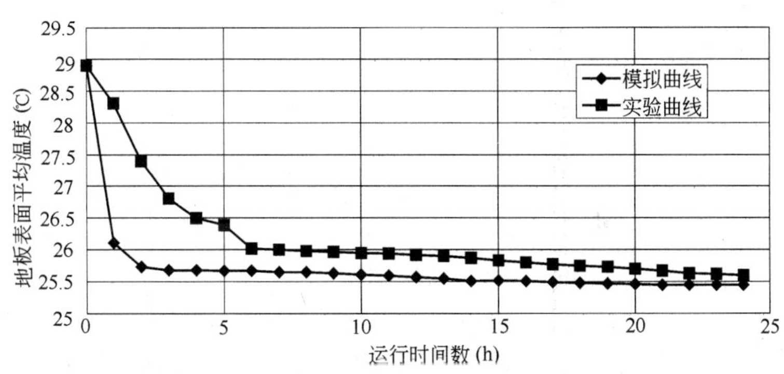

图 5　地板表面温度的模型计算值与实验值的对比

②实验中冷水机组的供水温度在起始阶段有一个慢慢冷却的过程，实测中地盘管供水温度经过 90 分钟才由初始状态下 30℃降至 18℃，这就大大延缓了地板表面的降温速度，而模型中的供水温度进行了合理的简化，从运行一开始即为固定值；③实验中人员的散热增加了室内负荷。

通过模型结果与实验结果的对比，可以认为对所建立的数学模型的求解结果与实验测试结果具有良好的一致性。

5 辐射地板动态模拟软件 FHC(Floor Heating and Cooling)的开发

基于 VB6.0 开发出来的辐射地板动态模拟软件 FHC 具有标准 Windows 风格界面，程序内部编制了大量模块，将整个系统有机的结合起来，实现了既分散又统一的系统集中管理。在程序中涉及的大量数据采用数组方式存入，一部分采用动态数组存储，在程序运行阶段占用内存，程序运行结束即释放内存，可以使计算机的内存得到充分利用。程序中的输入参数都可以以文件的形式保存在计算机磁盘上，文件的存取格式采用了随机文件，加快了数据存取速度，在程序中可以根据需要随机读写文件中的某个记录。

5.1 软件功能

(1) FHC 软件的一个显著特点是动态性，它可以逐时模拟地板表面温度、室内气温以及各壁面温度随供水温以及室外气象参数的变化值，同时还可以逐时计算地板表面向上、向下的传热量以及地板结构内能的变化量，另外在地板向上的传热量中还可以确定对流换热和辐射换热所占的百分比。

(2) FHC 软件的另一个显著特点是可变性，它的输入参数中地板构造层、水系统、盘管形式、房间建筑参数以及初始条件都是由用户自由输入的，因而是可以变化的，这就为分析不同扰量下地板的传热特性提供了可能，这也正是该软件优越于实验的所在。

(3) FHC 软件还考虑了室内空调系统的影响，当室内有空调系统除热时，只需输入空调系统的显热除热量，就可以模拟地板系统与空调系统联合作用下室内热环境的动态变化规律。

(4) FHC 软件结果输出在类似于 Excel 的电子表格中，该表格中的数据可以与 Excel 表格进行互相拷贝，同时它还可以直接输出为 Excel 文件格式，方便了在 Excel 中直接进行数据分析。

5.2 基本输入输出参数

(1) FHC 软件的基本输入参数为：

1) 系统参数：包括地板构造层(装饰层、埋管层、保温层、楼板层)的物性参数(厚度、导热系数、比热、密度)，盘管铺设参数(盘管管径、管间距、盘管长度、盘管导热系数)，水系统参数(供水温度、水流量)等。

2) 房间建筑尺寸参数：包括房间各围护结构的朝向、表面发射率、组成墙体的各层材料的物性参数(厚度、导热系数、比热、密度)，房间建筑的尺寸参数等。

3) 初始条件参数：包括室内余热部分(室外空气渗透量、空调显热除热量、水分蒸发耗热量、设备散热量、照明散热量、人体散热量等)和初始参数部分(地板表面初温、室内气温初温、各围护结构初温、门窗的传热系数、地板下表面的复合换热系数等)。

4) 其他参数：包括描述方面(建筑名称、建筑所在城市、房间名称等)和计算方面(模拟的起始时刻、模拟的时间数等)。

(2) FHC 软件的基本输出参数为：

1) 温度方面：包括地板表面温度的逐时值、室内气温的逐时值、各围护结构内表面温度的逐时值、回水温度的逐时值等。

2) 传热量方面：包括地板向上的传热量、地板向下的传热量、地板内能的变化量以及地板向上传热量中的对流部分和辐射部分的值，还有辐射部分所占的百分比。

5.3 软件界面

FHC 软件系统包括启动画面、主程序输入界面、结果输出界面。程序启动后，首先进入用户视野的是程序的启动界面，如图 6 所示。接下来就进入主程序界面，也就是程序的输入界面，如图 7 所示。在主程序界面上包含了如下信息：

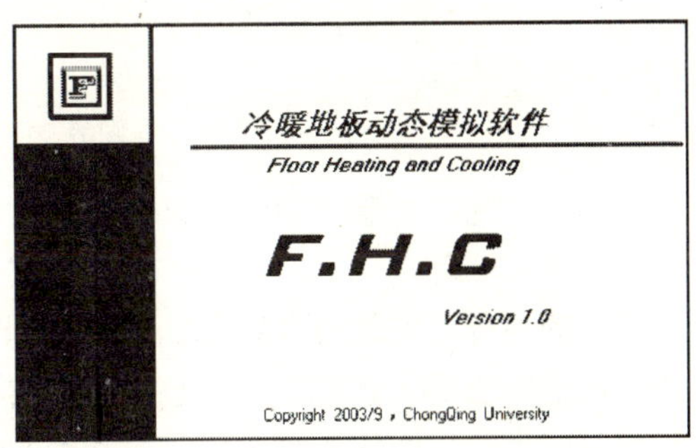

图 6 FHC 软件启动界面

(1) 标题栏：

标题栏是程序窗体顶部的水平条，它显示是应用程序的名称以及打开的文件的名称。

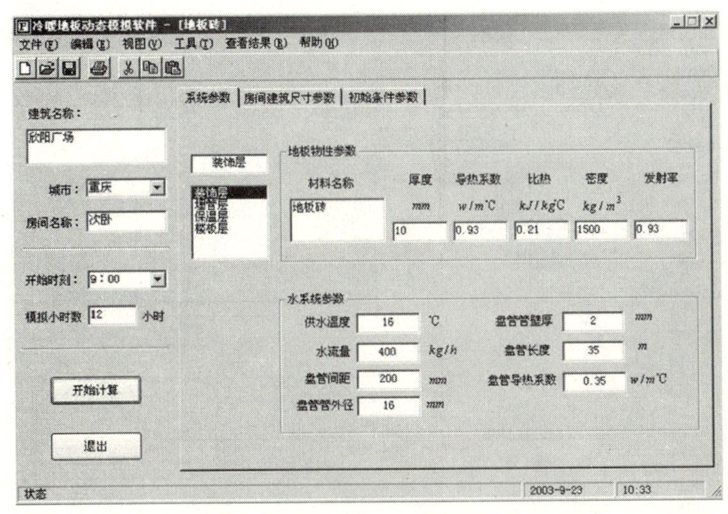

图 7 FHC 软件输入界面—系统参数

(2) 菜单栏：

在标题栏下方是程序的主菜单。主菜单共有 6 个菜单选项：【文件】、【编辑】、【视图】、【工具】、【查看结果】和【帮助】。每个菜单项含有若干个菜单命令，执行不同的操作，各个菜单的具体内容如图 8 所示。

(3) 工具栏

FHC 软件采用了 Windows 标准工具栏的图标按钮显示方法，包括新建、打开、保

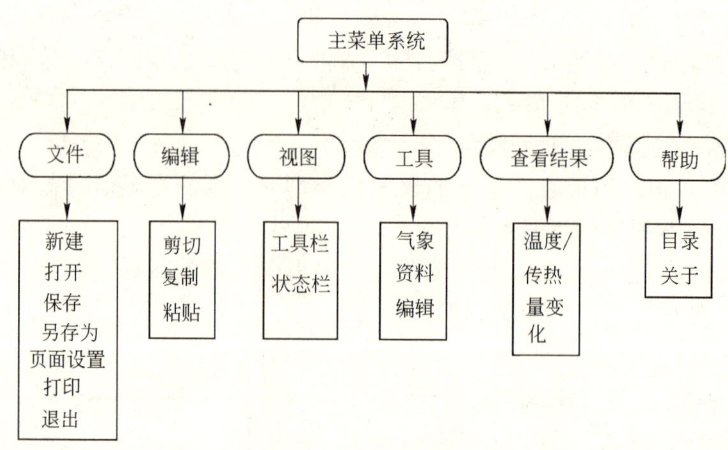

图 8 主菜单结构图

存、打印、剪切、复制、粘贴。这些标准按钮可以极大方便用户的操作。

（4）输入界面

程序的主界面即是输入界面，如图 7、图 9、图 10。在输入界面上，最左边一列是有关建筑工程信息以及计算选项，包括建筑名称、所在城市、房间名称以及模拟的开始时刻和模拟时间数；界面的右侧是程序的基本参数输入界面，由于涉及的输入参数较多，一面窗体很难容纳，所以采用了选项卡技术，选项卡最大的优点就是简洁明快，它可以在一个窗体位置上同时放置几个卡片，每个卡片之间互不影响，可以自由转换，极大方便操作。

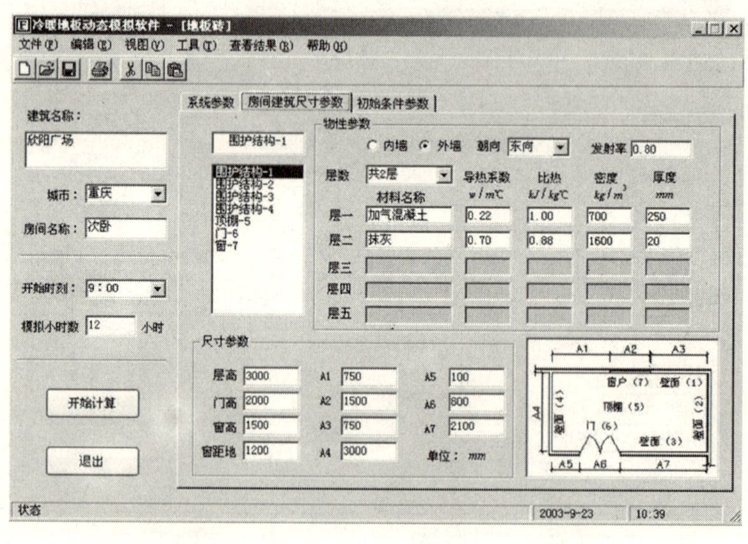

图 9 FHC 软件输入界面—房间建筑尺寸参数

输入界面按照参数的性质分了三大类：系统参数、房间建筑尺寸参数和初始条件参数，分别放在不同的选项卡上。这些参数包括地板构造各层物性参数、铺管的参数、水系统参数、房间各围护结构物性参数和建筑尺寸、室内余热、各壁面、室温的初始值，以及门窗的传热系数等。

（5）查看结果界面

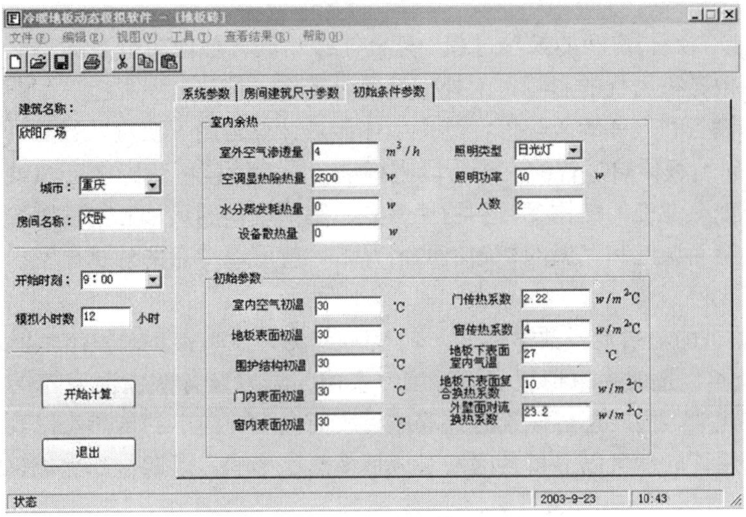

图 10　FHC 软件输入界面—初始条件参数

程序计算完毕后，单击查看结果菜单，即可进入查看结果界面。查看结果界面包括温度结果和传热量结果两个表格。前者罗列了地板表面温度、室温、各围护结构内表面温度以及回水温度从模拟时刻起的逐时值；后者则包括了相应时刻的地板向上、向下传热量和地板内能变化量。结果界面上的表格可以直接拷贝到 Excel 程序中，也可以直接输出为 Excel 格式文件。

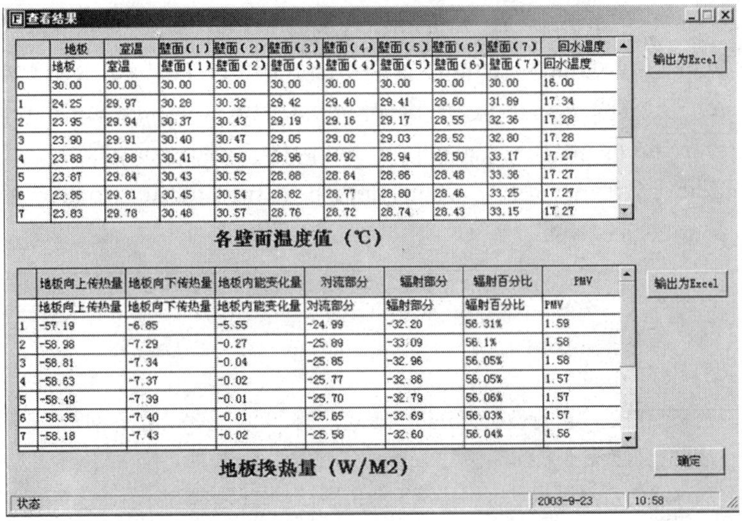

	地板	室温	壁面(1)	壁面(2)	壁面(3)	壁面(4)	壁面(5)	壁面(6)	壁面(7)	回水温度
	地板	室温	壁面(1)	壁面(2)	壁面(3)	壁面(4)	壁面(5)	壁面(6)	壁面(7)	回水温度
0	30.00	30.00	30.00	30.00	30.00	30.00	30.00	30.00	30.00	16.00
1	24.25	29.97	30.28	30.32	29.42	29.40	29.41	28.60	31.89	17.34
2	23.95	29.94	30.37	30.43	29.19	29.16	29.17	28.55	32.36	17.28
3	23.90	29.91	30.40	30.47	29.05	29.02	29.03	28.52	32.80	17.28
4	23.88	29.88	30.41	30.50	28.96	28.92	28.94	28.50	33.17	17.27
5	23.87	29.84	30.43	30.52	28.88	28.84	28.86	28.48	33.36	17.27
6	23.85	29.81	30.45	30.54	28.82	28.77	28.80	28.46	33.25	17.27
7	23.83	29.78	30.48	30.57	28.76	28.72	28.74	28.43	33.15	17.27

各壁面温度值（℃）

	地板向上传热量	地板向下传热量	地板内能变化量	对流部分	辐射部分	辐射百分比	PMV
	地板向上传热量	地板向下传热量	地板内能变化量	对流部分	辐射部分	辐射百分比	PMV
1	-57.19	-6.85	-5.55	-24.99	-32.20	56.31%	1.59
2	-58.98	-7.29	-0.27	-25.89	-33.09	56.1%	1.58
3	-58.81	-7.34	-0.04	-25.85	-32.96	56.05%	1.58
4	-58.63	-7.37	-0.02	-25.77	-32.86	56.05%	1.57
5	-58.49	-7.39	-0.01	-25.70	-32.79	56.06%	1.57
6	-58.35	-7.40	-0.01	-25.65	-32.69	56.03%	1.57
7	-58.18	-7.43	-0.02	-25.58	-32.60	56.04%	1.56

地板换热量（W/M2）

图 11　查看结果界面

6　小结

6.1　对已有的地板传热数学模型进行改进

(1) 在地板内部的传热过程中建立了非稳态导热方程，并运用有限差分法求解该微分方程；

(2) 在地板表面与房间的换热过程中建立了以房间各壁面为节点的热网方程组，并充分考虑了房间内各种内扰的影响：室外空气通过墙体的传热、室外空气的渗透得热、太阳辐射、照明、设备和人体的散热等，其中墙体传热计算运用了反应系数法；

(3) 将辐射地板传热过程的三个阶段进行耦合求解，就可达到在已知供水温度、水流量、盘管的长度、地板的结构参数和物性参数、围护结构的尺寸和物性参数、室外气象参数以及室内各壁面温度和气温的初始值的情况下，就可以动态的计算出室内各壁面温度和室温的逐时变化值。

6.2 借助于改进的辐射地板数学模型，以 VB6.0 为开发平台，开发出辐射地板动态模拟软件 FHC。它可以逐时模拟地板表面温度、室内气温以及各壁面温度随供水温以及室外气象参数的变化值，同时还可以逐时计算地板表面向上、向下的传热量以及地板结构内能的变化量。利用该软件可以分析各因素对地板换热的影响，以期达到优化辐射地板系统的目的。

A Radiant Floor Dynamic Simulation Software

Gao Diance[1] Ma Xiaowen[1] Lu Jun[2] Fu Xiangzhao[2]

([1] Shen Zhen Institute of Building Reseach, Shenzhen 518031 China;
[2] Faculty of Urban Construction & Environmental Engineering,
Chongqing University, Chongqing 400044 China)

Abstract: Based on the theory of finite difference method and response factor method, develop the radiant floor dynamic simulation software by VB6.0 computer language. This software can simulate the variation of floor temperature, indoor air temperature, wall surface temperature along with water supply temperature and outdoor weather parameter by hours. In addition, it also can compute floor heat transfer at two side surfaces and the variation of floor inner energy.

参考文献

[1] 王子介等. 地板辐射供冷可行性研究分析. 暖通空调，2002，32卷，6期，p56～58

[2] 付祥钊等. 冷暖地板换热性能分析方法. 暖通空调，2000，4期，p9～11

[3] 彦启森，赵庆珠. 建筑热过程. 中国建筑工业出版社，1986.12

[4] 章熙民，任泽霈，梅飞鸣. 传热学(第三版). 中国建筑工业出版社，1993.6

[5] 杨大地，涂光裕. 数值分析. 重庆大学出版社，1998.1

[6] 宋伟，吴建国等. 中文 Visual Basic 6.0 高级编程. 清华大学出版社，1999.4

建筑节能促进绿色建筑的健康发展

惠彦涛　鲁　奇　李辉山
（西安建筑科技大学管理学院，西安　710055）

摘　要：目前，能源问题已被列为人类面临的四大生存问题之一。据统计，在全世界的能源消耗中，建筑能源的消耗在人类总能源消耗量中所占比例甚高，达到了30%～40%。而且随着人们对建筑物各方面要求的不断提高，建筑能耗还在不断地增加。因此，推行建筑节能已成为我国发展绿色建筑、实现可持续发展的一项重要举措。文章给出了建筑能源消耗的现状及影响，对建筑节能的潜力进行了具体分析，指出了我国建筑节能工作中的一些特点和不足，提出了具体的进一步提高建筑节能效果的措施建议。

关键词：能源　建筑节能　绿色建筑　健康发展

随着我国经济的迅猛发展，对于能源的消耗与日俱增。2003年我国的能源消费量已占到全球消费总量的9.4%，相当于全球总产量30%的主要资源和原材料，成为能耗第二大国。在能源消耗中，建筑能耗上升势头很快，目前已接近发达国家水平，占到能源总消耗的1/3左右。21世纪头20年，是我国建筑业的鼎盛时期，2020年全国房屋建筑面积将接近2000年的2倍。目前我国建设高潮持续不断，每年建成的房屋面积高达16亿～20亿m^2，超过所有发达国家年建成建筑面积的总和。遗憾的是，不仅既有的近400亿m^2建筑中的99%为高耗能建筑，新建建筑中95%以上仍属于高耗能建筑。按照目前建筑能耗水平发展，到2020年，我国建筑能耗将达到10.89亿t标准煤，超过2000年的3倍。建筑节能已不只是经济问题，而成为国家的重大战略问题，已成为束缚我国绿色建筑发展的瓶颈。

1　建筑能耗现状及影响

1.1　建筑能耗约占社会总能耗的1/3

我国建筑能耗的总量逐年上升，在能源总消费量中所占的比例已从上世纪70年代末的10%，上升到近年的27.45%。而发达国家的建筑能耗一般占全国总能耗的33%左右。建设部科技司研究表明，随着城市化进程的加快和人民生活质量的改善，我国建筑耗能比重最终还将上升至35%左右。

1.2　高耗能建筑比例大，加剧能源危机

直到2002年末，我国节能建筑面积只有2.3亿m^2。目前，我国已建房屋有400亿m^2以上属于高耗能建筑，总量庞大，潜伏巨大能源危机。正如建设部有关负责人指出，仅到2000年末，我国建筑年消耗商品能源共计3.76亿t标准煤，占全社会终端能耗总量的

27.6%，而建筑用能的增加也多带来了25%的温室气体排放。因高耗能建筑比例大，单北方采暖地区每年就多耗标准煤1800万t，直接经济损失达70亿元，多排二氧化碳52万t。如果任由这种状况继续发展，到2020年，我国建筑耗能将达到1089亿t标准煤；到2020年，空调夏季高峰负荷将相当于10个三峡电站满负荷发电能力，这将会是一个十分惊人的数量。

1.3 能源匮乏

与当前发达国家建筑能耗已经大大降低的情况相比，我国单位建筑面积采暖能耗是发达国家标准的3倍以上，与发达国家存在较大的差距。而对于美国而言，全球石油资源的战略布局以及石油的开采区域和运输线路等关键点的调整工作已基本完成，我国却没有那样强有力的能源后盾支持，在这样的国情下，建筑节能水平的改善实际上应该比发达国家更为紧迫。

2 建筑节能潜力分析

2.1 中国建筑的特点决定了节能潜力很大

发达国家住宅中独户住宅、联户住宅很多，体形系数(建筑外表面积与其所包容的体积之比)较大，这类建筑能耗相对较高；而我国住宅以多层、高层公寓式建筑居多，体形系数较小，即冷天散热面积相对较小，在大体相同的围护结构保温条件下，以每平方米建筑面积计，大约只要用一半的采暖能源。但与发达国家对比，我们的外墙、屋顶单位面积能耗为它们的3～5倍，窗户能耗为它们的2～3倍。从节能的角度看，我国的建筑节能潜力很大。

2.2 积极贯彻节能标准，挖掘节能潜力

我国城乡建筑围护结构保温隔热和气密性能差、采暖空调系统的能源效率低，与发达国家不断提高的建筑节能要求相比，差距越拉越大。我国已经编制了居住建筑与公共建筑节能设计标准，该标准可使建筑节能50%。这些标准都已经通过充分的技术经济论证，符合我国国情，经过努力完全可以做得到的。然而，仍有不少地区无法真正落实这些节能标准，或者敷衍应付一下了事。如果这种情况不能从根本上改变，实现建筑节能目标就无从谈起。在2020年以前，只能希望首先把节能50%的标准贯彻执行好，新建建筑全部按节能标准建造，既有建筑有计划地每年按标准改造一批。在工作进展良好的条件下，2010年以后再逐步在各地实施节能更高的标准，节能成果就将更大了。

2.3 设备容量减小，可以提高经济效益

由于建筑能耗减少，即采暖空调照明负荷相应减少，因而所需设备容量减小，新建建筑可以减少这部分初始投资，改造既有建筑则可将节省下来的这部分负荷移用于其他建筑。在节能50%的情况下，设备容量相应减少50%，可见建筑节能的经济效益是十分显著的。

2.4 市场经济发展的需要

随着市场的发展，房地产市场已逐步进入了“买方”市场，重新考虑市场需求问题已经摆在每个开发商面前。事实上，北京、南京、上海、广州等城市，已经有多个节能建筑的成功范例，并受到市场的热烈追捧。只有不断创新才是创造市场商机的原动力。顺应建筑节能大潮，实现节能建筑的社会和经济效益双赢，为中国的可持续发展做出贡献，是每个有责任的开发商的必然之路。

3 我国目前建筑节能工作的特点

3.1 节能建筑面积比例甚小

我国到2002年只在城市中累计建成节能建筑2.3亿m^2（包括节能率30%及50%的建筑），仅占全国城市建筑面积的2.1%，其中约有一半建在北京和天津。因此，从总体上说，我国现在的建筑能耗水平与传统建筑相比变化并不大，有些地方甚至还不如以往的传统建筑。

3.2 建筑节能的重点是北方地区的采暖居住建筑

由于我国建筑能耗中占比重最大的是采暖能耗（占全国总能耗的13%），而且住宅室内热环境质量普遍较低，因此以采暖居住建筑作为建筑节能的突破口无疑是正确的决策。但近几年商用和公共空调建筑以及住宅空调的使用量迅速增加，相应的节能标准和节能措施显得有些滞后。

3.3 建筑节能标准主要针对新建建筑

出台的建筑节能标准大部分是设计标准或设计规范，缺乏对已建建筑的节能措施，也缺乏产品设备在运行工况下的节能标准，尤其缺乏建筑运行管理中合理用能的方法和措施。

3.4 建筑节能重点有所偏废

主要技术措施集中在改善维护结构的热工性能方面。然而建筑节能的重点是暖通空调和照明，重要任务是提高能源的转换效率。例如，如果我们将维护结构的保温性能提高一倍，大约能减少采暖负荷35W/m^2，而采暖系统的效率降低一个百分点，每燃烧1kg标准煤就要多损失81W。如果室内温度高出标准1℃，就会增加12%的能耗，在北方城市中因为房间过热而开窗睡觉的现象并不少见。

3.5 建筑节能标准执行不力

尽管我国已经有了初步的建筑节能法规标准的框架体系，但“执法”却十分不力。由于缺乏配套的节能评价指标体系，因此对建筑设计和产品的节能评审往往流于形式，也无法正确地评估建筑节能的目标是否实现。更由于没有经济制约手段，在市场经济条件下很难形成建筑节能的内在驱动力和自觉性。

3.6 部分标准制定不严密

我国的室内热舒适标准、制冷机标准中的制冷机的综合部分负荷值等，基本上是完全参照发达国家的标准制定的，这对各方面和国外有很大差别的中国人是否完全适用，并没有经过验证。

4 进一步提高建筑节能效果的措施建议

4.1 建立和健全建筑节能法规体系

其中的关键是全国人大对《节约能源法》进行补充修改，由国务院制定并颁布《建筑节能管理条例》，使建筑节能工作走上法制化的轨道，可依法推进和监督建筑节能工作。届时，建筑节能将不仅仅是一般提倡和号召，而成为各级政府、相关部门、单位和公民的执法行为。

4.2 强制性法规和优惠政策并用

为此，建议对不执行节能标准的新建（或改建、扩建）民用建筑工程恢复征收固定资产投资方向调节税，或者建筑能源消费税，税率仍为总投资额的5%；对节能建筑则实行零税率政策。同时，国家可以建立城市热网改造专项资金，以引导北方城市热网改造。鼓励生产和采用节能门窗和建筑保温(隔热)材料，经认证合格的节能门窗、建筑保温(隔热)材料减半征收产品增值税。制定推广节能空调、燃气空调和冷热电联产的具体措施；制定鼓励生产和使用建筑耗能器具的经济激励机制。

4.3 加强对建筑节能工作的组织领导

各级政府要把提高建筑能效的工作列入重要议事日程，纳入当地国民经济和发展计划，落实领导，落实管理机构，落实岗位和人员，逐级严格实行目标责任制考核，以确保预定目标的实现。

4.4 不断地开发各种新型节能材料

在各部分能耗中，围护结构(墙体、屋顶、外窗)是耗能量最大的部分，约占到76%左右，因此要不断地开发各种新型高效节能墙体材料、保温隔热材料和高性能建筑玻璃及其应用技术。同时要加强设备节能设计的研究和开发，提高能源的转换效率。

4.5 加强终端能耗管理

研究并建立我国节能建筑评定体系，选取试点城市和试点住宅小区，实施节能住宅建筑能耗性能评定；组织制定、修订和实施家用电器和照明器具的强制性能效标准，把好市场准入关，从源头上提高家用电器和采暖空调、照明器具的能源利用效率，限制并淘汰落后的高耗能设备；制定和实施超前性能效标准，积极引导生产企业主动提高产品能效，加快节能技术进步。进一步规范节能产品认证制度，扩大节能产品认证范围，提高节能产品的知名度和社会的接受程度。

5 结语

节约能源是我国的基本国策，建筑节能是国家节能工作的重要组成部分，发展绿色建筑是我国走可持续发展道路的重要体现。因此，我们要积极推进建筑节能与绿色建筑的发展，贯彻实施建筑节能法律法规，这对于建设事业走科技含量高、经济效益好、资源消耗低、环境污染少、人力资源优势得到充分发挥的新型工业化道路；贯彻落实党的十六届三中全会提出的“坚持以人为本，树立全面、协调、可持续的发展观，促进经济社会和人的全面发展”的科学发展观；按照减量化、再利用、资源化的原则，搞好资源综合利用，建设节约型社会，发展循环经济；实现建设事业健康、协调、可持续发展；全面建设小康社会进而实现现代化的宏伟目标，都具有重大而深远的意义。

Building Energy Efficiency Accelerates The Healthy Development of Green Architecture

Hui yantao　Lu qi　Li huishan

(Management Institute in Xian University of Architecture and Technology，Xi'an　710055　China)

Abstract: Today, energy sources issue has become one of the four survival issues for people. According to the statistics, among the total energy sources consumed in the world, the proportion of the construction energy sources consumed is much high, from 30% to 40%. And along with the continuously improving demand, construction energy sources consumed are still constantly increasing. So, pushing building energy efficiency has turned into one of the important acts for developing green architecture and realizing sustainable development. The paper presents the actuality and infection of the construction energy sources consumed, concretely analyzes the potential of the building energy efficiency, points out some characters and shortage in our building energy efficiency tasks, and advances some idiographic measures and advices for enhancing the effect of building energy efficiency.

参考文献

[1] 杨欣铭，陈军栋. 节能住宅设计技术经济分析. 河北建筑工程学院学报，2004.9

[2] 黎明地产. 节能住宅：可持续发展的绿色建筑. 城市开发，2004.11

[3] 龙惟定编著. 建筑节能与建筑能效管理. 北京：中国建筑工业出版社，2005

[4] 胡吉士，方子晋主编. 建筑节能与设计方法. 北京：中国计划出版社，2005

[5] 孙慧琴，刘军民. 绿色建筑大有作为. 建设科技，2004.19

INFRARED BUILDING ENERGY INSPECTION

D. R. Newport
(Institute of Infrared Thermography, Canada)

The need for identifying construction and energy related problems and to pinpoint areas of energy gain and loss is more important today than ever before. Inspections of buildings to determine these problems are difficult at best. This paper shows how the energy efficiency of the building envelope can be assessed with greater precision and identifies problems that were previously impossible to find using normal methods.

1 Introduction

Infrared Thermography used for building investigations is not new. It was used extensively in Canada during the 70's and continues to be used today. There are various applications of the technology for the analysis of the performance of building enclosures and facilities. The objective of thermography is to assess the efficiency of a building enclosure by locating surface temperature anomalies and identifying their causal mechanisms.

Performing infrared thermography inspections provides information that allows building owners to know exactly where their energy related problems are, thereby removing the guesswork to increasing building energy efficiency and occupant health and comfort level.

1.1 Energy Consumption

Energy consumption is usually divided into 3 sectors, industry, transport and residential/commercial sectors. Depending on the country, 40%~50% of energy usage is related to the last sector. With almost one half of the energy consumed used for comfort purposes, this sector is of great importance when the reduction of energy is considered. Increasing energy efficiency in the building sector must therefore be a priority.

For new buildings and structures this task if fairy simple, change the standards and codes and from a certain date on the structures will be more efficient. The difficulty arises with the requirement to ensure the buildings are being constructed to the energy efficiency standard.

Older buildings tend to be less energy-efficient and with only about 1%~2% of the existing building stock are renovated per year; there is an overwhelming need to upgrade existing buildings to reduce energy consumption. For existing buildings it is difficult to access the quality, current condition and effectiveness of the structure. Without a problem visibly showing up, it is difficult to find the weaknesses for the repair and upgrade. This presents a major problem.

2 Infrared Inspection and Testing

Defects in thermal insulation and airtightness in the external envelope of a building cause inconvenience such as uncomfortable interior temperatures, air leakages and moisture damage. It causes discomfort to the inhabitants of the building and also excessive consumption of energy. The best solution is to first, identify the problem, determine the cause of the anomaly and then to treat it accordingly. The main difficulty is to find suitable methods and equipment to initially identify the problem. Inspections and assessment of the building by visual means is usually ineffective, and chiefly relate to obvious faults and cosmetic defects. It is possible for example, to visually conceal extensive defects in workmanship, with serious consequences for the user. Often the problems, their causes and consequences-simply cannot be seen until after costly damage has been done. At that point the only recourse may be extensive, costly reconstruction.

Infrared Imaging is predictive and is the most cost effective & least intrusive of all building diagnostic procedures. Thermal defects and faults such as missing insulation, thermal bridges, air leakages, moisture and condensation will change the wall temperatures. With the use of infrared thermography surface temperature patterns are used as indirect indicators of subsurface anomalies. However, the interpretation of these characteristics and the formulation of solutions to the resulting problems require the skills of a trained building thermographer with an understanding of building science.

3 Conduction Losses

In designing the thermal resistance of the envelope, the aim is to achieve the desired indoor climate in the most economical way with regard to the specific outdoor climate. Experience has shown that lack of insulation, poor installation of insulation, air barrier and air sealing deficiencies reduces the thermal integrity of the envelope and gives rise to considerable higher rates of energy consumption of a building.

For new or old buildings, it is important they meet the new energy conservation requirements with regard to thermal insulation and airtightness and that any defects in construction which may occur should be identified and repaired.

Typical insulation defects:

—the insulation does not fill the entire space designed for it (cracks, holes, too thin insulation, settlement of insulation, shrinkage after installation, rigid insulation in the wrong place, etc)

Figure 1 The thermal image shows an example of a new house with a high level of energy efficiency. No thermal defects were identified during the inspection.

Construction and insulation standards have improved over the past decade. Many countries either have, or are working on, energy rating programs that conform with modern Environmental Energy Efficiency guidelines.

—incorrectly installed insulation
—HVAC installations through the insulation
—permeable insulation materials insufficiently protected against movement of air
—wet insulation

Figure 2 The Infrared Inspection clearly shows the extent of the energy loss problems with this house. This home is only 8 years old and the infrared camera shows areas of significant heat loss through the walls, roof, and the basement has no insulation whatsoever. An attic inspection revealed that there was no insulation in the ceiling.

The owner will now understand the extent of the energy loss and know exactly what must be done to rectify it. After the work is complete another inspection can be performed to evaluate the effectiveness and quality of the insulation contractors' work.

Walls in a building that are not adequately insulated are significant areas of energy loss or gain. The larger the differential between the inside and outside air temperatures, the lower the K-value of the envelope, the greater the heating or cooling loads. Figure 3 shows a void in the wall insulation between the window and ceiling.

Figure 3 The Infrared image identifies missing insulation between the ceiling and the window. The problem was evident in several wall areas of the same home. This can lead to more severe problems associated with moisture developing within the wall cavity.

The contractor neglected to put insulation in the wall cavity. It is easily identified with the infrared camera.

Installation of the insulation within the wall cavity is critical. It should fill the cavity and be tight against the warm side of the wall. If this is not installed in this way the effective insulation value will be reduced and has the possibility of becoming a pathway for convective air flow, further reducing the insulation value.

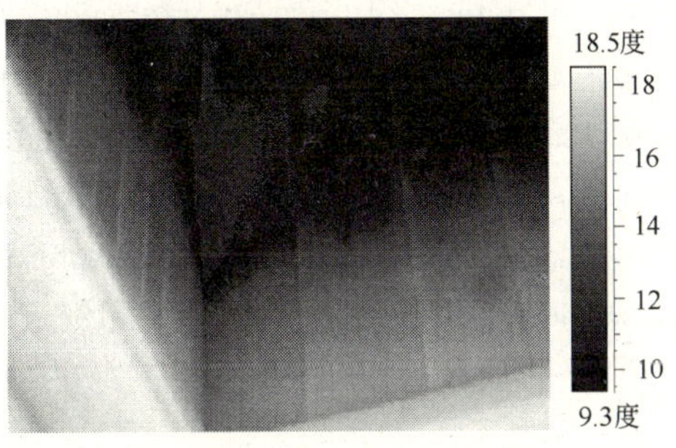

Figure 4　The Infrared image shows the effect of insulation improperly installed. The insulation is not in intimate contact with the inside wall. This will reduce the effective insulation value thereby increasing energy loss.

3.1 Thermal Bridges

Thermal bridges are paths of high conductivity that exist within an enclosure. They may be the framing members in insulated construction, the metal ties in cavity wall and panel constructions, certain structural members, metal window and door assemblies or any other element that penetrates a layer of the building envelope. Increased heat flow toward the exterior via thermal bridges results in temperature variations on the interior. Small temperature differences may not be serious, but add the energy loss up due to thermal bridging of a structure and significant losses may be revealed. More extreme differences may lead to condensation and subsequent deterioration.

Figure 5　Thermal image shows typical wood frame construction with thermal bridging between the warm interior and cold exterior at each 2×4 wall stud and in the corner. The k-value of the wood is higher than that of the insulation so there is increased energy loss here. Also note the corner where the construction has a double stud. To reduce energy loss the wood studs should be 2×6 which provides a longer heat transfer path and more insulation can be added. Or better, the framing studs should not make direct contact from the interior to the outside surfaces. The introduction of a thermal break or staggered studs would be ideal.

Figure 6　Excessive thermal bridging at the floor to wall attachment.

4 Convection Losses

Leakage of air through improperly sealed joints and junctions, and incorrectly installed or damaged air barrier appear to be the types of defects of greatest frequency. Air will flow quite easily through batt and loose fill types of insulation and through the joints between the sections of rigid insulation. Such defects give rise to unsatisfactory temperature distribution and air movements (draughts) in the living space and to local reductions in temperature on the building envelope risking condensation and deposition of dirt. The leakage pathway is often complex and, without infrared, extremely difficult to find.

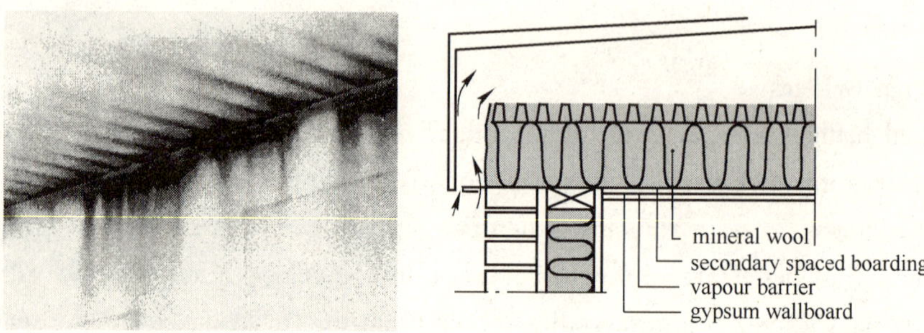

Figure 7 Thermal Image showing cold air infiltration at top plate of wall section. The entire house was poorly sealed and had excessively high heating bills.

Exfiltrating air is only identified from the outside of the building and has a reverse tone in its pattern (irregular flaring) to that of infiltrating air. The interpretation of exfiltration is slightly more complex than infiltration because of the many layers of material that the exfiltrating air has to pass through. If one is examining a composite construction (e. g. construction brick with veneer), the thermal pattern may not represent a significant surface temperature difference, even though the air leakage is severe.

Figure 8 Severe air exfiltration pattern on hotel building due to poor sealing at wall interface.

It was also found that vertical air gaps due to improper placing of insulation material near studs, electric cables, etc may exert a significant effect on the thermal insulation performance, owing to air movements in these gaps.

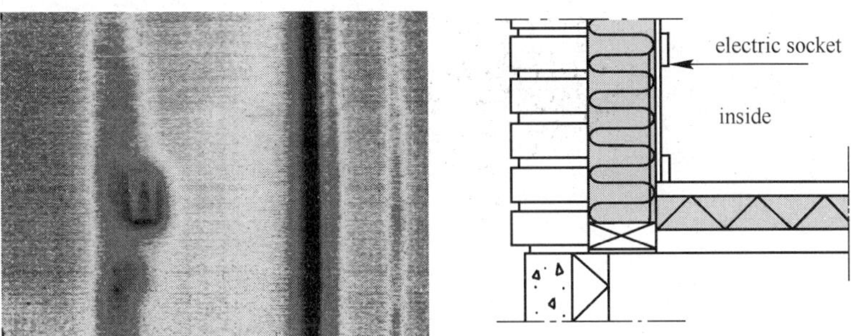

Figure 9 Insulation and airtightness defect due to incorrect placing of insulation material around electric cables and the tearing of the Polyethylene vapour retarder.

5 Moisture

Moisture is possibly the most important factor affecting the performance of a building enclosure. In the form of vapour, it can be a necessary and useful component of the air and of building materials. When it condenses to a liquid or a solid, however, it can create difficulties. Moisture in a building can originate from precipitation, condensation of vapour, or building materials which release moisture.

Vapor travels from high pressure to low pressure. Moisture-laden air is literally "pushed" through penetrations in the building envelope. The moisture could damage the insulation and even the framing members, but a continuous vapour retarder helps prevent this from happening.

Figure 10 Showing vapor retarder installed in the inside face of a cold climate building.

Defects in the inner or outer air barrier of the envelope involve the risk that movement of air will occur right through the envelope and cause moisture to collect. Consequently, in general practice, an air/ vapour retarder should be carefully installed on the warm side of the insulation to prevent the outward movement of water vapour, in a cold climate situation.

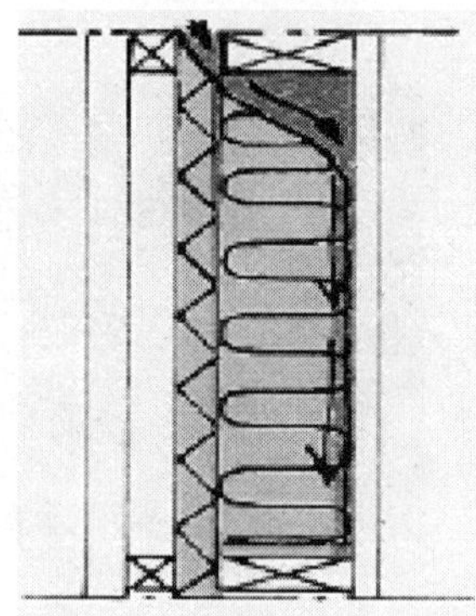

Figure 11 Improperly installed air/vapour retarder material. The material must be properly applied, overlapped and sealed to the wall structure.

As building designs and techniques produce tighter thermal envelopes, moisture (from leaks or condensation) has created more and more problems. The water can intrude through a small crack but it is then trapped between the relatively impermeable building materials. Unbonded areas between brick and mortar are usually the cause of both water and air leakage in brick walls. The use of an insufficient quantity of mortar promotes rain penetration since incompletely filled joints and poor bonding between brick and mortar produce openings through which rain can pass. Also, cracking of masonry, which may be a result of differential movements of the structural components of the buildings, produces openings through which rain may enter, as in Figure 13 below.

Figure 12

Buildings often have problems with water intruding into occupied apartments or offices from exterior sources. Because finding the source of water leakage paths using conventional methods are often unsuccessful. Water damage will continue and cause premature deterioration of finish materials, ruined equipment and furnishings. In worse case scenarios of prolonged moisture, mold and mildew can grow and affect indoor air quality.

Figure 13 Water leak entering the building. Infrared thermography was able to find the leak and identify the source. Even though there was no visible evidence of the leak.

The entry point of infiltrating water is often difficult to determine because water normally does not enter a building in a way that is predictable. For example, if there is an exterior facade with multiple entry points in close proximity, the result can be one large leak on the interior side. Another example is a single large leak area located on one side of a building that filters through to multiple interior places on the opposite side of the building.

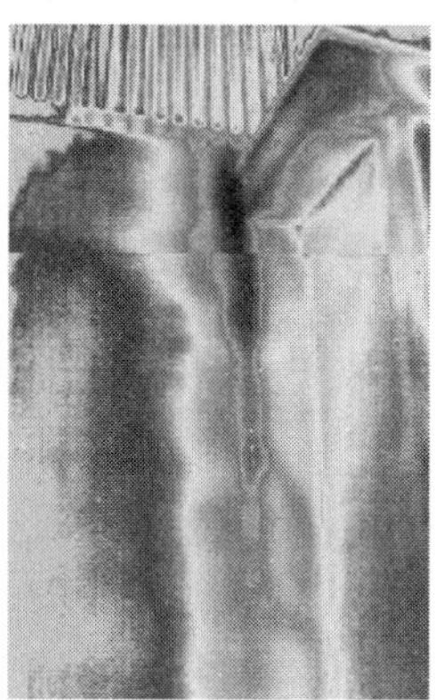

Figure 14 Water leaked between the concrete and stucco. Infrared thermography easily identified the location of the water, the extent of the damage and the problem source. In this area once again, there was no physical indications of moisture. Without the infrared inspection this problem could have continued for some time before it became visually apparent. The repair costs would have been substantially higher.

5.1 Moisture Detection-Key to the Identification of Mold Growth.

Mold, organic fumes or bacteria can play a significant role as either the result of, or culprit of a variety of issues. Mold typically occurs when porous building materials are exposed to condensation, leaks or where inadequate ventilation promotes ambient moisture saturation. Unattended mold infections are likely to result in a condition commonly known as dry rot, and damage can become quite extensive. An airborne mold spore spread has also become a matter of concern involving Indoor Air Quality (IAQ).

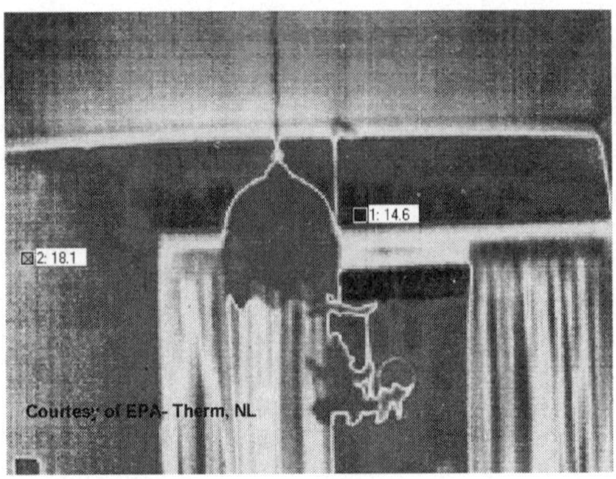

Figure 15 No vapour retarder was installed in this building. Moist vapour passed into the wall and condensed within the wall membrane causing serious moisture problem, leading to a serious mold condition. The occupants had to be removed from the dwelling because of the concern for health related problems.

6 Summary

Thermography has proven for over 30 years to be an efficient means of locating building energy deficiencies and effectively reduce operating and energy costs. Infrared thermography for investigative and predictive inspections of all aspects of small and large buildings is one of the most powerful additions to the arsenal available to combat high energy and maintenance costs. Infrared Thermography is an inspection tool that will continue to lend itself to more applications in every range of building investigation.

References

[1] Institute of Infrared Thermography; Building Thermography Certification Course Training Manual, 2005 Edition. www.infraredinstitute.com

[2] Colantonio, Antonio and Desroches, Garry: "Thermal patterns on solid masonry and cavity walls as a result of positive and negative building pressures", pp 176～187; Proc. Thermosense XXVII; SPIE Vol. 5782, March 2005.

Acknowledgement

The author would like to thank the Institute of Infrared Thermography for their continued support in my research and development projects and from the teaching staff for their critical reviews, as well as the encouragement and assistance from the administrative people.

D 建筑生态，材料与绿色建筑

原生态景观研究刍议
——如何营造更生态的 ners 地景和更清澈的 nars 水景

金元欢
(杭州思源环境与发展研究院，浙江大学 card 区域发展研究所，杭州 310027)

摘 要：在多年多学科交叉研究和实践的基础上，针对现在景观设计和水景治理方法中存在的问题，提出了原生态景观设计的基本概念、设计思路和原则，将原生态景观分成地景(即绿地生态景观系统)和水景即自然水景系统两类。认为原生态景观师法造化、生态设计和综合治理，可以营造出一个自然生动的绿地和水体景观，并能长期保持绿地的优美景观和水景的清澈秀美，养护成本很低。ners 和 nars 的研究成果，可以广泛应用于住宅小区、城市、园林、风景区等的景观设计、养护和治理中。
关键词：原生态景观 ners 地景 nars 水景 绿地生态景观系统 自然水景系统

综观目前景观设计，对景观的生态设计的理解往往存在偏差，尤其是在设计大面积的草坪、广场和水景、景观大道等方面，存在着养护成本偏高、不以人为本等不少问题。为此，本文提出了原生态景观设计的基本概念、设计思路和原则，将原生态景观分成地景即绿地生态景观系统(natural ecoscape restoration system 简称 ners)和水景即自然水景系统(natural aquascape restoration system 简称 nar)两类。所谓原生态设计，即在整个设计过程中贯彻一种生态与可持续环境的设计思想，不是简单的自然或绿化种植，而是促进维持自然系统必需的基本生态过程来恢复场地自然性的一种整体主义方法，使设计的景观更能让人亲近自然并实现景观的可持续发展。

1 ners 绿地生态景观系统

ners 绿地生态景观系统重视绿地景观建设和养护的经济性即低成本甚至免维护理念，强调观赏性与功能性的结合，注重本地植物群落的恢复和针叶和阔叶、常绿和落叶、乔木灌木草本的有机结合。应用光影、色彩、声音、质感等形式要素与地形、水体、植物、建筑、构筑物等形体要素创造景观环境，在深深理解大自然及其秩序、过程与形式的基础上，以一种艺术抽象的手段再现了自然的精神，而不是简单的移植或模仿。

在 ners 绿地生态景观设计中，可以考虑以下几点：

1.1 季相与地相

群落的外貌决定于群落的种类组成和层片构造。群落外貌常常随时间的推移而发生周期性的变化，这是群沿构造的另一重要特征。随着气候季节性交替，群落呈现不同的外貌，这就是季相。

温热地区四季分明，群落的季相变化十分显著，如在温带草原群落中，一年可有四或五个季相。草原群落中动物的季节性变化亦十分明显。生物群落的季节变化受环境条件(特别是气候)，周期性变化的制约，并与生物种的生活周期关联。群落的季节动态是群落本身内部的变化，并不影响整个群落的性质，有人称此为群落的内部动态。在中纬度及高纬度地区，气候的四季分明，群落的季节变化亦最明显。

亚热带和温带，四季分明，植物资源较为丰富，为表现植物季相提供了有利条件，故在植物配置上应充分利用本地植物资源，展现丰富多彩的四季景观。避免一季开花、一季萧瑟，偏枯偏荣的现象，尽量使春色早临，秋色晚去。在配置手法上，春、夏之花，秋之叶、果，冬之枝、干都是很好的素材，或以春花为主，或以秋实为主，组织得好可以体现本地鲜明的季相特点，使空间更具观赏性和艺术性。

地相(land facet 或 microchore)：由多个生态区所组成，每一地相内的各个生态区至少在某一地理因素(主要是地形)的影响下，在空间上出现一定的关系和分布格局。

从早春到深秋，景观的颜色由浅到深，但总的色相为绿色，在这个颜色的底色上，植物配置的色彩与季相对草坪空间景观与艺术效果的影响是相当显著的。景观中的色彩需要组织，要掌握好补色对比、冷色对比、中和色、近似色补色对比的应用造成的不同景观和意境，如在受光的草坪上种植大红的花木和花卉，能取得明快的对比效果。

树木的色彩体现在叶、花、果、干四个部位，设计时应最先考虑叶色的安排。除少数色叶树外，大多数树木的叶色都呈绿色，但随树种、四季时间的不同而有深、浅、浓、淡之分。据了解，很多落叶树的叶子在阳光透射下会形成光影闪烁的效果，叶背呈现出嫩黄色显得轻快活泼。落叶树的冬态也可持续较长时间。常绿树叶子浓密厚重，冬天其色彩过于醒目也会造成重点分散而影响整体布局。草坪上配置的植物宜选择花、果艳丽，花期(果期)较长的树种，如紫薇、石榴、月季、蔷薇、桃、杏、绣线菊类、樱花、海棠类、珍珠梅、迎春、连翘、棣棠、锦带花等。

草花或其他地被植物及石块等组成一些色彩艳丽、灵活多样的花丛、花境、模纹及小景，在草坪的边缘或中心，可以疏密相间、曲折有致地配置。在这类植物配置中，草花的栽植要根据草坪的空间形式、地形特点灵活掌握。另外要随季节及时更换品种，保证花开不断，常见的有国庆菊、一串红、万寿菊、太阳花、宝石蓝、鸡冠花、羽衣甘蓝、金盏菊等；模纹和图案植物色彩搭配一定要与草坪的绿色形成较鲜明的对比或较大的色度差，品种一般常用红叶小檗、金叶女贞、小叶黄杨等，而且要及时修剪，保证模纹、图案的清晰、整齐。

1.2 四维景观设计

绿色房产显然最应与绿色联系在一起。居住区绿地大致包括区内公共绿地(如老人休息绿地等)、专用绿地(如学校专用绿地等)、道路绿地、住宅组群绿地、宅旁绿地等类型。

绿色总与环境连在一起，江南的早春最是风景，我想也正是在于一个绿字。然而绿色的营造应考虑其四维特点，即时空的变化。

时间上应考虑当地的气候状况，使得绿色(花草树木)有春之绚烂、秋之金黄，顺应天时，景色随四季而变；或可侧重一面，突出一季：如可将楼盘建成为桃花源、葡萄沟、枇木巴林、杏花村、幽兰苑、荷花塘、修竹园、梅花弄、杨柳城等，将楼盘园林化，更具诗情画意。这方面绿城之桂花园系列大致可算个典范，桂花不仅秋季芳香四溢，也是杭州人喜爱的市花。

空间上须注意的是：墙面的绿化(我一直将其称为绿墙工程 GREEN PROJECT)，包括墙面和围墙的垂直绿化(如可种些常春藤、爬墙虎、凌霄、络石等)、阳台的修饰、窗台的点缀、屋顶花园等；利用自然地形如高地、缓坡、洼地、池塘等(或可人为建造)，使绿色有空间上的变化，其效果远大于单一的同面积的平地；萎叶或常绿以及乔木和灌木等空间布局，使之有高低起伏，远近不同，并要与建筑物轮廓相协调；还应强调的一点是“黄土不见天”的理念，即不要留出哪怕一小块裸地，其往往影响到整个绿色的总体效果，于细微处才能体现大智。

1.3 注重细节

如生态停车场、生态围墙、生态垃圾桶等等，丁细微处体现 ners 绿地生态景观系统的各个要素。

如以别墅原生态景观设计为例，可以考虑所谓 5S 要素：一是景观(scenery)：一般都是依山傍水建造，不仅有优美的自然景观，而且还要有私家花园，使居住环境与自然融为一体；二是逸事(story)：周边应有浓郁的人文气息和悠远的历史沉淀，充分体现社区人杰地灵之底蕴；三是阳光(sun)：由于别墅的建造大多选择在市郊或远郊，上有天下有地，或接景造物，或依山傍水，与大自然相得益彰，阳光更是相伴左右；四是运动(sport)：别墅作为一个极具个性化的顶尖的消费品，无论是在室内或室外，应有充足的自由空间和生活空间，让业主充分展现自我；五是溪流(stream)：自古人们就有靠水而居的习惯，水能给人带来灵气和财气。别墅的建造往往也因溪流或河渠而备显尊贵。从这些标准来看，别墅的特点应该集中体现在人文景观、自然环境上，这是区别于普通住宅产品的一个重要特点。

2 nars 自然水景系统

2.1 水景项目可能产生的问题

由于一般景观水体(如人工湖、池塘等)大多没有按自然水理设计，基本上是一个无自净能力的封闭系统，且其内部组构不合理，加上外来物质输入，随着时间的推移必将产生富营养化，最终水体变得浑浊不堪，后果严重者甚至会导致水体发黑、变臭，影响景观水体的美观，所以大多水景已完全没有自然中的那种美丽生动的景观了。

2.2 传统水景设计方法及存在问题

众所周知，目前水景问题不少，主要表现在两个方面：设计和治理。目前水景设计的问题主要有：设计与治理缺少同步考虑、防渗处理问题、驳坎与亲水和水生态系统等。

2.3 水景的原生态景观设计

我们知道，就自然界生态水景之循环过程中有四个基型态存在：流、落、滞、喷。水景也可以设计为上喷、下落、流动、静止，是乃师法造化。

水景原生态景观设计的基本原则即注重景观的外在观赏性、内部功能性、环境整体性、技术可靠性以及运行经济性。对水景的生态驳岸的设计，我们更愿意强调水景的生态设计，如水景的生态驳岸的设计，主要要考虑以下内容：

(1) 不同河道断面的选择

河道断面的处理和驳岸的处理有密切的关系。河道断面处理的关键是要设计一个能够常年保证有水的河道及能够应付不同水位、水量的河床，既能提供城市中理想的开敞空间环境，具有较好的亲水性、适于休闲游憩。如美国加州山裘斯格达鲁普河滨公园、维也纳的多瑙支河，即在河道断面处理上采取了多种形式，景观自然宜人。

(2) 生态驳岸的选择

- 自然型驳岸：

对于较陡的坡岸或冲蚀较严重的地段，不仅种植植被，还采用天然石材、木材护底，以增强堤岸抗洪能力。如在坡脚采用石笼、木桩或浆砌石块等护底，其上筑有一定坡度的土堤，斜坡种植植被。实行乔灌草相结合，固堤护岸。

- 台阶式人工自然驳岸：

对于防洪要求较高、而且腹地较小的河段，在必须建造重力式挡土墙时，也要采取台阶式分层处理。临水可种植芦苇、菖蒲等水生植物，可生长得繁茂葱绿。

2.4 传统水景治理方法及存在问题

目前水景治理的方式主要有：

(1) 物理方式

景观水体净化的物理方法引水换水、循环过滤、底泥疏浚、曝气充氧等有时虽一时有一定效果，难以大规模长期实施。

(2) 化学方法

化学方法主要指投加化学药剂来杀死藻类。虽一时效果较明显，但久而久之，水中会出现耐药的藻类，灭藻剂的效能会逐渐下降，导致投药间隔会越来越短，而投加量会越来越多，灭藻剂的品种也要频繁更换，对环境的污染也在不断地增加，而这种污染会影响我们的下一代。且投加的化学药剂，使得水生植物无法生长，也就无法营造生动美丽的水景。所以说用化学的方式处理水质，虽然是立竿见影的，但它的危害也是显而易见的。

(3) 微生物方法

在景观水水质恶化的时候，投加适当的适量的微生物(各类菌种)，加速水中污染物的分解，起到水质净化的作用。用微生物处理水质，必须定期进行微生物的筛选培育、保存、复壮等等一系列专业处理过程，而且不能保证水质状况长期处于良好的状态之中。

(4) 水生动植物系统修复

水生动植物系统在控制藻类的过度繁殖，对防止水体富营养化的发生起到的作用尚需进一步试验，若处理不当，反而会因此造成水体的污染和富营养化。许多景观水体养鱼类和种植一些水生植物后，鱼类的排泄物和水生植物的腐殖质落叶等，造成的大多是混浊、发臭的水体。

2.5 水景的治本方法——nars 自然水景系统

(1) 原理——师法造化，综合治理

水景治理其实是一个世界性的难题，大量科学试验表明，上述引水换水、循环过滤和底泥疏浚、投药等物理化学方法都是从污染后的结果再治理的，而没有从成因即水体富营养化生成机制上治理，故不用多久势必会引起重新污染。我们考虑的思路是截断造成污染的成因源头，从而来维持水景的清澈。

事实上，自然界是一个十分复杂的系统。要营造这样一个长期清澈、自然的水体景观，较为科学的方法应采用综合设计和治理的办法——即 nars（natural aquascape restoration system）自然水景系统。运用 nars 自然水景系统，可营造出清澈美丽生动的自然水景。

nars 的治理主要包括以下几个子系统：nars 底质综合治理、nars 水质综合治理、nars 微生物菌群、nars 水生动植物系统和其他措施(如突变事件而导致的水质变化等)。nars 水景系统是综合了各种方法的一种生态设计和综合治理技术。传统种种办法，均有利于水质的治理，但水体是一个十分复杂的系统。许多因素相互作用，可能一个因素就可以将水体污染，而治理绝不是单一方法即可，因为一个因素会使得许多因素随之而变。就我们掌握的资料来看，事实上目前无法找到一种单一方法治理好水景的技术。

nars 水景系统大多同时承担设计和治理，这样就避免了传统水景中先设计后施工再治理的弊病，也节省了大量的工程费用。以我们建造的多个 nars 水景为例(见后面附录)，它采用多种方法来加以设计和治理，并收集天然降雨，结合景观设计，动静相宜，建成一个建立具观赏性和参与性为一体的水景。近年来我们为国内众多知名楼盘和水景设计和营造了 nars 自然水景系统，也证明了该系统的可行性。

(2) nars 与传统过滤方式的比较

景观水体目前常用过滤等传统工业用水处理系统来治理，采用这种方法，要保持一定的水质需要的处理水量很大，加上药剂添加、电费、设备维修、设备损耗、过滤池清扫、人工管理等，每年维护费用也相当高昂。

传统过滤方式的关键问题可能还不单纯是费用问题，主要是技术上的不可行。因为目前这样的方式一般用在工业或民用水处理上，并不能真正解决景观水体的处理(因为一部分水体处理完毕后又汇流到整个污染的水体中；另一解决不可了的问题是水体藻类的繁殖速度可达数十小时翻倍，设备处理的水量很难承受。这就是为什么目前几乎所有的水景大多污染严重，景观不美的根本原因)。就我们掌握得资料看，目前为止尚无成功的案例(见表 1、图 1、图 2)。

各种治水方式比较 **表 1**

	水体的清澈透明度	水体的美丽度	治理和日常维护成本	适宜用场地	水景成功案例
机械过滤、换水	差或一般	差或一般	高	游泳池	无
化学处理、投药	好	单　调	很高	游泳池	无
nars 自然水景系统	好	优　美	低	景观水体	有

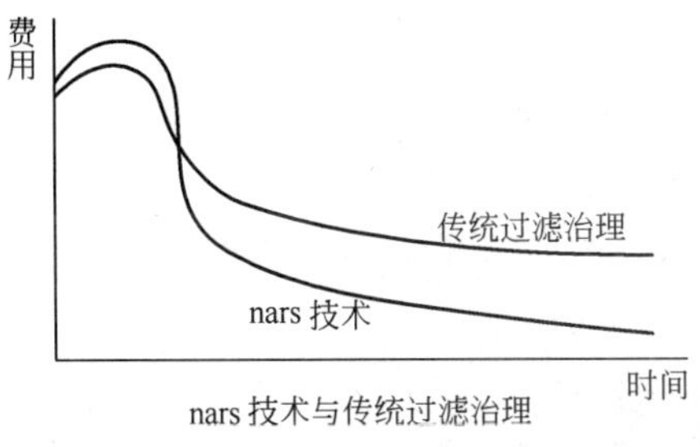

图 1　治理和维护成本比较示意图

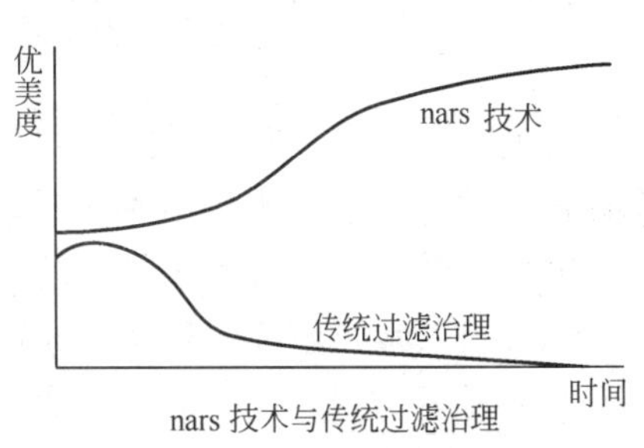

图 2　水景优美程度比较示意图

另一种方法是化学处理方式，由于其处理过程中添加的化学药剂，而使得一般水生景观植物无法生长，也无法达到房地产开发商希望的生动美丽的景观效果(此方法一般用在游泳池的水处理上)。

(3) 与传统治水的根本区别在哪里？——前期治本与后期治标

大量的科研成果证明，藻类一旦爆发，要控制它是很难的。传统的过滤并不能完全过滤干净或者成本太大。采用化学除藻剂或杀菌剂等，效果不佳，且马上藻类又会马上生长，成本太大且污染底质。靠换水等在 500m³ 以上水体时，并没有经济性，在 5000m³ 以上水体时，已经没有现实性：试想能用此方法治理太湖、滇池吗？如何换水？这么巨大的水源又来自哪里？——所以传统治水注重后期治标，应是导致现在水景治理效果到处不佳的主要原因(一般只能达到地表水环境质量标准Ⅴ类水质标准)。

因此，nars 理水更注重前期治本，破坏藻类爆发的条件，使得藻类不再爆发，从而达到水质的长期清澈(最终可达到地表水环境质量Ⅱ类甚至Ⅰ类水质标准)。这一理论已经为大量成功的案例验证。

(4) 实例：一个 nars 水景

这里以一个 nars 水景改造为例(图 3、图 4，表 2)，具体步骤及其景观效果大致如下：

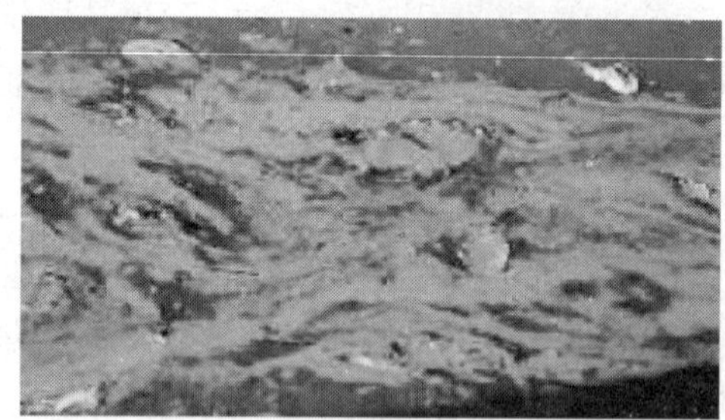

图 3 治理前(实景)

图 4 治理后(实景)

表 2 nars 实例有关指标比较

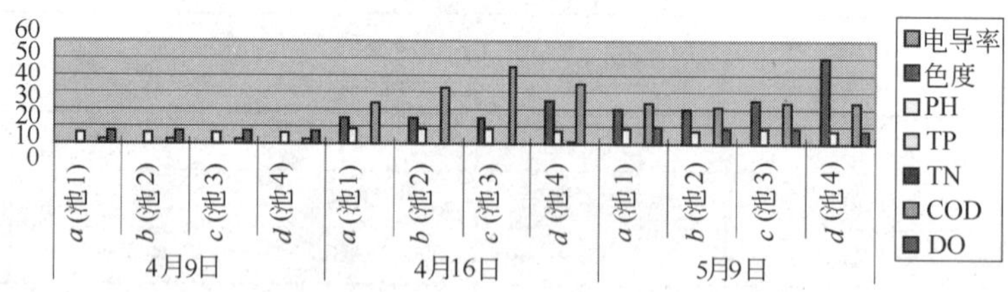

(1) 水清问题：用 nars 自然水景设计和治理底质水质，能使水体清澈自然，是为治本。

(2) 水美问题：为体现观赏性，通过养种些水生动植物(如鱼虾、水草等)，可营造出生动美丽的水岸、水面、水中、水底景观，是为治标。

(3) 维护问题：nars 自然水景可以保持长期稳定，清澈美丽，日常维护成本低廉。

(4) nars 自然水景系统的创新点或技术特点。

相比传统治水方式，nars 自然水景系统的创新点有：

- nars 注重前期治本，而传统治水方式注重的是后期治标；

- 师法造化，综合治理，治理效果佳，可以做到清澈见底；
- 不用换水，雨水利用，节约水资源；
- 微动力技术使得日常维护费用很低；
- 原生态设计使得水岸、水面、水中、水底景观生动美丽。

总之，运用生态景观设计的概念，nars 自然水景系统来设计和治理景观水体，可以广泛应用于住宅小区、城市、风景区等。既可以节水节电，还可营造一个美丽自然的生态景观和独特水景。有关具体技术、资料、常见问题等可参考网站 http：//www. nars. cn 及 nars@163. com。

Ecoscape：a solution to dynamic landscape design

Jin Yuanhuan，Ph. D
(Hangzhou Nars Institute for Environment And Development Research，
Zhejjiang University，Hangzhou 310027 China)

Abstract：based on the experiments and models of Ecoscape and aquascape studies in many years，two new solutions——and NARS (natural ecoscape restoration system) and nar (natural aquascape restoration system) technique have been worked out. With applying the NARS technique，we can design a dynamic living systems that are ecologically sustainable，mimicking the self-reliance，diversity，and resilience of natural systems. With applying the NARS technique，we make the water body of rivers and lakes very clear and beautiful，the polluted water body could be kept clear under a lower maintenance price. NARS could be used for water body treatment and landscape design in various fields such as urban planning，landscape design，residential quarters and scenic spots etc.

参考文献

[1] http：//www. nars. cn

[2] Jin Yuanhuan(金元欢)et al，1996，The golden waterway：its development and significance-study on the economic development along the Yangtze River Basin，in：proceedings on the 1996 International Conference on the coordination in the regional infrastructure development along the Yangtze River Basin，May 20～21，Fudan Univ.，105～110

[3] Jin Yunahuan(金元欢)，1991. Comprehensive research on the estuaries and tidal flats in Zhejiang. in：the proceedings of the 2nd international symposium on the coastal ocean space utilization，COSU-II，Long Beach，Cal.，USA，April 2～4

[4] Jin Yuanhuan(金元欢)，1991. Progress in the estuarine classifications，Marine Science Bulletin，(1)：32～45

[5] Ecosan-a holistic approach to material-flow- management insanitation，Christine Wemer，GTZ Germany，Ecosan - Closing the loop in wastewater management and sanitation，Proceedings of international symposium，p29～33

[6] Towards a recycling society，Ecological sanitation-closing the loop to food security，Dr. Steven A.

Esrey，(UNICEF)，Ecosan-Closing the loop in wastewater management and sanitation，Proceedings of international symposium，p34～44

[7] New development of ecosan in Germany and Europe，Prof. Dr. Ralf Otterpohl，TUHH，Germany，Ecosan-Closing the loop in wastewater management and sanitation，Proceedings of international symposium，p68～79

[8] Foreword，Ecosan-Closing the loop in wastewater management and sanitation，Proceedings of international symposium，GTZ ecosan-team

[9] 金元欢，1992，黄河口：作为一个特殊的过滤器，黄河三角洲研究，1(3～4)：16～24

[10] 金元欢，孙志林，1992. 中国河口盐淡水混合特征研究，北京，地理学报，47(2)：165～178

[11] 金元欢，1998，环境第一：房地产开发中的一个趋势，房地产信息，40(14)：23～24

[12] 金元欢，1998，关于绿色住宅的几个问题，杭州日报. 房地产专版，10.1

[13] 金元欢，王建宇，2001，杭州城市发展中几个值得忧虑的问题，城市规划，4：64～73

[14] 金元欢，2002，楼不在高，有水则灵——水景楼盘的兴起及其设计和治理，经济观察，8.19

绿色建筑室内空气质量控制策略及应用案例分析

李景广　韩继红　叶剑军　袁　静　谢晓东
（上海建筑科学研究院有限公司，上海　200032）

摘　要：创造健康、舒适、高效的室内环境是生态建筑体现“以人为本”的主要目标之一。室内空气质量是室内环境中的关键问题。本文阐述了在生态建筑规划设计、施工、验收和运行管理各阶段有效控制室内控制质量的关键技术和策略。并通过上海市建筑科学研究院生态办公楼的案例分析，详细介绍了室内空气质量控制策略和手段以及效果。

关键词：绿色建筑　室内空气质量　控制策略

1　研究背景

一个成年人平均呼吸次数为每分钟10～15次，每次需要0.5升空气，以平均70岁寿命来计算，每个人一生要用27万m^3空气。这些空气进入人体内，在总表面积为60～80m^2的肺泡里，经物理扩散进入体内交换。由于现代人平均80％～90％以上的时间在室内度过，可以想象在如此长的暴露时间，如此大的接触面积下，室内空气质量状况对人们身体健康影响何等巨大[1]。

关于室内空气质量问题的起因，人们最初认识到人体排出的二氧化碳、体臭是室内空气的污染物，并提出通过通风来解决气味问题[2,3]。20世纪70年代由于世界范围的节能，建筑物加强了气密性，减少了新风量。同时由于有机合成材料和新设备的广泛使用，使得室内空气污染源大大增加。在19世纪70年代美国环境署对于选定性有机性挥发物做了调查，发现由于室内污染源导致的有毒有机性挥发物（VOCs）对于人员的暴露影响远远高于由于工业污染导致的室外空气的影响[4]。而哈佛大学6城市的调查更进一步揭示了室内空气污染包括粒子、硫酸盐和氮氧化物等对于人员暴露的影响[5,6,7]。室内空气污染的严重性引发了以下三种病症：病态建筑综合症（SBS），与建筑有关的疾病（BRI），多种化学污染物过敏症（MCS）。

20世纪人类的平均寿命提高了30年，其中因为医疗医药发展的贡献仅仅是5年，而25年的提高是因为卫生和公共卫生设施的发展和提高，室内空气质量和饮用水一样都属于卫生范畴。对于发展中国家，除去营养不良和水质问题，室内空气质量是第三大危害人类生命的凶手，每年由于室内空气质量问题导致的呼吸传染病、肺癌、肺结核、哮喘等疾病，死亡人数超过200万人[8]。对于发达国家，由室内空气质量导致的过敏、空气传染、肺癌、SBS等造成了巨大的经济损失。据美国环境保护署（EPA）统计，美国每年因室内空气质量低劣造成的经济损失高达400亿美元。

中国疾病预防控制中心统计表明我国每年由于室内空气污染引起的超额死亡数可达11.1万人，超额门诊数22万人次，超额急诊数430万人。世界银行的一份研究报告表明，仅1995年，我国因室内空气污染危害健康所导致的经济损失按支付意愿价值估计就高达106亿美元。

同时，世界卫生组织(WHO)指出室内空气污染降低的收益不仅包括健康因素，同时也包括非健康因素，Fanger[9]、Clements-Croome[10]发现室内空气质量和工作效率具有显著的相关性，Kroner[11]调查发现良好的室内空气质量可以提高员工2%～16%的工作效率，Wargocki[12,13]在类似的试验中同样发现室内空气质量的提高和通风量的增加可以明显提高工作效率。

室内空气质量极大地影响着人们的生活质量、健康水平和生产效率，室内空气污染及由此引起的SBS、BRI等是当前国际上的普遍问题，也日益成为我国社会关注的焦点问题。如何避免室内空气污染问题在建筑的实施中出现已成为目前生态建筑室内环境的关键问题。

2 影响室内空气质量的主要因素及控制策略

影响室内空气质量的因素主要包括：室外空气污染、建筑材料和建筑装修材料、新风量、空调系统导致的霉菌、粒子等污染物、单区域和多区域的气流组织、家具家电办公设备、室内燃料燃烧和烹调油烟、家用化学品、吸烟、人员活动等。而建筑物的实现主要包括规划设计、施工、验收评估和运行管理阶段。在建筑物整个实现过程始终贯穿室内空气质量控制策略是解决目前室内空气污染问题中比较可行有效的手段。

2.1 规划设计阶段室内空气质量控制关键技术(见表1)

规划设计阶段室内空气质量控制关键技术 表1

建筑设计	通风系统	预评估
1. 建筑设计： 建筑物位置和周边建筑情况、房间功能和房间位置、建筑材料 2. 装饰设计： 墙体、天花板、地板装饰材料 3. 室内布置： 家具家电、办公设备位置和散发强度、净化设备位置和净化效率 4. 室内人员： 活动情况、吸烟情况	1. 系统形式 2. 气流组织 3. 净化方式和效率 4. 新风量和新风采集位置以及新风途径 5. 智能控制系统的仿真	1. 温度场 2. 速度场 3. 浓度场 4. 湿度场 5. 空气年龄场

2.2 施工阶段室内空气质量控制关键技术(见表2)

施工阶段室内空气质量控制关键技术 表2

工艺	过程
合理施工。不使用或者少使用含有甲醛等的材料	规范施工、洁净施工。避免材料与结构造成的尘埃等的积聚等

2.3 验收与运行管理阶段室内空气质量控制关键技术（见表 3）

验收与运行管理阶段室内空气质量控制关键技术　　表 3

室内空气质量	通风系统
1. 检测技术的应用 2. 室内空气质量评估体系的应用	1. 新风途径的检查 2. 通风系统的运行和维护 3. 智能控制系统的调试、运行

3 技术手段

为了实现良好的室内空气质量，在生态建筑的实现中从选址评估、设计优化、建筑材料筛选、空调系统选择、绿化配置以及自然通风等方面进行了相关工作，从而避免室内空气的污染，创造健康舒适的室内环境。

3.1 选址 IAQ 评估

生态楼的选址初期，为了了解室外空气污染状况，课题组于生态楼选址处的空气本底进行了测量，检测结果见表 4。

空气质量本底值检测结果汇总　　表 4

检测点	空气质量室外值检测结果	
	NO_x mg/m³	SO_2 mg/m³
GB 3095—1996 标准值	≤0.1	≤0.15
1	0.072	0.075
2	0.019	0.060
3	0.041	0.107
4	0.018	0.068
5	0.043	0.059
6	0.019	0.116
7	0.035	0.079
8	0.040	0.056
9	0.031	0.080

检测结果 NO_x 和 SO_2 符合 GB 3095—1996 标准[14]，生态楼的选址符合环境空气质量标准。

3.2 优化房间空间布置并满足通风量要求的设计优化

3.2.1 通风量模拟计算

为了评价室内外温差对自然通风效果的影响，设置室外温度为 20℃和 25℃（室内为 28℃）及室内外温差分别为 8℃和 3℃的情况下进行模拟，考察在不同情况下，典型区域室内通风量和换气次数的影响。表 5 给出了不同工况下典型房间通风参数的模拟结果值。

生态楼各层典型区域通风指标在不同温差情况下的模拟结果 **表 5**

室外温度	1楼				2楼		3楼	
	仪器自校室		生态建筑展示厅		第三测量室		仪器分析室	
	kg/s	ACH	kg/s	ACH	kg/s	ACH	kg/s	ACH
20℃	1.374	37.05	30.63	83.72	0.161	4.97	2.415	45.91
25℃	0.829	22.36	18.45	50.43	0.099	3.041	1.473	28.00

由表5中数据可以判断，随着室内外温差的降低，通风量和换气次数都减少。观察自然通风最小的2楼区域可以判断，如果将3次换气次数作为通风基本要求的话，室外25℃及室内外温差为3℃将是一个极限，也就是说当室内外温差大于或等于3℃时，可以满足自然通风的要求。在过渡季节可以满足室内各房间换气次数1次的要求。

3.2.2 优化房间空间布置

由于生态楼的实际功能是建筑环境实验楼，因此，有污染物释放的功能实验室的具体位置关系到生态楼里的空气质量现状。为了更好地控制污染物的释放及减少实验室污染物释放对空气质量的影响，我们采用CONTAMW对其室内污染和通风进行模拟(模拟结果如图1、2、3所示)，图中采用颜色代表污染物浓度的高低，红色代表高浓度，蓝色代表低浓度。由模拟结果有图可知，把可能有污染物释放的化学和微生物实验室放在三楼北面，则整个生态楼里建筑室内污染物浓度大致分布均匀，且处于相对较低的浓度水平；而把这些房间的浓度置于偏高，则发现这些处于高污染物浓度的房间，对周围的室内空间影响达到较小的程度。

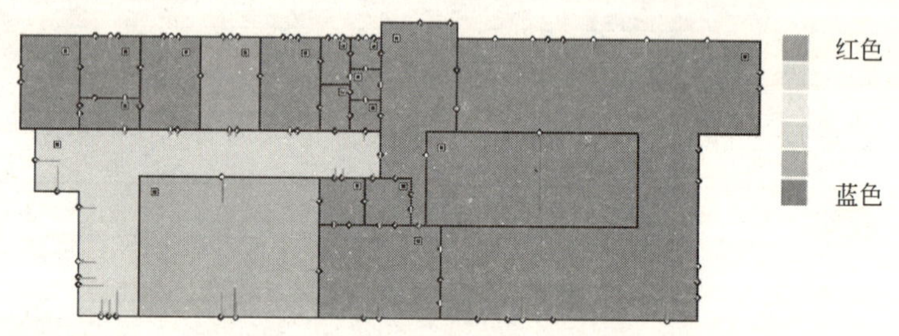

图1 生态办公楼1楼污染物模拟结果

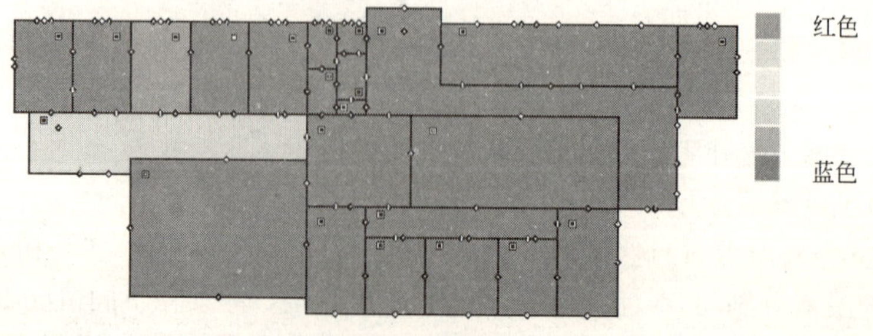

图2 生态办公楼2楼污染物模拟结果

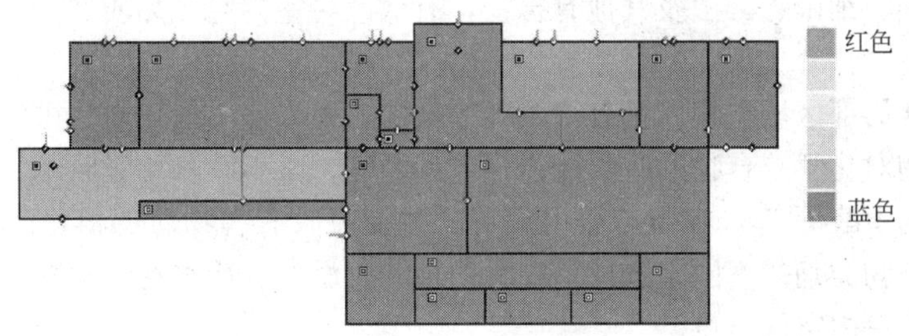

图 3　生态办公楼 3 楼污染物模拟结果

生态建筑楼部分区域污染影响模拟结果　　表 6

门窗状态	高→低			
开	2 楼西部楼梯区	1 楼实验设备区	1 楼左中厅	3 楼电梯前室
	0.339mg/m³	0.334mg/m³	0.156mg/m³	0.025mg/m³
关	2 楼西部楼梯区	3 楼北部楼梯区	1 楼实验设备区	1 楼左中厅
	0.349mg/m³	0.337mg/m³	0.206mg/m³	0.131mg/m³

由表 6 中数据可以得到，当某些功能房间由于工作的需要可能产生污染物的时候，它对其他房间的影响是相当小的，当 VOC 浓度在设定房间内为 1000ppm(2.33mg/m³)时，无论是门窗开关，受影响的区域最大浓度只有 0.35mg/m³，而标准为 0.6mg/m³。过渡季节自然通风的条件下，化学分析实验室、微生物实验室等有污染的区域，在热压驱动下，不会对周边房间产生污染危害。

3.3　生态楼建筑材料控制技术

建筑材料和装修材料的选择对室内空气质量起到重大作用，在设计及装修、使用生态楼的过程中，主要考虑生态楼内建筑材料的来源鉴定：采用清洁无害的绿色建材、适度装修、慎重装修、装修保证一次到位。

3.4　自然通风策略

对于有空调的生态楼来说，当伴有充足的室外空气进入后，70%的循环使用空气对人群不会造成不良的健康影响[15]。许多空调室内的污染物，如可吸入粒子、NO_x、CO、CO_2 以及甲醛等都可通过改善通风而加以降低。生态楼利用中庭顶的“老虎窗”拔风的自然通风方式来改善室内空气质量，通过室内二氧化碳传感器的监测，及时开关外窗以保证室内新风量和空气质量。

3.5　热湿分离的健康空调系统

在目前建筑中，空调系统普遍使用表冷器换热，通过降温将空气中的水分冷凝下来，以达到降温除湿的目的。潮湿的表冷器表面，是滋生大量霉菌的最好环境，同时也能造成其他病菌和病毒的聚集和滋生[16]。特别是在空调箱停机期间(如下班或放假)，在空调箱的密闭空间内，细菌浓度能骤然上升几十万倍[17]。这些微生物通过送风系统大量进入室内，会对室内的空气品质造成严重的影响。为了避免这样的问题，生态楼建筑中采用了热湿分离空调系统，这种热泵驱动的热、湿负荷独立控制的空调系统通过避免使用有凝结水的盘管，解决了目前空调系统中存在的霉菌滋生问题。同时通过除湿机内盐溶液的喷洒除去空气中的尘埃、细菌、霉菌及其他有害物，有效改善室内空气品质。

3.6 室内绿化

针对生态楼局部区域空气质量有待改善的情况，引入室内绿化对之进行调节改善，根据许多绿色植物对化学复合物的吸收能力强，如芦荟、常青藤等可吸收苯、龙舌兰、吊兰可吸收空气中的甲醛等[18]，在生态楼内放置一些能吸收有害物质的植物。引入的品种根据室内主要污染物类别的不同来选择室内绿化植物：吊兰、冷水花、常春藤、绿萝、富贵竹等，可以提高室内空气质量。

4 效果分析

4.1 空气质量现状

技术手段的最终目的是实现室内空气质量优良，通过对生态楼室内空气质量的现状分析，发现其是否产生问题以及导致的原因，并为将来的改善手段的应用提供基础数据。研究小组按照《民用建筑工程室内环境污染控制规范》GB 50325—2001 对部分重要指标和反映空气质量优劣的指标进行了检测。检测结果如表 7、8，结果表明生态楼室内空气质量符合国内相关标准。

甲醛检测结果　　表 7

房间编号	结果(mg/m^3)
101	0.08
1 楼会议室	0.09
204	0.08
205	0.09
209	0.07
208	0.07
210	0.07
211	0.05

苯和 TVOC 检测结果　　表 8

房间编号	苯	TVOC
生态大厅	0.011	0.421
205	0.007	0.393
213	0.007	0.398
210	0.007	0.391
209	0.009	0.134
214	0.006	0.329
305	0.004	0.269
208	0.004	0.068

4.2 房间通风量和功能房间布置合理性

4.2.1 测试工况

建筑运行工况：建筑各门和窗户关闭，一楼大厅开启机械通风，其余地区未开机械通风。室内温度为 28℃，室外温度 26℃。

4.2.2 实测结果

2 层办公室换气次数为 1.472，3 层实验室换气次数为 2.248。结果表明在最不利房间内通风量大于 1 次，整体建筑通风量满足设计目标。

同时根据测试结果，经过计算绘制图 4。由图中可知，无论是二层的污染还是 3 层的污染，对其他楼层的污染影响均不大于 10%。这一结论同样可以从设计阶段的模拟结果中得到，因此生态建筑的自然通风设计合理，室内功能房间分布恰当。

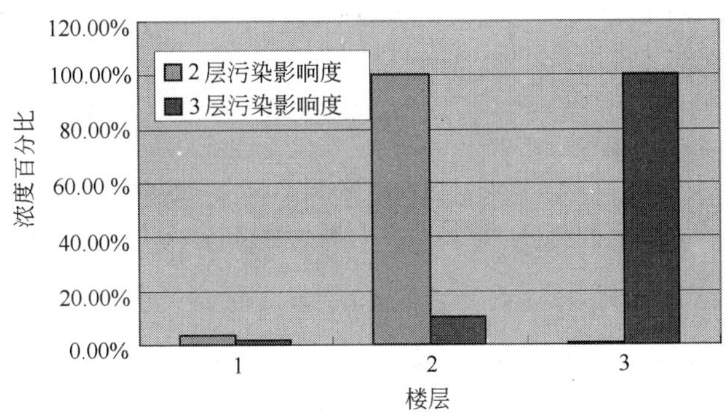

图 4　2、3 层污染对其他楼层的影响

5　展望

有效控制室内空气质量，在保证健康的前提之下进一步追求舒适和高效是发展生态建筑的一项艰巨的任务。在传统理念中，人们只重视使用中的室内空气问题，而忽略了规划设计、施工等过程的影响。室内空气质量的实现必须参照建筑周期方法来分析不同阶段的影响和采用不同的技术手段。本文通过上海市建筑科学研究院生态办公楼的案例分析，详细介绍了室内空气质量控制策略和手段以及效果。然而由于目前在室内空气质量研究领域仍然存在大量科学问题尚未得到解决，诸如污染物对人的影响规律、污染源种类、污染源特性、污染源的控制、污染物传播过程与传播过程中的反应、空气净化方法等问题，因此仍存在大量和艰巨的工作等待各界人士共同努力，共同构建人类安全、健康、舒适、高效的室内环境。

Control Strategies for Indoor Air Quality in Green Building and Case Analyses

Jingguang Li　Jihong Han　Jianjun Ye　Jing Yuan　Xiaodong Xie
(Shanghai Research Institute of Building Science，Shanghai　200032　China)

Abstract: Creating a healthy, comfortable and productive indoor environment is one of the most important tasks in green building. Indoor air quality is the key part for indoor environment. The paper discusses the important technical Strategies that are applied for green building during layout, design, construct, run and management phase to control the indoor air quality. And control strategies, technical method and result in detail are introduced by case, the model Eco-Building in Shanghai Research Institute of Building Science.

参考文献

[1]　姚志棋等. 环境卫生学(第二版). 北京：人民卫生出版社，1993

[2] Yaglou，C. P.，Riley，E. C. and Coggins，D. I.（1936）. Ventilation requirements. Heating，piping and air conditioning. ASHRAE Journal 42：65～76

[3] Yaglou，C. P. and Witheridge，W. N.（1937）. Ventilation requirements，Part 2. Heating，piping and air conditioning. ASHRAE Journal 43：1～4

[4] Wallace. L. R.（1987）. The Total Exposure Assessment Methodology（TEAM）Study：Summary and Analysis. Washington，DC：Office of Research and Development. US Environmental Protection Agency

[5] Speizer，F. E.，Ferris，B. G.，Jr.，Bishop，Y. M. and Spengler，J. D.（1980）. Respiratory disease rates and pulmonary function in children associated with NO_2 exposure. American Review of Respiratory Diseases 121：3～10

[6] Dockery，D. W.，Pope，C. A.，Ⅲ，Xu，X.，Spengler，J. D.，Ware，J. H.，Fay，M. E. et al.（1993）. An association between air pollution and mortality in six U. S. cities. New England Journal of Medicine 329(24)：1753～1759

[7] Ware，J. H.，Dockery，D. W. and Spiro，A.，Ⅲ.（1984）. Passive smoking，gas cooking，and respiratory health of children living in six cities. American Review of Respiratory Diseases 129：366～374.

[8] Jan Sundell，Indoor Environments And Health，SB04 Shanghai，2003

[9] Fanger，P. O.（1999）. Perceived air quality，sick building syndrome（SB）symptoms and productivity in an office with two different pollution loads，Proceedings of the 8th International Conference on Indoor Air Quality and Climate—Indoor Air'99，Edinburgh，Vol. 2

[10] Kroner，W. M. et al.（1997）. Employee productivity and the intelligent workplace，Proceedings of the International Conference on Creating the Productive Workplace，London，Vol. 2

[11] Clements-Croome，D. and Li，B.（1997）. Assessment of the influence of indoor environment on occupant productivity of occupants in offices，Proceedings of the Fifth International Conference on Healthy Buildings，Washington，Vol. 2：495～500

[12] Wargocki，P.，Lagercrantz，L.，Witterseh，T.，Sundell，J.，Wyon，D. P. and Fanger，P. O.. Subjective perceptions，symptom intensity and performance：a comparison of two independent studies，both changing similarly the pollution load in an office. Indoor Air 12，74～80

[13] Wargocki，P.，Wyon，D. P. and Fanger，P. O.. Call-centre operator performance with new and used filters at two outdoor air supply rates. Proceedings of Healthy Buildings 2003

[14] GB 3095—1996 环境空气质量标准

[15] Andrew K. Persily，A Modeling Study of Ventilation，IAQ and Energy Impacts of Residential Mechanical Ventilation

[16] 江亿. "非典"问题引起的对今后空调系统方式的思考. 暖通空调. 2003 暖通空调与 SARS 特集，4～7

[17] 沈晋明，邓伟鹏. 正确认识空调系统防范非典型肺炎. 暖通空调. 2003 暖通空调与 SARS 特集，16～19

[18] 胡海红，戴修道等. 室内绿化净化功能的探讨 [J]. 上海建设科技，1996，(2)：37～38

城市小区雨洪多功能调蓄设计案例分析

车　伍[1]　张　燕[1]　黄　宇[1]　李俊奇[1]
何建平[2]　刘　红[2]　孟光辉[2]　汪宏玲[2]
（[1] 北京建筑工程学院，北京　100044；[2] 北京市节水管理中心，北京　100036）

摘　要：城市的建设改变了原有的环境和地表条件，使得雨水滞留能力下降，洪峰流量增加，从而导致雨水资源流失和洪涝风险增大。多功能调蓄设施的运用不仅可以缓解城区的防洪压力，有效的利用雨水资源，同时可以高效率地利用城市宝贵的土地资源。本文以某住宅小区为例，分析了小区开发后带来的一些水环境问题，从雨水利用、水体规模、蓄洪能力、土地的有效利用等方面进行了分析，重点讨论了多功能雨洪调蓄设施在城市小区中应用的可行性及其优点。

关键词：城市雨水　多功能调蓄　雨水利用　小区

1　引言

随着城市的不断建设，城市大量的自然环境如绿地、农田等被建筑物、构筑物及硬化地面所取代，原有疏松透气的地表被混凝土等不透水材料所取代，阻断了雨水向土壤的渗透，造成雨水资源的流失、城市洪涝灾害增加等一系列的城市生态环境和社会问题。城市雨洪问题已成为城市可持续发展的重要制约因素之一。如何科学合理的利用城市雨水资源，解决水资源短缺问题，同时缓解城市防洪压力，近年来成为一个广受关注的重大课题。多功能雨洪调蓄设施是一种新的城市雨洪管理思路，它能将城市洪涝控制与雨水资源的综合利用相结合，同时又能高效地利用城市宝贵的土地资源，营造良好的城市生态环境。[1][2]

2　背景资料概述

2.1　原环境概况

该项目位于天津东丽湖湖畔。拟开发地块的用地性质大部分为闲置的沼泽荒地，外加少量耕地、水塘和宅地。该小区所处地理位置优越，地势平坦，自然环境优美，但近年来东丽湖水质已受到比较严重的污染。

2.2　发展后住区环境概况

该小区的建设分两期进行，一、二期规划用地见表 1。

一、二期规划用地构成表 **表 1**

用地性质		面积(万 m^2)	
		一期	二期
1. 居住用地		18.68	178.5
其中	住宅(私人住宅)	12.08(3.46)	101.11(42.83)
	公建	0.97	11.17
	公共绿地	0.94	20.60
	道路	3.89	45.62
2. 绿化带用地		1.42	16.88
3. 水系		14.00	14.50
4. 湿地和水渠		0.80	25.80
总计		34.10	235.68
综合径流系数		0.796	0.566

该小区在一期、二期分别修建面积为 14ha 和 14.5ha 的人工湖，小区内原有芦苇荡湿地面积 21.3ha。建成后一期和二期湖可连成整体，与四周的住宅构筑人与水和谐、自然的居住环境。

3 小区发展后存在的问题分析

3.1 水涝问题

小区建成后，不透水面积显著增加，径流系数由原来的 0.15 左右增至 0.57～0.80 左右，使得小区内雨水汇流的时间缩短，汇流总量和峰流量大幅度增加。按重现期为 5 年的一场降雨(降雨量 144mm)进行估算，开发后产生的汇流总量比开发前增加了约 4.5 倍。而且小区地势较低，小区外只有一条 *DN*600 市政规划排水管道，因此存在排洪压力和内涝隐患。

3.2 缺水问题

由于整个小区有大面积的水体，需要有充足的水源。而天津本身就是一个严重缺水的城市，如果利用自来水作为水体的补充水源不仅需要很大的运行成本，而且还面临越来越严格的用水限制。虽然小区附近的东丽湖水容量大，同样面临限制使用和水质污染的双重约束，也不可能将它作为小区长期而稳定的补水水源。因此，雨水资源成为该小区重要的补充水源。

3.3 水体规模问题

小区一期修建的水体水面面积较大，占一期总用地面积的 41.06%，具有较高的雨洪调蓄能力：当水位位于常水位时，利用常水位与溢流水位之间的调蓄空间计算，最大可容纳一次降雨量为 312mm 的降雨(相当于重现期大于 $P=20$ 年的降雨)。大规模的水体造成补水量也大：经过计算，即使充分利用雨水资源，每年湖体的补水量至少还需要 9万 m^3。

小区建成后，还存在其他问题，如：水环境的水质保障问题、湿地的维护问题等。

4 小区水环境的总体方案

针对上述问题，在进行二期湖体规划设计时充分考虑利用一期湖体过大的调蓄能力，减少整个水体的补水量，同时兼顾蓄洪、湖体景观及水生植物生长等的需要。

从补水、蓄洪、雨水利用等多因素综合考虑，通过水量平衡计算，确定二期湖适宜规模为 14.5ha。为了更好地维护湖体水质，设计两湖相通且高程一致，即湖底高程 0.5m，常水位 2.0m，溢流水位 2.45m。

小区水环境的总体方案见图 1。

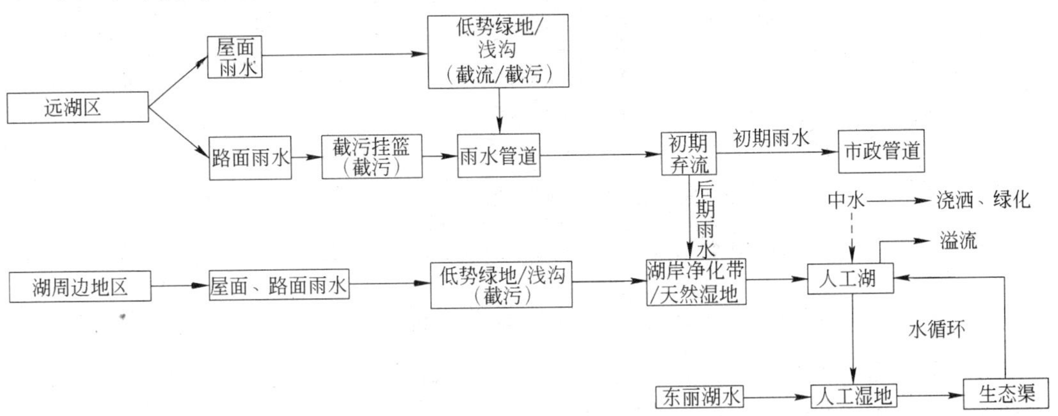

图 1 小区水环境总体方案框图

小区水体的第一次水源取自东丽湖，由于东丽湖水水质较差，因此先经过人工湿地处理后入湖，以后主要利用雨水作为水体的补充水源，雨量不足时才考虑从东丽湖补水。正常运行中湖水也可通过人工湿地和生态化的水渠循环净化。

小区雨水收集与截污采取的方案是：远离水体区域的雨水先通过低势绿地（植被浅沟）或截污挂篮截污净化进入雨水管道，在集中汇流处再经初期雨水弃流控制，后期雨水流入水体的湖滨净化带或天然芦苇荡湿地，最后进入水体。弃流后的雨水就近进入市政管线；水体周边区域的雨水通过低势绿地（植被浅沟）的截污处理后，直接通过水体的湖滨净化带或天然芦苇荡湿地流入水体。

在小区中利用人工湖体调蓄利用雨水不仅可以大幅度减少自来水的补充量，还能节省雨水储存池等设施的投资和减小小区的内涝。按多年平均降雨量计算，一期工程范围内年均可利用雨水量 6.7 万～23.5 万 m^3；二期建成后，整个小区年均可利用雨水量 39.6 万～138.9 万 m^3。

小区内建有中水站，但由于中水水质不利于水体的水质保障，因此设计方案使中水不进入水体，可直接用于小区的绿化及道路喷洒用水。当雨水量不足时，中水可作为水体的备用水源，适当补水。

5 应用多功能雨水调蓄设施的可行性分析

由于土地资源的稀缺及其价格的昂贵，综合考虑雨水利用、水涝问题、景观及土地有效利用率等多种因素，在小区内修建与人工湖调蓄相结合并兼具其他功能的多功能雨洪调蓄设施，有利于提高整个小区的防洪能力和土地的有效利用率，同时可以为小区创造更多亲水、自然优美的环境和休闲活动场所。

5.1 利用湖体的调蓄能力分析

根据天津市东丽区1986～2003年的降雨及蒸发量统计资料，可以得出年平均蒸发量为1700.5mm，丰水年降雨量为865.30mm，平水年降雨量为455.05mm，枯水年降雨量为246.82mm(见图2)。

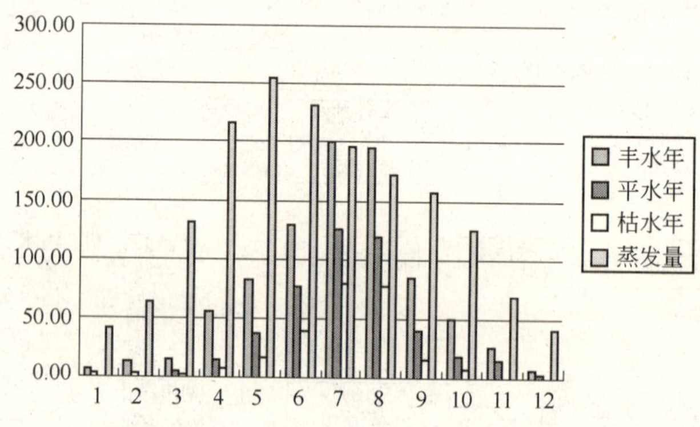

图2 东丽区多年平均蒸发量及不同年份的降雨量

在不考虑所有天然湿地调蓄能力的情况下，初步进行水量平衡分析，得出在全部雨水入湖和无补充水时整个湖的水深变化，见表2。

整个湖的水文变化 表2

不同年份	水位变化(m)	外排雨水量(m^3)	不同年份	水位变化(m)	外排雨水量(m^3)
丰水年	1.89～3.56	316350	枯水年	1.09～1.86	0
平水年	1.55～2.15	0			

由于建成后一期地面高程3.2m，二期最低地面高程3.75m，从表2中可以看出，如果排水不及时，在丰水年可能会造成小区积水，严重时产生小区内涝。因此，考虑在丰水年利用天然湿地调蓄雨水。

5.2 利用多功能雨洪调蓄设施的调蓄能力分析

在二期人工湖周边分布着较多的芦苇荡湿地，为净化处理水体水质，在二期湖的南面建造了人工湿地。人工湿地与二期水体周边保留的天然芦苇荡湿地组成了小区22.1ha的湿地面积。湿地的分布布置见图3。

天然湿地的底部高程为1.76m，最大调蓄高度0.69m(见图4)，充分利用湿地的调蓄能力，通过水量平衡计算得出一、二期整个湖体的水文变化理论数据见表3。

一、二期整个湖体水文变化 表3

不同年份	水位变化(m)	一次最大能容纳的降雨量(mm)	外排雨水量(m^3)
丰水年	1.85～2.97	174～0	169380
平水年	1.55～2.02	266～168	0
枯水年	1.09～1.83	384～182	0

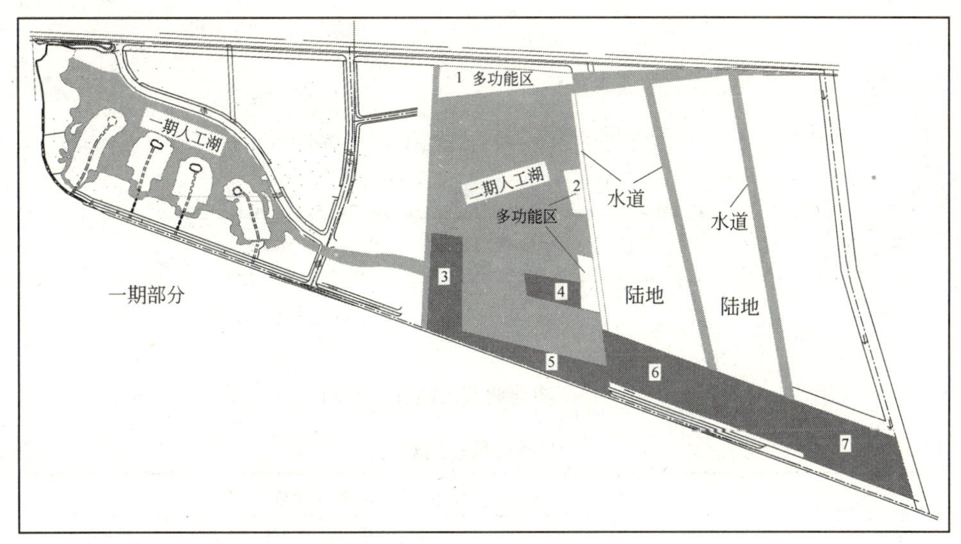

注：图中 1、2、3、4、6、7 区为天然芦苇荡湿地，5 区为人工湿地；一期以外的区域属二期部分

图 3　二期范围湿地分布平面图

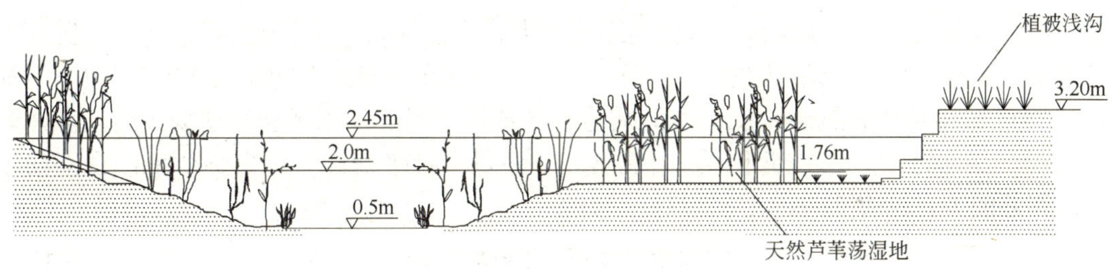

图 4　人工湖芦苇荡湿地高程示意图

大面积湿地不仅可以净化水质，还可以兼作洪峰时期的雨水滞留调蓄区，减少雨水的外排水量，有效缓解小区的防洪压力。

对比表 2 和表 3 可知，当充分利用整个天然湿地(21.3ha)的调蓄能力(0.69m)后，可明显降低最高水位，有效地缓解小区的防涝压力。

但是，大面积的湿地占据不少的土地资源，也加大了运行管理的难度，例如，当水位低于 1.76m 时，为了避免湿地长时间处于无水状态，影响湿地景观效果，要及时实施补水措施。

如果将二期湖体周边的天然芦苇荡湿地全部改建为其他多功能用途的雨洪调蓄设施，比如：运动场所、小区娱乐区、停车场等，可以提高小区土地的有效利用率，避免湿地不良运行带来的一些负面影响，又保证足够的蓄洪能力，还可以为人们提供更多丰富的休闲娱乐场地，创造良好的小区亲水环境，多功能区高程设计见图 5。多功能区只有出现多年一遇的暴雨且水位高于溢流水位时才发挥短时间的蓄洪功能，待湖水水位下降后，再通过水渠的控制闸将雨水排入湖内。

将湿地改建成多功能调蓄区后，增加了这片区域的使用频率，也解决了在水位低于 1.76m 时湿地的补水问题。具体数据见表 4。

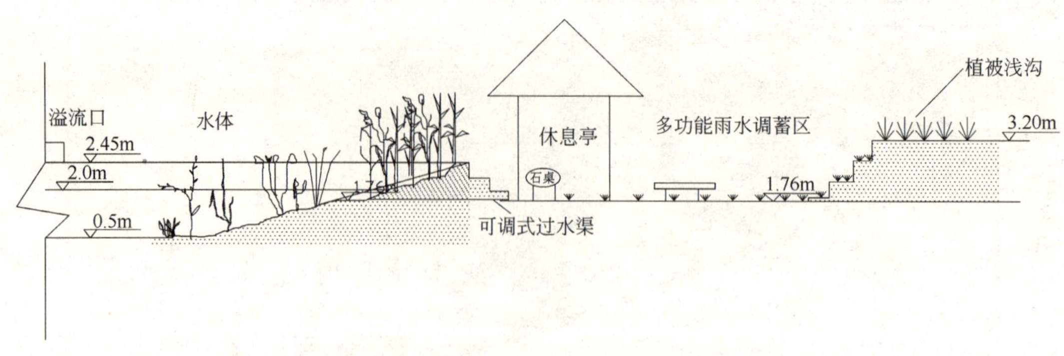

图5 多功能调蓄区高程示意图

土地利用率比较 **表4**

用地性质	一年中的使用频率		
	丰水年	平水年	枯水年
湿　　地	100%	75%	8.30%
多功能调蓄区	100%	100%	100%

注：湿地一年中的使用频率指的是一年中水位在1.76m以上的时间占全年的比例。

但湿地的大幅度减少会降低水体的净化能力，综合考虑小区水质维护和多功能区的建设成本，只将一部分湿地（大约35%）即77350m² 的面积改建成多功能调蓄区（即图3中的1、2号区域和4号的部分区域），可以将它们建成停车场、娱乐区和运动场。当只考虑剩余的65%的湿地与一、二期水体进行调蓄时，通过水量平衡计算得出整个水体在不同年份的调蓄能力，见表5。

不同年份整个水体的调蓄能力 **表5**

不同年份	一次能容纳的降雨量(mm)		
	最大	最小	平均
丰水年	167	0	83.5
平水年	254	161	207.5
枯水年	372	175	273.5

从表5中的数据可以看出，即使减少35%湿地的调蓄面积，整个水体还具备很高的调蓄能力。根据不同年份水体平均的调蓄能力，丰水年、平水年、枯水年分别可以调蓄1年一遇、10年一遇、20年一遇的暴雨。也就是说，多功能调蓄区在丰水年遇上大于1年一遇的暴雨时，才发挥雨水调蓄的功能，与水体、湿地共同构成雨水调蓄区；在平水年、枯水年分别遇上大于10年一遇、20年一遇的暴雨时，才用于雨水调蓄。

综上所述，将35%的湿地面积改建为多功能调蓄区，在保证整个小区具有较高的防洪能力条件下，又为小区的居民增加了更多的活动空间，提高了土地的有效利用率。

6 结束语

多功能雨洪调蓄设施不仅可用于城市小区，还可用于城市公园、滞洪区、城市低洼多涝区、绿化带等的许多地方。比如在北京总体规划(2004～2020)修编中规划2020年要建设7个万亩以上的绿色版块，新的规划中还进行了滞洪区的建设，规划建设公园45个，扩大水面1300hm^2(资料来源：北京规划展)。考虑到对于寸土寸金的北京而言，土地资源十分宝贵，可以利用7个万亩以上的绿色版块中的部分绿地，在合适的位置或条件下建成多功能雨洪调蓄设施，也可在立交桥和高速公路附近、在环绕城区的生态绿化带以及规划建设的45个公园中因地制宜地考虑采用多功能雨洪调蓄设施。从而实现充分发挥土地资源的效益和有效地利用城市雨水资源、控制洪涝、改善城市环境等一举多得的目的。

Case-Analysis of Stormwater Multi-Functional Detention Facilities in Uptown

Che Wu[1] Zhang Yan[1] Huang Yu[1] Li jun qi[1]
He jianping[2] Liu Hong[2] Meng guanghui[2] Wang hongling[2]
([1] Beijing Institute of Civil Engineering & Architecture, Beijing 100044 China;
[2] Beijing Water Conservation Office, Beijing 100036 China)

Abstract: Development of the city changed original environment and the land surface, reduced the detention ability of stormwater, increased the peak flow of stormwater and the risk of flooding in urban area, and caused the rainwater lost. Application of stormwater multi-functional detention facilities not only can mitigate flooding, but also can utilize effectively the rainwater and the land resources. This article takes a uptown as an example, analyzing some water environment problems with the developing, rainwater harvest, scale of the waterscape, stormwater detention ability, and the land utilization. Specially, feasibility of storm water multi-functional detention measures in uptown and its virtues is discussed.

参考文献

[1] 车伍，张燕，李俊奇，刘红，何建平，孟光辉，汪宏玲. 城市雨水多功能调蓄技术. 给水排水，2005，31(9)：25～29

[2] 雨水贮留浸透技术协会(1). エコロジカルポンド计画. 设计の手引き. 1998年，山海堂

[3] 车伍，李俊奇. 生态住宅小区雨水利用与水景观系统-案例分析. 城市环境与城市生态，2002.15(5)：34～36

偏高岭土替代硅灰做水泥混凝土掺合料研究

张忠伦　隋同波　文寨军　王　晶　范　磊

（中国建筑材料科学研究院，北京　100024）

摘　要： 本文研究了偏高龄土替代硅灰做水泥混凝土掺合料时水泥混凝土的力学性能和工作性能。研究表明，掺加偏高岭土的水泥早期强度与硅酸盐水泥相当，后期强度明显增高；水泥中掺入偏高岭土后，随掺量的增加，水泥早期强度略有增加，28d 强度明显提高，偏高岭土在水泥中的最佳掺量为 10%左右；用偏高龄土替代硅灰，在较低水胶比条件下制得的混凝土力学性能和工作性能优良。用偏高岭土可以满足高性能混凝土要求，偏高岭土是一种优良的混凝土掺合料。

关键词： 偏高岭土　水泥　混凝土

1　前言

高岭土——以高岭土族为主要成分的软质粘土，其主要成分是高岭石(结晶化学式为 $Al_4[Si_4O_{10}](OH)_8$，此外尚有石英与少量的云母)，是一种极为广泛的工业矿物。它资源丰富并具有一系列特殊的工艺性能：质软，粒细，白度高，不易磨蚀，导电、导热率低；有烧结性、可塑性及良好的化学惰性；优良的吸附性、分散性和易于加工，已被造纸、陶瓷、橡胶、塑料、涂料、耐火材料、化工生产等各个工业部门所采用，其新的用途也不断被发现。

高岭土是层状的铝硅酸盐粘土质矿物。破坏高岭土的有效结构，最常用的办法是在高温下焙烧，这也是一种最简单的活化方式。实验发现，活化方式不同，所得产物的性质差别很大，也直接影响最终的产品质量。高岭土活化的实质是破坏高岭土的晶体结构，使其中的硅铝活化，变成酸或碱的可溶组分。有关资料表明，高岭土经高温煅烧处理后，其晶体结构遭到破坏，变为一种类似无序结构的介稳状态结构，高岭土脱水温度区间在 509～636℃，而 900℃则开始出现重结晶的放热峰，这与 Swarmy 等的研究结果相吻合。当温度提高至 950℃以上后，偏高岭土就会重结晶转变为尖晶石，最后变为莫来石，则失去水化活性。

一般有两种活化方式：第一种为直接将高岭土加热至高温，焙烧一段时间；第二种是将高岭土和碱按一定的比例混合的高温焙烧法。高岭土经过 450～900℃脱水后，铝氧八面体结构发生较大的调整，成为不稳定的铝氧八面体网络，构成偏高岭土。

第一种活化法——“直接焙烧法”的微观分析表明，在约为 500℃时高岭土开始脱去其结构水，到 600℃左右完成，变成具有较高活性的偏高岭土。至约 1000℃时，有一放热峰，是高岭土晶体破坏后产生尖晶石的放热效应。

第二种活化方法——“加碱焙烧法”，将高岭土和碱按一定比例混合，经一定温度(约600℃)焙烧，高岭土的晶体结构完全破坏，石英和云母的衍射峰均消失，XRD谱图中未见衍射峰，烧结物为无定形相。说明加碱焙烧中碱起到了重要作用，它除了降低焙烧温度外，还与高岭土结构解体析出的硅铝氧化物反应生成可溶性的铝硅酸盐，其中的 Al_3+ 由原来的六配位变成了四配位。

偏高岭土可改善水泥的性能，如美国的 Pyrament 水泥中，由于使用了偏高岭土，使该水泥可全天候施工，特别适合于炎热天气以及低温条件下抢修、修补工程。在－20℃寒冷条件下施工时，强度也能较快增长，即使不用热力养护也可制备预制混凝土及预应力混凝土构件。研究表明，用高活性偏高岭土制备混凝土，可以较大幅度提高混凝土的早期和后期强度，特别是混凝土的后期强度。当偏高岭土掺量与硅灰相同时，其强度相当于或超过硅灰混凝土的强度，在与硅灰达到相同的流动度时，可减少高效减水剂用量。

2 试验研究

(1) 原材料

试验采用江苏苏州高龄土公司生产的一级高岭土进行煅烧研究。原材料成分如表 1 所示。

苏州高岭土化学成分表　　表 1

产　地	Loss	SiO_2	Al_2O_3	Fe_2O_3	CaO	MgO	SO_3
江苏苏州	13.56	47.29	36.34	0.85	0.10	0.49	0.83

(2) 试验过程

1) 不同煅烧温度时的偏高岭土水泥性能

分别取高岭土料 500g 试样置于快升温电炉中，分别在 600℃、700℃、800℃、900℃、1000℃，保温煅烧 2 小时后出炉，置于空气中自然冷却。所得偏高岭土经实验小球磨磨细，按 10%的掺量配入水泥中检验强度，胶砂比 1∶2.5，水灰比＝0.44，试体尺寸 4cm×4cm×16cm，试验结果见表 2、图 1 和图 2。

不同煅烧温度下的 MK 水泥的强度性能　　表 2

编　号	煅 烧 温 度	抗折/抗压强度(MPa)		
		3d	7d	28d
C-0	空白样、冀东 42.5 级水泥	7.3/42.7	7.9/54.2	8.6/62.0
C-1	未煅烧的高岭土	7.0/42.6	7.9/51.1	8.2/59.4
C-2	600℃	6.9/42.2	8.3/58.7	9.7/72.3
C-3	700℃	7.2/43.0	8.8/60.0	9.7/74.7
C-4	800℃	7.1/42.7	9.1/59.7	9.6/75.6
C-5	900℃	7.1/41.1	8.2/55.3	8.7/66.4
C-6	1000℃	6.9/39.7	8.1/51.0	9.4/68.3

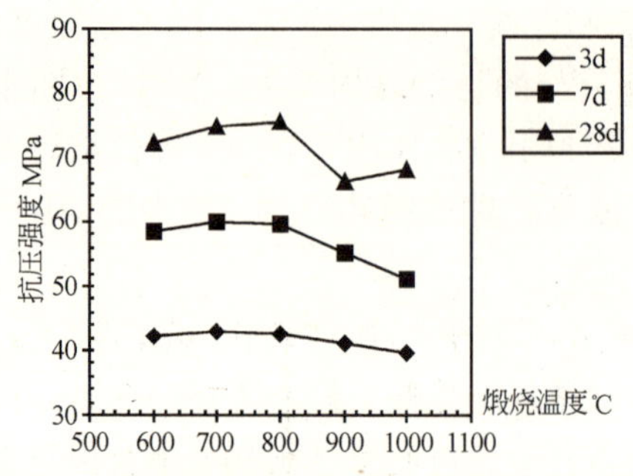

图 1 不同煅烧温度下 MK 水泥的强度性能

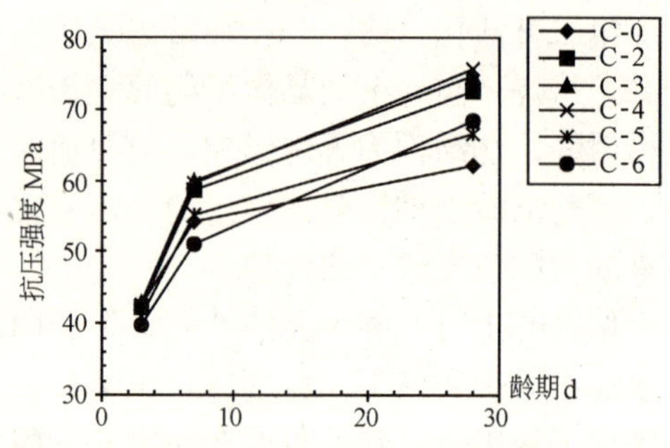

图 2 MK 掺合料对水泥性能影响

试验结果表明，未经煅烧的高岭土配入水泥中后，随龄期的增长，强度会出现下降。这主要是由于高岭土属层状结构的铝硅酸盐黏土矿物，未采取高温煅烧等活化激化措施，其水化活性较低或火山灰活性低，因而掺入水泥中会导致水泥强度的降低。而偏高岭土是高岭土经高温煅烧得到，其结构中铝氧八面体群发生较大的改变，成为不稳定的铝氧八面体网络，偏高岭土中原子排列不规则呈现热力学介稳状态，具有较强的火山灰活性。掺入水泥中后，3d 强度与空白样持平，7d 强度略好于空白样，28d 强度均高于空白样。

另外，从表 2、图 1 和图 2 可以看出，掺加偏高岭土的水泥早期强度与硅酸盐水泥相当，后期强度明显增高，经不同温度煅烧所得的偏高岭土所表现出的火山灰活性不同。当煅烧温度由 600℃提高至 800℃时，偏高岭土的活性略有提高，如 C-3、C-4 试样，其早期强度(3d、7d)略高于 C-2 试样，但 28d 强度约比 C-2 试样提高 3～5MPa。当煅烧温度提高至 900℃以上时，偏高岭土的活性反而降低，水泥的早期强度和后期强度都下降，尤其是 28d 强度下降幅度较大达 5～10MPa。因此，高活性的偏高岭土的适宜煅烧温度范围在 700℃～800℃之间，在这个温度范围内煅烧的偏高岭土作为水泥掺合料水泥的力学性能最佳。

2）偏高岭土混凝土性能试验

A. 原材料

水泥采用冀东 42.5 级硅酸盐水泥，常规物理性能列于表 3。粗集料采用河北产破碎石灰石，最大粒径为 25mm，连续级配，密度为 2.65kg/L。粗集料在使用前用水洗净，晾干并且剔除其中的针状颗粒。细集料选用北京普通河砂，细度模数为 2.8，密度为 2.7kg/L。细掺合料使用硅灰和细磨偏高岭土粉，其颗粒粒径分析结果见表 4。偏高岭土粉为苏州高岭土经 750℃煅烧后磨细制成。研究中选用内蒙古元宝山电厂Ⅰ级粉煤灰。外加剂选用天津产萘磺酸甲醛缩合物 UNF-5 高效减水剂，减水率≥20％。

冀东 42.5 级硅酸盐水泥物理性能 **表 3**

编 号	比表面积 m^2/kg	凝结时间(h∶min)		抗折/抗压强度(MPa)		
		初 凝	终 凝	3d	7d	28d
42.5 级硅酸盐水泥	355	2∶23	3∶14	6.8/38.4	7.6/51.3	8.3/61.5

偏高岭土和硅灰的粒径分布 **表 4**

粒径 / 名称	<1μm	<2μm	<4μm	<8μm	<16μm	<32μm	<64μm	<80μm	平均粒径 (μm)
偏高岭土	25.82	35.29	52.87	69.58	85.32	94.81	98.69	100	8.48
硅 灰	79.5	94.5	98.9	100	—	—	—	—	0.2

B. 混凝土试件的制备和试验方法

以常规混凝土生产工艺制作混凝土试件。使用卧轴式强制搅拌机，搅拌时间为 3min。加料顺序及方法为：先在搅拌机中加入粗细集料，然后加入胶凝材料，干拌适当时间，再加水搅拌均匀。试验中应预先将矿物掺合料、高效减水剂和水泥充分拌匀制成复合胶凝材料一起加入搅拌机中。混凝土出机后测定坍落度或扩展度，然后在振动台上振动成型各种试件。试件带模在室温下放置 1d，随后置于 20℃±3℃的水中养护至规定的龄期进行性能测试。混凝土拌合物试验按照《普通混凝土拌合物性能试验方法》GB/T 50080—2002 进行；力学性能按照《普通混凝土力学性能试验方法》GB/T 50081—2002 进行，长期性能按照《普通混凝土长期性能和耐久性试验方法》的规定进行。

C. 掺 MK 和 SF 混凝土的工作性

混凝土的坍落度是表征新拌混凝土性能的重要指标，它的大小直接影响混凝土的工作性能。另一方面，对于高性能混凝土而言，混凝土坍落度经时损失是其生产的重要控制指标，也是目前商品混凝土施工常遇的难题。为此，研究了在不同掺量硅灰和偏高岭土粉时混凝土的坍落度和混凝土经时损失，试验结果见表 5 和图 3。混凝土的配合比 1∶1.90∶2.84，胶凝材料用量为 385kg/m^3。

掺 MK 和 SF 混凝土的工作性 **表 5**

掺合料		胶凝材料用量 (kg/m^3)	*W/C*	高效减水剂		初始坍落度 (cm)	90min 坍落度 (cm)
编号及种类	掺量(%)			种类	掺量(%)		
0	0	385	0.42	UNF-5	1.0	22.3	18.6
MK-1	5	385	0.42	UNF-5	1.0	21.6	16.7
MK-2	10	385	0.42	UNF-5	1.0	20.0	15.1
MK-3	15	385	0.42	UNF-5	1.0	16.9	8.0
SF-1	5	385	0.42	UNF-5	1.0	21.2	16.0
SF-2	10	385	0.42	UNF-5	1.0	19.4	14.1
SF-3	15	385	0.42	UNF-5	1.0	15.1	5.5

从试验结果和图 3 可以明显看出，掺活性掺合料的混凝土和纯硅酸盐水泥混凝土的工作性略有差异，在相同水灰比时，掺活性掺合料混凝土的初始坍落度小且 90min 后坍落度损失要大一些。

从混凝土初始坍落度来看，随着掺合料掺量的增加，混凝土的初始坍落度降低，当掺量在 10%以内时，对混凝土的初始坍落度影响较小，但当掺合料掺量达到 15%时，混凝土的初始坍落度值明显减少。90min 后坍落度损失也显著增大，如 MK 和 SF 掺量达到 15%时，混凝土的坍落度保留值只有 5～10cm，基本上失去流动性能。相对而言，掺 MK

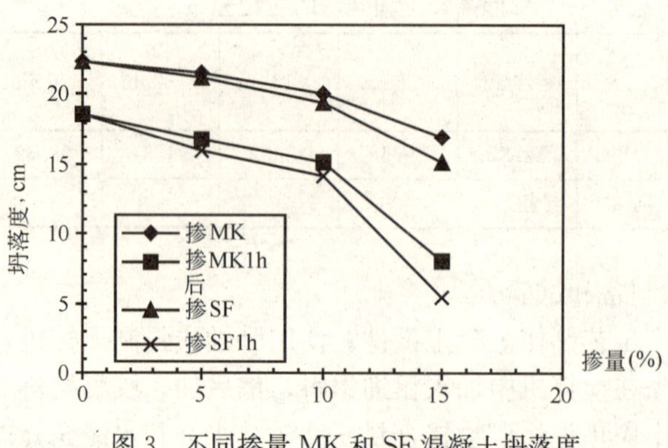

图 3 不同掺量 MK 和 SF 混凝土坍落度

混凝土的和易性稍优于掺 SF 混凝土。另外，从新拌混凝土状态看，与掺 MK 混凝土相比，掺 MK 混凝土的黏聚性较小，表现出较好的流动性能。

D. 偏高岭土的混凝土强度性能

用试烧偏高岭土配制混凝土，试体尺寸 10cm×10cm×10cm，高效减水剂采用 UNF-5，掺量 1.0%。配合比为 1∶1.90∶2.84，水泥用量 385kg。控制坍落度 19～21cm。试验结果见表 6、图 4 和图 5，MK 为偏高岭土，SF 为硅灰。

偏高岭土的混凝土强度性能 **表 6**

编 号	配 合 比	掺量(%)	W/C	坍落度(cm)	抗压强度(MPa)		
					3d	7d	28d
0	1∶1.90∶2.84	0	0.42	21	39.6	52.3	63.5
MK-1	1∶1.90∶2.84	5	0.42	22	40.4	58.2	69.3
MK-2	1∶1.90∶2.84	10	0.42	22	38.9	56.7	69.9
MK-3	1∶1.90∶2.84	15	0.43	18	37.8	53.9	70.9
SF-1	1∶1.90∶2.84	5	0.41	22	40.3	56.0	71.9
SF-2	1∶1.90∶2.84	10	0.41	21	39.2	53.3	71.2
SF-3	1∶1.90∶2.84	15	0.42	16	36.3	52.2	67.6

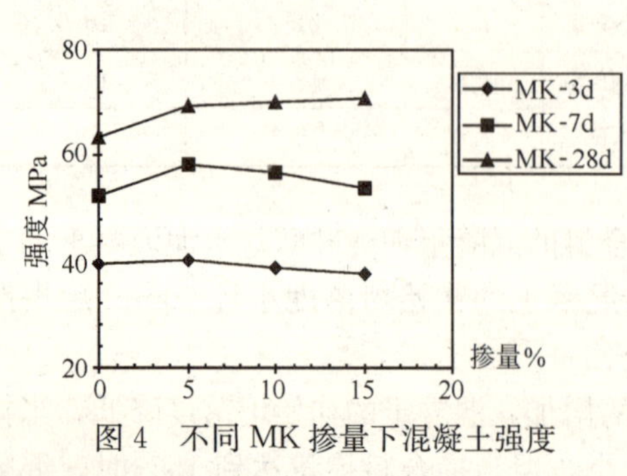

图 4 不同 MK 掺量下混凝土强度

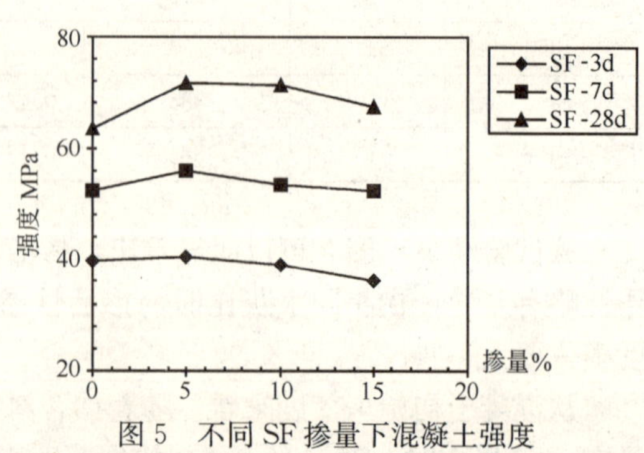

图 5 不同 SF 掺量下混凝土强度

试验结果表明，与空白混凝土相比，掺入偏高岭土和硅灰后，混凝土的早期强度与空白混凝土基本相当，而 28d 混凝土强度均有不同幅度的提高，随掺量的不同，混凝土的

28d 强度提高 4～9MPa。偏高岭土对混凝土性能的影响与硅灰相似，即随着偏高岭土掺量的增加，早期强度随掺量的增加略有下降，28d 强度增长。与空白混凝土相比，混凝土 28d 强度提高 5～10MPa。掺硅灰的混凝土强度发展规律与掺偏高岭土基本一致，但其混凝土 28d 强度随掺量的增加而降低，其适宜掺量为 5%～10%。掺入偏高岭土后，混凝土强度性能相当于掺硅灰的混凝土。

3 结论

高活性的偏高岭土的适宜煅烧温度范围在 700～800℃之间，在这个温度范围内煅烧的偏高岭土作为水泥掺合料水泥的力学性能最佳。用偏高岭土粉、硅灰作为辅助胶凝材配制高性能水泥混凝土，在相同条件下，掺高岭土的水泥砂浆性能在整体上略优于或相当于掺硅灰的水泥性能如水泥砂浆流动度、强度等。利用Ⅰ级粉煤灰-MK(或 SF)-水泥三元复合胶凝材料，掺入适宜的高效减水剂，在较低水胶比条件下均能制得具有优异强度性能和工作性能的混凝土，用偏高岭土取代硅灰配制高性能混凝土是可行的。在混凝土中掺入细磨的偏高岭土和优质粉煤灰，不仅可以节省水泥，而且可明显地改善混凝土的强度性能。单独掺入偏高岭土后会增大混凝土的黏聚性，当掺量超过 10%时，将明显降低混凝土拌合物的坍落度和增加坍落度损失。

Research Metakaolin as Active Admixture in Cement and Concrete Compared wth Silica Fume

Zhonglun Zhang　Tongbo Sui　Zhaijun Wen　Wang Jing　Fan Lei
(China Building Materials Academy, Beijing　100024　China)

Abstract: The workability and strength of cement prepared by metakaolin and silica fume has been researched. The results indicate that the strength of cement with metakaolin is come up to Portland cement in short term age and higher than Portland cement in long term age. When the Portland cement has been added metakaolin, the short term age strength increase less, but the long term strength increase greatly, and the best capacity add in cement is about 10%. The workability and strength of concrete under the low water ratio prepared by metakaolin and silica fume is good. The result shows that the metakaolin can meet the technical requirements of high-performance concrete and is a good kind of admixture for preparing high-performance concrete.

万科东丽湖生态实践

苏志刚　时　宇
（万科企业股份有限公司）

摘　要： 天津东丽湖项目是万科企业股份有限公司目前在国内开发的最大项目。东丽湖项目在生态住区的建设上从小区环境规划设计、能源与环境、室内环境质量、小区水环境、材料与资源五个方面着手，结合项目重点在于水环境、环境质量及节能等方面进行了20余项技术实施。2004年底建设部提出建设“节能省地型”住宅，万科东丽湖项目的生态实践与国家政策不谋而合。

关键词： 生态住区　东丽湖项目　节能　水环境

1　关于天津东丽湖项目

东丽湖是天津市七个自然生态保护区之一，湖面面积为 $7.3km^2$，是天津市周围最大的淡水水域，水面辽阔，水草丰美，属于典型的湿地生态系统。天津万科东丽湖项目占地4000多亩，位于东丽湖北侧，东丽湖旅游度假区内。东丽湖区域的空气质量远远好于东丽区的总体水平，同时也是天津市空气质量最好的区域之一。建设地块大部分为闲置荒地，土壤肥力低，盐碱含量高。

作为一个有着社会责任感及使命感的房地产开发企业，也由于东丽湖项目得天独厚的环境条件，万科将其作为集团第一个全方位的生态实践项目，并整合集团资源进行技术支持，创建一个生态住区。

俗话说：一方水土养一方人。在这样一个区域开发住宅项目，一方面通过统筹规划，合理布局，注重区域环境的保护，并创造良好的人文环境；另一方面，通过人工湖湿地环境的营造、雨水收集利用、东丽湖原状渗水渠的改造、改良盐碱地并形成较高的绿地率，对现有环境进行修复；构建出良好的生态环境。

具体而言，从小区环境规划设计、能源与环境、室内环境质量、小区水环境、材料与资源五个方面着手，结合项目拟定的技术措施多达200余项，经过合并归纳，最终整理出20余项应用技术进行实施，重点在于水环境、环境质量及节能。

东丽湖项目一期于2003年开始设计并在年底动工，目前已有一些生态技术得以“见庐山面目”。

2　建造生态住区

2.1　水环境系统

我们对水环境的理解是将住区作为一个整体来进行系统的考虑。水环境系统包括景观水环境、雨水利用、中水利用、给排水管网等。排水及给水系统在管网设计中属于常规内容，中水利用也比较成熟；但如何保障景观水环境的水质是一项疑难问题，而同时将雨水利用、中水利用、给排水等与景观水环境纳入到一个整体进行规划设计确是业界的新课题。

在实际的操作中，首先通过改造东丽湖北岸的原状渗水渠，形成循环通路，从而让景观湖循环起来，也为居住区营造了一条护城河；然后首次在北方引入高效垂直流人工湿地水质净化技术对景观湖水质进行处理，从投入运行以来半年的水质监测数据显示水质一直保持地表三类水标准，符合设计目标要求；能够满足将在今年 8 月份举行的“2005 年世界滑水锦标赛”的水质要求。目前景观湖的水量超过 20 万 m^3，没有采用一个 m^3 的自来水，第一次注水采用的是邻近的地表劣五类水源，经过人工湿地处理达到三类地表水后排到景观湖(图 1)。前段时间，建设部出台《住宅建筑规范》(征求意见稿)，其中明确“人工水体严禁采用自来水”的要求在东丽湖项目中已经提前达到了。

根据掌握的东丽湖地区的降雨量及蒸发量数据，对景观湖水的全年水位变化进行分析，并量化补水频率及补水量。补水的主要来源为雨水。合理利用自然资源，减少对生态的破坏是东丽湖项目建设的基本原则之一，东丽湖项目采用了较为合理的雨水利用措施。小区一期公建、住宅屋面及地面的雨水(图 2、3、4)通过渗透、排水系统收集基本上都得到了利用。

图 1　景观湖鸟瞰

图 2　情景大道雨水明渠

图 3　公建区雨水收集系统

图 4　住宅区雨水收集系统

对于一期的污水，将全部收集起来，通过中水处理系统后用于绿化浇灌及道路喷洒，实现污水零排放。为了形成均衡的水生态链系统，还在景观湖内放养鱼类、人工养殖水禽、在岸边种植挺水植物等。期望通过以上的措施逐渐营造出一个水生态环境圈(图5)，并与周边环境和谐统一。良好的生态水环境系统为小区亲水环境的形成创造了条件(图6)。

图5 生态水环境

2.2 环境质量

东丽湖项目地处市郊，住区空气环境优良，为了提高室内环境舒适度，在别墅及公寓内均设置了外墙式通风系统(图7)，使得业主尤其在冬季不必开窗也能呼吸到新鲜的室外空气；同时通过控制进出风量还能起到节约能源的作用。

图6 亲水环境

营造健康舒适的生活环境，声环境也是重要的衡量指标。东丽湖项目远离闹市区，附近没有明显的噪声源，为了更好的维护区内声环境，在住区与东丽之光市政道路之间，建设了生态景观防护林带，由堆坡、原生植物、草坪、围墙等共同构成(图8)。

图7 外墙式通风系统

图8 防风减噪林

为了事先全面了解小区在建设之后的小区环境状况，委托清华大学对规划设计进行了模拟评估。其中，对于风场环境的评价：小区内的风环境令人满意，小区内基本没有气流死角，容易形成良好的自然通风；对于“日照与采光”部分评述如下：“由于整个小区分布的均匀性和高度的一致性，冬季采暖日满足冬至日每户至少一小时日照要求，符合国家规范”。

东丽湖项目地处天津滨海地区，属于轻中度盐碱土，仅适合芦苇、香蒲等耐盐碱植物(图 9)，而不利于常规景观植物生长。东丽湖项目采用了盐碱地改造的专利技术，合理改造当地土壤资源，最大限度的减少了对当地自然环境的破坏，保证了住区内绿化环境的良好生长(图 10)。

图 9　盐碱地

图 10　滨湖公园

东丽湖项目盐碱地的改造符合“用进废退”原则；由于项目开发持续时间较长，需要有一个长期的观念。为了避免后期地块的土壤盐碱化，同时也为了提供后期绿化树木，在项目后期用地内建设了人工苗圃(图 11)。一期苗圃利用的是项目内土质较好的区域进行人工苗木的培育；苗圃还将逐步扩大。

图 11　苗圃

减少废弃物排放量是减少对自然环境破坏，保持环境质量的方式之一。东丽湖项目提倡垃圾分类收集处理，设计了垃圾分类收集箱，将有机垃圾、无机垃圾、有害垃圾分类收集；同时还将采用有机垃圾生化处理设备，对分类收集的有机垃圾进行处理，有效减少对外的清运量、减轻城市负担。

2.3 节能

一期由于容积率较低，大都为独体别墅以及低层连排公寓，在设计中严格控制围护结构的传热系数，提高墙体、屋面及窗的保温性能；特别是对于窗，由于设计的需要，在临院落、临湖部位开窗面积较大，所以通过综合控制各朝向的窗墙比以及采用性能更高的铝合金中空玻璃窗来实现节能(图 12)。

图 12 建筑外观

佛甲草种植屋面(图 13)在万科集团南方的项目中应用较多。经过了在天津水晶城项目的小规模试点成功之后，这次正式在公建平屋面上进行较大规模的实践，并成功越冬。佛甲草种植屋面具有施工简便、少维护的特点，不但能够丰富建筑的第五立面，而且对建筑节能能够做出贡献。

图 13 佛甲草种植屋面

除建筑节能采用各种技术手段实现之外，东丽湖项目一期还尝试在一些公共区域设置太阳能灯。太阳能属于清洁能源，将太阳能转换为电能并用于夜间照明，是有效利用自然能源的一大手段。

3 万科生态研发与实践

万科从 2001 年起开始对生态进行研究，作为发展商，不会停留在基础研究层面，需要侧重应用；因此，选择从单项技术层面进行系统实践作为突破口。以南京金色家园二期三滑道推拉百叶遮阳技术为例，从 2001 年底开始课题研发、到 2002 年落实到施工图设计、2003 年初与厂家共同研发产品成功、最后在施工完毕之后的 2004 年 6 月份间进行了实地检测，前后 3 年时间。数据显示课题研发阶段的模拟评估与实测结果基本吻合。类似的还有高效垂直流人工湿地技术在国内小区中的首次实践、在国内首次提出复合式厨房的概念，并将研发成果在多个项目进行实践、佛甲草种植屋面技术在住宅及公建平屋面上的应用等。单项技术的成功研发及实践，为东丽湖项目生态实践打下了坚实的基础。

在实践的过程中，逐步形成了万科的技术观：应用低技术，达到住宅的高舒适性。换言之，就是采用生态技术，从传统的“三高一低”(高消耗、高排放、高投入、低效益)转化为“三低一高”(低消耗、低排放、低投入、高效益)。

技术观对东丽湖项目的生态实践起到了较好的指导作用。不难发现，在东丽湖项目一期的生态实践中，采用的都是生态技术。从万科的技术观出发，实际上，东丽湖项目的生态住区实践主要围绕三个主题：一是创造健康、舒适的居住环境；二是与自然环境相融合；三是减少对地球资源与环境的负荷和影响。这也是万科在生态住区实践中秉承的理念，既有为业主的考虑，也有万科的社会责任感使然。

建设部部长汪光焘在《大力发展节能省地型住宅——贯彻中央经济工作会议精神的思考》一文中指出：“在住宅建设工作中，按照减量化、再利用、资源化的原则，搞好资源综合利用，大力抓好节能、节地、节水、节材工作，建设“节能省地型”住宅”。而在此指导思想下，建设部副部长仇保兴提出了“绿色建筑”的概念：指为人们提供健康、舒适、安全的居住、工作和活动的空间，同时在建筑全生命周期中实现高效率的利用资源(能源、水资源、土地、材料)、最低限度地影响环境的建筑物。

对比绿色建筑的概念与东丽湖项目生态住区实践围绕的三个主题，可以发现，两者不谋而合了。

The Ecological Practices in Dongli Lake Project Of Vanke

Su Zhigang　Shi Yu

(Vanke Co.，LTD)

Abstract: Tianjin Dongli Lake project is now the largest project of Vanke Company Ltd. in China. The construction of ecological residence of Dongli Lake is stressed on five aspects as conceptual design, energy & environment, quality of indoor climate, subzone aquatic environment, and material and resources. More than twenty techniques are applied to this project, which are aiming at aquatic environment design, environment quality improvement and energy economization. At the end of 2004, Ministry of Construction issued the policy of building house in terms of “land and energy saving”. The ecological practice in Dongli Lake project by Vanke happened to have the same idea with the government.

夏热冬冷地区主要外墙外保温系统的综合对比分析

王智宇[1]　樊先平[1]　闫成文[3]　王志平[3]　阮　华[3]　王小山[3]

（[1] 浙江大学材料科学与工程学系，浙江　310027；[2] 宁波大学建筑工程与环境学院，浙江　315000；[3] 宁波荣山新型材料有限公司，浙江　315000）

摘　要：膨胀聚苯板薄抹灰外墙外保温系统，胶粉聚苯颗粒外墙外保温系统和聚合物保温砂浆外墙外保温系统是当前我国夏热冬冷地区建筑节能所广泛使用的主流外墙外保温系统。本文从体系的性能指标，成本，安全性、环保消防性和适用范围等几个方面对三个系统的进行了全面的综合比较分析，并提出相关参考意见，以达到使用用户及研究设计人员正确设计与选用外墙外保温系统的目的。

关键词：膨胀聚苯板薄抹灰外墙外保温系统　胶粉聚苯颗粒外墙外保温系统　聚合物保温砂浆外墙外保温系统　综合对比分析

能源已成为人类发展的瓶颈，随着人们对生活品质不断追求，能源消耗在成倍地增长。能源是人类社会有限资源，而能源的滥用也造成人类生存环境的日益恶化。节能已成为21世纪人类发展主题。建筑节能往往是人们的盲点，一是对建筑能耗总量认识不足，在发达国家建筑能耗已达社会总能耗的30%～40%，我国也已达到20%～30%，二是对可节约空间认识不足，其他行业能耗如工业能耗可节约空间在10%以内，交通运输能耗可节空间更低，相比较而言，建筑能耗其可节约空间巨大，达50%以上。随着人们认识的提高，顺应时代的需要，近年来，建筑节能产品也呈现繁荣景象[1]。纵观国内夏热冬冷地区建筑节能市场，主要以膨胀聚苯板薄抹灰外墙外保温系统、胶粉聚苯颗粒外墙外保温系统、聚合物保温砂浆外墙外保温系统为主，本文针对三个保温系统的性能指标、成本、安全性、环保消防性和适用范围等方面作比较，为使用用户及研究设计人员提供参考意见。

1　主要外墙外保温系统构造

1.1　膨胀聚苯板薄抹灰外墙外保温系统构造

膨胀聚苯板薄抹灰外墙外保温系统构造见表1。系统要求基层墙面找平，平整度达到规范要求，用胶粘剂将膨胀聚苯板与基层墙体粘贴，辅以锚栓固定，当建筑物高度不超过20m时，也可采用单一粘贴固定方式。防护层为嵌埋有耐碱网布或热镀锌电焊网增强的聚合物抹面胶浆。以耐碱网格布为增强网的构造，为了增强抗冲击强度，首层需加一层加强耐碱玻纤网。饰面层只能是涂料。

膨胀聚苯板薄抹灰外墙外保温系统构造 **表 1**

外墙类型	外保温做法				
	找平层	胶粘层	保温层	防护层	外饰面
① 混凝土剪力墙、混凝土砌块墙、粘土砖墙、砌块等	② 普通水泥砂浆层	③ 胶粘剂层	④ 膨胀聚苯板保温层	⑤ 抹面胶浆复合耐碱网布或热镀锌电焊网	⑥ 涂料

1.2 胶粉聚苯颗粒外墙外保温系统构造

胶粉聚苯颗粒外墙外保温系统构造见表 2。本系统在基层墙体先涂一层界面剂，然后将胶粉聚苯颗粒浆料直接粉刷在墙面上，克服板材类保温材料在边角与凹凸表面难以施工的问题。保温层施工根据保温厚度不同分两遍或多遍施工，每遍施工厚度约 1cm。防护层为嵌埋有耐碱网布或热镀锌电焊网增强的抗裂砂浆。涂料饰面且建筑物高度不超过 30m 时，防护层可采用埋有耐碱网布为增强网，为了增强抗冲击强度，首层需加一层加强耐碱玻纤网。面砖饰面或建筑物高度超过 30m 时，防护层宜采用热镀锌电焊网为增强网，辅以锚栓固定。

胶粉聚苯颗粒外墙外保温系统构造 **表 2**

外墙类型	外保温做法			
	界面层	保温层	保护层	外饰面
① 混凝土剪力墙、混凝土砌块墙、粘土砖墙、砌块等	② 专用界面剂层	③ 胶粉聚苯颗粒保温层	④ 抗裂砂浆层复合耐碱网布或热镀锌电焊网	⑤ 涂料、面砖或其他重量不超过 20kg/m^2 的饰面材料

1.3 聚合物保温砂浆外墙外保温系统构造

聚合物保温砂浆外墙外保温系统构造见表 3。本系统在基层墙体涂一层界面剂，然后将聚合物保温砂浆浆料直接粉刷在墙面上，克服板材类保温材料在边角与凹凸表面难以施工的问题。保温层施工根据保温厚度不同分两遍或多遍施工，每遍施工厚度约 1cm。防护层为嵌埋有耐碱网布或热镀锌电焊网增强的抗裂砂浆。涂料饰面所有类型建筑及面砖饰面且建筑物高度不超过 35m 时，防护层可采用耐碱网布为增强网，面砖饰面且建筑物高度超过 35m 时或饰面材料超过 20kg/m^2，防护层宜采用热镀锌电焊网为增强网，辅以锚栓固定。

聚合物保温砂浆外墙外保温系统构造 **表 3**

外墙类型	外保温做法			
	界面层	保温层	保护层	外饰面
① 混凝土剪力墙、混凝土砌块墙、粘土砖墙、砌块等	② 专用界面剂层	③ 聚合物保温砂浆保温层	④ 抗裂砂浆层复合耐碱网布或热镀锌电焊网	⑤ 涂料、面砖或其他饰面材料

2 保温层材料性能对比分析

膨胀聚苯板，胶粉聚苯颗粒和聚合物保温砂浆等三种保温层材料的主要性能指标分别见表 4、表 5 和表 6。

膨胀聚苯板的主要性能指标 表 4

项 目	性能指标	项 目	性能指标
导热系数 W/(m·K)	≤0.041	垂直于板面方向的抗拉强度 MPa	≥0.10
蓄热系数 W/(m^2·K)	≥0.36	尺寸稳定性%	≤0.30
表观密度(kg/m^3)	18.0～22.0		

胶粉聚苯颗粒的主要性能指标 表 5

项 目	单 位	指 标	项 目	单 位	指 标
湿表观密度	kg/m^3	≤420	压剪粘结强度	kPa	≥50
干表观密度	kg/m^3	180～250	线形收缩率	%	≤0.3
导热系数	W/(m·K)	≤0.06	软化系数	—	≥0.5
蓄热系数	W/(m^2·K)	≥0.95	难燃性	—	B1 级
抗压强度	KPa	≥200			

聚合物保温砂浆的主要性能指标 表 6

项 目	单 位	指 标
外观	—	灰褐色、均匀、无结块含颗粒纤维粉状
堆积密度	kg/m^3	≤450
质量含水率	%	≤0.8
干制品密度	kg/m^3	≤520
导热系数	W/(m·K)	≤0.11
蓄热系数	W/(m^2·K)	≥3.5
抗压强度	MPa	≥2.5
压剪粘结强度	MPa	≥0.4
抗冻性(强度保留率)	%	≥75
干燥收缩值	mm/m	≤3.0

根据《夏热冬冷地区居住建筑节能设计标准》(JGJ 134—2001)和浙江省《居住建筑节能设计标准》(DB 33/1015—2003)的规定，有两个基本强制要求：一是外墙传热系数，二是热惰性指标：在热惰性指标 $D\geqslant3.0$ 时外墙传热系数 $K\leqslant1.5W/(m^2 \cdot K)$；热惰性指标 $3.0\geqslant D\geqslant2.5$ 时外墙传热系数 $K\leqslant1.0W/(m^2 \cdot K)$。

表 7 以夏热冬冷地区常见墙体为参考对象，综合考虑传热系数与热惰性指标两个基本要求，对三个保温系统达到国家标准所需的最低保温层厚度(括号中数据为在此最低保温层厚度下的墙体综合热惰性指标)进行了计算比较。

不同墙体材料最低保温层厚度(单位：mm)与相应的热惰性指标比较 **表7**

项　　目	膨胀聚苯板	胶粉聚苯颗粒	聚合物保温砂浆
混凝土多孔砖(240)	7(3.02)	26(3.01)	20(3.23)
KP1烧结多孔砖(240)	2(3.91)	5(3.62)	9(3.86)
钢筋混凝土墙(200)	31(2.83)	51(2.98)	48(3.59)
二排孔混凝土空心砌块(190)	26(2.84)	42(2.90)	32(3.21)
三排孔混凝土空心砌块(190)	25(2.82)	41(2.87)	30(3.15)
蒸压灰砂砖(240)	11(3.48)	19(3.31)	34(4.02)
轻集料混凝土空心砌块(240)	22(2.71)	36(2.72)	28(3.01)

计算条件为：膨胀聚苯板薄抹灰外墙外保温系统中抹面胶浆层厚度为3mm，内粉刷混合砂浆层厚度为20mm，修正系数1.1；胶粉聚苯颗粒外墙外保温系中抗裂防护层厚度为3mm，内粉刷混合砂浆层厚度为20mm，修正系数1.2；聚合物保温砂浆外墙外保温系统中抗裂防护层厚度为8mm，内粉刷混合砂浆层厚度为20mm，修正系数1.2。

由表7可以看出，因为要同时满足传热系数与热惰性指标，导热系数较低的胶粉聚苯颗粒材料在钢筋混凝土墙、二排孔混凝土空心砌块、三排孔混凝土空心砌块和轻集料混凝土空心砌块等墙体上所需使用的厚度甚至要大于具有较高导热系数的聚合物保温砂浆。因此在我们选择和使用保温体系的过程中，不能简单地选择低的传热系数，而必须全面综合考虑满足传热系数与热惰性指标两个要求：聚合物保温砂浆外墙外保温系统适宜于基础墙体为混凝土多孔砖、KP1烧结多孔砖、二排孔混凝土空心砌块、三排孔混凝土空心砌块和轻集料混凝土空心砌块的外墙外保温；胶粉聚苯颗粒外墙外保温系统适宜于混凝土多孔砖、KP1烧结多孔砖和蒸压灰砂砖的外墙外保温；膨胀聚苯板外墙外保温系统对上述所有墙体都比较适用。

为进一步对不同保温系统保温隔热性能开展综合对比分析，表8给出了在采用厚度为20mm的保温层时，各类墙体综合传热系数和热惰性指标的计算数据(括号中为热惰性指标计算数据，计算条件同上)。

不同墙体材料采用20mm厚保温层时墙体综合传热系数［W/(m^2·K)］及热惰性指标 **表8**

项　　目	膨胀聚苯板	胶粉聚苯颗粒	聚合物保温砂浆
混凝土多孔砖(240)	1.03(3.13)	1.27(2.93)	1.49(3.23)
KP1烧结多孔砖(240)	0.942(4.05)	1.14(3.84)	1.32(4.15)
混凝土剪力墙(200)	1.31(2.75)	1.73(2.54)	2.18(2.85)
二排孔混凝土空心砌块(190)	1.13(2.79)	1.42(2.59)	1.72(2.90)
三排孔混凝土空心砌块(190)	1.11(2.78)	1.39(2.57)	1.68(2.88)
蒸压灰砂砖(240)	1.15(3.55)	1.47(3.34)	1.78(3.65)
轻集料混凝土空心砌块(240)	1.03(2.69)	1.28(2.49)	1.51(2.80)

从表8可以看出，在采取相同的20mm保温层厚度情况下，聚合物保温砂浆外墙外保

温系统在混凝土多孔砖和KP1烧结多孔砖基层墙体能达标，胶粉聚苯颗粒外墙外保温系统在KP1烧结多孔砖和蒸压灰砂砖基层墙体能达标，而膨胀聚苯板外墙外保温系统在KP1烧结多孔砖、混凝土多孔砖和蒸压灰砂砖基层墙体能达标。

单纯从传热系数数据上分析，膨胀聚苯板和胶粉聚苯颗粒两种保温材料性能占优。但由于在夏热冬冷地区的夏天制冷所需能耗巨大，因此还必须综合考虑墙体的热惰性指标性能。热惰性指标大小反映了墙体的蓄热能力大小，具有高热惰性指标的墙体可以提高建筑物的居住舒适性，减少夏天空调开机时间。在综合考虑了热惰性指标后，传热系数稍高的聚合物保温砂浆由于热惰性指标较高，大大弥补了其在保温性能上面的欠缺问题。

3 保温系统的安全性对比分析

建筑是百年大计，安全应放在首位。在考虑节能的同时还应重点考虑保温系统的安全性。为有效解决常见的安全性问题，如空鼓、开裂和外饰面为面砖时的面砖脱落问题，传统外墙砂浆粉刷中设计要求中不允许外层粉刷砂浆层的强度高于内层强度。这是因为这样随着外层强度的逐渐提高，当外层强度大于内层强度，内外层界面结合力小于外层自身水平伸展力时，在界面上即会产生脱落，引起空鼓，而空鼓现象则导致外层砂浆层受到不同力场作用，特别是剪切力作用，进而产生开裂。

在建筑节能的各种外墙外保温系统中，外层的抗裂防护层强度必然高于内层的保温层强度，因此非常容易引发空鼓和开裂的安全性问题。为有效解决防护层空鼓、开裂问题，不同保温系统采取了不同的方案。

3.1 膨胀聚苯板薄抹灰外墙外保温系统

膨胀聚苯板薄抹灰外墙外保温系统为解决保护层防护层空鼓、开裂问题主要采取的措施是：

（1）采取薄抹灰系统，薄的防护层有利于减少防护层在强度形成时水平伸展力从而减少防护层与保温层之间的脱落。

（2）在防护层砂浆中加入聚合物，增加其柔性变形能力，提高其抗折力，同时提高防护层砂浆与保温层之间的粘结力。

（3）在防护层中夹耐碱玻璃纤维网格布或热镀锌电焊网，提高其抗折能力。

从材料力学角度，两种介质力学性能相差越大，则在两种介质界面上产生脱落剥离的可能性越大。膨胀聚苯板薄抹灰外墙外保温系统在传统无机脆性基层墙体材料上，粘结上一层柔性有机高分子保温材料，外表面再进行一层刚性防护砂浆层的施工，由于三层材料的强度和柔性相差大，安全性方面存在一定的隐患。尽管对粘结剂砂浆和防护砂浆进行聚合物改性提高了它们的柔性变形能力，但在许多实际工程中开裂、空鼓现象仍难避免，甚至经过几个冷暖期后出现了大面积脱落的现象。

即使对膨胀聚苯板薄抹灰外墙外保温系统采取了加强锚固的措施以后，安全性问题也不能彻底解决：《膨胀聚苯板薄抹灰外墙外保温系统》JG 149—2003中要求单个锚栓抗拉承载力≥0.3kN，根据一般施工工法，每m^2锚栓5～7个，则每m^2提供的承载力为2.1kN左右，与膨胀聚苯板自身0.1MPa的粘结/抗拉强度相比微不足道。

3.2 胶粉聚苯颗粒外墙外保温系统

胶粉聚苯颗粒外墙外保温系统所采取的技术措施与膨胀聚苯板薄抹灰外墙外保温系统相似。虽然胶粉聚苯颗粒的强度较膨胀聚苯板有所提高，但其≥0.25MPa的抗压强度和≥0.1MPa的抗拉强度与防护层砂浆仍相差甚远，安全性隐患难以避免。

国内一些公司甚至提供了的胶粉聚苯颗粒贴面砖外墙外保温系统，其具体做法如下：保温层做好之后，固化5～7天，然后铺热镀锌电焊网，用塑料锚栓固定，然后抹抗裂防护砂浆，等防护砂浆层固化后进行面砖施工。

首先，胶粉聚苯颗粒外墙外保温系统的自身粘结强度仅为≥0.1MPa，而单个塑料锚栓提供拉伸力为≥0.8kN，根据一般施工工法，每 m^2 锚栓数量为5～7个，按7个计算，则提供抗拉力为0.0056MPa，二者之和较《建筑工程饰面砖粘结强度检验标准》JGJ 110中规定的面砖平均拉伸粘结强度应≥0.4MPa的要求相差甚远。

其次，在这些保温系统的面砖拉拔强度测试中，仅切割到钢丝网面的拉拔强度测试结果仅能说明单个面砖脱落的可能性。而整个面砖外饰面系统是建立在非常薄弱之基础上，几个冷暖周期之后，由于热胀冷缩，或者由于地基沉降，或局部施工有问题，都会造成局部开裂、渗水，进而造成局部脱落，而由于保温层自身强度差，脱落面会迅速增加，将会导致大面积脱落，问题更加严重。

3.3 聚合物保温砂浆外墙外保温系统

聚合物保温砂浆外墙外保温系统为有效解决防护层空鼓和开裂问题主要采取的措施有：增加防护层砂浆中有机聚合物的掺入量，提高砂浆附着力，增加砂浆的柔性；在防护层中夹增强耐碱纤维网格布或者热镀锌电焊网，提高其抗裂能力，以防止其空鼓开裂。由于聚合物保温砂浆的强度与普通防护砂浆的强度接近，膨胀系数匹配，所以防护层砂浆易于满足所需的柔性和抗裂性匹配要求。

聚合物保温砂浆外墙外保温系统的整体设计思路兼顾了保温隔热性能和强度指标，其聚合物保温砂浆自身的抗压强度接近普通粉刷砂浆，拉伸粘结强度≥0.4MPa，完全满足《建筑工程饰面砖粘结强度检验标准》JGJ 110规定的面砖平均拉伸粘结强度的要求。其外墙外保温系统可直接粘贴面砖，是当前所有保温系统中粘贴面砖最安全的系统。

4 保温系统的环保消防性对比分析

三个保温系统中，胶粉聚苯颗粒外墙外保温系统和膨胀聚苯板外墙外保温系统采用的都是聚苯乙烯材质的发泡保温材料，在带来良好保温性能的同时，我们也应该认识到在建筑领域广泛、大量推广使用以聚苯乙烯材质为主的保温系统并非上选：聚苯乙烯属于白色垃圾，在自然界不可自然降解，大量使用对环境的影响恶劣；聚苯乙烯在高温下会释放出大量有毒烟气，这对于建筑物的消防安全性是致命的。

因此，目前世界各个国家对于这些采用了聚苯乙烯材质的保温系统的使用范围都是有着严格限制：在美国聚苯板保温系统的使用仅占17%；英国规定18m以上、德国规定20m以上的房屋必须使用不燃的无机质材料作为外墙保温材料[2]。

聚合物保温砂浆外墙外保温系统以无机质表面烧结膨胀珍珠岩为保温隔热的主体材料，因此其具有其他两个保温系统所不具备的环保性和消防安全性。

5 保温系统的成本与经济效益比较

下面以夏热冬冷地区最常见两种墙体类型 KP1 粘土多孔砖和混凝土剪力墙为例，对采用三种不同外墙外保温系统的成本与总体经济效益进行分析比较。

5.1 KP1 粘土多孔砖墙体

胶粉聚苯颗粒外墙外保温系统和聚合物保温砂浆外墙外保温系统以粘贴面砖系统为参考，膨胀聚苯板外墙外保温系统则以涂料饰面为参考，保温层厚度为 20mm，经市场调查，三个保温系统的综合成本分别见表 9、表 10 和表 11。

KP1 粘土多孔砖墙体/膨胀聚苯板薄抹灰外墙外保温系统成本分析　　表 9

序　号	构造层	厚度（mm）	材料名称	单　价（元/m^2）	备　注
1	胶粘剂层	—	胶　粘　剂	28.0	施工成本：～18 元/m^2 综合成本：(约)102.1 元/m^2
2	保温层	20	膨胀聚苯板	15	
3	抹面层	3	抹面胶浆	35.7	
			耐碱玻纤网格布	2.4	
			塑料锚栓	3	

KP1 粘土多孔砖墙体/胶粉聚苯颗粒外墙外保温系统成本分析　　表 10

序　号	构造层	厚度（mm）	材料名称	单　价（元/m^2）	备　注
1	界面剂层	—	界　面　剂	2.0	施工成本：～16 元/m^2 综合成本：(约)68.2 元/m^2
2	保温层	20	胶　粉　料	24	
			聚苯颗粒	10	
3	抗裂防护层	8	抗　裂　剂	5.6	
			热镀锌电焊网	7.6	
			塑料锚栓	3	

KP1 粘土多孔砖墙体/聚合物保温砂浆外墙外保温系统成本分析　　表 11

序　号	构造层	厚度（mm）	材料名称	单　价（元/m^2）	备　注
1	界面剂层	—	界　面　剂	1.3	施工成本：～13 元/m^2 综合成本：(约)49.1 元/m^2
2	保温层	20	聚合物保温砂浆	30	
3	抗裂防护层	8	抗　裂　剂	2.4	
			耐碱玻纤网格布	2.4	

从表8膨胀聚苯板薄抹灰外墙外保温系统、胶粉聚苯颗粒外墙外保温系统、聚合物保温砂浆外墙外保温系统三大系统保温20mm后总体传热系数分别为0.942、1.14、1.32，热惰性指标为4.05、3.84、4.15，单从传热系数膨胀聚苯板薄抹灰外墙外保温系统保温效果最佳，而从价格性能比较聚合物保温砂浆外墙外保温系统仍有优势。建筑节能实际上是一系统工程，需综合考虑各种影响因素，单一追求某一方面节能效果，而其他方面仍然保持原有水平，总体保温效果不一定理想。图1是以夏热冬冷地区其他保温条件不变前提下，利用清华DEST软件计算外墙传热系数对能耗的影响图。

从图1可以看出在其他方面节能措施不变情况下，单单改善墙体保温效果，当 K 值小于1.5 W/(m^2 · K)，K 值的变化对总体节能效果不明显。

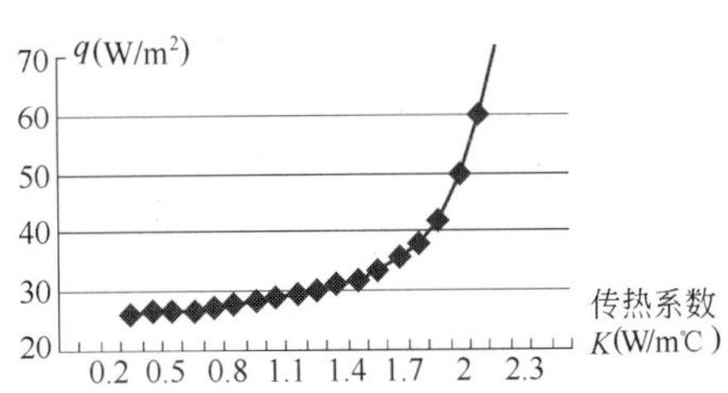

图1　外墙传热系数对能耗的影响

综合考虑了这几个保温系统的其他性能特点和在安全性与环保消防性方面的差异后，我们认为KP1粘土多孔砖墙体的首选是聚合物保温砂浆外墙外保温系统。

5.2　混凝土剪力墙体

由表7可知，膨胀聚苯板，胶粉聚苯颗粒和聚合物保温砂浆三个系统满足夏热冬冷地区节能设计标准的最小保温层厚度分别为31mm、51mm和48mm。为计算方便，将保温层厚度分别设计为35mm、55mm和50mm，以此得到的各个系统综合成本分别见表12、表13和表14。

混凝土剪力墙体/膨胀聚苯板薄抹灰外墙外保温系统成本分析　　表12

序　号	构造层	厚度(mm)	材料名称	单　价(元/m^2)	备　　注
1	胶粘剂层	—	胶　粘　剂	28.0	施工成本：~20元/m^2 综合成本：(约)115.6元/m^2
2	保温层	35	膨胀聚苯板	26.5	
3	抹面层	3	抹面胶浆	35.7	
			耐碱玻纤网格布	2.4	
			塑料锚栓	3	

混凝土剪力墙体/胶粉聚苯颗粒外墙外保温系统成本分析　　表13

序　号	构造层	厚度(mm)	材料名称	单　价(元/m^2)	备　　注
1	界面剂层	—	界　面　剂	2.0	施工成本：~22元/m^2 综合成本：(约)133.7元/m^2
2	保温层	55	胶　粉　料	66	
			聚苯颗粒	27.5	

续表

序 号	构造层	厚度 (mm)	材料名称	单 价 (元/m²)	备 注
3	抗裂防护层	8	抗 裂 剂	5.6	施工成本：～22 元/m² 综合成本：(约)133.7 元/m²
			热镀锌电焊网	7.6	
			塑 料 锚 栓	3	

混凝土剪力墙体/聚合物保温砂浆外墙外保温系统成本分析 **表 14**

序 号	构造层	厚度 (mm)	材料名称	单 价 (元/m²)	备 注
1	界面剂层	—	界 面 剂	1.3	施工成本：～ 18 元/m² 综合成本：(约)99.1 元/m²
2	保 温 层	50	聚合物保温砂浆	75	
3	抗裂防护层	8	抗 裂 剂	2.4	
			耐碱玻纤网格布	2.4	

无论是胶粉聚苯颗粒还是聚合物保温砂浆外墙外保温系统，由于保温粉刷层的厚度达到了 50mm 左右，在施工上难度加大，为保证质量，必须进行多遍粉刷施工，因此造成了施工费用一定的提升。

从安全角度出发，外粉刷的保温层太厚存在安全隐患，因此聚合物保温砂浆外墙外保温系统在混凝土剪力墙保温施工中提出了内外保温结合的施工方法，即在外墙的内、外两个面同时进行保温层的施工，保温层的厚度都控制在 3cm 以内，确保了施工可靠性与安全性。

由表 12、表 13 和表 14 的对比不难看出，在采用混凝土剪力墙体的情况下，膨胀聚苯板外墙外保温系统和聚合物保温砂浆外墙外保温系统的成本造价低，优势显著。在综合考虑安全性、施工性和环保消防性后，聚合物保温砂浆外墙外保温系统仍然是混凝土剪力墙保温施工的比较适当的选择。

6 结束语

当前，各种类型的建筑节能外墙保温产品和系统共存共发展是中国建筑节能工作的一个显著特色，这是源自于各种保温系统的各有所长和市场空间：

膨胀聚苯板薄抹灰外墙外保温系统的保温性能优势非常突出，尽管它在环保性、消防安全性和造价方面存在一定的弱点，但它仍然是目前具有最佳保温节能效果和施工可靠性的外墙外保温系统，尤其是在 65％或 75％的建筑节能标准全面执行以后；

胶粉聚苯颗粒外墙外保温系统保温性能有一定优势，但也存在类似膨胀聚苯板系统一些同样的问题，由于其方便的施工性能和良好的可推广性，使得其在目前的建筑节能市场上也占有重要份额；

聚合物保温砂浆外墙外保温系统保温性能在三个系统中相对差一点，但其在热容性、施工性、安全性、环保消防性和造价等方面的优势十分突出，显著优于上面两个保温系统，适用于面砖饰面的场合。它是根据我国夏热冬冷地区的气候气象特点、建筑节能标准、建筑特色和原材料供应等区域性特征而开发的针对性产品，在这个地区易于推广使用，市场份额最大。

The Comprehensive Contrast and Analysis of the Main External Thermal Insulation System in Exterior Wall in Hot Summer and Cold Winter Zone

Wang Zhiyu[1]　Fan Xianping[1]　Yan Chengwen[2]　Wang Zhiping[3]
Ruan Hua[3]　Wang Xiaoshan[3]
([1] State Key Laboratory of Silicon Materials, Department of Materials Science and Engineering, Zhejiang University, Zhejiang　310027, China
[2] The Faculty of Architectural Engineering and Environment Civil Engineering of Ningbo University, Zhejiang　315000, China [3] Ningbo Rongshan New material Co., Ltd, Zhejiang 315000, China)

Abstract: Presently expansion polyphenyl board thin-plastering, paste power polystyrene granule and polymer thermal insulation mortar exterior wall and exterior insulation system are adopted by Hot summer and cold winter zone of China for economizing energy architecturally. From performance parameter, safety, environmental toxicity in fire control and extent of applicability, this paper has comparatively analyzed in summary. And it shows relative reference opinion that can guide consumers and technologists to exactly choose and design exterior Wall and Exterior Insulation System.

参考文献

[1]　何勇智. 外墙外保温系统几个关键问题的探讨. 保温材料与建筑节能，2005，(5)：48～50
[2]　建设部科技发展促进中心，北京振利高新技术公司. 外墙外保温应用技术. 中国建筑工业出版社，2005

有效解决城市空间板结现象的城市生态措施
——城市建筑空间环境大面积植被化研究

戎 安
（北京建筑工程学院建筑系，北京 100044）
杨 红
（北京工业大学建筑与城市规划学院，北京 100022）

摘 要：运用城市生态工程原理，通过“城市建筑空间环境大面积植被化”理论研究并在此基础上开发相应的应用技术和工艺措施；通过城市规划设计，将开发出来的城市生态技术及措施运用到相应的城市工程项目中，有效地改善城市环境质量，控制城市环境向良性循环方向发展，重新恢复或再造生态良好的城市生存空间。它是生物科学与城市科学结合的产物，它是在城市化进程中的一场“生物科学—城市科学”革命；它是创建城市生态的一种有效途径。

关键词：城市化，城市板结现象，城市建筑空间环境大面积植被化。

城市水系及大气环境恶化的趋势依然严峻，城乡发展与资源、环境的矛盾越来越尖锐(略)。

——国家建设事业“十五”计划纲要序言

1 城市化与城市板结现象

伴随着高速城市化现象，大量的自然环境被城市建筑、道路等硬质人造物所代替。地球上自然的大地景观逐渐消失和淹没在柏油、混凝土和砖石的海洋中(图 1)，使城市变成了钢筋混凝土森林(图 2)。铺天盖地的城市硬化面层不仅导致“城市热岛”现象的形成(图 3*a*，3*b*)，还人为的割断了自然生态循环链。一方面雨水无法返回自然土地，严重破坏了土壤保水能力，致使城市的地表水面萎缩、地下浅层水源干枯；另一方面城市不断增加对地下水资源的开采，导致了地下水位的极度下降。大城市普遍存在水资源不足，日趋严重的用水供需失衡加剧了城市生存空间的潜在危机(表 1)。而城市排水管网负荷与日聚增，城市基础设施不堪重负，城市建设价格不断上涨。城市内部交通拥堵、噪

北京人均水资源量与全国、世界水平对比

资料来源：作者自绘 **表 1**

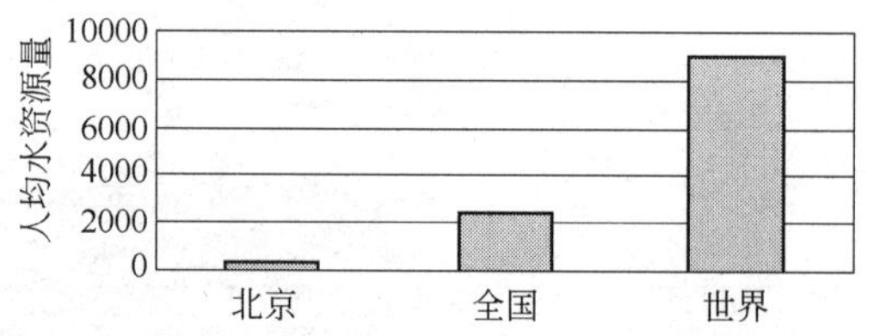

音、粉尘充斥，城市生态循环的自然调节力极度下降。城市机能效率逐渐低下……这一系列现代城市问题被统称为城市板结现象(Phaenomen der Agglomeration)(图 4)。

图 1　红外线航拍片显示：深圳从 1988 到 1996 的城市化过程中几乎丧失了绝大部分自然植被覆盖层，失去了自然环境的大部分的光和能力，也可以说城市生态的氧气库已被消耗殆尽

图 2　对土地资源的过度开发，高密度的城市建设使城市变成了钢筋混凝土的森林

Thermalbilder Atlanta

混凝土化的城市表面

True Color

白天城市热岛效应红外线测试图
白天：气温27℃，城市表面温度48℃

Thermal Day

夜晚城市热岛效应红外线测试图
夜晚：气温10~13℃，城市表面温度24℃

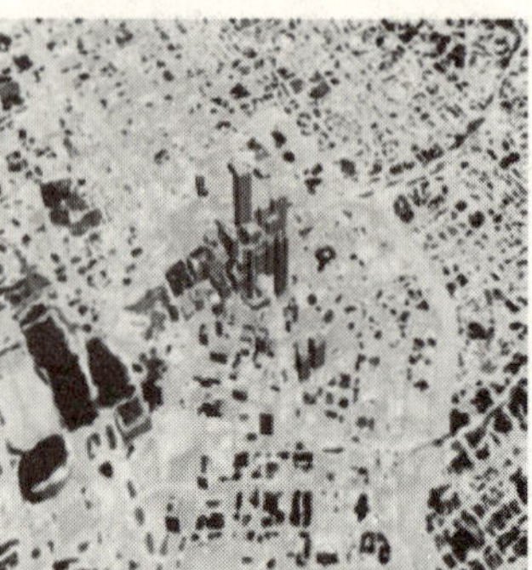

Thermal Night

11-12.05 97 Tageslufttemperatur 27℃,Oberflachentemperatur48℃,Nachtlufttemperaturen10~13℃,Oberflachentemperatur24℃.

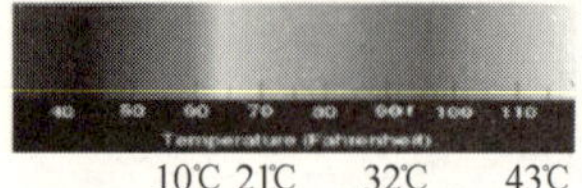

10℃ 21℃ 32℃ 43℃

(*a*)

Quelle:LandSat.NASA:Oberflachentemperaturen der Stadt Atlanta am 21.02.2000

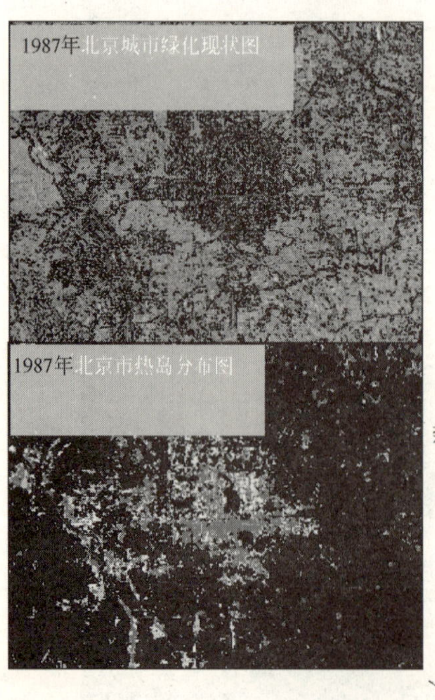

绿地不断减少

热岛效应加剧

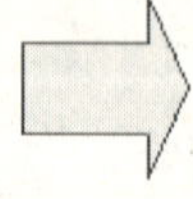

2001年北京城市绿化现状图

2001年北京市热岛分布图

(*b*)

图 3

(*a*)城市热岛现象；(*b*)北京市绿化及热岛分布图

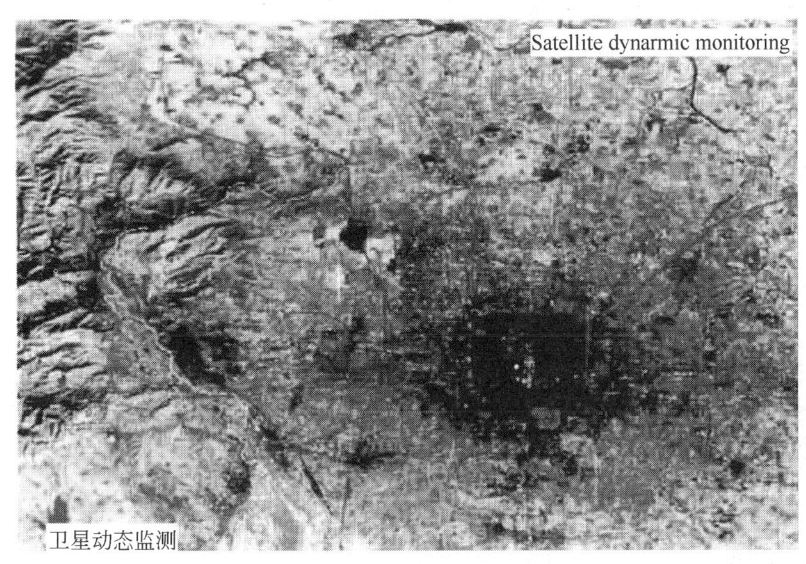

图 4　城市板结现象(Phaenomen der Agglomeration)

2　呼唤有效的城市生态工程措施

面对急速发展的城市化热潮中的城市板结现象，城市空间发展面临着几个亟待解决的环境问题：

◆ 城市空间板结——亟待软化；

◆ 城市环境污染——亟待净化；

◆ 城市景观混乱——亟待美化；

◆ 城市形态破碎——亟待整合；

◆ 城市物种消亡——亟待拯救。

由于我国土地资源有限，人口基数大，近十几年的高速发展，导致城市已建成的固有城市空间密度远远超过世界上同等大城市的密度。城市内部已开发建设的部分，不再可能开发出用作修复城市生态环境、代尝城市超负荷的环境负担和偿还城市“生态债务”的大片城市绿地。那么通过什么手段才能因地制宜和切实可行的恢复城市良性生态功能？如何才能克服“城市板结现象”这个城市可持续发展中的主要矛盾？

通过不断的研究，我们的回答是：“城市建筑空间环境大面积植被化”这一有效的城市生态工程措施堪当此重任！

3　有效的城市生态措施

在城市建筑环境中，如：大量的建筑物、构筑物，道路、桥梁、路轨，步行街道、广场等，运用生态理念和生物科技手段和工程措施对城市建筑环境的三维空间体部的表面层实施大面积植被化，尽可能多的在各种城市建筑环境表面覆盖植被层；例如：在城市建筑物的立面、屋面，城市公路的防噪板墙面、城市轨道交通的路基上、城市高架立交桥桥体表面，城市空间中的维护栅栏、隔断、围墙以及坡道；在所有城市建筑环境中的垂直的、水平的、斜向的多维空间中尽可能多的强化栽培各种植被覆盖层，利用植物生态特性和其特有的对环境的调节功能，来消解由于“城市板结现象”所带来的城市热岛效应，消解城市环境污染、缓和气候反常等一系列影响城市空间可持续发展的“大城市病”，这是世界先进国家采取的主要城市生态对策(图 5～7)。它可以起到：

图 5 植被化屋面

图 6 植被化公路防噪墙

- 软化城市空间；净化城市环境；
- 整合城市形态；协调城市景观；
- 丰富城市物种；美化城市生活的功效。

针对所确定的建筑环境，因地制宜的选用能够适应所在地城市气候的、土生土长的、具有较强洁净环境能力的、最易栽培、易成活、耐寒、耐旱、最少养护需求、四季都发生环境效应的植被物种。利用植被的光合作用，蓄水特性、滤水性能、吸尘能力、消解城市噪音等生态习性与功效，调节城市环境的辐射热、空气温度和湿度，调节对城市季风运动的影响改善城市生态环境和小气候。被植被化的城市建筑空间环境包括城市地面和建筑空间，可以连接城市郊区园野——产生新鲜空气和冷气流的生态库和由高密度硬化建材覆盖的城市中心热效应区，其植被覆盖层与城市绿地、水域空间系统相结合，整合构筑“城市生态廊道”和“城市冷桥”，为城市提供新鲜空气，消减城市热岛和温室效应对城市环境影响，修复城市业已被破坏的城市生态链。

图 7 植被化道轨轨基

4 城市建筑环境大面积植被化的城市生态基理和机能

4.1 生态型植被

瑞典生物学家杜尔松(Turcsson)认为，一种对某一特定生境发生基因型反应的个体群，它们在形态、生态特征和生理特性上均有稳定性并有遗传性是生态型植物。一个植物种分布区愈广，就常具有较多的生态型。生态型植被根据其所以形成的主要影响因子，可分为：气候、土壤、生物生态型等。

4.2 植被的光合作用特性

利用植被光合作用特性(表 2)，能够有效的吸收碳氧排放物、调节城市核心地区的碳化合物气体的浓度，为城市生产制造氧气，有效的控制城市环境温室效应，改善城市气候。

植物叶片光合作用(Photosynthese)❶ **表 2**

植物吸收	通过光合作用	生成植物物质基础	植物排放
$6CO_2$ + $12H_2O$ 二氧化碳 水	Strahlungsenergie → 消耗光辐射能量	$C_6H_{12}O_6$ 碳水化合物	$+6O_2+6H_2O$ 氧 水

4.3 植物光化学特性驱动植物泵作用

植被叶面的向阳面和背阳面有着明显的温差效应(表 3)，说明植被叶面光合作用中可以吸收大量辐射热，植物消耗掉许多热能用以驱动植物泵，植物泵可将大部分集水从根系输送到叶面，再通过叶面蒸发到空气中，在蒸发中带走热量调节空气温度和湿度。

植被叶面的阳面和阴面以及日光下的建筑表面温差测定❷ **表 3**

太阳下不同物体表面的温度比较(Comparison of temperature of different surfaces)			
表面 (Surface)	温度 Temperature(℃)	空气温度 Air temperature(℃)	太阳辐射 Solar Radiation (W/m^2)
树叶阴面(Leaves in Shadow)	30.5	31.2	430
阴影下的混凝土(Concrete in Shadow)	39.5	35.5	420
太阳下的绿色树叶表面(Green leaves in Sun)	35.2	35.5	980
太阳下的紫色树叶表面(Purple leaves in Sun)	37.1	35.5	980
太阳下的白色树叶表面(White leaves in Sun)	33.0	35.6	975
太阳下的枯死的树叶表面(Dying green leaves in Sun)	45.2	35.6	983
太阳下的混凝土地面(Concrete ground in Sun)	55.2	35.5	1010
太阳下的土地面(Soil ground in Sun)	43.4	35.3	1000
太阳下的砖墙面(Brick wall in Sun)	53.3	35.5	1015

❶ Kord Baeumer 著，《Allgemeiner Pflanzenbau》3. Auflage，Verlag Eugen Ulmer Stuttgart 1992 出版，SBN 3-8252-0018-3，P30，笔者翻译加工。

❷ 资料来源：Department of Architecture，The Chinese University of Hong Kong，Liao Zaiyi，Tsou Jinyeu. 笔者翻译并加工。

同时利用覆盖在屋面的植被层将屋面“防热”变为“消热”，将建筑环境外维护结构的单一维护性功能，转变为在维护的同时又具有光合作用的生态效益层。消解热岛现象，改善城市气候环境(表4)。

植被层吸收、反射辐射热和对屋面的影响值测定表 **表4**

随日光变化的植被化屋面的能量分布❶

【单位：光辐射或反射通量(μmol)/(mm·s)】

日进程	太阳总辐射量	非植化被屋面		植被化屋面			灌木群体		
		反射量	反射率	反射量	吸收量	吸收率	反射量	吸收量	吸收率
6	340	38	23.38	20.4	319.6	94.0			
7	615	55	22.33	41.5	573.5	93.2			
8	1473	119	20.21	90.3	1382.7	93.9	131.1	1341.9	91.1
9	2738	202	18.45	160.2	2577.8	94.2			
10	3063	232	18.94	165.1	2897.9	94.6	248.1	2814.9	91.9
11	3295	184	18.08	152.2	3142.8	95.4			
12	3645	258	17.72	180.1	3464.9	95.1	273.4	3371.6	92.5
13	2635	211	20.02	130.6	2530.0	96.0			
14	1665	144	21.56	99.4	1565.6	94.0	148.2	1516.8	91.1
15	1563	125	20.00	70.0	1493.0	95.5			
16	1015	85	20.96	48.4	966.6	95.2	75.1	939.9	92.6
17	453	38.6	21.28	21.7	431.3	95.2			
18	355	30	21.20	16.0	339.0	95.5			
平均值	1758.0	132.4	20.32	91.3	1666.7	94.8	144.2	1613.8	91.8

4.4 植被蓄水特性

落在植被化屋面上的雨水，50%～90%将被植被的根系吸收储存(表5)，剩余的小部分，通过檐口和落水管排出。这部分雨水一些可以直接返还自然地面补给地下水，另一些可以排给社区水面形成城市生态循环链。同时这个措施将大大缓解城市排水管网不足的压力，节约城市市政管网的建设投资。

屋顶植被化后1986～1988年间雨水贮存和雨水排放量统计表

(德国汉诺沃大学测定)❷ **表5**

基层厚度(cm)	植被物种形式	雨水贮蓄量/减少向市政管网排放的雨水量		
		春、秋(冷季节)	夏(热季节)	总时间阶段中
2	苔藓类植被	40～50(%)	55～65(%)	48～58(%)
4～6	苔藓类、草本、草类植被	45～50(%)	60～70(%)	52～62(%)
8～10	草本、草类植被	45～50(%)	65～75(%)	55～65(%)
14～16	草本、草类植被	50～60(%)	80～90(%)	65～75(%)

❶ 资料来源：广州市果树科学研究所，510406广州。刘恰．城市屋面绿化植被的能流反应。

❷ Jens Drefahl. Dach Begruenung 1995，P17 Tabelle5.1。笔者翻译并加工。

4.5 植被光热物理化学过程分析❶

由于植物群落和植株结构上的差异性和非均匀性和叶片结构、环境的不同有着显著的差异，叶片对光谱的吸收，反射和透射能力也随机改变，所以植被对太阳辐射能量的传递至今尚无令人满意推倒方法，下面就假设认为植被与周围空气进行的热、湿交换，主要由植物冠层显热和潜热二部分组成，而忽略光合作用等生化方式的能量传递[1]，给出植物冠层受太阳辐射最为简单的计算方法：

$$I_s = I_{pk} + I_{pd} \tag{1}$$

式中 I_s——太阳辐射强度，W/m²

I_{pk}——植被层的太阳净辐射能，即植被冠层显热和潜热之和，可表示为：

$$I_{pk} = J_c + LE_c \tag{2}$$

$$J_c = PC_p(t_c - t_a)/r_{ac} \tag{3}$$

$$LE = PC_p \cdot [P_b(t_c) - P_a]/\gamma(r_{ac} + r_s) \tag{4}$$

式中 t_c——植物冠层温度,℃；

t_a——室外空气温度,℃；

$P_b(t_c \cdot)$——冠层温度下的饱和水蒸气分压；

P_a——室外空气的水蒸气分压力。

室外空气的水蒸气分压力可表示为室外空气饱和水蒸气分压力与空气相对湿度的乘积，即：

$$P_a = \phi P_b(t_a)$$

r_{ac}——冠层的表面空气动力学阻力，s/m；

r_s——冠层的气孔阻力，s/m；

γ——热力学湿度常数，$\lambda = 67.5\text{Pa}/℃$。

冠层叶面与空气的换热阻力 r_{ac} 和 r_s 与植物的生态特性有关，通过测试确定。对于草坪，r_{ac} 在静风时，可取 $r_{ac} = 1.82\text{s/cm}$，但有风条件下，r_{ac} 随风速的平方根而减少，即 $r_{ac} = 1.82\sqrt{v}$。r_s 可取 0.19s/cm。

在一般温度范围内，饱和水蒸气分压可表示为温度 t 的线性关系[2]：

$$P_b(t) = 270t + 8350 \tag{5}$$

I_{pd}——穿透植被的太阳辐射强度，W/m²，它主要包括直接太阳辐射强度 IDN 和散射太阳辐射强度 I'_p，由公式(6)计算：

$$I_{pd} = \text{IDnexp}(-K_1 LAL) + \text{Ipexp}(-K_2 LAL) \tag{6}$$

式中 LAL——叶面积指数；表示单位截面积和高度 hL 的垂直圆柱体内的叶片(单面)面积的总和，若 $LAL=1\sim2$ 为疏稀冠层，$LAL=3\sim4$ 为中密度冠层，$LAL=5\sim6$ 为稠密度冠层。

K_1——为直接辐射消光系数，取 $K_1 = 0.5(1-\sigma)0.5/\sin\beta$；

K_2——为散射辐射消光系数，取 $K_2 = 0.8(1-\sigma)0.5$；

式中 σ——叶片的散射系数；

---

❶ 国家自然科学基金资助项目：南方［被动却除湿低能耗建筑研究(59878054)，引自杨红，冯雅. 植被热反应研究与测试。

β——太阳高度角。

5 植被化的环境功能

5.1 环境美学效果

植被层为城市建筑穿上了一层可调节城市气候的、可呼吸的绿色外装。这件外套成了建筑物的附加保温层，通过这个措施可以改善建筑保温隔热性能，同时减少城市能源消耗和因能耗所联生的环境污染问题。大面积植被覆盖层，可以为城市建筑穿上绿色的外衣，软化目前城市建筑“水泥化”所造成的城市形象的僵硬感，随着四季的变化，大面积的植被饰面，可以在大格调和谐统一的前提下，大大美化城市，城市的面貌将丰富多彩，城市空间将充满诗情画意(图 8～10)。

图 8 法兰克福生态村植被化屋面和墙面

图 9 建筑表面立体植被化细部近景

5.2 改善城市环境空气质量

大气中的 CO_2 的平均浓度为 320ppm，而在城市中由于人口密集和尾气排放等原因，CO_2 排放浓度可达 500～700ppm，局部地方尚高于此数。从卫生角度而言，当 CO_2 浓度达 500ppm 时，人的呼吸就会感到不舒适，如果达到 2000～6000ppm 时就会有明显受害症状，通常是头疼、耳鸣、血压增高、呕吐、脉搏过缓，而浓度达 10%以上则会造成死亡。

图 10 德国、法兰克福商业银行高层办公楼屋面植被化景观

植物是环境中 CO_2 和 O_2 的调节器，在光合作用中，植被吸收 CO_2 放出 O_2。每吸收 44g CO_2 可放出 32g O_2。虽然植物也进行呼吸作用，但在日间由光合作用所放出的 O_2 要比由呼吸作用所消耗的 O_2 量大 20 倍。一个体重 75kg 的成年人，每天呼吸需 O_2 量为 0.75kg，排出 CO_2 量为 0.9kg。通常每公顷森林每天可消耗 1000kg CO_2，放出

730kg O_2。所以每人若有 10m² 的树林，既可满足呼吸氧气的需要。生长良好的草坪，每 1m² 每小时可吸收 CO_2 1.5g，即约合 1ha 吸收 15kg，而每人每小时呼出 37.5g CO_2，所以每人有 50m² 草坪可以满足呼吸的平衡。根据 1966 年德国在柏林中心大公园所作实验的结果，得知每个居民需要绿地面积 30～40m²，才能满足呼吸的需要。

植物分泌杀菌素。城镇中闹市区空气里的细菌数比公园绿地中多 7 倍以上。公园绿地中细菌少的原圃之一是由于很多植物能分泌杀菌素。据计算 1ha 圆柏社于 24 小时内，能分泌出 30kg 杀菌素。

建筑空间环境表面的植被叶面可以吸附大气中 10%～20%的粉尘污染，同时还可以部分吸收空气中和雨水中所含的硝酸盐或其他有害物质，将被吸附和凝结的污染物作为自身营养利用和吸收。利用植被叶面吸附、凝结能力，消解部分城市大气中的有害物质、吸附城市粉尘，降解城市大气污染。

5.3 改善城市热环境

在强烈的光辐射热和严寒或酷暑气候条件下，对建筑物表面造成戏剧性的巨大温差变化，在特殊自然现象如冰雹的袭击下会严重的损害建筑外维护面特别是屋面，这些都会加速建筑材料特别是建筑防水层的老化，影响建筑质量和寿命。当它们置于植被化覆盖层保护下时，建筑材料的寿命将会大大增加，因此也可以大大减少一般性建筑维修费用。

5.4 降低噪音对城市生活的干扰

植被叶面分布的多方向性和它的发散性特点可以消散噪音 3dB；植被软质覆盖面与建筑外表面之间形成的夹层腔体效应可消耗 8dB 城市噪音；尤其在地处航道空中走廊和舞厅等娱乐设施处，强烈噪声源旁的建筑都有防噪、降噪要求，利用植被吸音效能，吸收部分城市噪音，可以有效降低噪音对城市生活的干扰，充分体现出其防噪消噪优越性。

5.5 再塑城市生态链、创造小生态环境

利用植物的物种群落关系，打造人工小生态环境，为城市小生物提供了可生存空间，昆虫和其他小生物、小动物，都可以在其所提供的的基床上建立自己的家园。“城市建筑空间环境大面积植被化”为他们的生存提供了空间，而他们又为飞禽鸟类提供了食物来源，弥合了生态圈内已断裂的食物链（图 11～14）。

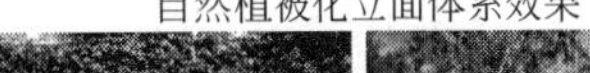

图 11 德国柏林市区住宅植被化环境

5.6 一种城市消防措施

植被叶面对飞星火种或强烈的辐射热所引起的火灾现象有一定的防护作用。故德国建筑规范(DIN 4102 第 7)将其也作为一种城市消防措施。

5.7 为城市建筑屋面的再次开发和利用提供条件

城市建筑屋面将被作为城市立体空间的休闲场所被继续开发成：空中苗圃、花园、菜园和屋顶植物园、屋顶咖啡、酒吧、游乐场等各种利用方式。为市民提供了多样化立体空间活动场地。

图 12 德国法兰克福郊区住宅区立体绿化状况

图 13 弗莱堡试验居住区住宅屋顶花园

6 城市建筑空间环境大面积植被化的规划设计原则

城市规划设计是指导城市建设行为，控制城市和谐有序发展的关键。在通常规划设计中：人均公共绿地面积、绿地率、绿化覆盖率等指标，对于宏观把握一个城市或地区的绿地规模具有积极意义，但是用它们控制城市生态效益进行规划设计，往往不能真实反映出生物量、生态效率以及生态功能等环境大面积植被化后所产生的实际生态效率。它们无法有效的评价、管理与控制城市生态效果，更不利于加强对生态环境修复所需工程措施的管理运行。针对指标系统的缺陷，城市建筑空间环境大面积植被化提出一套具体城市生态效益内涵和重要生态功能的规划设计、建设和管理措施和指标体系，它由三部分组成：

图 14 法兰克福居住区屋顶植物小群落

一、仍沿用原城市规划设计中所使用的绿地评价指标——绿地率及绿化覆盖率等；

二、绿量、绿量率是衡量城市绿地本身的生态效益及其绿化水平的指标；

三、植被容积率是将城市绿地建设与城市建筑空间环境大面积植被化建设结合起来的生态建设指数。

6.1 城市建筑空间环境植被容积率指标

借用城市规划设计中“建筑容积率”的概念，量化描述城市建筑环境大面积植被化度，我们引入了植被容积率的概念，它是城市建筑表面积上的植被化覆盖面积总量与总建筑用地面积的比值；它可将城市建筑空间环境大面积植被化量化并与城市生态的规划设计和建设、管理联系到一起，量化反映城市建筑环境中的生态效率总量。通常城市总规、控规、修规和相关城市设计中，针对绿地系统规划设计规定了绿地率指标(二维系统)。而从城市生态角度出发，建筑环境大面积植被化提出植被容积率概念，意在指导和控制规划设计中城市立体空间单位面积上植被总量及植被覆盖率的多少(三维系统)。其目的在于提高对城市建筑环境大面积植被化系统的控制和引导，规范城市建筑环境大面积植被化的建设责任与义务，提高城市生态环境的品质与效率。

6.2 建筑空间环境植被化的植被生态效率系数

建筑环境植被化的植被生态效率源于建筑屋顶植被化及建筑垂直面植被化的植被生态效率，由于其生态效益不如自然植被环境效益也不如人造的地面植被环境效益，参照国际惯例，其植被生态效率量计算应乘以修正系数，笔者建议其系数可用表 6 参考值。

表 6

建筑空间环境植被面积生态效益量调整系数(B-V)

植被化城市生态环境	代　　码	修正系数
自然植被覆盖面积的生态环境效益系数	N-V(Natural Vegetarieren)	1
人工植被化面积生态环境效益系数	A-V(Artificial Vegetarieren)	0.8
建筑屋顶面积植被化生态环境效益系数	B-V-H (Bauding Vegetarieren horizontal)	0.4
建筑垂直面积植被化生态环境效益系数	B-V-V(Bauding Vegetarieren vertical)	0.2

6.3 植被覆盖率(Vegetarieren Efficiency)

植被覆盖率是一个衡量建筑物表面植被化程度和植被生态系统的生态效益的指标，它可反映某建筑物单位面积上植被覆盖率的高低。它对建筑环境植被化的城市生态效率起关键性作用。

6.4 植被化率

建筑空间环境植被化率是城市建筑空间环境大面积植被化中的植被容积率与建筑容积率的比值，它将建筑环境植被化系统与建筑环境系统结合起来考虑。

6.5 绿量率(Green Efficiency)

同时相同面积上选用植物的绿量率也对城市生态效率有影响，相同面积上绿量率高的植被对环境影响效率也高。植物群落乔、灌、草合理搭配，物种的合理选择对总绿量的高低往往十分重要。

实施建筑空间环境大面积植被化系统的技术路线将以植被容积率为核心，建筑环境植被化率为评价标准，绿量率为技术保证，通过城市设计将城市建筑空间环境植被化系统落实到城市空间建设中。在操作中其控制体系可建立在 GIS 数字信息控制技术平台上，以 GS 系统支持和辅助城市建筑环境植被化系统的建构、实施、管理和评价。在城市系统规划设计和建设中应遵循以下原则：

6.5.1 保护性原则

最大限度的保护现有的自然生态植被存量，在规划设计中只要切实可行，就应该适当调整城市建筑空间布局，以保持和保护固有的原生态景观。

6.5.2 多样性原则

人工植物配置需考虑群落的多样性生态效应，组织群落系统。人工植被环境应由乔木、灌木、常绿与落叶、速生与慢生等相结合；乔灌木与地被、草皮相结合；适当点缀草花，多品种合理杂交，构成多层次的复合化、立体化植被结构层，以发挥各种植物的综合生态效应。

6.5.3 连续性原则

除了孤植等特殊造景需要，一般来讲，为了加强植被群落之间的联系，通过营造植物廊道，有助于物种的空间迁移。

7 结语

城市建筑环境大面积植被化可以改善城市中人类社会与自然、经济与自然、生物与生物以及生物与非生物环境之间的多元关系。它可以烘托城市文化氛围，创造和谐的生活氛围。它不但具有重要的生态价值，而且具有重要的社会和文化价值。它能够有效利用城市资源，保护环境，净化环境，亲和自然，美化环境，为城市生命种群生存提供良好的生态环境，为人们创造一种舒适、健康、安全、美好的城市生活空间。当然，面对城市化现象中如此错综复杂和多变的城市环境问题，不是仅靠一项举措就能够彻底扭转局面。城市问题的最终解决还要依靠人类理性的反思和发展科学技术来对城市环境综合治理，恢复和再生良性循环的城市生态。但无论如何 ，“城市建筑空间环境大面积植被化”提供了一个城市生态的可实施性方面重要的例证，它证明人们能够运用生态学观念和城市生态工程原理，通过城市设计，并运用在上述理论基础上，开发出来的相应技术措施，有效的控制和改善城市发展，重新恢复良好的城市环境质量，再造我们理想的城市生存空间。它是生物科学与建筑科学结合的产物，是在城市化进程中的一场“生物学一建筑学”革命，是创建城市生态的一种有效途径(图 15)。

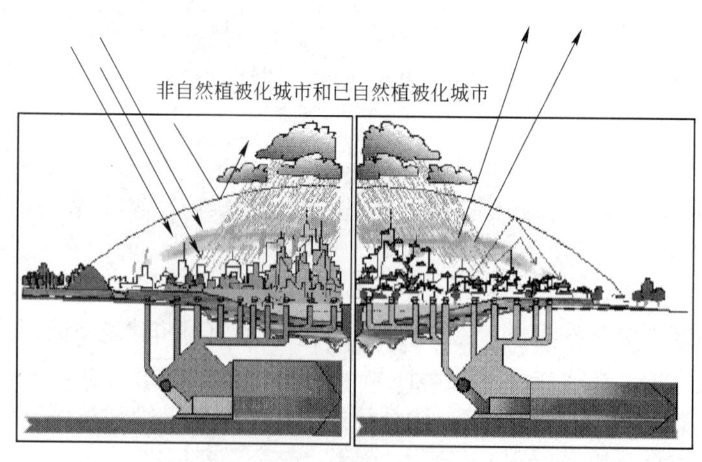

图 15 城市生态环境比较

本文在形成过程中得到德国友人、德国城市生态专家鲁道夫博士教授(Dr. Prof. Rodulf)的指点、帮助和支持，包括他的柏林城乡生态工程有限公司(GASP-Berlin)所组织的欧洲城市生态专业考察和他所提供的技术资料，特在此表示衷心感谢。

Large Area Vegetarieren of City-Architectural

An Rong
(Beijing Institute of Civil Engineering and Architecture, Beijing 100044 China)
Hong Yang
(Beijing University of Technology, Beijing 100022 China)

Abstract: "Large area vegetatieren of building" provides us an important illustration for the practicable of City-Ecology. It proves that people can use the concept of Ecology and the engineering principle of City Ecology, and can develop appropriate measures by considering of apply principle, by means of urban design, to control and improve the development of the city efficiently, to recover the good quantity of city environments, and to rebuild out ideal living space in the city. It is the product of the integration of Ecology and Architecture; it is an Ecology-Architecture revolution during the process of urbanlization; it is also an efficient way to create the City-Ecology.

参考文献

[1] Jens Drefahl. Dachbegruenung. M Rudolf Mueller, 1995. ISBN 3-481-00745-0

[2] H. Sukopp, R. Wittig. Stadtoekologie. Gustav Fischer, 1993. ISBN 3-437-26000-6

[3] James Wine. Gruene Architektur. Taschen Lektorat: Koeln London Madrid New York Paris Tokyo 1992～98. ISBN 3-8228-0811-3

[4] Kord Baeumer. Allgemeiner Pflanzenbau. 3. Auflage, Verlag Eugen Ulmer Stuttgart 1992, SBN 3-8252-0018-3

[5] 戎安. 德国城市建筑环境大面积植被化. 世界建筑, 150 期, 世界建筑杂志社, 2002

[6] 杨红, 冯雅, 刊名: 中国学术期刊文摘, 5/2001, P647 植被对城市环境的热反应. 中国学术期刊文摘, 中国科学技术协会学会部, 国家自然科学基金委计划局. 2001

[7] J. L. Monteith. Vegetation and the Atmosphere. Vol. I, Principles [M], London, Academic Press, 1975

[8] J. K. NAYAK. et, The relative performance of different approaches of the Passive cooling of roofs [J], Building and Enviroment, Vol. 17 NO2. PP145～161, 1982

[9] J. Goudiaam, 邹华武译. 作物微气象学: 模拟研究 [M]. 北京: 科学出版社, 1985

生态安全视野下的绿色建材

曹　伟
（厦门大学建筑学院，厦门　361005）

摘　要： 生态安全视野下的绿色建筑，首先必须考虑建筑材料的“4R”原则、本土观念以及生态建材所具有的系统工程特性。通过对国外生态建筑以及使用典型生态建材的木建筑的实例分析，研究生态建材的发展战略，试图探索一条适于我国的生态建材与生态建筑之道。

关键词： 生态建材　生态建筑　木材　纸屋　节能　“4R”原则　发展战略

1　生态建材

1.1　生态建材的特征

从城市生态安全的高度看，生态建材又称绿色建材、环保建材和健康建材等。它采用清洁生产技术、少用天然资源和能源、大量使用工业或城市固态废物生产的无毒害、无污染、无放射性、有利于环境保护和人体健康的建筑材料。它在材料的生产、使用、废弃和再生循环过程中以与生态环境相协调，满足最少资源和能源消耗，最小或无环境污染，最佳使用性能，最高循环再利用率要求设计生产的建筑材料。显然这样的环境协调性是一个相对和发展的概念。

生态建材与传统的建材相比，在有利于城市生态安全方面可归纳出 10 方面的基本特征：①其生产所用原料尽可能少用天然资源、大量使用尾渣、垃圾、废液等废弃物；②采用低能耗制造工艺和无污染环境的生产技术；③在产品配制或生产过程中，不得使用甲醛、卤化物溶剂或芳香族碳氢化合物；④产品的设计是以提高生活质量、改善生产环境、有利于生态安全为宗旨，即产品不仅不损害人体健康，而应有益于人体健康；⑤产品可循环或回收利用，无污染环境的废弃物；以减轻垃圾填埋的压力，节省自然资源；⑥避免使用能够产生温室效应、破坏臭氧层的化学物质的机构设备和绝缘材料；⑦购买本地生产的建筑材料；运输不仅需要消耗能量，同时会产生污染，因此应尽量购买当地生产的材料。体现建筑的乡土观念；⑧避免使用会释放污染物的材料；以免对身体健康及生态安全造成危害；⑨最大限度地减少加压处理木材的使用；⑩将包装废料减到最少；避免过分的包装。

1.2　生态建材注重全生命周期的安全性

生态建材与其他新型建材在概念上的主要不同在于生态建材是一个系统工程的概念，不能只看生产或使用过程中的某一环节。对材料环境协调性的评价取决于所考察的区间或所设定的边界。目前，国内外正在出现各种各样称之为生态建材的新型建筑材料，如利用废料或城市垃圾生产的“生态水泥”等。但如果没有系统工程的观点，设计生产的建筑材料有可能在一个方面反映出“绿色”面在其他方面则是“黑色”，评价时难免失之偏颇甚至误导。例如，高性能的陶瓷材料可能废弃后难以分解，建筑高分子材料常常难于降解，复合建筑材料因组成复杂也给再生利用带来难度；黏土陶粒混凝土砌块轻质、高强、热绝缘性和防火性能好，但其生产需要较高的能耗；塑钢门窗较钢窗和铝合金窗坚固、耐久，热绝缘性能更好，但它包含高的能源成本和废弃处理时将对环境产生严重的负担。为全面评价建筑材料的环境协调性能，需要采用生命周期评价（Life Cycle Assessment，简称LCA）。LCA方法是对材料整个生命周期中的环境污染、能源和资源消耗与资源影响大小的一种方法。对建筑材料而言，它还是一个正在研究和发展中的方法。

例如，一批用绿色建材——稻草为主要原材料的新型住房在上海市郊浦东北蔡建成。这种被称作“稻草别墅”（如图1、2）的住房外型美观、造价低、施工便捷，具有重量轻、体积小、保温隔音等优良性能。这类建筑的寿命周期内对环境可以说没有公害，因此，有益于城市生态安全的维护。

图1　外墙用无机玻璃钢装饰，内墙是稻草板

图2　美观、有利环保的稻草别墅

2　木构建筑材料的生态安全的特性

在所有的建筑材料中，木材的获取（包括制造、运输和供应）需要的能量最小，并且对环境所带来的负荷最小。进一步讲，木材还有很好的绝热特性，这也是它被当作低能耗房屋理想材料的原因。越是自然未经处理的木材，它的可循环利用的能力越强。建筑师、工程师以及森林工业界已经认识到：面对与日俱增的环境危机，有必要尝试恢复木构建筑，木屋属于具有高技术特性的低技术建筑。

今天使用木材，不仅是建筑师选择生态途径的重要方式，而且被认为是城市可持续发展的重要内容，当然只有森林管理得当，才能确保木材作为一个可持续的资源。然而，在我国使用木材作为主要建材，可能会引起不小的轰动效应。这一方面是因为我国的森林资源想对匮乏；另一方面，大量使用木材显然意味着森林的减少，甚至有人也认为使用木材作为建材，会导致生态环境的破坏，但它是可以再生的。而现实是，我国城市人均森林资源太少，这是无法实施这项有利于城市生态安全的策略。

2.1 结构性能——安全、节能、方便

木结构房屋具有很好的安全性。首先，木结构对于瞬间冲击和周期性疲劳破坏具有良好的延展性，即使有地震发生，木构架建筑仍然可以保持结构的稳定和完整；其次，在节能方面，木构架建筑具有良好的隔热保温性能。有研究表明：若达到同样的保温效果，木材需要的厚度是混凝土的1/15，是钢材的1/400。在使用同样的保温材料时，木结构比钢结构的保温性能好15%～70%。这意味着不仅建筑能耗的大大降低，而且居住环境质量将得到提高。再次，它具有施工周期短、布局与造型灵活以及维修和翻修方便的优势。

2.2 环境效益——低污染、可循环、环境友善

众所周知，建筑的一般寿命平均50年左右，那些被拆除的建筑物材料的处理，一直是困扰城市生态环境的一大问题。从表1主要建材生命周期对环境影响的比较中，我们可看出木材具有其他材料无可比拟的生态安全优势。建筑材料是影响建筑生命周期的重要因素，但建筑的设计、施工技术、使用与维护等因素的影响亦不可小视。

主要建材生命周期对环境影响比较 **表1**

材料分类	对水污染	能源消耗	温室效应	空气污染指数	固体废弃物
木　材	1	1	1	1	1
钢　材	120	1.9	1.47	1.44	1.37
水　泥	0.9	1.5	1.88	1.69	1.95

2.3 木构建筑的更新

如上所述，木构建筑对环境友善。因此，对它进行翻新以继续利用要比推翻拆除更符合可持续发展的思想。旧建筑的更新在欧洲比较普遍，这正体现了“4R”(Re-new，Recycle，Reuse，Reduce)原则在一些发达国家的重视。如德国的G·康哈蒂教授在外部结构不变的前提下，将一幢有200年历史的建筑更新为低能耗建筑，它的做法是在外墙填充轻质粘土砖保温材料，它由木屑和粘土混合后加压30分钟再自然干燥而成。用这种砖砌成50cm厚的外墙，其传热系数为0.27W/m·K，可达到较理想的节能效果。

2.4 生态特性

以美式木结构房屋为例，它几乎可以适应任何气候条件，通常省去了砖石和混凝土建筑所必需的大型建筑工程队。就房屋质量、能源效益、建筑简易性和建筑速度而言，美式木结构房屋是目前成本效益最好的。房屋的能源效益是由构成该房屋“外壳”的所有材料的保温选择性所决定的。房屋结构中除了使用木材外，美式木结构建筑还采用玻璃纤维或泡沫作为保温材料。同时可轻松达到砖混结构房屋所具有的隔音性能。美式木结构房屋在地震中一般极少受损失，这表明美式木屋不仅是生态的，而且也是安全的。

3 国外生态建材典型实例

3.1 生态超级纸屋

2000年汉诺威“世博会”的绝大部分博览馆都把木材当作建筑的主要材料，而日本馆则创造了建筑史上前所未有的最大规模的“纸造建筑”，当然它的室内与环保主题相关且减少二氧化碳排放量等内容展示不尽理想，但是将再生之后制造的纸管作为建筑材料使

用，而创造建筑的奇迹，还是受到所有参观者的高度赞赏。

该馆由青年纸建筑家西格卢·班(Shigeru Ban)设计，他从材料和结构的特性出发，来切合世博会的主题，关注资源和环保问题。同时注意到建筑的临时性，会后其大部分材料可以回收利用。它采用回收加工的纸建成拱形的结构，由 12.5cm 粗的纸管网状交叉而成，弧形屋面和墙身材料也是织物和纸膜。整个馆长 72m，宽 32m，最高处达 15.5m，面积 $3600m^2$(见图 3)。整个占地面积约为 $3600m^2$，全部使用再生纸管和纸板建造，建筑的框架全部是由在德国工厂特制的纸管构成，顶使用了纤维及防水纸质膜材料。由于担心安全问题，最后还是在某些地方，被迫接受德国当局的要求，加上一些金属支柱，这与西格卢当初的理想发生了冲突，既他想建造最大的、百分之百的纸建筑。白天，自然光经过半透明纸窗的过滤构成柔和、宜人的室内光环境；夜晚，纸窗又是神奇光影的“屏幕”。由于注重技术创造与革新，以及与自然高度融合的智慧，因而这座超级纸屋在博览会上获奖。

日本馆反映了日本人普遍具有的生态与环境意识，这种纸品构筑物的意义不仅仅在于环保。还因为它建造周期短，施工便利，为解决人类居住问题提供了一条捷径。

3.2 “人·自然·技术”

图 4 是由瑞士建筑家卒姆托(Peter Zumthor)设计，在所有与众不同的博览馆中，瑞士馆尤其不同。它将锯好的木材，以最简单的方式累积成架空木材壁，高度为 9m，做成纵横交叉面积为 $3000m^2$ 的迷宫。这个迷宫中有通道和中庭，在这个迷宫的木结构空隙中，萦绕着由瑞士传统乐器演奏的音乐，加之透进来的阳光，参观者的话语和脚步声，把人们从缭乱的视觉世界拉回到纯粹音色的世界。

图 3　日本馆内景(资料来源：设计选例．超级纸屋—日本馆．世界建筑，2002(11)：28)

图 4　瑞士馆内景(资料来源：设计选例．迷宫—瑞士展馆．世界建筑，2000(11)：32)

在木材夹壁的顶部，有长长伸出来的凹型金属顶，然而那是技术的抽象表现。卒姆托幽默地说："展期为153天，正好是从瑞士刚刚采伐的落叶松需要干燥的时间"。建筑构成方式为组装和拆借提供了最佳形式，在固定木材时没有使用固定角铁之类的建筑部件，所以博览会结束之后，那些木材可以轻易地重新利用。显然，卒姆托想通过这个建筑装置，体现对历史价值及自然生态的回归。我们可以体会到，其作品的力量正好是"人·自然·技术"这三者的合力，而采用这种策略的技术在这里确实实现了生态安全的效果。

3.3 以人为本的"绿色办公室"

最近美国国立资源保护委员会(NRDC)的总部所在地利用废旧回收物品的再生材料建造了一个"绿色办公室"。NRDC动工兴建的这个"绿色办公室"，以利于节能和保护环境及对人体不造成危害为原则。办公室的墙壁是由麦秆压成的，地板是由废玻璃制成的，办公桌是由废旧报纸和黄豆渣制成的。另外，它还设有很大的窗户，这样，办公室内非常明亮，从而可节约电力30%。

绿色办公室比通常建造的办公室可节省开支10%以上。NRDC官员指出，绿色办公室属节能性建筑，它不需投入太多的资金就可以建成一个明亮、漂亮而且有利于健康的办公室，它没有那些来自普通办公室的各种有害物质的污染。

3.4 健康建材

丹麦、芬兰、冰岛、挪威、瑞典等国于1989年实施了统一的北欧环境标志。丹麦为了促进绿色建材的发展，推出了"健康建材"（HMB)标准，规定所出售的建材产品在使用说明书上除了标出产品质量标准外，还必须标出健康指标。丹麦是实施健康住宅工程较早的国家，早在1984年底，就在阿尔胡斯(Arhus)市建成了"非过敏住宅建筑"示范工程。

瑞典已正式实施新的建筑法规，规定用于室内的建筑材料必须实行安全标签制。出口厂家已自觉在产品说明书上标出产品在4周和26周时，室内有机化合物空气浓度指导限值。瑞典最大的住宅银行于1995年宣布只向生态建筑开发商贷款。

英国是研究开发绿色建材较早的欧洲国家之一。早在1991年英国建筑研究院(BRE)曾对建筑材料及家具等室内用品对室内空气质量产生的有害影响进行了研究。

3.5 生态建材塑造生态美

图5是一个巨大的、舒展的屋顶结构，造型有力而优雅，建于汉诺威世界博览会的中心位置，被称为汉诺威世博会大屋顶。它是一个可容纳大量观众的开敞的户外空间，用于音乐演出、各种表演、艺术展示及其他活动，这个极富创意的木结构建筑由10个边长

图5 天人和一：汉诺威世博会大屋顶

40m、高度超过10m的大木伞组成，如同巨大的雕塑构成了一个水路网络和人工岛屿之上的复杂建筑组群的一部分。的确它像一把大雨伞在天气变化时为观众提供庇护。

通过将现代制造流程与传统工艺技术相结合，这个大尺度的建筑物由木质材料建成，采用一种透明的合成薄膜来抵抗气候的侵扰。因为它独特地使用了生态材料，木材作为原材料进行建造，在其有限的用色与变化的形式上，它是现代建筑设计在木结构建筑与面令人印象深刻的一个杰作。在美学与结构上达到了一种平衡，在材料的运用上可以说它达到了一种生态美的境界，在几何学上它简直是对称美的具体展现，在与环境的结合上它令人难以分出室内与室外，使得人与自然完美和谐、协调共生，如图6。

图6　鲲鹏展翅：汉诺威世博会大屋顶

4　我国生态建材的前景及其环境协调性分析

关于生态建材的发展方式和对环境协调性的改进，应进行建筑材料对环境协调性贡献大小的评价，并在产品改进，重新设计，功能创新和系统创新上充分考虑那些有利于城市生态安全的设计与技术策略。

事实上，环境协调性与使用性能之间并不总是能协调发展相互促进，这就要求生态建材的发展不能以过分牺牲使用性能为代价。但生态建材料使用性能的要求不一定都要高性能，而是指满足使用要求的优异性能或最佳使用性能。性能低的建筑材料势必影响耐久性和使用功能，如采用LCA方法评价，在生产环节中为节能利废而牺牲性能并不一定能提高材料的环境协调性。

与周围生态环境的协调包括：雨水利用，节水设备和系统，可再生性资源和建材，代替性的镶板，非使用性木材的有效利用，工程用木材，垃圾分类化，垃圾处理制肥器，高性能综合处理净化槽，建筑环境的健康和舒适化，利用地形的施工方法（与自然融合的宅地开发），建筑布局的技巧，雨水地下浸透的施工方法，自然型亲水空间，绿化的保护绿化的再生，建筑物绿化，小菜园庭园，鸟类生息庭园，建筑设计上的措施（采光、通风、换气），人所接触温度环境的考虑，温度调节材料的应用，低质热源辐射式空调，对不可见环境因素的考虑，考虑环境与健康类型的材料，考虑环境与健康类型的设备（替代氟里昂冷冻机和冷库），隔声楼面构造，对老年人和残疾人的考虑等。

从宏观来看，我国发展生态建材，现阶段的重点应放在引入资源和环境意识，采用高新技术对占主导地位的传统建筑材料进行环境协调化改造，尽快改善建材工业对资源能源的浪费和严重污染环境的状况。从我国的实际情况出发，结合国外生态建材的应用情况，我国制定生态建材的发展战略应立足于城市生态安全。主要考虑以下几点：

①建立建筑材料生命周期（LCA）的理论和方法；②以最低资源和能源消耗、最小环境污染代价生产传统建筑材料；③发展大幅度减少建筑能耗的建材制品；④开发具有高性

能长寿命的建筑材料，大幅度降低建筑工程的材料消耗和服务寿命；⑤发展具有改善居室生态环境和保健功能的建筑材料；⑥发展能替代生产能耗高、对环境污染大，对人体有毒有害的建筑材料；⑦开发工业废弃物再生资源化技术，利用工业废弃物生产优异性能的建筑材料；⑧发展能治理工业污染、净化修复环境或能扩大人类生存空间的新型建筑材料；⑨扩大可用原料和燃料范围，减少对优质、稀少或正在枯竭的重要原材料的依赖。

目前世界各国都在加强环保建材的研究，我国质量技术监督局也在组织专家制定建材产品的环保安全标准。现在，一些有眼光的建材厂家已纷纷给自己的产品打出了“环保牌”，在竞争激烈的市场上领先一步。如某涂料公司已生产出有空气净化功能的乳胶漆；有的地板商将“关注环保，关爱人生，回归自然，持续发展”作为企业理念。

结语

尽管上述列举的一些建筑应用绿色建材的实例，并不是城市建筑所具有的普遍性，但至少向人们显示：使用生态建材，回归自然，强调对人的安全无公害是由可持续建筑通向生态安全城市的必由之路。随着高技术、轻结构在建筑中的应用，建筑材料强调使用可再生、可循环、可重复使用、可降低污染的地方自然资源即生态建材将代表着一种方向、一种力量与一种趋势。同时生态建材也是城市生态安全的一个重要标志。

建筑从建造、运行到建筑拆除的全过程能源的节约，它既包括建筑材料的开采、运输、加工等能源的节约，也包括建筑运行中能源的节约，还包括建筑拆除能源的节约，以至建筑全寿命周期中节能、节地、节水、节材。而生态建材则是生态建筑之根本。

Green building material based on ecological security

Cao Wei

(School of Architecture, Xiamen University, Xiamen 361005 China)

Abstract: Based on ecological security, it was firstly considered that the "4R" principle, local provincialism and system engineering characteristic of ecological material in green building. The ecological material development strategy was researched, the path to ecological building material and ecological building were explored in the paper, by analyzing typical examples of typical ecological building including of wood construction buildings.

参考文献

[1] 曹伟. 城市生态安全导论. 中国建筑工业出版社，2004.9

更新装修观念　倡导绿色装修

李晓红　韩　勇　谭大可
（青岛理工大学艺术学院，青岛　266033）

摘　要： 随着“绿色”、“生态”、“可持续发展”的理念深入人心，建设绿色建筑成为民众广泛关注的焦点。而绿色建筑设计的准则和方法中很重要的两个方面便是：应用减轻环境负荷的建筑节能技术、创造健康舒适的室内环境。本文分几个方面就现今室内装修的观念更新问题进行探讨，倡导绿色装修观念，创造健康舒适的绿色室内环境。

关键词： 可持续发展　绿色建筑　绿色装修

1　引言——可持续发展、绿色建筑与绿色装修

近代以来，全球的环境问题引起了人们的普遍关注。人们在思考环境问题的同时，不得不反思人类的发展观。于是 1987 年有了联合国大会上的《我们共同的未来》的研究报告，提出了人类可持续发展的模式。进而里约会议又提出了 21 世纪的行动纲领——《21 世纪议程》，对可持续发展观进一步达成共识。

在上个世纪末，作为人类发展的人类住区，也随着人类发展观的转变而改变着自己的观念、模式，相继有人提出了“绿色建筑”、“生态建筑”的观点，制定了有关绿色生态建筑的指标体系，把节约能源、扩大使用非燃料能源（清洁能源）和可再生能源作为追求的目标之一。在“以人为本”的基础上，利用自然条件和人工手段来创造一个有利于人们舒适、健康的生活环境，同时又要控制对于自然资源的使用，实现向自然索取与回报之间的有效平衡。

绿色建筑设计的准则和方法中很重要的两个方面是：应用减轻环境负荷的建筑节能型技术和创造健康舒适的室内环境❶。因此，绿色室内环境成为绿色建筑的基本特征之一。绿色装修意在以适宜可行的方法对策，最大限度的利用自然资源的，节约能耗，创造健康舒适的绿色室内环境，其总目标亦是实现人与自然和谐共存、人类与地球可持续发展。

2　我国现今居住建筑及装修的现状

我国是一个能源储备量并不丰厚的国家。随着我国城市化进程的加速发展，加之我国的建筑耗能高、节能技术水平低的现状（据统计我国居住建筑每 m^2 能耗为相同气候条件下发达国家的 3 倍），全国建筑的能源利用效率，仅为发达国家的 1/3 左右。因此，在当

❶ 中建建筑承包公司编. 中国绿色建筑/可持续发展建筑国际研讨会论文集. 北京：中国建筑工业出版社，2001。

今的建筑中各种资源的节约问题变得日益紧迫。

随着中国经济快速发展和人民生活水平的提高，中国的室内装饰装修行业迅速发展。据统计，目前中国每年有装修需求的新房屋 9000 万套，加上二次装修和二手房装修，每年新装修的房屋将超过一亿套。如此巨大的装修工程量，势必耗费惊人的材料、能源等等。加之一些消费者盲目追求的过度装修、华而不实的豪华装修造成的材料、能源和资源的浪费惊人，且在环保装修方面也存在着诸多问题，由于设计、选材、家具等造成的室内环境污染也令人担忧。因此，更新装修观念，倡导绿色装修，创造宜人的绿色室内环境势在必行。

3 现今主要存在的装修观念问题

3.1 政府部门在建筑节能方面的推广力度不够

当前我国的有效的政策措施远远不够，环境法规过于宽泛和原则，与其他方面的政策不够配套，缺乏可实施性和可操作性，在创造绿色室内环境方面的专业性标准更是缺乏。一些政府部门还停留在“时尚意识”阶段，危机意识不够。加之关于绿色建筑主要是在建筑领域进行探讨、研究，并没有为广大群众所深知、熟知，自然谈不上绿色室内环境的营造问题。这对绿色建筑的整体进程十分不利。

3.2 大众的装修观念误区

不可否认，随着社会整体经济水平的提高，人们对生活舒适度的要求也逐渐提高。但大众的知识水平和审美理念普遍较低，再加上普遍存在的虚荣心和攀比心态，使大众很难以正常的心态去面对装饰装修，特别是存在着的生态意识差、环保观念片面等因素，使行业内存在着一个很奇怪的现象，一方面人们渴望拥有健康的空间，另一方面又想显现出富足和豪华。于是，大量的新建筑都开始采用暖通空调设备，装饰风格趋向浮躁和奢华，更加追求表面形式。

3.3 室内设计观念欠科学性

由设计不合理带来的各种能源浪费往往是潜在性的。许多室内设计中，因无视“绿色”而导致对环境的破坏、资源的浪费。总括起来有如下几个方面要予以考虑：设计缺乏真正意义上的“以人为本”、过多的设计、对建筑结构的再施工、过多的热负荷、选材不当造成室内的空气污染等等，盲目进行设计的堆积，从而误导消费者的“豪华型”主导消费心理，造成消费偏差，无论从哪个角度来讲都是有违于绿色装修观念。

4 当前整体装修观念的更新要求

正如郑曙旸老师所说：“……即使以绿色设计概念的最低限来衡量我们今天的室内环境设计，不是无所作为而是大有可为。关键在于我们的决策者、我们的设计师、我们的社会大众观念的转变。不少问题并非客观上的不能改变，而是我们主观上愿意不愿意改变。”❶ 这足以说明装修观念更新的重要性，只有先更新整体装修观念，才能从根本上加

❶ 郑曙旸．室内环境与绿色设计．中国建筑装饰．2002 年第 4 期。

快实施绿色装修的进程。

4.1　政府需进一步增强危机意识，加大推广力度

从长远来看，只有通过一定规范标准甚至法律手段，可持续发展的技术规划才能够转化为全体社会成员自觉或被迫遵循的行为。政府建设行政主管部门及技术监督部门给予进一步高度重视，制定相应的具体法规，从较高的程度上来管理和约束设计与施工部门，引导他们对“绿色”环境的实施做出应有的努力，使人们对健康空间的要求不断提高，在抓紧立法的同时，要狠抓法规、标准的执行，进一步整顿建筑装饰材料市场及整个装修市场。

4.2　大众舒适性概念、装修观念的更新及节约意识的强化

4.2.1　舒适性概念及装修观念的更新

早期的建筑对居住者的舒适性及健康要求的满足只是基于基本的温度和照明要求的满足。随着人类对自然界的认识程度不断的提升，反映在人们对居住环境的舒适性要求，便是在注重环境共生、生态与环境保护、可持续性发展等基础上，使室内环境具有适宜的温、湿度，必要的风速，新鲜的空气，充足的光线和不受周围环境的热、光辐射与噪声干扰等等。

装修观念的更新，首先应是本着经济节约的原则进行装修。我国迄今仍是一个经济相对落后的发展中国家，从某种角度上说也没有必要去倡导“奢华型”的装修。我们允许特定场合的“豪华”设计的存在，但不应为主导趋向。因此，推广节约型室内装饰装修观念成为主流，这与目前国家正在推广的节能型建筑，也是不谋而合的。

其次，应是本着健康环保的原则进行装修。最大限度的利用自然资源，包括光、热、水、空气等，最低限度地利用耗费常规能源达到热舒适的技术设备；对天然材料控制使用，大力倡导环保型复合材料，以“绿色设计”作为选用材料时的第一观念，更多使用可回收、可降解的建筑装饰材料。

4.2.2　节约意识的强化

现今正处加快建设节约型社会的大好时机，“全民节约，共同行动”成为各种宣传活动的主题。克服陋习小到随手关灯、随手关水龙头；大到设备选型、定制标准。贪大求全、讲排场、慕虚荣、喜炫耀成为滥用能源的病根。节约能源不仅是设计专业的准则，也应该成为全民的基本道德规范。

4.3　倡导绿色设计观念

4.3.1　注重人性化空间设计

引入绿色装修观念设计师应起到重要的引导作用，确立正确的观念与意识，其从设计自身来修正建设者的不良思想，可以引导社会更快地接受，使全社会逐渐认同符合生态原则的绿色装修。设计中注重人与自然的交流与沟通，强调室内、室外空间交融。对人性的关爱，也是绿色装修设计的基本出发点，室内设计中的见物不见人和以偏概全的手法是对人性关爱的漠视。主要体现在对老龄人群、残疾人群、特殊职业人群生活环境空间尺度的设计上。整体设计力求以简洁的手法，摈弃浮躁和奢华，崇尚简洁、自然，不仅要能实现其物质消费的需求，同时也要满足人们精神文明的需要，提倡各类符合生态保护与环境保护的各种设计方式，从风格、选材、施工方式等诸多方面统一考虑。如此将会成为今后室内设计流行趋势的主导。

4.3.2 设计中最大限度地利用自然资源

最大限度地利用自然资源的空间，也就是节能空间，这是降低建筑物整体成本最基本途径。建筑物的初始建造成本只占总成本的2%，而使用成本却占98%，充分利用自然资源能够有效地控制与降低使用成本，减少人类对大自然的索取。提高自然资源的利用效果，将有利于实现人类的可持续发展目标。

因此设计中，在建筑室内装修特征方面，力求功能合理，设计造型简化。通过有效的组织，最大限度地获得自然通风，提高室内空气质量，减少不必要的机械通风。在室内采光设计方面，充分利用天然光的照度变化进行设计，尽量减少不必要的人工照明。在室内采暖、隔热方面，提高门窗的密封性，减少传热损失和热辐射；尽而提高建筑围护结构保温隔热性能；提高屋内暖气的利用率，合理利用空调设备。

节约用水已经成为了一个全球性的话题，节水(water conservation)有两层含义：一是通常意义上的节约用水，倡导采用节水型家用设备。另一层含义是合理用水，即建筑中最重要的节水技术中水系统，提高了水的利用效率。目前在国际上受到广泛的认同。

装修后室内空气中有机物随时间变化❶ **表1**

污染物	通风一个月		通风三个月		通风六个月	
	A	B	A	B	A	B
甲醛	0.168～0.130	0.432～0.150	0.102～0.163	0.155～0.121	0.066～0.042	0.075～0.037
苯	0.055～0.034	0.235～0.178	0.024～0.017	0.102～0.090	0.009～0.005	0.039～0.017
甲苯	0.056～0.042	1.530～1.366	0.026～1.102	0.909～0.776	0.005～0.003	0.045～0.015
二甲苯	0.214～0.175	2.697～1.998	0.030～0.021	0.150～0.131	0.0018～0.0001	0.0024～0.0013
人体反映调查	有明显眼部和嗅觉刺激 B型长时间有胸闷感		轻微眼部和嗅觉刺激	明显眼部和嗅觉刺激	无明显感官反应	
GC/MS定性结果	庚烷、苯系物、乙酸丁脂、辛烷、莰烯、a-蒎烯等15种	庚烷、甲系物、乙酸丁脂、辛烷、莰烯、a-蒎烯、1-甲基氮杂环丁烷、2,4-二甲基庚烷、醛等28种	同一个月但其量有显著减少		苯系物、乙酸丁脂等	苯系物、乙酸丁脂、烷烃、烯烃等12种

注：A——120m² 环保型木地板，乳胶漆，合成板墙裙；
B——120m² 地板砖，壁纸，石膏吊顶，合成板墙裙，部分合成板新家具。

4.3.3 注重安全环保设计

经过SARS之后，人们重新认识生命与生存的含义、安全与健康的珍贵，认真思考人与自然、人与建筑环境和谐的关系，促使我们对前一时期的设计理念和装饰实践进行认真反思。设计中在材料的选用上注重材料的环保性能，即注重材料本身无危害自然环境的成分、不会向外发散有害物质以及材料的再生性。装修中尽量选用符合国家质量标准的装

❶ 王玲玲，多克辛等．室内空气污染现状与特征研究。北京：首届全国室内环境与健康研讨会论文集．中国环境科学出版社，2002。

饰材料。这对于保证室内空气的品质的优劣起着至关重要的作用。

据来自国外资料介绍，由于室内建筑、装饰装修等引起的不良建筑综合症(Sick Building Syndroms，缩写SBS)也称病态建筑综合症，主要是因为室内装饰装修的材料、家具及烧香、吸烟等所挥发出的污染物，在通风不良的情况下所致。又据德国德图公司提供的资料，室内空气污染按其比例来说：室内建筑材料污染占4%，室内通风不良占53%。因此选择环保的装饰材料和适宜地加强室内通风是改善室内空气质量的有效途径。

表1是对A、B两种不同装修类型的居室进行了污染物跟踪调查，时间为1～6月，并对人体进行了反应调查。可见，由于装修带来的有机污染物主要是甲醛、苯系物、醋酸醋类和烯烃类。这些有机物污染对人体健康有较大的危害，开窗通风，室温升高有利于挥发性有机物的减少。

5　结语

绿色，代表生命，代表健康和活力，是充满希望的颜色。绿色建筑与绿色装修已经成为今后建筑业的发展趋势，其发展首先需要在全社会建立相应的道德价值观和“绿色”的美学观，在社会观念上形成共识；其次应在业内围绕着“绿色”的室内理念进行全面的考虑，作为一种基本的设计方针贯穿到设计与施工的各个环节。让我们尽快在全社会树立绿色装修的消费观念，使之成为全社会的一致行动，共同装修出一个舒适健康的绿色之家。

Change the Conception Of Decoration & Sparkplug Green-Decoration

LI XiaoHong　HAN Yong　TAN DaKe

(Art college，Qingdao Technological University，Qingdao　266033　China)

Abstract: Deeply influenced by philosophies such as “Green”、“Ecosystems” and “Sustainable Development”，Constructing the Green Building is more and more widely focused by common people. In Green Building designing fields，two major principles/methods should be taken into consideration: Using Energy savable constructing technologies to reduce environmental workload; and creating healthy & comfortable indoor environment. This article will discuss，from multiple view points，changing the conception of decoration that is currently used in indoor decoration，and sparkplug the conception of Green-Decoration and create healthy & comfortable green environment.

参考文献

[1]　宋德萱编著. 节能建筑设计与技术. 上海：同济大学出版社，2003

[2]　西安建筑科技大学绿色建筑研究中心编. 绿色建筑. 北京：中国计划出版社，1999

[3]　陈衍庆，王玉荣主编. 建筑新技术. 北京：中国建筑工业出版社，2002

[4]　中建建筑承包公司. 中国绿色建筑/可持续发展建筑国际研讨会论文集. 北京：中国建筑工业出版社，2001

活性硅激发改性技术在新型节能建筑墙体围护结构中的应用

乌力吉

（上海英硕聚合物材料有限公司，上海 200233）

摘 要：通过对新型墙体材料围护结构的裂缝及渗漏等问题的研究，采用活性硅激发改性技术，研发了具有自组织功能的保温防水抗裂一体化的系列产品——封闭剂、填加剂、粘结砂浆、抹面砂浆、内墙抗裂粉刷材料等。经过在加气混凝土砌块、小型混凝土砌块等新型墙体材料围护结构上应用，取得了较好的效果。

关键词：新型墙体材料 活性硅 防水 抗裂 改性

前言

2002 年 6 月，建设部颁发了《建筑节能“十五”计划纲要》，提出在“十五”期间，要实现节能 2854 万 t 标准煤，减少向大气排放污染物数量：总悬浮颗粒物 10 万 t、二氧化硫 71.9 万 t、二氧化碳 6992 万 t 的目标。自 2000 年 10 月 1 日起实施的建设部第 76 号令《民用建筑节能管理规定》，严格地规定了工程建设单位、设计单位、施工单位未执行节能标准和设计规范的经济处罚办法，有力地推动了建筑节能事业的发展。显然，实施建筑节能已经成为国家大力发展经济的一项长远战略方针。

最近，胡锦涛总书记指出：要突出抓好节约能源、节约原材料和节约用水工作，大力发展循环经济，要发展节能省地型住宅和公共建筑。温家宝总理也强调要坚决贯彻节约优先的原则，把节能、节水、节地作为可持续发展的大政策，并多次批示、关注建筑领域的节能问题。这充分表明建筑节能已经得到了党中央、国务院的高度重视。

迄今为止，我国已建成节能建筑约 3 亿 m^2。北京、天津等大城市已率先编制并执行居住建筑节能 65％的标准，上海、武汉等地加强执行居住建筑节能设计标准的力度，并以政府办公建筑的节能为先导，引发了全国公共建筑节能的启动。实践证明，住宅建筑节能投入成本较低，所增加的成本一般可在 5 年左右收回，节能住宅已成为生态住宅打出的最响亮的牌，必定赢得大众的青睐。

建筑节能外围护结构的隔热保温，主要包括外墙体、门窗、屋顶、楼地面和户隔墙等建筑部位的隔热保温。其中，外墙体隔热保温的技术模式、材料质量、施工工艺，对节能建筑的节能投资和耐久性影响最大。在新型墙体材料推广过程中也不可避免地产生了一些问题，主要有以下几个方面：加气混凝土砌块、小型混凝土砌块、粉煤灰砌块、GRC 板，主要表现为由于温差交替收缩、干湿交替收缩等各种应力变形造成了墙体产生龟裂渗漏；

EPS/XPS 薄抹抗裂砂浆耐碱网格布增强外保温系统、聚合物砂浆复合聚苯外墙外保温系统、胶粉聚苯颗粒外保温系统等系统，主要问题有面层移位、空鼓、开裂、脱落的现象，成了上述外保温系统质量问题的通病；大模板内滞聚苯板与混凝土现浇做法主要有抹面砂浆开裂、脱落、聚苯板表面处理不佳造成粘结不牢，在拆模后出现错茬，造成后续施工的困难。这些问题严重地影响了新型墙体材料的推广应用。鉴于上述情况，作者长期对新型墙体材料围护结构的裂缝及渗漏等存在问题的研究，采用活性硅激发改性技术，研发了具有自组织功能的防水抗裂一体化的系列产品。经过在加气混凝土砌块、小型混凝土砌块、粉煤灰砌块、GRC 板等新型墙体材料围护结构上应用，得到了较好的效果。

1 活性硅激发改性技术

作者经长期研究发现，多数无机胶凝材料和绝热材料在形成过程中由高温向低温急冷转变时含有大量的玻璃态活性硅，其具有潜在的反应活性，其反应活性与存在的玻璃态硅的含量成正比。该研究课题关键的技术路线为选择适当的激发剂在无机绝热材料和胶凝材料中构成激发体系，使活性硅被激发，并在活化过程中加入含有有机硅烷高分子官能团的超渗透改性剂，生成有机无机共聚结构体，赋予被改性的无机材料各种功能即完成了分子水平的改性，使无机胶凝和绝热材料的防水性、抗裂性、黏结性等性能大大提高，这样由改性后的无机胶凝和绝热材料组合可以开发系列高性能的保温防水抗裂功能材料。现已利用该项技术开发了新型墙体材料围护结构防水保温抗裂系列产品。

2 新型墙体材料围护结构防水抗裂系列产品

2.1 抗裂防水封闭剂

该系列防水封闭剂是以活性硅改性复合高分子材料为基料，加入各种助剂等配制而成的高分子柔性透明防水封闭剂，具有防水、抗裂、无毒环保的功能。该系列封闭剂适用于以下几个方面：

2.1.1 可作为 GRC 板及加气混凝土、混凝土小型空心砌块抹灰的界面处理剂。

目前就国内而言，在新型墙体材料的推广应用中由于其吸水以及收缩等问题，造成表面抹灰产生龟裂，严重地影响推广应用。如在抹灰前用该封闭剂处理可防止其表面抹灰层龟裂，并提高砂浆的粘结强度。

2.1.2 可用于浴室及卫生间墙面的抗裂防水封闭，即在贴瓷砖之前刷涂二度封闭剂，待涂膜干燥后即可在其表面贴瓷砖。由于该封闭剂的涂膜与水泥砂浆有极好的亲合粘结性，并有一定的弹塑性，这样既有效地提高了墙面抗裂性、防水性，阻止卫生间的潮气渗入墙面并进而破坏其他相邻房间的装饰面，又提高了瓷砖粘结力。

2.1.3 可用于外墙涂料的抗裂底涂。

由于其具有一种神奇的自组织功能，可弥补混凝土墙面的龟裂，并有节约涂料用量的作用，也可用于旧墙面龟裂的修复。

2.1.4 可作为建筑外墙面砖界面抗裂防水封闭剂。即在贴外墙面砖前涂刷二度该封闭剂，待涂膜干燥后再贴面砖。这样可防止面砖空鼓造成雨水渗入墙体，既解决了外墙瓷砖墙面渗漏的难题，又提高了瓷砖的粘结力和抗温差引起变形造成瓷砖的脱落。

2.2 系列新型墙体材料围护结构粉刷砂浆抗裂填加剂

该产品是一种水泥砂浆增强剂，其主要原理是利用激发改性技术使硅酸盐水泥中的活性硅激发，使硅酸盐水泥的水化性和塑性等性能大大提高，从而在加入砂浆后，可大大提高和易性、抗裂性、防水性，同时增加水泥砂浆对GRC板及加气混凝土、混凝土小型空心砌块粘结力，解决抹灰浆面空鼓、面砖脱落、新老混凝土脱层问题，有效地解决新型墙体材料围护结构墙面的开裂渗漏问题。

2.3 系列新型墙体材料围护结构砌筑砂浆抗裂防水填加剂

该产品主要由活性硅激发改性剂改性的高分子聚合物组成。当其被加入到水泥砂浆时，激发剂可激活水泥的活性硅使其与高分子聚合物形成网状缔合结构，从而大大提高砌筑砂浆的塑性、和易性、粘结性，使加气混凝土、小型混凝土砌块、粉煤灰砌块等新型墙体材料的砌筑砂浆粘结性提高，有效地防止了墙体灰缝开裂渗漏。

2.4 新型墙体材料围护结构内外墙系列抗裂防水抹面砂浆

该产品是采用激发改性技术对无机硅酸盐等复合无机材料进行改性，使其活性硅与高分子材料在抹面砂浆内形成网状缔合物体系，使抹面砂浆的塑性、粘结强度大大提高，有效的解决了加气混凝土、小型混凝土砌块、粉煤灰砌块等新型墙体材料开裂的难题。

2.5 EPS、XPS外墙外保温粘结砂浆

该保温粘结砂浆由于采用活性硅激发剂，大大提高了粘结砂浆与EPS、XPS板及墙体的粘结强度，增加了砂浆的抗冲击性、和易性、施工性，解决了现有粘结砂浆与EPS、XPS板的粘结强度差的问题，同时使别墅等低层建筑的保温体系重装修成为现实。

2.6 EPS、XPS外墙外保温抗裂抹面砂浆

现有的保温抹面砂浆塑性差，抗冲击性能不佳。由于采用了活性硅激发技术，不仅提高了抹面砂浆的粘结性、施工性，同时解决了抗冲击性差的难题。

2.7 内墙抗裂粉刷材料

随着EPS、XPS外墙外保温技术的推广，内墙粉刷饰面开裂成为主要矛盾。我们将激发改性技术应用于内墙粉刷系统，研制出了内墙抗裂砂浆系列产品。

2.8 砌块抗裂防水封闭剂

目前，由于加气混凝土砌块、陶粒混凝土砌块、浮石混凝土砌块等新型建筑节能墙体砌块的推广，传统的粉刷饰面材料产生裂纹，使墙体产生渗漏，严重阻碍着上述产品的推广、应用。鉴于上述情况，采用激发改性技术开发了砌块抗裂防水封闭剂，基本解决了上述产品的粉刷饰面层的开裂、渗漏问题。

2.9 EPS、XPS界面处理剂

目前，聚苯乙烯发泡板与水泥砂浆的粘结强度低，成为影响EPS、XPS板外保温系统强度的主要因素。我们采用表面化学技术和激发改性技术研制开发了EPS、XPS板的界面处理剂，有效地解决了粘结砂浆、抹面砂浆与EPS、XPS的粘结问题。

3 新型墙体围护结构抗裂防水体系

3.1 非承重混凝土小砌块内外墙抗裂体系

该抗裂体系适用于轻骨料(如陶粒、膨胀珍珠岩、煤渣等)混凝土和蒸压加气混凝土(或泡沫混凝土)，组成如下：

第一步，该非承重混凝土小砌块的砌筑采用加入砌筑砂浆抗裂防水填加剂的专用砂浆。

第二步，采用上海英硕聚合物材料有限公司生产的抗裂防水封闭剂刷两度。

第三步，抹面采用围护结构内外墙系列抗裂专用抹面砂浆或加入上海英硕聚合物材料有限公司生产的粉刷砂浆抗裂填加剂的专用砂浆。

第四步，在填充墙与结构框架的结合部用专用抗碱玻纤网格布增强后抹面。

3.2 复合型保温板围护结构抗裂体系

复合型保温板围护结构包括玻璃纤维增强水泥珍珠岩空心隔墙板、玻璃纤维增强石膏水泥珍珠岩空心隔墙板、轻质陶粒隔墙板。该抗裂体系是由以下组成：

第一步，复合型保温板围护结构包括玻璃纤维增强水泥珍珠岩空心隔墙板、玻璃纤维增强石膏水泥珍珠岩空心隔墙板、轻质陶粒隔墙板。采用上海英硕聚合物材料有限公司生产的抗裂防水封闭剂刷两度。

第二步，抹面采用围护结构内外墙系列抗裂专用抹面砂浆或加入上海英硕聚合物材料有限公司生产的粉刷砂浆抗裂填加剂的专用砂浆。

第三步，在墙板与结构框架的结合部以及板缝用专用抗碱玻纤网格布增强后抹面。

3.3 EPS/XPS 薄抹抗裂砂浆耐碱网格布增强外保温系统

所谓外墙外保温，是指在垂直外墙的外表面上建造保温层，该外墙用砖石、空心砌块或混凝土建造，主要采用 EPS 和 XPS 做保温层。

3.3.1 保温层采用热阻值高，即导热系数小的高效保温材料，其导热系数一般小于 0.04W/m·K。保温层厚应根据设计计算，以满足节能标准对该地区墙体的保温要求。此外，保温材料的吸湿率要低，而粘结性能要好。

3.3.2 保温层的固定是采用粘结材料和塑料卡钉将保温层固结在基底上。粘结材料为事先预拌好的胶浆干粉料，故使用时不需添加其他任何物料，即无需再行配料，只需加适量水搅拌均匀就可使用，稳定可靠，操作方便。

3.3.3 面层保温层的表面层采用聚合物水泥胶浆抹面，厚度为 3mm 左右，内部包覆有加强材料(涂塑玻璃纤维方格网)，其作用是改善抹灰层的机械强度，保证其连续性，分散面层的收缩应力和温度应力，避免应力集中，防止面层出现裂纹。

3.3.4 装饰层如果需要，可以继续在面层上喷刷涂料，或粘贴陶瓷、石材等贴面材料，以增强墙面的装饰效果。

由于该体系采用了 EPS/XPS 高效保温材料，使墙面受环境温度变化影响较小，有效的减少了结构及围护结构砌体的变形开裂，由于上海英硕聚合物材料有限公司采用了激发改性技术开发的抗裂防水抹面砂浆，有效的防止了 EPS/XPS 薄抹抗裂砂浆耐碱网格布增强外保温系统的开裂。

因为活性硅广泛地存在于围护结构墙体材料及硅酸盐水泥中，通过活性硅激发改性技术可以激活无机保温材料和无机胶凝材料中的活性硅，从而大大提高了无机保温材料、无机胶凝材料的耐水性、粘结性、抗裂性等各种性能。例如，在 XPS 外保温系统中采用激发改性技术开发的粘结砂浆可以大大提高装饰层与 XPS 板之间的粘结强度，从而别墅外墙外保温用大理石、毛石重装修成为现实，解决了别墅外墙外保温重装修的难题。另外，采用砌块抗裂防水封闭剂有效地解决了加气混凝土砌块、陶粒混凝土砌块、小型混凝土砌块粉刷层开裂的问题。实践充分说明，活性硅激发改性技术可以广泛地应用于节能建筑，为新型建筑节能材料的推广应用找到了新的途径！

Application of active Silica Activation and Modification Technology in Energy Saving Space Enclosing Structure

Wu Liji

(Shanghai Insuring Polymer Material Co.，Ltd，Shanghai 200233 China)

Abstract: We have studied the problems of crack and leakage of new wall material space enclosing structure. Then we researched and developed a series of products with self organization、heat preservation、waterproof and anti-crack function-blocking agent、compound、adhesive mortar、rendering mortar、interior anti-crack rendering material on walls by means of activation and modification. We have applied them into concrete masonry block and aerated concrete block of new wall material space enclosing structure. It was very effective.

参考文献

[1] 涂逢祥. 建筑节能现状及发展趋势. 住宅科技，2004

[2] 加气混凝土砌块墙体裂缝的防治措施. 广东科技

[3] 汪冰. 加气混凝土墙面抹灰层空裂原因分析及处理. 建材工业信息，2005

[4] 陶有生. 加气混凝土建筑的节能效果分析. 新型建筑材料，2005

[5] 杨嗣信，吴琏. 论多层住宅结构体系问题. 施工技术，2004

[6] 2004 墙体革新与建筑节能力争新突破. 墙材革新与建筑节能，2004

[7] 江亿. 我国建筑能耗状况及有效的节能途径. 暖通空调，2005

[8] 曾珍，张雄. 建筑保温材料的发展. 上海建材，2005

[9] 曹同玉等. 聚合物乳液合成原理性能及应用

[10] 韩永奇. 关于推进我国墙体材料革新和推广节能建筑的思考. 砖瓦，2005

[11] 徐德恒等译. 功能性乳液·胶乳. 上海：上海科技大学出版社

[12] 石玉梅. 交联型乳胶漆，丙烯酸树脂及涂料论文集

[13] 轻质保温复合墙体材料的试验研究. 新型建筑材料，2004

[14] 大力推广新型墙体的重要配套材料：商品砂浆. 墙体革新建筑节能，2004

[15] 詹镇峰等. 聚合物水泥砂浆性能试验研究. 化学建材，2003

施工建筑垃圾管理调查

于震平　宋中南
（中国建筑工程总公司科技部，北京　100037）

摘　要：建筑垃圾的减量化和提高回收率是绿色建筑链条上的重要一环，对于保护环境，实现建筑业的可持续发展有着十分重要的意义。为此，首先应该对建筑垃圾的产生原因、数量、组成成分等进行调查统计分析，为治理建筑垃圾问题提供依据。本文对北京的五个建筑工地做了现场调查，了解了工地处理建筑垃圾的程序，统计了在处理建筑垃圾中的一些做法，和建筑垃圾的相对产出量，对产生建筑垃圾的原因作了归纳。认为施工单位提高使用与设计尺寸相符的定制材料以及提高建筑垃圾的回收率是今后应该加以重视的两个课题。

关键词：绿色施工　建筑垃圾　环境保护

1　引言

随着建设可持续发展社会的深入人心，绿色建筑也得到了空前的重视。政府有关部门，相继出台了一系列的举措，旨在贯彻落实科学发展观，促进节约资源、保护环境和实现建筑业的可持续发展，推动我国绿色建筑及其技术的健康发展。绿色建筑是一个综合的概念，不是仅局限于最终建筑产品本身，而是涵盖了规划、设计、施工、运行的整个建筑生命周期全过程。建筑施工是其整个过程中的重要一环，也就是目前提倡的绿色施工的理念。绿色施工涉及到很多方面，包含有丰富的内容，建筑垃圾的减量化和有效管理就是绿色施工中的重要内容之一。做好这方面的工作，对于保护环境，节约资源有着重要的意义。为了达到建筑垃圾的减量化和有效管理，首先应该了解施工现场建筑垃圾的产生情况。本文的目的就是提倡对施工现场的实际调查，为建立建筑垃圾的减量化措施和有效管理提供第一手资料。

2　国内外对建筑垃圾产生的研究

对建筑垃圾第一次大规模的调查报告是在 1976 年由英国建筑研究院（Building Research Establishment，U. K.）的 Skoyles 完成的[1]。其研究数据是基于 20 世纪 60 年代和 70 年代对 114 个建筑工地的调查。调查报告考察了 37 种建筑材料，提出了直接垃圾和间接垃圾，可避免垃圾和不可避免垃圾等概念。调查表明，建筑垃圾重量占设计建筑总量的 2%～15%。1992 年 Gavilan and Bernold[2] 对分散在美国四个地方的住房工地做了建筑垃圾产出量的调查。由于工程的规模小，主要对基础、主体结构和墙体饰面等做了仔细的统计。结果表明，建筑垃圾主要是由于在施工中的切割砖、木料、饰面等造成。而包装材料占了总建筑垃圾量的25%～30%。要减少建筑垃圾的产生应该综合考虑各种因素，而工人的熟练程度也是其因素之一。Bossink and Brouwers[3]在1993年和1994年间，对荷兰5个住房工程的7种建筑材料做了调查，计算了各种建材的浪费率，统计表明，建筑的浪费率以重量计为购买材料总数的1%～10%之间。浪费的原因主要是施工及其有关上游过程，如设计、采购和材料运输和储存中的不规范操作等。香港可利用的土地紧缺，对建筑垃圾的产生与管理一直十分重视。从1992年6月到1993年2月，香港理工大学和香港建筑协会（Hong Kong Polytechnic University and Hong Kong Construction Association)[4]对32个建筑工地做了监测调查，着重于了解建筑垃圾的产生，包括钢筋混凝土结构、砌体作业、抹灰和瓷砖贴面等。报告指出，其中建筑垃圾产生的一个主要原因是建筑承包商缺少对材料使用的控制，14个工地预拌混凝土的浪费为2.4%～26.5%，平均值为11%。为了减少建筑垃圾的产生，优化施工现场建筑垃圾的管理，Shen等[5]在2001年8月到11月对六个工地的建筑垃圾处理程序和做法作了调查，并提出了有利于建筑垃圾再利用和回收的优化操作程序。巴西对施工现场建筑垃圾的研究始于1989年Pinto对一栋18层住宅的10种建筑材料浪费的调查。调查表明，总的建筑垃圾重量占采购量的18%，费用的6%。1993年Picchi报道了对三处住宅建筑工地的调查结果，建筑垃圾重量占11%～17%，大概每万m^2的垃圾产出是950～1450t。Formoso等[6]在1992～1993年对5个相似的建筑工地，1996～1998年对69个建筑工地就各种建筑材料的浪费率及其原因进行了大型调查。结果表明，第一次调查中发现建筑材料的浪费率相当大，约占设计重量的27.6%，第二次调查得到的浪费率比第一次稍低。各国通过对工地现场建筑垃圾的综合调查，为实现建筑垃圾的减量化和回收打下了基础。

我国工程界对建筑垃圾的关注始于20世纪90年代，根据90年代的资料统计表明，砖混结构、全现浇混凝土结构等建筑每一万m^2的建筑施工会产生500～600t的建筑垃圾，则每年施工产生的建筑垃圾超过4000万吨[7]。在本文成文过程中所作的调查显示，目前每年施工产生的建筑垃圾数量应该远在4000万t以上。上海市容环境管理局在2005年春节期间(2月9日至2月16日)签准的建筑垃圾总量为886710t，据此大概推测全年的建筑垃圾总量约4000万t。根据北京市垃圾渣土管理处提供的数据，北京市2004年产生的建筑垃圾高达3000万t。另外，欧洲国家的一些统计资料表明，20世纪90年代初期建筑垃圾中施工产生的约占三分之一，建筑拆除的约占三分之二[3]。椐此推算，上海、北京每年建筑施工产生的建筑垃圾均在1000万t左右。这对城市垃圾的处理造成了很大的压力，也对环境造成了不利的影响。国内学者对于如何控制建筑垃圾的产生提出了不少好的建议[8,9]，对于建筑垃圾的利用也有不少论述[10]，但还少见对施工现场建筑垃圾产生的实际调查。本文的工作在于为建筑垃圾的减量化和综合利用的实施提供基础资料。

3 施工工地建筑垃圾管理现状

本次研究在北京地区调查了中国建筑工程总公司所属四个公司在施的五个工程，对建筑垃圾的处理情况作了初步了解。建筑垃圾的处理包含在工地环境管理计划中，项目组的环境管理因素一般包括有：噪声排放、施工现场扬尘、污水排放、夜间施工光污染、固体废弃物弃置、现场水电的浪费等。固体废弃物的管理目标是减少废弃物的产生，现场废弃物实行分类管理，提高回收利用量。

3.1 建筑垃圾处理程序

各工地建筑垃圾处理程序基本一致，大体如图1所示。

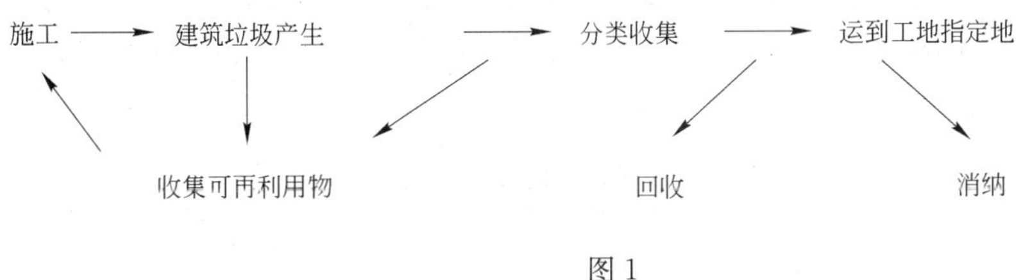

图1

3.2 工地建筑垃圾分类

工地建筑垃圾一般根据北京市的有关规定分为四类，即有毒有害可回收、有毒有害不可回收、无毒无害可回收、无毒无害不可回收。这些分类中的典型物品如表1所示。

建筑垃圾分类　　表1

分类名	有毒有害		无毒无害	
	可回收	不可回收	可回收	不可回收
典型品名	墨盒、电池、油漆桶	聚本板、聚脂板、废油漆涂料、保温材料	木材、钢材、空材料储存罐、废密目网、包装纸、塑料类等	碎砖瓦、碎混凝土块、碎石材、过期散装水泥、石膏板

3.3 废料的再利用

工地对于废弃材料缺少再利用的统筹计划，仅仅是比较随意地利用废旧材料，主要有以下几种再利用的形式：

- 利用短的钢筋头制作楼板钢筋的铁支撑、地锚拉环等。
- 木枋、木胶合板铺设办公区、道路防护棚的防护板、后浇带防护板。
- 每次浇筑剩余的混凝土浇筑女儿墙、构造柱、后浇带预制盖板等小型构件、硬化场区道路等。

3.4 建筑垃圾的回收

可回收的材料主要是钢材和木材等：

- 可回收材料中木料、木板、废纸由胶合板厂、造纸厂回收再利用。
- 废钢铁由项目组织，统一对社会钢铁制品工厂进行公开招标回收。
- 旧墨盒由制造商回收二次灌注墨粉，重复使用。
- 领取新电池时交回旧电池，最终由管理部统一移交公司处理。

3.5 不可回收建筑垃圾消纳

- 不可回收的建筑垃圾由有资质的垃圾渣土运输公司运送到垃圾消纳场。

4 施工现场建筑垃圾管理调查

本次调查选择的五个工程均是办公楼，建筑面积在 6 万到 15 万 m^2 不等，结构为钢混结构。承包商的施工管理水平基本上相差不大。

4.1 建筑垃圾处理的手段

各项目在垃圾处理的程序方面大同小异，但在具体的操作方面以及有效性等方面还是存在着差异。为了了解各项目的具体做法，优化垃圾建筑处理环节，设计了建筑垃圾处理的有利方式和不利方式表格，统计结果如表 2 和表 3 所示。

工地建筑垃圾管理中的不利方式 **表 2**

内容	工程编号				
	一	二	三	四	五
太多的建筑垃圾处理过程	□	△	△	▲	△
缺少垃圾分类过程	▲	▲	□	▲	△
没有考虑回收垃圾	▲	▲	□	□	▲
没有考虑重复利用	▲	▲	▲	□	▲
收集运送垃圾造成严重空气污染	△	△	▲		▲
运送垃圾造成严重的噪音污染	△	▲	▲		▲
垃圾运送中机械设备利用率低	△	△	□		▲
收集分散的垃圾耗时太多	△	▲	□	■	▲
允许更多的劳力收集垃圾而没有安全训练	▲	▲	▲		▲
垃圾处理人员之间缺少协调和监督	△	△	▲		▲
缺少机械设备应用使垃圾处理效率低	△	△	□		△
收集垃圾中广泛使用塑料袋	△	□	▲	▲	□
处理垃圾中不安全的操作	▲	▲	▲		▲
在工人中缺乏减少垃圾的宣传	△	▲	▲	▲	▲

注：■——认可；□——一般；△——稍微；▲——否认。

工地建筑垃圾管理中的有利方式 **表 3**

内容	工程编号				
	一	二	三	四	五
较少的建筑垃圾处理过程	□	□	■	■	■
适当的垃圾分类	■	■	□	■	■
考虑回收垃圾	□	■	■	□	■
考虑重复利用	■	□		▲	■
清洁的垃圾处理操作	□	■	□		■
避免二次垃圾处理(搬运)	□	□	■	■	■
运送垃圾中有效地使用机械工具	□	■	□		□
可见范围内较少的垃圾收集点	□	■	■		■
有很好的协调程序处理垃圾	△	□	□		■
从卖出可重复利用的垃圾中获得收入	□	■	■	■	□

续表

内　　容	工 程 编 号				
	一	二	三	四	五
在装备垃圾运送设备方面较少的投入	□	■	■	■	□
安全进行垃圾处理操作	□	■	■	■	■

注：■——认可；□——一般；△——稍微；▲——否认。

从调查结果可以看到，各项目在建筑垃圾的分类、回收、重复利用以及安全操作方面做得比较好，但是在减少垃圾处理过程、提高机械使用率、协调和监督垃圾处理，减少垃圾袋的使用等方面还有进一步改进的余地。

4.2　建筑材料浪费率统计

本研究选择了11种主要的建筑材料统计浪费率，以作为今后指导建筑垃圾减量化的依据。按重量计的材料浪费率根据下式求得：

$$浪费率=(采购量-设计用量-库存量)/设计用量 \quad (1)$$

由于各工地的进度不一，有些材料无法统计，所以在统计的意义上存在不足。各工程的统计结果如表4所示。

材料浪费率统计表　　　　**表4**

工程 材料名称	浪费率%(按重量计)					
	工程1	工程2	工程3	工程4	工程5	平均
1. 钢筋	1	2.5	2.2	2.5	2.0	2.04
2. 混凝土	0.1	0.2	0.15	3		0.94
3. 木料	100		5			
4. 砂浆	0.5	1	0.75			0.75
5. 各类装饰砖	1					
6. 各类装饰石材	30					
7. 墙体(板)	10	1				
砖			0.25			
8. 吊顶(矿棉板)	20					
9. 水暖管道	2					
10. 保温材料	30					
11. 电线	1					

从调查结果看，在结构施工阶段材料浪费率比较低，钢筋平均浪费2.04%，混凝土为0.94%。砂浆是0.75%。工程1木料100%的浪费是由于地下室采用了木模板，工程竣工后全部报废的缘故。Formoso等[11]在巴西调查了69个工地得到部分材料浪费率平均值以及Bossink等[2]在统计了荷兰5个工地得到的部分材料浪费率平均值见表5。

材料浪费率统计　　　　**表5**

材 料 名 称	材料浪费率%(以重量计)		
	巴西[6]	荷兰[3]	本调查
钢　筋	10.3		2.04
混凝土	9.5	3	0.94
砂　浆	59.8	10	0.75

通过比较可以看到，中建公司在北京地区 5 个工地的施工，在结构阶段材料浪费率是最小的。究其原因，很大一部分是由于采用了商品混凝土和商品砂浆，使浪费率大为降低。装修阶段的材料浪费率相对来说比较高，工程 1 饰面瓷砖 1%的浪费率是因为仅仅统计了马赛克，没有较大的切割浪费。装饰石材 30%的浪费率是石料在工厂备制过程中只有 70%的缘故。浪费率较大的墙体(10%)、吊顶(20%)以及保温材料(30%)大多是由于施工中的切割所造成的。

4.3 浪费原因

造成浪费的原因很多，研究者们已给出了各种不同的原因。Bossink[3] 在吸取前人经验的基础上，从设计、采购、材料运输、施工等不同方面归纳了 31 种造成材料浪费的原因。从本次调查结果看，在施工阶段主要有以下原因：

- 钢筋：配料时所产生的尾料；不合理的配料；采购超量。
- 混凝土：罐车内的余量，一般用来硬化场地；输送过程中的浪费，主要是输送泵堵塞、设备的冲洗；浇注时的溢出和散落；支模不合理造成的涨模剔凿和漏浆。
- 砂浆：落地灰清理不及时；开裂、剔凿进行的补灰；砌筑砌体时铺灰过厚；工地运输中的撒落。
- 墙体：石膏板墙体在切割中产生的废料；工厂定做的轻质隔墙在运输中的破损；设计变更拆改中的破损。
- 砌体：搬运时的损耗；不合理的组砌方法和操作方法；切割工具使用不当。
- 石材：基料在加工厂制备时的损耗。
- 保温棉：裁剪造成下脚料；裁剪不当；超量订购。

5 结论与建议

- 北京市每年施工产生的建筑垃圾约 1000 万 t，对环境的治理产生很大的压力。建筑垃圾的减量化具有很大的社会效率。
- 目前可回收的建筑垃圾主要有钢材、木材、包装箱等，而在建筑垃圾中占有很大比重的碎石(包括混凝土块、碎砖、碎装饰砖等)、石膏板、矿棉板、保温棉等材料同样可以重新加工利用，但目前还没有回收的单位。开发利用这部分材料，对建筑垃圾的减量化将起到十分重要的作用。
- 在建筑垃圾减量化方面，除了加强管理，严格执行环境管理体系外，应该重视材料的重复利用和回收工作。加强各种物资采购计划编制工作，尽量符合工程实际用量。
- 作为施工单位应尽量考虑按照设计规格在工厂定购相同尺寸的产品。如轻质墙体、石膏板墙体、矿棉板吊顶、石材饰面以及其他饰面等。也可开展装饰材料放大样工作，做到合理使用材料。而作为设计单位，设计中则应考虑到建筑模数，减少施工中可能发生的材料切割。
- 建议政府出台有关政策，规定生产经销石膏板、矿棉板、保温材料等的厂商，有义务回收这些材料的下脚料，经过加工处理重复利用；扶持回收建筑碎石的加工厂，使建筑碎石得以重复利用。只有使建筑垃圾的回收率大幅度提高，才能从根本上避免污染环境，从而实现可持续发展的战略。

Research Into Construction Waste Management

Yu Zhenping Song Zhongnan
(Science and technology Dep., China State Construction Engineering Corporation, Beijing 100037 China)

Abstract: Reduction of construction waste and increase of its recycling, a part of integrated chain management for green building, are very significance both for the environmental protection and for the sustainable construction. For to do that, it is first necessary to investigate and analyze the waste generation, quantity and component during construction, witch is a basic work. This paper investigated the waste handling process, the handling ways, the waste rates and the main causes of waste generation at five commercial building sites in Beijing. Suggestion of attention should be paid to both topics increasing use of materials to suit to design size and reducing materials to be cut and increasing construction waste recycling.

参考文献

[1] Skoyles, E. F. Material wastage: A misuse of resources. Building Research and Practice [R]. 1976, July/April, 232～243.

[2] Gavilan, R. M., and Bernold, L. E. Source evaluation of solid waste in building construction [J]. J. Constr. Eng. Manage., 1994, 120 (3): 536～552

[3] Bossink B. A. G. and Brouwers H. J. H. Construction waste: quantification and source evaluation [J]. J. Constr. Eng. Manag., 1996, 122(1): 55～60

[4] Hong Kong Polytechnic and Hong Kong Construction Association. Reduction of construction waste [R], 1993, Final report

[5] Shen, L. Y., Tam, V. W. Y., and Tam, C. M. etc. Mapping approach for examining waste management on construction sites [J], J. Constr. Eng. Manag., 2004, 130(4): 472～481

[6] Formoso, C. T., Soibelmam, L., Cesare, C. D., etc. Material waste in building industry: Main causes and prevention [J]. J. Constr. Eng. Manag., 2002, 128(4): 316～325

[7] 陆凯安. 我国建筑垃圾的现状与综合利用 [J]. 施工技术. 1999 年第 5 期

[8] 扬子江. 建筑垃圾对城市环境的影响及解决途径 [J]. 城市问题. 2003 年第 4 期: 60～63

[9] 郝建丽, 陈顺甜, 吴海燕等. 建筑废弃物的综合减量化措施 [J]. 重庆环境科学, 2003, 25(11): 10～12

[10] 谷显明. 建筑垃圾——可持续利用的再生资源 [J]. 中国环保产业. 2003 年第 8 期: 18～19

城市住宅小区水景观建设问题分析

车　伍[1]　程文静[1]　李海燕[1]　袁　望[2]

（[1] 北京建筑工程学院城建系，北京　100044；[2] 万科企业股份有限公司创研部，深圳　518049）

摘　要：目前全国各地的房地产开发商纷纷在可能的条件下建设城市住宅小区水景观项目，以改善小区环境并吸引购房者。规划模型里优美的水景成为小区商品房的一大卖点，但水景观建设的投资大、景观用水的紧张和水价的不断上涨，以及建成后常常出现的水质问题投诉及其运行维护投入都给开发商和物业管理者带来了很大的困扰。本文旨在分析与水景观建设相关的主要因素、探讨合理的水景观设计思路与设计流程。

关键词：水景观设计　水量平衡　水质保障

1　前言

进入 21 世纪以来，住宅小区注重于生态、人文特色建设，“城有水则秀，居有水增灵”，人工湖、人工溪流、喷泉水景等各种风格的水景观吸引了人们的目光，房价也因水而涨，但是水景观建设与运行仍面临重重问题。

水景观存在的最主要问题可归结为水景需水量较大，因水质恶化而采取的处理、清理、换水工作成本高，物业管理困难。水景观建设面临着如下几个困扰：一是，水资源的短缺使自来水的使用受到当地政策方面的限制，如自 2005 年 5 月 1 日起实施的《北京市节约用水办法》中规定，住宅小区景观环境用水应当使用雨水或者再生水，不得使用自来水，而且对自备井的控制也越来越严格，这就使水景观要面临水源问题；二是，即使可以使用自来水，自来水水价的持续上涨也使业主难以承受，物业管理压力增大；三是，水体换水清洗工作繁重，费时费力，且运行维护成本高，导致不少水体只能成为“节日水景”，长期闲置甚至废弃，结果是得不偿失。

水景观存在的上述问题已引起各界的普遍关注，本文针对具有一定规模、耗水量较大的人造水景或人工改造的水景，进行简要的讨论。

2　问题根源与分析

经过对大量住宅小区水景观的调研分析，探究水景问题的根源主要在于一些水景观设计、建设过于盲目草率，缺乏科学分析，设计观念存在误区，最终导致设计不合理、建设运行成本高。设计中出现的主要问题有以下几方面。

2.1 设计本身缺乏科学的统筹规划

水景观设计多是先在总体规划设计中进行景观效果的设计，而后再由不同的设计单位配合给水排水设计、土建设计、水质保障设计等，这样的分割设计缺乏统筹考虑，使水景观的功能、造价都受到影响。例如水体的水质保障措施，很多项目是在水景观设计甚至建成之后再配置水处理设备，而没有在设计的同时考虑水质保障措施和建立良好的水生态系统来增强水体的自净能力。

2.2 水体规模过大

水体规模过大是水景观建设的一大问题。特别是在北方大部分缺水地区，水资源紧缺，蒸发量大，庞大的水面虽然景观气势较好，但水体的蒸发损失量过大，再加上水费昂贵导致补水成本很高。水体的庞大也使水质保障难度加大，不论是水的循环还是水体清理或换水，运行费用和工作负荷都很高。

2.3 水体结构不合理

硬化的水体结构是小区水景观普遍采用的形式。为追求水景效果或贪图设计简单，很多水体构造了硬化的池底与池壁，缺乏生态功能，致使水体自净能力弱，水质难以保障，从而导致了水清无鱼⟶藻类疯长⟶水质恶化⟶高成本处理或补换清水的景观水恶性循环运行。据调查，一般硬化水体隔 10 天左右就需要清洗一次，夏季高温时换水的频率可能更快，给物业管理带来很大困难。另外大规模的硬化水体景色单调，缺乏生机，也不能达到良好的景观效果。

2.4 缺乏科学的开源节流

小区水景观的供水设计多采用自来水，造成水景观运行费用高，而且一旦自来水使用受限，景观水体便成为无源之水。而一些廉价的水资源，如雨水、中水，却未能被很好地利用。

2.5 水体水质保障措施不力

在小区清洁工作保障的前提下，由于水体生态系统的缺陷使得水质仍难以维持长久。一些大规模水景观通常要设置混凝过滤等水处理构筑物。这样虽能改善水质，但投资较大，运行费用也高，一些处理设备难以持续运行，长期闲置。另外当水体存在大量的藻类时，容易堵塞滤料，处理效果有限。因此配置传统的水处理设备成为小区水景观保障水质的惯用手段但无论从效果还是投资角度看，都非最佳解决方案。

3 水景观设计的关键环节

基于以上分析可以看出，水景观的设计问题涉及到多个重要环节，从大的方面讲可归结为“一点两面多环节”——首要的一点是从科学的设计理念和综合、开放的设计思路上整体把握水景观的规划，其次是从水量平衡和水质保障两方面作好水景的方案设计，而后是依靠新的技术体系对各技术环节进行合理的具体设计。

3.1 科学规划、统筹设计

生态住宅小区强调“节水、节能、节地”，水景观设计也应根据这三条原则，既要处理好景观水、给水、排水、雨水、地表水、地下水等水的各子系统的关系，也要妥善解决水环境与建筑、园林、道路等不同专业之间的关系，结合景观效果设计日标，运用先进的理念和技术，最终实现以合理的投资建设环境优美、运行经济、社会效益良好的水景项目。具体设计可从水景观的水量保障与水质保障两方面着手考虑。

3.2 水体水量平衡计算分析

水量平衡分析最主要目的就是合理地确定水景规模并解决水量保障问题，即对水体的输入项与输出项进行科学的全面分析。水体的水量输入项一般包括自来水或再生水补水量、收集利用的雨水径流量，水量输出量包括蒸发量、取用量以及渗漏损失量等。涉及到水体参数有水体的水面面积、有效水深和调蓄水深，汇流到水体的小区汇水面积以及径流系数，小区的补水、用水方案、水体结构及防洪能力等。

水量平衡计算需要在上述参数设定或计算的前提下，计算分析输入项与输出项及变化规律，确定水体合理的补水和防洪排涝方案，并检验参数设定的合理性，这些对于水景观设计都具有重大意义。

图1为水量平衡计算过程的示意说明。

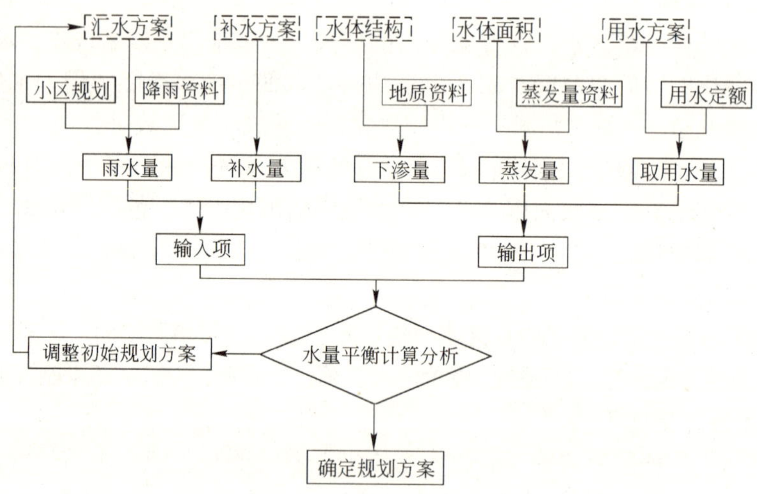

图1 水量平衡分析流程示意图

3.3 水体水质保障措施设计

3.3.1 加强水体生态系统的建设

水生态系统的建设可以模仿天然水环境，当地下水位高或渗透性很差时可考虑水体结构不防渗，并合理搭配水生动植物；如果不具备这样的条件而必须防渗时，同样可以合理地设计底层和驳岸的结构来构建水的生态系统，一般可以在防渗层之上铺设至少四十公分厚的沙土层，创造适宜水生物生存和栖息的环境。水体驳岸设计要将抗冲刷、抗风浪的能力与景观性、生态性的特点相结合，建设生态型驳岸等[3]，有时还要结合雨水的汇集设计缓冲带或净化区。

水体生态系统可选用的水生植物可大体分为三类，即挺水植物、浮水植物和沉水植物。挺水植物的根系置于水底淤泥，植体挺出水面，常见的有芦苇、茭白、菖蒲、灯心草、水葱、水芹等；浮水植物的植物体完全悬浮水面上或叶片悬浮在水面上，如凤眼莲、浮萍、菱、睡莲等；沉水植物完全沉没于水中，常见的有金鱼藻、苦草、菹草等。水生植物可以采用直接植于底层淤泥，或盆栽置入水体，或以生物浮岛的形式设置于水面，建议水面植物覆盖率达到30%左右。水体中可放养蚌类、鲢鳙等食藻鱼类、螺蛳、青蛙等动物，丰富生物种类，提高生物净化能力。

3.3.2　加强水的流动性避免水质恶化

正所谓流水不腐，增加水体的流动性可以为水体充氧，破坏水中藻类的生长环境，强化水体生态功能，同时潺潺水流也自有其韵味。促进水体流动的方法有借助水泵使水体循环流动、射流泵曝气以及跌水等。

3.3.3　强化水质净化处理

常规的混凝过滤等物化处理工艺的局限性如前文所述，因此在容积率较低或有自然条件可以利用的小区，可以优先考虑人工湿地处理方式或土地快速渗滤处理方式。这种系统的优点是对营养物有较好的去除效果，设备简易、运行简单，而且比较自然，可以和小区的景观设计结合，营造良好的生态，主要缺点是占地面积大，需要根据当地条件选择合适的植物，并随季节变化对植物和表层土壤进行必要的管理。

对一些规模较小的水体也可以采用直接加药进行净化，如北京某小区的硬化水景，水生生态系统脆弱，但采用投加漂白粉的净化措施后与不投药的情况相比，换水次数每年减少 3 到 4 次，节约了运行成本。常用的水处理药剂主要有硫酸铜、漂白粉等，但需要注意控制好投加剂量，避免带来副作用。

3.3.4　外源污染控制

控制外源污染进入水体是保障水质的重要措施。控制外源污染首先要保证动植物残体、生活垃圾等不滞留在水中，要求物业管理人员做好清洁工作；再者是要做好对补水水源水质的保障设计、水体周边径流污染控制设计等。

4　水景观设计流程建议

掌握上述水景观设计的关键环节后，还需要通过科学合理的设计流程来加以保证，避免建设赶进度或程序不当造成设计败笔。一般住宅小区水景观项目设计流程可按图 2 进行。

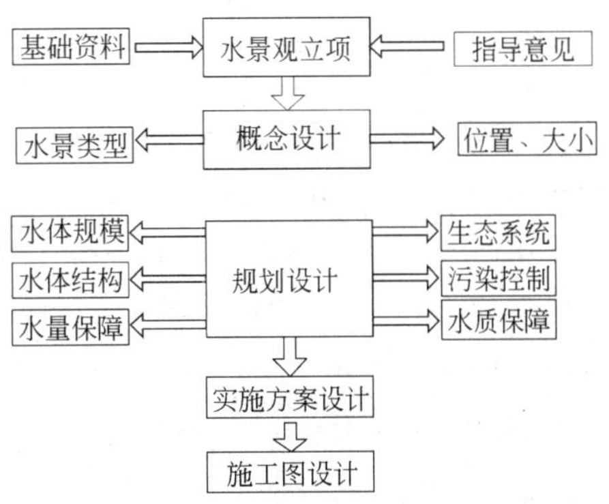

图 2　水景观项目设计流程图

项目资料收集主要包括建设场地的水文地质资料、气象资料、水资源和水环境状况资料、市政设施资料等尽可能详细的基础资料，根据这些资料有助于按实际情况对水景观的建设规模或目标有一个合理的初步意向。

作为小区总体规划的一部分，在水景观的立项阶段，开发商(可借助专家系统)应对水体的大致面积、水的生态性、经济性等提出具体的指导性意见和要求，避免对“水”的盲目性导致后续设计上的败笔或此后对总规做大的调整。立项后，规划设计者在小区的总体规划中对水景进行概念规划，给出水体的类型、位置和规模等。

在水景观规划设计阶段，该阶段设计者需要对水景观概念方案进行评估和方案的细化设计，主要工作内容包括水量平衡分析，水景的补水、雨水收集利用、再生水利用等方案设计，水景的面积、水深与防洪调蓄能力的调整，水体结构考虑，水生动植物选择与分布等生态系统设计，污染控制措施与水质保障设计等。显然，这是一个涉及到多学科专业的复杂的系统工程，并涉及到一些新的设计理念和技术，又直接关系到水景观实施的投资、运行费用和最后效果，因此，需要认真和反复地进行方案比选和调整，力争实现最优设计方案。该项工作最好由有经验并具有多学科综合规划设计能力的设计者或公司来完成。

当水景观规划设计完成之后，还应该形成水景观分项设计任务书，再由各专业设计公司分别进行实施方案或施工图设计。至此，水景观设计工作便圆满完成。

5 小结

住宅小区水景观建设是市场的需要，开发商在把握水景观建设先机的同时，也应注意科学设计、全面规划，思想先行，设计在后，依据资源循环、生态和可持续性的设计理念及相应的技术支持来保证水景观的生态效果和可持续的稳定运行，这是建设生态小区、走可持续发展道路的必然要求。另一方面，科学的设计方法与严谨的设计流程也是控制投资和保证经济实用的重要举措，特别是对于一些建设项目较多的大型房地产公司，更具有科学方法上的指导意义。

The Analysis on the Problem of Waterscape Construction in Urban Residential Subdivision

Che Wu[1] Cheng Wenjing[1] Li Haiyan[1] Yuanwang[2]

([1] Dept. of Urban Construction Engineering in Beijing Institute of Civil Engineering and Architecture, Beijing 100044; [2] Dept. of Innovation and Research in Vanke Co., Ltd, Shenzhen 518049 China)

Abstract: At present, waterscape construction in residential subdivision is quite popular in urban area of China. Though the planning model of waterscape is very beautiful and attractive, there are some trouble for the developer and the managers, which are the great investment in its construction, the lack of water resource, the increasing water price, as well as the water quality problem and its maintenance after built, etc.. This paper aims at analyzing the factors on waterscape plan and construction, and discussing reasonable idea and approach to the plan and design.

参考文献

[1] 李俊奇，车伍，汪宏玲. 雨水利用与生态小区. 给水排水. 2003.29(5)：14～17

[2] 车伍，李俊奇等. 雨污水利用和水环境修复案例分析. 中德水与发展国际研讨会文集. 水与发展. 德国卡尔斯鲁尔大学/浙江大学. 2003.12.4～8

[3] 王超，王沛芳. 城市水生态系统建设与管理. 北京：科学出版社. 2004.6：242～251

[4] 兰淑澄. 生态住区大面积景观水体、水质、水量保障. 2004 北京生态住宅水景观论坛. 全国工商联住宅产业商会/中国房地产报. 新北京楼市. 2004.11.18

室内可吸入颗粒物对 DNA 氧化性损伤能力的研究*

邵龙义[1]　李金娟[1]　赵厚银[2]　Jones T. P.[2]　李　慧[1]　Merolla L.[2]
（[1] 中国矿业大学（北京）煤炭资源教育部重点实验室及资源与地球科学系，
北京　100083；[2] Cardiff University，UK）

摘　要：本文使用质粒 DNA 评价法（plasmid DNA assay）和电感偶合等离子体质谱（ICP-MS）研究了北京市室内外 PM_{10} 的生物活性及其与微量元素的关系。本研究选择了 6 组样品分别代表室内和相应室外进行质粒 DNA 评价研究。结果表明，厨房内采集的 PM_{10} 的 TD50（50％的 DNA 被破坏时的 PM_{10} 的剂量）最低，为 45μg mL^{-1}，对 DNA 的损伤最大，表明具有最高的生物活性。而两个抽烟室内采集到的 PM_{10} 的 TD50 分别为 100 和 263μg mL^{-1}，具有较强的生物活性。结合 ICP-MS 得到的微量元素结果分析发现，PM_{10} 的水溶性 Zn 与 TD50 较其他元素相比具有更好的负相关性，说明水溶 Zn 可能造成 DNA 的氧化性损伤的主要原因。

关键词：质粒 DNA 评价　生物活性　微量元素　室内空气　可吸入颗粒物（PM_{10}）

近年来，PM_{10} 对人体的损伤效应引起了广泛的关注，流行病学研究结果表明 PM_{10} 浓度的增加与人们发病率及死亡率之间有显著的相关性[1,2]。关于颗粒物的肺损伤目前被广泛接受的观点是颗粒物表面的生物可利用的重金属离子会产生自由基，这些自由基对关键细胞的氧化性损伤可能是颗粒物导致肺损伤的主要原因[3~7]。Wilson 等[8]通过有过渡金属和没有过渡金属两种情况下的炭黑和超细炭粒子的自由基及致炎作用的研究，证明 PM_{10} 中的过渡金属对肺具有氧化性损伤。McNeilly 等[9]利用体外实验的方法证明水溶性重金属会导致炎症。Richards 等[10]研究表明导致肺损伤的是可溶的而非不可溶的锌；Adamson 等[11]利用活体实验研究发现，造成老鼠肺损伤的是水溶 Zn。王玉秋等[12]研究发现 Fe 是典型介导氧自由基过程的重金属。中国矿业大学（北京）环境与健康课题组的研究表明对 DNA 产生氧化性损伤的是颗粒物中的水溶性组分，其中水溶性的 Zn 元素可能是对 DNA 氧化性损伤的主要重金属元素[13,14]，并提出吸烟室内可吸入颗粒物比不吸烟室内可吸入颗粒物具有更强毒性[15]。本文利用质粒 DNA 评价法对居室室内 PM_{10} 样品的生物活性进行研究，并利用 ICP-MS 对 PM_{10} 的微量元素进行研究，最后判别不同类型室内 PM_{10} 的氧化性损伤能力强弱以及可能导致这种损伤能力的重元素类型。

* 教育部新世纪优秀人才计划项目（104028）及建设部科学技术项目部分内容。

1　样品采集及室内实验

1.1　样品采样

2003年7月在北京市2户(吸烟，非吸烟家庭)客厅、厨房(烹饪时)各采样3天，每天连续采样8小时以上。采样点位于中国矿业大学附近，由于条件限制，所以只在白天居民正常的生活条件下进行采样。同时在室外采集 PM_{10} 样品，避免雨天。采样仪空气流速为30L/min，采样头距地面高度为1.5m左右。随时记录室内外温度、湿度及风速。本次实验的采样头为Negretti(UK) PM_{10} 切割器，所用滤膜为直径47mm，孔径0.6μm聚碳酸酯滤膜(Millipore，UK)。

1.2　DNA损伤实验及电感耦合等离子体质谱(ICP-MS)实验

质粒DNA评价法是一种研究颗粒物表面的自由基对超螺旋DNA产生氧化性损伤的体外方法[13~16]。基本原理是颗粒物表面携带的自由基对超螺旋DNA产生氧化性损伤，最初的损伤是引起超螺旋DNA松弛；进一步的损伤表现为使DNA线化。这种损伤变化引起DNA在电泳仪中的电泳淌度的变化，进而将不同形态的DNA在琼脂糖凝胶中分离开，并在专门软件中进行定量分析，以此来评价颗粒物的生物活性(图1)。具体实验步骤参见文献[13]，首先将颗粒物从滤膜上分离出来后，使用超纯水将样品稀释成4～5个不同的浓度级别，分别与一定量的质粒DNA(ϕX174-RF DNA，Promega，London，UK)进行混合并使其充分反应，然后将混合溶液置入电泳槽凝胶中，使用电泳仪在30V电压下通电16小时，将不同形态的DNA分离开来，最后使用紫外凝胶成像系统(Synoptics LTM，Cambridge，UK)对凝胶成像，并使用Syngene Genetools软件(Synoptics Ltd.)和光密度

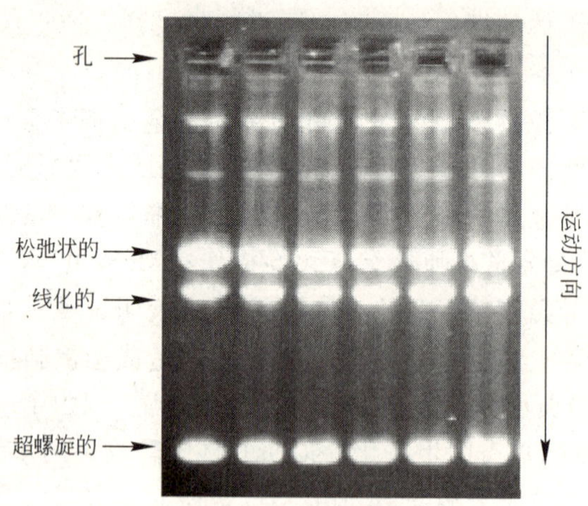

图1　质粒DNA评价法分析原理图

计对凝胶体中的超螺旋、松弛的及线化的等不同形态的DNA进行半定量分析，其中松弛和线化的DNA占总DNA的比例即代表样品的氧化性损伤能力或生物活性，其中一个重要的参数TD50，它代表造成50%的DNA损伤所需的样品的剂量(toxic dose)，TD50值越小，颗粒物生物活性越大。

使用 ICP-MS 进行颗粒物中微量重金属元素进行分析。对水溶部分微量元素的测量，是将滤膜放入 10mL HPLC 级水中，使用涡旋混合器震荡 6 小时；然后使用 0.2μm 孔径的聚碳酸酯滤膜(Millipore)过滤，即为水溶部分样品；从制备好的溶液中取出 1mL，与 0.5mL 的 50ppb 铑标准液混合，并加入 2%硝酸至 5mL；使用 ICP-MS(Perkin Elmer Elan 5000)进行化学成分分析。

2 结果与讨论

2.1 室内/外 PM_{10}的 DNA 氧化性损伤研究

本次选择六个样品中，No. 1 为厨房中采集到的 PM_{10}，其质量浓度为 205.9μg m^{-3}；No. 2 和 No. 3 为两个吸烟室内 PM_{10}，其质量浓度分别为 51.1μg m^{-3} 和 611.1μg m^{-3}；No. 4 和 No. 5 为非吸烟室内的 PM_{10}，质量浓度分别为 29.3 和 91.2μg m^{-3}；No. 6 为相应的室外 PM_{10}，其质量浓度为 38.8μg m^{-3}。六组样品对超螺旋 DNA 损伤的定量分析结果如图 2 所示，其 TD50 值见图 3。从图 3 可以看出，厨房中采集到的 PM_{10}样品（No. 1）对 DNA 的氧化性损伤最大，其 TD50 低达 45μg mL^{-1}，当其剂量为 75μg mL^{-1}时，就有 97.8%的 DNA 被破坏，而且当剂量再增加时，DNA 的破坏量略有增加（图 2）。吸烟的室内 PM_{10} 对 DNA 的损伤较非吸烟室内和室外 PM_{10} 的损伤大，TD50 为 100 和 263μg mL^{-1}；非吸烟室内 No. 4 样品在 250μg mL^{-1}的浓度下造成的损伤为 60%，而在 125μg mL^{-1}时，损伤为 17%，所以其 TD50 在 125～250μg mL^{-1}之间，而 No. 5 样品即使在浓度为 500μg mL^{-1}对 DNA 的损伤才达 38%，说明其生物活性较小；室外 No. 6 样品在 250μg mL^{-1}时对 DNA 的损伤达 30%，TD50 为 415μg mL^{-1}。很明显，厨房中采集到的 PM_{10}样品的 TD50 值最小，表明其生物活性最大，吸烟室内的样品次之，具较强的生物活性。笔者曾针对前四个样品的实验结果证明吸烟家庭的可吸入颗粒物比不吸烟家庭的可吸入颗粒物有更高的生物活性[15]，本次将新做的厨房可吸入颗粒物的实验结果加入数据组，结果显示厨房采集的可吸入颗粒物生物活性及毒性相对最高。

2.2 导致 PM_{10}的氧化性损伤的原因浅析

为说明引起质粒 DNA 损伤的主要原因，我们将 4 组样品的 TD50 值与这些样品的水溶性组分中的主要重金属元素 As、Cu、Ni、Pb、Zn 等的含量进行相关分析(图 4)(为了便于比较，No. 4 的 TD50 介于 125～250μg mL^{-1}，值取 200μg mL^{-1}；No. 5 的 TD50 取 600μg mL^{-1})。

从图 4 可以看出，虽然 As、Cu、Ni、Pb 在水溶组分中含量较高，但是它们与样品的 TD50 值并没有对应相关关系，说明这些元素可能不是造成质粒 DNA 损伤的主要元素。比较有意思的是 Zn 元素，从图 4 可看出水溶组分中 Zn 元素的含量与 TD50 值呈明显的负相关关系，即 Zn 含量越高，样品造成的 DNA 损伤越大。水溶性的 Zn 在 No. 2（吸烟室内)样品的含量最高，其 TD50 值也最小，这表明水溶性的 Zn 可能是 PM_{10}造成 DNA 损伤的重要重金属元素。

关于水溶性 Zn 的氧化性损伤能力，Richards 等[10]和 Adamson 等[11]都从对老鼠肺损伤的活体实验中证实。关于香烟烟雾对肺及 DNA 的氧化性损伤作用，已有许多学者论述[17～19]。结合本文实验，可以认为水溶性的 Zn 可能是导致 PM_{10}具有氧化性损伤的主要因

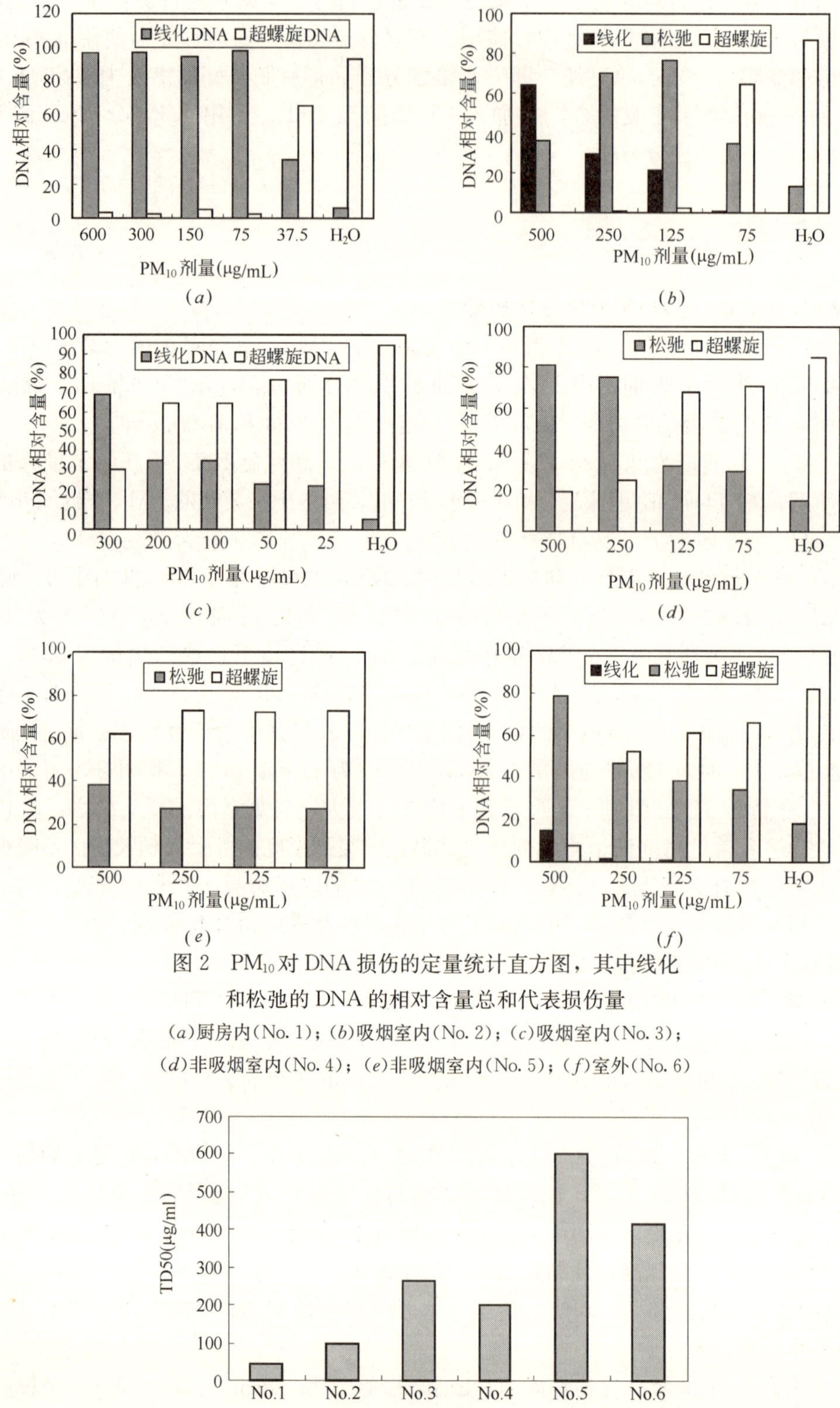

图 2 PM_{10}对 DNA 损伤的定量统计直方图，其中线化和松弛的 DNA 的相对含量总和代表损伤量

(*a*)厨房内(No. 1)；(*b*)吸烟室内(No. 2)；(*c*)吸烟室内(No. 3)；(*d*)非吸烟室内(No. 4)；(*e*)非吸烟室内(No. 5)；(*f*)室外(No. 6)

图 3 6 个 PM_{10}样品的 TD50 值（样品编号同图 2）

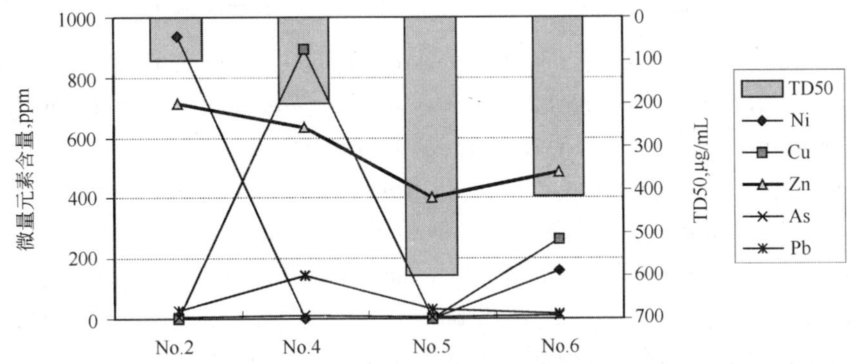

图4 北京市室内/外 PM_{10} 的生物活性与水溶性As、Cu、Ni、Pb、Zn等元素的关系(No.2、No.4-No.6样品与图2同)

素之一。

除Zn之外,Fe亦是一种具有较强氧化性损伤能力的重金属元素[12],但是在本次分析样品中,Fe与颗粒物的氧化性损伤能力的关系不甚明显。4个样品的水溶性的Fe的含量几乎都为0,说明在本次实验 PM_{10} 中的铁是以非水溶状态存在,因此不是造成DNA氧化性损伤的主要元素。

近来的研究[20]表明烹调油烟雾的各组分,包括油烟冷凝物、残留油、油烟颗粒物、油挥发性有机物均能诱导DNA氧化产生8-OhdG,造成氧化损伤。值得指出的是,吸烟造成的可吸入颗粒物的氧化性损伤力,除与重金属元素Zn有关外,还与烟草中的焦油含量关系密切,Landsberger等[21]应用中子活化分析室内吸烟产生的微量元素Cd、Zn是室内吸烟来源的主要污染物,吸烟的烟草烟雾(ETS)中含有4700多种化学物质,有较高浓度的焦油、苯、多环芳烃(PAHs)等,其中焦油中含有的许多微量元素,对人体产生极大危害[22,23]。吸烟室内可吸入颗粒物中的焦油含量及有机质与颗粒物氧化性损伤能力的关系还有待今后进一步探讨。

3 结论

(1)室内 PM_{10} 对超螺旋DNA的氧化性损伤略高于室外;在室内,厨房和吸烟居室 PM_{10} 对DNA的破坏性较大,厨房内 PM_{10} 的TD50最低,可低达45μg/mL,表明其氧化性损伤能力及生物活性最强。

(2) PM_{10} 中的水溶性Zn与TD50具有很好的负相关性,因此可能是引起 PM_{10} 对质粒DNA氧化性损伤的主要重金属元素。

Particle-induced oxidative capacities of residential indoor PM_{10}s

Shao Longyi[1], Li Jinjuan[1], Zhao Houyin[1], Jones T. P.[2],
Li hui[1], Merolla L.[2]

([1] Key Lab of Coal Resources of Chinese Ministry of Education and Department of Resources and Earth Sciences, China University of Mining and Technology,

Beijing 100083, China; [2] Cardiff University, UK)

Abstract: Plasmid DNA assay and ICP-MS were conducted to investigate the bioreactivity of PM_{10} and relationship between bioreactivity and trace elements of PM_{10} in Beijing air. Six PM_{10} samples were selected to represent the indoor and corresponding outdoor environments. Among six samples, the PM_{10} from the kitchen had the lowest TD50 (toxic dosage of PM_{10} causing 50% DNA to be damaged), being 45μg mL^{-1}, suggesting the highest bioreactivity. The TD50s of two PM_{10} samples from smoker's homes were 100 and 263μg mL^{-1}, indicating a relatively high bioreactivity. Trace elements are believed to be the main reasons for oxidative damage to plasmid DNA. The ICP-MS analysis combined with the DNA assay showed that water-soluble zinc had better relationship with TD50 values than other elements, implying that water-soluble zinc might play an important role in the damage of DNA.

参考文献

[1] 魏复盛，胡伟，吴国平，等．空气污染对儿童肺功能指标影响的初步分析．中国环境监测，2001，17(7)：61～65

[2] Schwartz J. What are people dying of on high air pollution days? Environmental Research, 1994, 64: 26～35

[3] Donaldson K, Beswick P H, Gilmour P S. Free radical activity associated with the surface of the particles: a unifying factor in determining biological activity. Toxicology Letter, 1996, 88: 293～298

[4] Costa D L and Dreher K L. Bioavailable transition metals in particulate matter mediate cardiopulmonary injury in healthy and compromised animal models. Environmental Health Perspectives, 1997, 105(Supp. 5): 1053～1060

[5] Dreher, K. L., Jascot, R. H., Lehmann, R., Richards, J. H., McGee, J. K. Soluble transition metals mediate residual oil fly ash induced acute lung injury. J. Toxicol. Environ. Health, 1997, 50: 285～305

[6] Pritchard, R. J., Ghio, A. J., Lehman, J. R., Winsett, D. W., Tepper, J. S., Park, P., Gilmour, M. I., Dreher, K. L., Costa, D. Oxidant generation and lung injury after particulate air pollutant exposure increase with the concentration of associated metals. Inhalation Toxicol, 1996, 8: 457～477

[7] 童永彭，倪新伯，张元勋，等．气溶胶自由基毒理学机制的研究．环境科学学报，2001，21(6)：654～659

[8] Wilson M. R., Lightbody J. H., Donaldson K, Sales J. and Stone V., Interactions between ultrafine particles and transition metals in vivo and in Vitro. Toxicology and Applied Pharmacology. 2002, 184, 172～179

[9] McNeilly J. D., Heal M. R., Beverland I. J., Howe A., Gibson M. D., Hibbs L. R., MacNee W., Donaldson K., Soluble transition metals cause the pro-inflammatory effects of welding fumes in vitro. Toxicology and Applied Pharmacology. 2004, 196: 95～107

[10] Richards R. J., Atkins J., Marrs T. C., Brown R. F. R., Masek L. C.. The biochemical and pathological changes produced by the intratracheal instillation of certain components of zinc-

hexachloroethane smoke. Toxicology Letter，1989，54：79～88

[11] Adamson I. Y. R.，Prieditis H.，Hedgecock C.，Vincent R.. Zinc is the toxic factor in the lung response to an atmospheric particulate sample. Toxicology and Applied Pharmacology，2000，166：111～119

[12] 王玉秋，张林，戴树桂，马永民．可吸入颗粒物上铁介导的活性氧产生及其对肺损伤的影响．环境科学进展 1998(增刊)，5：118～123

[13] 邵龙义、时宗波、李红等，2003，北京市大气中的 PM_{10} 的微区形貌及有害元素组成分析。北京市自然科学基金研究报告，137 页

[14] 时宗波，邵龙义，Jones T P，等．城市大气可吸入颗粒物的质粒 DNA 的氧化性损伤．科学通报，2004，49(1)：673～678

[15] 邵龙义、赵厚银、Jones T. P.、吕森林、Morella L.，2005，北京市室内 PM_{10} 对质粒 DNA 氧化性损伤研究，自然科学进展，15 卷 4 期，417～422 页

[16] Greenwell L L，Moreno T，Jones T P，Richards R J. Particle-induced oxidative damage is ameliorated by pulmonary antioxidants. Free Radical Biology & Medicine，2002，32(9)：898～905

[17] 孙咏梅，戴树桂，袭著革．香烟烟雾成分分析及其对 DNA 生物氧化能力研究．环境与健康杂志，2001，18(4)：203～207

[18] Kamp D W，Greenberger M J，Sbalchierro J S，et al. Cigarette smoke augments asbestos-induced alveolar epithelial cell injury：role of free radicals. Free Radical Biology & Medicine，1998，25(6)：728～739

[19] Jung M，Davis W P，Taatjes D J，et al. Asbestos and cigarette smoke cause increased DNA strand breaks and necrosis in bronchiolar epithelial cells in vivo. Free Radical Biology & Medicine，2000，28(8)：1295～1299

[20] 袭著革，李官贤，孙咏梅，晁福寰．烹调油烟雾诱导核酸氧化损伤及其标志物 8-羟基脱氧鸟苷的形成机制．环境与健康杂志，2003，20(5)：259～262

[21] Landsberger S，Wu D. The impact of heavy metals from environmental tobacco smoke on indoor air quality as determined by Compton suppression neutron activation analysis. The Science of the Total Environment，1995，173/174：323～337

[22] Evisken M P，Maistre C A，Newell G R. Health hazards of passive smoking. Annual Review of Public Health，1988，9(1)：47～70

[23] Higgins M. Risk factors associated with chronic obstructive lung disease. Ann：N，Y，Acad，Sci. 1991，624：7～17

193 聚氨酯彩色防水保温系统在住宅围护结构中的应用

孙生根　杨永巍　古小英
（上海市房地产科学研究院，上海　200031）

摘　要： 介绍了193聚氨酯彩色防水保温系统的组成结构，分析了与传统防水保温性价比，以及在屋面防水保温、外墙外保温工程的应用分析。

关键词： 防水保温　聚氨酯　纤维增强腻子　彩色防水涂膜　界面处理

1　前言

20世纪90年代以来，上海以前所未有的速度进行了大规模的住宅建设，极大地改善了居民的居住条件。然而，在大规模的住宅建设中，限于当时的建造标准与物质基础，住宅的防水保温存在严重不足，"保温材料不防水，防水材料不保温"这种状态一直延续至今，致使防水材料一旦老化损坏后，水渗至珍珠岩保温层内，形成一个"蓄水池"，给维修工作带来很大的不便。[1]

193聚氨酯彩色防水保温系统针对传统防水保温组合的缺陷，意欲在材料和施工工艺上将传统防水保温工艺的弊端得以解决，使防水保温材料有机统一。

2　193 聚氨酯彩色防水保温系统的组成

193聚氨酯彩色防水保温系统是由193彩色防水涂膜、193硬质聚氨酯泡沫塑料、193纤维增强抗裂腻子组成。[2]

2.1　193 彩色防水涂膜

一种以高分子聚合物乳胶和硅酸盐类材料为主要基料的防水涂层材料。常用颜色为灰色和米黄色。根据工程需要还可添加色浆，构成多种色彩。该材料有一定防水作用，涂刷在硬质聚氨酯表面可作为防紫外线保护层，并美化屋面环境。理化指标可参照表1。

193 彩色防水涂膜理化指标　　表1

试验项目		单　位	技术指标	
			Ⅰ型	Ⅱ型
固体含量		%	≥65	
干燥时间	表干时间	h	≤4	
	实干时间	h	≤8	

续表

试验项目		单　　位	技术指标	
			I 型	II 型
固体含量		%	≥65	
拉伸强度	无 处 理	MPa	≥1.2	≥1.8
	加热处理后保持率	%	≥80	≥80
	碱处理后保持率	%	≥70	≥80
	紫外线处理后保持率	%	≥80	—
断　　裂	无 处 理	%	≥200	—
伸 长 率	加热处理	%	≥150	—
	碱 处 理	%	≥140	—
	紫外线处理	%	≥150	—
低温柔性		φ10mm 棒	−10℃无裂纹	—
不透水性		0.3MPa，30min	不 透 水	不 透 水
潮湿基面粘结强度		MPa	≥0.5	≥1.0

注：I 型用于屋面工程，II 型用于墙体工程。

2.2 193 硬质聚氨酯泡沫塑料

由多元醇和异氰酸酯双组份材料组成。由于采用直接喷涂成型技术，产生闭孔率不低于 95%的硬泡沫化合物，使硬质聚氨酯层成为完整的不透水没有拼缝的整体。不仅具有优异的保温性，而且具有良好的防水性。物理性能指标可参照表 2。

硬质聚氨酯物理性能指标　　　　**表 2**

检测项目	单　　位	技术指标	
		H 型	M 型
密　　度	kg/m³	≥55	≥35
导热系数	W/(m·k)	≤0.027	≤0.025
吸 水 率	%	≤3	
抗压强度	MPa	≥0.3	≥0.15
抗拉强度	MPa	≥0.5	≥0.2
粘接强度	MPa	≥0.2	≥0.2
尺寸稳定性	%	≤1	
不透水性	0.3MPa，30min	不 透 水	
氧 指 数	%	26	

注：H 型硬质聚氨酯适用于上人屋面防水保温；
M 型硬质聚氨酯适用于墙体和不上人屋面防水保温。

2.3 193 纤维增强抗裂腻子

由高强化学纤维、有机高分子聚合物、硅酸盐材料等组成，具有抗裂保护及找平作用，在系统中起到保护硬质聚氨酯泡沫塑料的作用。物理性能指标可照表 3。

增强抗裂腻子物理性能指标 **表 3**

<table>
<tr><th colspan="3">检测项目</th><th>单 位</th><th>技术指标</th></tr>
<tr><td colspan="3">可操作时间</td><td>h</td><td>≤4</td></tr>
<tr><td rowspan="4">拉伸粘接强度</td><td rowspan="2">与水泥砂浆</td><td>原强度</td><td>MPa</td><td>≥0.6</td></tr>
<tr><td>耐 水</td><td>MPa</td><td>≥0.4</td></tr>
<tr><td rowspan="2">与硬质聚氨酯</td><td>原强度</td><td>MPa</td><td>≥0.2</td></tr>
<tr><td>耐 水</td><td>MPa</td><td>≥0.2</td></tr>
</table>

注：增强抗裂腻子与硬质聚氨酯之间有涂膜稀浆作界面处理。

3 屋面防水保温设计[3]

3.1 平屋面设计

平屋面基层找坡分结构找坡、轻集混凝土找坡(陶粒混凝土等)和水泥砂浆找坡三种。其中水泥砂浆找坡仅用于面积较小的平屋面。当基层(包括找坡层)平整度小于等于 10mm，可不加找平层。硬质聚氨酯与涂膜稀浆共同作用形成一道防水层。保护层分不上人屋面和上人屋面两种。

3.2 坡屋面设计

坡屋面分为彩色水泥瓦和彩陶瓦、多彩玻纤油毡瓦两种。坡屋面的基层平整度小于等于 10mm，可不加找平层。硬质聚氨酯与涂膜稀浆共同作用形成一道防水层。

3.3 保温层厚度选择

(1) 193 聚氨酯彩色防水保温系统中硬质聚氨酯泡沫塑料厚度的设计，应按所需的屋面传热系数(K[W/(m^2·K)])的大小而定，其最小应用厚度及其他节能要求应用厚度可参照表 4。

(2) 不需保温部位(如山墙、女儿墙及突出屋面的结构)的结构，硬质聚氨酯应用厚度不应小于 20mm。

(3) 屋面与山墙、女儿墙、天沟、檐沟以及突出屋面结构的连接处应用 1∶3 水泥砂浆或混凝土做成半径大于等于 80mm 的圆弧。

(4) 硬质聚氨酯层表面上必须设保护层，不上人屋面保护层为 2mm 厚防水涂膜及 3～5mm 厚增强抗裂腻子。上人屋面保护层为 40mm 厚细石混凝土，内配钢筋 Φ4@200mm 双向。

(5) 系统施工面积大于 40m^2，应对保护层进行分割，宜以 6m×6m 为分格上限，分格缝用防水涂膜或聚氨酯密封膏填补，分格缝宽宜大于等于 20mm。

(6) 质聚氨酯长度在 60m 以上时，应预留变形缝，缝用防水涂膜或聚氨酯密封膏填补，变形缝宽宜大于等于 20mm。

屋面保温硬质聚氨酯不同应用厚度(mm) **表 4**

项目		硬质聚氨酯应用厚度							
		$\delta=20$		$\delta=25$		$\delta=30$		$\delta=35$	
		K	D	K	D	K	D	K	D
陶粒混凝土找坡平均厚度 80mm $i=2\%$	不上人	0.95	2.92	0.81	3.00	0.71	3.09	0.63	3.17
	上　人	0.97	3.30	0.83	3.38	0.73	3.47	0.65	3.55
陶粒混凝土找坡平均厚度 160mm $i=5\%$	不上人	0.87	3.92	0.75	4.00	0.67	4.09	0.59	4.17
	上　人	0.89	4.30	0.77	4.38	0.69	4.47	0.62	4.55
结构找坡	不上人	1.04	1.92	0.87	2.00	0.76	2.09	0.67	2.17
	上　人	1.06	2.33	0.90	2.41	0.78	2.50	0.69	2.58

注：1. 表中硬质聚氨酯厚度以 δ 表示，单位为 mm。屋面的传热系数以 K 表示，单位为[W/(m² · k)]。热惰性指标以 D 表示。

2. 结构找坡屋面板厚度设定为 120 厚钢筋混凝土，内表面为 20 厚水泥砂浆。

3. 结构找坡，D 值小于 2.5，按《民用建筑热工设计规范》GB 50176—93 第 5.1.1 条验算，符合屋面隔热要求。

4. 硬质聚氨酯厚度可根据实际需求而计算，最大厚度可达 80mm。

5. K 值、D 值，根据《住宅建筑围护结构节能应用技术规程》DG/TJ 08—206—2002 表 4.7.2、表 4.8.1 计算而得。

6. 表中不上人屋面硬质聚氨酯导热系数 λ 取 0.025[W/(m · k)]，上人屋面硬质聚氨酯导热系数 λ 取 0.027[W/(m · k)]，并考虑系数 1.1。

7. 不上人屋面批嵌增强抗裂腻子，材料蓄热系数 S 取 11.37[W/(m² · k)]，厚度 δ 取 3mm，导热系数 λ 取 0.93[W/(m · k)]，则热阻 R 为 0.003(m² · k/W)，热惰性指标 D 为 0.03。

4 墙体防水保温设计

4.1 墙体防水保温设计

自墙体起，依次为 193 彩色防水涂膜稀浆、193 硬质聚氨酯泡沫塑料、193 彩色防水涂膜稀浆、193 纤维增强抗裂腻子及外墙装饰层，外墙装饰层一般选用弹性外墙防水涂料和面砖。基层平整度小于等于 10mm，可不加找平层。

4.2 保温层厚度选择

(1) 193 聚氨酯防水保温系统中硬质聚氨酯泡沫塑料厚度应按墙体所需的传热系数(K[W/(m² · K)])的大小而定，应用厚度可参照表 5。

(2) 墙面在离地高度 2.4m 范围内应增设耐碱玻璃纤维网格布。

(3) 若以外墙以防水涂料做装饰，宜采用弹性外墙防水涂料。

(4) 若以外墙用面砖做装饰层，粘接材料应选用专用面砖的粘接剂。

(5) 硬质聚氨酯的长、高度在 23m 以上，须预留变形缝。缝用聚氨酯密封膏填补。

墙体保温硬质聚氨酯不同应用厚度(mm)　　表 5

墙体材料与砌体厚度	硬质聚氨酯应用厚度								
	δ=15			δ=20			δ=25		
	K_p	K_m	D	K_p	K_m	D	K_p	K_m	D
240 多孔粘土砖	0.92	1.00	3.71	0.79	0.82	3.80	0.69	0.73	3.88
190 双排孔砌块	1.02	1.07	2.18	0.86	0.90	2.26	0.75	0.78	2.34
240 混凝土多孔砖	0.97	1.03	2.89	0.83	0.87	2.97	0.72	0.75	3.06
200 钢筋混凝土	1.21	1.22	2.51	1.00	1.00	2.59	0.85	0.85	2.67

注：1. 表中硬质聚氨酯厚度以 δ 表示，单位为 mm。外墙主体部位的传热系数以 K_p 表示，外墙的平均传热系数以 K_m 表示，单位为[W/(m²·k)]。热惰性指标以 D 表示。

2. 硬质聚氨酯厚度可根据实际需求而计算，最大厚度可达 60mm。

3. K_p 值、K_m 值、D 值，根据《住宅建筑围护结构节能应用技术规程》DG/TJ 08—206—2002 表附录 C 计算而得，导热系数 λ 取 0.025[W/(m·k)]。并考虑系数 1.1。

4. 不上人屋面批嵌增强抗裂腻子，材料蓄热系数 S 取 11.37[W/(m²·k)]，厚度 δ 取 3mm，导热系数 λ 取 0.93[W/(m·k)]，则热阻 R 为 0.003(m²·k/W)，热惰性指标 D 为 0.03。

5　193 聚氨酯彩色防水保温系统与传统保温综合性价比较

5.1　墙体防水保温(表 6)

193 彩色防水保温系统与传统保温性价比较表　　表 6

项　目	保温砂浆	聚苯板加聚合物砂浆夹网格布	193 系统
导热系数 W/m·k	0.070 左右	0.030～0.040	0.020～0.027
基层粘接材料粘接情况	无或直接抹灰直接粘接	粘接剂点粘接加螺栓固定	直接粘接
有无冷热桥效应	无	有	无
拼　缝	无	有	无
保温材料施工方法	抹灰、易施工	拼装、施工复杂	喷涂、易施工
保温材料施工厚度	最　厚	一　般	最　薄
保温防水性	吸水性较大	吸水率<4%，但拼缝及基础粘接存在空隙是渗水隐患	吸水率<3%，无拼缝、防水率优异
表面保护材料	外墙腻子等	聚合物砂浆夹网格布	193 纤维增强抗裂腻子
抗 裂 性	良	拼缝可能会出现裂缝	优
达到相同保温要求(K_p≤1.5)工程造价(元/m²)	80	75	85

5.2 屋面防水保温(表 7)

193 彩色防水保温系统与传统保温性价比较表 **表 7**

项　　目	珍珠岩保温加卷材或涂料	挤塑板加卷材或涂料	193 系统
导热系数 W/m·k	0.035～0.045	0.030～0.040	0.020～0.027
保温材料	过去常用材料，吸水性能较大、保温效果不明显	保温效果较佳，自身吸水率低但有拼缝，粘接施工、时间长久膨胀收缩易产生缝隙	新型保温材料，导热系数最低、易施工、现场喷涂完全无接缝，吸水率低
施工性能	与基层填砌的方法施工	采用粘接剂条状、点状粘接	现场直接喷涂与基层粘接最牢固
保温材料防水效果	无防水效果且会形成“蓄水池”	尽管材料本身不吸水，但因是粘接施工，存在诸多拼缝，与基层存在空隙达不到防水效果	因与基层是无空隙完全粘接，且是连续体，不仅具有优异的保温性能，且具有优良的防水性
防水材料	APP、SBS 卷材或黑色防水涂料等	APP、SBS 卷材或黑色防水涂料等	无机、有机复合材料，与基层粘接度高，施工最简便
砂浆与防水材料的粘接性	砂浆与防水材料难粘接会产生起壳现象	砂浆与防水材料难粘接会产生起壳现象	砂浆可直接与防水材料粘接不会产生起壳现象
防水保温综合性能	一旦防水层损坏，水渗至保温层维修困难，使用寿命 5～10 年	一旦防水层损坏，水沿保温板拼缝渗至基面，检修困难，使用寿命 5～10 年	与基础直接粘接，检修方便，使用寿命可达 25 年
保温材料价格(元/m^3)	300～400	800～1200	1800
工程造价(元/m^2)	60～70	80～90	75

6 结语

系统应用于屋面工程造价略高于苯板＋防水卷材组合，应用于墙面价格适中，适用于民用(包括住宅与公共建筑)和工业建筑的屋面工程防水保温，同样也适用于建筑物的外墙外保温及既有建筑的维修和改造。

The Use of 193 Colored Polyurethane Waterproof and Insulation System in Building Envelope

Sun Shenggen　Yang Yongwei　Guxiaoying
(Shanghai Real Estate Science Research Institute，Shanghai，200031　China)

Abstract: This article introduces the composition of 193 colour polyurethane waterproof insulation system, property-cost comparison with tradition waterproof insulation composite and its application in roofing waterproof insulation and exterior wall insulation projects.

参考文献

[1] 涂逢祥．建筑节能第 42 期．北京：中国建筑工业出版社
[2] 上海市建筑产品推荐性应用标准．193 聚氨酯彩色防水保温系统技术规程 DBJ/CT 022—2001
[3] 中国建筑业协会建筑节能专业委员会．建筑节能技术．北京：中国计划出版社，1996

绿色节能建筑的新材料
——稻草砌块

况青梅
（宜春学院土木系，江西宜春 336000）

摘　要： 稻草砌块是用干燥的水稻禾秆为原料的一种新型墙体建筑材料。该材料具有原料来源丰富，价格便宜，施工方便，保温性能好，耐腐蚀等特点。这是一种较便宜的绿色环保、可持续发展的红砖替代材料。

关键词： 稻草砌块　新型墙体材料　绿色节能

1　前言

当前我国正处于工业化和城镇化的快速发展时期，资源消耗多，能源短缺问题日益突出；建筑不仅对资源的消耗大，而且对环境影响大。据初步测算，住宅使用能耗占全国总能耗20％左右，若再考虑加上建材生产和建造的能耗，住宅总能耗约为37％左右。而传统的墙体材料——粘土砖不仅大量消耗土地资源，生产时释放的 CO_2 还破坏环境。为了在我国现代化建设中实现可持续发展战略，20 世纪 90 年代我国政府已将限制实心粘土砖的生产与使用、开发和推广新型墙体材料作为在相当长一个时期内的既定国策之一。

2　新型墙体材料的现状

2.1　根据国家建筑材料工业局的规划，我国新型墙体材料产量占全部墙体材料的百分比在 2000 年、2010 年、2030 年应分别达到 25％、40％和 60％。

2.2　经过十几年的努力，我国新型墙体材料取得很大发展。混凝土砌块、加气混凝土、新建烧结空心砖、水泥刨花板、PVC 外墙挂板、石膏板、纸面草板等新型墙体材料的产量在全部墙体材料中所占的比例越来越多。2000 年，全国新型墙体材料产量达 2100 亿块（标准砖），占墙体材料总量的 28％，比 1995 年提高 12％。实心粘土砖由 1995 年的 6300 亿块减少到 2000 年的 5400 亿块。累计节约土地近 40 万亩、节能 6000 万 t 标准煤、利用废渣 2 亿 t，实现新型墙材建筑面积 2 亿 m^3，节能建筑面积 7470 万 m^2。

2.3　但从全国来看，由于房地产的迅猛发展，许多中小城市和村镇仍在大量使用粘土砖。据预计 2005 年我国实心粘土砖的用量约 7500 亿块，其中新型墙体材料占 3000 亿块。小块实心粘土砖仍在墙体材料组成中占有绝对优势。究其原因，一是现在的新型墙材投资大，资源、产量供不应求，售价高；二是粘土砖的生产就地取材，成本低、技术简单，田间地头搭个大棚就可以大量生产，而不需建什么厂房。要想彻底取代红砖，就要找到与粘土同样资源丰富，取材容易，成本低，生产技术要求低，同时又符合节能、绿色、环保、可持续发展的原材料来生产墙体材料。

3 稻草是绿色可持续发展的材料

3.1 草纤维从人的生命起源就与人共存，人类从原始时期就利用草防寒过冬。历史上就有杜甫草堂的记载。稻草与粘土混合制作的土砖大量地用于建造“干打垒”房屋。从古至今，稻草作为一种建筑材料，在我国各个地方都有不同程度的应用，但一般用于建造简易房屋。随着高科技的发展，人类开始重新认识稻草的价值。利用稻草制造无毒、无污染、轻质、多功能的新型建筑材料，符合我国产业政策与可持续发展战略，有广阔的发展前景。

3.2 稻草是水稻除去稻穗和根部后的茎杆部分，俗称稻秆。水稻是人类主要食物之一，目前世界上至少有 1/ 3 的人以食用稻米为主。水稻的主要产地集中在亚洲，其中，我国年产稻谷约 2 亿 t，稻草年产量与之相当约 2 亿 t，占世界总产量的 35 %以上，居世界第一位。过去，稻草在我国农村除作燃料外，几乎全部作为废弃物浪费了；不仅没有得到利用，而且污染环境。

3.3 稻草是由大量的有机物(占 90%以上)和少量的无机物及水所组成的，其有机物的主要成分是纤维素类的碳水化合物，此外还有少量的粗蛋白质和粗脂肪。碳水化合物又由纤维素类物质和可溶性糖类组成。纤维素类物质是稻草细胞壁的主要成分，它包括纤维素、半纤维素和木质素等。不同品种的水稻品种，纤维素类物质的含量有不同，一般含纤维素 38.70%，半纤维素 18.30%，木质素 15.00%，粗蛋白质 4.10%和灰分 12.20%。无机盐由硅酸盐和其他少量微量元素组成，稻草中的硅酸盐含量较高达到 12%以上。纤维素纤维的相对密度为 1.2，抗拉强度为 300～500MPa，弹性模量为1.0×10^4MPa。

3.3.1 纤维素是由许多 b-D-葡萄糖分子以 b-(1-4)糖苷键相连而成直链。纤维素是植物细胞壁的主要结构成份，一般占植物体总重量的 1/3 左右。化学性能稳定，不溶于稀酸。在高温、高压和酸性条件下，可以水解成为葡萄糖。纤维素呈白色，比重 1.5kg/cm^3。由于纤维素分子含有羟基，在空气中能吸收水分，因而稻草的吸湿性强；但由于分子排列的规则性和大量氢键存在的缘故，纤维素不溶于水。纤维素在常温下是很稳定的，随着温度的升高，其抗张强度也相应的降低。在高温、光照和氧存在时，会发生氧化反应，使纤维素生成氧化纤维素，从而降低稻草的抗拉强度。

3.3.2 半纤维素 半纤维素是戊糖、己糖和多糖醛酸及其甲酯的缩合物，其主要成分是戊聚糖。一般不溶于热水，而溶于稀酸。在纤维素和木质素之间起联结作用。

3.3.3 木质素 木质素是一类酚酸多聚体混合物，它是由苯丙烷及其衍生物为基本单位构成的高分子芳香醇，它常常与半纤维素，纤维素镶嵌在一起，极不容易分开。木质素的功能是，在细胞壁中与其他成分一起形成复杂结构，防止微生物的侵袭；在细胞之间作为一种粘合剂起支架的作用；还可以缓和水通过细胞壁向内渗透。

3.3.4 无机盐 稻秆中无机盐含量都很低，但稻草细胞壁中二氧化硅的含量很高。

3.4 稻草在古代建筑中的应用

稻草在比较干燥的条件下，它的长纤维与木材一样是长寿的，金字塔里的稻草存放了几千年以后，它的强度还保留着现有稻草强度的60%；寺庙里佛像的制作就是用杉木捆绑稻草用泥巴封住，存放了几百年后不变质。寺庙中大雄宝殿的大柱都是将稻草捆扎在木头四周，然后用麻绳将稻草扎紧，外再用猪血老粉密封，外面涂上朱红大漆；几百年打开后，内部的稻草与新稻草的颜色也差不多，具有很好的抗拉强度。过去上海新式洋房采用稻草筋与粘土作为轻质墙的粉刷层，100年后，稻草筋粘土墙还是蛮好，这些洋房还在继续升值。客观存在的事实已经证明，稻草在干燥的条件下是长命的。这都应该归功于上述稻草的化学结构。美国稻草研究学会研究结论：因稻草内存在着大量的二氧化硅，老鼠、白蚁都不喜欢它。

3.5 稻草在现代建筑随着科学技术的进步，稻草的开发利用在国内外得到了迅速的发展，但与其年产量相比，利用率仍然很低。现在许多地方，稻草被一种称为草砖机的机器打压成块状，称做草捆(straw-bale)，以利于储存和运输，然后用于作饲料、包装等。发展应用于建筑便得到一种极好的墙体材料。

4 稻草砌块

4.1 稻草砌块，简称(俗称)草砖，是采用稻杆经过机械整理冲压挤击后用麻绳或铁丝打包而成大捆状的一种新型墙体材料。在国内，草砖外形尺寸多为高360mm、宽490mm、长300～1000mm的方捆，密度112～134kg/m^3，含水率小于15%。目前，仅在我国的内蒙古、黑龙江、辽宁、吉林等地冬季非常寒冷的地区农村少量的应用。积累了不少建筑技术经验，但可参考的研究数据很少。据美国能源部(DOE)有关草砖的资料显示，美国存在的草砖种类繁多，有方形草砖、圆形草砖等。其规格也不尽相同，如“三线草砖”的规格为0.58m×0.4m×1.1m，质量为34～39kg；“两线草砖”的规格为0.46m×0.36m×0.92m，质量为23～27kg。草砖的密度一般为83.2～132.8kg / m^3，经技术检测其承重压力为1,956kg/m^2。

4.2 稻草砌块生产

当草放进草砖机后，每一次的打压，会将草压成一层层“薄片”或者是将其压成捆。一块草砖，就是由一些“薄片”用铁丝或麻绳紧紧地捆在一起而组成的。草砖的高度和宽度，是由草砖机内部的压制空间来决定的。草砖的长度可以由制草砖的人员根据需要调节。如图1所示。

根据房屋设计的不同，有时需要用半块草砖，或者是特定长度的草砖，用来填满草砖与柱子之间或墙中间的空隙。只要将铁丝或麻绳剪断，取出适量的草砖薄片，然后再将铁丝或麻绳捆紧即可。

5 稻草砌块建造草砖房

5.1 草砖房的发展历史　19世纪90年代，人们无意间发现的打包机可将松散的农作物的秸秆打成可以用于建筑的大草块。稻草砌块最早用于建房是20世纪30年代在美国北部

图 1 草砖的生产

的草原区，当时由于木材资源稀缺且昂贵，刚到北方的移民要建房，而草原上只有大量的草。美国内布拉斯加人第一个用草块来建房，所建房屋至今仍然有效地为他们遮风挡雨。

初始的草砖房面积很小，是承重型的，正方形的房屋构成锥型屋顶。图一所展示的马丁/蒙哈特房屋分别摄于 1925 年和 1995 年，是这种草砖房结构的代表，该房屋至今还在已有 100 年。期间，由于交通条件的改善，木材资源逐渐丰富，草砖房建筑技术就开始衰退。20 世纪 80 年代以来，环境保护及节能需求日益高涨，促进了草砖建筑热在全球再度兴起。由于草砖房的节能性和环保性，草砖房建筑技术在美国、加拿大，德国、英国、澳大利亚等地又开始流行并大量推广应用，结构形式也多种多样。近年来用草砖盖房成为一种新兴的建筑业，在有些地方，其增长率几乎每年 30%。自 1999 年起我国 21 世纪议程管理中心环境无害化技术转移中心与安泽国际救援协会合作，在我国北方农村地区开始实施“节能草砖建房能力建设与示范工程”。几年来，我国北方 5 省 9 地共建立节能草砖示范民房 616 套和 3 所小学，深受广大农民群众的喜爱(图 2)。经过跟踪检测草专房节能达到 50%以上，人们称之“既原始又高效，既传统又现代”。

图 2 1998 年建于中国张北的草砖学校，摄于 2002 年
Straw-Bale School，Built 1998，Photographed 2002

5.2 草砖房的结构体系 房屋的结构一般分两类，一类是墙体传递水平荷载，称为自承

重结构形式；另一类是梁柱传递水平荷载，墙体只起围护作用，称为框架结构形式。草砖房也是如此，任何地方材料—木材、混凝土、砖、竹子、钢材等都能做它的框架结构。各种不同的草砖房建筑遍及全世界，每一种结构体系的建立都是基于当地的气候条件包括生态环境、可利用的材料、现有的建筑技术、建筑工人的技能，以及当地的建筑风格。

5.3 草砖墙自承重体系适合于小而简单的建筑(即草砖承受屋面体系的重量)。如图3所示。

图3 Straw-bale House，built 1994
Historic Photo and Present Day Photo
建于1904年的草砖房，历史照(左)与近期照(右)

5.4 框架结构适用面更广，因为框架结构可以先建楼、屋盖，然后再安砌草砖，这样可以保护草砖在施工中不受雨淋。框架结构体系可以使建筑设计更美观且可以采用低密度的草砖。如图4所示。

图4 2001年建于美国加利福尼亚州的草砖房

5.5 草砖房建筑经历了多种的结构测试，并且逐步规范化。目前，在美国有多种不同的草砖建筑规范。草砖房在美国已被广泛接受，并且正向世界各地传播。在加拿大、法国、澳大利亚、英国、荷兰、中国、蒙古、阿根廷、墨西哥和新西兰等国家都可以见到草砖房。但在国内结构方面的实验尚处于空白阶段，笔者将致力于草砖这种稻草砌块的结构研究。

6 草砖房的建筑节能效果

从1993年开始，草砖房的保温性能就公布于众了。根据测试结果，草砖墙的热阻值RSI至少是5.8。国外草砖墙保温性能的实验研究结果如表1所示。同济大学暖通空调及燃气研究所对中国辽宁省本溪市草砖房的全年采暖能耗进行了模拟计算。图5为供暖室内计算温度为16℃时模拟出的相同结构不同墙体材料的草砖房(Straw-Bale Building)与普通砖房(Brick Masonry Building)采暖季的动态热负荷曲线。由图5可见，草砖房动态热负荷曲线较普通砖房动态热负荷曲线明显偏下，前者显示了很好的节能特性。普通砖房的全年累计热负荷为11,401kW·h，而草砖房则为3,620kW·h，比普通砖房节省采暖能耗60%以上，折合成标煤达2t以上。中国环境无害化技术转移中心跟踪测试结果表明："在同一时间内，草砖房的每立方米能效比普通砖房高68%。"

国外草砖墙保温性能的实验研究结果 **表1**

Test 试验名称	Test Configuration 试验装置	Wall Thickness 墙厚（cm）	Bale Density 草砖密度(kg/m^3)	Thermal Resistance RSI 热阻值 ($m^2 \cdot K$)/W	Conductivity U- value 导热系数 ($W/m^2 \cdot K$)
McCabe，1993	Guarded Hot Plate method 保温板方法	57	133	11	0.09
Sandia Lab，1994	Thermal Probe 热工探测	46	83	7.74	0.129
Nova cotia，1995	Hot Plate in Existing Wall 墙内设置保温板	46	133	5.8	0.172
Oakridge Labs，1998	Guarded Hot Box 保温箱	58	107	5.8	0.172

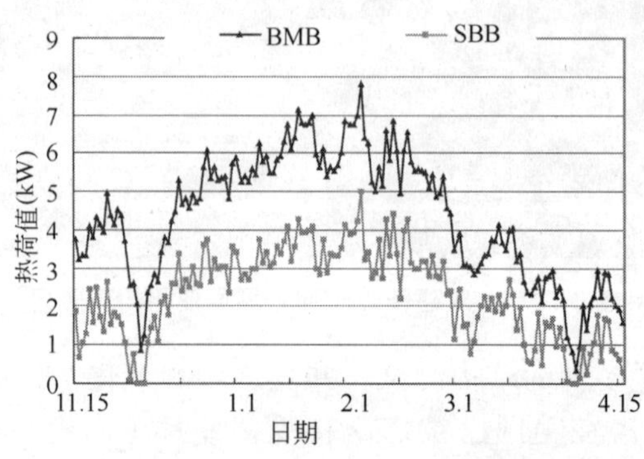

图5 采暖季动态热负荷曲线

7 结语

水稻是一年几生的植物，是一种可持续发展的农村经济作物；人类的发展和延续得益于它的果实。因此，利用稻草制成稻草砌块作为建筑的新型材料，是人类回归自然，保护环境，节约能源，延续人的生命进入一个良性生存链。发展循环经济、建设节约型社会需要人们重新认识稻草对人类的贡献。稻草砌块将是绿色节能建筑一种必不可少的墙体材料。

Green Economical Wall Material

——Straw-bale

Qingmei Kuang

(Department of Civil Engineering , Yichun Uinversity, Jiangxi 336000 China)

Abstract: This article has mainly introduced the research and development of complete sets of house which uses a new-type green wall material made of straw-bale. Straw-bale made up of dry crop straws . The straw-bale building can construct economic, easy-built, good-thermal Performance and instand brick. This kind of house accords with the strategic request for sustainable development of our country because its construction technology can utilize resources rationally, economize the energy, reduce the wastes, realize cleaner production , and protect the environment .

参考文献

[1] 徐至钧编著. 纤维混凝土技术及应用. 北京：中国建筑工业出版社，2003

[2] 李坚编著. 木材科学. 北京：高等教育出版社，2002.8

[3] 建设部人事教育司等边著译. MB 轻型房屋钢结构建筑体系. 北京：中国建筑工业出版社，2005.3

[4] [加拿大]Chris Magwood and Peter Mack. Staw Bale Building: How to Plan Design and Build with Straw. Gabriola Island Canada: new Island Publishers. 2000

[5] 大力推进住宅产业化，加快发展节能省地型住宅——刘志峰副部长在“节能省地生态—住宅产业可持续发展高峰论坛”上的讲话. http://www.cin.gov.cn/indus/speech/2005062106.htm.

[6] 陈君琛，郑金贵，杨菁，郑开斌，陈旺瑞 . 谷秆两用稻草栽培食用菌研究. 福建农业学报，2000，2

[7] 王婧，张旭. 草砖住宅的建筑节能性分析. 建筑材料学报，2005.8

[8] Ruppert, Jeff, and Grandsaert, Matt, “*A Compression Test of Plastered Straw-Bale Walls*”, 1999, University of Colorado, Boulder, Colorado Contact: Jeff Ruppert, jruppert2@uswest.net

[9] Faine, Michael, and Zhang, John “*A Pilot Study examining the Strength, Compressibility and Serviceability of Rendered Straw Bale Walls for Two Storey Load Bearing Construction*” University of Western Sydney, July 2002

[10] Donahue, Kevin, *Testing of Straw Bale Walls with Out-of-Plane Loads*, Report for Ecological Building Network Straw Bale Testing Program, September, 2003

[11] Aschheim, Mark, Ash, Cale, and Mar, David. *In-Plane Cyclical Tests of Plastered Straw Bale Wall Assemblies*, University of Illinois, 2003

[12] Simons, Bryce *Transverse Load Test and Small Scale E-119 Fire Test on Uncoated Straw Bale Wall Panels and Stucco Coated Wall Panels* SHB Agra Inc., 1993. Available from Natural Building Resources, 119 Main Street Kingston, NM 88042; ph 505-895-5652; www.strawbalecentral.com

[13] Haby, Guy and Fitch, William, *ASTM E84-98 Surface Burning Characteristics*, (*on*) *Straw Bale* Omega Point Laboratories, Inc., 2000 Available from Development Center for Appropriate Technology (DCAT) P.O. Box 27513, Tucson, AZ 85726-7513; ph 520-624-6628; www.azstarnet.com/~dcat

[14] Wimmer, Robert, Hohensinner, Hannes, Janisch, Luise and Drack, Manfred, *Wall Systems of Renewable Resources* (*"Wandsysteme aus nachwachsenden Rohstoffen"*). Gruppe Angepasste Technologie (GrAT) an der TU Wein (the Appropriate TechnologyGroup at Vienna Technological University), 2001

[15] Stone, Nehemiah *Thermal Performance of Straw Bale Wall Systems* Report for the Ecological Building Network, 2003

环保型绝热材料的开发与应用

尚建丽　桑国臣　王秀芬

（西安建筑科技大学材料学院，710055　西安）

摘　要：节能是绿色建筑具有的特征之一，开发环保型绝热材料则是实现建筑节能的重要途径。本文在对粉煤灰漂珠、准纳米材料形态的效应分析基础上，利用废弃的聚苯乙烯塑料颗粒，通过粉状材料和颗粒材料之间的优化级配设计，试验研究了环保型绝热材料的最佳比例和各种性能，工程实践应用表明，本文研制的绝热材料具有导热系数小，绝热性优良，力学性能不降低，施工性能稳定，无毒无放射污染等优点。

关键词：粉煤灰漂珠　准纳米　绝热机理　节能效应

纳米材料是指粒径在 1～100nm 范围内的粉体材料，由于粒径极小，所产生的尺寸效应和表面效应逐渐被人们所认识，目前在冶金、电子、磁性材料、传感器等领域得到了广泛应用。但是，针对量大面广的工程材料，若将高成本的纳米材料应用于此，所产生的性能效果会因经济成本的提高而受到抑制。因此，利用已有的工业废渣，进行纳米化处理，从而制备低成本、高性能节能建筑材料，具有重要的社会效益和经济效益。

粉煤灰漂珠是粉煤灰在形成过程中存在的一种特殊物质，虽然，从粒径尺度上达不到纳米级范畴，但针对建筑上广泛使用的保温材料，漂珠可谓准纳米级颗粒材料。

1　准纳米级漂珠的结构与性质

粉煤灰漂珠是煤粉在锅炉内燃烧，受湍流热空气作用高速自旋，形成的浑圆球状体。当裂解反应产生的氮、氢、二氧化碳等气体在熔融高温硅铝球内迅速膨胀，形成高真空的空心微珠，急冷下属于非晶态物质。在扫描电镜下观察其形态如图 1 所示。

从图中可以看出，其微观结构为玻璃态空心微珠，漂珠直径较小，断壁上有大量的纳米级气泡，具有质轻、分散性好的特点。

图 1　粉煤灰漂珠微观形貌

正是由于漂珠的特殊结构，使得其具有如下性质：

表面效应：漂珠的微小粒径，吸附能力强，可以吸附大量的活性分子，因此具有表面反应活性高。滚珠效应：浑圆的球状体，似为无数个滚珠，增加了材料润滑度。化学反应效应：漂珠本身具有活性反应度，与其他碱性或硫酸盐激发剂可产生化学反应，生成表面固相成分。阻热效应：空心漂珠中的空气在真空状态，导热性迅速降低，具有非常小的导热系数。

2 准纳米级漂珠的绝热机理

2.1 接触热阻机理

当导热过程发生在两个直接接触的漂珠固体材料之间时，由于漂珠固体表面不是理想平整的(微观结构)，所以直接接触的界面上容易出现点接触，导致准纳米空隙增加，准纳米空隙的存在则会给固体导热过程带来额外的阻力，将这种阻力称之为接触热阻，如图2 所示。

当界面上那些互不接触的界面空隙中充满气体时由于空气导热系数远小于固体的导热系数此时接触热阻的作用更为突出。因此，为了提高热阻，尽量增加分子间接触热阻。

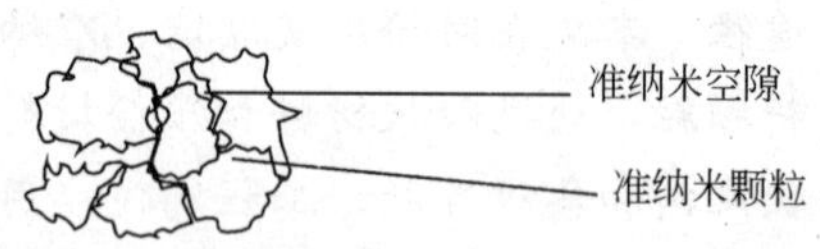

图 2 接触热阻示意图

根据这一原理，在复合建筑节能材料的制备中，添加适量准纳米级固体粉料，利用纳米颗粒尺寸小、表面积大，接触点迅速增多，大大提高固体导热过程中热阻的附加效用。

2.2 空气分子自由程机理

材料中热对流的产生主要来自块体材料内部孔隙和散粒材料空隙的空气存在，据资料介绍，空气中的氮气、氧气的自由程一般在 70nm 左右，也即当材料孔隙中绝大多数孔径较大时(大于 100nm)，材料导热过程中的空气热对流效应则不容忽略，显然，孔隙的存在降低了材料表观密度，是绝热材料实现的必备条件之一，但是孔隙尺寸增加势必会造成对流的增加，将会减弱阻热效果。因此，从减小孔隙尺寸入手可利用纳米孔这些大多数在 50nm～100nm 孔，可以基本消除对流传热；其次，掺入级配良好的微细纳米颗粒填充至大颗粒空隙内，使原有的空隙尺寸变为纳米空隙范围，从而减小热对流作用。

孔的存在与对流效应的关系在材料中随着空隙或孔隙的孔径由大变小而表现为强对流、弱对流、无对流、准真空的状态，在材料制备中，空气分子自由程的降低幅度达到准真空状态时，建筑绝热效果将成倍增加。

2.3 多界面反射机理

热辐射作为传热的一种形式，是分子、原子内部电子受激和振动时，产生交替变化的电场和磁场发出电磁波向空间传播的现象。与固相导热和空气对流换热不同的是，辐射换热不依赖于物体的接触，材料受有温差作用时发生高温物体辐射，当热射线投射在固体材料上，则发生部分被反射，随着纳米微粒的掺入、晶粒的超细化，导致结构内有序区域范围缩小，界面原子增多，反射几率增大；同时，固体材料结构不同，吸收能量不同，有些纳米材料具有很强的吸收中红外频段的特性，使得热辐射能量衰减幅度提高。纳米微粒可以将大块材料中连续能带分裂成分子能级，随粒径减小，能级间距增大，当辐射电场或磁场传播比平均能级间距小时，纳米微粒可以变成绝热体，有效地降低导热力。

3 准纳米建筑节能材料的配比方案设计

基于上述三方面机理，决定纳米节能材料具有优异绝热性的途径主要有两条：一是利用材料自身的纳米孔级的存在，以减小空气自由程产生的对流，同时，利用界面增加，提高反射率，消耗热量传递；二是以固相材料为载体，掺入准纳米材料，以增加接触热阻，减少固体空隙的对流，消耗热量传递。

本文在上述理论分析的基础上，利用粉煤灰漂珠这一特征效应进行了新型浆体绝热材料的研制及开发。通过大量试验摸索和工程实践证明，利用节能机理指导研发新型复合纳米节能墙体材料，可以获得较高节能指标的要求。

3.1 原材料的优选：在绝热机理及评价指标搭建的平台上，按以下层次对原材料进行了选择；见表1。

原材料的优选方案 **表1**

种　类	名　称	作　用
轻质材料	主相(废弃聚苯颗粒)辅相(膨胀珍珠岩颗粒)	降低密度、减小固相传导能力
改性材料	复合纤维、外加剂(减水剂、界面剂、粘结剂)	提高抗裂能力、提高保水性
粘结材料	无机胶凝材料、粉煤灰准纳米级漂珠	提高界面能、增加接触热阻、强度

3.2 原材料的组成设计

3.2.1 设计模式构思

浆体绝热材料的研制与开发：关键决定于以下两方面的解决：一是“质”，二是“量”。所谓“质”的方面，在于绝热材料的必备特性是轻，轻能削弱固相粒子传导的能力，轻才能使整体具有小的导热指标。本文选择的主相为废弃聚苯颗粒，辅相为优质珍珠岩，其主要性能指标见表2。

聚苯颗粒和膨胀珍珠岩的主要性能指标 **表2**

项　目	聚　苯　颗　粒		膨胀珍珠岩	
	技术指标	实测指标	技术指标	实测指标
堆积密度(kg/m³)	20～40	25	40～70	50
导热系数(W/m·k)	0.04～0.05	0.045	0.045～0.049	0.048
吸水率(%)	≤4	≤0.1	≤5	2.5
粒径范围(mm)	—	2～5	—	0.16～2

所谓“量”的方面，在于浆体绝热材料使用特性是必须满足一定的强度，且易于施工，由于强度来源于胶凝材料的水化反应，必然涉及到质轻的固相材料与粉状的粘结材料以及复合改性材料数量上的比例，本文则在两相交互作用方面进行了以下系统试验。

3.2.2 绝热骨料最佳比率

通过聚苯颗粒与膨胀珍珠岩不同比例组合试验，确定出空隙率最小匹配比率为1∶30。

测试数据如图 3。

从试验看出，聚苯颗粒与膨胀珍珠岩之间存在着一个最佳比例，寻找混合骨料最小空隙率，可以保证最大限度地发挥绝热材料的固相阻热效果。

3.2.3 骨料与胶凝材料的比率

在获得最佳骨料比率基础上，确定胶凝材料的数量比例是试验中保证强度，达到施工要求的重要内容，试验采用不同骨料比率对应的不同胶凝材料，按照沉入度均为 8cm 设计，测试结果见图 4。

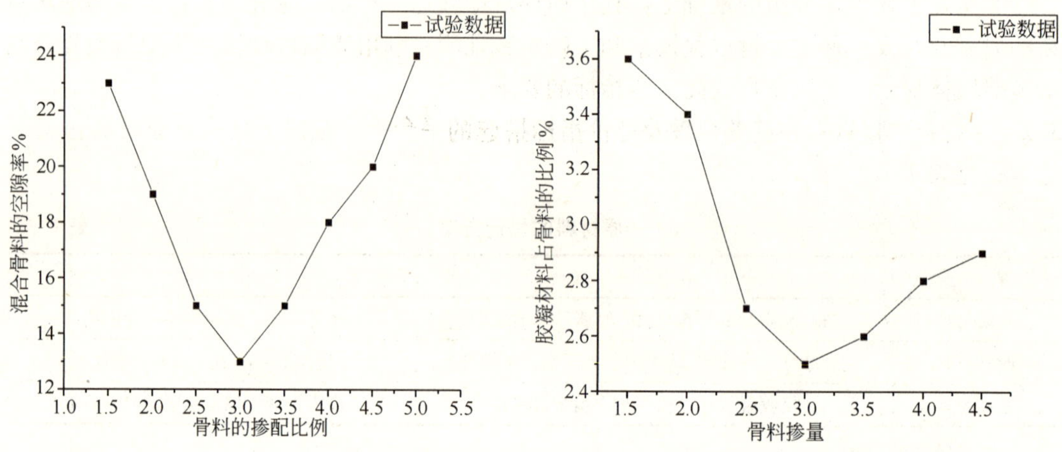

图 3 混合轻骨料空隙与骨料掺量关系

图 4 胶凝材料用量与骨料质量比关系

从图 4 可以看出，在最佳骨料比率附近出现了胶凝材料用量最小的趋势，说明此时确定的胶凝材料不是以填充为主，而是以粘结为主，最可能不破坏轻质骨料结构。

3.3 原材料的改性设计

按照前述的思路，尽管寻找出了轻质骨料和胶凝材料两大物相的数量匹配关系。但是两大相材料在堆积密度上出现了两个极端，轻质骨料的表观密度为 40～70kg/m^3，胶凝材料的表观密度为 1400～1600kg/m^3，相差极大的两种材料组合后，保证了保温性，但会影响强度。反之，照顾了强度，又会影响保温性。如何协调两者比例，是新型绝热材料研制的关键所在。本文利用漂珠准纳米级的粉粒（粒径小，质轻，具有活性反应性，抗压强度可达 10MPa 左右[3]）替代部分胶凝材料，可降低胶凝材料的堆积密度，但不降低其强度。同时，漂珠颗粒易于粘附在轻质骨料表面，还可以改善表面能，增加其分散性，试验表明，纤维、减水剂以及高分子聚合物在掺入漂珠颗粒后改性效果明显增加，表现在浆体流动性增加，硬化后强度不降低。

4 优化配比的试验过程及结果

将上述组份，按照正交设计方案 Lq(3^4)进行了系统和大量的试验，其结果及测试数据见表 3。

正交试验表 **表 3**

序号 \ 因素	混合绝热骨料 A	1#纤维 B	2#纤维 C	粘结材料 D	组合水平	测试值	
						导热系数 [W/(m^2·K)]	抗压强度 (MPa)
1	1	1	1	1	$A_1 B_1 C_1 D_1$	0.062	0.455
2	1	2	2	2	$A_1 B_2 C_2 D_2$	0.067	0.520
3	1	3	3	3	$A_1 B_3 C_3 D_3$	0.072	0.610
4	2	1	2	3	$A_2 B_1 C_2 D_3$	0.060	0.440
5	2	2	3	1	$A_2 B_2 C_3 D_1$	0.066	0.370
6	2	3	1	2	$A_2 B_3 C_1 D_2$	0.070	0.415
7	3	1	3	2	$A_3 B_1 C_3 D_2$	0.060	0.225
8	3	2	1	3	$A_3 B_2 C_1 D_3$	0.080	0.355
9	3	3	2	1	$A_3 B_3 C_2 D_1$	0.078	0.145

注：粘结材料是复合改性材料与漂珠材料的综合组份。

根据试验结果，按照设计要求，选择 $A_2 B_3 C_1 D_2$ 为优化组合配方。

5 准纳米级绝热材料的试验研究

按照上述优化配方，分别对粘结材料采用了两种对比实验，其一是不用漂珠替代水泥方案，另一种是用漂珠按一定比例替代水泥方案，分别进行了未改性和改性各项指标的测试，见表 4。

绝热材料性能的测试数据 **表 4**

序号	项　　目	指　　标	
		未改性的浆体保温材料	改性浆体保温材料
1	导热系数 [W/(m^2·K)]	0.081	0.065
2	粘结强度(kPa)	40	63
3	抗压强度(MPa)	0.40	0.45
4	抗折强度(MPa)	0.28	0.39
5	堆积密度(kg/m^3)	180	125
6	干表观密度(kg/m^3)	320	210
7	干燥收缩率(%)	0.59	0.48
8	抗冻性(质量损失≤5%；强度损失≤10%)	15 次冻融循环	20 次冻融循环

从表中数据可看出，掺入粉煤灰漂珠和外加剂、高分子聚合物的改性浆体保温材料，其性能明显优于未改性的浆体保温材料。

6 准纳米级绝热材料的工程应用

工程概况：西安市某高层住宅为框剪结构体系，地上 24 层，地下 2 层，外窗采用双层塑钢窗，屋面采用膨胀珍珠岩，冬季集中供暖，室外计算温度－9.6℃，室内计算温度16℃，其中外墙内保温采用了本实验研究的节能浆体保温材料，施工中和易性易于控制，涂抹性好，操作简单，经对单元用户进行抽检，采暖耗热量为 16.27W/m^2，采暖耗煤量 8.02kg/m^2，屋面传热系数 0.3W/m^2·k，外墙传热系数0.36W/m^2·k，户门传热系数 2.07W/m^2·k，窗户传热系数 2.67W/m^2·k，数据表明，各项技术指标达到了西安地区节能示范(65％)要求。

7 结论

新型浆体绝热材料的研究表明，利用废弃聚苯颗粒开发轻质保温材料，符合环保政策，将粉煤灰漂珠准纳米粉粒适量取代胶凝材料用量既解决了以往浆体绝热材料导热系数与强度之间的矛盾，又解决了难以在大宗材料中应用纳米材料的经济约束，准纳米级材料在绝热材料应用中有着良好的发展前景。

Development and Application of Environmental Protective Thermal Insulation Material

Shang Jian li Sang Guo chen Wang Xiu fen
(Xi'an university of architecture & technology, Xi'an 710055 China)

Abstract: Energy conservation is one of characteristics of green building, developing environmental protective thermal insulation material is an important path to realize building energy saving. In this paper, base on the effect analysis of quasi-nano float-bead fly ash, make use of waste polystyrene foam particle, design optimum size mix formula of powder and particulate material, study on the optimum formula of environmental protective thermal insulation material and get the optimum parameters. Proved by engineering practice, the thermal insulation material made in this paper is characterized by low heat conduction coefficient、good thermal insulation property、non-reduction of mechanical property、stable construction property、non-toxicity and non-radio pollution etc.

参考文献

[1] 王冲等．纳米颗粒材料在水泥基材料中应用的可行性研究．新型建材．2003.2
[2] 翟冠杰．粉煤灰漂珠的纳米结构及其传热机理研究．新型建材．2003.8
[3] 付晓茹等．粉煤灰漂珠的抗压强度及其影响因素研究．粉煤灰综合利用．2002.4
[4] 桑国臣．节能环保型保温材料的研制．硕士论文．2004.6

城市硬化地面铺装呼唤生态回归
——推广应用透水性铺装策略初探

王　波　　　　　　　　　　　　　　　　　　　　高建明
（厦门大学建筑与土木工程学院，厦门　361005）（东南大学材料学院，南京　210096）

摘　要： 推广应用透水性铺装，充分发挥透水性铺装所特有的生态环境优势，是改善城市生态环境的重要途径。城市硬化地面呼唤生态回归——透水性铺装。本文提出了透水性铺装推广应用的相关对策和建议。我国对透水性铺装的研究及应用尚处于起步阶段，但应用前景广阔。

关键词： 透水性铺装　生态回归　推广应用　策略

1　引言

广义的城市下垫面包括市区建筑物、构筑物和地面铺装。城市化地区作为人口高密度聚居和高强度生产经营活动地区，造成其下垫面的特殊性质。由于城市的发展，城市下垫面原有的自然环境如农田、牧场等发生了根本的变化，建筑物、构筑物高度集中，以水泥、沥青、砖石、陶瓦和金属板等坚硬密实、干燥而不透水的建筑材料，代替了原来疏松和植物覆盖的土壤或空旷的荒地。因此，城市化的重要特征之一就是原有的天然土壤植被不断被建筑物及非透水性硬化地面所取代，从而改变了自然土壤植被及下垫层的天然可渗透属性。这种改变是城市一系列环境问题的根源之一。透水性铺装的内部构造是由一系列与外部空气相连通的多孔结构形成骨架，同时又能满足路用强度和耐久性要求的地面铺装。透水性铺装生态环境方面的优势包括改善城市热、光环境、涵养地下水、水体净化、改善铺装地表土壤生态环境、吸声降噪、提高交通安全及城市防洪等生态环境效益。城市传统不透水硬化地面铺装呼唤生态回归——透水性铺装。针对推广应用透水性铺装急需解决的难题，结合中国国情，探索切实可行的政策措施，大力促进我国透水性铺装推广应用步伐，它是贯彻环境可持续发展战略，建设和谐社会的重要方面。

2　透水性铺装的应用推广工作急需解决的难题

透水性铺装的应用推广工作急需解决的难题包括以下几方面：

（1）公众对透水性铺装的环境效益认识不足，缺乏足够的了解。由于观念的问题，建筑规划及市政设计及管理者习惯于传统非透水性铺装的设计应用，对透水性铺装的可行性及生态环境效益持怀疑态度。

（2）在技术层面上，虽然国内陆续有几家透水性铺装的专业生产厂家生产出一些透水性铺装材料投放市场，与传统非透水性铺装相比，透水性铺装可供选择的种类以及路用性能指标均不同程度地存在着差距，透水性铺装的研究开发及产品质量还存在着很大的提升空间。另外，缺乏透水性铺装的相关设计及施工验收规范也是目前存在的突出问题。

（3）在市场开发方面，由于透水性铺装的生产工艺相对较为复杂，工艺配方还处于探索和保密状态，有些品种还处于小批量试生产，因此生产成本和施工成本相对较高，市场定价与传统非透水性铺装相比缺乏竞争力，这也是制约透水性铺装大面积推广的重要因素。

（4）透水性铺装的维护保养难度较大，需要专门的配套清洗设备，日常维护成本偏高。上述这些均不利于透水性铺装的推广应用。

3 实施透水性铺装改造及优化设计的原则措施

城市地面铺装实施透水性铺装改造及优化设计的原则包括：

（1）生态功能优先的原则——符合生态学原理，尊重自然规律，在生态功能优先的前提下，采用适宜的技术措施，优化地面铺装的材料与结构体系，充分挖掘透水性铺装的生态潜能和环境效益，维持包括城市地面铺装在内的城市自然生态系统的良性循环。

（2）因地制宜原则——城市有大小、经济水平高低、气候条件及地理环境差异，必须考虑城市的性质、规模、自然条件和城市环境现况，透水性地面铺装的制作、施工应尽可能利用当地原材料以降低成本。地面铺装的孔隙率及其下垫层透水体系应符合当地水文气候条件及植被特点，只有这样，才能建设起与城市特点相匹配的透水性铺装体系。

（3）整体优化原则——就是应着眼于城市生态环境整体，结合城市绿化配置、水体布局与地面铺装透水性改造进行优化整合，使其在实现城市生态环境可持续发展的共同目标下，将绿化配置、水体布局与地面铺装等要素的方综合环境效应，发挥最佳的生态环境的功能和作用。

（4）长远目标与近期目标结合的原则——城市地面铺装透水性改造及优化设计的实施，是一项复杂的系统工程，在规划设计时必须有超前意识，立足于长远的，高标准高起点的建设，营造和改造需要分期分批进行，合理确定城市地面铺装透水性改造的近期目标，使长期目标与近期目标很好地结合起来。

（5）科学性与可操作性结合的原则——城市地面铺装透水性改造及优化设计，应用科学的理论解决地面铺装建设的目标、框架结构、内容布局等原则问题。但为了提高其可操作性，对于部分重点的、近期的项目，应做到适应当前技术水平和施工条件的要求。推广应用透水性铺装的依据：一是，要以国家现行的法规、条例为依据。如《中华人民共和国城市规划法》、《中华人民共和国环境保护法》、《城市市政建设条例》等；二是，要与城市总体发展规划相协调衔接；三是，要与城市发展规模相适应，与区域性生态环境规划相协调；四是，要与城市生态环境现状相协调。

透水性铺装的施工具有自身的特点，其特殊性集中表现在如何保证透水性铺装本身及其下垫层的透水性要求上，因此施工过程中一定要确保透水性铺装砖缝的渗透要求，不能延用水泥沙浆勾缝形成刚性连接的传统做法，正确的做法应该是用细砂等透水性填充材料填缝处理以保证接缝的通透。下垫面材料的选用及施工应结合当地的实际，因地制宜地选用与该铺装面层透水性能相适应的砾石、粗砂等材料，形成良好的下垫层透水体系。透水性铺装面层与下垫层也应该注意避免用水泥砂浆砌筑形成的刚性不透水连接，应采用细砂干铺形成适于透水性要求的柔性连接。图 1 所示为透水性地砖砖缝构造示例。考虑到居民区、公园和街头广场绿化和美化的实际需求，选用实心板材铺路，但板材与板材之间要留出一定的泥土空隙，天然的草可在此处生长，这样的地面可以在满足路用要求的前提下，还有可观的绿化面积。图 2 所示为透水性铺装与地面绿化结合的示例。国内目前透水性铺装的施工质量不容乐观，很多名为透水性铺装实际并不透水或透水性很差。究其原因，一方面是施工单位对于透水性铺装的特殊要求缺乏足够的了解；另一方面是我国目前还没有制定出相关的透水性铺装施工验收规范，我国现行的地面铺装施工验收规范中很多内容不适合透水性铺装的特殊要求。相关职能部门应结合我国国情，尽快制定出相应的施工验收规范来指导透水性铺装的施工，满足透水性铺装透水体系的特殊要求，保证透水性铺装的施工质量。该项工作对于充分发挥透水性铺装生态环境效益具有至关重要的意义。

图 1　透水性地砖砖缝构造示例

图 2　透水性铺装与地面绿化结合的示例

通常透水性铺装的透水性能随使用年限的增加呈递减趋势。为确保该构造体系的透水要求，日常维护和保养工作也很重要。通常做法是定期使用喷射水枪或强力吸尘器等对透水性铺装进行清理。德国、日本等发达国家结合自身特点及各自的透水性铺装的应用情况，已经研制出相应的维护保养专用设备及产品。我国的透水性铺装的维护与保养的研究及相关配套产品开发也应逐步实施。

4　推广应用透水性铺装的相关策略

为了进一步做好透水性铺装的研究及推广工作，充分发挥透水性铺装在改善城市生态环境方面的效能，探索出一条适应我国市场经济的发展需要，实现最佳的经济效益和社会效益，形成具有良好可持续发展能力的透水性铺装产业化市场。这需要相关的材料研制开发、建筑规划设计、施工维护以及市政管理等各专业人员的共同努力，综合应用各专业知识，才能取得最佳效果。

4.1　加强透水性铺装生态环境效益的舆论宣传

应充分利用各种媒介及方式宣传透水性铺装生态环境效益，使公众明了非透水性铺装对生态环境的负面影响。针对公众普遍存在的“地面铺装理应是非透水性的”的思维定势及误解，通过科普宣传，改变人们对透水性铺装的可行性普遍持有的怀疑态度，使公众特别是市政工程设计以及决策人员树立应用推广透水性铺装的信心。

4.2　制定整体经济效益、社会效益最佳的透水性铺装专项规划

首先要调查我国城市现有非透水性铺装的现状，对不同地区城市硬化地面下垫层的地质结构以及当地的气候特色做出科学的评估，在此基础上，结合城市防洪排水以及雨水利用进行综合规划，因地制宜地制定出适应当地气候及地质条件的透水性铺装实施细则，编制城市透水性铺装的专项规划，并作为城镇建设的指导方针。透水性铺装专项规划应结合现有城市规划的相关内容，追求整体最佳的经济效益与社会效益，按照先局部后整体、先试行后推广的原则积极稳妥的进行。

4.3　建立创新激励机制

图3　多孔透水性混凝土球筏示例

透水性铺装的研究开发及推广是个利国利民的重要工作。在透水性铺装研究推广工作中，要大力培养创新人才，建立创新激励机制，营造创新氛围。近年来，随着改革开放的深入，相关技术人员积极走出国门、放眼世界，向透水性铺装应用推广技术发达的国家学习，通过学习大大缩短了研究开发的时间，缩小了与发达国家的差距。通过对国外先进技术的消化吸收，结合我国的地理环境、经济水平、技术水平等国情，积极开展透水性铺装新工艺、新设备(包括专用的透水性铺装日常维护设备)的研究，提出适应我国经济技术水平的新工艺，开发出性能更佳、费用更低的新设备，降低日常维护成本，为透水性铺装的推广普及和产业化提供保障。研究将城市现有不透水铺装改造成透水性铺装并结合附设雨水贮留设施的可行性技术方案，以及研究开发透水性铺装在净化水体方面的环境效能，这些方面都是透水性铺装今后值得关注的课题。图3所示为东南大学与日本合作研制开发的应用于护坡铺装及水体净化的多孔透水混凝土球筏，研究成果及示范工程已在上海黄浦江松浦水厂改造、镇江北固山滨水区、云南省罗时江滨水区推广应用，效果良好。

4.4　改善政府职能，加强宏观调控

我国市政设施建设过去一直由政府垄断经营，投资、生产、经营基本上由政府包办。随着改革开放的不断深入，这种经营体制已陷入比较困难的局面。政府部门应转变观念，进一步确定“政府引导、市场主导”的思路，按照“谁出资、谁决策、谁受益、谁担风险”的原则。在推广应用透水性铺装的初期，应加强政府部门的宏观调控，借鉴国外相关

经验，制定应用实施透水性铺装的相关优惠措施，让投资者能得到适宜的回报。大力拓展市政建设筹资渠道，多渠道筹措资金，打破行业垄断和地域限制，放宽投资准入，通过体制创新，引入竞争机制，积极拓展融资方式，探索推广BOT(建设一经营一转让)BLT(建设一租赁一转让)BOO(建设一拥有一经营)等多种方式，吸引社会资金、私人资金和境外资金投资城市地面铺装工程建设，为推广应用透水性铺装提供充足的资金。围绕透水性铺装产业化目标，制定相关法律、法规、政策、规范、标准等，通过地方行政手段或地方法律等方式，采取有力措施，落实以上政策。通过政府职能的改善和宏观调控，争取从公众思想到社会大环境都形成推广应用透水性铺装的新局面。

5 结语

城市透水性铺装改造是以透水性铺装材料开发及应用技术为支撑，以城市气候及生态环境为对象，结合材料、水文、地质等学科的相关理论，经过多学科的交叉协作，最终实施的市政工程措施。该技术设施的最终实施及推广是以环境、社会经济效益的统一为原则的。应该依托透水性铺装现有的材料开发、施工管理的技术水平为基础，结合地区气候特点、经济水平、资源状况以及当地水文地质条件，因地制宜地选择适宜的透水性铺装的相关技术参数及指标规格，以环境效益作为选择实施透水性铺装的重要参考依据。城市透水性铺装的产业化及其推广应用，其价值不但体现在相关行业的经济效益之中，还更多地体现为潜在的环境效益和社会效益方面。它对改善和提高城市生态环境质量、贯彻实施城市生态环境可持续发展等方面有着重要作用。综上所述，我国城市透水性铺装产业化潜力巨大，但任重道远，有众多的技术、意识、经济、管理问题有待解决。希望各专业人士携手合作，在透水性铺装技术及产品研发，项目的投资、建设、管理等多方面共同努力，保证透水性铺装产业化市场及推广应用健康有序的发展壮大。可以预计"透水性铺装"这一新生事物必将在祖国大地上生根、开花、结果。

Urban Pavement Calling On Ecological Regression

——Tactics of Generalization and Application Porous Pavement

Wang Bo

(Architectural and civil engineering college of Xiamen University, Xiamen 361005 China)

Gao Jianming

(Material college of Southeast University, Nanjing 210096 China)

Abstract: Generalizing and applying porous pavement, fully developing its ecology environmental benefit, it is an important measure to improve urban ecology environment. Urban pavement was calling on ecological regression. Some advice and measures was suggested. Even so the research and applying on porous pavement is still in its initial stage . Porous pavement's splendid future may be expected because of its various advantages.

参考文献

[1] 王 波，李 成. 城市生态物理环境与透水性铺装. 工业建筑. 2002(12). 29～31

[2] 杨士弘等编著. 城市生态环境学(第二版). 北京：科学出版社，2003

[3] 蒋剑虹. 城市污水资源产业化策略. 第46届IFHP世界大会大学生论坛论文集. 2002

[4] 王 波. 透水性铺装的生态环境效益. 全国首届博士生论坛(建筑分册)论文集. 2003

[5] 王 波，王 焱，高建明. 透水性铺装的透水体系，建筑技术. 2004年第7期

绿色居住小区的环境绿化

李晓红　郎小霞　蒋咏梅
（青岛理工大学艺术学院，青岛　266033）

摘　要： 本文就当今居住小区环境绿化中存在的许多一味注重绿化景观、而忽视其实用性的普遍现象，提出绿色住宅小区环境绿化的实用性设计原则，建议应用立体绿化技术，适宜可行的绿化方法，尽可能地加大绿化空间，节约土地资源，提高绿化数量和质量，真正提高住区的整体绿化环境，进一步推动我国绿色建筑的发展进程。

关键词： 绿色居住小区　环境绿化　实用性设计　立体绿化

引言

雅典宪章将人类的居住活动称为"城市的第一活动"，随着人们对生活质量要求的不断提高，以及对生存环境认识的逐渐深刻，全社会的环保意识在不断增强，营造绿色建筑、绿色居住小区正成为越来越多的开发商、建筑师追求的目标，成为人们向往的居住环境，而其中的居住区环境绿化系统更是衡量绿色居住小区的重要条件之一。

1　居住区环境绿化的作用

1.1　文化艺术方面

优美的小区绿化环境，可以为居民提供娱乐、欣赏、文化、休憩的场所。环境绿化丰富的色彩、婀娜的形态、淡淡的清香，配以景观设计的借景、对景，无不让人流连忘返，令人心旷神怡。可以说优美的小区绿化环境是居民不可缺少的精神食粮，这往往成为我们赏心悦目的表象，成为我们现今设计的重点。

1.2　生态健康性方面

小区绿化除了有创造空间、美化环境的作用外，设计中还要从生态的角度出发，注意它在生态健康性方面的作用。主要是指绿化有维持空气中的碳氧平衡，提高空气的清洁度、减少大气污染(吸尘、净化)、降温、增湿、降低噪音等方面的功效，这是我们设计中需要进一步加强的方面，也真正体现了绿化的重要实用性及创造住区微小气候的关键作用。

1.3　安全功能方面

主要是指住区绿化可以减少空气中放射性危害，植物通过对光的吸收和反射，可以减轻和消除眼部疲劳；多种植树木可以减轻台风的侵害、有效保持水土防洪水；住区的绿化场地也可以在受到地震、火灾时的避难场所。

植物叶片光合作用(Potosynthese)❶ **表 1**

植物吸收	通过光合作用	生成植物物质基础	植物排放
$6CO_2 + 12H_2O$ 二氧化碳 ＋水	Strahlungsenergie $\longrightarrow$ 光辐射能量	$C_6H_{12}O_6$ 碳水化合物	$+6O_2+6H_2O$ 氧＋水

2　当今住宅小区环境绿化中存在的误区

居住区环境是当前商品房竞争的热点之一，从开发商的角度分析，居住区绿化可以提升居住区档次。在寸土寸金的都市里，开发商想方设法增加绿化率，满足都市人生理上、心理上对绿色的需求以及对自然环境的向往。于是随之的出现的楼盘，在环境绿化方面可谓是把绿化中的花草树木、水、铺地等等全面地艺术化了，“以景观为中心”的开发模式蔚然成风，在楼盘广告中各以自己独特的环境绿化设计作为自己的主题，迎合了人们的购房心理，借此来推销自己，成为房地产商的卖点和追求利润的工具手段。

然而在这些炒作中，大多只是浮于形式外表，这些居住环境景观看起来很优美，但大多并没有从居住者切身的健康需要入手。在这些赏心悦目的绿化后面，是否思考过绿化其他方面的实用性呢？是否考虑环境绿化对于居住环境微小气候的营造问题呢？可以看到在我们当今的小区总体规划中，大多数小区缺乏的立体绿化规划配置，进行垂直绿化的少之又少。这普遍反映出对这方面的忽视。

3　绿色居住区环境绿化的构成和实用性设计原则

绿色居住区源于绿色建筑，绿色建筑体系即可持续发展人类住区。为了加强绿化，居住区绿化常以公共绿地为中心，以道路绿化为网格，以绿地带为纽带，以住宅庭院绿地为基础，点、线、面结合设计，由中心绿地、组团绿地、宅旁绿地、道路绿地、公共设施专用绿地、建筑及附属设施的垂直绿化、屋面绿化等构成。现今设计中常常更多地注重景观，而忽视实用性；注重平面，而忽视立面。由此，如何更好的“以人为本”考虑符合绿色居住区的环境绿化，日渐成为人们关心的问题。其主要侧重以下几个方面：

3.1　因地制宜，注重绿化的多样性

居住区绿化应适宜小区的地形地貌及气候条件，合理进行建筑平面布局；绿化的形式必须适宜居住区各种使用空间的功能、适宜小区不同类型使用者；植物的配置也必须“适地选树”，尽可能适应性强、有地方特色的本土植物。

3.2　艺术与生态合理结合

在考虑绿化生态健康作用的同时，注意绿化的艺术、景观效果，让小区的居住者拥有

❶ 摘自 Kord Baeumer. Allemeiner Pflanzenbau. 3 Auflage，Verlag Eugen Ulmer Stuttgart 1992，SBN3-8252-0018-3 P30。

诗情画意的健康的生活环境。

3.3 提高绿化的参与性

居住区的环境绿化只是用来欣赏的或者说只是一个装饰品，它需要居民的参与，一片大的草坪再加上“请勿践踏草地”给人的感受是冷冰冰的。这要求绿化设计考虑合理的实用功能，把地方特色的文化融入其中。

3.4 生态功能优先的原则

提倡植物的立体绿化，尽量用多层次绿化结构，避免或少用单层草地绿化的形式，多采用常绿阔叶树种。适当地增加流水、瀑布、喷泉等利用喷筒电效应，提高空气的清洁度。

3.5 多方位的绿化

提倡阳台绿化、屋顶花园的垂直绿化和室内绿化等多种绿化形式，即采用立体绿化技术，减少硬质铺地，加大绿化空间，珍惜土地资源，提高绿化量。

4 居住区环境绿化中的几点具体建议

4.1 提倡科学植树、生态铺地

建筑周围树木的布置位置往往能在一定程度上引导风的方向。如图 2 所示科学合理的植树，有利于建筑物的自然通风。当在建筑的长向迎风一侧种植树木时，如果树木在建筑的两端向外延伸，则可加强建筑内的通风效果。当在迎风一侧窗前种植一排低于窗台的灌木，且当灌木与窗的间距在 4.5～6m 时，可使吹进窗内的风的角度向下倾斜，有利于促进底层房间的通风效果(见图 3)[1]。

图 1

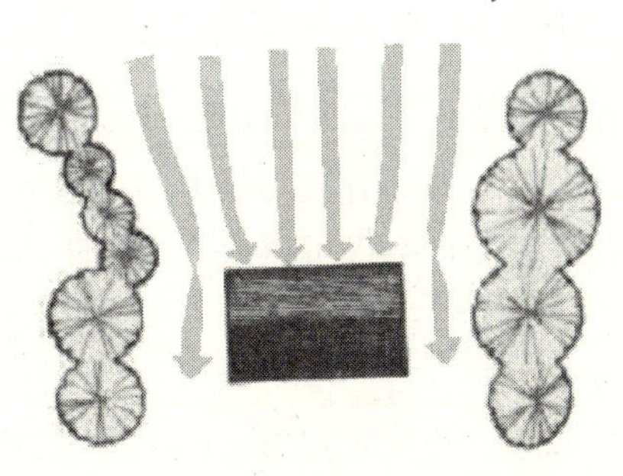

图 2

图 3

住宅周边的铺地吸收太阳辐射热后，由于自身的温度升高而对周围产生热辐射。所以在居住区绿化中尽可能减少硬质铺地，采用嵌草砖(透水的)铺地。这种透水的铺地比起车道、步道、停车场、游戏场、广场常变成不透水的硬质地面，如：停车场采用嵌草砖铺装(如图 1)，既节约了水泥材料，又增加了种植面积，美化了环境，又可以吸收利用雨水，

[1] 摘自林川等编著．小城镇住宅建筑节能设计与施工．北京：中国建材工业出版社．2004.1 P56-57。

减轻地表径流，减轻市政管网的压力。可以改善大地的渗透能力，对基地保水非常有利。同时，改变了铺地的物理热性，对太阳的热辐射能通过水分的蒸发而使得自身的温度较低，因此对周边住宅的热辐射降低。目前多以多孔质的植草砖、水泥砖、砌石来铺设地面，为相当普遍的透水铺面方式。

4.2 提倡垂直绿化

所谓垂直绿化，就是在建筑物或建筑小品上，利用植物形成立面绿化效果，多与建筑立面、栏杆灯柱、棚架、凉亭、拱门等结合，是提升小区绿化面积的途径之一。国外许多经济发达的国家，都十分注重这方面的发展，将垂直绿化作为建筑设计过程的一个组成部分，把植物作为附属于建筑的一种材料，并将植物自身的生长特性与建筑设计自身的规律有机结合，使其成为一个有机过程。

在居住区的绿化中，我们不仅仅只是将住宅周边环境“面”的绿化、小区内的道路“线”的绿化，还应该为建筑本身进行绿化。具体的方式可通过墙面绿化、阳台、露台绿化等立体绿化的手法，增加建筑“绿点”来提升住区整体的绿化量，不仅可以节约土地资源，又可以达到室内降温、缓和城市热岛效应的功效。

4.2.1 墙面的垂直绿化

对居住小区内住宅建筑进行垂直绿化，对美化立面环境、增加绿地面积和形成小区良好的生态环境有着重大的意义(图4)。此种垂直绿化主要应用在东西墙面，是防止“晨晒”和“西晒”的一种有效方法，能够更有效地利用植物的遮阳和蒸腾作用，缓和阳光对建筑的直射，间接地对室内居住空间降温隔热，显著调节其温度，起到降低房间热负荷的作用，并且降低墙体对周边环境的热辐射。攀缘植物生长速度快、覆盖面积大、绿化占地少，是进行垂直绿化主要材料。垂直绿化还可以按照人们的意图，为建筑物的立面进行遮挡和美化；同时可以减低墙面对噪音的反射，吸附灰尘，减少尘埃进入室内。如应用爬山虎，地锦等有吸附能力的材料不需任何支架，就可以绿化6层楼高的位置。小区内采用垂直绿化，不仅可以成为城市小区的重要景观，而且具有良好的生态效应。

图4

4.2.2 阳台、窗台绿化

在阳台上摆放盆花或设置花池种植植物，这本来是一种个人行为，但由于阳台在整个建筑中是最具公开性的地方，因此这种私人行为就产生了公众性的作用。如果对其进行适当的规范和管理，那么就不仅是一家一户得到了绿色，而是会使整个小区的绿化覆盖率大大提高。其实不光是居民住宅可以进行阳台绿化，点点滴滴，见缝插绿，所聚集起来的总绿量是很可观的。

阳台是住宅建筑立面重点绿化装饰的部位。由于阳台的空间有限，宜栽种攀援或蔓生植物。可以将多种种植方式如：种植槽、花箱、花盆等。窗台绿化可以进一步丰富绿化空间。当人们平视窗外时，可以欣赏到窗台的“小花园”，感受到接触自然的乐趣。最简单的窗台种植是将盆栽植物放置在窗台上，盆下用托盘垫住防止漏水。

4.2.3 屋顶绿化

屋顶绿化主要作用就是减少屋顶排水量和防止屋顶表面温度的激烈变化。因此其具有良好的隔热保暖作用，并能为居住者提供户外活动和交往空间。据研究发现夏季水泥屋面昼夜都向空气散热，而草坪屋面昼间向空气散热强度远小于水泥屋面，夜间冷却空气，具有减缓城市热岛的作用。这种对屋顶进行绿化的举措对扩大城市绿化、美化城市、调节城市气候、提高城市环境质量，也有着积极的作用(图5)。

图5

由于屋顶绿化的空间布局受到建筑固有平面的限制，施工又关系到相关工种的协同，其大面积的贯彻实施就需要有相对成熟的技术支持和良好的运行机制。在这个方面，德国一直处在世界的领先地位。其宝贵的实践经验值得我们研究和借鉴(见表2)。

屋顶植被化后1986～1988年间雨水储存和雨水排放量统计表(德国汉诺沃大学测定)❶ **表2**

基层厚度(cm)	植被物种形式	雨水储蓄量/减少向市政管网排放的雨水量		
		春、秋(冷季节)	夏(热季节)	总时间阶段中
2	苔藓类植被	40～50(%)	55～65(%)	48～58(%)
4～6	苔藓类、草本、草类植被	45～50(%)	60～70(%)	52～62(%)
8～10	草本、草类植被	45～50(%)	65～75(%)	55～65(%)
14～16	草本、草类植被	50～60(%)	80～90(%)	65～75(%)

4.3 大力提倡室内绿化

绿色植物对建筑室内环境的影响非常大，它可以改善室内环境与气候。据测定室内绿化可明显降低室内CO_2浓度，其效果与室内绿化量显著相关；对于飘尘含量降低作用明显，平均可达60%以上；室内绿化在一定程度上还可减少细菌数目；随着室内绿化时间的增长室内工作人员的警觉性和注意力也明显提高，进而提高人的反应能力；还可显著提高人的视力水平。另外，绿色植物色彩丰富艳丽，形态优美，作为室内装饰性陈设与许多价格昂贵的艺术品相比，更富于生机与活力，不仅使人赏心悦目，消除疲劳，还能愉悦心情，影响和改变人们的心态，减少焦躁与忧虑。目前在我国城市已很普遍，在普通的办公室、住宅内都有摆放盆花、盆景、插花，有的还设置室内景园等。

5 结语

绿色象征着生活、生命，象征着健康，正当跨入新世纪的历史时期，我国正在进行大规模的居住区建设，绿色一作为其主旋律，包含着广泛的内容。如何在建设过程中，更好

❶ 摘自Jens Drefahl. Dach Begruenugn. 1995. P17 Tabelle 5.1。

地利用自然条件、保护自然环境，以人为本，创建健康、无害、舒适的绿色环境，成为我们共同关注的切实问题。建设绿色居住小区是当前和今后我国住宅小区发展的必然选择。

The Environment Virescence of Green Residential Districot

LI Xiao Hong LANG Xiao Xia JIANG Yong Mei
(School of art，Qingdao Technological University，Qingdao 266033 China)

Abstract：Based on the general phenomena that the residential district only pay attention to make landscaping and neglect the practicability of environment virescence，this paper put forward the principle of practicable design and advise on the environment virescence of Green residential district that we should apply the technique of spatial virescence and doable methods to enlarge the green space，economize the ground resource，enhance the quantity and quality of virescence，veritably improve the whole environment virescence of residential quarter and impel the development course of our country Green building。

参考文献

［1］ 周浩明．张晓东编著．生态建筑—面向未来的建筑．南京：东南大学出版社 2002.3
［2］ 林川等编著．小城镇住宅建筑节能设计与施工．北京：中国建材工业出版社 2004.2
［3］ 中建建筑承包公司．中国绿色建筑/可持续发展建筑国际研讨会论文集．北京：中国建筑工业出版社 2001
［4］ 张涛．建筑结合竖向绿化设计．北京建筑工程学院硕士学位论文．2003.6

速生林木材自增强技术的开发和应用

马小冈

（深圳市高雅科技有限公司，深圳　518057）

摘　要：本文介绍了木材自增强技术的原理和特点，特别强调了自增强技术对于人工速生林木的适用性以及提高其利用价值的作用，简要分析了这项技术的应用前景。

关键词：自增强技术　人工速生林木材　木材密度

中国经济的蓬勃发展带动了房屋建筑、家具制造、室内装饰业的空前繁荣，刺激了对木材的大量需求。但我国的森林资源贫乏，缓解我国木材原料供应紧缺局面的有效途径之一是大力营造人工林，但随之而来的疑虑是人工林的材质能否满足市场需要的问题。以当前国内主要木材市场的供应行情来看，每 m^3 价格在 1500 元以下的普通木材主要有杨木、杉木、落叶松、泡桐等人工速生林木，但由于其质地疏松，除了用作人造板原料、低档木制品和纸浆原料以外，高性能高附加值的利用较少。

为了改善人工林木材的物理性能，我单位的科研人员进行了木材自增强技术的研究与开发，其第一代产品大幅提高了普通木材的密度和强度，可使原本用于火柴、一次性筷子的低档木材用于硬木地板、高档家具；第二代产品的市场价格、物理强度、节能特性等性价比处于同类产品的领先地位，其产品的最大优势在结构材；第三代产品使刨花、木屑具有混凝土的功能，是一种无缺陷高强度低成本的复合型木质材料，从而为废旧木材的回收利用找到了最佳途径。

1　木材自增强技术的原理与特点

1.1　原理

自增强技术(Auto-frettage)源于军工，其基本原理是通过超高压力突破材料的弹性极限，使材料产生塑性变形以达到向心性致密，使木材表面和深层整体强化，从而大幅提高木材的密度、强度等物理性能。自增强技术不仅适用于原木，圆木、方木、中小径材以及刨花、木屑，尤其适宜于材质较松软的人工林木材。

与现有的木材改性技术(压缩、树脂强化等)、人造板技术(胶合板、纤维板、刨花板等)、木质复合技术(塑合木、层积塑料等)相比，其区别主要表现在：木材的自增强技术不需要加温加湿蒸煮软化，更加趋显节能特色。

(1) 处理条件：依据木材学理论，木材的塑性与温度和含水率有直接关系，因此，木材的压缩需在 120 ℃以上高温条件下蒸煮软化，使木素产生充分的塑性流动，才能消除压缩木中的内应力，减少回弹性[1]。但高温高湿软化处理不可避免地会降低木材的强度。而用自增强方法处理木材时，木材的塑性变形在常温条件下产生，木材含水率越低，处理的效果就越好，处理后木材原有的纤维等主要组分的性质不会发生变化，改变的仅是木材的细胞孔隙。经中国林科院木材工业所对自增强技术处理杨木的显微检测："自增强杨木纤维没有明显压溃，但被挤压，细胞间隙变小，纤维直径变小，导管由圆形变成多角形。单位面积细胞增多，致使密度加大。非自增强杨木密度 0.40g/cm^3，自增强杨木密度 0.98g/cm^3。"[2]

(2) 处理手段：轴向加压方法的缺陷是容易造成木材开裂或者卷翘，而且质地不均匀，因此市场较少接受；目前比较广泛应用的平面压缩只适合于板材，不能压缩原木和圆形木材，并必须经过石蜡、树脂等化学药剂处理，否则板材受潮吸水后极易膨胀变形；而压缩木技术是仅从上下或左右两个方向施加压力挤压木材的管胞、导管等细胞孔隙，不仅容易造成木材开裂，而且因压缩应力的存在，致使压缩木产生回弹趋势。为弥补强度的降低，压缩木需经树脂等化学药物处理。而自增强技术对于木材的施压形式属于全方位，其表现就是木材体积致密性缩小，木材的天然纹理因致密化而纹理色泽加深。

1.2 技术特点

(1) 改进木材性能。经过自增强处理，可提高木材密度，改善木材的强度、硬度、阻燃、防腐等物理性能。经测算，应用该技术对泡桐、杉木和杨木进行处理，其气干密度可分别由处理前的 0.28、0.38、0.40g/cm^3 增加到 0.91、0.94、0.98g/cm^3。这项自增强技术对于木材密度提高极限可达到 1.2g/cm^3 左右。

(2) 工艺简单。与其他的强化方法相比，自增强技术可以直接对干燥后的原木进行处理，从设备启动到木材的后处理只用 2～3 道工序，30min 即可完成。

(3) 成本适中。与常规木材压缩方法相比，省却了高温、除水工艺，降低了生产成本。

(4) 原料广泛。尤其适合 30cm 以下的中小径原木和木屑等下脚料的加工。

(5) 实用性强。可以根据处理木材的不同用途，灵活调整工艺参数，获得不同密度、强度性能的木材，拓展人工林木材的利用途径。

2 自强木的应用

原木型自强木(自增强第一代)可使用于火柴、一次性筷子的低档木材用于硬木地板、高档家具。由于木材经自增强处理后会形成一种自紧现象，如果木材钻孔加进钢筋、钢管或 PVC 塑料管，可以使木材自动裹紧钢筋、钢管或 PVC 管，从而更进一步提高木材的强度。这种超高密度自强型"钢筋复合木"或中空结构的"塑管复合木"(自增强第二代)具有超强的承重力和强度，因此对于推广木质结构房屋和建造、修复中国古建具有重大意义。同时我们可以利用其芯材的特点以标准件插接组合，将木结构传统的卯榫连接、钉连接改革成钢结构螺栓连接或 PVC 套管式的插接，同时可以利用其空心特点插筋拉固、穿管走线、采暖通风、填充隔热、排水排烟等。由于具有快速组建和优良的抗震保温隔热性能，这种木结构建材又是抢险救灾临时建筑和野外营房的首选材料，可以说，用钢筋复合木能够制造最具抗震能力的房屋。同时根据这一特点，刨花木屑也可在制材时加进钢筋、

PVC管或金属管制成“钢筋刨花复合材”(自增强第三代),从而大大提高木材的利用率,扩展用途,降低生产成本。

经深圳市建设工程质量检测中心检测,杉木经过自增强技术处理后(自增强第一代),其破坏载荷是钢筋混凝土的1.86倍;杉木加钢筋经过自增强技术处理后(自增强第二代),其破坏载荷是钢筋混凝土的2.01倍;木屑加钢管经过自增强技术处理后(自增强第三代),其破坏载荷是钢管混凝土的1.97倍[3]。以刨花木屑为主的加筋的木材强化专利技术首次使刨花木屑具有混凝土的特性,例如,原始木屑材的抗弯强度和破坏载荷均低于水泥砌块,但是经过自增强处理后,其抗弯强度是水泥砌块的1.23倍,破坏载荷是水泥砌块1.12倍。木屑的强度能得到这样大幅提高,是废旧木材充分利用的重大突破。

木材的自增强专利技术特点表现为:第一,在常温条件下使木材产生塑性变形以达到向心性致密化;第二,将钢筋混凝土的概念引入木材,使木材或刨花木屑与钢筋紧密结合,达到超强化效果。木材的自增强市场特点表现为:它不仅符合建筑节能、生态、环保、抗震等要求,其综合成本低于传统的钢筋混凝土建材;尤其是它具有极强的市场竞争优势,以建材市场为例:进口的木结构建材(我国的木结构建材目前多以进口为主)市场价格为5500～6500元人民币/m^3左右,而我们的自增强产品市场价格为3500～4500元人民币/m^3左右,且其强度是进口同类产品的3倍以上,物理性能远远超过进口木结构建材,也就是说,无论是技术特征还是市场价格,我们的产品比进口产品都具有绝对优势。

3 结语

总之,该项技术及其产品可广泛应用于房屋的承重结构,提高房屋的抗震性能;应用于建筑的墙体、楼板,提高房屋的节能效应;应用于地板、家具,减少稀缺高档木材的进口需求。它将大幅提高人工速生林木的市场价值,并以“变废为宝”的方式利用木材的下脚料,这不仅符合国家提倡的充分利用可再生资源和废旧材料的循环经济的大政方针,而且可以为中国的木结构建筑开辟新的国内外市场。随着天然林木资源的日渐减少,以及国际绿色组织的环保压力日益加大,如何有效利用人工速生林木资源已成为人们共同研究开发的课题。

Development and Application of Auto-Frettage Technique for the Artificial Fast-Growth Wood

MA Xiaogang

(Shenzhen Gaoya Technology Co., Ltd. Shenzhen, Shenzhen, 518057 China)

Abstract: To increase the density and mechanical property for the artifical fast-growth wood, the author has made a development research on Auto-Frettage Technique for those materials. In this article, the principle and character of the technique are presented. It is pointed that this technique is special suitable for logs with diameter less than 30 cm. In the end, the application of this technique is analyzed.

参考文献

[1] 赵仁杰，喻云水．木质材料学．北京：中国林业出版社，2003
[2] 中国林业科学研究院木材工业研究所．“自增强杨木和非自增强杨木对比报告”．木编 2004-167 号
[3] 深圳市建设工程质量检测中心检验报告：1LX2005-3719．3720．3721

高性能塑料推拉窗

王治家

（大连爱必索门窗技术开发中心，大连 116033）

摘 要：我国门窗的气密水平低，导致的室内空气外泄，造成建筑能量大量的流失，提高建筑门窗的气密性是节能的首要课题。现有推拉窗作为建筑构件，虽然气密性最低，但是，其优异的力学结构和双杆件抗风压结构，比平开窗有更大的优越性。推拉窗的上述优异性能，是同时开发：门窗“保温性、抗风压、气密、水密、隔音性”等各项物理性能和经济性能，实现门窗的物理性、经济性同步提高的关键性资源。80 塑窗的缺点，是我们很少研究现代门窗理论，不能挖掘推拉窗的技术资源所致，非窗种不好。

关键词：焓 气密 K 当量 保温腔增倍结构 刚度增倍结构 气密行程机构

1 门窗的保温

室内的热空气，通过门窗进行：传导、热辐射、空气对流而向外界散失。下面，以实验数据结合理论推导，对门窗的传导辐射、气密流失的热量作定量分析，研讨门窗的保温性。

1.1 门窗的热传导、热辐射因素与热流失

《建筑外窗保温性能分级及检测方法》GB 8484—2002 的检测方法 5.6.4 条规定：“试件的开启缝应两面用胶带密封”。因此，此标准规定的“外窗保温系数 K 值”，仅仅是窗的“传导辐射系数 K 值”。

1.1.1 热工实验条件设定：开启缝隙贴胶带密封；境界条件设定：环境气温：暖室 20℃、冷室 0℃、冷热区的窗表面热阻按(jis)标准设定。

1.1.2 热工实验试件的使用环境的设定。

以北京 1 月为例(北京市观象台气象信息)：北京冬季 1 月份的平均气温：－3.7℃，平均相对湿度 44%，室内平均气温 20℃，平均相对湿度 60%，为试件的使用环境条件。

1.1.3 热工试验的热传导结果及使用状态下热量流失：见表 1 中的传导流失热部分。

1.2 门窗的气密因素与热流失

1.2.1 上述试件窗的气密检验结果：见表 1 中的气密流失热部分。

1.2.2 空气外泄与热量的流失：

(1) 空气外泄热流失的物理过程。

向室外泄露 $1m^3$ 暖湿空气，因气压的平衡，自然会由室外补充 $1m^3$ 干冷空气，干冷

空气被加热同时相对湿度降低，便吸收水分，促使室内水的自然蒸发，加上人为的烧水等活动的强制性挥发，最终自然蒸发加上强制性挥发量与暖湿气外泄量达到平衡的一个稳定的室内相对湿度。

（2）上述物理过程的北京1月地区，室内暖湿气外泄与热量流失的关系

A. 1kg 气温－3.7℃的干气，含："相对湿度 44％水蒸气"的冷空气的焓：

$$I_{a_1}=H_{a_1}+H_{v_1}=1.005\times(-3.7)+1.2166(2500.5+1.86\times(-3.7))/1000$$
$$=-3.7185+3.0337=-0.6847\text{kJ}$$

式中 H_{a_1}——1kg 气温－3.7℃的干气相对于零度的焓；

H_{v_1}——1kg 气温－3.7℃的干气中含的"－3.7℃相对湿度 44％"水蒸气相对于零度的焓。

B. 1kg 气温 20℃的干气，含："气温－3.7℃、相对湿度 44％的水蒸气"的暖空气的焓：

$$I_{a_2}=H_{a_2}+H_{v_2}=1.005\times(20)+1.2166(2500.5+1.86\times20)/1000=23.187(\text{kJ})$$

C. 1kg 气温－3.7℃的干气，含："相对湿度 44％的水蒸气"的冷空气，转变成："1kg 气温 20℃的干气，含相对湿度 60％的水蒸气"的室内暖湿空气吸收的水量：

$$G=14.68\times60\%-2.765\times44\%=7.5915(\text{g})$$

D. 室温 20℃，蒸发 7.5915g 水，消耗的热量：

$$Q_1=7.5915\times2453.3=18.624(\text{kJ})$$

E. 1kg 干 20℃空气含相对湿度 60％的水蒸气，流失的物理过程吸收的热量即流失的热量：

$$Q=I_{a_2}-I_{a_1}+Q_1=23.187-(-0.6847)+18.624=42.4957(\text{kJ})$$

F. 1kg 干空气含相对湿度 60％的水蒸气，20℃气温，标准大气压下的体积：

按分容积定律：

$$V=\Sigma(m/M——)\times R\times T/P=0.76866\text{m}^3$$

式中 m——干空气质量；

$m1$——水蒸气质量；

M——干空气摩尔质量；

$M1$——水蒸气摩尔质量；

R——摩尔气体常量；

T——开氏温度；

P——标准大气压。

结论：1m³ 室温 20℃相对湿度 60％的暖湿空气流失到到气温－3.7℃相对湿度 44％的室外损失热量：

$$Q=(1/0.76866)\times42.4957(\text{kJ})=55.285(\text{kJ})/\text{m}^3$$

1.2.3 窗的气密与热量流失及气密的保温 *K* 当量：

窗气密 *K* 当量＝缝长×气密量×55.285×1000J/(60 分×60 秒×2.25m²×23.7℃)

表 1

窗参数			传导流失热			气密流失热			
窗种	尺寸 (m)	玻璃 (m)	3 次检平均传热系数 [W/(m²·K)]	窗传热系数平均值 [W/(m²·K)]	北京 1 月份使用环境下传导辐射流失的热量	气密 (10Pa) (m³/h)	气密缝隙长 (m)	气密保温 K 当量值	窗流失热量 (W)
80 塑料推胶条	1.5×1.5	4+10+4	2.96 2.96 2.95	2.96	157.8W	1.1	6.95	2.2	117.4
80 塑料推毛条	1.5×1.5	4+10+4	2.96 2.96 2.95	2.96	157.8W	1.6	6.95	3.2	170.8
60 塑料平开窗	1.5×1.5 双扇	5+12+5	2.45 2.44 2.46	2.45	130.6W	0.3	8.06	0.696	37.1
60 塑料平开窗	1.5×1.5 双扇	5+12+5e	1.79 1.78 1.77	1.78	95W	0.3	8.06	0.696	37.1

结论：我国目前建筑保温节能的首要任务是：提高门窗气密性能。

1.3 推拉窗气密技术资源

由窗的气密检测可知：推拉窗的气密，由毛条充填式密封，实现如平开窗的密封行程机构压缩密封条，是推拉窗气密的技术资源。

1.4 换代推拉窗气密技术资源的开发

技术方案：滑动时撤销密封压紧力，关闭后施加密封压紧力。如图 1 所示的楔形块密封行程工作原理。

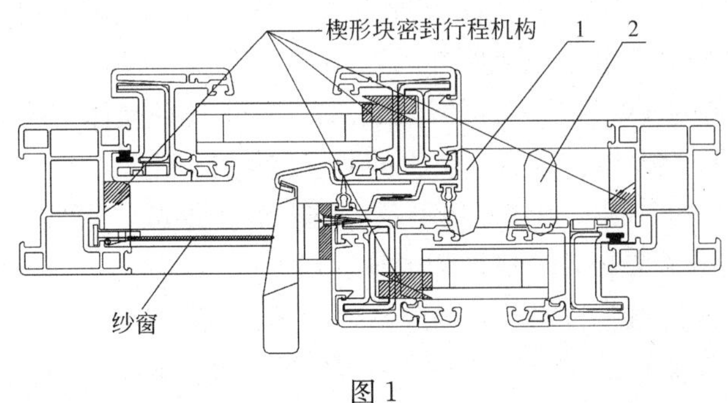

图 1

2 推拉塑窗的保温与推拉塑窗各部热传导

2.1 现行 80 推拉窗与型材相关的各部传导热流失与节能技术资源

2.1.1 通透玻璃部分资源。

现行 80 推拉窗通透玻璃部：玻璃的热阻结构：4mm＋10mm＋4mm，等效热阻：0.3334；热贯流率 2.9993(BL 实验条件：市内气温 20℃、表面热传达率 8.93(W/m^2·K)、室外气温 0℃、外表面热传达率 23.26(W/m^2·K)。中空玻璃结构是：4mm＋14mm＋4mm，等效热阻：0.3544；热贯流率 2.8217。1.5m×1.5m 推拉窗，在使用环境下，可开发热资源含量：Q=(2.999－2.8217)×0.62×1.29×2×23.7＝6.73W。两轨推拉窗，可最低成本开发此资源。

2.1.2 窗框结构部分的资源区域(图 2)。

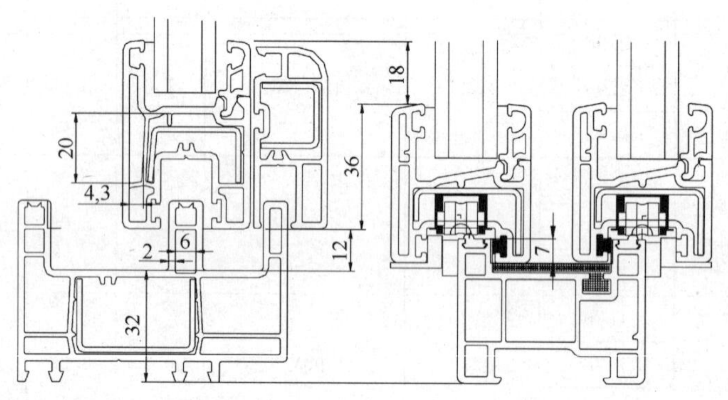

图 2

相对于三腔结构的框型材的胴体，高滑轨形成的 12mm 高的 2mm＋6mm＋2mm 的结构、扇形成的 20mm 高 4.3mm 厚的单腔保温结构及相对中空玻璃，型材框矮化后，形成的玻璃补偿区，是低热阻区。

2.2 换代推拉窗非玻璃部分的节能资源的挖掘技术

2.2.1 降低框型材的滑轨的高度，减小低热阻结构区域的散热面积。

(1) 解决降低轨高引入的窗扇的安装拆卸问题、窗的水密问题。

(2) 技术方案(图 3)：图 3 中 3 解决矮轨窗扇的安装拆卸问题；技术方案(图 1)：图 1 中 1、2 解决矮轨窗的水密问题(水密原理后述)。

2.2.2 增加窗扇的非抗风压三边的保温腔数，使推拉窗整体三腔保温。

(1) 解决增加扇型材保温腔数后，引入的窗扇抗风压问题。

(2) 技术方案(图 3)：图 3 中 2.8 扇型材的钢衬腔不缩减，扇框焊接前，在非抗风压的三边型材腔内组装分腔隔板，形成增加的加厚保温腔。

(3) 解决：组装式保温腔技术引入的腔室间的热对流问题。

(4) 技术方案：在非抗风压得三边型材腔内组装分腔隔板后，在腔的端口塞上发泡材质的端口塞，阻止腔室间的空气对流。

由技术方案的气密行程机构技术和矮轨技术以及上述三腔保温构成：保温腔增倍结构技术，使塑料推拉窗的杆件抵抗热传导性能接近三腔平开塑窗的杆件。节能效果如表 2。

节能效果：

(195.6－328)/328＝40.3％

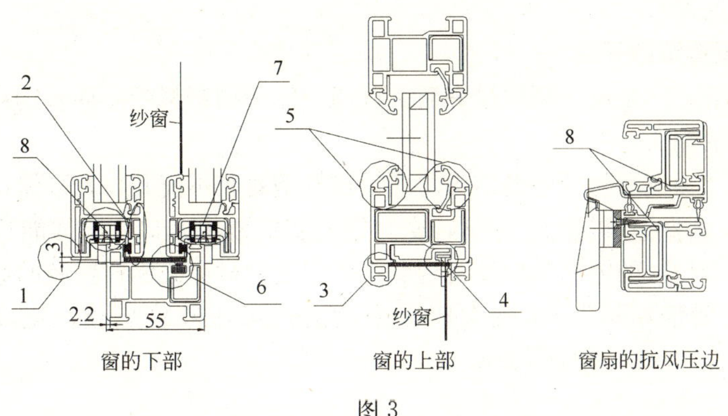

图 3

节能对比：标准窗，在北京 1 月份的环境下的热量损失　　表 2

	轨高区热资源		扇区资源		玻璃补偿区热资源		窗传导流失热	窗对流热流失	窗流失总热
散热面	0.06839		0.1981		0.0956				
参　数	K	散热	K	散热	K	散热			
80 塑推	3.27	5.3	3.24	15.21	3.24	7.34	157.8W	170.8W	328W
换代塑	2.557	4.14	2.557	12	2.99	6.79	152.88W	42.7W	195.6W

3　塑窗抗风压

3.1　塑窗的抗风压结构的基本理论

抗风压杆件的端的约束，只有“有效”的固定端约束，如：钢衬直接预埋或与预埋件焊接，才能使杆件刚性增倍，其他隔塑料型材的连接方式，都不能扩大钢衬的刚性。即：杆件端有微小变量，杆件就会产生大的挠度弯曲(图 4)。

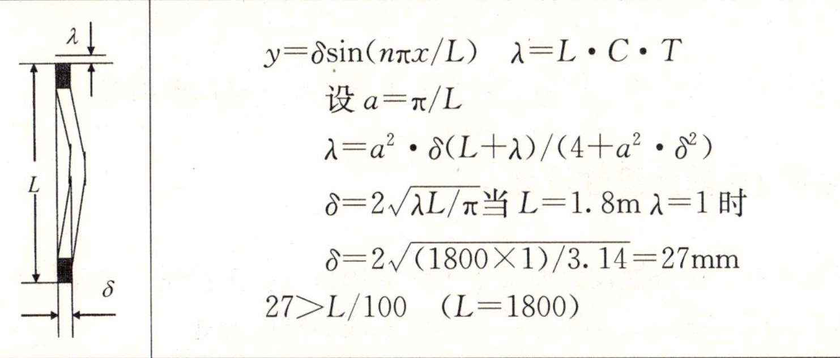

$$y=\delta\sin(n\pi x/L)\quad \lambda=L\cdot C\cdot T$$

设 $a=\pi/L$

$$\lambda=a^2\cdot\delta(L+\lambda)/(4+a^2\cdot\delta^2)$$

$\delta=2\sqrt{\lambda L/\pi}$ 当 $L=1.8\text{m}$ $\lambda=1$ 时

$$\delta=2\sqrt{(1800\times1)/3.14}=27\text{mm}$$

$$27>L/100\quad(L=1800)$$

图 4

3.2　现有 80 塑窗的抗风压的技术资源空间

如图 3 开发了推拉窗扇的抗风压技术资源，扇钢衬的惯性距：$I=2.61$；1.5m×

1.5m标准窗的安全风压达到3700Pa以上。并通过滑轮嵌入盒体内，组装到扇上的技术，解决钢衬的防锈问题。

3.3　中梃抗风压资源的开发

去掉框的内筋，作中梃，框焊接好后，开通中梃端的塑料壁，插入钢衬，留出预埋端或与预埋件焊接的端头。

此技术突破“塑窗型材及内筋结构对应设计及钢衬腔不能开口，以保持钢衬腔室封闭防锈的死原则”。用发泡聚氨酯密封材，密封钢衬腔的侵水开口，同样能达到：钢衬防锈的目的。塑窗中梃型材的内筋与框型材内筋不对应及钢衬突破钢衬壁腔的结构，配合密封材，构成“刚度增倍结构”，实现大跨距中梃的抗风压刚度增倍，挠度由 $f=5\omega L4/384EI$、缩小成：$f=\omega L4/384EI$。

4　塑料推拉窗的水密

4.1　塑料推拉窗的水密原理

窗扇滑轨间的槽，由密封块分割成：室内侧、室外侧。此处的水密原理：室内槽和室外槽通过密封块构成连通器。连通器外侧的水，由风压压入室内侧。窗扇的下侧附着的水，随气流和毛条的引导，越轨进入室内侧。

4.2　室内侧溢水条件

风压差扬起的水柱高，加上气泡引起浪涌高(或气泡的喷溅)之和大于轨高，则室内侵水。

4.3　塑料推拉门窗的水密的技术资源

现有推拉窗的“连通器结构”的存在，轨高的挡水方式不能与风压抗衡。因此，在连通器的底部开设阻止水进入室内侧的泄水壕，就是推拉窗的水密技术资源。

4.4　换代推拉窗的水密技术

(1) 开设泄水壕(图1中的1)。

(2) 在内扇滑轨的外侧，开设沿滑轨走向的泄水壕，形成无引导物、引导缝隙的空间结构。

(3) 框型材的滑轨槽的底部直接挤出泄水壕(图3中的6)，并开设泄水壕的排水孔(图1中的2)。

(4) 上述1.4的气密技术，使扇底附着的水滴无侵入室内的动力。

5　换代塑料推拉窗的经济性设计

对大众门窗来说，经济性设计取决于杆件的用材量。关键是：少用材料解决门窗抗风压的安全问题；强度问题；保障窗各结构功能实现的刚度问题。

5.1　矮轨框的技术手段，不仅降低了扇、框型材的延米重，而且框轨上直接安装玻璃压条，进一步降低成窗的用材量。上述3.3的“刚度增倍结构”的技术手段，实现一体窗的经济目的。

5.2　软边纱窗的技术手段(图3中的4)。不仅大幅度降低框型材的延米重，而且省掉纱窗框的用材。

5.3 框型材的大钢衬腔设计，在拼樘窗的结构中有抗风压的功能(图 5-80 推拉窗)，一体窗的技术下，窗框只有结构功能，安装连接功能。适当缩小框型材的断面尺寸，可提高其结构功能的刚度。宽由 80mm 缩小成 55mm；高由 52mm 缩小成 42mm，壁厚由 2.0mm 加厚到 2.2mm，大幅度提高型材断面的稳定刚度。以此技术手段，不但缩减型材的延米重、而且缩减框钢衬的延米重，更加提高推拉窗的经济性。

5.4 经济性设计技术资源挖掘的效果(表 3、图 5)。

表 3

窗种	换代		80 推拉		节省材料	
两室一厅住宅	型材(kg)	钢衬(kg)	型材(kg)	钢衬(kg)		
客厅 3×2.4(m)	28.49	25.0	50	47.39	型材	35.97%
主卧 1.8×1.7(m)	16.4	12.5	23.1	18.7		
辅卧、厨房 1.5×1.5(m)	24.1	20.3	34.6	25.64	钢衬	36.98%
合计	68.96	57.8	107.7	91.73		

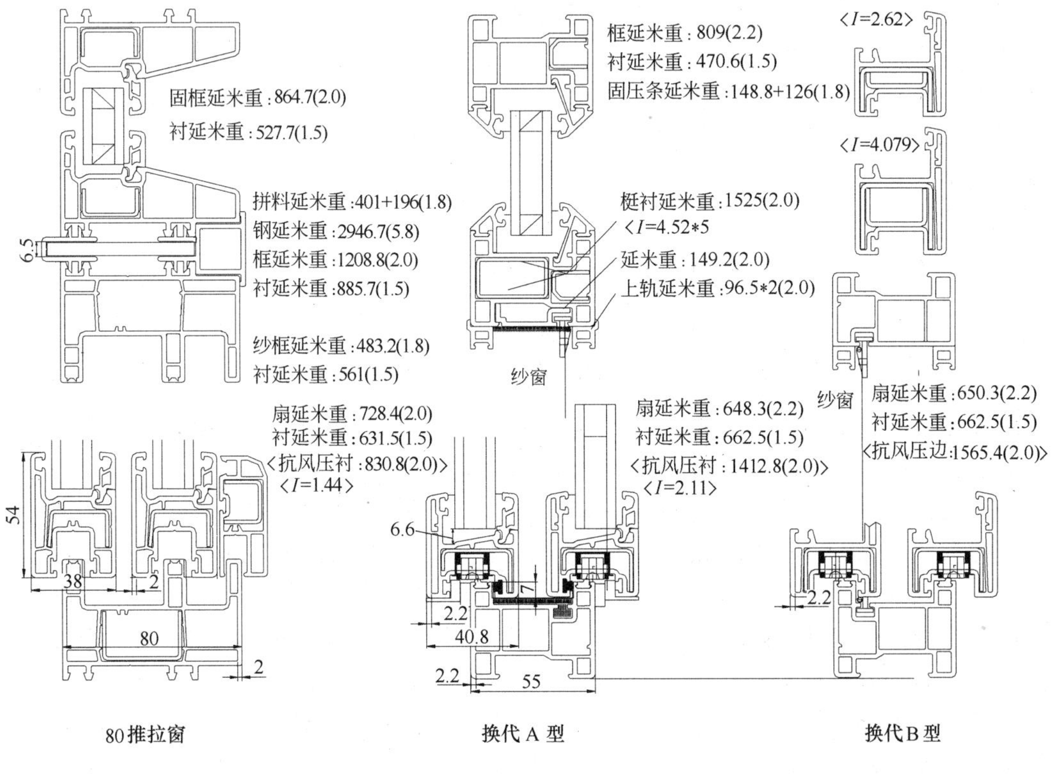

图 5

以两室一厅住宅的窗用材为比较基准，客厅窗：3m×2.4m；主卧窗：1.8m×1.7m；辅卧、厨房：1.5m×1.5m。

门窗虽然只有简单的结构，可由各行业生产好部件、五金，再由门窗厂组装成窗，再把窗应用于社会的每个人，就构成一个系统。

2004 年北京取缔 80 及以下规格塑料推拉窗，仿佛 82、88、90 不取缔；一年后、辽宁取缔 80 塑窗，取而代之是平开窗，人们渐渐认识到：推拉的 82、88、90 性能比 80 也没好多少。其实这两个行政指令的根本目的是："取缔所有材料、所有窗种、所有规格的达不到"使用性能要求、健康要求、环保要求、可持续性发展要求"的窗；非材料中的塑料、非窗种中的推拉、非规格中的 80"。看一下我们对简单结构门窗基本研究状况，就可以理解上述的行政指令的缺陷。

(1)《GB 8484—02 保温分级》5.6.4 条："要用胶带密封窗缝"，而且气密划分的标准《GB 7107—02 气密分级》不能应用于保温(因为：气密泄漏量的保温 K 当量与保温标准的保温 K 值，非相同数量级，不能和并应用划分窗保温等级)。

(2) 门窗气密水准划分作为标准，严重脱离现代门窗实际(密封条密封的现代窗，检测泄漏量超过 $3m^3/mh$，已经很少见，何况《GB 7107—02》规定的 $6m^3/mh$)。

(3)《GB 7106—86 抗风压》附录 A 规定：单层玻璃 $f<L/130$；中空玻璃 $f<L/180$ (玻利在 $L/100$ 的挠度下，容易产生碎裂，在 $L/125$ 的挠度下，通常的窗会产生残留变形。因此规定安全风压：$f<L/130$。$L/180$ 非玻璃碎裂的安全标准，作为："中空结构失效，玻璃漏气的基本风压条件下的耐疲劳标准"也许更合适一些)。《GB 7106—02》虽然取消附录 A 的内容，但是全国都在用 $f<L/180$，设计中空玻璃窗的安全抗风压，材料浪费可想而知。

总之，我国门窗的基本理论的研究，需要更多的人力物力投入，才能提高门窗整体系统技术和门窗的整体的社会效益。

High-performance plastic slide window

Wang Zhijia

(Dalian Abiso Window & Door Technology Development Center, Dalian 116033 China)

Abstract: The low air tightness of doors and windows in our country have resulted in the leakage of indoor air and great losses of energy in a building. The improvement of the air tightness is the key to energy conservation. The existing sliding window is of an enclosure structure in a building with the lowest air tightness, but its outstanding mechanical structure and wind-resisting double-bar structure are much superior to the casement window. They are the critical resources to simultaneously develop various physical and economic performances including the heat insulation, wind resistance, air tightness, water tightness and sound insulation for synchronous improvement of these performances. The defects of 80 plastic window is attributable to our failure to study the theory of modern doors and windows and tap the key technical resources of the sliding windows, not to the poor variety of window.

参考文献

[1] 专利申请《高气密性能推拉塑料门窗》. 公开号 CN 1683742A
[2] 专利申请《小遮光自洁滑槽推拉窗》. 公开号 CN 558078A
[3] 专利申请《两矮轨高性能框面内卷轴纱窗推拉塑窗》. 公开号 CN 1587625A
[4] 专利申请《两轨有纱窗可开启推拉窗》. 公开号 CN 1683741A
[5] 专利申请《任意转角高抗风压塑料门窗》. 公开号 CN 1605708A

INNOVATIVE ENERGY EFFICIENT AND ENVIRONMENT FRIENDLY BUILDING MATERIALS DEVELOPED IN INDIA

T. N. Gupta

(Former Executive Director,

Building Materials & Technology Promotion Council

Ministry of Urban Employment & Poverty Alleviation, Government of India

UNIDO Consultant in the area of

Energy Efficient & Eco Friendly Building

Materials Manufacturing Technologies for Low Cost Housing)

Abstract: Exponential growth in the construction works has been noticed, the world over, during recent several decades. The main reasons for such a high growth includes inter-alia rapidly rising population and high pace of economic development. The situation, thus, puts a heavy demand on building materials production sector. To meet this rising demand, traditional manufacturing methods can not be sustained with adversely affecting the environment, as these manufacturing technologies result in very high consumption rate of natural resources.

Scientific and technological researchers in India have developed a variety of innovative materials which can be manufactured by using wastes, by-products and residues from agricultural and industrial operations. Such wastes have been causing pollution problems. Recycling of these renewable resources with newly developed manufacturing processes reduced environmental pollution on one hand and conserves energy on the other-because researches carried out in India, have established that recycling of wastes consume much less energy than used in traditional manufacturing processes.

This paper presents energy requirements of traditional building materials and quantity of savings that can be achieved in energy and non-renewable natural resources by recycling of wastes and by-products from agriculture and industry and natural fibres. Such technologies help in producing alternate building materials which can effectively substitute traditional materials/products both in quality and performance.

Keywords: Energy-Conservation Environmental-Protection Recycling of Wastes for Green Building Materials

1 Background

Unprecedented pace of economic development, the world over, is leading to increased construction activities and higher and higher rate of exploitation of natural resources. Man has been ignoring ecological consequences and causing substantial damage to the atmospheric chemistry which in turn adversely impact the natural life support systems. While scientific and technological advancements in construction and building materials industry do offer a large variety of options to meet the needs and aspirations of fast progressing societies, the rapidly rising quantum of construction of buildings and manufacturing of building materials needed, continue to affect the environment through high resource consumption, environmental disruption, chemical pollution and deforestation. Construction activity, particularly the manufacturing of energy intensive materials, significantly contributes to high carbon dioxide emissions. An estimated 8 to 20 percent of these emissions in different countries are due to construction and building materials production activities. Process of conversion of natural raw material resources may vary from simple handmade tools to most sophisticated industrial production methods and manufacturing technologies.

Life becoming more entrenched in the economic activities and with meteoric rise of population, dependence on traditional materials and production methods could not sustain pace of building activity without detriment to natural environment and quality of life in human settlements. This led to vigorous efforts on the part of scientists and technologists for improving performance and durability of materials, reducing consumption of raw material resources and energy.

With increasing inter-disciplinary appreciation and continuing scientific efforts it has become possible to relate the engineering properties of materials to their typical structural variables such as the molecular weight, degree of molecular orientation, crystallinity and phase morphology. Indepth knowledge of properties not only helped in developing new products, but also in specifying materials with predetermined criteria for thermal comfort, structural strength, safety against forces of natural hazards, durability and cost to meet needs buildings required by different socio-economic segments in any society - developed or developing.

2 Selection Criteria for Building Materials

The complete range of materials having bearing on design and development of housing and buildings can be broadly classified as: stones, clays, metals, polymers, ceramics, inorganic glasses, wood and composites. Though the mechanical, electrical and thermal properties, performance through service life in varied environmental and other aggressive conditions and aspects of durability, reliability, quality and aesthetics are very important for choice of materials for desired applications, the signal parameter for final selection, however, remains the trade-off between cost and availability on one hand and knowledge about performance on the other. Innovations in materials sector have provided ample opportunities to introduce new products that would enable reducing strains on the mineral resources and non-renewable sources of a country. The principal factors affecting a country's materials production and usage are consumption of energy, natural non-renewable resources and structure of economy. All these factors coupled with national policies and programmes for housing, buildings and infrastructure development govern the demand of materials, potential for investment in R&D, and domestic industrial production or for import.

3 Innovative Low Energy-Consumption and Environment Friendly Building Materials

In view of the rapid increase in the volume and complexity of building works, extensive research and development has been undertaken in India for developing innovative green materials. Such efforts have mainly focussed in the following areas:

ⅰ. reduction of consumption of natural non-renewable raw material resources

ⅱ. development of environment friendly building materials and composites by recycling of wastes, by-products residues and natural fibres from agricultural and industrial operations/manufacturing

ⅲ. development of energy-efficient manufacturing technologies for reducing energy consumption in production of building materials and reducing embodied energy in buildings

Now efforts are being made to evolve an 'Eco-Rating Tool' to guide the choice of materials that will maximize the efficiency of critical resources. Extensive efforts are also being made for promoting use of passive solar systems to reduce operational energy in buildings.

4 Use of Industrial and Agricultural Wastes in Production of Building Materials

There is a well recognized group of industrial wastes which has found wide application, in many countries, as alternatives to clays, limestone and various other argillaceous and siliceous materials in the production of bricks, tiles, cement, concrete, glass and ceramics, with considerable advantages of not only conservation of natural resources but also saving of precious energy as some of the wastes have their own intrinsic fuel value. Steel and aluminium and a variety of synthetic materials are now considered well established substitutes for traditional timbers in buildings. The developing countries must be selective to adopt all such technologies in which industrial, mining and mineral wastes constitute a raw material resource for building materials.

Besides, inorganic industrial wastes, there are many natural fibres and agro-industrial wastes which have great potential for their transformation into structural components for roofing, panelling, partition and ceiling, replacing high energy-intensive materials produced by high temperature sintering, melting and fiberisation. Developed countries might use high and energy-costing glass fibre, kevlar, ceramic fibres and polymeric fibres but where there is serious fuel shortages, emphasis should definitely be on the utilisation of natural fibres particularly in developing countries.

Table 1 provides comparative energy requirements of different building materials, Table 2 shows a list of well-established industrial, mining and mineral wastes and by-products and their use in production of building materials as developed in India. This table also shows approximate values of savings of important minerals, and Table 3 gives energy saving in the manufacture of alternative building materials.

Table 4 indicates a list of important natural fibres and allied agro-industrial wastes and the names of some well accepted commercially produced building boards and sheets made from them. This table also indicates the saving of timber and the binder used in the materials which get substituted by organic-fibre based composite sheets and panels.

Comparative energy requirements of building materials **Table 1**

Material	Primary energy requirement MJ/kg
Very high energy	
Aluminium	200-250
Plastics	50-100
Copper	100+
Stainless steel	100
High energy	
Steel	20-60
Lead, Zinc	25+
Glass	12-25
Cement	5-8
Plaster board	8-10
Medium energy	
Lime	3-5
Clay bricks & tiles	2-7
Concrete	
In-situ	0.8-1.5
Blocks	0.8-3.5
Precast	1.5-8
Sand-lime brick	0.8-1.2
Timber(sawn)	0.1-5
Low energy	
Sand, aggregate	<0.5
Fly ash, RHA, volcanic ash	<0.5
Soil	<0.5
Adobe	<0.2

Source: *Building Materials in India*: 50 *Years - A Commemorative Volume*, *editedby* T. N. *Gupta*, *published by Building Materials & Technology Promotion Council* (*BMTPC*), *New Delhi*, *India*, 1998.

Industrial, mining, mineral wastes and their application in building materials as alternate to traditional materials and natural resources **Table 2**

S. No.	Industrial waste/by-product	Application in building materials as alternative to traditional materials	Traditional material saved fully or partl (20%~30%)	Natural resources saved fully or partly (10%~20%)
1.	Blast furnace slags (i) air cooled (ii) foamed (iii) granulated	• Dense aggregate in concrete or road • Light wt. Aggregate for concrete • Portland-slag cement super sulphate cement	• Rock, stone • Traditional light wt. (high energy) • Ordinary portland cement (OPC) sulphate resisting portland cement	• Stone • Clay, slate, shale • Limestone, clay
2.	Ferro-alloys and other metallurgical slags	• Pozzolana - metallurgical masonry cement	Lime pozzolana	• Clays (for pozzolana) • Limestone
3.	Flyash (Puverised fuel ash)	i) Portland - pozzolana ii) Concrete filler iii) Sintered light wt. aggregate iv) Lime-flyash calcium silicate brick v) Cellular concrete (flyash-lime) vi) Clay-flyash brick vii) Stabilisation in roads, mines, lagoons etc.	• OPC • Fine aggregate • Other L. W. Aggregate • Sand-lime brick • Cement - sand based cellular concrete • Burnt-clay brick • Road materials & other fillers • Limestone, clay • Sand, crushed stone • Clay, shale, slate	• Limestone, clay • Sand, crushed stone • Clay, shale, slate • Sand, lime • Cement, sand • Clay • Traditional road material clay
4.	Byproduct gypsum (from fertilizer, hydro fluoric acid, boric acid)	• Gypsum for cement • Gypsum plaster & blocks • Gypsum plaster fibrous board • Special cements	Mineral gypsum plaster and blocks from mineral gypsum. Sulphate-resisting portland cement.	• Mineral gypsum • Mineral gypsum • Clay & limestone (for OPC)
5.	Lime sludges from acetylene, sugar, paper & fertiliser industries	• Raw meal component in cement • Lime pozzolana mixture, (L. P.) • Building lime • Maonsry cement	Raw meal in cement Traditional L. P. Lime from limestone Limestone based masonry cement	• Limestone & clay • Limestone • Limestone • Limestone

续表

S. No.	Industrial waste/by-product	Application in building materials as alternative to traditional materials	Traditional material saved fully or partl (20%～30%)	Natural resources saved fully or partly (10%～20%)
6.	Red mud (from alumina in aluminium)	• Cement raw meal • Bricks and tiles • Sintered aggregate	Farraginous mailer High strength brick Stone and other aggregates	• Oxides of iron • Clay, feldspar • Clay, shale, slate
7.	Mine tailings (from zinc, copper, gold, iron mines)	• Filler in concrete • Calcium silicate bricks • Cellular concrete • Tailing-clay brick • Masonry cement (tailing + cement)	Fine aggregates Sand (in sand-lime brick) Ground sand Clay bricks Limestone-cement based	• Sand • Sand • Sand • Clay • Limestone

Source: *Building Materials in India* : 50 *Years - A Commemorative Volume*, *edited by* T. N. *Gupta*, *published by Building Materials & Technology Promotion Council* (BMTPC), *New Delhi*, *India*, 1998.

Energy saving through use of industrial wastes **Table 3**

S. No.	Building Material	Composition	Material Compared	% Energy saving
1.	Portland pozzolana cement	75% OPC 25% Flyash	100% OPC	20
2.	Portland blast furnace slag cement	60% OPC 40% B. F. Slag	100% OPC	30
3.	Masonry cement	50% OPC 50% Tailings/waste chalk	100% (Masonry cement) (50% OPC + 50% Limestone)	20
4.	Lime-pozzolana mixture	25% Acetylene gas lime 75% Flyash	25% Lime 75% Calcined brick	75
5.	Calcium silicate brick	90% FA Tailings 10% Lime (waste source)	Burnt clay brick	40
6.	Burnt-brick	75% clay 25% Flyash	Burnt clay brick	15

Source: *Building Materials in India* : 50 *Years - A Commemorative Volume*, *edited by* T. N. *Gupta*, *published by Building Materials & Technology Promotion Council* (BMTPC), *New Delhi*, *India*, 1998.

Resource and energy saving through use of natural fibres and agro-wastes in building materials

Table 4

S. No.	Waste and source	Commercial product using natural fibre &agro-waste	Traditional resource fully or partly saved	Energy %
1.	Coir fibre (coir industry)	Coir-fibre-cement roofing sheet & panels	Asbestos	10
2.	Rick husk (Rice mill)	Rick-husk-cement building board	Resin (PR or UF) bonded particle board timber	20
3.	Ground nut hulls (Oil mills)	Ground nut-hull-cement building board	Resin-bonded particle board timber	20
4.	Jute fibre (Jute mills)	Jute-fibre-polymer bonded panel; door and window	Timber, metal	10
5.	Cotton waste (Textile mills)	Cotton-lint-cement bonded board	Gypsum, timber	25
6.	Bagasse (Sugar mills)	Bagasse-polymer-bonded boards	Timber fibres (in insulation board)	30
7.	Corn cobs (Corn mill)	Corn cobs-cement bonded boards	Timber, polymer	40
8.	Sisal fibre (Sisal plant)	Sisal fibre - polymer/cement bonded roofing sheet, door, window	Asbestos fibre Timber	20～15
9.	Rice straw & Wheat straw (Farms)	Compressed and paper covered board	Timber Polymer	40
10.	Banana fibre (Banana plant)	Banana fibre + cotton pulp/paper pulp and polymer insulation boards	Timber Traditional light weight mineral viz. vermiculite or mica	25

Source: *Building Materials in India* : 50 *Years - A Commemorative Volume*, *edited by T. N. Gupta*, *published by Building Materials & Technology Promotion Council* (*BMTPC*), *New Delhi*, *India*, 1998.

There are nearly a thousand natural fibres and grasses which are derived from renewable plant sources possessing potential for making fibre-reinforced composites. It is sometimes argued that natural fibres and husks have a very good fuel value, almost equal to normal bituminous coal(20%～40% as content)and therefore, may not be available for making composite building materials. But the fact remains that cost of these fibres would never exceed the market price of timber fibre-the traditional material for production of building boards, or of asbestos fibre which is often imported. Recycling of industrial and agro-wastes as substitute in traditional building materials industries vis-ἀ-vis the energy derived by their incineration should be choice in a particular situation.

There have been continuous efforts in making the industries aware to the various advantages in the utilisation of the wastes and by-products. There are additional applications which have great values in improving the environment and quality of life of the people living around the points of generation of the industrial wastes. Newer applications are filling the empty mines, making stabilised land for buildings and also for growing forests as well as selected agricultural products.

5 Energy in recycling of municipal wastes

Out of numerous types of municipal wastes only steel, aluminium, waste paper, glass, rubber, plastics and wood products are considered here. In all these materials there are three main approaches: (ⅰ) direct compression and use as a building material in the form of block or sheet, (ⅱ) palletisation and incineration for heat energy, and (ⅲ) recycling into the main product or into ore its major by-products. Since the naturally occurring resource minerals or vegetable fibres as exploited for producing the above mentioned materials at the primary level, there is considerable saving of energy both at the mining and processing stage and later at the manufacturing stage.

The recycled material could be considered as secondary material in which all the chemical and metallurgical processes are not required. For example, production of primary steel requires winning and processing of iron ore and its sintering, blast furnace reduction and further smelting and refining whereas the secondary steel produced from steel-scraps does not require the basic ore sintering operations. It is, therefore, highly important for energy-deficient developing countries to exploit all kinds of municipal wastes for usable products and effect saving of energy. This, however, should not encourage dumping of wastes containing highly toxic substances in the developing countries.

Table 5 gives the energy break-up, in the recycling of some wastes, which enables direct comparison of the energy required in primary and secondary (recycled) materials.

Energy saving in recycling of some wastes **Table 5**

S. No.	Material	Total energy primary product (MJ/kg)	Total energy in recycled product (MJ/kg)	Saving %
1.	Steel sheet	333.50	21.00	30.00
2.	Aluminium sheet	235.00	11.00	94.00
3.	Paper	28.00	25.00	10.30
4.	Rubber	100.00	56.00	44.00
5.	Plastics (PVC)	54.10	14.50	80.00
6.	Glass	23.00	17.00	26.00
7.	Wood products	3.00	2.00	33.00

Among the typical examples given in Table 5 wastes of rubber, plastics and wood products could be used for both as fuel as recycled(secondary)material whereas glass and metals could be used only for making recycled materials. The calorific value of gas produced by incineration of the above wastes(organic)ranges between 10～15MJ/kg.

A few aspects are, however, important for successful planning for saving of minerals by utilisation of wastes.

a) The waste to be returned/recycled must be easily identifiable.

b) The collected wastes materials are relatively uncontaminated.

c) The consumer motivation is necessary, through waste collection bins.

d) The collection could be through properly trained labour who could identify and sort out various wastes.

6 Concluding Remarks

As the level of technology increases, costs can increase by a factor of two to five or even higher for several other options. The issue of demand management of energy is crucial to developing countries where the shortages are expected to continue. The activities that the utilities at national or state level agencies can undertake to play an effective role for promoting demand management of energy in the construction sector particularly in the materials sub-sectors, could be under the following three phases:

The first phase should relate to creating public awareness on energy efficiency and the availability of energy efficient building materials and other products for application in buildings.

The second phase should cover setting up of industry level efficiency consultative groups for continuous interaction amongst the involved actors and also should evolve targets for improving energy efficiency in the manufacturing sector of building materials.

The third phase should deal with demonstration projects where both governmental and private sectors should invest suitable resources for bringing about energy efficiency improvements in manufacturing of environment-friendly building materials.

The multi-lateral institutions such as UNIDO, UNCHS should be requested to fund the demand management programmes in building materials sector which is fast becoming a major source of energy consumption in most developing countries. The Global Environment Facility which is expected to finance agreed increment cost for emission reduction could also be source of funding for these projects.

Construction Industry Development Council(CIDC)and Building Materials & Technology Promotion Council(BMTPC)of India are also making extensive efforts for promoting design concepts and manufacturing technologies for Green and Energy Efficient Building and New Technologies in India and several other countries.

UNIDO and Government of India, Ministry of Urban Employment & Poverty Alleviation, are already implementing a Cooperation Programme in the area of Capacity Building and Transfer of Manufacturing Technologies for Eco-friendly and Energy-efficient Building Materials for Low Cost Housing in the countries in African, Latin American and Asian regions.

印度创新型节能环境友好建材

T. N. Gupta
(Former Executive Director,
Building Materials & Technology Promotion Council
Ministry of Urban Employment & Poverty Alleviation, Government of India
UNIDO Consultant in the area of
Energy Efficient & Eco Friendly Building
Materials Manufacturing Technologies for Low Cost Housing)

摘　要：最近几十年来，建筑工程的迅猛增长速度已经在全世界范围内引起了关注。究其原因，除了一些客观因素，主要还是因为世界人口的飞速膨胀和经济的高速发展。在这种情况下，建筑材料需求巨大。而传统的建筑材料生产方式又是以消耗大量自然资源为基础的，所以根本无法在不影响环境的前提下持续发展以满足建材的巨大需求。

印度的科技研究学者已经研制出多种创新型材料，生产这些材料所需要的原材料则是废弃物、工业和农业生产过程中的副产品和剩余物，而这些原材料是造成严重环境污染的重要原因。利用最新开发的生产工艺将这些可利用资源进行循环使用不但可以减轻环境污染，同时可以节省能源。因为在印度进行的研究中已经确定：废弃物的循环利用所消耗的能量要远远低于传统生产方式。

本文介绍了传统生产工艺的能源需求以及循环利用废弃物、工农业副产品和天然纤维所能节省的能源和不可再生资源。通过质量和性能的比较，利用这些技术生产出的建筑材料将是传统材料最有力的替代者。

HYBRID NATURAL/ MECHANICAL ENVIRONMENTAL DESIGN STRATEGIES FOR LOW ENERGY BUILDINGS ACROSS A SPECTRUM OF CLIMATES

C. A. SHORT[1,2]

[1] Professor of Architecture Department of Architecture University of Cambridge, 1-5 Scroope Terrace, Cambridge, CB2 1PK United Kingdom

[2] Short and Associates, Chartered Architects 24A Marshalsea Road, Borough, London, SE1 1HF United Kingdom

Abstract: Natural ventilation and passive cooling are promoted as important strategies to reduce carbon dioxide emissions associated with the built environment, but evidence is emerging that natural means alone will not provide comfort all year round even in the centre of temperate zone cities such as London. The design of such buildings can be enhanced by the parallel incorporation of low energy supplementary equipment with judicious controls to achieve relatively very good performance. This paper describes two such built projects in their full context. Thereis potentially direct applicability across the various climate zones of China.

Keywords: natyral ventilation mechanical ventilation hybrid climate

Introduction

There is broad international recognition that worldwide temperatures are rising, and that this phenomenon may be attributed to global warming, induced by an increase in the concentration of greenhouse gases, notably carbon dioxide(DTI, 2003). The increase in carbon dioxide concentration can be correlated with the increase in energy consumption as a result of human endeavour measured as traded energy(Lomborg, 2001).

An important strategy for reducing emissions of carbon dioxide associated with the built environment is to adopt an integrated and holistic design approach exploiting natural ventilation to cool non-domestic buildings, whilst reducing the internal heat loads which might necessitate cooling, and by naturally lighting as much of the interior usable floorspace as possible (Short/Thomas, 2003). However even in the urban heat island of city centres such as London, within a Temperate Climate Zone, it has recently emerged that natural ventilation needs to be mechanically supplemented to maintain comfort conditions. But this can be achieved whilst retaining the low energy characteristics of the design. This paper describes a series of designs and completed buildings which employ this strategy, more particularly in deep-plan configurations, which may be more economic to build and operate and offer more flexibility in space planning.

Case Study 1

A downdraught-cooled Institutional building in the centre of the London heat island The School of Slavonic and East European Studies, Bloomsbury

Functional Requirements

The School of Slavonic and East European Studies(SSEES) was founded in October 1915, and is currently located on several sites in Bloomsbury. The intent is to concentrate all of the constituent elements of the School onto a single site; its academics, research assistants, post-doctoral and teacher fellows, its language unit, its administrators and its important library, holding 280,000 books, 1,000 periodical titles and an extensive film and video collection. It serves 600 undergraduates and 150 graduate students. The School has acquired an almost embassy-like function over the decades, representing the East European community as a bloc in London. Insofar as EU Accession for a significant number of these countries is scheduled for mid 2004, the significance of this Institution is likely to intensify. The author won the Europe wide competition for the new building in early 2002 (Fig 1). Construction commenced in December 2003 with completion due in mid 2005.

Density of occupation

A gross floor area of 3750m^2 is under construction to accommodate 50 academics and 20 administrators within their own cellular spaces, 15 library staff, and, at any time between 9.00am and 7.00pm, 200 library readers and 50 graduate students and research assistants, approximately one person per 11.2m^2. However, the site is so cramped the building will rise to seven storeys with a deep section front to back, some 26m^2 (Fig 2). The SSEES Library is a high grade academic research centre and the majority of its readers will remain at their desks for the whole day, and inhabit the library over extended visits to London.

Fig. 1 Queens Building view across concourse(Peter Cook)

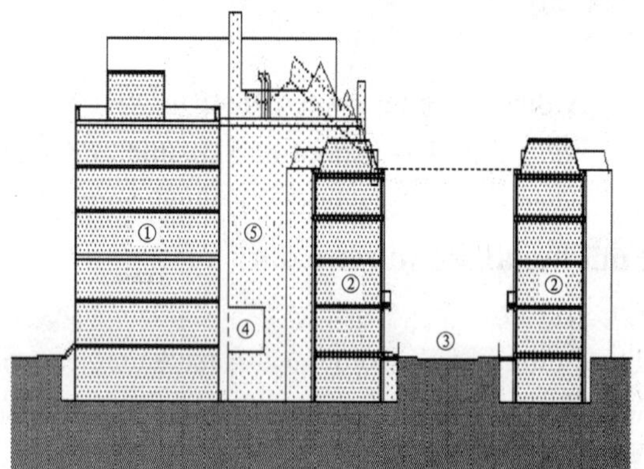

Fig. 2 SSEES, the completed building (Short & Associates)

Physical location

The SSEES site is 600m from the epicentre of the London urban heat island, which appears to be over the British Museum (Watkins et al. , 2002). It enjoys other city centre characteristics: high levels of low frequency traffic noise, and neighbouring users both sensitive to noise and pollution and productive of both! Pollution is a particular issue in London during periods of summer temperature inversion when the incidence of asthma attacks rises.

Security is an important requirement implying that the exterior be sealed, with the additional instruction to dissuade rough sleepers from colonising cavities formed in the street front, such as air intakes.

The site sits in the angle of the University College chemistry block with the requirement to minimise the impact of the new building on the natural lighting of the chemistry laboratories behind. Delivery vehicles will continue to service the chemistry loading dock. The building has a Mews 'drive-through' on its upper ground floor to facilitate this through traffic and has to accommodate the noise and fire risk of vehicles manoeuvring within the site .

There are stringent planning and zoning requirements within Bloomsbury's vestigial eighteenth and nineteenth century Georgian landscape, demanding set comice heights, window arrangements, external materials and an overall silhouette to match the lost Georgian town houses. This Conservation-driven policy raised complex issues relating to the provision of stack ventilation and not least achieving the required height for effective exhaust. Another section of the Borough's legislation strongly encourages low-energy design, citing natural ventilation and passive cooling as useful tactics. The two policies have yet to be fully synchronised(CBC 1994).

Climate

The site enjoys the temperate climate of the UK with the added complexity of the superimposition of a heat island effect generated by the city centre. Watkins et al. (2002) report an annual mean heat island of 2K for 1999～2000. As London warms, Watkins et al. deduce that more cooling and less heating will be required, resulting in a positive increase in CO_2 emissions of 4.7% by 2050 and a 7.2% increase by 2080. Their records of an August night in 1999 show a 5K elevation in night-time temperatures for the site compared to the urban fringe. An important consequence is the potentially reduced benefit of night purging.

This effect is superimposed upon the observed trend for summers in the UK to become warmer. CIBSE has recently published a Design Summer Year(DSY) (CIBSE 2002) data file, which recommends that naturally ventilated buildings be designed such that no more than 5% of working hours exceed 25℃. A simple calculation yields a figure of 150 hours duration as a maximum above 25℃. However, the DSY for Heathrow airport at the Western edge of London, in effect the 1989 weather data, yields 267 hours in which ambient dry bulb temperatures exceed 25℃, some 9% of occupied hours, not taking into account the urban heat island effect of the city centre.

Using these criteria and adding the disbenefit of the local heat island, a wholly natural unassisted ventilation strategy will not be viable. A basic calculation reveals the London DSY applied to the unassisted Coventry Library scheme yields 966 hours above 25℃, equivalent to 32% of working hours. However, the Coventry building is not experiencing this level of discomfort by a wide margin, so one might conclude that Coventry is considerably cooler than London or in someway the application of the new UK Design Summer Year and the more onerous new criterion in concert are not productive in promoting low energy building solutions, being overly pessimistic given particular types of heavy construction for building envelopes.

Nonetheless, it is clearly prudent to provide the SSEES building with some form of mechanical cooling to be applied as required in the most energy efficient way, by avoiding fan power as a minimum measure.

Environmental Design Strategy

i) *Natural and assisted ventilation and cooling*

The SSEES building is different from other naturally ventilated buildings because it has downward airflow in its atrium and upward airflow in the stacks during the summer and, to some extent, during the mid-season periods when downdraught natural and cooled supply air is admitted(Short et al, 2004). The flows through different floors are coupled together and this can lead to multiple flow regimes and potentially to stagnation of flow (Woods et al, 2003). The two upper floors were detached from the system as a consequence of the findings of physical water and computational modelling.

A triangular lightwell punctures all floors at the centre of the plan . The plan is zoned, for operational purposes, into a 'core area' which is generally open plan for bookstacks, reader spaces or research bases, and a 'perimeter zone' which may be cellularised into individual or shared offices, as on the upper School floors, or open or part-cellularised, as it is on the lower Library floors. It is not quite a 'mat-building' plan form but could be developed into one.

The plan is enclosed by a variant of the double facade. To the public street front, rooms are buffered by the principal stair hall 1. 8m in width, heavily constructed in load-bearing brickwork. The plan is divided laterally so that to the street side all spaces exhaust into the stair hall which employs an operable ventilating parapet device at its head which, coincidentally, re-introduces the cornice line of the residual half of the Georgian terrace which lines Taviton Street. The floor plate to the radiused, more private side retracts from the enclosing envelope to entrap stack voids related to the potential radial subdivision of the plan into academics' offices. Supply air passes across the partition lines at low level, either through open slots where a high level of acoustic separation is not essential, or through lateral transfer ducts, acoustically treated, where privacy is a priority. This device was used in the Coventry Library between teaching spaces and the main library floor.

Air passes up into the lightwell via a plenum sized to supply the winter fresh air requirement. The top of the lightwell is developed into a double-layered steel-framed wheel, ringed by top-hung openable windows. Translucent ethylene-tetra-flouro-ehtylene (ETFE) cushions are stretched across the upper and lower layers forming a void, which can be continuously vented in summer to flush out solar gains which would compromise the capacity of the atrium to maintain a reservoir of cooled air. It is tilted to bring its public street side presence within the acceptable planning/conservation silhouette, which is not ideal. Cooling coils ring the aperture of the atrium wheel. In peak summer conditions, the perimeter windows and the ring of dampers above the coils are opened by BEMS controls, allowing air to pass across the coils and directly into the lightwell. Chilled water is fed into the coils derived from absorption chillers driven off the UCL campus Combined Heat and Power circuit and the building goes into downdraught cooling mode. It is essential in this mode that the dampers at its base are tightly sealed. Air is drawn off from the reservoir of cooled air contained within the lightwell into each floor through conventional bottom hung windows.

Air moves across the floors in all directions, passing across partition lines as required to perimeter exhausts. The lower ground floor has a full height dedicated stack. Floors ground, one, two and three are coupled in pairs to perimeter stacks around the radiused side, while ground, one and two on the street side vent into the stairwell. The third floor on the street side vents directly into dedicated stacks, as does the fourth floor. On the radiused side of the building, the fourth and fifth floors drain into dedicated stacks. The terminations are an adaptation in terracotta of the Lanchester Library parallel split tube pattern in aluminium, designed to be robust to wind direction to prevent downdraughting.

ii) *Seasonal modes of operation* (Short et al, 2004)

1) *Winter condition; heating in place as required, unassisted ventilation*

Air enters the base of the building from opposing orientations in sufficient quantity to satisfy the occupancy fresh air requirement. It is introduced into the base of the lightwell through dampers and heating coils. Finer control of space temperatures is achieved by locating heating elements at the point of entry into the occupied spaces at each floor level, as at the Lanchester Library Coventry (Cook et al, 1999) (Short, 2003). The amount of heating delivered can be controlled closely on a space-by-space basis, taking into account the varying levels of heat gain and occupancy across the building. Air passes across the floors and is exhausted as describe through dedicated stacks. The building is in its pure buoyancy driven mode.

2) *Mid Season Condition; unassisted ventilation*

Air is admitted via the low level plenum at a rate determined by the occupants' fresh air requirements. However, the free areas provided only allow for this base requirement. Air may also be introduced simultaneously from the top of the lightwell to boost mass flow rates for ventilation cooling. The same top and bottom intake strategy will be exploited to achieve purging of the unoccupied structure during mid-season and summer nights.

3) *Summer Condition; downdraught cooling mode*

The building is configured by the control system into downdraught cooling mode. Absorption chillers driven off the University's CHP circuit feed chilled water to cooling coils at the lightwell head. Incoming ambient air is cooled and drops down, filling the sealed lightwell, before being released into the connecting floors. The Building Energy Management System controls the opening sizes into each floor based on internal temperature and CO_2 readings. This mode may also be used in particularly warm spells when higher night-time temperatures are maintained as a result of the heat island effect, to enhance night-cooling of the building.

Natural lighting and solar control

Maximising natural light penetration and usage is a critical component of the formula to reduce internal heat gains. The site is surrounded by buildings on three sides, but has a limited southerly aspect. However, the proposal adopts a D-shaped plan, presenting a half cylinder form to the rear optimising the available daylight on both the surrounding building elevations and the rear elevation of the SSEES building. The relative efficiency of this radial geometry was established through simulation at the Institute of Energy and Sustainable Development using Radiance software.

The section through the central lightwell is configured to admit as much view of the sky as possible from its base, the cooling wheel develops a 'funnel' shape, and unlike the Coventry supply lightwells, heating coils at the base of the SSEES lightwell are retracted within the plenum. This enables maximum daylight penetration into the lower ground floor library reading area. The dampers at this point play a crucial role in all operating modes and high operating performance is required to avoid leakage.

The 'building within a building' strategy provides more than adequate buffering for direct solar gain, but perhaps with the penalty of reduced daylight factors within the occupied perimeter spaces.

Predicted Performance

Dynamic thermal simulations were conducted by IESD Leicester for a typical section of the building encompassing core and perimeter spaces. The simulations were based on full occupancy and maximum use of personal computers between 9am and 5pm (weekdays) and 70% of these values between 5pm and 10pm (weekdays). During weekends (10am- 5pm), 70% of the full occupancy was assumed. An airflow network was also embedded into the thermal model to represent the natural ventilation flow from the lightwell, through the occupied zones into the perimeter stacks.

Without cooling, simulation results for the zone with the highest heat gains predicted 966 hours above 25℃ (32% of working hours) in line with the Coventry experiment. When the atrium is cooled to 23℃ the model predicts peak temperatures in the perimeter zones varying between 25. 5℃ to 28℃, the second floor perimeter zone being the hottest. The computer model revealed no evidence of stagnation or backflow in the cooling mode. However, flow reversal occurred when there were clear openings at the base of the atrium.

However, physical water modelling conducted at the BP institute at Cambridge University revealed a real possibility of stalling and reversal of flow on the fourth and fifth floors if connected directly into shared stacks, particularly when the building is in downdraught cooling mode and air moving towards the stacks is below ambient temperature (Woods et al, 2003).

The physical water modelling involved construction of a Perspex model of the building, scaled by a factor of 1 : 100, as well as a simplified model to illustrate key principles. The models were immersed in a reservoir of 5% saline water, and fresh water was added to different floors in the model building to simulate the generation of heat associated with the occupants. The pre-cooling at the top of the atrium was modelled by the injection of 10% saline fluid at this point. By dying the input fluids the flow patterns into the floor space and stacks may be visualised (Woods *et al.*, 2003).

In the summer pre-cooling mode, with high external temperatures, the greater elevation of the stacks to the rear of the building compared to the elevation of the atrium, led to stalling when the fluid within the stacks was sufficiently dense (cool) compared to the exterior fluid. By contrast, the same ventilation strategy proved extremely efficient with external temperatures compared to the desired interior temperatures. The pre-cooled inflow through the atrium and relatively warm outflow through the stacks then enhances the net pressure head driving the flow, increasing the ventilation rate.

Project Learnings

Hybrid natural/mechanical environmental strategies can be surprisingly efficient, even in city centres, with a comprehensive control system. Conservation issues need not deter their implementation.

A further hypothetical experiment was conducted, to model the performance of the down-draught cooling strategy out of London and further north. Under the regime of the Manchester Design Summer Year cooling is only required for 400 hours, 13% of working hours. Further north in Scotland using the Edinburgh Design Summer Year, cooling is theoretically only required for 119 hours, 4% of the working time. Certainly in climates equivalent to that of Scotland a wholly unassisted natural ventilation strategy is likely to be successful and perhaps only the new, more onerous UK criteria prevent such a strategy being widely applicable across the UK outside the centre of London.

Case Study 2

An integrated hybrid low energy design strategy for a college library in a continental climate Judson College Academic Center, Chicago, Illinois

Functional requirements

The College, requires a new 7500m^2 combined Library, Art & Design, and Architecture teaching building. A low energy design strategy has secured $7.5 million of Federal 'Earth Stewardship' grant aid, and $100,000 of research funding to aid the development of the energy aspects of the design from the Illinois State's 'Clean Energy Programme'. This paper concentrates on the design of the deep plan library component within the centre.

The librarians, as at the Lanchester Library Coventry, required a simple square floor plate. For Judson this becomes 108 feet by 108 feet, in sixteen 27.5 foot square bays. The plans envisage an open-access circulatory book collection growing to 100,000 volumes and a reference collection of 12,000 volumes, with workspace for up to 250 readers and up to 15 staff, computer work stations to be co-located in a computer laboratory with a limited number of catalogue terminals within the main library space.
Security is an important issue, as much in terms of personal safety, requiring close surveillance of entry points, as to protect book stock.

Density of occupation

The scheme envisages up to 250 readers, and 15 staff in 40,000 square feet (3,700m^2), approximately 1 person per 150 square feet, or 1 person per 14m^2. The occupancy density will be relatively low with implications for the fresh air requirement.

Physical location

Judson College lies 64km west of Chicago, Illinois, in the Fox River Valley on a parkland site on the eastern edge of the prairies. It lies west of O'Hare International Airport and the climate modification effect of Lake Michigan is mitigated. To the northwest of the site is a dual carriage highway leading to the I-90 Interstate further to the north. Traffic noise is clearly audible on the site which could be described as being on the 'urban fringe'.

Climate

The climate is classified as continental, 'a cool temperate rain climate with severe winters'. Winter temperatures frequently dip below −20℃ but summer highs exceed 35℃ with clear skies and high humidity at the seasonal changes. During both seasons there can be large and sudden temperature swings. It would appear that a high proportion of the Chinese mainland enjoys a similar climate. The inclusion of an operating mode employing full mechanical air conditioning with humidity control would seem to be essential to combat extreme low humidity in winter and high summer humidities, given the environmental expectations of north American educators, administrators and students. However, the profile of annual temperature variations set against annual heating and cooling demand reveals that thermal comfort could potentially be maintained for much of the spring and autumn (Fall) seasons using natural ventilation.

The Environmental Design Strategy

The design challenge is to reduce the duration of operation of the mechanical system during the year by reducing the cooling and heating load imposed on it, to integrate the mechanical system with a passive ventilation strategy to avoid duplication of air distribution paths (and thus maintain cost effectiveness), to design an energy efficient mechanical system and to utilise, wherever practicable, renewable energy- either by active or passive means.

Ventilation

Natural ventilation is a viable strategy in the mid season periods, provided the building comprises:

- a compact form, which reduces the exposed surface area.
- an air-tight and well insulated external envelope.
- night-time ventilation coupled with sufficient thermal mass exposed to the interior
- appropriate areas of shaded external glazing to give adequate natural light on the working plane with minimal solar gain.

Such a strategy should reduce the frequency of summertime operation of the air conditioning system quite considerably, and reduce the cooling loads imposed on it.

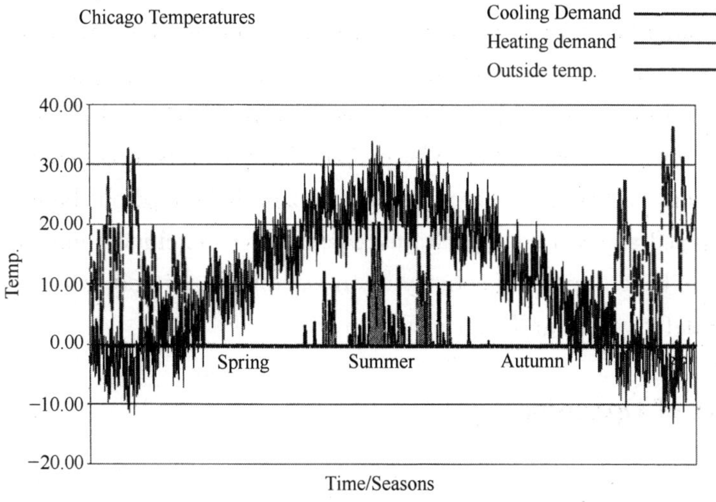

Predicted demand for artificial environmental heating and cooling in the Chicago area

Fig. 3 Judson College, graph of annual temperature variations and potential heating and cooling demand (IESD, Leicester)

Fig 3 shows a diagrammatic cross-section through the deep plan library portion of the plan. Air is introduced at high level through a radial array of air intakes, to respond to all

potential wind directions. Dampers admit air through a fan driven duct to the enthalpy wheel, or through an unassisted route bypassing the mechanical equipment.
The natural ventilation flows will be driven by the stack effect. Computational Fluid Dynamics simulations by IESD Leicester were used to test proposed designs. The airflow paths are mirrored by the mechanical system, thereby simplifying controls, avoiding the duplication of air distribution systems and maintaining clarity of operation for the building's users and operators. The large openings necessary for the natural system reduce the loads on the fans, when operating in mechanical mode.
Fresh air via either route is admitted into each floor through a central atrium. The air enters each floor at low level, either onto the floor plate, or within a raised floor from which it discharges into the space through floor registers. This is a buoyancy-driven ventilation scheme in which a clear stratification should develop in each space.
Floor to ceiling heights will be 11 feet. Air is exhausted into perimeter ducts embedded in the external wall construction. The warmed air is either diverted to recharge the enthalpy wheel, or freely discharges into the rooftop stacks, or both. The terminations are static, lifted 10 feet above roof level out of the turbulence zone, and offset so that fine snow would have to track some distance laterally to infiltrate their interior.

Natural Lighting and Solar Control

The use of daylight to reduce artificial lighting loads minimising direct solar gain in summer, whilst admitting winter sun to warm and enliven the interiors, are important design considerations. Simulations with the Radiance (Ward Larson et al., 1998) programme enabled the glazing areas and shading to be refined and the resulting levels of internal illumination to be predicted. The proposed external wall construction envisages a deep elevation (some 1.2m/4 ft.) in which the glazing is located on the internal face. The openings are splayed and variedorientation, calculated to exclude summer solar radiation level. The reveals have a high reflectivity in an attempt to mitigate the loss of high angle sky view, potentially very significant in winter.

Seasonal Modes of Operation

ⅰ) *Winter Mode*
At night temperatures are allowed to fall to 15 to 18℃. Before the occupied day commences the spaces are warmed to 72℉. Solar shading precautions do reduce passive solar gains but render the building less likely to overheat in the early afternoon and call for cooling.
ⅱ) *Summer Mode*
The daytime cooling set-point is 27℃(80.6℉). The upper acceptable moisture content is

set at 10g/kg to give some tolerance around the ASHRAE 55 standard. Cooling is switched off at 8pm. Temperatures start to rise but night venting commences when the building is unoccupied. By 2.00am cooler air floods the building and heat transfer occurs with the high thermal mass elements of the superstructure. By the start of the following day the spaces are already cooled to 79°F, the cooling system merely fights casual gains. Lighting gains are reduced by the provision of natural lighting. Night venting also reduces the build-up of temperatures through the weekend when the plant in a conventional US building would be unlikely to be employed for economy reasons, building up higher temperatures for the Monday morning start-up.

iii) *Mid Season Mode*

The building controls switch the building to passive mode when external temperatures rise above 6℃ (42.8°F) and the moisture content of the air lies between 12g/kg and 3.0g/kg. As the air handling unit ceases to operate airflow is driven by internal heat gains. The passive mode can operate in heating mode, supply air being heated by the coils at the plenum inlet, and in ambient supply mode on warmer days and to provide night purging to cool the exposed thermal mass of the ceilings and walls.

Predicted Performance

Fig. 4 summarises a comparative study of the energy demands, using ESP-r (ESRU, 1995) of firstly a building of the size and shape of the Judson proposal specified as a US standard building, secondly a building which is well shaded from summer solar gains and adequately daylit, and thirdly the same environmentally conscious proposal night purged with cooler night air.

The standard US building has unshaded glazing, internal blinds and a suspended ceiling masking the floor structure from the internal occupied volume. The Winter heating set point is taken as 71°F and the summer cooling set point as 77°F, probably rather warm as compared to the expectation of an air conditioned interior held by the American public.

The immediate solution reduces cooling energy demand by reducing solar gains and heat derived from artificial lighting. It also eases the cooling set point to 79°F, envisaging some behavioural response. Winter heat loads rise a little as solar gains are reduced by the shading geometry.

The final proposal incorporates all of the above, introduces high floor to ceiling heights, removes the hung ceilings, exposes the thermal mass of the floor slabs and night vents the interior. The consequence is a reduction in overall predicted energy demand of 53% by manipulating the architectural design. The use of renewable energy sources will reduce this further.

The renewable sources proposed are:

i) an enthalpy wheel to recover both sensible and latent heat

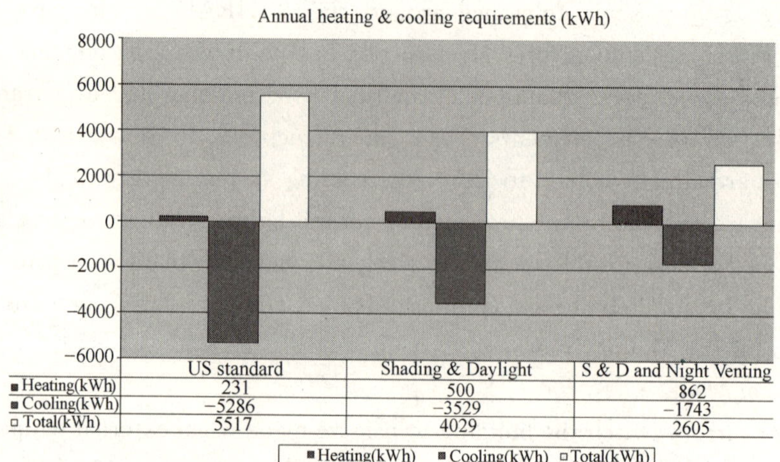

	US standard	Shading & Daylight	S & D and Night Venting
■ Heating(kWh)	231	500	862
■ Cooling(kWh)	−5286	−3529	−1743
□ Total(kWh)	5517	4029	2605

Fig. 4 Judson College, predictions of proportional energy savings given certain design measures, control of solar gain, night ventilation (IESD)

ⅱ) a geothermal array (linked to heat pumps for both winter heating and summer cooling)

ⅲ) roof top photovoltaic panels, funded in part by the Illinois 'Clean Energy Campaign'.

Project Learnings

- Close study of regional and local climate data will reveal more precisely the duration of conditions which a natural ventilation scheme can and cannot fully ameliorate.
- The definition of these intolerable periods also depends on established comfort criteria. Comfort, social and cultural expectations, and social responsibility need to be harmonised in order to achieve a clear brief.
- Purely architectural measures, relating to the design of the fabric of the building, can reduce energy demand by 50% in Illinois type harsh continental climates. This in a region where bioclimatic architecture is normally thought to be unrealistic. It would appear that much of northern and eastern China enjoys broadly similar conditions.

Conclusion

There is an opportunity for building owners and their design teams in China to avoid many of the energy-consuming mistakes made in northwest Europe and north America. It may be that an assumption has arisen that low energy, sustainable design as developed through the second half of the twentieth century is appropriate only in temperate climates. This paper proposes that this is not the case, and that viable energy-efficient design strategies can be evolved for a wide range of climate types. Certainly the high diurnal temperature ranges which seem to be recorded across much of China would seem to offer opportunities for the exploitation of stack driven natural ventilation and passive night cooling, mechanically assisted, when necessary, by innovative hybrid mechanical means.

The paper also promotes the advantages of using locally available materials, particularly if they offer a degree of thermal mass. It makes the case that international commercial and intellectual credibility need not be expressed solely through all-glass building envelopes, an unhappy convention developed in the West. Regional architectures often offer ingenious strategies to deflect climatic extremes in particularised, expressive and meaningful architectural vocabularies.

Britain is adopting more stringent comfort criteria in the context of a warming climate. We are extremely interested to read of recommendations for much less restrictive comfort criteria in China. For example, JGJ 134—2001, the Building Code for Residential Buildings in Hot Summer/Cold Winter Climate Zones ANSI/ASHRAE(2004), advises an internal temperature range of 16-18℃ in winter, and a peak of 26-28℃ in summer. This makes a very interesting comparison with the UK CIBSE Recommendation, as referred to earlier in this paper. It would seem to be most important that the built environment community in China consider any reduction in the temperature ranges described by these criteria very carefully. The potential implications for national energy consumption could be very considerable.

不同气候下的低能耗环境建筑设计混合式自然和机械通风策略

C. A. SHORT[1,2]

[1] Professor of Architecture Department of Architecture University of Cambridge, 1-5 Scroope Terrace, Cambridge, CB2 1PK United Kingdom

[2] Short and Associates, Chartered Architects 24A Marshalsea Road, Borough, London, SE1 1HF United Kingdom

摘　要：自然通风和被动式制冷在建筑设计中是减少碳化物排放的重要手段，但是现实中仅自然通风方法不能满足全年的热舒适要求，即使是像伦敦这样的温和地区的城市也是如此。设计这样的建筑可以辅助于低能耗机械设备，通过恰当的控制手段使建筑达到良好的性能。本文介绍了两个项目的设计方案，对中国不同气候区的建筑设计也有潜在的直接应用价值。

References

[1] ANSI/ASHRAE (2004). Thermal conditions for human occupancy. American National Standards Institute(ANSI), American Society of Heating, Refrigeration and Air-conditioning Engineers (ASHRAE), Standard 55-2004, 170 pp

[2] CBC (Camden Borough Council) (1994). *Supplementary Planning Guidance*

[3] CIBSE (2002) *Guide J: Weather, solar and illuminance data*. Chartered Institution of Building Services Engineers, pp. 8.1-8.6 and p. 1-4

[4] Cook, M.J., Lomas, K.J. & Eppel, H. (1999). Design and operating concept for an innovative naturally ventilated library. *Proceedings of the CIBSE* (1999) *National Conference*. Harrogate, UK, pp. 500-507

[5] DTI (UK Department of Trade and Industry) (2003). Our energy future-creating a low carbon economy. Energy White paper, pp. 80

[6] ESRU (1995). ESP-r User Guide Version 8 Series. Energy Systems Research Unit, University of Strathclyde, Glasgow

[7] JGN (2001). Design Standard for Energy Efficiency of Residential Buildings in Hot Summer and Cold Winter Zone, JGN 134-2001, Industrial Standard of PRC, China Architecture and Building Press, Beijing

[8] Holmes, M. J. & Hitchin, E. R. (1978). An example year for the calculation of energy demand in buildings. *Building Services Engineer*, 45, pp. 186-190

[9] Lomborg, B. (2001). *The Skeptical Environmentalist: Measuring the State of the Real World*. Cambridge, U. K.: Cambridge University Press

[10] Rickaby, P. (1991). The art of energy: Peake Short's Malta brewery. *Architecture Today*, 14, pp. 34-43

[11] Short, C. A. (2003). Sustainable design in an urban context: three case studies. In Thomas, R. (Ed.) *Sustainable Urban Design*. Spon Press

[12] Ward Larson G. and Shakepeare R. (1998). *Rendering with Radiance: The Art and Science of Lighting Visualisation*. Morgan Kaufman, San Francisco

[13] Short, C. A., Lomas, K. & Woods, A. (2004). *Design strategy for low-energy ventilation and cooling within an urban heat island*. Building Research & Information, (May-June 2004) 32(3), 187-20

[14] Watkins, R., Palmer, J., Koloktroni, M. & Littlefair, P. (2002). The balance of the annual heating and cooling demand within the London urban heat island. *Building Service Engineering Research Technology*, 23(4), pp. 207-213

[15] Woods, A., Torwong, C. & Short, C. A. (2003). Reversing flow in a naturally ventilated building with multiple stacks. Proceedings of the Plea 2003 Conference. Pontificia Universidad Catolica de Chile

[16] Woods, A., Short, C. A. & Gladstone, C. (2003). Stack driven ventilation with pre-cooled inflow from a central atrium. Proceedings of the Plea 2003 Conference. Pontificia Universidad Catolica de Chile

ALTERNATIVE ENERGY EFFICIENT BUILDING MATERIALS FOR LOW COST HOUSING

Dr. J. S. Chauhan
(Prof. & Head
Civil Engineering Department
Samrat Ashok Technological Institute (Engg. College)
Vidisha (M. P.)India-464001)

Abstract: The significance of cost effective, energy efficient and long lasting housing and Civil structures in developing country like India can not be denied. The majority of Construction activities in the country are mainly based on age old technologies. Now it is time to develop innovative, alternative, energy efficient building materials by effectively recycling industrial wastes, utilizing natural resources as major raw materials to reduce the vulnerability and protect the environment. Most significant aspects of commercial acceptance of innovative & energy efficient building materials and technologies are cost effectiveness, better quality, ease of adoptability, energy efficiency and environment friendliness. To increase the safety of buildings against natural disasters, new green building materials are essentially to be developed with a provision to resist these impacts.

The present paper highlights that the nature of emerging green building materials will be dictated by choice of materials, energy conservation, national plan about cost effective & energy efficient building materials. The focus has to be on valorization and recycling of industrial wastes, waste utilization and development of green building materials and its uses.

INTRODUCTION

Building materials account for 60%~70% (Table 1) of the total cost of construction. Due to large-scale construction programmes in the country, the demand for conventional building materials like cement, steel, bricks and timber has outstripped their supply. Moreover, the exponential population growth and the existing housing shortage have made the situation even more alarming. There is a general shortage of conventional building materials like cement, steel, bricks and timber in different parts of the country. In 2003, there was an estimated shortage of 55,000 million bricks, 17 million tonnes of cement, 285 million tonnes of stones, 2.7 million tonnes of steel and 13 million cum. of timber (Table 2). It is thus quite obvious that the present available stock of building materials in the country is not in a position to meet the ever-growing demand of housing. In the present situation, the country has very little option but to rely increasingly on locally available cost-effective building materials and components. The R&D efforts undertaken in the country by various research organisations like CBRI, Roorkee; SERC, Madras; National Council for Cement and Building Materials (NCB), Ballabgarh; Central Fuel Research Institute (CFRI), Dhanbad; NBO and BMTPC have led to development of various low cost, low energy consuming building materials using several industrial and agricultural wastes. There is considerable potential for exploitation of the agro-industrial wastes for their commercial production and large-scale application in construction programmes.

Average cost break-up of building construction **Table 1**

Materials (67%)		Labour (33%)		Component-wise (100%)	
Cement	17%	Mason's wages	12%	Foundation	10%
Iron & steel	9%	Carpenter's wages	6%	Walls	30%
Bricks	15%	Unskilled labour	15%	Roofs	25%
Timber	12%			Doors & Window	15%
Sand	6%			Flooring	10%
Aggregate	8%			Finishing	10%

Table 2

No. S.	Particulars	Cement Million Ton	Aggre-gate Million	Bricks Million Cum.	Steel Million Ton	Timber Million Cum.	Houses (No.) Million
1.	DEMAND Urban Areas						
(a)	New Construction	94.53	416.30	159458	11.76	24.93	8.46
(b)	Upgradation Rural Areas	4.87	10.72	3082	0.07	0.64	1.09
(c)	New Construction	29.25	121.97	118875	2.08	3.33	9.51
(d)	Upgradation	3.33	6.95	5081	0.03	0.19	2.17
2.	AVAILABILITY	114.93	270.35	231598	11.24	15.75	—
3.	SHORTAGE	17.05	285.59	54898	2.70	13.34	—

The building materials and components are being produced by using a wide variety of processes ranging from labour intensive to capital intensive. Continuing dependence of building industry on conventional, energy intensive building materials is adversely effecting the construction costs. To meet the shortages of building materials appropriate investment in Research and Development during past three to four decades have resulted in a large number of innovative building materials and construction techniques, prefab systems etc. In order to promote these low cost building material technologies through large scale adoption and production in different regions, BMTPC has been set up for strengthening technology transfer mechanism in the country.

The Housing Policy envisages the need to promote use of standardized low-cost components in the housing and building construction.

Availability of finance required for construction of houses is a big problem in the way of increasing housing stock. In order to tackle this problem Govt. of India provides institutional finances. Housing Development & Finance Corporation(HDFC)and Housing Promotion Finance Corporation were established to cater to needs of finance exclusively for housing to Individuals, groups and corporate enterprises. Many commercial banks also provide finance for housing.

The government envisages a more active role for commercial banks with their country wide branch network in provision of housing finance. Housing and Urban Development Corporation(HUDCO) plays a very vital role by providing finance for building material industries and housing.

The most important development in strengthening the housing finance system has been establishment of National Housing Bank(NHB) as an apex housing finance institution.

In order to encourage housing, Government of India has offered Income Tax incentives. For encouraging building material industries in the country. The housing finance institutions are considering to form a consortium to extend loan and equity participation to the entrepreneurs. The existing system of taxes, levies on production of building components and systems has been rationalized.

CHOICE OF MATERIALS

The policy should also promote use of appropriate building materials for mass housing constructions. The matrix to guide the choice of building materials for low-cost housing is shown in Fig. 1.

Special emphasis in national policy would have to be laid on prevention of environmental degradation, energy conservation, and ecological balance.

Prevention of Environmental Degradation

Production of building materials is undertaken mostly in a decentralized manner all over the country, particularly in small-scale and cottage industries. Appropriate measures have, therefore, to be devised to prevent pollution of environment due to the process of producing these building materials. For example, burning of bricks in the vicinity of

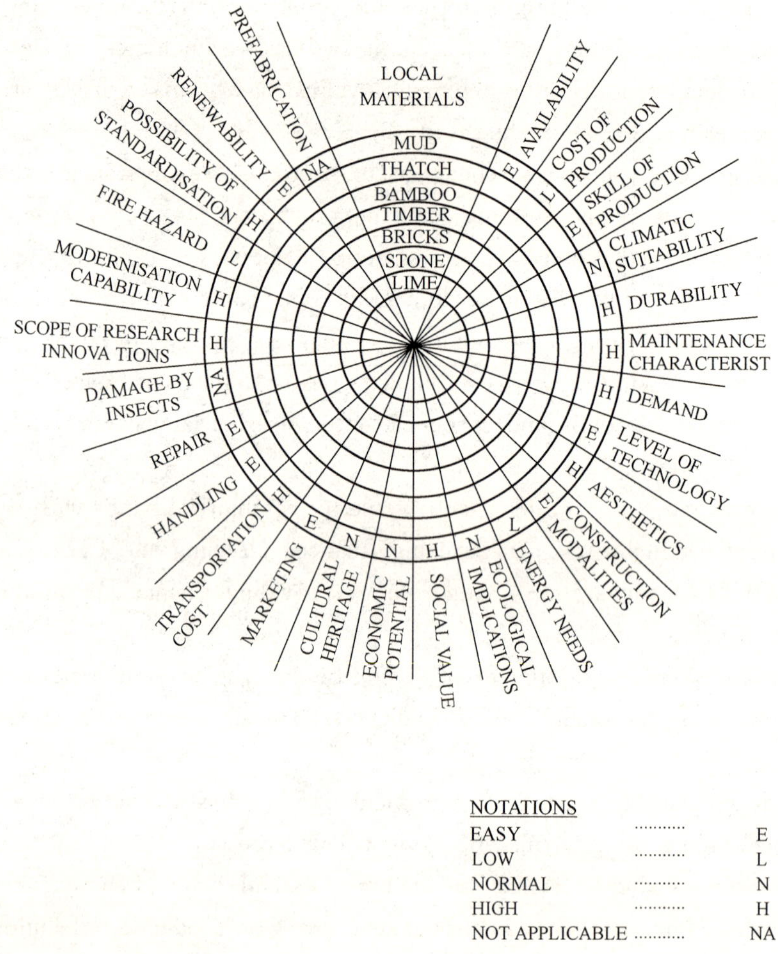

Fig. 1 Matrix to guide choice of building materials in low-cost housing

mangroves damages plant life, and digging of soil for brick making causes collection of waste water in cess pools creating unhygienic conditions and erosion of good agricultural soils.

Energy Conservation

Production of durable materials requires use of energy in significant proportion. In light of severe constraints with regard to availability of conventional energy resources and their high cost, it is necessary to adopt measures which result in energy conservation and also to use non-conventional energy resources for production of building materials to the extent possible.

Ecological Balance

Only such types of building materials and products should be used for construction of houses which do not disturb ecological balance. For example, excessive quarrying of limestone for lime burning or cement manufacturing and extraction of timber has denuded forest wealth and destroyed the ecology of hilly regions.

NATIONAL PLAN

Hitherto, the production and supply of building materials for housing has been almost entirely left to the private sector particularly in respect of indigenous materials like mud, bamboos, reeds, and local timbers and traditional materials like burnt clay bricks, lime and stone.

Due to several constraints it has not been possible for the building materials industry in the private sector to augment its production to meet the demand. As a result there is a great demand for indigenous and traditional building materials and some materials are in short supply and their prices have risen considerably.

ADVANCE PLANNING

To meet the requirements of construction programmes, the availability and production of building materials of different kinds from in-digenous resources would have to be considerably stepped up. Emphasis has to be laid on increasing the production of new building materials which can be used as alternative materials effecting saving in consumption of costly and scarce materials like cement, steel, and timber. New production units have to be installed in different parts of the country based on the latest advances in technology.

As a matter of fact the augmentation of production of building materials required for mass housing programmes should be undertaken well in advance so that the required types of building materials are easily available at reasonable cost.

INDIGENOUS BUILDING MATERIALS

Indigenous building materials have been used extensively for building houses at affordable cost. They are easily available locally at little or no cost and local people are conversant with the technology of using them for construction of houses largely through self-help.

The performance of the houses built with indigenous building materials is suited to the geoclimatic conditions. However, frequent repairs and reconstruction are required. Moreover, large-scale damage and destruction takes place due to natural disasters like heavy rainfall and floods, earthquakes and land slides, strong winds and cyclones, and fires.

It is, therefore, incumbent on builders to achieve economic optimisation in the use of indigenous building materials by application of modern science and technology, so that more durable and liveable houses can be constructed at lowest possible cost. This is also necessary to ensure sustainable development without causing environmental degradation and ecological imbalance, which have become matters of over riding significance.

Some notable progress has been made in some developing countries in the improved use of indigenous building materials for low-cost housing and is briefly mentioned here, along with its potential for tackling the massive problem of shelter for the people at affordable cost.

INORGANIC MATERIALS

Inorganic materials include commonly used materials like soils and laterite, burnt clay bricks and tiles, building lime and clay pozzolana, stones, and sands. As a result of research work done by CBRI, Roorkee, India and similar research organisations in other developing countries improved use of these materials for construction of low-cost houses has been made for achieving economy. Some of these are as given below:

1. Techniques of soil stabilisation using materials such as lime, cement, and emulsified bitumen to make the walls stronger and durable, with better dimensional stability.

2. Water-resistant mud plaster using bitumen emulsion for application on soil-based walls and roofs to prevent erosion caused by rainfall.

3. Good quality laterite and lime bricks for construction of walls.

4. Better quality burnt clay bricks and tiles for construction of walls and roof.

5. Good quality building lime and clay pozzolana and also ready to use lime and lime pozzolana mixtures.

6. Stone spalls with cement concrete to produce economical stone blocks for masonry work.

7. Sand-lime bricks and blocks that are strong and durable.

ORGANIC MATERIALS

Organic materials include widely used materials such as grasses and leaves, husks and straws, reeds and bamboos, and wood and wood products which are vulnerable to early decay and insect attack. Improved techniques for extending the service life of these materials have been developed which will lead to economic optimization in the use of organic materials are as follows :

1. Application of chemical treatment for preservation and extending the service life of grass thatch and palm leaves.

2. Use of rice husk with cement to form roofing sheets.

3. Production of small roofing tiles using straw mixed with mud.

4. Reed boards formed by compressing a small bunch of reeds to gather and wiring them tightly.

5. Use of bamboo treated with preservatives and also bamboo-reinforced cement concrete construction.

6. Use of secondary species of timber after preservative treatment and solar kilns for seasoning of timber.

SUBSTITUTE FOR SCARCE MATERIALS

Research findings of various research institutions like CBRI, NCB and CFRI, etc. have led to introduction of number of alternative building materials like hydrated lime, asphaltic corrugated sheets, cellular concrete, flyash bricks, plastic pipes, secondary species of timber, pozzolana cement and ferrocement overhead tanks, etc. The NBO had proposed feasibility reports for setting up plants for manufacture of such building products. Application of these substitute materials have resulted in the saving of scarce materials like cement and steel. Some of the alternative materials which are promoted by CETDEC, SATI, Vidisha for low cost housing are listed below.

S. No.	Name of Component	Size (mm)	Cost per Components (Rs.)
1	Cement Concrete Blocks	400×100×200	9.00
	Cement Concrete Blocks	400×150×200	12.00
	Cement Concrete Blocks	400×200×200	16.50
2	Cement Concrete Hollow Blocks	300×150×150	8.50
	Cement Concrete Hollow Blocks	400×150×200	11.50
	Cement Concrete Hollow Blocks	400×200×200	15.00
3	Stone masonary Blocks	300×150×150	8.00
4	Soil stabilized blocks	250×130×150	2.00 to 3.00
5	Precast R. C. C. Beam	300×150×150	172/RMT
6	Precast R. C. C. Channel	3500×300×60	130/RMT
7	Precast R. C. C. Plank	1500×300×60	98.00
8	Precast R. C. C. Lintel	1500×200×75	80.00
9	R. C. C. Chaukhat	900×2100	300 to 340

续表

S. No.	Name of Component	Size (mm)	Cost per Components (Rs.)
10	Funicular shell roof	900×900×25	120.00
11	Filler slab		50.00
12	Ferrocement water Tank		3.00 to 2.00/lit.
13	Ferrocement Tree Guard		225/each
14	Ferrocement Urinal		4500/each
15	Ferrocement Bench		800/each
16	Ferrocement Chair		110/each
17	Ferrocement Sign Board		110/each
18	Ferrocement wash basin		200/each
19	Ferrocement sink		250/each
20	Ferrocement U-Shape drain		75/RMT
21	Ferrocement Manhole cover		50/sqft.
22	Ferrocement shell roof		35/sq. ft.
23	M. C. R. Tiles		5-7/No.

供低成本住房使用的新颖节能建材

Dr. J. S. Chauhan
(Prof. & Head
Civil Engineering Department
Samrat Ashok Technological Institute (Engg. College)
Vidisha (M. P.) India - 464001)

摘　要：发展节约成本、高能效、耐久型住宅和城市建筑，对于印度等发展中国家来说，作用是不能否认的。目前国内的现存建筑大多是用古老的建筑工艺建造的。现在是开发新型建筑的时候了，即通过有效地回收利用工业废物、提高建材的能效、使用天然材料作为主要的建筑原料来减少环境冲击保护环境。创新、节能型建筑材料和技术的商业兑现中最主要的问题是如何保证产品的成本效率、高质量、适用性广泛、节能且环境友好等性能。为增强建筑物抵御自然灾害的能力，很有必要研发新型绿色建筑材料以抗御冲击。

本文强调，通过材料选择、能源保护、国家有关节约成本、高能效建筑材料规划，将规定绿色建材呈现的自然特性。文章将重点论述工业废物限价和回收、废料利用、开发绿色建材，及绿色建材应用。

E 房地产发展与绿色建筑

绿色建筑的实践与探索
——建设部科技示范工程(绿建·北秀蓝湾)

刘晓天　张　宬　赵健伟　刘　宏　周和进
(依柯尔绿色建筑技术研究发展(北京)有限公司，北京　100835)

摘　要：本文介绍了建设部绿色建筑示范项目——绿建·北秀蓝湾在节能、节水、绿色建材、生态环保等方面的实践，提出应关注建设项目背景，明确项目定位，采取适宜技术，创建平民化的宜居建筑，并为绿色建筑的发展探索出了一条行之有效的道路。

关键词：绿色建筑　城市背景　价值取向　宜居　适宜技术

引言

中国社会现在正经历着前所未有的高速经济发展，正在最大化地将自然资源转化为供人类消费的物质产品，满足人类日益增长的物质需要。其中与我们息息相关的建筑活动所消耗的自然资源大约占到全球总消耗量的 40%；发电产生的温室气体占全球温室气体排放的 37%，而其中 48%直接用于建筑的能量供应；同时，建筑产生的相关空气、电磁污染占环境污染总量的 34%。到 2020 年我国还要建造约 300 亿 m^2 的建筑，在这样一个生态环境日益恶化、能源日益匮乏、资源浪费、人口数量日益膨胀的时期，建筑自身以什么样的状况运行和建筑行业走一条什么样的路都将成为制约着我国经济快速发展和威胁着我们生存的关键因素。经过 SARS、禽流感等一系列事件的冲击，人们对生存环境的关注远远超过了以往任何时期。我们的发展无论是当前还是未来，都不能以生态环境的日趋恶化为代价，才能可持续地进行下去。

20 世纪 60 年代，美籍意大利建筑师保罗·索勒瑞(Paola Soleri)把生态学(Ecology)和建筑学(Architecture)两词合并为“Arology”，首次提出了著名的“绿色建筑”的概念，将绿色思想引入建筑领域，这是建筑界对人类可持续发展战略所采取的积极回应，也必将成为未来建筑的主导趋势。

1　绿建·北秀蓝湾项目背景与概况

1.1　项目背景

浙江省是我国能源最缺乏的省份之一，其能源消费总量的 96%左右依靠外省调入或进口，2005 年浙江拉限电高达 49.29 万条次，已成为我国最缺电的省份。浙江能源、土地等生产要素匮乏对经济发展的制约也越来越明显，“块状经济”的稳定性也开始受到挑战。然而，日益增多的高耗能建筑正极大地消耗着浙江本就匮乏的资源，如何解决其与环境和能源的关系，开始被越来越多的业内人士和建筑的最终使用者所关注。随着“全民建设节约型社会”的开展，建筑行业开始了对绿色建筑的探求。2003年12月浙江省建设厅正式发布《浙江省居住建筑节能设计标准》，提出居住建筑节能50%的要求，又率全国之先于2005年6月立项编写《浙江省绿色建筑标准》，给予绿色建筑发展一个明确的指导方向。绿建·北秀蓝湾就是浙江省探索绿色建筑发展道路上第一个大胆的尝试。

1.2 价值取向

绿色建筑不仅仅关注的是物质上的创造，更需要关注建筑、人、社会和自然生态环境之间的关系。

价值取向一：尽可能节约资源

浙江省经济比较发达，但是能源紧缺。在这样的环境下，要求我们必须对建筑能源消耗和资源消耗进行严格控制，而要实现这一目的，必须约束从规划、设计、施工、运行直至拆除各个阶段的自然资源消耗，实现建筑全生命周期内的可持续发展。但是我们又不反对采用适宜的超前技术来提高资源能源的使用效率，减少不必要的浪费，提高建筑的宜居程度。

价值取向二：适宜的使用空间

曾经有一段时间，过大的空间要求、超豪华的装修、夸张的高科技堆砌成为人们一时选择建筑的时尚，这使得我国已建的建筑中有90%成为高耗能建筑。但是随着人们的节约意识越来越强，过度舒适、过度浪费的建筑空间已经不再时尚，人们越来越注重建筑的自然美，提倡用适宜的材料、技术来为人们提供适宜的居住空间和沟通场所，满足人类“居”的要求。

价值取向三：体现本土文化

建筑作为一种载体，使得社会文化在建筑中延续，巩固社会文化的稳定性与永续性，促进人类对自身居住观念的不断进步。在一个能见证中国5000年文明史的地域里，绿建·北秀蓝湾作为绿色建筑示范，就必须考虑到建筑本身对良渚文化展现与传承，统合建筑与社会的持续发展。

1.3 项目概况

绿建·北秀蓝湾位于杭州市余杭区良渚镇，属于城乡结合的地区，是四千多年前良渚文化的发祥地，有着得天独厚的环境优势。

小区的南面与104新国道以50m绿化带隔离，东临东西大道，北临二号干道，西临三号支路和一条内河，面积为44113m^2，由21幢建筑组成，共分为A、B两个区。全区共570户，建筑面积80230.5m^2，地下建筑面积6953m^2，容积率1.81，绿地率35.5%，见图1。

本项目由杭州绿建房地产有限公司开发，由依柯尔绿色建筑研究中心(北京)有限公司负责整体策划。

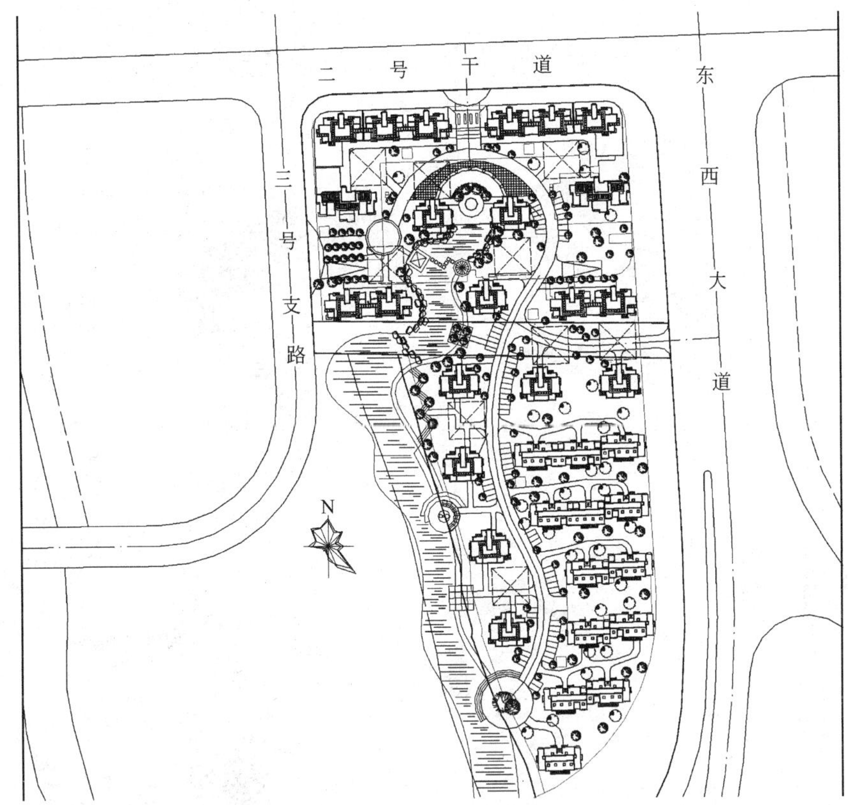

图 1　绿建·北秀蓝湾工程总平面布置图

2　绿建·北秀蓝湾项目的适宜技术集成

2.1　节能技术

浙江省是夏热冬冷地区，其采暖降温措施均由民众自行选择，缺乏科学指导。而其冬季采暖、夏季降温基本上都是以电暖器、空调为主，能源消耗极大，且多是电能这种高品位能源，仅 2004 年 1～9 月份浙江省全社会用电量就达 546.83 亿 kWh。电力资源的紧缺将成为浙江省经济快速发展的最大瓶颈，随着城市规模的日益扩大，用电压力必将越来越大，因此建筑节能工作势在必行。

提高空调设备节能性能、加强围护结构保温性能、合理确定室内外照明等等都是建筑节能的有效措施。通过实施建筑节能，可以减小建筑总的用电负荷，那么与之相配套的电力设备、设施和管线也将减少。如果把这笔节省下来的资金用于物业管理和物业基金，那最终受益的必将是众多业主。

2.1.1　建筑围护结构节能技术

(1) 目标与方法

在达到浙江省《居住建筑节能设计标准》DB 33/1015—2003 节能 50%的基础上，通过改变墙体材料和外墙、屋面保温材料的材质和厚度，增加外窗的外遮阳设施，改善外窗的传热性和气密性等有效措施，使得建筑物节能达到 60%。现以 A 区 7 号楼为例说明(图 2～

图 5、表 2)。

(2) 7 号楼改造前后的外围护结构的传热系数和保温状况见表 1。

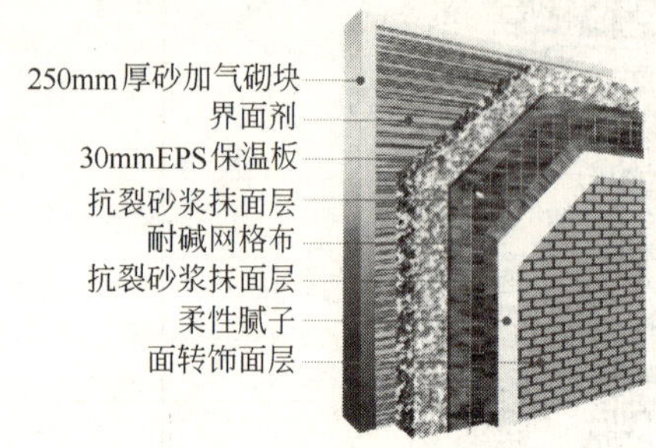

图 2　外墙保温构造示意图

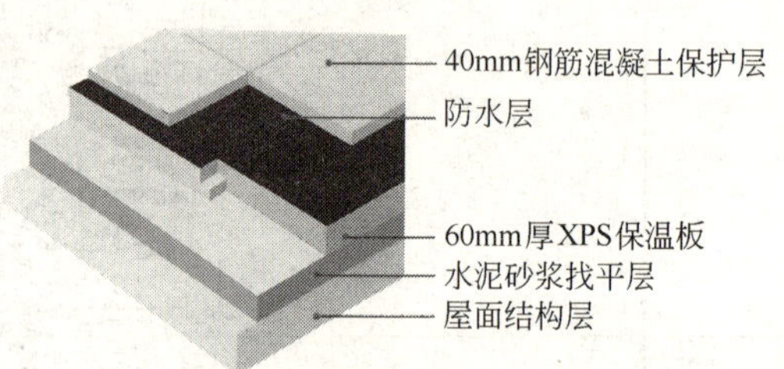

图 3　屋面保温构造示意图

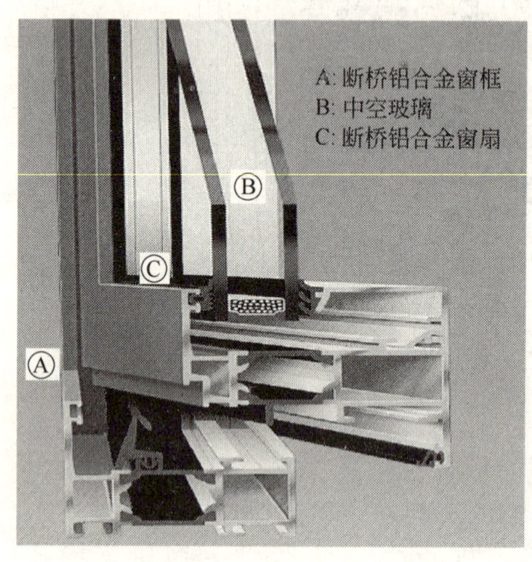

图 4　铝合金断桥中空窗示意图

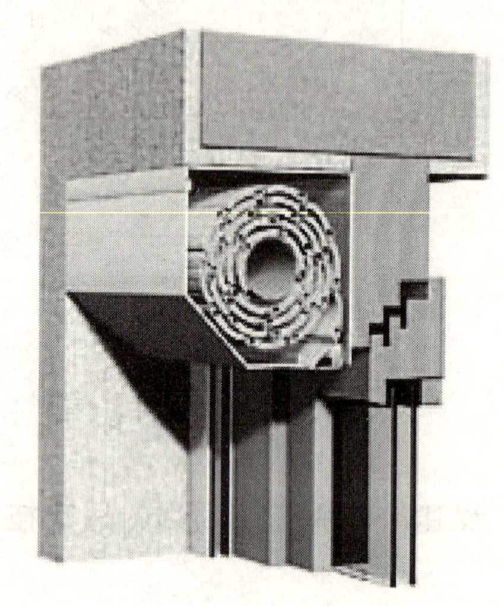

图 5　铝合金卷帘外遮阳

7 号楼外围护结构构造表　　**表 1**

围护结构	外墙		屋顶		外窗		外遮阳
	平均传热系数[W/(m²·K)]	保温状况	平均传热系数[W/(m²·K)]	保温状况	平均传热系数[W/(m²·K)]	配置	遮阳系数遮阳形式及部位
优化前	2.67	240 黏土实心砖+20mm 厚水泥砂浆	3.06	120mm 现浇混凝土+20mm 水泥砂浆找平层+40mm 混凝土保护层	5.81	单玻塑料推拉窗	无
优化后	0.74	250 砂加气砌块+30mm 厚膨胀聚苯	0.44	120mm 现浇混凝土+60mm 挤塑聚苯+40mm 混凝土保护层	2.6	铝合金断桥框中空窗 5(白玻)+12(空气)+5(白玻)	东、西、南向铝合金卷帘外遮阳、遮阳系数 0.2

(3) 建筑全年能耗计算

计算条件：室内计算温度：冬季为18℃，夏季为26℃；室外计算温度：夏季计算温度35.7℃，冬季室外计算温度−4℃；采暖和空调时，换气次数为1.0次/h；采暖、空调设备为家用空气源热泵空调器，空调额定能效比取2.3，采暖额定能效比取1.9；室内照明得热为每平方米每天0.0141kWh，室内其他得热平均强度为4.3W/m^2。

A区7号楼建筑概况 **表2**

城市名称：	杭州(北纬=30.20°，东经=120.20°海拔=41.70m)		
建筑名称	7号楼	朝　向	南　向
建筑形状	点形建筑	体形系数	0.34
建筑面积(地上)	2910.40m^2	建筑面积(地下)	0.00m^2
建筑体积(地上)	8452.70m^3	建筑体积(地下)	0.00m^3
建筑表面积	2849.80m^2	层　数	11
建筑物高度	36.10m		
建筑窗墙比	北向：0.34　西向：0.30　南向：0.34　东向：0.30		

经过中国建筑科学研究院编制的《夏热冬冷地区居住建筑能耗分析软件》计算，参照建筑采暖年耗电量：38.33 (kWh/m^2)，空调年耗电量：22.69(kWh/m^2)，空调采暖年耗电量之和：61.02 (kWh/m^2)。7号楼的实际年节能综合指标如表3。

7号楼优化后年耗电量指标 **表3**

HDD18 (℃·D)	采暖年耗电量指标 Eh(kWh/m^2)	CDD26 (℃·D)	空调年耗电量指标 Ec(kWh/m^2)
1647	29.68	196	18.33

经对比可以看出，优化后的采暖年耗电量和空调年耗电量之和较现行标准降低了约22%。

2.1.2　新型地源热泵空调技术

(1) 计算目标与方法

B区建筑为高标准节能住宅，适当地采用新技术和新工艺，其节能目标为更低能耗。在原建筑的基础上，采取如下措施：减小建筑的体形系数，适当降低窗墙比，取消建筑北向的突窗，改变外墙、屋面保温材料的材质和厚度，增加外遮阳设施，改善外窗的传热性和气密性。

采用DeST动态建筑能耗模拟软件，对B区的10号楼全年逐时的采暖和空调能耗进行了模拟分析(图6)，见表4～表7。

(2) 建筑全年逐时能耗计算

建筑概况见表4，外围护结构的传热系数和保温状况见表5。杭州室外的逐日最高温度、平均温度和最低温度见图7，月平均温度见图8。杭州最高气温为37°C，最低气温为−2.4°C；最热月的平均温度为28.2°C，最冷月的平均温度为5.2°C。日平均和月平均的含湿量分别见图9和图10。建筑物基本负荷和新风负荷见表8、9。建筑物基本负荷曲线见图11。

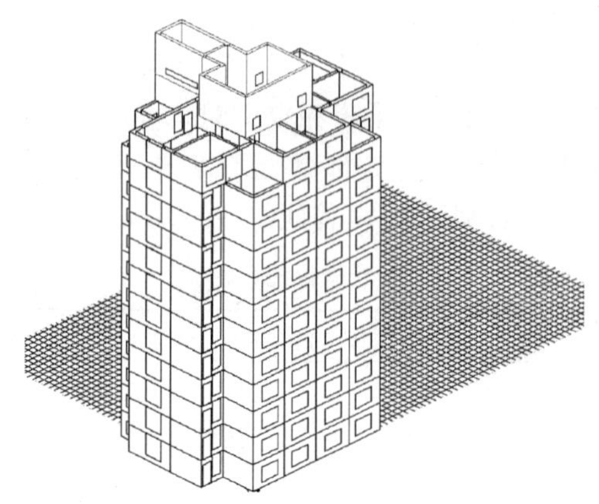

图 6 10 号楼 DeST 计算模拟图

B 区 10 号楼建筑概况 **表 4**

城市名称：	杭州(北纬=30.20°，东经=120.20°海拔=41.70m)		
建筑名称	10 号楼	朝向	南向
建筑形状	点状建筑	体形系数	0.33
建筑面积(地上)	3047.40m^2	建筑面积(地下)	0.00m^2
建筑体积(地上)	8676.82m^3	建筑体积(地下)	0.00m^3
建筑表面积	2815.84m^2	层数	11
建筑物高度	36.10m		
建筑窗墙比	北向：0.20 西向：0.14 南向：0.30 东向：0.14		

B 区 10 号楼外围护结构构造表 **表 5**

围护结构	外墙		屋顶		外窗		外遮阳
	平均传热系数[W/(m^2·K)]	保温状况	平均传热系数[W/(m^2·K)]	保温状况	平均传热系数[W/(m^2·K)]	配置	遮阳系数遮阳形式及部位
	0.24	250 砂加气砌块+70mm厚硬泡聚氨酯	0.20	120mm 现浇混凝土+120mm 硬泡聚氨酯+40mm混凝土保护层	2.0	铝合金断桥中空低辐射窗(6 白玻+12+6Low—e)	东、南、西、北向铝合金卷帘外遮阳、遮阳系数 0.2

建筑室内环境控制参数 **表 6**

夏季		冬季	
温度(℃)	相对湿度(%)	温度(℃)	相对湿度(%)
26	55	20	

室内新风分配 **表7**

	主卧室	客房	餐厅
最多人数(人)	2	2	3
灯光、设备(W/m²)	10	15	6
最低新风量(m³/人)	30	30	30

注：在本模拟中，冬季不考虑人员、灯光的作息情况。

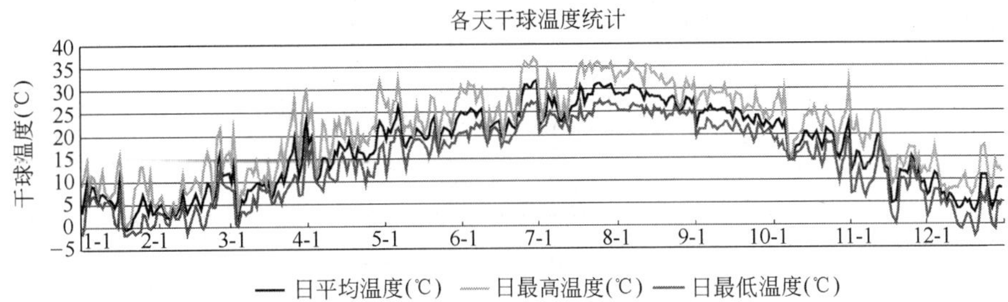

图7 杭州室外逐日温度

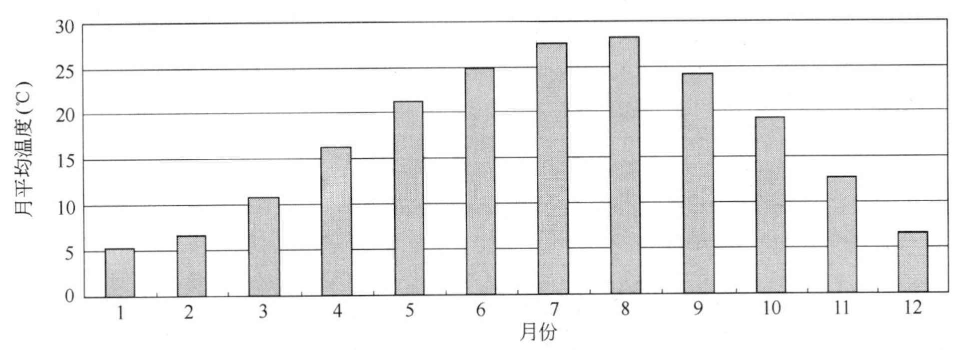

图8 杭州室外月平均温度

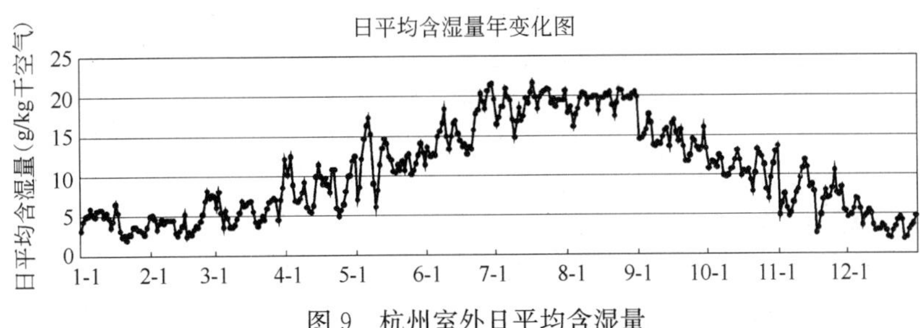

图9 杭州室外日平均含湿量

建筑物基本负荷 **表8**

名称	计算结果
采暖期累计耗热量(kW·h)	14785.68
空调期累计耗冷量(kW·h)	105695.15
单位建筑面积峰值热负荷(W/m²)	8.93
单位建筑面积峰值冷负荷(W/m²)	24.35

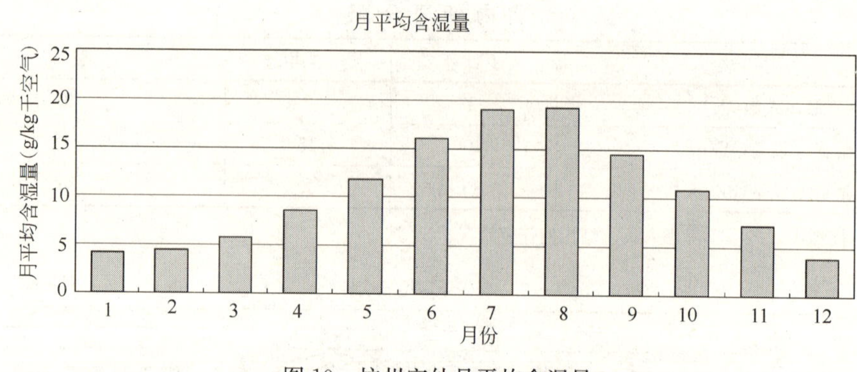

图 10 杭州室外月平均含湿量

新风负荷 **表 9**

名称	计算结果
采暖期累计耗热量（kW·h）	45349.13
空调期累计耗冷量（kW.h）	147398.31
单位建筑面积峰值热负荷（W/m²）	23.01
单位建筑面积峰值冷负荷（W/m²）	39.14

注：表 8 和表 9 中采暖期为 12 月 31 日～2 月 11 日，空调供冷期为 5 月 15 日～10 月 15 日。采暖期累计耗热量的物理意义为：采暖期的累计耗热量等于采暖期内的所有小时的叠加；空调期累计耗冷量的物理意义为：空调供冷期的累计耗热量等于空调期内的所有小时的叠加；表 8 中所有负荷均指建筑物本身的负荷，即除新风机带来的负荷之外的所有显热负荷。

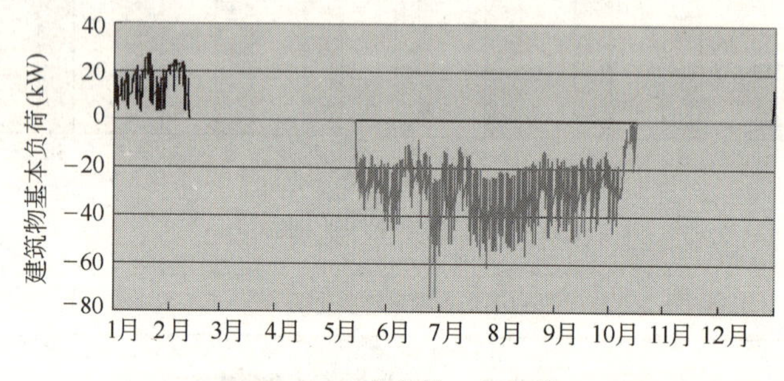

图 11 建筑物基本负荷曲线

（3）地源热泵空调系统

地源热泵空调系统组成为：冷热源（分户式土壤热泵机组＋地下埋管换热器系统）＋平面式辐射末端系统（毛细管）＋独立除湿新风系统三部分。现以 B 区 10 号楼为例说明。

1）土壤源热泵技术。

土壤热泵系统采用地下埋管形式，利用地下浅层土壤温度常年保持在 12～20℃左右的特点，通过地下埋管管内的介质循环与土壤进行闭式热交换达到供冷供热目的。夏季通过热泵将建筑内的热量转移到地下，对建筑进行降温；冬季通过热泵将大地中的低位热能提高品位对建筑供暖。最终根据建筑逐时负荷和实际的热响应实验实测土壤地质参数为依据，计算确定能源井为 40 口，深度为 100m。跃层选用 1 台制冷量 14kW，制热量 8.4kW，输入制冷功率 2.2kW，输入制热功率 3.0kW 的土壤热泵机组；标准层选用 1 台制冷量 10kW，制热量 6kW，输入制冷功率 1.5kW，输入制热功率 2.0kW 的土壤热泵机组。

2）高舒适度平面式冷暖辐射空调技术(图 12)。

图 12　毛细管平面辐射形式

图 13　置换式新风示意图

此种技术主要是以毛细管低温辐射的形式进行采暖制冷，室内温度均匀柔和，无温度死角、无噪声、无吹风感，是目前国际上公认的舒适度最高的空调形式，是一种典型的舒适、健康、节能的空调末端形式。在自然界中，利用充满水的毛细管进行传热的例子很多，例如植物的叶脉或人的皮下组织都是这种原理。由于毛细管换热的速度极快，因此换热效率会更高。毛细管辐射系统就是利用这种“仿生学”原理开发出来的新型产品。土壤热泵机组直接配套，与后者组合使用可达到最佳的节能效果。土壤热泵经过整个冬季的运行便会在地下土壤中形成一个天然的“廉价冷源”，夏季供冷初期可不启动制冷机，直接利用土壤中的冷源对房间进行制冷，基本实现夏季空调“零能耗”。

根据冷热负荷计算指标，进行毛细管选型和布置。本项目中辐射末端均选用直接采用特制砂浆粘贴在顶棚上的毛细管席来供冷及供暖。采用吊顶敷设(局部房间布置在墙面上)毛细管的方式，整个空调系统根据使用功能的不同进行环路划分。由空调机组提供冷热水：夏季供回水温度 18～21℃；冬季供水温度 28～32℃。分集水器设在每户的卫生间及储藏室内。经计算 B 区 1 号楼和 10 号楼毛细管面积总计为：1852.47m^2。

3）独立置换式送风及湿度控制技术(图 13)。

新风经除湿处理后以置换送风的方式送入空调房间，用以保证房间的高空气品质，保证室内人员时刻呼吸到新鲜空气；除湿机针对平面空调制冷系统设计，它只将房间湿度控制在 30％～60％之间，而不会改变平面空调辐射制冷系统的房间温度。下送上回置换式新风管露出新风机房后沿墙角暗敷至设计位置，与外墙内预埋风管对接，新风自上而下输送，至设计位置由踢脚上新风口送出。

根据负荷计算结果，每户选用 1 台新风量 400m^3/h(跃层 500m^3/h)的全热回收器和 1 台新风量 400m^3/h(跃层 500m^3/h)的除湿新风机组。由空调机组提供冷热水：夏季供回水温度 18～21℃；冬季供水温度 32℃。经过除湿设备处理后的新风经管道进入静压箱，从静压箱分出支风管经地板送进各房间，每个房间设置一个条形风口。各房间与走廊之间设一个消声回风口，每户在客厅设集中回风口，用风管引入全热回收器，经过全热交换器进行能量交换后排到室外。

4）室内温湿度控制。

在各户每个房间安装1个温度控制器，通过控制每个房间支路上的电动阀门通断，控制各房间的温度。温控器具有冬夏转换开关和温度设定功能。每个房间安装1个露点探测器，当房间内有结露现象时，自动关闭各房间支管上的电动阀门作为防结露保护。温控开关同时控制分户水泵的运行，温控开关打开时水泵运行，全部关闭时水泵停止运行。

5）新型空调系统的全年运行费用模拟。

计算依据：电价0.53元/kWh；制冷时间150d；供暖时间90d；每天运行18小时；负荷系数0.7；空调同时使用系数0.7；空调使用面积4000m²。模拟运行费用见表10、表11。

夏季运行费用 **表10**

负荷百分比	100%	75%	50%	25%
运行天数	10	50	50	40
运行费用(万元)	0.670	2.513	1.676	0.670
合计(万元)	5.529			
折算运行费用(元/m²)	13.8			

冬季运行费用 **表11**

负荷百分比	100%	75%	50%	25%
运行天数	10	20	30	30
运行费用(万元)	0.725	1.087	1.087	0.544
合计(万元)	3.443			
折算运行费用(元/m²)	8.607			

2.1.3 全循环生活热水技术

建筑选用1台制热量155kW，输入功率36kW，出水温度50℃的污水源热泵机组来实现生活热水供应。先收集洗浴水盥洗水等优质杂排水，然后通过热泵机组将水中难以直接利用的低品位能量提取出来，作为生活热水的热源，实现取热和排污两不误。而且采用每天24小时全循环方式，因为每个龙头都是循环管路上的一个节点，使用者无论何时打开热水龙头都能直接使用到热水，而不会有滞留的凉水。

2.1.4 太阳能庭院灯

太阳能庭院灯不需要架设输电线路或挖沟铺设电缆，省去了配电柜及路灯之间的配线工序，不用专人管理和控制，可方便地安装在小区道路旁、室外停车场等地方，而且没有线损，不消耗常规电力能源，节能环保。

根据杭州地区的太阳能总辐射量、直接辐射量、散射辐射量、年平均气温与最高、最低气温和最长连续阴雨天数等资料计算出需要安装200W庭院灯约120盏，两灯间距15m。与普通庭院灯相比，不仅在整个寿命周期内每盏太阳能庭院灯可节约5000元，而且其不消耗化石燃料，无二氧化碳、二氧化硫等有害气体的排放，清洁无污染。

2.2 再生水技术

2.2.1 项目再生水使用概况

为了增加水的重复利用率以实现节水的目的，绿建·北秀蓝湾示范小区采用室内污废分流以便于收集优质杂排水(主要为盥洗水、洗浴水)作为再生水源，并配套建设了再生水处理、回用设施，将处理后的再生水用于冲厕、绿化、景观、道路浇洒等用水，且水质符合《城市污水再生利用 城市生活杂用水水质标准》GB/T 18920—2002 中的相关要求。

2.2.2 水量计算与经济分析

本社区有 570 户，绿化面积为 15660m^2，硬质路面面积为 9000m^2，景观水池容积为 2300m^3，建筑总面积为 80230.5m^2。

浇洒道路用水 13.5m^3/d、绿化用水 31.4m^3/d、回用冲厕用水 108m^3/d，再生水总用量 153m^3/d，可以收集到的优质杂排水为 184m^3/d，所以可收集的杂用水水量完全满足再生水水源的水量要求，故再生水处理设施得以充分运转。而且，再生水成本约合 1.10 元/t，远低于杭州地区 1.65 元/t 的自来水价格。

考虑到杭州的气候条件(雨水比较多)，把场地内和建筑屋面雨水经过处理作为人工湖的雨水景观水源。考虑初期入住率的影响，可以将雨水的一部分作为再生水水源。

2.2.3 再生水系统的供水保障技术措施

供水保障技术措施：主要设备选用进口或国产优质品牌并均有备用，可保证系统连续不间断运行；回用水池上方设置了自来水应急补给系统，以保证用户用水在最不利的情况下仍能得以保证。

2.2.4 节水型器具和设施(图 14～图 16)

图 14 微喷浇灌

图 15 挂墙式坐便器

图 16 节水龙头

项目全面采用节水器具，如节水型便器、环保节水浴缸和洗浴龙头等，可以在生活机能丝毫不受影响的情况下，每人节约日常用水量 20%。另外，采用微喷等节水灌溉技术和措施，提高户外用水效率。

2.2.5 项目节水的综合意义

整个浙江省地下水资源一般(储量约 200～299 亿 m^3)，目前冲厕、绿化浇灌和道路清洗等需要大量自来水，因此节水方面的工作不能忽视。随着杭州城市规模的日益扩大，用水方面的压力也将越来越大，节水势在必行。目前全国大部分城市的自来水供给需要政府补贴，北京的水价将可能提高到 6 元/t，杭州居民生活用水经价格调整听证后拟增加 0.27 元/t。节水不仅关系到全社会的资源节约和可持续发展，也关系到普通百姓的直接利益。通过再生水的收集和利用，不仅可以减轻市政排水的压力，也可以减少小区的污水排放增容费。将这笔节省下来的费用用于物业管理和物业基金，最终受益的还是众多业主。

2.3　绿色建材

为从源头上控制建筑的绿色度，工程中使用了许多新型的绿色建材。

住宅建筑的墙体材料全部采用砂加气砌块。这种材料是由石英砂、水泥和其他废料蒸压而成，与常规的多孔黏土砖比起来具有轻质、保温性能好等显著特点。其组成材料的生产能耗很低，而且回收的材料容易再利用，可以成为混凝土的细骨料。

门窗和遮阳材料全部采用铝合金材料。这些材料容易回收，不会像PVC材料难以回收，且如果处理不当，还会对环境产生很大危害。

保温材料采用新的施工技术，加长保温系统的使用寿命，从而减少这些难以回收材料的更换次数，在建筑全生命周期内高效使用。

工程中的钢筋混凝土采用高强度混凝土和水泥、再生混凝土、高强钢筋，这样可以减小梁柱截面，节省其用量，提高耐久性。

装饰材料是利用废弃建筑材料和矿山尾料生产的再造石，用作小区内的建筑立面和建筑小品装饰。

小区垃圾桶等公共卫生设施均由回收的无毒害固体废弃物制成。

2.4　绿色景观

2.4.1　屋顶绿化(图17、图18)

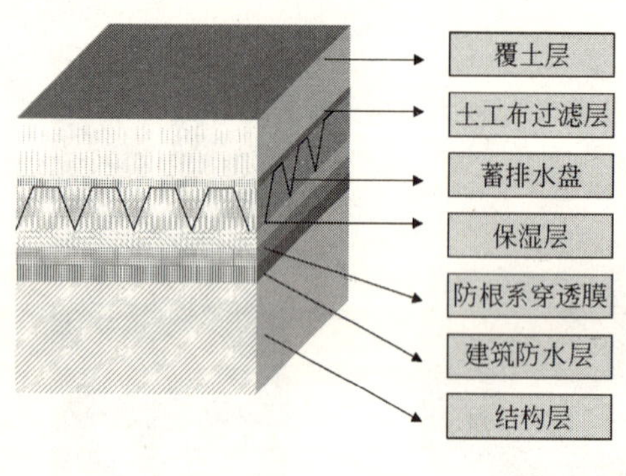

图17　屋顶绿化构造图

图18　屋顶绿化实景

小区屋顶绿化主要分布在项目的1～6号楼的平屋顶部分，17、18号楼之间会所的房顶，20、21号楼的裙房房顶，总共的屋顶绿化面积为1767.7m^2。

绿化设计的具体实施中主要考虑了杭州地区的降雨量和台风等因素，最终确定的屋顶绿化的植物选择原则是：

(1) 选择适宜在杭州生长的本土植物，包括宿根草本、灌木、地被植物；

(2) 保证在植物生长良好的同时，起到绿化美观屋面、隔热保温及净化空气的作用；

(3) 实现植物的免维护或少维护；

(4) 考虑花期、叶形、叶色的搭配，选择金叶过路黄＋欧亚活血丹＋佛甲草＋金叶景天＋凹叶景天的低矮草本配植模式。

2.4.2　地表绿化

本工程在切合当地自然植被规律、注重生物多样性的基础上，通过不同的植物配置方式，充分利用植物的特性，为小区提供良好的生态环境。在小区东、北、西三侧，为了发挥植被的降噪、滞尘、吸收有害气体等生态功能，植物与建筑之间的距离定为植物高度的5～10倍。在小区南侧，利用植物廊道的通风、降温与增湿的功能，结合植物的疏密高低等不同的形态和配置方式，合理地将新鲜的空气输送给每一栋建筑。在中心生活区——良渚文化广场、生态水景、活动娱乐区等处，充分发挥植物的降温、放氧、杀菌、降噪、除尘、消纳污染、涵养水源、净化水体的生态功能，选用芳香植物、传统名花、色叶树种，如香樟、桂花、枫香、广玉兰、无患子等，并充分融入中国的优秀文化内涵，创造一个舒适空间(图 19)。

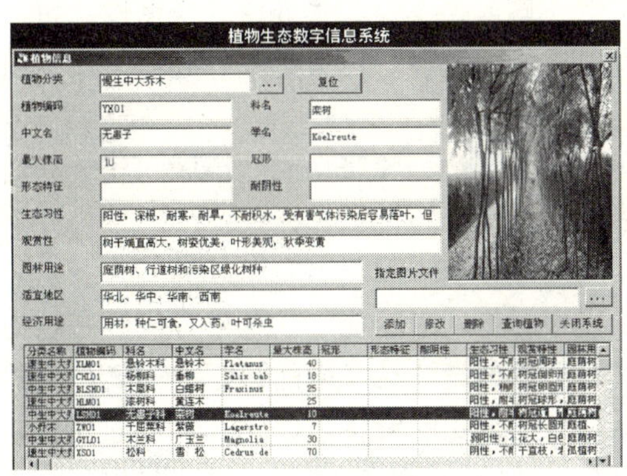

图 19　无患子的生态绿化计算

小区选择高大适宜的落叶乡土树种如榉树创建林荫广场，更大限度地发挥广场的使用程度，使人们能够在即使是夏季中午也可以在其中活动休息。选择高大适宜的乡土树种对地面停车场进行遮荫绿化，减少夏季对车体的暴晒。

小区结合原有的地形，营建流动的生态水景，防止“水体富营养化”的发生，通过植物净化池等湿地进行雨水的收集和净化，尽量降低雨水的排放。

2.4.3　水景设计(图 20、图 21)

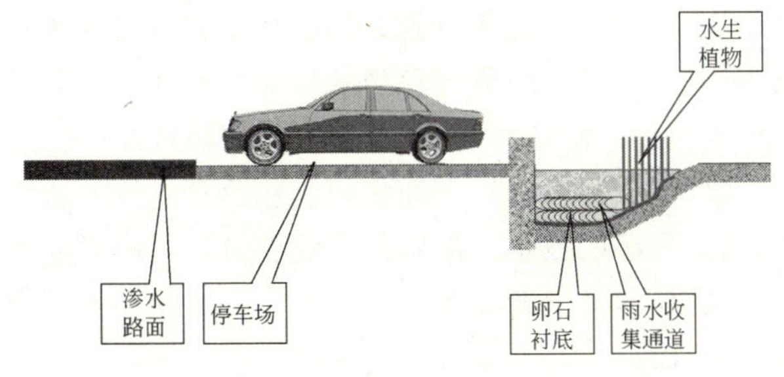

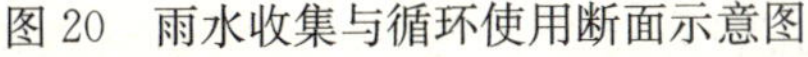
图 20　雨水收集与循环使用断面示意图

图 21　渗水路面

整个小区的水景设计与雨水收集和循环利用有效结合。以小区内道路为雨水收集的分水岭，落在绿化带内的雨水直接渗透到导流暗渠内流入收集水景中；落在道路上的雨水由渗透路面直接渗透一并收集；落在屋面上的雨水，一部分经过种植屋面蓄水后(每平方米可以蓄水 20L)流入落水管，一部分直接进入落水管，而后汇集到水景中。最后经过道路两侧的停车位旁边的雨水通道把小区南部各个水景联系起来，由太阳能水泵使其循环，确保整个水景内的水为活水，而且也能保证其水质良好。

2.5 适宜环境

热环境：通过改善建筑围护结构的性能和小区植物配置，调节室外环境的温度湿度，可以在不使用任何机械设备采暖和制冷的手段下，使室内温度比室外低1～2℃，确保良好的室内热舒适环境。

风环境：杭州地区春夏季的主导风向为东风或东南风，秋冬季主导风向为北风或西北风，因此通过调整建筑体形和园区内的植物搭配起到引风、阻风和导风的作用，强化春夏秋季的自然通风，减弱冬季的冷风侵袭，从而改善建筑室内在开窗下的风环境。

声环境：通过使用中空玻璃窗，不仅改善外窗的气密性，也可以起到隔声的效果，可以在室外吵闹的环境下室内依然很安静。白天卧室、起居室在关窗的情况下不大于45dB，夜间不大于35dB。

人文环境：小区广场内文化雕塑和建筑小品、小区内建筑立面的装饰材料和太阳能庭院灯上无不体现着对良渚文化的展现与传承。

2.6 智能化系统

小区的智能化系统工程为即将到来的住宅信息化时代做好准备，为将来的数字化应用打下良好的软硬件基础。整个系统设计结合当前智能化小区系统的优点，并考虑到今后各系统的发展方向，采用具有高性价比的设备，以达到和超过国内同类小区的智能化要求。整个小区智能化建设实现对小区内的物业实施统一的监控和和集成管理，并在此基础上允许逐步建立包括综合物业管理，社区服务等内容的应用服务系统，做到更好地服务于业主。

本小区的弱电系统主要由安保监控系统(包括闭路电视监控、防盗报警、电子巡更、家居报警)、车辆管理系统、小区室外背景音乐广播系统、电子公告系统、有线电视、电话和宽带系统等部分组成。

3 结束语

通过对绿建·北秀蓝湾绿色工程的探索，我们深刻地体会到，目前在中国发展绿色建筑还存在着许多困难和亟待解决的问题。首先，国家和地方绿色建筑标准及其评估体系的缺乏，使我们只能一边探索项目，一边结合项目实际和国内外已有的绿色建筑或者可持续建筑的评价标准完成浙江省自己的绿色建筑标准。再者，对于新建建筑和已有建筑，由于政府缺乏财政税收等经济激励政策，单纯依靠简单的技术“堆砌”试验项目和用户以及开发商的自发行为难以把绿色建筑推向市场，这样包括节能、节水、节地、节材和环保等指标得不到落实。

绿建·北秀蓝湾项目的绿色建筑实践，是我们面对市场迈出的既艰难又准确的第一步。说它艰难是因为我们是“第一个吃螃蟹的人”，说它准确是因为我们把准了绿色建筑“跳动”的脉搏，探索了一条行之有效的绿色建筑发展道路。因此，我们要发展绿色建筑就必须尽快制定并不断完善我国的绿色建筑的相关标准与评定体系，而且国家应该出台相关的激励政策，来扶持那些真正愿意而且也正在努力推动中国绿色建筑发展的企业。

绿建·北秀蓝湾项目的探索与实践过程中，依柯尔绿色建筑研究中心(北京)有限公司投入了大量的精力，调查杭州的实际情况，咨询了大量多方面专家，最终完成了绿建·北秀蓝湾的策划与实施方案。在此我们衷心地感谢建设部科技司、浙江省建设厅、中国建筑科学研究院、上海建筑科学研究院和绿建房地产开发有限公司对建设部绿色建筑示范项目绿建·北秀蓝湾的大力支持与帮助，以及对本项目给予过帮助的单位个人表示真诚的谢意。

Practice and Research for Green Building

——The Technology Demonstration Project of Ministry of Construction

(Lvjian-Beixiu Lanwan)

Liu Xiaotian　Zhang Cheng　Zhao Jianwei　Liu Hong　Zhou Hejin

(Eco Green Building Technologies Research and Development Co., Ltd (Beijing), Beijing　100835　China)

Abstract: This paper introduces the demonstration project named Lvjian-Beixiu Lanwan for green building of Ministry of Construction, about the practice of energy efficiency, water efficiency, ecology , environmental protection . It points out we should concern the project background, understand the project orientation, take the appropriate technologies , establish the appropriate habitation, and search an effective way for green building development.

参考文献

[1] 中华人民共和国建设部科学技术司和《智能与绿色建筑文集》编委会. 智能与绿色建筑文集. 北京：中国建筑工业出版社，2005. 3

[2] 中国建筑科学研究院. 绿色建筑技术文集. 北京：中国建筑工业出版社，2005

[3] [美] Davida Gissen. BIG & GREEN-TOWARD SUSTAINABLE ARCHITECTURE IN THE 21ST CENTURY . Princeton Architectural Press. December 1, 2002

[4] 仇保兴. 中国城市化进程中的城市规划变革. 上海：同济大学出版社，2005

推广绿色建筑与房地产业的可持续发展

王洪波[1]　刘长滨[2]
（[1] 哈尔滨工业大学管理学院，哈尔滨　150001；[2] 北京建筑工程学院经济管理与人居环境研究所，北京　100044）

摘　要：房地产业作为国民经济的支柱产业，其快速发展对于国家的经济和社会繁荣都有着重要的意义。但是，房地产业的发展也不是越快越好，同样面临着可持续发展问题，以及与社会和自然的和谐。在能源和资源紧缺、环境问题突出的当前，探讨房地产业的可持续发展具有重要意义。本文分析了房地产业发展过程中受到的各种制约，并阐述了绿色建筑的概念和内涵，提出推广绿色建筑是解决房地产业可持续发展的必然选择。

关键词：绿色建筑　房地产业　可持续发展

1　引言

可持续发展的概念是在1987年世界环境与发展委员会公布的报告——《我们共同的未来》中被第一次提出的，是“指既满足当代人的需要，又不危害后代人满足其需要的能力的发展”。可持续发展最初的含义侧重于“代际公平原则”，强调资源在各代人之间的分配。此后，学者们陆续补充了“代内公平”、“生活质量与环境保护相协调”等原则，逐步将可持续的思想发展成为一种理论，并得到了全世界的认同。我国在1992年的“联合国环境与发展大会”以后开始积极研究中国的可持续发展问题，并制定了国家级的可持续发展纲要——《中国21世纪议程》。后来，可持续发展被定为我国国民经济和社会发展的战略目标。

房地产业是进行房地产投资、开发、经营、管理的行业，是固定资产投资的重要组成部分。尽管房地产业在旧中国已经有了相对较高的发达程度，但在新中国则是一个新的产业部门，虽然起步较晚，但发展十分迅速。由于任何行业的发展都离不开房地产业，可以说房地产是国民经济发展的一个基本生产要素，房地产业已成为我国国民经济的支柱产业。

对房地产业的研究大多侧重于房地产业的发展与国民经济之间的关系，但从另一个角度来看，房地产业的开发活动占用土地，直接和间接地消耗各种资源，对资源和能源需求以及周围自然环境都具有重要影响。根据有关研究结果，房地产开发建造及使用过程所产生的二氧化碳占全社会二氧化碳排放量的1/3，对环境造成巨大破坏。所以房地产业的发展不是个单纯的经济问题，而是个综合性问题，它在国民经济中的地位越重要，它的活动与环境等方面的关系就越密切。当前全世界都面临着能源和资源紧缺，以及环境污染等严峻问题的挑战，在这种情况下，研究房地产业的可持续发展问题就显得十分重要而迫切。

2 房地产业可持续发展应考虑的因素

近几年我国的房地产开发投资连续多年增长保持在19%以上，年增长额保持在固定资产投资增长额的30%左右，直接和间接拉动GDP增长在2个百分点左右。但是在快速发展的同时，房地产业也遇到了一些难题，甚至出现了瓶颈，主要反映在土地、资金、资源等要素上。2004年以来增速有所放缓，于是人们担心房地产业能否持续地快速发展?

应该看到，在过去的房地产业活动中，偏重于考虑它对经济的影响，而忽略了对资源及环境等方面的影响，致使房地产业的活动与人类社会的可持续发展失去了应有的联系。我国正处于经济快速发展时期，以房地产业拉动国民经济发展是我国的一项重要措施。但是，房地产业的发展也并不是越快越好，如果房地产业发展过于超前，就会受到资源和环境的制约，同时也可能产生通货膨胀，影响金融系统的运行，对社会造成严重的影响。因此，从能源、资源及环境的角度研究房地产业的可持续发展问题，对房地产业乃至整个国民经济的发展协调都具有重要意义。以下分析了房地产业可持续发展应考虑的几方面因素。

2.1 能源方面

目前我国的经济规模占世界的第七位，但是能源消耗总量到2003年已经达到了16.8亿t标准煤，居全球第二，占全世界能源消费的10%。这充分说明了我国的能耗非常高。能源已经是制约我国经济发展的重要因素之一，同时也是制约房地产发展的瓶颈。房地产业开发建造过程所使用的建材生产能耗占全社会终端能耗的16.7%，而开发的房屋在使用过程的能耗占全社会终端能耗的27.5%，因此，节能是房地产业可持续发展的首要问题。

2.2 土地资源方面

由于土地是房地产的载体，因此房地产业的可持续发展应该考虑土地资源的可持续利用。根据《中华人民共和国土地管理法》的划分方法，按用途将土地分为农用地、建设用地和未利用地三类，而我国绝大部分未利用的土地属于不可利用的范围，有数据显示，尚可利用的未利用土地存量为73.9万km^2(占国土总面积的7.7%)，其中宜建设用地仅为0.88万km^2，占尚可利用的未利用土地存量的1.19%。

然而近几年房地产开发的规模、品位和质量不断提升。商品住房供给中出现了一种片面追求高档次、大套型的倾向，每套150m^2以上的房子越建越多，花园别墅动辄300m^2以上。与此同时土地利用效率却越来越低，尤其是占用耕地越来越多。土地供给与需求的矛盾愈加突出，因此，房地产开发应转变粗放式观念，集约利用土地资源才是其可持续发展之路。

2.3 建筑材料方面

在建筑材料的利用方面同样应贯彻可持续发展思想。目前我国商品住宅开发面积越来越大，装修越来越豪华，但是与提高住宅科技含量并非同步，材料资源的循环利用率低，导致资源的极大浪费。我国住宅建设的物耗水平与发达国家相比，钢材消耗高出10%～25%，每立方米混凝土要多消耗水泥80kg。建材工业是资源耗费最大、破坏土地最多、大气污染最严重的行业之一，仅每年生产水泥和石灰排放的CO_2量就达6亿t之多，建材在生产过程中还要消耗大量的能源和排放SO_2、NOx等有害气体。同时由于我国房地产开发建设用材中相当一部分来自于自然环境，如木材和过去常用的红砖。它们存量很少，再生速度缓慢。可见，节约利用材料和使用替代品是房地产用材的可持续利用应考虑的主要措施。

2.4 环境质量方面

房地产业的发展中应注意改善室内的环境质量，室内环境的无害化问题是人们关心已久的问题。建筑物的室内空气污染严重时，如空气温度、湿度、一氧化碳浓度、二氧化碳浓度、粉尘和挥发性有机物等物质的含量超出人体所能适应的范围，使用者将时常遭受头痛、疲劳、恶心等病态反应的困扰，影响使用者的身体健康，降低了生活质量。改善环境质量方面还应该包括在房地产开发中减少建筑粉尘和建筑噪声，采取高效的建筑节能措施。

2.5 自然生态方面

房地产业发展也不能忽视生物保护。房地产开发占用了土地，使原生动、植物的生存环境受到影响，甚至从根本上破坏了其生存环境，使它们无法在当地继续生存。生物的多样性是可持续发展的重要支持力量，我国过去的房地产开发忽视保护生物的多样性，应该在以后的开发活动中加以注重。

通过从能源、资源、环境的角度分析我国当前的房地产业发展，可以看出，为了实现其可持续发展，必须注重节能、节地、节材，以及改善环境质量和保护环境等。这正是绿色建筑所体现的设计原则，可以说推广绿色建筑就是房地产业的可持续发展之路。

3 解决对策：推广绿色建筑

3.1 绿色建筑的含义

绿色建筑是指在建筑的全寿命周期内，最大限度地节约资源(节能、节地、节水、节材)、保护环境和减少污染，为人们提供健康、适用和高效的使用空间，与自然和谐共生的建筑。

尽管很早就制定了可持续建筑的一些原则，但是绿色建筑的概念直到20世纪80年代末期才有了定义并进入全球建筑事业中。“绿色建筑”的产生和发展，是将绿色思想引入建筑领域的结果。这是建筑界对人类可持续发展战略所采取的积极回应，作为一种设计规则和建筑环境性能的衡量标准，绿色建筑的发展势头迅猛。原因在于地球资源的有限性，以及人与自然环境和谐共生的需求，绿色建筑恰恰是这一设计理念的体现。绿色建筑设计要求合理处理节能、节地、节水、节材、环保及满足建筑功能之间的辩证关系，从建筑的全寿命周期综合核算效益和成本，注重考虑环境因素，确保施工过程及运营阶段对环境的影响最低，为人们提供健康、舒适、低耗、无害空间，同时尽可能将拆除的材料循环利用，将拆除后又对环境的危害降到最低。

分析绿色建筑的概念，可以看出，绿色建筑包含了三层含义，第一是节约资源，这个资源包括了能源、土地资源、水资源和材料；第二是创造舒适的环境，从使用者的舒适度需求出发，保障使用安全的同时，降低环境污染，改善室内环境质量，努力创造优美和谐的环境；第三是强调与自然界的和谐共存，充分利用建筑场地周边的自然条件，尽量保留和合理利用现有适宜的地形、地貌、植被和自然水系，尽可能减少对自然环境的负面影响，如减少有害气体和废弃物的排放，减少对生态环境的破坏等，最大限度地保护自然环境。

绿色建筑的理念充分考虑了房地产开发建造及使用活动与资源和环境的和谐、可持续发展。改变粗放式、盲目追求高档次和视觉享受的错误观念，以科学的发展观统领房地产业发展，推广绿色建筑才是房地产业可持续发展的必然选择。

3.2 绿色建筑的设计原则

绿色建筑的设计原则要求适应于可持续发展的要求，即可持续发展在建筑设计中的反映，贴切而直观地表达了可持续发展的概念与内涵。其原则包括全寿命设计、资源经济和较低费用、适用与健康的环境、资源节约与综合利用、减轻环境负荷等，具体体现在以下方面：节地、节能、节水、节材、室内环境质量及环境保护等。

3.2.1 节地

建设开发用地应适度密集，适当提高公共建筑的建筑密度，住宅建筑立足创造适宜居住环境为依据合理确定建筑密度和容积率。同时强调土地的集约化利用，充分利用周边的配套公共建筑设施，合理规划用地，加强对耕地的保护，可采取高效利用土地的措施，如开发利用地下空间，采用新型结构体系与高强轻质结构材料，提高建筑空间的使用率。

3.2.2 节能

房地产开发建设充分利用自然条件，如合理考虑建筑朝向和楼距，利用自然通风和天然采光，减少空调和照明能耗，并注重提高建筑围护结构的保温隔热性能，采用由高效保温材料制成的复合墙体和屋面及密封保温隔热性能好的门窗，采用有效的遮阳设施，以减少建筑的使用能耗。还采用高效建筑供能、用能系统和设备，提高用能效率。同时根据自然条件开发利用可再生能源，以减少常规能源的消耗。

3.2.3 节水

房地产开发应考虑节水措施，采用节水系统、节水器具和设备，提高用水效率。按高质高用、低质低用的原则，将生活用水、景观用水和绿化用水等按用水水质要求分别提供、梯级处理回用。还可以根据当地水资源状况，因地制宜地制定中水回用、雨水集流利用等节水方案，充分利用水资源。

3.2.4 节材

建筑材料占房地产开发总投资的 50%以上，建筑物的功能是通过合理选择建筑材料和施工来完成的，绿色建筑的内涵大多通过建筑材料来体现。绿色建筑从节约能源和材料的角度出发，希望尽可能实现重复使用和循环使用的建筑材料。这有助于减少废弃物的产生，对于降低自然生态系统的破坏、减少环境污染都具有重要意义。一方面，应采用高性能、低材耗、耐久性好的新型建筑体系，选用可循环、可回用和可再生的建材，做到节约用材并可循环利用。另一方面，减少建材在全寿命周期中的能源消耗，选用可降解、对环境污染少的建材，做到节能并环保。

3.2.5 室内环境质量

室内环境质量包括光环境、热环境、声环境和空气品质等。在设计时，尽量考虑自然采光、自然通风的问题。这样不但能改善室内环境质量，且可通过节约大量能源减少对整个大环境的污染。在热工性能方面，应考虑优化建筑外围护结构，防止因外围护结构内表面温度过高过低、透过玻璃进入室内的太阳辐射热等引起的不舒适感。还应使用可改善室内空气质量的新型装饰装修材料，避免由装修材料引起的空气质量下降。

3.2.6 自然环境保护

绿色建筑还考虑了与周围环境的协调，设计中，尽量减少建筑产生的废水、废气、废物的排放，利用园林绿化和建筑外部设计减少热岛效应，并考虑减少建筑外立面和室外照明引起的光污染，以及采用雨水回渗措施，维持土壤水生态系统的平衡，降低环境负荷，保护自然生态多样性。

总之，绿色建筑的设计包含了节约能源和资源，以及环境保护等措施，解决了房地产业可持续发展中遇到的瓶颈，是房地产业可持续发展的必然选择。

4 结论

过去单纯就房地产业的经济问题来讨论房地产业发展的思路，不是解决我国房地产业的可持续发展的途径。房地产业的可持续发展应该同时兼顾房地产业与环境、社会和国民经济的协调发展关系，即综合考虑才是解决房地产业可持续发展问题的根本途径。资源节约与环境保护已成为全球的发展趋势，作为消耗资源的重要行业来说，房地产业更应该贯彻可持续发展的观念，大力推广绿色建筑。

The Spread Of Green Buildings & Sustainable Development Of Real Estate Industry

Wang Hongbo[1] Liu Changbin[2]

([1] School of Management, Harbin Institute of Technology, Harbin 150001China,;
[2] Research Institute of Economy Management and Habitation environment, Beijing Institute of Civil Engineering and Architecture, Beijing 100044 China)

Abstract: Real estate industry serves as a pillar-industry of national economy whose rapid development plays an important role in the prosperity of state economy and society. However, it doesn't make sense that real estate industry should develop as fast as possible. It also confronts the issue of realizing sustainable development and keeping harmonious relation with society and nature. Therefore, discussion on the sustainable development of real estate industry is full of significance when taking account for the today's shortage of energy and resources and outstanding problems of environment. The paper analyzes many restrict factors that real estate industry confronts in the process of its development, illustrates the concept and meaning of green buildings , comes up with a conclusion that the spread of green buildings is the inevitable choice to realize the sustainable development of real estate industry.

参考文献

[1] 胡志刚. 用科学发展观统领房地产开发. 中国房地产，2005(4)
[2] 杨同利，刘长滨，刘琳. 我国房地产业可持续发展对策研究. 城市开发，2000(12)
[3] 赵苏，杨合. 绿色建筑与可持续发展. 低温建筑技术，2003(5)
[4] 泰佑国. 中国国情下的绿色建筑. 中外建筑，2005(3)
[5] 游娜. 探索建筑新概念：生态建筑与绿色建筑. 中国房地信息，2003(7)

循环经济与住宅产业

——以3R原则促进住宅产业化

林　驹

（上海市房地产科学研究院，上海　200031）

摘　要：本文提出住宅产业化必须走循环经济的道路，坚持3R原则，改变传统的住宅建造模式。本文首次将住宅生命周期的概念引入，以3R原则围绕住宅生命周期建设资源节约型住宅，即将住宅的生命周期分为四个阶段：设计阶段、建设阶段、使用阶段、老化阶段，从整个周期来整体进行考虑。最后本文阐述了以3R原则促进住宅产业化的措施，从政策层面来保证产业化的实施。

关键词：住宅产业化　循环经济　3R原则　资源节约　全生命周期

进入21世纪后，人类面临着能源、资源、环境、生态与工业化和城镇化加快、经济快速增长的矛盾，这一矛盾已成为瓶颈约束。特别是在住宅的生命周期中，需要消耗大量的能源、资源，因此，如何节约能源资源，提高使用效率，保护和改善环境，缓解能源资源短缺的矛盾，实现住宅产业的可持续发展，需要我们进行长期不懈的探索与实践，本文将从循环经济和住宅产业的相互关系着手，探讨以循环经济3R原则实现住宅产业化的途径。

1　住宅产业化必须走循环经济之路

工业化和城市化是人类社会发展的规律，也是人类社会实现现代化的必经之路和艰巨的历史性任务。怎样突破资源能源环境的瓶颈约束，实现可持续发展已变得越来越重要。

循环经济要求以“3R原则”为经济活动的行为准则。第一，减量化原则(reduce)。第二，再使用原则(reuse)。第三，再循环原则(recycle)。

对循环经济理论的认识和实践在我国也有一个不断深入的过程，2004年提出从不同的空间规模：即城市、区域、国家层面等方面大力发展循环经济。

住宅产业化其本质是将目前住宅产业中以劳动力密集、资本密集型的外延式增长转变成技术密集的内涵式增长，以达到全面提升住宅的综合性能和提高人民群众居住水平，促进住宅产业的可持续发展。

发达国家住宅产业化发展历程大致经历了三个阶段：第一阶段是产业化形成的初期阶段，解决的重点是建立工业化生产(建造)体系；第二阶段是推进产业化的发展期，解决的重点是提高产品(住宅)的质量和性能价格比；第三阶段是产业化发展的成熟期(发达国家目前正处在这个阶段)，解决的重点是进一步降低住宅的物耗和对环境的负荷，发展资源循环型住宅。

我国是发展中国家，住宅产业在整个国民经济中所占的地位十分显著，特别对资源的占用、能源的消耗、环境的负荷是相当大的。因此我国目前的住宅产业还处于发展的第一和第二阶段，即产业化形成的初期阶段和推进产业化的发展期。

正因为住宅产业在整个生命周期中要消耗大量的资源能源，对环境也造成了极大的影响；要达到全面提升住宅的综合性能和提高人民居住生活水平，促进住宅产业的可持续发展，住宅无疑是发展循环经济最大的载体。我们必须以循环经济理论促进住宅产业，以3R原则实现住宅产业化。

2 以3R原则建设资源节约型住宅的前景

循环经济的3R原则在住宅产业领域中主要体现在“省地、节能、节水、节材”等方面。只有大力发展可持续住宅、推广建筑节能、完善可持续住宅关键技术支撑体系，研究开发推广应用建筑新材料、新技术、新产品、新工艺、提高住宅的科技含量，才能实现住宅产业的可持续发展。以下我们进行了分析，这些分析将对上述观点有一个充分的印证。

我国的土地资源特别是耕地严重匮乏，土地资源十分宝贵。通过住宅规划设计中对荒地、废地、滩涂的合理有效利用，以及推广新型墙体应用等措施都能有效地节省土地。

低容积率的住宅建筑会占用更多的土地资源，依据新出台的上海市普通住宅标准，容积率小于1.0的住宅为非普通商品房。每降低0.1的容积率，占用1万m^2的土地将少建1000m^2的住宅建筑，那么上海新建配套商品房两个1000万m^2，建筑容积率从1.0降至0.9，将多占土地222.2万m^2(折合3333.0亩)。

另外，以使用新型墙体为国家节约土地资源和节能为例：采用新型墙材之前，每1m^3墙材耗用578块标准黏土砖，建造10万m^2的住宅墙材约为2万m^3，折合标准黏土砖1156万块，按每亩耕地生产标准黏土砖60.6万块，使用新型墙体材料可节省烧砖用地19.08亩[1]。上海将新建配套商品房两个1000万m^2，以此推算，仅推广新墙体一项就能保护耕地3816亩。

建筑节能是以3R原则促进住宅产业化的关键技术，主要包括住宅单体结构节能和运营设备节能两方面。通过完善、应用围护结构保温隔热技术；提高空调采暖系统的效率；利用可再生能源(如太阳能、风能)与清洁能源；电能的削峰填谷等多种能源互补的途径，可以有效地推进建筑节能。

上海属于夏热冬冷地区，根据标准，节能效率应达50%，则上海地区建筑物节能综合指标限值为55.06kWh/m^2。目前上海既有住宅节能效率达到50%的住宅建筑数量很少，住宅建筑物节能综合指标较高，而且参差不齐，据保守估算建筑物节能综合指标平均值达70kWh/m^2以上。试想上海将新建配套商品房两个1000万m^2若能采用各种先进的建筑节能技术，使住宅节能效率达到50%，建筑物节能综合指标限值降至55.06kWh/m^2，那么每年能节约能耗29880万kWh，折合成标准煤[2]为12.25万t。

水是生命之源，我国2/3的城市供水不足，其中1/6的城市严重缺水。依据上海市生态型住宅小区技术实施细则，通过采用这一系列的节水措施住宅小区的节水率可达20%以上。按每户三口之家计，人均日生活用水定额为230L[3]，若节水率为20%，每户年平均节约用水50.37t。上海将新建的配套商品房为两个1000万m^2，约20万户，那么将节水1007.4万t。

全面推进住宅装修的工业化生产方式，全装修房是一个切入口。全装修房中的设计一体化，施工一体化，管理一体化能有效地节约建材，避免二次装修。保守估算住户装修时重新布线100m，50%的住户拆除墙体5m^2（折合1.2m^3），上海将新建的配套商品房两个1000万m^2，住户约20万户（每100m^2为一单元），那么实行全装修房将减少建筑垃圾12万m^3，节约电线2000万m。

建筑垃圾的回收和循环再利用能够保护资源，降低对环境的影响。有资料表明建筑废料的回收率英国为48%（其中中高级利用占6%）；日本为65%（其中中高级利用占20%），是回收利用较好的国家；新加坡为63%。我国建筑废料的回收利用较低，据统计每年仅施工建设所产生排放的建筑垃圾超过亿吨，若有效措施将其减量化和再利用，将节省大量资金和资源。

上述数据只是初略的和局部的，以小见大，可见以3R原则建设资源节约型住宅潜力巨大，前景广阔。

3 以3R原则围绕住宅生命周期建设资源节约型住宅

以循环经济的3R原则围绕住宅生命周期转变住宅产业的增长方式，建设资源节约型住宅，将实现住宅产业的可持续发展。

住宅的生命周期分为四个阶段，即：设计阶段、建设阶段、使用阶段、老化阶段。

3.1 设计阶段

1. 设计的合理性

在区位选址上要尽可能省地，鼓励使用无毒无害的废弃土地、滩涂作为住宅用地。在景观设计、绿化设计、水景设计时也应考虑资源、能源的最少消耗，将对原有环境的破坏减少到最小程度。

2. 设计的耐久性

首先要在住宅设计时引入混凝土结构的耐久性理念，其次必须注重耐久性研究成果的应用，进而延长住宅的生命周期。

3. 设计的可改造性

在住宅小区的设计之初，考虑生命周期轮回中住宅小区的可改造性。在不改变原有结构体系的前提下，使设计的住宅具有可改造性，这样有利于资源的节约。

3.2 建设阶段

住宅小区的建设阶段对资源的占用，能源的消耗，环境的影响程度等在整个住宅的生命周期里所占的比例很大。因此在这一阶段应依据循环经济的3R原则，注重资源节约型住宅技术集成。

住宅主体建筑是住宅技术、材料、部品的集成，因此要考虑性能与价格的合理比值，应使住宅的性能质量，诸如安全性、耐久性、适用性、建筑节能等成为住宅的亮点，使住宅的科技含量成为住宅的附加值，这样才能促进住宅产业现代化。

3.3 使用阶段

1. 减少建筑使用阶段废弃物的数量及再利用

推广居住区生活垃圾分类的收集和处置，有机垃圾生化处理，无机垃圾综合利用技术，做到居住区生活垃圾的减量化、资源化与无害化并循环利用。

2. 使用节水器具节约水资源，提高水的回用率

要普及使用节水器具，提高用水效率，减少小区管网的渗透率，节约水资源。大力推广雨水就地渗透和雨水收集综合利用等技术，以达到水体自净的目的。

3. 注重住宅使用的维护保养

加强住宅使用阶段的维修及保养，制定合理的住宅维护计划，及时排除影响住宅安全和使用的隐患，用较少的投入使住宅的生命周期得以延长。

4. 全生命周期使用住宅

很多住宅未到生命周期就被拆除，造成了人力、财力和资源的极大浪费。因此严禁非法拆除尚处生命周期内的住宅是节约资源、能源和保护环境的有效措施。

3.4 老化阶段

我们知道住宅是有生命周期的。住宅的设计基准周期为 50 年。一般住宅的结构寿命可以达到 70 年。住宅的老化有双重意义：住宅主体结构的老化和居住功能的老化。因此住宅老化阶段应从延长结构寿命和改善居住功能方面，综合提高住宅的安全性、适用性及耐久性，以 3R 原则延续住宅的生命周期。

1. 对老化住宅进行鉴定，实施加固

应对住宅建筑的安全性进行评估，保证安全的住宅建筑继续使用；对存在安全隐患但未到生命周期的住宅建筑制定经济、科学的结构加固措施，有效地提高住宅的安全性，从而达到延年益寿的目的。

2. 改造既有住宅，提升功能质量

通过改造，提高既有住宅的品质和综合性能，延长住宅的生命周期，使原来居住者的居住条件和居住环境得以改善。这些都是低投入高产出的措施，能有效改善人们的居住条件，达到全生命周期使用住宅的目的。

3. 既有建筑节能改造

采用先进的外墙外保温技术既能提高住宅的节能效率，使房屋冬暖夏凉，提高居住的热舒适性，对住宅主体结构起到了保护作用，从而有效地提高了主体结构的耐久性能，起到了增加住宅使用寿命的作用。

4. 协调区域环境，完善城市功能

随着城市的发展和历史的变迁，在旧小区拆除改建过程中，我们必须遵循 3R 原则，在改造过程中充分利用原有的有价值的建筑材料，对小区内具有历史文化价值并可继续使用的建筑物进行保护和利用，将改造过程中对周边环境的影响降到最低限度。

5. 注重住宅的生命轮回，提高居住水平

当旧的住宅居住小区生命周期已到，建设新的居住小区就成为必然。这样我们又将经历前面所提到的住宅生命周期了。在拆旧建新过程中，值得重视的是：要充分保护利用原小区内资源，在拆除原有建筑时采用新的施工技术，使对原来的生态破坏最小。

4 保证以 3R 原则促进住宅产业化的措施

（1）加大科技投入，加强对基础性、关键性、全局性和战略性问题的研究，为实行住宅产业化构筑技术支撑。

（2）构建住宅产业基础信息体系

要根据本市住宅产业投资、资源消耗、能耗及构成等情况，逐步建立健全有关基础数据的采集统计与分析系统。

（3）加强政府管理与扶持

政府部门应制定相关的政策法规、强制措施和激励机制，特别是要建立基于市场的财税、信贷、保险等经济政策保障。建立基于市场为主的推进机制。“政府导向、全社会参与、投资者受益”，充分发挥各实施主体的积极性。

5 结论

综上所述，笔者认为：

住宅产业要做到可持续发展，遵循循环经济 3R 原则是必然选择，得出以下结论：

（1）发展循环经济，住宅产业必须先行。住宅产业要可持续发展，遵循循环经济的 3R 原则，是惟一途径。

（2）必须加大科技投入，全面提高住宅产业的科技含量，通过科技创新，用先进的科技手段为实现住宅产业化提供保障。

（3）必须依托政府的监管和扶持，以有效的管理机制和体制保证住宅产业化实施。

（4）必须结合城市规划，从整个城市总体布局的角度来进行住宅小区住宅生命周期轮回中的改造。从发展循环经济的角度出发，坚持 3R 原则，对不同性质的旧住宅小区采用不同的方式进行改造或重建。

Circulatory Economy and House Industrialization

——Promote House Industrialization By “3R” Principles

Lin Ju

(Shanghai Real Estate Science Research Institute, Shanghai 200031 China)

Abstract: This article points out that construction industrialization must follow circulatory economy, hold on “3R” principles, change the traditional construct pattern . It induct the conception of life cycle of construction. Dividing the life cycle of construction into four phases, include design, construct, use and aging. In the end, it expatiates the measure of promoting construction industrialization by “3R” principles.

参考文献

［1］ 相关数据参考．涂逢祥主编．建筑节能(42)，2004

［2］ 电耗折煤．1kWh 电相当于 0.41kg 标准煤．全国集中供热锅炉网．“九五”期间供暖能耗调查报告，2002

［3］ 参照 DGJ 08-20-2001《住宅设计标准》

绿色建筑在我国的市场化发展分析

杨建荣　韩继红　安　宇
（上海市建筑科学研究院有限公司，上海　200032）

摘　要： 通过分析绿色建筑所包含的理论研究、工程实践和政府作用，结合我国的建筑业发展特点，指出我国的绿色建筑发展虽处于起步阶段，但市场化发展将成为必然，且我国近年来的大型工程项目也为绿色建筑的发展提供了良好契机。同时指出了在绿色建筑市场化发展过程中各参与方应该注意几个方面的问题。

关键词： 绿色建筑　市场化　发展

1　引言

目前我国正处在城市化高速发展的过程中。为适应城镇人口飞速增加的需求和继续改善人民生活水平的需要，在 2020 年前我国每年城镇新建建筑的总量将持续保持在 10 亿 m^2/年左右，到 2020 年新增城镇民用建筑面积将为 100～150 亿 m^2。按照目前建筑能耗水平，则需要每年增加 1.4 亿 t 标煤/年用于采暖，每年增加 4000～4500 亿度用电量[1]。如此巨大的建设量，在建造和运行中将要消耗大量的土地、材料、水资源，形成的室内环境优劣也将对人们的生活质量产生影响。

由于社会发展直接或间接对环境造成影响，近 30 年来国际上提出了可持续发展的概念，即在发展过程中不仅考虑某个单体建筑或工程的功能需要，更要兼顾未来资源利用的可持续性。具体到建筑和建设的快速化增长过程中，对品质和消耗之间的平衡提出了较高的要求，为此有的评估导则把这两者的比值作为评判项目优劣的指标[2]。

绿色建筑或生态建筑的概念和发展模式正是在这样的大背景下产生。根据我国《绿色建筑评价标准》（送审稿）的定义，绿色建筑是指在建筑的全寿命周期内，最大限度地节约资源（节能、节地、节水、节材）、保护环境和减少污染，为人们提供健康、适用和高效的使用空间，与自然和谐共生的建筑[3]。在全世界范围内，虽然绿色建筑的概念和评价体系层出不穷，但从社会化发展的角度审视，绿色建筑仍然是一个努力的方向，人们仍然处于通过一个个实践来摸索其市场化、社会化道路的阶段。

近年来在我国建设总量的驱动下，人们对绿色建筑的研究业已全面展开。同时建筑的特点是其实践性，而其大面积的推广离不开市场化的操作，这同时对政府、政策、技术、资金等都提出了新的要求。相对而言，国外对绿色建筑的研究更早期就得到了关注和开展。为此通过分析国外的绿色建筑实践之路，结合近年来我国的绿色建筑发展，可以对我国未来的建设过程提供借鉴。

2 国外绿色建筑发展

绿色建筑相关内容包含多个方面。对于有借鉴意义的国外绿色建筑发展，有三个方面的内容值得引起关注，即绿色建筑理论、绿色建筑实践，以及政府行动。

绿色建筑的理论包含了多方面的内容。在对建筑和建设领域的直接指导上，绿色建筑评估体系无疑具有最大的代表性。自 20 世纪 90 年代英国建筑科学研究院(BRE)的 BREEAM 评估体系开始，影响较大绿色建筑评估体系还包括美国绿色建筑理事会(USG-BC)的 LEED 评估体系、国际可持续建筑环境促进会(iiSBE)的 GBTool 评价工具、日本“建筑物综合环境评价委员会”开发的 CASBEE 工具等。

前期的评价体系较多关注各单项指标的要求，例如节能、室内环境舒适等方面；后期的评价体系则将各指标间的整体平衡纳入考虑，如 CASBEE 工具中就创造性地提出了用品质(Q)与代价(L)的比值作为衡量标尺的概念。除了日本之外，国外较有影响的其他评价体系的出台并不具有太强的政府背景，而是由行业协会、机构等自发提出和逐步改进。这些评估工具之所以引起学术界和管理部门的关注，很大程度上是由于它们各自具有特色的市场化操作之路，即这些评价工具被用来提供咨询服务、进行建设项目过程优化和后期评估、绿色度等级标识等的工作。

对发达国家而言，他们具有良好的节约资源的意识和较发达的技术基础，在适当的引导(如市场化的绿色建筑导向)下，绿色建筑得到了很好地探索和实践。在英国 BRE 的生态环境楼和 Integer 生态住宅样板房等具有示范和研究价值的项目之后，出现了许多的实际工程项目。例如，在以节约全球闻名地北欧(包括德国在内的)地区，出现了德国太阳能技术应用园区(图 1)、丹麦的低能耗建筑(图 2)和瑞典 Bo01 生态小区(图 3)等工程应用。在澳大利亚悉尼的奥运村、体育场馆中也增加了建筑新技术和新材料的应用。这些实际工程往往并非简单套用其他国家的评估或技术导则而成，而是结合本国、当地的情况进行了合理建设，这种思路和过程也为我国未来的建筑可持续发展提供了榜样。

图 1 德国的太阳能示范园区

图 2 丹麦斯科特帕肯低能耗建筑

在大多数西方国家，土地和资本具有高度私有化、建筑技术领域具有高度市场化和行业化特点，这使得绿色建筑相关理论和实践的各个分支比较发散，通过设备、系统等的可

图 3 瑞典 Bo01 生态小区

靠性和先进性造就了建筑品质的总体水平提高。这种情况下，政府层面的大部分行动则具有另外的特点：并非在绿色建筑的旗帜下提供导向或标准，而是通过各种专业、行业标准来界定，绿色建筑的推动靠的是市场化运作。美国政府即是采用这种方式，在各行业的博弈和妥协之后制订相关的节能、通风的标准，由此改善建筑的基本性能。而 Leed 评估体系的作用则是在市场中引导各种资源整合和优化，由此产生不同的绿色建筑理念和实践。

相对而言，虽然同为发达国家，日本政府则不仅制订了单项的建筑和建设标准，也同时提出了综合的绿色建筑导向，直接在环境友好、绿色建筑的旗帜下综合调控市场行为，例如 CASBEE 便是由政府主导建立、完善和运作，同完全市场化的 Leed 评估体系进行推动的模式截然不同。无论何种模式，都体现了各自国家的特点，即市场化的优势或者政府较强的控制能力，这为我国的绿色建筑发展过程中政府作用的发挥同样具有很好的借鉴意义。

3 我国绿色建筑的研究与实践

1994 年我国发表了“中国 21 世纪议程”，同时开始启动“国家重大科技产业工程——2000 年小康型城乡住宅科技产业工程”。1996 年又发表了“中华人民共和国人类住区发展报告”，对进一步改善和提高居住环境质量提出了更高的要求和保证措施。因此，到目前为止，我国的绿色建筑整体探索只有几年的时间，在理论、技术和工程实践方面尚处于起步阶段。

绿色的概念本身带有可持续发展的意义，要强调实事求是和因地制宜。在中国发展绿色建筑、制订标准和进行评估必须考虑中国国情，在生态技术策略上要考虑因地制宜，在传统技术策略、中低技术策略、高新技术策略中采用适宜技术策略。在绿色建筑的相关评价工具开发方面，2001 年建设部住宅产业化促进中心制订的《绿色生态住宅小区建设要点与技术导则》和全国工商联制订的《中国生态住宅技术评估手册》是这一类工作的开始。2003 年以来，北京、上海、浙江等地区也分别制订了绿色建筑评价标准或生态住宅的实施细则，使得绿色建筑的理念在这些经济较发达的地区率先得到宣传和推广。2005年，我国出台了《绿色建筑评价标准》的国家标准，在国家层面上正式吹响了绿色建筑发展的号角。

同发达国家不同的是，我国的各级政府在绿色建筑的推动上起到了很大的作用。这些推动体现在两个方面：一是借助具有较大影响的项目开展，例如2008年的北京奥运、2010年的上海世博会等就对我国的绿色建筑评价体系建立和建筑新技术应用提供了很好的契机；另一方面是政府直接支持各种评价体系的开发，例如我国“十五”科技攻关中的绿色建筑评估体系、2005年刚刚编写而成的《绿色建筑评价标准》等均是在政府的直接支持下完成。这种模式是我国体制的反映，在今后的发展中将进一步延续。

伴随着绿色建筑理论完善的是示范工程开展。2004年和2005年，由上海市建筑科学研究院和北京清华大学分别负责的“上海生态建筑示范楼”（图4）和“清华大学超低能耗示范楼”（图5）等示范项目分别建成，并成为我国绿色建筑技术展示、教育基地和后续研发平台。

(*a*) 生态办公示范楼　　(*b*) 生态住宅示范楼

图4　上海生态建筑示范楼

图5　清华大学超低能耗示范楼

在绿色建筑的实践和市场化方面，虽然也出现了科技部大楼、锋尚国际公寓、上海万科朗润园等实际工程，在绿色建筑技术的真正应用上踏出了坚实的一步，但整体性较强的引导缺乏使得我国绿色建筑的基础理论薄弱，市场化还处于起步阶段。

4 我国的绿色建筑市场化分析

绿色建筑的市场化既要考虑对全国范围乃至全球范围的环境影响，如节约各种资源，也必须在不违背我国国情、在技术经济合理的情况下进行。同时，建筑的特点的包含了多个方面的参与者(stakeholder)，包括政府、开发商、技术人员和使用者。只有在各方面共同参与、全面努力的前提下，才能保证建筑在全寿命周期内的绿色性能。

对比国内外的绿色建筑发展与实践，可以看出发达国家基本已经跨过了初期示范工程的阶段。借助于他们坚实的科技基础和工业基础，发达国家已将许多较成熟的建筑新技术自然、和谐地给予了实际应用，在建筑师和工程师的合作上已进入初步融合的阶段。这些工程的经验也反哺至理论的深入，为此发达国家的各种绿色建筑评价工具都带有较强的国家特色和操作性。

而发展中国家在这个发展浪潮中尚处于起步阶段。虽然理论研究、工程示范等工作已全面开展，但由于后者的缺乏使得理论同实际相距较大，评价体系或技术导则的指导性较弱；而后者由于缺乏具有当地特色的合理理论支撑，目前仍处于借助国外的概念进行实验的阶段。无论在东南亚还是我国，这两年来出现了较多的“零能耗”或低能耗建筑，但这种目标下的产物是否适应了当地特点仍然需要大量的研究投入和总结。为此在巨大的建筑市场上，绿色建筑项目尚是凤毛麟角。

在绿色建筑的市场化发展问题上，我国具有同其他国家相似的问题，但也不乏自己的特点。

第一，在我国的社会和经济发展态势下，绿色建筑的市场化将成为必然趋势，且存在良好契机。

我国大力宣传绿色建筑的理念，引导建设市场增加绿色建筑的技术含量是我国国情的现实和未来的需要。在我国当前每年超过 10 亿 m^2 的城镇新建建筑总量的前提下，资源的大量消耗不可避免，这其中包括土地资源、能源、水资源、材料资源等等，而建筑之外的社会发展同样对上述资源提出了持续性需求。相对于其他国家，人口众多的中国将面临更加严峻的压力。为此，如何满足连续性需求和大量消耗之间的矛盾，这本身就是可持续发展所包含的最根本问题。

“十一五”期间，我国的建设高潮中还包括奥运会、世博会这样的大型项目，都同建筑具有最为直接的关系。这些项目的出现推动了绿色建筑理论在我国从萌芽到快速发展的过程，但这些项目的实施和完善更为这些理论提供了示范和检验的大平台，对我国建筑业的持续发展提供了机遇，具有重要意义。

第二，我国绿色建筑成熟技术的改善进程需加快。

同国外相比，我国当前在建筑技术的研发和应用上都存在不小的差距。这些差距包含几个方面，一是对国外新技术的盲目引进，未充分考虑其使用背景和我国使用者的生活方式。例如近年来在我国出现的区域供冷模式，就是因为忽视了我国居民的生活模式而遇到了系统效率低、运行费过高的现象。二是技术的实际开发与应用，包括节能新技术的开发、水资源回用技术等方面。而建筑师在节地方面设计手段、建筑运营管理水平等方面的技术提高更具有极大的空间。

第三，政府的法律法规和相关政策函需出台。

我国的国情决定了在绿色建筑发展方面，我国政府不可能走美国那种完全自由市场化的道路，而是应该借鉴日本在引导可持续建筑发展上的操作方式，由政府主导进行发展，因此政策的系统性、引导性具有更为重要的意义。

由于绿色建筑的发展很大程度上对一个国家、地区的常用意义更大，在一栋建筑、一个小区的层面上并非总能很快同时体现社会效益和经济效益，为此政府有必要通过相关措施的出台平衡这一矛盾。具体而言，建筑新技术应用一般意味着投资增加，这对于开发商而言是最为敏感的一方面，因此政府有必要出台相应的奖励、税收减免等政策引导市场的正确方向，通过建立工程或城市级的示范，由点及面推动全社会的实际参与。而这些政策必须在仔细调查研究和论证的基础上制订，为此同时也对相关政策的系统性研究提出了新的要求。

第四，市场化的技术支持和相关管理模式需加快建设。

我国已出台了《绿色建筑评价标准》，但横亘在标准执行面前的问题是如何实施。从前面的分析不难看出，发达国家的技术专业化是其重要特点，为此促生了大量独立于设计单位之外的咨询机构出现。在我国面临如此高速的发展压力，而设计单位的技术储备和经验无法短时间跟上的情况下，由专业公司开展系统的技术支撑服务将成为必然。我国的开发商目前对新技术的应用已表现出浓厚的兴趣，且大部分项目的定位已超越了市场炒作的初级阶段，为此对技术服务提出了大量要求。虽然我国经济技术较发达地区的一些建筑研究机构已开始从事这样的工作，在技术上成功策划了一些建筑项目的开发，但总量上仍然极小。国外的很多大型咨询公司正是通过这种方式获得发展壮大，目前他们大多都已进入我国的建筑市场，而在这一轮的发展大潮中充分占领我国的自身市场，及时发展壮大我们的建筑技术知识产权，同样是我国绿色建筑市场化大潮中的重要方面。

另一方面，为防止绿色建筑市场化过程中炒作概念的现象发生，则需要加大公平的评估机构建设，在土地使用、节约能源、水和材料资源等分项技术的把关上出台公证、合理的结论，杜绝不良现象的发生。

总之，从某种意义上讲，我国的绿色建筑发展相对于其他国家更具有紧迫性。在充分认识我国发展趋势的前提下，认真借鉴国内外绿色建筑发展中的有益经验是绿色建筑市场化良性发展的重要方面。同时，对发展绿色建筑必要性的认识不仅仅政府需要，普通老百姓也只有在拥有这些基本认识的前提下才能约束各种浪费行为，最终真正实现建筑业的可持续发展。

Analysis on Development of Green Building development in China

Yang Jianrong　Han Jihong　An Yu

(Shanghai Research Institute of Building Sciences, Shanghai　200032　China)

Abstract: Green building is a challenge but also an opportunity for China's sustainable development. This paper presents the green building related theory, application and government actions in the western countries, and the relevant aspects in China are analyzed afterwards. It concludes that China will also experience the similar mechanization process with the opportunity of several big projects. In the end, the actions that different stakeholders should take are presented for the successful development of green building mechanization in China.

参考文献

[1] 江亿. 我国建筑耗能状况及有效的节能途径 . 暖通空调 . 2005. 35(5). 30～40

[2] Japan Sustainable Building Consortium. Comprehensive Assessment System for Building environmental Efficiency(CASBEE). 2003

[3] 中华人民共和国国家标准《绿色建筑评价标准》(待颁布)

建设绿色建筑，做可持续发展的房地产企业

戴海锋

（万科企业股份有限公司建筑研究中心，深圳　518049）

摘　要：在可持续发展的背景下，可持续发展型企业是以后企业的发展主导潮流。中国的房地产开发企业不仅将迎接日益成熟的市场需求所带来的机遇，还将面临着资源高效利用和环境保护所带来的挑战。笔者认为在新形势下，房地产企业需要建设绿色建筑，实现可持续发展型企业的转型，并以此作为彰显房地产企业社会责任感的手段之一。

关键词：可持续发展型企业　房地产企业　绿色建筑

1　转型中的可持续发展型企业

1.1　可持续发展的背景

可持续发展(sustainable development)在1987年世界环境与发展委员会的《我们共同的未来》报告中第一次明确给出定义，其概念是指既满足当代人的需要，又不对后代人满足其需要的能力构成危害的发展，概念提出后得到了国际社会的广泛共识。

中国政府在1992年就开始了关于可持续发展的研究工作，2002年发布了《中华人民共和国可持续发展国家报告》，报告阐明了中国政府对可持续发展的认识和行动，而在2003年发布了《中国21世纪初可持续发展行动纲要》更是提出了中国可持续发展的指导思想、目标与原则，规定了重点领域，提出了实现目标的保障措施。中国“十一五”规划中也明确了“全面贯彻落实科学发展观 ”，“坚持节约发展、清洁发展、安全发展，实现可持续发展”的战略思路。中国香港特别行政区政府经过2年多的公众咨询也在2005年发布了《香港首个可持续发展策略》报告。目前，可持续发展概念已经纳入到许多国家和地区发展的行动当中。

1.2　主导的可持续发展型企业

然而，可持续发展并不只是国家或地区政府层面专用的概念，国际上已有不少企业以此为目标并制定自身的战略目标。例如杜邦公司(Dupont)把可持续发展作为公司目标：“杜邦正致力于一种既增加股东和社会价值，又减少对环境的影响的可持续增长。”为此，杜邦公司制定了包括科学、知识和生产力三项内容的策略。宝洁公司(Procter & Gamble)把可持续发展与他们公司的宗旨结合起来，同时把可持续发展看作为一个发掘60亿人需求和赢得顾客信任的契机。为此，宝洁公司成立了专门的行动部门，并就其工作每年制定可持续发展报告总结成果和问题并计划下一年的工作目标。太古集团旗下的香港太古地产在每年发布《环境、健康、安全报告》中认为他们房地产开发业务“最重要的是，要以可持续

发展的方式经营业务，在环境、社会责任和经济三方面取得可以接受的平衡”。

2005 年，毕马威(KPMG)公司发布了三年一次的企业责任全球调查报告。报告内容调查了全球 1600 多家领先企业的责任报告，其中包括全球 500 强企业的 250 家(G250)和 16 个国家的前 100 强企业(N100)。调查报告中显示，企业责任报告从 1993 年开始逐年增加，并在过去三年中有显著的增长，2005 年有 52%G250 和 33% (N100)发表了独立的企业责任报告(如果把包含了企业责任信息的年终财务报告算在内，比例分别为 64% 和 41%)(见图 1)，比 2002 年分别增长了 7% 和 10%。KPMC 还发现报告的形式从原先纯粹的环境报告变成了可持续发展报告(包括社会，环境和经济)。从报告中我们可以看到国际先进的企业正遵循着可持续发展原则，追求着社会、环境和经济效益的均衡：采用绿色技术、进行绿色管理、开发绿色产品、开展绿色销售、构建着企业可持续发展新模式。

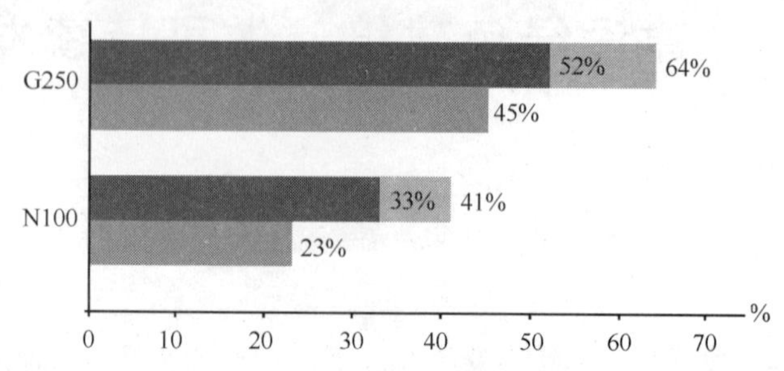

图 1　全球 500 强企业的 250 家和 16 个国家的前 100 强企业发布社会责任报告的比例[1]

企业的可持续发展模式转型不仅为企业创造了良好的形象，更为企业创造出新产品和提供无限商机。《中国 21 世纪议程》曾指出：“对人体和环境无害的绿色产品随着可持续发展进程的深入，日益成为今后产品生产的主导方向。”例如，海尔开发的 A＋低能耗标准海尔冰箱，因率先达到欧盟规定的能耗最高标准，德国、荷兰、比利时、卢森堡等国家政府规定，购买达到 A＋能耗标准的海尔冰箱政府将补贴 100 欧元。为此，海尔不仅在欧洲名声大振，同时带来了巨大的效益。西安开米股份有限公司，在公司成立之时从产品的选择一直到整个市场选择都把环境作为第一位来考虑。其在 20 世纪 90 年代中期创建了中国第一个环保型洗涤剂品牌，目前已成为国内最大的环保无磷液体专业生产企业。又如，近年来中国的低排量汽车已成为了市场的主导，在 2005 年 1 到 6 月的销售就占了近 60% 的市场份额，于是，现在越来越多的汽车厂家已经开始转向低排量汽车的制造。因此，与其说可持续发展型企业是以后的发展主导潮流，还不如说未来的企业将围绕着绿色产品这个主导市场团团转。

2　面临转型的中国房地产企业

2.1　持续增长的房屋需求

我国的房屋施工面积从 1985 年开始就一直在增长(见图 2)，2003 年施工面积达到近 26 亿 m^2，人均面积也从 1978 年的 6.7m^2 增加到 2003 年末的 22.8m^2。与此同时，我国目前的城镇化进程正在处于快速发展时期。有研究预测从现在到 2030 年我国的城镇化速率平均每年为 1～1.3 个百分点。现在我国的城镇化率为 40%左右，按照城镇化进程城镇化过程诺瑟姆 S 型曲线模型(见图 3)，我国的城镇化进程进入了高速度增长期。这意味着中国每年将约有 1200～1500 万人口从农村转移到城市，为城镇房地产市场源源不断地输入客户群体。虽然 2003 年以来各界关于房地产泡沫和房价过高的声音不绝于耳，政府也在 2005 年出台了系列的宏观调控措施，然而从销售量持续大幅增长却表明市场需求依然旺盛，居民的购房热情不减。

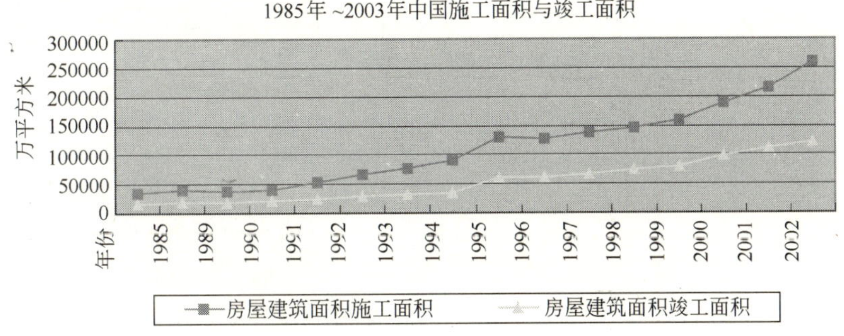

图 2　1985～2003 年中国施工面积与竣工面积

但是，我们也看到在房地产行业一片欣欣向荣的景象下，越来越多的竞争者开始进入了房地产开发市场。1986 年全国房地产开发企业仅有 1991 家，到 2002 年已增加到 32618 家左右，从业人员也从 1997 年的 68.3 万人，上升到 2002 年的 113.4 万人(见图 4、图 5)。有分析认为，未来中国房地产业竞争将继续加剧。经过前几年的高速发展，目前正处在一个结构性转变的关键时期。这个时期内房地产企业需要加速形成自身的核心竞争能力，以应对新一轮的洗牌。随着消费者购房行为成熟，房地产的概念炒作将逐渐失去效力。相反产品的品牌、质量成为主要的竞争力。过去几年房地产市场中曾出现过各种标新立异的概念，其满足了高收入购房者对房屋品质和舒适程度以及文化品位和时尚的需求。目前市场重心开始向中低收入者转移，这些消费群体是务实型的，对产品要求比较理性，即适于安身居住、功能配套、设施齐全、高性价比等。这就要求开发商在销售策略下功夫以外，还要实实在在地在产品开发方面做出相应改进。

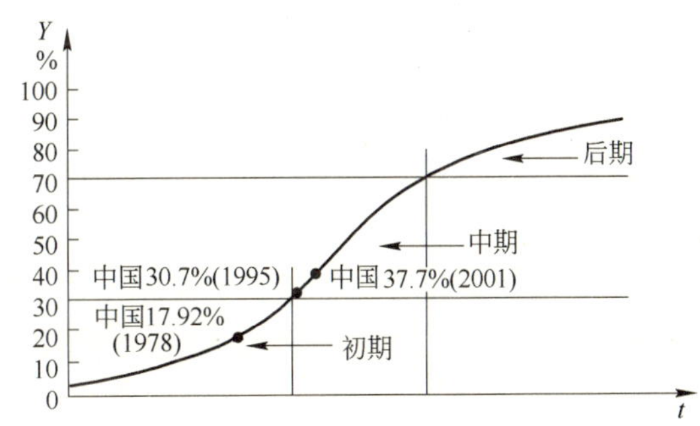

图 3　中国处于城镇化进程模型的点(诺瑟姆 S 型曲线模型)[2]

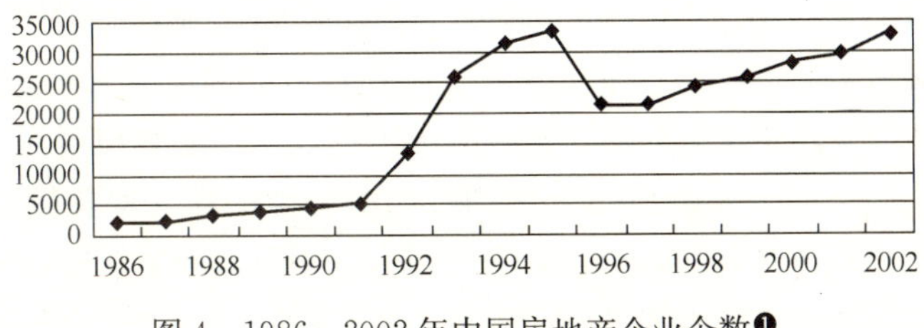

图 4　1986～2002 年中国房地产企业个数❶

❶　资料来源：国家信息中心中经网。

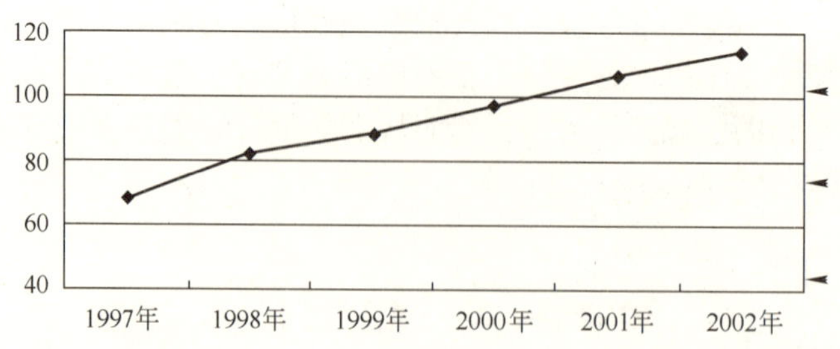

图 5 1997～2002 年房地产业从业人数❶

2.2 中国房地产面临的环境问题

中国的房地产开发企业不仅将迎接日益成熟的市场需求所带来的机遇，还将面临着资源高效利用和环境保护所带来的挑战。2005 年《中国房地产发展报告》中谈关于中国房地产业发展过程中存在的主要问题时，就提出了房地产开发中的短期行为忽视生态环境的保护与建设，与其他“土地资源浪费流失”、“商品房空置量加大”、“缺乏稳定资金来源”三个问题并行。众所周知，建筑是资源和能源的消耗大户。人类从自然界所获得的 50%以上的物质原料，用来建造各类建筑及其附属设施，这些建筑在建造与使用过程中又消耗了全球能源的 50%左右。中国是一个能源、土地、水、原材料等资源严重短缺的国家，同时又是一个资源利用效率很低的国家，于是在生产过程中给自然环境造成了很大的破坏。在我国化石能源资源探明储量中，90%以上是煤炭，人均储量也仅为世界平均水平的 1/2；人均石油储量和天然气储量仅为世界平均水平的 11%和 4.5%，人均耕地只有世界人均耕地的 1/3，水资源仅是世界人均占有量的1/4。与此同时，我国单位建筑面积能耗却是发达国家的 2～3 倍以上，实心粘土砖每年毁田 12 万亩。物耗水平与发达国家相比，钢材消耗高出 10%～25%，每拌和 1 立方米混凝土要多消耗水泥 80 公斤，卫生洁具的耗水量高出 30%以上，而污水回用率仅为发达国家的 25%。在 2003 年由于环境污染和生态破坏造成的损失已占 GDP 的 15%。然而，在我国每年城乡新建房屋建筑面积近 20 亿 m^2，其中 80%以上为高耗能建筑；既有建筑近400 亿 m^2,95%以上是高能耗建筑。所以对资源的高效利用和环境保护对于房地业来说显得尤为紧迫。

于是，中国从上到下在各个领域开展了“节约”行动。中共中央“十一五”规划中提出“建设资源节约型、环境友好型社会”要求，将“全面推行清洁生产，形成低投入、低消耗、低排放和高效率的节约型增长方式”。国务院发布了《国务院关于做好建设节约型社会重点工作通知》，该通知是针对建设节约型社会发布今明两年的工作重点，经国家发改委(成立了部级的工作协调小组)把工作拆分，共有 110 多项具体工作，已经分配到国家 25 个部委上了。略统计，与房地产业务相关的工作近有 30 项。另外，建设部也在《关于发展节能省地型住宅和公共建筑的指导意见》提出建设“节能省地型住宅”总体目标，即到 2020 年，我国住宅和公共建筑建造和使用的能源资源消耗水平要接近或达到现阶段中等发达国家的水平。为达到该目标，建设部已制定了相关的政策法规在全国各地区强制性地推行建筑节能，各地区节水条例也纷纷出台。

❶ 资料来源：国家信息中心中经网。

3 建设绿色建筑，实现企业转型

在目前面临的机遇和挑战下，笔者认为房地产开发商应向可持续发展型企业转型，以应对更广更深远的挑战，而建设绿色建筑是房地产企业转型的直接手段。绿色建筑指为人们提供健康、舒适、安全的居住、工作和活动的空间，同时在建筑全生命周期中实现高效率的利用资源，最低限度地影响环境的建筑物。首先，从可持续发展的思路来看，绿色建筑首先是满足人们的居住、工作和活动的需求，同时要高效利用资源和保护环境，为下一代的可再发展留下空间，实现企业以致社会的可持续性。再者，绿色建筑要给客户提供的是健康、舒适、安全的空间，这要求开发商从建筑的规划设计、技术措施、建造工艺和材料等方面进行创新，以提升的室外内环境质量，为客户提供超出他们期望的建筑品质。最后，绿色建筑要求开发者关注建筑全生命周期，也就是由建材生产到建筑物规划设计、施工、使用、管理及拆除等系列过程的资源利用效率和对环境的影响。那么开发商需要对目前的开发流程管理以及技术运用进行反思和变革，并在这个过程中不断的创新以实现企业产品开发的转型。目前，已有部分房地产企业正在转型中，例如有“社区综合开发，走绿色地产之路”，有“建立健康住宅产品体系和研发体系”，有通过“建筑与景观、景观与人、人与建筑三者之间亲密沟通、和谐统一，体现‘环境＋质量＋文化’的地产品牌”，有实现“恒温恒湿，创造中国未来居住环境新标准”等等的目标和行动。比起以前的概念性炒作，这些房地产企业有自身的方法论也更有实际的行动去落实。从他们的一些实践看来，虽然初期建造成本会提高，但经过精心的论证，分析和选用，客户最终还是会认可其给带来的品质上的提升，并且给与了相应的回报。

现在越来越多人认为企业社会责任终将会成为企业未来竞争的重要因素。《财富》和《福布斯》在企业排名评比时，都加上了“社会责任”标准。面对着目前中国的环境情况和未来建筑开发量，房地产企业作为实施的主体是有不可逃避的责任。综上所述，中国房地产企业需要制定企业发展策略，建设绿色建筑，实现企业可持续发展的转型，彰显房地产企业社会责任竞争力。

Building Green Architecture, To be A Sustainable Real Estate Enterprise

Dai Haifeng

(Vanke Architecture Research Center, Shenzhen 518049 China)

Abstract: In the background of sustainable development, sustainable enterprise will be the mainstream in the future. The real estate developer enterprises will face not only the opportunities which come from the gradually maturating real estate market, but also the challenges which come from the high-efficient resource utilization and environment protection. The author think that in the new situation, the real estate enterprise need to build the green architecture, and change their models to the sustainable enterprises and, therefore, take it as one mean to show their social responsibilities.

参考文献

[1] KPMG，KPMG International Survey of Corporate Responsibility Reporting 2005，The Netherlands：University of Amsterdam：2005
(Report download webpage：http：//www. kpmg. com/Rut2000 _ prod/Documents/9/Survey2005. pdf)

[2] 仇保兴，推行绿色建筑加快资源节约型社会建设(第五届中国城市住宅研讨会论文集)，北京：中国建筑工业出版社，2005

[3] 牛凤瑞，房地产蓝皮书-中国房地产发展报告 NO. 2，北京：社会科学文献出版社，2005

[4] 魏锋，管理新视界丛书-绿色壁垒攻略，武汉：湖北人民出版社，2005

建设节约型建筑行业

胡玉银
（上海建工(集团)总公司，上海　200120）

摘　要： 建设节约型建筑行业是建设节约型社会的有机组成部分，也是一项系统工程。通过对建设节约型建筑行业必要性的论述，指出了健全技术管理和增强技术创新是建设节约型社会的必由途径，并对其具体内容进行了论述。

关键词： 节约型建筑行业　节能建筑工业化　绿色施工技术及规程

1　引言

21世纪的今天，实施生态环境保障、资源循环利用的可持续发展战略已经成为人类的共识，也已经成为当今世界发展的主题。

建筑活动是人类对自然资源、环境影响最大的活动之一。一般讲，建筑活动使用了人类所使用的自然资源总量的40%，能源总量的40%。而造成的建筑垃圾也占人类活动产生的垃圾总量的40%(见图1)。对我国而言，坚持可持续发展尤为重要。我国人口数量巨大，发展任务迫切，年建筑量世界排名第一，资源消耗总量增长迅速，而许多资源的人均拥有量居世界平均水平以下。因此，在建筑领域中，遵循可持续发展原则，将对我国实现可持续发展发挥极其重要的作用。

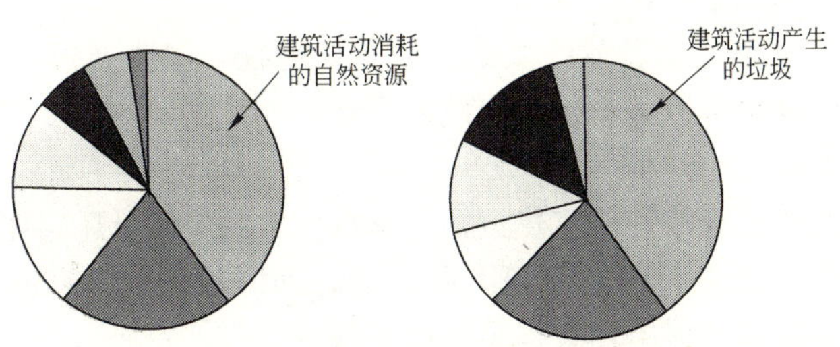

图1　建筑活动消耗自然资源的情况

建筑施工企业是资源消耗大户，在建设节约型社会的进程中担任了重要的社会责任。日本建筑业同行环保意识非常强，除了企业本身的业绩外，还定期发布环保工作年报，公布企业环保工作方针、主要任务和工作成效，值得我们借鉴。

2　节约型建筑行业的重点工作内容

可以从两个方面开展建设节约型社会的工作：一是，健全技术管理；二是，增强技术创新。

建设节约型社会是一项系统工程，需要社会全员参与、协同努力，因此必须高度重视制度建设，通过严密的技术规程规范各方行为。制作技术规程是科技人员义不容辞的责任。

2004年国家科技部和上海市在征集世博科技专项研究课题时，我们就提出了研究和制定绿色施工技术规程的建议，得到有关领导和专家的肯定。现在该课题已经列入今年启动的世博重大课题专项。今后我们将根据绿色施工技术规程编制的需要，选择部分施工企业和项目进行合作与示范，争取在2006年底完成编制工作并申请上海市地方标准。

绿色施工技术规程内容大致包括：

- 研究建设单位(业主)和施工单位的相关责任，制定相应管理要求。
- 研究现场环境管理的人员、制度与资金保障等相关要求，并制定相应规范。
- 研究在施工过程中的各个阶段对自然环境和相邻环境的影响，制定相应的施工现场管理措施和手段。尽量减少对工地及其周边的动植物资源、地形地貌、地下水位，以及现存文物、地方特色资源等带来的破坏。
- 研究施工各个环节中，可能出现粉尘及其他污物污染的情况，从每个细节上制定相应的措施和应达到的标准。如车辆运输时的遮盖措施；搅拌站防尘和封尘的措施和设施等等。
- 研究现场垃圾及渣土要临时放置及处置措施和要求，研究回收利用的方法和相关管理措施。
- 研究施工临时用水及降水等的处置管理措施及应达到的指标和要求，增大水资源利用率。
- 研究制定噪声污染以及光污染等的控制措施和手段，并制定相应控制标准和管理规范等等。

除了绿色施工技术规程之外，我们还准备改造现有的施工工艺标准，与时俱进，努力将建设节约型社会的概念融入这些技术标准中。

增强技术创新力度也是建设节约型社会的重要途径，主要通过在节能、资源再生利用、建筑改造、微创施工等方面的技术突破来实现。

(1) 节能建筑工业化

节能建筑在近几年已经成为研究热点，但是在技术的应用上还存在不足。现在的建筑节能水平是不到30%，近期目标是节能达到50%，而国外发达国家的平均水准已经到达65%～70%。所以在建筑节能上还有很多工作要做。

节能建筑工业化的内容大体包括：

1) 建筑设计研究。

(a) 房型与维护相互关系设计研究。

(b) 维护系统外观设计。

2) 设计技术研究。

(a) 预制维护结构体系研究(图2)。

(b) 维护体系保温性能研究。

(c) 维护体系抗震性能研究。

图 2　预制节能墙板

(*d*) 维护体系抗渗性能研究。

(*e*) 维护系统节点体系研究。

(*f*) 高精度维护体系和低精度结构连接研究。

(*g*) 维护体系设计规程研究。

3) 构件制作技术研究。

(*a*) 高精度模具体系研究。

(*b*) 构件制作工艺研究。

(*c*) 测量定位技术研究。

(*d*) 配件集成技术研究。

(*e*) 预埋件高精度定位技术研究。

(*f*) 构件涂装工艺研究。

(*g*) 构件运输与保护技术研究。

(*h*) 高精度构件及保温材料的研究。

(*i*) 维护系统耐久性与永久结构耐久性协调研究。

(*j*) 高精度维护体系构件制作技术规程研究。

4) 施工技术研究。

(*a*) 维护体系高精度连接件定位技术研究。

(*b*) 维护体系吊装装配工艺研究。

(*c*) 维护体系吊装设备研究。

(*d*) 维护体系质量现场检测技术研究。

(*e*) 施工技术规程研究。

5) 技术经济研究

(2) 地源热泵技术(图 3)

地源热泵技术是利用地下常温土壤或地下水温度相对稳定的特性，通过深埋于建筑物周围的管路系统或地下水与建物内部完成热交换的装置。冬季它代替锅炉从土壤中取热，向建筑物供暖；夏季它代替普通空调向土壤排热给建筑物制冷。地热泵供暖时 70％以上

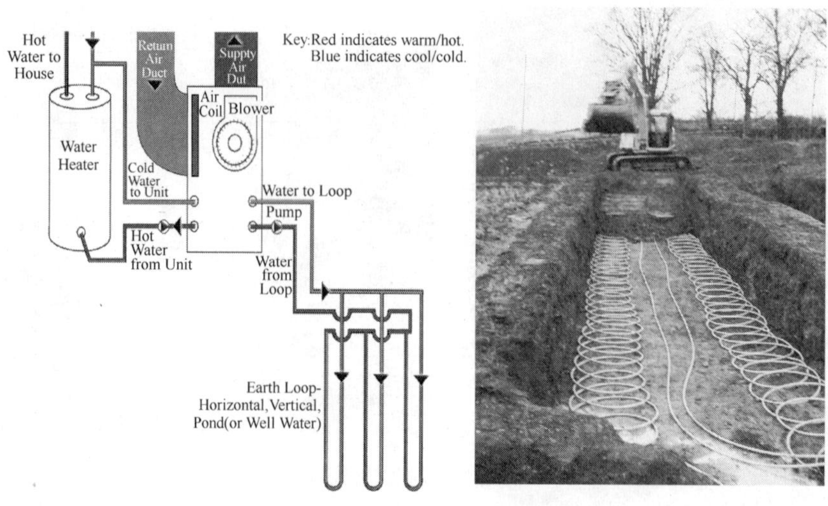

图 3　地源热泵技术

的能源来源于土壤中的能量，30％以下的能源来源于电能，所以用它替代冬季采暖锅炉，至少可减排温室气体 70％以上。同时，它还能供应生活热水，被称为 21 世纪的“绿色空调技术”。

开发利用低品位能源空调系统如地埋管式土壤源热泵作为建筑物供暖空调的冷热源，即有节能、环保的效益，又符合了生态建筑的发展趋势。其研究内容包括：

1）研究地源热泵的设计方法和系统参数的确定。

2）埋管换热器的布置和传热模型研究。

3）塑料管、防冻液和回填材料的研发。

4）土壤热物性的研究。

5）地源热泵成套安装技术和标准工艺工法研究。

6）减少环境影响的施工方法研究。

7）地源热泵安装的相应规范的制定。

(3) 建筑改造技术(图 4)

从世界范围内来看，现代大多数发达国家已经完成了城市开发建设，建筑行业的重点转换到旧城的改造、保护上。他们通过对现有建筑的改造，使其重新焕发活力，同时保留历史和文化的魅力。但国外这些改建扩建工程和我国有着较大的不同，主要是这些国家的建筑物通常年代久远、工程不大；而我国特别是上海、北京等城市，由于土地资源日趋紧张，面临的则是大量的高层和超高层建筑的改建和扩建的问题。在改建和扩建过程中，对现有建筑物采取加层、扩大平面面积、改变结构形式并辅以结构加固、基础托换等措施，

图 4　建筑改造技术

使其满足新的使用功能，达到节约土地资源的目的。这对设计、施工都提出了新的挑战。

建筑改造技术的主要内容包括：

1）结构设计及施工过程目标计算模型的研究。

2）改造过程监测技术的研究。

3）新旧桩基、桩土、新旧结构材料共同作用的研究。

4）超高层建筑加固设计及施工方法的研究。

5）改建过程中噪声、尘土等污染控制的研究。

6）拆除、置换施工技术研究。

（4）建筑垃圾循环利用技术

对于在施工过程中产生的渣土，通常要运输到指定地点埋置。这不仅耗费了大量的运输和处置费用，占用了处置土地，也使渣土没有了一点利用价值。因此使渣土变废为宝，不仅可以解决上述问题，而且还能产生良好的经济和社会效应。更为关键的是渣土的循环再利用本身就是绿色施工技术的一个必然要求。

对于渣土的循环再利用，相关的研究还比较少，我们拟以以下技术措施解决这个问题：绿化用土，回填还耕，制渣土砖及道路用渣土。我们将在已有研究的基础上，建立专门的研究小组，开展渣土循环利用的技术和设备。

世博会建设过程中，将牵涉数量巨大的土方工程，会产生大量的渣土，这个问题的解决将极大地保护环境和实现材料的重复利用。

（5）建筑用水循环利用技术

建筑业是用水大户，建筑工程的施工对城市地下水位的影响乃至对整个城市水资源的影响可谓不容忽视。

解决这个问题一方面靠节水，而另一方面则必须解决施工用水及降水的循环再利用问题。施工过程中牵涉到的水主要有两类，一是，外部引入水，二是，由于采用降水等措施而新产生的水。

不同的水有不同的处理办法：对于外部引入水，如施工过程中的生活用水等需要在节水用具和净化再利用设备上进行相关开发和研究，而对于新产生的水，则需要采用回灌等技术措施，尽可能的减小降水等对地面沉降的影响。从另一个角度需要优化设计，尽量减少降水量，开发现场水循环技术，增大水的利用率。

（6）地下工程非开挖技术

21 世纪是地下空间开发的黄金时代，无论是土地资源的有效利用，还是城市环境的改善，都对城市地下空间开发提出了前所未有的高要求。传统的明挖方法，对环境的影响大，而且在闹市区经常让市政管线改线，不但影响市民生活，而且造成极大的浪费。

因此，建筑行业开发了多种地下空间开挖方法，包括暗挖法、盖挖法、沉管法、盾构法等等。

3 结语

建设节约型社会对工程技术人员而言是严峻挑战，许多技术难关有待大家攻克。建设节约型社会对工程技术人员而言更是良好机遇，科技创新的需求更加迫切，科技创新的舞台更加广阔。让我们增强社会责任感，加大科技创新力度，为建设节约型施工企业提供有力的科技支撑！

Developing the Saving Type Construction Trade

Hu Yuyin

(Shanghai Construction (Group) General Co., Shanghai 200120 China)

Abstract: Developing the saving type construction trade is one organic compositional part of developing the saving type society. It is pointed out that technology management and technology original innovation are the most important things in developing the saving type construction trade.

我国绿色房地产评价体系的误区与改进探讨

周建亮　孙碧襄
（青岛理工大学，青岛　266520）

摘　要：随着我国建设循环经济和可持续发展的要求，绿色建筑推广逐渐成为政府建设主管部门工作的重点。论文在梳理当前国内绿色建筑评价体系研究的基础上，首先探讨了我国绿色建筑评价体系的缺陷和不足：一是，在绿色理念上，把绿色评价仅仅等同于对建筑自身系统的考察，缺乏从更大的社会系统层面上进行评价；二是，在评价对象上，基本等同于环境方面的评价，忽视了社会因子和经济因子方面的考量；三是，在评价主体上，注重专业机构的定量评定，忽视了消费者的需求和使用者的评价。其次，论文分析了由于评价体系的不足，可能会带来的绿色房地产推行的三种误区。最后，论文在对绿色建筑评价不足和推动误区分析的基础上，汲取了当前国内外绿色建筑评价指标的研究成果，建立了绿色房地产评价体系的层次分析基础模型。

关键词：绿色建筑　绿色房地产评价　层次分析

1　我国绿色建筑评价体系概述

绿色建筑，也有人称之为生态建筑、可持续建筑，“是指在建筑的全寿命周期内，最大限度地节约资源(节能、节地、节水、节材)、保护环境和减少污染，为人们提供健康、适用和高效的使用空间，与自然和谐共生的建筑。”❶。现阶段我国推行或者正在讨论中的与绿色建筑有关的标准和评价体系大致如下：

（1）2001年5月，国家建设部颁布了《绿色生态住宅建设要点与技术导则》，该导则将绿色生态住宅小区的环境性能分为九大系统，即能源系统、水环境系统、气环境系统、声环境系统、光环境系统、热环境系统、绿化系统、废弃物管理与处置系统、绿色建材，分别对这九大系统的建设要点和技术要求提出了明确要求。

（2）2001年10月，国家住宅与居住环境工程中心推出了健康住宅的建设标准。健康住宅的建设要点和评价标准包括：人居环境的健康性、自然环境的亲和性、居住环境的保护、健康环境的保障。

（3）2003年12月，为配合北京2008年奥运会倡导的“绿色奥运”理念，清华大学、中国建筑科学研究院、北京市建筑设计研究院等专业机构组成的课题组公布了“绿色奥运建筑评估体系”。这套评估体系旨在对绿色建筑进行全程监督管理，涉及项目招标、设计、

❶ 建设部住宅产业化促进中心．绿色建筑评价标准(征求意见稿)．2005年10月．

施工、调试及运行管理等环节，并分别从环境、能源、水资源、材料、室内环境等方面进行评估。

2 我国绿色建筑评价的不足及其在房地产业的应用误区分析

从绿色建筑的定义表述来看，它强调了“以人为本”，追求“人—建筑—自然”三者的和谐统一，指出了绿色建筑的最终目的在于为居住人或者使用人提供一个“健康、适用和高效的使用空间”，但就目前绿色建筑的评价标准来看，至少存在着以下不足：

第一，更多对建筑物的自身系统方面进行了绿色节能效率的考察，忽视了从更大的社会系统层面上对建筑物的整体效率进行评价。在一栋建筑绿色得分甚高的情况下，只能说符合指标的绿色建筑仅仅拥有了绿色的物品属性，但是这栋建筑假如长期空置，其绿色并未对社会造成贡献，反而占据了资源，因此，从整体的角度来说，无人居住和使用的建筑其绿色得分为零。

第二，更多的注重了对建筑环境因子方面的评价，忽视了社会因子和经济因子方面的考量。从以上绿色建筑的评价标准来看，各个评价标准的权重基本都倾向与对环境效益的评估，注重了绿色建筑“健康”的一面，但是对绿色建筑 Z 的“适用性”（社会效益评定）和“高效性”（经济效益评定）考虑不足。

第三，更多的注重了专业机构或者专业人士对建筑设计与建设阶段绿色与否方面的定量评定，忽视了使用人在房屋居住和使用中的定量分析和定性评价。尽管绿色建筑的定义涵盖了建筑的全生命周期，但是从评价体系的指标设计来看，更多的是着重于对建筑设计阶段的调节和建筑建设阶段的控制，但是，必须注意的是，设计阶段的某些理想值和实际使用中常常有一定差距。正如澳大利亚的建筑环境评估体系 NABERS 的完成者威尔夫妇特别强调的“对建筑实践的研究永远超过对其的估计”。

具体到房地产业而言，尤其是商品住宅和商业建筑类的房地产，评价体系的缺陷和不足常常会对绿色房地产的推广带来意想不到的负面引导作用，因此，在加大绿色建筑的理论研究和实际推动的前提下，必须防止以下误区：

误区一：谨防忽视需求、盲目追求绿色的误区。

由于绿色建筑的建造成本要比普通建筑要高 10%～15%左右，在目前普通建筑的房价已经超出了购房者的能力的情况下，又如何奢求消费者去购买更高价格的绿色住宅呢？房子是要住的，一味的追求房地产的绿色指标如节能、绿化等，造成建筑使用者入住成本的大幅提高，最终的结果只能是造成房屋空置、开发商的资金积压。没有得到有效居住或者利用的建筑，再好、再高的绿色指标也是徒然，而且从整个社会运行成本来说，它是非绿色甚至是黑色的建筑。绿色建筑的评定和推广更应是从消费者或者最终用户角度出发，以个人居住成本和社会外部成本两项经济指标评价房地产业的绿色实现程度，以需求推动绿色建筑、绿色房地产业的快速发展。

误区二：谨防形象、标志工程的误区。

当前，随着我国经济社会发展的转型，政府对绿色建筑的推广极为重视。但是，政府的推动作用往往是一把双刃剑，极有可能造成开发商为了迎合政府政绩的需要，而脱离大多数居民的需求，建造极少数的标志型天价楼盘。在目前房地产价格已经远远超出普通民众购买力的前提下，这些“标志性楼盘”作为试验绿色的指标探索固然需要，但是应当考虑到建造一个100%绿色的空置小区，不如建造10个节能10%的准绿色小区，尤其在我国城市化推动房地需求旺盛而购买力不足的情况下，更为需要。

误区三：谨防利益短视，建筑伪绿色包装的误区。

一般说来，开发商是把短期经济效益放在首位，“绿色建筑”、“节能住宅”和“生态住宅”的标签，对开发商进行楼盘促销时的诱惑力更大。但当获取这些标志认可的建造成本过高时，难免会出现为绿色认证而弄虚作假的可能。因此，在大力推行绿色建筑评估的同时，如何加强对评估过程和评估结果的监督，同样也是摆在政府面前的一道难题。

3　若干国外绿色建筑综合评价体系比较

目前国际上发展较成熟的绿色建筑综合评价系统有英国BREEAM、美国LEEDTM、加拿大的GBC和日本的CASBEE等，它们科学的架构和良好的运作，使之成为各国建立新型绿色建筑评估体系的重要参考，参见表1。

表1

	BREEAM英国	LEED美国	GBTool加拿大	CASBEE日本
版本	1990年初版 2005年最新版	1996年初版 2004年最新版	1998年初版 2005年最新版	2002年初版 2005年最新版
评估项目	1. 管理 2. 健康与舒适性 3. 能耗 4. 交通 5. 水耗 6. 材料 7. 土地利用 8. 生态 9. 污染	1. 场地可持续性 2. 用水的利用率 3. 耗能与大气 4. 材料与资源保护 5. 室内空气质量 6. 创新与设计和施工	1. 资源消耗 2. 环境负荷 3. 室内环境 4. 服务设施质量 5. 经济性 6. 管理 7. 出入与交通	Q 建筑物的质量 Q1 室内环境 Q2 服务设施质量 Q3 占地内的室外环境 L 环境负荷 L1 能源 L2 资源与材料 L3 占地以外的环境 Q/L 建筑环境效益

注：引自：中国土木工程协会：2005年赴日建筑住宅考察报告。

此外，其他一些国家开发具有鲜明特色的评估系统也具有一定的借鉴价值，如法国的ESCALE、南非的SBAT和澳大利亚的NABERS等。这些评价体系大都基于层次结构模型，综合了多种定性定量指标，建立了多层次的带权递阶结构，通过逐层累加评分得到建筑物的绿色分值认定。但是，从其指标的设计可以看出，一方面，多数指标主要是与环境相关，经济与社会指标仍然欠缺；另一方面，评价系统的导向大都重点放在了设计和建造的引导上，对使用阶段的建筑的实际“绿色”表现如何，也缺少相应的考虑。

4　绿色房地产评价体系的改进模型探讨

对以出售和出租为目的的房地产业来说，与其他建筑如公共建筑等相比，其最大的不同在于市场机制的隐性主导作用。如图1所示绿色房地产的供求曲线，卖方即开发商的主要供给目的在于追求最大化的利润，绿色建筑在追求自身“绿色”实施过程中，必然会将原本属于外部成本的部分成本转嫁入开发商的内部成本中，造成开发商利润的降低；而买方的需求线对价格的敏感程度即需求弹性则受到多方面因素的影响。这些因素可以简单分为物质要素、自然要素和社会要素。在初期，购房者的需求主要是受到物质要素的影响，即购房是为了满足自己的基本居住条件；在中期，随着经济发展到一定水平，人们在居住的物质需求基本得到满足以后，将会扩大对自然要素的需求，购房者对未来居住区的设施、绿化、空气等环境要素的要求会越来越高；在未来，随着经济社会的进一步发展，社会因素对购房者的需求的影响将逐步加大，居住区将在市场机制的作用下，以小区区位、住宅特点、教育条件和就业条件以及居住者身份等形成不同特色的居住社区。在我国当前阶段，应当说大多数房地产需求还处于对物质要素的需求阶段，需求弹性相对较大。房地产绿色评价不仅不能脱离对绿色建筑的供给与需求分析，评价模型中也应当包含对绿色房地产供给和需求两方面的指标因素。

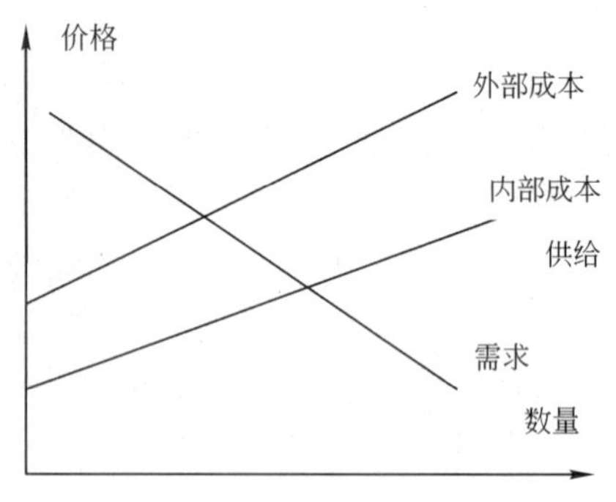

图1　绿色建筑的供给曲线

基于以上分析，在汲取我国现有研究进展的基础上，并综合了国外相关绿色建筑评价研究的成果，以层次分析法为主要技术路线，建立的绿色房地产评价改进模型基本框架如表2。

表2

指标层A	因素层B	分因素C
经济指标	投资成本	一次性建设投入
	改建成本	建筑用途或内外布局改变的便易性及其可能成本
	使用效益	房屋入住率、房屋使用度、人均居住面积等
	居住开支	水、电、气、暖等生活必需消费支出
	管　　理	设计管理、施工管理和物业管理规范性
社会指标	室内舒适度	日照、通风、室温、隔音和视野等定量指标与主观感受
	小区便利度	小区周围银行、学校、商店、邮政和运动场所等的可达性(步行/自行车/汽车)
	设施完善程度	小区内外公交、出入交通、通讯、卫生等设施的配套程度
	居民管理参与度	居民对小区环境、社交等公共事务的参与程度
	健康生活保障	社区安全程度、康体设施、医疗保健设施、社区老人设施、健康行动、危机处置等
环境指标	水环境	参照现有的绿色建筑环境评价指标
	气环境	
	废弃物	
	能　　源	
	建筑材料	
	环境整体负荷	

该绿色房地产评价改进模型的应用要点包括：

（1）评估系统既包括适用于建筑全过程的一般性指标，也包含了部分仅仅适用于设计、施工或者使用等某一阶段的特殊性指标，以期达到评估工具应有的设计引导功能、施工监督功能和使用评价功能的目的。

（2）评估面向规划设计、绿色施工、居住使用部分或者全部阶段，围绕拟建建筑、竣工建筑或者交付使用建筑的经济、社会和环境 3 大类 16 小项进行单独评分，在通过权重系统得出总分，划分为优秀、良好、合格、不合格四个等级。建议在项目开发期间首先对设计和施工阶段进行预评分，待项目交付使用后再进行综合评定，作为房地产绿色与否的最后评估结果。

（3）采取定性与定量、评测与实测相结合的方式。量化标准值的选取以国家、地方有关法规标准或科学的研究成果为依据，若缺少相关标准，又无法提供科学的研究成果，可采取与同类型参考建筑相比较的方式给予评分。

（4）制定相应的财税激励政策，根据总分评出建筑绿色等级，采取建筑成本补贴、购买成本补贴等手段，减少开发商的建造成本和消费者的购买成本，以促进绿色房地产的供应和扩大绿色房地产的需求。

（5）评估认证机构必须具有权威性，建议由政府部门推动，建立相应的行业协会。由该机构负责组织评估过程、给予评估认证、进行系统研究，不断改进和更新版本。

Study on the Mistakens of Green Building Assessment and Its Model Improvement of Green Real Estate in China

Zhou Jianliang　Sun Bixiang

(Qingdao Technology University, Qingdao　266520　China)

Abstract: As our country builds the requests of recycle economy and sustainable development, the green building is popularized and its application becomes government's focal work point gradually. At first, the thesis is on the basis of the present domestic research on the assessment system of green building, have probed into three major defects of the assessment systems: the first defect is the idea on the green concept. The assessment system takes much more considering to the building itself, while less consideration in the higher and larger social system; the second defect is assessment objects. It makes the green assessment only equal with the assessment to the building environment or building energy-conservation, lacking the investigation from economic angle and social angle; the third defect is the makeup of the assessment group. The assessment system ignores both consumers 'demand and users' feeling. Secondly, the paper discussed three mistakes in developing green real estate caused by the defects of assessment system. Finally, basing on both fully analysis to the three mistakes and deriving from the current research results, the paper has established a hierarchy analytic foundation model of the green real estate assessment system.

参考文献

[1] 杨宁，董聪，江见鲸．绿色建筑综合评估系统分析与探索．工程力学增刊 2003：631～635
[2] 绿色奥运建筑课题组．绿色奥运建筑评估体系．北京：中国建筑工业出版社．2003
[3] 建设部住宅产业化促进中心．绿色生态住宅建设要点与技术导则．2001.5
[4] 国家住宅与居住环境工程中心．健康住宅建设技术要点．2001.10

开发商建绿色建筑既是机遇更是责任

张在东　史　勇　陈亚君
（北京锋尚房地产开发有限公司，北京　100089）

摘　要：论文指出开发商对绿色建筑的理解对项目最终是否能建成绿色建筑起主导作用，结合北京锋尚国际公寓开发实践，系统阐述了绿色建筑在一个具体项目中的构成要素。指出建绿色建筑是开发商的机遇和责任。

关键词：开发商　绿色建筑　机遇　责任

1　起主导作用的开发商如何认识绿色建筑

在现行开发模式中，从项目立项、规划设计、施工安装及精装修和园林景观，开发商都在起主导作用，所有参与项目的人员都是“建议”，只有开发商是最后“决策”，很多有识的科研和设计人员，给开发商提供了很多绿色建筑的建议，但多数被开发商否决。开发商是否具备绿色建筑的概念或者对绿色建筑理解的不同程度，将决定一个项目的最终结果，鼓励引导开发商迅速补上绿色建筑这一课是当务之急。目前开发商队伍中的决策人物对绿色建筑的理解参差不齐，有的甚至是很著名的发展商认为“绿色建筑就是多种些树”、“南方地区不需要保温隔热”；也有的发展商认为绿色建筑要浪费很大的成本，市场买房人没这种需求；更多的发展商被市场上充斥的“节能建筑”、“健康住宅”、“生态建筑”和“绿色建筑”等耀眼光环名词的说法说糊涂了，搞不清楚绿色建筑的真正概念，于是就自下结论“绿色建筑也不过是在卖房子时可以利用的一个噱头而已”。澄清、改变一些开发商的糊涂观念，进而使他们认识到这是一个难得的历史机遇，是保证最终建成绿色建筑的重要因素；要透过政策、法规和媒体社会舆论，帮助开发商建立起正确的观念，要他们清楚建绿色建筑既是机遇也是责任。

2　绿色建筑投放市场后的反应

2.1　对绿色建筑的理解

“绿色”概念，也就是可持续发展(Sustainability)概念，首先要强调与周围自然环境相融合，要利用最少的资源(排放最少的污染)创造出高尚的健康、舒适居住环境。也就是说绿色建筑要因地制宜，适合中国国情和地方特点，包括地方文化保护等；同时绿色建筑一定是节能的和智能的。传统的节能建筑和绿色建筑有些方面是冲突的，绿色建筑更强调节能和舒适的辩证关系，特别强调室内空气环境质量，绿色建筑不一定豪华，但它一定是低能耗和健康舒适的。具体说就是要做到节能、节地、节水、节材，特别要注意的是，这个节约不单指狭义的节约(Save)，而是讲投入产出比例、循环经济的广义的节约(Efficiency)。

2.2 市场反应

北京锋尚国际公寓 2002 年 5 月推向市场后，立即引起市场的轩然大波，售价是同地段其他商品房的二倍，在热销期 3 个多月时间里，300 多套房子一销而空，来自社会各界的参观人员达 5 万人次，媒体进行了 1300 多次报道。自 2003 年 3 月入住后，一切使用正常，累计收到业主各种形式的表扬信 100 多封，租金是同地段其他商品房的三倍，证明这种节能环保的绿色建筑是有市场的。当然这样好的市场反应，也同市场营销中提出的“告别空调暖气时代”口号有很大关系。

3 绿色建筑要素在锋尚的系统集成

3.1 总体规划

北京锋尚国际公寓位于北京中关村地区苏州桥附近，属纯住宅小区，占地面积约 2.6ha，总建筑面积约 10 万 m^2，容积率 3.0。园区建筑由 4 幢塔楼、2 幢板楼、会所及地下停车场组成，所有建筑以小区内最大的水面积和绿地中心呈围合式布局，不仅满足了人们依山傍水而居的亲自然情结，也强调了这里的中心感和凝聚力，很自然地构成了纯居住的理想环境。业主和来访客人的车辆均由园区入口进入地下车库，避免了汽车噪声、废气对居民生活的干扰，同时也增添了老人和儿童在园区活动的安全感。

3.2 单体设计

住宅单体从属于整个建筑群错落有致的空间形态。其高度、体量、造型均满足营造最佳整体效果的需要。每幢建筑均因地制宜，根据所处位置、景观、日照、交通等多种因素，进行单独设计，避免生硬套用。小区内住宅户型设计突出与环境对话的特点，满足住宅的适用性和私密性。设计中充分研究社会生活模式的变化对住宅设计提出的新要求，使其具有一定超前性。考虑到居家办公的需要，大部分住户均设置书房，并配以电子网络接口。

3.3 绿化系统

园区环境经过精心设计，绿荫间散落的水面、蜿蜒曲折的山坡和无路牙的砾石小路亲切、自然、温馨的气息在这里处处弥漫。锋尚小区的实际绿化覆盖率达到 63%。

锋尚内有一个人工湖，不同于一般的水泥砌筑水池，湖底和四周的土壤都进行了特殊改良，湖边的边缘与草地自然过渡。夏季湖中有鱼虾和水生植物，冬天湖面结冰后成为天然溜冰场；人工湖还特别设计了浅水池，夏季用于儿童嬉水，冬季则是露出铺满鹅卵石的景观小品。园区道路与绿地自然过渡、衔接、人工湖与自然草坡有机结合，相映成趣；社区内相当部分的成树均采用整体移栽手法，常绿植物和天然草坡品种各异、搭配合理，构成了园区绿化的主体。

3.4 道路与停车

小区内道路系统结构清晰，分级明确。可同时满足居民出入、消防、救护和社区服务的要求。

小区采用人车完全分离的交通组织模式，机动车主要通道全在地下与地下车库有机结合，使园区散步的老人和游戏的儿童有较强的安全感。小区内提供了近1：1的机动车停车位。地下停车场均匀地布置在庭院地下，使用方便，利于管理。此外，室外还设置了近1：1自行车停车位。

3.5 智能系统

小区从功能上把智能化包括：安全防范智能化，居民生活智能化和物业管理智能化。

3.5.1 安防系统——社区围墙和出入口处，多区域设置全天候工作的红外线对射控测器和电视监控探头，保证社区内24小时全方位的安全。

3.5.2 消防系统——由消防弱电系统和消防水系统两部分组成；消防弱电系统光电感烟控测器位置——住宅卧室、起居室及地下室、电梯间、走廊等公共部分。

3.5.3 门禁对讲——园区大门、楼宇入口和电梯共设有三道门禁，专门身份识别系统可自动识别业主手中的IC卡，有效避免了闲杂人员进入园区内，增加业主们的安全感。

3.5.4 户内紧急呼叫——各户起居室和主卧室内安装有紧急求助按钮，如遇紧急情况可直接按下监控中心报警，监控中心就会自动显示具体房间，随即在第一时间组织支援。

3.5.5 可燃气体探测器——住宅厨房设可燃气体探测，当发生可燃气体漏立即报警。

3.5.6 手动火灾报警按钮及消防报警电话插口——设于住宅楼各层、各区。

3.5.7 IC卡交费系统——社区内自来水、电、燃气的计量和缴费均通过IC卡系统完成，免去人工抄表、核算金额的麻烦。热水则采取远传抄表系统进行计量收费。

3.6 环境质量和废水排放系统

3.6.1 采用新型高性能草种和透气透水型室外铺装材料和施工技术，使人行道可同时长矮草。

3.6.2 垃圾处理系统由中央吸尘、食物垃圾处理器和可回收分类垃圾周转箱三部分构成。无需转运，避免二次搬运垃圾污染。

3.6.3 卫生间采用隐蔽式水箱和同层排水系统，彻底改变了传统的地排水方式所带来的噪声大、管线穿楼板、卫生洁具以后不能改变位置、易漏水及楼上楼下的纠纷，并且维修简单易行、消除了卫生间的卫生死角。

3.6.4 小区设置了中水系统，将洗浴废水作为中水水源，经处理后用于园林景观的人工湖补水、园林绿化用水和洗车、冲洗道路等用水。生活污水采用室内分流，室外合流的排水系统，其流量600t/日，排入埋地式无动力生化污水处理装置经达标后排入市政管道。

3.7 防噪声系统

3.7.1 420mm厚外墙系统和断热铝合金窗、外遮阳卷帘，可有效防止室外噪声传入。

3.7.2 楼板总厚度为250mm，地板下设置的隔音垫，能有效隔绝楼层间噪声的传递。

3.7.3 采用具有良好的隔音措施的同层排水系统，能有效消除水流撞击管壁声及对下层住户的噪声。

3.8 建筑节能成套技术系统

3.8.1 混凝土制冷采暖子系统

首次在国内高层住宅应用了混凝土制冷采暖系统。混凝土制冷采暖系统同传统的空气调节，采暖和制冷方式有着很大的不同。它的目标是将建筑物内部用很少的能耗而长时间均匀地保持在高舒适度的范围，不仅节约能源而且舒适和健康。该系统的节能效果也给每位住户都带来了实际上的好处，节约了采暖制冷费用的支出，同时也减少了采暖期的燃煤消耗和 CO_2 排放。该系统单独的投入并不大，但对建筑整体的保温隔热水平提出了更高的要求，并在设计、选材、施工等诸多环节上均须加以整体考虑，即它是一个系统工程。

3.8.2 健康新风子系统

新鲜空气由卧室和起居室墙边的地面送出，将污浊空气由厨房和卫生间的顶部排风口排出。其优点首先是略低于室内温度的新鲜空气因密度大而自然沉降于房间下方并流向各处而形成“新风湖”，因为人体是新鲜空气所能接触到的最热的热源，所以它顺着人体爬升，将新鲜空气直接送到口、鼻处；其次是因人体呼出的气体是室内较热的气体，所以不存在呼出的污浊气体与新鲜空气交叉混合的可能，对人体健康十分有利；其三是它利用排出气体的热或冷与将要送入室内的新风进行热量交换，以减少新风预热或预冷的能耗。

3.8.3 外窗子系统

断热铝合金窗和低辐射(Low-E)玻璃。Low-E 玻璃能阻止室内热量的外泄，冬季能将白天的照射到室内的太阳能量留在室内，用最少的能源消耗得到最温暖的室内温度，提高室内舒适质量。考虑到北京夏季较高强度的太阳辐射，外窗上安装了外遮阳卷帘，从而使室内温度尽量少受太阳能辐射热能的影响。

3.8.4 屋面及地下子系统

屋面采用 200mm 厚的外保温板，女儿墙采用与外墙相同的保温板完全包裹，外墙外保温板沿外墙深入地下 1.5m(北京地区的冻土层为 0.8m)，根绝来自地下的湿冷和来自屋顶的曝晒。

4 建绿色建筑是发展商的机遇和责任

中国正处在城市化和工业化的高速发展期，房地产业面临一个崭新的课题，发展绿色建筑是大势所趋。如何应对这一契机？是该迅速落实行动的时候了。我们与欧洲先进国家的差距，不在材料和技术上，而是在观念上。北京锋尚 5 年前开始建设，现已正常运行 2 年多的时间，可比冬季采暖能耗已降到 12.5W/m^2，每年每平方米可节约 4.9kg 标准煤，两套住宅一年就能节省 1t 煤。证明了欧洲的低能耗住宅材料和技术在中国完全可以实现。中国目前能源紧张状态，“不是能源少了，是能源浪费太多”。统计数据表明，目前中国现有建筑总面积 400 多亿 m^2，预计到 2020 年还将新增 300 亿 m^2；我国每年在房地产方面的投资约为 1 万亿元人民币(商品房)，还不包括农村的建筑，在城市基础设施方面的投资为 5000 亿元人民币。建筑在建造和使用过程中直接消耗的能源占全社会总能源的 30%，使用建材的生产能耗占 16.7%，用水占城市用水的 47%，使用钢材占全国用钢量的 30%，水泥占 25%，据预测，到 2020 年，如果达到绿色建筑的标准，每年可节约 3.35 亿 t 标准煤，仅空调节能就达 8000 万 kW，节水 47 亿 t。从以上数据可以看出，是继续建高能耗建筑还是建绿色建筑，意义十分重大。2005 年 2 月 28 日《中华人民共和国可再生能源法》颁布，为低能耗向零能耗建筑发展创造了法律环境。我国化石能源资源探明储量中，90％以上是煤炭，人均储量仅为世界平均水平的 1/2；人均石油储量仅为世界平均水平的 11％；天然气仅为 4.5％，因此，不但要鼓励建低能耗建筑，而且要鼓励建零能耗建筑，目前南京锋尚正在做零能耗的实践。高能耗建筑不仅仅是多浪费了能源，而且更为严重的是这些能源燃烧后产生的温室气体排放给地球造成的影响。作为在建设过程中起决策作用的发展商，不但要考虑机遇，更要考虑到你的社会历史责任，否则无法面对子孙后代。开发商要团结、组织参与建设的研究、设计和施工单位，开创出一条既适合我国国情，又具有自主知识产权的绿色建筑体系。

Green Building is Developers' Opportunity & Responsibility

Zhang Zaidong Shi Yong Chen Yajun

(Beijing Fengshang Real Estate Co. Ltd, Beijing 100089 China)

Abstract: This paper shows the comprehensive role played by developers in the development and operation of Green Buildings. Developers play a decisive role in determining whether the development can successfully achieve the designation of Green Building. Using examples and lessons from the current practice of Beijing International Apartments, it sets forth the components of Green Building with a representative project already in practice. The paper shows the Green Building is the developers' ongoing opportunity and responsibility.

发展绿色建筑——中国房地产业的光明之路

惠彦涛　李　瑞
（西安建筑科技大学管理学院，西安　710055）

摘　要：近年来，中国住宅市场正处以蓬勃发展阶段，绿色建筑生态智能住宅已经开始走入我们的生活，人居环境已成为全世界共同关注的课题，由此，绿色建筑正逐步走进我国的房地产业。本文在详细分析我国当前房地产业发展状况及存在问题的基础上，提出了在我国的房地产业中发展绿色建筑势必在行。并从绿色建筑政策标准的制定和完善，行政监管的加强和四节一保的强制执行，关键技术的研究与引进，地方政府参与积极性的调动、可再生能源的利用等方面提出了在房地产业中发展绿色建筑的主要对策，阐述了中国房地产业发展绿色建筑的重大意义。

关键词：绿色建筑　和谐相关　可持续发展

伴随着时代的发展，建筑业发生了日新月异的变化，诸如智能建筑、节能建筑等不断出现。近年来，绿色建筑开始在中国出现并越来越受到重视。早在 1992 年联合国环境与发展大会上就比较明确地提出了绿色建筑这一概念。当时全球的科学家和社会各界认识到，日益快速发展的经济给我们的环境带来了巨大的压力，怎样利用我们有限的资源，尽可能减少对环境的影响，取得社会经济的可持续发展，成为全人类关心的问题。由于建筑业对环境的巨大影响，能够为人们提供健康、舒适、安全的居住、工作和活动的空间，同时在建筑全生命周期(物料生产、建筑规划、设计、施工、运营维护及拆除、回用过程)中实现高效率地利用资源、能源、土地、水、材料，最大限度降低对环境影响的绿色建筑理念受到了广泛的重视与推广。中国正处在城镇化高速发展的起步阶段，是目前建筑量最大的国家。关系到十三亿中国人居住问题的房地产业如何避免走发达国家对节能建筑进行二次改造的弯路，能否及时普及和推广绿色建筑成为我国能否大量节约能源和资源进行可持续发展的重要因素。

1　中国房地产业的发展状况

近年来，我国房地产业正处于蓬勃发展阶段，我国每年新建约 20 亿 m^2 建筑，现有建筑面积达到 440 多亿 m^2，未来 20 年内，我国还需建造 400 亿 m^2 的新建筑，相当于中国数千年文明史积累的总建筑量。通过近 20 年的发展，中国房地产业的发展状况有以下几个方面：

1.1　地区差异大

我国是一个幅员辽阔的国家，地区差异十分巨大。从南到北，从东到西，气候条件、自然资源、生活水平与社会习俗等都有很大的差异。各地在结合当地历史文化传统，汲取先人在建筑上积累的智慧的基础上，修建了大量随着地区差异呈现出不同的风貌，富有地方特色的建筑。同时根据各地经济发展水平，房屋的功能、设备、装修、造型方面存在较大差异，总体上经济较发达地区，在修建的住房中车库、贮藏室等设置较全，热水、暖气、空调设备齐全；装修标准较高；造型上较多引进诸如“欧美风情”、“现代城”等风格的建筑样式，建筑物的造型较丰富。而经济欠发达地区，房屋的功能设置较简单，设备安装较少，装修上主要利用当地材料，造型简洁。如湖南湘西地区的新建住房中很多仍采取二层木式斜坡顶住房，充分利用了当地丰富的木材资源，又符合当地的气候条件，经济和环境效果都很好。

1.2 房地产业开始推广绿色建筑

从 20 世纪 80 年代开始，中国房地产业经历了两次大的发展高峰，第一次从 80 年代中期到 90 年代初期，房地产业得到了迅猛发展，这一时期修建的建筑，从资源消耗到环境影响方面基本达不到绿色建筑的要求。修建的住房一方面满足了居民的住房消费需求，另一方面由于缺乏长远规划和整体布局，单纯追求数量盲目建房，造成了很多住宅因为缺乏基本生活条件或者质量较差而无人居住，加上大量的烂尾楼工程，造成了物质资源的极大浪费。第二次高峰是从 90 年代末至今，这一时期的建筑在规划和布局上有了较大改进，同时也出现了不少智能建筑和节能建筑，使建筑本身更趋向于满足人的舒适性、观赏性要求。这一时期建筑节能工作得到了逐步推进，取得了较大成绩，与此同时，伴随着可持续发展观念在全世界的认同，绿色建筑理念在我国渐渐得到了重视，众多的科研院所、大专院校启动了绿色建筑技术研究，并在一些城市的住宅小区、农村住宅进行了绿色建筑实践。尤其是 2002 年底，对绿色奥运建筑标准和评估体系进行了研究，系统地提出了绿色建筑所涉及的内容重点，建立了科学的绿色奥运建筑评估体系。中国房地产业开始对绿色建筑进行研究和探索。

1.3 新旧建筑共存的局面将长期存在

建筑尤其是住宅属于长期的消费品，不可能进行频繁地更新。我国房地产业通过几十年的发展，建造了大量住房，加上今后每年都会建造大量的住房，因此必然形成新建建筑与原有建筑，耗能高的建筑与节能建筑，普通建筑与智能建筑等各种建筑形态共存的局面。同时面临着对已建建筑延长使用或装修期限；对存在安全问题的建筑进行加固改造；对报废建筑进行拆除并对建筑垃圾通过回收等处理方法变废为宝，并减少对自然环境的影响等一系列的问题。如果能够处理好这些问题，就能够大量地节约能源和资源。

2 房地产业存在的问题

在高速发展的房地产业后面也暴露出一些亟需解决的问题。

2.1 建筑耗能大，运行和维护费用高

现有建筑中，95%以上是高能建筑，我国平均单位建筑面积能耗是发达国家的 2～3 倍。相反，我国的能源储量极其匮乏，人均煤炭储量为世界平均水平的 1/3，人均石油为 11%，天然气仅为 4.5%，人均耕地为世界平均耕地的 1/3，水资源占有量只有 1/4，因此我们更要节约能源，按照可持续发展的规划充分利用能源。而绿色建筑和高能建筑相

比，能够降低能耗70%～80%，节能效果特别显著。另外房地产商在开发实践中，为了降低造价，追求商业上的最大利润，往往采取降低材料、设备性能的方法，居民在购房时看起来减少了费用，但在实际使用中运行和维护费用高昂。城镇居民在水、电、煤气、暖气等住房消费上所花的比例占家庭收入的18%左右，超出了一些低收入家庭的经济承受能力。因此降低能耗，减少运行和维护成本成为房地产业亟待解决的问题。

2.2 建筑、装修材料污染严重，室内环境差

随着生活水平的提高和住房条件的改善，住房和办公公共场所的装修程度越来越高，各种新型建筑、装修材料不断涌现，由材料引发的甲醛、氨、苯、氡和可挥发的有机化合物等有毒物质超标而造成的室内环境污染纠纷不断。加上现有住宅结构上趋向于封闭，在设计上与自然环境隔离。建筑内部与外部缺乏有效连通的方式，无法对气候变化自动进行适应调节，因此室内环境品质(即空气质量，温度、湿度，舒适感，自然光照明，隔噪声等)大大降低，室内环境往往不利于健康，无法满足人们对住房的舒适性、健康性的要求。

2.3 二次装修，加重自然环境污染

新建住宅二次装修问题是现阶段住房供给体系带来的问题。刚建好的新房子，在住进去之前一般都要进行重复装修，拆除一些尚未使用过的建筑构件、设备，造成了极大的材料、人工、经济上的浪费。同时在拆除和重新装修过程中产生大量的建筑垃圾，加重对当地的环境污染。有的因为对重要结构部位进行改造或拆除，对房屋的整体安全构成了安全隐患。无法满足人们对住房的安全性要求。

2.4 追求奢华，没有充分利用当地资源

不少地方建筑一味追求豪华气派，放着当地便宜实用的材料不用，不惜从外地甚至从国外运来各种高档材料，造成了资源和能源的浪费；在建筑风貌上追捧某个风格，并通过商品化的生产技术加以复制，一方面缺乏本地文化特征，一方面又因为采用标准化的生产过程，造成了很多地方新建的建筑外貌上大同小异，形成“千城一面”的现象。如很多地方的政府机构，建筑风格上大多外墙采用高档石材加玻璃幕墙的造型，造价高昂但有时与当地建筑风格难以融合，综合效果并不好。

3 房地产业实施绿色建筑需要采取的主要对策

事实说明，要实现房地产业的可持续发展，就必须推进绿色建筑的发展。绿色建筑在我国刚刚起步，还处于试点、发展阶段。如何实事求是、因地制宜地实施绿色建筑，是影响今后中国房地产业发展方向的关键。

3.1 尽快建立、完善适应中国国情的绿色建筑标准和政策

我国绿色建筑起步晚，基础差，理论研究不足，工程实践少。建设部虽已先后颁布针对三个气候区的节能标准，初步形成了比较完善的民用建筑节能标准，但工业建筑的节能标准尚未出台，公共建筑的节能标准也才刚刚颁布。而关于建筑节能、节地、节水、节材和环境保护的综合性标准体系，还没有建立。我国的绿色建筑评价标准和技术导则尚处于研究阶段，我们在制定标准时要考虑到中国的实际情况，我国的国土与资源有限，人口数量巨大，不能照搬国外的形式。需要尽快建立、完善适应中国国情的绿色建筑标准和政策。

3.2 加强全程监管，强制执行四节和环保标准

要在建筑的立项、规划、设计、审图、施工、监理、检测、竣工验收、核准销售、维护使用等各个环节加强监管，层层把关，做到不符合绿色建筑标准的不批准、不立项、不施工，在根本上控制不达标建筑的建造。在资源、能源消耗的各个领域强制执行包括节能、节地、节水、节材和环境保护等方面的标准规范，并不断制定、完善新的内容更加宽泛的四节标准和技术规范。鼓励各地根据本地的原材料、风俗习惯和居住条件，开展多种形式、多种途径的建筑创新。推行使用当地材料，规定建筑材料运输范围，减少材料运输环节中的能源消耗。结合当地特点，广泛利用太阳能、风能、地热能、沼气等可再生能源，减少能耗和对环境的污染，达到最高效率地利用当地资源、能源，最低限度地影响环境的目标。

3.3 加强技术研究和材料开发，系统引进国外先进技术

我国的绿色建筑关键技术和设备十分缺乏，需要组织科技力量，加强对绿色建筑关键技术的研究，开发出符合绿色建筑标准的，具有自主知识产权的重要设备和材料。同时一些发达国家在推行绿色建筑方面已建立了非常完善的技术体系、零部件的供应体系和评价体系，及时、系统地引进发达国家在这方面取得的成熟技术和体系并结合中国国情加以创新、发展，对我们少走弯路，采取成本最低的办法来加快绿色建筑的进程起到强有力的助推作用。

3.4 调动地方政府的积极性，合理利用当地资源、能源，创造与自然和谐相关的绿色建筑

推行绿色建筑必须调动地方政府参预的主动性和创造性，才能有效地进行分类指导。把节能与绿色建筑列入各级政府的工作目标，出台相应激励政策，例如：将绿色建筑的推广作为评选国家园林城市的条件之一等来鼓励地方政府大力发展绿色建筑。根据各地气候条件和发展程度的差异，利用当地政府和建筑师了解当地风土人情的优势，发展富有地方历史文化特征的绿色建筑。尽可能地使用当地材料，规定建筑材料运输范围，减少材料运输环节中的能源消耗。结合当地特点，广泛利用太阳能、风能、地热能、沼气等可再生能源，减少能耗和对环境的污染，达到最高效率地利用当地资源、能源，最低限度地影响环境的目标。

3.5 对老建筑进行改造，普及绿色建筑

我国现有的400多亿建筑物中，95%以上是高能建筑，节能潜力巨大，如果把这些建筑物改造按绿色建筑的标准进行改造，将节省大量的能源。这对于我国能源短缺的状况会有极大的改善。发达国家是在工业化、城镇化高速发展之后的后工业化时期才开始绿色建筑进程，因此有大量的老建筑需要进行改造，他们通过多年的实践，形成了一整套成熟的技术。我们可以在借鉴这些成熟技术的基础上，结合实际情况，因地制宜地制定出符合中国国情的改造技术，对数量庞大的老建筑进行改造，增加绿色建筑的数量，普及绿色建筑。

4 结束语

中国住宅与房地产业发展绿色建筑，是围绕着为人们提供健康、舒适、安全的居住、

工作和活动的空间，同时高效率地利用资源、最低限度地影响环境而展开的，最终受益的仍是广大民众，因此它离不开全民的参与。通过我们共同的智慧、经验和努力，我国的建设事业一定会走上一条资源节约、环境和谐、持续发展的光明大道！

Develop the Green Building——the Bright Road of the Chinese Real Estate Industry

Hui Yantao Li Rui

(Management College of Xi'an University of Architecture and Technology, Xi'an 710055 China)

Abstract: In recent years, the Chinese residence market has being in its prosperous developing stage, green building ecology intelligence residence his came to our life already. Residence environment has become the common topic that people pay attention to all over the world. Hence, green building is penetrating to Chinese real estate gradually. This paper puts forward the urgency of developing green building in our country's real estate, on the basic of analying the industry's present developing condition and existed problems in detail. And from the espect of the situation of China, the establishment and perfect of green building policy standards strengthen of the administration surveillance, compulsory execution of the "four savings and one protecting" Key technique's researching and bringing in, stimulating the local government's participation aggressive, using of the recycling etc. Puts forward to the main counterplan of developing green building in Chinese real estate industry, and narratives the important meaning of it.

参考文献

[1] 中国建筑材料科学研究院. 绿色建材与建材绿色化. 北京：化学工业出版社，2003

[2] 西安建筑科技大学绿色建筑建筑研究所. 绿色建筑. 北京：中国计划出版社，1999

[3] 中国室内装饰协会室内环境监测工作委员会. 绿色家装全攻略. 北京：机械工业出版社，2004

[4] 绿色奥运建筑研究课题组. 绿色奥运建筑实施指南. 北京：中国建筑工业出版社，2004

[5] 于思远主编. 房地产. 住房改革运作全书. 北京：中国建材出版社，1998

城市"层峡"地带的热环境部分影响因素浅析
——以香港石峡尾商住区为例

王　振[1]　王　萍[1]　赵　逵[1]　管凯雄[1]　李保峰[1]　卢兆明[2]

([1] 华中科技大学建筑与城市规划学院，武汉　430074；[2] 香港城市大学建筑系，香港　999077)

摘　要： 城市"层峡"地带是最为重要的城市活动场所和行为空间之一，不同的城市空间形态和建筑布局往往影响该地带的微环境气候。本文以香港商住区的城市"层峡"地带的城市热环境为例，结合现场测试和CFD软件数字技术模拟城市"层峡"地带不同的尺度、界面以及风源状况对城市热环境的影响，提出了关于城市"层峡"地带热环境研究的方法及其微气候环境改善的技术策略。

关键词： 街道"层峡"　CFD　城市空间　热环境

1　前言

伴随中国快速城市化进程，人类赖以生存的城市建设环境发生了巨大的变化。这些变化在交通、农业、工程和机械等许多领域得到了很好的研究。然而，城市化对气候带来的影响并未在城市设计师、规划师、建筑师和工程师中得到应有的认识和重视。许多建筑和城市设计项目与发展循环经济、建设节约型社会以及保护环境、建立可持续发展的目标背道而驰。

对于城市建成环境，城市规划师和建筑师可以关注什么，或者说什么值得关注？技术(如空调、电梯和汽车)的应用(且不说个人消耗的能源)，保留了人在室内环境里的体面，但谁为排向室外的热量和废气买单？街道和开放空间又交给了谁？合理的规划仅仅是经济层面的规划？是交通层面的规划？还是管理层面的规划？规划不是某一方面的消费品，植于气候和环境的规划是达到和谐规划的基石，是其他单方面规划的首要约束。那么，在基于环境和气候的城市规划和设计当中，在无人买单的城市街道和开放空间环境当中我们可以做些什么？具有高度城市化的香港，地处于亚热带热湿气候区域，经过城市高速发展，其人口稠密，建筑密集，工业集中。在这样一个发达的城市中，城市的街道"层峡"随处可见。街道"层峡"在英文中称为"STREET CANYON"，包括城市街道两边的建筑及其形成的狭长街道空间。它是最为重要的城市活动场所和行为空间之一，而不同的城市空间形态和建筑布局往往影响该地带的微环境气候。因此研究城市街道"层峡"的空间形态，建筑方式对其热环境的影响对分析城市建成环境变化的成因、改善城市热环境、提高城市设计质量具有重要的意义。笔者于 2005 年 6～9 月，在香港九龙塘的石硖尾地区[1]首先对不同城市空间形态进行了温湿度的测试，并对于重点研究的街道“层峡”地带利用 CFD 软件模拟对影响街道“层峡”热环境的因素进行了分析。

2 现场测试

从香港城市大学北出口向南，经南山邨，往西折向大坑西邨，绕过一座山丘就进入石硖尾邨，而继续南行则是结构清晰的城市商住混合区。这条路线上有三个典型区域值得关注，按建设时间来看是城大校区(独体集中式，若干连续合院，室内完全依赖空调)，大坑西邨片(自由分散式，成片绿化和围合空间，常有行人驻留)，石硖尾邨片(严格网格式，建筑直压道路红线，形成典型的城市“层峡”，白天行人少停留)。在上述路线的三个典型区域(如图 1)，设置基本测试点，采集相关气候环境的同步数据(主要是温湿度和风向)。其中城大校区(A 点)，南山邨片(B 点)，石硖尾邨片(C 点)。

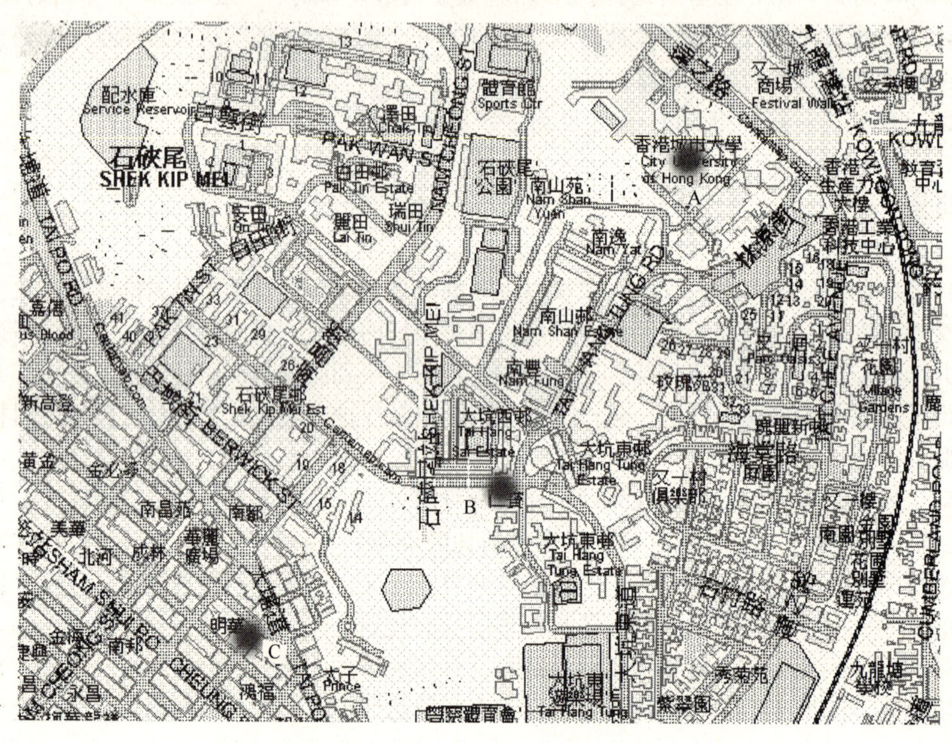

图 1 所选三个典型区域的基本环境测试点的设置

如图 2，从整体上来看，三个点的测试温度五天日最高温度的最低值均在 34℃以上，而五天最高湿度的最低值也在 85%以上，说明香港确实是典型的热湿气候城市。另外，在城大六楼室外天井朝西南的 A 点温度始终最高，太阳直接辐射和大量空调对外排热是

[1] 香港石硖尾地区曾于 1953 年曾经发生大火，无数房屋毁于一旦，上万人无家可归。火灾后政府立即着手在石硖尾建设“徙置区”。至此，成片的方格规划和林立的简易公屋伴随了石硖尾 50 多年。见龙炳颐先生编撰的《香港古今建筑》。

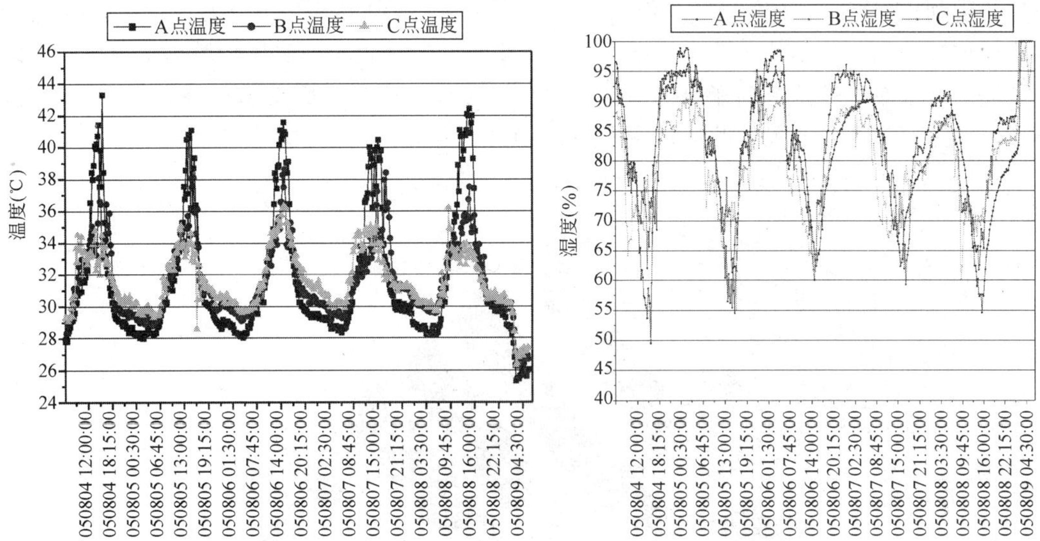

图 2　所选三个典型区域的基本环境测试点持续 5 天的温湿度曲线图

重要原因，而湿度较高说明天井形成的上下气流作用仍然不足以带来较为干燥的空气。在石硖尾邨黄竹街“层峡”地段朝东南设置的 C 点温度(与 A 点最高温差达 8℃)和湿度都是最低，这说明城市的街道“层峡”效应确实存在，但究竟是建筑密集区域相互遮挡，屏蔽太阳辐射的大部分作用，还是层峡通风带走热量，需要进一步分析。至于位于大坑西邨十字路口西北朝向的 B 点，由于中午以后的低纬度太阳直射仍能直接影响到其热环境，所以温度较 C 点高，同时由于处于主导风向(香港 8 月份风向主要为东风，或东偏南风)的背风向，其湿度也保持较高的水平。

图 3　石硖尾片区中的典型地段

由于 C 点因为其独特的城市结构环境和其测试数据反映出来的可继续深入的问题，现把 C 点所属的石硖尾片区中某典型地段(如图 3)作为 CFD 软件模拟分析热湿地区典型的城市街道“层峡”地段城市空间布局与气候热环境的关系考察的对象。

3　街道“层峡”的模拟

3.1　前期调研数据和计算机模拟技术

3.1.1　模拟之前期准备

作为城市街道“层峡”热环境研究的深入，需要如下之手段其中之一或更多。其一，实验室之风洞实验。其二，各种实验仪器的实测，并以此为基础的理论解析。其三，计算机之模拟计算。鉴于实验条件和更大现场操作的不便，目前仅从计算机之模拟计算的角度锲入。现利用 CFX 5.7 模拟石硖尾片区城市“层峡”地段的热环境。石硖尾片区作为计算模拟地段范围，包括长沙湾道、南昌街、界限街和大埔道等石硖尾片城市干道范围以内的区块。其中主干道两旁的建筑也囊括在内。计算模拟所需的数据包括建筑层数，高度，体形，道路以及巷弄的尺寸等通过调研采取(图 4、图 5)。

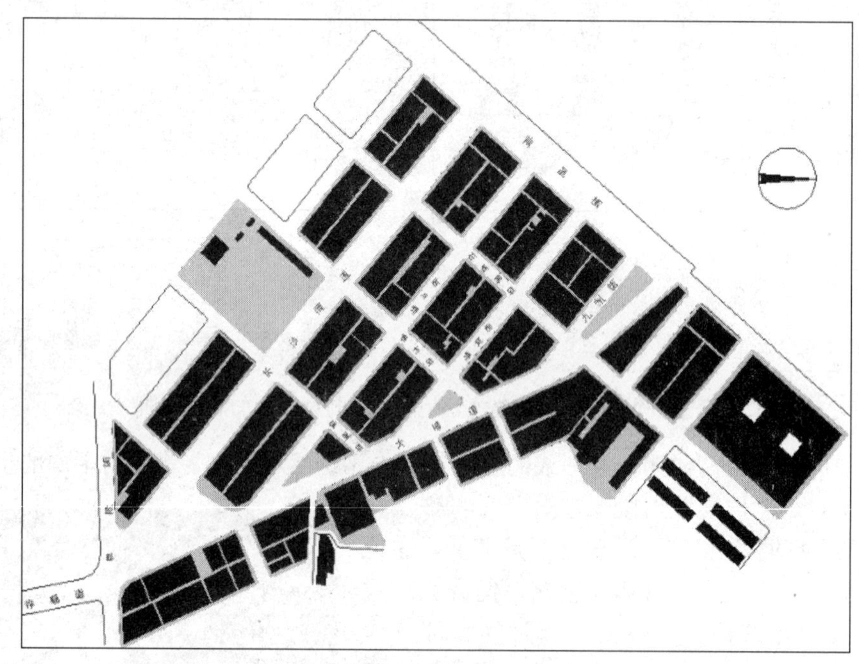

图 4 调研地段的城市 BLOCKS 结构图

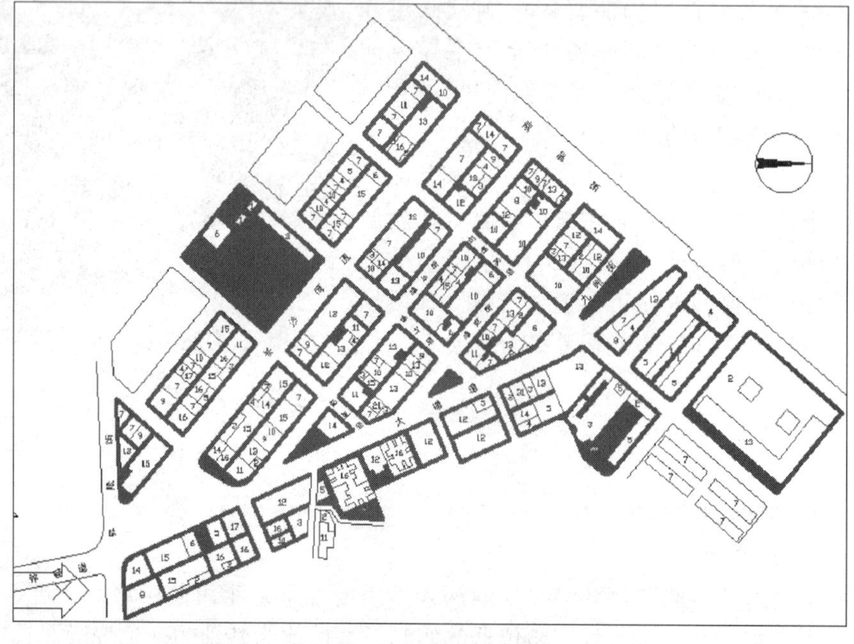

图 5 调研地段的建筑层数和巷弄结构图

2001年香港全年气候数据 **表1**

月份	平均气压(Pa)	最高气温(℃)	平均气温(℃)	最低气温(℃)	平均露点(℃)	平均相对湿度(%)	平均云量(%)	总日照时间(h)	总雨量(mm)	盛行风向(度)	平均风速(km/h)
一月	1017.5	23.3	17.3	9.8	13.3	78	68	112.2	47.6	060	24.9
二月	1016.9	26.0	17.6	10.5	13.7	79	66	127.9	10.9	050	22.1
三月	1014.9	27.9	20.6	14.3	16.2	77	72	140.4	56.5	060	20.5
四月	1012.0	31.8	22.4	15.0	19.7	85	86	79.0	133.0	070	21.6
五月	1007.5	32.2	26.7	22.9	23.3	82	79	149.5	162.0	080	17.4
六月	1005.6	33.2	27.6	23.7	24.8	85	81	114.0	1083.6	220	20.2
七月	1004.0	33.2	28.1	24.3	25.0	84	77	136.3	656.4	100	25.0
八月	1005.4	34.0	29.2	24.3	25.2	80	70	207.9	318.9	090	17.2
九月	1007.4	33.8	28.1	24.6	23.1	75	60	162.3	563.3	360	19.2
十月	1014.7	31.2	26.1	22.0	20.7	73	61	205.9	10.7	080	26.9
十一月	1018.6	28.3	21.5	13.1	14.2	65	43	206.4	4.3	070	26.3
十二月	1022.0	26.2	18.3	8.9	12.9	72	62	143.0	44.6	060	26.4
全年	1012.2	34.0	23.6	8.9	19.3	78	69	1784.8	3091.8	070	22.3

注：由 www.gb.weather.gov.hk 提供。

根据2001年香港政府气象网站提供的香港全年基础数据(见表1)，香港8月平均气温29.2℃平均相对湿度80%盛行东风和东偏南风，平均风速4.78m/s，平均气压1005.4P等。以上种种数据作为模拟研究热湿地区香港街道“层峡”地带的边界环境特征。

3.1.2 关于CFX模拟

计算模拟的全过程基于CFX 5.7对城市风热环境的处理。CFX 5.7软件是基于计算流体动力学(CFD)技术开发的流体动力学分析专用模块。因为城市模型的复杂性和计算机的性能要求，城市“层峡”地段整体模拟和分区分块模拟结合进行，保证计算面片要素控制在百万左右，以提高运算能力和速度。前期模拟使用ANSYS ICEM CFD 5.0建模，对于独立建筑和简单建筑组合的域使用非结构化四面体(Tetra)网格操作，对于整体城市空间的域使用结构化六面体(Hexa)网格操作。中后期使用CFX-5.7 for Windows进行定义、运算、求解以及后处理。

3.2 模拟模型

图6显示为ANSYS ICEM CFD 5.0模拟模型。软件模拟了在不同风速(1.5m/s、4.78m/s和7.5m/s)和不同界面(混凝土界面、绿化界面)条件下对不同尺度的街道“层峡”内热环境的影响。其中三种不同风速代表了本地区正常条件下的典型最低、平均和最高风速，而两种常用界面(建筑物表面和地面)旨在说明针对街道“层峡”热环境的建筑界面设计的选择。模拟以香港夏季实测数据和常年气象数据为依据，其中设定为香港常年主导风——东偏南风为风源，空气气流平均温度为29.2℃，平均气压为1005.4P。

3.3 模拟分析

经过模拟，我们发现城市街道“层峡”的热环境是一个非常复杂的热作用场，它跟周边的边界条件如入口空气气流、界面材料以及街道“层峡”尺寸密切相关。

3.3.1 风速的影响

图7中(*a*)、(*b*)和(*c*)分别是1.5m/s、4.78m/s和7.5m/s等不同风速，并视路面和

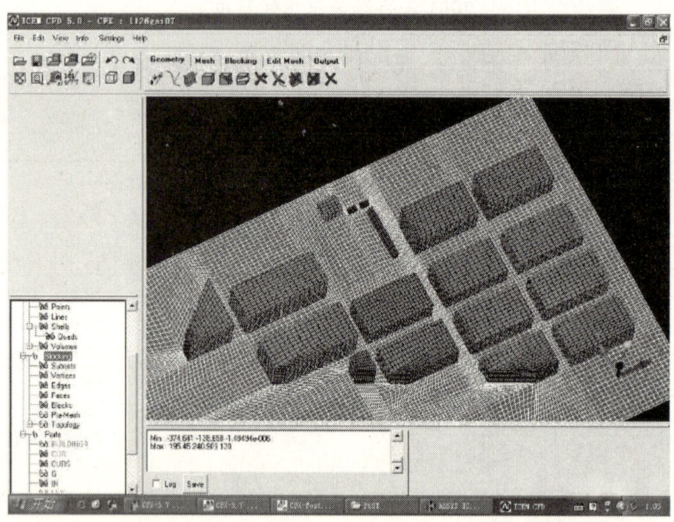

图 6　ANSYS ICEM CFD 5.0 界面以及调研地段的模拟模型

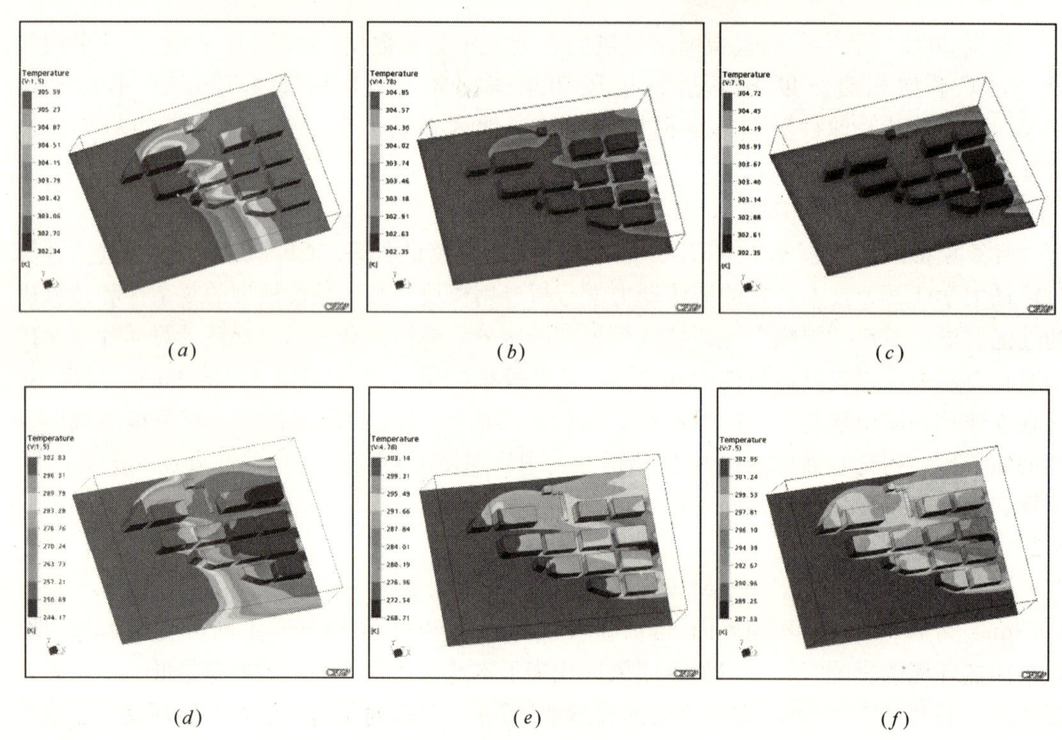

图 7　所模拟街道“层峡”地段的 3D 温度场示意图

建筑为理想界面(自身无热作用)的条件下的温度场示意；图 7 中(d)、(e)和(f)分别是 1.5m/s、4.78m/s 和 7.5m/s 等不同风速，考虑路面和建筑为物质界面(自身有传热作用)的条件下的温度场示意。

如图 7 所示的两组对比模拟试验当中，在路面和建筑为理想界面的条件下，街道两边的建筑以及街道地面仅仅对街道“层峡”所形成的狭长风道起到导风的作用，所以伴随主导风源风速的提高，街道“层峡”内空气龄极大降低，街道“层峡”内空气携热的能力也显著加强。然而在路面和建筑为物质界面(模拟混凝土界面)的条件下，如果界面本身的热传导作用带来的热量，无法及时通过空气流动带走，那么将在街道“层峡”内滞留。所以，在第二组模拟当中，1.5m/s 的低风速相反在一定程度上阻止了街道“层峡”内部和外部较高的热交换过程，其温度场的整体温度最低；而风速达到 7.5m/s 的时候，相比较 4.78m/s 的风速，街道“层峡”内部更容易通过空气携热的能力带走一部分外界和内部产生的大量热量，其温度场也较好一些。不过要注意的是第一组模拟实验的温度场局部最高温度要高于第二组模拟实验。

3.3.2 界面的影响

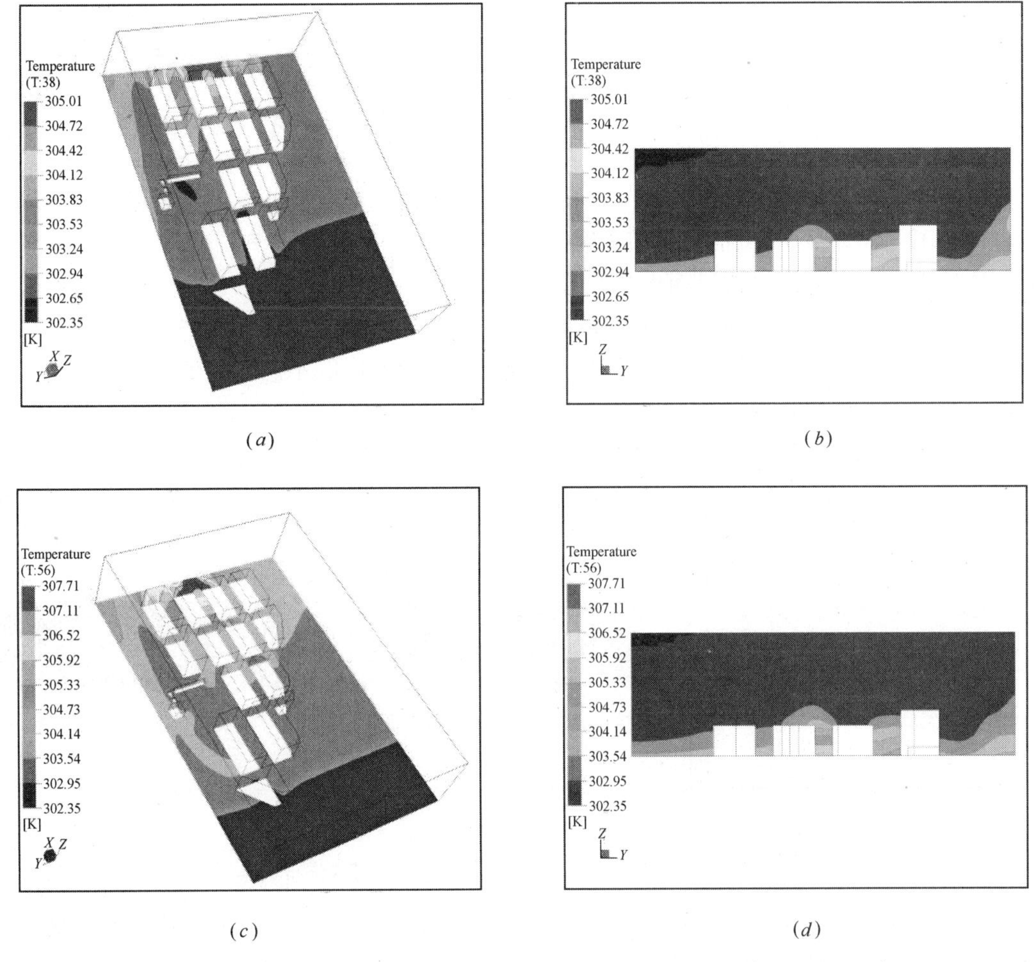

图 8 所模拟街道“层峡”地段的 3D 温度场示意图和温度切面分布图

图 8 中(*a*)和(*b*)表示的是绿化路面；(*c*)和(*d*)表示的是全混凝土路面。

图 9 中(*a*)表示的是理想界面；(*b*)表示的是路面和建筑物为全混凝土界面。

从两组不同道路界面的对比模拟试验(如图 8)中可以看出，在同样风速条件下和相同街道“层峡”内，绿化路面同全混凝土路面相比较，其街道“层峡”范围内最高温差为 2.7℃，近路面空气温度也普遍低 0.6℃左右，而且其温度场分布更为均匀，“层峡”内热

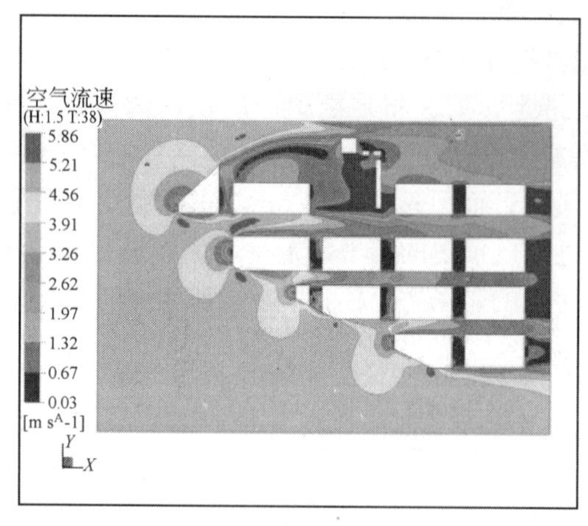

(a)

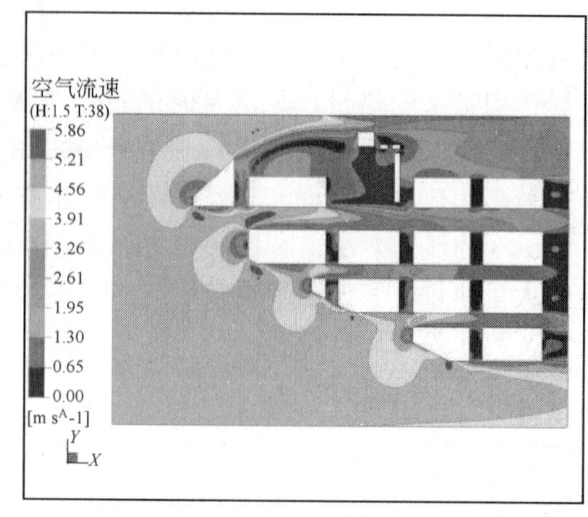

(b)

图 9 所模拟街道“层峡”地段的空气流场平面分布图

环境也更为理想。不同地表覆盖界面在模拟试验中出现的温度的差异，主要是因为不同界面材料的不同物理性质(包括对太阳辐射的吸收率、发射率、传导率和比热等)。同样，街道“层峡”范围内的建筑物界面也会改善街道“层峡”微环境气候。也由此可见，在太阳辐射下街道“层峡”的不同覆盖界面通过其与环境的热交换作用对街道“层峡”内的热环境有重要的影响。另外，从图 9 不难看出，街道“层峡”内的不同覆盖界面对于其空气流场并没有太多明显的影响。

3.3.3 街道“层峡”尺寸的影响

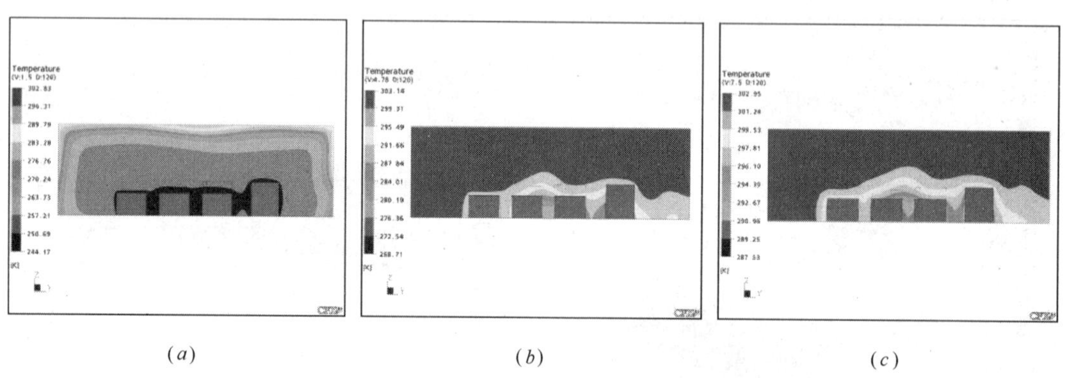

(a) (b) (c)

图 10 所模拟街道“层峡”地段的温度切面分布图

图 10 中(a)、(b)和(c)分别是 1.5m/s、4.78m/s 和 7.5m/s 等不同风速条件下靠近街道“层峡”出口的温度场切面示意。

在所模拟的街道“层峡”范围内沿主导风方向主要有两种街道，分别是 $H/W=1$ 的城市干道(以长沙湾道为代表，见图 4、图 5)和 $H/W=2$(以福华街为代表，见图 4、图 5)的一般道路。从图 10 可以清楚看到不同街道“层峡”的温度轮廓线。在较低 H/W 比率的长沙湾道，当风源风速达到 4.78m/s 以上，“层峡”内热空气积极上扬；当风源风速较低时，热空气下沉并向街道“层峡”内部甚至道路中央集聚。而在较高 H/W 比率的福华街，伴随风源风速的变化，其街道“层峡”内的温度场变化相对较为平缓。

4 结论

一系列的实测和CFD软件模拟已经说明了入口空气气流、界面材料以及街道“层峡”尺寸在城市街道“层峡”内热环境的形成当中有着及其重要的作用。在一定的覆盖界面情况下，较高的风源风速并不能明显地影响街道“层峡”内的热环境，同时街道“层峡”内的不同覆盖界面对于其空气流场也没有太多明显的影响。但是路面和建筑物表面的绿化处理却能极大改善街道“层峡”内的微气候环境，并且可以结合街道“层峡”的 H/W 比率对“层峡”内的微气候环境进行更进一步的利用和控制。

注：本论文得到国家自然科学基金项目“夏热冬冷地区基于节能及气候适应性的城市设计策略研究”资助，项目批准号为50578067。

A Study on Some Factors Influenced the Thermal Environment of the Street Canyon

——Citing a mixed residential and commercial area in Shixiawei of Hong Kong

Wang Zhen[1] Wang Ping[1] Zhao Kui[1] Guang Kaixiong[1]
Li Baofeng[1] Lu Zhaoming[2]
([1] School of Architecture & Urban Planning HUST, WuHan 430074 China;
[2] Department of Building & Construction, City University of Hong Kong, Hong Kong 999077 China)

Abstract: The street canyon in urban is one of the most important place of activities and public space of cities. Different forms of architecture and urban space affect its microenvironment of climate. After the test of fieldwork and the application of Computational Fluid Dynamics (CFD) technique for the thermal environment of the street canyon in a mixed residential and commercial area in Hong Kong, this paper study the thermal environment of the street canyon influenced with some factors such as the different geometry of canyon, materials of surface and wind speed, and proposes one study method about the microenvironment of the street canyon and some technique strategies to improve the microenvironment of the street canyon.

参考文献

[1] M. Rohinton Emmanuel 著. An Urban Approach to Climate-Sensitive Design. London and New York: Spon Press, 2005

[2] A. K. M. Chu, R. C. W. Kwok, K. N. Yu, 2005. Study of pollution dispersion in urban areas using Computational Fluid Dynamics (CFD) and Geographic Information System (GIS), Environmental Modelling & Software 273~277